EDITION

4

PREALGEBRA

ALAN S. TUSSY
CITRUS COLLEGE

R. DAVID GUSTAFSON
ROCK VALLEY COLLEGE

DIANE R. KOENIG
ROCK VALLEY COLLEGE

D0580314

BROOKS/COLE
CENGAGE Learning

Australia • Brazil • Japan • Korea • Mexico • Singapore • Spain • United Kingdom • United States

BROOKS/COLE
CENGAGE Learning™

Prealgebra, **Fourth Edition**
Alan S. Tussy, R. David Gustafson, Diane R. Koenig

Publisher: Charlie Van Wagner

Senior Developmental Editor: Danielle Derbenti

Senior Development Editor for Market Strategies: Rita Lombard

Assistant Editor: Stefanie Beeck

Editorial Assistant: Jennifer Cordoba

Media Editors: Heleny Wong, Guanglei Zhang

Marketing Manager: Gordon Lee

Marketing Assistant: Angela Kim

Marketing Communications Manager: Katy Malatesta

Content Project Manager: Jennifer Risden

Creative Director: Rob Hugel

Art Director: Vernon Boes

Print Buyer: Linda Hsu

Rights Acquisitions Account Manager, Text: Mardell Glinksi-Schultz

Rights Acquisitions Account Manager, Image: Don Schlotman

Production Service: Graphic World Inc.

Text Designer: Diane Beasley

Photo Researcher: Bill Smith Group

Illustrators: Lori Heckelman; Graphic World Inc.

Cover Designer: Terri Wright

Cover Image: Background: © JTB Photo/Masterfile; Y Button: © Art Parts/Fotosearch RF

Compositor: Graphic World Inc.

For product information and technology assistance, contact us at **Cengage Learning Customer & Sales Support, 1-800-354-9706**

For permission to use material from this text or product, submit all requests online at **www.cengage.com/permissions**

Further permissions questions can be e-mailed to **permissionrequest@cengage.com**

Library of Congress Control Number: 2010920734

ISBN-13: 978-1-4390-4431-5

ISBN-10: 1-4390-4431-7

Brooks/Cole
20 Davis Drive
Belmont, CA 94002-3098
USA

Cengage Learning is a leading provider of customized learning solutions with office locations around the globe, including Singapore, the United Kingdom, Australia, Mexico, Brazil, and Japan. Locate your local office at **www.cengage.com/global**

Cengage Learning products are represented in Canada by Nelson Education, Ltd.

To learn more about Brooks/Cole, visit **www.cengage.com/brookscole**

Purchase any of our products at your local college store or at our preferred online store **www.cengagebrain.com**

Printed in the United States of America
2 3 4 5 6 7   14 13 12 11

To three good friends,

Jennifer,

Danielle,

and

Charlie

ALAN S. TUSSY

R. DAVID GUSTAFSON

DIANE R. KOENIG

CONTENTS

Comstock Images/Getty Images

© OJO Images Ltd/Alamy

CHAPTER **3**

The Language of Algebra 225

CHAPTER 4

Fractions and Mixed Numbers 303

CHAPTER 5

Decimals 443

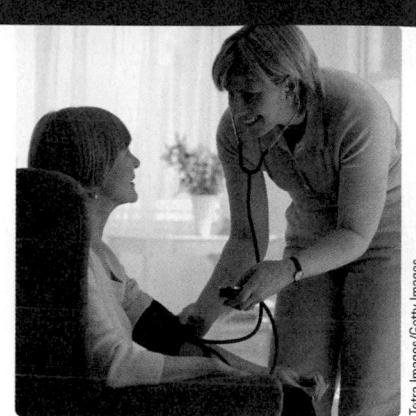

Tetra Images/Getty Images

CHAPTER 6

Ratio, Proportion, and Measurement 557

Nick White/Getty Images

Ariel Skelley/Getty Images

Kim Steele/Photodisc/Getty Images

CHAPTER 9

An Introduction to Geometry   819

CHAPTER **10**

Exponents and Polynomials   949

APPENDIXES

PREFACE

Prealgebra, Fourth Edition, is more than a simple upgrade of the third edition. Substantial changes have been made to the worked example structure, the *Study Sets,* and the pedagogy. Throughout the revision process, our objective has been to ease teaching challenges and meet students' educational needs.

Mathematics, for many of today's developmental math students, is like a foreign language. They have difficulty translating the words, their meanings, and how they apply to problem solving. With these needs in mind (and as educational research suggests), our fundamental goal is to have students read, write, think, and speak using the *language of algebra.* Instructional approaches that include vocabulary, practice, and well-defined pedagogy, along with an emphasis on reasoning, modeling, communication, and technology skills have been blended to address this need.

The most common question that students ask as they watch their instructors solve problems and as they read the textbook is … *Why?* The new fourth edition addresses this question in a unique way. Experience teaches us that it's not enough to know *how* a problem is solved. Students gain a deeper understanding of algebraic concepts if they know *why* a particular approach is taken. This instructional truth was the motivation for adding a **Strategy** and **Why** explanation to the solution of each worked example. The fourth edition now provides, on a consistent basis, a concise answer to that all-important question: *Why?*

These are just two of several reasons we trust that this revision will make this course a better experience for both instructors and students.

NEW TO THIS EDITION

- **New Chapter Openers**
- **New Worked Example Structure**
- **New Calculation Notes in Examples**
- **New Five-Step Problem-Solving Strategy**
- **New Study Skills Workshop Module**
- **New Language of Algebra, Success Tip, and Caution Boxes**
- **New Chapter Objectives**
- **New Guided Practice and Try It Yourself Sections in the Study Sets**
- **New Chapter Summary and Review**
- **New Study Skills Checklists**

Chapter Openers That Answer the Question: When Will I Use This?

Instructors are asked this question time and again by students. In response, we have written chapter openers called *From Campus to Careers.* This feature highlights vocations that require various algebraic skills. Designed to inspire career exploration, each includes job outlook, educational requirements, and annual earnings information. Careers presented in the openers are tied to an exercise found later in the *Study Sets.*

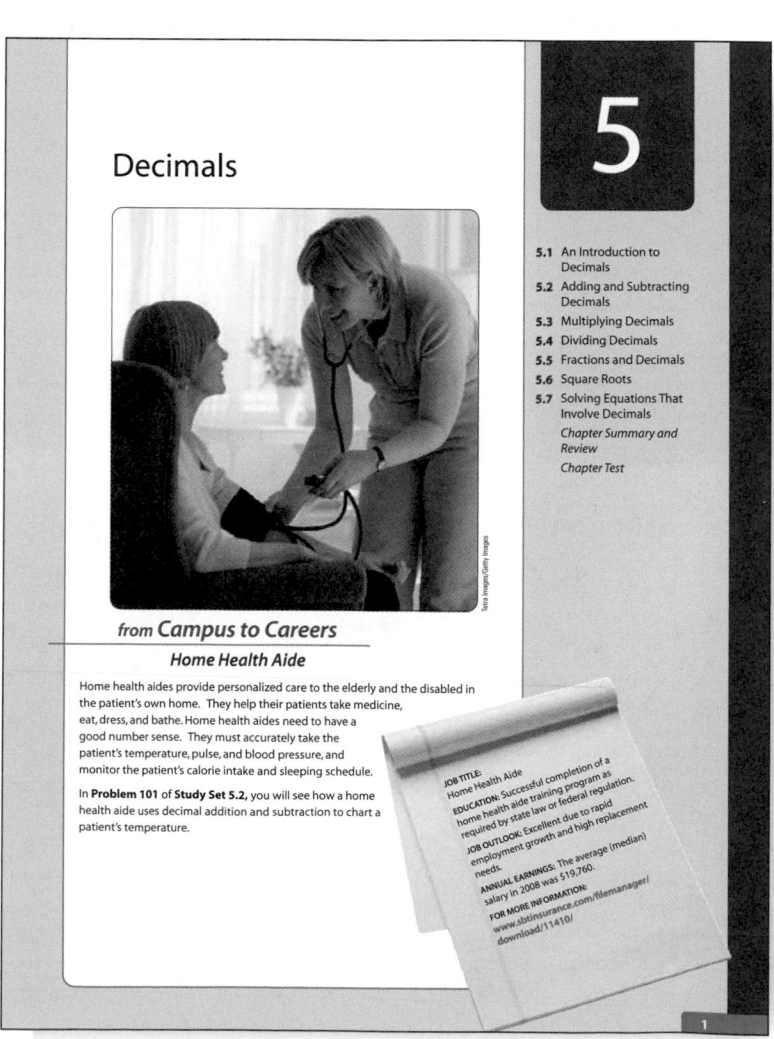

Decimals

5

5.1 An Introduction to Decimals
5.2 Adding and Subtracting Decimals
5.3 Multiplying Decimals
5.4 Dividing Decimals
5.5 Fractions and Decimals
5.6 Square Roots
5.7 Solving Equations That Involve Decimals
Chapter Summary and Review
Chapter Test

from *Campus to Careers*
Home Health Aide

Home health aides provide personalized care to the elderly and the disabled in the patient's own home. They help their patients take medicine, eat, dress, and bathe. Home health aides need to have a good number sense. They must accurately take the patient's temperature, pulse, and blood pressure, and monitor the patient's calorie intake and sleeping schedule.

In **Problem 101** of **Study Set 5.2,** you will see how a home health aide uses decimal addition and subtraction to chart a patient's temperature.

JOB TITLE: Home Health Aide
EDUCATION: Successful completion of a home health aide training program as required by state law or federal regulation.
JOB OUTLOOK: Excellent due to rapid employment growth and high replacement needs.
ANNUAL EARNINGS: The average (median) salary in 2008 was $19,760.
FOR MORE INFORMATION: www.abtinsurance.com/filemanager/download/11410/

Examples That Tell Students Not Just How, But WHY

Why? That question is often asked by students as they watch their instructor solve problems in class and as they are working on problems at home. It's not enough to know *how* a problem is solved. Students gain a deeper understanding of the algebraic concepts if they know *why* a particular approach was taken. This instructional truth was the motivation for adding a *Strategy* and *Why* explanation to each worked example.

Examples That Offer Immediate Feedback

Each worked example includes a *Self Check*. These can be completed by students on their own or as classroom lecture examples, which is how Alan Tussy uses them. Alan asks selected students to read aloud the *Self Check* problems as he writes what the student says on the board. The other students, with their books open to that page, can quickly copy the *Self Check* problem to their notes. This speeds up the note-taking process and encourages student participation in his lectures. It also teaches students how to read mathematical symbols. Each *Self Check* answer is printed adjacent to the corresponding problem in the *Annotated Instructor's Edition* for easy reference. *Self Check* solutions can be found at the end of each section in the student edition before each *Study Set*.

Examples That Ask Students to Work Independently

Each worked example ends with a *Now Try* problem. These are the final step in the learning process. Each one is linked to a similar problem found within the *Guided Practice* section of the *Study Sets*.

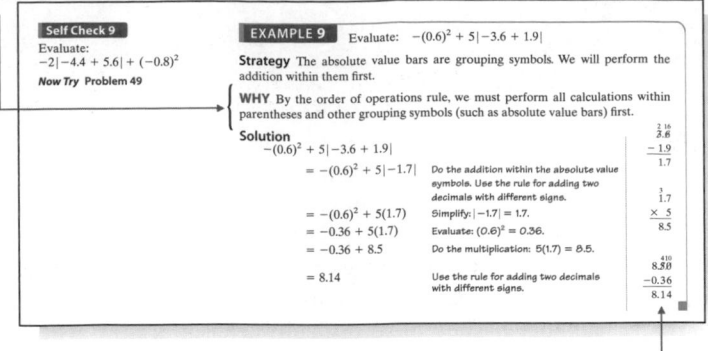

Examples That Show the Behind-the-Scenes Calculations

Some steps of the solutions to worked examples in *Prealgebra* involve arithmetic calculations that are too complicated to be performed mentally. In these instances, we have shown the actual computations that must be made to complete the formal solution. These computations appear directly to the right of the author notes and are separated from them by a thin, gray rule. The necessary addition, subtraction, multiplication, or division (usually done on scratch paper) is placed at the appropriate stage of the solution where such a computation is required. Rather than simply list the steps of a solution horizontally, making no mention of how the numerical values within the solution are obtained, this unique feature will help answer the often-heard question from a struggling student, "How did you get that answer?" It also serves as a model for the calculations that students must perform independently to solve the problems in the Study Sets.

Emphasis on Problem-Solving

New to *Prealgebra*, the five-step problem-solving strategy guides students through applied worked examples using the Analyze, Form, Solve, State, and Check process. This approach clarifies the thought process and mathematical skills necessary to solve a wide variety of problems. As a result, students' confidence is increased and their problem-solving abilities are strengthened.

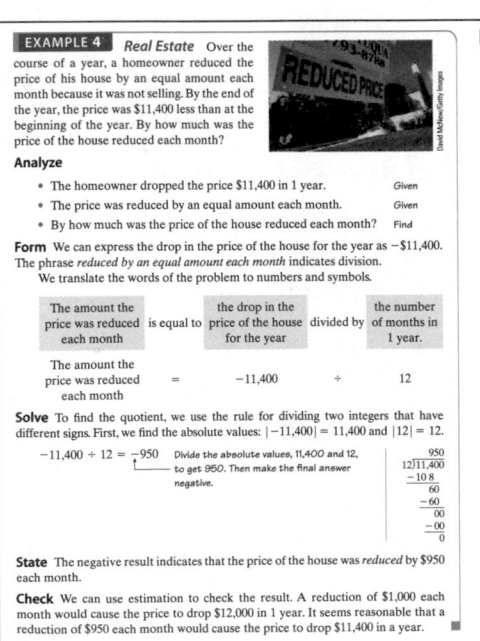

Strategy for Problem Solving

1. **Analyze the problem** by reading it carefully. What information is given? What are you asked to find? What vocabulary is given? Often, a diagram or table will help you visualize the facts of the problem.
2. **Form a plan** by translating the words of the problem to numbers and symbols.
3. **Solve the problem** by performing the calculations.
4. **State the conclusion** clearly. Be sure to include the units (such as feet, seconds, or pounds) in your answer.
5. **Check the result.** An estimate is often helpful to see whether an answer is reasonable.

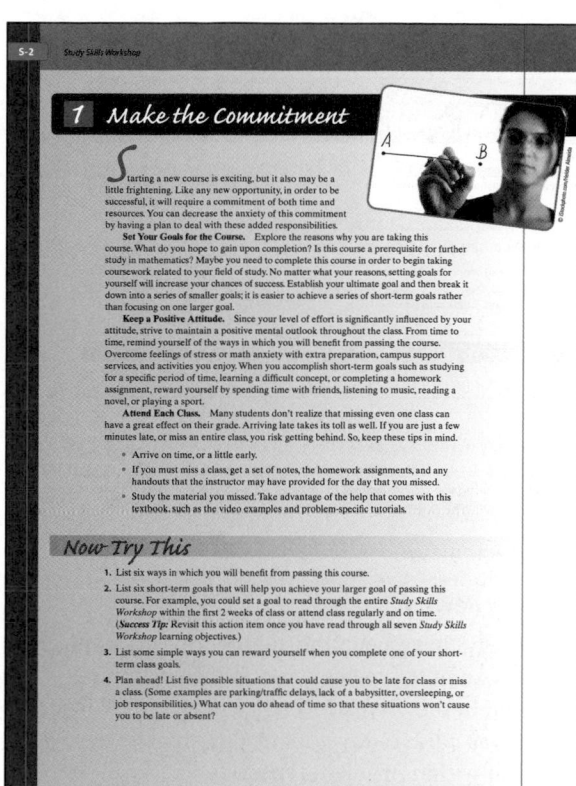

Emphasis on Study Skills

Prealgebra begins with a *Study Skills Workshop* module. Instead of simple, unrelated suggestions printed in the margins, this module contains one-page discussions of study skills topics followed by a *Now Try This* section offering students actionable skills, assignments, and projects that will impact their study habits throughout the course.

The Language of Algebra   The word *decimal* comes from the Latin word *decima,* meaning a tenth part.

Integrated Focus on the Language of Algebra

Language of Algebra boxes draw connections between mathematical terms and everyday references to reinforce the language of algebra approach that runs throughout the text.

Guidance When Students Need It Most

Appearing at key teaching moments, *Success Tips* and *Caution* boxes improve students' problem-solving abilities, warn students of potential pitfalls, and increase clarity.

Success Tip   In the newspaper example, we found a *part of a part* of a page. Multiplying proper fractions can be thought of in this way. When taking a *part of a part* of something, the result is always smaller than the original part that you began with.

Caution!   In Example 5, it was very helpful to prime factor and simplify when we did (the third step of the solution). If, instead, you find the product of the numerators and the product of the denominators, the resulting fraction is difficult to simplify because the numerator, 126, and the denominator, 420, are large.

$$\frac{2}{3} \cdot \frac{9}{14} \cdot \frac{7}{10} = \frac{2 \cdot 9 \cdot 7}{3 \cdot 14 \cdot 10} = \frac{126}{420}$$

Factor and simplify at this stage, before multiplying in the numerator and denominator.

Don't multiply in the numerator and denominator and then try to simplify the result. You will get the same answer, but it takes much more work.

Useful Objectives Help Keep Students Focused

Each section begins with a set of numbered *Objectives* that focus students' attention on the skills that they will learn. As each objective is discussed in the section, the number and heading reappear to the reader to remind them of the objective at hand.

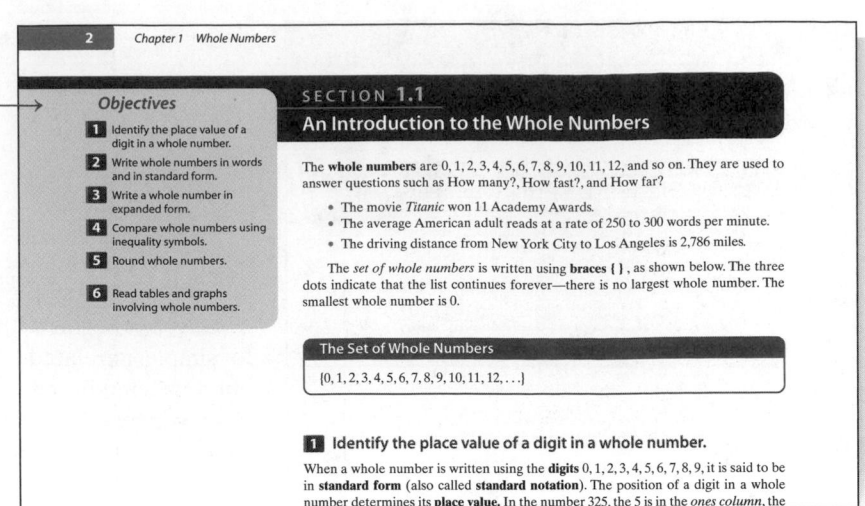

2   Chapter 1   Whole Numbers

Objectives

1. Identify the place value of a digit in a whole number.
2. Write whole numbers in words and in standard form.
3. Write a whole number in expanded form.
4. Compare whole numbers using inequality symbols.
5. Round whole numbers.
6. Read tables and graphs involving whole numbers.

SECTION 1.1
An Introduction to the Whole Numbers

The **whole numbers** are 0, 1, 2, 3, 4, 5, 6, 7, 8, 9, 10, 11, 12, and so on. They are used to answer questions such as How many?, How fast?, and How far?

• The movie *Titanic* won 11 Academy Awards.
• The average American adult reads at a rate of 250 to 300 words per minute.
• The driving distance from New York City to Los Angeles is 2,786 miles.

The *set of whole numbers* is written using **braces { }**, as shown below. The three dots indicate that the list continues forever—there is no largest whole number. The smallest whole number is 0.

The Set of Whole Numbers

{0, 1, 2, 3, 4, 5, 6, 7, 8, 9, 10, 11, 12, . . .}

1 Identify the place value of a digit in a whole number.

When a whole number is written using the **digits** 0, 1, 2, 3, 4, 5, 6, 7, 8, 9, it is said to be in **standard form** (also called **standard notation**). The position of a digit in a whole number determines its **place value**. In the number 325, the 5 is in the *ones column*, the

Thoroughly Revised Study Sets

The *Study Sets* have been thoroughly revised to ensure that every example type covered in the section is represented in the *Guided Practice* problems. Particular attention was paid to developing a gradual level of progression within problem types.

Guided Practice Problems

All of the problems in the *Guided Practice* portion of the *Study Sets* are linked to an associated worked example or objective from that section. This feature promotes student success by referring them to the proper worked example(s) or objective(s) if they encounter difficulties solving homework problems.

GUIDED PRACTICE

Perform each operation and simplify, if possible. See Example 1.

17. $\frac{4}{9} + \frac{1}{9}$

18. $\frac{3}{7} + \frac{1}{7}$

19. $\frac{3}{8} + \frac{1}{8}$

20. $\frac{7}{12} + \frac{1}{12}$

21. $\frac{11}{15} - \frac{7}{15}$

22. $\frac{10}{21} - \frac{5}{21}$

23. $\frac{11}{20} - \frac{3}{20}$

24. $\frac{7}{18} - \frac{5}{18}$

Subtract and simplify, if possible. See Example 2.

25. $-\frac{11}{5} - \left(-\frac{8}{5}\right)$

26. $-\frac{15}{9} - \left(-\frac{11}{9}\right)$

27. $-\frac{7}{21} - \left(-\frac{2}{21}\right)$

28. $-\frac{21}{25} - \left(-\frac{9}{25}\right)$

Perform the operations and simplify, if possible. See Example 3.

29. $\frac{19}{40} - \frac{3}{40} - \frac{1}{40}$

30. $\frac{11}{24} - \frac{1}{24} - \frac{7}{24}$

31. $\frac{13}{33} + \frac{1}{33} + \frac{7}{33}$

32. $\frac{21}{50} + \frac{1}{50} + \frac{13}{50}$

49. $\frac{1}{6} + \frac{5}{8}$

50. $\frac{7}{12} + \frac{3}{8}$

51. $\frac{4}{9} + \frac{5}{12}$

52. $\frac{1}{9} + \frac{5}{6}$

Subtract and simplify, if possible. See Example 9.

53. $\frac{9}{10} - \frac{3}{14}$

54. $\frac{11}{12} - \frac{11}{30}$

55. $\frac{11}{12} - \frac{7}{15}$

56. $\frac{7}{15} - \frac{5}{12}$

Determine which fraction is larger. See Example 10.

57. $\frac{3}{8}$ or $\frac{5}{16}$

58. $\frac{5}{6}$ or $\frac{7}{12}$

59. $\frac{4}{5}$ or $\frac{2}{3}$

60. $\frac{7}{9}$ or $\frac{4}{5}$

61. $\frac{7}{9}$ or $\frac{11}{12}$

62. $\frac{3}{8}$ or $\frac{5}{12}$

63. $\frac{23}{20}$ or $\frac{7}{6}$

64. $\frac{19}{15}$ or $\frac{5}{4}$

Add and simplify, if possible. See Example 11.

Try It Yourself

To promote problem recognition, the *Study Sets* now include a collection of *Try It Yourself* problems that *do not* link to worked examples. These problem types are thoroughly mixed, giving students an opportunity to practice decision making and strategy selection as they would when taking a test or quiz.

TRY IT YOURSELF

Perform each operation.

69. $-\frac{1}{12} - \left(-\frac{5}{12}\right)$

70. $-\frac{1}{16} - \left(-\frac{15}{16}\right)$

71. $\frac{4}{5} + \frac{2}{3}$

72. $\frac{1}{4} + \frac{2}{3}$

73. $\frac{12}{25} - \frac{1}{25} - \frac{1}{25}$

74. $\frac{7}{9} + \frac{1}{9} + \frac{1}{9}$

75. $-\frac{7}{20} - \frac{1}{5}$

76. $-\frac{5}{8} - \frac{1}{3}$

77. $-\frac{7}{16} + \frac{1}{4}$

78. $-\frac{17}{20} + \frac{4}{5}$

79. $\frac{11}{12} - \frac{2}{3}$

80. $\frac{2}{3} - \frac{1}{6}$

81. $\frac{2}{3} + \frac{4}{5} + \frac{5}{6}$

82. $\frac{3}{4} + \frac{2}{5} + \frac{3}{10}$

83. $\frac{9}{20} - \frac{1}{30}$

84. $\frac{5}{6} - \frac{3}{10}$

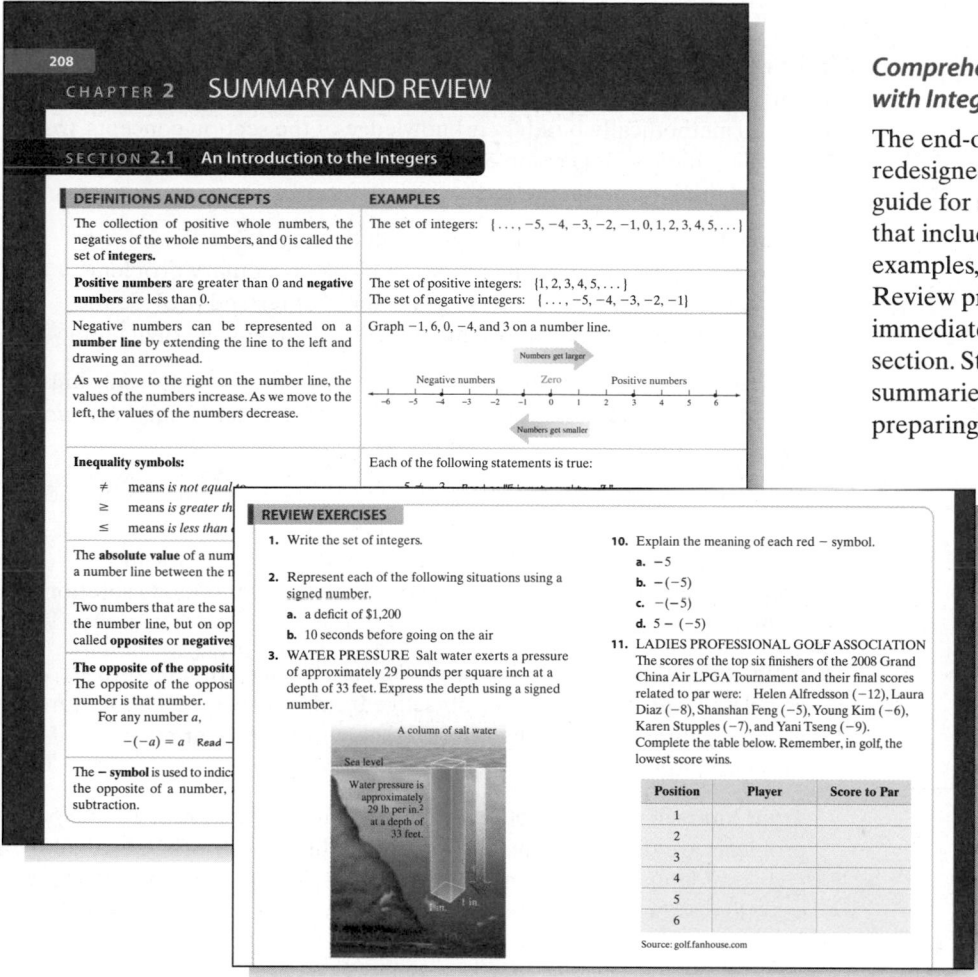

Comprehensive End-of-Chapter Summary with Integrated Chapter Review

The end-of-chapter material has been redesigned to function as a complete study guide for students. New chapter summaries that include definitions, concepts, and examples, by section, have been written. Review problems for each section immediately follow the summary for that section. Students will find the detailed summaries a very valuable study aid when preparing for exams.

Study Skills That Point Out Common Student Mistakes

In Chapter 1, we have included four *Study Skills Checklists* designed to actively show students how to effectively use the key features in this text. Subsequent chapters include one checklist just before the *Chapter Summary and Review* that provides another layer of preparation to promote student success. These *Study Skills Checklists* warn students of common errors, giving them time to consider these pitfalls before taking their exam.

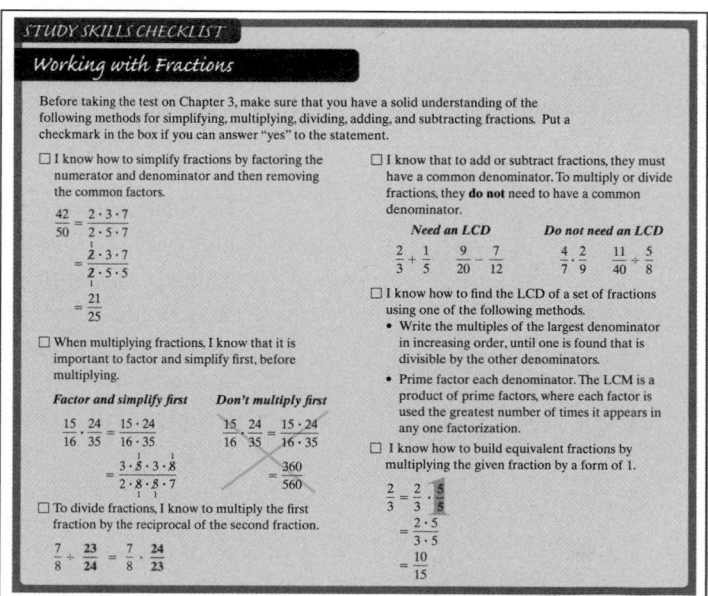

TRUSTED FEATURES

- **Study Sets** found in each section offer a multifaceted approach to practicing and reinforcing the concepts taught in each section. They are designed for students to methodically build their knowledge of the section concepts, from basic recall to increasingly complex problem solving, through reading, writing, and thinking mathematically.

  Vocabulary—Each *Study Set* begins with the important *Vocabulary* discussed in that section. The fill-in-the-blank vocabulary problems emphasize the main concepts taught in the chapter and provide the foundation for learning and communicating the language of algebra.

  Concepts—In *Concepts,* students are asked about the specific subskills and procedures necessary to successfully complete the *Guided Practice* and *Try It Yourself* problems that follow.

  Notation—In *Notation,* the students review the new symbols introduced in a section. Often, they are asked to fill in steps of a sample solution. This strengthens their ability to read and write mathematics and prepares them for the *Guided Practice* problems by modeling solution formats.

  Guided Practice—The problems in *Guided Practice* are linked to an associated worked example or objective from that section. This feature promotes student success by referring them to the proper examples if they encounter difficulties solving homework problems.

  Try It Yourself—To promote problem recognition, the *Try It Yourself* problems are thoroughly mixed and are *not* linked to worked examples, giving students an opportunity to practice decision-making and strategy selection as they would when taking a test or quiz.

  Applications—The *Applications* provide students the opportunity to apply their newly acquired algebraic skills to relevant and interesting real-life situations.

  Writing—The *Writing* problems help students build mathematical communication skills.

  Review—The *Review* problems consist of randomly selected problems from previous chapters. These problems are designed to keep students' successfully mastered skills up-to-date before they move on to the next section.

- **Detailed Author Notes** that guide students along in a step-by-step process appear in the solutions to every worked example.

- **Think It Through** features make the connection between mathematics and student life. These relevant topics often require algebra skills from the chapter to be applied to a real-life situation. Topics include tuition costs, student enrollment, job opportunities, credit cards, and many more.

- **Chapter Tests,** at the end of every chapter, can be used as preparation for the class exam.

- **Cumulative Reviews** follow the end-of-chapter material and keep students' skills current before moving on to the next chapter. Each problem is linked to the associated section from which the problem came for ease of reference. The final *Cumulative Review* is often used by instructors as a Final Exam Review.

- **Using Your Calculator** is an optional feature (formerly called *Calculator Snapshots*) that is designed for instructors who wish to use calculators as part of the instruction in this course. This feature introduces keystrokes and shows how scientific and graphing calculators can be used to solve problems. In the *Study Sets,* icons are used to denote problems that may be solved using a calculator.

CHANGES TO THE TABLE OF CONTENTS

Based on feedback from colleagues and users of the third edition, the following changes have been made to the table of contents in an effort to further streamline the text and make it even easier to use.

- The Chapter 1 topics have been expanded and reorganized:

    1.1 *An Introduction to the Whole Numbers* (expanded coverage of rounding and integrated estimation)

    1.2 *Adding and Subtracting Whole Numbers* (integrated estimation)

    1.3 *Multiplying Whole Numbers* (integrated estimation; now covered in its own section)

    1.4 *Dividing Whole Numbers* (integrated estimation; now covered in its own section)

    1.5 *Prime Factors and Exponents*

    1.6 *The Least Common Multiple and the Greatest Common Factor* (new section)

    1.7 *Order of Operations*

    1.8 *Solving Equations Using Addition and Subtraction*

    1.9 *Solving Equations Using Multiplication and Division*

- In Chapter 2, *The Integers,* there is added emphasis on problem-solving.

- The Chapter 3 topics have been heavily revised and reorganized for an improved introduction to the language of algebra that is consistent with our approach taken in the other books of our series.

    3.1 *Algebraic Expressions*

    3.2 *Evaluating Algebraic Expressions and Formulas*

    3.3 *Simplifying Algebraic Expressions and the Distributive Property*

    3.4 *Combining Like Terms*

    3.5 *Simplifying Expressions to Solve Equations*

    3.6 *Using Equations to Solve Application Problems*

- In Chapter 4, *Fractions and Mixed Numbers,* the topics of the least common multiple are revisited as this applies to fractions and there is an added emphasis on problem-solving.

- The concept of estimation is integrated into Section 5.4, *Dividing Decimals.* Also, there is an added emphasis on problem-solving.

- The chapter *Ratio, Proportion, and Measurement* has been moved up to precede the chapter *Percent* so that proportions can be used to solve percent problems.

- Section 7.2, *Solving Percent Problems Using Percent Equations and Proportions,* has two separate objectives, giving instructors a choice in approach.

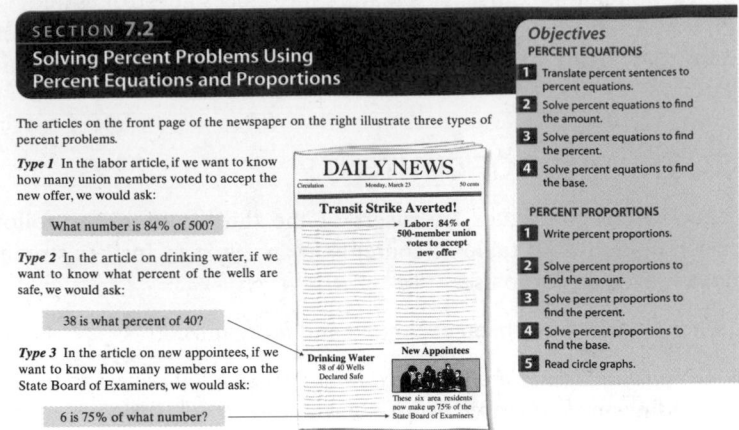

- Section 7.4, *Estimation with Percent,* is new and continues with the integrated estimation we include throughout the text.

- Chapter 8, *Graphs and Statistics,* is new to this edition:

    8.1 *Reading Graphs and Tables*

    8.2 *Mean, Median, and Mode*

    8.3 *Equations in Two Variables; The Rectangular Coordinate System* (formerly located in the chapter on exponents and polynomials)

    8.4 *Graphing Linear Equations* (formerly located in the chapter on exponents and polynomials)

- The Chapter 9 topics have been reorganized and expanded:

    9.1 *Basic Geometric Figures; Angles*

    9.2 *Parallel and Perpendicular Lines*

    9.3 *Triangles*

    9.4 *The Pythagorean Theorem*

    9.5 *Congruent Triangles and Similar Triangles*

    9.6 *Quadrilaterals and Other Polygons*

    9.7 *Perimeters and Areas of Polygons*

    9.8 *Circles*

    9.9 *Volume*

GENERAL REVISIONS AND OVERALL DESIGN

- We have edited the prose so that it is even more clear and concise.
- Strategic use of color has been implemented within the new design to help the visual learner.
- Added color in the solutions highlights key steps and improves readability.
- We have updated much of the data and graphs and have added scaling to all axes in all graphs.
- We have added more real-world applications.
- We have included more problem-specific photographs and improved the clarity of the illustrations.

INSTRUCTOR RESOURCES

Print Ancillaries

Instructor's Resource Binder (0-538-73675-5)
Maria H. Andersen, *Muskegon Community College*
NEW! Each section of the main text is discussed in uniquely designed *Teaching Guides* containing instruction tips, examples, activities, worksheets, overheads, assessments, and solutions to all worksheets and activities.

Complete Solutions Manual (0-538-79886-6)
Nathan G. Wilson, *St. Louis Community College at Meramec*
The *Complete Solutions Manual* provides worked-out solutions to all of the problems in the text.

Annotated Instructor's Edition (1-4390-4866-5)
The *Annotated Instructor's Edition* provides the complete student text with answers next to each respective exercise. New to this edition: Teaching Examples have been added for each worked example.

Electronic Ancillaries

Enhanced WebAssign

Instant feedback and ease of use are just two reasons why WebAssign is the most widely used homework system in higher education. WebAssign's homework delivery system allows you to assign, collect, grade, and record homework assignments via the web. Personal Study Plans provide diagnostic quizzing for each chapter that identifies concepts that students still need to master, and directs them to the appropriate review material. And now, this proven system has been enhanced to include links to textbook sections, video examples, and problem-specific tutorials. For further utility, students will also have the option to purchase an online multimedia eBook of the text. Enhanced WebAssign is more than a homework system—it is a complete learning system for math students. Contact your local representative for ordering details.

Solution Builder
Easily build solution sets for homework or exams using *Solution Builder's* online solutions manual. Visit www.cengage.com/solutionbuilder

PowerLecture with ExamView® (0-538-45207-2)
This CD-ROM provides the instructor with dynamic media tools for teaching. Create, deliver, and customize tests (both print and online) in minutes with *ExamView® Computerized Testing Featuring Algorithmic Equations*. Easily build solution sets for homework or exams using *Solution Builder's* online solutions manual. Microsoft® PowerPoint® lecture slides, figures from the book, and Test Bank (in electronic format) are also included on this CD-ROM.

Text Specific Videos (0-538-79884-X)
Rena Petrello, *Moorpark College*
These 10- to 20-minute problem-solving lessons cover nearly every learning objective from each chapter in the Tussy/Gustafson/Koenig text. Recipient of the "Mark Dever Award for Excellence in Teaching," Rena Petrello presents each lesson using her experience teaching online mathematics courses. It was through this online teaching experience that Rena discovered the lack of suitable content for online instructors, which caused her to develop her own video lessons—and ultimately create this video project. These videos have won four awards: two Telly Awards, one Communicator Award, and one Aurora Award (an international honor). Students will love the additional guidance and support when they have missed a class or when they are preparing for an upcoming quiz or exam. The videos are available for purchase as a set of DVDs or online via CengageBrain.com.

STUDENT RESOURCES

Print Ancillaries

Student Solutions Manual (0-538-49377-1)

Nathan G. Wilson, *St. Louis Community College at Meramec*
The *Student Solutions Manual* provides worked-out solutions to the odd-numbered problems in the text.

Electronic Ancillaries

Enhanced WebAssign

Get instant feedback on your homework assignments with Enhanced WebAssign (assigned by your instructor). Personal Study Plans provide diagnostic quizzing for each chapter that identifies concepts that you still need to master, and directs you to the appropriate review material. This online homework system is easy to use and includes helpful links to textbook sections, video examples, and problem-specific tutorials. For further ease of use, purchase an online multimedia eBook via WebAssign.

Website *www.cengage.com/math/tussy*

Visit us on the web for access to a wealth of learning resources.

ACKNOWLEDGMENTS

We want to express our gratitude to all those who helped with this project: Steve Odrich, Mary Lou Wogan, Paul McCombs, Maria H. Andersen, Sheila Pisa, Laurie McManus, Alexander Lee, Ed Kavanaugh, Karl Hunsicker, Cathy Gong, Dave Ryba, Terry Damron, Marion Hammond, Lin Humphrey, Doug Keebaugh, Robin Carter, Tanja Rinkel, Bob Billups, Jeff Cleveland, Jo Morrison, Sheila White, Jim McClain, Paul Swatzel, Matt Stevenson, Carole Carney, Joyce Low, Rob Everest, David Casey, Heddy Paek, Ralph Tippins, Mo Trad, Eagle Zhuang, and the Citrus College library staff (including Barbara Rugeley) for their help with this project. Your encouragement, suggestions, and insight have been invaluable to us.

We would also like to express our thanks to the Cengage Learning editorial, marketing, production, and design staff for helping us craft this new edition: Charlie Van Wagner, Danielle Derbenti, Gordon Lee, Rita Lombard, Greta Kleinert, Stefanie Beeck, Jennifer Cordoba, Angela Kim, Maureen Ross, Heleny Wong, Jennifer Risden, Vernon Boes, Diane Beasley, and Carol O'Connell and Graphic World.

Additionally, we would like to say that authoring a textbook is a tremendous undertaking. A revision of this scale would not have been possible without the thoughtful feedback and support from the following colleagues listed below. Their contributions to this edition have shaped this revision in countless ways.

Alan S. Tussy
R. David Gustafson
Diane R. Koenig

Advisory Board

J. Donato Fortin, *Johnson and Wales University*
Geoff Hagopian, *College of the Desert*
Jane Wampler, *Housatonic Community College*
Mary Lou Wogan, *Klamath Community College*
Kevin Yokoyama, *College of the Redwoods*

Reviewers

Darla Aguilar, *Pima Community College*
Sheila Anderson, *Housatonic Community College*
David Behrman, *Somerset Community College*
Michael Branstetter, *Hartnell College*
Joseph A. Bruno, Jr., *Community College of Allegheny County*
Joy Conner, *Tidewater Community College*
Ruth Dalrymple, *Saint Philip's College*
John D. Driscoll, *Middlesex Community College*
LaTonya Ellis, *Bishop State Community College*
Steven Felzer, *Lenoir Community College*
Rhoderick Fleming, *Wake Technical Community College*
Heather Gallacher, *Cleveland State University*
Kathirave Giritharan, *John A. Logan College*
Marilyn Green, *Merritt College and Diablo Valley College*
Joseph Guiciardi, *Community College of Allegheny County*
Deborah Hanus, *Brookhaven College*
A.T. Hayashi, *Oxnard College*
Susan Kautz, *Cy-Fair College*
Sandy Lofstock, *Saint Petersburg College–Tarpon Springs*
Mikal McDowell, *Cedar Valley College*
Gregory Perkins, *Hartnell College*
Euguenia Peterson, *City Colleges of Chicago–Richard Daley*
Carol Ann Poore, *Hinds Community College*
Christopher Quarles, *Shoreline Community College*
George Reed, *Angelina College*
John Squires, *Cleveland State Community College*
Sharon Testone, *Onondaga Community College*
Bill Thompson, *Red Rocks Community College*
Donna Tupper, *Community College of Baltimore County–Essex*
Andreana Walker, *Calhoun Community College*
Jane Wampler, *Housatonic Community College*
Mary Young, *Brookdale Community College*

Focus Groups

David M. Behrman, *Somerset Community College*
Eric Compton, *Brookdale Community College*
Nathalie Darden, *Brookdale Community College*
Joseph W. Giuciardi, *Community College of Allegheny County*
Cheryl Hobneck, *Illinois Valley Community College*
Todd J. Hoff, *Wisconsin Indianhead Technical College*
Jack Keating, *Massasoit Community College*
Russ Alan Killingsworth, *Seattle Pacific University*
Lynn Marecek, *Santa Ana College*
Lois Martin, *Massasoit Community College*
Chris Mirbaha, *The Community College of Baltimore County*
K. Maggie Pasqua, *Brookdale Community College*
Patricia C. Rome, *Delgado Community College*
Patricia B. Roux, *Delgado Community College*
Rebecca Rozario, *Brookdale Community College*
Barbara Tozzi, *Brookdale Community College*
Arminda Wey, *Brookdale Community College*
Valerie Wright, *Central Piedmont Community College*

Reviewers of Previous Editions

Cedric E. Atkins, *Mott Community College*
William D. Barcus, *SUNY, Stony Brook*
Kathy Bernunzio, *Portland Community College*
Linda Bettie, *Western New Mexico University*
Girish Budhwar, *United Tribes Technical College*
Sharon Camner, *Pierce College–Fort Steilacoom*
Robin Carter, *Citrus College*
John Coburn, *Saint Louis Community College–Florissant Valley*
Sally Copeland, *Johnson County Community College*
Ann Corbeil, *Massasoit Community College*
Ben Cornelius, *Oregon Institute of Technology*
Carolyn Detmer, *Seminole Community College*
James Edmondson, *Santa Barbara Community College*
David L. Fama, *Germanna Community College*
Maggie Flint, *Northeast State Technical Community College*
Charles Ford, *Shasta College*
Barbara Gentry, *Parkland College*
Kathirave Giritharan, *John A. Logan College*
Michael Heeren, *Hamilton College*
Laurie Hoecherl, *Kishwaukee College*
Judith Jones, *Valencia Community College*
Therese Jones, *Amarillo College*
Joanne Juedes, *University of Wisconsin–Marathon County*
Dennis Kimzey, *Rogue Community College*
Monica C. Kurth, *Scott Community College*
Sally Leski, *Holyoke Community College*
Sandra Lofstock, *St. Petersberg College–Tarpon Springs Center*
Elizabeth Morrison, *Valencia Community College*
Jan Alicia Nettler, *Holyoke Community College*
Marge Palaniuk, *United Tribes Technical College*
Scott Perkins, *Lake-Sumter Community College*
Angela Peterson, *Portland Community College*
Jane Pinnow, *University of Wisconsin–Parkside*
J. Doug Richey, *Northeast Texas Community College*
Angelo Segalla, *Orange Coast College*
Eric Sims, *Art Institute of Dallas*
Lee Ann Spahr, *Durham Technical Community College*
Annette Squires, *Palomar College*
John Strasser, *Scottsdale Community College*
June Strohm, *Pennsylvania State Community College–Dubois*
Rita Sturgeon, *San Bernardino Valley College*
Stuart Swain, *University of Maine at Machias*
Celeste M. Teluk, *D'Youville College*
Jo Anne Temple, *Texas Technical University*
Sharon Testone, *Onondaga Community College*
Marilyn Treder, *Rochester Community College*
Sven Trenholm, *Herkeimer County Community College*
Thomas Vanden Eynden, *Thomas More College*
Stephen Whittle, *Augusta State University*
Mary Lou Wogan, *Klamath Community College*

ABOUT THE AUTHORS

Alan S. Tussy

Alan Tussy teaches all levels of developmental mathematics at Citrus College in Glendora, California. He has written nine math books—a paperback series and a hardcover series. A creative and visionary teacher who maintains a keen focus on his students' greatest challenges, Alan Tussy is an extraordinary author, dedicated to his students' success. Alan received his Bachelor of Science degree in Mathematics from the University of Redlands and his Master of Science degree in Applied Mathematics from California State University, Los Angeles. He has taught up and down the curriculum from Prealgebra to Differential Equations. He is currently focusing on the developmental math courses. Professor Tussy is a member of the American Mathematical Association of Two-Year Colleges.

R. David Gustafson

R. David Gustafson is Professor Emeritus of Mathematics at Rock Valley College in Illinois and coauthor of several best-selling math texts, including Gustafson/Frisk's *Beginning Algebra, Intermediate Algebra, Beginning and Intermediate Algebra: A Combined Approach, College Algebra,* and the Tussy/Gustafson developmental mathematics series. His numerous professional honors include Rock Valley Teacher of the Year and Rockford's Outstanding Educator of the Year. He earned a Master of Arts from Rockford College in Illinois, as well as a Master of Science from Northern Illinois University.

Diane R. Koenig

Diane Koenig received a Bachelor of Science degree in Secondary Math Education from Illinois State University in 1980. She began her career at Rock Valley College in 1981, when she became the Math Supervisor for the newly formed Personalized Learning Center. Earning her Master's Degree in Applied Mathematics from Northern Illinois University, Ms. Koenig in 1984 had the distinction of becoming the first full-time woman mathematics faculty member at Rock Valley College. In addition to being nominated for AMATYC's Excellence in Teaching Award, Diane Koenig was chosen as the Rock Valley College Faculty of the Year by her peers in 2005, and, in 2006, she was awarded the NISOD Teaching Excellence Award as well as the Illinois Mathematics Association of Community Colleges Award for Teaching Excellence. In addition to her teaching, Ms. Koenig has been an active member of the Illinois Mathematics Association of Community Colleges (IMACC). As a member, she has served on the board of directors, on a state-level task force rewriting the course outlines for the developmental mathematics courses, and as the association's newsletter editor.

APPLICATIONS INDEX

Examples that are applications are shown with boldface page numbers.
Exercises that are applications are shown with lightface page numbers.

Study Skills Workshop

OBJECTIVES

1 Make the Commitment

2 Prepare to Learn

3 Manage Your Time

4 Listen and Take Notes

5 Build a Support System

6 Do Your Homework

7 Prepare for the Test

SUCCESS IN YOUR COLLEGE COURSES requires more than just mastery of the content. The development of strong study skills and disciplined work habits plays a crucial role as well. Good note-taking, listening, test-taking, team-building, and time management skills are habits that can serve you well, not only in this course, but throughout your life and into your future career. Students often find that the approach to learning that they used for their high school classes no longer works when they reach college. In this Study Skills Workshop, we will discuss ways of improving and fine-tuning your study skills, providing you with the best chance for a successful college experience.

1 Make the Commitment

Starting a new course is exciting, but it also may be a little frightening. Like any new opportunity, in order to be successful, it will require a commitment of both time and resources. You can decrease the anxiety of this commitment by having a plan to deal with these added responsibilities.

Set Your Goals for the Course.   Explore the reasons why you are taking this course. What do you hope to gain upon completion? Is this course a prerequisite for further study in mathematics? Maybe you need to complete this course in order to begin taking coursework related to your field of study. No matter what your reasons, setting goals for yourself will increase your chances of success. Establish your ultimate goal and then break it down into a series of smaller goals; it is easier to achieve a series of short-term goals rather than focusing on one larger goal.

Keep a Positive Attitude.   Since your level of effort is significantly influenced by your attitude, strive to maintain a positive mental outlook throughout the class. From time to time, remind yourself of the ways in which you will benefit from passing the course. Overcome feelings of stress or math anxiety with extra preparation, campus support services, and activities you enjoy. When you accomplish short-term goals such as studying for a specific period of time, learning a difficult concept, or completing a homework assignment, reward yourself by spending time with friends, listening to music, reading a novel, or playing a sport.

Attend Each Class.   Many students don't realize that missing even one class can have a great effect on their grade. Arriving late takes its toll as well. If you are just a few minutes late, or miss an entire class, you risk getting behind. So, keep these tips in mind.

- Arrive on time, or a little early.
- If you must miss a class, get a set of notes, the homework assignments, and any handouts that the instructor may have provided for the day that you missed.
- Study the material you missed. Take advantage of the help that comes with this textbook, such as the video examples and problem-specific tutorials.

Now Try This

1. List six ways in which you will benefit from passing this course.

2. List six short-term goals that will help you achieve your larger goal of passing this course. For example, you could set a goal to read through the entire *Study Skills Workshop* within the first 2 weeks of class or attend class regularly and on time. (*Success Tip:* Revisit this action item once you have read through all seven *Study Skills Workshop* learning objectives.)

3. List some simple ways you can reward yourself when you complete one of your short-term class goals.

4. Plan ahead! List five possible situations that could cause you to be late for class or miss a class. (Some examples are parking/traffic delays, lack of a babysitter, oversleeping, or job responsibilities.) What can you do ahead of time so that these situations won't cause you to be late or absent?

2 Prepare to Learn

Many students believe that there are two types of people—those who are good at math and those who are not—and that this cannot be changed. This is not true! You can increase your chances for success in mathematics by taking time to prepare and taking inventory of your skills and resources.

Discover Your Learning Style.   Are you a visual, verbal, or auditory learner? The answer to this question will help you determine how to study, how to complete your homework, and even where to sit in class. For example, visual-verbal learners learn best by reading and writing; a good study strategy for them is to rewrite notes and examples. However, auditory learners learn best by listening, so listening to the video examples of important concepts may be their best study strategy.

Get to Know Your Textbook and Its Resources.   You have made a significant investment in your education by purchasing this book and the resources that accompany it. It has been designed with you in mind. Use as many of the features and resources as possible in ways that best fit your learning style.

Know What Is Expected.   Your course syllabus maps out your instructor's expectations for the course. Read the syllabus completely and make sure you understand all that is required. If something is not clear, contact your instructor for clarification.

Organize Your Notebook.   You will definitely appreciate a well-organized notebook when it comes time to study for the final exam. So let's start now! Refer to your syllabus and create a separate section in the notebook for each chapter (or unit of study) that your class will cover this term. Now, set a standard order within each section. One recommended order is to begin with your class notes, followed by your completed homework assignments, then any study sheets or handouts, and, finally, all graded quizzes and tests.

Now Try This

1. To determine what type of learner you are, take the *Learning Style Survey* at http://www.metamath.com/multiple/multiple_choice_questions.html. You may also wish to take the *Index of Learning Styles Questionnaire* at http://www.engr.ncsu.edu/learningstyles/ilsweb.html, which will help you determine your learning type and offer study suggestions by type. List what you learned from taking these surveys. How will you use this information to help you succeed in class?

2. Complete the *Study Skills Checklists* found at the end of sections 1–4 of Chapter 1 in order to become familiar with the many features that can enhance your learning experience using this book.

3. Read through the list of Student Resources found in the Preface of this book. Which ones will you use in this class?

4. Read through your syllabus and write down any questions that you would like to ask your instructor.

5. Organize your notebook using the guidelines given above. Place your syllabus at the very front of your notebook so that you can see the dates over which the material will be covered and for easy reference throughout the course.

3 Manage Your Time

© iStockphoto.com/Yiannos Ioannou

Now that you understand the importance of attending class, how will you make time to study what you have learned while attending? Much like learning to play the piano, math skills are best learned by practicing a little every day.

Make the Time.   In general, 2 hours of independent study time is recommended for every hour in the classroom. If you are in class 3 hours per week, plan on 6 hours per week for reviewing your notes and completing your homework. It is best to schedule this time over the length of a week rather than to try to cram everything into one or two marathon study days.

Prioritize and Make a Calendar.   Because daily practice is so important in learning math, it is a good idea to set up a calendar that lists all of your time commitments, as well as the time you will need to set aside for studying and doing your homework. Consider how you spend your time each week and prioritize your tasks by importance. During the school term, you may need to reduce or even eliminate certain nonessential tasks in order to meet your goals for the term.

Maximize Your Study Efforts.   Using the information you learned from determining your learning style, set up your blocks of study time so that you get the most out of these sessions. Do you study best in groups or do you need to study alone to get anything done? Do you learn best when you schedule your study time in 30-minute time blocks or do you need at least an hour before the information kicks in? Consider your learning style to set up a schedule that truly suits your needs.

Avoid Distractions.   Between texting and social networking, we have so many opportunities for distraction and procrastination. On top of these, there are the distractions of TV, video games, and friends stopping by to hang out. Once you have set your schedule, honor your study times by turning off any electronic devices and letting your voicemail take messages for you. After this time, you can reward yourself by returning phone calls and messages or spending time with friends after the pressure of studying has been lifted.

Now Try This

1. Keep track of how you spend your time for a week. Rate each activity on a scale from 1 (not important) to 5 (very important). Are there any activities that you need to reduce or eliminate in order to have enough time to study this term?

2. List three ways that you learn best according to your learning style. How can you use this information when setting up your study schedule?

3. Download the *Weekly Planner Form* from www.cengage.com/math/tussy and complete your schedule. If you prefer, you may set up a schedule in Google Calendar (calendar.google.com), www.rememberthemilk.com, your cell, or your email system. Many of these have the ability to set up useful reminders and to-do lists in addition to a weekly schedule.

4. List three ways in which you are most often distracted. What can you do to avoid these distractions during your scheduled study times?

4 Listen and Take Notes

© iStockphoto.com/Jacob Wackerhausen

Make good use of your class time by listening and taking notes. Because your instructor will be giving explanations and examples that may not be found in your textbook, as well as other information about your course (test dates, homework assignments, and so on), it is important that you keep a written record of what was said in class.

Listen Actively.   Listening in class is different from listening in social situations because it requires that you be an *active* listener. Since it is impossible to write down everything that is said in class, you need to exercise your active listening skills to learn to write down what is *important*. You can spot important material by listening for cues from your instructor. For instance, pauses in lectures or statements from your instructor such as "This is really important" or "This is a question that shows up frequently on tests" are indications that you should be paying special attention. Listen with a pencil (or highlighter) in hand, ready to record or highlight (in your textbook) any examples, definitions, or concepts that your instructor discusses.

Take Notes You Can Use.   Don't worry about making your notes really neat. After class you can rework them into a format that is more useful to you. However, you should organize your notes as much as possible as you write them. Copy the examples your instructor uses in class. Circle or star any key concepts or definitions that your instructor mentions while explaining the example. Later, your homework problems will look a lot like the examples given in class, so be sure to copy each of the steps in detail.

Listen with an Open Mind.   Even if there are concepts presented that you feel you already know, keep tuned in to the presentation of the material and look for a deeper understanding of the material. If the material being presented is something that has been difficult for you in the past, listen with an open mind; your new instructor may have a fresh presentation that works for you.

Avoid Classroom Distractions.   Some of the same things that can distract you from your study time can distract you, and others, during class. Because of this, be sure to turn off your cell phone during class. If you take notes on a laptop, log out of your email and social networking sites during class. In addition to these distractions, avoid getting into side conversations with other students. Even if you feel you were only distracted for a few moments, you may have missed important verbal or body language cues about an upcoming exam or hints that will aid in your understanding of a concept.

Now Try This

1. Before your next class, refer to your syllabus and read the section(s) that will be covered. Make a list of the terms that you predict your instructor will think are most important.

2. During your next class, bring your textbook and keep it open to the sections being covered. If your instructor mentions a definition, concept, or example that is found in your text, highlight it.

3. Find at least one classmate with whom you can review notes. Make an appointment to compare your class notes as soon as possible after the class. Did you find differences in your notes?

4. Go to www.cengage.com/math/tussy and read the *Reworking Your Notes* handout. Complete the action items given in this document.

5 Build a Support System

© iStockphoto.com/Chris Schmidt

Have you ever had the experience where you understand everything that your instructor is saying in class, only to go home and try a homework problem and be completely stumped? This is a common complaint among math students. The key to being a successful math student is to take care of these problems before you go on to tackle new material. That is why you should know what resources are available outside of class.

Make Good Use of Your Instructor's Office Hours.   The purpose of your instructor's office hours is to be available to help students with questions. Usually these hours are listed in your syllabus and no appointment is needed. When you visit your instructor, have a list of questions and try to pinpoint exactly where in the process you are getting stuck. This will help your instructor answer your questions efficiently.

Use Your Campus Tutoring Services.   Many colleges offer tutorial services for free. Sometimes tutorial assistance is available in a lab setting where you are able to drop in at your convenience. In some cases, you need to make an appointment to see a tutor in advance. Make sure to seek help as soon as you recognize the need, and come to see your tutor with a list of identified problems.

Form a Study Group.   Study groups are groups of classmates who meet outside of class to discuss homework problems or study for tests. Get the most out of your study group by following these guidelines:

- Keep the group small—a maximum of four committed students. Set a regularly scheduled meeting day, time, and place.
- Find a place to meet where you can talk and spread out your work.
- Members should attempt all homework problems before meeting.
- All members should contribute to the discussion.
- When you meet, practice verbalizing and explaining problems and concepts to each other. The best way to really learn a topic is by teaching it to someone else.

Now Try This

1. Refer to your syllabus. Highlight your instructor's office hours and location. Next, pay a visit to your instructor during office hours this week and introduce yourself. (***Success Tip:*** Program your instructor's office phone number and email address into your cell phone or email contact list.)

2. Locate your campus tutoring center or math lab. Write down the office hours, phone number, and location on your syllabus. Drop by or give them a call and find out how to go about making an appointment with a tutor.

3. Find two to three classmates who are available to meet at a time that fits your schedule. Plan to meet 2 days before your next homework assignment is due and follow the guidelines given above. After your group has met, evaluate how well it worked. Is there anything that the group can do to make it better next time you meet?

4. Download the *Support System Worksheet* at www.cengage.com/math/tussy. Complete the information and keep it at the front of your notebook following your syllabus.

6 Do Your Homework

© iStockphoto.com/djordje zivaljevic

Attending class and taking notes are important, but the only way that you are really going to learn mathematics is by completing your homework. Sitting in class and listening to lectures will help you to place concepts in short-term memory, but in order to do well on tests and in future math classes, you want to put these concepts in long-term memory. When completed regularly, homework assignments will help with this.

Give Yourself Enough Time.   In Objective 3, you made a study schedule, setting aside 2 hours for study and homework for every hour that you spend in class. If you are not keeping this schedule, make changes to ensure that you can spend enough time outside of class to learn new material.

Review Your Notes and the Worked Examples from Your Text.   In Objective 4, you learned how to take useful notes. Before you begin your homework, review or rework your notes. Then, read the sections in your textbook that relate to your homework problems, paying special attention to the worked examples. With a pencil in hand, work the *Self Check* and *Now Try* problems that are listed next to the examples in your text. Using the worked example as a guide, solve these problems and try to understand each step. As you read through your notes and your text, keep a list of anything that you don't understand.

Now Try Your Homework Problems.   Once you have reviewed your notes and the textbook worked examples, you should be able to successfully manage the bulk of your homework assignment easily. When working on your homework, keep your textbook and notes close by for reference. If you have trouble with a homework question, look through your textbook and notes to see if you can identify an example that is similar to the homework question. See if you can apply the same steps to your homework problem. If there are places where you get stuck, add these to your list of questions.

Get Answers to Your Questions.   At least one day before your assignment is due, seek help with the questions you have been listing. You can contact a classmate for assistance, make an appointment with a tutor, or visit your instructor during office hours.

Now Try This

1. Review your study schedule. Are you following it? If not, what changes can you make to adhere to the rule of 2 hours of homework and study for every hour of class?

2. Find five homework problems that are similar to the worked examples in your textbook. Were there any homework problems in your assignment that didn't have a worked example that was similar? (**Success Tip:** Look for the *Now Try* and *Guided Practice* features for help linking problems to worked examples.)

3. As suggested in this Objective, make a list of questions while completing your homework. Visit your tutor or your instructor with your list of questions and ask one of them to work through these problems with you.

4. Go to www.cengage.com/math/tussy and read the *Study and Memory Techniques* handout. List the techniques that will be most helpful to you in your math course.

7 Prepare for the Test

Taking a test does not need to be an unpleasant experience. Use your time management, organization, and these test-taking strategies to make this a learning experience and improve your score.

Make Time to Prepare.   Schedule at least four daily 1-hour sessions to prepare specifically for your test.

Four days before the test: Create your own study sheet using your reworked notes. Imagine you could bring one $8\frac{1}{2} \times 11$ sheet of paper to your test. What would you write on that sheet? Include all the key definitions, rules, steps, and formulas that were discussed in class or covered in your reading. Whenever you have the opportunity, pull out your study sheet and review your test material.

Three days before the test: Create a sample test using the in-class examples from your notes and reading material. As you review and work these examples, make sure you understand how each example relates to the rules or definitions on your study sheet. While working through these examples, you may find that you forgot a concept that should be on your study sheet. Update your study sheet and continue to review it.

Two days before the test: Use the *Chapter Test* from your textbook or create one by matching problems from your text to the example types from your sample test. Now, with your book closed, take a timed trial test. When you are done, check your answers. Make a list of the topics that were difficult for you and review or add these to your study sheet.

One day before the test: Review your study sheet once more, paying special attention to the material that was difficult for you when you took your practice test the day before. Be sure you have all the materials that you will need for your test laid out ahead of time (two sharpened pencils, a good eraser, possibly a calculator or protractor, and so on). The most important thing you can do today is get a good night's rest.

Test day: Review your study sheet, if you have time. Focus on how well you have prepared and take a moment to relax. When taking your test, complete the problems that you are sure of first. Skip the problems that you don't understand right away, and return to them later. Bring a watch or make sure there will be some kind of time-keeping device in your test room so that you can keep track of your time. Try not to spend too much time on any one problem.

Now Try This

1. Create a study schedule using the guidelines given above.

2. Read the *Preparing for a Test* handout at www.cengage.com/math/tussy.

3. Read the *Taking the Test* handout at www.cengage.com/math/tussy.

4. After your test has been returned and scored, read the *Analyzing Your Test Results* handout at www.cengage.com/math/tussy.

5. Take time to reflect on your homework and study habits after you have received your test score. What actions are working well for you? What do you need to improve?

6. To prepare for your final exam, read the *Preparing for Your Final Exam* handout at www.cengage.com/math/tussy. Complete the action items given in this document.

Whole Numbers

Comstock Images/Getty Images

from *Campus to Careers*

Landscape Designer

Landscape designers make outdoor places more beautiful and useful. They work on all types of projects. Some focus on yards and parks, others on land around buildings and highways. The training of a landscape designer should include botany classes to learn about plants; art classes to learn about color, line, and form; and mathematics classes to learn how to take measurements and keep business records.

In **Problem 104** of **Study Set 1.5,** you will see how a landscape designer uses division to determine the number of pine trees that are needed to form a windscreen for a flower garden.

JOB TITLE:
Landscape designer

EDUCATION: A bachelor's degree in landscape design. Most states require a license.

JOB OUTLOOK: Excellent

ANNUAL EARNINGS: Salaries range from $45,000–$70,000.

FOR MORE INFORMATION
www.ashs.org/careers/profiles/
landscape.lasso

Objectives

1 Identify the place value of a digit in a whole number.

2 Write whole numbers in words and in standard form.

3 Write a whole number in expanded form.

4 Compare whole numbers using inequality symbols.

5 Round whole numbers.

6 Read tables and graphs involving whole numbers.

SECTION 1.1

An Introduction to the Whole Numbers

The **whole numbers** are 0, 1, 2, 3, 4, 5, 6, 7, 8, 9, 10, 11, 12, and so on. They are used to answer questions such as How many?, How fast?, and How far?

- The movie *Titanic* won 11 Academy Awards.
- The average American adult reads at a rate of 250 to 300 words per minute.
- The driving distance from New York City to Los Angeles is 2,786 miles.

The *set of whole numbers* is written using **braces { }**, as shown below. The three dots indicate that the list continues forever—there is no largest whole number. The smallest whole number is 0.

The Set of Whole Numbers

{0, 1, 2, 3, 4, 5, 6, 7, 8, 9, 10, 11, 12, . . .}

---

1 Identify the place value of a digit in a whole number.

When a whole number is written using the **digits** 0, 1, 2, 3, 4, 5, 6, 7, 8, 9, it is said to be in **standard form** (also called **standard notation**). The position of a digit in a whole number determines its **place value.** In the number 325, the 5 is in the *ones column*, the 2 is in the *tens column*, and the 3 is in the *hundreds column*.

```
              Tens column
Hundreds column ┐ │ ┌ Ones column
                3 2 5
```

To make large whole numbers easier to read, we use commas to separate their digits into groups of three, called **periods.** Each period has a name, such as *ones, thousands, millions, billions,* and *trillions.* The following **place-value chart** shows the place value of each digit in the number 2,691,537,557,000, which is read as:

Two trillion, six hundred ninety-one billion, five hundred thirty-seven million, five hundred fifty-seven thousand

In 2007, the federal government collected a total of $2,691,537,557,000 in taxes.
(Source: Internal Revenue Service.)

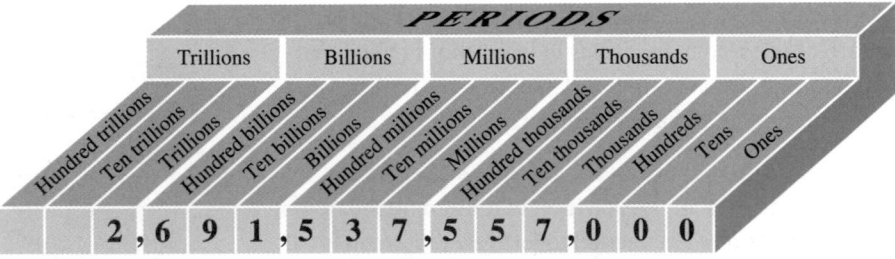

Each of the 5's in 2,691,537,557,000 has a different place value because of its position. The place value of the red 5 is 5 *hundred millions.* The place value of the blue 5 is 5 *hundred thousands*, and the place value of the green 5 is 5 *ten thousands*.

The Language of Algebra   As we move to the left in the chart, the place value of each column is 10 times greater than the column directly to its right. This is why we call our number system the *base-10 number system.*

EXAMPLE 1 *Airports* Hartsfield-Jackson Atlanta International Airport is the busiest airport in the United States, handling 89,379,287 passengers in 2007. (Source: Airports Council International–North America)

a. What is the place value of the digit 3?
b. Which digit tells the number of millions?

Strategy We will begin in the ones column of 89,379,287. Then, moving to the left, we will name each column (ones, tens, hundreds, and so on) until we reach the digit 3.

WHY It's easier to remember the names of the columns if you begin with the smallest place value and move to the columns that have larger place values.

Solution

a. 89,379,287   Say, "Ones, tens, hundreds, thousands, ten thousands, hundred
                        thousands" as you move from column to column.

3 hundred thousands is the place value of the digit 3.

b. 89,379,287

The digit 9 is in the millions column.

> **The Language of Algebra** Each of the worked examples in this textbook includes a *Strategy* and *Why* explanation. A **strategy** is a plan of action to follow to solve the given problem.

2 Write whole numbers in words and in standard form.

Since we use whole numbers so often in our daily lives, it is important to be able to read and write them.

Reading and Writing Whole Numbers

> To write a whole number in words, start from the left. Write the number in each period followed by the name of the period (except for the *ones period*, which is not used). Use commas to separate the periods.
>
> To read a whole number out loud, follow the same procedure. The commas are read as slight pauses.

> **The Language of Algebra** The word *and* should not be said when reading a whole number. It should only be used when reading a mixed number such as $5\frac{1}{2}$ (five *and* one-half) or a decimal such as 3.9 (three *and* nine-tenths).

EXAMPLE 2 Write each number in words:
a. 63   **b.** 499   **c.** 89,015   **d.** 6,070,534

Strategy For the larger numbers in parts c and d, we will name the periods from right to left to find the *greatest* period.

WHY To write a whole number in words, we must give the name of each period (except for the ones period). Finding the largest period helps to start the process.

Solution

a. 63 is written: *sixty-three*.   Use a hyphen to write whole numbers from 21 to 99 in
                                        words (except for 30, 40, 50, 60, 70, 80, and 90).

b. 499 is written: *four hundred ninety-nine*.

Self Check 1

CELL PHONES In 2007, there were 255,395,600 cellular telephone subscribers in the United States. (Source: International Telecommunication Union)
a. What is the place value of the digit 2?
b. Which digit tells the number of hundred thousands?

Now Try Problem 23

Self Check 2

Write each number in words:
a. 42
b. 798
c. 97,053
d. 23,000,017

Now Try Problems 31, 33, and 35

c. Thousands   Ones   Say the names of the periods, working from right to left.

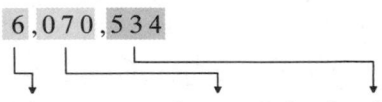

89 , 015

Eighty-nine **thousand**, **fifteen**   *We do not use a hyphen to write numbers between 1 and 20, such as 15. The ones period is not written.*

d. Millions Thousands Ones   Say the names of the periods, working from right to left.

6,070,534

Six **million**, seventy **thousand**, five **hundred** thirty-four.   *The ones period is not written.*

Caution!   Two numbers, 40 and 90, are often misspelled: write *forty* (*not fourty*) and *ninety* (*not ninty*).

Self Check 3

Write each number in standard form:
a. *Two hundred three thousand, fifty-two*
b. *Nine hundred forty-six million, four hundred sixteen thousand, twenty-two*
c. *Three million, five hundred seventy-nine*

Now Try **Problems 39 and 45**

EXAMPLE 3   Write each number in standard form:

a. *Twelve thousand, four hundred seventy-two*
b. *Seven hundred one million, thirty-six thousand, six*
c. *Forty-three million, sixty-eight*

Strategy We will locate the commas in the written-word form of each number.

WHY When a whole number is written in words, commas are used to separate periods.

Solution

a. Twelve thousand , four hundred seventy-two

12, 472

b. Seven hundred one million , thirty-six thousand , six

701,036,006

c. Forty-three million , sixty-eight   *The written-word form does not mention the thousands period.*

43,000,068   *If a period is not named, three zeros hold its place.*

Success Tip   Four-digit whole numbers are sometimes written without a comma. For example, we may write 3,911 or 3911 to represent three thousand, nine hundred eleven.

---

3   **Write a whole number in expanded form.**

In the number 6,352, the digit 6 is in the thousands column, 3 is in the hundreds column, 5 is in the tens column, and 2 is in the ones (or units) column. The meaning of 6,352 becomes clear when we write it in **expanded form** (also called **expanded notation**).

$$6{,}352 = 6 \text{ thousands} + 3 \text{ hundreds} + 5 \text{ tens} + 2 \text{ ones}$$

or

$$6{,}352 = \quad 6{,}000 \quad + \quad 300 \quad + \quad 50 \quad + \quad 2$$

EXAMPLE 4 Write each number in expanded form:
a. 85,427 **b.** 1,251,609

Self Check 4
Write 708,413 in expanded form.
Now Try Problems 49, 53, and 57

Strategy Working from left to right, we will give the place value of each digit and combine them with + symbols.

WHY The term *expanded form* means to write the number as an addition of the place values of each of its digits.

Solution

a. The expanded form of 85,427 is:

   8 ten thousands +**5** thousands +**4** hundreds +**2** tens +**7** ones

   which can be written as:

   80,000 + 5,000 + 400 + 20 + 7

b. The expanded form of 1,251,609 is:

   1 million + **2** hundred thousands + **5** ten thousands + **1** thousand + **6** hundreds + **0** tens + **9** ones

   Since 0 tens is zero, the expanded form can also be written as:

   1 million + **2** hundred thousands + **5** ten thousands + **1** thousand + **6** hundreds + **9** ones

   which can be written as:

   1,000,000 + 200,000 + 50,000 + 1,000 + 600 + 9

4 Compare whole numbers using inequality symbols.

Whole numbers can be shown by drawing points on a **number line.** Like a ruler, a number line is straight and has uniform markings. To construct a number line, we begin on the left with a point on the line representing the number 0. This point is called the **origin.** We then move to the right, drawing equally spaced marks and labeling them with whole numbers that increase in value. The arrowhead at the right indicates that the number line continues forever.

A number line

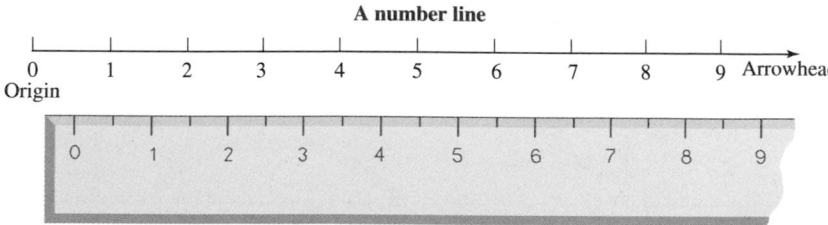

Using a process known as **graphing,** we can represent a single number or a set of numbers on a number line. **The graph of a number** is the point on the number line that corresponds to that number. *To graph a number* means to locate its position on the number line and highlight it with a heavy dot. The graphs of 5 and 8 are shown on the number line below.

As we move to the right on the number line, the numbers increase in value. Because 8 lies to the right of 5, we say that 8 is greater than 5. The **inequality symbol** > ("is greater than") can be used to write this fact:

   8 > 5   Read as "8 is greater than 5."

Since 8 > 5, it is also true that 5 < 8. We read this as "5 is less than 8."

Inequality Symbols

> means *is greater than*

< means *is less than*

Success Tip To tell the difference between these two inequality symbols, remember that they always point to the smaller of the two numbers involved.

8 > 5   5 < 8

└──Points to the──┘
smaller number

Self Check 5

Place an < or an > symbol in the box to make a true statement:

a. 12 ☐ 4

b. 7 ☐ 10

Now Try Problems 59 and 61

EXAMPLE 5 Place an < or an > symbol in the box to make a true statement: **a.** 3 ☐ 7   **b.** 18 ☐ 16

Strategy To pick the correct inequality symbol to place between a pair of numbers, we need to determine the position of each number on the number line.

WHY For any two numbers on a number line, the number to the *left* is the smaller number and the number to the *right* is the larger number.

Solution

a. Since 3 is to the left of 7 on the number line, we have 3 < 7.

b. Since 18 is to the right of 16 on the number line, we have 18 > 16.

5 Round whole numbers.

When we don't need exact results, we often round numbers. For example, when a teacher with 36 students orders 40 textbooks, he has rounded the actual number to the *nearest ten,* because 36 is closer to 40 than it is to 30. We say 36, rounded to the nearest 10, is 40. This process is called **rounding up.**

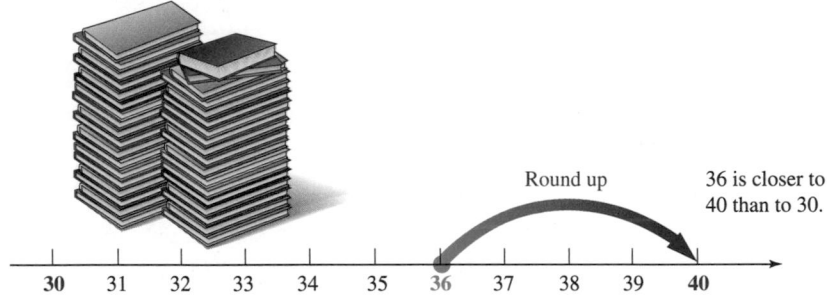

Round up   36 is closer to 40 than to 30.

30 31 32 33 34 35 36 37 38 39 40

When a geologist says that the height of Alaska's Mount McKinley is "about 20,300 feet," she has rounded to the *nearest hundred,* because its actual height of 20,320 feet is closer to 20,300 than it is to 20,400. We say that 20,320, rounded to the nearest hundred, is 20,300. This process is called **rounding down.**

20,320 is closer to 20,300 than 20,400.
Round down

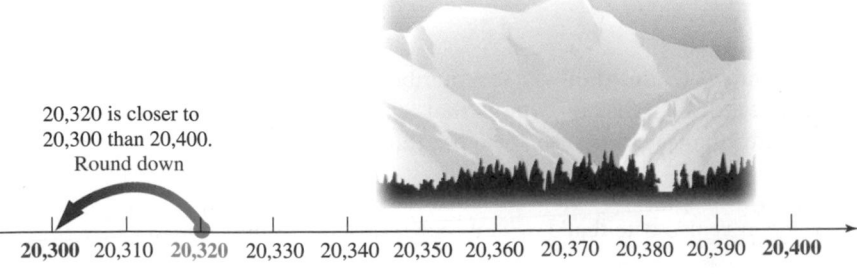

20,300 20,310 20,320 20,330 20,340 20,350 20,360 20,370 20,380 20,390 20,400

> **The Language of Algebra**   When we round a whole number, we are finding an *approximation* of the number. An *approximation* is close to, but not the same as, the exact value.

To round a whole number, we follow an established set of rules. To round a number to the nearest ten, for example, we locate the **rounding digit** in the tens column. If the **test digit** to the right of that column (the digit in the ones column) is 5 or greater, we *round up* by increasing the tens digit by 1 and replacing the test digit with 0. If the test digit is less than 5, we *round down* by leaving the tens digit unchanged and replacing the test digit with 0.

EXAMPLE 6   Round each number to the nearest ten: **a.** 3,761 **b.** 12,087

Strategy We will find the digit in the tens column and the digit in the ones column.

WHY To round to the nearest ten, the digit in the tens column is the rounding digit and the digit in the ones column is the test digit.

Solution

a. We find the rounding digit in the tens column, which is 6. Then we look at the test digit to the right of 6, which is the 1 in the ones column. Since $1 < 5$, we round down by leaving the 6 unchanged and replacing the test digit with 0.

┌─ Rounding digit: tens column   ┌─ Keep the rounding digit: Do not add 1.

3,761   3,761

└─ Test digit: 1 is less than 5.   └─ Replace with 0.

Thus, 3,761 rounded to the nearest ten is 3,760.

b. We find the rounding digit in the tens column, which is 8. Then we look at the test digit to the right of 8, which is the 7 in the ones column. Because 7 is 5 or greater, we round up by adding 1 to 8 and replacing the test digit with 0.

┌─ Rounding digit: tens column   ┌─ Add 1.

12,087   12,087

└─ Test digit: 7 is 5 or greater.   └─ Replace with 0.

Thus, 12,087 rounded to the nearest ten is 12,090.

A similar method is used to round numbers to the nearest hundred, the nearest thousand, the nearest ten thousand, and so on.

Rounding a Whole Number

1. To round a number to a certain place value, locate the **rounding digit** in that place.
2. Look at the **test digit,** which is directly to the right of the rounding digit.
3. If the test digit is 5 or greater, round up by adding 1 to the rounding digit and replacing all of the digits to its right with 0.

   If the test digit is less than 5, replace it and all of the digits to its right with 0.

EXAMPLE 7   Round each number to the nearest hundred:

a. 18,349   **b.** 7,960

Strategy We will find the rounding digit in the hundreds column and the test digit in the tens column.

Self Check 6

Round each number to the nearest ten:
a. 35,642
b. 9,756

Now Try Problem 63

Self Check 7

Round 365,283 to the nearest hundred.

Now Try Problems 69 and 71

WHY To round to the nearest hundred, the digit in the hundreds column is the rounding digit and the digit in the tens column is the test digit.

Solution

a. First, we find the rounding digit in the hundreds column, which is 3. Then we look at the test digit 4 to the right of 3 in the tens column. Because $4 < 5$, we round down and leave the 3 in the hundreds column. We then replace the two rightmost digits with 0's.

┌─ Rounding digit: hundreds column   ┌─ Keep the rounding digit: Do not add 1.
18,349   18,349
  └─ Test digit: 4 is less than 5.   └─ Replace with 0's.

Thus, 18,349 rounded to the nearest hundred is 18,300.

b. First, we find the rounding digit in the hundreds column, which is 9. Then we look at the test digit 6 to the right of 9. Because 6 is 5 or greater, we round up and increase 9 in the hundreds column by 1. Since the 9 in the hundreds column represents 900, increasing 9 by 1 represents increasing 900 to 1,000. Thus, we replace the 9 with a 0 and add 1 to the 7 in the thousands column. Finally, we replace the two rightmost digits with 0's.

                                            ┌─ Add 1. Since 9 + 1 = 10, write 0 in this
┌─ Rounding digit: hundreds column   │   column and carry 1 to the next column.
                                     7+1 0
7,960   7,960
  └─ Test digit: 6 is 5 or greater.   └─ Replace with 0s.

Thus, 7,960 rounded to the nearest hundred is 8,000.

> **Caution!** To round a number, use *only* the test digit directly to the right of the rounding digit to determine whether to round up or round down.

U.S. CITIES Round the elevation of Denver:
a. to the nearest hundred feet
b. to the nearest thousand feet

Now Try **Problems 75 and 79**

EXAMPLE 8 *U.S. Cities* In 2007, Denver was the nation's 26th largest city. Round the 2007 population of Denver shown on the sign to:
a. the nearest thousand
b. the nearest hundred thousand

Strategy In each case, we will find the rounding digit and the test digit.

WHY We need to know the value of the test digit to determine whether we round the population up or down.

Solution

a. The rounding digit in the thousands column is 8. Since the test digit 3 is less than 5, we round down. To the nearest thousand, Denver's population in 2007 was 588,000.

b. The rounding digit in the hundred thousands column is 5. Since the test digit 8 is 5 or greater, we round up. To the nearest hundred thousand, Denver's population in 2007 was 600,000.

6 Read tables and graphs involving whole numbers.

The following table is an example of the use of whole numbers. It shows the number of women members of the U.S. House of Representatives for the years 1997–2007.

Year	Number of women members
1997	51
1999	56
2001	60
2003	59
2005	67
2007	71

Source: www.ergd.org/
HouseOfRepresentatives

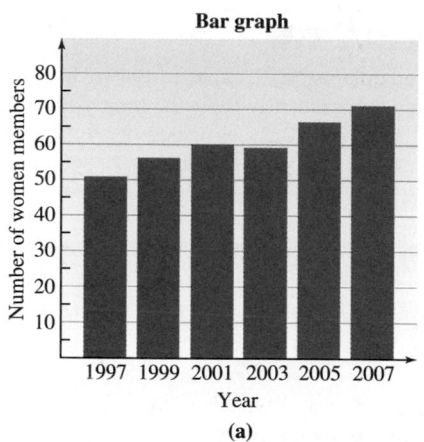

(a)

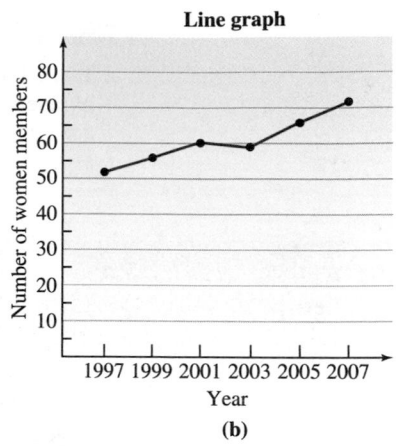

(b)

In figure (a), the information in the table is presented in a **bar graph.** The *horizontal* scale is labeled "Year" and units of 2 years are used. The *vertical* scale is labeled "Number of women members" and units of 10 are used. The bar directly over each year extends to a height that shows the number of women members of the House of Representatives that year.

> **The Language of Algebra**   *Horizontal* is a form of the word *horizon.* Think of the sun setting over the *horizon. Vertical* means in an upright position. Pro basketball player LeBron James' *vertical* leap measures more than 49 inches.

Another way to present the information in the table is with a **line graph.** Instead of using a bar to represent the number of women members, we use a dot drawn at the correct height. After drawing data points for 1997, 1999, 2001, 2003, 2005, and 2007, the points are connected to create the line graph in figure (b).

THINK IT THROUGH   *Re-entry Students*

"A re-entry student is considered one who is the age of 25 or older, or those students that have had a break in their academic work for 5 years or more. Nationally, this group of students is growing at an astounding rate."

Student Life and Leadership Department, University Union, Cal Poly University, San Luis Obispo

Some common concerns expressed by adult students considering returning to school are listed below in Column I. Match each concern to an encouraging reply in Column II.

Column I	Column II
1. I'm too old to learn.	**a.** Many students qualify for some type of financial aid.
2. I don't have the time.	**b.** Taking even a single class puts you one step closer to your educational goal.
3. I didn't do well in school the first time around. I don't think a college would accept me.	**c.** There's no evidence that older students can't learn as well as younger ones.
4. I'm afraid I won't fit in.	**d.** More than 41% of the students in college are older than 25.
5. I don't have the money to pay for college.	**e.** Typically, community colleges and career schools have an open admissions policy.

Source: Adapted from *Common Concerns for Adult Students,* Minnesota Higher Education Services Office

STUDY SKILLS CHECKLIST

Get to Know Your Textbook

Congratulations. You now own a state-of-the-art textbook that has been written especially for you. The following checklist will help you become familiar with the organization of this book. Place a check mark in each box after you answer the question.

☐ Turn to the **Table of Contents** on page v. How many chapters does the book have?

☐ Each chapter of the book is divided into **sections.** How many sections are there in Chapter 1, which begins on page 1?

☐ Learning **Objectives** are listed at the start of each section. How many objectives are there for Section 1.2, which begins on page 15?

☐ Each section ends with a **Study Set.** How many problems are there in Study Set 1.2, which begins on page 29?

☐ Each chapter has a **Chapter Summary and Review.** Which column of the Chapter 1 Summary found on page 114 contains examples?

☐ How many review problems are there for Section 1.1 in the **Chapter 1 Summary and Review,** which begins on page 114?

☐ Each chapter has a **Chapter Test.** How many problems are there in the Chapter 1 Test, which begins on page 132?

☐ Each chapter (except Chapter 1) ends with a **Cumulative Review.** Which chapters are covered by the Cumulative Review which begins on page 221?

Answers: 10, 9, 122, the right, 14, 48, 1–2

SECTION 1.1   STUDY SET

VOCABULARY

Fill in the blanks.

1. The numbers 0, 1, 2, 3, 4, 5, 6, 7, 8, and 9 are the _____.
2. The set of _____ numbers is {0, 1, 2, 3, 4, 5, …}.
3. When we write five thousand eighty-nine as 5,089, we are writing the number in _____ form.
4. To make large whole numbers easier to read, we use commas to separate their digits into groups of three, called _____.
5. When 297 is written as 200 + 90 + 7, we are writing 297 in _____ form.
6. Using a process called *graphing*, we can represent whole numbers as points on a _____ line.

7. The symbols > and < are _____ symbols.
8. If we _____ 627 to the nearest ten, we get 630.

CONCEPTS

9. Copy the following place-value chart. Then enter the whole number 1,342,587,200,946 and fill in the place value names and the periods.

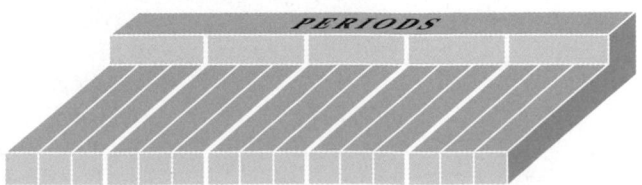

PERIODS

10. a. Insert commas in the proper positions for the following whole number written in standard form: 5467010

b. Insert commas in the proper positions for the following whole number written in words: *seventy-two million four hundred twelve thousand six hundred thirty-five*

11. Write each number in words.

   a. 40   **b.** 90

   c. 68   **d.** 15

12. Write each number in standard form.

   a. 8 ten thousands + 1 thousand + 6 hundreds + 9 tens + 2 ones

   b. 900,000 + 60,000 + 5,000 + 300 + 40 + 7

Graph the following numbers on a number line.

13. 1, 3, 5, 7

14. 0, 2, 4, 6, 8

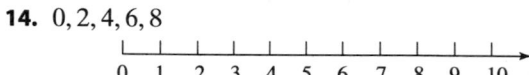

15. 2, 4, 5, 8

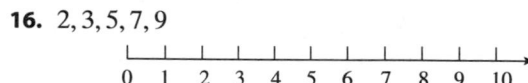

16. 2, 3, 5, 7, 9

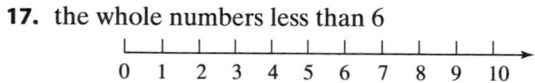

17. the whole numbers less than 6

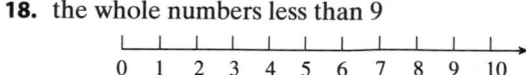

18. the whole numbers less than 9

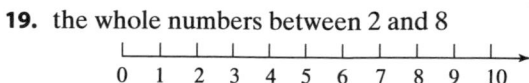

19. the whole numbers between 2 and 8

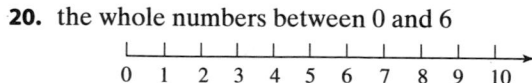

20. the whole numbers between 0 and 6

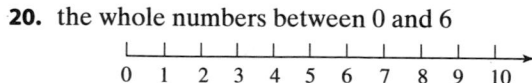

▌ NOTATION

Fill in the blanks.

21. The symbols { }, called _____, are used when writing a set.

22. The symbol > means ___ _____ _____, and the symbol < means ___ ____ _____.

▌ GUIDED PRACTICE

Find the place values. **See Example 1.**

23. Consider the number 57,634.

   a. What is the place value of the digit 3?

   b. What digit is in the thousands column?

   c. What is the place value of the digit 6?

   d. What digit is in the ten thousands column?

24. Consider the number 128,940.

   a. What is the place value of the digit 8?

   b. What digit is in the hundreds column?

   c. What is the place value of the digit 2?

   d. What digit is in the hundred thousands column?

25. WORLD HUNGER On the website Freerice.com, sponsors donate grains of rice to feed the hungry. As of October 2008, there have been 47,167,467,790 grains of rice donated.

   a. What is the place value of the digit 1?

   b. What digit is in the billions place?

   c. What is the place value of the 9?

   d. What digit is in the ten billions place?

26. RECYCLING It is estimated that the number of beverage cans and bottles that were *not* recycled in the United States from January to October of 2008 was 102,780,365,000.

   a. What is the place value of the digit 7?

   b. What digit is in the ten thousands place?

   c. What is the place value of the digit 2?

   d. What digit is in the ten billions place?

Write each number in words. **See Example 2.**

27. 93   **28.** 48

29. 732   **30.** 259

31. 154,302

32. 615,019

33. 14,432,500

34. 104,052,005

35. 970,031,500,104

36. 5,800,010,700

37. 82,000,415

38. 51,000,201,078

Write each number in standard form. **See Example 3.**

39. Three thousand, seven hundred thirty-seven

40. Fifteen thousand, four hundred ninety-two

41. Nine hundred thirty

42. Six hundred forty

43. Seven thousand, twenty-one

44. Four thousand, five hundred

45. Twenty-six million, four hundred thirty-two

46. Ninety-two billion, eighteen thousand, three hundred ninety-nine

***Write each number in expanded form.* See Example 4.**

47. 245   **48.** 518

49. 3,609   **50.** 3,961

51. 72,533

52. 73,009

53. 104,401

54. 570,003

55. 8,403,613

56. 3,519,807

57. 26,000,156

58. 48,000,061

***Place an < or an > symbol in the box to make a true statement.* See Example 5.**

59. a. 11 ☐ 8   **b.** 29 ☐ 54

60. a. 410 ☐ 609   **b.** 3,206 ☐ 3,231

61. a. 12,321 ☐ 12,209   **b.** 23,223 ☐ 23,231

62. a. 178,989 ☐ 178,898 **b.** 850,234 ☐ 850,342

***Round to the nearest ten.* See Example 6.**

63. 98,154   **64.** 26,742

65. 512,967   **66.** 621,116

***Round to the nearest hundred.* See Example 7.**

67. 8,352   **68.** 1,845

69. 32,439   **70.** 73,931

71. 65,981   **72.** 5,346,975

73. 2,580,952   **74.** 3,428,961

***Round each number to the nearest thousand and then to the nearest ten thousand.* See Example 8.**

75. 52,867   **76.** 85,432

77. 76,804   **78.** 34,209

79. 816,492   **80.** 535,600

81. 296,500   **82.** 498,903

▌ TRY IT YOURSELF

83. Round 79,593 to the nearest . . .

   a. ten   **b.** hundred

   c. thousand   **d.** ten thousand

84. Round 5,925,830 to the nearest . . .

   a. thousand   **b.** ten thousand

   c. hundred thousand   **d.** million

85. Round $419,161 to the nearest . . .

   a. $10   **b.** $100

   c. $1,000   **d.** $10,000

86. Round 5,436,483 ft to the nearest . . .

   a. 10 ft   **b.** 100 ft

   c. 1,000 ft   **d.** 10,000 ft

Write each number in standard notation.

87. 4 ten thousands + 2 tens + 5 ones

88. 7 millions + 7 tens + 7 ones

89. 200,000 + 2,000 + 30 + 6

90. 7,000,000,000 + 300 + 50

91. Twenty-seven thousand, five hundred ninety-eight

92. Seven million, four hundred fifty-two thousand, eight hundred sixty

93. Ten million, seven hundred thousand, five hundred six

94. Eighty-six thousand, four hundred twelve

▌ APPLICATIONS

95. GAME SHOWS On *The Price is Right* television show, the winning contestant is the person who comes closest to (without going over) the price of the item up for bid. Which contestant shown below will win if they are bidding on a bedroom set that has a suggested retail price of $4,745?

96. PRESIDENTS The following list shows the ten youngest U.S. presidents and their ages (in years/days) when they took office. Construct a two-column table that presents the data in order, beginning with the youngest president.

J. Polk 49 yr/122 days	U. Grant 46 yr/236 days
G. Cleveland 47 yr/351 days	J. Kennedy 43 yr/236 days
W. Clinton 46 yr/154 days	F. Pierce 48 yr/101 days
M. Filmore 50 yr/184 days	B. Obama 47 yr/169 days
J. Garfield 49 yr/105 days	T. Roosevelt 42 yr/322 days

97. MISSIONS TO MARS The United States, Russia, Europe, and Japan have launched Mars space probes. The graph shows the success rate of the missions, by decade.

a. Which decade had the greatest number of successful or partially successful missions? How many?

b. Which decade had the greatest number of unsuccessful missions? How many?

c. Which decade had the greatest number of missions? How many?

d. Which decade had no successful missions?

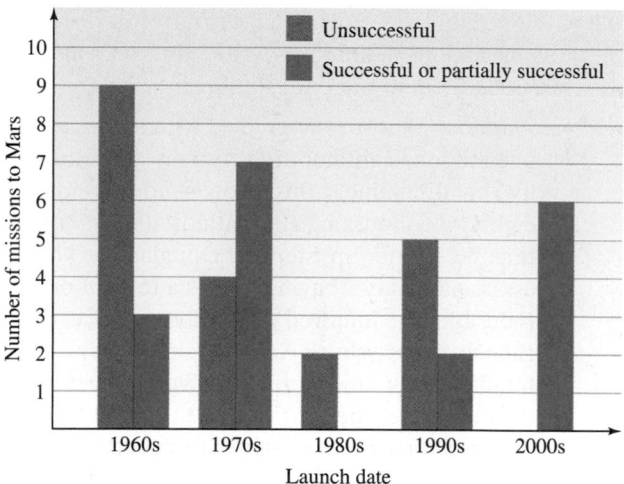

Source: The Planetary Society

98. SPORTS The graph shows the maximum recorded ball speeds for five sports.

a. Which sport had the fastest recorded maximum ball speed? Estimate the speed.

b. Which sport had the slowest maximum recorded ball speed? Estimate the speed.

c. Which sport had the second fastest maximum recorded ball speed? Estimate the speed.

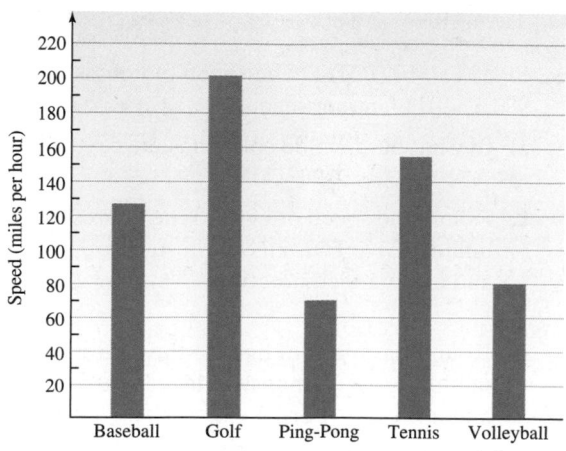

99. COFFEE Complete the bar graph and line graph using the data in the table.

Starbucks Locations

Year	Number
2000	3,501
2001	4,709
2002	5,886
2003	7,225
2004	8,569
2005	10,241
2006	12,440
2007	15,756

Source: Starbucks Company

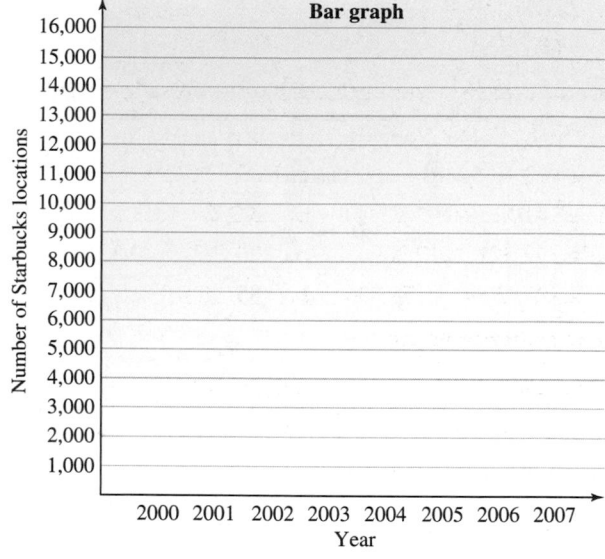

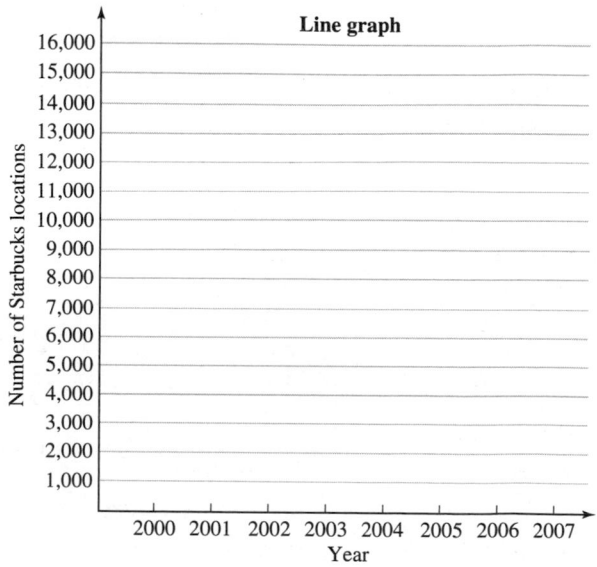

100. ENERGY RESERVES Complete the bar graph and line graph using the data in the table.

**Natural Gas Reserves, 2008
Estimates (in Trillion Cubic Feet)**

United States	211
Venezuela	166
Canada	58
Argentina	16
Mexico	14

Source: *Oil and Gas Journal*, August 2008

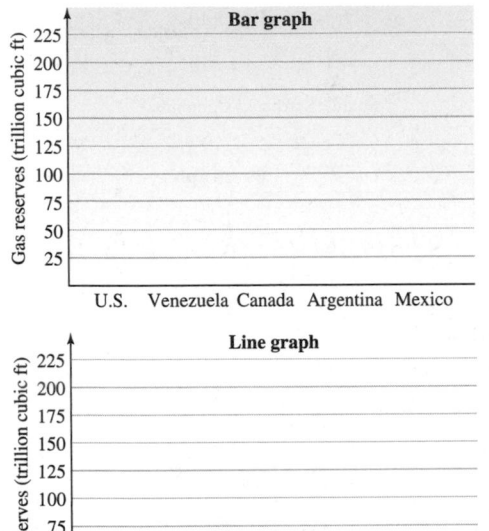

101. CHECKING ACCOUNTS Complete each check by writing the amount in words on the proper line.

a.

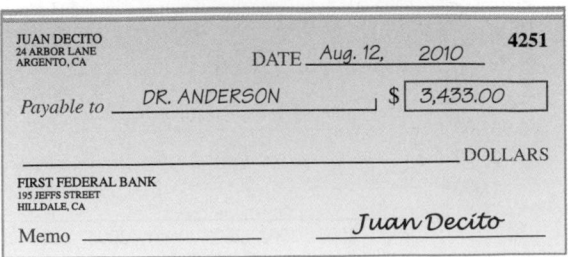

DON SMITH
1234 MILL STREET
HILLDALE, CA   7155
DATE _March 9, 2010_
Payable to _Davis Chevrolet_ $ 15,601.00
_____ DOLLARS
FIRST FEDERAL BANK
195 JEFFS STREET
HILLDALE, CA
Memo _____   _Don Smith_

b.

JUAN DECITO
24 ARBOR LANE
ARGENTO, CA   4251
DATE _Aug. 12, 2010_
Payable to _DR. ANDERSON_ $ 3,433.00
_____ DOLLARS
FIRST FEDERAL BANK
195 JEFFS STREET
HILLDALE, CA
Memo _____   _Juan Decito_

102. ANNOUNCEMENTS One style used when printing formal invitations and announcements is to write all numbers in words. Use this style to write each of the following phrases.

a. This diploma awarded this 27th day of June, 2005.

b. The suggested contribution for the fundraiser is $850 a plate, or an entire table may be purchased for $5,250.

103. COPYEDITING Edit this excerpt from a history text by circling all numbers written in words and rewriting them in standard form using digits.

Abraham Lincoln was elected with a total of one million, eight hundred sixty-five thousand, five hundred ninety-three votes—four hundred eighty-two thousand, eight hundred eighty more than the runner-up, Stephen Douglas. He was assassinated after having served a total of one thousand, five hundred three days in office. Lincoln's Gettysburg Address, a mere two hundred sixty-nine words long, was delivered at the battle site where forty-three thousand, four hundred forty-nine casualties occurred.

104. READING METERS The amount of electricity used in a household is measured in kilowatt-hours (kwh). Determine the reading on the meter shown below. (When the pointer is between two numbers, read the *lower* number.)

Thousands of kwh	Hundreds of kwh	Tens of kwh	Units of kwh

105. SPEED OF LIGHT The speed of light is 983,571,072 feet per second.

a. In what place value column is the 5?

b. Round the speed of light to the nearest ten million. Give your answer in standard notation and in expanded notation.

c. Round the speed of light to the nearest hundred million. Give your answer in standard notation and in written-word form.

106. CLOUDS Graph each cloud type given in the table at the proper altitude on the vertical number line below.

Cloud type	Altitude (ft)
Altocumulus	21,000
Cirrocumulus	37,000
Cirrus	38,000
Cumulonimbus	15,000
Cumulus	8,000
Stratocumulus	9,000
Stratus	4,000

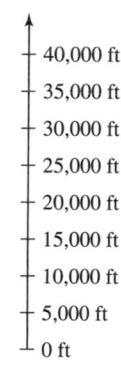

40,000 ft
35,000 ft
30,000 ft
25,000 ft
20,000 ft
15,000 ft
10,000 ft
5,000 ft
0 ft

WRITING

107. Explain how you would round 687 to the nearest ten.

108. The houses in a new subdivision are priced "in the low 130s." What does this mean?

109. A million is a thousand thousands. Explain why this is so.

110. Many television infomercials offer the viewer creative ways to make a six-figure income. What is a six-figure income? What is the smallest and what is the largest six-figure income?

111. What whole number is associated with each of the following words?

duo   decade   zilch   a grand   four score

dozen   trio   century   a pair   nil

112. Explain what is wrong by reading 20,003 as *twenty thousand and three.*

SECTION 1.2
Adding and Subtracting Whole Numbers

Objectives

1 Add whole numbers.

2 Use properties of addition to add whole numbers.

3 Estimate sums of whole numbers.

4 Solve application problems by adding whole numbers.

5 Find the perimeter of a rectangle and a square.

6 Subtract whole numbers.

7 Check subtractions using addition.

8 Estimate differences of whole numbers.

9 Solve application problems by subtracting whole numbers.

10 Evaluate expressions involving addition and subtraction.

Addition and subtraction of whole numbers is used by everyone. For example, to prepare an annual budget, an accountant adds separate line item costs. To determine the number of yearbooks to order, a principal adds the number of students in each grade level. To find the sale price of an item, a store clerk subtracts the discount from the regular price.

1 Add whole numbers.

Addition is the process of finding the total of two (or more) numbers. It can be illustrated using a number line, as shown below. For example, to compute $4 + 5$, we begin at 0 and draw an arrow 4 units long, extending to the right. This represents 4. From the tip of that arrow, we draw another arrow 5 units long, also extending to the right. The second arrow points to 9. This result corresponds to the addition fact $4 + 5 = 9$.

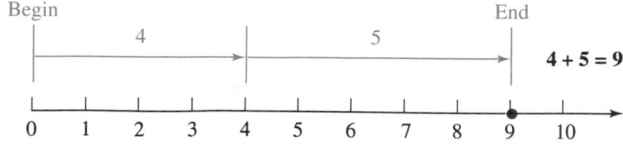

We can write this addition problem in **horizontal** or **vertical form** using an **addition symbol +,** which is read as "plus." The numbers that are being added are called **addends** and the answer is called the **sum** or **total.**

Horizontal form

4   +   5   =   9

Addend   Addend   Sum

We read each form as
"4 plus 5 equals (or is) 9."

Vertical form

4 ⟵ Addend
+ 5 ⟵ Addend
9 ⟵ Sum

To add whole numbers that are less than 10, we rely on our understanding of basic addition facts. For example,

$$2 + 3 = 5, \qquad 6 + 4 = 10, \qquad \text{and} \qquad 9 + 7 = 16$$

To add whole numbers that are greater than 10, we can use vertical form by stacking them with their corresponding place values lined up. Then we simply add the digits in each corresponding column. If an addition of the digits in any place value column produces a sum that is greater than 9, we must **carry.**

Self Check 1

Add: 675 + 1,497 + 1,527

Now Try **Problems 27 and 31**

EXAMPLE 1   Add: 9,835 + 692 + 7,275

Strategy We will write the numbers in vertical form so that corresponding place value columns are lined up. Then we will add the digits in each column, watching for any sums that are greater than 9.

WHY If the sum of the digits in any column is more than 9, we must carry.

Solution
We write the addition in vertical form, so that the corresponding digits are lined up. Each step of this addition is explained separately. Your solution need only look like the *last* step.

$$
\begin{array}{r}
{}^{1}\\
9,8\ 3\ \mathbf{5}\\
6\ 9\ \mathbf{2}\\
+\ 7,2\ 7\ \mathbf{5}\\
\hline
\mathbf{2}
\end{array}
$$

Add the digits in the ones column: 5 + 2 + 5 = 12. Write 2 in the ones column of the answer and carry 1 to the tens column.

$$
\begin{array}{r}
{}^{2}\ {}^{1}\\
9,8\ \mathbf{3}\ 5\\
6\ \mathbf{9}\ 2\\
+\ 7,2\ \mathbf{7}\ 5\\
\hline
\mathbf{0}\ 2
\end{array}
$$

Add the digits in the tens column: 1 + 3 + 9 + 7 = 20. Write 0 in the tens column of the answer and carry 2 to the hundreds column.

$$
\begin{array}{r}
{}^{1}\ {}^{2}\ {}^{1}\\
9,\mathbf{8}\ 3\ 5\\
\mathbf{6}\ 9\ 2\\
+\ 7,\mathbf{2}\ 7\ 5\\
\hline
\mathbf{8}\ 0\ 2
\end{array}
$$

Add the digits in the hundreds column: 2 + 8 + 6 + 2 = 18. Write 8 in the hundreds column of the answer and carry 1 to the thousands column.

$$
\begin{array}{r}
{}^{1}\ {}^{2}\ {}^{1}\\
\mathbf{9},8\ 3\ 5\\
6\ 9\ 2\\
+\ \mathbf{7},2\ 7\ 5\\
\hline
\mathbf{17},8\ 0\ 2
\end{array}
$$

Add the digits in the thousands column: 1 + 9 + 7 = 17. Write 7 in the thousands column of the answer. Write 1 in the ten thousands column.

Your solution should look like this:

$$
\begin{array}{r}
{}^{1}\ {}^{2}{}^{1}\\
9,835\\
692\\
+\ 7,275\\
\hline
17,802
\end{array}
$$

The sum is 17,802.

Success Tip In Example 1, the digits in each place value column were added from *top to bottom*. To check the answer, we can instead add from *bottom to top*. Adding down or adding up should give the same result. If it does not, an error has been made and you should re-add. You will learn why the two results should be the same in Objective 2, which follows.

First add
top to
bottom

$$
\begin{array}{r}
17,802 \\
9,835 \\
692 \\
+\ 7,275 \\
\hline
17,802
\end{array}
$$

To check,
add
bottom
to top

2 Use properties of addition to add whole numbers.

We have used a number line to find that $4 + 5 = 9$. If we add 4 and 5 in the opposite order, we see on the number line below that the result is the same:

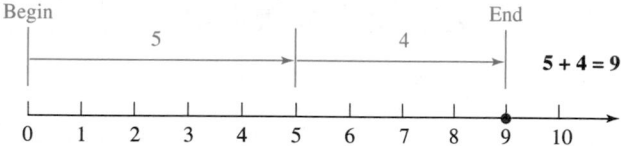

This example illustrates that the order in which we add two numbers does not affect the result. This property is called the **commutative property of addition.** To state the commutative property of addition in a compact form, we can use *variables*.

Variables

A **variable** is a letter (or a symbol) that is used to stand for a number.

We now use the variables *a* and *b* to state the communtative property of addition.

Commutative Property of Addition

The order in which whole numbers are added does not change their sum.
    For any whole number *a* and *b*,

$$a + b = b + a$$

The Language of Algebra *Commutative* is a form of the word *commute*, meaning to go back and forth. *Commuter* trains take people to and from work.

To find the sum of three whole numbers, we add two of them and then add the sum to the third number. In the following examples, we add $3 + 4 + 7$ in two ways. We will use the grouping symbols ( ), called **parentheses,** to show this. It is standard practice to perform the operations within the parentheses first. The steps of the solutions are written in horizontal form.

> ***The Language of Algebra***   In the following example, read $(3 + 4) + 7$ as "The *quantity* of 3 plus 4," pause slightly, and then say "plus 7." We read $3 + (4 + 7)$ as, "3 plus the *quantity* of 4 plus 7." The word *quantity* alerts the reader to the parentheses that are used as grouping symbols.

Method 1: Group 3 and 4

$(3 + 4) + 7 = 7 + 7$ Because of the parentheses, add 3 and 4 first to get 7. Then add 7 and 7 to get 14.

$= 14$

Method 2: Group 4 and 7

$3 + (4 + 7) = 3 + 11$ Because of the parentheses, add 4 and 7 first to get 11. Then add 3 and 11 to get 14.

$= 14$

———— Same result ————

Either way, the answer is 14. This example illustrates that changing the grouping when adding numbers doesn't affect the result. This property is called the **associative property of addition.**

Associative Property of Addition

The way in which whole numbers are grouped does not change their sum.
For any whole numbers a, b, and c,

$$(a + b) + c = a + (b + c)$$

> ***The Language of Algebra***   *Associative* is a form of the word *associate*, meaning to join a group. The WNBA (Women's National Basketball *Association*) is a group of 14 professional basketball teams.

Sometimes, an application of the associative property can simplify a calculation.

Self Check 2

Find the sum: $(139 + 25) + 75$

Now Try Problem 35

EXAMPLE 2   Find the sum: $98 + (2 + 17)$

Strategy We will use the associative property to group 2 with 98.

WHY It is helpful to regroup because 98 and 2 are a pair of numbers that are easily added.

Solution
We will write the steps of the solution in horizontal form.

$98 + (2 + 17) = (98 + 2) + 17$ Use the associative property of addition to regroup the addends.

$= 100 + 17$ Do the addition within the parentheses first.

$= 117$

Whenever we add 0 to a whole number, the number is unchanged. This property is called the **addition property of 0.**

Addition Property of 0

The sum of any whole number and 0 is that whole number.
For any whole number a,

$$a + 0 = a \quad \text{and} \quad 0 + a = a$$

3 Estimate sums of whole numbers.

Estimation is used to find an approximate answer to a problem. Estimates are helpful in two ways. First, they serve as an accuracy check that can find errors. If an answer does not seem reasonable when compared to the estimate, the original problem should be reworked. Second, some situations call for only an approximate answer rather than the exact answer.

There are several ways to estimate, but the objective is the same: Simplify the numbers in the problem so that the calculations can be made easily and quickly. One popular method of estimation is called **front-end rounding.**

EXAMPLE 3 Use front-end rounding to estimate the sum:
3,714 + 2,489 + 781 + 5,500 + 303

Strategy We will use front-end rounding to approximate each addend. Then we will find the sum of the approximations.

WHY Front-end rounding produces addends containing many 0's. Such numbers are easier to add.

Solution

Each of the addends is rounded to its *largest place value* so that all but its first digit is zero. Then we add the approximations using vertical form.

3,714	$\longrightarrow$	4,000	*Round to the nearest thousand.*
2,489	$\longrightarrow$	2,000	*Round to the nearest thousand.*
781	$\longrightarrow$	800	*Round to the nearest hundred.*
5,500	$\longrightarrow$	6,000	*Round to the nearest thousand.*
+ 303	$\longrightarrow$	+ 300	*Round to the nearest hundred.*
		13,100	

The estimate is 13,100.

If we calculate 3,714 + 2,489 + 781 + 5,500 + 303, the sum is exactly 12,787. Note that the estimate is close: It's just 313 more than 12,787. This illustrates the tradeoff when using estimation: The calculations are easier to perform and they take less time, but the answers are not exact.

Success Tip Estimates can be greater than or less than the exact answer. It depends on how often rounding up and rounding down occurs in the estimation.

Self Check 3

Use front-end rounding to estimate the sum:

6,780
3,278
566
4,230
+ 1,923

Now Try Problem 37

4 Solve application problems by adding whole numbers.

Since application problems are almost always written in words, the ability to understand what you read is very important.

The Language of Algebra Here are some key words and phrases that are often used to indicate addition:

gain	*increase*	*up*	*forward*	*rise*	*more than*
total	*combined*	*in all*	*in the future*	*altogether*	*extra*

Self Check 4

AIRLINE ACCIDENTS The numbers of accidents involving U.S. airlines for the years 2000 through 2007 are listed in the table below. Find the total number of accidents for those years.

Year	Accidents
2000	56
2001	46
2002	41
2003	54
2004	30
2005	40
2006	33
2007	26

Now Try Problem 99

EXAMPLE 4

Sharks The graph on the right shows the number of shark attacks worldwide for the years 2000 through 2007. Find the total number of shark attacks for those years.

Strategy We will carefully read the problem looking for a key word or phrase.

WHY Key words and phrases indicate which arithmetic operation(s) should be used to solve the problem.

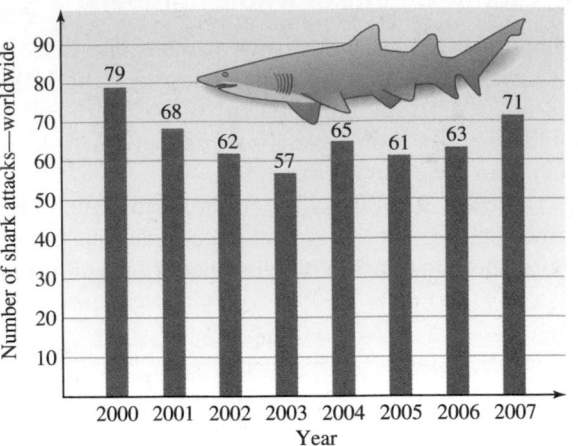

Source: University of Florida

Solution

In the second sentence of the problem, the key word *total* indicates that we should add the number of shark attacks for the years 2000 through 2007. We can use vertical form to find the sum.

$$
\begin{array}{r}
\overset{5\,3}{79} \\
68 \\
62 \\
57 \\
65 \\
61 \\
63 \\
+\ 71 \\
\hline
526
\end{array}
$$

Add the digits, one column at a time, working from right to left. To simplify the calculations, we can look for groups of two or three numbers in each column whose sum is 10.

The total number of shark attacks worldwide for the years 2000 through 2007 was 526.

> **The Language of Algebra** To solve the application problems, we must often *translate* the words of the problem to numbers and symbols. To *translate* means to change from one form to another, as in *translating* from Spanish to English.

EXAMPLE 5   ***Endangered Eagles*** In 1963, there were only 487 nesting pairs of bald eagles in the lower 48 states. By 2007, the number of nesting pairs had increased by 9,302. Find the number of nesting pairs of bald eagles in 2007. (Source: U.S. Fish and Wildlife Service)

Strategy We will carefully read the problem looking for key words or phrases.

WHY Key words and phrases indicate which arithmetic operations should be used to solve the problem.

Solution

The phrase *increased by* indicates addition. With that in mind, we translate the words of the problem to numbers and symbols.

| The number of nesting pairs in 2007 | is equal to | the number of nesting pairs in 1963 | increased by | 9,302. |

| The number of nesting pairs in 2007 | = | 487 | + | 9,302 |

Use vertical form to perform the addition:

```
  9,302
+   487
  9,789
```

Many students find vertical form addition easier if the number with the larger amount of digits is written on top.

In 2007, the number of nesting pairs of bald eagles in the lower 48 states was 9,789. ∎

5 Find the perimeter of a rectangle and a square.

Figure (a) below is an example of a four-sided figure called a **rectangle.** Either of the longer sides of a rectangle is called its **length** and either of the shorter sides is called its **width.** Together, the length and width are called the **dimensions** of the rectangle. For any rectangle, opposite sides have the same measure.

When all four of the sides of a rectangle are the same length, we call the rectangle a **square.** An example of a square is shown in figure (b).

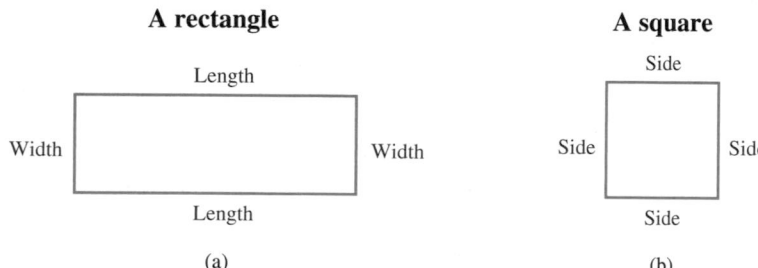

A rectangle (a)

A square (b)

The distance around a rectangle or a square is called its **perimeter.** To find the perimeter of a rectangle, we add the lengths of its four sides.

The perimeter of a rectangle = length + length + width + width

To find the perimeter of a square, we add the lengths of its four sides.

The perimeter of a square = side + side + side + side

The Language of Algebra When you hear the word *perimeter,* think of the distance around the "rim" of a flat figure.

EXAMPLE 6 *Money* Find the perimeter of the dollar bill shown below.

Width = 65 mm

Length = 156 mm

mm stands for millimeters

Self Check 5

MAGAZINES In 2005, the monthly circulation of *Popular Mechanics* magazine was 1,210,126 copies. By 2007, the circulation had increased by 24,199 copies per month. What was the monthly circulation of *Popular Mechanics* magazine in 2007? (Source: *The World Almanac Book of Facts,* 2009)

Now Try Problem 97

Self Check 6

BOARD GAMES A Monopoly game board is a square with sides 19 inches long. Find the perimeter of the board.

Now Try Problems 41 and 43

Strategy   We will add two lengths and two widths of the dollar bill.

WHY   A dollar bill is rectangular-shaped, and this is how the perimeter of a rectangle is found.

Solution
We translate the words of the problem to numbers and symbols.

The perimeter of the dollar bill	is equal to	the length of the dollar bill	plus	the length of the dollar bill	plus	the width of the dollar bill	plus	the width of the dollar bill.

The perimeter of the dollar bill	=	156	+	156	+	65	+	65

Use vertical form to perform the addition:

$$
\begin{array}{r}
\overset{\scriptstyle 2\,2}{156} \\
156 \\
65 \\
+\ 65 \\
\hline
442
\end{array}
$$

The perimeter of the dollar bill is 442 mm.

To see whether this result is reasonable, we estimate the answer. Because the rectangle is about 160 mm by 70 mm, its perimeter is approximately $160 + 160 + 70 + 70$, or 460 mm. An answer of 442 mm is reasonable.

6   Subtract whole numbers.

Subtraction is the process of finding the difference between two numbers. It can be illustrated using a number line, as shown below. For example, to compute $9 - 4$, we begin at 0 and draw an arrow 9 units long, extending to the right. From the tip of that arrow, we draw another arrow 4 units long, but extending to the left. (This represents taking away 4.) The second arrow points to 5, indicating that $9 - 4 = 5$.

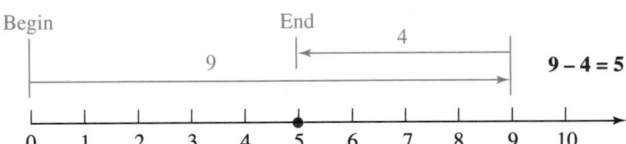

We can write this subtraction problem in **horizontal** or **vertical form** using a **subtraction symbol −,** which is read as "minus." We call the number from which another number is subtracted the **minuend.** The number being subtracted is called the **subtrahend,** and the answer is called the **difference.**

> ***The Language of Algebra***   The prefix *sub* means *below,* as in *sub*marine or *sub*way. Notice that in vertical form, the *sub*trahend is written below the minuend.

To subtract two whole numbers that are less than 10, we rely on our understanding of basic subtraction facts. For example,

$$6 - 3 = 3, \qquad 7 - 2 = 5, \qquad \text{and} \qquad 9 - 8 = 1$$

To subtract two whole numbers that are greater than 10, we can use vertical form by stacking them with their corresponding place values lined up. Then we simply subtract the digits in each corresponding column.

EXAMPLE 7   Subtract 235 from 6,496.

Strategy We will translate the sentence to mathematical symbols and then perform the subtraction. We must be careful when translating the instruction to subtract one number *from* another number.

WHY The order of the numbers in the sentence must be reversed when we translate to symbols.

Solution
Since 235 is the number to be subtracted, it is the subtrahend.

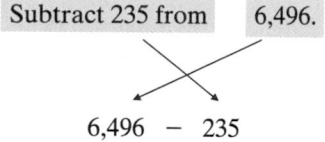

$$6{,}496 \ - \ 235$$

To find the difference, we write the subtraction in vertical form and subtract the digits in each column, working from right to left.

$$
\begin{array}{r}
6{,}496 \\
-\ \ 235 \\
\hline
6{,}261
\end{array}
$$

    ∟ *Bring down the 6 in the thousands column.*

When 235 is subtracted from 6,496, the difference is 6,261.

Self Check 7

Subtract 817 from 1,958.

Now Try **Problem 49**

> *Caution!*   When subtracting two numbers, it is important that we write them in the correct order, because subtraction is *not* commutative. For instance, in Example 2, if we had incorrectly translated "*Subtract 235 from 6,496*" as $235 - 6{,}496$, we see that the difference is not 6,261. In fact, the difference is not even a whole number.

If the subtraction of the digits in any place value column requires that we subtract a larger digit from a smaller digit, we must **borrow** or **regroup**. Some subtractions require borrowing from two (or more) place value columns.

EXAMPLE 8   Subtract: $9{,}927 - 568$

Strategy We will write the subtraction in vertical form and subtract as usual. In each column, we must watch for a digit in the subtrahend that is greater than the digit directly above it in the minuend.

WHY If a digit in the subtrahend is greater than the digit above it in the minuend, we need to borrow (regroup) to subtract in that column.

Self Check 8

Subtract: $6{,}734 - 356$

Now Try **Problem 53**

Solution

We write the subtraction in vertical form, so that the corresponding digits are lined up. Each step of this subtraction is explained separately. Your solution should look like the last step.

$$
\begin{array}{r}
9,927 \\
-568 \\
\end{array}
$$

Since 8 in the ones column of 568 is greater than 7 in the ones column of 9,927, we cannot immediately subtract. To subtract in that column, we must regroup by borrowing 1 ten from 2 in the tens column. In this process, we use the fact that 1 ten = 10 ones.

$$
\begin{array}{r}
^{1\ 17} \\
9,9\cancel{2}\cancel{7} \\
-568 \\
\hline
9
\end{array}
$$

Borrow 1 ten from 2 in the tens column and change the 2 to 1. Add the borrowed 10 to the digit 7 in the ones column of the minuend to get 17. Then subtract in the ones column: $17 - 8 = 9$.

Since 6 in the tens column of 568 is greater than 1 in the tens column directly above it, we cannot immediately subtract. To subtract in that column, we must regroup by borrowing 1 hundred from 9 in the hundreds column. In this process, we use the fact that 1 hundred = 10 tens.

$$
\begin{array}{r}
^{11} \\
^{8\ 1\ 17} \\
9,\cancel{9}\cancel{2}\cancel{7} \\
-568 \\
\hline
59
\end{array}
$$

Borrow 1 hundred from 9 in the hundreds column and change the 9 to 8. Add the borrowed 10 to the digit 1 in the tens column of the minuend to get 11. Then subtract in the tens column: $11 - 6 = 5$.

Complete the solution by subtracting in the hundreds column ($8 - 5 = 3$) and bringing down the 9 in the thousands column.

$$
\begin{array}{r}
^{11} \\
^{8\ \cancel{1}\ 17} \\
9,\cancel{9}\cancel{2}\cancel{7} \\
-568 \\
\hline
9,359
\end{array}
$$

Your solution should look like this:

$$
\begin{array}{r}
^{11} \\
^{8\ \cancel{1}\ 17} \\
9,\cancel{9}\cancel{2}\cancel{7} \\
-568 \\
\hline
9,359
\end{array}
$$

The difference is 9,359.

The borrowing process is more difficult when the minuend contains one or more zeros.

Self Check 9

Subtract: $65,304 - 1,445$

Now Try Problem 57

EXAMPLE 9   Subtract: $42,403 - 1,675$

Strategy We will write the subtraction in vertical form. To subtract in the ones column, we will borrow from the hundreds column of the minuend 42,403.

WHY Since the digit in the tens column of 42,403 is 0, it is not possible to borrow from that column.

Solution

We write the subtraction in vertical form so that the corresponding digits are lined up. Each step of this subtraction is explained separately. Your solution should look like the *last* step.

$$
\begin{array}{r}
42,403 \\
-1,675 \\
\end{array}
$$

Since 5 in the ones column of 1,675 is greater than 3 in the ones column of 42,403, we cannot immediately subtract. It is not possible to borrow from the digit 0 in the tens column of 42,403. We can, however, borrow from the hundreds column to regroup in the tens column, as shown below. In this process, we use the fact that 1 hundred = 10 tens.

$$\overset{\overset{3\ 10}{}}{42,\cancel{4}\cancel{0}3}$$
$$-\ 1,675$$

Borrow 1 hundred from 4 in the hundreds column and change the 4 to 3. Add the borrowed 10 to the digit 0 in the tens column of the minuend to get 10.

Now we can borrow from the 10 in the tens column to subtract in the ones column.

$$\overset{\overset{9}{3\ 10\ 13}}{42,\cancel{4}\cancel{0}\cancel{3}}$$
$$\underline{-\ 1,675}$$
$$8$$

Borrow 1 ten from 10 in the tens column and change the 10 to 9. Add the borrowed 10 to the digit 3 in the ones column of the minuend to get 13. Then subtract in the ones column: 13 − 5 = 8.

Next, we perform the subtraction in the tens column: $9 - 7 = 2$.

$$\overset{\overset{9}{3\ 10\ 13}}{42,\cancel{4}\cancel{0}\cancel{3}}$$
$$\underline{-\ 1,675}$$
$$28$$

To subtract in the hundreds column, we borrow from the 2 in the thousands column. In this process, we use the fact that 1 thousand = 10 hundreds.

$$\overset{\overset{13\ 9}{1\ \ \ 3\ 10\ 13}}{42,\cancel{4}\cancel{0}\cancel{3}}$$
$$\underline{-\ 1,675}$$
$$728$$

Borrow 1 thousand from 2 in the thousands column and change the 2 to 1. Add the borrowed 10 to the digit 3 in the hundreds column of the minuend to get 13. Then subtract in the hundreds column: 13 − 6 = 7.

Complete the solution by subtracting in the thousands column ($1 - 1 = 0$) and bringing down the 4 in the ten thousands column.

$$\overset{\overset{13\ 9}{1\ \ \ 3\ 10\ 13}}{42,\cancel{4}\cancel{0}\cancel{3}}$$
$$\underline{-\ 1,6\ 7\ 5}$$
$$40,7\ 2\ 8$$

> Your solution should look like this:
>
> $$\overset{\overset{13\ 9}{1\ \ \ 3\ 10\ 13}}{42,\cancel{4}\cancel{0}\cancel{3}}$$
> $$\underline{-\ 1,675}$$
> $$40,728$$

The difference is 40,728.

7 Check subtractions using addition.

Every subtraction has a **related addition statement.** For example,

$9 - 4 = 5$	because	$5 + 4 = 9$
$25 - 15 = 10$	because	$10 + 15 = 25$
$100 - 1 = 99$	because	$99 + 1 = 100$

These examples illustrate how we can check subtractions. If a subtraction is done correctly, *the sum of the difference and the subtrahend will always equal the minuend:*

Difference + subtrahend = minuend

> **The Language of Algebra**   To describe the special relationship between addition and subtraction, we say that they are **inverse operations.**

EXAMPLE 10   Check the following subtraction using addition:

$$3,682$$
$$\underline{-1,954}$$
$$1,728$$

Strategy We will add the difference (1,728) and the subtrahend (1,954) and compare that result to the minuend (3,682).

Self Check 10

Check the following subtraction using addition:

$$9,784$$
$$\underline{-4,792}$$
$$4,892$$

Now Try Problem 61

WHY If the sum of the difference and the subtrahend gives the minuend, the subtraction checks.

Solution

The subtraction to check   *Its related addition statement*

$$
\begin{array}{r}
3{,}682 \\
-1{,}954 \\
\hline
1{,}728
\end{array}
\qquad
\begin{array}{l}
\text{difference} \\
+\,\text{subtrahend} \\
\hline
\text{minuend}
\end{array}
\qquad
\begin{array}{r}
\overset{1}{}\overset{1}{1}{,}728 \\
+1{,}954 \\
\hline
3{,}682
\end{array}
$$

Since the sum of the difference and the subtrahend is the minuend, the subtraction is correct.

8 Estimate differences of whole numbers.

Estimation is used to find an approximate answer to a problem.

Self Check 11

Estimate the difference:
$64{,}259 - 7{,}604$

Now Try Problem 65

EXAMPLE 11   Estimate the difference: $89{,}070 - 5{,}431$

Strategy We will use front-end rounding to approximate the 89,070 and 5,431. Then we will find the difference of the approximations.

WHY Front-end rounding produces whole numbers containing many 0's. Such numbers are easier to subtract.

Solution

Both the minuend and the subtrahend are rounded to their *largest place value* so that all but their first digit is zero. Then we subtract the approximations using vertical form.

$$
\begin{array}{rcl}
89{,}070 & \rightarrow & 90{,}000 \qquad \text{Round to the nearest ten thousand.}\\
-\;5{,}431 & \rightarrow & -\;5{,}000 \qquad \text{Round to the nearest thousand.}\\
& & \overline{85{,}000}
\end{array}
$$

The estimate is 85,000. If we calculate $89{,}070 - 5{,}431$, the difference is exactly 83,639. Note that the estimate is close: It's only 1,361 more than 83,639.

9 Solve application problems by subtracting whole numbers.

To answer questions about *how much more* or *how many more*, we use subtraction.

Self Check 12

ELEPHANTS An average male African elephant weighs 13,000 pounds. An average male Asian elephant weighs 11,900 pounds. How much more does an African elephant weigh than an Asian elephant?

Now Try Problem 105

EXAMPLE 12   *Horses*   Radar, the world's largest horse, weighs 2,540 pounds. Thumbelina, the world's smallest horse, weighs 57 pounds. How much more does Radar weigh than Thumbelina? (Source: *Guinness Book of World Records, 2008*)

Strategy We will carefully read the problem, looking for a key word or phrase.

WHY Key words and phrases indicate which arithmetic operation(s) should be used to solve the problem.

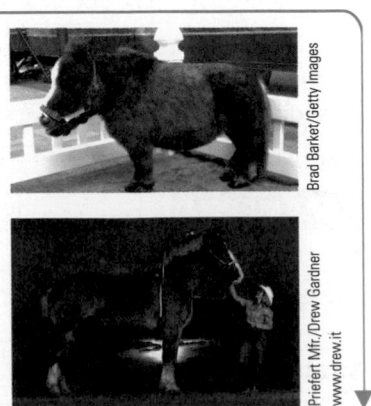

Brad Barker/Getty Images

Priefert Mfr./Drew Gardner
www.drew.it

Solution

In the second sentence of the problem, the phrase *How much more* indicates that we should subtract the weights of the horses. We translate the words of the problem to numbers and symbols.

The number of pounds more that Radar weighs	is equal to	the weight of Radar	minus	the weight of Thumbelina.
The number of pounds more that Radar weighs	=	2,540	−	57

Use vertical form to perform the subtraction:

$$
\begin{array}{r}
\overset{\overset{13}{\scriptstyle 4\ 3\ 10}}{2,\cancel{5}\cancel{4}\cancel{0}} \\
-\ \ \ \ 57 \\
\hline
2,483
\end{array}
$$

Radar weighs 2,483 pounds more than Thumbelina.

The Language of Algebra   Here are some more key words and phrases that often indicate subtraction:

loss	*decrease*	*down*	*backward*	*fell*	*less than*	*fewer*
reduce	*remove*	*debit*	*in the past*	*remains*	*declined*	*take away*

EXAMPLE 13   *Radio Stations*   In 2005, there were 773 oldies radio stations in the United States. By 2007, there were 62 less. How many oldies radio stations were there in 2007? (Source: *The M Street Radio Directory*)

Strategy We will carefully read the problem, looking for a key word or phrase.

WHY Key words and phrases indicate which arithmetic operations should be used to solve the problem.

Solution

The key phrase 62 *less* indicates subtraction. We translate the words of the problem to numbers and symbols.

The number of oldies radio stations in 2007	is	the number of oldies radio stations in 2005	less 62.
The number of oldies radio stations in 2007	=	773	− 62

Use vertical form to perform the subtraction

$$
\begin{array}{r}
773 \\
-\ \ 62 \\
\hline
711
\end{array}
$$

In 2007, there were 711 oldies radio stations in the United States.

Self Check 13

HEALTHY DIETS When Jared Fogle began his reduced-calorie diet of Subway sandwiches, he weighed 425 pounds. With dieting and exercise, he eventually dropped 245 pounds. What was his weight then?

Now Try Problem 111

10 Evaluate expressions involving addition and subtraction.

In arithmetic, numbers are combined with the operations of addition, subtraction, multiplication, and division to create **expressions.** For example,

$$15 + 6, \quad 873 - 99, \quad 6,512 \times 24, \quad \text{and} \quad 42 \div 7$$

are expressions.

Expressions can contain more than one operation. That is the case for the expression $27 - 16 + 5$, which contains addition *and* subtraction. To **evaluate** (find the value of) expressions written in horizontal form that involve addition and subtraction, we perform the operations as they occur *from left to right.*

Self Check 14

Evaluate: $75 - 29 + 8$

Now Try **Problem 71**

EXAMPLE 14   Evaluate: $27 - 16 + 5$

Strategy We will perform the subtraction first and add 5 to that result.

WHY The operations of addition and subtraction must be performed as they occur from left to right.

Solution

We will write the steps of the solution in horizontal form.

$$27 - 16 + 5 = \mathbf{11} + 5 \qquad \text{Working left to right, do the subtraction first: } 27 - 16 = 11.$$
$$= 16 \qquad \text{Now do the addition.}$$

> *Caution!* When making the calculation in Example 14, we must perform the subtraction first. If the addition is done first, we get the incorrect answer 6.
>
> $$27 - 16 + 5 = 27 - 21$$
> $$= 6$$

Using Your CALCULATOR **The Addition and Subtraction Keys**

Calculators are useful for making lengthy calculations and checking results. They should not, however, be used until you have a solid understanding of the basic arithmetic facts. This textbook ***does not*** require you to have a calculator. Ask your instructor if you are allowed to use a calculator in the course.

The *Using Your Calculator* feature explains the keystrokes for an inexpensive scientific calculator. If you have any questions about your specific model, see your user's manual.

To check the result in Example 6 using a scientific calculator, we can use the addition key $\boxed{+}$.

156 $\boxed{+}$ 156 $\boxed{+}$ 65 $\boxed{+}$ 65 $\boxed{=}$ $\boxed{442}$

On some calculator models, the $\boxed{\text{Enter}}$ key is pressed instead of the $\boxed{=}$ for the result to be displayed.

We can use a scientific calculator to check the result in Example 9 using the subtraction key $\boxed{-}$.

42403 $\boxed{-}$ 1675 $\boxed{=}$ $\boxed{40728}$

ANSWERS TO SELF CHECKS

1. 3,699 **2.** 239 **3.** 16,600 **4.** The total number of accidents for 2000–2007 was 326. **5.** The monthly circulation in 2007 was 1,234,325. **6.** 76 in. **7.** 1,141 **8.** 6,378 **9.** 63,859 **10.** The subtraction is incorrect. **11.** 52,000 **12.** An African elephant weighs 1,100 lb more than an Asian elephant. **13.** After the dieting and exercise program, Jared weighed 180 lb. **14.** 54

STUDY SKILLS CHECKLIST

Learning From the Worked Examples

The following checklist will help you become familiar with the example structure in this book. Place a check mark in each box after you answer the question.

☐ Each section of the book contains worked **Examples** that are numbered. How many worked examples are there in Section 1.3, which begins on page 34?

☐ Each worked example contains a **Strategy**. Fill in the blanks to complete the following strategy for Example 3 on page 4: We will locate the commas in the written-word ____ __ ____ _____.

☐ Each Strategy statement is followed by an explanation of **Why** that approach is used. Fill in the blanks to complete the following Why for Example 3 on page 4: When a whole number is written in words, commas are ____ __ _____ _____.

☐ Each worked example has a **Solution**. How many lettered parts are there to the Solution in Example 3 on page 4?

☐ Each example uses red **Author notes** to explain the steps of the solution. Fill in the blanks to complete the first author note in the solution of Example 3 on page 19: Round to the _____ _____.

☐ After reading a worked example, you should work the Self Check problem. How many **Self Check** problems are there for Example 5 on page 6?

☐ At the end of each section, you will find the **Answers to Self Checks**. What is the answer to Self Check problem 2 on page 28?

☐ After completing a Self Check problem, you can **Now Try** similar problems in the Study Sets. For Example 7 on page 23, which Study Set problem is suggested?

Answers: 11, form of each number, used to separate periods, 3, nearest thousand, 2, 239, 49

SECTION **1.2**   **STUDY SET**

VOCABULARY

Fill in the blanks.

1. In the addition problem shown below, label each *addend* and the *sum*.

$$10 \quad + \quad 15 \quad = \quad 25$$

2. When using the vertical form to add whole numbers, if the addition of the digits in any one column produces a sum greater than 9, we must _____.

3. The _____ property of addition states that the order in which whole numbers are added does not change their sum.

4. The _____ property of addition states that the way in which whole numbers are grouped does not change their sum.

5. To see whether the result of an addition is reasonable, we can round the addends and _____ the sum.

6. The words *rise, gain, total,* and *increase* are often used to indicate the operation of _____. The words *fall, lose, reduce,* and *decrease* often indicate the operation of _____.

7. The figure below on the left is an example of a _____. The figure on the right is an example of a _____.

8. Label the *length* and the *width* of the rectangle below. Together, the length and width of a rectangle are called its _____.

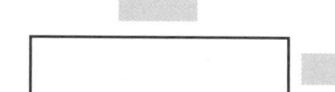

9. When all the sides of a rectangle are the same length, we call the rectangle a _____ .

10. The distance around a rectangle is called its

_____ .

11. In the subtraction problem shown below, label the *minuend, subtrahend,* and the *difference.*

$$25 \quad - \quad 10 \quad = \quad 15$$

12. If the subtraction of the digits in any place value column requires that we subtract a larger digit from a smaller digit, we must _____ or *regroup.*

13. Every subtraction has a _____ addition statement. For example,

$$7 - 2 = 5 \text{ because } 5 + 2 = 7$$

14. To *evaluate* an expression such as $58 - 33 + 9$ means to find its _____ .

CONCEPTS

15. Which property of addition is shown?

   a. $3 + 4 = 4 + 3$

   b. $(3 + 4) + 5 = 3 + (4 + 5)$

   c. $(36 + 58) + 32 = 36 + (58 + 32)$

   d. $319 + 507 = 507 + 319$

16. a. Use the commutative property of addition to complete the following:

   $19 + 33 =$ _____

   b. Use the associative property of addition to complete the following:

   $3 + (97 + 16) =$ _____

17. The subtraction $7 - 3 = 4$ is related to the addition statement ☐ + ☐ = ☐.

18. The operation of _____ can be used to check the result of a subtraction: If a subtraction is done correctly, *the _____ of the difference and the subtrahend will always equal the minuend.*

19. To *evaluate* (find the value of) an expression that contains both addition and subtraction, we perform the operations as they occur from _____ to _____ .

20. To answer questions about *how much more* or *how many more,* we can use _____ .

NOTATION

21. Fill in the blanks. The symbols ( ) are called _____ . It is standard practice to perform the operations within them _____ .

22. Which expression is the correct translation of the sentence: *Subtract 30 from 83.*

$$83 - 30 \quad \text{or} \quad 30 - 83$$

23. Complete the solution to find the sum.

$$12 + (15 + 2) = 12 + \boxed{}$$
$$= \boxed{}$$

24. Fill in the blanks to complete the solution:

$$36 - 11 + 5 = \boxed{} + 5$$
$$= \boxed{}$$

GUIDED PRACTICE

Add. See Example 1.

25. $25 + 13$

26. $47 + 12$

27. $\begin{array}{r} 406 \\ + 283 \\ \hline \end{array}$

28. $\begin{array}{r} 213 \\ + 751 \\ \hline \end{array}$

29. $156 + 305$

30. $647 + 138$

31. $4{,}301 + 789 + 3{,}847$

32. $5{,}576 + 649 + 1{,}922$

Apply the associative property of addition to find the sum. See Example 2.

33. $(13 + 8) + 12$

34. $(19 + 7) + 13$

35. $94 + (6 + 37)$

36. $92 + (8 + 88)$

Use front-end rounding to estimate the sum. See Example 3.

37. $686 + 789 + 12{,}233 + 24{,}500 + 5{,}768$

38. $404 + 389 + 11{,}802 + 36{,}902 + 7{,}777$

39. $567{,}897 + 23{,}943 + 309{,}900 + 99{,}113$

40. $822{,}365 + 15{,}444 + 302{,}417 + 99{,}010$

Find the perimeter of each rectangle or square. See Example 6.

41.  32 feet (ft) 12 ft

42. 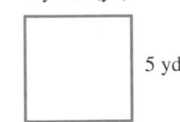 127 meters (m) 91 m

43. 17 inches (in.) 17 in.

44. 5 yards (yd) 5 yd

45.  94 mi (miles) 94 mi

46.  56 ft (feet) 56 ft

47.

87 cm (centimeters)

6 cm

48.

77 in. (inches)

76 in.

Subtract. See Example 7.

49. 347 from 7,989   **50.** 283 from 9,799

51. 405 from 2,967   **52.** 304 from 1,736

Subtract. See Example 8.

53. 8,746 − 289   **54.** 7,531 − 276

55.   6,961   **56.**   4,823
      − 478   − 667

Subtract. See Example 9.

57. 54,506 − 2,829   **58.** 69,403 − 4,635

59.   48,402   **60.**   39,506
      − 3,958   − 1,729

Check each subtraction using addition. See Example 10.

         298   469
61. −175   **62.** −237
         123   132

        4,539   2,698
63. −3,275   **64.** −1,569
        1,364   1,129

Estimate each difference. See Example 11.

65. 67,219 − 4,076   **66.** 45,333 − 3,410

67. 83,872 − 27,281   **68.** 74,009 − 37,405

Evaluate each expression. See Example 14.

69. 35 − 12 + 6   **70.** 47 − 23 + 4

71. 574 + 47 − 13   **72.** 863 + 39 − 11

TRY IT YOURSELF

Perform the operations.

        8,539   5,799
73. + 7,368   **74.** + 6,879

        3,430   2,470
75. −   529   **76.** −   863

77. 51,246 + 578 + 37 + 4,599

78. 4,689 + 73,422 + 26 + 433

79. 633 − 598 + 30   **80.** 600 − 497 + 60

81. (45 + 16) + 4   **82.** 7 + (63 + 23)

83. 20,007 − 78   **84.** 70,006 − 48

85. 852 − 695 + 40   **86.** 397 − 348 + 65

        632   423
87. +347   **88.** +570

89.   15,700   **90.**   35,600
       − 15,397   − 34,799

91. 16,427 increased by 13,573

92. 13,567 more than 18,788

93. Subtract 1,249 from 50,009.

94. Subtract 2,198 from 20,020.

APPLICATIONS

95. DIMENSIONS OF A HOUSE Find the length of the house shown in the blueprint.

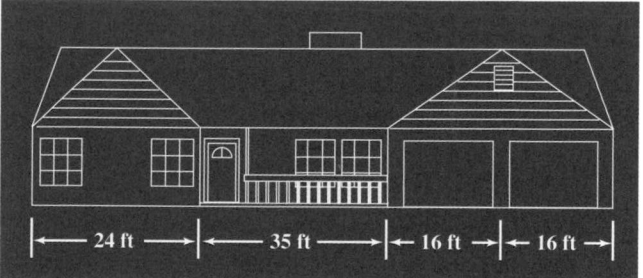

96. FAST FOOD Find the total number of calories in the following lunch from McDonald's: Big Mac (540 calories), small French fries (230 calories), Fruit 'n Yogurt Parfait (160 calories), medium Coca-Cola Classic (210 calories).

97. EBAY In July 2005, the eBay website was visited at least once by 61,715,000 people. By July 2007, that number had increased by 18,072,000. How many visitors did the eBay website have in July 2007? (Source: *The World Almanac and Book of Facts*, 2006, 2008)

98. BRIDGE SAFETY The results of a 2007 report of the condition of U.S. highway bridges is shown below. Each bridge was classified as either *safe*, *in need of repair*, or *should be replaced*. Complete the table.

Number of safe bridges	Number of bridges that need repair	Number of outdated bridges that should be replaced	Total number of bridges
445,396	72,033	80,447	

Source: *Bureau of Transportation Statistics*

99. WEDDINGS The average wedding costs for 2007 are listed in the table below. Find the total cost of a wedding.

Clothing/hair/makeup	$2,293
Ceremony/music/flowers	$4,794
Photography/video	$3,246
Favors/gifts	$1,733
Jewelry	$2,818
Transportation	$361
Rehearsal dinner	$1,085
Reception	$12,470

Source: tickledpinkbrides.com

100. CANDY The graph below shows U.S. candy sales in 2007 during four holiday periods. Find the sum of these seasonal candy sales.

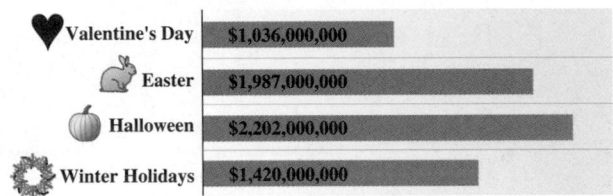

Valentine's Day	$1,036,000,000
Easter	$1,987,000,000
Halloween	$2,202,000,000
Winter Holidays	$1,420,000,000

Source: National Confectioners Association

101. FLAGS To decorate a city flag, yellow fringe is to be sewn around its outside edges, as shown. The fringe is sold by the inch. How many inches of fringe must be purchased to complete the project?

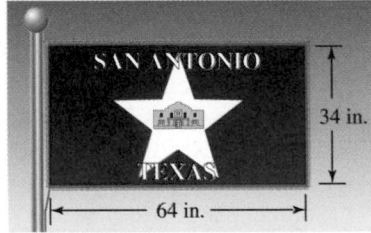

SAN ANTONIO
TEXAS

34 in.

64 in.

102. DECORATING A child's bedroom is rectangular in shape with dimensions 15 feet by 11 feet. How many feet of wallpaper border are needed to wrap around the entire room?

103. BOXING How much padded rope is needed to make a square boxing ring, 24 feet on each side?

104. FENCES A square piece of land measuring 209 feet on all four sides is approximately one *acre*. How many feet of chain link fencing are needed to enclose a piece of land this size?

105. WORLD RECORDS The world's largest pumpkin weighed in at 1,689 pounds and the world's largest watermelon weighed in at 269 pounds. How much more did the pumpkin weigh? (Source: *Guinness Book of World Records*, 2008)

106. BULLDOGS See the graph below. How many more bulldogs were registered in 2007 as compared to 2000?

Number of new bulldogs registered with the American Kennel Club

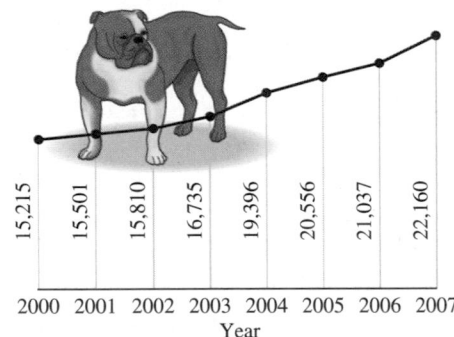

15,215	15,501	15,810	16,735	19,396	20,556	21,037	22,160
2000	2001	2002	2003	2004	2005	2006	2007

Year

Source: American Kennel Club

107. MILEAGE Find the distance (in miles) that a trucker drove on a trip from San Diego to Houston using the odometer readings shown below.

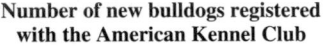

7	0	1	5	4

Truck odometer reading leaving San Diego

7	1	6	4	9

Truck odometer reading arriving in Houston

108. DIETS Use the bathroom scale readings shown below to find the number of pounds that a dieter lost.

January   October

109. MAGAZINES In 2007, *Reader's Digest* had a circulation of 9,322,833. By what amount did this exceed *TV Guide*'s circulation of 3,288,740?

110. TRANSPLANTS See the graph below. Find the decrease in the number of patients waiting for a liver transplant from:

    a. 2001 to 2002   **b.** 2007 to 2008

Waiting list for liver transplants

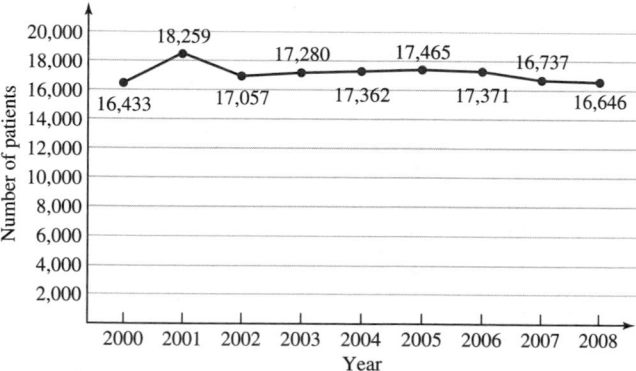

Source: U.S. Department of Health and Human Services

111. JEWELRY Gold melts at about 1,947°F. The melting point of silver is 183°F lower. What is the melting point of silver?

112. READING BLUEPRINTS Find the length of the motor on the machine shown in the blueprint.

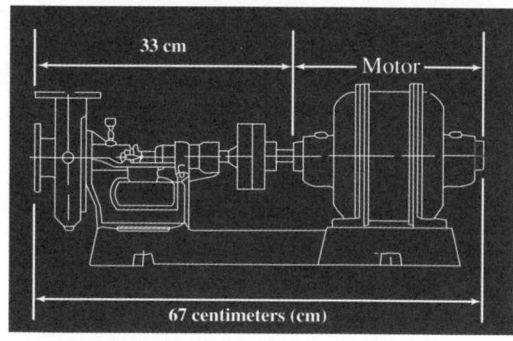

Refer to the teachers' salary schedule shown in the next column. To use this table, note that a fourth-year teacher (Step 4) in Column 2 makes $42,209 per year.

113. a. What is the salary of a teacher on Step 2/Column 2?

    b. How much more will that teacher make next year when she gains 1 year of teaching experience and moves down to Step 3 in that column?

114. a. What is the salary of a teacher on Step 4/Column 1?

    b. How much more will that teacher make next year when he gains 1 year of teaching experience and takes enough coursework to move over to Column 2?

Teachers' Salary Schedule
ABC Unified School District

Years teaching	Column 1	Column 2	Column 3
Step 1	$36,785	$38,243	$39,701
Step 2	$38,107	$39,565	$41,023
Step 3	$39,429	$40,887	$42,345
Step 4	$40,751	$42,209	$43,667
Step 5	$42,073	$43,531	$44,989

WRITING

115. In this section, it is said that estimation is a *tradeoff*. Give one benefit and one drawback of estimation.

116. A student added three whole numbers top to bottom and then bottom to top, as shown below. What do the results in red indicate? What should the student do next?

$$\begin{array}{r} \mathbf{1{,}689} \\ 496 \\ 315 \\ +\ \ 788 \\ \hline 1{,}599 \end{array}$$

117. Explain why the operation of subtraction is not commutative.

118. Explain how addition can be used to check subtraction.

REVIEW

119. Write each number in expanded notation.

    a. 3,125

    b. 60,037

120. Round 6,354,784 to the nearest…

    a. ten

    b. hundred

    c. ten thousand

    d. hundred thousand

121. Round 5,370,645 to the nearest…

    a. ten

    b. ten thousand

    c. hundred thousand

122. Write 72,001,015

    a. in words

    b. in expanded notation

Objectives

1 Multiply whole numbers by one-digit numbers.

2 Multiply whole numbers that end with zeros.

3 Multiply whole numbers by two- (or more) digit numbers.

4 Use properties of multiplication to multiply whole numbers.

5 Estimate products of whole numbers.

6 Solve application problems by multiplying whole numbers.

7 Find the area of a rectangle.

SECTION 1.3
Multiplying Whole Numbers

Multiplication of whole numbers is used by everyone. For example, to double a recipe, a cook multiplies the amount of each ingredient by two. To determine the floor space of a dining room, a carpeting salesperson multiplies its length by its width. An accountant multiplies the number of hours worked by the hourly pay rate to calculate the weekly earnings of employees.

1 Multiply whole numbers by one-digit numbers.

In the following display, there are 4 rows, and each of the rows has 5 stars.

5 stars in each row

We can find the total number of stars in the display by adding: $5 + 5 + 5 + 5 = 20$.

This problem can also be solved using a simpler process called **multiplication.** Multiplication is repeated addition, and it is written using a **multiplication symbol ×**, which is read as "times." Instead of *adding* four 5's to get 20, we can multiply 4 and 5 to get 20.

Repeated addition   **Multiplication**

$5 + 5 + 5 + 5$ $=$ $4 \times 5 = 20$ Read as "4 times 5 equals (or is) 20."

We can write multiplication problems in **horizontal** or **vertical form.** The numbers that are being multiplied are called **factors** and the answer is called the **product.**

Horizontal form   **Vertical form**

$$4 \quad \times \quad 5 \quad = \quad 20$$

Factor   Factor   Product

$$\begin{array}{r} 5 \leftarrow \text{Factor} \\ \times\ 4 \leftarrow \text{Factor} \\ \hline 20 \leftarrow \text{Product} \end{array}$$

A **raised dot ·** and **parentheses ( )** are also used to write multiplication in horizontal form.

Symbols Used for Multiplication

Symbol		Example
$\times$	times symbol	4×5
$\cdot$	raised dot	$4 \cdot 5$
( )	parentheses	$(4)(5)$ or $4(5)$ or $(4)5$

Recall that a variable is a letter that stands for a number. We often multiply a variable by another number or multiply a variable by another variable. When we do this, we don't need to use a symbol for multiplication.

$5a$ means $5 \cdot a$, ab means $a \cdot b$ and xyz means $x \cdot y \cdot z$

> **Caution!**   In this book, we seldom use the $\times$ symbol, because it can be confused with the letter x.

To multiply whole numbers that are less than 10, we rely on our understanding of basic multiplication facts. For example,

$$2 \cdot 3 = 6, \qquad 8(4) = 32, \qquad \text{and} \qquad 9 \times 7 = 63$$

To multiply larger whole numbers, we can use vertical form by stacking them with their corresponding place values lined up. Then we make repeated use of basic multiplication facts.

EXAMPLE 1   Multiply: 8×47

Strategy We will write the multiplication in vertical form. Then, working right to left, we will multiply each digit of 47 by 8 and carry, if necessary.

WHY This process is simpler than treating the problem as repeated addition and adding eight 47's.

Solution
To help you understand the process, each step of this multiplication is explained separately. Your solution need only look like the *last* step.

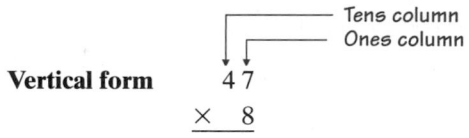

We begin by multiplying 7 by 8.

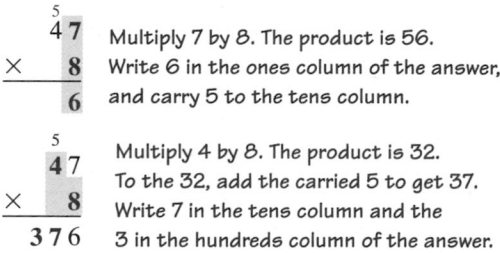

Multiply 7 by 8. The product is 56. Write 6 in the ones column of the answer, and carry 5 to the tens column.

Multiply 4 by 8. The product is 32. To the 32, add the carried 5 to get 37. Write 7 in the tens column and the 3 in the hundreds column of the answer.

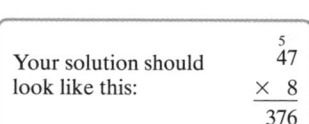

Your solution should look like this:

The product is 376.

---

2 **Multiply whole numbers that end with zeros.**

An interesting pattern develops when a whole number is multiplied by 10, 100, 1,000, and so on. Consider the following multiplications involving 8:

$8 \cdot 10 = 80$ There is one zero in 10. The product is 8 with one 0 attached.

$8 \cdot 100 = 800$ There are two zeros in 100. The product is 8 with two 0's attached.

$8 \cdot 1,000 = 8,000$ There are three zeros in 1,000. The product is 8 with three 0's attached.

$8 \cdot 10,000 = 80,000$ There are four zeros in 10,000. The product is 8 with four 0's attached.

These examples illustrate the following rule.

Multiplying a Whole Number by 10, 100, 1,000, and So On

To find the product of a whole number and 10, 100, 1,000, and so on, attach the number of zeros in that number to the right of the whole number.

EXAMPLE 2   Multiply: **a.** $6 \times 1,000$ **b.** $45 \cdot 100$ **c.** $912(10,000)$

Strategy For each multiplication, we will identify the factor that ends in zeros and count the number of zeros that it contains.

WHY Each product can then be found by attaching that number of zeros to the other factor.

Solution
a. $6 \times 1,000 = 6,000$ Since 1,000 has three zeros, attach three 0's after 6.
b. $45 \cdot 100 = 4,500$ Since 100 has two zeros, attach two 0's after 45.
c. $912(10,000) = 9,120,000$ Since 10,000 has four zeros, attach four 0's after 912.

We can use an approach similar to that of Example 2 for multiplication involving any whole numbers that end in zeros. For example, to find $67 \cdot 2,000$, we have

$$67 \cdot \mathbf{2,000} = 67 \cdot \mathbf{2} \cdot \mathbf{1,000}$$ Write 2,000 as $2 \cdot 1,000$.
$$= 134 \cdot 1,000$$ Working left to right, multiply 67 and 2 to get 134.
$$= 134,000$$ Since 1,000 has three zeros, attach three 0's after 134.

This example suggests that to find $67 \cdot 2,000$ we simply multiply 67 and 2 and attach three zeros to that product. This method can be extended to find products of two factors that *both* end in zeros.

EXAMPLE 3   Multiply: **a.** $14 \cdot 300$ **b.** $3,500 \cdot 50,000$

Strategy We will multiply the nonzero leading digits of each factor. To that product, we will attach the sum of the number of trailing zeros in the factors.

WHY This method is faster than the standard vertical form multiplication of factors that contain many zeros.

Solution
a. The factor **300** has two trailing zeros.

$14 \cdot 300 = 4,200$ Attach two 0's after 42.
Multiply 14 and 3 to get 42.

b. The factors **3,500** and **50,000** have a total of six trailing zeros.

$3,500 \cdot 50,000 = 175,000,000$ Attach six 0's after 175.
Multiply 35 and 5 to get 175.

Success Tip Calculations that you cannot perform in your head should be shown outside the steps of your solution.

3 Multiply whole numbers by two- (or more) digit numbers.

EXAMPLE 4   Multiply: $23 \cdot 436$

Strategy We will write the multiplication in vertical form. Then we will multiply 436 by 3 and by 20, and add those products.

WHY Since $23 = 3 + 20$, we can multiply 436 by 3 and by 20, and add those products.

Solution

Each step of this multiplication is explained separately. Your solution need only look like the *last* step.

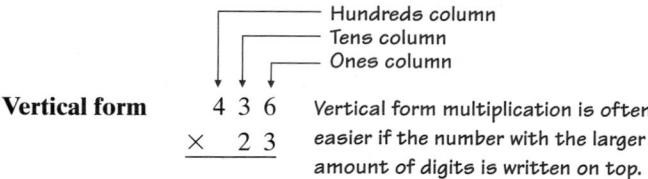

We begin by multiplying 436 by 3.

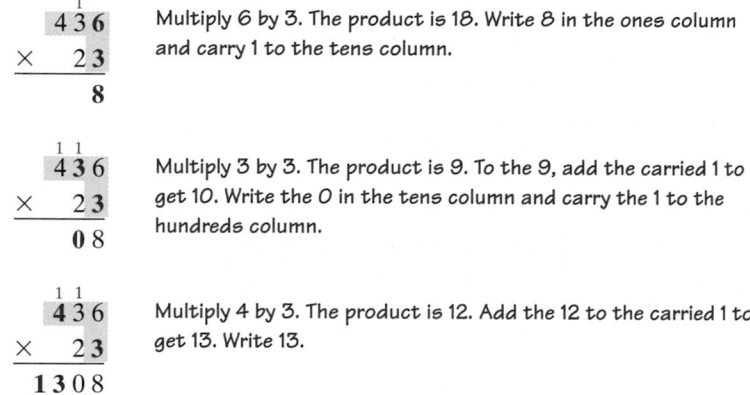

Multiply 6 by 3. The product is 18. Write 8 in the ones column and carry 1 to the tens column.

Multiply 3 by 3. The product is 9. To the 9, add the carried 1 to get 10. Write the 0 in the tens column and carry the 1 to the hundreds column.

Multiply 4 by 3. The product is 12. Add the 12 to the carried 1 to get 13. Write 13.

We continue by multiplying 436 by 2 tens, or 20. If we think of 20 as 2 · 10, then we simply multiply 436 by 2 and attach one zero to the result.

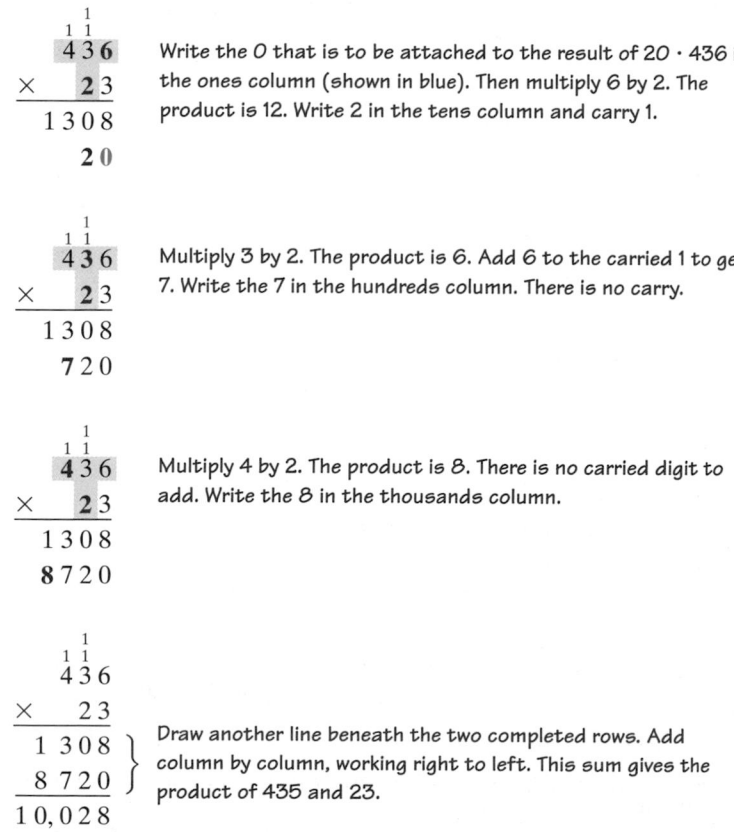

Write the 0 that is to be attached to the result of 20 · 436 in the ones column (shown in blue). Then multiply 6 by 2. The product is 12. Write 2 in the tens column and carry 1.

Multiply 3 by 2. The product is 6. Add 6 to the carried 1 to get 7. Write the 7 in the hundreds column. There is no carry.

Multiply 4 by 2. The product is 8. There is no carried digit to add. Write the 8 in the thousands column.

Draw another line beneath the two completed rows. Add column by column, working right to left. This sum gives the product of 435 and 23.

The product is 10,028.

> *The Language of Algebra*   In Example 4, the numbers **1,308** and **8,720** are called **partial products.** We added the partial products to get the answer, 10,028. The word *partial* means *only a part*, as in a *partial* eclipse of the moon.

$$\begin{array}{r} 436 \\ \times\quad 23 \\ \hline 1\,308 \\ 8\,720 \\ \hline 10,028 \end{array}$$

When a factor in a multiplication contains one or more zeros, we must be careful to enter the correct number of zeros when writing the partial products.

Self Check 5

Multiply:
a. 706(351)
b. 4,004(2,008)

Now Try   Problem 41

EXAMPLE 5   Multiply: **a.** 406 · 253   **b.** 3,009(2,007)

Strategy   We will think of 406 as 6 + 400 and 3,009 as 9 + 3,000.

WHY   Thinking of the multipliers (406 and 3,009) in this way is helpful when determining the correct number of zeros to enter in the partial products.

Solution
We will use vertical form to perform each multiplication.

a. Since 406 = 6 + 400, we will multiply 253 by 6 and by 400, and add those partial products.

$$\begin{array}{r} 253 \\ \times\quad 406 \\ \hline 1\,518 \\ 101\,200 \\ \hline 102,718 \end{array}$$ ← 6 · 253
← 400 · 253. Think of 400 as 4 · 100 and simply multiply 253 by 4 and attach two zeros (shown in blue) to the result.

The product is 102,718.

b. Since 3,009 = 9 + 3,000, we will multiply 2,007 by 9 and by 3,000, and add those partial products.

$$\begin{array}{r} 2,007 \\ \times\quad 3,009 \\ \hline 18\,063 \\ 6\,021\,000 \\ \hline 6,039,063 \end{array}$$ ← 9 · 2,007
← 3,000 · 2,007. Think of 3,000 as 3 · 1,000 and simply multiply 2,007 by 3 and attach three zeros (shown in blue) to the result.

The product is 6,039,063.

4   Use properties of multiplication to multiply whole numbers.

Have you ever noticed that two whole numbers can be multiplied in either order because the result is the same? For example,

$$4 \cdot 6 = 24 \quad \text{and} \quad 6 \cdot 4 = 24$$

This example illustrates the **commutative property of multiplication.**

> **Commutative Property of Multiplication**
>
> The order in which whole numbers are multiplied does not change their product.
>   For any whole numbers *a* and *b*,
>
> $$a \cdot b = b \cdot a \quad \text{or, more simply,} \quad ab = ba$$

Whenever we multiply a whole number by 0, the product is 0. For example,

$$0 \cdot 5 = 0, \qquad 0 \cdot 8 = 0, \qquad \text{and} \qquad 9 \cdot 0 = 0$$

Whenever we multiply a whole number by 1, the number remains the same. For example,

$$3 \cdot 1 = 3, \qquad 7 \cdot 1 = 7, \qquad \text{and} \qquad 1 \cdot 9 = 9$$

These examples illustrate the multiplication properties of 0 and 1.

Multiplication Properties of 0 and 1

The product of any whole number and 0 is 0.
The product of any whole number and 1 is that whole number.
   For any whole number a,

$$a \cdot 0 = 0 \qquad \text{and} \qquad 0 \cdot a = 0$$
$$a \cdot 1 = a \qquad \text{and} \qquad 1 \cdot a = a$$

Success Tip   If one (or more) of the factors in a multiplication is 0, the product will be 0. For example,

$$16(27)(0) = 0 \qquad \text{and} \qquad 109 \cdot 53 \cdot 0 \cdot 2 = 0$$

To multiply three numbers, we first multiply two of them and then multiply that result by the third number. In the following examples, we multiply $3 \cdot 2 \cdot 4$ in two ways. The parentheses show us which multiplication to perform first. The steps of the solutions are written in horizontal form.

The Language of Algebra   In the following example, read $(3 \cdot 2) \cdot 4$ as "The *quantity* of 3 times 2," pause slightly, and then say "times 4." Read $3 \cdot (2 \cdot 4)$ as "3 times the *quantity* of 2 times 4." The word *quantity* alerts the reader to the parentheses that are used as grouping symbols.

Method 1: Group $3 \cdot 2$	**Method 2: Group $2 \cdot 4$**
$(3 \cdot 2) \cdot 4 = 6 \cdot 4$ Multiply 3 and 2 to get 6.	$3 \cdot (2 \cdot 4) = 3 \cdot 8$ Then multiply 2 and 4 to get 8.
$= 24$ Multiply 6 and 4 to get 24.	$= 24$ Then multiply 3 and 8 to get 24.

Same result

Either way, the answer is 24. This example illustrates that changing the grouping when multiplying numbers doesn't affect the result. This property is called the **associative property of multiplication.**

Associative Property of Multiplication

The way in which whole numbers are grouped does not change their product.
   For any whole numbers a, b, and c,

$$(a \cdot b) \cdot c = a \cdot (b \cdot c) \qquad \text{or, more simply,} \qquad (ab)c = a(bc)$$

Sometimes, an application of the associative property can simplify a calculation.

Self Check 6
Find the product: $(23 \cdot 25) \cdot 4$
Now Try Problem 45

EXAMPLE 6   Find the product: $(17 \cdot 50) \cdot 2$

Strategy We will use the associative property to group 50 with 2.

WHY It is helpful to regroup because 50 and 2 are a pair of numbers that are easily multiplied.

Solution
We will write the solution in horizontal form.

$$(17 \cdot 50) \cdot 2 = 17 \cdot (50 \cdot 2) \qquad \text{Use the associative property of multiplication to regroup the factors.}$$
$$= 17 \cdot 100 \qquad \text{Do the multiplication within the parentheses first.}$$
$$= 1{,}700 \qquad \text{Since 100 has two zeros, attach two 0's after 17.}$$

5   Estimate products of whole numbers.

Estimation is used to find an approximate answer to a problem.

Self Check 7
Estimate the product: $74 \cdot 488$
Now Try Problem 51

EXAMPLE 7   Estimate the product: $59 \cdot 334$

Strategy We will use front-end rounding to approximate the factors 59 and 334. Then we will find the product of the approximations.

WHY Front-end rounding produces whole numbers containing many 0's. Such numbers are easier to multiply.

Solution
Both of the factors are rounded to their *largest place value* so that all but their first digit is zero.

$$
\begin{array}{ll}
& \text{Round to the nearest ten.} \\
59 \cdot 334 & \qquad\qquad 60 \cdot 300 \\
& \text{Round to the nearest hundred.}
\end{array}
$$

To find the product of the approximations, $60 \cdot 300$, we simply multiply 6 by 3, to get 18, and attach 3 zeros. Thus, the estimate is 18,**000**.

    If we calculate $59 \cdot 334$, the product is exactly 19,706. Note that the estimate is close: It's only 1,706 less than 19,706.

6   Solve application problems by multiplying whole numbers.

Application problems that involve repeated addition are often more easily solved using multiplication.

Self Check 8
DAILY PAY In 2008, the average U.S. construction worker made $22 per hour. At that rate, how much money was earned in an 8-hour workday? (Source: Bureau of Labor Statistics)
Now Try Problem 86

EXAMPLE 8   *Daily Pay*   In 2008, the average U.S. manufacturing worker made $18 per hour. At that rate, how much money was earned in an 8-hour workday? (Source: Bureau of Labor Statistics)

Strategy To find the amount earned in an 8-hour workday, we will multiply the hourly rate of $18 by 8.

WHY For each of the 8 hours, the average manufacturing worker earned $18. The amount earned for the day is the sum of eight 18's: $18 + 18 + 18 + 18 + 18 + 18 + 18 + 18$. This repeated addition can be calculated more simply by multiplication.

Solution
We translate the words of the problem to numbers and symbols.

The amount earned in an 8-hr workday | is equal to | the rate per hour | times | 8 hours.

The amount earned in
an 8-hr workday $=$ 18 $\cdot$ 8

Use vertical form to perform the multiplication:

$$\begin{array}{r} \overset{6}{1}8 \\ \times\ \ 8 \\ \hline 144 \end{array}$$

In 2008, the average U.S. manufacturing worker earned \$144 in an 8-hour workday. ∎

We can use multiplication to count objects arranged in patterns of neatly arranged rows and columns called **rectangular arrays**.

> ***The Language of Algebra***   An *array* is an orderly arrangement. For example, a jewelry store might display a beautiful *array* of gemstones.

EXAMPLE 9   *Pixels*   Refer to the illustration at the right. Small dots of color, called *pixels*, create the digital images seen on computer screens. If a 14-inch screen has 640 pixels from side to side and 480 pixels from top to bottom, how many pixels are displayed on the screen?

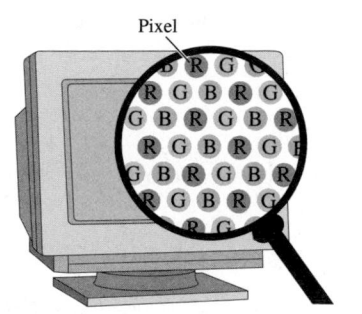
Pixel

Self Check 9

PIXELS If a 17-inch computer screen has 1,024 pixels from side to side and 768 from top to bottom, how many pixels are displayed on the screen?

Now Try Problem 93

Strategy We will multiply 640 by 480 to determine the number of pixels that are displayed on the screen.

WHY The pixels form a rectangular array of 640 rows and 480 columns on the screen. Multiplication can be used to count objects in a rectangular array.

Solution
We translate the words of the problem to numbers and symbols.

The number of pixels on the screen | is equal to | the number of pixels in a row | times | the number of pixels in a column.

The number of pixels
on the screen $=$ 640 $\cdot$ 480

To find the product of 640 and 480, we use vertical form to multiply 64 and 48 and attach two zeros to that result.

$$\begin{array}{r} 48 \\ \times\ 64 \\ \hline 192 \\ 2\ 880 \\ \hline 3{,}072 \end{array}$$

Since the product of 64 and 48 is 3,072, the product of 64**0** and 48**0** is 307,2**00**. The screen displays 307,200 pixels.

> **The Language of Algebra** Here are some key words and phrases that are often used to indicate multiplication:
>
> double   triple   twice   of   times

Self Check 10

INSECTS Leaf cutter ants can carry pieces of leaves that weigh 30 times their body weight. How much can an ant lift if it weighs 25 milligrams?

Now Try Problem 99

EXAMPLE 10 *Weight Lifting* In 1983, Stefan Topurov of Bulgaria was the first man to lift three times his body weight over his head. If he weighed 132 pounds at the time, how much weight did he lift over his head?

Strategy To find how much weight he lifted over his head, we will multiply his body weight by 3.

WHY We can use multiplication to determine the result when a quantity increases in size by 2 *times*, 3 *times*, 4 *times*, and so on.

Solution
We translate the words of the problem to numbers and symbols.

| The amount he lifted over his head | was | 3 | times | his body weight. |

$$\text{The amount he lifted over his head} = 3 \cdot 132$$

Use vertical form to perform the multiplication:

$$\begin{array}{r} 132 \\ \times\ \ 3 \\ \hline 396 \end{array}$$

Stefan Topurov lifted 396 pounds over his head.

> **Using Your CALCULATOR**   **The Multiplication Key: Seconds in a Year**
>
> There are 60 seconds in 1 minute, 60 minutes in 1 hour, 24 hours in 1 day, and 365 days in 1 year. We can find the number of seconds in 1 year using the multiplication key $\boxed{\times}$ on a calculator.
>
> 60 $\boxed{\times}$ 60 $\boxed{\times}$ 24 $\boxed{\times}$ 365 $\boxed{=}$ $\boxed{31536000}$
>
> One some calculator models, the $\boxed{\text{ENTER}}$ key is pressed instead of the $\boxed{=}$ for the result to be displayed.
>   There are 31,536,000 seconds in 1 year.

---

7 **Find the area of a rectangle.**

One important application of multiplication is finding the area of a rectangle. The **area of a rectangle** is the measure of the amount of surface it encloses. Area is measured in square units, such as square inches (written *in.*^{2}) or square centimeters (written *cm*^{2}), as shown below.

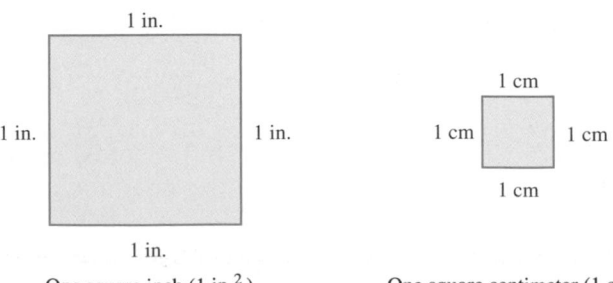

One square inch (1 in.^{2})   One square centimeter (1 cm^2)

The rectangle in the figure below has a length of 5 centimeters and a width of 3 centimeters. Since each small square region covers an area of one square centimeter, each small square region measures 1 cm². The small square regions form a rectangular pattern, with 3 rows of 5 squares.

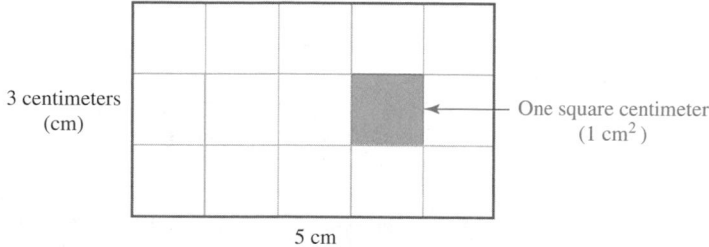

3 centimeters (cm)

One square centimeter (1 cm²)

5 cm

Because there are 5 · 3, or 15, small square regions, the area of the rectangle is 15 cm². This suggests that the area of any rectangle is the product of its length and its width.

Area of a rectangle = length · width

By using the letter A to represent the area of the rectangle, the letter l to represent the length of the rectangle, and the letter w to represent its width, we can write this **formula** in simpler form. Letters (or symbols), such as A, l, and w, that are used to represent numbers are called **variables.**

Area of a Rectangle

The area, A, of a rectangle is the product of the rectangle's length, l, and its width, w.

$$\text{Area} = \text{length} \cdot \text{width} \quad \text{or} \quad A = l \cdot w \quad \text{or} \quad A = lw$$

EXAMPLE 11   *Gift Wrapping*   When completely unrolled, a long sheet of gift wrapping paper has the dimensions shown below. How many square feet of gift wrap are on the roll?

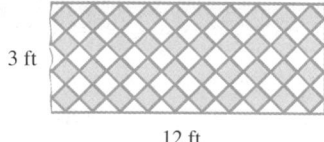

3 ft

12 ft

Strategy We will substitute 12 for the length and 3 for the width in the formula for the area of a rectangle.

WHY To find the number of square feet of paper, we need to find the area of the rectangle shown in the figure.

Solution

$A = lw$ This is the formula for the area of a rectangle.

$\quad = 12 \cdot 3$ Replace the length l with 12 and the width w with 3.

$\quad = 36$ Do the multiplication.

There are 36 square feet of wrapping paper on the roll. This can be written in more compact form as 36 ft².

Self Check 11

ADVERTISING The rectangular posters used on small billboards in the New York subway are 59 inches wide by 45 inches tall. Find the area of a subway poster.

Now Try **Problems 53 and 55**

> **Caution!** Remember that the perimeter of a rectangle is the distance around it and is measured in units such as inches, feet, and miles. The area of a rectangle is the amount of surface it encloses and is measured in square units such as in.^{2}, ft^2, and mi^2.

ANSWERS TO SELF CHECKS

1. 324 **2. a.** 9,000 **b.** 2,500 **c.** 875,000 **3. a.** 13,500 **b.** 21,700,000
4. 12,024 **5. a.** 247,806 **b.** 8,040,032 **6.** 2,300 **7.** 35,000 **8.** $176
9. 786,432 **10.** 750 milligrams **11.** 2,655 in.^{2}

STUDY SKILLS CHECKLIST

Getting the Most from the Study Sets

The following checklist will help you become familiar with the Study Sets in this book. Place a check mark in each box after you answer the question.

☐ Answers to the odd-numbered **Study Set** problems are located in the appendix on page A-21. On what page do the answers to Study Set 1.3 appear?

☐ Each Study Set begins with **Vocabulary** problems. How many Vocabulary problems appear in Study Set 1.3?

☐ Following the Vocabulary problems, you will see **Concepts** problems. How many Concepts problems appear in Study Set 1.3?

☐ Following the Concepts problems, you will see **Notation** problems. How many Notation problems appear in Study Set 1.3?

☐ After the Notation problems, **Guided Practice** problems are given which are linked to similar

examples within the section. How many Guided Practice problems appear in Study Set 1.3?

☐ After the Guided Practice problems, **Try It Yourself** problems are given and can be used to help you prepare for quizzes. How many Try It Yourself problems appear in Study Set 1.3?

☐ Following the Try It Yourself problems, you will see **Applications** problems. How many Applications problems appear in Study Set 1.3?

☐ After the Applications problems in Study Set 1.3, how many **Writing** problems are given?

☐ Lastly, each Study Set ends with a few **Review** problems. How many Review problems appear in Study Set 1.3?

Answers: A-21 6, 6, 4, 40, 24, 26, 2, 2

SECTION 1.3   STUDY SET

▌ VOCABULARY

Fill in the blanks.

1. In the multiplication problem shown below, label each *factor* and the *product*.

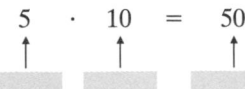

5   ·   10   =   50

2. Multiplication is _____ addition.

3. The _____ property of multiplication states that the order in which whole numbers are multiplied does not change their product. The _____ property of multiplication states that the way in which whole numbers are grouped does not change their product.

4. Letters that are used to represent numbers are called _____.

5. If a square measures 1 inch on each side, its area is 1 _____ inch.

6. The ____ of a rectangle is a measure of the amount of surface it encloses.

▌ CONCEPTS

7. a. Write the repeated addition 8 + 8 + 8 + 8 as a multiplication.

   b. Write the multiplication 7 · 15 as a repeated addition.

8. a. Fill in the blank: A rectangular _____ of red squares is shown below.

  b. Write a multiplication statement that will give the number of red squares shown below.

9. a. How many zeros do you attach to the right of 25 to find 25 · 1,000?

  b. How many zeros do you attach to the right of 8 to find 400 · 2,000?

10. a. Using the variables x and y, write a statement that illustrates the commutative property of multiplication.

  b. Using the variables x, y, and z, write a statement that illustrates the associative property of multiplication.

11. Determine whether the concept of *perimeter* or that of *area* should be applied to find each of the following.

  a. The amount of floor space to carpet

  b. The number of inches of lace needed to trim the sides of a handkerchief

  c. The amount of clear glass to be tinted

  d. The number of feet of fencing needed to enclose a playground

12. Perform each multiplication.

  a. $1 \cdot 25$ **b.** $62(1)$
  c. $10 \cdot 0$ **d.** $0(4)$

NOTATION

13. Write three symbols that are used for multiplication.

14. What does ft^2 mean?

15. Write the formula for the area of a rectangle using variables.

16. Write each multiplication in simpler form.

  a. $8 \cdot x$ **b.** $a \cdot b$

GUIDED PRACTICE

Multiply. See Example 1.

17. 15×7 **18.** 19×9
19. 34×8 **20.** 37×6

Perform each multiplication without using pencil and paper or a calculator. See Example 2.

21. $37 \cdot 100$ **22.** $63 \cdot 1,000$
23. 75×10 **24.** $88 \times 10,000$

25. $107(10,000)$ **26.** $323(100)$
27. $512(1,000)$ **28.** $673(10)$

Multiply. See Example 3.

29. $68 \cdot 40$ **30.** $83 \cdot 30$
31. $56 \cdot 200$ **32.** $222 \cdot 500$
33. $130(3,000)$ **34.** $630(7,000)$
35. $2,700(40,000)$ **36.** $5,100(80,000)$

Multiply. See Example 4.

37. $73 \cdot 128$ **38.** $54 \cdot 173$
39. $64(287)$ **40.** $72(461)$

Multiply. See Example 5.

41. $602 \cdot 679$ **42.** $504 \cdot 729$

43. $3,002(5,619)$ **44.** $2,003(1,376)$

Apply the associative property of multiplication to find the product. See Example 6.

45. $(18 \cdot 20) \cdot 5$ **46.** $(29 \cdot 2) \cdot 50$

47. $250 \cdot (4 \cdot 135)$ **48.** $250 \cdot (4 \cdot 289)$

Estimate each product. See Example 7.

49. $86 \cdot 249$ **50.** $56 \cdot 631$
51. $215 \cdot 1,908$ **52.** $434 \cdot 3,789$

Find the area of each rectangle or square. See Example 11.

53.

54.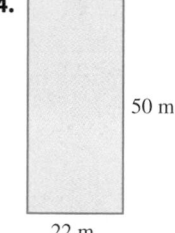

55.

56. 20 cm, 20 cm

TRY IT YOURSELF

Multiply.

57. $\begin{array}{r} 213 \\ \times\ 7 \end{array}$ **58.** $\begin{array}{r} 863 \\ \times\ 9 \end{array}$

59. $34,474 \cdot 2$ **60.** $54,912 \cdot 4$

61. 99
× 77

62. 73
× 59

63. 44(55)(0)

64. 81 · 679 · 0 · 5

65. 53 · 30

66. 20 · 78

67. 754
× 59

68. 846
× 79

69. (2,978)(3,004)

70. (2,003)(5,003)

71. 916
× 409

72. 889
× 507

73. 25 · (4 · 99)

74. (41 · 5) · 20

75. 4,800 × 500

76. 6,400 × 700

77. 2,779
× 128

78. 3,596
× 136

79. 370 · 450

80. 280 · 340

APPLICATIONS

81. BREAKFAST CEREAL A cereal maker advertises "Two cups of raisins in every box." Find the number of cups of raisins in a case of 36 boxes of cereal.

82. SNACKS A candy warehouse sells large four-pound bags of M & M's. There are approximately 180 peanut M & M's per pound. How many peanut M & M's are there in one bag?

83. NUTRITION There are 17 grams of fat in one Krispy Kreme chocolate-iced, custard-filled donut. How many grams of fat are there in one dozen of those donuts?

84. JUICE It takes 13 oranges to make one can of orange juice. Find the number of oranges used to make a case of 24 cans.

85. BIRDS How many times do a hummingbird's wings beat each minute?

65 wingbeats
per second

86. LEGAL FEES Average hourly rates for lead attorneys in New York are $775. If a lead attorney bills her client for 15 hours of legal work, what is the fee?

87. CHANGING UNITS There are 12 inches in 1 foot and 5,280 feet in 1 mile. How many inches are there in a mile?

88. FUEL ECONOMY Mileage figures for a 2009 Ford Mustang GT convertible are shown in the table.

   a. For city driving, how far can it travel on a tank of gas?

   b. For highway driving, how far can it travel on a tank of gas?

© Car Culture/Corbis

Fuel tank capacity	16 gal
Fuel economy (miles per gallon)	15 city/23 hwy

89. WORD COUNT Generally, the number of words on a page for a published novel is 250. What would be the expected word count for the 308-page children's novel *Harry Potter and the Philosopher's Stone*?

90. RENTALS Mia owns an apartment building with 18 units. Each unit generates a monthly income of $450. Find her total monthly income.

91. CONGRESSIONAL PAY The annual salary of a U.S. House of Representatives member is $169,300. What does it cost per year to pay the salaries of all 435 voting members of the House?

92. CRUDE OIL The United States uses 20,730,000 barrels of crude oil per day. One barrel contains 42 gallons of crude oil. How many gallons of crude oil does the United States use in one day?

93. WORD PROCESSING A student used the *Insert Table* options shown when typing a report. How many entries will the table hold?

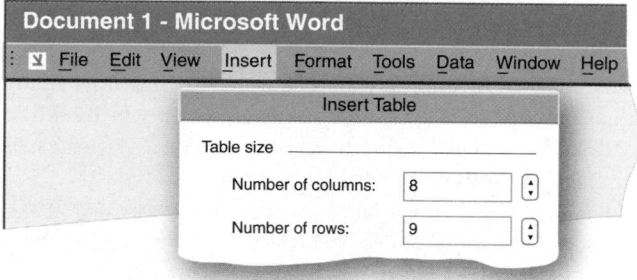

Document 1 - Microsoft Word

File Edit View Insert Format Tools Data Window Help

Insert Table

Table size

Number of columns: 8

Number of rows: 9

94. BOARD GAMES A checkerboard consists of 8 rows, with 8 squares in each row. The squares alternate in color, red and black. How many squares are there on a checkerboard?

95. ROOM CAPACITY A college lecture hall has 17 rows of 33 seats each. A sign on the wall reads, "Occupancy by more than 570 persons is prohibited." If all of the seats are taken, and there is one instructor in the room, is the college breaking the rule?

96. ELEVATORS There are 14 people in an elevator with a capacity of 2,000 pounds. If the average weight of a person in the elevator is 150 pounds, is the elevator overloaded?

97. KOALAS In one 24-hour period, a koala sleeps 3 times as many hours as it is awake. If it is awake for 6 hours, how many hours does it sleep?

98. FROGS Bullfrogs can jump as far as ten times their body length. How far could an 8-inch-long bullfrog jump?

99. TRAVELING During the 2008 Olympics held in Beijing, China, the cost of some hotel rooms was 33 times greater than the normal charge of $42 per night. What was the cost of such a room during the Olympics?

100. ENERGY SAVINGS An ENERGY STAR light bulb lasts eight times longer than a standard 60-watt light bulb. If a standard bulb normally lasts 11 months, how long will an ENERGY STAR bulb last?

© Image copyright Jose Gill, 2009. Used under license from Shutterstock.com

101. PRESCRIPTIONS How many tablets should a pharmacist put in the container shown in the illustration?

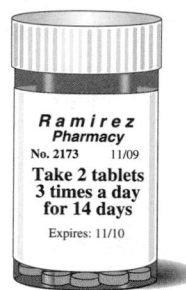

Ramirez Pharmacy
No. 2173   11/09
Take 2 tablets
3 times a day
for 14 days
Expires: 11/10

102. HEART BEATS A normal pulse rate for a healthy adult, while resting, can range from 60 to 100 beats per minute.

   a. How many beats is that in one day at the lower end of the range?

   b. How many beats is that in one day at the upper end of the range?

103. WRAPPING PRESENTS When completely unrolled, a long sheet of wrapping paper has the dimensions shown. How many square feet of gift wrap are on the roll?

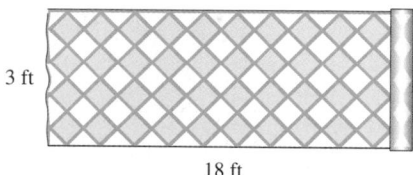

3 ft

18 ft

104. POSTER BOARDS A rectangular-shaped poster board has dimensions of 24 inches by 36 inches. Find its area.

105. WYOMING The state of Wyoming is approximately rectangular-shaped, with dimensions 360 miles long and 270 miles wide. Find its perimeter and its area.

106. COMPARING ROOMS Which has the greater area, a rectangular room that is 14 feet by 17 feet or a square room that is 16 feet on each side? Which has the greater perimeter?

WRITING

107. Explain the difference between 1 foot and 1 square foot.

108. When two numbers are multiplied, the result is 0. What conclusion can be drawn about the numbers?

REVIEW

109. Find the sum of 10,357, 9,809, and 476.

110. DISCOUNTS A radio, originally priced at $367, has been marked down to $179. By how many dollars was the radio discounted?

Objectives

1. Write the related multiplication statement for a division.
2. Use properties of division to divide whole numbers.
3. Perform long division (no remainder).
4. Perform long division (with a remainder).
5. Use tests for divisibility.
6. Divide whole numbers that end with zeros.
7. Estimate quotients of whole numbers.
8. Solve application problems by dividing whole numbers.

SECTION **1.4**
Dividing Whole Numbers

Division of whole numbers is used by everyone. For example, to find how many 6-ounce servings a chef can get from a 48-ounce roast, he divides 48 by 6. To split a $36,000 inheritance equally, a brother and sister divide the amount by 2. A professor divides the 35 students in her class into groups of 5 for discussion.

1 Write the related multiplication statement for a division.

To divide whole numbers, think of separating a quantity into equal-sized groups. For example, if we start with a set of 12 stars and divide them into groups of 4 stars, we will obtain 3 groups.

A set of 12 stars.

There are 3 groups of 4 stars.

We can write this division problem using a **division symbol** $\div$, a **long division symbol** $\overline{)}$, or a **fraction bar** $-$. We call the number being divided the **dividend** and the number that we are dividing by is called the **divisor.** The answer is called the **quotient.**

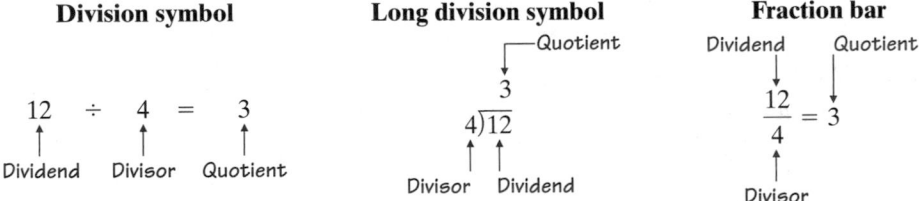

We read each form as "12 divided by 4 equals (or is) 3."

Recall from Section 1.3 that multiplication is repeated addition. Likewise, division is repeated subtraction. To divide 12 by 4, we ask, "How many 4's can be subtracted from 12?"

$$\begin{array}{r} 12 \\ -\ 4 \end{array}\Big\} \text{ Subtract 4 one time.}$$

$$\begin{array}{r} 8 \\ -\ 4 \end{array}\Big\} \text{ Subtract 4 a second time.}$$

$$\begin{array}{r} 4 \\ -\ 4 \end{array}\Big\} \text{ Subtract 4 a third time.}$$

$$\overline{0}$$

Since exactly three 4's can be subtracted from 12 to get 0, we know that $12 \div 4 = 3$.

Another way to answer a division problem is to think in terms of multiplication. For example, the division $12 \div 4$ asks the question, "What must I multiply 4 by to get 12?" Since the answer is 3, we know that

$$12 \div 4 = 3 \text{ because } 3 \cdot 4 = 12$$

We call $3 \cdot 4 = 12$ the **related multiplication statement** for the division $12 \div 4 = 3$. In general, to write the related multiplication statement for a division, we use:

$$\text{Quotient} \cdot \text{divisor} = \text{dividend}$$

EXAMPLE 1 Write the related multiplication statement for each division.

a. $10 \div 5 = 2$ **b.** $6\overline{)24}$ with quotient 4 **c.** $\dfrac{21}{3} = 7$

Strategy We will identify the quotient, the divisor, and the dividend in each division statement.

WHY A related multiplication statement has the following form:
Quotient · divisor = dividend.

Solution

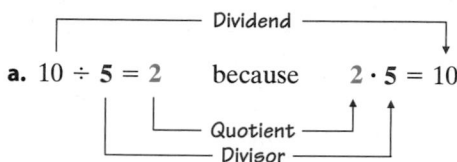

a. $10 \div 5 = 2$ because $2 \cdot 5 = 10$.

b. $6\overline{)24}$ with quotient 4 because $4 \cdot 6 = 24$.   4 is the quotient, 6 is the divisor, and 24 is the dividend.

c. $\dfrac{21}{3} = 7$ because $7 \cdot 3 = 21$.   7 is the quotient, 3 is the divisor, and 21 is the dividend.

> **The Language of Algebra** To describe the special relationship between multiplication and division, we say that they are **inverse operations.**

2 Use properties of division to divide whole numbers.

Recall from Section 1.3 that *the product of any whole number and 1 is that whole number.* We can use that fact to establish two important properties of division. Consider the following examples where a whole number is divided by 1:

$8 \div 1 = \mathbf{8}$ because $\mathbf{8} \cdot 1 = 8$.

$1\overline{)4}$ with quotient $\mathbf{4}$ because $\mathbf{4} \cdot 1 = 4$.

$\dfrac{20}{1} = \mathbf{20}$ because $\mathbf{20} \cdot 1 = 20$.

These examples illustrate that *any whole number divided by 1 is equal to the number itself.*

Consider the following examples where a whole number is divided by itself:

$6 \div 6 = \mathbf{1}$ because $\mathbf{1} \cdot 6 = 6$.

$9\overline{)9}$ with quotient $\mathbf{1}$ because $\mathbf{1} \cdot 9 = 9$.

$\dfrac{35}{35} = \mathbf{1}$ because $\mathbf{1} \cdot 35 = 35$.

These examples illustrate that *any nonzero whole number divided by itself is equal to 1.*

Properties of Division

Any whole number divided by 1 is equal to that number.
Any nonzero whole number divided by itself is equal to 1.
    For any whole number a,

$$\frac{a}{1} = a \qquad \text{and} \qquad \frac{a}{a} = 1 \qquad (\text{where } a \neq 0) \; \text{Read} \neq \text{as "is not equal to."}$$

Self Check 1

Write the related multiplication statement for each division.
a. $8 \div 2 = 4$
b. $7\overline{)56}$ with quotient 8
c. $\dfrac{36}{4} = 9$

Now Try Problems 19 and 23

Recall from Section 1.3 that *the product of any whole number and 0 is 0.* We can use that fact to establish another property of division. Consider the following examples where 0 is divided by a whole number:

$$0 \div 2 = \mathbf{0} \text{ because } \mathbf{0} \cdot 2 = 0.$$

$$\overset{\mathbf{0}}{7\overline{)0}} \text{ because } \mathbf{0} \cdot 7 = 0.$$

$$\frac{0}{42} = \mathbf{0} \text{ because } \mathbf{0} \cdot 42 = 0.$$

These examples illustrate that *0 divided by any nonzero whole number is equal to 0.*

We cannot divide a whole number by 0. To illustrate why, we will attempt to find the quotient when 2 is divided by 0 using the related multiplication statement shown below.

Division statement	*Related multiplication statement*
$\dfrac{2}{0} = ?$	$? \cdot 0 = 2$

There is no number that gives
2 when multiplied by 0.

Since $\dfrac{2}{0}$ does not have a quotient, we say that division of 2 by 0 is *undefined.* Our observations about division of 0 and division by 0 are listed below.

Division with Zero

1. Zero divided by any nonzero number is equal to 0.
2. Division by 0 is undefined.

For any nonzero whole number a,

$$\frac{0}{a} = 0 \quad \text{and} \quad \frac{a}{0} \text{ is undefined}$$

---

3 **Perform long division (no remainder).**

A process called **long division** can be used to divide larger whole numbers.

Self Check 2

Divide using long division: 2,968 ÷ 4. Check the result.

Now Try Problem 31

EXAMPLE 2   Divide using long division: 2,514 ÷ 6. Check the result.

Strategy We will write the problem in long-division form and follow a four-step process: **estimate, multiply, subtract,** and **bring down.**

WHY The repeated subtraction process would take too long to perform and the related multiplication statement (? · 6 = 2,514) is too difficult to solve.

Solution
To help you understand the process, each step of this division is explained separately. Your solution need only look like the *last* step.

We write the problem in the form $6\overline{)2514}$. The quotient will appear above the long division symbol. Since 6 will not divide 2,

$$6\overline{)2514}$$

we divide 25 by 6.

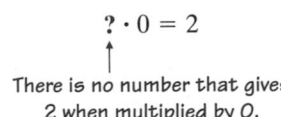

Ask: "How many times will 6 divide 25?" We [estimate] that 25 ÷ 6 is about 4, and write the 4 in the hundreds column above the long division symbol.

Next, we multiply 4 and 6, and subtract their product, 24, from 25, to get 1.

$$
\begin{array}{r}
4 \\
6\overline{)2514} \\
-24 \\
\hline
1
\end{array}
$$

Now we bring down the next digit in the dividend, the 1, and again estimate, multiply, and subtract.

$$
\begin{array}{r}
41 \\
6\overline{)2514} \\
-24\downarrow \\
\hline
\mathbf{11} \\
-6 \\
\hline
5
\end{array}
$$

Ask: "How many times will 6 divide 11?" We estimate that 11 ÷ 6 is about 1, and write the 1 in the tens column above the long division symbol. Multiply 1 and 6, and subtract their product, 6, from 11, to get 5.

To complete the process, we bring down the last digit in the dividend, the 4, and estimate , multiply , and subtract one final time.

$$
\begin{array}{r}
419 \\
6\overline{)2514} \\
-24 \\
\hline
11 \\
-6 \\
\hline
\mathbf{54} \\
-54 \\
\hline
0
\end{array}
$$

Ask: "How many times will 6 divide 54?" We estimate that 54 ÷ 6 is 9, and we write the 9 in the ones column above the long division symbol. Multiply 9 and 6, and subtract their product, 54, from 54, to get 0.

Your solution should look like this:

$$
\begin{array}{r}
419 \\
6\overline{)2514} \\
-24 \\
\hline
11 \\
-6 \\
\hline
54 \\
-54 \\
\hline
0
\end{array}
$$

To check the result, we see if the product of the quotient and the divisor equals the dividend.

$$
\begin{array}{r}
{\scriptstyle 1\,5} \\
419 \leftarrow\text{Quotient} \\
\times\quad 6 \leftarrow\text{Divisor} \\
\hline
2,514 \leftarrow\text{Dividend}
\end{array}
\qquad 6\overline{)2514}
$$

The check confirms that 2,514 ÷ 6 = 419.

The Language of Algebra In Example 2, the long division process ended with a 0. In such cases, we say that the divisor divides the dividend *exactly*.

We can see how the long division process works if we write the names of the place-value columns above the quotient. The solution for Example 2 is shown in more detail on the next page.

$$\begin{array}{r} 419 \\ 6{\overline{\smash{\big)}\,2514}} \\ -2400 \\ \hline 114 \\ -60 \\ \hline 54 \\ -54 \\ \hline 0 \end{array}$$

Here, we are really subtracting 400 · 6, which is 2,400, from 2,514. That is why the 4 is written in the hundreds column of the quotient.

Here, we are really subtracting 10 · 6, which is 60, from 114. That is why the 1 is written in the tens column of the quotient.

Here, we are subtracting 9 · 6, which is 54, from 54. That is why the 9 is written in the ones column of the quotient.

The extra zeros (shown in the steps highlighted in red and blue) are often omitted.

We can use long division to perform divisions when the divisor has more than one digit. The estimation step is often made easier if we approximate the divisor.

Self Check 3

Divide using long division:

$57{\overline{\smash{\big)}\,45{,}885}}$

Now Try Problem 35

EXAMPLE 3   Divide using long division: $48{\overline{\smash{\big)}\,33{,}888}}$

Strategy We will follow a four-step process: **estimate, multiply, subtract,** and **bring down.**

WHY This is how long division is performed.

Solution
To help you understand the process, each step of this division is explained separately. Your solution need only look like the *last* step.

Since 48 will not divide 3, nor will it divide 33, we divide 338 by 48.

$$\begin{array}{r} 6 \\ 48{\overline{\smash{\big)}\,33888}} \end{array}$$

Ask: "How many times will 48 divide 338?" Since 48 is almost 50, we can estimate the answer to that question by thinking 33 ÷ 5 is about 6, and we write the 6 in the hundreds column of the quotient.

$$\begin{array}{r} 6 \\ 48{\overline{\smash{\big)}\,33888}} \\ -288 \\ \hline 50 \end{array}$$

Multiply 6 and 48, and subtract their product, 288, from 338 to get 50. Since 50 is greater than the divisor, 48, the estimate of 6 for the hundreds column of the quotient is *too small*. We will erase the 6 and increase the estimate of the quotient by 1 and try again.

$$\begin{array}{r} 7 \\ 48{\overline{\smash{\big)}\,33888}} \\ -336 \\ \hline 2 \end{array}$$

Change the estimate from 6 to 7 in the hundreds column of the quotient. Multiply 7 and 48, and subtract their product, 336, from 338 to get 2. Since 2 is less than the divisor, we can proceed with the long division.

$$\begin{array}{r} 70 \\ 48{\overline{\smash{\big)}\,33888}} \\ -336 \\ \hline 28 \\ -0 \\ \hline 28 \end{array}$$

Bring down the 8 from the tens column of the dividend. Ask: "How many times will 48 divide 28?" Since 28 cannot be divided by 48, write a 0 in the tens column of the quotient. Multiply 0 and 48, and subtract their product, 0, from 28 to get 28.

$$\begin{array}{r} 705 \\ 48{\overline{\smash{\big)}\,33888}} \\ -336 \\ \hline 28 \\ -0 \\ \hline 288 \\ -240 \\ \hline 48 \end{array}$$

Bring down the 8 from the ones column of the dividend. Ask: "How many times will 48 divide 288?" We can estimate the answer to that question by thinking 28 ÷ 5 is about 5, and we write the 5 in the ones column of the quotient. Multiply 5 and 48, and subtract their product, 240, from 288 to get 48. Since 48 is equal to the divisor, the estimate of 5 for the ones column of the quotient is *too small*. We will erase the 5 and increase the estimate of the quotient by 1 and try again.

Caution!   If a difference at any time in the long division process is greater than or equal to the divisor, the estimate made at that point should be increased by 1, and you should try again.

$$
\begin{array}{r}
706 \\
48\overline{)\;33888} \\
-336 \\
\hline
28 \\
-0 \\
\hline
288 \\
-288 \\
\hline
0
\end{array}
$$

Change the estimate from 5 to 6 in the ones column of the quotient.

Multiply 6 and 48, and subtract their product, 288, from 288 to get 0. Your solution should look like this.

The quotient is 706. Check the result using multiplication.

4 Perform long division (with a remainder).

Sometimes, it is not possible to separate a group of objects into a whole number of equal-sized groups. For example, if we start with a set of 14 stars and divide them into groups of 4 stars, we will have 3 groups of 4 stars and 2 stars left over. We call the left over part the **remainder.**

A set of 14 stars.

There are 3 groups of 4 stars.   There are 2 stars left over.

In the next long division example, there is a remainder. To check such a problem, we add the remainder to the product of the quotient and divisor. The result should equal the dividend.

(Quotient · divisor) + remainder = dividend   Recall that the operation within the parentheses must be performed first.

EXAMPLE 4   Divide: $23\overline{)832}$. Check the result.

Strategy We will follow a four-step process: **estimate, multiply, subtract,** and **bring down.**

WHY This is how long division is performed.

Solution
Since 23 will not divide 8, we divide 83 by 23.

$$
\begin{array}{r}
4 \\
23\overline{)\;832}
\end{array}
$$

Ask: "How many times will 23 divide 83?" Since 23 is about 20, we can estimate the answer to that question by thinking 8 ÷ 2 is 4, and we write the 4 in the tens column of the quotient.

$$
\begin{array}{r}
4 \\
23\overline{)\;832} \\
-92 \\
\hline
\end{array}
$$

Multiply 4 and 23, and write their product, 92, under the 83. Because 92 is greater than 83, the estimate of 4 for the tens column of the quotient is *too large*. We will erase the 4 and decrease the estimate of the quotient by 1 and try again.

Self Check 4

Divide: $34\overline{)792}$. Check the result.

Now Try Problem 39

$$
\begin{array}{r}
3 \\
23\,\overline{)\,832} \\
-69 \\
\hline
14
\end{array}
$$

Change the estimate from 4 to 3 in the tens column of the quotient. Multiply 3 and 23, and subtract their product, 69, from 83, to get 14.

$$
\begin{array}{r}
3 \\
23\,\overline{)\,832} \\
-69\!\downarrow \\
\hline
142
\end{array}
$$

Bring down the 2 from the ones column of the dividend.

$$
\begin{array}{r}
37 \\
23\,\overline{)\,832} \\
-69 \\
\hline
142 \\
-161 \\
\hline
\end{array}
$$

Ask: "How many times will 23 divide 142?" We can estimate the answer to that question by thinking $14 \div 2$ is 7, and we write the 7 in the ones column of the quotient. Multiply 7 and 23, and write their product, 161, under 142. Because 161 is greater than 142, the estimate of 7 for the ones column of the quotient is *too large*. We will erase the 7 and decrease the estimate of the quotient by 1 and try again.

$$
\begin{array}{r}
36 \\
23\,\overline{)\,832} \\
-69 \\
\hline
142 \\
-138 \\
\hline
4
\end{array}
$$

Change the estimate from 7 to 6 in the ones column of the quotient. Multiply 6 and 23, and subtract their product, 138, from 142, to get 4.

←   The remainder

The quotient is 36, and the remainder is 4. We can write this result as 36 R 4.

To check the result, we multiply the divisor by the quotient and then add the remainder. The result should be the dividend.

Check: Quotient   Divisor   Remainder
$$
(36 \;\cdot\; 23) \;+\; 4 \;=\; 828 + 4
$$
$$
= 832 \leftarrow \text{Dividend}
$$

Since 832 is the dividend, the answer 36 R 4 is correct.

Self Check 5

Divide: $\dfrac{28{,}992}{629}$

Now Try Problem 43

EXAMPLE 5   Divide: $\dfrac{13{,}011}{518}$

Strategy We will write the problem in long-division form and follow a four-step process: **estimate, multiply, subtract,** and **bring down.**

WHY This is how long division is performed.

Solution
We write the division in the form: $518\,\overline{)\,13011}$. Since 518 will not divide 1, nor 13, nor 130, we divide 1,301 by 518.

$$
\begin{array}{r}
2 \\
518\,\overline{)\,13011} \\
-1036 \\
\hline
265
\end{array}
$$

Ask: "How many times will 518 divide 1,301?" Since 518 is about 500, we can estimate the answer to that question by thinking $13 \div 5$ is about 2, and we write the 2 in the tens column of the quotient. Multiply 2 and 518, and subtract their product, 1,036, from 1,301, to get 265.

$$\begin{array}{r} 25 \\ 518\overline{)13011} \\ -1036 \\ \hline \mathbf{2651} \\ -2590 \\ \hline 61 \end{array}$$

Bring down the 1 from the ones column of the dividend. Ask: "How many times will 518 divide 2,651?" We can estimate the answer to that question by thinking 26 ÷ 5 is about 5, and we write the 5 in the ones column of the quotient. Multiply 5 and 518, and subtract their product, 2,590, from 2,651, to get a remainder of 61.

The result is 25 R 61. To check, verify that $(25 \cdot 518) + 61$ is 13,011.

5 Use tests for divisibility.

We have seen that some divisions end with a 0 remainder and others do not. The word *divisible* is used to describe such situations.

Divisibility

One number is **divisible** by another if, when dividing them, we get a remainder of 0.

Since $27 \div 3 = 9$, with a 0 remainder, we say that *27 is divisible by 3*. Since $27 \div 5 = 5$ R 2, we say that *27 is not divisible by 5*.

There are tests to help us decide whether one number is divisible by another.

Tests for Divisibility

A number is divisible by

- 2 if its last digit is divisible by 2.
- 3 if the sum of its digits is divisible by 3.
- 4 if the number formed by its last two digits is divisible by 4.
- 5 if its last digit is 0 or 5.
- 6 if it is divisible by 2 and 3.
- 9 if the sum of its digits is divisible by 9.
- 10 if its last digit is 0.

There are tests for divisibility by a number other than 2, 3, 4, 5, 6, 9, or 10, but they are more complicated. See problems 109 and 110 of Study Set 1.4 for some examples.

EXAMPLE 6   Is 534,840 divisible by:

a. 2   **b.** 3   **c.** 4   **d.** 5   **e.** 6   **f.** 9   **g.** 10

Strategy We will look at the last digit, the last two digits, and the sum of the digits of each number.

WHY The divisibility rules call for these types of examination.

Solution

a. 534,840 is divisible by 2, because its last digit **0** is divisible by 2.

b. 534,840 is divisible by 3, because the sum of its digits is divisible by 3.

$$5 + 3 + 4 + 8 + 4 + 0 = 24 \quad \text{and} \quad 24 \div 3 = 8$$

Self Check 6

Is 73,311,435 divisible by:

a. 2   **b.** 3   **c.** 5
d. 6   **e.** 9   **f.** 10

Now Try **Problems 49 and 53**

c. 534,8**40** is divisible by 4, because the number formed by its last two digits is divisible by 4.

$$40 \div 4 = 10$$

d. 534,84**0** divisible by 5, because its last digit is 0 or 5.

e. 534,840 is divisible by 6, because it is divisible by 2 and 3. (See parts a and b.)

f. 534,840 is not divisible by 9, because the sum of its digits is not divisible by 9. There is a remainder.

$$24 \div 9 = 2 \text{ R } 6$$

g. 534,84**0** is divisible by 10, because its last digit is 0.

6 Divide whole numbers that end with zeros.

There is a shortcut for dividing a dividend by a divisor when both end with zeros. We simply *remove the ending zeros in the divisor and remove the same number of ending zeros in the dividend.*

Self Check 7

Divide: **a.** 50 ÷ 10
b. 62,000 ÷ 100
c. 12,000 ÷ 1,500

Now Try **Problems 55 and 57**

EXAMPLE 7   Divide: **a.** 80 ÷ 10   **b.** 47,000 ÷ 100   **c.** $350\overline{)9{,}800}$

Strategy We will look for ending zeros in each divisor.

WHY If a divisor has ending zeros, we can simplify the division by removing the same number of ending zeros in the divisor and dividend.

Solution

There is one zero in the divisor.

a. $80 \div 10 = 8 \div 1 = 8$

Remove one zero from the dividend and the divisor, and divide.

There are two zeros in the divisor.

b. $47{,}000 \div 100 = 470 \div 1 = 470$

Remove two zeros from the dividend and the divisor, and divide.

c. To find

$$350\overline{)9{,}800}$$

we can drop *one zero* from the divisor and the dividend and perform the division $35\overline{)980}$.

$$
\begin{array}{r}
28 \\
35\overline{)980} \\
-70 \\
\hline
280 \\
-280 \\
\hline
0
\end{array}
$$

Thus, 9,800 ÷ 350 is 28.

7 Estimate quotients of whole numbers.

To estimate quotients, we use a method that approximates both the dividend and the divisor so that they divide easily. There is one rule of thumb for this method: If possible, round both numbers up or both numbers down.

EXAMPLE 8   Estimate the quotient: 170,715 ÷ 57

Strategy We will round the dividend and the divisor up and find 180,000 ÷ 60.

WHY The division can be made easier if the dividend and the divisor end with zeros. Also, 6 divides 18 exactly.

Solution

The dividend is approximately

170,715 ÷ 57   180,000 ÷ 60 = 3,000   To divide, drop one zero from 180,000
         The divisor is approximately   and from 60 and find 18,000 ÷ 6.

The estimate is 3,000.

If we calculate 170,715 ÷ 57, the quotient is exactly 2,995. Note that the estimate is close: It's just 5 more than 2,995.

Self Check 8
Estimate the quotient:
33,642 ÷ 42

Now Try Problem 59

8 Solve application problems by dividing whole numbers.

Application problems that involve forming equal-sized groups can be solved by division.

EXAMPLE 9 *Managing a Soup Kitchen* A soup kitchen plans to feed 1,990 people. Because of space limitations, only 144 people can be served at one time. How many group seatings will be necessary to feed everyone? How many will be served at the last seating?

Strategy We will divide 1,990 by 144.

WHY Separating 1,990 people into equal-sized groups of 144 indicates division.

Solution
We translate the words of the problem to numbers and symbols.

The number of group seatings	is equal to	the number of people to be fed	divided by	the number of people at each seating.
The number of group seatings	=	1,990	÷	144

Use long division to find 1,990 ÷ 144.

```
        13
144)1,990
   -144
    550
   -432
    118
```

The quotient is 13, and the remainder is 118. This indicates that fourteen group seatings are needed: 13 full-capacity seatings and one partial seating to serve the remaining 118 people.

Self Check 9

MOVIE TICKETS On a Saturday, 3,924 movie tickets were purchased at an IMAX theater. Each showing of the movie was sold out, except for the last. If the theater seats 346 people, how many times was the movie shown on Saturday? How many people were at the last showing?

Now Try Problem 91

> **The Language of Algebra**   Here are some key words and phrases that are often used to indicate division:
>
> | split equally | distributed equally | how many does each |
> |:---:|:---:|:---:|
> | goes into | per | how much extra (remainder) |
> | shared equally | among | how many left (remainder) |

Self Check 10

TOURING   A rock band will take a 275-day world tour and spend the same number of days in each of 25 cities. How long will they stay in each city?

Now Try **Problem 97**

EXAMPLE 10   *Timeshares*   Every year, the 73 part-owners of a timeshare resort condominium get use of it for an equal number of days. How many days does each part-owner get to stay at the condo? (Use a 365-day year.)

Strategy We will divide 365 by 73.

WHY Since the part-owners get use of the condo for an equal number of days, the phrase *"How many days does each"* indicates division.

Solution
We translate the words of the problem to numbers and symbols.

The number of days each part-owner gets to stay at the condo	is equal to	the number of days in a year	divided by	the number of part-owners.
The number of days each part-owner gets to stay at the condo	=	365	÷	73

Use long division to find 365 ÷ 73.

$$\begin{array}{r} 5 \\ 73\overline{)365} \\ -365 \\ \hline 0 \end{array}$$

Each part-owner gets to stay at the condo for 5 days during the year.

Using Your CALCULATOR
The Division Key

Bottled water
A beverage company production run of 604,800 bottles of mountain spring water will be shipped to stores on pallets that hold 1,728 bottles each. We can find the number of full pallets to be shipped using the division key ÷ on a calculator.

604800 ÷ 1728 =   350

On some calculator models, the ENTER key is pressed instead of = for the result to be displayed.

The beverage company will ship 350 full pallets of bottled water.

STUDY SKILLS CHECKLIST

Get the Most from Your Textbook

The following checklist will help you become familiar with some useful features in this book. Place a check mark in each box after you answer the question.

☐ Locate the **Definition** for divisibility on page 55 and the Order of Operations Rules on page 85. What color are these boxes?

☐ Find the **Caution** box on page 34 and the **Language of Algebra** box on page 39. What color is used to identify these boxes?

☐ Each chapter begins with **From Campus to Careers** (see page 225). Chapter 3 gives information on how to become a school guidance counselor. On what page does a related problem appear in Study Set 3.4?

☐ Locate the **Study Skills Workshop** at the beginning of your text beginning on page S-1. How many Objectives appear in the Study Skills Workshop?

Answers: Green, Red, 225, 7

SECTION 1.4   STUDY SET

VOCABULARY

Fill in the blanks.

1. In the three division problems shown below, label the *dividend*, *divisor*, and the *quotient*.

$$12 \div 4 = 3$$

$$4\overline{)12} \qquad \dfrac{12}{4} = 3$$

2. We call $5 \cdot 8 = 40$ the related _____ statement for the division $40 \div 8 = 5$.

3. The problem $6\overline{)246}$ is written in _____-division form.

4. If a division is not exact, the leftover part is called the _____.

5. One number is _____ by another number if, when we divide them, the remainder is 0.

6. Phrases such as *split equally* and *how many does each* indicate the operation of _____.

CONCEPTS

7. a. Divide the objects below into groups of 3. How many groups of 3 are there?

• • • • • • • • • • • • • • • • • • • •

b. Divide the objects below into groups of 4. How many groups of 4 are there? How many objects are left over?

* * * * * * * * * * * * * * * * * * * *

8. Tell whether each statement is true or false.

a. Any whole number divided by 1 is equal to that number.

b. Any nonzero whole number divided by itself is equal to 1.

c. Zero divided by any nonzero number is undefined.

d. Division of a number by 0 is equal to 0.

Fill in the blanks.

9. Divide, if possible.

a. $\dfrac{25}{25} =$ **b.** $\dfrac{6}{1} =$

c. $\dfrac{100}{0}$ is _____   **d.** $\dfrac{0}{12} =$

10. To perform long division, we follow a four-step process: _____, _____, _____, and _____ _____ .

11. Find the *first* digit of each quotient.

   a. 5)$\overline{1147}$ **b.** 9)$\overline{587}$

   c. 23)$\overline{7501}$ **d.** 16)$\overline{892}$

12. a. Quotient · divisor = _____

   b. (Quotient · divisor) + _____ = dividend

13. To check whether the division 9)$\overline{333}$ is correct, we use multiplication:

$$\begin{array}{r} 37 \\ \times\ \ 9 \\ \hline \quad \end{array}$$

14. a. A number is divisible by if its last digit is divisible by 2.

   b. A number is divisible by 3 if the _____ of its digits is divisible by 3.

   c. A number is divisible by 4 if the number formed by its last _____ digits is divisible by 4.

15. a. A number is divisible by 5 if its last digit is ▢ or ▢ .

   b. A number is divisible by 6 if it is divisible by ▢ and ▢ .

   c. A number is divisible by 9 if the _____ of its digits is divisible by 9.

   d. A number is divisible by ▢ if its last digit is 0.

16. We can simplify the division 43,800 ÷ 200 by removing two _____ from the dividend and the divisor.

❚ NOTATION

17. Write three symbols that can be used for division.

18. In a division, 35 R 4 means "a quotient of 35 and a _____ of 4."

❚ GUIDED PRACTICE

Fill in the blanks. See Example 1.

19. 9)$\overline{45}$ (with 5 above) because ▢ · ▢ = ▢ .

20. $\frac{54}{6}$ = 9 because ▢ · ▢ = ▢ .

21. 44 ÷ 11 = 4 because ▢ · ▢ = ▢ .

22. 120 ÷ 12 = 10 because ▢ · ▢ = ▢ .

Write the related multiplication statement for each division. See Example 1.

23. 21 ÷ 3 = 7   **24.** 32 ÷ 4 = 8

25. $\frac{72}{12}$ = 6   **26.** 15)$\overline{75}$ (with 5 above)

Divide using long division. Check the result. See Example 2.

27. 96 ÷ 6   **28.** 72 ÷ 4

29. $\frac{87}{3}$ **30.** $\frac{98}{7}$

31. 2,275 ÷ 7   **32.** 1,728 ÷ 8

33. 9)$\overline{1,962}$ **34.** 5)$\overline{1,635}$

Divide using long division. Check the result. See Example 3.

35. 62)$\overline{31,248}$ **36.** 71)$\overline{28,613}$

37. 37)$\overline{22,274}$ **38.** 28)$\overline{19,712}$

Divide using long division. Check the result. See Example 4.

39. 24)$\overline{951}$ **40.** 33)$\overline{943}$

41. 999 ÷ 46   **42.** 979 ÷ 49

Divide using long division. Check the result. See Example 5.

43. $\frac{24,714}{524}$ **44.** $\frac{29,773}{531}$

45. 178)$\overline{3,514}$ **46.** 164)$\overline{2,929}$

If the given number is divisible by 2, 3, 4, 5, 6, 9, or 10, enter a checkmark ✓ in the box. See Example 6.

	Divisible by ⟶	2	3	4	5	6	9	10
47.	2,940							
48.	5,850							
49.	43,785							
50.	72,954							
51.	181,223							
52.	379,157							
53.	9,499,200							
54.	6,653,100							

Use a division shortcut to find each quotient. See Example 7.

55. 700 ÷ 10   **56.** 900 ÷ 10

57. 450)$\overline{9,900}$ **58.** 260)$\overline{9,100}$

Estimate each quotient. See Example 8.

59. 353,922 ÷ 38   **60.** 237,621 ÷ 55

61. 46,080 ÷ 933   **62.** 81,097 ÷ 419

TRY IT YOURSELF

Divide.

63. $\dfrac{25{,}950}{6}$ **64.** $\dfrac{23{,}541}{7}$

65. $54 \div 9$ **66.** $72 \div 8$

67. $273 \div 31$ **68.** $295 \div 35$

69. $\dfrac{64{,}000}{400}$ **70.** $\dfrac{125{,}000}{5{,}000}$

71. 745 divided by 7   **72.** 931 divided by 9

73. $29\overline{)14{,}761}$ **74.** $27\overline{)10{,}989}$

75. $539{,}000 \div 175$ **76.** $749{,}250 \div 185$

77. $75 \div 15$ **78.** $96 \div 16$

79. $212\overline{)5{,}087}$ **80.** $214\overline{)5{,}777}$

81. $42\overline{)1{,}273}$ **82.** $83\overline{)3{,}363}$

83. $89{,}000 \div 1{,}000$ **84.** $930{,}000 \div 1{,}000$

85. $\dfrac{57}{8}$ **86.** $\dfrac{82}{9}$

APPLICATIONS

87. TICKET SALES A movie theater makes a $4 profit on each ticket sold. How many tickets must be sold to make a profit of $2,500?

88. RUNNING Brian runs 7 miles each day. In how many days will Brian run 371 miles?

89. DUMP TRUCKS A 15-cubic-yard dump truck must haul 405 cubic yards of dirt to a construction site. How many trips must the truck make?

90. STOCKING SHELVES After receiving a delivery of 288 bags of potato chips, a store clerk stocked each shelf of an empty display with 36 bags. How many shelves of the display did he stock with potato chips?

91. LUNCH TIME A fifth grade teacher received 50 half-pint cartons of milk to distribute evenly to his class of 23 students. How many cartons did each child get? How many cartons were left over?

92. BUBBLE WRAP A furniture manufacturer uses an 11-foot-long strip of bubble wrap to protect a lamp when it is boxed and shipped to a customer. How many lamps can be packaged in this way from a 200-foot-long roll of bubble wrap? How many feet will be left on the roll?

93. GARDENING A metal can holds 640 fluid ounces of gasoline. How many times can the 68-ounce tank of a lawnmower be filled from the can? How many ounces of gasoline will be left in the can?

94. BEVERAGES A plastic container holds 896 ounces of punch. How many 6-ounce cups of punch can be served from the container? How many ounces will be left over?

95. LIFT SYSTEMS If the bus shown below weighs 58,000 pounds, how much weight is on each jack?

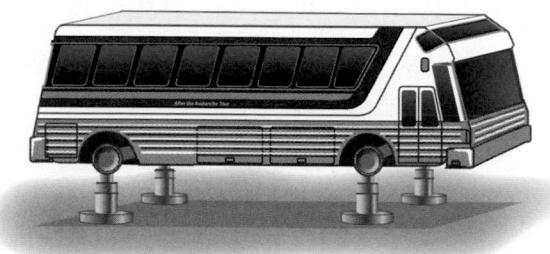

96. LOTTERY WINNERS In 2008, a group of 22 postal workers, who had been buying Pennsylvania Lotto tickets for years, won a $10,282,800 jackpot. If they split the prize evenly, how much money did each person win?

97. TEXTBOOK SALES A store received $25,200 on the sale of 240 algebra textbooks. What was the cost of each book?

98. DRAINING POOLS A 950,000-gallon pool is emptied in 20 hours. How many gallons of water are drained each hour?

99. MILEAGE A tour bus has a range of 700 miles on one tank (140 gallons) of gasoline. How far does the bus travel on one gallon of gas?

100. WATER MANAGEMENT The Susquehanna River discharges 1,719,000 cubic feet of water into Chesapeake Bay in 45 seconds. How many cubic feet of water is discharged in one second?

101. ORDERING SNACKS How many *dozen* doughnuts must be ordered for a meeting if 156 people are expected to attend, and each person will be served one doughnut?

102. TIME A *millennium* is a period of time equal to one thousand years. How many decades are in a millennium?

103. VOLLEYBALL A total of 216 girls are going to play in a city volleyball league. How many girls should be put on each team if the following requirements must be met?

- All the teams are to have the same number of players.
- A reasonable number of players on a team is 7 to 10.
- For scheduling purposes, there must be an even number of teams (2, 4, 6, 8, and so on).

104. A landscape designer intends to plant pine trees 12 feet apart to form a windscreen along one side of a flower garden, as shown below. How many trees are needed if the length of the flower garden is 744 feet?

from Campus to Careers
Landscape Designer

Comstock Images/Getty Images

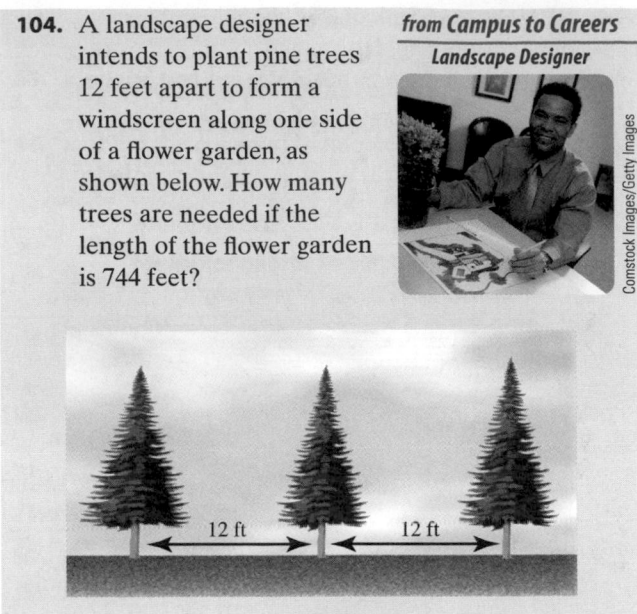

12 ft   12 ft

105. ENTRY-LEVEL JOBS The typical starting salaries for 2008 college graduates majoring in nursing, marketing, and history are shown below. Complete the last column of the table.

College major	Yearly salary	Monthly salary
Nursing	$52,128	
Marketing	$43,464	
History	$35,952	

Source: CNN.com/living

106. POPULATION To find the **population density** of a state, divide its population by its land area (in square miles). The result is the number of people per square mile. Use the data in the table to approximate the population density for each state.

State	2008 Population*	Land area* (square miles)
Arizona	6,384,000	114,000
Oklahoma	3,657,000	69,000
Rhode Island	1,100,000	1,000
South Carolina	4,500,000	30,000

Source: Wikipedia   *approximation

WRITING

107. Explain how $24 \div 6$ can be calculated by repeated subtraction.

108. Explain why division of 0 is possible, but division by 0 is impossible.

109. DIVISIBILTY TEST FOR 7 Use the following rule to show that 308 is divisible by 7. Show each of the steps of your solution in writing.

> Subtract twice the units digit from the number formed by the remaining digits. If that result is divisible by 7, then the original number is divisible by 7.

110. DIVISIBILTY TEST FOR 11 Use the following rule to show that 1,848 is divisible by 11. Show each of the steps of your solution in writing.

> Start with the digit in the one's place. From it, subtract the digit in the ten's place. To that result, add the digit in the hundred's place. From that result, subtract the digit in the thousands place, and so on. If the final result is a number divisible by 11, the original number is divisible by 11.

REVIEW

111. Add: $2,903 + 378$

112. Subtract: $2,903 - 378$

113. Multiply: $2,903 \times 378$

114. DISCOUNTS A car, originally priced at $17,550, is being sold for $13,970. By how many dollars has the price been decreased?

SECTION **1.5**
Prime Factors and Exponents

In this section, we will discuss how to express whole numbers in factored form. The procedures used to find the factored form of a whole number involve multiplication and division.

1 Factor whole numbers.

The statement $3 \cdot 2 = 6$ has two parts: the numbers that are being multiplied and the answer. The numbers that are being multiplied are called *factors,* and the answer is the *product.* We say that 3 and 2 are factors of 6.

Factors

Numbers that are multiplied together are called **factors.**

EXAMPLE 1 Find the factors of 12.

Strategy We will find all the pairs of whole numbers whose product is 12.

WHY Each of the numbers in those pairs is a factor of 12.

Solution
The pairs of whole numbers whose product is 12 are:

$1 \cdot 12 = 12, \quad 2 \cdot 6 = 12, \quad$ and $\quad 3 \cdot 4 = 12$

In order, from least to greatest, the factors of 12 are 1, 2, 3, 4, 6, and 12.

Success Tip In Example 1, once we determine the pair 1 and 12 are factors of 12, any remaining factors must be *between* 1 and 12. Once we determine that the pair 2 and 6 are factors of 12, any remaining factors must be *between* 2 and 6. Once we determine that the pair 3 and 4 are factors of 12, any remaining factors of 12 must be *between* 3 and 4. Since there are no whole numbers between 3 and 4, we know that all the possible factors of 12 have been found.

In Example 1, we found that **1, 2, 3, 4, 6,** and **12** are the factors of 12. Notice that each of the factors divides 12 exactly, leaving a remainder of 0.

$$\frac{12}{1} = 12 \qquad \frac{12}{2} = 6 \qquad \frac{12}{3} = 4 \qquad \frac{12}{4} = 3 \qquad \frac{12}{6} = 2 \qquad \frac{12}{12} = 1$$

In general, if a whole number is a factor of a given number, it also divides the given number exactly.

When we say that 3 is a factor of 6, we are using the word *factor* as a noun. The word *factor* is also used as a verb.

Factoring a Whole Number

To **factor** a whole number means to express it as the product of other whole numbers.

Objectives

1 Factor whole numbers.

2 Identify even and odd whole numbers, prime numbers, and composite numbers.

3 Find prime factorizations using a factor tree.

4 Find prime factorizations using a division ladder.

5 Use exponential notation.

6 Evaluate exponential expressions.

Self Check 1
Find the factors of 20.
Now Try Problems 21 and 27

Self Check 2
Factor 18 using   **a.** two factors
b. three factors
Now Try Problems 39 and 45

EXAMPLE 2   Factor 40 using   **a.** two factors   **b.** three factors

Strategy We will find a pair of whole numbers whose product is 40 and three whole numbers whose product is 40.

WHY To *factor* a number means to express it as the product of two (or more) numbers.

Solution

a. To factor 40 using two factors, there are several possibilities.

$$40 = 1 \cdot 40, \qquad 40 = 2 \cdot 20, \qquad 40 = 4 \cdot 10, \qquad \text{and} \qquad 40 = 5 \cdot 8$$

b. To factor 40 using three factors, there are several possibilities. Two of them are:

$$40 = 5 \cdot 4 \cdot 2 \qquad \text{and} \qquad 40 = 2 \cdot 2 \cdot 10$$

Self Check 3
Find the factors of 23.
Now Try Problem 49

EXAMPLE 3   Find the factors of 17.

Strategy We will find all the pairs of whole numbers whose product is 17.

WHY Each of the numbers in those pairs is a factor of 17.

Solution
The only pair of whole numbers whose product is 17 is:

$$1 \cdot 17 = 17$$

Therefore, the only factors of 17 are 1 and 17.

2 **Identify even and odd whole numbers, prime numbers, and composite numbers.**

A whole number is either *even* or *odd*.

Even and Odd Whole Numbers

If a whole number is divisible by 2, it is called an **even** number.

If a whole number is not divisible by 2, it is called an **odd** number.

The even whole numbers are the numbers

0, 2, 4, 6, 8, 10, 12, 14, 16, 18, …

The odd whole numbers are the numbers

1, 3, 5, 7, 9, 11, 13, 15, 17, 19, …

The three dots at the end of each list shown above indicate that there are infinitely many even and infinitely many odd whole numbers.

The Language of Algebra   The word *infinitely* is a form of the word *infinite*, meaning *unlimited*.

In Example 3, we saw that the only factors of 17 are 1 and 17. Numbers that have only two factors, 1 and the number itself, are called **prime numbers.**

Prime Numbers

A **prime number** is a whole number greater than 1 that has only 1 and itself as factors.

The prime numbers are the numbers:

2, 3, 5, 7, 11, 13, 17, 19, 23, 29, 31, 37, 41, 43, 47, 53, 59, 61, 67, 71, 73, 79, 83, 89, 97, 101, …

There are infinitely many prime numbers.

Note that the only even prime number is 2. Any other even whole number is divisible by 2, and thus has 2 as a factor, in addition to 1 and itself. Also note that not all odd whole numbers are prime numbers. For example, since 15 has factors of 1, 3, 5, and 15, it is not a prime number.

The set of whole numbers contains many prime numbers. It also contains many numbers that are not prime.

Composite Numbers

The **composite numbers** are whole numbers greater than 1 that are *not* prime.

The composite numbers are the numbers

4, 6, 8, 9, 10, 12, 14, 15, 16, 18, …

There are infinitely many composite numbers.

Caution!   The numbers 0 and 1 are neither prime nor composite, because neither is a whole number greater than 1.

EXAMPLE 4   **a.** Is 37 a prime number?   **b.** Is 45 a prime number?

Strategy We will determine whether the given number has only 1 and itself as factors.

WHY If that is the case, it is a prime number.

Solution

a. Since 37 is a whole number greater than 1 and its only factors are 1 and 37, it is prime. Since 37 is not divisible by 2, we say it is an odd prime number.

b. The factors of 45 are 1, 3, 5, 9, 15, and 45. Since it has factors other than 1 and 45, 45 is *not* prime. It is an odd composite number.

Self Check 4

a. Is 39 a prime number?

b. Is 57 a prime number?

Now Try **Problems 53 and 57**

3   Find prime factorizations using a factor tree.

Every composite number can be formed by multiplying a specific combination of prime numbers. The process of finding that combination is called **prime factorization.**

Prime Factorization

To find the **prime factorization** of a whole number means to write it as the product of only prime numbers.

One method for finding the prime factorization of a number is called a **factor tree.** The factor trees shown below are used to find the prime factorization of 90 in two ways.

1. Factor 90 as $9 \cdot 10$.

2. Neither 9 nor 10 are prime, so we factor each of them.

3. The process is complete when only prime numbers appear at the bottom of all branches.

```
      90
     /  \
    9    10
   /\    /\
  3  3  2  5
```

1. Factor 90 as $6 \cdot 15$.

2. Neither 6 nor 15 are prime, so we factor each of them.

3. The process is complete when only prime numbers appear at the bottom of all branches.

```
      90
     /  \
    6    15
   /\    /\
  2  3  3  5
```

Either way, the prime factorization of 90 contains one factor of 2, two factors of 3, and one factor of 5. Writing the factors in order, from least to greatest, the **prime-factored form** of 90 is $2 \cdot 3 \cdot 3 \cdot 5$. It is true that no other combination of prime factors will produce 90. This example illustrates an important fact about composite numbers.

Fundamental Theorem of Arithmetic

Any composite number has exactly one set of prime factors.

Self Check 5

Use a factor tree to find the prime factorization of 126.

Now Try Problems 61 and 71

EXAMPLE 5   Use a factor tree to find the prime factorization of 210.

Strategy We will factor each number that we encounter as a product of two whole numbers (other than 1 and itself) until *all the factors involved are prime.*

WHY The prime factorization of a whole number contains only prime numbers.

Solution

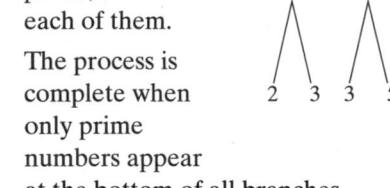

Factor 210 as $7 \cdot 30$. (The resulting prime factorization will be the same no matter which two factors of 210 you begin with.) Since 7 is prime, circle it. That branch of the tree is completed.

Since 30 is not prime, factor it as $5 \cdot 6$. (The resulting prime factorization will be the same no matter which two factors of 30 you use.) Since 5 is prime, circle it. That branch of the tree is completed.

Since 6 is not prime, factor it as $2 \cdot 3$. Since 2 and 3 are prime, circle them. All the branches of the tree are now completed.

The prime factorization of 210 is $7 \cdot 5 \cdot 2 \cdot 3$. Writing the prime factors in order, from least to greatest, we have $210 = 2 \cdot 3 \cdot 5 \cdot 7$.

Check: Multiply the prime factors. The product should be 210.

$$2 \cdot 3 \cdot 5 \cdot 7 = 6 \cdot 5 \cdot 7 \quad \text{Write the multiplication in horizontal form.}$$
$$\text{Working left to right, multiply 2 and 3.}$$
$$= 30 \cdot 7 \quad \text{Working left to right, multiply 6 and 5.}$$
$$= 210 \quad \text{Multiply 30 and 7. The result checks.}$$

Caution! Remember that there is a difference between the *factors* and the *prime factors* of a number. For example,

> The factors of 15 are: 1, 3, 5, 15
>
> The prime factors of 15 are: 3 and 5

4   Find prime factorizations using a division ladder.

We can also find the prime factorization of a whole number using an inverted division process called a **division ladder.** It is called that because of the vertical "steps" that it produces.

Success Tip   The divisibility rules found in Section 1.5 are helpful when using the division ladder method. You may want to review them at this time.

EXAMPLE 6   Use a division ladder to find the prime factorization of 280.

Strategy We will perform repeated divisions by prime numbers until the final quotient is itself a prime number.

WHY If a prime number is a factor of 280, it will divide 280 exactly.

Solution
It is helpful to begin with the *smallest prime,* 2, as the first trial divisor. Then, if necessary, try the primes 3, 5, 7, 11, 13, ... in that order.

Step 1 The prime number 2 divides 280 exactly.

The result is 140, which is not prime. Continue the division process.

$$2\underline{)280}$$
$$140$$

Step 2 Since 140 is even, divide by 2 again.

The result is 70, which is not prime. Continue the division process.

$$2\underline{)280}$$
$$2\underline{)140}$$
$$70$$

Step 3 Since 70 is even, divide by 2 a third time. The result is 35, which is not prime.

Continue the division process.

$$2\underline{)280}$$
$$2\underline{)140}$$
$$2\underline{)70}$$
$$35$$

Step 4 Since neither the prime number 2 nor the next greatest prime number 3 divide 35 exactly, we try 5. The result is 7, which is prime. We are done.

The prime factorization of 280 appears in the left column of the division ladder: $2 \cdot 2 \cdot 2 \cdot 5 \cdot 7$. Check this result using multiplication.

$$2\underline{)280}$$
$$2\underline{)140}$$
$$2\underline{)70}$$
$$5\underline{)35}$$
$$7 \leftarrow \text{Prime}$$

Caution! In Example 6, it would be incorrect to begin the division process with

$$4\underline{)280}$$
$$70$$

because 4 is not a prime number.

Self Check 6

Use a division ladder to find the prime factorization of 108.

Now Try Problems 63 and 73

5 Use exponential notation.

In Example 6, we saw that the prime factorization of 280 is $2 \cdot 2 \cdot 2 \cdot 5 \cdot 7$. Because this factorization has three factors of 2, we call 2 a *repeated factor*. We can use **exponential notation** to write $2 \cdot 2 \cdot 2$ in a more compact form.

Exponent and Base

An **exponent** is used to indicate repeated multiplication. It tells how many times the **base** is used as a factor.

The exponent is 3.

$$\underbrace{2 \cdot 2 \cdot 2}_{\text{Repeated factors}} = \underset{\uparrow}{2^3} \qquad \text{Read } 2^3 \text{ as "2 to the third power" or "2 cubed."}$$

The base is 2.

The prime factorization of 280 can be written using exponents: $2 \cdot 2 \cdot 2 \cdot 5 \cdot 7 = 2^3 \cdot 5 \cdot 7$.

In the **exponential expression** 2^3, the number 2 is the base and 3 is the exponent. The expression itself is called a **power of 2.**

Self Check 7

Write each product using exponents:

a. $3 \cdot 3 \cdot 7$

b. $5(5)(7)(7)$

c. $2 \cdot 2 \cdot 2 \cdot 3 \cdot 3 \cdot 5$

Now Try Problems 77 and 81

EXAMPLE 7 Write each product using exponents:

a. $5 \cdot 5 \cdot 5 \cdot 5$ **b.** $7 \cdot 7 \cdot 11$ **c.** $2(2)(2)(2)(3)(3)(3)$

Strategy We will determine the number of repeated factors in each expression.

WHY An exponent can be used to represent repeated multiplication.

Solution

a. The factor 5 is repeated 4 times. We can represent this repeated multiplication with an exponential expression having a base of 5 and an exponent of 4:

$$5 \cdot 5 \cdot 5 \cdot 5 = 5^4$$

b. $7 \cdot 7 \cdot 11 = 7^2 \cdot 11$ 7 is used as a factor 2 times.

c. $2(2)(2)(2)(3)(3)(3) = 2^4(3^3)$ 2 is used as a factor 4 times, and 3 is used as a factor 3 times.

---

6 Evaluate exponential expressions.

We can use the definition of exponent to **evaluate** (find the value of) exponential expressions.

Self Check 8

Evaluate each expression:

a. 9^2 **b.** 6^3

c. 3^4 **d.** 12^1

Now Try Problem 89

EXAMPLE 8 Evaluate each expression:

a. 7^2 **b.** 2^5 **c.** 10^4 **d.** 6^1

Strategy We will rewrite each exponential expression as a product of repeated factors, and then perform the multiplication. This requires that we identify the base and the exponent.

WHY The exponent tells the number of times the base is to be written as a factor.

Solution

We can write the steps of the solutions in horizontal form.

a. $7^2 = 7 \cdot 7$ Read 7^2 as "7 to the second power" or "7 squared." The base is 7 and the exponent is 2. Write the base as a factor 2 times.

$= 49$ Multiply.

b. $2^5 = 2 \cdot 2 \cdot 2 \cdot 2 \cdot 2$ Read 2^5 as "2 to the 5th power." The base is 2 and the exponent is 5. Write the base as a factor 5 times.

$= 4 \cdot 2 \cdot 2 \cdot 2$ Multiply, working left to right.

$= 8 \cdot 2 \cdot 2$

$= 16 \cdot 2$

$= 32$

c. $10^4 = 10 \cdot 10 \cdot 10 \cdot 10$ Read 10^4 as "10 to the 4th power." The base is 10 and the exponent is 4. Write the base as a factor 4 times.

$= 100 \cdot 10 \cdot 10$ Multiply, working left to right.

$= 1,000 \cdot 10$

$= 10,000$

d. $6^1 = 6$ Read 6^1 as "6 to the first power." Write the base 6 once.

Caution!   Note that 2^5 means $2 \cdot 2 \cdot 2 \cdot 2 \cdot 2$. It does not mean $2 \cdot 5$. That is, $2^5 = 32$ and $2 \cdot 5 = 10$.

EXAMPLE 9   The prime factorization of a number is $2^3 \cdot 3^4 \cdot 5$. What is the number?

Strategy To find the number, we will evaluate each exponential expression and then do the multiplication.

WHY The exponential expressions must be evaluated first.

Solution
We can write the steps of the solutions in horizontal form.

$2^3 \cdot 3^4 \cdot 5 = 8 \cdot 81 \cdot 5$ Evaluate the exponential expressions: $2^3 = 8$ and $3^4 = 81$.

$= 648 \cdot 5$ Multiply, working left to right.

$= 3,240$ Multiply.

$2^2 \cdot 3^4 \cdot 5$ is the prime factorization of 3,240.

$$\begin{array}{r} 81 \\ \times\ 8 \\ \hline 648 \end{array}$$

$$\begin{array}{r} {\scriptstyle 2\ 4} \\ 648 \\ \times\ 5 \\ \hline 3,240 \end{array}$$

Success Tip Calculations that you cannot perform in your head should be shown outside the steps of your solution.

Self Check 9
The prime factorization of a number is $2 \cdot 3^3 \cdot 5^2$. What is the number?

Now Try Problems 93 and 97

Using Your CALCULATOR   **The Exponential Key: Bacteria Growth**

At the end of 1 hour, a culture contains two bacteria. Suppose the number of bacteria doubles every hour thereafter. Use exponents to determine how many bacteria the culture will contain after 24 hours.

We can use a table to help model the situation. From the table, we see a pattern developing: The number of bacteria in the culture after 24 hours will be 2^{24}.

Time	Number of bacteria
1 hr	$2 = 2^1$
2 hr	$4 = 2^2$
3 hr	$8 = 2^3$
4 hr	$16 = 2^4$
24 hr	$? = 2^{24}$

We can evaluate this exponential expression using the exponential key $\boxed{y^x}$ on a scientific calculator ($\boxed{x^y}$ on some models).

   2 $\boxed{y^x}$ 24 $\boxed{=}$ $\boxed{\text{16777216}}$

On a graphing calculator, we use the carat key $\boxed{\wedge}$ to raise a number to a power.

   2 $\boxed{\wedge}$ 24 $\boxed{\text{ENTER}}$ $\boxed{\text{16777216}}$

Since $2^{24} = 16{,}777{,}216$, there will be 16,777,216 bacteria after 24 hours.

ANSWERS TO SELF CHECKS

1. 1, 2, 4, 5, 10, and 20   **2. a.** $1 \cdot 18, 2 \cdot 9,$ or $3 \cdot 6$ **b.** Two possibilities are $2 \cdot 3 \cdot 3$ and $1 \cdot 2 \cdot 9$ **3.** 1 and 23   **4. a.** no   **b.** no   **5.** $2 \cdot 3 \cdot 3 \cdot 7$ **6.** $2 \cdot 2 \cdot 3 \cdot 3 \cdot 3$ **7. a.** $3^2 \cdot 7$ **b.** $5^2(7^2)$ **c.** $2^3 \cdot 3^2 \cdot 5$ **8. a.** 81   **b.** 216   **c.** 81   **d.** 12   **9.** 1,350

SECTION 1.5   STUDY SET

VOCABULARY

Fill in the blanks.

1. Numbers that are multiplied together are called _____.

2. To _____ a whole number means to express it as the product of other whole numbers.

3. A _____ number is a whole number greater than 1 that has only 1 and itself as factors.

4. Whole numbers greater than 1 that are not prime numbers are called _____ numbers.

5. To prime factor a number means to write it as a product of only _____ numbers.

6. An exponent is used to represent _____ multiplication. It tells how many times the _____ is used as a factor.

7. In the exponential expression 6^4, the number 6 is the _____, and 4 is the _____.

8. We can read 5^2 as "5 to the second power" or as "5 _____." We can read 7^3 as "7 to the third power" or as "7 _____."

CONCEPTS

9. Fill in the blanks to find the pairs of whole numbers whose product is 45.

   $1 \cdot \boxed{} = 45$ $3 \cdot \boxed{} = 45$ $5 \cdot \boxed{} = 45$

The factors of 45, in order from least to greatest, are: $\boxed{}, \boxed{}, \boxed{}, \boxed{}, \boxed{}, \boxed{}$

10. Fill in the blanks to find the pairs of whole numbers whose product is 28.

   $1 \cdot \boxed{} = 28$ $2 \cdot \boxed{} = 28$ $4 \cdot \boxed{} = 28$

The factors of 28, in order from least to greatest, are: $\boxed{}, \boxed{}, \boxed{}, \boxed{}, \boxed{}, \boxed{}$

11. If 4 is a factor of a whole number, will 4 divide the number exactly?

12. Suppose a number is divisible by 10. Is 10 a factor of the number?

13. a. Fill in the blanks: If a whole number is divisible by 2, it is an _____ number. If it is not divisible by 2, it is an _____ number.

   b. List the first 10 even whole numbers.

   c. List the first 10 odd whole numbers.

14. a. List the first 10 prime numbers.

   b. List the first 10 composite numbers.

15. Fill in the blanks to prime factor 150 using a factor tree.

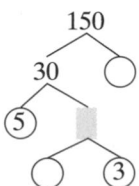

The prime factorization of 150 is $\boxed{} \cdot \boxed{} \cdot \boxed{} \cdot \boxed{}$.

16. Which of the whole numbers, 1, 2, 3, 4, 5, 6, 7, 8, 9, and 10, could be at the top of this factor tree?

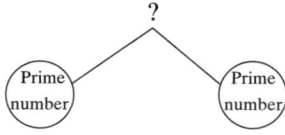

17. Fill in the blanks to prime factor 150 using a division ladder.

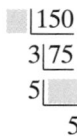

The prime factorization of 150 is ▓ · ▓ · ▓ · ▓.

18. **a.** When using the division ladder method to find the prime factorization of a number, what is the first divisor to try?

  b. If 2 does not divide the given number exactly, what other divisors should be tried?

▌ NOTATION

19. For each exponential expression, what is the base and the exponent?

  a. 7^6 **b.** 15^1

20. Consider the expression $2 \cdot 2 \cdot 2 \cdot 3 \cdot 3$.

  a. How many repeated factors of 2 are there?

  b. How many repeated factors of 3 are there?

▌ GUIDED PRACTICE

Find the factors of each whole number. List them from least to greatest. **See Example 1.**

21. 10	**22.** 6
23. 40	**24.** 75
25. 18	**26.** 32
27. 44	**28.** 65
29. 77	**30.** 81
31. 100	**32.** 441

Factor each of the following whole numbers using two factors. Do not use the factor 1 in your answer. **See Example 2.**

33. 8	**34.** 9
35. 27	**36.** 35
37. 49	**38.** 25
39. 20	**40.** 16

Factor each of the following whole numbers using three factors. Do not use the factor 1 in your answer. **See Example 2**

41. 30	**42.** 28
43. 63	**44.** 50
45. 54	**46.** 56
47. 60	**48.** 64

Find the factors of each whole number. **See Example 3.**

49. 11	**50.** 29
51. 37	**52.** 41

Determine whether each of the following numbers is a prime number. **See Example 4.**

53. 17	**54.** 59
55. 99	**56.** 27
57. 51	**58.** 91
59. 43	**60.** 83

Find the prime factorization of each number. Use exponents in your answer, when it is helpful. **See Examples 5 and 6.**

61. 30	**62.** 20
63. 39	**64.** 105
65. 99	**66.** 400
67. 162	**68.** 98
69. 64	**70.** 243
71. 147	**72.** 140
73. 220	**74.** 385
75. 102	**76.** 114

Write each product using exponents. **See Example 7.**

77. $2 \cdot 2 \cdot 2 \cdot 2 \cdot 2$ **78.** $3 \cdot 3 \cdot 3 \cdot 3 \cdot 3 \cdot 3$

79. $5 \cdot 5 \cdot 5 \cdot 5$ **80.** $9 \cdot 9 \cdot 9$

81. $4(4)(8)(8)(8)$ **82.** $12(12)(12)(16)$

83. $7 \cdot 7 \cdot 7 \cdot 9 \cdot 9 \cdot 7 \cdot 7 \cdot 7 \cdot 9$

84. $6 \cdot 6 \cdot 6 \cdot 5 \cdot 5 \cdot 6 \cdot 6 \cdot 6$

Evaluate each exponential expression. **See Example 8.**

85. a. 3^4	**b.** 4^3	**86. a.** 5^3	**b.** 3^5
87. a. 2^5	**b.** 5^2	**88. a.** 4^5	**b.** 5^4
89. a. 7^3	**b.** 3^7	**90. a.** 8^2	**b.** 2^8
91. a. 9^1	**b.** 1^9	**92. a.** 20^1	**b.** 1^{20}

The prime factorization of a number is given. What is the number? See Example 9.

93. $2 \cdot 3 \cdot 3 \cdot 5$

94. $2 \cdot 2 \cdot 2 \cdot 7$

95. $7 \cdot 11^2$

96. $2 \cdot 3^4$

97. $3^2 \cdot 5^2$

98. $3^3 \cdot 5^3$

99. $2^3 \cdot 3^3 \cdot 13$

100. $2^3 \cdot 3^2 \cdot 11$

APPLICATIONS

101. **PERFECT NUMBERS** A whole number is called a **perfect number** when the sum of its factors that are less than the number equals the number. For example, 6 is a perfect number, because $1 + 2 + 3 = 6$. Find the factors of 28. Then use addition to show that 28 is also a perfect number.

102. **CRYPTOGRAPHY** Information is often transmitted in code. Many codes involve writing products of large primes, because they are difficult to factor. To see how difficult, try finding two prime factors of 7,663. (*Hint:* Both primes are greater than 70.)

103. **LIGHT** The illustration shows that the light energy that passes through the first unit of area, 1 yard away from the bulb, spreads out as it travels away from the source. How much area does that energy cover 2 yards, 3 yards, and 4 yards from the bulb? Express each answer using exponents.

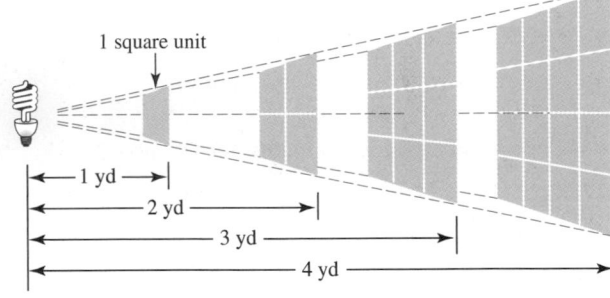

104. **CELL DIVISION** After 1 hour, a cell has divided to form another cell. In another hour, these two cells have divided so that four cells exist. In another hour, these four cells divide so that eight exist.

   a. How many cells exist at the end of the fourth hour?

   b. The number of cells that exist after each division can be found using an exponential expression. What is the base?

   c. Find the number of cells after 12 hours.

WRITING

105. Explain how to check a prime factorization.

106. Explain the difference between the *factors* of a number and the *prime factors* of a number. Give an example.

107. Find 1^2, 1^3, and 1^4. From the results, what can be said about any power of 1?

108. Use the phrase *infinitely many* in a sentence.

REVIEW

109. **MARCHING BANDS** When a university band lines up in eight rows of fifteen musicians, there are five musicians left over. How many band members are there?

110. **U.S. COLLEGE COSTS** In 2008, the average yearly tuition cost and fees at a private four-year college was $25,143. The average yearly tuition cost and fees at a public four-year college was $6,585. At these rates, how much less are the tuition costs and fees at a public college over four years? (Source: The College Board)

Objectives

1. Find the LCM by listing multiples.
2. Find the LCM using prime factorization.
3. Find the GCF by listing factors.
4. Find the GCF using prime factorization.

SECTION **1.6**

The Least Common Multiple and the Greatest Common Factor

As a child, you probably learned how to count by 2's and 5's and 10's. Counting in that way is an example of an important concept in mathematics called *multiples*.

1 Find the LCM by listing multiples.

The **multiples** of a number are the products of that number and 1, 2, 3, 4, 5, and so on.

EXAMPLE 1   Find the first eight multiples of 6.

Strategy We will multiply 6 by 1, 2, 3, 4, 5, 6, 7, and 8.

WHY The *multiples of a number* are the products of that number and 1, 2, 3, 4, 5, and so on.

Solution
To find the multiples, we proceed as follows:

$6 \cdot 1 = 6$ This is the first multiple of 6.
$6 \cdot 2 = 12$
$6 \cdot 3 = 18$
$6 \cdot 4 = 24$
$6 \cdot 5 = 30$
$6 \cdot 6 = 36$
$6 \cdot 7 = 42$
$6 \cdot 8 = 48$ This is the eighth multiple of 6.

The first eight multiples of 6 are 6, 12, 18, 24, 30, 36, 42, and 48.

The first eight multiples of 3 and the first eight multiples of 4 are shown below. The numbers highlighted in red are *common multiples* of 3 and 4.

$3 \cdot 1 = 3$ $4 \cdot 1 = 4$
$3 \cdot 2 = 6$ $4 \cdot 2 = 8$
$3 \cdot 3 = 9$ $4 \cdot 3 = \mathbf{12}$
$3 \cdot 4 = \mathbf{12}$ $4 \cdot 4 = 16$
$3 \cdot 5 = 15$ $4 \cdot 5 = 20$
$3 \cdot 6 = 18$ $4 \cdot 6 = \mathbf{24}$
$3 \cdot 7 = 21$ $4 \cdot 7 = 28$
$3 \cdot 8 = \mathbf{24}$ $4 \cdot 8 = 32$

If we extend each list, it soon becomes apparent that 3 and 4 have infinitely many common multiples.

The common multiples of 3 and 4 are: **12, 24, 36, 48, 60, 72,** …

Because 12 is the smallest number that is a multiple of both 3 and 4, it is called the **least common multiple (LCM)** of 3 and 4. We can write this in compact form as:

LCM $(3, 4) = 12$ Read as "The least common multiple of 3 and 4 is 12."

The Least Common Multiple (LCM)

The **least common multiple** of two whole numbers is the smallest common multiple of the numbers.

We have seen that the LCM of 3 and 4 is 12. It is important to note that 12 is divisible by both 3 and 4.

$$\frac{12}{3} = 4 \quad \text{and} \quad \frac{12}{4} = 3$$

This observation illustrates an important relationship between divisibility and the least common multiple.

Self Check 1
Find the first eight multiples of 9.
Now Try Problems 17 and 85

The Least Common Multiple (LCM)

The **least common multiple (LCM)** of two whole numbers is the smallest whole number that is divisible by both of those numbers.

When finding the LCM of two numbers, writing both lists of multiples can be tiresome. From the previous definition of LCM, it follows that we need only list the multiples of the larger number. The LCM is simply *the first multiple of the larger number that is divisible by the smaller number.* For example, to find the LCM of 3 and 4, we observe that

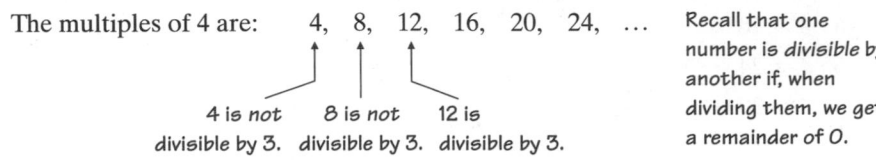

The multiples of 4 are: 4,   8,   12,   16,   20,   24,   …

4 is not divisible by 3.   8 is not divisible by 3.   12 is divisible by 3.

Recall that one number is *divisible* by another if, when dividing them, we get a remainder of 0.

Since 12 is the first multiple of 4 that is divisible by 3, the LCM of 3 and 4 is 12. As expected, this is the same result that we obtained using the two-list method.

Finding the LCM by Listing the Multiples of the Largest Number

To find the least common multiple of two (or more) whole numbers:

1. Write multiples of the largest number by multiplying it by 1, 2, 3, 4, 5, and so on.

2. Continue this process until you find the first multiple of the larger number that is divisible by each of the smaller numbers. That multiple is their LCM.

Self Check 2

Find the LCM of 8 and 10.

Now Try Problem 25

EXAMPLE 2   Find the LCM of 6 and 8.

Strategy We will write the multiples of the larger number, 8, until we find one that is divisible by the smaller number, 6.

WHY The LCM of 6 and 8 is the smallest multiple of 8 that is divisible by 6.

Solution

The 1st multiple of 8: $8 \cdot 1 = 8$ ← 8 is not divisible by 6. (When we divide, we get a remainder of 2.) Since 8 is not divisible by 6, find the next multiple.

The 2nd multiple of 8: $8 \cdot 2 = 16$ ← 16 is not divisible by 6. Find the next multiple.

The 3rd multiple of 8: $8 \cdot 3 = 24$ ← 24 is divisible by 6. This is the LCM.

The first multiple of 8 that is divisible by 6 is 24. Thus,

LCM $(6, 8) = 24$ Read as "The least common multiple of 6 and 8 is 24."

We can extend this method to find the LCM of three whole numbers.

Self Check 3

Find the LCM of 3, 4, and 8.

Now Try Problem 35

EXAMPLE 3   Find the LCM of 2, 3, and 10.

Strategy We will write the multiples of the largest number, 10, until we find one that is divisible by both of the smaller numbers, 2 and 3.

WHY The LCM of 2, 3, and 10 is the smallest multiple of 10 that is divisible by 2 and 3.

Solution

The 1st multiple of 10: $10 \cdot 1 = 10$ ← 10 is divisible by 2, but not by 3. Find the next multiple.

The 2nd multiple of 10: $10 \cdot 2 = 20$ ← 20 is divisible by 2, but not by 3. Find the next multiple.

The 3rd multiple of 10: $10 \cdot 3 = 30$ ← 30 is divisible by 2 and by 3. It is the LCM.

The first multiple of 10 that is divisible by 2 and 3 is 30. Thus,

LCM $(2, 3, 10) = 30$ Read as "The least common multiple of 2, 3, and 10 is 30." ■

2 Find the LCM using prime factorization.

Another method for finding the LCM of two (or more) whole numbers uses prime factorization. This method is especially helpful when working with larger numbers. As an example, we will find the LCM of 36 and 54. First, we find their prime factorizations:

$36 = 2 \cdot 2 \cdot 3 \cdot 3$ Factor trees (or division ladders) can be used to find the prime factorizations.

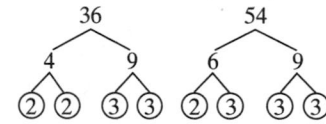

$54 = 2 \cdot 3 \cdot 3 \cdot 3$

The LCM of 36 and 54 must be divisible by 36 and 54. If the LCM is divisible by 36, it must have the prime factors of 36, which are $2 \cdot 2 \cdot 3 \cdot 3$. If the LCM is divisible by 54, it must have the prime factors of 54, which are $2 \cdot 3 \cdot 3 \cdot 3$. The smallest number that meets both requirements is

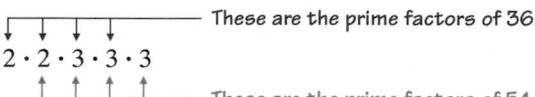

These are the prime factors of 36.
$2 \cdot 2 \cdot 3 \cdot 3 \cdot 3$
These are the prime factors of 54.

To find the LCM, we perform the indicated multiplication:

LCM $(36, 54) = 2 \cdot 2 \cdot 3 \cdot 3 \cdot 3 = 108$

Caution!   The LCM $(36, 54)$ is not the product of the prime factorization of 36 and the prime factorization of 54. That gives an incorrect answer of 2,052.

LCM $(36, 54) = 2 \cdot 2 \cdot 3 \cdot 3 \cdot 2 \cdot 3 \cdot 3 \cdot 3 = 1,944$

The LCM should contain all the prime factors of 36 and all the prime factors of 54, but the prime factors that 36 and 54 have in common are not repeated.

The prime factorizations of 36 and 54 contain the numbers 2 and 3.

$36 = 2 \cdot 2 \cdot 3 \cdot 3$ $54 = 2 \cdot 3 \cdot 3 \cdot 3$

We see that

- The greatest number of times the factor 2 appears in any one of the prime factorizations is twice and the LCM of 36 and 54 has 2 as a factor twice.
- The greatest number of times that 3 appears in any one of the prime factorizations is three times and the LCM of 36 and 54 has 3 as a factor three times.

These observations suggest a procedure to use to find the LCM of two (or more) numbers using prime factorization.

Finding the LCM Using Prime Factorization

To find the least common multiple of two (or more) whole numbers:

1. Prime factor each number.

2. The LCM is a product of prime factors, where each factor is used the greatest number of times it appears in any one factorization.

Self Check 4

Find the LCM of 18 and 32.

Now Try Problem 37

EXAMPLE 4   Find the LCM of 24 and 60.

Strategy We will begin by finding the prime factorizations of 24 and 60.

WHY To find the LCM, we need to determine the greatest number of times each prime factor appears in any one factorization.

Solution

Step 1 Prime factor 24 and 60.

$24 = 2 \cdot 2 \cdot 2 \cdot 3$ Division ladders (or factor trees) can be used to find the prime factorizations.

$60 = 2 \cdot 2 \cdot 3 \cdot 5$

$$
\begin{array}{ll}
2\,|\,24 & 2\,|\,60 \\
2\,|\,12 & 2\,|\,30 \\
2\,|\,6 & 3\,|\,15 \\
\quad 3 & \quad 5
\end{array}
$$

Step 2 The prime factorizations of 24 and 60 contain the prime factors 2, 3, and 5. To find the LCM, we use each of these factors the greatest number of times it appears in any one factorization.

- We will use the factor 2 three times, because 2 appears three times in the factorization of 24. Circle $2 \cdot 2 \cdot 2$, as shown below.

- We will use the factor 3 once, because it appears one time in the factorization of 24 and one time in the factorization of 60. When the number of times a factor appears are equal, circle either one, but not both, as shown below.

- We will use the factor 5 once, because it appears one time in the factorization of 60. Circle the 5, as shown below.

$$24 = \boxed{2 \cdot 2 \cdot 2} \cdot ③$$
$$60 = 2 \cdot 2 \cdot 3 \cdot ⑤$$

Since there are no other prime factors in either prime factorization, we have

Use 2 three times.
Use 3 one time.
Use 5 one time.

$$\text{LCM } (24, 60) = 2 \cdot 2 \cdot 2 \cdot 3 \cdot 5 = 120$$

Note that 120 is the smallest number that is divisible by both 24 and 60:

$$\frac{120}{24} = 5 \quad \text{and} \quad \frac{120}{60} = 2$$

In Example 4, we can express the prime factorizations of 24 and 60 using exponents. To determine the greatest number of times each factor appears in any one factorization, we circle the factor with the greatest exponent.

$$24 = \boxed{2^3} \cdot \boxed{3^1}$$

  The greatest exponent on the factor 2 is 3.
  The greatest exponent on the factor 3 is 1.

$$60 = 2^2 \cdot 3^1 \cdot \boxed{5^1}$$

  The greatest exponent on the factor 5 is 1.

The LCM of 24 and 60 is

$$2^3 \cdot 3^1 \cdot 5^1 = 8 \cdot 3 \cdot 5 = 120 \qquad \text{Evaluate: } 2^3 = 8.$$

EXAMPLE 5   Find the LCM of 28, 42, and 45.

Strategy We will begin by finding the prime factorizations of 28, 42, and 45.

WHY To find the LCM, we need to determine the greatest number of times each prime factor appears in any one factorization.

Solution

Step 1 Prime factor 28, 42, and 45.

$$28 = \boxed{2 \cdot 2} \cdot 7 \quad \text{This can be written as} \boxed{2^2} \cdot 7^1.$$
$$42 = 2 \cdot 3 \cdot \boxed{7} \quad \text{This can be written as } 2^1 \cdot 3^1 \cdot \boxed{7^1}.$$
$$45 = \boxed{3 \cdot 3} \cdot \boxed{5} \quad \text{This can be written as} \boxed{3^2} \cdot \boxed{5}.$$

Step 2 The prime factorizations of 28, 42, and 45 contain the prime factors 2, 3, 5, and 7. To find the LCM (28, 42, 45), we use each of these factors the greatest number of times it appears in any one factorization.

- We will use the factor 2 two times, because 2 appears two times in the factorization of 28. Circle $2 \cdot 2$, as shown above.
- We will use the factor 3 twice, because it appears two times in the factorization of 45. Circle $3 \cdot 3$, as shown above.
- We will use the factor 5 once, because it appears one time in the factorization of 45. Circle the 5, as shown above.
- We will use the factor 7 once, because it appears one time in the factorization of 28 and one time in the factorization of 42. You may circle either 7, but only circle one of them.

Since there are no other prime factors in either prime factorization, we have

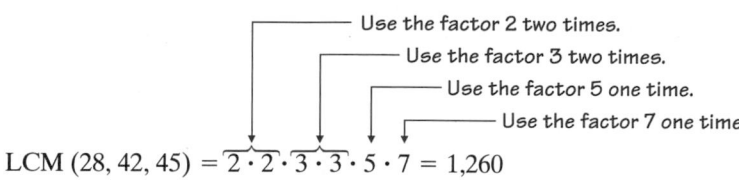

  Use the factor 2 two times.
  Use the factor 3 two times.
  Use the factor 5 one time.
  Use the factor 7 one time.

$$\text{LCM (28, 42, 45)} = 2 \cdot 2 \cdot 3 \cdot 3 \cdot 5 \cdot 7 = 1{,}260$$

If we use exponents, we have

$$\text{LCM (28, 42, 45)} = 2^2 \cdot 3^2 \cdot 5 \cdot 7 \qquad = 1{,}260$$

Either way, we have found that the LCM (28, 42, 45) = 1,260. Note that 1,260 is the smallest number that is divisible by 28, 42, and 45:

$$\frac{1{,}260}{4} = 315 \qquad \frac{1{,}260}{42} = 30 \qquad \frac{1{,}260}{45} = 28$$

Self Check 5
Find the LCM of 45, 60, and 75.

Now Try Problem 45

EXAMPLE 6   *Patient Recovery* Two patients recovering from heart surgery exercise daily by walking around a track. One patient can complete a lap in 4 minutes. The other can complete a lap in 6 minutes. If they begin at the same time and at the same place on the track, in how many minutes will they arrive together at the starting point of their workout?

Self Check 6

AQUARIUMS A pet store owner changes the water in a fish aquarium every 45 days and he changes the pump filter every 20 days. If the water and filter are changed on the same day, in how many days will they be changed again together?

Now Try Problem 87

Strategy We will find the LCM of 4 and 6.

WHY Since one patient reaches the starting point of the workout every 4 minutes, and the other is there every 6 minutes, we want to find the least common multiple of those numbers. At that time, they will both be at the starting point of the workout.

Solution

To find the LCM, we prime factor 4 and 6, and circle each prime factor the greatest number of times it appears in any one factorization.

$4 = \boxed{2 \cdot 2}$ Use the factor 2 two times, because 2 appears two times in the factorization of 4.

$6 = 2 \cdot \boxed{3}$ Use the factor 3 once, because it appears one time in the factorization of 6.

Since there are no other prime factors in either prime factorization, we have

$$\text{LCM } (4, 6) = 2 \cdot 2 \cdot 3 = 12$$

The patients will arrive together at the starting point 12 minutes after beginning their workout.

3 Find the GCF by listing factors.

We have seen that two whole numbers can have common multiples. They can also have *common factors*. To explore this concept, let's find the factors of 26 and 39 and see what factors they have in common.

To find the factors of 26, we find all the pairs of whole numbers whose product is 26. There are two possibilities:

$$1 \cdot 26 = 26 \qquad 2 \cdot 13 = 26$$

Each of the numbers in the pairs is a factor of 26. From least to greatest, the factors of 26 are 1, 2, 13, and 26.

To find the factors of 39, we find all the pairs of whole numbers whose product is 39. There are two possibilities:

$$1 \cdot 39 = 39 \qquad 3 \cdot 13 = 39$$

Each of the numbers in the pairs is a factor of 39. From least to greatest, the factors of 39 are 1, 3, 13, and 39. As shown below, the *common factors* of 26 and 39 are 1 and 13.

1, 2, 13, 26   These are the factors of 26.
1, 3, 13, 39   These are the factors of 39.

Because 13 is the largest number that is a factor of both 26 and 39, it is called the **greatest common factor (GCF)** of 26 and 39. We can write this in compact form as:

GCF (26, 39) = 13   Read as "The greatest common factor of 26 and 39 is 13."

The Greatest Common Factor (GCF)

The **greatest common factor** of two whole numbers is the largest common factor of the numbers.

Self Check 7

Find the GCF of 30 and 42.

Now Try Problem 49

EXAMPLE 7   Find the GCF of 18 and 45.

Strategy We will find the factors of 18 and 45.

WHY Then we can identify the largest factor that 18 and 45 have in common.

Solution

To find the factors of 18, we find all the pairs of whole numbers whose product is 18. There are three possibilities:

$$1 \cdot 18 = 18 \qquad 2 \cdot 9 = 18 \qquad 3 \cdot 6 = 18$$

To find the factors of 45, we find all the pairs of whole numbers whose product is 45. There are three possibilities:

$$1 \cdot 45 = 45 \qquad 3 \cdot 15 = 45 \qquad 5 \cdot 9 = 45$$

The factors of 18 and 45 are listed below. Their common factors are circled.

Factors of 18:   1,   2,   3,   6,   9,   18

Factors of 45:   1,   3,   5,   9,   15,   45

The common factors of 18 and 45 are 1, 3, and 9. Since 9 is their largest common factor,

GCF (18, 45) = 9   *Read as "The greatest common factor of 18 and 45 is 9."*

In Example 7, we found that the GCF of 18 and 45 is 9. Note that 9 is the greatest number that divides 18 and 45.

$$\frac{18}{9} = 2 \qquad \frac{45}{9} = 5$$

In general, the greatest common factor of two (or more) numbers is the largest number that divides them exactly. For this reason, the greatest common factor is also known as the **greatest common divisor (GCD)** and we can write GCD (18, 45) = 9.

4 Find the GCF using prime factorization.

We can find the GCF of two (or more) numbers by listing the factors of each number. However, this method can be lengthy. Another way to find the GCF uses the prime factorization of each number.

Finding the GCF Using Prime Factorization

To find the greatest common factor of two (or more) whole numbers:

1. Prime factor each number.
2. Identify the common prime factors.
3. The GCF is a product of all the common prime factors found in Step 2. If there are no common prime factors, the GCF is 1.

EXAMPLE 8   Find the GCF of 48 and 72.

Strategy We will begin by finding the prime factorizations of 48 and 72.

WHY Then we can identify any prime factors that they have in common.

Solution

Step 1 Prime factor 48 and 72.

$$48 = 2 \cdot 2 \cdot 2 \cdot 2 \cdot 3$$
$$72 = 2 \cdot 2 \cdot 2 \cdot 3 \cdot 3$$

Self Check 8
Find the GCF of 36 and 60.
Now Try Problem 57

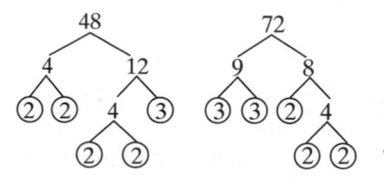

Step 2 The circling on the previous page shows that 48 and 72 have four common prime factors: Three common factors of 2 and one common factor of 3.

Step 3 The GCF is the product of the circled prime factors.

GCF (48, 72) = 2 · 2 · 2 · 3 = 24

Self Check 9

Find the GCF of 8 and 25.

Now Try Problem 61

EXAMPLE 9   Find the GCF of 8 and 15.

Strategy We will begin by finding the prime factorizations of 8 and 15.

WHY Then we can identify any prime factors that they have in common.

Solution
The prime factorizations of 8 and 15 are shown below.

8 = 2 · 2 · 2
15 = 3 · 5

Since there are no common factors, the GCF of 8 and 15 is 1. Thus,

GCF (8, 15) = 1   Read as "The greatest common factor of 8 and 15 is 1."

Self Check 10

Find the GCF of 45, 60, and 75.

Now Try Problem 67

EXAMPLE 10   Find the GCF of 20, 60, and 140.

Strategy We will begin by finding the prime factorizations of 20, 60, and 140.

WHY Then we can identify any prime factors that they have in common.

Solution
The prime factorizations of 20, 60, and 140 are shown below.

20 = 2 · 2 · 5
60 = 2 · 2 · 3 · 5
140 = 2 · 2 · 5 · 7

The circling above shows that 20, 60, and 140 have three common factors: two common factors of 2 and one common factor of 5. The GCF is the product of the circled prime factors.

GCF (20, 60, 140) = 2 · 2 · 5 = 20   Read as "The greatest common factor of 20, 60, and 140 is 20."

Note that 20 is the greatest number that divides 20, 60, and 140 exactly.

$$\frac{20}{20} = 1 \qquad \frac{60}{20} = 3 \qquad \frac{140}{20} = 7$$

Self Check 11

SCHOOL SUPPLIES A bookstore manager wants to use some leftover items (36 markers, 54 pencils, and 108 pens) to make identical gift packs to donate to an elementary school.

a. What is the greatest number of gift packs that can be made?   *(continued)*

EXAMPLE 11   Bouquets A florist wants to use 12 white tulips, 30 pink tulips, and 42 purple tulips to make as many identical arrangements as possible. Each bouquet is to have the same number of each color tulip.

a. What is the greatest number of arrangements that she can make?

b. How many of each type of tulip can she use in each bouquet?

Strategy We will find the GCF of 12, 30, and 42.

WHY Since an equal number of tulips of each color will be used to create the identical arrangements, division is indicated. The greatest common factor of three numbers is the largest number that divides them exactly.

Solution

a. To find the GCF, we prime factor 12, 30, and 42, and circle the prime factors that they have in common.

$$12 = 2 \cdot 2 \cdot 3$$
$$30 = 2 \cdot 3 \cdot 5$$
$$42 = 2 \cdot 3 \cdot 7$$

The GCF is the product of the circled numbers.

$$\text{GCF } (12, 30, 42) = 2 \cdot 3 = 6$$

The florist can make 6 identical arrangements from the tulips.

b. To find the number of white, pink, and purple tulips in each of the 6 arrangements, we divide the number of tulips of each color by 6.

White tulips: Pink tulips: Purple tulips:

$$\frac{12}{6} = 2 \qquad \frac{30}{6} = 5 \qquad \frac{42}{6} = 7$$

Each of the 6 identical arrangements will contain 2 white tulips, 5 pink tulips, and 7 purple tulips.

b. How many of each type of item will be in each gift pack?

Now Try Problem 93

ANSWERS TO SELF CHECKS

1. 9, 18, 27, 36, 45, 54, 63, 72   **2.** 40   **3.** 24   **4.** 288   **5.** 900   **6.** 180 days   **7.** 6   **8.** 12
9. 1   **10.** 15   **11. a.** 18 gift packs   **b.** 2 markers, 3 pencils, 6 pens

---

SECTION 1.6   STUDY SET

VOCABULARY

Fill in the blanks.

1. The _____ of a number are the products of that number and 1, 2, 3, 4, 5, and so on.

2. Because 12 is the smallest number that is a multiple of both 3 and 4, it is the ____ _____ _____ of 3 and 4.

3. One number is _____ by another if, when dividing them, we get a remainder of 0.

4. Because 6 is the largest number that is a factor of both 18 and 24, it is the _____ _____ _____ of 18 and 24.

CONCEPTS

5. a. The LCM of 4 and 6 is 12. What is the smallest whole number divisible by 4 and 6?

b. Fill in the blank: In general, the LCM of two whole numbers is the _____ whole number that is divisible by both numbers.

6. a. What are the common multiples of 2 and 3 that appear in the list of multiples shown in the next column?

b. What is the LCM of 2 and 3?

Multiples of 2	Multiples of 3
$2 \cdot 1 = 2$	$3 \cdot 1 = 3$
$2 \cdot 2 = 4$	$3 \cdot 2 = 6$
$2 \cdot 3 = 6$	$3 \cdot 3 = 9$
$2 \cdot 4 = 8$	$3 \cdot 4 = 12$
$2 \cdot 5 = 10$	$3 \cdot 5 = 15$
$2 \cdot 6 = 12$	$3 \cdot 6 = 18$

7. a. The first six multiples of 5 are 5, 10, 15, 20, 25, and 30. What is the first multiple of 5 that is divisible by 4?

b. What is the LCM of 4 and 5?

8. Fill in the blanks to complete the prime factorization of 24.

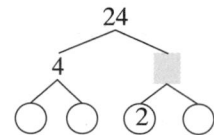

9. The prime factorizations of 36 and 90 are:

$$36 = 2 \cdot 2 \cdot 3 \cdot 3$$
$$90 = 2 \cdot 3 \cdot 3 \cdot 5$$

What is the greatest number of times

a. 2 appears in any one factorization?

b. 3 appears in any one factorization?

c. 5 appears in any one factorization?

d. Fill in the blanks to find the LCM of 36 and 90:

$$\text{LCM} = \boxed{} \cdot \boxed{} \cdot \boxed{} \cdot \boxed{} \cdot \boxed{} = \boxed{}$$

10. The prime factorizations of 14, 70, and 140 are:

$$14 = 2 \cdot 7$$
$$70 = 2 \cdot 5 \cdot 7$$
$$140 = 2 \cdot 2 \cdot 5 \cdot 7$$

What is the greatest number of times

a. 2 appears in any one factorization?

b. 5 appears in any one factorization?

c. 7 appears in any one factorization?

d. Fill in the blanks to find the LCM of 14, 70, and 140:

$$\text{LCM} = \boxed{} \cdot \boxed{} \cdot \boxed{} \cdot \boxed{} = \boxed{}$$

11. The prime factorizations of 12 and 54 are:

$$12 = 2^2 \cdot 3^1$$
$$54 = 2^1 \cdot 3^3$$

What is the greatest number of times

a. 2 appears in any one factorization?

b. 3 appears in any one factorization?

c. Fill in the blanks to find the LCM of 12 and 54:

$$\text{LCM} = 2^{\boxed{}} \cdot 3^{\boxed{}} = \boxed{}$$

12. The factors of 18 and 45 are shown below.

Factors of 18:   1, 2, 3, 6, 9, 18

Factors of 45:   1, 3, 5, 9, 15, 45

a. Circle the common factors of 18 and 45.

b. What is the GCF of 18 and 45?

13. The prime factorizations of 60 and 90 are:

$$60 = 2 \cdot 2 \cdot 3 \cdot 5$$
$$90 = 2 \cdot 3 \cdot 3 \cdot 5$$

a. Circle the common prime factors of 60 and 90.

b. What is the GCF of 60 and 90?

14. The prime factorizations of 36, 84, and 132 are:

$$36 = 2 \cdot 2 \cdot 3 \cdot 3$$
$$84 = 2 \cdot 2 \cdot 3 \cdot 7$$
$$132 = 2 \cdot 2 \cdot 3 \cdot 11$$

a. Circle the common factors of 36, 84, and 132.

b. What is the GCF of 36, 84, and 132?

NOTATION

15. a. The abbreviation for the greatest common factor is _____.

b. The abbreviation for the least common multiple is _____.

16. a. We read LCM $(2, 15) = 30$ as "The _____ _____ multiple ___ 2 and 15 ___ 30."

b. We read GCF $(18, 24) = 6$ as "The _____ _____ factor ___ 18 and 24 ___ 6."

GUIDED PRACTICE

Find the first eight multiples of each number. **See Example 1.**

17. 4   **18.** 2

19. 11   **20.** 10

21. 8   **22.** 9

23. 20   **24.** 30

Find the LCM of the given numbers. **See Example 2.**

25. 3, 5   **26.** 6, 9

27. 8, 12   **28.** 10, 25

29. 5, 11   **30.** 7, 11

31. 4, 7   **32.** 5, 8

Find the LCM of the given numbers. **See Example 3.**

33. 3, 4, 6   **34.** 2, 3, 8

35. 2, 3, 10   **36.** 3, 6, 15

Find the LCM of the given numbers. **See Example 4.**

37. 16, 20   **38.** 14, 21

39. 30, 50   **40.** 21, 27

41. 35, 45   **42.** 36, 48

43. 100, 120   **44.** 120, 180

Find the LCM of the given numbers. **See Example 5.**

45. 6, 24, 36   **46.** 6, 10, 18

47. 5, 12, 15   **48.** 8, 12, 16

Find the GCF of the given numbers. **See Example 7.**

49. 4, 6   **50.** 6, 15

51. 9, 12   **52.** 10, 12

Find the GCF of the given numbers. **See Example 8.**

53. 22, 33   **54.** 14, 21

55. 15, 30   **56.** 15, 75

57. 18, 96   **58.** 30, 48

59. 28, 42   **60.** 63, 84

Find the GCF of the given numbers. **See Example 9.**

61. 16, 51   **62.** 27, 64

63. 81, 125   **64.** 57, 125

Find the GCF of the given numbers. **See Example 10.**

65. 12, 68, 92   **66.** 24, 36, 40

67. 72, 108, 144   **68.** 81, 108, 162

▌TRY IT YOURSELF

Find the LCM and the GCF of the given numbers.

69. 100, 120   **70.** 120, 180

71. 14, 140   **72.** 15, 300

73. 66, 198, 242   **74.** 52, 78, 130

75. 8, 9, 49   **76.** 9, 16, 25

77. 120, 125   **78.** 98, 102

79. 34, 68, 102   **80.** 26, 39, 65

81. 46, 69   **82.** 38, 57

83. 50, 81   **84.** 65, 81

▌APPLICATIONS

85. OIL CHANGES Ford has officially extended the oil change interval for 2007 and newer cars to every 7,500 miles. (It used to be every 5,000 miles). Complete the table below that shows Ford's new recommended oil change mileages.

1st oil change	2nd oil change	3rd oil change	4th oil change	5th oil change	6th oil change
7,500 mi					

86. ATMs An ATM machine offers the customer cash withdrawal choices in multiples of $20. The minimum withdrawal is $20 and the maximum is $200. List the dollar amounts of cash that can be withdrawn from the ATM machine.

87. NURSING A nurse is instructed to check a patient's blood pressure every 45 minutes and another is instructed to take the same patient's temperature every 60 minutes. If both nurses are in the patient's room now, how long will it be until the nurses are together in the room once again?

88. BIORHYTHMS Some scientists believe that there are natural rhythms of the body, called *biorhythms,* that affect our physical, emotional, and mental cycles. Our physical biorhythm cycle lasts 23 days, the emotional biorhythm cycle lasts 28 days, and our mental biorhythm cycle lasts 33 days. Each biorhythm cycle has a high, low and critical zone. If your three cycles are together one day, all at their lowest point, in how many more days will they be together again, all at their lowest point?

89. PICNICS A package of hot dogs usually contains 10 hot dogs and a package of buns usually contains 12 buns. How many packages of hot dogs and buns should a person buy to be sure that there are equal numbers of each?

90. WORKING COUPLES A husband works for 6 straight days and then has a day off. His wife works for 7 straight days and then has a day off. If the husband and wife are both off from work on the same day, in how many days will they both be off from work again?

91. DANCE FLOORS A dance floor is to be made from rectangular pieces of plywood that are 6 feet by 8 feet. What is the minimum number of pieces of plywood that are needed to make a square dance floor?

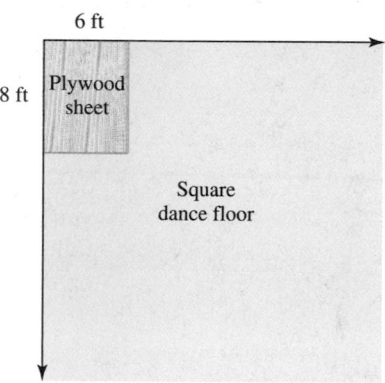

92. BOWLS OF SOUP Each of the bowls shown below holds an exact number of *full ladles* of soup.

   a. If there is no spillage, what is the greatest-size ladle (in ounces) that a chef can use to fill all three bowls?

   b. How many ladles will it take to fill each bowl?

12 ounces   21 ounces   18 ounces

93. ART CLASSES Students in a painting class must pay an extra art supplies fee. On the first day of class, the instructor collected $28 in fees from several students. On the second day she collected $21 more from some different students, and on the third day she collected an additional $63 from some other students.

    a. What is the most the art supplies fee could cost a student?

    b. Use your answer from part a to determine how many students paid the art supplies fee each day.

94. SHIPPING A toy manufacturer needs to ship 135 brown teddy bears, 105 black teddy bears, and 30 white teddy bears. They can pack only one type of teddy bear in each box, and they must pack the same number of teddy bears in each box. What is the greatest number of teddy bears they can pack in each box?

▌WRITING

95. Explain how to find the LCM of 8 and 28 using prime factorization.

96. Explain how to find the GCF of 8 and 28 using prime factorization.

97. The prime factorization of 12 is $2 \cdot 2 \cdot 3$ and the prime factorization of 15 is $3 \cdot 5$. Explain why the LCM of 12 and 15 is *not* $2 \cdot 2 \cdot 3 \cdot 3 \cdot 5$.

98. How can you tell by looking at the prime factorizations of two whole numbers that their GCF is 1?

▌REVIEW

Perform each operation.

99. $9,999 + 1,111$ **100.** $10,000 - 7,989$

101. $305 \cdot 50$ **102.** $2,100 \div 105$

Objectives

1 Use the order of operations rule.

2 Evaluate expressions containing grouping symbols.

3 Find the mean (average) of a set of values.

SECTION **1.7**
Order of Operations

Recall that numbers are combined with the operations of addition, subtraction, multiplication, and division to create **expressions.** We often have to **evaluate** (find the value of) expressions that involve more than one operation. In this section, we introduce an order of operations rule to follow in such cases.

1 Use the order of operations rule.

Suppose you are asked to contact a friend if you see a Rolex watch for sale while you are traveling in Europe. While in Switzerland, you find the watch and send the following text message, shown on the left. The next day, you get the response shown on the right from your friend.

You sent this
message.

You get this
response.

Something is wrong. The first part of the response (No price too high!) says to buy the watch at any price. The second part (No! Price too high.) says not to buy it, because it's too expensive. The placement of the exclamation point makes us read the two parts of the response differently, resulting in different meanings. When reading a mathematical statement, the same kind of confusion is possible. For example, consider the expression

$2 + 3 \cdot 6$

We can evaluate this expression in two ways. We can add first, and then multiply. Or we can multiply first, and then add. However, the results are different.

$2 + 3 \cdot 6 = 5 \cdot 6$ Add 2 and 3 first. | $2 + 3 \cdot 6 = 2 + 18$ Multiply 3 and 6 first.
$\qquad = 30$ Multiply 5 and 6. | $\qquad = 20$ Add 2 and 18.

—————— Different results ——————

If we don't establish a uniform order of operations, the expression has two different values. To avoid this possibility, we will always use the following order of operations rule.

Order of Operations

1.   Perform all calculations within parentheses and other grouping symbols following the order listed in Steps 2–4 below, working from the innermost pair of grouping symbols to the outermost pair.

2.   Evaluate all exponential expressions.

3.   Perform all multiplications and divisions as they occur from left to right.

4.   Perform all additions and subtractions as they occur from left to right.
When grouping symbols have been removed, repeat Steps 2–4 to complete the calculation.

If a fraction bar is present, evaluate the expression above the bar (called the **numerator**) and the expression below the bar (called the **denominator**) separately. Then perform the division indicated by the fraction bar, if possible.

It isn't necessary to apply all of these steps in every problem. For example, the expression $2 + 3 \cdot 6$ does not contain any parentheses, and there are no exponential expressions. So we look for multiplications and divisions to perform and proceed as follows:

$2 + 3 \cdot 6 = 2 + 18$ Do the multiplication first.
$\qquad = 20$ Do the addition.

EXAMPLE 1   Evaluate: $2 \cdot 4^2 - 8$

Strategy We will scan the expression to determine what operations need to be performed. Then we will perform those operations, one at a time, following the order of operations rule.

WHY If we don't follow the correct order of operations, the expression can have more than one value.

Solution
Since the expression does not contain any parentheses, we begin with Step 2 of the order of operations rule: Evaluate all exponential expressions. We will write the steps of the solution in horizontal form.

Self Check 1
Evaluate: $4 \cdot 3^3 - 6$
Now Try Problem 19

$$2 \cdot \mathbf{4}^2 - 8 = 2 \cdot \mathbf{16} - 8 \qquad \text{Evaluate the exponential expression: } 4^2 = 16.$$
$$= 32 - 8 \qquad\qquad \text{Do the multiplication: } 2 \cdot 16 = 32.$$
$$= 24 \qquad\qquad\quad \text{Do the subtraction.}$$

$$\begin{array}{r} \overset{1}{16} \\ \times\ 2 \\ \hline 32 \end{array} \qquad \begin{array}{r} \overset{2\ 12}{\cancel{32}} \\ -\ 8 \\ \hline 24 \end{array}$$

> **Success Tip**   Calculations that you cannot perform in your head should be shown outside the steps of your solution.

Self Check 2

Evaluate: $60 - 2 \cdot 3 + 22$

Now Try   Problem 23

EXAMPLE 2   Evaluate: $80 - 3 \cdot 2 + 16$

Strategy   We will perform the multiplication first.

WHY   The expression does not contain any parentheses, nor are there any exponents.

Solution

We will write the steps of the solution in horizontal form.

$$80 - \mathbf{3 \cdot 2} + 16 = 80 - \mathbf{6} + 16 \qquad \text{Do the multiplication: } 3 \cdot 2 = 6.$$
$$= 74 + 16 \qquad\qquad \text{Working from left to right, do the}$$
$$\qquad\qquad\qquad\qquad\quad \text{subtraction: } 80 - 6 = 74.$$
$$= 90 \qquad\qquad\qquad \text{Do the addition.}$$

$$\begin{array}{r} \overset{1}{74} \\ +\ 16 \\ \hline 90 \end{array}$$

> **Caution!**   In Example 2, a common mistake is to forget to work from left to right and *incorrectly* perform the addition before the subtraction. This error produces the wrong answer, 58.
>
> $$80 - 3 \cdot 2 + 16 = 80 - 6 + 16$$
> $$= 80 - 22$$
> $$= 58$$
>
> Remember to perform additions and subtractions *in the order in which they occur*. The same is true for multiplications and divisions.

Self Check 3

Evaluate: $144 \div 9 + 4(2)3$

Now Try   Problem 27

EXAMPLE 3   Evaluate: $192 \div 6 - 5(3)2$

Strategy   We will perform the division first.

WHY   Although the expression contains parentheses, there are no calculations to perform *within* them. Since there are no exponents, we perform multiplications and divisions as they are occur from left to right.

Solution

We will write the steps of the solution in horizontal form.

$$\mathbf{192 \div 6} - 5(3)2 = \mathbf{32} - 5(3)2 \qquad \text{Working from left to right, do the}$$
$$\qquad\qquad\qquad\qquad\qquad\quad \text{division: } 192 \div 6 = 32.$$
$$= 32 - 15(2) \qquad \text{Working from left to right, do the}$$
$$\qquad\qquad\qquad\qquad \text{multiplication: } 5(3) = 15.$$
$$= 32 - 30 \qquad\quad \text{Complete the multiplication: } 15(2) = 30.$$
$$= 2 \qquad\qquad\quad \text{Do the subtraction.}$$

$$\begin{array}{r} 32 \\ 6\overline{)192} \\ -\ 18 \\ \hline 12 \\ -\ 12 \\ \hline 0 \end{array}$$

We will use the five-step problem solving strategy introduced in Section 1.6 and the order of opertions rule to solve the following application problem.

EXAMPLE 4   *Long-Distance Calls*

The rates that Skype charges for overseas landline calls from the United States are shown on the right. A newspaper editor in Washington, D.C., made a 60-minute call to Canada, a 45-minute call to Panama, and a 30-minute call to Vietnam. What was the total cost of the calls?

Landline calls	
All rates are per minute.	
Afghanistan	41¢
Canada	2¢
Haiti	28¢
Panama	12¢
Russia	6¢
Vietnam	38¢
Includes tax	

Analyze

- The 60-minute call to Canada costs 2 cents per minute.   Given
- The 45-minute call to Panama costs 12 cents per minute.   Given
- The 30-minute call to Vietnam costs 38 cents per minute.   Given
- What is the total cost of the calls?   Find

Form   We translate the words of the problem to numbers and symbols. Since the word *per* indicates multiplication, we can find the cost of each call by multiplying the length of the call (in minutes) by the rate charged per minute (in cents). Since the word *total* indicates addition, we will add to find the total cost of the calls.

The total cost of the calls	is equal to	the cost of the call to Canada	plus	the cost of the call to Panama	plus	the cost of the call to Vietnam.
The total cost of the calls	=	60(2)	+	45(12)	+	30(38)

Solve   To evaluate this expression (which involves multiplication and addition), we apply the order of operations rule.

$$\begin{aligned}
\text{The total cost of the calls} &= 60(2) + 45(12) + 30(38) &\text{The units are cents.}\\
&= 120 + 540 + 1{,}140 &\text{Do the multiplication first.}\\
&= 1{,}800 &\text{Do the addition.}
\end{aligned}$$

$$\begin{array}{r} \overset{1}{1}20 \\ 540 \\ + 1{,}140 \\ \hline 1{,}800 \end{array}$$

State   The total cost of the overseas calls is 1,800¢, or $18.00.

Check   We can check the result by finding an estimate using front-end rounding. The total cost of the calls is approximately 60(2¢) + 50(10¢) + 30(40¢) = 120¢ + 500¢ + 1,200¢ or 1,820¢. The result of 1,800¢ seems reasonable.

Self Check 4

LONG-DISTANCE CALLS   A newspaper reporter in Chicago made a 90-minute call to Afghanistan, a 25-minute call to Haiti, and a 55-minute call to Russia. What was the total cost of the calls?

Now Try Problem 105

2   Evaluate expressions containing grouping symbols.

Grouping symbols determine the order in which an expression is to be evaluated. Examples of grouping symbols are parentheses ( ), brackets [ ], braces { }, and the fraction bar —.

EXAMPLE 5   Evaluate each expression: **a.** 12 − 3 + 5   **b.** 12 − (3 + 5)

Strategy   To evaluate the expression in part a, we will perform the subtraction first. To evaluate the expression in part b, we will perform the addition first.

WHY   The similar-looking expression in part b is evaluated in a different order because it contains parentheses. Any operations within parentheses must be performed first.

Self Check 5

Evaluate each expression:

a. 20 − 7 + 6

b. 20 − (7 + 6)

Now Try Problem 33

Solution

a. The expression does not contain any parentheses, nor are there any exponents, nor any multiplication or division. We perform the additions and subtractions as they occur, from left to right.

$$12 - 3 + 5 = 9 + 5 \quad \text{Do the subtraction: } 12 - 3 = 9.$$
$$= 14 \quad \text{Do the addition.}$$

b. By the order of operations rule, we must perform the operation within the parentheses first.

$$12 - (3 + 5) = 12 - 8 \quad \text{Do the addition: } 3 + 5 = 8. \text{ Read as "12 minus the quantity of 3 plus 5."}$$
$$= 4 \quad \text{Do the subtraction.}$$

> **The Language of Algebra** When we read the expression $12 - (3 + 5)$ as "12 minus the *quantity* of 3 plus 5," the word *quantity* alerts the reader to the parentheses that are used as grouping symbols.

Self Check 6

Evaluate: $(1 + 3)^4$

Now Try Problem 35

EXAMPLE 6 Evaluate: $(2 + 6)^3$

Strategy We will perform the operation within the parentheses first.

WHY This is the first step of the order of operations rule.

Solution
$$(2 + 6)^3 = 8^3 \quad \text{Read as "The cube of the quantity of 2 plus 6." Do the addition.}$$
$$= 512 \quad \text{Evaluate the exponential expression: } 8^3 = 8 \cdot 8 \cdot 8 = 512.$$

$$\begin{array}{r} \overset{3}{64} \\ \times 8 \\ \hline 512 \end{array}$$

Self Check 7

Evaluate: $50 - 4(12 - 5 \cdot 2)$

Now Try Problem 39

EXAMPLE 7 Evaluate: $5 + 2(13 - 5 \cdot 2)$

Strategy We will perform the multiplication within the parentheses first.

WHY When there is more than one operation to perform within parentheses, we follow the order of operations rule. Multiplication is to be performed before subtraction.

Solution
We apply the order of operations rule within the parentheses to evaluate $13 - 5 \cdot 2$.

$$5 + 2(13 - 5 \cdot 2) = 5 + 2(13 - 10) \quad \text{Do the multiplication within the parentheses.}$$
$$= 5 + 2(3) \quad \text{Do the subtraction within the parentheses.}$$
$$= 5 + 6 \quad \text{Do the multiplication: } 2(3) = 6.$$
$$= 11 \quad \text{Do the addition.}$$

Some expressions contain two or more sets of grouping symbols. Since it can be confusing to read an expression such as $16 + 6(4^2 - 3(5 - 2))$, we use a pair of **brackets** in place of the second pair of parentheses.

$$16 + 6[4^2 - 3(5 - 2)]$$

If an expression contains more than one pair of grouping symbols, we always begin by working within the **innermost pair** and then work to the **outermost pair.**

Innermost parentheses

$$16 + 6[4^2 - 3(5 - 2)]$$

Outermost brackets

The Language of Algebra   Multiplication is indicated when a number is next to a parenthesis or a bracket. For example,

$$16 + 6[4^2 - 3(5 - 2)]$$

Multiplication   Multiplication

EXAMPLE 8   Evaluate: $16 + 6[4^2 - 3(5 - 2)]$

Self Check 8
Evaluate:
$130 - 7[2^2 + 3(6 - 2)]$
Now Try Problem 43

Strategy We will work within the parentheses first and then within the brackets. Within each set of grouping symbols, we will follow the order of operations rule.

WHY By the order of operations, we must work from the *innermost* pair of grouping symbols to the *outermost*.

Solution

$16 + 6[4^2 - 3(\mathbf{5} - \mathbf{2})] = 16 + 6[4^2 - 3(\mathbf{3})]$	Do the subtraction within the parentheses.
$= 16 + 6[16 - 3(3)]$	Evaluate the exponential expression: $4^2 = 16$.
$= 16 + 6[16 - 9]$	Do the multiplication within the brackets.
$= 16 + 6[7]$	Do the subtraction within the brackets.
$= 16 + 42$	Do the multiplication: $6[7] = 42$.
$= 58$	Do the addition.

Caution!   In Example 8, a common mistake is to *incorrectly* add 16 and 6 instead of *correctly* multiplying 6 and 7 first. This error produces a wrong answer, 154.

$$16 + 6[4^2 - 3(5 - 2)] = 16 + 6[4^2 - 3(3)]$$
$$= 16 + 6[16 - 3(3)]$$
$$= 16 + 6[16 - 9]$$
$$= 16 + 6[7]$$
$$= 22[7]$$
$$= 154$$

EXAMPLE 9   Evaluate: $\dfrac{2(13) - 2}{3(2^3)}$

Self Check 9
Evaluate: $\dfrac{3(14) - 6}{2(3^2)}$
Now Try Problem 47

Strategy We will evaluate the expression above and the expression below the fraction bar separately. Then we will do the indicated division, if possible.

WHY Fraction bars are grouping symbols. They group the numerator and denominator. The expression could be written $[2(13) - 2] \div [3(2^3)]$.

Solution

$$\frac{2(13) - 2}{3(2^3)} = \frac{26 - 2}{3(8)} \qquad$$ In the numerator, do the multiplication.
In the denominator, evaluate the exponential expression within the parentheses.

$$= \frac{24}{24} \qquad$$ In the numerator, do the subtraction.
In the denominator, do the multiplication.

$$= 1 \qquad$$ Do the division indicated by the fraction bar: $24 \div 24 = 1$. ▪

3 Find the mean (average) of a set of values.

The **mean** (sometimes called the **arithmetic mean** or **average**) of a set of numbers is a value around which the values of the numbers are grouped. It gives you an indication of the "center" of the set of numbers. To find the mean of a set of numbers, we must apply the order of operations rule.

Finding the Mean

To find the mean (average) of a set of values, divide the sum of the values by the number of values.

NFL DEFENSIVE LINEMEN The weights of the 2008–2009 New York Giants starting defensive linemen were 273 lb, 305 lb, 317 lb, and 265 lb. What was their mean (average) weight? (Source: nfl.com/New York Giants depth chart)

Now Try Problems 51 and 113

EXAMPLE 10 *NFL Offensive Linemen* The weights of the 2008–2009 New York Giants starting offensive linemen are shown below. What was their mean (average) weight?

© Larry French/Getty Images

| Left tackle #66 D. Diehl 319 lb | Left guard #69 R. Seubert 310 lb | Center #60 S. O'Hara 302 lb | Right guard #76 C. Snee 317 lb | Right tackle #67 K. McKenzie 327 lb |

(Source: nfl.com/New York Giants depth chart)

Strategy We will add 327, 317, 302, 310, and 319 and divide the sum by 5.

WHY To find the mean (average) of a set of values, we divide the sum of the values by the number of values.

Solution
Since there are 5 weights, divide the sum by 5.

$$\text{Mean} = \frac{327 + 317 + 302 + 310 + 319}{5}$$

$$= \frac{1{,}575}{5} \qquad$$ In the numerator, do the addition.

$$= 315 \qquad$$ Do the indicated division: $1{,}575 \div 5$.

$$\begin{array}{r} \overset{2}{327} \\ 317 \\ 302 \\ 310 \\ + \ 319 \\ \hline 1{,}575 \end{array}$$

$$\begin{array}{r} 315 \\ 5\overline{)1{,}575} \\ -15 \\ \hline 7 \\ -5 \\ \hline 25 \\ -25 \\ \hline 0 \end{array}$$

In 2008–2009, the mean (average) weight of the starting offensive linemen on the New York Giants was 315 pounds.

Using Your CALCULATOR   **Order of Operations and Parentheses**

Calculators have the rules for order of operations built in. A left parenthesis key $\boxed{(}$ and a right parenthesis key $\boxed{)}$ should be used when grouping symbols, including a fraction bar, are needed. For example, to evaluate $\frac{240}{20-5}$, the parentheses keys must be used, as shown below.

$$240 \;\boxed{\div}\; \boxed{(}\; 20 \;\boxed{-}\; 5 \;\boxed{)}\; \boxed{=} \qquad\qquad \boxed{16}$$

On some calculator models, the $\boxed{\text{ENTER}}$ key is pressed instead of $\boxed{=}$ for the result to be displayed.

If the parentheses are not entered, the calculator will find $240 \div 20$ and then subtract 5 from that result, to produce the wrong answer, 7.

THINK IT THROUGH   *Education Pays*

"Education does pay. It has a high rate of return for students from all racial/ethnic groups, for men and for women, and for those from all family backgrounds. It also has a high rate of return for society."

The College Board, Trends in Higher Education *Series*

Attending school requires an investment of time, effort, and sacrifice. Is it all worth it? The graph below shows how average weekly earnings in the U.S. increase as the level of education increases. Begin at the bottom of the graph and work upward. Use the given clues to determine each of the missing weekly earnings amounts.

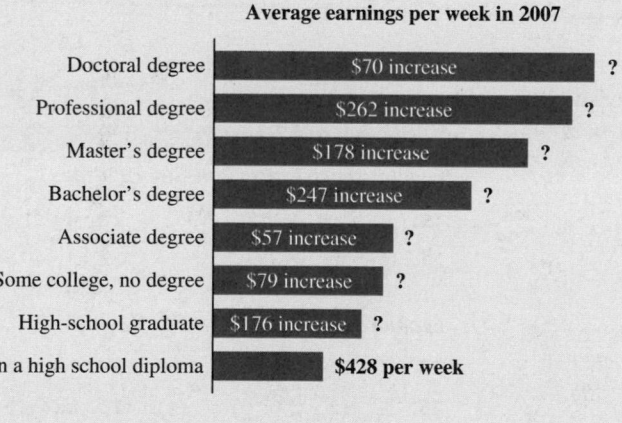

Average earnings per week in 2007

Doctoral degree	$70 increase ?
Professional degree	$262 increase ?
Master's degree	$178 increase ?
Bachelor's degree	$247 increase ?
Associate degree	$57 increase ?
Some college, no degree	$79 increase ?
High-school graduate	$176 increase ?
Less than a high school diploma	**$428 per week**

(Source: Bureau of Labor Statistics, Current Population Survey)

ANSWERS TO SELF CHECKS

1. 102   **2.** 76   **3.** 40   **4.** 4,720¢ = $47.20   **5. a.** 19   **b.** 7   **6.** 256   **7.** 42   **8.** 18   **9.** 2   **10.** 290 lb

SECTION 1.7   STUDY SET

VOCABULARY

Fill in the blanks.

1. Numbers are combined with the operations of addition, subtraction, multiplication, and division to create _____ .

2. To *evaluate* the expression $2 + 5 \cdot 4$ means to find its _____ .

3. The grouping symbols ( ) are called _____ , and the symbols [ ] are called _____ .

4. The expression above a fraction bar is called the _____ . The expression below a fraction bar is called the _____ .

5. In the expression $9 + 6[8 + 6(4 - 1)]$, the parentheses are the _____ most grouping symbols and the brackets are the _____ most grouping symbols.

6. To find the _____ of a set of values, we add the values and divide by the number of values.

CONCEPTS

7. List the operations in the order in which they should be performed to evaluate each expression. *You do not have to evaluate the expression.*

   a. $5(2)^2 - 1$

   b. $15 + 90 - (2 \cdot 2)^3$

   c. $7 \cdot 4^2$

   d. $(7 \cdot 4)^2$

8. List the operations in the order in which they should be performed to evaluate each expression. *You do not have to evaluate the expression.*

   a. $50 + 8 - 40$

   b. $50 - 40 + 8$

   c. $16 \cdot 2 \div 4$

   d. $16 \div 4 \cdot 2$

9. Consider the expression $\dfrac{5 + 5(7)}{(5 \cdot 20 - 8^2) - 28}$. In the numerator, what operation should be performed first? In the denominator, what operation should be performed first?

10. To find the mean (average) of 15, 33, 45, 12, 6, 19, and 3, we add the values and divide by what number?

NOTATION

11. In the expression $\dfrac{60 - 5 \cdot 2}{5 \cdot 2 + 40}$, what symbol serves as a grouping symbol? What does it group?

12. Use brackets to write $2(12 - (5 + 4))$ in clearer form.

Fill in the blanks.

13. We read the expression $16 - (4 + 9)$ as "16 minus the _____ of 4 plus 9."

14. We read the expression $(8 - 3)^3$ as "The cube of the _____ of 8 minus 3."

Complete each solution to evaluate the expression.

15. $7 \cdot 4 - 5(2)^2 = 7 \cdot 4 - 5\left(\boxed{}\right)$

    $= 28 - \boxed{}$

    $= \boxed{}$

16. $2 + (5 + 6 \cdot 2) = 2 + \left(5 + \boxed{}\right)$

    $= 2 + \boxed{}$

    $= \boxed{}$

17. $[4(2 + 7)] - 4^2 = \left[4\left(\boxed{}\right)\right] - 4^2$

    $= \boxed{} - 4^2$

    $= 36 - \boxed{}$

    $= \boxed{}$

18. $\dfrac{12 + 5 \cdot 3}{3^2 - 2 \cdot 3} = \dfrac{12 + \boxed{}}{\boxed{} - 6}$

    $= \dfrac{\boxed{}}{3}$

    $= \boxed{}$

GUIDED PRACTICE

Evaluate each expression. See Example 1.

19. $3 \cdot 5^2 - 28$

20. $4 \cdot 2^2 - 11$

21. $6 \cdot 3^2 - 41$

22. $5 \cdot 4^2 - 32$

Evaluate each expression. See Example 2.

23. $52 - 6 \cdot 3 + 4$

24. $66 - 8 \cdot 7 + 16$

25. $32 - 9 \cdot 3 + 31$

26. $62 - 5 \cdot 8 + 27$

Evaluate each expression. See Example 3.

27. $192 \div 4 - 4(2)3$

28. $455 \div 7 - 3(4)5$

29. $252 \div 3 - 6(2)6$

30. $264 \div 4 - 7(4)2$

Evaluate each expression. See Example 5.

31. a. $26 - 2 + 9$

    b. $26 - (2 + 9)$

32. a. $37 - 4 + 11$

    b. $37 - (4 + 11)$

33. a. $51 - 16 + 8$

    b. $51 - (16 + 8)$

34. a. $73 - 35 + 9$

    b. $73 - (35 + 9)$

Evaluate each expression. See Example 6.

35. $(4 + 6)^2$ **36.** $(3 + 4)^2$

37. $(3 + 5)^3$ **38.** $(5 + 2)^3$

Evaluate each expression. See Example 7.

39. $8 + 4(29 - 5 \cdot 3)$ **40.** $33 + 6(56 - 9 \cdot 6)$

41. $77 + 9(38 - 4 \cdot 6)$ **42.** $162 + 7(47 - 6 \cdot 7)$

Evaluate each expression. See Example 8.

43. $46 + 3[5^2 - 4(9 - 5)]$

44. $53 + 5[6^2 - 5(8 - 1)]$

45. $81 + 9[7^2 - 7(11 - 4)]$

46. $81 + 3[8^2 - 7(13 - 5)]$

Evaluate each expression. See Example 9.

47. $\dfrac{2(50) - 4}{2(4^2)}$ **48.** $\dfrac{4(34) - 1}{5(3^2)}$

49. $\dfrac{25(8) - 8}{6(2^3)}$ **50.** $\dfrac{6(31) - 26}{4(2^3)}$

Find the mean (average) of each list of numbers. See Example 10.

51. 6, 9, 4, 3, 8   **52.** 7, 1, 8, 2, 2

53. 3, 5, 9, 1, 7, 5   **54.** 8, 7, 7, 2, 4, 8

55. 19, 15, 17, 13   **56.** 11, 14, 12, 11

57. 5, 8, 7, 0, 3, 1   **58.** 9, 3, 4, 11, 14, 1

TRY IT YOURSELF

Evaluate each expression, if possible.

59. $(8 - 6)^2 + (4 - 3)^2$ **60.** $(2 + 1)^2 + (3 + 2)^2$

61. $2 \cdot 3^4$ **62.** $3^3 \cdot 5$

63. $7 + 4 \cdot 5$ **64.** $10 - 2 \cdot 2$

65. $(7 - 4)^2 + 1$ **66.** $(9 - 5)^3 + 8$

67. $\dfrac{10 + 5}{52 - 47}$ **68.** $\dfrac{18 + 12}{61 - 55}$

69. $5 \cdot 10^3 + 2 \cdot 10^2 + 3 \cdot 10^1 + 9$

70. $8 \cdot 10^3 + 0 \cdot 10^2 + 7 \cdot 10^1 + 4$

71. $20 - 10 + 5$ **72.** $80 - 5 + 4$

73. $25 \div 5 \cdot 5$ **74.** $6 \div 2 \cdot 3$

75. $150 - 2(2 \cdot 6 - 4)^2$ **76.** $760 - 2(2 \cdot 3 - 4)^2$

77. $190 - 2[10^2 - (5 + 2^2)] + 45$

78. $161 - 8[6(6) - 6^2] + 2^2(5)$

79. $2 + 3(0)$ **80.** $5(0) + 8$

81. $\dfrac{(5 - 3)^2 + 2}{4^2 - (8 + 2)}$ **82.** $\dfrac{(4^3 - 2) + 7}{5(2 + 4) - 7}$

83. $4^2 + 3^2$ **84.** $12^2 + 5^2$

85. $3 + 2 \cdot 3^4 \cdot 5$ **86.** $3 \cdot 2^3 \cdot 4 - 12$

87. $60 - \left(6 + \dfrac{40}{2^3}\right)$ **88.** $7 + \left(5^3 - \dfrac{200}{2}\right)$

89. $\dfrac{(3 + 5)^2 + 2}{2(8 - 5)}$ **90.** $\dfrac{25 - (2 \cdot 3 - 1)}{2 \cdot 9 - 8}$

91. $(18 - 12)^3 - 5^2$ **92.** $(9 - 2)^2 - 3^3$

93. $30(1)^2 - 4(2) + 12$

94. $5(1)^3 + (1)^2 + 2(1) - 6$

95. $16^2 - \dfrac{25}{5} + 6(3)4$ **96.** $15^2 - \dfrac{24}{6} + 8(2)(3)$

97. $\dfrac{3^2 - 2^2}{(3 - 3)^2}$ **98.** $\dfrac{5^2 + 17}{4 - 2^2}$

99. $3\left(\dfrac{18}{3}\right) - 2(2)$ **100.** $2\left(\dfrac{12}{3}\right) + 3(5)$

101. $4[50 - (3^3 - 5^2)]$ **102.** $6[15 + (5 \cdot 2^2)]$

103. $80 - 2[12 - (5 + 4)]$

104. $15 + 5[12 - (2^2 + 4)]$

APPLICATIONS

Write an expression to solve each problem and evaluate it.

105. SHOPPING At the supermarket, Carlos is buying 3 cases of soda, 4 bags of tortilla chips, and 2 bottles of salsa. Each case of soda costs $7, each bag of chips costs $4, and each bottle of salsa costs $3. Find the total cost of the snacks.

106. BANKING When a customer deposits cash, a teller must complete a currency count on the back of the deposit slip. In the illustration, a teller has written the number of each type of bill to be deposited. What is the total amount of cash being deposited?

Currency count, for financial use only		
24	x 1's	
—	x 2's	
6	x 5's	
10	x 10's	
12	x 20's	
2	x 50's	
1	x 100's	
	TOTAL $	

107. DIVING The scores awarded to a diver by seven judges as well as the degree of difficulty of his dive are shown on the next page. Use the two-step process shown on the next page to calculate the diver's overall score.

Step 1 Throw out the lowest score and the highest score.

Step 2 Add the sum of the remaining scores and multiply by the degree of difficulty.

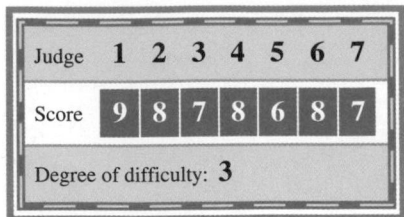

Judge	1	2	3	4	5	6	7
Score	9	8	7	8	6	8	7

Degree of difficulty: **3**

108. WRAPPING GIFTS How much ribbon is needed to wrap the package shown if 15 inches of ribbon are needed to make the bow?

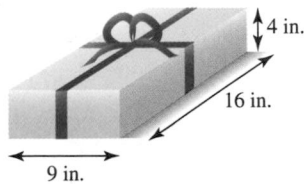

4 in.

16 in.

9 in.

109. SCRABBLE Illustration (a) shows part of the game board before and illustration (b) shows it after the words *brick* and *aphid* were played. Determine the scoring for each word. (*Hint:* The number on each tile gives the point value of the letter.)

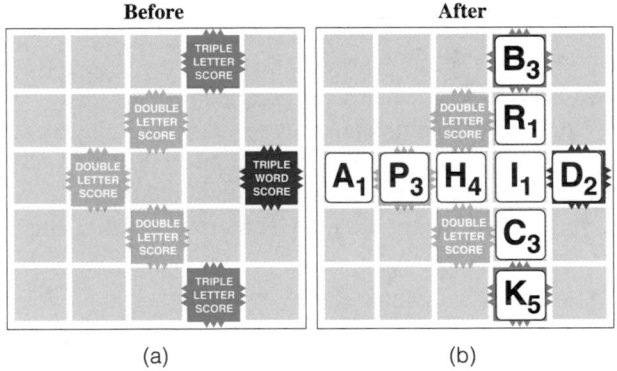

(a)   (b)

110. THE GETTYSBURG ADDRESS Here is an excerpt from Abraham Lincoln's Gettysburg Address:

> Fourscore and seven years ago, our fathers brought forth on this continent a new nation, conceived in liberty, and dedicated to the proposition that all men are created equal.

Lincoln's comments refer to the year 1776, when the United States declared its independence. If a score is 20 years, in what year did Lincoln deliver the Gettysburg Address?

111. PRIME NUMBERS Show that 87 is the sum of the squares of the first four prime numbers.

112. A 27-foot-long by 19-foot-wide rectangular garden is one feature of a landscape design for a community park. A concrete walkway is to run through the garden and will occupy 125 square feet of space. How many square feet are left for planting in the garden?

from Campus to Careers
Landscape Designer

113. CLIMATE One December week, the high temperatures in Honolulu, Hawaii, were 75°, 80°, 83°, 80°, 77°, 72°, and 86°. Find the week's mean (average) high temperature.

114. GRADES In a science class, a student had test scores of 94, 85, 81, 77, and 89. He also overslept, missed the final exam, and received a 0 on it. What was his test average (mean) in the class?

115. ENERGY USAGE See the graph below. Find the mean (average) number of therms of natural gas used per month for the year 2009.

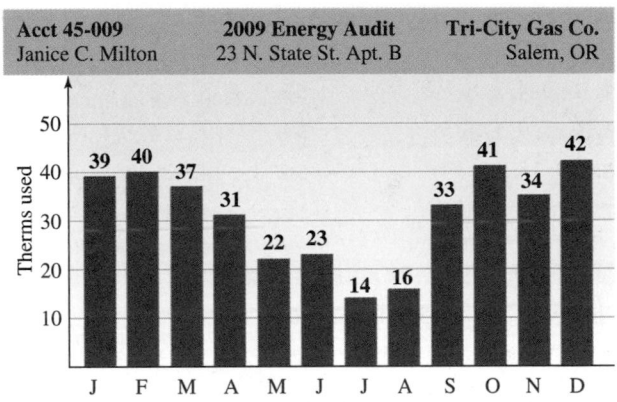

Acct 45-009	2009 Energy Audit	Tri-City Gas Co.
Janice C. Milton	23 N. State St. Apt. B	Salem, OR

Therms used: 39, 40, 37, 31, 22, 23, 14, 16, 33, 41, 34, 42

J F M A M J J A S O N D

116. COUNTING NUMBERS What is the average (mean) of the first nine counting numbers: 1, 2, 3, 4, 5, 6, 7, 8, and 9?

117. FAST FOODS The table shows the sandwiches Subway advertises on its 6 grams of fat or less menu. What is the mean (average) number of calories for the group of sandwiches?

6-inch subs	Calories
Veggie Delite	230
Turkey Breast	280
Turkey Breast & Ham	295
Ham	290
Roast Beef	290
Subway Club	330
Roasted Chicken Breast	310
Chicken Teriyaki	375

(Source: Subway.com/NutritionInfo)

118. TV RATINGS The table below shows the number of viewers* of the 2008 Major League Baseball World Series between the Philadelphia Phillies and the Tampa Bay Rays. How large was the average (mean) audience?

Game 1	Wednesday, Oct. 22	14,600,000
Game 2	Thursday, Oct. 23	12,800,000
Game 3	Saturday, Oct. 25	9,900,000
Game 4	Sunday, Oct. 26	15,500,000
Game 5 (suspended in 6th inning by rain)	Monday, Oct. 27	13,200,000
Game 5 (conclusion of game 5)	Wednesday, Oct. 29	19,800,000

* Rounded to the nearest hundred thousand
(Source: The Nielsen Company)

AP Images

119. YOUTUBE A YouTube video contest is to be part of a kickoff for a new sports drink. The cash prizes to be awarded are shown below.

a. How many prizes will be awarded?

b. What is the total amount of money that will be awarded?

c. What is the average (mean) cash prize?

> **YouTube Video Contest**
>
> **Grand prize: Disney World vacation plus $2,500**
>
> Four 1st place prizes of $500
> Thirty-five 2nd place prizes of $150
> Eighty-five 3rd place prizes of $25

120. SURVEYS Some students were asked to rate their college cafeteria food on a scale from 1 to 5. The responses are shown on the tally sheet.

a. How many students took the survey?

b. Find the mean (average) rating.

Poor				Fair		Excellent
1		2	3		4	5
IIII		I	THL		I	IIII

WRITING

121. Explain why the order of operations rule is necessary.

122. What does it mean when we say to do all additions and subtractions *as they occur from left to right?* Give an example.

123. Explain the error in the following solution:

Evaluate:

$$8 + 2[6 - 3(9 - 8)] = 8 + 2[6 - 3(1)]$$
$$= 8 + 2[6 - 3]$$
$$= 8 + 2(3)$$
$$= 10(3)$$
$$= 30$$

124. Explain the error in the following solution:

Evaluate:

$$24 - 4 + 16 = 24 - 20$$
$$= 4$$

REVIEW

Write each number in words.

125. 254,309

126. 504,052,040

SECTION 1.8

Solving Equations Using Addition and Subtraction

The first seven sections of this textbook have been devoted to an in-depth study of whole-number arithmetic. It's now time to begin the move toward *algebra*. **Algebra** is the language of mathematics. It is the result of contributions from many cultures over thousands of years. The word *algebra* comes from the title of the book *Ihm Al-jabr wa'l muqābalah,* written by an Arabian mathematician around A.D. 800. In this section, we will introduce one of the most powerful concepts in algebra, the *equation*.

1 Determine whether a number is a solution.

An **equation** is a statement that two expressions are equal. All equations contain an $=$ symbol. An example is $x + 5 = 15$. The equal symbol $=$ separates the equation into two parts: The expression $x + 5$ is the **left side** and 15 is the **right side**. The letter x is the **variable** (or the **unknown**). The sides of an equation can be reversed, so we can write $x + 5 = 15$ or $15 = x + 5$.

- An equation can be true: $6 + 3 = 9$

- An equation can be false: $2 + 4 = 7$

- An equation can be neither true nor false. For example, $x + 5 = 15$ is neither true nor false because we don't know what number x represents.

An equation that contains a variable is made true or false by substituting a number for the variable. If we substitute 10 for x in $x + 5 = 15$, the resulting equation is true: $10 + 5 = 15$. If we substitute 1 for x, the resulting equation is false: $1 + 5 = 15$. A number that makes an equation true when substituted for the variable is called a **solution** and it is said to **satisfy** the equation. Therefore, 10 is a solution of $x + 5 = 15$, and 1 is not.

> **The Language of Algebra**   To *substitute* means to put or use in place of another, as with a *substitute* teacher. In the previous example, we *substituted* 10 for x in $x + 5 = 15$.

Self Check 1

Is 8 a solution of $x + 17 = 25$?

Now Try Problem 17

EXAMPLE 1   Is 18 a solution of $x - 3 = 15$?

Strategy We will substitute 18 for x in the equation and evaluate the left side.

WHY If a true statement results, 18 is a solution of the equation. If we obtain a false statement, 18 is not a solution.

Solution

$$x - 3 = 15 \quad \text{This is the given equation.}$$
$$\mathbf{18} - 3 \stackrel{?}{=} 15 \quad \text{Substitute 18 for } x. \text{ Read } \stackrel{?}{=} \text{ as "is possibly equal to."}$$
$$15 = 15 \quad \text{On the left side, do the subtraction.}$$

$$\begin{array}{r} 18 \\ -\ 3 \\ \hline 15 \end{array}$$

Since $15 = 15$ is a true statement, 18 is a solution of $x - 3 = 15$. ∎

> **The Language of Algebra**   It is important to know the difference between an *equation* and *expression*. An equation contains an $=$ symbol and an expression does not.

EXAMPLE 2   Is 23 a solution of $32 = y + 10$?

Strategy We will substitute 23 for y in the equation and evaluate the right side.

WHY If a true statement results, 23 is a solution of the equation. If we obtain a false statement, 23 is not a solution.

Solution

$32 = y + 10$	This is the given equation.	23
$32 \stackrel{?}{=} 23 + 10$	Substitute 23 for y.	+10
$32 = 33$	On the right side, do the addition.	33

Since $32 = 33$ is a false statement, 23 is not a solution of $32 = y + 10$.

2 Use the addition property of equality.

Since the solution of an equation is usually not given, we must develop a process to find it. This process is called *solving the equation*.

To **solve an equation** means to find all values of the variable that make the equation true. To develop an understanding of how to solve equations, refer to the scales shown on the right.

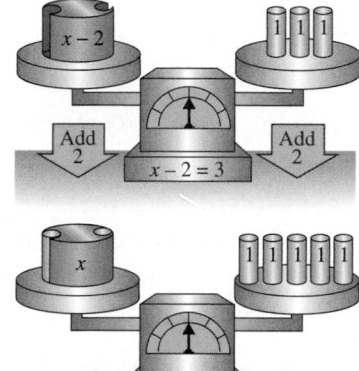

The first scale represents the equation $x - 2 = 3$. The scale is in balance because the weights on the left side and right side are equal. To find x, we must add 2 to the left side. To keep the scale in balance, we must also add 2 to the right side. After doing this, we see from the second scale that x is balanced by 5. Therefore, x must be 5. We say that we have solved the equation $x - 2 = 3$ and that the solution is 5.

In this example, we solved $x - 2 = 3$ by transforming it to a simpler *equivalent equation, $x = 5$.*

Equivalent Equations

Equations with the same solutions are called **equivalent equations.**

The Language of Algebra We *solve* equations. An expression can be *evaluated* (or *simplified*), but never *solved.*

The procedure that we used with the scales suggests the following property of equality.

Addition Property of Equality

Adding the same number to both sides of an equation does not change its solution.
    For any numbers $a, b,$ and $c,$

$$\text{if } a = b, \text{then } a + c = b + c$$

When we use this property, the resulting equation is *equivalent to the original one.* We will now show how it is used to solve $x - 2 = 3$ algebraically.

Self Check 3

Solve $x - 10 = 33$ and check the result.

Now Try Problem 25

EXAMPLE 3   Solve: $x - 2 = 3$

Strategy We will use the addition property of equality to isolate the variable x on the left side of the equation.

WHY To solve the original equation, we want to find a simpler equivalent equation of the form x **= a number**, whose solution is obvious.

Solution

$$x - 2 = 3 \qquad \text{This is the equation to solve.}$$
$$x - 2 + 2 = 3 + 2 \qquad \text{To isolate x, undo the subtraction of 2 by adding 2 to both sides.}$$
$$x = 5 \qquad \text{On the left side, adding 2 undoes the subtraction of 2 and leaves x. On the right side, do the addition: } 3 + 2 = 5.$$

Since 5 is obviously the solution of the equivalent equation $x = 5$, the solution of the original equation, $x - 2 = 3$, is also 5.

To check this result, we substitute 5 for x in the original equation and simplify.

Check: $x - 2 = 3$ This is the original equation.
$$5 - 2 \overset{?}{=} 3 \qquad \text{Substitute 5 for x.}$$
$$3 = 3 \qquad \text{On the left side, do the subtraction.}$$

Since $3 = 3$ is a true statement, 5 is the solution of $x - 2 = 3$.

> *The Language of Algebra*   We solve equations by writing a series of steps that result in an equivalent equation of the form
>
> $$x = a\ number \qquad \text{or} \qquad a\ number = x$$
>
> We say the variable is *isolated* on one side of the equation. *Isolated* means alone or by itself.

Self Check 4

Solve $75 = b - 38$ and check the result.

Now Try Problem 29

EXAMPLE 4   Solve: $19 = y - 7$

Strategy We will use the addition property of equality to isolate the variable y on the right side of the equation.

WHY To solve the original equation, we want to find a simpler equivalent equation of the form **a number** $= y$, whose solution is obvious.

Solution

$$19 = y - 7 \qquad \text{This is the equation to solve.}$$
$$19 + 7 = y - 7 + 7 \qquad \text{To isolate y, undo the subtraction of 7 by adding 7 to both sides.}$$
$$26 = y \qquad \text{On the left side, do the addition: } 19 + 7 = 26. \text{ On the right side, adding 7 undoes the subtraction of 7 and leaves y.}$$

$$\begin{array}{r} \overset{1}{1}9 \\ +\ 7 \\ \hline 26 \end{array}$$

> *Success Tip*   Calculations that you cannot perform in your head should be shown outside the steps of your solution.

Since 26 is obviously the solution of the equivalent equation $26 = y$, the solution of the original equation, $19 = y - 7$, is also 26.

To check this result, we substitute 26 for y in the original equation and simplify.

Check: $19 = y - 7$ This is the original equation.
$$19 \overset{?}{=} 26 - 7 \qquad \text{Substitute 26 for y.}$$
$$19 = 19 \qquad \text{On the right side, do the subtraction.}$$

Since $19 = 19$ is a true statement, 26 is the solution of $19 = y - 7$.

Success Tip Perhaps you are more comfortable by first reversing the sides of equations like that of Example 4 before attempting to solve them:

$$19 = y - 7 \quad \text{can be rewritten as} \quad y - 7 = 19$$

That step is fine; however, when solving equations, it is not necessary that the variable be isolated on the left side of the equation.

3 Use the subtraction property of equality.

To introduce another property of equality, consider the first scale shown on the right, which represents the equation $x + 3 = 5$. The scale is in balance because the weights on the left and right sides are equal. To find x, we need to remove 3 from the left side. To keep the scale in balance, we must also remove 3 from the right side. After doing this, we see from the second scale that x is balanced by 2. Therefore, x must be 2. We say that we have solved the equation $x + 3 = 5$ and that the solution is 2. This example illustrates the following property of equality.

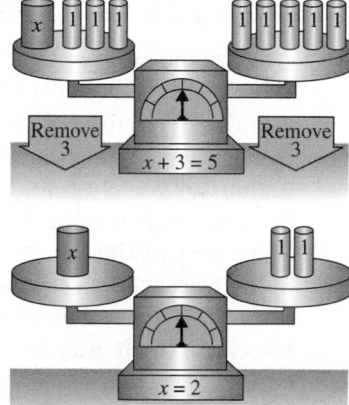

Subtraction Property of Equality

Subtracting the same number from both sides of an equation does not change its solution.

    For any numbers a, b, and c,

$$\text{if } a = b, \text{ then } a - c = b - c$$

When we use this property, the resulting equation is equivalent to the original one.

EXAMPLE 5   Solve: $x + 3 = 5$

Strategy We will use the subtraction property of equality to isolate the variable x on the left side of the equation.

WHY To solve the original equation, we want to find a simpler equivalent equation of the form $x = \textbf{a number}$, whose solution is obvious.

Solution

$x + 3 = 5$	This is the equation to solve.
$x + 3 - 3 = 5 - 3$	To isolate x, undo the addition of 3 by subtracting 3 from both sides.
$x = 2$	On the left side, subtracting 3 undoes the addition of 3 and leaves x. On the right side, do the subtraction: $5 - 3 = 2$.

We check by substituting 2 for x in the original equation and simplifying. If 2 is the solution, we will obtain a true statement.

Check: $x + 3 = 5$ This is the original equation.

         $2 + 3 \overset{?}{=} 5$ Substitute 2 for x.

           $5 = 5$ On the left side, do the addition.

Since the resulting equation $5 = 5$ is true, 2 is the solution of $x + 3 = 5$.

Self Check 5

Solve $m + 7 = 14$ and check the result.

Now Try Problem 33

4 Use equations to solve application problems.

The key to problem solving is to understand the problem and then to develop a plan for solving it. The following list of steps provides a good strategy to follow.

> ***The Language of Algebra***   A **strategy** is a careful plan or method. For example, a businessman might develop a new advertising *strategy* to increase sales or a long distance runner might have a *strategy* to win a marathon.

> ### Strategy for Problem Solving
>
> 1. **Analyze the problem** by reading it carefully to understand the given facts. What information is given? What are you asked to find? What vocabulary is given? Often, a diagram will help you visualize the facts of the problem.
> 2. **Form an equation** by picking a variable to represent the quantity to be found. Key words or phrases can be helpful. Finally, translate the words of the problem into an equation.
> 3. **Solve the equation.**
> 4. **State the conclusion clearly.** Be sure to include the units (such as feet, seconds, or pounds) in your answer.
> 5. **Check the result** using the original wording of the problem, not the equation that was formed in step 2 from the words.

We will now use this five-step strategy to solve application problems. The purpose of the following examples is to help you learn the strategy, even though you can probably solve the problems without it. If you learn how to use the strategy now, you will gain valuable problem-solving experience that will pay off later in the course when you are asked to solve more difficult problems.

Self Check 6

GASOLINE STORAGE A tank currently contains 1,325 gallons of gasoline. If 450 gallons were pumped from the tank earlier, how many gallons did it originally contain?

Now Try Problems 38 and 75

EXAMPLE 6   *Small Businesses*   Last year a hairstylist lost 17 customers who moved away. If she now has 73 customers, how many did she have originally?

Analyze

- She lost 17 customers.   *Given*
- She now has 73 customers.   *Given*
- How many customers did she originally have?   *Find*

> ***Caution!***   Unlike an arithmetic approach, you *do not* have to determine whether to add, subtract, multiply, or divide at this stage. Simply translate the words of the problem to mathematical symbols to form an equation that describes the situation. Then solve the equation.

Form

We can let c = the original number of customers. To form an equation involving c, we look for a key word or phrase in the problem.

Key phrase: *moved away*   **Translation:** *subtraction*

Now we translate the words of the problem into an equation.

This is called the verbal model.

The original number of customers	minus	17	is equal to	the number of customers she now has.
c	$-$	17	$=$	73

Solve

$$c - 17 = 73 \qquad \text{We need to isolate } c \text{ on the left side.}$$

$$c - 17 + 17 = 73 + 17 \qquad \text{To isolate } c, \text{ add 17 to both sides}$$
$$\text{to undo the subtraction of 17.}$$

$$c = 90 \qquad \text{Do the addition.}$$

$$\begin{array}{r} \overset{1}{7}3 \\ + 17 \\ \hline 90 \end{array}$$

State

She originally had 90 customers.

Check

If the hairstylist originally had 90 customers, and we decrease that number by the 17 that moved away, we should obtain the number of customers she now has.

$$\begin{array}{r} \overset{8\,10}{9\,\cancel{0}} \\ -\ 17 \\ \hline 73 \end{array} \quad \text{This is the number of customers the hairstylist now has.}$$

The result, 90, checks.

> *Caution!*   Check the result using the original wording of the problem, not by substituting it into the equation. Why? The equation may have been solved correctly, but the danger is that you may have formed it incorrectly. ∎

EXAMPLE 7   *Mortgages*   Sue wants to buy a house that costs $87,000. Since she has only $15,000 for a down payment, she will have to borrow some money by taking a mortgage. How much will she have to borrow?

Analyze

- The house costs $87,000.   Given
- Sue has $15,000 for a down payment.   Given
- How much money does she need to borrow?   Find

Form

We can let x = the amount of money that she needs to borrow. To form an equation involving x, we look for a key word or phrase in the problem.

Key phrase: *borrow some additional money*   **Translation:**   *addition*

Now we translate the words of the problem into an equation.

The amount Sue now has	plus	the amount she borrows	is equal to	the total cost of the house.
15,000	+	x	=	87,000

Solve

$$15,000 + x = 87,000 \qquad \text{We need to isolate } x$$
$$\text{on the left side.}$$

$$15,000 + x - 15,000 = 87,000 - 15,000 \qquad \text{To isolate } x, \text{ subtract}$$
$$\text{15,000 from both sides}$$
$$\text{to undo the addition}$$
$$\text{of 15,000.}$$

$$\begin{array}{r} 87,000 \\ -\ 15,000 \\ \hline 72,000 \end{array}$$

$$x = 72,000 \qquad \text{Do the subtraction.}$$

Self Check 7

STUDENT LOANS A student has saved $1,500 to pay for his first year of college. How much money will he have to borrow if books, tuition, and expenses for first-year students are estimated to total $3,750?

Now Try Problems 38 and 73

State

Sue must borrow $72,000.

Check

If Sue has $15,000 and we add the amount of money she needs to borrow, we should obtain the cost of the house.

$15,000
+$72,000
$87,000   *This is the cost of the house.*

The result, $72,000, checks.

ANSWERS TO SELF CHECKS

1. yes   **2.** no   **3.** 43   **4.** 113   **5.** 7   **6.** The tank originally contained 1,775 gallons of gasoline.   **7.** The student needs to borrow $2,250.

SECTION 1.8   STUDY SET

VOCABULARY

Fill in the blanks.

1. An _____ is a statement indicating that two expressions are equal. All equations contain an [ ] symbol.

2. A number that makes an equation true when substituted for the variable is called a _____ of the equation. Such numbers are said to _____ the equation.

3. To _____ an equation means to find all values of the variable that make the equation true.

4. To solve an equation, we _____ the variable on one side of the equal symbol.

5. Equations with the same solutions are called _____ equations.

6. To _____ the solution of an equation, we substitute the value for the variable in the original equation and determine whether the result is a true statement.

CONCEPTS

7. Given: $x + 6 = 12$

a. What is the left side of the equation?

b. Is this equation true or false?

c. Is 5 a solution of this equation?

d. Does 6 satisfy the equation?

8. Tell whether each of the following is an equation.

a. $x - 3$ **b.** $m + 12 = 40$

c. $7 < 8$ **d.** $18 > 0$

9. Fill in the blanks.

a. The addition property of equality: Adding the _____ number to both sides of an equation does not change its solution.

b. If $a = b$, then $a + c = b +$ [ ].

10. Fill in the blanks.

a. The subtraction property of equality: Subtracting the same number from _____ sides of an equation does not change its solution.

b. If $a = b$, then $a - c = b -$ [ ].

11. Fill in the blanks.

a. To solve $x - 8 = 24$, we _____ 8 to both sides of the equation.

b. To solve $x + 4 = 11$, we _____ 4 from both sides of the equation.

12. Simplify each expression.

a. $x + 7 - 7$ **b.** $y - 2 + 2$

NOTATION

Complete each solution to solve the equation. Check the result.

13. $x - 5 = 45$

$x - 5 +$ [ ] $= 45 +$ [ ]

$x =$ [ ]

Check: $x - 5 = 45$

[ ] $- 5 \stackrel{?}{=} 45$

[ ] $= 45$ True

[ ] is the solution.

14. $y + 11 = 12$

$y + 11 - \boxed{} = 12 - \boxed{}$

$y = \boxed{}$

Check: $y + 11 = 12$

$\boxed{} + 11 \stackrel{?}{=} 12$

$\boxed{} = 12$ True

$\boxed{}$ is the solution.

15. What does the symbol $\stackrel{?}{=}$ mean?

16. If you solve an equation and obtain $50 = x$, can you write $x = 50$?

▍GUIDED PRACTICE

Check to determine whether the given number is a solution of the equation. See Example 1.

17. Is 1 a solution of $x + 2 = 3$?

18. Is 4 a solution of $x + 2 = 6$?

19. Is 7 a solution of $a - 7 = 0$?

20. Is 16 a solution of $x - 8 = 8$?

Check to determine whether the given number is a solution of the equation. See Example 2.

21. Is 40 a solution of $50 = y - 8$?

22. Is 5 a solution of $16 = 10 + c$?

23. Is 2 a solution of $1 = x + 2$?

24. Is 4 a solution of $8 = x + 1$?

Solve each equation and check the result. See Example 3.

25. $x - 7 = 3$ **26.** $y - 11 = 7$

27. $a - 20 = 50$ **28.** $z - 31 = 60$

Solve each equation and check the result. See Example 4.

29. $1 = b - 2$ **30.** $0 = t - 1$

31. $19 = n - 42$ **32.** $17 = m - 16$

Solve each equation and check the result. See Example 5.

33. $x + 9 = 12$ **34.** $x + 3 = 9$

35. $y + 7 = 12$ **36.** $c + 11 = 22$

In Exercises 37 and 38, fill in the blanks to complete each solution.

37. HISTORY A 1,700-year-old scroll is 425 years older than the clay jar in which it was found. How old is the jar? See Example 6.

Analyze
- The scroll is _____ years old.   *Given*
- The scroll is _____ years older than the jar.   *Given*
- How old is the ___?   *Find*

Form Let $x =$ the age of the ___. Now we look for a key word or phrase in the problem.

Key phrase: *older than*   **Translation:** _____

Now we translate the words of the problem into an equation.

The age of the scroll	is	425 years	plus	the age of the jar.
$\boxed{}$	$=$	425	$+$	$\boxed{}$

Solve

$\boxed{} = 425 + x$

$1,700 - \boxed{} = 425 + x - \boxed{}$

$\boxed{} = x$

State The jar is _____ years old.

Check If the jar is 1,275 years old, and if we add 425 years to its age, we should get the age of the scroll.

$\overset{1\ 1}{1{,}275}$
$+\ \ 425$

$\boxed{}$ *This is the age of the scroll.*

The result checks.

38. BANKING After a student wrote a $1,500 check to pay for a car, he had a new balance of $750 in his account. What was the account balance before he wrote the check?

Analyze
- The student wrote a _____ check.   *Given*
- The new balance in the account was ____.   *Given*
- What was the _____ before he wrote the check?   *Find*

Form Let $x =$ the account _____ before he wrote the check. Now we look for a key word or phrase in the problem.

Key phrase: *wrote a check*   **Translation:** _____

Now we translate the words of the problem into an equation.

The account balance before the check	minus	the amount of the check	is equal to	the new balance.
$\boxed{}$	$-$	1,500	$=$	$\boxed{}$

Solve

$$\boxed{} - 1,500 = 750$$

$$x - 1,500 + \boxed{} = 750 + \boxed{}$$

$$x = \boxed{}$$

State The account balance before he wrote the check was

_____.

Check If the old balance was $2,250, and if we subtract the $1,500 check from it, we should get the new balance.

$$\overset{1\ 12}{\$2,\!\cancel{2}50}$$
$$-\$1,500$$
$$\$\boxed{}$$ This is the new balance in the account.

The result checks.

■ TRY IT YOURSELF

Solve each equation and check the result.

39. $s + 55 = 100$ **40.** $n + 37 = 200$

41. $x - 4 = 0$ **42.** $c - 3 = 0$

43. $y - 7 = 6$ **44.** $a - 2 = 4$

45. $70 = x - 5$ **46.** $66 = b - 6$

47. $312 = x - 428$ **48.** $113 = x - 307$

49. $x - 117 = 222$ **50.** $y - 27 = 317$

51. $t + 19 = 28$ **52.** $s + 45 = 84$

53. $23 + x = 33$ **54.** $34 + y = 34$

55. $5 = 4 + c$ **56.** $41 = 23 + x$

57. $99 = r + 43$ **58.** $92 = r + 37$

59. $512 = 428 + x$ **60.** $513 = 307 + x$

61. $x + 117 = 222$ **62.** $y + 38 = 321$

63. $3 + x = 7$ **64.** $4 + b = 8$

65. $y - 5 = 7$ **66.** $z - 9 = 23$

67. $4 + a = 12$ **68.** $5 + x = 13$

69. $x - 13 = 34$ **70.** $x - 23 = 19$

■ APPLICATIONS

Let a variable represent the unknown quantity. Then write and solve an equation to answer the question.

71. FAST FOOD The franchise fee and start-up costs for a Pizza Hut restaurant are $316,500. If a woman has $68,500 to invest, how much money will she need to borrow to open her own Pizza Hut restaurant? (Source: yumfranchises.com)

72. PARTY INVITATIONS Three of Mia's party invitations were lost in the mail, but 59 were delivered. How many invitations did she send?

73. HIP HOP *Forbes* magazine estimates that in 2008, Shawn "Jay-Z" Carter earned $82 million. If this was $68 million less than Curtis "50 Cent" Jackson's earnings, how much did 50 Cent earn in 2008?

74. GOLF CLUBS A man wants to buy a new set of golf clubs for $345. How much more money does he need if he now has $317?

75. HEARING PROTECTION The sound intensity of a jet engine is 110 decibels. If an airplane mechanic wears earplugs when working near a jet, she only experiences 81 decibels of sound intensity. By how many decibels do the earplugs reduce the noise level?

76. HELP WANTED From the following ad from the classifed section of a newspaper, determine the value of the beneft package. ($45 K means $45,000.)

★ACCOUNTS PAYABLE★
2-3 yrs exp as supervisor. Degree a +. High vol company. Good pay, $45K & xlnt benefits; total compensation worth $52K. Fax resume.

77. POWER OUTAGES The electrical system in a building automatically shuts down when the meter shown reads 85. By how much must the current reading shown below increase to cause the system to shut down?

78. VIDEO GAMES After a week of playing Sega's *Sonic Adventure,* a boy scored 11,053 points in one game—an improvement of 9,485 points over the very first time he played. What was the score for his first game?

79. AUTO REPAIRS A woman paid $29 less to have her car repaired at a muffler shop than she would have paid at a gas station. If she paid $190 at the muffler shop, what was the gas station going to charge her?

80. RIDING BUSES A man had to wait 20 minutes for a bus today. Three days ago, he had to wait 15 minutes longer than he did today, because several buses passed by without stopping. How long did he wait three days ago?

81. HIT RECORDS The oldest artist to have a number 1 single was 67-year-old Louis Armstrong, with his version of *Hello Dolly*. He was 55 years older than the youngest artist to have a number 1 single, Jimmy Boyd, who sang *I Saw Mommy Kissing Santa Claus*. How old was Jimmy Boyd when he had the number 1 song? (Source: *The Top 10 of Everything*, 2000)

82. REBATES The price of a new Honda Civic was advertised in a newspaper as $15,305*. A note at the bottom of the ad read, "*Reflects $1,550 factory rebate." What was the car's original sticker price?

WRITING

83. Explain what it means for a number to *satisfy* an equation.

84. Explain how to tell whether a number is a solution of an equation.

85. Explain what the pair of figures on page 97 are trying to show.

86. Think of a number. Add 8 to it. Now subtract 8 from that result. Explain why we will always obtain the original number.

87. When solving equations, we *isolate* the variable. Write a sentence in which the word *isolate* is used in a different context.

88. What do you find to be the most difficult step of the five-step problem solving strategy? Explain why it is.

89. Unlike an arithmetic approach, you *do not* have to determine whether to add, subtract, multiply, or divide to solve the application problems in this section. That decision is made for you when you solve the equation that mathematically describes the situation. Explain.

90. What does the word *translate* mean?

REVIEW

91. Round 325,784 to the nearest ten.

92. Evaluate: 1^5

93. Evaluate: $2 \cdot 3^2 \cdot 5$

94. **a.** Represent $4 + 4 + 4$ as a multiplication.
   b. Represent $4 \cdot 4 \cdot 4$ using an exponential expression.

95. Evaluate: $8 - 2(2^2 - 1) + 1^3$

96. Write 1,055 in words.

SECTION 1.9
Solving Equations Using Multiplication and Division

Objectives

1 Use the multiplication property of equality.

2 Use the division property of equality.

3 Use equations to solve application problems.

In the previous section, we solved simple equations such as

$$x - 2 = 3 \quad \text{and} \quad x + 8 = 11$$

by using the addition and subtraction properties of equality. In this section, we will learn how to solve equations such as

$$\frac{x}{3} = 25 \quad \text{and} \quad 2x = 6$$

by using the multiplication and division properties of equality.

1 Use the multiplication property of equality.

To introduce a third property of equality, consider the first scale shown on the right, which represents the equation $\frac{x}{3} = 25$. The scale is in balance because the weights on the left side and right side are equal. To find x, we must triple (multiply by 3) the weight on the left side. To keep the scale in balance, we must also triple the weight on the right side. After doing this, we see in the second scale that x is balanced by 75. Therefore, x must be 75.

The procedure that we just used suggests the following property of equality.

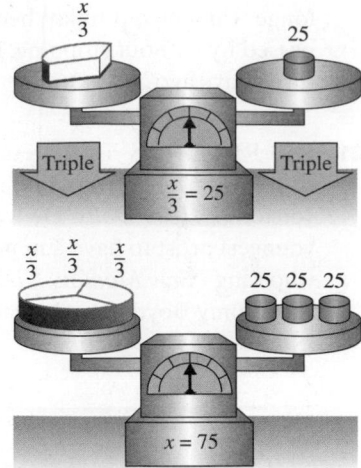

Multiplication Property of Equality

Multiplying both sides of an equation by the same nonzero number does not change its solution.

For any numbers a, b, and c, where c is not 0,

if $a = b$, then $ca = cb$

When we use this property, the resulting equation is equivalent to the original one. We will now show how it is used to solve $\frac{x}{3} = 25$ algebraically.

Self Check 1

Solve $\dfrac{x}{12} = 24$ and check the result.

Now Try Problem 13

EXAMPLE 1   Solve: $\dfrac{x}{3} = 25$

Strategy We will use the multiplication property of equality to isolate the variable x on the left side of the equation.

WHY To solve the original equation, we want to find a simpler equivalent equation of the form $x =$ **a number**, whose solution is obvious.

Solution

$$\frac{x}{3} = 25 \qquad \text{This is the equation to solve.}$$

$$3 \cdot \frac{x}{3} = 3 \cdot 25 \qquad \text{To isolate } x, \text{ undo the division by 3 by multiplying both sides by 3.}$$

$$x = 75 \qquad \text{On the left side, when x is divided by 3 and that quotient is then multiplied by 3, the result is x. Multiplication by 3 undoes division by 3. On the right side, do the multiplication: } 3 \cdot 25 = 75.$$

$$\begin{array}{r} \overset{1}{25} \\ \times\ 3 \\ \hline 75 \end{array}$$

Since 75 is obviously the solution of $x = 75$, the solution of the original equation, $\frac{x}{3} = 25$, is also 75.

Check: $\dfrac{x}{3} = 25 \qquad \text{This is the original equation.}$

$$\frac{75}{3} \overset{2}{=} 25 \qquad \text{Substitute 75 for x.}$$

$$25 = 25 \qquad \text{On the left side, do the division: } 75 \div 3 = 25.$$

$$\begin{array}{r} 25 \\ 3\overline{)75} \\ -6 \\ \hline 15 \\ -15 \\ \hline 0 \end{array}$$

Since $25 = 25$ is true statement, 75 is the solution of $\frac{x}{3} = 25$.

EXAMPLE 2 Solve: $84 = \dfrac{n}{16}$

Strategy We will use the multiplication property of equality to isolate the variable n on the right side of the equation.

WHY To solve the original equation, we want to find a simpler equivalent equation of the form **a number** $= n$, whose solution is obvious.

Solution

$$84 = \dfrac{n}{16} \qquad \text{This is the equation to solve.}$$

$$16 \cdot 84 = 16 \cdot \dfrac{n}{16} \qquad \begin{array}{l}\text{To isolate } n\text{, undo the division by 16 by multiplying} \\ \text{both sides by 16.}\end{array}$$

$$1{,}344 = n \qquad \begin{array}{l}\text{On the left side, do the multiplication: } 16 \cdot 84 = 1{,}344. \text{ On the} \\ \text{right side, when } n \text{ is divided by 16 and that quotient is then} \\ \text{multiplied by 16, the result is } n.\end{array}$$

$$\begin{array}{r} 84 \\ \times\ 16 \\ \hline 504 \\ 840 \\ \hline 1{,}344 \end{array}$$

To check this result, we substitute 1,344 for n in the original equation

Check: $84 = \dfrac{n}{16} \qquad$ This is the original equation.

$84 \overset{?}{=} \dfrac{\mathbf{1{,}344}}{16} \qquad$ Substitute 1,344 for n.

$84 = 84 \qquad$ On the right side, do the division.

$$\begin{array}{r} 84 \\ 16\overline{)1{,}344} \\ -128 \\ \hline 64 \\ -64 \\ \hline 0 \end{array}$$

Since $84 = 84$ is a true statement, 1,344 is the solution of $84 = \frac{n}{16}$.

Success Tip Calculations that you cannot perform in your head should be shown outside the steps of your solution.

---

2 Use the division property of equality.

To introduce a fourth property of equality, consider the first scale shown on the right, which represents the equation $2x = 6$. The scale is in balance because the weights on the left and right sides are equal. To find x, we need to split the amount of weight on the left side in half (divide by 2). To keep the scale in balance, we must split the amount of weight in half on the right side. After doing this, we see in the second scale that x is balanced by 3. Therefore, x must be 3. We say that we have solved the equation $2x = 6$ and that the solution is 3. This example illustrates the following property of equality.

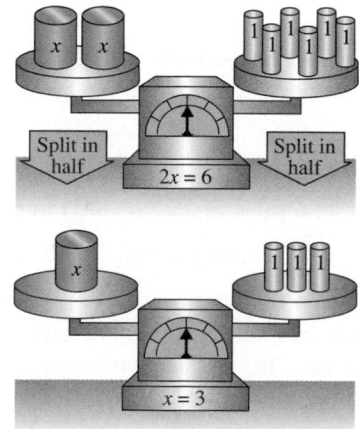

Division Property of Equality

Dividing both sides of an equation by the same nonzero number does not change its solution.

For any numbers a, b, and c, where c is not 0,

$$\text{if } a = b, \text{ then } \frac{a}{c} = \frac{b}{c}$$

When we use this property, the resulting equation is equivalent to the original one. We will now show how it is used to solve $2x = 6$ algebraically.

Self Check 3

Solve $17x = 153$ and check the result.

Now Try Problem 23

EXAMPLE 3   Solve: $2x = 6$

Strategy We will use the division property of equality to isolate the variable x on the left side of the equation.

WHY To solve the original equation, we want to find a simpler equivalent equation of the form $x = $ **a number**, whose solution is obvious.

Solution   Recall that $2x = 6$ means $2 \cdot x = 6$. To isolate x on the left side of the equation, we undo the multiplication by 2 by dividing both sides of the equation by 2.

$2x = 6$ This is the equation to solve.

$\dfrac{2x}{2} = \dfrac{6}{2}$ Divide both sides by 2.

$x = 3$ When x is multiplied by 2 and that product is then divided by 2, the result is x. On the right side, do the division: $6 \div 2 = 3$.

To check this result, we substitute 3 for x in $2x = 6$.

Check: $2x = 6$ This is the original equation.

$2 \cdot 3 \overset{?}{=} 6$ Substitute 3 for x.

$6 = 6$ On the left side, do the multiplication: $2 \cdot 3 = 6$.

Since $6 = 6$ is a true statement, 3 is the solution of $2x = 6$.

3 Use equations to solve application problems.

As before, we can use equations to solve application problems. Remember that the purpose of these early examples is to help you learn the strategy, even though you can probably solve the problems without it.

Self Check 4

CLASSICAL MUSIC A woodwind quartet (four musicians) was hired to play at an art exhibit. If each musician made $85 for the performance, what fee did the quartet charge?

Now Try Problem 25

EXAMPLE 4   *Entertainment Costs*   A five-piece band worked on New Year's Eve. If each player earned $120, what fee did the band charge?

Analyze
- There were 5 players in the band.   Given
- Each player made $120.   Given
- What fee did the band charge?   Find

Form
We can let $f = $ the band's fee. To form an equation, we look for a key word or phrase. In this case, we find it in the analysis of the problem. If each player earned the same amount ($120), the band's fee must have been divided into 5 equal parts.

Key phrase: *divided into 5 equal parts*   Translation: *division*

Now we translate the words of the problem into an equation.

The band's fee	divided by	the number of players in the band	is	each person's share.
f	$\div$	5	=	120

Solve

$$\dfrac{f}{5} = 120 \qquad \text{We need to isolate } f \text{ on the left side.}$$

$$5 \cdot \dfrac{f}{5} = 5 \cdot 120 \qquad \text{To isolate } f, \text{ multiply both sides by 5 to undo the division by 5.}$$

$$f = 600 \qquad \text{Do the multiplication.}$$

$$\begin{array}{r} \overset{1}{120} \\ \times\ \ 5 \\ \hline 600 \end{array}$$

State

The band's fee was $600.

Check

If the band's fee was $600, and we divide it into 5 equal parts, we should get the amount that each player earned.

$$\begin{array}{r} 120 \quad \longleftarrow \text{This is the amount each band member earned.}\\ 5\overline{)600} \\ \underline{-5} \\ 10 \\ \underline{-10} \\ 00 \\ \underline{-0} \\ 0 \end{array}$$

The result, $600, checks.

EXAMPLE 5   *Traffic Fines*   For speeding in a construction zone, a motorist had to pay a fine of $592. The violation occurred on a highway posted with signs like the one shown on the right. What would the fine have been if such signs were not posted?

> **TRAFFIC FINES DOUBLED IN CONSTRUCTION ZONE**

Analyze

- For speeding, the motorist was fined $592.   *Given*
- The fine was double what it would normally have been.   *Given*
- What would the fine have been, had the sign not been posted?   *Find*

Form

We can let f = the amount that the fine would normally have been. To form an equation, we look for a key word or phrase in the problem or analysis.

Key word: *double*   **Translation:** *multiply by 2*

Now we translate the words of the problem into an equation.

Two	times	the normal speeding fine	is	the new fine.
2	$\cdot$	f	=	592

Self Check 5

SPEED READING A speed reading course claims it can teach a person to read four times faster. After taking the course, a student can now read 700 words per minute. If the company's claims are true, what was the student's reading rate before taking the course?

Now Try Problem 26

Solve

$2f = 592$ We need to isolate f on the left side.

$\dfrac{2f}{2} = \dfrac{592}{2}$ To isolate f, divide both sides by 2 to undo the multiplication by 2.

$f = 296$ Do the division.

```
      296
   2)592
    - 4
     19
    - 18
      12
    - 12
       0
```

State

The fine would normally have been $296.

Check

If the normal fine was $296, and we double it, we should get the new fine.

```
  11
 296
×  2
 592   This is the new fine.
```

The result, $296, checks.

ANSWERS TO SELF CHECKS

1. 288   **2.** 1,020   **3.** 9   **4.** The quartet charged $340.00 for the performance.   **5.** The student used to read 175 words per minute.

SECTION 1.9   STUDY SET

VOCABULARY

Fill in the blanks.

1. To _____ an equation means to find all values of the variable that make the equation true.

2. A number that makes an equation true when substituted for the variable is called a _____ of the equation.

3. To solve an equation, we _____ the variable on one side of the equal symbol.

4. In this section, we used the multiplication and division properties of _____ to solve equations.

CONCEPTS

Fill in the blanks.

5. **a.** The multiplication property of equality: Multiplying both sides of an equation by the _____ nonzero number does not change its solution.

   b. If $a = b$, then $ca = $ ⬜ .   (provided c is not 0)

6. **a.** The division property of equality: Dividing both sides of an equation by the _____ nonzero number does not change its solution.

   b. If $a = b$, then $\dfrac{a}{c} = \dfrac{b}{c}$.   (provided c is not 0)

7. **a.** If we multiply x by 6 and then divide that product by 6, the result is ⬜ .

   b. If we divide x by 8 and then multiply that quotient by 8, the result is ⬜ .

8. Simplify each expression.

   a. $9 \cdot \dfrac{x}{9}$ **b.** $\dfrac{6y}{6}$

9. Fill in the blanks.

   a. To solve $\dfrac{x}{5} = 10$, we _____ both sides of the equation by 5.

   b. To solve $5x = 10$, we _____ both sides of the equation by 5.

   c. To solve $x - 5 = 10$, we _____ 5 to both sides of the equation.

   d. To solve $x + 5 = 10$, we _____ 5 from both sides of the equation.

10. Use a check to determine whether the given number is a solution of the equation.

   a. Is 8 a solution of $16 = 8t$?

   b. Is 2 a solution of $16 = \dfrac{t}{8}$?

NOTATION

Complete each solution to solve the equation. Check the result.

11. $\dfrac{x}{5} = 9$

$\boxed{} \cdot \dfrac{x}{5} = \boxed{} \cdot 9$

$x = \boxed{}$

Check: $\dfrac{x}{5} = 9$

$\dfrac{\boxed{}}{5} \overset{?}{=} 9$

$\boxed{} = 9$ True

$\boxed{}$ is the solution.

12. $3x = 12$

$\dfrac{3x}{\boxed{}} = \dfrac{12}{\boxed{}}$

$x = \boxed{}$

Check: $3x = 12$

$3 \cdot \boxed{} \overset{?}{=} 12$

$\boxed{} = 12$ True

$\boxed{}$ is the solution.

GUIDED PRACTICE

Solve each equation and check the result. **See Example 1.**

13. $\dfrac{x}{7} = 2$ **14.** $\dfrac{x}{12} = 4$

15. $\dfrac{y}{14} = 3$ **16.** $\dfrac{y}{13} = 5$

Solve each equation and check the result. **See Example 2.**

17. $16 = \dfrac{x}{24}$ **18.** $22 = \dfrac{x}{18}$

19. $31 = \dfrac{t}{11}$ **20.** $33 = \dfrac{m}{19}$

Solve each equation and check the result. **See Example 3.**

21. $3x = 3$ **22.** $5x = 5$

23. $9z = 90$ **24.** $3z = 60$

In Exercises 25 and 26, fill in the blanks to complete each solution.

25. THE NOBEL PRIZE In 1998, three Americans, Louis Ignarro, Robert Furchgott, and Fred Murad, were awarded the Nobel Prize for Medicine. They shared the prize money equally. If each person received $318,500, what was the amount of the Nobel Prize cash award? **See Example 4.**

Analyze

- __ people shared the cash award equally.   *Given*
- Each person received _____.   *Given*
- What was the _____ of the Nobel Prize cash award?   *Find*

Form

Let $x =$ the _____ of the Nobel Prize cash award. Now we look for a key word or phrase in the problem.

Key phrase: *shared the prize money equally*
Translation: division

Now we translate the words of the problem into an equation.

The Nobel Prize cash award	divided by	the number of prize winners	is equal to	each person's share.
$\boxed{}$	$\div$	3	$=$	$\boxed{}$

Solve

$\dfrac{x}{3} = 318{,}500$

$\boxed{} \cdot \dfrac{x}{3} = \boxed{} \cdot 318{,}500$

$x = \boxed{}$

State

The amount of the Nobel Prize cash award was _____.

Check

If we divide the Nobel Prize cash award by 3, we should get the amount each winner received.

$\dfrac{\$955{,}500}{3} = \boxed{}$ This is the amount each winner received.

The result checks.

26. THE STOCK MARKET An investor has seen the value of his stock double in the last 12 months. If the current value of his stock is $274,552, what was its value one year ago?

Analyze

- The value of the stock _____ in 12 months. Given

- The current value of the stock is _____. Given

- What was the _____ of the stock one year ago? Find

Form

We can let x = the _____ of the stock one year ago (in dollars). We now look for a key phrase in the problem.

Key phrase: *double*   **Translation:** _____ by 2

Now we translate the words of the problem into an equation.

2	times	the value of the stock one year ago	is equal to	the current value of the stock.

$$2 \cdot \boxed{} = \boxed{}$$

Solve

$$2x = \boxed{}$$

$$\frac{2x}{\boxed{}} = \frac{274,552}{\boxed{}}$$

$$x = \boxed{}$$

State

The value of the stock one year ago was _____ .

Check If we multiply the value that the stock had one year ago by 2, we should get its current value.

$$\begin{array}{r} \$137,276 \\ \times 2 \\ \hline \boxed{} \end{array}$$ This is the current value of the stock.

The result checks.

TRY IT YOURSELF

Solve each equation and check the result.

27. $100 = 100x$

28. $35 = 35y$

29. $\dfrac{a}{15} = 5$

30. $\dfrac{b}{25} = 5$

31. $16 = 8r$

32. $44 = 11m$

33. $21s = 210$

34. $155 = 31x$

35. $\dfrac{c}{1,000} = 3$

36. $\dfrac{d}{100} = 11$

37. $1 = \dfrac{x}{50}$

38. $1 = \dfrac{x}{25}$

39. $7 = \dfrac{t}{7}$

40. $4 = \dfrac{m}{4}$

41. $7x = 21$

42. $13x = 52$

43. $172 = 43t$

44. $288 = 96t$

45. $\dfrac{d}{20} = 201$

46. $\dfrac{x}{60} = 106$

47. $417 = \dfrac{t}{3}$

48. $259 = \dfrac{y}{7}$

49. $170y = 5,100$

50. $190y = 7,600$

51. $\dfrac{t}{3} = 47$

52. $\dfrac{d}{9} = 83$

53. $34y = 204$

54. $18y = 162$

APPLICATIONS

Let a variable represent the unknown quantity. Then write and solve an equation to answer the question.

55. SPEED READING An advertisement for a speed reading program claimed that successful completion of the course could triple a person's reading rate. After taking the course, Alicia can now read 399 words per minute. If the company's claims are true, what was her reading rate before taking the course?

56. PHYSICAL EDUCATION A high school PE teacher had the students in her class form three-person teams for a basketball tournament. Thirty-two teams participated in the tournament. How many students were in the PE class?

57. COST OVERRUNS Lengthy delays and skyrocketing costs caused a rapid-transit construction project to go over budget by a factor of 10. The final audit showed the project costing $540 million. What was the initial cost estimate?

58. STAMPS Large sheets of commemorative stamps honoring Marilyn Monroe are to be printed. On each sheet, there are 112 stamps, with 8 stamps per row. How many rows of stamps are on a sheet?

59. SPREADSHEETS The grid shown below is a computerized spreadsheet. The rows are labeled with numbers, and the columns are labeled with letters. Each empty box of the grid is called a *cell*. Suppose a certain project calls for a spreadsheet with 294 cells, using columns A through F. How many rows will need to be used?

Book 1					
File **Edit** **View** **Insert** **Format** **Tools**					
A	**B**	**C**	**D**	**E**	**F**
1					
2					
3					
4					
5					
6					
7					
8					
\Sheet 1 \Sheet 2 \Sheet 3 \Sheet 4 \Sheet 5/					

60. LOTTO WINNERS The grocery store employees listed below pooled their money to buy $120 worth of lottery tickets each week, with the understanding they would split the prize equally if they happened to win. One week they did have the winning ticket and won $480,000. What was each employee's share of the winnings?

Sam M. Adler	Ronda Pellman	Manny Fernando
Lorrie Jenkins	Tom Sato	Sam Lin
Kiem Nguyen	H. R. Kinsella	Tejal Neeraj
Virginia Ortiz	Libby Sellez	Alicia Wen

61. ANIMAL SHELTERS The number of phone calls to an animal shelter quadrupled after the evening news showed a segment explaining the services the shelter offered. Before the publicity, the shelter received 8 calls a day. How many calls did the shelter receive each day after being featured on the news?

62. OPEN HOUSES The attendance at an elementary school open house was only half of what the principal had expected. If 120 people visited the school that evening, how many had she expected to attend?

63. GRAVITY The weight of an object on Earth is 6 times greater than what it is on the moon. The situation shown to the right took place on Earth. If it took place on the moon, what weight would the scale register?

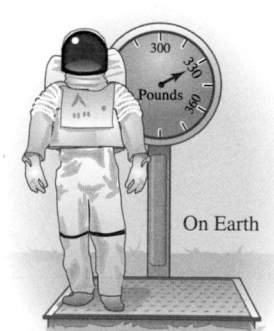

On Earth

64. INFOMERCIALS The number of orders received each week by a company selling skin care products increased fivefold after a Hollywood celebrity was added to the company's infomercial. After adding the celebrity, the company received about 175 orders each week. How many orders were received each week before the celebrity took part?

65. LIFE SPAN The average life span of an Amazon parrot is 104 years. That is thirteen times longer than the average life span of a Guinea pig. Find the average life span of a Guinea pig. (Source: petdoc.com)

66. CHILI HEAT SCALE In 1912, a chemist by the name of Wilbur Scoville developed a method to measure the heat level of chili peppers. For example, the heat rating on the Scoville scale for a habanero chili is 320,000 units. That is forty times greater than heat rating of a jalapeño chili. What is the Scoville rating for a jalapeño chili? (Source: ushotstuff.com)

WRITING

67. Explain what the pair of figures on page 106 are trying to show.

68. Draw a pair of figures like those on page 107. Explain what the figures illustrate.

69. What does it mean to *solve an equation*?

70. Think of a number. Double it. Now divide it by 2. Explain why you always obtain the original number.

REVIEW

71. Find the perimeter of a rectangle with sides measuring 8 cm and 16 cm.

72. Find the area of a rectangle with sides measuring 23 inches and 37 inches.

73. Find the prime factorization of 120.

74. Find the prime factorization of 150.

75. Evaluate: $3^2 \cdot 2^3$

76. Evaluate: $5 + 6 \cdot 3$

77. Divide, if possible: $\dfrac{0}{12}$

78. Divide, if possible: $\dfrac{50}{0}$

DEFINITIONS AND CONCEPTS	EXAMPLES
The set of **whole numbers** is {0, 1, 2, 3, 4, 5, …}. When a whole number is written using the **digits** 0, 1, 2, 3, 4, 5, 6, 7, 8, 9, it is said to be in **standard form**.	Some examples of whole numbers written in standard form are: 2, 16, 530, 7,894, and 3,201,954
The position of a digit in a whole number determines its **place value**. A place-value chart shows the place value of each digit in the number. To make large whole numbers easier to read, we use commas to separate their digits into groups of three, called **periods**.	 The place value of the digit 7 is 7 ten millions. The digit 4 tells the number of thousands.
To **write a whole number in words,** start from the left. Write the number in each period followed by the name of the period (except for the *ones period,* which is not used). Use commas to separate the periods. **To read a whole number out loud,** follow the same procedure. The commas are read as slight pauses.	Millions Thousands Ones 2 , 5 6 8 , 0 1 9 Two **million,** five hundred sixty-eight **thousand,** nineteen
To change from the **written-word form of a number to standard form,** look for the commas. Commas are used to separate periods.	Six billion , forty-one million , two hundred eight thousand , thirty-six 6,041,208,036
To write a number in **expanded form (expanded notation)** means to write it as an addition of the place values of each of its digits.	The expanded form of 32,159 is: $30,000 \; + \; 2,000 \; + \; 100 \; + \; 50 \; + \; 9$
Whole numbers can be shown by drawing points on a **number line.**	The graphs of 3 and 7 are shown on the number line below. 0 1 2 3 4 5 6 7 8
Inequality symbols are used to compare whole numbers: > means *is greater than* < means *is less than*	$9 > 8$ and $2,343 > 762$ $1 < 2$ and $9,000 < 12,453$

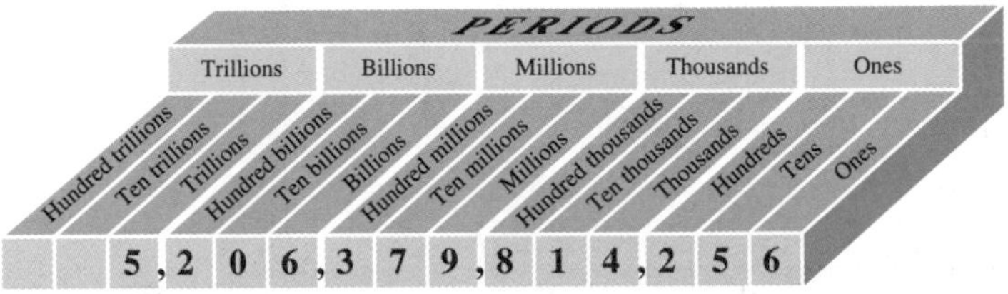

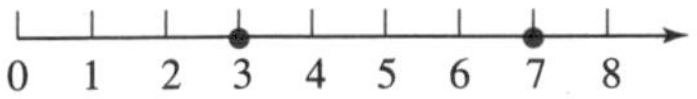

When we don't need exact results, we often **round** numbers.

Rounding a Whole Number

1. To round a number to a certain place value, locate the **rounding digit** in that place.

2. Look at the **test digit,** which is directly to the right of the rounding digit.

3. If the test digit is 5 or greater, round up by adding 1 to the rounding digit and replacing all of the digits to its right with 0.

If the test digit is less than 5, replace it and all of the digits to its right with 0.

Round 9,842 to the nearest ten.

Rounding digit: tens column

9,842

Test digit: Since 2 is less than 5, leave the rounding digit unchanged and replace the test digit with 0.

Thus, 9,842 rounded to the nearest ten is 9,840.

Round 63,179 to the nearest hundred.

Rounding digit: hundreds column

63,179

Test digit: Since 7 is 5 or greater, add 1 to the rounding digit and replace all the digits to its right with 0.

Thus, 63,179 rounded to the nearest hundred is 63,200.

Whole numbers are often used in **tables, bar graphs,** and **line graphs.**

See page 9 for an example of a table, a bar graph, and a line graph.

REVIEW EXERCISES

Consider the number 41,948,365,720.

1. **a.** Which digit is in the ten thousands column?

   b. Which digit is in the hundreds column?

   c. What is the place value of the digit 1?

   d. Which digit tells the number of millions?

2. Write each number in words.

   a. 97,283

   b. 5,444,060,017

3. Write each number in standard form.

   a. Three thousand, two hundred seven

   b. Twenty-three million, two hundred fifty-three thousand, four hundred twelve

4. Write $60,000 + 1,000 + 200 + 4$ in standard form.

Write each number in expanded form.

5. 570,302

6. 37,309,154

Graph the following numbers on a number line.

7. 0, 2, 8, 10

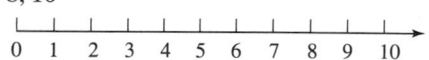

8. The whole numbers between 3 and 7.

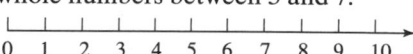

Place an < or an > symbol in the box to make a true statement.

9. 9 ☐ 7

10. 301 ☐ 310

11. Round 2,507,348

    a. to the nearest hundred

    b. to the nearest ten thousand

    c. to the nearest ten

    d. to the nearest million

12. Round 969,501

    a. to the nearest thousand

    b. to the nearest hundred thousand

13. **CONSTRUCTION** The following table lists the number of building permits issued in the city of Springsville for the period 2001–2008.

Year	2001	2002	2003	2004	2005	2006	2007	2008
Building permits	12	13	10	7	9	14	6	5

   a. Construct a bar graph of the data.

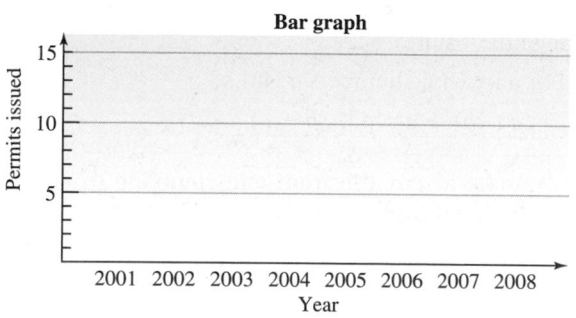

b. Construct a line graph of the data.

Line graph

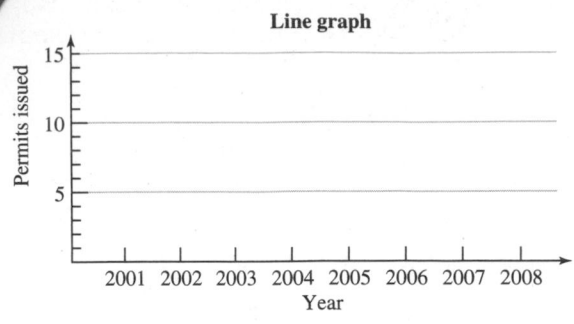

14. GEOGRAPHY The names and lengths of the five longest rivers in the world are listed below. Write them in order, beginning with the longest.

Amazon (South America)	4,049 mi
Mississippi-Missouri (North America)	3,709 mi
Nile (Africa)	4,160 mi
Ob-Irtysh (Russia)	3,459 mi
Yangtze (China)	3,964 mi

(Source: geography.about.com)

SECTION 1.2   Adding and Subtracting Whole Numbers

DEFINITIONS AND CONCEPTS	EXAMPLES
To **add whole numbers,** think of combining sets of similar objects. **Vertical form:** Stack the addends. Add the digits in the ones column, the tens column, the hundreds column, and so on. **Carry** when necessary.	Add: $10,892 + 5,467 + 499$ Carrying $\overset{1\ 2\ 1}{10,892}$ ← Addend $5,467$ ← Addend $+\ \ \ 499$ ← Addend $16,858$ ← Sum To check, add bottom to top
A **variable** is a letter (or symbol) that stands for a number.	Variables: x, a, and y
Commutative property of addition: The order in which whole numbers are added does not change their sum. For any whole numbers a and b, $a + b = b + a$	$6 + 5 = 5 + 6$ By the commutative property, the sum is the same.
Associative property of addition: The way in which whole numbers are grouped does not change their sum. For any whole numbers a and b, $(a + b) + c = a + (b + c)$	$(17 + 5) + 25 = 17 + (5 + 25)$ By the associative property, the sum is the same.
To estimate a sum, use **front-end rounding** to approximate the addends. Then add.	Estimate the sum: $7,219 \rightarrow 7,000$ Round to the nearest thousand. $592 \rightarrow 600$ Round to the nearest hundred. $+3,425 \rightarrow +3,000$ Round to the nearest thousand. $10,600$ The estimate is 10,600.

To solve the application problems, we must often *translate* the **key words** and **phrases** of the problem to numbers and symbols. Some key words and phrases that are often used to indicate addition are:

gain	*increase*	*up*	*forward*
rise	*more than*	*total*	*combined*
in all	*in the future*	*extra*	*altogether*

Translate the words to numbers and symbols:

VACATIONS There were 4,279,439 visitors to Grand Canyon National Park in 2006. The following year, attendance increased by 134,229. How many people visited the park in 2007?

The phrase *increased by* indicates addition:

The number of visitors to the park in 2007 = 4,279,439 + 134,229

The distance around a rectangle or a square is called its **perimeter.**

Perimeter of a rectangle = length + length + width + width

Perimeter of a square = side + side + side + side

Find the perimeter of the rectangle shown below.

15 ft

10 ft

Perimeter = 15 + 15 + 10 + 10   Add the two lengths and the two widths.

= 50

The perimeter of the rectangle is 50 feet.

To **subtract whole numbers,** think of taking away objects from a set.

Vertical form: Stack the numbers. Subtract the digits in the ones column, the tens column, the hundreds column, and so on. **Borrow** when necessary.

To **check:** Difference + subtrahend = minuend

Subtract: 4,957 − 869

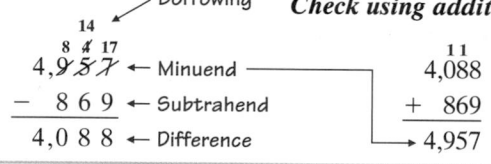

Check using addition:

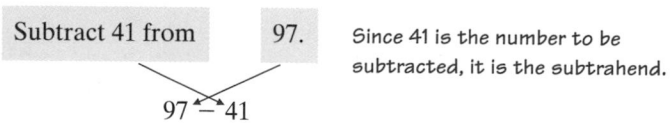

Be careful when translating the instruction to subtract one number *from* another number. The order of the numbers in the sentence must be reversed when we translate to symbols.

Translate the words to numbers and symbols:

Subtract 41 from   97.   Since 41 is the number to be subtracted, it is the subtrahend.

97 − 41

Every subtraction has a **related addition statement.**

10 − 3 = 7   because   7 + 3 = 10

To estimate a difference, use **front-end rounding** to approximate the minuend and subtrahend. Then subtract.

Estimate the difference:

59,033 →	60,000	Round to the nearest ten thousand.
− 4,124 →	− 4,000	Round to the nearest thousand.
	56,000	

The estimate is 56,000.

of the **key words** and **phrases** that are
en used to indicate subtraction are:

loss	decrease	down	backward
fell	less than	fewer	reduce
remove	debit	in the past	remains
declined			take away

To answer questions about *how much more* or
how many more, we use subtraction.

WEIGHTS OF CARS A Chevy Suburban weighs 5,607 pounds and a
Smart Car weighs 1,852 pounds. How much heavier is the Suburban?

The phrase *how much heavier* indicates subtraction:

$$
\begin{array}{r}
5,607 \quad \text{Weight of the Suburban} \\
-1,852 \quad \text{Weight of the Smart Car} \\
\hline
3,755
\end{array}
$$

The Suburban weighs 3,755 pounds more than the Smart Car.

To **evaluate** (find the value of) expressions
that involve addition and subtraction written
in **horizontal form,** we perform the operations
as they occur *from left to right.*

Evaluate: $75 - 23 + 9$

$75 - 23 + 9 = 52 + 9$ Working left to right, do the subtraction first.

$\qquad\qquad\quad = 61$ Now do the addition.

REVIEW EXERCISES

Add.

15. $27 + 436$

16. $4 + (36 + 19)$

17. $\begin{array}{r} 5,345 \\ +\ 655 \\ \hline \end{array}$

18. $2 + 1 + 38 + 3 + 6$

19. $4,447 + 7,478 + 676$

20. $\begin{array}{r} 32,812 \\ 65,034 \\ +54,323 \\ \hline \end{array}$

21. Use front-end rounding to estimate the sum.

$\qquad 615 + 789 + 14,802 + 39,902 + 8,098$

22. a. Use the commutative property of addition to
complete the following:

$\qquad 24 + 61 = $ ▢

b. Use the associative property of addition to
complete the following:

$\qquad 9 + (91 + 29) = $ ▢

23. AIRPORTS The nation's three busiest airports in
2007 are listed below. Find the total number of
passengers passing through those airports.

Airport	Total passengers
Hartsfield-Jackson Atlanta	89,379,287
Chicago O'Hare	76,177,855
Los Angeles International	61,896,075

Source: Airports Council International–North America

24. Add from bottom to top to check the sum. Is it
correct?

$$
\begin{array}{r}
1,291 \\
859 \\
345 \\
+\ 226 \\
\hline
1,821
\end{array}
$$

25. What is the sum of three thousand seven hundred
six and ten thousand nine hundred fifty-five?

26. What is 451,775 more than 327,891?

27. CAMPAIGN SPENDING In the 2004 U.S.
presidential race, candidates spent $717,900,000. In
the 2008 presidential race, spending increased by
$606,800,000 over 2004. How much was spent by the
candidates on the 2008 presidential race? (Source:
Center for Responsive Politics)

28. Find the perimeter of the rectangle shown below.

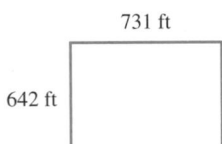

731 ft

642 ft

Subtract.

29. $148 - 87$

30. Subtract 10,218 from 10,435.

31. $750 - 259 + 14$

32. $\begin{array}{r} 7,800 \\ -5,725 \\ \hline \end{array}$

33. Check the subtraction using addition.

$$
\begin{array}{r}
8,017 \\
-6,949 \\
\hline
1,168
\end{array}
$$

34. Fill in the blank: $20 - 8 = 12$ because [   ].

35. Estimate the difference: $181{,}232 - 44{,}810$

36. LAND AREA Use the data in the table below to determine how much larger the land area of Russia is compared to that of Canada.

Country	Land area (square miles)
Russia	6,592,115
Canada	3,551,023

(Source: *The World Almanac, 2009*)

37. BANKING A savings account contains $12,975. If the owner makes a withdrawal of $3,800 and later deposits $4,270, what is the new account balance?

38. SUNNY DAYS In the United States, the city of Yuma, Arizona, typically has the most sunny days per year—about 242. The city of Buffalo, New York, typically has 188 days less than that. How many sunny days per year does Buffalo have?

SECTION 1.3   Multiplying Whole Numbers

DEFINITIONS AND CONCEPTS	EXAMPLES
Multiplication of whole numbers is repeated addition but with different notation.	Repeated addition: The sum of four 6's Multiplication $6 + 6 + 6 + 6 \quad = \quad 4 \times 6 \quad = \quad 24$
To write multiplication, we use a times symbol $\times$, a raised dot $\cdot$, and parentheses ( ).	$4 \times 6 \qquad\qquad 4 \cdot 6 \qquad\qquad 4(6)$ or $(4)(6)$ or $(4)6$
When multiplying a variable by a number or a variable by a variable, we can omit the symbol for multiplication.	$3x$ means $3 \cdot x$ and ab means $a \cdot b$
Vertical form: Stack the factors. If the bottom factor has more than one digit, multiply in steps to find the partial products. Then add them to find the product.	Multiply: $24 \cdot 163$ $163 \leftarrow$ Factor $\underline{\times\ 24} \leftarrow$ Factor $652 \leftarrow$ Partial product: $4 \cdot 163$ $\underline{3260} \leftarrow$ Partial product: $20 \cdot 163$ $3{,}912 \leftarrow$ Product
To find the **product of a whole number and 10, 100, 1,000, and so on,** attach the number of zeros in that number to the right of the whole number. This rule can be extended to multiply any two whole numbers that end in zeros.	Multiply: $8 \cdot 1{,}000 = 8{,}000$ Since 1,000 has three zeros, attach three 0's after 8. $43(10{,}000) = 430{,}000$ Since 10,000 has four zeros, attach four 0's after 43. $160 \cdot 20{,}000 = 3{,}200{,}000$ 160 and 20,000 have a total of five trailing zeros. Attach five 0's after 32. Multiply 16 and 2 to get 32.

Multiplication Properties of 0 and 1
 	
 The product of any whole number and 0 is 0.
     For any whole number a,
        $a \cdot 0 = 0$ and $0 \cdot a = 0$
	$0 \cdot 9 = 0$ and $3(0) = 0$

The product of any whole number and 1 is that whole number.
     For any whole number a,
        $a \cdot 1 = a$ and $1 \cdot a = a$
	$15 \cdot 1 = 15$ and $1(6) = 6$

Commutative property of multiplication: The order in which whole numbers are multiplied does not change their product.
 	
    For any whole numbers a and b,
 	
      $ab = ba$
 	
Associative property of multiplication: The way in which whole numbers are grouped does not change their product.
 	
    For any whole numbers a and b,
 	
      $(ab)c = a(bc)$
	$5 \cdot 9 = 9 \cdot 5$
	
	By the commutative property, the product is the same.
	
	$(3 \cdot 7) \cdot 10 = 3 \cdot (7 \cdot 10)$
	
	By the associative property, the product is the same.

To **estimate** a product, use **front-end rounding** to approximate the factors. Then multiply.	To estimate the product for $74 \cdot 873$, find $70 \cdot 900$.
	
	          ⌐— Round to the nearest ten —⌐
	$74 \cdot 873$ $70 \cdot 900$
	            ⌐— Round to the nearest hundred —⌐

Application problems that involve **repeated addition** are often more easily solved using multiplication.	HEALTH CARE A doctor's office is open 210 days a year. Each day the doctor sees 25 patients. How many patients does the doctor see in 1 year? This **repeated addition** can be calculated by multiplication: The number of patients seen each year $= \;\; 25 \cdot 210$
We can use multiplication to count objects arranged in rectangular patterns of neatly arranged rows and columns called **rectangular arrays.** Some **key words** and **phrases** that are often used to indicate multiplication are: *double triple twice of times*	CLASSROOMS A large lecture hall has 16 rows of desks and there are 12 desks in each row. How many desks are in the lecture hall? The **rectangular array** of desks indicates multiplication: The number of desks in the lecture hall $= \;\; 16 \cdot 12$

The **area of a rectangle** is the measure of the amount of surface it encloses. Area is measured in square units, such as square inches (written in.^{2}) or square centimeters (written cm^2).
 	
    Area of a rectangle $=$ length $\cdot$ width
 	
                or
 	
            $A = lw$
 	
Letters (or symbols) that are used to represent numbers are called **variables.**
	Find the area of the rectangle shown below.
	
	                25 in.
	        [rectangle] 4 in.
	
	  $A = lw$
	    $= 25 \cdot 4$ Replace the length l with 25 and the width w with 4.
	    $= 100$ Multiply.
	
	The area of the rectangle is 100 square inches, which can be written in more compact form as 100 in.^{2}

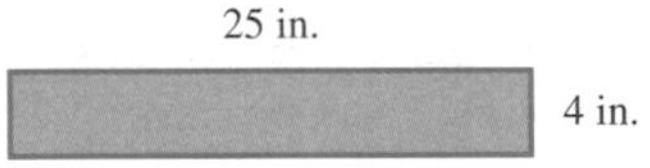

REVIEW EXERCISES

Multiply.

39. 47×9

40. $5 \cdot (7 \cdot 6)$

41. $72 \cdot 10,000$

42. $157 \cdot 59$

43. $\begin{array}{r} 5,624 \\ \times\ 281 \end{array}$

44. $502 \cdot 459$

45. Estimate the product: $6,891 \cdot 438$

46. a. Write the repeated addition $7 + 7 + 7 + 7 + 7$ as a multiplication.

   b. Write $2 \cdot t$ in simpler form.

   c. Write $m \cdot n$ in simpler form.

47. Find each product:

   a. $8 \cdot 0$ **b.** $7 \cdot 1$

48. What property of multiplication is shown?

   a. $2 \cdot (5 \cdot 7) = (2 \cdot 5) \cdot 7$

   b. $100(50) = 50(100)$

Find the area of the rectangle and the square.

49.

8 cm

4 cm

50.

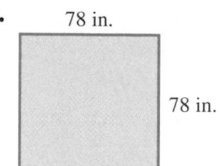

78 in.

78 in.

51. SLEEP The National Sleep Foundation recommends that adults get from 7 to 9 hours of sleep each night.

   a. How many hours of sleep is that in one year using the smaller number? (Use a 365-day year.)

   b. How many hours of sleep is that in one year using the larger number?

52. GRADUATION For a graduation ceremony, the graduates were assembled in a rectangular 22-row and 15-column formation. How many members are in the graduating class?

53. PAYCHECKS Sarah worked 12 hours at $9 per hour, and Santiago worked 14 hours at $8 per hour. Who earned more money?

54. SHOPPING There are 12 eggs in one dozen, and 12 dozen in one gross. How many eggs are in a shipment of 100 gross?

SECTION 1.4   Dividing Whole Numbers

DEFINITIONS AND CONCEPTS	EXAMPLES
To **divide whole numbers,** think of separating a quantity into equal-sized groups. To write division, we can use a division symbol $\div$, a long division symbol $\overline{)}$, or a fraction bar $-$.	Dividend Divisor $\downarrow$ $\quad$ $\downarrow$ $8 \div 2 = 4$ $\qquad 2\overline{)8}^{\,4} \qquad \dfrac{8}{2} = 4$ $\uparrow$ Quotient
Another way to answer a division problem is to think in terms of multiplication and write a **related multiplication statement.**	$8 \div 2 = 4$ because $4 \cdot 2 = 8$
A process called **long division** can be used to divide whole numbers. Follow a four-step process: • Estimate • Multiply • Subtract • Bring down	Divide: $8,317 \div 23$ Quotient $361 \text{ R } 14$ Divisor → $23\overline{)8,317}$ ← Dividend $\underline{-69\downarrow}$ $1\,41$ $\underline{-1\,38\downarrow}$ 37 $\underline{-23}$ 14 — Remainder

To **check** the result of a division, we multiply the divisor by the quotient and add the remainder. The result should be the dividend.	For the division on the previous page, the result checks. Quotient · divisor remainder $(361 \cdot 23) + 14 = 8{,}303 + 14$ $= 8{,}317 \leftarrow$ Dividend

Properties of Division

Any whole number divided by 1 is equal to that number.

For any whole number a,
$$\frac{a}{1} = a.$$

$$\frac{4}{1} = 4 \qquad \text{and} \qquad \frac{58}{1} = 58$$

Any nonzero whole number divided by itself is equal to 1.

For any nonzero whole number a,
$$\frac{a}{a} = 1$$

$$\frac{9}{9} = 1 \qquad \text{and} \qquad \frac{103}{103} = 1$$

Division with Zero

Zero divided by any nonzero number is equal to 0.

Division by 0 is undefined.

For any nonzero whole number a,
$$\frac{0}{a} = 0 \qquad \text{and} \qquad \frac{a}{0} \text{ is undefined}$$

$$\frac{0}{7} = 0 \qquad \text{and} \qquad \frac{0}{23} = 0$$

$$\frac{7}{0} \text{ is undefined} \qquad \text{and} \qquad \frac{2{,}190}{0} \text{ is undefined}$$

There are **divisibility tests** to help us decide whether one number is divisible by another. They are listed on page 55.	Is 21,507 divisible by 3? 21,507 is divisible by 3, because the sum of its digits is divisible by 3. $$2 + 1 + 5 + 0 + 7 = 15 \qquad \text{and} \qquad 15 \div 3 = 5$$
There is a shortcut for **dividing a dividend by a divisor when both end with zeros.** We simply *remove the ending zeros in the divisor and remove the same number of ending zeros in the dividend.*	Divide: $$64{,}000 \div 1{,}600 = 640 \div 16$$ Remove two zeros from the dividend and the divisor, and divide.
To **estimate quotients,** we use a method that approximates both the dividend and the divisor so that they divide easily.	Estimate the quotient for $154{,}908 \div 46$ by finding $150{,}000 \div 50$. The dividend is approximately $154{,}908 \div 46 \qquad 150{,}000 \div 50$ The divisor is approximately
Application problems that involve **forming equal-sized groups** can be solved by division. Some **key words** and **phrases** that are often used to indicate division: split equally distributed equally shared equally how many does each how many left (remainder) per how much extra (remainder) among	BRACES An orthodontist offers his patients a plan to pay the \$5,400 cost of braces in 36 equal payments. What is the amount of each payment? The phrase *36 equal payments* indicates division: The amount of each payment $= 5{,}400 \div 36$

REVIEW EXERCISES

Divide, if possible.

55. $\dfrac{72}{4}$

56. $1{,}443 \div 39$

57. $68\overline{)20{,}876}$

58. $21\overline{)405}$

59. $\dfrac{0}{10}$

60. $\dfrac{165}{0}$

61. $127\overline{)5{,}347}$

62. $1{,}482{,}000 \div 3{,}900$

63. Write the related multiplication statement for $160 \div 4 = 40$.

64. Use a check to determine whether the following division is correct.

$$45\text{ R }6$$
$$7\overline{)320}$$

65. Is 364,545 divisible by 2, 3, 4, 5, 6, 9, or 10?

66. Estimate the quotient: $210{,}999 \div 53$

67. TREATS If 745 candies are distributed equally among 45 children, how many will each child receive? How many candies will be left over?

68. PURCHASING A county received an $850,000 grant to purchase some new police patrol cars. If a fully equipped patrol car costs $25,000, how many can the county purchase with the grant money?

SECTION **1.5**   Prime Factors and Exponents

DEFINITIONS AND CONCEPTS	EXAMPLES
Numbers that are multiplied together are called **factors.** To **factor** a whole number means to express it as the product of other whole numbers. If a whole number is a factor of a given number, it also *divides the given number exactly.*	The pairs of whole numbers whose product is 6 are: $1 \cdot 6 = 6$ and $2 \cdot 3 = 6$ From least to greatest, the factors of 6 are 1, 2, 3, and 6. Each of the factors of 6 divides 6 exactly (no remainder): $\dfrac{6}{1} = 6 \quad \dfrac{6}{2} = 3 \quad \dfrac{6}{3} = 2 \quad \dfrac{6}{6} = 1$
If a whole number is divisible by 2, it is called an **even** number. If a whole number is not divisible by 2, it is called an **odd** number.	Even whole numbers: 0, 2, 4, 6, 8, 10, 12, 14, 16, 18, … Odd whole numbers: 1, 3, 5, 7, 9, 11, 13, 15, 17, 19, …
A **prime number** is a whole number greater than 1 that has only 1 and itself as factors. There are infinitely many prime numbers.	Prime numbers: 2, 3, 5, 7, 11, 13, 17, 19, 23, 29, 31, …
The **composite numbers** are whole numbers greater than 1 that are *not* prime. There are infinitely many composite numbers.	Composite numbers: 4, 6, 8, 9, 10, 12, 14, 15, 16, 18, …

To find the **prime factorization** of a whole number means to write it as the product of only prime numbers.

A **factor tree** and a **division ladder** can be used to find prime factorizations.

Use a *factor tree* to find the prime factorization of 30.

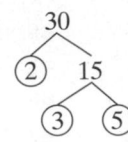

Factor each number that is encountered as a product of two whole numbers (other than 1 and itself) until all the factors involved are prime.

The prime factorization of 30 is $2 \cdot 3 \cdot 5$.

Use a *division ladder* to find the prime factorization of 70.

$$\begin{array}{r} 2\,\lfloor 70 \\ 5\,\lfloor 35 \\ \hline 7 \end{array}$$

Perform repeated divisions by prime numbers until the final quotient is itself a prime number.

The prime factorization of 70 is $2 \cdot 5 \cdot 7$.

An **exponent** is used to indicate repeated multiplication. It tells how many times the **base** is used as a factor.

$$\underbrace{2 \cdot 2 \cdot 2 \cdot 2}_{\text{Repeated factors}} = 2^{\overset{\text{Exponent}}{4}}$$

Base

2^4 is called an exponential expression.

We can use the definition of exponent to **evaluate** (find the value of) exponential expressions.

Evaluate: 7^3

$$\begin{aligned} 7^3 &= 7 \cdot 7 \cdot 7 & &\text{Write the base 7 as a factor 3 times.} \\ &= 49 \cdot 7 & &\text{Multiply, working left to right.} \\ &= 343 & &\text{Multiply.} \end{aligned}$$

Evaluate: $2^2 \cdot 3^3$

$$\begin{aligned} 2^2 \cdot 3^3 &= 4 \cdot 27 & &\text{Evaluate the exponential expressions first.} \\ &= 108 & &\text{Multiply.} \end{aligned}$$

REVIEW EXERCISES

Find all of the factors of each number. List them from least to greatest.

69. 18

70. 75

71. Factor 20 using two factors. *Do not use the factor 1 in your answer.*

72. Factor 54 using three factors. *Do not use the factor 1 in your answer.*

Tell whether each number is a prime number, a composite number, or neither.

73. a. 31

b. 100

c. 1

d. 0

e. 125

f. 47

Tell whether each number is an even or an odd number.

74. a. 171

b. 214

c. 0

d. 1

Find the prime factorization of each number. Use exponents in your answer, when helpful.

75. 42

76. 75

77. 220

78. 140

Write each expression using exponents.

79. $6 \cdot 6 \cdot 6 \cdot 6$

80. $5(5)(5)(13)(13)$

Evaluate each expression.

81. 5^3

82. 11^2

83. $2^4 \cdot 7^2$

84. $2^2 \cdot 3^3 \cdot 5^2$

SECTION 1.6   The Least Common Multiple and the Greatest Common Factor

DEFINITIONS AND CONCEPTS	EXAMPLES
The **multiples** of a number are the products of that number and 1, 2, 3, 4, 5, and so on.	Multiples of 2: 2, 4, **6**, 8, 10, **12**, 14, 16, **18**, 20, 22, **24**, … Multiples of 3: 3, **6**, 9, **12**, 15, **18**, 21, **24**, 27, … The common multiples of 2 and 3 are: 6, 12, 18, 24, 30, …
The **least common multiple (LCM)** of two whole numbers is the smallest common multiple of the numbers. The **LCM** of two whole numbers is the *smallest* whole number that is divisible by both of those numbers.	The least common multiple of 2 and 3 is 6, which is written as: LCM (2, 3) = 6. $\dfrac{6}{2} = 3$ and $\dfrac{6}{3} = 2$

To **find the LCM** of two (or more) whole numbers **by listing:**
 	
 1. Write multiples of the largest number by multiplying it by 1, 2, 3, 4, 5, and so on.
 2. Continue this process until you find the *first* multiple of the larger number that is divisible by each of the smaller numbers. That multiple is their LCM.
	Find the LCM of 3 and 5.
	 Multiples of 5: 5, 10, 15, 20, 25, …
	     Not divisible Not divisible Divisible by 3.
	     by 3. by 3.
	 Since 15 is the first multiple of 5 that is divisible by 3, the LCM (3, 5) = 15.

To **find the LCM** of two (or more) whole numbers **using prime factorization:** 1. Prime factor each number. 2. The LCM is a product of prime factors, where each factor is used the greatest number of times it appears in any one factorization.	Find the LCM of 6 and 20. $6 = 2 \cdot ③$ *The greatest number of times 3 appears is once.* $20 = (2 \cdot 2) \cdot ⑤$ *The greatest number of times 2 appears is twice.* *The greatest number of times 5 appears is once.* *Use the factor 2 two times.* *Use the factor 3 one time.* *Use the factor 5 one time.* LCM (6, 20) = $2 \cdot 2 \cdot 3 \cdot 5 = 60$
The **greatest common factor (GCF)** of two (or more) whole numbers is the largest common factor of the numbers.	The factors of 18: 1, 2, 3, 6, 9, 18 The factors of 30: 1, 2, 3, 5, 6, 10, 15, 30 The common factors of 18 and 30 are 1, 2, 3, and 6. The greatest common factor of 18 and 30 is 6, which is written as: GCF (18, 30) = 6.
The greatest common factor of two (or more) numbers is the *largest* whole number that divides them exactly.	$\dfrac{18}{6} = 3$ and $\dfrac{30}{6} = 5$
To **find the GCF** of two (or more) whole numbers using **prime factorization:** 1. Prime factor each number. 2. Identify the common prime factors. 3. The GCF is a product of all the common prime factors found in Step 2. If there are no common prime factors, the GCF is 1.	Find the GCF of 36 and 60. $36 = 2 \cdot 2 \cdot 3 \cdot 3$ *36 and 60 have two common factors* $60 = 2 \cdot 2 \cdot 3 \cdot 5$ *of 2 and one common factor of 3.* The GCF is the product of the circled prime factors. GCF (36, 60) = $2 \cdot 2 \cdot 3 = 12$

REVIEW EXERCISES

85. Find the first ten multiples of 9.

86. a. Find the common multiples of 6 and 8 in the lists below.

>   Multiples of 6: 6, 12, 18, 24, 30, 36, 42, 48, 54 ...
>   Multiples of 8: 8, 16, 24, 32, 40, 48, 56, 64, 72 ...

b. Find the common factors of 6 and 8 in the lists below.

>   Factors of 6: 1, 2, 3, 6
>   Factors of 8: 1, 2, 4, 8

Find the LCM of the given numbers.

87. 4, 6   **88.** 3, 4

89. 9, 15   **90.** 12, 18

91. 18, 21   **92.** 24, 45

93. 4, 14, 20   **94.** 21, 28, 42

Find the GCF of the given numbers.

95. 8, 12   **96.** 9, 12

97. 30, 40   **98.** 30, 45

99. 63, 84   **100.** 112, 196

101. 48, 72, 120 **102.** 88, 132, 176

103. MEETINGS The Rotary Club meets every 14 days and the Kiwanis Club meets every 21 days. If both clubs have a meeting on the same day, in how many more days will they again meet on the same day?

104. FLOWERS A florist is making flower arrangements for a 4th of July party. He has 32 red carnations, 24 white carnations, and 16 blue carnations. He wants each arrangement to be identical.

a. What is the greatest number of arrangements that he can make if every carnation is used?

b. How many of each type of carnation will be used in each arrangement?

SECTION 1.7   Order of Operations

DEFINITIONS AND CONCEPTS	EXAMPLES

To **evaluate** (find the value of) expressions that involve more than one operation, use the order-of-operations rule.

Order of Operations

1. Perform all calculations within parentheses and other grouping symbols following the order listed in Steps 2–4 below, working from the innermost pair of grouping symbols to the outermost pair.

2. Evaluate all exponential expressions.

3. Perform all multiplications and divisions as they occur from left to right.

4. Perform all additions and subtractions as they occur from left to right.

When grouping symbols have been removed, repeat Steps 2–4 to complete the calculation.

If a fraction bar is present, evaluate the expression above the bar (called the **numerator**) and the expression below the bar (called the **denominator**) separately. Then perform the division indicated by the fraction bar, if possible.

Evaluate: $10 + 3[2^4 - 3(5 - 2)]$

Work within the *innermost* parentheses first and then within the *outermost* brackets.

$$10 + 3[2^4 - 3(5 - 2)] = 10 + 3[2^4 - 3(3)]$$ Subtract within the parentheses.

$$= 10 + 3[16 - 3(3)]$$ Evaluate the exponential expression within the brackets: $2^4 = 16$.

$$= 10 + 3[16 - 9]$$ Multiply within the brackets.

$$= 10 + 3[7]$$ Subtract within the brackets.

$$= 10 + 21$$ Multiply: $3[7] = 21$.

$$= 31$$ Do the addition.

Caution!   A common error is to incorrectly add 10 and 3 in Step 5 of the solution.

$$= 10 + 3[7]$$

~~$= 13[7]$~~   Multiply before adding.

~~$= 91$~~

Evaluate: $\dfrac{3^3 + 8}{7(15 - 14)}$

Evaluate the expressions above and below the fraction bar separately.

$\dfrac{3^3 + 8}{7(15 - 14)} = \dfrac{27 + 8}{7(1)}$ In the numerator, evaluate the exponential expression. In the denominator, subtract.

$= \dfrac{35}{7}$ In the numerator, add. In the denominator, multiply.

$= 5$ Divide.

The **mean**, or **average**, of a set of numbers is a value around which the values of the numbers are grouped.

To **find the mean (average)** of a set of values, divide the sum of the values by the number of values.

Find the mean (average) of the test scores 74, 83, 79, 91, and 73.

$\text{Mean} = \dfrac{74 + 83 + 79 + 91 + 73}{5}$ Since there are 5 scores, divide by 5.

$= \dfrac{400}{5}$ Do the addition in the numerator.

$= 80$ Divide.

The mean (average) test score is 80.

REVIEW EXERCISES

Evaluate each expression.

105. $3^2 + 12 \cdot 3$

106. $35 - 5 \cdot 3 + 3$

107. $(6 \div 2 \cdot 3)^2 \cdot 3$

108. $(35 - 5 \cdot 3) \div 5$

109. $2^3 \cdot 5 - 4 \div 2 \cdot 4$

110. $8 \cdot (5 - 4 \div 2)^2$

111. $2 + 3\left(\dfrac{100}{10} - 2^2 \cdot 2\right)$

112. $4(4^2 - 5 \cdot 3 + 2) - 4$

113. $\dfrac{4(6) - 6}{2(3^2)}$

114. $\dfrac{6 \cdot 2 + 3 \cdot 7}{5^2 - 2(7)}$

115. $7 + 3[3^3 - 10(4 - 2)]$

116. $5 + 2\left[\left(2^4 - 3 \cdot \dfrac{8}{2}\right) - 2\right]$

Find the arithmetic mean (average) of each set of test scores.

117.

Test	1	2	3	4
Score	80	74	66	88

118.

Test	1	2	3	4	5
Score	73	77	81	0	69

SECTION 1.8   Solving Equations Using Addition and Subtraction

DEFINITIONS AND CONCEPTS	EXAMPLES
An **equation** is a statement indicating that two expressions are equal. All equations contain an = symbol. The equal symbol separates an equation into two parts: the left side and the right side.	Equations: $x + 4 = 10$ $y - 7 = 15$ $\dfrac{x}{3} = 9$ $6x = 42$
A number that makes an equation a true statement when substituted for the variable is called a **solution** of the equation.	Use a check to determine whether 6 is a solution of $x + 4 = 10$. **Check:** $x + 4 = 10$ $\qquad 6 + 4 \overset{?}{=} 10$ Substitute 6 for x. $\qquad\quad 10 = 10$ On the left side, do the addition. Since the resulting statement $10 = 10$ is true, 6 is the solution.
Equivalent equations have the same solutions.	$x - 2 = 6$ and $x = 8$ are equivalent equations because they have the same solution, 8.
To **solve an equation,** isolate the variable on one side of the equation by undoing the operation performed on it using a property of equality. The **addition property of equality:** Adding the same number to both sides of an equation does not change its solution. If $a = b$, then $a + c = b + c$	Solve: $x - 8 = 12$ We can use the addition property of equality to isolate x on the left side of the equation. $x - 8 = 12$ This is the equation to solve. $x - 8 + 8 = 12 + 8$ Undo the subtraction of 8 by adding 8 to both sides. $\qquad x = 20$ On the left side, adding 8 undoes the subtraction of 8 and leaves x. On the right side add: 12 + 8 = 20. The solution is 20. Check this result by substituting 20 for x in the original equation.
The **subtraction property of equality:** Subtracting the same number from both sides of an equation does not change its solution. If $a = b$, then $a - c = b - c$	Solve: $59 = y + 31$ We can use the subtraction property of equality to isolate y on the right side of the equation. $59 = y + 31$ This is the equation to solve. $59 - 31 = y + 31 - 31$ Undo the addition of 31 by subtracting 31 from both sides. $28 = y$ On the left side, do the subtraction: 59 − 31 = 28. On the right side, subtracting 31 undoes the addition of 31 and leaves y. The solution is 28. Check this result by substituting 28 for y in the original equation.

To solve application problems, use the five-step problem-solving strategy.

1. **Analyze the problem:** What information is given? What are you asked to find?

2. **Form an equation:** Pick a variable to represent the numerical value to be found. Translate the words of the problem into an equation.

3. **Solve the equation.**

4. **State the conclusion** clearly. Be sure to include the units (such as feet, seconds, or pounds) in your answer.

5. **Check the result:** Use the original wording of the problem, not the equation that was formed in step 2 from the words.

AUTO REPAIRS   A man paid $34 less for a new set of tires at a gas station than he would have paid for the same tires at a car dealer. If he paid $356 at the gas station, what was the car dealer going to charge him for the tires?

Analyze

• He paid $356 for the tires at the gas station.   Given

• The gas station charged $34 less than what the car dealer would have charged.   Given

• What would the car dealer have charged for the tires?   Find

Form   Let x = the amount the car dealer would have charged for the tires.

Key phrase: *$34 less*   **Translation:** subtraction

Now we translate the words of the problem into an equation.

The amount the car dealer would have charged	minus	$34	is equal to	the amount the gas station charged.
x	$-$	34	$=$	356

Solve

$$x - 34 = 356$$ We need to isolate x on the left side.

$$x - 34 + 34 = 356 + 34$$ To isolate x, undo the subtraction of 34 by adding 34 to both sides.

$$x = 390$$ Do the addition.

State   The car dealer would have charged $390 for the tires.

Check   If the car dealer was going to charge $390 for the tires, and if we subtract the $34 from that cost, we should get the amount the gas station charged.

$$\begin{array}{r} \overset{810}{\$3\cancel{9}\cancel{0}} \\ -\ \$\ \ 34 \\ \hline \$\ 356 \end{array}$$ This is what the gas station charged for the tires.

The result checks.

REVIEW EXERCISES

Use a check to determine whether the given number is a solution of the equation.

119. Is 5 a solution of $x + 2 = 13$?

120. Is 4 a solution of $x - 3 = 1$?

Solve each equation and check the result.

121. $x - 7 = 2$ **122.** $x - 11 = 20$

123. $225 = y - 115$ **124.** $101 = p - 32$

125. $x + 9 = 18$ **126.** $b + 12 = 26$

127. $175 = p + 55$ **128.** $212 = m + 207$

Let a variable represent the unknown quantity. Then write and solve an equation to answer the question.

129. FINANCING   A newly married couple made a $25,500 down payment on a $122,750 house. How much did they need to borrow?

130. HEALTH CARE   After moving his office, a doctor lost 13 patients. If he had 172 patients left, how many did he have originally?

DEFINITIONS AND CONCEPTS	EXAMPLES
The **multiplication property of equality:** Multiplying both sides of an equation by the same nonzero number does not change its solution. If $a = b$, then $ca = cb$ (provided $c \neq 0$)	Solve: $\dfrac{m}{5} = 32$ We can use the multiplication property of equality to isolate m on the left side. $\dfrac{m}{5} = 32$ This is the equation to solve. $5 \cdot \dfrac{m}{5} = 5 \cdot 32$ Undo the division by 5 by multiplying both sides by 5. $m = 160$ On the left side, multiplying both sides by 5 undoes the division by 5 and leaves m. On the right side, multiply: $5 \cdot 32 = 160$. The solution is 160. Verify this by substituting 160 into the original equation.
The **division property of equality:** Dividing both sides of an equation by the same nonzero number does not change its solution. If $a = b$, then $\dfrac{a}{c} = \dfrac{b}{c}$ (provided $c \neq 0$)	Solve: $17 = 17c$ We can use the division property of equality to isolate c on the right side. $17 = 17c$ This is the equation to solve. $\dfrac{17}{17} = \dfrac{17c}{17}$ Undo the multiplication by 17 by dividing both sides by 17. $1 = c$ On the left side, divide: $17 \div 17 = 1$. On the right side, dividing both sides by 17 undoes the multiplication by 17 and leaves c. The solution is 1. Verify this by substituting 1 into the original equation.
To solve application problems, use the **five-step problem-solving strategy.**	CONSTRUCTION DELAYS Because of bad weather and a labor stoppage, the final cost of a construction project was three times greater than the original estimate. Upon completion, the project cost $126 million. What was the original cost estimate? *Analyze* • The final cost of the construction project was 3 times greater than the estimate. Given • The completed project cost $126 million. Given • What was the original cost estimate? Find *Form* Let x = the original cost estimate (in millions of dollars). We now look for a key phrase in the problem. **Key phrase:** *three times* **Translation:** multiplication

Now we translate the words of the problem into an equation. The units are millions of dollars.

3	times	the original cost	is equal to	the final cost.
3	·	x	=	126

Solve

$$3x = 126 \qquad \text{We need to isolate x on the left side.}$$

$$\frac{3x}{3} = \frac{126}{3} \qquad \text{To isolate x, undo the multiplication by 3 by dividing both sides by 3.}$$

$$x = 42 \qquad \text{Do the division.}$$

State

The original cost estimate for the project was $42 million.

Check

If we multiply the original estimate $42 million by 3, we should get the final cost.

$$\begin{array}{r} \$42 \\ \times\ 3 \\ \hline \$126 \end{array} \quad \text{This is the final cost in millions of dollars.}$$

The result checks.

REVIEW EXERCISES

Solve each equation and check the result.

131. $3x = 12$

132. $15y = 45$

133. $105 = 5r$

134. $224 = 16q$

135. $\dfrac{x}{7} = 3$

136. $\dfrac{a}{3} = 12$

137. $15 = \dfrac{s}{21}$

138. $25 = \dfrac{d}{17}$

Let a variable represent the unknown quantity. Then write and solve an equation to answer the question.

139. INFOMERCIALS The number of orders received by a company selling juicers doubled the week after a sports celebrity was added to the company's infomercial. If the company received 364 orders that week, how many did they receive the week before?

140. BIRTHDAY PRESENTS Four sisters split the cost of a gold chain that they were giving to their mother as a birthday present. How much did the chain cost if each sister's share was $32?

1. a. The set of _____ numbers is {0, 1, 2, 3, 4, 5, … }.

  b. The symbols > and < are _____ symbols.

  c. The _____ of a rectangle is a measure of the amount of surface it encloses.

  d. The grouping symbols ( ) are called _____, and the symbols [ ] are called _____.

  e. A _____ number is a whole number greater than 1 that has only 1 and itself as factors.

  f. An _____ is a statement indicating that two expressions are equal.

  g. A number that makes an equation true when substituted for the variable is called a _____ of the equation.

  h. In this chapter, we used the addition, subtraction, multiplication, and division properties of _____ to solve equations.

2. Graph the whole numbers less than 7 on a number line.

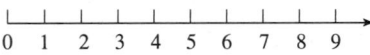

3. Consider the whole number 402,198.

  a. What is the place value of the digit 1?

  b. What digit is in the ten thousands column?

4. a. Write 7,018,641 in words.

  b. Write "one million, three hundred eighty-five thousand, two hundred sixty-six" in standard form.

  c. Write 92,561 in expanded form.

5. Place an < or an > symbol in the box to make a true statement.

  a. 15 ☐ 10   **b.** 1,247 ☐ 1,427

6. Round 34,759,841 to the …

  a. nearest million

  b. nearest hundred thousand

  c. nearest thousand

7. THE NHL The table below shows the number of teams in the National Hockey League at various times during its history. Use the data to complete the bar graph.

Year	1960	1970	1980	1990	2000	2008
Number of teams	6	14	21	21	28	30

Source: www.rauzulusstreet.com

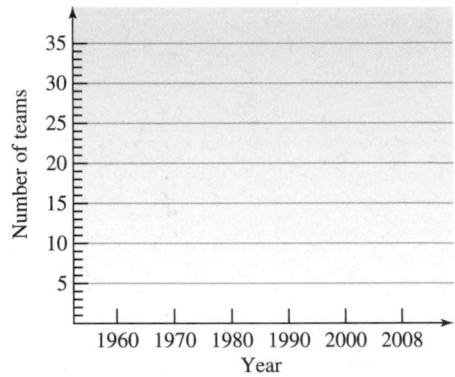

8. Subtract 287 from 535. Show a check of your result.

9. Add: 136,231
         82,574
       + 6,359

10. Subtract: 4,521
         −3,579

11. Multiply: 53
         × 8

12. Multiply: 74 · 562

13. Divide: $6\overline{)432}$

14. Divide: 8,379 ÷ 73. Show a check of your result.

15. Find the product of 23,000 and 600.

16. Find the quotient of 125,000 and 500.

17. Use front-end rounding to estimate the difference: 49,213 − 7,198

18. A rectangle is 327 inches wide and 757 inches long. Find its perimeter.

19. Find the area of the square shown below.

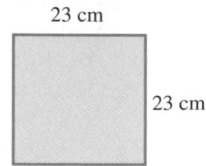

23 cm

23 cm

20. a. Find the factors of 12.

   b. Find the first six multiples of 4.

   c. Write $5 + 5 + 5 + 5 + 5 + 5 + 5 + 5$ as a multiplication.

21. Find the prime factorization of 1,260.

22. TOSSING A COIN During World War II, John Kerrich, a prisoner of war, tossed a coin 10,000 times and wrote down the results. If he recorded 5,067 heads, how many tails occurred? (Source: *Figure This!*)

23. P.E. CLASSES In a physical education class, the students stand in a rectangular formation of 8 rows and 12 columns when the instructor takes attendance. How many students are in the class?

24. FLOOR SPACE The men's, women's, and children's departments in a clothing store occupy a total of 12,255 square feet. Find the square footage of each department if they each occupy the same amount of floor space.

25. MILEAGE The fuel tank of a Hummer H3 holds 23 gallons of gasoline. How far can a Hummer travel on one tank of gas if it gets 18 miles per gallon on the highway?

26. What property is illustrated by each statement?

   a. $18 \cdot (9 \cdot 40) = (18 \cdot 9) \cdot 40$

   b. $23{,}999 + 1 = 1 + 23{,}999$

27. Perform each operation, if possible.

   a. $15 \cdot 0$ **b.** $\dfrac{0}{15}$

   c. $\dfrac{8}{8}$ **d.** $\dfrac{8}{0}$

28. Find the LCM of 15 and 18.

29. Find the LCM of 8, 9, and 12.

30. Find the GCF of 30 and 54.

31. Find the GCF of 24, 28, and 36.

32. STOCKING SHELVES Boxes of rice are being stacked next to boxes of instant mashed potatoes on the same bottom shelf in a supermarket display. The boxes of rice are 8 inches tall and the boxes of instant potatoes are 10 inches high.

   a. What is the shortest height at which the two stacks will be the same height?

   b. How many boxes of rice and how many boxes of potatoes will be used in each stack?

33. Is 521,340 divisible by 2, 3, 4, 5, 6, 9, or 10?

34. GRADES A student scored 73, 52, 95, and 70 on four exams and received 0 on one missed exam. Find his mean (average) exam score.

Evaluate each expression.

35. $9 + 4 \cdot 5$

36. $3^4 \cdot 10 - 2(6)(4)$

37. $20 + 2[4^2 - 2(6 - 2^2)]$

38. $\dfrac{3^3 - 2(15 - 14)^2}{33 - 9 + 1}$

39. Use a check to determine whether 3 is a solution of the equation $x + 13 = 16$.

40. Explain what it means to *solve* an equation.

Solve each equation and check the result.

41. $100 = x + 1$

42. $y - 12 = 18$

43. $5m = 55$

44. $\dfrac{q}{3} = 27$

Let a variable represent the unknown quantity. Then write and solve an equation to answer the question.

45. PARKING After many student complaints, a college decided to triple the number of parking spaces on campus by constructing a parking structure. That increase will bring the total number of spaces up to 6,240. How many parking spaces does the college have at this time?

46. HEARING PROTECTION When a sound technician at a rock concert wears ear plugs, the sound intensity that he experiences from a heavy metal band is only 73 decibels. If the ear plugs reduce the sound intensity by 41 decibels, what is the actual sound intensity of the band?

47. DISCUSSION GROUPS A sociology professor had the students in her class split up into six-person discussion groups. If there were exactly twelve discussion groups of that size, how many students were in the class?

48. KITCHEN REMODELING A woman wants to have her kitchen remodeled. If she has saved $12,500, and the project costs $27,250, how much money does she need to borrow?

The Integers

© OJO Images Ltd/Alamy

from *Campus to Careers*

Personal Financial Advisor

Personal financial advisors help people manage their money and teach them how to make their money grow. They offer advice on how to budget for monthly expenses, as well as how to save for retirement. A bachelor's degree in business, accounting, finance, economics, or statistics provides good preparation for the occupation. Strong communication and problem-solving skills are equally important to achieve success in this field.

In **Problem 90** of **Study Set 2.2,** you will see how a personal financial planner uses integers to determine whether a duplex rental unit would be a money-making investment for a client.

JOB TITLE:
Personal Financial Advisor

EDUCATION: Must have at least a bachelor's degree. Some states require a certificate or license.

JOB OUTLOOK: Excellent—Jobs are projected to grow by 41% over the next decade.

ANNUAL EARNINGS: In 2007, average yearly earnings were $89,220.

FOR MORE INFORMATION:
http://www.collegeboard.com/csearch/majors_careers/profiles/careers/101000.html

Objectives

1. Define the set of integers.

2. Graph integers on a number line.

3. Use inequality symbols to compare integers.

4. Find the absolute value of an integer.

5. Find the opposite of an integer.

An Introduction to the Integers

We have seen that whole numbers can be used to describe many situations that arise in everyday life. However, we cannot use whole numbers to express temperatures below zero, the balance in a checking account that is overdrawn, or how far an object is below sea level. In this section, we will see how negative numbers can be used to describe these three situations as well as many others.

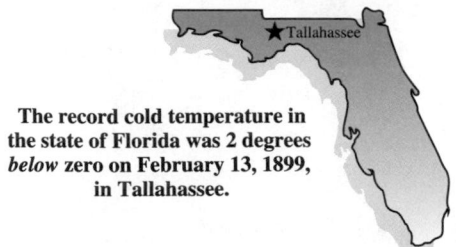

The record cold temperature in the state of Florida was 2 degrees *below* zero on February 13, 1899, in Tallahassee.

A check for $500 was written when there was only $450 in the account. The checking account is *overdrawn*.

The American lobster is found off the East Coast of North America at depths as much as 600 feet *below* sea level.

1 Define the set of integers.

To describe a temperature of 2 degrees above zero, a balance of $50, or 600 feet above sea level, we can use numbers called **positive numbers.** All positive numbers are greater than 0, and we can write them with or without a **positive sign** $+$.

In words	In symbols	Read as
2 degrees above zero	$+2$ or 2	positive two
A balance of $50	$+50$ or 50	positive fifty
600 feet above sea level	$+600$ or 600	positive six hundred

To describe a temperature of 2 degrees below zero, $50 overdrawn, or 600 feet below sea level, we need to use negative numbers. **Negative numbers** are numbers less than 0, and they are written using a **negative sign** $-$.

In words	In symbols	Read as
2 degrees below zero	-2	negative two
$50 overdrawn	-50	negative fifty
600 feet below sea level	-600	negative six hundred

Together, positive and negative numbers are called **signed numbers.**

Positive and Negative Numbers

Positive numbers are greater than 0. **Negative numbers** are less than 0.

Caution! Zero is neither positive nor negative.

The collection of positive whole numbers, the negatives of the whole numbers, and 0 is called the set of **integers** (read as "in-ti-jers").

The Set of Integers

$$\{\ldots, -5, -4, -3, -2, -1, 0, 1, 2, 3, 4, 5, \ldots\}$$

The three dots on the right indicate that the list continues forever—there is no largest integer. The three dots on the left indicate that the list continues forever—there is no smallest integer. The set of **positive integers** is $\{1, 2, 3, 4, 5, \ldots\}$ and the set of **negative integers** is $\{\ldots, -5, -4, -3, -2, -1\}$.

The Language of Algebra Since every whole number is an integer, we say that the set of whole numbers is a **subset** of the integers.

The set of integers → $\{\ldots, -5, -4, -3, -2, -1, \underbrace{0, 1, 2, 3, 4, 5, \ldots}\}$

The set of whole numbers

2 Graph integers on a number line.

In Section 1.1, we introduced the number line. We can use an extension of the number line to learn about negative numbers.

Negative numbers can be represented on a number line by extending the line to the left and drawing an arrowhead. Beginning at the origin (the 0 point), we move to the left, marking equally spaced points as shown below. As we move to the right on the number line, the values of the numbers increase. As we move to the left, the values of the numbers decrease.

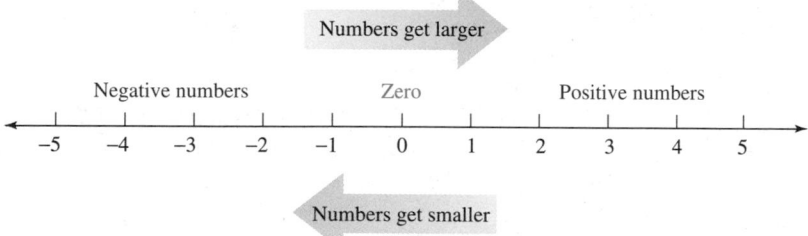

The thermometer shown on the next page is an example of a vertical number line. It is scaled in degrees and shows a temperature of $-10°$. The time line is an example of a horizontal number line. It is scaled in units of 500 years.

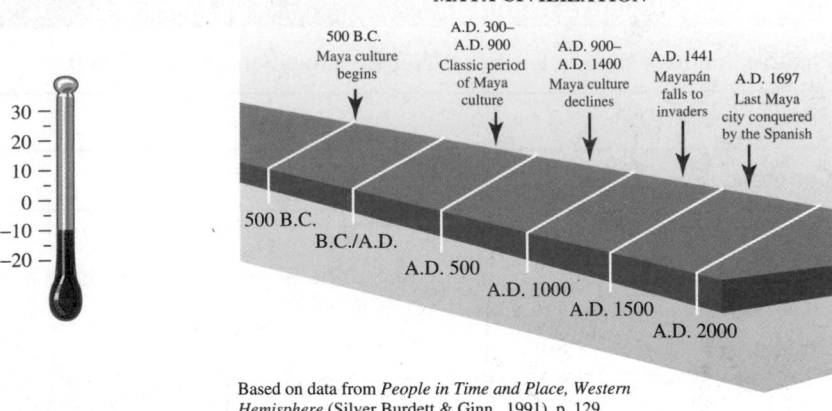

MAYA CIVILIZATION

Based on data from *People in Time and Place, Western Hemisphere* (Silver Burdett & Ginn., 1991), p. 129

A vertical number line   A horizontal number line

Graph −4, −2, 1, and 3 on a number line.

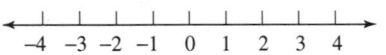

Now Try Problem 23

EXAMPLE 1   Graph −3, 2, −1, and 4 on a number line.

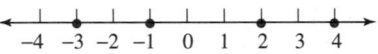

Strategy We will locate the position of each integer on the number line and draw a bold dot.

WHY To *graph a number* means to make a drawing that represents the number.

Solution
The position of each negative integer is to the left of 0. The position of each positive integer is to the right of 0.

By extending the number line to include negative numbers, we can represent more situations using bar graphs and line graphs. For example, the following bar graph shows the net income of the Eastman Kodak Company for the years 2000 through 2007. Since the net income in 2004 was positive $556 million, the company made a *profit*. Since the net income in 2005 was −$1,362 million, the company had a *loss*.

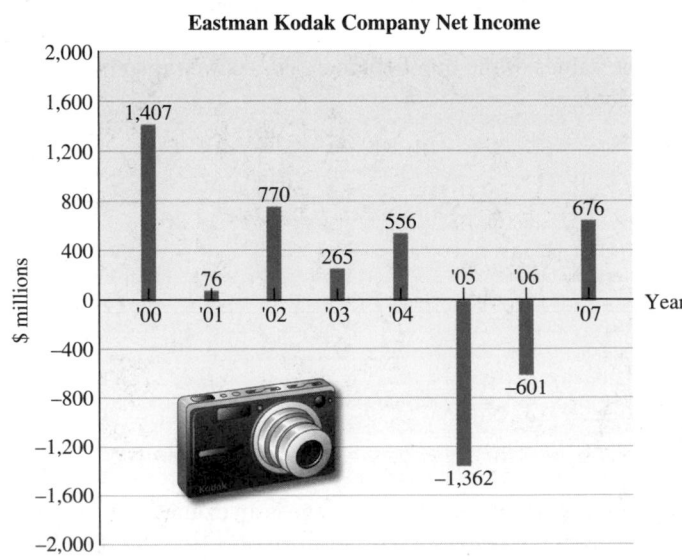

Eastman Kodak Company Net Income

Source: morningstar.com

> **The Language of Algebra** *Net* refers to what remains after all the deductions (losses) have been accounted for. **Net income** is a term used in business that often is referred to as the *bottom line*. Net income indicates what a company has earned (or lost) in a given period of time (usually 1 year).

THINK IT THROUGH   *Credit Card Debt*

"The most dangerous pitfall for many college students is the overuse of credit cards. Many banks do their best to entice new card holders with low or zero-interest cards."

Gary Schatsky, certified financial planner

Which numbers on the credit card statement below are actually debts and, therefore, could be represented using negative numbers?

		Account Summary	
Previous Balance	New Purchases	Payments & Credits	New Balance
$4,621	$1,073	$2,369	$3,325

04/21/10 Billing Date	05/16/10 Date Payment Due	$67 Minimum payment

BANK STAR   Periodic rates may vary.
See reverse for explanation and important information.
Please allow sufficient time for mail to reach Bank Star.

---

3 Use inequality symbols to compare integers.

Recall that the symbol $<$ means "is less than" and that $>$ means "is greater than." The figure below shows the graph of the integers -2 and 1. Since -2 is to the left of 1 on the number line, $-2 < 1$. Since $-2 < 1$, it is also true that $1 > -2$.

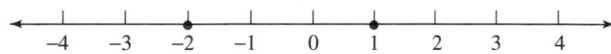

EXAMPLE 2   Place an $<$ or an $>$ symbol in the box to make a true statement.   **a.** 4 ⬚ -5 **b.** -8 ⬚ -7

Strategy To pick the correct inequality symbol to place between the pair of numbers, we will determine the position of each number on the number line.

WHY For any two numbers on a number line, the number to the *left* is the smaller number and the number on the *right* is the larger number.

Solution
a. Since 4 is to the right of -5 on the number line, $4 > -5$.
b. Since -8 is to the left of -7 on the number line, $-8 < -7$.

Self Check 2

Place an $<$ or an $>$ symbol in the box to make a true statement.

a. 6 ⬚ -6

b. -11 ⬚ -10

Now Try Problems 31 and 35

> **The Language of Algebra** Because the symbol $<$ requires one number to be strictly less than another number and the symbol $>$ requires one number to be strictly greater than another number, mathematical statements involving the symbols $<$ and $>$ are called *strict inequalities*.

There are three other commonly used inequality symbols.

Inequality Symbols

$\neq$ means *is not equal to*

$\geq$ means *is greater than or equal to*

$\leq$ means *is less than or equal to*

$-5 \neq -2$	Read as "-5 is not equal to -2."
$-6 \leq 10$	Read as "-6 is less than or equal to 10." This statement is true, because $-6 < 10$.
$12 \leq 12$	Read as "12 is less than or equal to 12." This statement is true, because $12 = 12$.
$-15 \geq -17$	Read as "-15 is greater than or equal to -17." This statement is true, because $-15 > -17$.
$-20 \geq -20$	Read as "-20 is greater than or equal to -20." This statement is true, because $-20 = -20$.

Self Check 3

Tell whether each statement is true or false.

a. $-17 \geq -15$

b. $-35 \leq -35$

c. $-2 \geq -2$

d. $-61 \leq -62$

Now Try Problems 41 and 45

EXAMPLE 3   Tell whether each statement is true or false.

a. $-9 \geq -9$ **b.** $-1 \leq -5$ **c.** $-27 \geq 6$ **d.** $-32 \leq -32$

Strategy We will determine if either the strict inequality or the equality that the symbols $\leq$ and $\geq$ allow is true.

WHY If either is true, then the given statement is true.

Solution

a. $-9 \geq -9$ This statement is true, because $-9 = -9$.

b. $-1 \leq -5$ This statement is false, because neither $-1 < -5$ nor $-1 = -5$ is true.

c. $-27 \geq 6$ This statement is false, because neither $-27 > 6$ nor $-27 = 6$ is true.

d. $-32 \leq -31$ This statement is true, because $-32 < -31$. ■

---

4 Find the absolute value of an integer.

Using a number line, we can see that the numbers 3 and -3 are both a distance of 3 units away from 0, as shown below.

The **absolute value** of a number gives the distance between the number and 0 on the number line. To indicate absolute value, the number is inserted between two vertical bars, called the **absolute value symbol.** For example, we can write $|-3| = 3$. This is read as "The absolute value of negative 3 is 3," and it tells us that the distance between -3 and 0 on the number line is 3 units. From the figure, we also see that $|3| = 3$.

Absolute Value

The **absolute value** of a number is the distance on the number line between the number and 0.

Caution! Absolute value expresses distance. The absolute value of a number is always positive or 0. It is never negative.

EXAMPLE 4   Find each absolute value: **a.** $|8|$ **b.** $|-5|$ **c.** $|0|$

Strategy We need to determine the distance that the number within the vertical absolute value bars is from 0 on a number line.

WHY The absolute value of a number is the distance between 0 and the number on a number line.

Solution

a. On the number line, the distance between 8 and 0 is 8. Therefore,

$$|8| = 8$$

b. On the number line, the distance between -5 and 0 is 5. Therefore,

$$|-5| = 5$$

c. On the number line, the distance between 0 and 0 is 0. Therefore,

$$|0| = 0$$

Self Check 4

Find each absolute value:

a. $|-9|$

b. $|4|$

Now Try Problems 47 and 49

5 Find the opposite of an integer.

Opposites or Negatives

Two numbers that are the same distance from 0 on the number line, but on opposite sides of it, are called **opposites** or **negatives.**

The figure below shows that for each whole number on the number line, there is a corresponding whole number, called its *opposite*, to the left of 0. For example, we see that 3 and -3 are opposites, as are -5 and 5. Note that 0 is its own opposite.

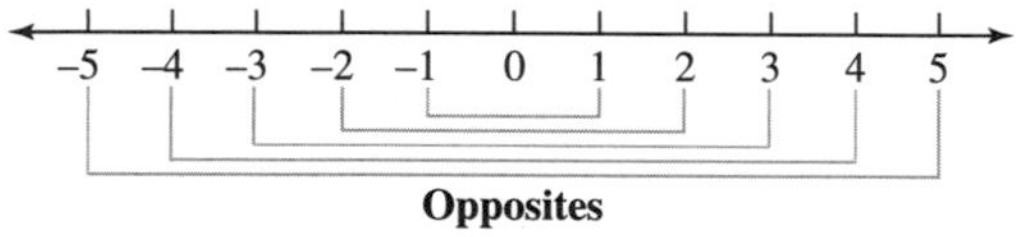

Opposites

To write the opposite of a number, a $-$ symbol is used. For example, the opposite of 5 is -5 (read as "negative 5"). Parentheses are needed to express the opposite of a negative number. The opposite of -5 is written as $-(-5)$. Since 5 and -5 are the same distance from 0, the opposite of -5 is 5. Therefore, $-(-5) = 5$. This illustrates the following rule.

The Opposite of the Opposite Rule

The opposite of the opposite (or negative) of a number is that number.
For any number a,

$$-(-a) = a \qquad \text{Read } -a \text{ as "the opposite of } a."$$

Number	Opposite	
57	-57	Read as "negative fifty-seven."
-8	$-(-8) = 8$	Read as "the opposite of negative eight is eight."
0	$-0 = 0$	Read as "the opposite of 0 is 0."

The concept of opposite can also be applied to an absolute value. For example, the opposite of the absolute value of -8 can be written as $-|-8|$. Think of this as a two-step process, where the absolute value symbol serves as a grouping symbol. Find the absolute value first, and then attach a $-$ sign to that result.

First, find the absolute value.

$$-|-8| = -8$$

Then attach a $-$ sign.

Read as "the opposite of the absolute value of negative eight is negative eight."

Simplify each expression:

a. $-(-1)$

b. $-|4|$

c. $-|-99|$

Now Try Problems 55, 65, and 67

EXAMPLE 5   Simplify each expression: **a.** $-(-44)$ **b.** $-|11|$ **c.** $-|-225|$

Strategy We will find the opposite of each number.

WHY In each case, the $-$ symbol is written outside the grouping symbols means "the opposite of."

Solution

a. $-(-44)$ means the opposite of -44. Since the opposite of -44 is 44, we write

$$-(-44) = 44$$

b. $-|11|$ means the opposite of the absolute value of 11. Since $|11| = 11$, and the opposite of 11 is -11, we write

$$-|11| = -11$$

c. $-|-225|$ means the opposite of the absolute value of -225. Since $|-225| = 225$, and the opposite of 225 is -225, we write

$$-|-225| = -225$$

The $-$ symbol is used to indicate a negative number, the opposite of a number, and the operation of subtraction. The key to reading the $-$ symbol correctly is to examine the context in which it is used.

Reading the $-$ Symbol

-12	Negative twelve	A $-$ symbol directly in front of a number is read as "negative."
$-(-12)$	The opposite of negative twelve	The first $-$ symbol is read as "the opposite of" and the second as "negative."
$12 - 5$	Twelve minus five	Notice the space used before and after the $-$ symbol. This indicates subtraction and is read as "minus."

ANSWERS TO SELF CHECKS

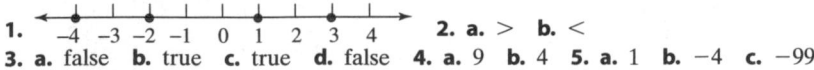

1. [number line from -4 to 4]   **2. a.** $>$ **b.** $<$
3. a. false   **b.** true   **c.** true   **d.** false   **4. a.** 9   **b.** 4   **5. a.** 1   **b.** -4 **c.** -99

SECTION 2.1   STUDY SET

VOCABULARY

Fill in the blanks.

1. _____ numbers are greater than 0 and _____ numbers are less than 0.

2. $\{\ldots, -5, -4, -3, -2, -1, 0, 1, 2, 3, 4, 5, \ldots\}$ is called the set of _____.

3. To _____ an integer means to locate it on the number line and highlight it with a dot.

4. The symbols $>$ and $<$ are called _____ symbols.

5. The _____ _____ of a number is the distance between the number and 0 on the number line.

6. Two numbers that are the same distance from 0 on the number line, but on opposite sides of it, are called _____.

CONCEPTS

7. Represent each of these situations using a signed number.

   a. $225 overdrawn

   b. 10 seconds before liftoff

   c. 3 degrees below normal

   d. A deficit of $12,000

   e. A 1-mile retreat by an army

8. Represent each of these situations using a signed number, and then describe its opposite in words.

   a. A trade surplus of $3 million

   b. A bacteria count 70 more than the standard

   c. A profit of $67

   d. A business $1 million in the "black"

   e. 20 units over their quota

9. Determine what is wrong with each number line.

   a.
   b.
   c.
   d.

10. a. If a number is less than 0, what type of number must it be?

    b. If a number is greater than 0, what type of number must it be?

11. On the number line, what number is

    a. 3 units to the right of -7?

    b. 4 units to the left of 2?

12. Name two numbers on the number line that are a distance of

    a. 5 away from -3.

    b. 4 away from 3.

13. a. Which number is closer to -3 on the number line: 2 or -7?

    b. Which number is farther from 1 on the number line: -5 or 8?

14. Is there a number that is both greater than 10 and less than 10 at the same time?

15. a. Express the fact $-12 < 15$ using an $>$ symbol.

    b. Express the fact $-4 > -5$ using an $<$ symbol.

16. Fill in the blank: The opposite of the _____ of a number is that number. For any number a, $-(-a) = $ ▢.

17. Complete the table by finding the opposite and the absolute value of the given numbers.

Number	Opposite	Absolute value
-25		
39		
0		

18. Is the absolute value of a number always positive?

NOTATION

19. Translate each phrase to mathematical symbols.

    a. The opposite of negative eight

    b. The absolute value of negative eight

    c. Eight minus eight

    d. The opposite of the absolute value of negative eight

20. a. Write the set of integers.

    b. Write the set of positive integers.

    c. Write the set of negative integers.

21. Fill in the blanks.

    a. We read ≥ as "is _____ than or _____ to."

    b. We read ≤ as "is ____ than or _____ to."

22. Which of the following expressions contains a minus sign?

    15 − 8   −(−15)   −15

GUIDED PRACTICE

Graph the following numbers on a number line. See Example 1.

23. −3, 4, 3, 0, −1

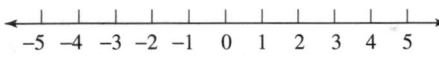

24. 2, −4, 5, 1, −1

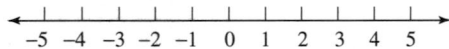

25. The integers that are less than 3 but greater than −5

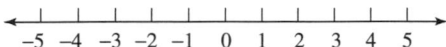

26. The integers that are less than 4 but greater than −3

27. The opposite of −3, the opposite of 5, and the absolute value of −2

28. The absolute value of 3, the opposite of 3, and the number that is 1 less than −3

29. 2 more than 0, 4 less than 0, 2 more than negative 5, and 5 less than 4

30. 4 less than 0, 1 more than 0, 2 less than −2, and 6 more than −4

Place an < or an > symbol in the box to make a true statement. See Example 2.

31. −5 ☐ 5

32. 0 ☐ −1

33. −12 ☐ −6

34. −7 ☐ −6

35. −10 ☐ −17

36. −11 ☐ −20

37. −325 ☐ −532

38. −401 ☐ −104

Tell whether each statement is true or false. See Example 3.

39. −15 ≤ −14

40. −77 ≤ −76

41. 210 ≥ 210

42. 37 ≥ 37

43. −1,255 ≥ −1,254

44. −6,546 ≥ −6,465

45. 0 ≤ −8

46. −6 ≤ −6

Find each absolute value. See Example 4.

47. |9|

48. |12|

49. |−8|

50. |−1|

51. |−14|

52. |−85|

53. |180|

54. |371|

Simplify each expression. See Example 5.

55. −(−11)

56. −(−1)

57. −(−4)

58. −(−9)

59. −(−102)

60. −(−295)

61. −(−561)

62. −(−703)

63. −|20|

64. −|143|

65. −|6|

66. −|0|

67. −|−253|

68. −|−11|

69. −|−0|

70. −|97|

TRY IT YOURSELF

Place an < or an > symbol in the box to make a true statement.

71. |−12| ☐ −(−7)

72. |−50| ☐ −(−40)

73. −|−71| ☐ −|−65|

74. −|−163| ☐ −|−150|

75. −(−343) ☐ −(−161)

76. −(−999) ☐ −(−998)

77. −|−30| ☐ −|−(−8)|

78. −|−100| ☐ −|−(−88)|

Write the integers in order, from least to greatest.

79. 82, −52, 52, −22, 12, −12

80. 49, −9, 19, −39, 89, −49

Fill in the blanks to continue each pattern.

81. 5, 3, 1, −1, ☐, ☐, ☐, . . .

82. 4, 2, 0, −2, ☐, ☐, ☐, . . .

APPLICATIONS

83. HORSE RACING In the 1973 Belmont Stakes, *Secretariat* won by 31 lengths over second place finisher, *Twice a Prince*. Some experts call it the greatest performance by a thoroughbred in the history of racing. Express the position of *Twice a Prince* compared to *Secretariat* as a signed number. (Source: ezinearticles.com)

© Bettmann/Corbis

84. NASCAR In the NASCAR driver standings, negative numbers are used to tell how many points behind the leader a given driver is. Jimmie Johnson was the leading driver in 2008. The other drivers in the top ten were Greg Biffle (-217), Clint Bowyer (-303), Jeff Burton (-349), Kyle Busch (-498), Carl Edwards (-69), Jeff Gordon (-368), Denny Hamlin (-470), Kevin Harvick (-276), and Tony Stewart (-482). Use this information to rank the drivers in the table below.

AP Images

2008 NASCAR Final Driver Standings

Rank	Driver	Points behind leader
1	Jimmie Johnson	Leader
2		
3		
4		
5		
6		
7		
8		
9		
10		

(Source: NASCAR.com)

85. FREE FALL A boy launches a water balloon from the top of a building, as shown in the next column. At that instant, his friend starts a stopwatch and keeps track of the time as the balloon sails above

the building and then falls to the ground. Use the number line to estimate the position of the balloon at each time listed in the table below.

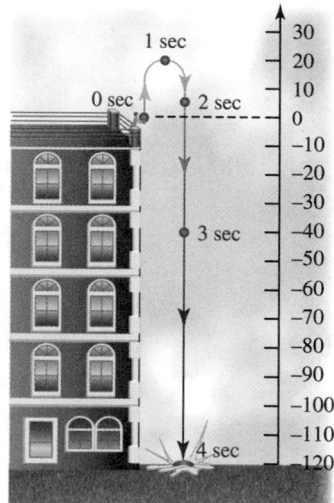

Time	Position of balloon
0 sec	
1 sec	
2 sec	
3 sec	
4 sec	

86. CARNIVAL GAMES At a carnival shooting gallery, players aim at moving ducks. The path of one duck is shown, along with the time it takes the duck to reach certain positions on the gallery wall. Use the number line to estimate the position of the duck at each time listed in the table below.

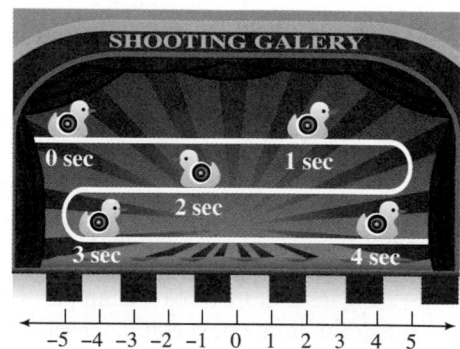

Time	Position of duck
0 sec	
1 sec	
2 sec	
3 sec	
4 sec	

87. TECHNOLOGY The readout from a testing device is shown. Use the number line to find the height of each of the peaks and the depth of each of the valleys.

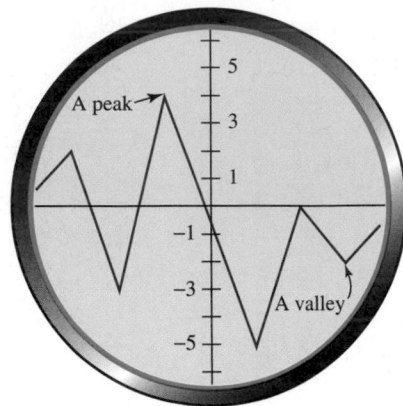

88. FLOODING A week of daily reports listing the height of a river in comparison to flood stage is given in the table. Complete the bar graph shown below.

Flood Stage Report

Sun.	2 ft below
Mon.	3 ft over
Tue.	4 ft over
Wed.	2 ft over
Thu.	1 ft below
Fri.	3 ft below
Sat.	4 ft below

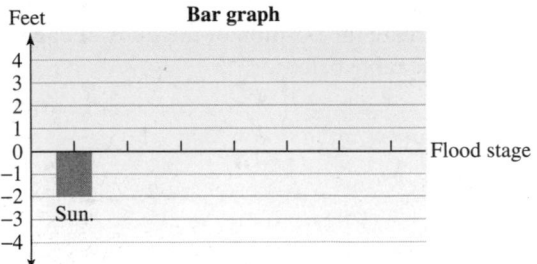

89. GOLF In golf, *par* is the standard number of strokes considered necessary on a given hole. A score of −2 indicates that a golfer used 2 strokes less than par. A score of +2 means 2 more strokes than par were used. In the graph in the next column, each golf ball represents the score of a professional golfer on the 16th hole of a certain course.

a. What score was shot most often on this hole?

b. What was the best score on this hole?

c. Explain why this hole appears to be too easy for a professional golfer.

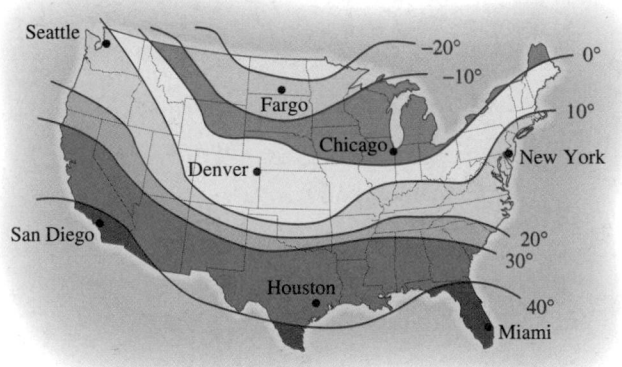

90. PAYCHECKS Examine the items listed on the following paycheck stub. Then write two columns on your paper—one headed "positive" and the other "negative." List each item under the proper heading.

Tom Dryden Dec. 09	Christmas bonus	$100
Gross pay $2,000	**Reductions**	
Overtime $300	Retirement	$200
Deductions	**Taxes**	
Union dues $30	Federal withholding	$160
U.S. Bonds $100	State withholding	$35

91. WEATHER MAPS The illustration shows the predicted Fahrenheit temperatures for a day in mid-January.

a. What is the temperature range for the region including Fargo, North Dakota?

b. According to the prediction, what is the warmest it should get in Houston?

c. According to this prediction, what is the coldest it should get in Seattle?

92. INTERNET COMPANIES The graph on the next page shows the net income of Amazon.com for the years 1998–2007. (Source: Morningstar)

a. In what years did Amazon suffer a loss? Estimate each loss.

b. In what year did Amazon first turn a profit? Estimate it.

c. In what year did Amazon have the greatest profit? Estimate it.

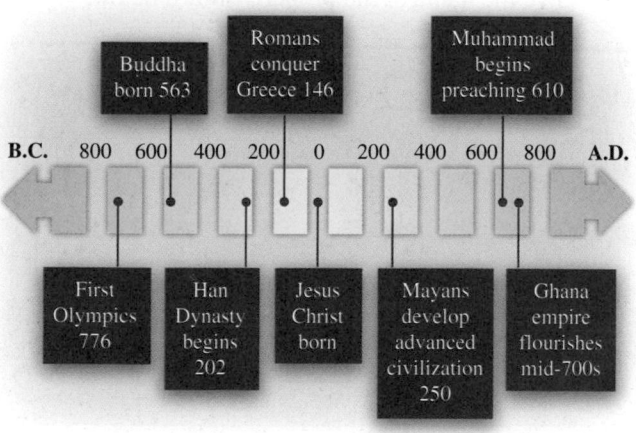

93. **HISTORY** Number lines can be used to display historical data. Some important world events are shown on the time line below.

a. What basic unit is used to scale this time line?

b. What can be thought of as positive numbers?

c. What can be thought of as negative numbers?

d. What important event distinguishes the positive from the negative numbers?

94. **ASTRONOMY** Astronomers use an inverted vertical number line called the *apparent magnitude scale* to denote the brightness of objects in the sky. The brighter an object appears to an observer on Earth, the more negative is its apparent magnitude. Graph each of the following on the scale to the right.

- Visual limit of binoculars +10
- Visual limit of large telescope +20
- Visual limit of naked eye +6
- Full moon −12
- Pluto +15
- Sirius (a bright star) −2
- Sun −26
- Venus −4

95. **LINE GRAPHS** Each thermometer in the illustration gives the daily high temperature in degrees Fahrenheit. Use the data to complete the line graph below.

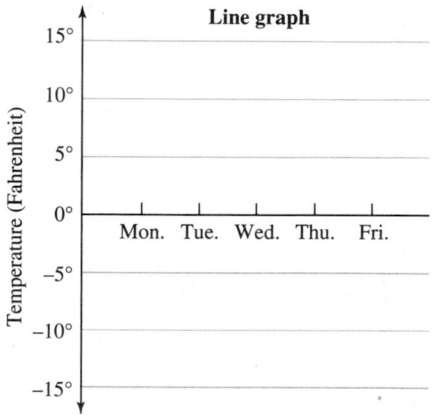

96. GARDENING The illustration shows the depths at which the bottoms of various types of flower bulbs should be planted. (The symbol ″ represents inches.)

a. At what depth should a tulip bulb be planted?

b. How much deeper are hyacinth bulbs planted than gladiolus bulbs?

c. Which bulb must be planted the deepest? How deep?

Ground level

−1″   *Anemone*
−2″   *Sparaxis*
Ranunculus
−3″
Narcissus
−4″   *Freesia*
−5″
Gladiolus
−6″   *Hyacinth*
−7″   *Tulip*
−8″
Daffodil
−9″
−10″
−11″   **Planting Chart**

WRITING

97. Explain the concept of *the opposite of a number.*

98. What real-life situation do you think gave rise to the concept of a negative number?

99. Explain why the absolute value of a number is never negative.

100. Give an example of the use of the number line that you have seen in another course.

101. DIVING Divers use the terms *positive buoyancy*, *neutral buoyancy*, and *negative buoyancy* as shown. What do you think each of these terms means?

102. GEOGRAPHY Much of the Netherlands is low-lying, with half of the country below sea level. Explain why it is not under water.

103. Suppose integer *A* is greater than integer *B*. Is the opposite of integer *A* greater than integer *B*? Explain why or why not. Use an example.

104. Explain why −11 is less than −10.

REVIEW

105. Round 23,456 to the nearest hundred.

106. Evaluate: $19 - 2 \cdot 3$

107. Subtract 2,081 from 2,842.

108. Divide 346 by 15.

109. Give the name of the property shown below:

$$(13 \cdot 2) \cdot 5 = 13 \cdot (2 \cdot 5)$$

110. Write *four times five* using three different symbols.

Objectives

1 Add two integers that have the same sign.

2 Add two integers that have different signs.

3 Perform several additions to evaluate expressions.

4 Identify opposites (additive inverses) when adding integers.

5 Solve application problems by adding integers.

SECTION 2.2
Adding Integers

An amazing change in temperature occurred in 1943 in Spearfish, South Dakota. On January 22, at 7:30 A.M., the temperature was −4 degrees Fahrenheit. Strong warming winds suddenly kicked up and, in just 2 minutes, the temperature rose 49 degrees! To calculate the temperature at 7:32 A.M., we need to add 49 to −4.

$$-4 + 49$$

To perform this addition, we must know how to add positive and negative integers. In this section, we develop rules to help us make such calculations.

> **The Language of Algebra**   In 1724, Daniel Gabriel *Fahrenheit,* a German scientist, introduced the temperature scale that bears his name. The United States is one of the few countries that still use this scale. The temperature -4 degrees Fahrenheit can be written in more compact form as $-4°F$.

1 Add two integers that have the same sign.

We can use the number line to explain addition of integers. For example, to find $4 + 3$, we begin at 0 and draw an arrow 4 units long that points to the right. It represents positive 4. From the tip of that arrow, we draw a second arrow, 3 units long, that points to the right. It represents positive 3. Since we end up at 7, it follows that $4 + 3 = 7$.

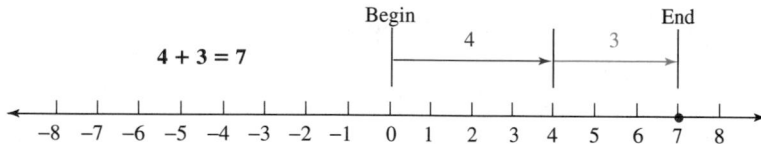

To check our work, let's think of the problem in terms of money. If you had $4 and earned $3 more, you would have a total of $7.

To find $-4 + (-3)$ on a number line, we begin at 0 and draw an arrow 4 units long that points to the left. It represents -4. From the tip of that arrow, we draw a second arrow, 3 units long, that points to the left. It represents -3. Since we end up at -7, it follows that $-4 + (-3) = -7$.

Let's think of this problem in terms of money. If you lost $4 ($-4$) and then lost another $3 ($-3$), overall, you would have lost a total of $7 ($-7$).

Here are some observations about the process of adding two numbers that have the same sign on a number line.

- The arrows representing the integers point in the same direction and they build upon each other.
- The answer has the same sign as the integers that we added.

These observations illustrate the following rules.

> ### Adding Two Integers That Have the Same (Like) Signs
>
> 1. To add two positive integers, add them as usual. The final answer is positive.
> 2. To add two negative integers, add their absolute values and make the final answer negative.

> **The Language of Algebra**   When writing additions that involve integers, write negative integers within parentheses to separate the negative sign − from the plus symbol +.
>
> $9 + (-4)$ ~~$9 + -4$~~   and $-9 + (-4)$ ~~$-9 + -4$~~

Self Check 1

Add:

a. $-7 + (-2)$

b. $-25 + (-48)$

c. $-325 + (-169)$

Now Try Problems 19, 23, and 27

EXAMPLE 1   Add: **a.** $-3 + (-5)$ **b.** $-26 + (-65)$ **c.** $-456 + (-177)$

Strategy We will use the rule for adding two integers that have the *same sign*.

WHY In each case, we are asked to add two negative integers.

Solution

a. To add two negative integers, we add the absolute values of the integers and make the final answer negative. Since $|-3| = 3$ and $|-5| = 5$, we have

$$-3 + (-5) = -8$$

Add their absolute values, 3 and 5, to get 8. Then make the final answer negative.

b. Find the absolute values: $|-26| = 26$ and $|-65| = 65$

$$-26 + (-65) = -91$$

Add their absolute values, 26 and 65, to get 91. Then make the final answer negative.

$$\begin{array}{r} \overset{1}{26} \\ +65 \\ \hline 91 \end{array}$$

c. Find the absolute values: $|-456| = 456$ and $|-177| = 177$

$$-456 + (-177) = -633$$

Add their absolute values, 456 and 177, to get 633. Then make the final answer negative.

$$\begin{array}{r} \overset{11}{456} \\ +177 \\ \hline 633 \end{array}$$

> **Success Tip**   Calculations that you cannot perform in your head should be shown outside the steps of your solution.

> **The Language of Algebra**   Two negative integers, as well as two positive integers, are said to have *like* signs.

2 Add two integers that have different signs.

To find $4 + (-3)$ on a number line, we begin at 0 and draw an arrow 4 units long that points to the right. This represents positive 4. From the tip of that arrow, we draw a second arrow, 3 units long, that points to the left. It represents -3. Since we end up at 1, it follows that $4 + (-3) = 1$.

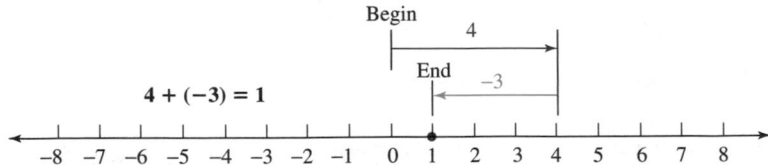

In terms of money, if you won $4 and then lost $3 ($-3$), overall, you would have $1 left.

To find $-4 + 3$ on a number line, we begin at 0 and draw an arrow 4 units long that points to the left. It represents -4. From the tip of that arrow, we draw a second arrow, 3 units long, that points to the right. It represents positive 3. Since we end up at -1, it follows that $-4 + 3 = -1$.

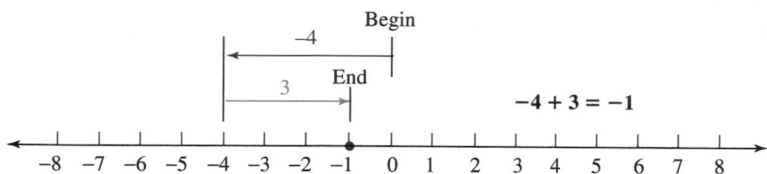

In terms of money, if you lost \$4 (-4) and then won \$3, overall, you have lost \$1 (-1).

Here are some observations about the process of adding two integers that have different signs on a number line.

- The arrows representing the integers point in opposite directions.
- The longer of the two arrows determines the sign of the answer. If the longer arrow represents a positive integer, the sum is positive. If it represents a negative integer, the sum is negative.

These observations suggest the following rules.

Adding Two Integers That Have Different (Unlike) Signs

To add a positive integer and a negative integer, subtract the smaller absolute value from the larger.

1. If the positive integer has the larger absolute value, the final answer is positive.

2. If the negative integer has the larger absolute value, make the final answer negative.

EXAMPLE 2   Add: $5 + (-7)$

Strategy We will use the rule for adding two integers that have different signs.

WHY The addend 5 is positive and the addend -7 is negative.

Solution

Step 1 To add two integers with different signs, we first subtract the smaller absolute value from the larger absolute value. Since $|5|$, which is 5, is smaller than $|-7|$, which is 7, we begin by subtracting 5 from 7.

$$7 - 5 = 2$$

Step 2 Since the negative number, -7, has the larger absolute value, we attach a negative sign $-$ to the result from step 1. Therefore,

$$5 + (-7) = -2$$

— Make the final answer negative.

Self Check 2

Add: $6 + (-9)$

Now Try Problem 31

The Language of Algebra   A positive integer and a negative integer are said to have *unlike* signs.

Self Check 3
Add:

a. $7 + (-2)$

b. $-53 + 39$

c. $-506 + 888$

Now Try Problems 33, 35, and 39

EXAMPLE 3   Add: **a.** $8 + (-4)$ **b.** $-41 + 17$ **c.** $-206 + 568$

Strategy We will use the rule for adding two integers that have different signs.

WHY In each case, we are asked to add a positive integer and a negative integer.

Solution

a. Find the absolute values: $|8| = 8$ and $|-4| = 4$

$$8 + (-4) = 4$$ Subtract the smaller absolute value from the larger: $8 - 4 = 4$. Since the positive number, 8, has the larger absolute value, the final answer is positive.

b. Find the absolute values: $|-41| = 41$ and $|17| = 17$

$$-41 + 17 = -24$$ Subtract the smaller absolute value from the larger: $41 - 17 = 24$. Since the negative number, -41, has the larger absolute value, make the final answer negative.

$$\begin{array}{r} \overset{311}{\cancel{4}\cancel{1}} \\ -17 \\ \hline 24 \end{array}$$

c. Find the absolute values: $|-206| = 206$ and $|568| = 568$

$$-206 + 568 = 362$$ Subtract the smaller absolute value from the larger: $568 - 206 = 362$. Since the positive number, 568, has the larger absolute value, the answer is positive.

$$\begin{array}{r} 568 \\ -206 \\ \hline 362 \end{array}$$

> *Caution!* Did you notice that the answers to the addition problems in Examples 2 and 3 were found using subtraction? This is the case when the addition involves two integers that have *different signs*.

THINK IT THROUGH   *Cash Flow*

"College can be trial by fire — a test of how to cope with pressure, freedom, distractions, and a flood of credit card offers. It's easy to get into a cycle of overspending and unnecessary debt as a student."

Planning for College, Wells Fargo Bank

If your income is less than your expenses, you have a *negative* cash flow. A negative cash flow can be a red flag that you should increase your income and/or reduce your expenses. Which of the following activities can increase income and which can decrease expenses?

- Buy generic or store-brand items.
- Get training and/or more education.
- Use your student ID to get discounts at stores, events, etc.
- Work more hours.
- Turn a hobby or skill into a money-making business.
- Tutor young students.
- Stop expensive habits, like smoking, buying snacks every day, etc
- Attend free activities and free or discounted days at local attractions.
- Sell rarely used items, like an old CD player.
- Compare the prices of at least three products or at three stores before buying.

Based on the *Building Financial Skills* by National Endowment for Financial Education.

3 Perform several additions to evaluate expressions.

To evaluate expressions that contain several additions, we make repeated use of the rules for adding two integers.

EXAMPLE 4 Evaluate: $-3 + 5 + (-12) + 2$

Strategy Since there are no calculations within parentheses, no exponential expressions, and no multiplication or division, we will perform the additions, working from the left to the right.

WHY This is step 4 of the order of operations rule that was introduced in Section 1.7.

Solution

$-3 + 5 + (-12) + 2 = \mathbf{2} + (-12) + 2$ Use the rule for adding two integers that have different signs: $-3 + 5 = 2$.

$= -10 + 2$ Use the rule for adding two integers that have different signs: $2 + (-12) = -10$.

$= -8$ Use the rule for adding two integers that have different signs.

Self Check 4
Evaluate:
$-12 + 8 + (-6) + 1$
Now Try Problem 43

The properties of addition that were introduced in Section 1.2, *Adding and Subtracting Whole Numbers,* are also true for integers.

Commutative Property of Addition

The order in which integers are added does not change their sum.

Associative Property of Addition

The way in which integers are grouped does not change their sum.

Another way to evaluate an expression like that in Example 4 is to use these properties to reorder and regroup the integers in a helpful way.

EXAMPLE 5 Use the commutative and/or associative properties of addition to help evaluate the expression: $-3 + 5 + (-12) + 2$

Strategy We will use the commutative and/or associative properties of addition so that we can add the positives and add the negatives separately. Then we will add those results to obtain the final answer.

WHY It is easier to add integers that have the same sign than integers that have different signs. This approach lessens the possibility of an error, because we only have to add integers that have different signs once.

Solution

$-3 + 5 + (-12) + 2$

$= -3 + (-12) + 5 + 2$ Use the commutative property of addition to reorder the integers.

Negatives   Positives
$= [-3 + (-12)] + (5 + 2)$ Use the associative property of addition to group the negatives and group the positives.

Self Check 5
Use the commutative and/or associative properties of addition to help evaluate the expression:
$-12 + 8 + (-6) + 1$
Now Try Problem 45

$= -15 + 7$ *Use the rule for adding two integers that have the same sign twice. Add the negatives within the brackets. Add the positives within the parentheses.*

$= -8$ *Use the rule for adding two integers that have different signs. This is the same result as in Example 4.*

Self Check 6

Evaluate:
$(-6 + 8) + [10 + (-17)]$

Now Try Problem 47

EXAMPLE 6   Evaluate: $[-21 + (-5)] + (-17 + 6)$

Strategy We will perform the addition within the brackets and the addition within the parentheses first. Then we will add those results.

WHY By the order of operations rule, we must perform the calculations within the grouping symbols first.

Solution Use the rule for adding two integers that have the same sign to do the addition within the brackets and the rule for adding two integers that have different signs to do the addition within the parentheses.

$$[-21 + (-5)] + (-17 + 6) = -26 + (-11) \quad \text{Add within each pair of grouping symbols.}$$

$$= -37 \quad \text{Use the rule for adding two integers that have the same sign.}$$

4   Identify opposites (additive inverses) when adding integers.

Recall from Section 1.2 that when 0 is added to a whole number, the whole number remains the same. This is true for any number. For example, $-5 + 0 = -5$ and $0 + (-43) = -43$. Because of this, we call 0 the **additive identity.**

> **The Language of Algebra**   *Identity* is a form of the word *identical,* meaning the same. You have probably seen *identical* twins.

Addition Property of 0

The sum of any number and 0 is that number.
     For any number a

$$a + 0 = a \quad \text{and} \quad 0 + a = a$$

There is another important fact about the operation of addition and 0. To illustrate it, we use the number line below to add 6 and its opposite, -6. Notice that $6 + (-6) = 0$.

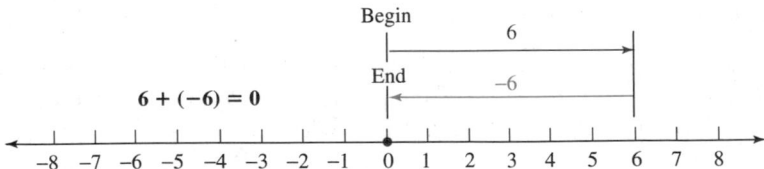

If the sum of two numbers is 0, the numbers are said to be **additive inverses** of each other. Since $6 + (-6) = 0$, we say that 6 and -6 are additive inverses. Likewise, -7 is the additive inverse of 7, and 51 is the additive inverse of -51.

We can now classify a pair of integers such as 6 and -6 in three ways: as opposites, negatives, or additive inverses.

Addition Property of Opposites

The sum of a number and its opposite (additive inverse) is 0.
 For any number a,

$$a + (-a) = 0 \qquad \text{and} \qquad -a + a = 0$$

 At certain times, the addition property of opposites can be used to make addition of several integers easier.

EXAMPLE 7 Evaluate: $12 + (-5) + 6 + 5 + (-12)$

Strategy Instead of working from left to right, we will use the commutative and associative properties of addition to add *pairs of opposites*.

WHY Since the sum of an integer and its opposite is 0, it is helpful to identify such pairs in an addition.

Solution

$$12 + (-5) + 6 + 5 + (-12) = 0 + 0 + 6 \qquad \text{Locate pairs of opposites and add them to get 0.}$$

$$= 6 \qquad \text{The sum of any integer and 0 is that integer.}$$

Self Check 7

Evaluate:
$8 + (-1) + 6 + (-8) + 1$

Now Try Problem 51

5 Solve application problems by adding integers.

Since application problems are almost always written in words, the ability to understand what you read is very important. Recall from Chapter 1 that words and phrases such as *gained, increased by,* and *rise* indicate addition.

EXAMPLE 8 *Record Temperature Change* At the beginning of this section, we learned that at 7:30 A.M. on January 22, 1943, in Spearfish, South Dakota, the temperature was −4°F. The temperature then rose 49 degrees in just 2 minutes. What was the temperature at 7:32 A.M.?

Strategy We will carefully read the problem looking for a key word or phrase.

WHY Key words and phrases indicate what arithmetic operations should be used to solve the problem.

Solution The phrase *rose 49 degrees* indicates addition. With that in mind, we translate the words of the problem to numbers and symbols.

The temperature at 7:32 A.M.	was	the temperature at 7:30 A.M.	plus	49 degrees.
The temperature at 7:32 A.M.	=	−4	+	49

To find the sum, we will use the rule for adding two integers that have different signs. First, we find the absolute values: $|-4| = 4$ and $|49| = 49$.

$$-4 + 49 = 45 \qquad \text{Subtract the smaller absolute value from the larger absolute value: } 49 - 4 = 45. \text{ Since the positive number, 49, has the larger absolute value, the final answer is positive.}$$

At 7:32 A.M., the temperature was 45°F.

Self Check 8

TEMPERATURE CHANGE On the morning of February 21, 1918, in Granville, North Dakota, the morning low temperature was −33°F. By the afternoon, the temperature had risen a record 83 degrees. What was the afternoon high temperature in Granville? (Source: *Extreme Weather* by Christopher C. Burt)

Now Try Problem 83

Using Your CALCULATOR   Entering Negative Numbers

Canada is the largest U.S. trading partner. To calculate the 2007 U.S. trade balance with Canada, we add the $249 billion worth of U.S. exports *to* Canada (considered positive) to the $317 billion worth of U.S. imports *from* Canada (considered negative). We can use a calculator to perform the addition: $249 + (-317)$.

We do not have to do anything special to enter a positive number. Negative numbers are entered using either **direct** or **reverse entry,** depending on the type of calculator you have.

To enter -317 using reverse entry, press the change-of-sign key $\boxed{+/-}$ *after entering* 317. To enter -317 using direct entry, press the negative key $\boxed{(-)}$ *before* entering 317. In either case, note that $\boxed{+/-}$ and the $\boxed{(-)}$ keys are different from the subtraction key $\boxed{-}$.

Reverse entry: 249 $\boxed{+}$ 317 $\boxed{+/-}$ $\boxed{=}$

Direct entry: 249 $\boxed{+}$ $\boxed{(-)}$ 317 $\boxed{\text{ENTER}}$ $\boxed{\qquad -68}$

In 2007, the United States had a trade balance of $-\$68$ billion with Canada. Because the result is negative, it is called a trade *deficit.*

ANSWERS TO SELF CHECKS

1. a. -9 **b.** -73 **c.** -494 **2.** -3 **3. a.** 5 **b.** -14 **c.** 382 **4.** -9 **5.** -9 **6.** -5
7. 6 **8.** 50°F

SECTION 2.2   STUDY SET

VOCABULARY

Fill in the blanks.

1. Two negative integers, as well as two positive integers, are said to have the same or _____ signs.

2. A positive integer and a negative integer are said to have different or _____ signs.

3. When 0 is added to a number, the number remains the same. We call 0 the additive _____.

4. Since $-5 + 5 = 0$, we say that 5 is the additive _____ of -5. We can also say that 5 and -5 are _____.

5. _____ property of addition: The order in which integers are added does not change their sum.

6. _____ property of addition: The way in which integers are grouped does not change their sum.

CONCEPTS

7. **a.** What is the absolute value of 10? What is the absolute value of -12?

b. Which number has the larger absolute value, 10 or -12?

c. Using your answers to part a, subtract the smaller absolute value from the larger absolute value. What is the result?

8. **a.** If you lost $6 and then lost $8, overall, what amount of money was lost?

b. If you lost $6 and then won $8, overall, what amount of money have you won?

Fill in the blanks.

9. To add two integers with unlike signs, _____ their absolute values, the smaller from the larger. Then attach to that result the sign of the number with the _____ absolute value.

10. To add two integers with like signs, add their _____ values and attach their common _____ to the sum.

11. **a.** Is the sum of two positive integers always positive?

    b. Is the sum of two negative integers always negative?

    c. Is the sum of a positive integer and a negative integer always positive?

    d. Is the sum of a positive integer and a negative integer always negative?

12. Complete the table by finding the additive inverse, opposite, and absolute value of the given numbers.

Number	Additive inverse	Opposite	Absolute value
19			
−2			
0			

13. **a.** What is the sum of an integer and its additive inverse?

    b. What is the sum of an integer and its opposite?

14. **a.** What number must be added to −5 to obtain 0?

    b. What number must be added to 8 to obtain 0?

NOTATION

Complete each solution to evaluate the expression.

15. $-16 + (-2) + (-1) = \boxed{} + (-1)$

    $= \boxed{}$

16. $-8 + (-2) + 6 = \boxed{} + 6$

    $= \boxed{}$

17. $(-3 + 8) + (-3) = \boxed{} + (-3)$

    $= \boxed{}$

18. $-5 + [2 + (-9)] = -5 + (\boxed{})$

    $= \boxed{}$

GUIDED PRACTICE

Add. See Example 1.

19. $-6 + (-3)$
20. $-2 + (-3)$
21. $-5 + (-5)$
22. $-8 + (-8)$
23. $-51 + (-11)$
24. $-43 + (-12)$
25. $-69 + (-27)$
26. $-55 + (-36)$
27. $-248 + (-131)$
28. $-423 + (-164)$
29. $-565 + (-309)$
30. $-709 + (-187)$

Add. See Examples 2 and 3.

31. $-8 + 5$
32. $-9 + 3$
33. $7 + (-6)$
34. $4 + (-2)$

35. $20 + (-42)$
36. $-18 + 10$
37. $71 + (-23)$
38. $75 + (-56)$
39. $479 + (-122)$
40. $589 + (-242)$
41. $-339 + 279$
42. $-704 + 649$

Evaluate each expression. See Examples 4 and 5.

43. $9 + (-3) + 5 + (-4)$
44. $-3 + 7 + (-4) + 1$
45. $6 + (-4) + (-13) + 7$
46. $8 + (-5) + (-10) + 6$

Evaluate each expression. See Example 6.

47. $[-3 + (-4)] + (-5 + 2)$
48. $[9 + (-10)] + (-7 + 9)$
49. $(-1 + 34) + [16 + (-8)]$
50. $(-32 + 13) + [5 + (-14)]$

Evaluate each expression. See Example 7.

51. $23 + (-5) + 3 + 5 + (-23)$
52. $41 + (-1) + 9 + 1 + (-41)$
53. $-10 + (-1) + 10 + (-6) + 1$
54. $-14 + (-30) + 14 + (-9) + 9$

TRY IT YOURSELF

Add.

55. $-2 + 6 + (-1)$
56. $4 + (-3) + (-2)$
57. $-7 + 0$
58. $0 + (-15)$
59. $24 + (-15)$
60. $-4 + 14$
61. $-435 + (-127)$
62. $-346 + (-273)$
63. $-7 + 9$
64. $-3 + 6$
65. $2 + (-2)$
66. $-10 + 10$
67. $2 + (-10 + 8)$
68. $(-9 + 12) + (-4)$
69. $-9 + 1 + (-2) + (-1) + 9$
70. $5 + 4 + (-6) + (-4) + (-5)$
71. $[6 + (-4)] + [8 + (-11)]$
72. $[5 + (-8)] + [9 + (-15)]$
73. $(-4 + 8) + (-11 + 4)$
74. $(-12 + 6) + (-6 + 8)$
75. $-675 + (-456) + 99$
76. $-9,750 + (-780) + 2,345$
77. Find the sum of $-6, -7,$ and $-8.$
78. Find the sum of $-11, -12,$ and $-13.$
79. $-2 + [789 + (-9,135)]$
80. $-8 + [2,701 + (-4,089)]$
81. What is 25 more than -45?
82. What is 31 more than -65?

APPLICATIONS

Use signed numbers to solve each problem.

83. RECORD TEMPERATURES The lowest recorded temperatures for Michigan and Minnesota are shown below. Use the given information to find the highest recorded temperature for each state.

State	Lowest temperature	Highest temperature
Michigan	Feb. 9, 1934: −51°F	July 13, 1936: 163°F warmer than the record low
Minnesota	Feb. 2, 1996: −60°F	July 6, 1936: 174°F warmer than the record low

(Source: *The World Almanac Book of Facts*, 2009)

84. ELEVATIONS The lowest point in the United States is Death Valley, California, with an elevation of −282 feet (282 feet below sea level). Mt. McKinley (Alaska) is the highest point in the United States. Its elevation is 20,602 feet higher than Death Valley. What is the elevation of Mt. McKinley? (Source: *The World Almanac Book of Facts*, 2009)

85. SUNKEN SHIPS Refer to the map below.

  a. The German battleship *Bismarck,* one of the most feared warships of World War II, was sunk by the British in 1941. It lies on the ocean floor 15,720 feet below sea level off the west coast of France. Represent that depth using a signed number.

  b. In 1912, the famous cruise ship *Titanic* sank after striking an iceberg. It lies on the North Atlantic ocean floor, 3,220 feet higher than the *Bismarck*. At what depth is the *Titanic* resting?

86. JOGGING A businessman's lunchtime workout includes jogging up ten stories of stairs in his high-rise office building. He starts the workout on the fourth level below ground in the underground parking garage.

  a. Represent that level using a signed number.

  b. On what story of the building will he finish his workout?

87. FLOODING After a heavy rainstorm, a river that had been 9 feet under flood stage rose 11 feet in a 48-hour period.

  a. Represent that level of the river before the storm using a signed number.

  b. Find the height of the river after the storm in comparison to flood stage.

88. ATOMS An atom is composed of protons, neutrons, and electrons. A proton has a positive charge (represented by +1), a neutron has no charge, and an electron has a negative charge (−1). Two simple models of atoms are shown below.

  a. How many protons does the atom in figure (a) have? How many electrons?

  b. What is the net charge of the atom in figure (a)?

  c. How many protons does the atom in figure (b) have? How many electrons?

  d. What is the net charge of the atom in figure (b)?

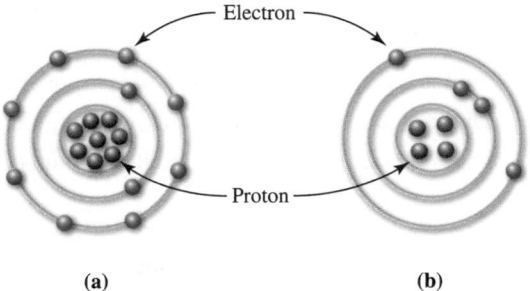

(a)   (b)

89. CHEMISTRY The three steps of a chemistry lab experiment are listed here. The experiment begins with a compound that is stored at −40°F.

  Step 1 Raise the temperature of the compound 200°.

  Step 2 Add sulfur and then raise the temperature 10°.

  Step 3 Add 10 milliliters of water, stir, and raise the temperature 25°.

  What is the resulting temperature of the mixture after step 3?

90. Suppose as a personal financial advisor, your clients are considering purchasing income property. You find a duplex apartment unit that is for sale and learn that the maintenance costs, utilities, and taxes on it total $900 per month. If the current owner receives monthly rental payments of $450 and $380 from the tenants, does the duplex produce a positive cash flow each month?

from Campus to Careers
Personal Financial Advisor

© OJO Images Ltd/Alamy

91. HEALTH Find the point total for the six risk factors (shown with blue headings) on the medical questionnaire below. Then use the table at the bottom of the form (under the red heading) to determine the risk of contracting heart disease for the man whose responses are shown.

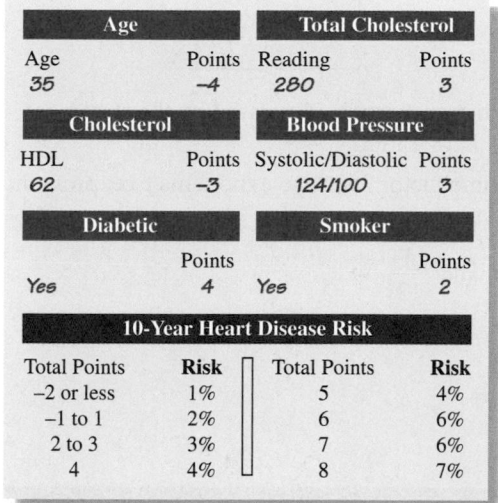

Age		Total Cholesterol	
Age	Points	Reading	Points
35	−4	280	3
Cholesterol		**Blood Pressure**	
HDL	Points	Systolic/Diastolic	Points
62	−3	124/100	3
Diabetic		**Smoker**	
	Points		Points
Yes	4	Yes	2

10-Year Heart Disease Risk			
Total Points	**Risk**	Total Points	**Risk**
−2 or less	1%	5	4%
−1 to 1	2%	6	6%
2 to 3	3%	7	6%
4	4%	8	7%

Source: National Heart, Lung, and Blood Institute

92. POLITICAL POLLS Six months before a general election, the incumbent senator found himself trailing the challenger by 18 points. To overtake his opponent, the campaign staff decided to use a four-part strategy. Each part of this plan is shown below, with the anticipated point gain.

  Part 1 Intense TV ad blitz: gain 10 points

  Part 2 Ask for union endorsement: gain 2 points

  Part 3 Voter mailing: gain 3 points

  Part 4 Get-out-the-vote campaign: gain 1 point

With these gains, will the incumbent overtake the challenger on election day?

93. MILITARY SCIENCE During a battle, an army retreated 1,500 meters, regrouped, and advanced 3,500 meters. The next day, it advanced 1,250 meters. Find the army's net gain.

94. AIRLINES The graph in the next column shows the annual net income for Delta Air Lines during the years 2004–2007.

  a. Estimate the company's total net income over this span of four years in millions of dollars.

  b. Express your answer from part a in billions of dollars.

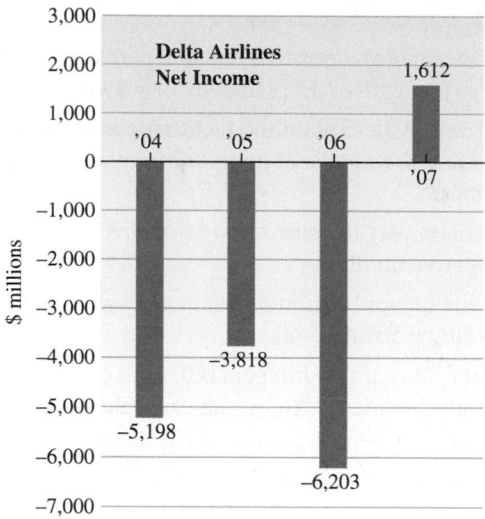

(Source: *The Wall Street Journal*)

95. ACCOUNTING On a financial balance sheet, debts (considered negative numbers) are written within parentheses. Assets (considered positive numbers) are written without parentheses. What is the 2009 fund balance for the preschool whose financial records are shown below?

ABC Preschool Balance Sheet, June 2009

Fund	Balance $
Classroom supplies	$5,889
Emergency needs	$927
Holiday program	($2,928)
Insurance	$1,645
Janitorial	($894)
Licensing	$715
Maintenance	($6,321)
BALANCE	?

96. SPREADSHEETS Monthly rain totals for four counties are listed in the spreadsheet below. The −1 entered in cell B1 means that the rain total for Suffolk County for a certain month was 1 inch *below* average. We can analyze this data by asking the computer to perform various operations.

	A	B	C	D	E	F
1	Suffolk	−1	−1	0	+1	+1
2	Marin	0	−2	+1	+1	−1
3	Logan	−1	+1	+2	+1	+1
4	Tipton	−2	−2	+1	−1	−3
5						

File Edit View Insert Format Tools Data Window Help

Book 1

  a. To ask the computer to add the numbers in cells B1, B2, B3, and B4, we type SUM(B1:B4). Find this sum.

  b. Find SUM(F1:F4).

WRITING

97. Is the sum of a positive and a negative number always positive? Explain why or why not.

98. How do you explain the fact that when asked to *add* -4 and 8, we must actually *subtract* to obtain the result?

99. Explain why the sum of two negative numbers is a negative number.

100. Write an application problem that will require adding -50 and -60.

101. If the sum of two integers is 0, what can be said about the integers? Give an example.

102. Explain why the expression $-6 + -5$ is not written correctly. How should it be written?

REVIEW

103. a. Find the perimeter of the rectangle shown below.

    b. Find the area of the rectangle shown below.

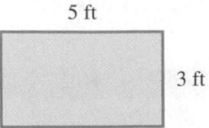

5 ft

3 ft

104. What property is illustrated by the statement $5 \cdot 15 = 15 \cdot 5$?

105. Prime factor 250. Use exponents to express the result.

106. Divide: $\dfrac{144}{12}$

Objectives

1 Use the subtraction rule.

2 Evaluate expressions involving subtraction and addition.

3 Solve application problems by subtracting integers.

SECTION 2.3

Subtracting Integers

In this section, we will discuss a rule that is helpful when subtracting signed numbers.

1   Use the subtraction rule.

The subtraction problem $6 - 4$ can be thought of as taking away 4 from 6. We can use a number line to illustrate this. Beginning at 0, we draw an arrow of length 6 units long that points to the right. It represents positive 6. From the tip of that arrow, we draw a second arrow, 4 units long, that points to the left. It represents taking away 4. Since we end up at 2, it follows that $6 - 4 = 2$.

Begin

6

End

4

6 − 4 = 2

−4 −3 −2 −1   0   1   2   3   4   5   6   7

Note that the illustration above also represents the *addition* $6 + (-4) = 2$. We see that

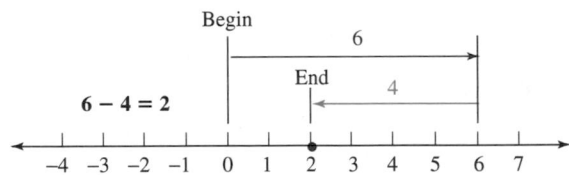

Subtracting 4 from 6 . . .   is the same as . . .   adding the opposite of 4 to 6.

$$6 - 4 = 2 \qquad\qquad\qquad 6 + (-4) = 2$$

The results are the same.

This observation suggests the following rule.

Rule for Subtraction

To subtract two numbers, add the first number to the opposite (additive inverse) of the number to be subtracted.

For any numbers a and b,

$$a - b = a + (-b)$$

Put more simply, this rule says that *subtraction is the same as adding the opposite.*

After rewriting a subtraction as addition of the opposite, we then use one of the rules for the addition of signed numbers discussed in Section 2.2 to find the result.

You won't need to use this rule for every subtraction problem. For example, $6 - 4$ is obviously 2; it does not need to be rewritten as adding the opposite. But for more complicated problems such as $-6 - 4$ or $3 - (-5)$, where the result is not obvious, the subtraction rule will be quite helpful.

EXAMPLE 1 Subtract and check the result:

a. $-6 - 4$ **b.** $3 - (-5)$ **c.** $7 - 23$

Strategy To find each difference, we will apply the rule for subtraction: Add the first integer to the opposite of the integer to be subtracted.

WHY It is easy to make an error when subtracting signed numbers. We will probably be more accurate if we write each subtraction as addition of the opposite.

Solution

a. We read $-6 - 4$ as "negative six *minus* four." Thus, the number to be subtracted is 4. Subtracting 4 is the same as adding its opposite, -4.

Change the subtraction to addition.

$$-6 - 4 \;=\; -6 + (-4) = -10$$ Use the rule for adding two integers with the same sign.

Change the number being subtracted to its opposite.

To check, we add the *difference,* -10, and the *subtrahend,* 4. We should get the *minuend,* -6.

Check: $-10 + 4 = -6$ The result checks.

Caution! Don't forget to write the opposite of the number to be subtracted within parentheses if it is negative.

$$-6 - 4 = -6 + (-4)$$

b. We read $3 - (-5)$ as "three *minus* negative five." Thus, the number to be subtracted is -5. Subtracting -5 is the same as adding its opposite, 5.

Add . . .

$$3 - (-5) \;=\; 3 + 5 = 8$$

. . . the opposite

Check: $8 + (-5) = 3$ The result checks.

Self Check 1
Subtract and check the result:

a. $-2 - 3$
b. $4 - (-8)$
c. $6 - 85$

Now Try Problems 21, 25, and 29

c. We read $7 - 23$ as "seven *minus* twenty-three." Thus, the number to be subtracted is 23. Subtracting 23 is the same as adding its opposite, -23.

Add . . .

$$7 - 23 \quad = \quad 7 + (-23) = -16 \qquad \text{Use the rule for adding two integers with different signs.}$$

. . . the opposite

Check: $-16 + 23 = 7$ The result checks.

Caution! When applying the subtraction rule, *do not change* the first number.

$$-6 - 4 = -6 + (-4) \qquad 3 - (-5) = 3 + 5$$

Self Check 2
a. Subtract -10 from -7.
b. Subtract -7 from -10.
Now Try Problem 33

EXAMPLE 2   **a.** Subtract -12 from -8.   **b.** Subtract -8 from -12.

Strategy We will translate each phrase to mathematical symbols and then perform the subtraction. We must be careful when translating the instruction to subtract one number *from* another number.

WHY The order of the numbers in each word phrase must be reversed when we translate it to mathematical symbols.

Solution

a. Since -12 is the number to be subtracted, we reverse the order in which -12 and -8 appear in the sentence when translating to symbols.

Subtract -12 from -8

$-8 - (-12)$ Write -12 within parentheses.

To find this difference, we write the subtraction as addition of the opposite:

Add . . .

$$-8 - (-12) = -8 + 12 = 4 \qquad \text{Use the rule for adding two integers with different signs.}$$

. . . the opposite

b. Since -8 is the number to be subtracted, we reverse the order in which -8 and -12 appear in the sentence when translating to symbols.

Subtract -8 from -12

$-12 - (-8)$ Write -8 within parentheses.

To find this difference, we write the subtraction as addition of the opposite:

Add . . .

$$-12 - (-8) = -12 + 8 = -4 \qquad \text{Use the rule for adding two integers with different signs.}$$

. . . the opposite

The Language of Algebra   When we change a number to its opposite, we say we have *changed* (or *reversed*) its sign.

Remember that any subtraction problem can be rewritten as an equivalent addition. We just add the opposite of the number that is to be subtracted. Here are four examples:

- $4 - 8 = 4 + (-8) = -4$
- $4 - (-8) = 4 + 8 = 12$
- $-4 - 8 = -4 + (-8) = -12$
- $-4 - (-8) = -4 + 8 = 4$

Any subtraction can be written as addition of the opposite of the number to be subtracted.

2 Evaluate expressions involving subtraction and addition.

Expressions can involve repeated subtraction or combinations of subtraction and addition. To evaluate them, we use the order of operations rule discussed in Section 1.9.

EXAMPLE 3 Evaluate: $-1 - (-2) - 10$

Strategy This expression involves two subtractions. We will write each subtraction as addition of the opposite and then evaluate the expression using the order of operations rule.

WHY It is easy to make an error when subtracting signed numbers. We will probably be more accurate if we write each subtraction as addition of the opposite.

Solution We apply the rule for subtraction twice and then perform the additions, working from left to right. (We could also add the positives and the negatives separately, and then add those results.)

$-1 - (-2) - 10 = -1 + 2 + (-10)$ Add the opposite of −2, which is 2. Add the opposite of 10, which is −10.

$= 1 + (-10)$ Work from left to right. Add −1 + 2 using the rule for adding integers that have different signs.

$= -9$ Use the rule for adding integers that have different signs.

Self Check 3
Evaluate: $-3 - 5 - (-1)$
Now Try Problem 37

EXAMPLE 4 Evaluate: $-80 - (-2 - 24)$

Strategy We will consider the subtraction within the parentheses first and rewrite it as addition of the opposite.

WHY By the order of operations rule, we must perform all calculations within parentheses first.

Solution
$-80 - (-2 - 24) = -80 - [-2 + (-24)]$ Add the opposite of 24, which is −24. Since −24 must be written within parentheses, we write −2 + (−24) within brackets.

$= -80 - (-26)$ Within the brackets, add −2 and −24. Since only one set of grouping symbols is now needed, we can write the answer, −26, within parentheses.

$= -80 + 26$ Add the opposite of −26, which is 26.

$= -54$ Use the rule for adding integers that have different signs.

$$\begin{array}{r}\overset{7\,10}{8\!\!\!/0}\\-26\\\hline54\end{array}$$

Self Check 4
Evaluate: $-72 - (-6 - 51)$
Now Try Problem 49

EXAMPLE 5 Evaluate: $-(-6) + (-18) - 4 - (-51)$

Strategy This expression involves one addition and two subtractions. We will write each subtraction as addition of the opposite and then evaluate the expression.

Self Check 5
Evaluate:
$-(-3) + (-16) - 9 - (-28)$
Now Try Problem 55

WHY It is easy to make an error when subtracting signed numbers. We will probably be more accurate if we write each subtraction as addition of the opposite.

Solution We apply the rule for subtraction twice. Then we will add the positives and the negatives separately, and add those results. (By the commutative and associative properties of addition, we can add the integers in any order.)

$$-(-6) + (-18) - 4 - (-51)$$
$$= 6 + (-18) + (-4) + 51$$ Simplify: $-(-6) = 6$. Add the opposite of 4, which is -4, and add the opposite of -51, which is 51.
$$= (6 + 51) + [(-18) + (-4)]$$ Reorder the integers. Then group the positives together and group the negatives together.
$$= 57 + (-22)$$ Add the positives and add the negatives.
$$= 35$$ Use the rule for adding integers that have different signs.

3 **Solve application problems by subtracting integers.**

Subtraction finds the *difference* between two numbers. When we find the difference between the maximum value and the minimum value of a collection of measurements, we are finding the **range** of the values.

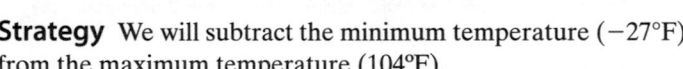

$$\text{Range} = \text{maximum value} - \text{minimum value}$$

EXAMPLE 6   *The Windy City* The record high temperature for Chicago, Illinois, is 104°F. The record low is −27°F. Find the temperature range for these extremes. (Source: *The World Almanac and Book of Facts,* 2009)

Strategy We will subtract the minimum temperature (−27°F) from the maximum temperature (104°F).

WHY The *range* of a collection of data indicates the spread of the data. It is the difference between the maximum and minimum values.

Solution We apply the rule for subtraction and add the opposite of −27.

$$104 - (-27) = 104 + 27$$ 104° is the highest temperature and −27° is the lowest.
$$= 131$$

The temperature range for these extremes is 131°F.

Things are constantly changing in our daily lives. The amount of money we have in the bank, the price of gasoline, and our ages are examples. In mathematics, the operation of subtraction is used to measure change. To find the **change** in a quantity, we subtract the earlier value from the later value.

$$\text{Change} = \text{later value} - \text{earlier value}$$

The five-step problem-solving strategy introduced in Section 1.6 can be used to solve more complicated application problems.

EXAMPLE 7

Water Management On Monday, the water level in a city storage tank was 16 feet above normal. By Friday, the level had fallen to a mark 14 feet below normal. Find the change in the water level from Monday to Friday.

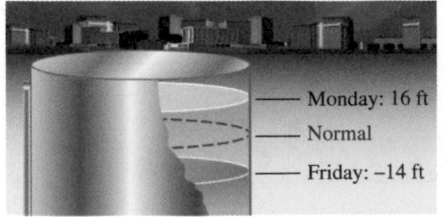

Analyze It is helpful to list the given facts and what you are to find.

- On Monday, the water level was 16 feet above normal.   *Given*
- On Friday, the water level was 14 feet below normal.   *Given*
- What was the change in the water level?   *Find*

Form To find the change in the water level, we *subtract the earlier value from the later value.* The water levels of 16 feet above normal (the earlier value) and 14 feet below normal (the later value) can be represented by 16 and −14.

We translate the words of the problem to numbers and symbols.

The change in the water level	is equal to	the later water level (Friday)	minus	the earlier water level (Monday).
The change in the water level	=	−14	−	16

Solve We can use the rule for subtraction to find the difference.

$$-14 - 16 = -14 + (-16) \qquad \text{Add the opposite of 16, which is } -16.$$
$$= -30 \qquad\qquad \text{Use the rule for adding integers with the same sign.}$$

State The negative result means the water level *fell* 30 feet from Monday to Friday.

Check If we represent the change in water level on a horizontal number line, we see that the water level fell 16 + 14 = 30 units. The result checks.

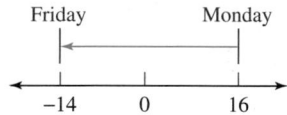

Self Check 7

CRUDE OIL On Wednesday, the level of crude oil in a storage tank was 5 feet above standard capacity. Thursday, after a large refining session, the level fell to a mark 76 feet below standard capacity. Find the change in the crude oil level from Wednesday to Thursday.

Now Try Problem 103

Using Your CALCULATOR   Subtraction with Negative Numbers

The world's highest peak is Mount Everest in the Himalayas. The greatest ocean depth yet measured lies in the Mariana Trench near the island of Guam in the western Pacific. To find the range between the highest peak and the greatest depth, we must subtract:

$$29{,}035 - (-36{,}025)$$

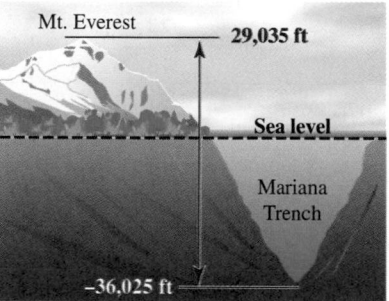

To perform this subtraction on a calculator, we enter the following:

Reverse entry: 29035 $\boxed{-}$ 36025 $\boxed{+/-}$ $\boxed{=}$

Direct entry: 29035 $\boxed{-}$ $\boxed{(-)}$ 36025 $\boxed{\text{ENTER}}$ $\boxed{65060}$

The range is 65,060 feet between the highest peak and the lowest depth. (We could also write $29{,}035 - (-36{,}025)$ as $29{,}035 + 36{,}025$ and then use the addition key $\boxed{+}$ to find the answer.)

ANSWERS TO SELF CHECKS

1. a. −5 **b.** 12 **c.** −79 **2. a.** 3 **b.** −3 **3.** −7 **4.** −15 **5.** 6 **6.** 125°F
7. The crude oil level fell 81 ft.

SECTION 2.3   STUDY SET

VOCABULARY

Fill in the blanks.

1. -8 is the _____ (or _____ inverse) of 8.

2. When we change a number to its opposite, we say we have *changed* (or *reversed*) its _____.

3. To evaluate an expression means to find its _____.

4. The difference between the maximum and the minimum value of a collection of measurements is called the _____ of the values.

CONCEPTS

Fill in the blanks.

5. To subtract two integers, add the first integer to the _____ (additive inverse) of the integer to be subtracted.

6. Subtracting is the same as _____ the opposite.

7. Subtracting 3 is the same as adding ▢. Subtracting -6 is the same as adding ▢.

8. For any numbers a and b, $a - b = a +$ ▢.

9. We can find the _____ in a quantity by subtracting the earlier value from the later value.

10. After rewriting a subtraction as addition of the opposite, we then use one of the rules for the _____ of signed numbers discussed in the previous section to find the result.

11. In each case, determine what number is being subtracted.

    a. $-7 - 3$ **b.** $1 - (-12)$

12. Fill in the blanks to rewrite each subtraction as addition of the opposite of the number being subtracted.

    a. $2 - 7 = 2 +$ ▢

    b. $2 - (-7) = 2 +$ ▢

    c. $-2 - 7 = -2 +$ ▢

    d. $-2 - (-7) = -2 +$ ▢

13. Apply the rule for subtraction and fill in the three blanks.

    $3 - (-6) = 3$ ▢ ▢ $=$ ▢

14. Use addition to check this subtraction: $14 - (-2) = 12$. Is the result correct?

NOTATION

15. Write each phrase using symbols.

    a. negative eight minus negative four

    b. negative eight subtracted from negative four

16. Write each phrase in words.

    a. $7 - (-2)$

    b. $-2 - (-7)$

Complete each solution to evaluate each expression.

17. $1 - 3 - (-2) = 1 + ($ ▢ $) + 2$

    $\qquad\qquad\quad = -2 +$ ▢

    $\qquad\qquad\quad =$ ▢

18. $-6 + 5 - (-5) = -6 + 5 +$ ▢

    $\qquad\qquad\qquad =$ ▢ $+ 5$

    $\qquad\qquad\qquad =$ ▢

19. $(-8 - 2) - (-6) = [-8 + ($ ▢ $)] - (-6)$

    $\qquad\qquad\qquad\; =$ ▢ $- (-6)$

    $\qquad\qquad\qquad\; = -10 +$ ▢

    $\qquad\qquad\qquad\; =$ ▢

20. $-(-5) - (-1 - 4) =$ ▢ $- [-1 + ($ ▢ $)]$

    $\qquad\qquad\qquad\;\; = 5 - ($ ▢ $)$

    $\qquad\qquad\qquad\;\; = 5 +$ ▢

    $\qquad\qquad\qquad\;\; =$ ▢

GUIDED PRACTICE

Subtract. See Example 1.

21. $-4 - 3$	**22.** $-4 - 1$
23. $-5 - 5$	**24.** $-7 - 7$
25. $8 - (-1)$	**26.** $3 - (-8)$
27. $11 - (-7)$	**28.** $10 - (-5)$
29. $3 - 21$	**30.** $8 - 32$
31. $15 - 65$	**32.** $12 - 82$

Perform the indicated operation. See Example 2.

33. **a.** Subtract -1 from -11.

    b. Subtract -11 from -1.

34. **a.** Subtract -2 from -19.

    b. Subtract -19 from -2.

35. **a.** Subtract -41 from -16.

    b. Subtract -16 from -41.

36. **a.** Subtract -57 from -15.

    b. Subtract -15 from -57.

Evaluate each expression. See Example 3.

37. $-4 - (-4) - 15$	**38.** $-3 - (-3) - 10$
39. $10 - 9 - (-8)$	**40.** $16 - 14 - (-9)$

41. $-1 - (-3) - 4$ **42.** $-2 - 4 - (-1)$
43. $-5 - 8 - (-3)$ **44.** $-6 - 5 - (-1)$

Evaluate each expression. See Example 4.

45. $-1 - (-4 - 6)$ **46.** $-7 - (-2 - 14)$
47. $-42 - (-16 - 14)$ **48.** $-45 - (-8 - 32)$
49. $-9 - (6 - 7)$ **50.** $-13 - (6 - 12)$
51. $-8 - (4 - 12)$ **52.** $-9 - (1 - 10)$

Evaluate each expression. See Example 5.

53. $-(-5) + (-15) - 6 - (-48)$
54. $-(-2) + (-30) - 3 - (-66)$
55. $-(-3) + (-41) - 7 - (-19)$
56. $-(-1) + (-52) - 4 - (-21)$

 Use a calculator to perform each subtraction. See Using Your Calculator.

57. $-1,557 - 890$ **58.** $20,007 - (-496)$
59. $-979 - (-44,879)$ **60.** $-787 - 1,654 - (-232)$

TRY IT YOURSELF

Evaluate each expression.

61. $5 - 9 - (-7)$ **62.** $6 - 8 - (-4)$
63. Subtract -3 from 7. **64.** Subtract 8 from -2.
65. $-2 - (-10)$ **66.** $-6 - (-12)$
67. $0 - (-5)$ **68.** $0 - 8$
69. $(6 - 4) - (1 - 2)$ **70.** $(5 - 3) - (4 - 6)$
71. $-5 - (-4)$ **72.** $-9 - (-1)$
73. $-3 - 3 - 3$ **74.** $-1 - 1 - 1$
75. $-(-9) + (-20) - 14 - (-3)$
76. $-(-8) + (-33) - 7 - (-21)$
77. $[-4 + (-8)] - (-6) + 15$
78. $[-5 + (-4)] - (-2) + 22$
79. Subtract -6 from -10.
80. Subtract -4 from -9.
81. $-3 - (-3)$ **82.** $-5 - (-5)$
83. $-8 - [4 - (-6)]$ **84.** $-1 - [5 - (-2)]$
85. $4 - (-4)$ **86.** $-3 - 3$
87. $(-6 - 5) - 3 + (-11)$ **88.** $(-2 - 1) - 5 + (-19)$

APPLICATIONS

Use signed numbers to solve each problem.

89. SUBMARINES A submarine was traveling 2,000 feet below the ocean's surface when the radar system warned of a possible collision with another sub. The captain ordered the navigator to dive an additional 200 feet and then level off. Find the depth of the submarine after the dive.

90. SCUBA DIVING A diver jumps from his boat into the water and descends to a depth of 50 feet. He pauses to check his equipment and then descends an additional 70 feet. Use a signed number to represent the diver's final depth.

91. GEOGRAPHY Death Valley, California, is the lowest land point in the United States, at 282 feet below sea level. The lowest land point on the Earth is the Dead Sea, which is 1,348 feet below sea level. How much lower is the Dead Sea than Death Valley?

92. HISTORY Two of the greatest Greek mathematicians were Archimedes (287–212 B.C.) and Pythagoras (569–500 B.C.).

   a. Express the year of Archimedes' birth as a negative number.

   b. Express the year of Pythagoras' birth as a negative number.

   c. How many years apart were they born?

93. AMPERAGE During normal operation, the ammeter on a car reads $+5$. If the headlights are turned on, they lower the ammeter reading 7 amps. If the radio is turned on, it lowers the reading 6 amps. What number will the ammeter register if they are both turned on?

94. GIN RUMMY After a losing round, a card player must deduct the value of each of the cards left in his hand from his previous point total of 21. If face cards are counted as 10 points, what is his new score?

95. FOOTBALL A college football team records the outcome of each of its plays during a game on a stat sheet. Find the net gain (or loss) after the third play.

Down	Play	Result
1st	Run	Lost 1 yd
2nd	Pass—sack!	Lost 6 yd
Penalty	Delay of game	Lost 5 yd
3rd	Pass	Gained 8 yd

96. ACCOUNTING Complete the balance sheet below. Then determine the overall financial condition of the company by subtracting the total debts from the total assets.

Walker Corporation Balance Sheet 2010				
Assets				
Cash	$11	1	0	9
Supplies	7	8	6	2
Land	67	5	4	3
Total assets	$			
Debts				
Accounts payable	$79	0	3	7
Income taxes	20	1	8	1
Total debts	$			

97. OVERDRAFT PROTECTION A student forgot that she had only $15 in her bank account and wrote a check for $25, used an ATM to get $40 cash, and used her debit card to buy $30 worth of groceries. On each of the three transactions, the bank charged her a $20 overdraft protection fee. Find the new account balance.

98. CHECKING ACCOUNTS Michael has $1,303 in his checking account. Can he pay his car insurance premium of $676, his utility bills of $121, and his rent of $750 without having to make another deposit? Explain.

99. TEMPERATURE EXTREMES The highest and lowest temperatures ever recorded in several cities are shown below. List the cities in order, from the largest to smallest range in temperature extremes.

Extreme Temperatures

City	Highest	Lowest
Atlantic City, NJ	106	−11
Barrow, AK	79	−56
Kansas City, MO	109	−23
Norfolk, VA	104	−3
Portland, ME	103	−39

100. EYESIGHT *Nearsightedness*, the condition where near objects are clear and far objects are blurry, is measured using negative numbers. *Farsightedness*, the condition where far objects are clear and near objects are blurry, is measured using positive numbers. Find the range in the measurements shown in the next column.

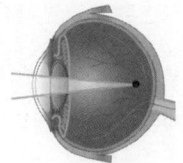

Nearsighted	Farsighted
−2	+4

101. FREEZE DRYING To make freeze-dried coffee, the coffee beans are roasted at a temperature of 360°F and then the ground coffee bean mixture is frozen at a temperature of −110°F. What is the temperature range of the freeze-drying process?

102. WEATHER Rashawn flew from his New York home to Hawaii for a week of vacation. He left blizzard conditions and a temperature of −6°F, and stepped off the airplane into 85°F weather. What temperature change did he experience?

103. READING PROGRAMS In a state reading test given at the start of a school year, an elementary school's performance was 23 points below the county average. The principal immediately began a special tutorial program. At the end of the school year, retesting showed the students to be only 7 points below the average. How did the school's reading score change over the year?

104. LIE DETECTOR TESTS On one lie detector test, a burglar scored −18, which indicates deception. However, on a second test, he scored −1, which is inconclusive. Find the change in his scores.

WRITING

105. Explain what is meant when we say that subtraction is the same as addition of the opposite.

106. Give an example showing that it is possible to subtract something from nothing.

107. Explain how to check the result: $-7 - 4 = -11$

108. Explain why students don't need to change every subtraction they encounter to an addition of the opposite. Give some examples.

REVIEW

109. a. Round 24,085 to the nearest ten.
  b. Round 5,999 to the nearest hundred.

110. List the factors of 20 from least to greatest.

111. It takes 13 oranges to make one can of orange juice. Find the number of oranges used to make 12 cans.

112. a. Find the LCM of 15 and 18.
  b. Find the GCF of 15 and 18.

SECTION 2.4
Multiplying Integers

Objectives

1 Multiply two integers that have different signs.

2 Multiply two integers that have the same sign.

3 Perform several multiplications to evaluate expressions.

4 Evaluate exponential expressions that have negative bases.

5 Solve application problems by multiplying integers.

Multiplication of integers is very much like multiplication of whole numbers. The only difference is that we must determine whether the answer is positive or negative.

When we multiply two nonzero integers, they either have different signs or they have the same sign. This means that there are two possibilities to consider.

1 Multiply two integers that have different signs.

To develop a rule for multiplying two integers that have different signs, we will find $4(-3)$, which is the product of a positive integer and negative integer. We say that the signs of the factors are *unlike*. By the definition of multiplication, $4(-3)$ means that we are to add -3 four times.

$$4(-3) = (-3) + (-3) + (-3) + (-3) \quad \text{Write } -3 \text{ as an addend four times.}$$
$$= -12 \quad \text{Use the rule for adding two integers that have the same sign.}$$

The result is negative. As a check, think in terms of money. If you lose \$3 four times, you have lost a total of \$12, which is written $-\$12$. This example illustrates the following rule.

Multiplying Two Integers That Have Different (Unlike) Signs

To multiply a positive integer and a negative integer, multiply their absolute values. Then make the final answer negative.

EXAMPLE 1 Multiply:
a. $7(-5)$ **b.** $20(-8)$ **c.** $-93 \cdot 16$ **d.** $-34(1,000)$

Strategy We will use the rule for multiplying two integers that have different (unlike) signs.

WHY In each case, we are asked to multiply a positive integer and a negative integer.

Solution
a. Find the absolute values: $|7| = 7$ and $|-5| = 5$.

$7(-5) = -35$ Multiply the absolute values, 7 and 5, to get 35. Then make the final answer negative.

b. Find the absolute values: $|20| = 20$ and $|-8| = 8$.

$20(-8) = -160$ Multiply the absolute values, 20 and 8, to get 160. Then make the final answer negative.

c. Find the absolute values: $|-93| = 93$ and $|16| = 16$.

$-93 \cdot 16 = -1,488$ Multiply the absolute values, 93 and 16, to get 1,488. Then make the final answer negative.

$$\begin{array}{r} 93 \\ \times\ 16 \\ \hline 558 \\ 930 \\ \hline 1,488 \end{array}$$

d. Recall from Section 1.4, to find the product of a whole number and 10, 100, 1,000, and so on, *attach the number of zeros in that number to the right of the whole number.* This rule can be extended to products of integers and 10, 100, 1,000, and so on.

$-34(1,000) = -34,000$ Since 1,000 has three zeros, attach three 0's after -34. ∎

Self Check 1
Multiply:
a. $2(-6)$
b. $30(-4)$
c. $-75 \cdot 17$
d. $-98(1,000)$

Now Try Problems 21, 25, 29, and 31

> **Caution!**   When writing multiplication involving signed numbers, do not write a negative sign − next to a raised dot · (the multiplication symbol). Instead, use parentheses to show the multiplication.
>
> $6(-2)$ ~~6 · −2~~   and $-6(-2)$ ~~−6 · −2~~

2 Multiply two integers that have the same sign.

To develop a rule for multiplying two integers that have the same sign, we will first consider 4(3), which is the product of two positive integers. We say that the signs of the factors are *like*. By the definition of multiplication, 4(3) means that we are to add 3 four times.

$$4(3) = 3 + 3 + 3 + 3 \quad \text{Write 3 as an addend four times.}$$
$$= 12 \qquad\qquad\quad \text{The result is 12, which is a positive number.}$$

As expected, the result is positive.

To develop a rule for multiplying two negative integers, consider the following list, where we multiply −4 by factors that decrease by 1. We know how to find the first four products. Graphing those results on a number line is helpful in determining the last three products.

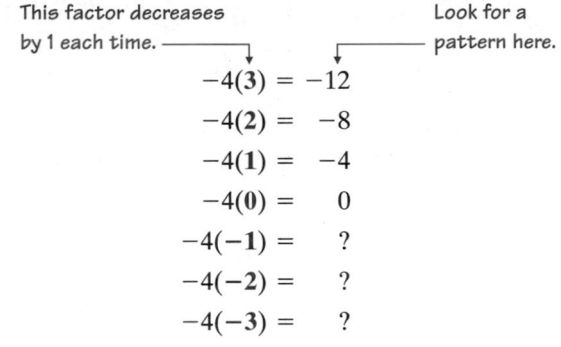

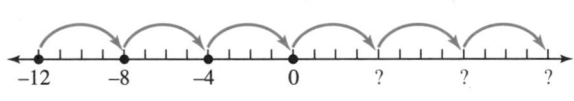

A graph of the products

From the pattern, we see that the product increases by 4 each time. Thus,

$$-4(-1) = 4, \qquad -4(-2) = 8, \qquad \text{and} \qquad -4(-3) = 12$$

These results illustrate that *the product of two negative integers is positive*. As a check, think of it as losing four debts of $3. This is equivalent to gaining $12. Therefore, $-4(-\$3) = \$12$.

We have seen that the product of two positive integers is positive, and the product of two negative integers is also positive. Those results illustrate the following rule.

> ### Multiplying Two Integers That Have the Same (Like) Signs
>
> To multiply two integers that have the same sign, multiply their absolute values. The final answer is positive.

EXAMPLE 2 Multiply:

a. $-5(-9)$ **b.** $-8(-10)$ **c.** $-23(-42)$ **d.** $-2,500(-30,000)$

Strategy We will use the rule for multiplying two integers that have the same (like) signs.

WHY In each case, we are asked to multiply two negative integers.

Solution

a. Find the absolute values: $|-5| = 5$ and $|-9| = 9$.

$-5(-9) = 45$ Multiply the absolute values, 5 and 9, to get 45.
The final answer is positive.

b. Find the absolute values: $|-8| = 8$ and $|-10| = 10$.

$-8(-10) = 80$ Multiply the absolute values, 8 and 10, to get 80.
The final answer is positive.

c. Find the absolute values: $|-23| = 23$ and $|-42| = 42$.

$-23(-42) = 966$ Multiply the absolute values, 23 and 42, to get 966.
The final answer is positive.

$$\begin{array}{r} 42 \\ \times\ 23 \\ \hline 126 \\ 840 \\ \hline 966 \end{array}$$

d. We can extend the method discussed in Section 1.4 for multiplying whole-number factors with trailing zeros to products of integers with trailing zeros.

$-2,500(-30,000) = 75,000,000$ Attach six 0's after 75.

Multiply -25 and -3 to get 75.

We now summarize the multiplication rules for two integers.

Self Check 2

Multiply:

a. $-9(-7)$

b. $-12(-2)$

c. $-34(-15)$

d. $-4,100(-20,000)$

Now Try Problems 33, 37, 41, and 43

Multiplying Two Integers

To multiply two nonzero integers, multiply their absolute values.

1. The product of two integers that have the same (*like*) signs is positive.

2. The product of two integers that have different (*unlike*) signs is negative.

Using Your CALCULATOR   Multiplication with Negative Numbers

At Thanksgiving time, a large supermarket chain offered customers a free turkey with every grocery purchase of $200 or more. Each turkey cost the store $8, and 10,976 people took advantage of the offer. Since each of the 10,976 turkeys given away represented a loss of $8 (which can be expressed as $-\$8$), the company lost a total of $10,976(-\$8)$. To perform this multiplication using a calculator, we enter the following:

Reverse entry: 10976 ✕ 8 +/− =   ⎡ −87808 ⎤

Direct entry: 10976 ✕ (−) 8 ENTER   ⎡ −87808 ⎤

The negative result indicates that with the turkey giveaway promotion, the supermarket chain lost $87,808.

---

3 **Perform several multiplications to evaluate expressions.**

To evaluate expressions that contain several multiplications, we make repeated use of the rules for multiplying two integers.

Self Check 3

Evaluate each expression:

a. $3(-12)(-2)$

b. $-1(9)(-6)$

c. $-4(-5)(8)(-3)$

Now Try Problems 45, 47, and 49

EXAMPLE 3 Evaluate each expression:

a. $6(-2)(-7)$ **b.** $-9(8)(-1)$ **c.** $-3(-5)(2)(-4)$

Strategy Since there are no calculations within parentheses and no exponential expressions, we will perform the multiplications, working from the left to the right.

WHY This is step 3 of the order of operations rule that was introduced in Section 1.9.

Solution

a. $6(-2)(-7) = -12(-7)$ Use the rule for multiplying two integers that have different signs: $6(-2) = -12$.

$$= 84$$ Use the rule for multiplying two integers that have the same sign.

$$\begin{array}{r} \overset{1}{12} \\ \times 7 \\ \hline 84 \end{array}$$

b. $-9(8)(-1) = -72(-1)$ Use the rule for multiplying two integers that have different signs: $-9(8) = -72$.

$$= 72$$ Use the rule for multiplying two integers that have the same sign.

c. $-3(-5)(2)(-4) = 15(2)(-4)$ Use the rule for multiplying two integers that have the same sign: $-3(-5) = 15$.

$$= 30(-4)$$ Use the rule for multiplying two integers that have the same sign: $15(2) = 30$.

$$= -120$$ Use the rule for multiplying two integers that have different signs.

The properties of multiplication that were introduced in Section 1.3, *Multiplying Whole Numbers,* are also true for integers.

Properties of Multiplication

Commutative property of multiplication: The order in which integers are multiplied does not change their product.

Associative property of multiplication: The way in which integers are grouped does not change their product.

Multiplication property of 0: The product of any integer and 0 is 0.

Multiplication property of 1: The product of any integer and 1 is that integer.

Another approach to evaluate expressions like those in Example 3 is to use the properties of multiplication to reorder and regroup the factors in a helpful way.

Self Check 4

Use the commutative and/or associative properties of multiplication to evaluate each expression from Self Check 3 in a different way:

a. $3(-12)(-2)$

b. $-1(9)(-6)$

c. $-4(-5)(8)(-3)$

Now Try Problems 45, 47, and 49

EXAMPLE 4 Use the commutative and/or associative properties of multiplication to evaluate each expression from Example 3 in a different way:

a. $6(-2)(-7)$ **b.** $-9(8)(-1)$ **c.** $-3(-5)(2)(-4)$

Strategy When possible, we will use the commutative and/or associative properties of multiplication to multiply pairs of negative factors.

WHY The product of two negative factors is positive. With this approach, we work with fewer negative numbers, and that lessens the possibility of an error.

Solution

a. $6(-2)(-7) = 6(14)$ Multiply the last two negative factors to produce a positive product: $-7(-2) = 14$.

$$= 84$$

$$\begin{array}{r} \overset{2}{14} \\ \times 6 \\ \hline 84 \end{array}$$

b. $-9(8)(-1) = 9(8)$ Multiply the negative factors to produce a positive product: $-9(-1) = 9$.

        $= 72$

c. $-3(-5)(2)(-4) = 15(-8)$ Multiply the first two negative factors to produce a positive product. Multiply the last two factors.

           $= -120$ Use the rule for multiplying two integers that have different signs.

$$\begin{array}{r} \overset{4}{15} \\ \times 8 \\ \hline 120 \end{array}$$

EXAMPLE 5   Evaluate: **a.** $-2(-4)(-5)$ **b.** $-3(-2)(-6)(-5)$

Strategy When possible, we will use the commutative and/or associative properties of multiplication to multiply pairs of negative factors.

WHY The product of two negative factors is positive. With this approach, we work with fewer negative numbers, and that lessens the possibility of an error.

Solution

a. Note that this expression is the product of three (an odd number) negative integers.

     $-2(-4)(-5) = 8(-5)$ Multiply the first two negative factors to produce a positive product.

            $= -40$ The product is negative.

b. Note that this expression is the product of four (an even number) negative integers.

     $-3(-2)(-6)(-5) = 6(30)$ Multiply the first two negative factors and the last two negative factors to produce positive products.

               $= 180$ The product is positive.

Example 5, part a, illustrates that a product is negative when there is an odd number of negative factors. Example 5, part b, illustrates that a product is positive when there is an even number of negative factors.

Multiplying an Even and an Odd Number of Negative Integers

The product of an even number of negative integers is positive.
The product of an odd number of negative integers is negative.

---

4 **Evaluate exponential expressions that have negative bases.**

Recall that exponential expressions are used to represent repeated multiplication. For example, 2 to the third power, or 2^3, is a shorthand way of writing $2 \cdot 2 \cdot 2$. In this expression, the *exponent* is 3 and the base is *positive* 2. In the next example, we evaluate exponential expressions with bases that are negative numbers.

EXAMPLE 6   Evaluate each expression: **a.** $(-2)^4$ **b.** $(-5)^3$ **c.** $(-1)^5$

Strategy We will write each exponential expression as a product of repeated factors and then perform the multiplication. This requires that we identify the base and the exponent.

WHY The exponent tells the number of times the base is to be written as a factor.

Self Check 5

Evaluate each expression:

a. $-1(-2)(-5)$

b. $-2(-7)(-1)(-2)$

Now Try Problems 53 and 57

Self Check 6

Evaluate each expression:

a. $(-3)^4$

b. $(-4)^3$

c. $(-1)^7$

Now Try Problems 61, 65, and 67

Solution

a. We read $(-2)^4$ as "negative two raised to the fourth power" or as "the fourth power of negative two." Note that the exponent is even.

$(-2)^4 = \mathbf{(-2)(-2)(-2)(-2)}$ Write the base, −2, as a factor 4 times.

$\quad\quad\quad = 4(4)$ Multiply the first two negative factors and the last two negative factors to produce positive products.

$\quad\quad\quad = 16$ The result is positive.

b. We read $(-5)^3$ as "negative five raised to the third power" or as "the third power of negative five," or as " negative five, cubed." Note that the exponent is odd.

$(-5)^3 = \mathbf{(-5)(-5)(-5)}$ Write the base, −5, as a factor 3 times.

$\quad\quad\quad = \mathbf{25}(-5)$ Multiply the first two negative factors to produce a positive product.

$\quad\quad\quad = -125$ The result is negative.

$$\begin{array}{r} \overset{2}{2}5 \\ \times\,5 \\ \hline 125 \end{array}$$

c. We read $(-1)^5$ as "negative one raised to the fifth power" or as "the fifth power of negative one." Note that the exponent is odd.

$(-1)^5 = \mathbf{(-1)(-1)(-1)(-1)(-1)}$ Write the base, −1, as a factor 5 times.

$\quad\quad\quad = \mathbf{1(1)}(-1)$ Multiply the first and second negative factors and multiply the third and fourth negative factors to produce positive products.

$\quad\quad\quad = -1$ The result is negative.

In Example 6, part a, −2 was raised to an even power, and the answer was positive. In parts b and c, −5 and −1 were raised to odd powers, and, in each case, the answer was negative. These results suggest a general rule.

Even and Odd Powers of a Negative Integer

When a negative integer is raised to an even power, the result is positive.
When a negative integer is raised to an odd power, the result is negative.

Although the exponential expressions $(-3)^2$ and -3^2 look similar, they are not the same. We read $(-3)^2$ as "negative 3 squared" and -3^2 as "the opposite of the square of three." When we evaluate them, it becomes clear that they are not equivalent.

$(-3)^2 = (-3)(-3)$ Because of the parentheses, the base is −3. The exponent is 2.

$\quad\quad = 9$

$-3^2 = -(3 \cdot 3)$ Since there are no parentheses around −3, the base is 3. The exponent is 2.

$\quad\quad = -9$

Different results

Caution!   The base of an exponential expression *does not include* the negative sign unless parentheses are used.

$$-7^3 \quad\quad\quad\quad\quad (-7)^3$$

Positive base: 7   Negative base: −7

EXAMPLE 7 Evaluate: -2^2

Self Check 7
Evaluate: -4^2
Now Try Problem 71

Strategy We will rewrite the expression as a product of repeated factors, and then perform the multiplication. We must be careful when identifying the base. It is 2, not -2.

WHY Since there are no parentheses around -2, the base is 2.

Solution

$-2^2 = -(2 \cdot 2)$ Read as "the opposite of the square of two."

$\quad\ = -4$ Do the multiplication within the parentheses to get 4. Then write the opposite of that result.

Using Your CALCULATOR   **Raising a Negative Number to a Power**

We can find powers of negative integers, such as $(-5)^6$, using a calculator. The keystrokes that are used to evaluate such expressions vary from model to model, as shown below. You will need to determine which keystrokes produce the positive result that we would expect when raising a negative number to an even power.

5 $\boxed{+/-}$ $\boxed{y^x}$ 6 $\boxed{=}$ Some calculators don't require the parentheses to be entered.

$\boxed{(}$ 5 $\boxed{+/-}$ $\boxed{)}$ $\boxed{y^x}$ 6 $\boxed{=}$ Other calculators require the parentheses to be entered.

$\boxed{(}$ $\boxed{(-)}$ 5 $\boxed{)}$ $\boxed{\wedge}$ 6 $\boxed{\text{ENTER}}$ $\boxed{15625}$

From the calculator display, we see that $(-5)^6 = 15{,}625$.

---

5 **Solve application problems by multiplying integers.**

Problems that involve repeated addition are often more easily solved using multiplication.

EXAMPLE 8   *Oceanography*

Scientists lowered an underwater vessel called a *submersible* into the Pacific Ocean to record the water temperature. The first measurement was made 75 feet below sea level, and more were made every 75 feet until it reached the ocean floor. Find the depth of the submersible when the 25th measurement was made.

Emory Kristof/National Geographic/Getty Images

Self Check 8
GASOLINE LEAKS To determine how badly a gasoline tank was leaking, inspectors used a drilling process to take soil samples nearby. The first sample was taken 6 feet below ground level, and more were taken every 6 feet after that. The 14th sample was the first one that did not show signs of gasoline. How far below ground level was that?
Now Try Problem 97

Analyze

- The first measurement was made 75 feet below sea level.   *Given*
- More measurements were made every 75 feet.   *Given*
- What was the depth of the submersible when it made the 25th measurement?   *Find*

Form If we use negative numbers to represent the depths at which the measurements were made, then the first was at -75 feet. The depth (in feet) of the submersible when the 25th measurement was made can be found by adding -75 twenty-five times. This repeated addition can be calculated more simply by multiplication.

We translate the words of the problem to numbers and symbols.

The depth of the submersible when it made the 25th measurement	is equal to	the number of measurements made	times	the amount it was lowered each time.
The depth of the submersible when it made the 25th measurement	=	25	·	(-75)

Solve To find the product, we use the rule for multiplying two integers that have different signs. First, we find the absolute values: $|25| = 25$ and $|-75| = 75$.

$$25(-75) = -1,875$$

Multiply the absolute values, 25 and 75, to get 1,875. Since the integers have different signs, make the final answer negative.

$$\begin{array}{r} 75 \\ \times\,25 \\ \hline 375 \\ 1\;500 \\ \hline 1,875 \end{array}$$

State The depth of the submersible was 1,875 feet below sea level ($-1,875$ feet) when the 25th temperature measurement was taken.

Check We can use estimation or simply perform the actual multiplication again to see if the result seems reasonable.

ANSWERS TO SELF CHECKS

1. a. -12 **b.** -120 **c.** $-1,275$ **d.** $-98,000$ **2. a.** 63 **b.** 24 **c.** 510 **d.** $82,000,000$
3. a. 72 **b.** 54 **c.** -480 **4. a.** 72 **b.** 54 **c.** -480 **5. a.** -10 **b.** 28 **6. a.** 81
b. -64 **c.** -1 **7.** -16 **8.** 84 ft below ground level (-84 ft)

SECTION 2.4   STUDY SET

VOCABULARY

Fill in the blanks.

1. In the multiplication problem shown below, label each *factor* and the *product*.

$$-5 \quad \cdot \quad 10 \quad = \quad -50$$

2. Two negative integers, as well as two positive integers, are said to have the same signs or _____ signs.

3. A positive integer and a negative integer are said to have different signs or _____ signs.

4. _____ property of multiplication: The order in which integers are multiplied does not change their product.

5. _____ property of multiplication: The way in which integers are grouped does not change their product.

6. In the expression $(-3)^5$, the _____ is -3, and 5 is the _____.

CONCEPTS

Fill in the blanks.

7. Multiplication of integers is very much like multiplication of whole numbers. The only difference is that we must determine whether the answer is _____ or _____.

8. When we multiply two nonzero integers, they either have _____ signs or _____ sign.

9. To multiply a positive integer and a negative integer, multiply their absolute values. Then make the final answer _____.

10. To multiply two integers that have the same sign, multiply their absolute values. The final answer is _____.

11. The product of two integers with _____ signs is negative.

12. The product of two integers with _____ signs is positive.

13. The product of any integer and 0 is   .

14. The product of an even number of negative integers is _____ and the product of an odd number of negative integers is _____.

15. Find each absolute value.
   a. $|-3|$ **b.** $|12|$

16. If each of the following expressions were evaluated, what would be the *sign* of the result?
   a. $(-5)^{13}$ **b.** $(-3)^{20}$

NOTATION

17. For each expression, identify the base and the exponent.
   a. -8^4 **b.** $(-7)^9$

18. Translate to mathematical symbols.
   a. negative three times negative two
   b. negative five squared
   c. the opposite of the square of five

Complete each solution to evaluate the expression.

19. $-3(-2)(-4) = \boxed{}(-4)$

$= \boxed{}$

20. $(-3)^4 = (-3)(-3)(-3)\boxed{}$

$= \boxed{}(9)$

$= \boxed{}$

GUIDED PRACTICE

Multiply. See Example 1.

21. $5(-3)$ **22.** $4(-6)$
23. $9(-2)$ **24.** $5(-7)$
25. $18(-4)$ **26.** $17(-8)$
27. $21(-6)$ **28.** $39(-3)$
29. $-45 \cdot 37$ **30.** $-42 \cdot 24$
31. $-94 \cdot 1,000$ **32.** $-76 \cdot 1,000$

Multiply. See Example 2.

33. $(-8)(-7)$ **34.** $(-9)(-3)$
35. $-7(-1)$ **36.** $-5(-1)$
37. $-3(-52)$ **38.** $-4(-73)$
39. $-6(-46)$ **40.** $-8(-48)$
41. $-59(-33)$ **42.** $-61(-29)$
43. $-60,000(-1,200)$ **44.** $-20,000(-3,200)$

Evaluate each expression. See Examples 3 and 4.

45. $6(-3)(-5)$ **46.** $9(-3)(-4)$
47. $-5(10)(-3)$ **48.** $-8(7)(-2)$
49. $-2(-4)(6)(-8)$ **50.** $-3(-5)(2)(-9)$
51. $-8(-3)(7)(-2)$ **52.** $-9(-3)(4)(-2)$

Evaluate each expression. See Example 5.

53. $-4(-2)(-6)$ **54.** $-4(-6)(-3)$
55. $-3(-9)(-3)$ **56.** $-5(-2)(-5)$
57. $-1(-3)(-2)(-6)$ **58.** $-1(-4)(-2)(-4)$
59. $-9(-4)(-1)(-4)$ **60.** $-6(-3)(-6)(-1)$

Evaluate each expression. See Example 6.

61. $(-3)^3$ **62.** $(-6)^3$
63. $(-2)^5$ **64.** $(-3)^5$
65. $(-5)^4$ **66.** $(-7)^4$
67. $(-1)^8$ **68.** $(-1)^{10}$

Evaluate each expression. See Example 7.

69. $(-7)^2$ and -7^2
70. $(-5)^2$ and -5^2
71. $(-12)^2$ and -12^2
72. $(-11)^2$ and -11^2

TRY IT YOURSELF

Evaluate each expression.

73. $6(-5)(2)$ **74.** $4(-2)(2)$
75. $-8(0)$ **76.** $0(-27)$
77. $(-4)^3$ **78.** $(-8)^3$
79. $(-2)10$ **80.** $(-3)8$
81. $-2(-3)(3)(-1)$ **82.** $5(-2)(3)(-1)$

83. Find the product of -6 and the opposite of 10.
84. Find the product of the opposite of 9 and the opposite of 8.

85. $-6(-4)(-2)$ **86.** $-3(-2)(-3)$
87. $-42 \cdot 200,000$ **88.** $-56 \cdot 10,000$
89. -5^4 **90.** -2^4
91. $-12(-12)$ **92.** $-5(-5)$
93. $(-1)^6$ **94.** $(-1)^5$
95. $(-1)(-2)(-3)(-4)(-5)$
96. $(-10)(-8)(-6)(-4)(-2)$

APPLICATIONS

Use signed numbers to solve each problem.

97. SUBMARINES As part of a training exercise, the captain of a submarine ordered it to descend 250 feet, level off for 5 minutes, and then repeat the process several times. If the sub was on the ocean's surface at the beginning of the exercise, find its depth after the 8th dive.

98. BUILDING A PIER A *pile driver* uses a heavy weight to pound tall poles into the ocean floor. If each strike of a pile driver on the top of a pole sends it 6 inches deeper, find the depth of the pole after 20 strikes.

Image Source/Getty Images

99. MAGNIFICATION A mechanic used an electronic testing device to check the smog emissions of a car. The results of the test are displayed on a screen.

   a. Find the high and low values for this test as shown on the screen.

   b. By switching a setting, the picture on the screen can be magnified. What would be the new high and new low if every value were doubled?

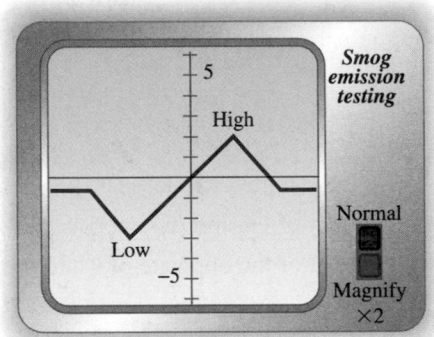

100. LIGHT Sunlight is a mixture of all colors. When sunlight passes through water, the water absorbs different colors at different rates, as shown.

   a. Use a signed number to represent the depth to which red light penetrates water.

   b. Green light penetrates 4 times deeper than red light. How deep is this?

   c. Blue light penetrates 3 times deeper than orange light. How deep is this?

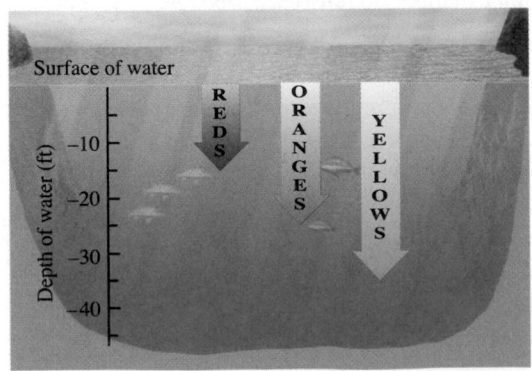

101. JOB LOSSES Refer to the bar graph. Find the number of jobs lost in . . .

   a. September 2008 if it was about 6 times the number lost in April.

   b. October 2008 if it was about 9 times the number lost in May.

   c. November 2008 if it was about 7 times the number lost in February.

   d. December if it was about 6 times the number lost in March.

2008 U.S. Monthly Net Job Losses

Source: Bureau of Labor Statistics

102. RUSSIA The U.S. Census Bureau estimates that Russia's population is decreasing by about 700,000 per year because of high death rates and low birth rates. If this pattern continues, what will be the total decline in Russia's population over the next 30 years? (Source: About.com)

103. PLANETS The average surface temperature of Mars is −81°F. Find the average surface temperature of Uranus if it is four times colder than Mars. (Source: *The World Almanac and Book of Facts*, 2009)

104. CROP LOSS A farmer, worried about his fruit trees suffering frost damage, calls the weather service for temperature information. He is told that temperatures will be decreasing approximately 5 degrees every hour for the next five hours. What signed number represents the total change in temperature expected over the next five hours?

105. TAX WRITE-OFF For each of the last six years, a businesswoman has filed a $200 depreciation allowance on her income tax return for an office computer system. What signed number represents the total amount of depreciation written off over the six-year period?

106. EROSION A levee protects a town in a low-lying area from flooding. According to geologists, the banks of the levee are eroding at a rate of 2 feet per year. If something isn't done to correct the problem, what signed number indicates how much of the levee will erode during the next decade?

107. DECK SUPPORTS After a winter storm, a homeowner has an engineering firm inspect his damaged deck. Their report concludes that the original foundation poles were not sunk deep enough, by a factor of 3. What signed number represents the depth to which the poles should have been sunk?

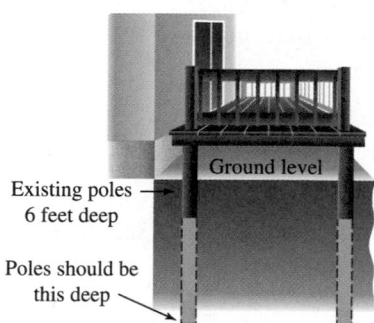

Existing poles
6 feet deep

Ground level

Poles should be
this deep

108. DIETING After giving a patient a physical exam, a physician felt that the patient should begin a diet. The two options that were discussed are shown in the following table.

	Plan #1	Plan #2
Length	10 weeks	14 weeks
Daily exercise	1 hr	30 min
Weight loss per week	3 lb	2 lb

a. Find the expected weight loss from Plan 1. Express the answer as a signed number.

b. Find the expected weight loss from Plan 2. Express the answer as a signed number.

c. With which plan should the patient expect to lose more weight? Explain why the patient might not choose it.

109. ADVERTISING The paid attendance for the last night of the 2008 Rodeo Houston was 71,906. Suppose a local country music radio station gave a sports bag, worth $3, to everyone that attended. Find the signed number that expresses the radio station's financial loss from this giveaway.

110. HEALTH CARE A health care provider for a company estimates that 75 hours per week are lost by employees suffering from stress-related or preventable illness. In a 52-week year, how many hours are lost? Use a signed number to answer.

WRITING

111. Explain why the product of a positive number and a negative number is negative, using $5(-3)$ as an example.

112. Explain the multiplication rule for integers that is shown in the pattern of signs below.

$$(-)(-) = +$$
$$(-)(-)(-) = -$$
$$(-)(-)(-)(-) = +$$
$$(-)(-)(-)(-)(-) = -$$
$$\vdots$$

113. When a number is multiplied by -1, the result is the opposite of the original number. Explain why.

114. A student claimed, "A positive and a negative is negative." What is wrong with this statement?

REVIEW

115. List the first ten prime numbers.

116. ENROLLMENT The number of students attending a college went from 10,250 to 12,300 in one year. What was the increase in enrollment?

117. Divide: $175 \div 4$

118. What does the symbol $<$ mean?

SECTION 2.5
Dividing Integers

Objectives

1 Divide two integers.

2 Identify *division of 0* and *division by 0*.

3 Solve application problems by dividing integers.

In this section, we will develop rules for division of integers, just as we did earlier for multiplication of integers.

1 Divide two integers.

Recall from Section 1.4 that every division has a related multiplication statement. For example,

$$\frac{6}{3} = 2 \qquad \text{because} \qquad 2(3) = 6$$

and

$$\frac{20}{5} = 4 \qquad \text{because} \qquad 4(5) = 20$$

We can use the relationship between multiplication and division to help develop rules for dividing integers. There are four cases to consider.

Case 1: A positive integer divided by a positive integer
From years of experience, we already know that the result is positive. Therefore, *the quotient of two positive integers is positive.*

Case 2: A negative integer divided by a negative integer
As an example, consider the division $\frac{-12}{-2} = ?$. We can find ? by examining the related multiplication statement.

Related multiplication statement

$?(-2) = -12$

This must be *positive 6* if the product is to be negative 12.

Division statement

$\frac{-12}{-2} = ?$

So the quotient is *positive 6.*

Therefore, $\frac{-12}{-2} = 6$. This example illustrates that *the quotient of two negative integers is positive.*

Case 3: A positive integer divided by a negative integer
Let's consider $\frac{12}{-2} = ?$. We can find ? by examining the related multiplication statement.

Related multiplication statement

$?(-2) = 12$

This must be -6 if the product is to be positive 12.

Division statement

$\frac{12}{-2} = ?$

So the quotient is -6.

Therefore, $\frac{12}{-2} = -6$. This example illustrates that *the quotient of a positive integer and a negative integer is negative.*

Case 4: A negative integer divided by a positive integer
Let's consider $\frac{-12}{2} = ?$. We can find ? by examining the related multiplication statement.

Related multiplication statement

$?(2) = -12$

This must be -6 if the product is to be -12.

Division statement

$\frac{-12}{2} = ?$

So the quotient is -6.

Therefore, $\frac{-12}{2} = -6$. This example illustrates that *the quotient of a negative integer and a positive integer is negative.*

We now summarize the results from the previous examples and note that they are similar to the rules for multiplication.

Dividing Two Integers

To divide two integers, divide their absolute values.

1.   The quotient of two integers that have the same (*like*) signs is positive.
2.   The quotient of two integers that have different (*unlike*) signs is negative.

EXAMPLE 1   Divide and check the result:

a. $\dfrac{-14}{7}$ b. $30 \div (-5)$ c. $\dfrac{176}{-11}$ d. $-24{,}000 \div 600$

Strategy We will use the rule for dividing two integers that have different (unlike) signs.

WHY Each division involves a positive and a negative integer.

Solution

a. Find the absolute values: $|-14| = 14$ and $|7| = 7$.

$$\dfrac{-14}{7} = {-2} \quad \text{Divide the absolute values, 14 by 7, to get 2.}$$
Then make the final answer negative.

To check, we multiply the *quotient*, -2, and the *divisor*, 7. We should get the *dividend*, -14.

Check: $-2(7) = -14$ The result checks.

b. Find the absolute values: $|30| = 30$ and $|-5| = 5$.

$$30 \div (-5) = {-6} \quad \text{Divide the absolute values, 30 by 5, to get 6.}$$
Then make the final answer negative.

Check: $-6(-5) = 30$ The result checks.

c. Find the absolute values: $|176| = 176$ and $|-11| = 11$.

$$\dfrac{176}{-11} = {-16} \quad \text{Divide the absolute values, 176 by 11, to get 16.}$$
Then make the final answer negative.

$$\begin{array}{r} 16 \\ 11\overline{)176} \\ -11 \\ \hline 66 \\ -\ 66 \\ \hline 0 \end{array}$$

Check: $-16(-11) = 176$ The result checks.

d. Recall from Section 1.5, that if a divisor has ending zeros, we can simplify the division by removing the same number of ending zeros in the divisor and dividend.

There are two zeros in the divisor.

$$-24{,}000 \div 60\widetilde{0} \; = \; -240 \div 6 \; = \; {-40} \quad \text{Divide the absolute values, 240 by 6, to get 40.}$$
Remove two zeros from the dividend and the divisor, and divide.   Then make the final answer negative.

Check: $-40(600) = -24{,}000$ Use the original divisor and dividend in the check.

EXAMPLE 2   Divide and check the result:

a. $\dfrac{-12}{-3}$ b. $-48 \div (-6)$ c. $\dfrac{-315}{-9}$ d. $-200 \div (-40)$

Strategy We will use the rule for dividing two integers that have the same (like) signs.

WHY In each case, we are asked to find the quotient of two negative integers.

Solution

a. Find the absolute values: $|-12| = 12$ and $|-3| = 3$.

$$\dfrac{-12}{-3} = 4 \quad \text{Divide the absolute values, 12 by 3, to get 4.}$$
The final answer is positive.

Check: $4(-3) = -12$ The result checks.

Self Check 1

Divide and check the result:

a. $\dfrac{-45}{5}$

b. $28 \div (-4)$

c. $\dfrac{336}{-14}$

d. $-18{,}000 \div 300$

Now Try Problems 13, 15, 21, and 27

Self Check 2

Divide and check the result:

a. $\dfrac{-27}{-3}$

b. $-24 \div (-4)$

c. $\dfrac{-301}{-7}$

d. $-400 \div (-20)$

Now Try Problems 33, 37, 41, and 43

b. Find the absolute values: $|-48| = 48$ and $|-6| = 6$.

$-48 \div (-6) = 8$ Divide the absolute values, 48 by 6, to get 8.
The final answer is positive.

Check: $8(-6) = -48$ The result checks.

c. Find the absolute values: $|-315| = 315$ and $|-9| = 9$.

$$\frac{-315}{-9} = 35$$ Divide the absolute values, 315 by 9, to get 35.
The final answer is positive.

$$\begin{array}{r} 35 \\ 9\overline{)315} \\ -27 \\ \hline 45 \\ -45 \\ \hline 0 \end{array}$$

Check: $35(-9) = -315$ The result checks.

d. We can simplify the division by removing the same number of ending zeros in the divisor and dividend.

There is one zero in the divisor.

$-200 \div (-40) = -20 \div (-4) = 5$ Divide the absolute values, 20 by 4, to get 5. The final answer is positive.

Remove one zero from the dividend and the divisor, and divide.

Check: $5(-40) = -200$ The result checks.

2 Identify *division of 0* and *division by 0*.

To review the concept of division of 0, we consider $\frac{0}{-2} = ?$. We can attempt to find ? by examining the related multiplication statement.

Related multiplication statement

$(?)(-2) = 0$

This must be 0 if the product is to be 0.

Division statement

$\frac{0}{-2} = ?$

So the quotient is 0.

Therefore, $\frac{0}{-2} = 0$. This example illustrates that *the quotient of 0 divided by any non-zero number is 0.*

To review division by 0, let's consider $\frac{-2}{0} = ?$. We can attempt to find ? by examining the related multiplication statement.

Related multiplication statement

$(?)0 = -2$

There is no number that gives −2 when multiplied by 0.

Division statement

$\frac{-2}{0} = ?$

There is no quotient.

Therefore, $\frac{-2}{0}$ does not have an answer and we say that $\frac{-2}{0}$ is undefined. This example illustrates that *the quotient of any nonzero number divided by 0 is undefined.*

Division with 0

1. If 0 is divided by any nonzero number, the quotient is 0.
   For any nonzero number a,
   $$\frac{0}{a} = 0$$

2. Division of any nonzero number by 0 is undefined.
   For any nonzero number a,
   $$\frac{a}{0} \text{ is undefined}$$

EXAMPLE 3   Divide, if possible: **a.** $\dfrac{-4}{0}$ **b.** $0 \div (-8)$

Strategy In each case, we need to determine if we have division *of* 0 or division *by* 0.

WHY *Division of 0* by a nonzero integer is defined, and the answer is 0. However, *division of a nonzero integer by 0* is undefined; there is no answer.

Solution

a. $\dfrac{-4}{0}$ is undefined.   *This is division by 0.*

b. $0 \div (-8) = 0$ because $0(-8) = 0.$ *This is division of 0.*

Self Check 3
Divide, if possible:
a. $\dfrac{-12}{0}$ **b.** $0 \div (-6)$

Now Try Problems 45 and 47

3 Solve application problems by dividing integers.

Problems that involve forming equal-sized groups can be solved by division.

EXAMPLE 4   *Real Estate* Over the course of a year, a homeowner reduced the price of his house by an equal amount each month because it was not selling. By the end of the year, the price was $11,400 less than at the beginning of the year. By how much was the price of the house reduced each month?

Self Check 4

SELLING BOATS The owner of a sail boat reduced the price of the boat by an equal amount each month, because there were no interested buyers. After 8 months, and a $960 reduction in price, the boat sold. By how much was the price of the boat reduced each month?

Now Try Problem 81

Analyze

- The homeowner dropped the price $11,400 in 1 year.   *Given*
- The price was reduced by an equal amount each month.   *Given*
- By how much was the price of the house reduced each month?   *Find*

Form We can express the drop in the price of the house for the year as $-\$11,400$. The phrase *reduced by an equal amount each month* indicates division. We translate the words of the problem to numbers and symbols.

The amount the price was reduced each month	is equal to	the drop in the price of the house for the year	divided by	the number of months in 1 year.
The amount the price was reduced each month	$=$	$-11,400$	$\div$	12

Solve To find the quotient, we use the rule for dividing two integers that have different signs. First, we find the absolute values: $|-11,400| = 11,400$ and $|12| = 12$.

$-11,400 \div 12 = -950$ *Divide the absolute values, 11,400 and 12, to get 950. Then make the final answer negative.*

$$\begin{array}{r} 950 \\ 12\overline{)11,400} \\ -10\ 8 \\ \hline 60 \\ -60 \\ \hline 00 \\ -00 \\ \hline 0 \end{array}$$

State The negative result indicates that the price of the house was *reduced* by $950 each month.

Check We can use estimation to check the result. A reduction of $1,000 each month would cause the price to drop $12,000 in 1 year. It seems reasonable that a reduction of $950 each month would cause the price to drop $11,400 in a year.

> ### *Using Your CALCULATOR*   Division with Negative Numbers
>
> The Bureau of Labor Statistics estimated that the United States lost 162,000 auto manufacturing jobs (motor vehicles and parts) in 2008. Because the jobs were lost, we write this as $-162,000$. To find the average number of manufacturing jobs lost each month, we divide: $\frac{-162,000}{12}$. We can use a calculator to perform the division.
>
> Reverse entry: 162000 $\boxed{+/-}$ $\boxed{\div}$ 12 $\boxed{=}$
>
> Direct entry: 162000 $\boxed{\div}$ $\boxed{(-)}$ 12 $\boxed{\text{ENTER}}$ $\boxed{-13500}$
>
> The average number of auto manufacturing jobs lost each month in 2008 was 13,500.

ANSWERS TO SELF CHECKS

1. a. -9 **b.** -7 **c.** -24 **d.** -60 **2. a.** 9   **b.** 6   **c.** 43   **d.** 20   **3. a.** undefined
b. 0   **4.** The price was reduced by \$120 each month.

SECTION 2.5   STUDY SET

VOCABULARY

Fill in the blanks.

1. In the division problems shown below, label the *dividend, divisor,* and *quotient.*

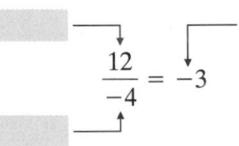

$$12 \;\div\; (-4) \;=\; -3$$

$$\frac{12}{-4} = -3$$

2. The related _____ statement for $\dfrac{-6}{3} = -2$ is $-2(3) = -6$.

3. $\dfrac{-3}{0}$ is division ____ 0 and $\dfrac{0}{-3} = 0$ is division ____ 0.

4. Division of a nonzero integer by 0, such as $\dfrac{-3}{0}$, is _____.

CONCEPTS

5. Write the related multiplication statement for each division.

   a. $\dfrac{-25}{5} = -5$ **b.** $-36 \div (-6) = 6$ **c.** $\dfrac{0}{-15} = 0$

6. Using multiplication, check to determine whether $-720 \div 45 = -12$.

7. Fill in the blanks.
To divide two integers, divide their absolute values.

   a. The quotient of two integers that have the same (*like*) signs is _____.

   b. The quotient of two integers that have different (*unlike*) signs is _____.

8. If a divisor has ending zeros, we can simplify the division by removing the same number of ending zeros in the divisor and dividend. Fill in the blank: $-2,400 \div 60 = -240 \div$ ▢

9. Fill in the blanks.

   a. If 0 is divided by any nonzero integer, the quotient is ▢.

   b. Division of any nonzero integer by 0 is _____.

10. What operation can be used to solve problems that involve forming equal-sized groups?

11. Determine whether each statement is always true, sometimes true, or never true.

   a. The product of a positive integer and a negative integer is negative.

   b. The sum of a positive integer and a negative integer is negative.

   c. The quotient of a positive integer and a negative integer is negative.

12. Determine whether each statement is always true, sometimes true, or never true.

   a. The product of two negative integers is positive.

   b. The sum of two negative integers is negative.

   c. The quotient of two negative integers is negative.

GUIDED PRACTICE

Divide and check the result. **See Example 1.**

13. $\dfrac{-14}{2}$

14. $\dfrac{-10}{5}$

15. $\dfrac{-20}{5}$

16. $\dfrac{-24}{3}$

17. $36 \div (-6)$

18. $36 \div (-9)$

19. $24 \div (-3)$

20. $42 \div (-6)$

21. $\dfrac{264}{-12}$

22. $\dfrac{364}{-14}$

23. $\dfrac{702}{-18}$

24. $\dfrac{396}{-12}$

25. $-9,000 \div 300$

26. $-12,000 \div 600$

27. $-250,000 \div 5,000$

28. $-420,000 \div 7,000$

Divide and check the result. **See Example 2.**

29. $\dfrac{-8}{-4}$

30. $\dfrac{-12}{-4}$

31. $\dfrac{-45}{-9}$

32. $\dfrac{-81}{-9}$

33. $-63 \div (-7)$

34. $-21 \div (-3)$

35. $-32 \div (-8)$

36. $-56 \div (-7)$

37. $\dfrac{-400}{-25}$

38. $\dfrac{-490}{-35}$

39. $\dfrac{-651}{-31}$

40. $\dfrac{-736}{-32}$

41. $-800 \div (-20)$

42. $-800 \div (-40)$

43. $-15,000 \div (-30)$

44. $-36,000 \div (-60)$

Divide, if possible. **See Example 3.**

45. a. $\dfrac{-3}{0}$

b. $\dfrac{0}{-3}$

46. a. $\dfrac{-5}{0}$

b. $\dfrac{0}{-5}$

47. a. $\dfrac{0}{-24}$

b. $\dfrac{-24}{0}$

48. a. $\dfrac{0}{-32}$

b. $\dfrac{-32}{0}$

TRY IT YOURSELF

Divide, if possible.

49. $-36 \div (-12)$

50. $-45 \div (-15)$

51. $\dfrac{425}{-25}$

52. $\dfrac{462}{-42}$

53. $0 \div (-16)$

54. $0 \div (-6)$

55. Find the quotient of -45 and 9.

56. Find the quotient of -36 and -4.

57. $-2,500 \div 500$

58. $-52,000 \div 4,000$

59. $\dfrac{-6}{0}$

60. $\dfrac{-8}{0}$

61. $\dfrac{-19}{1}$

62. $\dfrac{-9}{1}$

63. $-23 \div (-23)$

64. $-11 \div (-11)$

65. $\dfrac{40}{-2}$

66. $\dfrac{35}{-7}$

67. $9 \div (-9)$

68. $15 \div (-15)$

69. $\dfrac{-10}{-1}$

70. $\dfrac{-12}{-1}$

71. $\dfrac{-888}{37}$

72. $\dfrac{-456}{24}$

73. $\dfrac{3,000}{-100}$

74. $\dfrac{-60,000}{-1,000}$

75. Divide 8 by -2.

76. Divide -16 by -8.

Use a calculator to perform each division.

77. $\dfrac{-13,550}{25}$

78. $\dfrac{3,876}{-19}$

79. $\dfrac{27,778}{-17}$

80. $\dfrac{-168,476}{-77}$

APPLICATIONS

Use signed numbers to solve each problem.

81. LOWERING PRICES A furniture store owner reduced the price of an oak table an equal amount each week, because it was not selling. After six weeks, and a $210 reduction in price, the table was purchased. By how much was the price of the table reduced each week?

82. TEMPERATURE DROP During a five-hour period, the temperature steadily dropped 20°F. By how many degrees did the temperature change each hour?

83. SUBMARINES In a series of three equal dives, a submarine is programmed to reach a depth of 3,030 feet below the ocean surface. What signed number describes how deep each of the dives will be?

84. GRAND CANYON A mule train is to travel from a stable on the rim of the Grand Canyon to a camp on the canyon floor, approximately 5,500 feet below the rim. If the guide wants the mules to be rested after every 500 feet of descent, how many stops will be made on the trip?

85. CHEMISTRY During an experiment, a solution was steadily chilled and the times and temperatures were recorded, as shown in the illustration below. By how many degrees did the temperature of the solution change each minute?

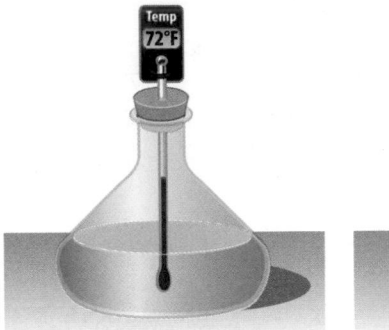 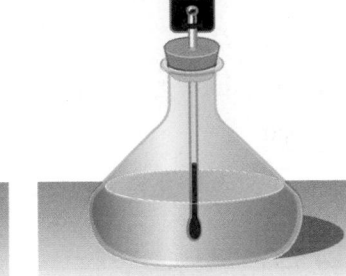

Beginning of experiment	End of experiment
8:00 A.M.	8:06 A.M.

86. OCEAN EXPLORATION The Mariana Trench is the deepest part of the world's oceans. It is located in the North Pacific Ocean near the Philippines and has a maximum depth of 36,201 feet. If a remote-controlled vessel is sent to the bottom of the trench in a series of 11 equal descents, how far will the vessel descend on each dive? (Source: marianatrench.com)

87. BASEBALL TRADES At the midway point of the season, a baseball team finds itself 12 games behind the league leader. Team management decides to trade for a talented hitter, in hopes of making up at least half of the deficit in the standings by the end of the year. Where in the league standings does management expect to finish at season's end?

88. BUDGET DEFICITS A politician proposed a two-year plan for cutting a county's $20-million budget deficit, as shown. If this plan is put into effect, how will the deficit change in two years?

	Plan	Prediction
1st year	Raise taxes, drop failing programs	Will cut deficit in half
2nd year	Search out waste and fraud	Will cut remaining deficit in half

89. MARKDOWNS The owner of a clothing store decides to reduce the price on a line of jeans that are not selling. She feels she can afford to lose $300 of projected income on these pants. By how much can she mark down each of the 20 pairs of jeans?

90. WATER STORAGE Over a week's time, engineers at a city water reservoir released enough water to lower the water level 105 feet. On average, how much did the water level change each day during this period?

91. THE STOCK MARKET On Monday, the value of Maria's 255 shares of stock was at an all-time high. By Friday, the value had fallen $4,335. What was her per-share loss that week?

92. CUTTING BUDGETS In a cost-cutting effort, a company decides to cut $5,840,000 from its annual budget. To do this, all of the company's 160 departments will have their budgets reduced by an equal amount. By how much will each department's budget be reduced?

WRITING

93. Explain why the quotient of two negative integers is positive.

94. How do the rules for multiplying integers compare with the rules for dividing integers?

95. Use a specific example to explain how multiplication can be used as a check for division.

96. Explain what it means when we say that division by 0 is undefined.

97. Explain the division rules for integers that are shown below using symbols.

$$\frac{+}{+} = + \qquad \frac{-}{-} = + \qquad \frac{-}{+} = - \qquad \frac{+}{-} = -$$

98. Explain the difference between *division of 0* and *division by 0*.

REVIEW

99. Evaluate: $5^2\left(\dfrac{2 \cdot 3^2}{6}\right)^2 - 7(2)$

100. Find the prime factorization of 210.

101. The statement $(4 + 8) + 10 = 4 + (8 + 10)$ illustrates what property?

102. Is $17 \geq 17$ a true statement?

103. Does $8 - 2 = 2 - 8$?

104. Sharif has scores of 55, 70, 80, and 75 on four mathematics tests. What is his mean (average) score?

SECTION 2.6
Order of Operations and Estimation

In this chapter, we have discussed the rules for adding, subtracting, multiplying, and dividing integers. Now we will use those rules in combination with the order of operations rule from Section 1.9 to evaluate expressions involving more than one operation.

1 Use the order of operations rule.

Recall that if we don't establish a uniform order of operations, an expression such as $2 + 3 \cdot 6$ can have more than one value. To avoid this possibility, always use the following rule for the order of operations.

Order of Operations

1. Perform all calculations within parentheses and other grouping symbols in the following order listed in Steps 2–4 below, working from the innermost pair of grouping symbols to the outermost pair.
2. Evaluate all the exponential expressions.
3. Perform all multiplications and divisions as they occur from left to right.
4. Perform all additions and subtractions as they occur from left to right.

When grouping symbols have been removed, repeat Steps 2–4 to complete the calculation.

If a fraction bar is present, evaluate the expression above the bar (called the **numerator**) and the expression below the bar (the **denominator**) separately. Then perform the division indicated by the fraction bar, if possible.

We can use this rule to evaluate expressions involving integers.

EXAMPLE 1   Evaluate: $-4(-3)^2 - (-2)$

Strategy We will scan the expression to determine what operations need to be performed. Then we will perform those operations, one at a time, following the order of operations rule.

WHY If we don't follow the correct order of operations, the expression can have more than one value.

Solution Although the expression contains parentheses, there are no calculations to perform *within* them. We begin with step 2 of the order of operations rule: Evaluate all exponential expressions.

$$-4(\mathbf{-3})^2 - (-2) = -4(\mathbf{9}) - (-2)$$ Evaluate the exponential expression: $(-3)^2 = 9$.

$$= -36 - (\mathbf{-2})$$ Do the multiplication: $-4(9) = -36$.

$$= -36 + \mathbf{2}$$ If it is helpful, use the subtraction rule: Add the opposite of -2, which is 2.

$$= -34$$ Do the addition.

Self Check 1
Evaluate: $-5(-2)^2 - (-6)$
Now Try Problem 13

Evaluate:
$4(9) + (-4)(-3)(-2)$

Now Try Problem 17

EXAMPLE 2   Evaluate: $12(3) + (-5)(-3)(-2)$

Strategy We will perform the multiplication first.

WHY There are no operations to perform within parentheses, nor are there any exponents.

Solution

$$12(3) + (-5)(-3)(-2) = 36 + (-30)$$ Working from left to right, do the multiplications.

$$= 6$$ Do the addition.

Evaluate: $45 \div (-5)3$

Now Try Problem 21

EXAMPLE 3   Evaluate: $40 \div (-4)5$

Strategy This expression contains the operations of division and multiplication. We will perform the divisions and multiplications as they occur from left to right.

WHY There are no operations to perform within parentheses, nor are there any exponents.

Solution

$$40 \div (-4)5 = -10 \cdot 5$$ Do the division first: $40 \div (-4) = -10$.

$$= -50$$ Do the multiplication.

> ***Caution!***   In Example 3, a common mistake is to forget to work from left to right and incorrectly perform the multiplication first. This produces the wrong answer, -2.
>
> $$40 \div (-4)5 = 40 \div (-20)$$
> $$= -2$$

Evaluate: $-3^2 - (-3)^2$

Now Try Problem 25

EXAMPLE 4   Evaluate: $-2^2 - (-2)^2$

Strategy There are two exponential expressions to evaluate and a subtraction to perform. We will begin with the exponential expressions.

WHY Since there are no operations to perform within parentheses, we begin with step 2 of the order of operations rule: Evaluate all exponential expressions.

Solution Recall from Section 2.4 that the values of -2^2 and $(-2)^2$ are not the same.

$$-2^2 - (-2)^2 = -4 - 4$$ Evaluate the exponential expressions: $-2^2 = -(2 \cdot 2) = -4$ and $(-2)^2 = -2(-2) = 4$.

$$= -4 + (-4)$$ If it is helpful, use the subtraction rule: Add the opposite of 4, which is -4.

$$= -8$$ Do the addition.

2 Evaluate expressions containing grouping symbols.

Recall that **parentheses** ( ), **brackets** [ ], **absolute value symbols** | |, and the **fraction bar** — are called **grouping symbols.** When evaluating expressions, we must perform all calculations within parentheses and other grouping symbols first.

EXAMPLE 5 Evaluate: $-15 + 3(-4 + 7 \cdot 2)$

Strategy We will begin by evaluating the expression $-4 + 7 \cdot 2$ that is within the parentheses. Since it contains more than one operation, we will use the order of operations rule to evaluate it. We will perform the multiplication first and then the addition.

WHY By the order of operations rule, we must perform all calculations within the parentheses first following the order listed in Steps 2–4 of the rule.

Solution

$$-15 + 3(-4 + \mathbf{7 \cdot 2}) = -15 + 3(-4 + \mathbf{14})$$ *Do the multiplication within the parentheses: $7 \cdot 2 = 14$.*

$$= -15 + 3(10)$$ *Do the addition within the parentheses: $-4 + 14 = 10$.*

$$= -15 + 30$$ *Do the multiplication: $3(10) = 30$.*

$$= 15$$ *Do the addition.*

Expressions can contain two or more pairs of grouping symbols. To evaluate the following expression, we begin within the innermost pair of grouping symbols, the parentheses. Then we work within the outermost pair, the brackets.

Innermost pair

$$67 - 5[-1 + (2 - 8)^2]$$

Outermost pair

EXAMPLE 6 Evaluate: $67 - 5[-1 + (2 - 8)^2]$

Strategy We will work within the parentheses first and then within the brackets. Within each pair of grouping symbols, we will follow the order of operations rule.

WHY We must work from the *innermost* pair of grouping symbols to the *outermost*.

Solution

$$67 - 5[-1 + (\mathbf{2 - 8})^2]$$

$$= 67 - 5[-1 + (\mathbf{-6})^2]$$ *Do the subtraction within the parentheses: $2 - 8 = -6$.*

$$= 67 - 5[-1 + 36]$$ *Evaluate the exponential expression within the brackets.*

$$= 67 - 5[35]$$ *Do the addition within the brackets: $-1 + 36 = 35$.*

$$= 67 - \mathbf{175}$$ *Do the multiplication: $5(35) = 175$.*

$$= 67 + (\mathbf{-175})$$ *If it is helpful, use the subtraction rule: Add the opposite of 175, which is -175.*

$$= -108$$ *Do the addition.*

$\begin{array}{r} \overset{2}{3}5 \\ \times\ 5 \\ \hline 175 \end{array}$

$\begin{array}{r} ^{6\,1}5 \\ 17\overset{}{5} \\ -67 \\ \hline 108 \end{array}$ ■

Success Tip Any arithmetic steps that you cannot perform in your head should be shown outside of the horizontal steps of your solution.

Self Check 7

Evaluate: $-\left[8 - \left(3^3 + \dfrac{90}{-9}\right)\right]$

Now Try Problem 37

EXAMPLE 7   Evaluate: $-\left[1 - \left(2^4 + \dfrac{66}{-6}\right)\right]$

Strategy We will work within the parentheses first and then within the brackets. Within each pair of grouping symbols, we will follow the order of operations rule.

WHY We must work from the *innermost* pair of grouping symbols to the *outermost*.

Solution

$$-\left[1 - \left(\mathbf{2^4} + \dfrac{66}{-6}\right)\right] = -\left[1 - \left(\mathbf{16} + \dfrac{66}{-6}\right)\right]$$ Evaluate the exponential expression within the parentheses: $2^4 = 16$.

$$= -\left[1 - \left(16 + (-11)\right)\right]$$ Do the division within the parentheses: $66 \div (-6) = -11$.

$$= -[1 - 5]$$ Do the addition within the parentheses: $16 + (-11) = 5$.

$$= -[-4]$$ Do the subtraction within the brackets: $1 - 5 = -4$.

$$= 4$$ The opposite of -4 is 4.

Self Check 8

Evaluate: $\dfrac{-9 + 6(-4)}{28 - (-5)^2}$

Now Try Problem 41

EXAMPLE 8   Evaluate: $\dfrac{-20 + 3(-5)}{21 - (-4)^2}$

Strategy We will evaluate the expression above and the expression below the fraction bar separately. Then we will do the indicated division, if possible.

WHY Fraction bars are grouping symbols that group the numerator and the denominator. The expression could be written $[-20 + 3(-5)] \div [21 - (-4)^2]$.

Solution

$$\dfrac{-20 + \mathbf{3(-5)}}{21 - (-4)^2} = \dfrac{-20 + (\mathbf{-15})}{21 - 16}$$ In the numerator, do the multiplication: $3(-5) = -15$. In the denominator, evaluate the exponential expression: $(-4)^2 = 16$.

$$= \dfrac{-35}{5}$$ In the numerator, add: $-20 + (-15) = -35$. In the denominator, subtract: $21 - 16 = 5$.

$$= -7$$ Do the division indicated by the fraction bar.

3 Evaluate expressions containing absolute values.

Earlier in this chapter, we found the absolute values of integers. For example, recall that $|-3| = 3$ and $|10| = 10$. We use the order of operations rule to evaluate more complicated expressions that contain absolute values.

Self Check 9

Evaluate each expression:

a. $|(-6)(5)|$

b. $|-3 + 96|$

Now Try Problem 45

EXAMPLE 9   Evaluate each expression: **a.** $|-4(3)|$ **b.** $|-6 + 1|$

Strategy We will perform the calculation within the absolute value symbols first. Then we will find the absolute value of the result.

WHY Absolute value symbols are grouping symbols, and by the order of operations rule, all calculations within grouping symbols must be performed first.

Solution

a. $|-4(3)| = |-12|$ Do the multiplication within the absolute value symbol: $-4(3) = -12$.

$$= 12$$ Find the absolute value of -12.

b. $|-6 + 1| = |-5|$ Do the addition within the absolute value symbol: $-6 + 1 = -5$.

$$= 5$$ Find the absolute value of -5.

> **The Language of Algebra**   Multiplication is indicated when a number is outside and next to an absolute value symbol. For example,
>
> $$8 - 4|-6 - 2| \quad \text{means} \quad 8 - 4 \cdot |-6 - 2|$$

EXAMPLE 10   Evaluate: $8 - 4|-6 - 2|$

Strategy   The absolute value bars are grouping symbols. We will perform the subtraction within them first.

WHY   By the order of operations rule, we must perform all calculations within parentheses and other grouping symbols (such as absolute value bars) first.

Solution

$8 - 4|-6 - 2| = 8 - 4|-6 + (-2)|$ If it is helpful, use the subtraction rule within the absolute value symbol: Add the opposite of 2, which is −2.

$\qquad\qquad = 8 - 4|-8|$ Do the addition within the absolute value symbol: −6 + (−2) = −8.

$\qquad\qquad = 8 - 4(8)$ Find the absolute value: |−8| = 8.

$\qquad\qquad = 8 - 32$ Do the multiplication: 4(8) = 32.

$\qquad\qquad = 8 + (-32)$ If it is helpful, use the subtraction rule: Add the opposite of 32, which is −32.

$\qquad\qquad = -24$ Do the addition.

$$\begin{array}{r} \overset{2}{\cancel{3}}\overset{1}{2} \\ -8 \\ \hline 24 \end{array}$$

Self Check 10

Evaluate: $7 - 5|-1 - 6|$

Now Try Problem 49

4   Estimate the value of an expression.

Recall that the idea behind estimation is to simplify calculations by using rounded numbers that are close to the actual values in the problem. When an exact answer is not necessary and a quick approximation will do, we can use estimation.

EXAMPLE 11   *The Stock Market*

The change in the Dow Jones Industrial Average is announced at the end of each trading day to give a general picture of how the stock market is performing. A positive change means a good performance, while a negative change indicates a poor performance. The week of October 13–17, 2008, had some record changes, as shown below. Round each number to the nearest ten and estimate the net gain or loss of points in the Dow that week.

EIGHTFISH/Getty Images

Self Check 11

THE STOCK MARKET For the week of December 15–19, 2008, the Dow Jones Industrial Average performance was as follows, Monday: −63, Tuesday: +358, Wednesday: −98, Thursday: −219, Friday: −27. Round each number to the nearest ten and estimate the net gain or loss of points in the Dow for that week. (Source: finance.yahoo.com)

Now Try Problems 53 and 97

Strategy   To estimate the net gain or loss, we will round each number to the nearest ten and *add* the approximations.

+936	−78	−733	+402	−123
Monday Oct. 13, 2008 (largest 1-day increase)	Tuesday Oct. 14, 2008	Wednesday Oct. 15, 2008 (second-largest 1-day decline)	Thursday Oct. 16, 2008 (tenth-largest 1-day increase)	Friday Oct. 17, 2008

Source: finance.yahoo.com

WHY The phrase *net gain or loss* refers to what remains after all of the losses and gains have been combined (added).

Solution To nearest ten:

936 rounds to 940   −78 rounds to −80   −733 rounds to −730
402 rounds to 400   −123 rounds to −120

To estimate the net gain or loss for the week, we add the rounded numbers.

$$\mathbf{940} + (\mathbf{-80}) + (\mathbf{-730}) + \mathbf{400} + (\mathbf{-120})$$

$$= \mathbf{1{,}340} + (\mathbf{-930}) \quad \text{Add the positives and the negatives separately.}$$

$$= 410 \quad\quad\quad\quad\quad\quad \text{Do the addition.}$$

$$\begin{array}{r} \overset{13}{1{,}\cancel{3}40} \\ -930 \\ \hline 410 \end{array}$$

The positive result means there was a net gain that week of approximately 410 points in the Dow.

ANSWERS TO SELF CHECKS

1. −14 **2.** 12 **3.** −27 **4.** −18 **5.** 48 **6.** 25 **7.** 9 **8.** −11 **9. a.** 30 **b.** 93
10. −28 **11.** There was a net loss that week of approximately 50 points.

SECTION 2.6 STUDY SET

VOCABULARY

Fill in the blanks.

1. To evaluate expressions that contain more than one operation, we use the _____ of operations rule.

2. Absolute value symbols, parentheses, and brackets are types of _____ symbols.

3. In the expression $-9 + 2[-5 - 6(-3 - 1)]$, the parentheses are the _____ most grouping symbols and the brackets are the _____ most grouping symbols.

4. In situations where an exact answer is not needed, an approximation or _____ is a quick way of obtaining a rough idea of the size of the actual answer.

CONCEPTS

5. List the operations in the order in which they should be performed to evaluate each expression. *You do not have to evaluate the expression.*

 a. $5(-2)^2 - 1$

 b. $15 - 3 + (-5 \cdot 2)^3$

 c. $4 + 2(-7 - 3)$

 d. $-2 \cdot 3^2$

6. Consider the expression $\dfrac{5 + 5(7)}{2 + (4 - 8)}$. In the numerator, what operation should be performed first? In the denominator, what operation should be performed first?

NOTATION

7. Give the name of each grouping symbol: ( ), [ ], | |, and —.

8. What operation is indicated?

$$-2 + 9\overset{\downarrow}{|}8 - (-2 + 4)|$$

Complete each solution to evaluate the expression.

9. $-8 - 5(-2)^2 = -8 - 5(\boxed{})$

$$= -8 - \boxed{}$$

$$= -8 + (\boxed{})$$

$$= \boxed{}$$

10. $2 + (5 - 6 \cdot 2) = 2 + (5 - \boxed{})$

$$= 2 + [5 + (\boxed{})]$$

$$= 2 + (\boxed{})$$

$$= \boxed{}$$

11. $-9 + 5[-4 \cdot 2 + 7] = -9 + 5[\boxed{} + 7]$

$$= -9 + 5[\boxed{}]$$

$$= -9 + (\boxed{})$$

$$= \boxed{}$$

12. $\dfrac{|-9 + (-3)|}{9 - 6} = \dfrac{|\boxed{}|}{3}$

$$= \dfrac{\boxed{}}{3}$$

$$= \boxed{}$$

GUIDED PRACTICE

Evaluate each expression. See Example 1.

13. $-2(-3)^2 - (-8)$ **14.** $-6(-2)^2 - (-9)$

15. $-5(-4)^2 - (-18)$ **16.** $-3(-5)^2 - (-24)$

Evaluate each expression. See Example 2.

17. $9(7) + (-6)(-2)(-4)$

18. $9(8) + (-2)(-5)(-7)$

19. $8(6) + (-2)(-9)(-2)$

20. $7(8) + (-3)(-6)(-2)$

Evaluate each expression. See Example 3.

21. $30 \div (-5)2$ **22.** $50 \div (-2)5$

23. $60 \div (-3)4$ **24.** $120 \div (-4)3$

Evaluate each expression. See Example 4.

25. $-6^2 - (-6)^2$ **26.** $-7^2 - (-7)^2$

27. $-10^2 - (-10)^2$ **28.** $-8^2 - (-8)^2$

Evaluate each expression. See Example 5.

29. $-14 + 2(-9 + 6 \cdot 3)$

30. $-18 + 3(-10 + 3 \cdot 7)$

31. $-23 + 3(-15 + 8 \cdot 4)$

32. $-31 + 6(-12 + 5 \cdot 4)$

Evaluate each expression. See Example 6.

33. $77 - 2[-6 + (3 - 9)^2]$

34. $84 - 3[-7 + (5 - 8)^2]$

35. $99 - 4[-9 + (6 - 10)^2]$

36. $67 - 5[-6 + (4 - 7)^2]$

Evaluate each expression. See Example 7.

37. $-\left[4 - \left(3^3 + \dfrac{22}{-11}\right)\right]$

38. $-\left[1 - \left(2^3 + \dfrac{40}{-20}\right)\right]$

39. $-\left[50 - \left(5^3 + \dfrac{50}{-2}\right)\right]$

40. $-\left[12 - \left(2^5 + \dfrac{40}{-4}\right)\right]$

Evaluate each expression. See Example 8.

41. $\dfrac{-24 + 3(-4)}{42 - (-6)^2}$ **42.** $\dfrac{-18 + 6(-2)}{52 - (-7)^2}$

43. $\dfrac{-38 + 11(-2)}{69 - (-8)^2}$ **44.** $\dfrac{-36 + 8(-2)}{85 - (-9)^2}$

Evaluate each expression. See Example 9.

45. a. $|-6(2)|$ **b.** $|-12 + 7|$

46. a. $|-4(9)|$ **b.** $|-15 + 6|$

47. a. $|15(-4)|$ **b.** $|16 + (-30)|$

48. a. $|12(-5)|$ **b.** $|47 + (-70)|$

Evaluate each expression. See Example 10.

49. $16 - 6|-2 - 1|$ **50.** $15 - 6|-3 - 1|$

51. $17 - 2|-6 - 4|$ **52.** $21 - 9| - 3 - 1|$

Estimate the value of each expression by rounding each number to the nearest ten. See Example 11.

53. $-379 + (-13) + 287 + (-671)$

54. $-363 + (-781) + 594 + (-42)$

Estimate the value of each expression by rounding each number to the nearest hundred. See Example 11.

55. $-3,887 + (-5,806) + 4,701$

56. $-5,684 + (-2,270) + 3,404 + 2,689$

TRY IT YOURSELF

Evaluate each expression.

57. $(-3)^2 - 4^2$ **58.** $-7 + 4 \cdot 5$

59. $3^2 - 4(-2)(-1)$ **60.** $2^3 - 3^3$

61. $|-3 \cdot 4 + (-5)|$ **62.** $|-8 \cdot 5 - 2 \cdot 5|$

63. $(2 - 5)(5 + 2)$ **64.** $-3(2)^2 4$

65. $6 + \dfrac{25}{-5} + 6 \cdot 3$ **66.** $-5 - \dfrac{24}{6} + 8(-2)$

67. $\dfrac{-6 - 2^3}{-2 - (-4)}$ **68.** $\dfrac{-6 - 6}{-2 - 2}$

69. $-12 \div (-2)2$ **70.** $-60(-2) \div 3$

71. $-16 - 4 \div (-2)$ **72.** $-24 + 4 \div (-2)$

73. $-|2 \cdot 7 - (-5)^2|$ **74.** $-|8 \div (-2) - 5|$

75. $|-4 - (-6)|$ **76.** $|-2 + 6 - 5|$

77. $(7 - 5)^2 - (1 - 4)^2$ **78.** $5^2 - (-9 - 3)$

79. $-1(2^2 - 2 + 1^2)$ **80.** $(-7 - 4)^2 - (-1)$

81. $\dfrac{-5 - 5}{1^4 + 1^5}$ **82.** $\dfrac{-7 - (-3)}{2 - 2^2}$

83. $-50 - 2(-3)^3(4)$ **84.** $(-2)^3 - (-3)(-2)(4)$

85. $-6^2 + 6^2$

86. $-9^2 + 9^2$

87. $3\left(\dfrac{-18}{3}\right) - 2(-2)$

88. $2\left(\dfrac{-12}{3}\right) + 3(-5)$

89. $2|1 - 8| \cdot |-8|$

90. $2(5) - 6(|-3|)^2$

91. $\dfrac{2 + 3[5 - (1 - 10)]}{|2(-8 + 2) + 10|}$

92. $\dfrac{11 + (-2 \cdot 2 + 3)}{|15 + (-3 \cdot 4 - 8)|}$

93. $-2 + |6 - 4^2|$

94. $-3 - 4|6 - 7|$

95. $\dfrac{-4(-5) - 2}{3 - 3^2}$

96. $\dfrac{(-6)^2 - 1}{-(2^2 - 3)}$

APPLICATIONS

97. THE STOCK MARKET For the week of January 5–9, 2009, the Dow Jones Industrial Average performance was as follows, Monday: -74, Tuesday: $+61$, Wednesday: -227, Thursday: -27, Friday: -129. Round each number to the nearest ten and estimate the net gain or loss of points in the Dow for that week. (Source: finance.yahoo.com)

98. STOCK MARKET RECORDS Refer to the tables below. Round each of the record Dow Jones point gains and losses to the nearest hundred and then add all ten of them. There is an interesting result. What is it?

5 Greatest Dow Jones Daily Point Gains

Rank	Date	Gain
1	10/13/2008	+936
2	10/28/2008	+889
3	11/13/2008	+553
4	11/21/2008	+494
5	9/30/2008	+485

5 Greatest Dow Jones Daily Point Losses

Rank	Date	Loss
1	9/29/2008	−778
2	10/15/2008	−733
3	12/1/2008	−680
4	10/9/2008	−679
5	10/22/2008	−514

(Source: Dow Jones Indexes)

99. TESTING In an effort to discourage her students from guessing on multiple-choice tests, a professor uses the grading scale shown in the table in the next column. If unsure of an answer, a student does best to skip the question, because incorrect responses are penalized very heavily. Find the test score of a student who gets 12 correct and 3 wrong and leaves 5 questions blank.

Response	Value
Correct	+3
Incorrect	−4
Left blank	−1

100. SPREADSHEETS The table shows the data from a chemistry experiment in spreadsheet form. To obtain a result, the chemist needs to add the values in row 1, double that sum, and then divide that number by the smallest value in column C. What is the final result of these calculations?

	A	B	C	D
1	12	−5	6	−2
2	15	4	5	−4
3	6	4	−2	8

101. BUSINESS TAKEOVERS Six investors are taking over a poorly managed company, but first they must repay the debt that the company built up over the past four quarters. (See the graph below.) If the investors plan equal ownership, how much of the company's total debt is each investor responsible for?

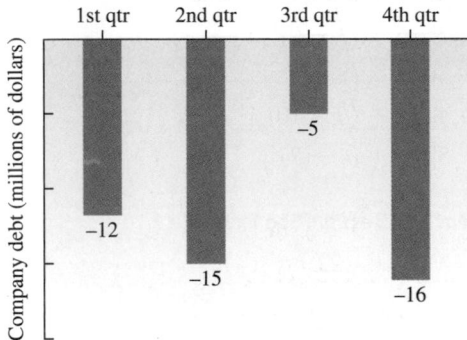

102. DECLINING ENROLLMENT Find the drop in enrollment for each Mesa, Arizona, high school shown in the table below. Express each drop as a negative number. Then find the mean (average) drop in enrollment for these four schools.

High school	2008 enrollment	2009 enrollment	Drop
Mesa	2,683	2,573	
Red Mountain	2,754	2,662	
Skyline	1,948	1,875	
Westwood	2,257	2,192	

(Source: azcentral.com)

103. THE FEDERAL BUDGET See the graph below. Suppose you were hired to write a speech for a politician who wanted to highlight the improvement in the federal government's finances during the 1990s. Would it be better for the politician to talk about the mean (average) budget deficit/surplus for the last half of the decade, or for the last four years of that decade? Explain your reasoning.

U.S. Budget Deficit/Surplus
($ billions)

Deficit	Year	Surplus
−164	1995	
−107	1996	
−22	1997	
	1998	+70
	1999	+123

104. SCOUTING REPORTS The illustration below shows a football coach how successful his opponent was running a "28 pitch" the last time the two teams met. What was the opponent's mean (average) gain with this play?

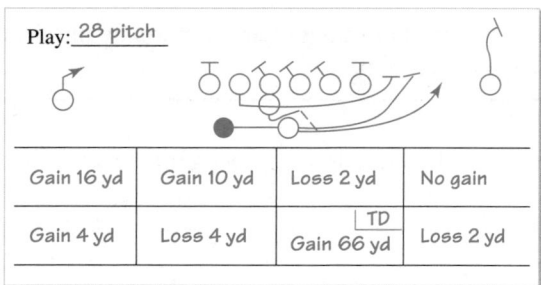

Play: 28 pitch			
Gain 16 yd	Gain 10 yd	Loss 2 yd	No gain
Gain 4 yd	Loss 4 yd	TD Gain 66 yd	Loss 2 yd

105. ESTIMATION Quickly determine a reasonable estimate of the exact answer in each of the following situations.

  a. A scuba diver, swimming at a depth of 34 feet below sea level, spots a sunken ship beneath him. He dives down another 57 feet to reach it. What is the depth of the sunken ship?

  b. A dental hygiene company offers a money-back guarantee on its tooth whitener kit. When the kit is returned by a dissatisfied customer, the company loses the $11 it cost to produce it, because it cannot be resold. How much money has the company lost because of this return policy if 56 kits have been mailed back by customers?

  c. A tram line makes a 7,891-foot descent from a mountaintop in 18 equal stages. How much does it descend in each stage?

106. ESTIMATION Quickly determine a reasonable estimate of the exact answer in each of the following situations.

  a. A submarine, cruising at a depth of −175 feet, descends another 605 feet. What is the depth of the submarine?

  b. A married couple has assets that total $840,756 and debts that total $265,789. What is their net worth?

  c. According to pokerlistings.com, the top five online poker losses as of January 2009 were $52,256; $52,235; $31,545; $28,117; and $27,475. Find the total amount lost.

WRITING

107. When evaluating expressions, why is the order of operations rule necessary?

108. In the rules for the order of operations, what does the phrase *as they occur from left to right* mean?

109. Explain the error in each evaluation below.

  a. $80 \div (-2)4 = 80 \div (-8)$
  $$= -10$$

  b. $-1 + 8|4 - 9| = -1 + 8|-5|$
  $$= 7|-5|$$
  $$= 35$$

110. Describe a situation in daily life where you use estimation.

REVIEW

111. On the number line, what number is

  a. 4 units to the right of −7?

  b. 6 units to the left of 2?

112. Is 834,540 divisible by: **a.** 2   **b.** 3   **c.** 4   **d.** 5   **e.** 6   **f.** 9   **g.** 10

113. ELEVATORS An elevator has a weight capacity of 1,000 pounds. Seven people, with an average weight of 140 pounds, are in it. Is it overloaded?

114. **a.** Find the LCM of 12 and 44.

  b. Find the GCF of 12 and 44.

Objectives

1 Use one property of equality to solve equations.

2 Solve equations involving −*x*.

3 Use more than one property of equality to solve equations.

4 Use equations to solve application problems involving integers.

SECTION 2.7
Solving Equations That Involve Integers

In this section, we revisit the topic of solving equations. The equations that we will solve involve negative numbers, and some of the solutions are negative numbers as well.

1 Use one property of equality to solve equations.

Recall that **to solve an equation** means to find all the values of the variable that make the equation true. In Chapter 1, we used the following properties of equality to solve equations involving whole numbers.

Properties of Equality

Addition Property of Equality: Adding the same number to both sides of an equation does not change its solution.

Subtraction Property of Equality: Subtracting the same number from both sides of an equation does not change its solution.

Multiplication Property of Equality: Multiplying both sides of an equation by the same nonzero number does not change its solution.

Division Property of Equality: Dividing both sides of an equation by the same nonzero number does not change its solution.

These properties are also used to solve equations involving integers.

Self Check 1

Solve $x + (-3) = -12$ and check the result.

Now Try Problem 17

EXAMPLE 1   Solve: $x + (-8) = -10$

Strategy We will use a property of equality to isolate the variable on one side of the equation.

WHY To solve the original equation, we want to find a simpler equivalent equation of the form $x = $ **a number**, whose solution is obvious.

Solution We will use the addition property of equality to isolate x on the left side of the equation. We can undo the addition of -8 by adding 8 to both sides.

$$x + (-8) = -10 \quad \text{This is the equation to solve.}$$
$$x + (-8) + 8 = -10 + 8 \quad \text{Add 8 to both sides.}$$
$$x + 0 = -2 \quad \text{On the left side, the sum of a number and its opposite is zero: } (-8) + 8 = 0. \text{ On the right side add: } -10 + 8 = -2.$$
$$x = -2 \quad \text{On the left side, the sum of any number and 0 is that number: } x + 0 = x.$$

To check, we substiute -2 for x in the original equation and simplify.

$$x + (-8) = -10 \quad \text{This is the original equation.}$$
$$-2 + (-8) \stackrel{?}{=} -10 \quad \text{Substitute } -2 \text{ for x.}$$
$$-10 = -10 \quad \text{On the left side, do the addition.}$$

Since the resulting statement $-10 = -10$ is true, -2 is the solution of $x + (-8) = -10$. ∎

Success Tip From Example 1, we see that to undo addition, we can *add the opposite* of the number that is added to the variable.

EXAMPLE 2   Solve: $t + 16 = -8$

Strategy We will use a property of equality to isolate the variable on one side of the equation.

WHY To solve the original equation, we want to find a simpler equation of the form $t = $ **a number**, whose solution is obvious.

Solution We will use the subtraction property of equality to isolate t on the left side of the equation. We can undo the addition of 16 by subtracting 16 from both sides.

$$t + 16 = -8 \qquad \text{This is the equation to solve.}$$

$$t + 16 - \mathbf{16} = -8 - \mathbf{16} \qquad \text{Subtract 16 from both sides.}$$

$$t + 0 = -8 + (-16) \qquad \begin{array}{l} \text{On the left side, } 16 - 16 = 0. \text{ On the} \\ \text{right side, write the subtraction as} \\ \text{addition of the opposite.} \end{array}$$

$$\begin{array}{r} \overset{1}{16} \\ +\ 8 \\ \hline 24 \end{array}$$

$$t = -24 \qquad \begin{array}{l} \text{On the left side, the sum of any number and 0 is that} \\ \text{number: } t + 0 = t. \text{ On the right side, do the addition.} \end{array}$$

Check:

$$t + 16 = -8 \qquad \text{This is the original equation.}$$

$$-24 + 16 \overset{?}{=} -8 \qquad \text{Substitute } -24 \text{ for } t.$$

$$-8 = -8 \qquad \text{On the left side, do the addition.}$$

$$\begin{array}{r} \overset{1\,14}{2\!\!\!/4} \\ -16 \\ \hline 8 \end{array}$$

Since the resulting statement $-8 = -8$ is true, -24 is the solution of $t + 16 = -8$. ∎

Self Check 2

Solve $c + 4 = -3$ and check the result.

Now Try Problem 21

EXAMPLE 3   Solve: $-3 + 7 = h + 11(-2)$

Strategy We will begin by performing the addition on the left side of the equation and the multiplication on the right side.

WHY The expressions on each side of the equation should be simplified before we use any properties of equality.

Solution

$$-3 + 7 = h + 11(-2) \qquad \text{This is the equation to solve.}$$

$$4 = h + (-22) \qquad \begin{array}{l} \text{On the left side, do the addition: } -3 + 7 = 4. \text{ On the} \\ \text{right side, do the multiplication: } 11(-2) = -22. \end{array}$$

Now we use the addition property of equality to isolate h on the right side of the equation.

$$4 + \mathbf{22} = h + (-22) + \mathbf{22} \qquad \begin{array}{l} \text{To isolate } h, \text{ undo the addition of } -22 \text{ by adding 22} \\ \text{to both sides.} \end{array}$$

$$26 = h \qquad \text{Simplify each side: } 4 + 22 = 26 \text{ and } (-22) + 22 = 0.$$

Check:

$$-3 + 7 = h + 11(-2) \qquad \text{This is the original equation.}$$

$$-3 + 7 \overset{?}{=} 26 + 11(-2) \qquad \text{Substitute 26 for } h.$$

$$4 \overset{?}{=} 26 + (-22) \qquad \text{On the left side, add. On the right side, multiply.}$$

$$4 = 4 \qquad \text{On the right side, do the addition.}$$

Since the resulting statement $4 = 4$ is true, 26 is the solution.

Self Check 3

Solve $-2 + 8 = y + 3(-4)$ and check the result.

Now Try Problem 25

Solve each equation and check the result:

a. $-7k = 28$

b. $-40 = -8k$

Now Try Problem 29

EXAMPLE 4   Solve: **a.** $-3y = 15$ **b.** $-16 = -4y$

Strategy We will use a property of equality to isolate the variable on one side of the equation.

WHY To solve each of the original equations, we want to find a simpler equivalent equation of the form $y = $ **a number** or **a number** $= y$, whose solution is obvious.

Solution

a. Recall that $-3y$ indicates multiplication: $-3 \cdot y$. We must undo the multiplication of y by -3. To do this, we use the division property of equality and divide both sides of the equation by -3.

$$-3y = 15 \qquad \text{This is the equation to solve.}$$

$$\frac{-3y}{-3} = \frac{15}{-3} \qquad \text{Divide both sides by } -3.$$

$$y = -5 \qquad \begin{array}{l}\text{On the left side, } -3 \text{ times } y, \text{ divided by } -3, \text{ is } y. \text{ On the right side, do}\\ \text{the division: } 15 \div (-3) = -5.\end{array}$$

Check:

$$-3y = 15 \qquad \text{This is the original equation.}$$

$$-3(\mathbf{-5}) \overset{?}{=} 15 \qquad \text{Substitute } -5 \text{ for } y.$$

$$15 = 15 \qquad \text{On the left side, do the multiplication: } -3(-5) = 15.$$

Since the resulting statement $15 = 15$ is true, -5 is the solution of $-3y = 15$.

b. $-16 = -4y$ This is the equation to solve.

$$\frac{-16}{-4} = \frac{-4y}{-4} \qquad \text{To isolate } y, \text{ undo the multiplication by } -4, \text{ by dividing both sides by } -4.$$

$$4 = y \qquad \begin{array}{l}\text{On the left side, do the division: } -16 \div (-4) = 4.\\ \text{On the right side, } -4 \text{ times } y, \text{ divided by } -4 \text{ is } y.\end{array}$$

Check the result to verify that 4 is the solution.

Solve $\dfrac{t}{-3} = 4$ and check the result.

Now Try Problem 33

EXAMPLE 5   Solve: $\dfrac{x}{-5} = -10$

Strategy We will use a property of equality to isolate the variable on one side of the equation.

WHY To solve the original equation, we want to find a simpler equivalent equation of the form $x = $ **a number**, whose solution is obvious.

Solution In this equation, x is being divided by -5. To undo this division, we use the multiplication property of equality and multiply both sides of the equation by -5.

$$\frac{x}{-5} = -10 \qquad \text{This is the equation to solve.}$$

$$-5\left(\frac{x}{-5}\right) = -5(-10) \qquad \text{Multiply both sides by } -5.$$

$$x = 50 \qquad \begin{array}{l}\text{On the left side, when } x \text{ is divided by } -5 \text{ and then}\\ \text{multiplied by } -5, \text{ the result is } x. \text{ On the right side,}\\ \text{do the multiplication: } -5(-10) = 50.\end{array}$$

Check:

$$\frac{x}{-5} = -10 \qquad \text{This is the original equation.}$$

$$\frac{50}{-5} \overset{?}{=} -10 \qquad \text{Substitute 50 for x.}$$

$$-10 = -10 \qquad \text{On the left side, do the division: } 50 \div (-5) = -10.$$

Since the resulting statement $-10 = -10$ is true, 50 is the solution of $\frac{x}{-5} = -10$. ■

2 Solve equations involving −x.

Recall from Chapter 1 that we don't need to write a multiplication symbol when multiplying a variable by a number. For example,

$$5a \quad \text{means} \quad 5 \cdot a, \qquad -9m \quad \text{means} \quad -9 \cdot m, \qquad \text{and} \qquad -1x \quad \text{means} \quad -1 \cdot x$$

A simpler way to write the last expression, $-1x$, is $-x$. When we examine what each notation means, it becomes clear why this is true.

$$-1x \quad = \quad -x$$

This means multiply the value of x by −1.   This means find the opposite of the value of x.

We can use the fact that $-1x = -x$ to solve equations that involve the expression $-x$.

EXAMPLE 6   Solve: $-x = 3$

Strategy The variable x is not isolated, because there is a $-$ sign in front of it. Since the term $-x$ has an understood coefficient of -1, the equation can be written as $-1x = 3$. We need to select a property of equality and use it to isolate the variable on one side of the equation.

WHY To find the solution of the original equation, we want to find a simpler equivalent equation of the form $x = \textbf{a number}$, whose solution is obvious.

Solution To isolate x, we can either multiply or divide both sides by -1.

Multiply both sides by −1:

$$-x = 3 \qquad \text{The equation to solve}$$

$$-1x = 3 \qquad \text{Write } -x \text{ as } -1x.$$

$$(-1)(-1x) = (-1)3$$

$$1x = -3 \qquad \text{On the left, } (-1)(-1) = 1.$$

$$x = -3 \qquad 1x = x$$

Divide both sides by −1:

$$-x = 3 \qquad \text{The equation to solve}$$

$$-1x = 3 \qquad \text{Write } -x \text{ as } -1x.$$

$$\frac{-1x}{-1} = \frac{3}{-1}$$

$$1x = -3 \qquad \text{On the left side, } \frac{-1}{-1} = 1.$$

$$x = -3 \qquad 1x = x$$

Check: $-x = 3$ This is the original equation.

$$-(-3) \overset{?}{=} 3 \qquad \text{Substitute } -3 \text{ for x.}$$

$$3 = 3 \qquad \text{On the left side, the opposite of } -3 \text{ is 3.}$$

Since the statement $3 = 3$ is true, -3 is the solution of $-x = 3$. ■

Self Check 6

Solve $-h = -17$ and check the result.

Now Try Problem 37

3   Use more than one property of equality to solve equations.

In the previous examples, each equation was solved by using a single property of equality. Sometimes we must use two (or more) properties of equality to solve more complicated equations. For example, on the left side of $2x + 6 = 10$, the variable x is multiplied by 2, and then 6 is added to that product. To solve the equation, we use the order of operations rule in reverse. First, we isolate the variable term $2x$ by undoing the addition of 6. Then isolate the variable x by undoing the multiplication by 2.

$$2x + 6 = 10 \qquad \text{This is the equation to solve.}$$

$$2x + 6 - 6 = 10 - 6 \qquad \text{To undo the addition of 6, subtract 6 from both sides.}$$

$$2x = 4 \qquad \text{Do the subtractions.}$$

$$\frac{2x}{2} = \frac{4}{2} \qquad \text{To undo the multiplication by 2, divide both sides by 2.}$$

$$x = 2 \qquad \text{Do the division.}$$

The solution is 2.

The Language of Algebra   In the example above, we subtracted 6 from both sides to *isolate the variable term*, $2x$. Then we divided both sides by 2 to *isolate the variable*, x.

$$\underset{\underset{\text{The variable term}}{\uparrow}}{2x} + 6 = 10$$

Self Check 7

Solve $-6b - 1 = 11$ and check the result.

Now Try   Problem 41

EXAMPLE 7   Solve: $-4x - 5 = 15$

Strategy   First we will use a property of equality to isolate the *variable term* on one side of the equation. Then we will use a second property of equality to isolate the *variable* itself.

WHY   To solve the original equation, we want to find a simpler equivalent equation of the form $x = \textbf{a number}$, whose solution is obvious.

Solution   On the left side of the equation, x is multiplied by -4, and then 5 is subtracted from that product. To solve the equation, we undo the operations in the opposite order.

- To isolate the variable term, $-4x$, we add 5 to both sides to undo the subtraction of 5.

- To isolate the variable, x, we divide both sides by -4 to undo the multiplication by -4.

$$-4x - 5 = 15 \qquad \text{This is the equation to solve.}$$

$$-4x - 5 + 5 = 15 + 5 \qquad \text{Use the addition property of equality: Add 5 to both sides to isolate } -4x.$$

$$-4x = 20 \qquad \text{Do the additions: } -5 + 5 = 0 \text{ and } 15 + 5 = 20. \text{ Now we want to isolate } x.$$

$$\frac{-4x}{-4} = \frac{20}{-4} \qquad \text{Use the division property of equality: Divide both sides by } -4 \text{ to isolate } x.$$

$$x = -5 \qquad \text{Do the division.}$$

Check:

$$-4x - 5 = 15 \qquad \text{This is the original equation.}$$
$$-4(\mathbf{-5}) - 5 \overset{?}{=} 15 \qquad \text{Substitute } -5 \text{ for } x.$$
$$20 - 5 \overset{?}{=} 15 \qquad \text{On the left side, do the multiplication: } -4(-5) = 20.$$
$$15 = 15 \qquad \text{On the left side, do the subtraction.}$$

Since the resulting statement $15 = 15$ is true, -5 is the solution of $-4x - 5 = 15$. ∎

EXAMPLE 8   Solve: $-1 = 2 - 3p$

Strategy First we will use a property of equality to isolate the *variable term* on one side of the equation. Then we will use a second property of equality to isolate the *variable* itself.

WHY To solve the original equation, we want to find a simpler equivalent equation of the form **a number** $= p$, whose solution is obvious.

Solution On the right side of the equation, p is multiplied by -3, and then 2 is added to that product. Think of $2 - 3p$ as $2 + (-3p)$.

- To isolate the variable term, $-3p$, we subtract 2 from both sides to undo the addition of 2.

- To isolate the variable, p, we divide both sides by -3 to undo the multiplication by -3.

$$-1 = 2 \boxed{-\ 3p} \qquad \text{This is the equation to solve.}$$
$$-1 \mathbf{-2} = 2 - 3p \mathbf{-2} \qquad \begin{array}{l}\text{Use the subtraction property of equality: Subtract 2}\\ \text{from both sides to isolate } -3p.\end{array}$$
$$-3 = \boxed{-3p} \qquad \begin{array}{l}\text{On the right side, do the subtraction: } 2 - 2 = 0.\\ \text{On the left side do the subtraction: } -1 - 2 = -3.\end{array}$$
$$\frac{-3}{\mathbf{-3}} = \frac{-3p}{\mathbf{-3}} \qquad \begin{array}{l}\text{Use the division property of equality: Divide both sides}\\ \text{by } -3 \text{ to isolate } p.\end{array}$$
$$1 = p \qquad \text{Do the division.}$$

Check this result in the *original* equation to verify that 1 is the solution. ∎

> **Caution!** In Example 8, a common error is to forget to write the $-$ symbol in front of $3p$ after subtracting 2 from both sides of the equation.
>
> $$-1 \mathbf{-2} = 2 - 3p \mathbf{-2}$$
> $$-3 = -3p$$
> $$\uparrow$$
> Don't forget to write the $-$ symbol.

EXAMPLE 9   Solve: $\dfrac{y}{-2} - 6 = -43$

Strategy First we will use a property of equality to isolate the *variable term* on one side of the equation. Then we will use a second property of equality to isolate the *variable* itself.

WHY To solve the original equation, we want to find a simpler equivalent equation of the form $y = $ **a number**, whose solution is obvious.

Self Check 8

Solve $-34 = 6 - 8k$ and check the result.

Now Try Problem 45

Self Check 9

Solve $\dfrac{m}{-8} - 10 = -74$ and check the result.

Now Try Problem 49

Solution On the left side of the equation, y is divided by -2, and 6 is subtracted from the quotient.

- To isolate the variable term, $\dfrac{y}{-2}$, we add 6 to both sides to undo the subtraction of 6.

- To isolate the variable, y, we multiply both sides by -2 to undo the division by -2.

$$\boxed{\dfrac{y}{-2}} - 6 = -43 \qquad \text{This is the equation to solve.}$$

$$\dfrac{y}{-2} - 6 + \mathbf{6} = -43 + \mathbf{6} \qquad \begin{array}{l}\text{Use the addition property of equality:}\\ \text{Add 6 to both sides to isolate } \dfrac{y}{-2}.\end{array}$$

$$\boxed{\dfrac{y}{-2}} = -37 \qquad \begin{array}{l}\text{Do the addition: } -6 + 6 = 0 \text{ and}\\ -43 + 6 = -37.\end{array}$$

$$\mathbf{-2}\left(\dfrac{y}{-2}\right) = \mathbf{-2}(-37) \qquad \begin{array}{l}\text{Use the multiplication property of equality:}\\ \text{Multiply both sides by } -2 \text{ to isolate } y.\end{array}$$

$$y = 74 \qquad \text{Do the multiplication.}$$

$$\begin{array}{r} \overset{3\;13}{\cancel{4}\cancel{3}} \\ -6 \\ \hline 37 \end{array}$$

$$\begin{array}{r} \overset{1}{3}7 \\ \times 2 \\ \hline 74 \end{array}$$

Check:

$$\dfrac{y}{-2} - 6 = -43 \qquad \text{This is the original equation.}$$

$$\dfrac{74}{-2} - 6 \overset{?}{=} -43 \qquad \text{Substitute 74 for } y.$$

$$-37 - 6 \overset{?}{=} -43 \qquad \text{On the left side, do the division: } 74 \div (-2) = -37.$$

$$-43 = -43 \qquad \text{On the left side, do the subtraction.}$$

$$\begin{array}{r} 37 \\ 2\overline{)74} \\ -6 \\ \hline 14 \\ -14 \\ \hline 0 \end{array}$$

$$\begin{array}{r} \overset{1}{3}7 \\ + 6 \\ \hline 43 \end{array}$$

Since the resulting statement $-43 = -43$ is true, 74 is the solution. ∎

4 Use equations to solve application problems involving integers.

In Chapter 1, we used the concepts of variable and equation to solve application problems involving whole numbers. We will now use a similar approach to solve problems involving integers. Like Chapter 1, we will follow the five-step problem-solving strategy of analyze, form, solve, state, and check.

> **The Language of Algebra** As you read the application problems, watch for the following words and phrases. They often indicate negative numbers.
>
> | | | | | | |
> |---|---|---|---|---|---|
> | behind | below | before | deficit | debt | drop |
> | in the red | overdrawn | under | loss | B.C. | |

Self Check 10

FAST FOOD In 2008, Wendy's International (the hamburger restaurant chain) lost $480 million. The year before, the company made a modest profit. If the company lost a total of $464 million over this two-year span, how much profit did Wendy's make in 2007? (Source: wikinvest.com)

Now Try Problem 91

EXAMPLE 10 *Home Entertainment* In 2007, TiVo, Inc., suffered a loss due to large operating expenses and ended the year $32 million in the red. In 2008, the company did much better and made a large profit. If the company made a total of $72 million over this two-year span, how much profit did TiVo make in 2008? (Source: wikinvest.com)

Analyze

- In 2007, TiVo lost $32 million.   *Given*
- The company made a total of $72 million in 2007 and 2008.   *Given*
- How much profit did TiVo make in 2008?   *Find*

Form We will let x = the profit that TiVo made in 2008. If we work in terms of millions of dollars, we can represent the loss in 2007 using the negative number $-\$32$, and the total amount made in 2007 and 2008 can be represented by the positive number \$72.

The key word *total* suggests addition. Now we translate the words of the problem to numbers and symbols.

The loss in 2007	plus	the profit in 2008	equals	the total amount made in 2007 and 2008.
-32	$+$	x	$=$	72

Solve $-32 + x = 72$

$-32 + x + \mathbf{32} = 72 + \mathbf{32}$ To isolate x on the left side, add 32 to both sides.

$$
\begin{array}{r} 72 \\ +32 \\ \hline 104 \end{array}
$$

$x = 104$ Do the addition. The units are millions of dollars.

State In 2008, TiVo made a profit of \$104 million.

Check We can check the result using estimation with front-end rounding.

$$-\$30 \text{ million} \quad + \quad \$100 \text{ million} \quad = \quad \$70 \text{ million}$$

Approximate loss in 2007	Approximate profit in 2008	Approximate total for 2007 and 2008

Since the approximate two-year total of \$70 million is close to the actual total of \$72 million, the result seems reasonable.

ANSWERS TO SELF CHECKS

1. -9 **2.** -7 **3.** 18 **4. a.** -4 **b.** 5 **5.** -12 **6.** 17 **7.** -2 **8.** 5 **9.** 512
10. Wendy's made a profit of \$16 million in 2007.

SECTION 2.7   STUDY SET

VOCABULARY

Fill in the blanks.

1. To _____ an equation means to find all the values of the variable that make the equation true.

2. In the equation $3x + 1 = 10$, we call $3x$ the _____ term.

3. To _____ the solution of an equation, we substitute the value for the variable in the original equation and determine whether the result is a true statement.

4. Words such as *debt, overdrawn,* and *loss* are often used to indicate a _____ number.

CONCEPTS

5. What operation is performed on the variable x?
   a. $-2x = -100$
   b. $-6 + x = -9$
   c. $\dfrac{x}{-5} = 2$
   d. $-20 = x - 4$

6. What operations are performed on the variable x?
   a. $-4x - 1 = 11$
   b. $-1 = -28 + 9x$
   c. $\dfrac{x}{-6} - 3 = 9$

7. What step should be used to isolate the *variable* on one side of the equation?

a. $x + (-9) = 14$

b. $-32 = -8x$

8. What step should be used to isolate the *variable term* on one side of the equation?

a. $-11x + 3 = -19$

b. $-6 + \dfrac{h}{-3} = -14$

Fill in the blanks.

9. The addition property of equality: Adding the _____ number to both sides of an equation does not change its solution.

10. The multiplication property of _____: Multiplying both sides of an equation by the same nonzero number does not change its solution.

11. It takes two steps to solve the equation
$$4x + 10 = -6$$

- To isolate the variable term $4x$, we undo the addition of 10 by _____ 10 from both sides.
- To isolate the variable x, we undo the multiplication by 4 by _____ both sides by 4.

12. To solve $-x = 6$, we can multiply or divide both sides of the equation by ▢.

NOTATION

Complete each solution to solve the equation. Then check the result.

13.
$$y + (-7) = -16 + 3$$
$$y + (-7) = \boxed{}$$
$$y + (-7) + \boxed{} = -13 + \boxed{}$$
$$y = \boxed{}$$

Check:
$$y + (-7) = -16 + 3$$
$$\boxed{} + (-7) \overset{?}{=} -13$$
$$\boxed{} = -13 \quad \text{True}$$

The solution is ▢.

14.
$$-31 = -4y + 1$$
$$-31 - \boxed{} = -4y + 1 - \boxed{}$$
$$\boxed{} = -4y$$
$$\dfrac{-32}{\boxed{}} = \dfrac{-4y}{\boxed{}}$$
$$\boxed{} = y$$

Check:
$$-31 = -4y + 1$$
$$-31 \overset{?}{=} -4(\boxed{}) + 1$$
$$-31 \overset{?}{=} \boxed{} + 1$$
$$-31 = \boxed{} \quad \text{True}$$

The solution is ▢.

15. a. What does $-10x$ mean?

b. What does $\dfrac{x}{-8}$ mean?

16. Fill in the blank: $-x = \boxed{} x$

GUIDED PRACTICE

Solve each equation and check the result. **See Example 1.**

17. $x + (-3) = -12$

18. $y + (-1) = -4$

19. $m + (-6) = -1$

20. $r + (-12) = -2$

Solve each equation and check the result. **See Example 2.**

21. $y + 20 = -4$

22. $s + 18 = -10$

23. $t + 19 = -33$

24. $x + 17 = -32$

Solve each equation and check the result. **See Example 3.**

25. $-7 + 9 = x + 5(-3)$

26. $-1 + 7 = x + 2(-9)$

27. $-6 + 3 = f + 2(-4)$

28. $-10 + 4 = t + 3(-3)$

Solve each equation and check the result. **See Example 4.**

29. $-2s = 16$

30. $-3t = 9$

31. $-25 = -5t$

32. $-60 = -6m$

Solve each equation and check the result. **See Example 5.**

33. $\dfrac{t}{-3} = -9$

34. $\dfrac{w}{-4} = -5$

35. $\dfrac{x}{-7} = 11$

36. $\dfrac{s}{-9} = 9$

Solve each equation and check the result. **See Example 6.**

37. $-x = 14$

38. $-m = 32$

39. $-y = -58$

40. $-x = -73$

Solve each equation and check the result. **See Example 7.**

41. $-5x - 9 = 11$

42. $-6x - 4 = 44$

43. $-11y - 1 = 87$

44. $-12y - 9 = 39$

Solve each equation and check the result. **See Example 8.**

45. $-22 = 8 - 3x$

46. $-60 = 3 - 7x$

47. $49 = 4 - 5t$

48. $21 = 15 - 6n$

Solve each equation and check the result. **See Example 9.**

49. $\dfrac{x}{-2} - 6 = -9$

50. $\dfrac{a}{-5} - 7 = -16$

51. $\dfrac{y}{4} + 5 = -8$

52. $\dfrac{r}{2} + 5 = -13$

TRY IT YOURSELF

Solve each equation and check the result.

53. $-21 = 4h - 5$

54. $-22 = 7l - 8$

55. $-9h = -3(-3)$

56. $-6k = -2(-3)$

57. $0 = \dfrac{y}{8}$

58. $0 = \dfrac{h}{7}$

59. $-5 - 6 - 5x = 4$

60. $-7 - 5 - 7x = 16$

61. $-15 = -k$

62. $-4 = -p$

63. $\dfrac{h}{-6} + 4 = 5$

64. $\dfrac{p}{-3} + 3 = 8$

65. $r - (-7) = -1 - 6$

66. $x - (-1) = -4 - 3$

67. $h - 8 = -9$

68. $x - 1 = -7$

69. $2x + 3(0) = -6$

70. $3x - 4(0) = -12$

71. $-x = 8$

72. $-y = 12$

73. $0 = y + 9$

74. $0 = t + 5$

75. $-1 = -8 + \dfrac{h}{-2}$

76. $-5 = 4 + \dfrac{g}{-4}$

77. $34 = 4 - 5x$

78. $21 = 15 - 6x$

79. $t - 4 = -8 - (-2)$

80. $r - 1 = -3 - (-4)$

81. $5 - t = 500$

82. $4 - r = 300$

83. $4 = -3x + (-2)$

84. $15 = -2x + (-11)$

85. $-2(4) = \dfrac{t}{-6} + 1$

86. $-2(5) = \dfrac{y}{-3} + 3$

87. $2y + 8 = -6$

88. $5y + 1 = -9$

APPLICATIONS

Complete each solution.

89. SHARKS During a research project, a diver inside a shark cage made the first observations at a depth of 120 feet below sea level. For a second set of observations, the cage was raised to a depth of 75 feet below sea level. How many feet was the cage raised between observations?

Analyze

We can represent depths below sea level using negative numbers.

- The first observations were at a depth of ☐ ft.   *Given*

- The second set of observations were at a depth of ☐ ft.   *Given*

- How many ___ was the cage raised?   *Find*

Form

Let x = the number of feet the cage was ___ between observations.

The key word *raised* suggests ___ . We now translate the words of the problem into an equation.

The first depth of the cage	plus	the amount the cage was raised	is equal to	the second depth of the cage.
-120	$+$	☐	$=$	☐

Solve

$$-120 + \boxed{} = -75$$
$$-120 + x + \boxed{} = -75 + \boxed{}$$
$$x = \boxed{}$$

State

The shark cage was raised ☐ feet.

Check

If we add the number of feet that the cage was raised to the first depth, we should get the second depth.

$$-120 \text{ ft} + 45 \text{ ft} = \boxed{} \text{ ft}$$

The result checks.

90. PROFITS AND LOSSES In its first year of business, a plant nursery suffered a loss due to frost damage, ending the year $11,500 in the red. In the second year, it made a sizable profit. If the nursery made a total of $32,000 the first two years in business, how much profit was made the second year?

Analyze

We can represent a loss using a ___ number and a profit using a positive number.

- The first year loss was $-$ ☐ .   *Given*

- The total amount made the first two years in business was ☐ .   *Given*

- How much profit was made the ___ year?   *Find*

Form

Let x = the ___ made the second year.

The key word *total* suggests ___ . We now translate the words of the problem into an equation.

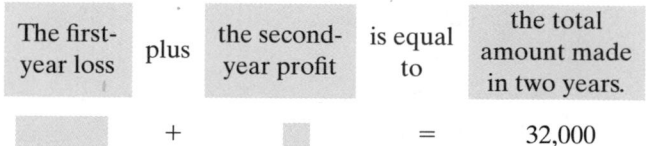

The first-year loss	plus	the second-year profit	is equal to	the total amount made in two years.
☐	$+$	☐	$=$	32,000

Solve

$$-11,500 + \boxed{} = 32,000$$
$$-11,500 + x + \boxed{} = 32,000 + \boxed{}$$
$$x = \boxed{}$$

State

The business made a profit of $\boxed{}$ the second year.

Check

If we add the second-year profit to the first-year loss, we should get the total amount made in two years.

$$-\$11,500 + \$43,500 = \boxed{}$$

The result checks.

In each of the following problems, let a variable represent the unknown quantity. Then write and solve an equation to answer the question.

91. FOOTWEAR TRENDS Because of tough economic times and cheap knock-offs from competitors, Crocs, Inc. (a shoe manufacturer), lost $185 million in 2008. Just one year before, the company made a very large profit. If the company lost a total of $17 million in this two-year span, how much profit did Crocs make in 2007? (Source: wikinvest.com)

92. AIRLINES In 2008, Jet Blue Airways lost $76 million. In 2007, the company made a modest profit. If the company lost a total of $58 million in this two-year span, how much profit did Jet Blue make in 2007? (Source: wikinvest.com)

93. MARKET SHARE After its first year of business, a manufacturer of smoke detectors found its market share 43 points behind the industry leader. Five years later, it trailed the leader by only 9 points. How many points of market share did the company gain over this five-year span?

94. POLLS Six months before an election, a political candidate was 31 points behind in the polls. Two days before the election, polls showed that his support had skyrocketed; he was now only 2 points behind. How much support had he gained over the six-month period?

95. CHECKING ACCOUNTS After he made deposits of $95 and $65, a student's account was still $15 overdrawn. What was his checking account balance before the deposit?

96. WEATHER FORECASTS The weather forecast for Fairbanks, Alaska, warned listeners that the daytime high temperature of 2° below zero would drop to a nighttime low of 28° below. By how many degrees did the temperature fall overnight?

97. FOOTBALL STATISTICS Most football teams keep track of how many yards their offense gains or loses by rushing (running) and by passing the ball during a game. Then they combine those two numbers to find the total yards gained (or lost). The chart below shows the statistics for a game in 1943 between the Detroit Lions and the Chicago Cardinals in which Detroit set the NFL record for *fewest rushing yards in a game*. Incredibly, the Lions still won the game 7-0. Find the number of yards Detroit had rushing that day. (Source: pro-football-reference.com)

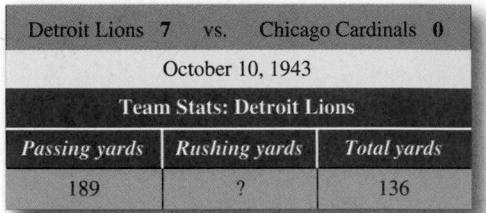

Detroit Lions **7**	vs.	Chicago Cardinals **0**
October 10, 1943		
Team Stats: Detroit Lions		
Passing yards	*Rushing yards*	*Total yards*
189	?	136

98. ROLLER COASTERS The end of a roller-coaster ride consists of a steep plunge from a peak 145 feet above ground level. The car then comes to a screeching halt in a cave that is 25 feet below ground level. How many feet does the roller coaster drop at the end of the ride shown in the illustration below?

99. THE ROMAN EMPIRE Historians usually date the beginning of the Roman Empire as 27 B.C. The date given for the fall of the Roman Empire is 476 A.D. For how many years did the Roman Empire last?

100. HISTORY The Roman–Persian wars were a series of conflicts between the Greco-Roman world and two Iranian empires that began in 92 B.C. and finally concluded in 627 A.D. For how many years did the Roman–Persian wars last?

101. AIRLINES Refer to the graph below. Find the 2009 second quarter net income for Continental Airlines.

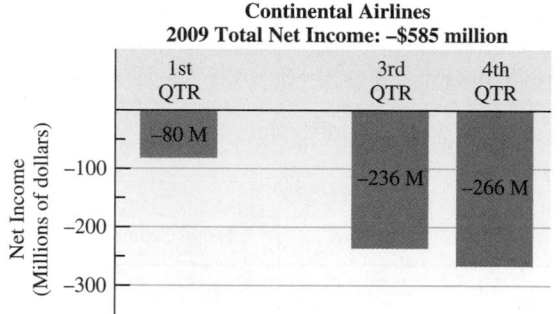

Continental Airlines
2009 Total Net Income: –$585 million

Source: Wikinvest.com

102. MERCURY The freezing point of Mercury is $-38°$ F. By how many degrees must it be heated to reach its boiling point, which is $674°$ F?

WRITING

103. Explain why the variable is not isolated in the equation $-x = 10$.

104. Explain the two-step process to solve the equation $-3x + 6 = -9$. What properties of equality are used?

REVIEW

105. Write the repeated multiplication that 5^6 represents.

106. How can the addition $2 + 2 + 2 + 2 + 2$ be represented using multiplication?

107. Perform the division, if possible: $\dfrac{0}{8}$

108. What are the first five prime numbers?

109. Subtract: $10,000 - 782$

110. Divide: $54)\overline{2,303}$

111. Add: $23 + 234 + 2,345 + 23,456$

112. Multiply: $1,000 \cdot 409$

STUDY SKILLS CHECKLIST

Do You Know the Basics?

The key to mastering the material in Chapter 2 is to know the basics. Put a checkmark in the box if you can answer "yes" to the statement.

☐ I understand order on the number line:

$-4 < -3$ and $-15 > -20$

☐ I know how to add two integers that have the same sign.

- The sum of two positive numbers is *positive*.

$4 + 5 = 9$

- The sum of two negative numbers is *negative*.

$-4 + (-5) = -9$

☐ I know how to add two integers that have different signs.

- If the positive integer has the larger absolute value, the sum is positive.

$-7 + 11 = 4$

- If the negative integer has the larger absolute value, the sum is negative.

$12 + (-20) = -8$

☐ I know how to use the subtraction rule: *Subtraction is the same as addition of the opposite.*

$-2 - (-7) = -2 + 7 = 5$

and

$-9 - 3 = -9 + (-3) = -12$

☐ I know that the rules for multiplying and dividing two integers are the same.

- Like signs: positive result

$(-2)(-3) = 6$ and $\dfrac{-15}{-3} = 5$

- Unlike signs: negative result

$2(-3) = -6$ and $\dfrac{-15}{3} = -5$

☐ I know the meaning of a $-$ symbol:

$-(-6) = 6$ $-|-6| = -6$

CHAPTER 2   SUMMARY AND REVIEW

DEFINITIONS AND CONCEPTS	EXAMPLES
The collection of positive whole numbers, the negatives of the whole numbers, and 0 is called the set of **integers.**	The set of integers: $\{\ldots, -5, -4, -3, -2, -1, 0, 1, 2, 3, 4, 5, \ldots\}$
Positive numbers are greater than 0 and **negative numbers** are less than 0.	The set of positive integers: $\{1, 2, 3, 4, 5, \ldots\}$ The set of negative integers: $\{\ldots, -5, -4, -3, -2, -1\}$
Negative numbers can be represented on a **number line** by extending the line to the left and drawing an arrowhead. As we move to the right on the number line, the values of the numbers increase. As we move to the left, the values of the numbers decrease.	Graph $-1, 6, 0, -4,$ and 3 on a number line. Numbers get larger Negative numbers Zero Positive numbers $-6\ -5\ -4\ -3\ -2\ -1\ 0\ 1\ 2\ 3\ 4\ 5\ 6$ Numbers get smaller
Inequality symbols: $\neq$ means *is not equal to* $\geq$ means *is greater than or equal to* $\leq$ means *is less than or equal to*	Each of the following statements is true: $5 \neq -3$ Read as "5 is not equal to -3." $4 \geq -6$ Read as "4 is greater than or equal to -6." $-2 \leq -2$ Read as "-2 is less than or equal to -2."
The **absolute value** of a number is the distance on a number line between the number and 0.	Find each absolute value: $\lvert 12 \rvert = 12$ $\lvert -9 \rvert = 9$ $\lvert 0 \rvert = 0$
Two numbers that are the same distance from 0 on the number line, but on opposite sides of it, are called **opposites** or **negatives.**	The opposite of 4 is -4. The opposite of -77 is 77. The opposite of 0 is 0.
The opposite of the opposite rule The opposite of the opposite (or negative) of a number is that number. For any number a, $-(-a) = a$ Read $-a$ as "the opposite of a."	Simplify each expression: $-(-6) = 6$ $-\lvert 8 \rvert = -8$ $-\lvert -26 \rvert = -26$
The **− symbol** is used to indicate a negative number, the opposite of a number, and the operation of subtraction.	-2 $-(-4)$ $6 - 1$ negative 2 the opposite of negative four six minus one

REVIEW EXERCISES

1. Write the set of integers.

2. Represent each of the following situations using a signed number.

  a. a deficit of $1,200

  b. 10 seconds before going on the air

3. WATER PRESSURE Salt water exerts a pressure of approximately 29 pounds per square inch at a depth of 33 feet. Express the depth using a signed number.

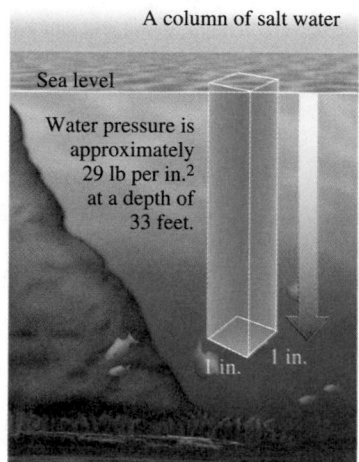

A column of salt water

Sea level

Water pressure is approximately 29 lb per in.^{2} at a depth of 33 feet.

1 in. 1 in.

4. Graph the following integers on a number line.

  a. $-3, 0, 4, -1$

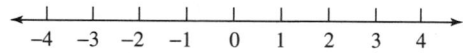

    -4 -3 -2 -1 0 1 2 3 4

  b. the integers greater than -3 but less than 4

    -4 -3 -2 -1 0 1 2 3 4

5. Place an $<$ or an $>$ symbol in the box to make a true statement.

  a. $0 \;\square\; -7$ **b.** $-20 \;\square\; -19$

6. Tell whether each statement is true or false.

  a. $-17 \geq -16$ **b.** $-56 \leq -56$

7. Find each absolute value.

  a. $|5|$ **b.** $|-43|$ **c.** $|0|$

8. a. What is the opposite of 8?

  b. What is the opposite of -8?

  c. What is the opposite of 0?

9. Simplify each expression.

  a. $-|12|$

  b. $-(-12)$

  c. -0

10. Explain the meaning of each red $-$ symbol.

  a. -5

  b. $-(-5)$

  c. $-(-5)$

  d. $5 - (-5)$

11. LADIES PROFESSIONAL GOLF ASSOCIATION The scores of the top six finishers of the 2008 Grand China Air LPGA Tournament and their final scores related to par were: Helen Alfredsson (-12), Laura Diaz (-8), Shanshan Feng (-5), Young Kim (-6), Karen Stupples (-7), and Yani Tseng (-9). Complete the table below. Remember, in golf, the lowest score wins.

Position	Player	Score to Par
1		
2		
3		
4		
5		
6		

Source: golf.fanhouse.com

12. FEDERAL BUDGET The graph shows the U.S. government's deficit/surplus budget data for the years 1980–2007.

  a. When did the first budget surplus occur? Estimate it.

  b. In what year was there the largest surplus? Estimate it.

  c. In what year was there the greatest deficit? Estimate it.

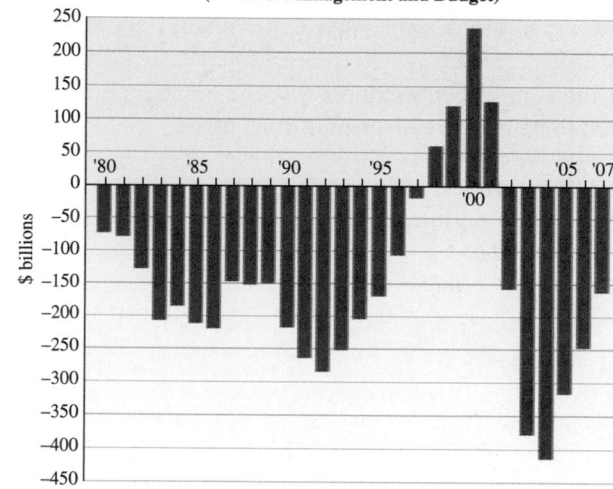

Federal Budget Deficit/Surplus
(Office of Management and Budget)

(Source: U.S. Bureau of the Census)

SECTION 2.2   Adding Integers

DEFINITIONS AND CONCEPTS	EXAMPLES
Adding two integers that have the same (like) signs 1. To add two positive integers, add them as usual. The final answer is positive. 2. To add two negative integers, add their absolute values and make the final answer negative.	Add: $-5 + (-10)$ Find the absolute values: $\|-5\| = 5$ and $\|-10\| = 10$. $\qquad -5 + (-10) = -15$ Add their absolute values, 5 and 10, to get 15. Then make the final answer negative.
Adding two integers that have different (unlike) signs To add a positive integer and a negative integer, subtract the smaller absolute value from the larger. 1. If the positive integer has the larger absolute value, the final answer is positive. 2. If the negative integer has the larger absolute value, make the final answer negative.	Add: $-7 + 12$ Find the absolute values: $\|-7\| = 7$ and $\|12\| = 12$. $\qquad -7 + 12 = 5$ Subtract the smaller absolute value from the larger: $12 - 7 = 5$. Since the positive number, 12, has the larger absolute value, the final answer is positive. Add: $-8 + 3$ Find the absolute values: $\|-8\| = 8$ and $\|3\| = 3$. $\qquad -8 + 3 = -5$ Subtract the smaller absolute value from the larger: $8 - 3 = 5$. Since the negative number, -8, has the larger absolute value, make the final answer negative.
To **evaluate expressions** that contain several additions, we make repeated use of the rules for adding two integers.	Evaluate: $-7 + 1 + (-20) + 1$ Perform the additions working left to right. $\mathbf{-7 + 1} + (-20) + 1 = \mathbf{-6} + (-20) + 1$ $= -26 + 1$ $= -25$
We can use the **commutative** and **associative properties of addition** to *reorder* and *regroup* addends.	Another way to evaluate this expression is to add the negatives and add the positives separately. Then add those results. $\qquad\qquad\qquad\qquad\quad$ Negatives $\qquad$ Positives $-7 + 1 + (-20) + 1 = [\mathbf{-7 + (-20)}] + (\mathbf{1 + 1})$ $= \mathbf{-27} + \mathbf{2}$ $= -25$
Addition property of 0 The sum of any number and 0 is that number. $\quad$ For any number a, $\qquad a + 0 = a \qquad$ and $\qquad 0 + a = a$	$-2 + 0 = -2 \qquad$ and $\qquad 0 + (-25) = -25$
If the sum of two numbers is 0, the numbers are said to be **additive inverses** of each other.	3 and -3 are *additive inverses* because $3 + (-3) = 0$.
Addition property of opposites The sum of an integer and its opposite (additive inverse) is 0. $\quad$ For any number a, $\qquad a + (-a) = 0 \qquad$ and $\qquad -a + a = 0$	$4 + (-4) = 0 \qquad$ and $\qquad -712 + 712 = 0$

At certain times, the **addition property of opposites** can be used to make addition of several integers easier.	Evaluate: $14 + (-9) + 8 + 9 + (-14)$ Locate pairs of opposites and add them to get 0. $$14 + (-9) + 8 + 9 + (-14) = 0 + 0 + 8$$ $$= 8$$ The sum of any integer and 0 is that integer.

REVIEW EXERCISES

Add.

13. $-6 + (-4)$ **14.** $-3 + (-6)$

15. $-28 + 60$ **16.** $93 + (-20)$

17. $-8 + 8$ **18.** $73 + (-73)$

19. $-1 + (-4) + (-3)$ **20.** $3 + (-2) + (-4)$

21. $[7 + (-9)] + (-4 + 16)$

22. $(-2 + 11) + [(-5) + 4]$

23. $-4 + 0$

24. $0 + (-20)$

25. $-2 + (-1) + (-76) + 1 + 2$

26. $-5 + (-31) + 9 + (-9) + 5$

27. Find the sum of $-102, 73$, and -345.

28. What is 3,187 more than -59?

29. What is the additive inverse of each number?

    a. -11 **b.** 4

30. a. Is the sum of two positive integers always positive?

    b. Is the sum of two negative integers always negative?

    c. Is the sum of a positive integer and a negative integer always positive?

    d. Is the sum of a positive integer and a negative integer always negative?

31. DROUGHT During a drought, the water level in a reservoir fell to a point 100 feet below normal. After a lot of rain in April it rose 16 feet, and after even more rain in May it rose another 18 feet.

    a. Express the water level of the reservoir before the rainy months as a signed number.

    b. What was the water level after the rain?

32. TEMPERATURE EXTREMES The world record for lowest temperature is $-129°$ F. It was set on July 21, 1983, in Antarctica. The world record for highest temperature is an amazing $265°$ F warmer. It was set on September 13, 1922, in Libya. Find the record high temperature. (Source: *The World Almanac Book of Facts*, 2009)

SECTION 2.3   Subtracting Integers

DEFINITIONS AND CONCEPTS	EXAMPLES

The **rule for subtraction** is helpful when subtracting signed numbers.
 	
To subtract two integers, add the first integer to the opposite of the integer to be subtracted.
 	
Subtracting is the same as adding the opposite.
 	
    For any numbers a and b,
    $a - b = a + (-b)$
	Subtract: $3 - (-5)$
	
	Add . . .
	$$3 - (-5) = 3 + 5 = 8$$ Use the rule for adding two integers with the same sign.
	. . . the opposite
	
	Check using addition: $8 + (-5) = 3$

After rewriting a subtraction as addition of the opposite, use one of the rules for the addition of signed numbers discussed in Section 2.2 to find the result.	Subtract: $-3 - 5 = -3 + (-5) = -8$ Add the opposite of 5, which is -5. $-4 - (-7) = -4 + 7 = 3$ Add the opposite of -7, which is 7.
Be careful when translating the instruction to subtract one number *from* another number.	Subtract -6 from -9. $-9 - (-6)$ The number to be subtracted is -6.

Expressions can involve repeated subtraction or combinations of subtraction and addition. To evaluate them, we use the **order of operations rule** discussed in Section 1.7.

Evaluate: $43 - (-6 - 15)$

$-43 - (-6 - 15) = -43 - [-6 + (-15)]$ Within the parentheses, add the opposite of 15, which is −15.

$= -43 - [-21]$ Within the brackets, add −6 and −15.

$= -43 + 21$ Add the opposite of −21, which is 21.

$= -22$ Use the rule for adding integers that have different signs.

When we find the difference between the maximum value and the minimum value of a collection of measurements, we are finding the **range** of the values.

Range = maximum value − minumum value

GEOGRAPHY The highest point in the United States is Mt. McKinley at 20,230 feet. The lowest point is −282 feet at Death Valley, California. Find the range between the highest and lowest points.

Range $= 20,320 - (-282)$

$= 20,320 + 282$ Add the opposite of −282, which is 282.

$= 20,602$ Do the addition.

The range between the highest point and lowest point in the United States is 20,602 feet.

To find the **change** in a quantity, we subtract the *earlier value* from the *later value*.

Change = later value − earlier value

SUBMARINES A submarine was traveling at a depth of 165 feet below sea level. The captain ordered it to a new position of only 8 feet below the surface. Find the change in the depth of the submarine.

We can represent 165 feet below sea level as −165 feet and 8 feet below the surface as −8 feet.

Change of depth $= -8 - (-165)$ Subtract the earlier depth from the later depth.

$= -8 + 165$ Add the opposite of −165, which is 165.

$= 157$ Use the rule for adding integers that have different signs.

The change in the depth of the submarine was 157 feet.

REVIEW EXERCISES

33. Fill in the blank: Subtracting an integer is the same as adding the _____ of that integer.

34. Write each phrase using symbols.

  a. negative nine minus negative one.

  b. negative ten subtracted from negative six

Subtract.

35. $5 - 8$

36. $-9 - 12$

37. $-4 - (-8)$

38. $-8 - (-2)$

39. $-6 - 106$

40. $-7 - 1$

41. $0 - 37$

42. $0 - (-30)$

Evaluate each expression.

43. $12 - 2 - (-6)$

44. $-16 - 9 - (-1)$

45. $-9 - 7 + 12$

46. $-5 - 6 + 33$

47. $1 - (2 - 7)$

48. $-12 - (6 - 10)$

49. $-70 - [(-6) - 2]$

50. $89 - [(-2) - 12]$

51. $-(-5) + (-28) - 2 - (-100)$

52. a. Subtract 27 from −50.

  b. Subtract −50 from 27.

Use signed numbers to solve each problem.

53. MINING Some miners discovered a small vein of gold at a depth of 150 feet. This encouraged them to continue their exploration. After descending another 75 feet, they came upon a much larger find. Use a signed number to represent the depth of the second discovery.

54. RECORD TEMPERATURES The lowest and highest recorded temperatures for Alaska and Virginia are shown. For each state, find the range between the record high and low temperatures.

Alaska	Virginia
Low: $-80°$ Jan. 23, 1971	Low: $-30°$ Jan. 22, 1985
High: $100°$ June 27, 1915	High: $110°$ July 15, 1954

55. POLITICS On July 20, 2007, a CNN/Opinion Research poll had Barack Obama trailing Hillary Clinton in the South Carolina Democratic Presidential Primary race by 16 points. On January 26, 2008, Obama finished 28 points ahead of Clinton in the actual primary. Find the point change in Barack Obama's support.

56. OVERDRAFT FEES A student had a balance of $255 in her checking account. She wrote a check for rent for $300, and when it arrived at the bank she was charged an overdraft fee of $35. What is the new balance in her account?

SECTION 2.4   Multiplying Integers

DEFINITIONS AND CONCEPTS	EXAMPLES
Multiplying two integers that have different (unlike) signs To multiply a positive integer and a negative integer, multiply their absolute values. Then make the final answer negative.	Multiply: $6(-8)$ Find the absolute values: $\lvert 6 \rvert = 6$ and $\lvert -8 \rvert = 8$. $6(-8) = -48$ Multiply the absolute values, 6 and 8, to get 48. Then make the final answer negative.
Multiplying two integers that have the same (like) signs To multiply two integers that have the same sign, multiply their absolute values. The final answer is positive.	Multiply: $-2(-7)$ Find the absolute values: $\lvert -2 \rvert = 2$ and $\lvert -7 \rvert = 7$. $-2(-7) = 14$ Multiply the absolute values, 2 and 7, to get 14. The final answer is positive.
To **evaluate expressions** that contain several multiplications, we make repeated use of the rules for multiplying two integers.	Evaluate $-5(3)(-6)$ in two ways. Perform the multiplications, working left to right. $$-5(3)(-6) = -15(-6)$$ $$= 90$$
Another approach to evaluate expressions is to use the **commutative** and/or **associative properties of multiplication** to reorder and regroup the factors in a helpful way.	First, multiply the pair of negative factors. $-5(3)(-6) = 30(3)$ Multiply the negative factors to produce a positive product. $$= 90$$
Multiplying an even and an odd number of negative integers The product of an even number of negative integers is positive. The product of an odd number of negative integers is negative.	Four negative factors: $-5(-1)(-6)(-2) = 60$ ← positive Five negative factors: $-2(-4)(-3)(-1)(-5) = -120$ ← negative
Even and odd powers of a negative integer When a negative integer is raised to an even power, the result is positive. When a negative integer is raised to an odd power, the result is negative.	Evaluate: $(-3)^4 = (-3)(-3)(-3)(-3)$ The exponent is even. $\qquad\qquad = 9(9)$ Multiply pairs of integers. $\qquad\qquad = 81$ The answer is positive. Evaluate: $(-2)^3 = (-2)(-2)(-2)$ The exponent is odd. $\qquad\qquad = -8$ The answer is negative.

Although the exponential expressions $(-6)^2$ and -6^2 look similar, they are not the same. The bases are different.

Evaluate: $(-6)^2$ and -6^2

Because of the parentheses, the base is -6. The exponent is 2.

$$(-6)^2 = (-6)(-6)$$
$$= 36$$

Since there are no parentheses around -6, the base is 6. The exponent is 2.

$$-6^2 = -(6 \cdot 6)$$
$$= -36$$

Application problems that involve **repeated addition** are often more easily solved using multiplication.

CHEMISTRY A chemical compound that is normally stored at 0°F had its temperature lowered 8°F each hour for 6 hours. What signed number represents the change in temperature of the compound after 6 hours?

$$-8 \cdot 6 = -48$$ Multiply the change in temperature each hour by the number of hours.

The change in temperature of the compound is -48°F.

REVIEW EXERCISES

Multiply.

57. $7(-2)$

58. $(-8)(47)$

59. $-23(-14)$

60. $-5(-5)$

61. $-1 \cdot 25$

62. $(6)(-34)$

63. $-4,000(17,000)$

64. $-100,000(-300)$

65. $(-6)(-2)(-3)$

66. $-4(-3)(-3)$

67. $(-3)(4)(2)(-5)$

68. $(-1)(10)(10)(-1)$

69. Find the product of -15 and the opposite of 30.

70. Find the product of the opposite of 16 and the opposite of 3.

71. DEFICITS A state treasurer's prediction of a tax shortfall was two times worse than the actual deficit of $130 million. The governor's prediction of the same shortfall was even worse—three times the amount of the actual deficit. Complete the labeling of the vertical axis of the graph in the next column to show the two incorrect predictions.

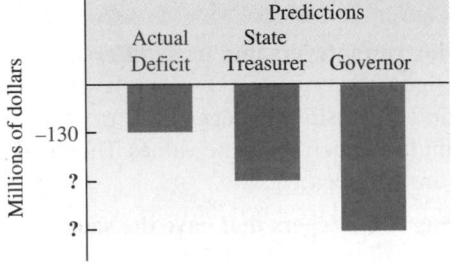

72. MINING An elevator is used to lower coal miners from the ground level entrance to various depths in the mine. The elevator stops every 45 vertical feet to let off miners. At what depth do the miners work who get off the elevator at the 12th stop?

Evaluate each expression.

73. $(-5)^3$

74. $(-2)^5$

75. $(-8)^4$

76. $(-4)^4$

77. When $(-17)^9$ is evaluated, will the result be positive or negative?

78. Explain the difference between -9^2 and $(-9)^2$ and then evaluate each expression.

SECTION 2.5   Dividing Integers

DEFINITIONS AND CONCEPTS	EXAMPLES

Dividing two integers
To divide two integers, divide their absolute values.

1. The quotient of two integers that have the same (*like*) signs is positive.

2. The quotient of two integers that have different (*unlike*) signs is negative.

To **check division** of integers, multiply the *quotient* and the *divisor*. You should get the *dividend*.

Divide: $\dfrac{-21}{-7}$

Find the absolute values: $|-21| = 21$ and $|-7| = 7$.

$$\frac{-21}{-7} = 3$$ Divide the absolute values, 21 by 7, to get 3. The final answer is positive.

Check: $3(-7) = -21$ The result checks.

Divide: $-54 \div 9$

Find the absolute values: $|-54| = 54$ and $|9| = 9$.

$$-54 \div 9 = -6 \quad \text{Divide the absolute values, 54 by 9, to get 6.}$$
$$\text{Then make the final answer negative.}$$

Check: $-6(9) = -54$ The result checks.

Division with 0

If 0 is divided *by* any nonzero number, the quotient is 0.

For any nonzero number *a*,

$$\frac{0}{a} = 0$$

$$\frac{0}{-8} = 0 \qquad\qquad 0 \div (-20) = 0$$

Division *of* any nonzero number by 0 is undefined.

For any nonzero number *a*,

$$\frac{a}{0} \text{ is undefined.}$$

$$\frac{-2}{0} \text{ is undefined.} \qquad -6 \div 0 \text{ is undefined.}$$

Problems that involve forming **equal-sized groups** can be solved by division.

USED CAR SALES The price of a used car was reduced each day by an equal amount because it was not selling. After 7 days, and a $1,050 reduction in price, the car was finally purchased. By how much was the price of the car reduced each day?

$$\frac{-1,050}{7} = -150 \quad \text{Divide the change in the price of the car by the number of days the price was reduced.}$$

The negative result indicates that the price of the car was *reduced* by $150 each day.

REVIEW EXERCISES

79. Fill in the blanks: We know that $\dfrac{-15}{5} = -3$ because

$$\boxed{} \left(\boxed{} \right) = \boxed{}.$$

80. Check using multiplication to determine whether $-152 \div (-8) = 18$.

Divide, if possible.

81. $\dfrac{25}{-5}$

82. $\dfrac{-14}{7}$

83. $-64 \div (-8)$

84. $72 \div (-9)$

85. $\dfrac{-10}{-1}$

86. $\dfrac{-673}{-673}$

87. $-150,000 \div 3,000$

88. $-24,000 \div (-60)$

89. $\dfrac{-1,058}{-46}$

90. $-272 \div 16$

91. $\dfrac{0}{-5}$

92. $\dfrac{-4}{0}$

93. Divide -96 by 3.

94. Find the quotient of -125 and -25.

95. PRODUCTION TIME Because of improved production procedures, the time needed to produce an electronic component dropped by 12 minutes over the past six months. If the drop in production time was uniform, how much did it change each month over this period of time?

96. OCEAN EXPLORATION The Puerto Rico Trench is the deepest part of the Atlantic Ocean. It has a maximum depth of 28,374 feet. If a remote-controlled unmanned submarine is sent to the bottom of the trench in a series of 6 equal dives, how far will the vessel descend on each dive? (Source: marianatrench.com)

SECTION 2.6 Order of Operations and Estimation

DEFINITIONS AND CONCEPTS	EXAMPLES

Order of operations

1. Perform all calculations within parentheses and other grouping symbols following the order listed in Steps 2–4 below, working from the innermost pair of grouping symbols to the outermost pair.

2. Evaluate all exponential expressions.

3. Perform all multiplications and divisions as they occur from left to right.

4. Perform all additions and subtractions as they occur from left to right.

When grouping symbols have been removed, repeat Steps 2–4 to complete the calculation.

If a fraction bar is present, evaluate the expression above the bar (called the **numerator**) and the expression below the bar (called the **denominator**) separately. Then perform the division indicated by the fraction bar, if possible.

Evaluate: $-3(-5)^2 - (-40)$

$$-3(-5)^2 - (-40) = -3(25) - (-40) \quad \text{Evaluate the exponential expression.}$$
$$= -75 - (-40) \quad \text{Do the multiplication.}$$
$$= -75 + 40 \quad \text{Use the subtraction rule: Add the opposite of } -40.$$
$$= -35 \quad \text{Do the addition.}$$

Evaluate: $\dfrac{-6 + 4(-2)}{16 - (-3)^2}$

$$\frac{-6 + 4(-2)}{16 - (-3)^2} = \frac{-6 + (-8)}{16 - 9} \quad \begin{array}{l}\text{In the numerator, do the multiplication.}\\ \text{In the denominator, evaluate the exponential expression.}\end{array}$$
$$= \frac{-14}{7} \quad \begin{array}{l}\text{In the numerator, do the addition.}\\ \text{In the denominator, do the subtraction.}\end{array}$$
$$= -2 \quad \text{Do the division.}$$

Absolute value symbols are grouping symbols, and by the order of operations rule, all calculations within grouping symbols must be performed first.

Evaluate: $10 - 2|-8 + 1|$

$$10 - 2|-8 + 1| = 10 - 2|-7| \quad \begin{array}{l}\text{Do the addition within the absolute value symbol.}\end{array}$$
$$= 10 - 2(7) \quad \text{Find the absolute value of } -7.$$
$$= 10 - 14 \quad \text{Do the multiplication.}$$
$$= -4 \quad \text{Do the subtraction.}$$

When an exact answer is not necessary and a quick approximation will do, we can use **estimation.**

Estimate the value of $-56 + (-67) + 89 + (-41) + 14$ by rounding each number to the nearest ten.

$$-60 + (-70) + 90 + (-40) + 10$$
$$= -170 + 100 \quad \begin{array}{l}\text{Add the positives and the negatives separately.}\end{array}$$
$$= -70 \quad \text{Do the addition.}$$

REVIEW EXERCISES

Evaluate each expression.

97. $2 + 4(-6)$

98. $7 - (-2)^2 + 1$

99. $65 - 8(9) - (-47)$

100. $-3(-2)^3 - 16$

101. $-2(5)(-4) + \dfrac{|-9|}{3^2}$

102. $-4^2 + (-4)^2$

103. $-12 - (8 - 9)^2$

104. $7|-8| - 2(3)(4)$

105. $-4\left(\dfrac{15}{-3}\right) - 2^3$

106. $-20 + 2(12 - 5 \cdot 2)$

107. $-20 + 2[12 - (-7 + 5)^2]$

108. $8 - 6|-3 \cdot 4 + 5|$

109. $\dfrac{2 \cdot 5 + (-6)}{-3 - 1^5}$

110. $\dfrac{3(-6) - 11 + 1}{4^2 - 3^2}$

111. $-\left[1 - \left(2^3 + \dfrac{100}{-50}\right)\right]$

112. $-\left[45 - \left(5^3 + \dfrac{100}{-4}\right)\right]$

113. Round each number to the nearest hundred to estimate the value of the following expression:
$-4,471 + 7,935 + 2,094 + (-3,188)$

114. Find the mean (average) of $-8, 4, 7, -11, 2, 0, -6,$ and $-4.$

SECTION 2.7   Solving Equations That Involve Integers

DEFINITIONS AND CONCEPTS	EXAMPLES
To **solve an equation** means to find all the values of the variable that make the equation true. To isolate the variable on one side of the equation, we use: **1.** Addition property of equality **2.** Subtraction property of equality **3.** Multiplication property of equality **4.** Division property of equality	Solve: $x + (-3) = 8$ $x + (-3) + 3 = 8 + 3$ To isolate x, undo the addition of −3 by adding 3 to both sides. $x + 0 = 11$ On the left side, the sum of a number and its opposite is 0. $x = 11$ On the left side, the sum of any number and 0 is that number. **Check:** $x + (-3) = 8$ This is the original equation. $11 + (-3) \stackrel{?}{=} 8$ Substitute 11 for x. $8 = 8$ True Since the resulting statement $8 = 8$ is true, the solution is 11.
The expressions on each side of an equation should be simplified before using any properties of equality to isolate the variable.	Solve: $-19 + 4 = y + 2(3)$ $-15 = y + 6$ On the left side, do the addition. On the right side, do the multiplication. $-15 - 6 = y + 6 - 6$ To isolate y, undo the addition of 6 by subtracting 6 from both sides. $-21 = y$ Do the subtraction. Check the result in the *original* equation to verify that -21 is the solution.
The notation $-x$ means $-1x$.	Solve: $-x = 14$ We can multiply or divide both sides by -1 to isolate x. $-1x = 14$ Write −x as −1x. $\dfrac{-1x}{-1} = \dfrac{14}{-1}$ To isolate x, undo the multiplication by −1 by dividing both sides by −1. $x = -14$ Do the division. Check the result in the *original* equation to verify that -14 is the solution.
Sometimes we must use two (or more) properties of equality to solve more complicated equations.	Solve: $-6x + 2 = -10$ To solve the equation, we use the order of operations rule in reverse. • To isolate the variable term $-6x$, subtract 2 from both sides to undo the addition of 2. • To isolate the variable x, divide both sides by -6 to undo the multiplication by -6. $-6x + 2 - 2 = -10 - 2$ Subtract 2 from both sides to isolate −6x. $-6x = -12$ Do the subtraction. $\dfrac{-6x}{-6} = \dfrac{-12}{-6}$ Divide both sides by −6 to isolate x. $x = 2$ Do the division. Check the result in the *original* equation to verify that 2 is the solution.

We can use the concepts of variable and equation to solve application problems involving integers.

The following words are often used to indicate negative numbers.

behind	below	before	deficit
debt	drop	under	loss
in the red	B.C.	overdrawn	

BANKING After a student made a deposit of $165, his checking account was still $38 overdrawn. What was his checking account balance before the deposit?

Analyze
An *overdrawn* account balance can be represented by a negative number.
- A deposit of $165 was made.   Given
- After the deposit, the account balance was $-\$38$.   Given
- What was the account balance before the deposit?   Find

Form Let x = the account balance before the deposit. The word *deposit* indicates addition. Now we translate the words of the problem to numbers and symbols.

The balance before the deposit	plus	the deposit	is equal to	the new balance.
x	$+$	165	$=$	-38

Solve
$$x + 165 = -38$$
$$x + 165 - 165 = -38 - 165$$
$$x = -203$$

$$\begin{array}{r} \overset{1\ 1}{165} \\ +\ 38 \\ \hline 203 \end{array}$$

State His checking account balance before the deposit was $-\$203$.

Check If we add the deposit to the original balance, we should get the new balance.
$$-\$203 + \$165 = -\$38$$

$$\begin{array}{r} \overset{9}{\cancel{2}}\overset{11}{\cancel{0}}\overset{0\ 13}{\cancel{3}} \\ -\ 165 \\ \hline 38 \end{array}$$

The result checks.

REVIEW EXERCISES

Solve each equation and check the result.

115. $x + (-16) = -6$ **116.** $-y = -32$

117. $x + 8 = -42$ **118.** $-20 + 4 = a + 3(-7)$

119. $-84 = -2t$ **120.** $\dfrac{n}{5} + (-2) = -7$

121. $\dfrac{s}{3} = -1$ **122.** $-11y - 10 = 67$

123. $16 - 6n = 22$ **124.** $\dfrac{n}{-2} - 27 = -27$

125. $-9 - 3 - 20x = -52$

126. $15 = -13b + (-11)$

In Exercises 127 and 128, let a variable represent the unknown quantity. Then write and solve an equation to answer the question.

127. FINANCIAL STATEMENTS In 2008, Foot Locker (a chain of athletic shoes stores) lost $80 million. In 2007, the company made a modest profit of $50 million, and in 2006, they made a very large profit, as well. If the company made a total of $223 million in this three-year span, how much profit did Foot Locker make in 2006? (Source: wikinvest.com)

128. POLITICS Eight weeks before an election, a political candidate was 32 points behind in the polls. On election day, she narrowly lost the race by 3 points. How much support had she gained over the eight-week period?

1. Fill in the blanks.

   a. $\{\ldots, -5, -4, -3, -2, -1, 0, 1, 2, 3, 4, 5, \ldots\}$ is called the set of _____.

   b. The symbols $>$ and $<$ are called _____ symbols.

   c. The _____ _____ of a number is the distance between the number and 0 on the number line.

   d. Two numbers that are the same distance from 0 on the number line, but on opposite sides of it, are called _____.

   e. In the expression $(-3)^5$, the _____ is -3 and 5 is the _____.

   f. To _____ an equation means to find all the values of the variable that make the equation true.

   g. To _____ the solution of an equation, we substitute the value for the variable in the original equation and determine whether the result is a true statement.

2. Insert one of the symbols $>$ or $<$ in the blank to make the statement true.

   a. -8 ___ -9 b. -213 ___ 123 c. -5 ___ 0

3. Tell whether each statement is true or false.

   a. $19 \geq 19$ b. $-(-8) = 8$
   c. $-|-2| > |6|$ d. $-7 + 0 = 0$
   e. $-5(0) = 0$

4. SCHOOL ENROLLMENT According to the projections in the table, which high school will face the greatest shortage of classroom seats in the year 2020?

   High Schools with Shortage of Classroom Seats by 2020

Lyons	-669
Tolbert	$-1,630$
Poly	$-2,488$
Cleveland	-350
Samuels	-586
South	$-2,379$
Van Owen	$-1,690$
Twin Park	-462
Heywood	$-1,004$
Hampton	-774

5. Graph the following numbers on a number line: $-3, 4, -1,$ and 3

   $$\begin{array}{c} \longleftarrow\!|\!\!-\!\!|\!\!-\!\!|\!\!-\!\!|\!\!-\!\!|\!\!-\!\!|\!\!-\!\!|\!\!-\!\!|\!\!-\!\!|\!\!-\!\!|\!\!-\!\!|\!\!\longrightarrow \\ {\scriptstyle -5\ -4\ -3\ -2\ -1\ \ 0\ \ 1\ \ 2\ \ 3\ \ 4\ \ 5} \end{array}$$

6. Add.

   a. $-6 + 3$ b. $-72 + (-73)$
   c. $8 + (-6) + (-9) + 5 + 1$
   d. $(-31 + 12) + [3 + (-16)]$
   e. $-24 + (-3) + 24 + (-5) + 5$

7. Subtract.

   a. $-7 - 6$ b. $-7 - (-6)$
   c. $82 - (-109)$ d. $0 - 15$
   e. $-60 - 50 - 40$

8. Multiply.

   a. $-10 \cdot 7$ b. $-4(-73)$
   c. $-4(2)(-6)$ d. $-9(-3)(-1)(-2)$
   e. $-20,000(1,300)$

9. Write the related multiplication statement for $\dfrac{-20}{-4} = 5$.

10. Divide and check the result.

    a. $\dfrac{-32}{4}$ b. $24 \div (-3)$

    c. $-54 \div (-6)$ d. $\dfrac{408}{-12}$

    e. $-560,000 \div 7,000$

11. a. What is 15 more than -27?
    b. Subtract -19 from -1.
    c. Divide -28 by -7.
    d. Find the product of 10 and the opposite of 8.

12. a. What property is shown: $-3 + 5 = 5 + (-3)$
    b. What property is shown: $-4(-10) = -10(-4)$
    c. Fill in the blank: Subtracting is the same as _____ the opposite.

13. Divide, if possible.

    a. $\dfrac{-21}{0}$ b. $\dfrac{-5}{1}$

    c. $\dfrac{0}{-6}$ d. $\dfrac{-18}{-18}$

14. Evaluate each expression:

    a. $(-4)^2$ b. -4^2

Evaluate each expression.

15. $4 - (-3)^2 - (-6)$ **16.** $-18 \div 2 \cdot 3$

17. $-3 + \left(\dfrac{-16}{4}\right) - 3^3$

18. $94 - 3[-7 + (5 - 8)^2]$

19. $\dfrac{4(-6) - 4^2 + (-2)}{-3 - 4 \cdot 1^5}$

20. $6(-2 \cdot 6 + 5 \cdot 4)$

21. $21 - 9|-3 - 4 + 2|$

22. $-\left[2 - \left(4^3 + \dfrac{20}{-5}\right)\right]$

23. CHEMISTRY In a lab, the temperature of a fluid was reduced 6°F per hour for 12 hours. What signed number represents the change in temperature?

24. GAMBLING On the first hand of draw poker, a player won the chips shown on the left. On the second hand, he lost the chips shown on the right. Determine his net gain or loss for the first two hands. The dollar value of each colored poker chip is shown.

		Value
Won	Lost	= \$1
		= \$5
		= \$10
		= \$25
		= \$100

25. GEOGRAPHY The lowest point on the African continent is the Qattarah Depression in the Sahara Desert, 436 feet below sea level. The lowest point on the North American continent is Death Valley, California, 282 feet below sea level. Find the difference in these elevations.

26. TRAMS A tram line makes a 5,250-foot descent from a mountaintop to the base of the mountain in 15 equal stages. How much does it descend in each stage?

27. CARD GAMES After the first round of a card game, Tommy had a score of 8. When he lost the second round, he had to deduct the value of the cards left in his hand from his first-round score. (See the illustration.) What was his score after two rounds of the game? For scoring, face cards (Kings, Queens, and Jacks) are counted as 10 points and aces as 1 point.

28. BANK TAKEOVERS Before three investors can take over a failing bank, they must repay the losses that the bank had over the past three quarters. If the investors plan equal ownership, how much of the bank's total losses is each investor responsible for?

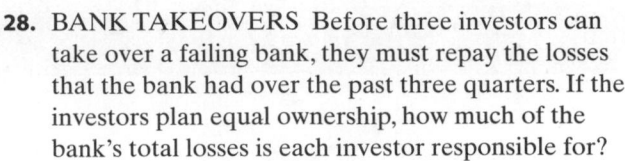

Bank Losses

Solve each equation and check the result.

29. $\dfrac{x}{-4} = 10$

30. $-10 = 6 - x$

31. $c - (-7) = -8$

32. $-6x = 0$

33. $3x + (-7) = -11 + (-11)$

34. $-a = 38$

35. $-5 = -6s + 7$

36. $\dfrac{x}{-2} + 3 = (-2)(-6)$

Let a variable represent the unknown quantity. Then write and solve an equation to answer each question.

37. BANKING After making deposits of \$125 and \$100, a student's account was still \$19 overdrawn. What was her account balance before the deposits?

38. ELEVATORS The weight of the passengers on board an elevator as it traveled from the first to the second floor was 165 pounds *under* capacity. When the doors opened at the second floor, no one exited, and several people entered . The weight of the passengers in the elevator was then 85 pounds *over* capacity.   What was the weight of the people that boarded the elevator on the second floor?

1. Consider the number 7,326,549. [Section 1.1]

   a. What is the place value of the digit 7?

   b. Which digit is in the hundred thousands column?

   c. Round to the nearest hundred.

   d. Round to the nearest ten thousand.

2. BIDS A school district received the bids shown in the table for electrical work. If the lowest bidder wins, which company should be awarded the contract? [Section 1.1]

Citrus Unified School District Bid 02-9899
Cabling and Conduit Installation

Datatel	$2,189,413
Walton Electric	$2,201,999
Advanced Telecorp	$2,175,081
CRF Cable	$2,174,999
Clark & Sons	$2,175,801

3. NUCLEAR POWER The table gives the number of nuclear power plants operating in the United States for selected years. Complete the bar graph using the given data. [Section 1.1]

Year	1978	1983	1988	1993	1998	2003	2008
Plants	70	81	109	110	104	104	104

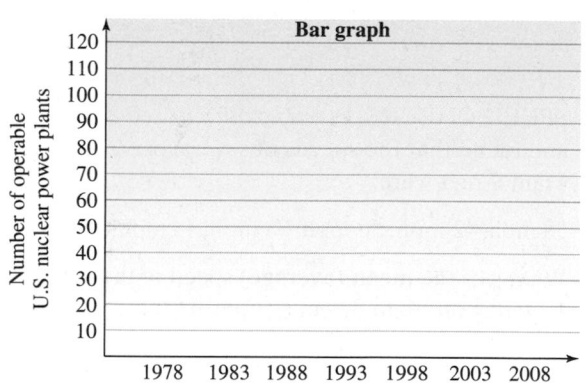

Source: allcountries.org and *The World Almanac and Book of Facts*, 2009

4. THREAD COUNT The thread count of a fabric is the sum of the number of horizontal and vertical threads woven in one square inch of fabric. One square inch of a bed sheet is shown below. Find the thread count. [Section 1.2]

Horizontal count
180 threads

Vertical count
180 threads

Add. [Section 1.2]

5. $1,237 + 68 + 549$

6.
$$\begin{array}{r} 8,907 \\ 2,345 \\ 7,899 \\ + 5,237 \end{array}$$

Subtract. [Section 1.2]

7. $6,375 - 2,569$

8.
$$\begin{array}{r} 5,369 \\ - 685 \end{array}$$

9.
$$\begin{array}{r} 39,506 \\ - 1,729 \end{array}$$

10. Subtract 304 from 1,736. [Section 1.2]

11. Check the subtraction below using addition. Is it correct? [Section 1.2]

$$\begin{array}{r} 469 \\ - 237 \\ \hline 132 \end{array}$$

12. SHIPPING FURNITURE In a shipment of 147 pieces of furniture, 27 pieces were sofas, 55 were leather chairs, and the rest were wooden chairs. Find the number of wooden chairs. [Section 1.2]

Multiply. [Section 1.3]

13. $435 \cdot 27$

14. $\begin{array}{r} 9,183 \\ \times\ 602 \\ \hline \end{array}$

15. $3,100 \cdot 7,000$

16. PACKAGING There are 3 tennis balls in one can, 24 cans in one case, and 12 cases in one box. How many tennis balls are there in one box? [Section 1.3]

17. GARDENING Find the perimeter and the area of the rectangular garden shown below. [Section 1.3]

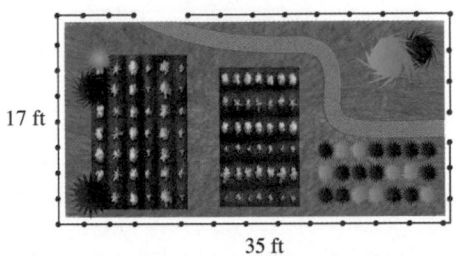

17 ft

35 ft

18. PHOTOGRAPHY The photographs below are the same except that different numbers of *pixels* (squares of color) are used to display them. The number of pixels in each row and each column of the photographs are given. Find the total number of pixels in each photograph. [Section 1.3]

5 pixels

5 pixels

12 pixels

12 pixels

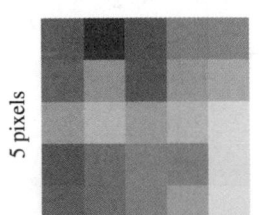

100 pixels

100 pixels

© iStockphoto.com/Aldo Murillo

Divide. [Section 1.4]

19. $\dfrac{701}{8}$

20. $1,261 \div 97$

21. $38\overline{)17,746}$

22. $350\overline{)9,800}$

23. Check the division below using multiplication. Is it correct? [Section 1.4]

$9\overline{)1,962} = 218$

24. GARDENING A metal can holds 320 fluid ounces of gasoline. How many times can the 30-ounce tank of a lawnmower be filled from the can? How many ounces of gasoline will be left in the can? [Section 1.4]

25. BAKING A baker uses 4-ounce pieces of bread dough to make dinner rolls. How many dinner rolls can he make from 15 pounds of dough? (*Hint:* There are 16 ounces in one pound.) [Section 1.4]

26. List the factors of 18, from least to greatest. [Section 1.5]

27. Identify each number as a prime number, a composite number, or neither. Then identify it as an even number or an odd number. [Section 1.5]

a. 17

b. 18

c. 0

d. 1

28. Find the prime factorization of 504. Use exponents to express your answer. [Section 1.5]

29. Write the expression $11 \cdot 11 \cdot 11 \cdot 11$ using an exponent. [Section 1.5]

30. Evaluate: $5^2 \cdot 7$ [Section 1.5]

31. Find the LCM of 8 and 12. [Section 1.6]

32. Find the LCM of 3, 6, and 15. [Section 1.6]

33. Find the GCF of 30 and 48. [Section 1.6]

34. Find the GCF of 81, 108, and 162. [Section 1.6]

Evaluate each expression. [Section 1.7]

35. $16 + 2[14 - 3(5 - 4)^2]$

36. $264 \div 4 - 7(4)2$

37. $\dfrac{4^2 - 2 \cdot 3}{2 + (3^2 - 3 \cdot 2)}$

38. SPEED CHECKS A traffic officer used a radar gun and found that the speeds of several cars traveling on Main Street were:

38 mph, 42 mph, 36 mph, 38 mph, 48 mph, 44 mph

What was the mean (average) speed of the cars traveling on Main Street? [Section 1.7]

39. Use a check to determine whether 6 is a solution of the equation $x - 2 = 4$. [Section 1.8]

40. Tell whether each of the following is an equation. [Section 1.8]

a. $d + 4$

b. $a - 11 = 19$

c. $4 < 5$

d. $\dfrac{x}{6} + 12$

Solve each equation and check the result. [Sections 1.8 and 1.9]

41. $50 = x + 37$

42. $a - 12 = 41$

43. $5p = 135$

44. $\dfrac{y}{8} = 3$

Let a variable represent the unknown quantity. Then write and solve an equation to answer the question.

45. FRANCHISES Dunkin' Donuts would have to open up 22,828 more shops to match the number of Subway stores. If there are 31,663 Subway stores, how many Dunkin' Donuts shops are there? (Sources: dunkindonuts.com and subway.com, 2008 data) [Section 1.8]

46. STADIUMS The May Day Stadium in Pyongyang, North Korea, has the largest nonracing stadium capacity in the world: 150,000 people. This is exactly twice the capacity for a football game at Arizona State's Sun Devil Stadium, in Phoenix, Arizona. What is the capacity of Sun Devil Stadium? (Sources: stubpass.com and worldstadiums.com [Section 1.8]

47. Graph the following integers on a number line. [Section 2.1]

a. $-2, -1, 0, 2$

```
←――┼――┼――┼――┼――┼――┼――┼――→
   -3  -2  -1   0   1   2   3
```

b. The integers greater than -4 but less than 2

```
←――┼――┼――┼――┼――┼――┼――┼――→
   -4  -3  -2  -1   0   1   2
```

48. Find the sum of $-11, 20, -13,$ and 1. [Section 2.2]

Use signed numbers to solve each problem.

49. LIE DETECTOR TESTS A burglar scored -18 on a polygraph test, a score that indicates deception. However, on a second test, he scored $+3$, a score that is uncertain. Find the change in the scores. [Section 2.3]

50. BANKING A student has $48 in his checking account. He then writes a check for $105 to purchase books. The bank honors the check, but charges the student an additional $22 service fee for being overdrawn. What is the student's new checking account balance? [Section 2.3]

51. CHEMISTRY The *melting point* of a solid is the temperature range at which it changes state from solid to liquid. The melting point of helium is seven times colder than the melting point of mercury. If the melting point of mercury is $-39°$ Celsius (a temperature scale used in science), what is the melting point of helium? (Source: chemicalelements.com) [Section 2.4]

52. BUYING A BUSINESS When 12 investors decided to buy a bankrupt company, they agreed to assume equal shares of the company's debt of $660,000. How much debt was each investor responsible for? [Section 2.5]

Evaluate each expression. [Section 2.6]

53. $5 + (-3)(-7)(-2)$

54. $-2[-6(5 \cdot 1^3) - 5]$

55. $\dfrac{10 - (-5)}{1 - 2 \cdot 3}$

56. $\dfrac{3(-6) - 10}{3^2 - 4^2}$

57. $3^4 + 6(-12 + 5 \cdot 4)$

58. $15 - 2|-3 - 4|$

59. $2\left(\dfrac{-12}{3}\right) + 3(-5)$

60. $-9^2 + (-9)^2$

61. $-\left|\dfrac{45}{-9} - (-9)\right|$

62. $\dfrac{-4(-5) - 2}{3 - 3^2}$

For Exercises 55 and 56, quickly determine a reasonable estimate of the exact answer. [Section 2.6]

63. CAMPING Hikers make a 1,150-foot descent into a canyon in 12 stages. How many feet do they descend in each stage?

64. RECALLS An automobile maker has to recall 19,250 cars because they have a faulty engine mount. If it costs $195 to repair each car, how much of a loss will the company suffer because of the recall?

Solve each equation and check the result. [Section 2.7]

65. $m + (-6) = -1$

66. $4 - 5t = 49$

67. $-1 + 7 = x + 2(-9)$

68. $\dfrac{r}{2} + 5 = -13$

Let a variable represent the unknown quantity. Then write and solve an equation to answer the question. [Section 2.7]

69. BANKING After she made deposits of $255 and $395, a business owner's account was still $85 overdrawn. What was the account balance before the deposit?

70. ALCOHOL The freezing point of ethanol alcohol is $-173°$F. By how many degrees must it be heated to reach its boiling point, which is $173°$F? (Source: about.com)

The Language of Algebra

© iStockphoto.com/Dejan Ljamić

from *Campus to Careers*

Broadcasting

It takes many people behind the scenes at radio and television stations to make what we see and hear over the airwaves possible. There are a wide variety of job opportunities in broadcasting for talented producers, directors, writers, editors, audio and video engineers, lighting technicians, and camera operators. These jobs require skills in business and marketing, programming and scheduling, operating electronic equipment, and the mathematical ability to analyze ratings and data.

In **Problem 49** of **Study Set 3.6**, you will see how a television producer determines the amount of commercial time and program time he should schedule for a 30-minute time slot.

JOB TITLE: Broadcasting

EDUCATION: Broadcasting jobs in large markets are usually offered to individuals who have a degree.

JOB OUTLOOK: Employment in broadcasting is expected to increase about 9 percent over the 2006–2016 period.

ANNUAL EARNINGS: Ranges from a low of $25,000 for an entry-level position to $70,000 or more for top positions.

FOR MORE INFORMATION: http://www.bls.gov/oco/cg/cgs017.htm

SECTION **3.1**

Algebraic Expressions

In Chapter 1, we introduced the following strategy for solving application problems.

1. Analyze the problem.
2. Form an equation.
3. Solve the equation.
4. State the conclusion.
5. Check the result.

To successfully form an equation in step 2 of the strategy, we must be able to translate English words and phrases into mathematical symbols.

1 Translate word phrases to algebraic expressions.

Recall that a **variable** is a letter (or symbol) that stands for a number. When we combine variables and numbers using arithmetic operations, the result is an *algebraic expression*.

> ### Algebraic Expressions
>
> Variables and/or numbers can be combined with the operations of addition, subtraction, multiplication, and division to create **algebraic expressions.**

> ***The Language of Mathematics***   We often refer to *algebraic expressions* as simply *expressions*.

Here are some examples of algebraic expressions.

$4a + 7$ This expression is a combination of the numbers 4 and 7, the variable *a*, and the operations of multiplication and addition.

$\dfrac{10 - y}{3}$ This expression is a combination of the numbers 10 and 3, the variable *y*, and the operations of subtraction and division.

Algebraic expressions can contain two (or more) variables.

$15mn(2m)$ This expression is a combination of the numbers 15 and 2, the variables *m* and *n*, and the operation of multiplication.

In order to solve application problems, which are almost always given in words, we must translate those words into mathematical symbols. The following tables show how key words and phrases can be translated into algebraic expressions.

Addition	
the sum of a and 8	$a + 8$
4 plus c	$4 + c$
16 added to m	$m + 16$
4 more than t	$t + 4$
20 greater than F	$F + 20$
T increased by r	$T + r$
exceeds y by 35	$y + 35$

Subtraction	
the difference of 23 and P	$23 - P$
550 minus h	$550 - h$
18 less than w	$w - 18$
7 decreased by j	$7 - j$
M reduced by x	$M - x$
12 subtracted from L	$L - 12$
5 less f	$5 - f$

Caution!   Be careful when translating subtraction. Order is important. For example, when a translation involves the phrase *less than*, note how the terms are reversed.

18 less than w

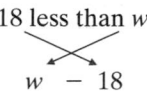

$w \quad - \quad 18$

Multiplication	
the product of 4 and x	$4x$
20 times B	$20B$
twice r	$2r$
double the amount a	$2a$
triple the profit P	$3P$
three-fourths of m*	$\frac{3}{4}m$

* This translation is discussed in more detail in Chapter 4.

Division	
the quotient of R and 19	$\frac{R}{19}$
s divided by d	$\frac{s}{d}$
k split into 4 equal parts	$\frac{k}{4}$
the ratio of c to d*	$\frac{c}{d}$

* This translation is discussed in more detail in Chapter 6.

Caution!   Be careful when translating division. As with subtraction, order is important. For example, s divided by d is *not* written $\frac{d}{s}$.

EXAMPLE 1   Write each phrase as an algebraic expression:

a. twice the profit P

b. 5 less than the capacity c

c. the product of the weight w and 2,000, increased by 300

Strategy We will begin by identifying any key words or phrases.

WHY Key words or phrases can be translated to mathematical symbols.

Solution

a. Key word: twice   **Translation:** multiplication by 2

The algebraic expression is: $2P$.

b. Key phrase: *less than*   **Translation:** subtraction

Sometimes thinking in terms of specific numbers makes translating easier. Suppose the capacity was 100. Then 5 *less than* 100 would be $100 - 5$. If the capacity is c, then we need to make c 5 less. The algebraic expression is: $c - 5$.

Caution! $5 < c$ is the translation of the statement 5 *is less than* the capacity c and not 5 *less than* the capacity c.

c. Key phrase: *product of*   **Translation:** multiplication

Key phrase: *increased by*   **Translation:** addition

In the given wording, the comma after 2,000 means w is first multiplied by 2,000; then 300 is added to that product. The algebraic expression is: $2{,}000w + 300$.

Self Check 1

Write each phrase as an algebraic expression:
a. 80 less than the total t
b. $\frac{2}{3}$ of the time T
c. the difference of twice a and 15, squared

Now Try Problems 15, 17, and 23

2 Write algebraic expressions to represent unknown quantities.

To solve application problems, we let a variable stand for an unknown quantity. We can use the translation skills just discussed to describe any other unknown quantities in the problem by using algebraic expressions.

Self Check 2

COMMUTING   It takes Val *m* minutes to get to work if she drives her car. If she takes the bus, her travel time exceeds this by 15 minutes. Write an algebraic expression that represents the time (in minutes) that it takes her to get to work by bus.

Now Try   Problem 43

EXAMPLE 2   *Banking*   Javier deposited *d* dollars in his checking account. He deposited $500 more than that in his savings account. Write an algebraic expression that represents the amount that he deposited in the savings account.

Strategy We will carefully read the problem, looking for a key word or key phrase.

WHY Then we can translate the key word (or phrase) to mathematical symbols to represent the unknown amount that Javier deposited in the savings account.

Solution The deposit that Javier made to the savings account was $500 *more than* the *d* dollars he deposited in his checking account.

   Key phrase: *more than*   **Translation:**   add

The number of dollars he deposited in the savings account was $d + 500$.

When solving application problems, we are rarely told which variable to use. We must decide what the unknown quantities are and how to represent them using variables.

Self Check 3

CLOTHING SALES   The sale price of a sweater is $20 less than the regular price. Choose a variable to represent one price. Then write an algebraic expression that represents the other price.

Now Try   Problem 55

EXAMPLE 3   *Sports Memorabilia*
The value of the baseball card shown on the right is 4 times that of the football card. Choose a variable to represent the value of one card. Then write an algebraic expression that represents the value of the other card.

Strategy There are two unknowns—the value of the baseball card and the value of the football card. We will let *v* = the value of the football card.

WHY The words of the problem tell us that the value of the baseball card is related to (based on) the value of the football card.

Solution The baseball card's value is *4 times* that of the football card.

   Key phrase: 4 *times*   **Translation:**   multiply by 4

Therefore, 4*v* is the value of the baseball card.

Caution!   A variable is used to represent an unknown number. Therefore, in the previous example, it would be incorrect to write, "Let *v* = football card," because the football card is not a number. We need to write, "Let *v* = the *value* of the football card."

EXAMPLE 4   *Swimming*   A pool is to be sectioned into eight equally wide swimming lanes. Write an algebraic expression that represents the width of each lane.

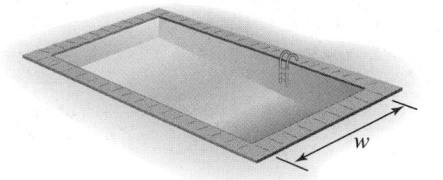

Strategy There are two unknowns—the width of the pool and the width of each lane. We will begin by letting w = the width of the pool (in feet), as shown in the illustration.

WHY The width of each lane is related to (based on) the width of the pool.

Solution The width of the pool is sectioned into *eight equally wide lanes*.

   Key phrase: *eight equally wide lanes*   **Translation:** division by 8

Therefore, the width of each lane is $\frac{w}{8}$ feet.

EXAMPLE 5   *Enrollments*   Second semester enrollment in a nursing program was 32 more than twice that of the first semester. Choose a variable to represent the enrollment for one of the semesters. Then write an algebraic expression that represents the enrollment for the other semester.

Strategy There are two unknowns—the enrollment for the first semester and the enrollment for the second semester. We will begin by letting x = the enrollment for the first semester.

WHY The second-semester enrollment is related to (based on) the first-semester enrollment.

Solution

   Key phrase: *more than*   **Translation:** addition
   Key phrase: *twice that*   **Translation:** multiplication by 2

The second semester enrollment was $2x + 32$.

EXAMPLE 6   *Painting*   A 10-inch-long paintbrush has two parts: a handle and bristles. Choose a variable to represent the length of one of the parts. Then write an algebraic expression to represent the length of the other part.

Strategy There are two approaches. We can let h = the length of the handle or we can let b = the length of the bristles.

WHY Both the length of the handle and the length of the bristles are unknown.

Self Check 4

LOTTOS The payoff for a winning lottery ticket is to be split equally among fifteen friends. Write an algebraic expression that represents each person's share of the prize (in dollars).

Now Try Problem 57

Self Check 5

ELECTIONS In an election, the incumbent received 55 fewer votes than three times the challenger's votes. Choose a variable to represent the number of votes received by one candidate. Then write an algebraic expression that represents the number of votes received by the other.

Now Try Problem 59

Self Check 6

SCHOLARSHIPS Part of a $900 donation to a college went to the scholarship fund, the rest to the building fund. Choose a variable to represent the amount donated to one of the funds. Then write an expression that represents the amount donated to the other fund.

Now Try Problem 63

Somos/Veer/Getty Images

Solution Refer to the first drawing on the right. If we let h = the length of the handle (in inches), then the length of the bristles is $10 - h$.

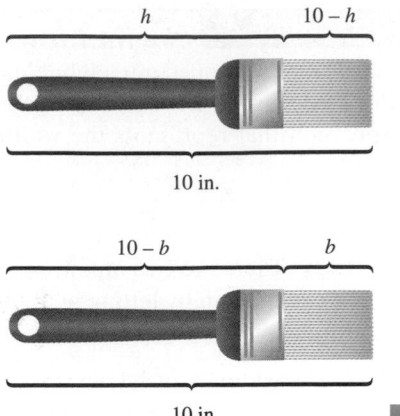

Now refer to the second drawing. If we let b = the length of the bristles (in inches), then the length of the handle is $10 - b$.

Sometimes we must analyze the wording of a problem carefully to detect hidden operations.

Self Check 7

FAMOUS BILLS Bill Cosby was born 9 years before Bill Clinton. Bill Gates was born 9 years after Bill Clinton. Write algebraic expressions to represent the ages of each of these famous men. (Source: celebritybirthdaylist.com)

Now Try Problem 67

EXAMPLE 7   *Engineering*   The Golden Gate Bridge was completed 28 years before the Houston Astrodome was opened. The CN Tower in Toronto was built 10 years after the Astrodome. Write algebraic expressions to represent the ages (in years) of each of these engineering wonders. (Source: Wikipedia)

Strategy There are three unknowns—the ages of the Golden Gate Bridge, the Astrodome, and the CN tower. We will begin by letting x = the age of the Astrodome (in years).

WHY The ages of the Golden Gate Bridge and the CN Tower are both related to (based on) the age of the Astrodome.

Solution Reading the problem carefully, we find that the Golden Gate Bridge was built 28 years before the dome, so its age is more than that of the Astrodome.

Key phrase: *more than*   **Translation:** add

In years, the age of the Golden Gate Bridge is $x + 28$.

The CN Tower was built 10 years after the dome, so its age is less than that of the Astrodome.

Key phrase: *less than*   **Translation:** subtract

In years, the age of the CN Tower is $x - 10$.
The results are summarized in the table at the right.

Engineering feat	Age
Astrodome	x
Golden Gate Bridge	$x + 28$
CN Tower	$x - 10$

EXAMPLE 8   *Packaging*   Write an algebraic expression that represents the number of eggs in d dozen.

Strategy First, we will determine how many eggs are in 1 dozen, 2 dozen, and 3 dozen.

WHY There are no key words or phrases in the problem. It will be helpful to consider some specific cases to determine which operation (addition, subtraction, multiplication, or division) is called for.

Solution If we calculate the number of eggs in 1 dozen, 2 dozen, and 3 dozen (as shown in the table below), a pattern becomes apparent.

Number of dozen	Number of eggs
1	$12 \cdot 1 = 12$
2	$12 \cdot 2 = 24$
3	$12 \cdot 3 = 36$
d	$12 \cdot d = 12d$

We multiply the number of dozen
by 12 to find the number of eggs.

If d = the number of dozen eggs, the number of eggs is $12 \cdot d$, or, more simply, $12d$. ∎

Self Check 8

Complete the table. Then use that information to write an algebraic expression that represents the number of yards in f feet.

Number of feet	Number of yards
3	
6	
9	
f	

Now Try Problems 71 and 75

ANSWERS TO SELF CHECKS

1. a. $t - 80$ **b.** $\frac{2}{3}T$ **c.** $(2a - 15)^2$ **2.** $m + 15$ **3.** p = the regular price of the sweater (in dollars); $p - 20$ = the sale price of the sweater (in dollars)

4. x = the lottery payoff (in dollars); $\frac{x}{15}$ = each person's share (in dollars)

5. x = the number of votes received by the challenger; $3x - 55$ = the number of votes received by the incumbent **6.** s = the amount donated to the scholarship fund (in dollars); $900 - s$ = the amount donated to the building fund (in dollars) **7.** x = the age of Bill Clinton; $x + 9$ = the age of Bill Cosby; $x - 9$ = the age of Bill Gates **8.** $1, 2, 3, \frac{f}{3}$

SECTION 3.1   STUDY SET

VOCABULARY

Fill in the blanks.

1. A _____ is a letter (or symbol) that stands for a number.

2. Variables and/or numbers can be combined with the operations of addition, subtraction, multiplication, and division to create algebraic _____.

3. Phrases such as *increased by* and *more than* indicate the operation of _____. Phrases such as *decreased by* and *less than* indicate the operation of _____.

4. The word *product* indicates the operation of _____. The word *quotient* indicates the operation of _____.

CONCEPTS

5. **a.** Write an algebraic expression that is a combination of the number 10, the variable x, and the operation of addition.

   b. Write an algebraic expression that is a combination of the numbers 3 and 2, the variable t, and the operations of multiplication and subtraction.

6. The illustration below shows the commute to work (in miles) for two men, Mr. Lamb and Mr. Lopez, who work in the same office.

   a. Who lives farther from the office?

   b. How much farther?

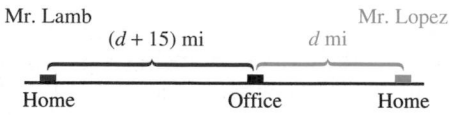

7. Match each algebraic expression to the correct phrase.

   a. $c + 2$ **i.** twice c

   b. $2 - c$ **ii.** c increased by 2

   c. $c - 2$ **iii.** c less than 2

   d. $2c$ **iv.** 2 less than c

8. Fill in the blank to complete the translation.

   a. 16 less than m

$$\boxed{} \; - \; \boxed{}$$

   b. 16 is less than m

        16 $\boxed{}$ m

9. CUTLERY The knife shown below is 12 inches long. Write an algebraic expression that represents the length of the blade (in inches).

10. The following table shows the ages of three family members.

   a. Who is the youngest person shown in the table?

   b. Who is the oldest person listed in the table?

   c. On whose age are the ages in the table based?

	Age (years)
Matthew	x
Sarah	$x - 8$
Joshua	$x + 2$

11. Complete the table. Then fill in the blank.

Number of decades	Number of years
1	
2	
3	
d	

We _____ the number of decades by 10 to find the number of years.

12. Complete the table. Then fill in the blank.

Number of inches	Number of feet
12	
24	
36	
i	

We _____ the number of inches by 12 to find the number of feet.

NOTATION

13. Write each algebraic expression in simpler form.

   a. $x \cdot 8$ **b.** $5(t)$ **c.** $10 \div g$

14. Consider the phrase:

       the product of 5 and w increased by 30

   Insert a comma in the phrase so that it translates to $5w + 30$.

GUIDED PRACTICE

Translate each phrase to an algebraic expression. If no variable is given, use x as the variable. **See Example 1.**

15. The sum of the length l and 15

16. The difference of a number and 10

17. The product of a number and 50

18. Three-fourths of the population p

19. The ratio of the amount won w and lost l

20. The tax t added to c

21. P increased by two-thirds of p

22. 21 less than the total height h

23. The square of k, minus 2,005

24. s subtracted from S

25. 1 less than twice the attendance a

26. J reduced by 500

27. 1,000 split n equal ways

28. Exceeds the cost c by 25,000

29. 90 more than twice the current price p

30. 64 divided by the cube of y

31. 3 times the total of 35, h, and 300

32. Decrease x by -17

33. 680 fewer than the entire population p

34. Triple the number of expected participants

35. The product of d and 4, decreased by 15

36. The quotient of y and 6, cubed

37. Twice the sum of 200 and t

38. The square of the quantity 14 less than x

39. The absolute value of the difference of a and 2

40. The absolute value of a, decreased by 2

41. One-tenth of the distance d

42. Double the difference of x and 18

Write an algebraic expression that represents the unknown quantity. **See Example 2.**

43. GARDENING The height of a hedge was f feet before a gardener cut 2 feet off the top. Write an algebraic expression that represents the height of the trimmed hedge (in feet).

44. SHOPPING A married couple needed to purchase 21 presents for friends and relatives on their holiday gift list. If the husband purchased g presents, write an algebraic expression that represents the number of presents that the wife needs to buy.

45. PACKAGING A restaurant owner purchased s six-packs of cola. Write an algebraic expression that represents the number of cans that this would be.

46. NOISE The highest decibel reading during a rock concert was only 5 decibels shy of that of a jet engine. If a jet engine is normally j decibels, write an algebraic expression that represents the decibel reading for the concert.

47. SUPPLIES A pad of yellow legal paper contains p pages. If a lawyer uses 15 pages every day, write an algebraic expression that represents the number of days that one pad will last.

48. ACCOUNTING The projected cost c (in dollars) of a freeway was too low by a factor of 10. Write an algebraic expression that represents the actual cost of the freeway (in dollars).

49. RECYCLING A campus ecology club collected t tons of newspaper. A Boy Scout troop then contributed an additional 2 tons. Write an algebraic expression that represents the number of tons of newspaper that were collected by the two groups.

50. GRADUATION A graduating class of x people took buses that held 40 students each to an all-night graduation party. Write an algebraic expression that represents the number of buses that were needed to transport the class.

51. STUDYING A student will devote h hours to study for a government final exam. She wants to spread the studying evenly over a four-day period. Write an algebraic expression that represents the number of hours that she should study each day.

52. BASEBALL TEAMS After all c children complete a Little League tryout, the league officials decide that they have enough players for 8 teams of equal size. Write an algebraic expression that represents the number of players that will be on each team.

53. SCOTCH TAPE Suppose x inches of tape have been used off the roll shown below. Write an algebraic expression that represents the number of inches of tape that are left on the roll.

54. MODELING A model's skirt is x inches long. The designer then lets the hem down 2 inches. Write an algebraic expression that represents the length of the altered skirt (in inches).

In Problems 55–58, there are two unknowns. See Examples 3 and 4.

55. GEOMETRY The length of a rectangle is 6 inches longer than its width. Choose a variable to represent one of the unknown dimensions of the rectangle. Then write an algebraic expression that represents the other dimension.

Length

Width

56. PLUMBING The smaller pipe shown below takes three times longer to fill the tank than does the larger pipe. Choose a variable to represent one of the unknown times it takes to fill the tank. Then write an algebraic expression that represents the other time.

57. TRUCK REPAIR The truck radiator shown below was full of coolant. Then three quarts of coolant were drained from it. Choose a variable to represent one of the unknown amounts of coolant in the radiator. Then write an algebraic expression that represents the other amount.

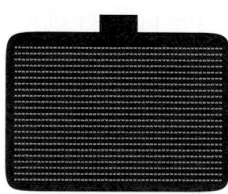

58. SALE PRICES During a sale, the regular price of a CD was reduced by $2. Choose a variable to represent one of the unknown prices of the CD. Then write an algebraic expression that represents the other price.

In Problems 59–62, there are two unknowns. **See Example 5.**

59. GEOGRAPHY Alaska is much larger than Vermont. To be exact, the area of Alaska is 380 square miles more than 50 times that of Vermont. Choose a variable to represent one area. Then write an algebraic expression that represents the other area.

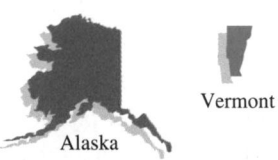

Vermont

Alaska

60. ROAD TRIPS On the second part of her trip, Tamiko drove 20 miles less than three times as far as the first part. Choose a variable to represent the number of miles driven on one part of her trip. Then write an algebraic expression that represents the number of miles driven on the other part.

61. DESSERTS The number of calories in a slice of pie is 100 more than twice the calories in a scoop of ice cream. Choose a variable to represent the number of calories in one type of dessert. Then write an algebraic expression that represents the number of calories in the other type.

62. WASTE DISPOSAL A waste disposal tank buried in the ground holds 15 gallons less than four times what a tank mounted on a truck holds. Choose a variable to represent the number of gallons that one type of tank holds. Then write an algebraic expression that represents the number of gallons the other tank holds.

In Problems 63–66, two approaches are used to represent the unknowns. **See Example 6.**

63. LANDSCAPING

   a. Let b represent the height of the birch tree (in feet) that is shown in the next column. Write an algebraic expression that represents the height of the elm tree (in feet).

   b. Let e stand for the height of the elm tree (in feet). Write an algebraic expression that represents the height of the birch tree (in feet).

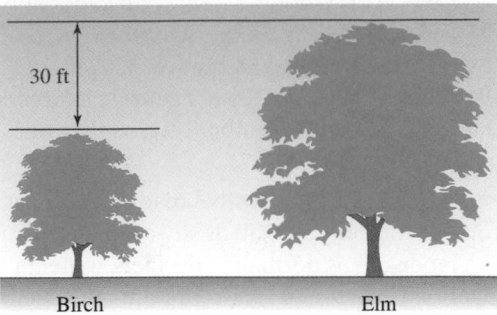

30 ft

Birch   Elm

64. BUILDING MATERIALS

   a. Let b = the length of the beam shown below (in feet). Write an algebraic expression that represents the length of the pipe.

   b. Let p = the length of the pipe (in feet). Write an algebraic expression that represents the length of the beam.

15 ft

65. MARINE SCIENCE

   a. Let s represent the length (in feet) of the great white shark shown below. Write an algebraic expression that represents the length (in feet) of the orca (killer whale).

   b. Let w represent the length (in feet) of the orca (killer whale) shown below. Write an algebraic expression that represents the length (in feet) of the great white shark.

11 ft

Great white shark

Orca (killer whale)

66. WEIGHTS AND MEASURES

   a. Refer to the scale shown on the next page. Which mixture is heavier, A or B? How much heavier is it?

b. Let *a* represent the weight (in ounces) of mixture A. Write an algebraic expression that represents the weight (in ounces) of mixture B.

c. Let *b* represent the weight (in ounces) of mixture B. Write an algebraic expression that represents the weight (in ounces) of mixture A.

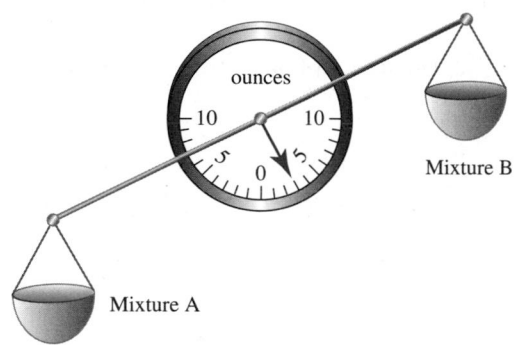

In Problems 67–70, there are three unknowns. **See Example 7.**

67. INVENTIONS The digital clock was invented 11 years before the automatic teller machine (ATM). The camcorder was invented 15 years after the ATM. Write algebraic expressions to represent the ages (in years) of each of these inventions. (Source: Wikipedia)

68. FAMOUS TOMS Tom Petty was born 6 years before Tom Hanks. Tom Cruise was born 6 years after Tom Hanks. Write algebraic expressions to represent the ages of each of these celebrities. (Source: celebritybirthdaylist.com)

69. NEW YORK ARCHITECTURE The Woolworth Building was completed 18 years before the Empire State Building. The United Nations Building was completed 21 years after the Empire State Building. Write algebraic expressions to represent the ages of each of these buildings. (Source: emporis.com)

70. CHILDREN'S BOOKS *The Tale of Peter Rabbit* was first published 24 years before *Winnie-the-Pooh*. *The Cat in the Hat* was first published 31 years after *Winnie-the-Pooh*. Write algebraic expressions to represent the ages of each of these books. (Source: Wikipedia)

Use a table to help answer Problems 71–78. **See Example 8.**

71. Write an algebraic expression that represents the number of seconds in *m* minutes.

72. Write an algebraic expression that represents the number of minutes in *h* hours.

73. Write an algebraic expression that represents the number of inches in *f* feet.

74. Write an algebraic expression that represents the number of feet in *y* yards.

75. Write an algebraic expression that represents the number of centuries in *y* years.

76. Write an algebraic expression that represents the number of decades in *y* years.

77. Write an algebraic expression that represents the number of dozen eggs in *e* eggs.

78. Write an algebraic expression that represents the number of days in *h* hours.

TRY IT YOURSELF

Translate each algebraic expression into words. (Answers may vary.)

79. $\frac{3}{4}r$

80. $\frac{2}{3}d$

81. $t - 50$

82. $c + 19$

83. xyz

84. $10ab$

85. $2m + 5$

86. $2s - 8$

87. A man sleeps *x* hours per day. Write an algebraic expression that represents

   a. the number of hours that he sleeps in a week.

   b. the number of hours that he sleeps in a year (non–leap year).

88. A store manager earns *d* dollars an hour. Write an algebraic expression that represents

    a. the amount of money he will earn in an 8-hour day.

    b. the amount of money he will earn in a 40-hour week.

89. A secretary earns an annual salary of *s* dollars. Write an algebraic expression that represents

    a. her salary per month.

    b. her salary per week.

90. Write an algebraic expression that represents the number of miles in *f* feet. (Hint: There are 5,280 feet in one mile.)

APPLICATIONS

91. ELECTIONS In 1960, John F. Kennedy was elected President of the United States with a popular vote only 118,550 votes more than that of Richard M. Nixon. Choose a variable to represent the number of votes received by one candidate. Then write an algebraic expression that represents the number of votes received by the other candidate.

92. THE BEATLES According to music historians, sales of the Beatles' second most popular single, *Hey Jude,* trail the sales of their most popular single, *I Want to Hold Your Hand,* by 2,000,000 copies. Choose a variable to represent the number of copies sold of one song. Then write an algebraic expression that represents the number of copies sold of the other song.

93. COMPUTER COMPANIES IBM was founded 80 years before Apple Computer. Dell Computer Corporation was founded 9 years after Apple. Let *x* represent the age (in years) of one of the companies. Write algebraic expressions to represent the ages (in years) of the other two companies.

94. VEHICLE WEIGHTS Refer to the illustration below. The car is 1,000 pounds lighter than the van. Choose a variable to represent the weight (in pounds) of one of the vehicles. Then write an algebraic expression that represents the weight (in pounds) of the other vehicle.

95. WITH/AGAINST THE WIND On a flight from Dallas to Miami, a jet airliner, which can fly 500 mph in still air, has a tail wind of *x* mph. The tail wind increases the speed of the jet. On the return flight to Dallas, the airliner flies into a head wind of the same strength. The head wind decreases the speed of the jet. Use this information to complete the table.

Wind conditions	Speed of jet (mph)
In still air	
With the tail wind	
Against the head wind	

96. SUB SANDWICHES Refer to the illustration below. Write an algebraic expression that represents the length (in inches) of the second piece of the sandwich.

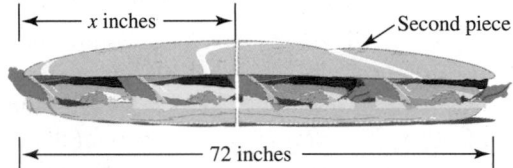

97. SAVINGS ACCOUNTS A student inherited $5,000 and deposits *x* dollars in American Savings. Write an algebraic expression that represents the amount of money (in dollars) left to deposit in a City Mutual account.

$5,000

American Savings
$*x*

City Mutual
$?

98. a. MIXING SOLUTIONS Solution 1 is poured into solution 2. Write an algebraic expression that represents the number of ounces in the mixture.

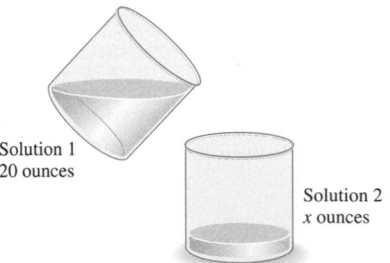

Solution 1
20 ounces

Solution 2
x ounces

b. SNACKS Cashews were mixed with *p* pounds of peanuts to make 100 pounds of a mixture. Write an algebraic expression that represents the number of pounds of cashews that were used.

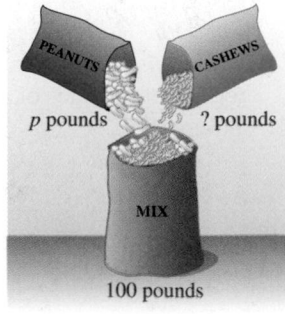

p pounds

? pounds

MIX

100 pounds

99. Explain how variables are used in this section.

100. Explain the difference between the phrases *greater than* and *is greater than*.

101. Suppose in an application problem you were asked to find the unknown height of a building. Explain what is wrong with the following start.

Let x = building

102. What is an algebraic expression?

REVIEW

103. Find the sum: $-5 + (-6) + 1$

104. Evaluate: $2 + 3(-3)$

105. Solve: $-x = 4$

106. Write the related multiplication statement for $\dfrac{-18}{2} = -9$.

107. Write the set of integers.

108. Represent a *deficit* of $1,200 using a signed number.

109. Subtract: $-3 - 2$

110. Evaluate: $(-5)^3$

SECTION 3.2

Evaluating Algebraic Expressions and Formulas

Recall that an **algebraic expression** is a combination of variables and numbers with the operation symbols of addition, subtraction, multiplication, and division. In this section, we will be replacing the variables in algebraic expressions with numbers. Then, using the rule for the order of operations, we will *evaluate* each expression. We will also study formulas. Like algebraic expressions, formulas involve variables.

Objectives

1 Evaluate algebraic expressions.

2 Use formulas from business to solve application problems.

3 Use formulas from science to solve application problems.

4 Find the mean (average) of a set of values.

1 Evaluate algebraic expressions.

EXAMPLE 1 *Plumbing* The manufacturer's instructions for installing a kitchen garbage disposal are shown below.

a. Choose a variable to represent the length of one of the pieces of pipe (A, B, or C). Then write algebraic expressions to represent the lengths of the other two pieces.

b. Suppose model #201 is being installed. Find the length of each piece of pipe that is needed to connect the disposal to the drain line.

Self Check 1

Refer to Example 1. Suppose model #101 is being installed. Find the length of each piece of pipe that is needed to connect the disposal to the drain line.

Now Try Problem 11

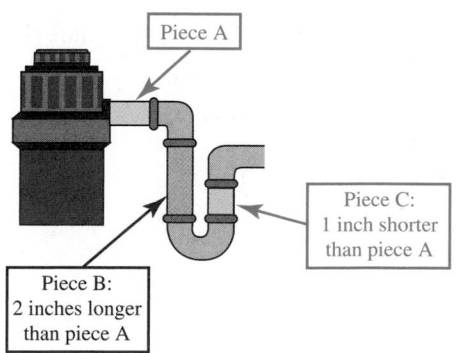

Piece A

Piece C:
1 inch shorter
than piece A

Piece B:
2 inches longer
than piece A

Model	Length of piece A
#101	2 inches
#201	3 inches
#301	4 inches

Strategy There are three lengths of pipe to represent. We will begin by letting x = the length (in inches) of piece A.

WHY The lengths of the other pieces are related to (based on) the length of piece A.

Solution

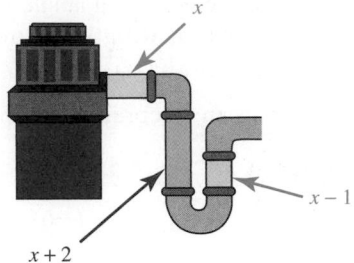

a. Since the instructions call for piece B to be *2 inches longer* than piece A, and the length of piece A is represented by x, we have:

$x + 2$ = the length of piece B (in inches)

Since piece C is to be *1 inch shorter* than piece A,

$x - 1$ = the length of piece C (in inches)

The illustration on the right shows the algebraic expressions that can be used to represent the length of each piece of pipe.

b. If model #201 is being installed, the table tells us that piece A should be 3 inches long. We can find the lengths of the other two pieces of pipe by replacing x with 3 in each of the algebraic expressions.

To find the length of piece B:	**To find the length of piece C:**
Replace x with 3.	Replace x with 3.
$x + 2 = 3 + 2$	$x - 1 = 3 - 1$
$= 5$	$= 2$
Piece B should be 5 inches long.	Piece C should be 2 inches long.

When we substitute given numbers for each of the variables in an algebraic expression and apply the order of operations rule, we are **evaluating the expression.** In the previous example, we say that we *substituted* 3 for x to evaluate the algebraic expressions $x + 2$ and $x - 1$.

> **Caution!** When replacing a variable with its numerical value, we must often write the replacement number within parentheses to convey the proper meaning.

Self Check 2

Evaluate each expression for $y = 5$:

a. $5y - 4$

b. $\dfrac{-y - 15}{2}$

Now Try Problems 15 and 17

EXAMPLE 2   Evaluate each expression for $x = 3$:

a. $2x - 1$ **b.** $\dfrac{-x - 15}{6}$

Strategy We will replace x with the given value of the variable and evaluate the expression using the order of operations rule.

WHY To *evaluate an algebraic expression* means to find its numerical value, once we know the value of its variable.

Solution

a. $2x - 1 = 2(3) - 1$ Substitute 3 for x. Use parentheses.

$\qquad\qquad = 6 - 1$ Do the multiplication first: 2(3) = 6.

$\qquad\qquad = 5$ Do the subtraction.

b. $\dfrac{-x-15}{6}=\dfrac{-(3)-15}{6}$ Substitute 3 for x. Use parentheses. Don't forget to write the − sign in front of (3).

$$=\dfrac{-3-15}{6}$$ Simplify: −(3) = −3.

$$=\dfrac{-3+(-15)}{6}$$ If it is helpful, write the subtraction of 15 as addition of the opposite of 15.

$$\begin{array}{r} 15 \\ +\,3 \\ \hline 18 \end{array}$$

$$=\dfrac{-18}{6}$$ Do the addition: −3 + (−15) = −18.

$$=-3$$ Do the division.

EXAMPLE 3   Evaluate each expression for $a = -2$:

a. $4a^2 - 3a$ **b.** $-a + 3(1 + a)$ **c.** $a^3 - 5$

Strategy We will replace each a in the expression with the given value of the variable and evaluate the expression using the order of operations rule.

WHY To *evaluate an algebraic expression* means to find its numerical value, once we know the value of its variable.

Solution

a. $4a^2 - 3a = 4(-2)^2 - 3(-2)$ Substitute −2 for each a. Use parentheses.

$$= 4(4) - 3(-2)$$ Evaluate the exponential expression: $(-2)^2 = 4$.

$$= 16 - (-6)$$ Do each multiplication.

$$= 16 + 6$$ If it is helpful, write the subtraction of −6 as addition of the opposite of −6.

$$\begin{array}{r} \overset{1}{1}6 \\ +\,6 \\ \hline 22 \end{array}$$

$$= 22$$ Do the addition.

b. $-a + 3(1 + a) = -(-2) + 3[1 + (-2)]$ Substitute −2 for each a. Use parentheses. Don't forget to write the − sign in front of (−2). Since another pair of grouping symbols are now needed, write brackets around $1 + (-2)$.

$$= -(-2) + 3(-1)$$ Do the addition within the brackets.

$$= 2 + (-3)$$ Simplify: −(−2) = 2. Do the multiplication: 3(−1) = −3.

$$= -1$$ Do the addition.

c. $a^3 - 5 = (-2)^3 - 5$ Substitute −2 for a. Use parentheses.

$$= -8 - 5$$ Evaluate the exponential expression: $(-2)^3 = -8$.

$$= -8 + (-5)$$ If it is helpful, write the subtraction of 5 as addition of the opposite of 5.

$$= -13$$ Do the addition.

Self Check 3

Evaluate each expression for $t = -3$:

a. $4t^2 - 2t$

b. $-t + 2(t + 1)$

c. $t^3 + 16$

Now Try Problems 19, 21, and 23

To evaluate algebraic expressions containing two or more variables, we need to know the value of each variable.

Self Check 4

Evaluate each expression for
$r = -1$ and $s = 5$:

a. $(5rs + 4s)^2$

b. $|-8s - 2r|$

Now Try Problems 27 and 29

EXAMPLE 4   Evaluate each expression for $h = -1$ and $g = 5$:
a. $(8hg + 6g)^2$ **b.** $|-5g - 7h|$

Strategy We will replace each h and g in the expression with the given value of the variable and evaluate the expression using the order of operations rule.

WHY To *evaluate an expression* means to find its numerical value, once we know the values of its variables.

Solution

a. $(8hg + 6g)^2 = [8(-1)(5) + 6(5)]^2$ Substitute −1 for h and 5 for g. Use parentheses. Since another pair of grouping symbols are now needed, write brackets around 8(−1)(5) + 6(5).

$= (-40 + 30)^2$ Do the multiplication within the brackets.

$= (-10)^2$ Do the addition within the parentheses.

$= 100$ Evaluate the exponential expression: $(-10)^2 = 100$.

b. $|-5g - 7h| = |-5(5) - 7(-1)|$ Substitute −1 for h and 5 for g. Use parentheses.

$= |-25 - (-7)|$ Do the multiplication within the absolute value symbols.

$= |-25 + 7|$ If it is helpful, write the subtraction of −7 as the addition of the opposite of −7.

$= |-18|$ Do the addition.

$= 18$ Find the absolute value of −18.

2 Use formulas from business to solve application problems.

A **formula** is an equation that is used to state a relationship between two or more variables. Formulas are used in many fields: economics, physical education, biology, automotive repair, and nursing, just to name a few. In this section, we will consider several formulas from business, science, and mathematics.

A formula to find the sale price

If a car that usually sells for $22,850 is discounted $1,500, you can find the **sale price** using the formula

| Sale price | = | original price | − | discount |

Using the variables s to represent the sale price, p the original price, and d the discount, we can write this formula as

$$s = p - d$$

To find the sale price of the car, we substitute 22,850 for p, 1,500 for d, and evaluate the right side of the equation.

$$s = p - d$$ This is the sale price formula.

$$= 22{,}850 - 1{,}500$$ Substitute 22,850 for p and 1,500 for d.

$$= 21{,}350$$ Do the subtraction.

$$\begin{array}{r} 22{,}850 \\ -\ 1{,}500 \\ \hline 21{,}350 \end{array}$$

The sale price of the car is $21,350.

A formula to find the retail price

To make a profit, a merchant must sell a product for more than he paid for it. The price at which he sells the product, called the **retail price,** is the *sum* of what the item cost him and the markup.

$$\boxed{\text{Retail price}} \ = \ \boxed{\text{cost}} \ + \ \boxed{\text{markup}}$$

Using the variables r to represent the retail price, c the cost, and m the markup, we can write this formula as

$$\boxed{r = c + m}$$

As an example, suppose that a store owner buys a lamp for $35 and then marks up the cost $20 before selling it. We can find the retail price of the lamp using this formula.

$$r = c + m$$ This is the retail price formula.

$$= 35 + 20$$ Substitute 35 for c and 20 for m.

$$= 55$$ Do the addition.

The retail price of the lamp is $55.

A formula to find profit

The **profit** a business makes is the *difference* of the revenue (the money it takes in) and the costs.

$$\boxed{\text{Profit}} \ = \ \boxed{\text{revenue}} \ - \ \boxed{\text{costs}}$$

Using the variables p to represent the profit, r the revenue, and c the costs, we have the formula

$$\boxed{p = r - c}$$

EXAMPLE 5   *Films*   It cost Universal Studios about $523 million to make and distribute the film *Jurassic Park*. If the studio has received approximately $920 million to date in worldwide revenue from the film, find the profit the studio has made on this movie. (Source: swivel.com)

Strategy To find the profit, we will substitute the given values in the formula $p = r - c$ and evaluate the expression on the right side of the equation.

WHY The variable p represents the unknown profit.

Self Check 5

FILMS It cost Paramount Pictures about $394 million to make and distribute the film *Forrest Gump*. If the studio has received approximately $679 million to date in worldwide revenue from the film, find the profit the studio has made on this movie. (Source: swivel.com)

Now Try Problem 35

Solution The studio has received $920 million in revenue *r* and the cost *c* to make and distribute the movie was $523 million. To find the profit *p*, we proceed as follows.

$p = r - c$ This is the formula for profit.

$\quad = 920 - 523$ Substitute 920 for *r* and 523 for *c*. The units are millions of dollars.

$\quad = 397$ Do the subtraction.

$$\begin{array}{r} \overset{11}{8}\overset{}{\cancel{9}}\overset{10}{2\cancel{0}} \\ -523 \\ \hline 397 \end{array}$$

Universal Studios has made $397 million in profit on the film *Jurassic Park*.

3 Use formulas from science to solve application problems.

A formula to find the distance traveled

If we know the rate (speed) at which we are traveling and the time we will be moving at that rate, we can find the distance traveled using the formula

$$\text{Distance} = \text{rate} \cdot \text{time}$$

Using the variables *d* to represent the distance, *r* the rate, and *t* the time, we have the formula

$$d = rt$$

Self Check 6

SPEED LIMITS Nevada's speed limit for trucks on rural interstate highways is 75 mph. How far would a truck travel in 3 hours at that speed?

Now Try Problem 39

EXAMPLE 6 *Interstate Speed Limits* Three state speed limits for trucks are shown below. At each of these speeds, how far would a truck travel in 3 hours?

Oregon SPEED LIMIT 55 TRUCKS   Michigan SPEED LIMIT 60 TRUCKS   Virginia SPEED LIMIT 65 TRUCKS

Strategy To find the distance traveled, we will substitute the given values in the formula $d = rt$ and evaluate the expression on the right side of the equation.

WHY The variable *d* represents the unknown distance traveled.

Solution To find the distance traveled by a truck in Oregon, we write

$d = rt$ This is the formula for distance traveled.

$\quad = 55(3)$ Substitute: 55 mph is the rate *r* and 3 hours is the time *t*.

$\quad = 165$ Do the multiplication. The units of the answer are miles.

$$\begin{array}{r}\overset{1}{5}5\\ \times\ 3\\ \hline 165\end{array}$$

At 55 mph, a truck would travel 165 miles in 3 hours. We can use a table to display the calculations for each state.

	r	·	t	=	d
Oregon	55		3		165
Michigan	60		3		180
Virginia	65		3		195

$$\begin{array}{r}60\\ \times\ 3\\ \hline 180\end{array}\qquad \begin{array}{r}\overset{1}{6}5\\ \times\ 3\\ \hline 195\end{array}$$

This column gives the distance traveled, in miles.

> **Caution!** When using $d = rt$ to find distance, make sure that the units are similar. For example, if the rate is given in miles per hour, the time must be expressed in hours.

A formula for converting degrees Fahrenheit to degrees Celsius

Electronic message boards in front of some banks flash two temperature readings. This is because temperature can be measured using the Fahrenheit or the Celsius scales. The **Fahrenheit scale** is used in the American system of measurement, and temperatures are measured in degrees Fahrenheit, written °F. The **Celsius scale** is used in the metric system, and temperatures are measured in degrees Celsius, written °C. The two scales are shown on the thermometers to the right. This should help you to see how the two scales are related. There is a formula to convert a Fahrenheit reading F to a Celsius reading C.

$$C = \frac{5(F - 32)}{9} \;*$$

Later we will see that there is a formula to convert a Celsius reading to a Fahrenheit reading.

*An alternate form of this formula is
$C = \frac{5}{9}(F - 32)$.

Celsius scale   **Fahrenheit scale**

Water boils — 100°C / 210°F
Normal body temperature
Room temperature
Water freezes — 0°C / 30°F

> **The Language of Algebra** In 1724, Daniel Gabriel *Fahrenheit*, a German scientist, introduced the temperature scale that bears his name. The Celsius scale was invented in 1742 by Swedish astronomer Anders *Celsius*.

EXAMPLE 7 **Heating** The thermostat in an office building was set at 77°F. Convert this setting to degrees Celsius.

Strategy To find the temperature in degrees Celsius, we will substitute the given Fahrenheit temperature in the formula $C = \dfrac{5(F - 32)}{9}$ and evaluate the expression on the right side of the equation.

WHY The variable C represents the temperature in degrees Celsius.

Self Check 7

SATURN Change −283°F, the temperature on Saturn, to degrees Celsius.

Now Try Problem 43

Solution

$$C = \frac{5(F - 32)}{9}$$ This is the formula for temperature conversion.

$$= \frac{5(77 - 32)}{9}$$ Substitute the Fahrenheit temperature, 77, for F.

$$= \frac{5(45)}{9}$$ Do the subtraction within the parentheses first: $77 - 32 = 45$.

$$= \frac{225}{9}$$ Do the multiplication: $5(45) = 225$.

$$= 25$$ Do the division.

$$
\begin{array}{r}
77 \\
-\ 32 \\
\hline
45
\end{array}
\qquad
\begin{array}{r}
\overset{2}{4}5 \\
\times\ \ 5 \\
\hline
225
\end{array}
$$

$$
\begin{array}{r}
25 \\
9\overline{)225} \\
-18 \\
\hline
45 \\
-45 \\
\hline
0
\end{array}
$$

The thermostat is set at 25°C.

A formula to find the distance an object falls

The distance an object falls (in feet) when it is dropped from a height is related to the time (in seconds) that it has been falling by the formula

> Distance fallen   =   16 · (time)^2

Using the variables d to represent the distance and t the time, we have

$$\boxed{d = 16t^2}$$

Self Check 8

FREEFALL Find the distance a rock fell in 3 seconds if it was dropped over the edge of the Grand Canyon.

Now Try Problem 47

EXAMPLE 8   *Balloon Rides*   Find the distance a camera fell in 6 seconds if it was dropped overboard by a vacationer taking a hot-air balloon ride.

Strategy To find the distance the camera fell, we will substitute the given time in the formula $d = 16t^2$ and evaluate the expression on the right side of the equation.

WHY The variable d represents the distance fallen.

Solution

$$d = 16t^2$$ This is the formula for distance fallen.

$$= 16(6)^2$$ The camera fell for 6 seconds. Substitute 6 for t.

$$= 16(36)$$ Evaluate the exponential expression: $6^2 = 36$.

$$= 576$$ Do the multiplication.

$$
\begin{array}{r}
36 \\
\times\ 16 \\
\hline
216 \\
360 \\
\hline
576
\end{array}
$$

The camera fell 576 feet.

4   Find the mean (average) of a set of values.

A formula to find the mean (average)

The **mean,** or **average,** of a set of numbers is a value around which the numbers are grouped. To find the mean, we divide the *sum* of all the values by the *number* of values. Writing this as a formula, we get

> Mean   = $\dfrac{\text{sum of the values}}{\text{number of values}}$

Using the variables S to represent the sum and n the number of values, we have

$$\boxed{\text{Mean} = \frac{S}{n}}$$

EXAMPLE 9   *Response Time*   A police department recorded the length of time between incoming 911 calls and the arrival of a police unit at the scene. The response times (in minutes) for one 24-hour period are listed below. Find the mean (average) response time.

Response times

5 min   3 min   6 min   2 min   7 min   4 min   3 min   2 min

Strategy We will count the number of response times and calculate their sum.

WHY To find the mean of a set of values, we divide the sum of the values by the number of values.

Solution There are 8 response times. To find their sum, it is helpful to look for groups of numbers that add to 10.

$$5 + 3 + 6 + 2 + 7 + 4 + 3 + 2 = 32 \qquad 5 + 3 + 2 = 10$$
$$6 + 4 = 10$$
$$7 + 3 = 10$$

Now we use the formula to find the mean.

$$\text{Mean} = \frac{S}{n} \qquad \text{This is the formula to find the mean (average).}$$

$$= \frac{32}{8} \qquad \begin{array}{l} \text{Substitute 32 for } S, \text{ the sum of the response times.} \\ \text{Substitute 8 for } n, \text{ the number of response times.} \end{array}$$

$$= 4 \qquad \text{Do the division.}$$

The mean response time was 4 minutes.

Self Check 9

WEB TRAFFIC The number of hits a website received each day for one week are listed below.
Mon: 392, Tues: 931,
Wed: 842, Thurs: 566,
Fri: 301, Sat: 103, Sun: 43
Find the mean (average) number of hits each day.

Now Try Problem 51

THINK IT THROUGH   *Study Time*

"Your success in school is dependent on your ability to study effectively and efficiently. The results of poor study skills are wasted time, frustration, and low or failing grades."

Effective Study Skills, Dr. Bob Kizlik, 2004

For a course that meets for h hours each week, the formula $H = 2h$ gives the suggested number of hours H that a student should study the course outside of class each week. If a student expects difficulty in a course, the formula can be adjusted upward to $H = 3h$. Use the formulas to complete the table on the right.

If a course meets for:	Suggested study time (hours per week)	Expanded study time (hours per week)
2 hours per week		
3 hours per week		
4 hours per week		
5 hours per week		

ANSWERS TO SELF CHECKS

1. piece B: 4 in.; piece C: 1 in.   **2. a.** 21   **b.** −10   **3. a.** 42   **b.** −1   **c.** −11
4. a. 25   **b.** 38   **5.** Paramount Pictures has made $285 million in profit on the movie *Forrest Gump.*   **6.** 225 mi   **7.** −175°C   **8.** 144 ft   **9.** 454 hits

SECTION 3.2   STUDY SET

VOCABULARY

Fill in the blanks.

1. An algebraic _____ is a combination of variables, numbers, and the operation symbols for addition, subtraction, multiplication, and division.

2. When we substitute 5 for x in the algebraic expression $7x + 10$ and apply the order of operations rule, we are _____ the expression.

3. To evaluate $a^2 + 10a + 1$ for $a = -3$, we _____ -3 for a and apply the order of operations rule.

4. A _____ is an equation that states a relationship between two or more variables.

5. Temperature can be measured using the Fahrenheit or _____ scale.

6. To find the _____ (or average) of a set of values, we divide the sum of the values by the number of values.

CONCEPTS

7. Use variables to write the formula that relates each of the quantities listed below.
   a. Sale price, original price, discount
   b. Profit, revenue, costs
   c. Retail price, cost, markup

8. Use variables to write the formula that relates each of the quantities listed below.
   a. Distance, rate, time
   b. Celsius temperature, Fahrenheit temperature
   c. The distance an object falls when dropped, time
   d. Mean, number of values, sum of values

NOTATION

9. Complete the solution. Evaluate the expression for $a = 5$.
$$9a - a^2 = 9(\quad) - (5)^2$$
$$= 9(5) - \boxed{}$$
$$= \boxed{} - 25$$
$$= 20$$

10. Fill in the blanks. The symbol °F stands for degrees _____ and the symbol °C stands for degrees _____.

GUIDED PRACTICE

In Problems 11–14, write algebraic expressions to represent the three unknowns and then evaluate each of them for the given value of the variable. **See Example 1.**

11. PLAYGROUND EQUIPMENT The plans for building a children's swing set are shown below.
   a. Choose a variable to represent the length (in inches) of one part of the swing set. Then write algebraic expressions that represent the lengths (in inches) of the other two parts.
   b. If the builder chooses to have part 1 be 60 inches long, how long should parts 2 and 3 be?

Part 3: crossbar.
This is to be 16 inches longer than part 1.

Part 2: brace.
This is to be 40 inches less than part 1.

Part 1: leg

12. ART DESIGN A television studio art department plans to construct two sets of decorations out of plywood, using the plan shown below.
   a. Choose a variable to represent the height (in inches) of one piece of plywood. Then write algebraic expressions that represent the heights (in inches) of the other two pieces.
   b. Designers will make the first set of three pieces for the foreground. Piece A will be 15 inches high. How high should pieces B and C be?
   c. Designers will make another set of three pieces for the background. Piece A will be 30 inches high. How high should pieces B and C be?

Piece C–three times as high as piece A

Piece B–twice as high as piece A

Piece A

13. VEHICLE WEIGHTS An H2 Hummer weighs 340 pounds less than twice a Honda Element. A Smart Fortwo car weighs 1,720 pounds less than a Honda Element.

   a. Choose a variable to represent the weight (in pounds) of one car. Then write algebraic expressions that represent the weights (in pounds) of the other two cars.

   b. If the weight of the Honda Element is 3,370 pounds, find the weights of the other two cars.

14. BATTERIES An AAA-size battery weighs 53 grams less than a C-size battery. A D-size battery weighs 5 grams more than twice a C-size battery.

   a. Choose a variable to represent the weight (in grams) of one size battery. Then write algebraic expressions that represent the weights (in grams) of the other two batteries.

   b. If the weight of a C-size battery is 65 grams, find the weights of the other two batteries.

Evaluate each expression for the given value of the variable. See Example 2.

15. $10x - 3$ for $x = 3$ **16.** $4a - 2$ for $a = 9$

17. $\dfrac{-n - 1}{3}$ for $n = 11$ **18.** $\dfrac{-b - 2}{7}$ for $b = 5$

Evaluate each expression for the given value of the variable. See Example 3.

19. $3x^2 - 2x$ for $x = -2$ **20.** $4n^2 - 5n$ for $n = -3$

21. $-y + 3(1 + y)$ for $y = -10$

22. $-b + 6(2 + b)$ for $b = -8$

23. $h^3 - 24$ for $h = -3$ **24.** $t^3 - 30$ for $h = -4$

25. $n^4 + n^2$ for $n = -1$ **26.** $d^4 + d^3$ for $n = -2$

Evaluate each expression for the given values of the variables. See Example 4.

27. $(2ab + 4b)^2$ for $a = -5$ and $b = 2$

28. $(3xy + 2y)^2$ for $x = -4$ and $y = 3$

29. $|-6r - 8s|$ for $r = -11$ and $s = 9$

30. $|-7t - 10x|$ for $t = -12$ and $x = 15$

Use the correct formula to solve each problem. See Objective 2 and Example 5.

31. SPORTING GOODS Find the sale price of a pair of skis that usually sells for $200 but is discounted $35.

32. OFFICE FURNISHINGS If a desk chair that usually sells for $199 is discounted $38, what is the sale price of the chair?

33. CLOTHING STORES A store owner buys a pair of pants for $125 and marks them up $65 for sale. What is the retail price of the pants?

34. SNACKS It costs a snack bar owner 20 cents to make a snow cone. If the markup is 50 cents, what is the retail price of a snow cone?

35. SMALL BUSINESSES On its first night of business, a pizza parlor brought in $445. The owner estimated his costs that night to be $295. What was the profit?

36. FLORISTS For the month of June, a florist's cost of doing business was $3,795. If June revenues totaled $5,115, what was her profit for the month?

37. FUNDRAISERS A school carnival brought in revenues of $13,500 and had costs of $5,300. What was the profit?

38. PRICING A shopkeeper marks up the cost of every item she carries by the amount she paid for the item. If a fan costs her $27, what does she charge for the fan?

Use the correct formula to solve each problem. See Example 6.

39. AIRLINES Find the distance covered by a jet if it travels for 3 hours at 550 mph.

40. ROAD TRIPS Find the distance covered by a car traveling 60 miles per hour for 5 hours.

41. HIKING A hiker can cover 12 miles per day. At that rate, how far will the hiker travel in 8 days?

42. TURTLES A turtle can walk 250 feet per minute. At that rate, how far can a turtle walk in 5 minutes?

Use the correct formula to convert each Fahrenheit temperature to a Celsius temperature. See Example 7.

43. $59°F$ **44.** $113°F$

45. $-4°F$ **46.** $-22°F$

Use the correct formula to solve each problem. See Example 8.

47. FREE FALL Find the distance a ball has fallen 2 seconds after being dropped from a tall building.

48. SIGHTSEEING A visitor to the Grand Canyon accidently dropped her sunglasses over the edge. It took 9 seconds for the sunglasses to fall directly to the bottom of the canyon. How far above the canyon bottom was she standing?

49. BRIDGE REPAIR A steel worker dropped his wrench while tightening a cable on the top of a bridge. It took 4 seconds for the wrench to fall straight to the ground. How far above ground level was the man working?

50. LIGHTHOUSES An object was dropped from the top of the Tybee Island Lighthouse (located near Savanna, Georgia). It took 3 seconds for the object to hit the ground. How tall is the lighthouse?

Use the correct formula to find each mean (average). **See Example 9.**

51. BOWLING Find the mean score for a bowler who rolled scores of 254, 225, and 238.

52. YAHTZEE A player had scores of 288, 192, 264, and 124 at a Yahtzee tournament. What was his mean score?

53. FISHING The weights of each of the fish caught by those on a deep-sea fishing trip are listed below. What was the mean weight?

23 lb   18 lb   37 lb   11 lb   18 lb   26 lb   42 lb   25 lb

54. GRADES Find the mean score of the following test scores: 76, 83, 79, 91, 0, 73.

TRY IT FOR YOURSELF

Evaluate each expression for the given value(s) of the variable(s).

55. $\dfrac{x-8}{2}$ for $x = -4$ **56.** $\dfrac{-10+y}{-4}$ for $y = -6$

57. $-p$ for $p = -4$ **58.** $-j$ for $j = -9$

59. $2(p + 9) + 2p$ for $p = -12$

60. $3(r - 20) + 2r$ for $r = 15$

61. $x^2 - x - 7$ for $x = -5$ **62.** $a^2 + 3a - 9$ for $a = -3$

63. $\dfrac{x-y}{a-b}$ for $x = -1, y = 8, a = 6,$ and $b = 3$

64. $\dfrac{m-n}{c-d}$ for $m = -20, n = -40, c = -5,$ and $d = -10$

65. $\dfrac{-b^2 + 3b}{2b + 1}$ for $b = -4$ **66.** $\dfrac{-a^2 + 5a}{2a + 12}$ for $a = -3$

67. $\dfrac{24 + k}{3k}$ for $k = 3$ **68.** $\dfrac{4 - h}{h - 4}$ for $h = -1$

69. $(x - a)^2 + (y - b)^2$ for $x = -2, y = 1, a = 5,$ and $b = -3$

70. $2a^2 + 2ab + b^2$ for $a = -5$ and $b = -1$

71. $|6 - x|$ for $x = 50$ **72.** $|3c - 1|$ for $c = -1$

73. $-2|x| - 7$ for $x = -7$ **74.** $|x^2 - 7^2|$ for $x = 7$

75. $2 - [10 - x(5h - 1)]$ for $x = -2$ and $h = 2$

76. $1 - [8 - c(2k - 7)]$ for $c = -3$ and $k = 4$

77. $b^2 - 4ac$ for $b = -3, a = 4,$ and $c = -1$

78. $3r^2h$ for $r = 4$ and $h = 2$

79. $\dfrac{x}{y + 10}$ for $x = 30$ and $y = -10$

80. $\dfrac{e}{3f + 24}$ for $e = 24$ and $f = -8$

81. $\dfrac{50 - 6s}{-t}$ for $s = 5$ and $t = 4$

82. $\dfrac{7v - 5r}{-r}$ for $v = 8$ and $r = 4$

83. $5rs^2t$ for $r = 2, s = -3,$ and $t = -3$

84. $-3bk^2t$ for $b = -5, k = -2,$ and $t = -3$

85. $\dfrac{|a^2 - b^2|}{2a - b}$ for $a = -2$ and $b = -5$

86. $\dfrac{-|2x - 3y + 10|}{-3 - y}$ for $x = 0$ and $y = -4$

APPLICATIONS

87. ACCOUNTING Refer to the financial statement for Avon Products, Inc., shown below. Find the operating profit for the year ending January 2008 and the year ending January 2009.

Annual Financials: Income Statement (All dollar amounts in millions)

	Year ending Jan. '08	Year ending Jan. '09
Total revenues	9,939	10,690
Cost of goods sold	3,773	3,946
Operating profit		

(Source: *Business Week*)

88. CONSTRUCTING TABLES Complete the table below by finding the distance traveled in each instance.

	Rate (mph)	·	time (hr)	=	distance (mi)
Bike	12		4		
Walking	3		2		
Car	3		x		

89. DASHBOARDS The illustration below shows part of a dashboard. Explain what each of the three instruments measures. What is the formula that mathematically relates these measurements?

90. SPREADSHEETS A store manager wants to use a spreadsheet to post the prices of items on sale. If column B in the following table lists the regular price and column C lists the discount, write a formula using column names to have a computer find the sale price to print in column D. Then fill in column D with the correct sale price.

	A	B	C	D
1	Bath towel set	$25	$5	
2	Pillows	$15	$3	
3	Comforter	$53	$11	

91. THERMOMETERS A thermometer manufacturer wishes to scale a thermometer in both degrees Celsius and degrees Fahrenheit. Find the missing Celsius degree measures in the illustration.

92. DEALER MARKUPS A car dealer marks up the cars he sells $500 above factory invoice (that is, $500 over what it costs him to purchase the car from the factory).

a. Complete the following table.

Model	Factory invoice ($)	Markup ($)	Price ($)
Minivan	25,600		
Pickup	23,200		
Convertible	x		

b. Write a formula that represents the price p of a car if the factory invoice is f dollars.

93. FALLING OBJECTS See the table below. First, find the distance in feet traveled by a falling object in 1, 2, 3, and 4 seconds. Enter the results in the middle column. Then find the distance the object traveled over each time interval and enter it in the right column.

Time falling	Distance traveled (ft)	Time intervals
1 sec		Distance traveled from 0 sec to 1 sec
2 sec		Distance traveled from 1 sec to 2 sec
3 sec		Distance traveled from 2 sec to 3 sec
4 sec		Distance traveled from 3 sec to 4 sec

94. DISTANCE TRAVELED

a. When in orbit, the space shuttle travels at a rate of approximately 17,250 miles per hour. How far does it travel in one day?

b. The speed of light is approximately 186,000 miles per second. How far will light travel in 1 minute?

c. The speed of a sound wave in air is about 1,100 feet per second at normal temperatures. How far does it travel in half a minute?

95. ENERGY USAGE The number of therms of natural gas that were used each month by a household are listed below. Find the mean number of therms the household used per month that year.

January: 39	May: 22	September: 33
February: 41	June: 23	October: 41
March: 37	July: 16	November: 35
April: 34	August: 16	December: 47

96. CUSTOMER SATISFACTION As customers were leaving a restaurant, they were asked to rate the service they had received. Good service was rated with a 5, fair service with a 3, and poor service with a 1. The tally sheet compiled by the questioner is shown below. What was the restaurant's average score on this survey?

Type of service	Point value	Number
Good	5	ЖЖЖ ЖЖЖ ЖЖЖ
Fair	3	ЖЖ Ж|||
Poor	1	Ж||||

WRITING

97. Explain the error in the student's work shown below.

Evaluate $-a + 3a$ for $a = -6$.

$$-a + 3a = -6 + 3(-6)$$
$$= -6 + (-18)$$
$$= -24$$

98. Explain how we can use a stopwatch to find the distance traveled by a falling object.

99. Write a definition for each of these business words: *revenue, markup,* and *profit.*

100. What is a formula?

101. In this section we *substituted* a number for a variable. List some other uses of the word *substitute* that you encounter in everyday life.

102. Temperature can be measured using the Fahrenheit or the Celsius scale. How do the scales differ?

103. Show the misunderstanding that occurs if we don't write parentheses around -8 when evaluating the expression $2x + 10$ for $x = -8$.

$$2x + 10 = 2 - 8 + 10$$
$$= -6 + 10$$
$$= 4$$

104. Explain why the following instruction is incomplete.
Evaluate the algebraic expression $3a^2 - 4$.

105. What occupation might use a formula that finds:
   a. target heart rate after a workout
   b. gas mileage of a car
   c. age of a fossil
   d. equity in a home
   e. dose to administer
   f. cost-of-living index

106. A car travels at a rate of 65 mph for 15 minutes. What is wrong with the following thinking?

$$d = rt$$
$$= 65(15)$$
$$= 975$$

The car travels 975 miles in 15 minutes.

REVIEW

107. Which of these are prime numbers? 9, 15, 17, 33, 37, 41

108. How can this repeated multiplication be rewritten in simpler form? $2 \cdot 2 \cdot 2 \cdot 2 \cdot 2$

109. Evaluate: $|-2 + (-5)|$

110. Multiply: $-3(-2)(4)$

111. In the equation $\frac{x}{3} = -4$, what operation is performed on the variable?

112. Is -6 a solution of $2t - 3 = 15$? Explain.

113. Subtract: $-3 - (-6)$

114. Which is undefined: division of 0 or division by 0?

SECTION 3.3
Simplifying Algebraic Expressions and the Distributive Property

Objectives
1 Simplify products.

2 Use the distributive property.

3 Distribute a factor of -1.

In algebra, we frequently replace one algebraic expression with another that is equivalent and simpler in form. That process, called *simplifying an algebraic expression,* often involves the use of one or more properties of real numbers.

1 Simplify products.

The commutative and associative properties of multiplication can be used to simplify certain products. For example, let's simplify $8(4x)$.

$8(4x) = 8 \cdot (4 \cdot x)$ Rewrite $4x$ as $4 \cdot x$.

$\quad\quad = (8 \cdot 4) \cdot x$ Use the associative property of multiplication to group 4 with 8.

$\quad\quad = 32x$ Do the multiplication within the parentheses.

We have found that $8(4x) = 32x$. We say that $8(4x)$ and $32x$ are **equivalent expressions** because for each value of x, they represent the same number. For example, if $x = 10$, both expressions have a value of 320. If $x = -3$, both expressions have a value of -96.

If $x = 10$

$8(4x) = 8[4(10)]$ $32x = 32(10)$
$\quad\quad = 8(40)$ $\quad\quad = 320$
$\quad\quad = 320$

same result

If $x = -3$

$8(4x) = 8[4(-3)]$ $32x = 32(-3)$
$\quad\quad = 8(-12)$ $\quad\quad = -96$
$\quad\quad = -96$

same result

> **Success Tip**   By the commutative property of multiplication, we can change the *order* of factors. By the associative property of multiplication, we can change the *grouping* of factors.

EXAMPLE 1   Simplify: **a.** $2 \cdot 7x$ **b.** $-12t(-6)$

Strategy We will use the commutative and associative properties of multiplication to reorder and regroup the factors in each expression.

WHY We want to group all of the numerical factors of an expression together so that we can find their product.

Solution
a. $2 \cdot 7x = (2 \cdot 7)x$ Use the associative property of multiplication to group the numerical factors together.

$\quad\quad = 14x$ Do the multiplication within the parentheses: $2 \cdot 7 = 14$.

b. $-12t(-6) = -12(-6)t$ Use the commutative property of multiplication to change the order of the factors.

$\quad\quad = [-12(-6)]t$ Use the associative property of multiplication to group the numbers together. Use brackets to show this.

$\quad\quad = 72t$ Do the multiplication within the brackets: $-12(-6) = 72$.

$$\begin{array}{r}\overset{1}{12}\\ \times\ 6\\ \hline 72\end{array}$$

Self Check 1
Simplify:

a. $4 \cdot 8r$

b. $-3y(-5)$

Now Try Problems 21 and 25

Self Check 2

Simplify:

a. $-7k \cdot 5t$

b. $2(-3d)(4a)$

Now Try Problems 29 and 33

EXAMPLE 2   Simplify: **a.** $-4m \cdot 5n$ **b.** $2(-4z)(6y)$

Strategy We will use the commutative and associative properties of multiplication to reorder and regroup the factors in each expression.

WHY We want to group all of the numerical factors of an expression together so that we can find their product.

Solution

a. $-4m \cdot 5n = (-4 \cdot 5)(m \cdot n)$ Group the numbers and variables separately, using the commutative and associative properties of multiplication.

$\qquad\qquad = -20mn$ Do the multiplication within the parenthese: $-4 \cdot 5 = -20$. Write $m \cdot n$ as mn.

b. $2(-4z)(6y) = [2(-4)(6)](z \cdot y)$ Use the commutative and associative properties to reorder and regroup the factors. Use brackets to show this.

$\qquad\qquad = -48zy$ Do the multiplication within the brackets: $2(-4)(6) = -48$. Write $z \cdot y$ as zy.

$\qquad\qquad = -48yz$ Standard practice is to write variable factors in alphabetical order: $zy = yz$.

The Language of Algebra   Be careful when using the words *simplify* and *solve*. In mathematics, we *simplify expressions* and we *solve equations*.

---

2 Use the distributive property.

Another property that is often used to simplify algebraic expressions is the **distributive property.** To introduce it, we will evaluate $4(5 + 3)$ in two ways.

Method 1	*Method 2*
Use the order of operations:	*Distribute the multiplication:*
$4(5 + 3) = 4(8)$	$4(5 + 3) = 4(5) + 4(3)$
$\qquad\qquad = 32$	$\qquad\qquad = 20 + 12$
	$\qquad\qquad = 32$

Each method gives a result of 32. This observation suggests the following property.

The Distributive Property

For any numbers a, b, and c,

$$a(b + c) = ab + ac$$

The Language of Algebra   To *distribute* means to give from one to several. You have probably *distributed* candy to children coming to your door on Halloween.

To illustrate one use of the distributive property, let's consider the expression $5(x + 3)$. Since we are not given the value of x, we cannot add x and 3 within the parentheses. However, we can distribute the multiplication by the factor of 5 that is outside the parentheses to x and to 3 and add those products.

$$5(x + 3) = 5(x) + 5(3) \qquad \text{Distribute the multiplication by 5.}$$
$$= 5x + 15 \qquad \text{Do the multiplication.}$$

In the expression $5(x + 3)$, we say that there are two *terms* within the parentheses, x and 3. In general, a **term** is a product or quotient of numbers and/or variables. A single number or variable is also a term. Some examples of terms are:

$$4, \qquad -22, \qquad y, \qquad -6r, \qquad x^2, \qquad \text{and} \qquad 15ab$$

We will discuss terms in more detail in the next section.

Since subtraction is the same as adding the opposite, the distributive property also holds for subtraction.

The Distributive Property

For any numbers a, b, and c,

$$a(b - c) = ab - ac$$

EXAMPLE 3 Multiply: **a.** $3(x + 7)$ **b.** $6(5x - 1)$

Strategy In each case, we will distribute the multiplication by the factor *outside* the parentheses over each term *within* the parentheses.

WHY In each case, we cannot simplify the expression within the parentheses. To multiply, we must use the distributive property.

Solution

a. We read $3(x + 7)$ as "three times the *quantity* of x plus seven." The word *quantity* alerts us to the grouping symbols in the expression.

$$3(x + 7) = 3 \cdot x + 3 \cdot 7 \qquad \text{Distribute the multiplication by 3.}$$
$$= 3x + 21 \qquad \text{Do the multiplication. Try to go to this step immediately.}$$

Caution! A common mistake is to forget to distribute the multiplication over each of the terms within the parentheses.

$$3(x + 7) = 3x + 7$$

b. $6(5x - 1) = 6 \cdot 5x - 6 \cdot 1 \qquad \text{Distribute the multiplication by 6.}$
$$= 30x - 6 \qquad \text{Do the multiplication. Try to go to this step immediately.} \qquad \blacksquare$$

Self Check 3

Multiply:

a. $5(h + 4)$

b. $9(2a - 3)$

Now Try Problems 37 and 41

The Language of Algebra Formally, it is called the *distributive property of multiplication over addition*. When we use it to write a product, such as $3(x + 7)$, as a sum, $3x + 21$, we say that we have *removed* or *cleared* the parentheses.

Self Check 4

Multiply:

a. $-4(6y + 8)$

b. $-7(2 - 8m)$

c. $-10(-9r - 5)$

d. $-1(x - 3)$

Now Try Problems 45, 49, 53, and 57

EXAMPLE 4   Multiply: **a.** $-3(4x + 2)$ **b.** $-9(3 - 2t)$

c. $-6(-3y - 8)$ **d.** $-1(t - 9)$

Strategy In each case, we will distribute the multiplication by the factor *outside* the parentheses over each term *within* the parentheses.

WHY In each case, we cannot simplify the expression within the parentheses. To multiply, we must use the distributive property.

Solution

a. $-3(4x + 2) = -3(4x) + (-3)(2)$ Distribute the multiplication by -3.

$= -12x + (-6)$ Do the multiplication.

$= -12x - 6$ Write the answer in simpler form. Adding -6 is the same as subtracting 6. Try to go to this step immediately.

b. $-9(3 - 2t) = -9(3) - (-9)(2t)$ Distribute the multiplication by -9.

$= -27 - (-18t)$ Do the multiplication.

$= -27 + 18t$ Write the answer in simpler form. Add the opposite of $-18t$. Try to go to this step immediately.

c. $-6(-3y - 8) = -6(-3y) - (-6)(8)$ Distribute the multiplication by -6.

$= 18y - (-48)$ Do the multiplication.

$= 18y + 48$ Write the result in simpler form. Add the opposite of -48. Try to go to this step immediately.

Another approach is to write the subtraction within the parentheses as addition of the opposite. Then we distribute the multiplication by -6 over the addition.

$-6(-3y - 8) = -6[-3y + (-8)]$ Add the opposite of 8.

$= -6(-3y) + (-6)(-8)$ Distribute the multiplication by -6.

$= 18y + 48$ Do the multiplication.

d. $-1(t - 9) = -1(t) - (-1)(9)$ Distribute the multiplication by -1.

$= -t - (-9)$ Do the multiplication.

$= -t + 9$ Write the result in simpler form. Add the opposite of -9. Try to go to this step immediately.

Notice that distributing the multiplication by -1 *changes the sign* of each term within the parentheses.

Success Tip It is common practice to write answers in simplified form. For instance, the answer to Example 4, part a, is expressed as $-12x - 6$ because it involves fewer symbols than $-12x + (-6)$. For the same reason, the answer to Example 4, part b, is given as $-27 + 18t$ instead of $-27 - (-18t)$.

> **Caution!** The distributive property does not apply to every expression that contains parentheses—only those where multiplication is distributed over addition (or subtraction). For example, to simplify $6(5x)$, we do not use the distributive property.
>
> | **Correct** | **Incorrect** |
> |---|---|
> | $6(5x) = (6 \cdot 5)x = 30x$ | $6(5x) = 30 \cdot 6x = 180x$ |

The distributive property can be extended to several other useful forms. Since multiplication is commutative, we have:

$$(b + c)a = ba + ca \qquad\qquad (b - c)a = ba - ca$$

For situations in which there are more than two terms within parentheses, we have:

$$a(b + c + d) = ab + ac + ad \qquad a(b - c - d) = ab - ac - ad$$

EXAMPLE 5 Multiply: **a.** $(5 + 3r)7$ **b.** $(4 - x)2$ **c.** $2(a - 3b)8$
d. $-6(-3x - 6y + 8)$

Strategy We will multiply each term within the parentheses by the factor (or factors) outside the parentheses.

WHY In each case, we cannot simplify the expression within the parentheses. To multiply, we use the distributive property.

Solution

a. $(5 + 3r)\mathbf{7} = (5)7 + (3r)\mathbf{7}$ Distribute the multiplication by 7.

$\qquad\quad = 35 + 21r$ Do the multiplication. Try to go to this step immediately.

b. $(4 - x)\mathbf{2} = (4)2 - (x)\mathbf{2}$ Distribute the multiplication by 2.

$\qquad\quad = 8 - 2x$ Do the multiplication.

c. This expression contains 3 factors.

$2(a - 3b)\mathbf{8} = 2 \cdot \mathbf{8}(a - 3b)$ Use the commutative property of multiplication to reorder the factors.

$\qquad\quad = 16(a - 3b)$ Multiply 2 and 8 to get 16.

$\qquad\quad = 16(a) - 16(3b)$ Distribute the multiplication by 16.

$\qquad\quad = 16a - 48b$ Do the multiplication.

d. There are three terms within the parentheses.

$\mathbf{-6}(-3x - 6y + 8)$

$\quad = \mathbf{-6}(-3x) - (\mathbf{-6})(6y) + (\mathbf{-6})(8)$ Distribute the multiplication by −6.

$\quad = 18x - (-36y) + (-48)$ Do the multiplication.

$\quad = 18x + 36y - 48$ Write the answer in simplest form. Try to go to this step immediately.

Self Check 5

Multiply:

a. $(8 + 7x)5$

b. $(5 - c)3$

c. $4(m - 6n)2$

d. $-2(-7c - 4d + 1)$

Now Try Problems 61, 67, 69, and 75

3 Distribute a factor of −1.

We can use the distributive property to find the opposite of a sum. For example, to find $-(x + 10)$, we interpret the — symbol as a factor of −1, and proceed as follows:

$$-(x + 10) = -\mathbf{1}(x + 10) \qquad \text{Replace the — symbol with −1.}$$
$$= -\mathbf{1}(x) + (-\mathbf{1})(10) \qquad \text{Distribute the multiplication by −1.}$$
$$= -x + (-10) \qquad \text{Do the multiplication.}$$
$$= -x - 10 \qquad \text{Write the answer in simplest form.}$$

In general, we have the following property.

The Opposite of a Sum

The opposite of a sum is the sum of the opposites.
    For any numbers a and b,

$$-(a + b) = -a + (-b)$$

Self Check 6

Simplify: $-(-5x + 18)$

Now Try Problem 77

EXAMPLE 6   Simplify: $-(-9s - 3)$

Strategy We will multiply each term within the parentheses by −1.

WHY The — outside the parentheses represents a factor of −1 that is to be distributed.

Solution

$$-(-9s - 3) = -\mathbf{1}(-9s - 3) \qquad \begin{array}{l}\text{Replace the — symbol in front} \\ \text{of the parentheses with −1.}\end{array}$$
$$= -\mathbf{1}(-9s) - (-\mathbf{1})(3) \qquad \text{Distribute the multiplication by −1.}$$
$$= 9s - (-3) \qquad \text{Do the multiplication.}$$
$$= 9s + 3 \qquad \begin{array}{l}\text{Write the answer in simplest form. Try to go} \\ \text{to this step immediately}\end{array}$$

Success Tip   After working several problems like Example 6, you will notice that it is not necessary to show each of the steps. The result can be obtained very quickly by *changing the sign of each term within the parentheses and dropping the parentheses.*

ANSWERS TO SELF CHECKS

1. a. $32r$ **b.** $15y$ **2. a.** $-35kt$ **b.** $-24ad$ **3. a.** $5h + 20$ **b.** $18a - 27$
4. a. $-24y - 32$ **b.** $-14 + 56m$ **c.** $90r + 50$ **d.** $-x + 3$ **5. a.** $40 + 35x$
b. $15 - 3c$ **c.** $8m - 48n$ **d.** $14c + 8d - 2$ **6.** $5x - 18$

SECTION 3.3   STUDY SET

VOCABULARY

Fill in the blanks.

1. To _____ the expression $5(6x)$ means to write it in simpler form: $5(6x) = 30x$.

2. $5(6x)$ and $30x$ are _____ expressions because for each value of x, they represent the same number.

3. In the expression $2(x + 8)$, there are two _____ within the parentheses, x and 8.

4. To perform the multiplication $2(x + 8)$, we use the _____ property.

5. When we use the distributive property to write a product, such as $7(4y + 3)$, as the sum $28y + 21$, we say we have _____ or *cleared* the parentheses.

6. We call $-(c + 9)$ the _____ of a sum.

CONCEPTS

7. **a.** Fill in the blanks to simplify the expression.

   $$4(9t) = (\;\;\; \cdot 9)t = \;\;\; t$$

   b. What property did you use in part a?

8. **a.** Fill in the blanks to simplify the expression.

   $$-6y \cdot 2 = \;\;\; \cdot 2 \cdot y = \;\;\; y$$

   b. What property did you use in part a?

9. State the distributive property using the variables x, y, and z.

10. Fill in the blanks.

    a. $2(x + 4) = 2x \;\;\; 8$
    b. $2(x - 4) = 2x \;\;\; 8$
    c. $-2(x + 4) = -2x \;\;\; 8$
    d. $-2(-x - 4) = 2x \;\;\; 8$

11. Fill in the blanks: Distributing multiplication by -1 changes the ____ of each term within the parentheses.

    $$-(x + 10) = \;\;\; (x + 10) = -x \;\;\; 10$$

12. For each of the following expressions, determine whether the distributive property applies. Write *yes* or *no*.

    a. $3(5t)$ **b.** $3(t + 5)$
    c. $5(3 \cdot t)$ **d.** $(3t)5$
    e. $(3)(-t)5$ **f.** $(5 - t)3$

13. **a.** Simplify: $6(4x)$
    b. Remove parentheses: $6(4 + x)$

14. Explain what the arrows are illustrating.

$$\overset{\curvearrowright}{-9(y - 7)}$$

NOTATION

Complete each solution.

15. $-5 \cdot 7n = (\;\;\; \cdot 7)n$
    $ = \;\;\; n$

16. $6y(-9) = 6(\;\;\;)y$
    $ = [6(\;\;\;)]y$
    $ = -54y$

17. $9(5y - 4) = \;\;\;(5y) - \;\;\;(4)$
    $ = \;\;\; - 36$

18. $4(2a + b - 1) = 4(\;\;\;) + 4(\;\;\;) - \;\;\;(1)$
    $ = \;\;\; + 4b - 4$

19. Write each expression in simpler form, using fewer mathematical symbols.

    a. $-(-x)$
    b. $x - (-5)$
    c. $10y + (-15)$
    d. $5 \cdot x$

20. In each expression, determine what number is to be distributed.

    a. $6(x - 2)$ **b.** $(t + 1)(-5)$
    c. $(a + 24)8$ **d.** $-(z - 16)$

GUIDED PRACTICE

Simplify. **See Example 1.**

21. $2 \cdot 6x$	**22.** $4 \cdot 7b$
23. $5 \cdot 8y$	**24.** $12 \cdot 6t$
25. $-10t(-10)$	**26.** $-8k(-6)$
27. $-15a(-3)$	**28.** $-11n(-9)$

Simplify. **See Example 2.**

29. $-7x \cdot 9y$	**30.** $-13a \cdot 2b$
31. $-4r \cdot 4s$	**32.** $-7x \cdot 7y$
33. $2(-5x)(3y)$	**34.** $4(-3y)(4z)$
35. $5r(2)(-3b)$	**36.** $4d(5)(-3e)$

Multiply. **See Example 3.**

37. $4(x + 1)$	**38.** $5(y + 3)$
39. $7(b + 2)$	**40.** $8(k + 7)$
41. $9(3e - 3)$	**42.** $10(7t - 2)$
43. $3(2q - 7)$	**44.** $6(3p - 1)$

Multiply. See Example 4.

45. $-2(3h + 5)$

46. $-5(7t + 3)$

47. $-10(4y + 6)$

48. $-9(2t + 9)$

49. $-8(2q - 4)$

50. $-2(22x - 1)$

51. $-5(7g - 1)$

52. $-7(3p - 8)$

53. $-4(-5s - 3)$

54. $-6(-3d - 1)$

55. $-6(-15t - 9)$

56. $-4(-5d - 6)$

57. $-1(x - 5)$

58. $-1(y - 1)$

59. $-1(5d + 8)$

60. $-1(6w + 2)$

Multiply. See Example 5.

61. $(4d + 7)6$

62. $(8r + 2)7$

63. $(3q + 20)7$

64. $(30x + 12)3$

65. $(-4 - d)6$

66. $(-9 - j)5$

67. $(t - 12)9$

68. $(x - 25)6$

69. $2(4t - 3)3$

70. $3(9m - 2)2$

71. $4(3h + 1)5$

72. $4(2w + 1)6$

73. $-3(-3z - 3x + 5)$

74. $-10(-5e - 4a + 6)$

75. $-8(2a + 4b - 6)$

76. $-9(3r + 6s - 9)$

Simplify. See Example 6.

77. $-(-3w - 4)$

78. $-(-4y - 6)$

79. $-(-18x - 19)$

80. $-(-50n - 100)$

81. $-(x + 3)$

82. $-(5 + y)$

83. $-(4t + 5)$

84. $-(8x + 4)$

TRY IT YOURSELF

Perform the indicated operations.

85. $(-13c - 3)(-6)$

86. $(-10s - 11)(-2)$

87. $(4s)3$

88. $(9j)7$

89. $5(-7q)$

90. $-7(5t)$

91. $6(-6c + 7)$

92. $9(-9d + 3)$

93. $3(3x - 7y + 2)$

94. $5(4 - 5r + 8s)$

95. $-5 \cdot 8h$

96. $-8 \cdot 4d$

97. $5 \cdot 8c \cdot 2$

98. $3 \cdot 6j \cdot 2$

99. $2(3t + 2)8$

100. $3(2q + 1)9$

101. $(-1)(-2e)(-4)$

102. $(-1)(-5t)(-1)$

103. $-(5x - 4y + 1)$

104. $-(6r - 5f + 1)$

Each expression is the result of an application of the distributive property. What was the original algebraic expression?

105. $2(4x) + 2(5)$

106. $3(3y) + 3(7)$

107. $-3(4y) + (-3)(2)$

108. $-5(11s) + (-5)(11t)$

109. $3(4) - 3(7t) - 3(5s)$

110. $2(7y) + 2(8x) - 2(4)$

111. $-4(5) - 3x(5)$

112. $-8(7) - (4s)(7)$

WRITING

113. Explain what it means to simplify an algebraic expression. Give an example.

114. Explain how to apply the distributive property. Give an example.

115. Use the word *distribute* in a sentence that describes a situation from everyday life.

116. Explain why the distributive property applies to $2(3 + x)$ but does not apply to $2(3x)$.

117. Explain the mistake: $5(6x + 2) = 30x + 2$

118. The distributive property can be demonstrated using the following illustration.

    a. Fill in the blanks: Two groups of 6 plus three groups of 6 is ▢ groups of 6.

      Therefore,

$$▢ \cdot 2 + ▢ \cdot 3 = 6(▢ + ▢)$$

    b. Draw a diagram that illustrates $5 \cdot 4 + 5 \cdot 6 = 5(4 + 6)$.

REVIEW

119. Evaluate: $|-6 + 1|$

120. Subtract: $-1 - (-4)$

121. Identify the operation associated with each word: *product, quotient, difference, sum.*

122. What steps are used to find the mean (average) of a set of values?

123. Insert the proper inequality symbol: -6 ▢ -7

124. Fill in the blank: To factor a number means to express it as the _____ of other whole numbers.

125. Which of the following involve area: carpeting a room, fencing a yard, walking around a lake, painting a wall?

126. Write seven squared and seven cubed.

SECTION 3.4
Combining Like Terms

In this section, we will show how the distributive property can be used to simplify algebraic expressions that involve addition and subtraction. We will also review the concept of perimeter and write the formulas for the perimeter of a rectangle and a square using variables.

1 Identify terms and coefficients of terms.

Addition symbols separate expressions into parts called *terms*. For example, the expression $x + 8$ has two terms.

$$\underset{\text{First term}}{x} \quad + \quad \underset{\text{Second term}}{8}$$

Since subtraction can be written as addition of the opposite, the expression $a^2 - 3a - 9$ has three terms.

$$a^2 - 3a - 9 = \underset{\text{First term}}{a^2} \quad + \quad \underset{\text{Second term}}{(-3a)} \quad + \quad \underset{\text{Third term}}{(-9)}$$

In general, a **term** is a product or quotient of numbers and/or variables. A single number or variable is also a term. Examples of terms are:

$$4, \quad y, \quad 6r, \quad -w^3, \quad 7x^5, \quad \frac{3}{n}, \quad -15ab^2$$

> ***Caution!*** By the commutative property of multiplication, $r6 = 6r$ and $-15b^2a = -15ab^2$. However, when writing terms, we usually write the numerical factor first and the variable factors in alphabetical order.

EXAMPLE 1   Identify the terms of each expression:
a. $3x^2 + 5x + 8$ **b.** $-24rs$ **c.** $a - 5 - 3a + 10$

Strategy We will locate the addition symbols in each expression.

WHY Addition symbols separate expressions into terms.

Solution

a. After locating the addition symbols in the given expression, we see that it has three terms: $3x^2, 5x,$ and 8.

$$\underset{\text{First term}}{3x^2} \quad + \quad \underset{\text{Second term}}{5x} \quad + \quad \underset{\text{Third term}}{8}$$

b. Since the given expression does not contain any addition symbols, it has only one term, $-24rs$.

c. When we write each subtraction in the expression $a - 5 - 3a + 10$ as addition of the opposite, we see that it has four terms: $a, -5, -3a,$ and 10.

$$a + (-5) + (-3a) + 10$$

Self Check 1

Identify the terms of each expression:

a. $12y^2 + y + 10$

b. $-4ab$

c. $9 - m - 6m + 12$

Now Try **Problems 23 and 27**

It is important to be able to distinguish between the *terms* of an expression and the *factors* of a term.

Self Check 2

Is *b* used as a *factor* or a *term* in each expression?
a. $-27b$
b. $5a + b$

Now Try **Problems 31 and 33**

EXAMPLE 2   Is *m* used as a *factor* or a *term* in each expression?

a. $m + 6$ **b.** $8m$

Strategy We will begin by determining whether *m* is involved in an addition or a multiplication.

WHY Addition symbols separate expressions into *terms*. A *factor* is a number being multiplied.

Solution

a. Since *m* is added to 6, *m* is a term of $m + 6$.

b. Since *m* is multiplied by 8, *m* is a factor of $8m$.

The numerical factor of a term is called the **coefficient** of the term. For instance, the term $6r$ has a coefficient of 6 because $6r = 6 \cdot r$. The coefficient of $-15ab^2$ is -15 because $-15ab^2 = -15 \cdot ab^2$. More examples are shown below.

A term such as 4, that consists of a single number, is called a **constant term.**

Term	Coefficient	
$8y^2$	8	
$9pq$	9	
$-78m$	-78	
$-2b^2$	-2	
x	1	Because x = 1x
$-y$	-1	Because −y = −1y
27	27	The coefficient of a constant term is that constant.

Notice that when there is no number in front of a variable, the coefficient is understood to be 1. For example, the coefficient of the term *x* is 1. If there is only a negative (or opposite) sign in front of the variable, the coefficient is understood to be -1. Therefore, $-y$ can be thought of as $-1y$.

> **The Language of Algebra**   Terms such as *x* and *y* have *implied* coefficients of 1. *Implied* means suggested without being precisely expressed.

Self Check 3

Identify the coefficient of each term in the expression:
$p^3 - 12p^2 + 3p - 4$

Now Try **Problems 35 and 41**

EXAMPLE 3   Identify the coefficient of each term in the expression: $7x^2 - x + 6$

Strategy We will begin by writing the subtraction as addition of the opposite. Then we will determine the numerical factor of each term.

WHY Addition symbols separate expressions into terms.

Solution If we write $7x^2 - x + 6$ as $7x^2 + (-x) + 6$, we see that it has three terms: $7x^2$, $-x$, and 6. The numerical factor of each term is its coefficient.

The coefficient of $7x^2$ is **7** because $7x^2$ means $\mathbf{7} \cdot x^2$.

The coefficient of $-x$ is **-1** because $-x$ means $\mathbf{-1} \cdot x$.

The coefficient of the constant 6 is 6.

2 Identify like terms.

Before we can discuss methods for simplifying algebraic expressions involving addition and subtraction, we need to introduce some new vocabulary.

Like Terms

Like terms are terms containing exactly the same variables raised to exactly the same powers. Any constant terms in an expression are considered to be like terms. Terms that are not like terms are called **unlike terms.**

Here are several examples.

Like terms	*Unlike terms*	
$4x$ and $7x$	$4x$ and $7y$	The variables are not the same.
$-10p^2$ and $25p^2$	$-10p$ and $25p^2$	Same variable, but different powers.
$8c^3d$ and c^3d	$8c^3d$ and c^3	The variables are not the same.

Success Tip   When looking for like terms, don't look at the coefficients of the terms. Consider only the variable factors of each term. If two terms are like terms, only their coefficients may differ.

EXAMPLE 4   Identify the like terms in each expression:

a. $7r + 5 + 3r$ **b.** $6x^4 - 6x^2 - 6x$ **c.** $-17m^3 + 3 - 2 + m^3$

Strategy First, we will identify the terms of the expression. Then we will look for terms that contain the same variables raised to exactly the same powers.

WHY If two terms contain the same variables raised to the same powers, they are like terms.

Solution

a. $7r + 5 + 3r$ contains the like terms $7r$ and $3r$.

b. Since the exponents on x are different, $6x^4 - 6x^2 - 6x$ contains no like terms.

c. $-17m^3 + 3 - 2 + m^3$ contains two pairs of like terms: $-17m^3$ and m^3 are like terms, and the constant terms, 3 and -2, are like terms.

Self Check 4

Identify the like terms in each expression:

a. $2x - 2y + 7y$

b. $5p^2 - 12 + 17p^2 + 2$

Now Try Problems 43 and 47

3 Combine like terms.

To add or subtract objects, they must be similar. For example, fractions that are to be added must have a common denominator. When adding decimals, we align columns to be sure to add tenths to tenths, hundredths to hundredths, and so on. The same is true when working with terms of an algebraic expression. They can be added or subtracted only if they are like terms.

This expression can be simplified because it contains like terms.

$$3x + 4x$$

This expression *cannot* be simplified because its terms are not like terms.

$$3x + 4y$$

Recall that the distributive property can be written in the following forms:

$$(b + c)a = ba + ca \qquad (b - c)a = ba - ca$$

We can use these forms of the distributive property in reverse to simplify a sum or difference of like terms. For example, we can simplify $3x + 4x$ as follows:

$$3x + 4x = (3 + 4)x \quad \text{Use the form: } ba + ca = (b + c)a.$$
$$= 7x \qquad\quad \text{Do the addition within the parentheses.}$$

> **Success Tip**   Just as 3 apples plus 4 apples is 7 apples,
>
> $$3x + 4x = 7x$$

We can simplify $15m^2 - 9m^2$ in a similar way:

$$15m^2 - 9m^2 = (15 - 9)m^2 \quad \text{Use the form: } ba - ca = (b - c)a.$$
$$= 6m^2 \qquad\qquad \text{Do the subtraction within the parentheses.}$$

> **The Language of Algebra**   Simplifying a sum or difference of like terms is called *combining like terms*. In each example above, we say that we combined like terms.

These examples suggest the following general rule.

Combining Like Terms

Like terms can be combined by adding or subtracting the coefficients of the terms and keeping the same variables with the same exponents.

Self Check 5

Simplify, if possible:

a. $3x + 5x$

b. $-6y + (-6y) + 9y$

c. $4s^4 - 2s^4$

d. $4a - 2$

e. $10r + 6r - 9r$

Now Try Problems 51, 55, 59, 63, and 67

EXAMPLE 5   Simplify by combining like terms, if possible:

a. $2x + 9x$ **b.** $-8p + (-2p) + 4p$ **c.** $5s^3 - 3s^3$ **d.** $4w + 6$ **e.** $8a + 2a - 3a$

Strategy We will use the distributive property in reverse to add (or subtract) the coefficients of the like terms. We will keep the same variables raised to the same powers.

WHY To *combine like terms* means to add or subtract the like terms in an expression.

Solution

a. Since $2x$ and $9x$ are like terms with the common variable x, we can combine them.

$$2x + 9x = 11x \quad \text{Think: } (2 + 9)x = 11x.$$

b. $-8p + (-2p) + 4p = -6p \quad \text{Think: } [-8 + (-2) + 4]p = -6p.$

c. $5s^3 - 3s^3 = 2s^3 \quad \text{Think: } (5 - 3)s^3 = 2s^3.$

d. Since $4w$ and 6 are not like terms, they cannot be combined. The expression $4w + 6$ does not simplify.

e. $8a + 2a - 3a = 7a \quad \text{Think: } (8 + 2 - 3)a = 7a.$

EXAMPLE 6   Simplify by combining like terms:

a. $16t - 15t$ **b.** $16t^2 - t^2$ **c.** $15t - 16t$ **d.** $16t + t$

Strategy As we combine like terms, we must be careful when working with terms such as t and $-t$.

WHY Coefficients of 1 and -1 are usually not written.

Solution

a. $16t - 15t = t$ Think: $(16 - 15)t = 1t = t$.

b. $16t^2 - t^2 = 15t^2$ Think: $16t^2 - 1t^2 = (16 - 1)t^2 = 15t^2$.

c. $15t - 16t = -t$ Think: $(15 - 16)t = -1t = -t$.

d. $16t + t = 17t$ Think: $16t + 1t = (16 + 1)t = 17t$.

Self Check 6
Simplify:
a. $9h - h$ **b.** $9h + h$
c. $9h - 8h$ **d.** $8h - 9h$
Now Try Problems 71 and 77

EXAMPLE 7   Simplify: $6a^2 + 54a - 4a - 36$

Strategy First, we will identify any like terms in the expression. Then we will use the distributive property in reverse to combine them.

WHY To *simplify* an expression we use properties of real numbers to write an equivalent expression in simpler form.

Solution We can combine the like terms that involve the variable a.

$6a^2 + \mathbf{54a} - \mathbf{4a} - 36 = 6a^2 + \mathbf{50a} - 36$ Think: $(54 - 4)a = 50a$.

Self Check 7
Simplify: $7y^2 + 21y - 2y - 6$
Now Try Problem 79

EXAMPLE 8   Simplify: $4(x + 5) - 5 - (2x - 4)$

Strategy First, we will remove the parentheses. Then we will identify any like terms and combine them.

WHY To *simplify* an expression we use properties of real numbers, such as the distributive property, to write an equivalent expression in simpler form.

Solution Here, the distributive property is used both *forward* (to remove parentheses) and in *reverse* (to combine like terms).

$4(x + 5) - 5 - (2x - 4) = 4(x + 5) - 5 - \mathbf{1}(2x - 4)$ Replace the $-$ symbol in front of $(2x - 4)$ with -1.

$= \mathbf{4x} + 20 - 5 - \mathbf{2x} + 4$ Distribute the multiplication by 4 and -1.

$= \mathbf{2x} + 19$ Think: $(4 - 2)x = 2x$.
Think: $(20 - 5 + 4) = 19$.

Self Check 8
Simplify:
$6(3y - 1) + 2 - (-3y + 4)$
Now Try Problem 83

4   Find the perimeter of a rectangle and square.

To develop the formula for the perimeter of a rectangle, we let l = the length of the rectangle and w = the width of the rectangle, as shown on the right. Then

$P = l + w + l + w$ The perimeter is the distance around the rectangle.

$= 2l + 2w$ Combine like terms: $l + l = 2l$ and $w + w = 2w$.

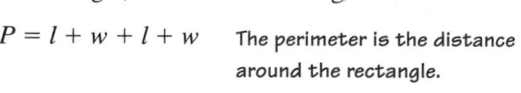

The Formula for the Perimeter of a Rectangle

The **perimeter P of a rectangle** with length l and width w is given by

$$P = 2l + 2w$$

To develop the formula for the perimeter of a square, we let s = the length of a side of the square, as shown on the right. Then

$P = s + s + s + s$ Add the lengths of the four sides.

$= 4s$ Combine like terms. Recall that $s = 1s$.

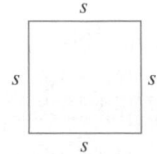

The Formula for the Perimeter of a Square

The **perimeter of a square** with sides of length s is given by

$$P = 4s$$

Self Check 9

ENERGY CONSERVATION   Refer to the figure in Example 9. Find the cost to weatherstrip around the door and window if the door is 8 feet tall and 3 feet wide and the window is 5 feet long and 3 feet high.

Now Try   Problem 119

EXAMPLE 9   *Energy Conservation*

Refer to the figure to the right. Find the cost to weatherstrip around the front door and the window of the house if the material costs 20¢ a foot.

Analyze

- The door is in the shape of a rectangle.   Given
- The window is in the shape of a square.   Given
- The weatherstripping material costs 20¢ a foot.   Given
- What will it cost to weatherstrip around the front door and window?   Find

Form

Let P = the total perimeter, and translate the words of the problem into an equation.

The total perimeter	is	the perimeter of the door	plus	the perimeter of the window.
P	$=$	$2l + 2w$	$+$	$4s$ Use the formulas for the perimeter of a rectangle and a square.

Solve

$P = 2l + 2w + 4s$ This is a formula for the combined perimeter.

$= 2(\mathbf{7}) + 2(\mathbf{3}) + 4(\mathbf{3})$ Substitute 7 for *l*, 3 for *w*, and 3 for *s*.

$= 14 + 6 + 12$ Do the multiplication.

$= 32$ Do the addition.

$$\begin{array}{r} \overset{1}{1}4 \\ 6 \\ +12 \\ \hline 32 \end{array}$$

The total perimeter is 32 feet. At 20¢ a foot, the total cost will be $(32 \cdot 20)$¢.

$$\begin{array}{r} 32 \\ \times\ 20 \\ \hline 640 \end{array}$$

State

It will cost 640¢ or $6.40 to weatherstrip around the front door and window.

Check

We can check the result by estimation. The perimeter is approximately 30 feet, and $30 \cdot 20 = 600$¢, which is $6. The answer, $6.40, seems reasonable. ∎

ANSWERS TO SELF CHECKS

1. a. $12y^2, y, 10$ **b.** $-4ab$ **c.** $9, -m, -6m, 12$ **2. a.** factor   **b.** term
3. $1, -12, 3, -4$ **4. a.** $-2y$ and $7y$ **b.** $5p^2$ and $17p^2$; -12 and 2 **5. a.** $8x$ **b.** $-3y$
c. $2s^4$ **d.** does not simplify   **e.** $7r$ **6. a.** $8h$ **b.** $10h$ **c.** h **d.** $-h$
7. $7y^2 + 19y - 6$ **8.** $21y - 8$ **9.** 760¢ or $7.60

SECTION 3.4   STUDY SET

VOCABULARY

Fill in the blanks.

1. A _____ is a number or a product of a number and one or more variables. A single number or variable is also a _____.

2. A _____ is a number being multiplied.

3. In the term $5t$, 5 is called the _____.

4. A term that consists of a single number is called a _____ term.

5. Terms such as x and y have an _____ coefficient of 1.

6. Terms with exactly the same variables raised to exactly the same powers are called _____ terms.

7. When we write $9x + x$ as $10x$, we say that we have _____ like terms.

8. The _____ of a rectangle is the distance around it.

CONCEPTS

Fill in the blanks.

9. The expression $5x - 10 + 8x$ has ▢ terms. The second term is ▢. The coefficient of the third term is ▢.

10. The expression $2a - 12 + 5a + 15$ has ▢ terms. The third term is ▢. The coefficient of the first term is ▢.

11. The term $8m$ has a coefficient of 8 because $8m = ▢ \cdot ▢$.

12. Just as 5 pencils plus 6 pencils is 11 _____, $5x + 6x = 11▢$.

13. Are the given pair of terms *like* or *unlike* terms?
    a. $6a$ and $6b$
    b. $5x^2$ and $5x^3$
    c. $3mn$ and $3m^2n$
    d. 15 and 16

14. What exponent must appear in each box to have like terms?
    a. $6x^2, 3x^▢$ c. $7a^3, 21a^▢$
    b. $-8h^5, -5h^▢$ d. $25n^4, -15n^▢$

NOTATION

Complete each solution to simplify the expression.

15. $2x + 3x = (▢ + ▢)x$
    $ = ▢x$

16. $16w^2 - 12w^2 = (▢ - ▢)w^2$
    $ = ▢w^2$

17. $2(x - 1) + 3x = 2x - ▢ + 3x$
    $ = ▢ - 2$

18. $-3(1 - b) - b = -3 + ▢ - b$
    $ = ▢ - 3$

19. In the formula $P = 2l + 2w$,
    a. what does P represent?
    b. what does $2l$ mean?
    c. what does $2w$ mean?

20. In the formula $P = 4s$,
    a. what does P represent?
    b. what does $4s$ mean?

21. Determine whether each statement is true or false.

    a. $x = 1x$ **c.** $100yx = 100xy$

    b. $-y = -1y$ **d.** $7x - x = 7$

22. Fill in the blank:

$$y^2 - 7y - 3 = y^2 + (\quad) + (\quad)$$

GUIDED PRACTICE

Identify the terms of each expression. **See Example 1.**

23. $3x^2 - 9x + 4$ **24.** $y^2 - 12y + 6$

25. $5 + 5t - 8t - 1$ **26.** $3x - y - 5x + y$

27. $-35a$ **28.** $-7t$

29. $9mn - 6n$ **30.** $3rs - 2r$

Determine whether x is used as a factor or as a term. **See Example 2.**

31. a. $x + 12$ **b.** $7x$

32. a. $12x + 12y - 6$ **b.** $x - 36y$

33. a. $5x(-10)$ **b.** $8 + x - z$

34. a. $100 + x + z$ **b.** $-xz$

Identify the coefficient of each term in the expression. **See Example 3.**

35. $5x^2 + x - 12$ **36.** $9y^2 + y - 8$

37. $a^3 - 27$ **38.** $b^3 - 64$

39. $xy - x + y + 10$

40. $mn + m - n - 4$

41. $-a^2 + 6b^2 - a + 5$

42. $-8x^3 - 4x^2 + 3x + 1$

Identify the like terms in each expression. **See Example 4.**

43. $8x + 7 + 2x$ **44.** $9y + 12 + 11y$

45. $5y^2 + 5y - 5$ **46.** $2m^2 - 2m + 2$

47. $-3k^3 + 6k + k^3 - 3k$

48. $-r^4 - 2r^3 - 9r^4 + 5r^3$

49. $12a - 8 + 15a + 1$

50. $33t + 4 + 18t - 9$

Simplify by combining like terms, if possible. **See Example 5.**

51. $6t + 9t$ **52.** $7r + 5r$

53. $20b + 30b$ **54.** $18c + 12c$

55. $-5x + (-6x) + 2x$ **56.** $-8m + (-6m) + 7m$

57. $-5d + (-9d) + 10d$ **58.** $-4a + (-12a) + 11a$

59. $5s^2 - 3s^2$ **60.** $8y^2 - 5y^2$

61. $3e^3 - 17e^3$ **62.** $2s^3 - 14s^3$

63. $h - 7$ **64.** $j - 8$

65. $14z - 8z + 2z$ **66.** $9w - 3w + 8w$

67. $53a + 6a - 21a$ **68.** $72n + 8n - 35n$

69. $2x + 2y$ **70.** $5a - 5b$

Simplify by combining like terms. **See Example 6.**

71. $10s - 9s$ **72.** $7q - 6q$

73. $40a^2 - a^2$ **74.** $13z^2 - z^2$

75. $6m - 7m$ **76.** $4h - 5h$

77. $14r + r$ **78.** $21w + w$

Simplify. **See Example 7.**

79. $5x^2 + 19x - 3x + 6$ **80.** $2b^2 - 6b + 12b + 1$

81. $y^2 - 8y - 2y - 4$ **82.** $n^2 - 4n - 7n - 3$

Simplify. **See Example 8.**

83. $5(m + 2) - 8 - (3m - 1)$

84. $7(r + 1) - 9 - (2r - 4)$

85. $4(x - 1) - 2 - (x + 5)$

86. $10(x - 1) - 6 - (x + 8)$

Use the formulas from this section to find the perimeter of each figure. **See Example 9.**

87. A rectangle with length 16 feet and width 7 feet

88. A rectangle with length 24 inches and width 11 inches

89. A square with a side 37 yards long

90. A square with a side 98 miles long

TRY IT YOURSELF

Simplify each expression, if possible.

91. $-3x^3 - 4x^3$ **92.** $-7y^4 - 9y^4$

93. $-4(-4y + 5) + 4 - 6(y + 2)$

94. $-3(-6y - 8) - 15 - 4(5 - y)$

95. $6t + 9 + 5t + 3$ **96.** $5x + 3 + 5x + 4$

97. $x + x + x + x$ **98.** $s - s - s$

99. $3t - (t - 8)$ **100.** $6n - (4n + 1)$

101. $5(2x)(-5)$ **102.** $2(-3x)(3)$

103. $2a + 2b$ **104.** $9y - 9$

105. $4x^2 - 3x - 7 + 4x^2 - 2 - x$

106. $2a^2 + 8 - a - 5 + 5a^2 - 9a$

107. $-6s + 6s$ **108.** $19c + (-19c)$

109. $-4r + 8R + 2R - 3r + R$

110. $12a - A - a - 8A - a$

111. $0 - 2y^3$ **112.** $0 - 7x^4$

113. $5(3 - 2s) + 4(2 - 3s) + 19s$

114. $7(y + 1) + 8(2y + 3) + 12y$

APPLICATIONS

115. a. COMMUTING The illustration below shows the distances (in miles) that two men live from the office where they both work. Write an algebraic expression that represents the *total distance* that the two men live from the office.

b. BOTANY Write an algebraic expression that represents the sum of the heights (in feet) of the two trees shown below.

116. THE RED CROSS In 1891, Clara Barton founded the Red Cross. Its symbol is a white flag bearing a red cross. If each side of the cross has length x, write an algebraic expression that represents the perimeter of the cross.

117. PING-PONG Write an algebraic expression that represents the perimeter of the Ping-Pong table in feet.

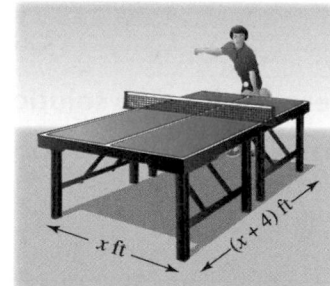

118. SEWING Write an algebraic expression that represents the length (in cm) of the blue trim needed to outline a pennant with the given side lengths.

119. MOBILE HOMES The design of a mobile home calls for a thin strip of stained pine around the outside of all four exterior sides, as shown in brown below. If the strip costs 80¢ a running foot, how much will be spent on the pine used for the trim? (*Hint:* the left and right sides have the same design, as do the front and back of the mobile home.)

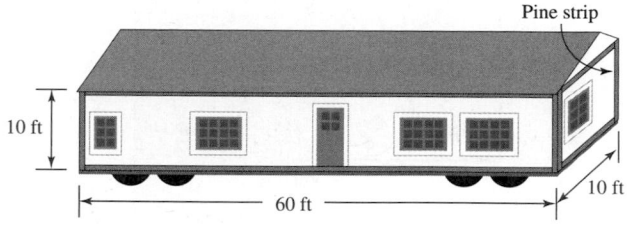

120. LANDSCAPING A landscape architect has designed a planter surrounding two birch trees, as shown below. The planter is to be outlined with redwood edging in the shape of a rectangle and two squares. If the material costs 17¢ a running foot, how much will the redwood cost for this project?

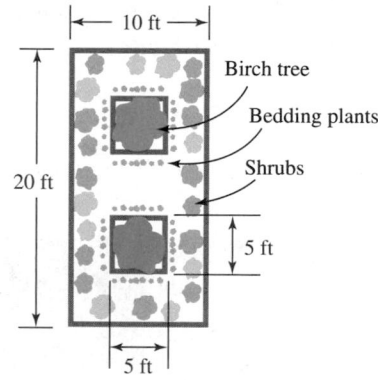

121. PARTY PREPARATIONS The appropriate size of a dance floor for a given number of dancers can be determined from the table shown. Find the perimeter of each of the dance floors listed.

Slow dancers	Fast dancers	Size of floor (in feet)
8	5	9 × 9
14	9	12 × 12
22	15	15 × 15
32	20	18 × 18
50	30	21 × 21

122. COASTAL DRILLING The map shows an area of the California coast where oil drilling is planned. Use the scale to estimate the lengths of the sides of the area highlighted on the map. Then find its perimeter.

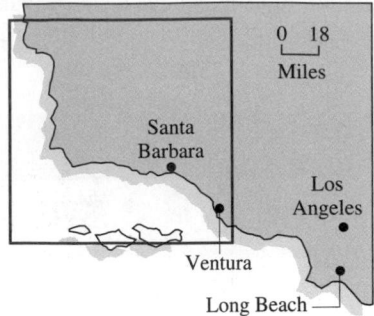

123. Explain what it means for two terms to be like terms.

124. Explain what it means to say that the coefficient of x is an *implied* (or *understood*) 1.

125. Explain the difference between a term and a factor. Give some examples.

126. When simplifying an algebraic expression, some students use underlining, as shown below. What purpose does the underlining serve?

$$\underline{3y} + \underline{4} + \underline{5y} + \underline{8}$$

■ REVIEW

127. Solve: $-4t - 3 = -11$

128. Find the prime factorization of 100. Use exponents in your answer.

129. Evaluate: $\left| -5^2 + (-3)^2 \right|$

130. A store manager earns d dollars an hour. Write an algebraic expression that represents

    a. the amount of money he will earn in an 8-hour day.

    b. the amount of money he will earn in a 40-hour week.

Objectives

1 Determine whether a number is a solution.

2 Combine like terms to solve equations.

3 Solve equations that have variable terms on both sides.

4 Use the distributive property to solve equations.

5 Apply a strategy to solve equations.

SECTION 3.5
Simplifying Expressions to Solve Equations

We must often simplify algebraic expressions to solve equations. Sometimes it will be necessary to combine like terms in order to isolate the variable on one side of the equation. At other times, it will be necessary to apply the distributive property to write an equation in a form that can be solved. In this section, we will discuss both of these situations.

1 Determine whether a number is a solution.

Recall that a number that makes an equation true when substituted for the variable is called a **solution** and is said to **satisfy** the equation.

Self Check 1

Is 25 a solution of
$10 - x = 35 - 2x$?

Now Try **Problem 19**

EXAMPLE 1   Is 9 a solution of $3y - 1 = 2y + 7$?

Strategy We will substitute 9 for each y in the equation and evaluate the expression on the left side and the expression on the right side separately.

WHY If a true statement results, 9 is a solution of the equation. If we obtain a false statement, 9 is not a solution.

Solution

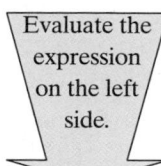 Evaluate the expression on the left side.

$$3y - 1 = 2y + 7$$
$$3(9) - 1 \overset{?}{=} 2(9) + 7 \qquad \text{Read } \overset{?}{=} \text{ as "is}$$
$$27 - 1 \overset{?}{=} 18 + 7 \qquad \text{possibly equal to."}$$
$$26 = 25$$

Evaluate the expression on the right side.

Since $26 = 25$ is false, 9 is *not* a solution of $3y - 1 = 2y + 7$. ■

2 Combine like terms to solve equations.

Recall that **like terms** are terms containing exactly the same variables raised to exactly the same powers. Like terms can appear on the left side of an equation, on the right side of an equation, or on both sides. When asked to solve such equations, we should combine the like terms first before using a property of equality.

EXAMPLE 2   Solve: $7x - 4x = 15$

Strategy We will begin by combining the like terms on the left side of the equation.

WHY It is best to simplify the algebraic expressions on each side of an equation before using a property of equality.

Solution

$$7x - 4x = 15 \qquad \text{This is the equation to solve.}$$
$$3x = 15 \qquad \text{Combine like terms: } 7x - 4x = 3x.$$
$$\frac{3x}{3} = \frac{15}{3} \qquad \text{To isolate x, undo the multiplication by 3 by dividing both sides by 3.}$$
$$x = 5 \qquad \text{Do the division.}$$

To check, we substitute 5 for x in the original equation and evaluate the left side.

$$7x - 4x = 15 \qquad \text{This is the original equation.}$$
$$7(5) - 4(5) \overset{?}{=} 15 \qquad \text{Substitute 5 for x.}$$
$$35 - 20 \overset{?}{=} 15 \qquad \text{Do the multiplication.}$$
$$15 = 15 \qquad \text{Do the subtraction.}$$

Since the statement $15 = 15$ is true, 5 is the solution of $7x - 4x = 15$. ■

Self Check 2

Solve: $8r - 6r = 16$. Check the result.

Now Try Problem 23

EXAMPLE 3   Solve: $100 + 248 = t - 20 + t$

Strategy We will begin by combining the like terms on the left side and on the right side of the equation.

WHY It is best to simplify the algebraic expressions on each side of an equation before using a property of equality.

Self Check 3

Solve: $150 + 5 = d - 1 + 3d$. Check the result.

Now Try Problem 27

Solution

$100 + 248 = t - 20 + t$	This is the equation to solve.	$\begin{array}{r} 348 \\ + 20 \\ \hline 368 \end{array}$
$348 = 2t - 20$	Combine like terms on each side of the equation: $100 + 248 = 348$ and $t + t = 2t$.	
$348 + 20 = 2t - 20 + 20$	To isolate $2t$, undo the subtraction of 20 by adding 20 to both sides.	$\begin{array}{r} 184 \\ 2\overline{)368} \\ -2 \\ \hline 16 \\ -16 \\ \hline 08 \\ -8 \\ \hline 0 \end{array}$
$368 = 2t$	Do the addition.	
$\dfrac{368}{2} = \dfrac{2t}{2}$	To isolate t, undo the multiplication by 2 by dividing both sides by 2.	
$184 = t$	Do the division: $368 \div 2 = 184$.	

Verify that 184 is the solution by substituting it into the original equation. ∎

3 Solve equations that have variable terms on both sides.

When solving an equation, if variables appear on both sides, we can use the addition (or subtraction) property of equality to get all variable terms on one side and all constant terms on the other.

Self Check 4

Solve: $30 + 6n = 4n - 2$

Now Try Problem 31

EXAMPLE 4   Solve: $3x - 15 = 4x + 36$

Strategy There are variable terms ($3x$ and $4x$) on both sides of the equation. We will eliminate $3x$ from the left side of the equation by subtracting $3x$ from both sides.

WHY To solve for x, all the terms containing x must be on the same side of the equation.

Solution

$3x - 15 = 4x + 36$	This is the equation to solve. There are variable terms (in blue) on both sides of the equation.
$3x - 15 - 3x = 4x + 36 - 3x$	Subtract $3x$ from both sides to isolate the variable term on the right side.
$-15 = x + 36$	Combine like terms: $3x - 3x = 0$ and $4x - 3x = x$. Now we want to isolate the variable, x.
$-15 - 36 = x + 36 - 36$	To undo the addition of 36, subtract 36 from both sides. This isolates x.
$-51 = x$	Do the subtraction.

$\begin{array}{r} \overset{1}{1}5 \\ + 36 \\ \hline 51 \end{array}$

Check:

$3x - 15 = 4x + 36$	This is the original equation.
$3(-51) - 15 \overset{?}{=} 4(-51) + 36$	Substitute -51 for x.
$-153 - 15 \overset{?}{=} -204 + 36$	Do the multiplications.
$-168 = -168$	True

$\begin{array}{rr} 51 & 51 \\ \times 3 & \times 4 \\ \hline 153 & 204 \end{array}$

$\begin{array}{rr} & \overset{9}{1}\overset{}{1}\overset{}{0}14 \\ 153 & 20\not{4} \\ + 15 & - 36 \\ \hline 168 & 168 \end{array}$

The solution is -51. ∎

Success Tip   In Example 4, we could have eliminated $4x$ from the right side by subtracting $4x$ from both sides:

$3x - 15 - 4x = 4x + 36 - 4x$

$-x - 15 = 36$ Note that the coefficient of x is negative.

However, it is usually easier to isolate the variable term on the side that will result in a *positive* coefficient.

EXAMPLE 5   Solve: $-9 - 2a - 4a = 5a + 2$

Strategy We will begin by combining the like terms on the left side of the equation.

WHY It is best to simplify the algebraic expressions on each side of an equation before using a property of equality.

Solution

$-9 - 2a - 4a = 5a + 2$ This is the equation to solve.

$-9 - 6a = 5a + 2$ Combine like terms: $-2a - 4a = -6a$.

There are variable terms (highlighted in blue) on both sides of the equation. Either we can subtract $5a$ from both sides to isolate a on the left side, or we can add $6a$ to both sides to isolate a on the right side. We will add $6a$ to both sides. That way, the coefficient of the resulting variable term on the right side will be positive.

$-9 - 6a + 6a = 5a + 2 + 6a$ Add 6a to both sides to isolate the variable term on the right side.

$-9 = 11a + 2$ Combine like terms: $-6a + 6a = 0$ and $5a + 6a = 11a$. Now we want to isolate the variable term, 11a.

$-9 - 2 = 11a + 2 - 2$ To isolate 11a, undo the addition of 2 by subtracting 2 from both sides.

$-11 = 11a$ Do the subtraction. Now we want to isolate the variable, a.

$\dfrac{-11}{11} = \dfrac{11a}{11}$ To isolate a, undo the multiplication by 11 by dividing both sides by 11.

$-1 = a$ Do the division.

Verify that the solution is -1 by substituting it into the original equation.

Self Check 5

Solve: $72 - 8d - 5d = 12d - 3$. Check the result.

Now Try Problem 35

---

4   Use the distributive property to solve equations.

At times, we must use the distributive property to remove parentheses when solving an equation.

EXAMPLE 6   Solve: $3(6x + 15) = 45$

Strategy We will use the distributive property on the left side of the equation.

WHY This will remove the parentheses and make it easier to see which properties of equality should be used to isolate x on the left side.

Solution

$3(6x + 15) = 45$ This is the equation to solve.

$3(6x) + 3(15) = 45$ Distribute the multiplication by 3.

$18x + 45 = 45$ Do the multiplication.

$18x + 45 - 45 = 45 - 45$ To isolate the variable term, 18x, undo the addition of 45 by subtracting 45 from both sides.

$18x = 0$ Do the subtraction.

$\dfrac{18x}{18} = \dfrac{0}{18}$ To isolate x, undo the multiplication by 18 by dividing both sides by 18.

$x = 0$ Do the division.

$$\begin{array}{r} \overset{1}{15} \\ \times\,3 \\ \hline 45 \end{array}$$

Verify that the solution is 0 by substituting it into the original equation.

Self Check 6

Solve: $7(5b + 5) = -70$. Check the result.

Now Try Problem 39

5 Apply a strategy to solve equations.

The previous examples suggest the following strategy for solving equations. You won't always have to use all four steps to solve a given equation. If a step doesn't apply, skip it and go to the next step.

Stategy for Solving Equations

1. **Simplify each side of the equation:** Use the distributive property to remove parentheses, and then combine like terms on each side.
2. **Isolate the variable term on one side:** Add (or subtract) to get the variable term on one side of the equation and a number on the other using the addition (or subtraction) property of equality.
3. **Isolate the variable:** Multiply (or divide) to isolate the variable using the multiplication (or division) property of equality.
4. **Check the result:** Substitute the possible solution for the variable in the *original* equation to see if a true statement results.

Self Check 7

Solve:
$6(5x - 30) - 2x = 8(x + 50)$

Now Try Problem 43

EXAMPLE 7   Solve: $3(4x - 80) + 6x = 2(x + 40)$

Strategy We will follow the steps of the equation-solving strategy to solve the equation.

WHY This is the most efficient way to solve an equation.

Solution

$$3(4x - 80) + 6x = 2(x + 40)$$ This is the equation to solve.

$$12x - 240 + 6x = 2x + 80$$ Distribute the multiplication by 3 and by 2.

$$18x - 240 = 2x + 80$$ On the left side, combine like terms: $12x + 6x = 18x$. There are variable terms on both sides.

$$18x - 240 - 2x = 2x + 80 - 2x$$ To eliminate the term $2x$ on the right side, subtract $2x$ from both sides.

$$16x - 240 = 80$$ Combine like terms on each side: $18x - 2x = 16x$ and $2x - 2x = 0$.

$$16x - 240 + 240 = 80 + 240$$ To isolate the variable term, $16x$, on the left side, add 240 to both sides to undo the subtraction of 240.

$$\begin{array}{r} \overset{1}{2}40 \\ + \ 80 \\ \hline 320 \end{array}$$

$$16x = 320$$ Do the addition on each side: $-240 + 240 = 0$ and $80 + 240 = 320$. Now we want to isolate the variable, x.

$$\frac{16x}{16} = \frac{320}{16}$$ To isolate x on the left side, divide both sides by 16 to undo the multiplication by 16.

$$\begin{array}{r} 20 \\ 16\overline{)320} \\ -\ 32 \\ \hline 00 \\ -\ 0 \\ \hline 0 \end{array}$$

$$x = 20$$ Do the division.

To check, we substitute 20 for x in the original equation and evaluate each side.

$$3(4x - 80) + 6x = 2(x + 40)$$ This is the original equation.

$$3[4(20) - 80] + 6(20) \stackrel{?}{=} 2(20 + 40)$$ Substitute 20 for each x.

$$3[80 - 80] + 120 \stackrel{?}{=} 2(60)$$

$$3[0] + 120 \stackrel{?}{=} 120$$

$$120 = 120$$ True

Since the statement $120 = 120$ is true, 20 is the solution of $3(4x - 80) + 6x = 2(x + 40)$. ∎

SECTION 3.5   STUDY SET

VOCABULARY

Fill in the blanks.

1. To _____ an equation means to find all values of the variable that make the equation a true statement.

2. A number that makes an equation true when substituted for the variable is called a _____ and is said to *satisfy* the equation.

3. To _____ a solution means to substitute that value into the original equation to see whether a true statement results.

4. The equation $6x + 1 = 2x + 7$ has variable _____ on both sides.

5. When solving equations, _____ the expressions that make up the left and right sides of the equation before using properties of equality to isolate the variable.

6. When we write the expression $2y + 3 + 6y$ as $8y + 3$, we say we have _____ _____ terms.

CONCEPTS

7. a. Circle the variable terms.

   $5x + 3x = 8$ $5t = 3t + 8$ $7 = 5h + 3h - 1$

  b. Which equation has variable terms on both sides?

8. To solve $6k = 5k - 18$, we need to eliminate $5k$ from the right side. To do this, what should we subtract from both sides?

9. Perform only the first step in solving each equation. **You do not have to solve the equation.**

  a. $2x + 4x = 36$

  b. $5(x + 1) = 15$

  c. $7x - 5 + 4x = x + 4 + x$

  d. $-3(x - 4) = 2(x + 1)$

10. Consider the equation $2x - 8 = -4x - 14$.

  a. To solve this equation by isolating x on the left side, what should we add to both sides?

  b. To solve this equation by isolating x on the right side, what should we subtract from both sides?

11. Fill in the blanks.

$$6 - (d - 4) = 8$$
$$6 - \blacksquare(d - 4) = 8$$
$$6 - \blacksquare + \blacksquare = 8$$

12. Fill in the blanks to complete the strategy for solving equations.

  Step 1. _____ each side of the equation.

  Step 2. Isolate the variable _____ on one side.

  Step 3. Isolate the _____.

  Step 4. _____ the result.

13. a. Simplify: $3t - t - 8$

  b. Solve: $3t - t = -8$

  c. Evaluate: $3t - t - 8$ for $t = -2$

  d. Check: Is 5 a solution of $3t - t = -8$?

14. a. Simplify: $2(x + 1) - 4$

  b. Solve: $2(x + 1) = -4$

  c. Evaluate: $2(x + 1)$ for $x = -1$

  d. Check: Is 3 a solution of $2(x + 1) = -4$?

NOTATION

Complete each solution to solve the equation.

15. $5x - 2x = -27$

$$\blacksquare = -27$$
$$\frac{3x}{\blacksquare} = \frac{-27}{\blacksquare}$$
$$x = \blacksquare$$

Check:

$$5x - 2x = -27$$
$$5(\blacksquare) - 2(\blacksquare) \overset{?}{=} -27$$
$$\blacksquare - (-18) \overset{?}{=} -27$$
$$-45 + \blacksquare \overset{?}{=} -27$$
$$\blacksquare = -27 \quad \text{True}$$

The solution is $\blacksquare$.

16.
$$8y - 6 = -2 + 10y$$
$$8y - 6 - \boxed{} = -2 + 10y - \boxed{}$$
$$-6 = -2 + \boxed{}$$
$$-6 + \boxed{} = -2 + 2y + \boxed{}$$
$$-4 = \boxed{}$$
$$\frac{-4}{\boxed{}} = \frac{2y}{\boxed{}}$$
$$\boxed{} = y$$

Check:
$$8y - 6 = -2 + 10y$$
$$8(\boxed{}) - 6 \stackrel{?}{=} -2 + 10(\boxed{})$$
$$-16 - 6 \stackrel{?}{=} -2 + (\boxed{})$$
$$\boxed{} = -22 \quad \text{True}$$

The solution is $\boxed{}$.

17.
$$5(x - 9) = 5$$
$$5x - 5(\boxed{}) = 5$$
$$5x - \boxed{} = 5$$
$$5x - 45 + \boxed{} = 5 + \boxed{}$$
$$\boxed{} = 50$$
$$\frac{5x}{\boxed{}} = \frac{50}{\boxed{}}$$
$$x = \boxed{}$$

Check:
$$5(x - 9) = 5$$
$$5(\boxed{} - 9) \stackrel{?}{=} 5$$
$$5(\boxed{}) \stackrel{?}{=} 5$$
$$\boxed{} = 5 \quad \text{True}$$

The solution is $\boxed{}$.

18. $-4(-1 - x) = 16$
$$4 + \boxed{} = 16$$
$$4 + 4x - \boxed{} = 16 - \boxed{}$$
$$4x = \boxed{}$$
$$\frac{4x}{\boxed{}} = \frac{12}{\boxed{}}$$
$$x = \boxed{}$$

Check:
$$-4(-1 - x) = 16$$
$$-4(-1 - \boxed{}) \stackrel{?}{=} 16$$
$$-4(\boxed{}) \stackrel{?}{=} 16$$
$$\boxed{} = 16 \quad \text{True}$$

The solution is $\boxed{}$.

GUIDED PRACTICE

Use a check to determine whether the given number is a solution of the equation. See Example 1.

19. Is 3 a solution of $5f + 8 = 4f + 11$?

20. Is 5 a solution of $3r + 8 = 5r - 2$?

21. Is 12 a solution of $2(x - 1) = 33$?

22. Is 8 a solution of $-6(x + 4) = -40$?

Solve each equation. Check the result. See Example 2.

23. $3x + 6x = 54$ **24.** $4c + 4c = 16$

25. $6x - 3x = 9$ **26.** $12b - 10b = 6$

Solve each equation. Check the result. See Example 3.

27. $250 + 350 = m - 12 + m$

28. $213 + 190 = x - 3 + x$

29. $255 + 275 = a + 16 + a$

30. $170 + 180 = m + 26 + m$

Solve each equation. Check the result. See Example 4.

31. $3s - 1 = 4s + 7$ **32.** $6v - 2 = 7v + 13$

33. $x - 14 = 2x - 10$ **34.** $2x - 27 = 3x - 20$

Solve each equation. Check the result. See Example 5.

35. $-16 - 3r - 5r = 2r + 4$

36. $-29 - x - 6x = 2x + 7$

37. $60 - a - 6a = 2a - 3$

38. $19 - 4t - 7t = t - 5$

Solve each equation. Check the result. See Example 6.

39. $2(6x + 7) = 14$ **40.** $9(3y + 2) = 18$

41. $8(9b + 5) = -32$ **42.** $5(11n + 5) = -30$

Solve each equation. Check the result. See Example 7.

43. $3(x - 4) + 3x = 2(x + 10)$

44. $9(w - 1) + 7w = 5(w + 7)$

45. $6(2j + 6) + 4j = 4(j - 30)$

46. $4(9h + 2) + 8h = 4(h - 18)$

TRY IT YOURSELF

Solve each equation. Check the result.

47. $-16 = 2(t + 2)$ **48.** $-10 = 5(y - 7)$

49. $-7 + 5r = 83 - 10r$ **50.** $-20 + t = 44 - 7t$

51. $T + T - 17 = 57$ **52.** $r + r - 15 = 95$

53. $-15 = 5 + 5(2x + 10)$ **54.** $-1 = 2 + 3(4x + 7)$

55. $60 = 3v - 5v$ **56.** $28 = x - 3x$

57. $9q + 3(q - 7) = 18 - q$

58. $q + 6(q - 4) = 24 - q$

59. $5 - (7 - y) = -5$ **60.** $10 - (5 - x) = 40$

61. $50a - 1 = 60a - 101$ **62.** $25y - 2 = 75y - 202$

63. $-(4 - c) = -3$ **64.** $-(6 - 2x) = -8$

65. $-20 - 8 = -m + 2m$ **66.** $-100 - 20 = -p + 4p$

67. $x + x + 6 = 90$ **68.** $c + c + 1 = 51$

69. $8 + 4(2x - 2) = -16 + 4x$

70. $3 + 3(2x - 1) = -18 + 3x$

71. $1,500 = b + 30 + b$ **72.** $8,000 = h + 100 + h$

73. $7x = 3x + 8$ **74.** $4x = 2x + 14$

75. $100 - y = 100 + y$ **76.** $-60 + z = -60 - z$

77. $2(4y + 8) + 3y = -3(2 - 3y)$

78. $3(7 - y) = 3(2y + 1)$

79. $t + 5t + 3t - 40 = 14t$ **80.** $5r - 24 = r + 5r + 2r$

81. $-2(9 - 3s) - (5s + 2) = -25$

82. $4(x - 5) - 3(12 - x) = 7$

83. $25 + 4j = 9j$ **84.** $36 + 5j = 9j$

85. $-3(3 - 2w) = -9$ **86.** $-4(5t + 2) = -8$

87. $-4(12) + 12t = 16t$ **88.** $-4(7) + 7t = 21t$

89. $-5g - 40 = -15g$ **90.** $-20s - 20 = -40s$

91. $4(p - 2) = 0$ **92.** $10(4s - 4) = 0$

93. $16 - (x + 3) = -13$ **94.** $10 - (w + 4) = -12$

WRITING

95. Explain the error in the work shown below.

Solve: $2x = 4x - x$

$2x = 4$

$\dfrac{2x}{2} = \dfrac{4}{2}$

$x = 2$

96. To solve $3x - 4 = 5x + 1$, one student began by subtracting $3x$ from both sides. Another student solved the same equation by first subtracting $5x$ from both sides. Will the students get the same solution? Explain why or why not.

97. Explain the error in the following solution.

Solve: $2x + 4 = 30$

$\dfrac{2x}{2} + 4 = \dfrac{30}{2}$

$x + 4 = 15$

$x + 4 - 4 = 15 - 4$

$x = 11$

98. Consider $3x = 2x + 9$. Why is it necessary to eliminate one of the variable terms in order to solve for x?

99. What does it mean to *solve an equation*?

100. Explain how to determine whether a number is a solution of an equation.

REVIEW

101. Subtract: $-7 - 9$

102. Which of the following numbers are *not* factors of 28? $4, 6, 7, 8$

103. Evaluate: $\dfrac{-8 + 2}{-2 + 4}$

104. Translate to mathematical symbols: 4 less than x

105. Simplify: $-(-5)$

106. Using x and y, illustrate the commutative property of addition.

107. What is the sign of the product of two negative integers?

108. Complete the table.

Term	Coefficient
$6m$	
$-75t$	
w	
$4bh$	

Objectives

1 Solve application problems to find one unknown.

2 Solve application problems to find two unknowns.

3 Solve number-value problems.

Using Equations to Solve Application Problems

The skills that we have studied in this chapter can now be used to solve more complicated application problems. Once again, we will use the five-step problem-solving strategy as an outline for each solution.

1 **Solve application problems to find one unknown.**

Self Check 1

SERVICE CLUBS To become a member of a service club, students at one college must complete 72 hours of volunteer service by working 4-hour shifts at the tutoring center. If a student has already volunteered 48 hours, how many more 4-hour shifts must she work to meet the service requirement for membership in the club?

Now Try Problem 21

EXAMPLE 1 *Volunteer Service Hours* To receive a degree in child development, students at one college must complete 135 hours of volunteer service by working 3-hour shifts at a local preschool. If a student has already volunteered 87 hours, how many more 3-hour shifts must she work to meet the service requirement for her degree?

Analyze

- Students must complete 135 hours of volunteer service.   *Given*
- Students work 3-hour shifts.   *Given*
- A student has already completed 87 hours of service.   *Given*
- How many more 3-hour shifts must she work?   *Find*

Form

Let x = the number of shifts needed to complete the service requirement. Since each shift is 3 hours long, multiplying 3 by the number of shifts will give the number of additional hours the student needs to volunteer.

The number of hours she has already completed	plus	3 times	the number of shifts yet to be completed	is	the number of hours required.
87	+	3 ·	x	=	135

Solve

$$87 + 3x = 135$$ *We need to isolate x on the left side.*

$$87 + 3x - 87 = 135 - 87$$ *To isolate the variable term 3x, subtract 87 from both sides to undo the addition of 87.*

$$3x = 48$$ *Do the subtraction.*

$$\frac{3x}{3} = \frac{48}{3}$$ *To isolate x, divide both sides by 3 to undo the multiplication by 3.*

$$x = 16$$ *Do the division.*

$$\begin{array}{r} \overset{12}{\cancel{1}}\overset{15}{3\!\!\!/ 5} \\ -\ 8\ 7 \\ \hline 4\ 8 \end{array}$$

$$\begin{array}{r} 16 \\ 3\overline{)48} \\ -\ 3 \\ \hline 18 \\ -\ 18 \\ \hline 0 \end{array}$$

State

The student needs to complete 16 more 3-hour shifts of volunteer service.

Check

The student has already completed 87 hours. If she works 16 more shifts, each 3 hours long, she will have 16 · 3 = 48 more hours. Adding the two sets of hours, we get:

$$\begin{array}{r} 87 \\ +\ 48 \\ \hline 135 \end{array}$$ ← This is the total number of hours needed.

The result, 16, checks.

EXAMPLE 2 *Attorney's Fees* In return for her services, an attorney and her client split the jury's cash award equally. After paying her assistant $1,000, the attorney ended up making $10,000 from the case. What was the amount of the award?

Analyze

- The attorney and client split the award equally. *Given*
- The attorney's assistant was paid $1,000. *Given*
- The attorney made $10,000. *Given*
- What was the amount of the award? *Find*

Form

Let x = the amount of the award. Two key phrases in the problem help us form an equation.

Key phrase: *split the award equally* **Translation:** divide by 2
Key phrase: *paying her assistant $1,000* **Translation:** subtract $1,000

Now we translate the words of the problem into an equation.

The award split in half	minus	the amount paid to the assistant	is	the amount the attorney makes.
$\frac{x}{2}$	$-$	1,000	$=$	10,000

Solve

$$\frac{x}{2} - 1,000 = 10,000 \quad \text{We need to isolate x on the left side.}$$

$$\frac{x}{2} - 1,000 + \mathbf{1,000} = 10,000 + \mathbf{1,000} \quad \begin{array}{l}\text{To isolate the variable term } \frac{x}{2},\\ \text{add 1,000 to both sides to}\\ \text{undo the subtraction of 1,000.}\end{array}$$

$$\frac{x}{2} = 11,000 \quad \text{Do the addition.}$$

$$2 \cdot \frac{x}{2} = 2 \cdot 11,000 \quad \begin{array}{l}\text{To isolate the variable x,}\\ \text{multiply both sides by 2}\\ \text{to undo the division by 2.}\end{array} \quad \begin{array}{r}11,000\\ \times\quad 2\\ \hline 22,000\end{array}$$

$$x = 22,000 \quad \text{Do the multiplication.}$$

State

The amount of the award was $22,000.

Check

If the award of $22,000 is split in half, the attorney's share is $11,000. If $1,000 is paid to her assistant, we subtract to get:

$$\begin{array}{r}\$11,000\\ -\ 1,000\\ \hline \$10,000\end{array} \leftarrow \text{This is what the attorney made.}$$

The result, $22,000, checks.

Self Check 2

YARD SALES A husband and wife split the money equally that they made on a yard sale. The husband gave $75 of his share to charity, leaving him with $210. How much money did the couple make at their yard sale?

Now Try Problem 22

2 **Solve application problems to find two unknowns.**

When solving application problems, we usually let the variable stand for the quantity we are asked to find. In the next two examples, each problem contains a second unknown quantity. We will look for a key word or phrase in the problem to help us describe it using an algebraic expression.

Self Check 3

CIVIL SERVICE A candidate for a position with the IRS scored 15 points higher on the written part of the civil service exam than he did on his interview. If his combined score was 155, what were his scores on the interview and on the written part?

Now Try Problem 23

EXAMPLE 3   *Civil Service*   A candidate for a position with the FBI scored 12 points higher on the written part of the civil service exam than she did on her interview. If her combined score was 92, what were her scores on the interview and on the written part of the exam?

Analyze

- She scored 12 points higher on the written part than on the interview.   Given
- Her combined score was 92.   Given
- What were her scores on the interview and on the written part?   Find

Form

Since we are told that her score on the written part was related to her score on the interview, we let $x =$ her score on the interview.

There is a second unknown quantity—her score on the written part of the exam. We look for a key phrase to help us decide how to represent that score using an algebraic expression.

Key phrase: 12 points *higher* on the written part than on the interview

Translation: add 12 points to the interview score

So $x + 12 =$ her score on the written part of the exam. Now we translate the words of the problem into an equation.

The score on the interview	plus	the score on the written part	is	the overall score.
x	$+$	$x + 12$	$=$	92

Solve

$x + x + 12 = 92$	We need to isolate x on the left side.
$2x + 12 = 92$	On the left side, combine like terms: x + x = 2x.
$2x + 12 - \mathbf{12} = 92 - \mathbf{12}$	To isolate the variable term, 2x, subtract 12 from both sides to undo the addition of 12.
$2x = 80$	Do the subtraction.
$\dfrac{2x}{\mathbf{2}} = \dfrac{80}{\mathbf{2}}$	To isolate the variable x, divide both sides by 2 to undo the multiplication by 2.
$x = 40$	Do the division. This is her score on the interview.

To find the second unknown, we substitute 40 for x in the expression that represents her score on the written part.

$$x + 12 = \mathbf{40} + 12$$
$$= 52 \qquad \text{This is her score on the written part.}$$

State

Her score on the interview was 40 and her score on the written part was 52.

Check

Her score of 52 on the written part was 12 points higher than her score of 40 on the interview. Also, if we add the two scores, we get:

$$\begin{array}{r} 40 \\ + \ 52 \\ \hline 92 \end{array}$$ ⟵ This is her combined score.

The results, 40 and 52, check.

EXAMPLE 4 *Playgrounds* After receiving a donation of 400 feet of chain link fencing, the staff of a preschool decided to use it to enclose a playground that is rectangular. Find the length and the width of the playground if the length is three times the width.

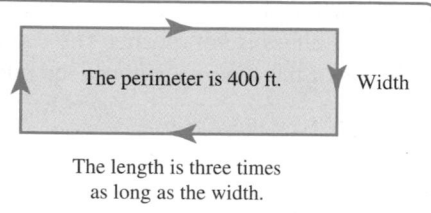

The perimeter is 400 ft.   Width

The length is three times as long as the width.

Self Check 4

CRIME SCENES Police used 800 feet of yellow tape to fence off a rectangular-shaped lot for an investigation. Fifty less feet of tape was used for each width as for each length. Find the length and the width of the lot.

Now Try Problem 24

Analyze

- The perimeter is 400 ft.   *Given*
- The length is three times as long as the width.   *Given*
- What is the length and what is the width of the rectangle?   *Find*

Form

Since we are told that the length is related to the width, we will let w = the width of the playground. There is a second unknown quantity: the length of the playground. We look for a key phrase to help us decide how to represent it using an algebraic expression.

Key phrase: length is *three times* the width   **Translation:** multiply width by 3

So $3w$ = the length of the playground.

The formula for the perimeter of a rectangle is $P = 2l + 2w$. In words, we can write

$2 \cdot$	the length of the playground	plus	$2 \cdot$	the width of the playground	is	the perimeter.
$2 \cdot$	$3w$	$+$	$2 \cdot$	w	$=$	400

Solve

$2 \cdot 3w + 2w = 400$ We need to isolate w on the left side.

$6w + 2w = 400$ Do the multiplication: $2 \cdot 3w = 6w$.

$8w = 400$ On the left side, combine like terms: $6w + 2w = 8w$.

$\dfrac{8w}{8} = \dfrac{400}{8}$ To isolate w, divide both sides by 8 to undo the multiplication by 8.

$w = 50$ Do the division. This is the width.

$$
\begin{array}{r}
50 \\
8\overline{)400} \\
-40 \\
\hline
00 \\
-0 \\
\hline
0
\end{array}
$$

To find the second unknown, we substitute 50 for w in the expression that represents the length of the playground.

$3w = 3(\mathbf{50})$ Substitute 50 for w.

$= 150$ This is the length of the playground.

State

The width of the playground is 50 feet and the length is 150 feet.

Check

If we add two lengths and two widths, we get $2(150) + 2(50) = 300 + 100 = 400$. Also, the length (150 ft) is three times the width (50 ft). The results check.

3 Solve number-value problems.

Some problems deal with quantities that have a value. In these problems, we must distinguish between the *number of* and the *value of* the unknown quantity. For example, to find the value of 3 quarters, we multiply the number of quarters by the value (in cents) of one quarter. Therefore, the value of 3 quarters is $3 \cdot 25$ cents = 75 cents.

The same approach must be taken if the number is unknown. For example, the value of d dimes is not d cents. The value of d dimes is $d \cdot 10$ cents $= 10d$ cents. For problems of this type, we will use the relationship

$$\text{Number} \cdot \text{value} = \text{total value}$$ This is the number-value formula.

Self Check 5

INVESTING   A T-bill (Treasury bill) is worth $10,000. Find the value of:

a. two T-bills

b. x T-bills

c. $(x + 3)$ T-bills

Now Try Problem 27

EXAMPLE 5   *Pricing*   Delicious apples sell for 89 cents a pound. Find the cost of:
a. 5 pounds of apples   **b.** p pounds of apples   **c.** $(p - 2)$ pounds of apples

Strategy In each case, we will multiply the number of pounds of apples by their value (89 cents a pound) to find the total cost.

WHY Since *value* and *cost* are similar concepts, we can use the number-value formula in this situation.

Solution

a.   Total value $=$ **number** $\cdot$ **value**   This is the number-value formula.

$\qquad\qquad\quad\; = \quad 5 \quad \cdot \quad 89$ Substitute 5 for the number of pounds and 89 for the value (cost per pound) of the apples.

$\qquad\qquad\quad\; = 445$ Do the multiplication.

$$\begin{array}{r} \overset{4}{8}9 \\ \times\; 5 \\ \hline 445 \end{array}$$

The cost of 5 pounds of apples is 445 cents, or $4.45.

b. Since the number of pounds of apples is unknown (p pounds), we cannot calculate the total cost as in part a. We can only represent it using an algebraic expression.

$\qquad$ Total value $=$ **number** $\cdot$ **value**   This is the number-value formula.

$\qquad\qquad\quad\; = \quad p \quad \cdot \quad 89$ Substitute p for the number of pounds and 89 for the value (cost per pound) of the apples.

$\qquad\qquad\quad\; = 89p$ It is standard practice to write the numerical factor, 89, in front of the variable factor.

The cost of p pounds of apples can be represented by the algebraic expression $89p$ cents.

c. Since the number of pounds of apples is unknown, $(p - 2)$ pounds, we cannot calculate the total cost, as in part a. We can only represent it using an algebraic expression.

$\qquad$ Total value $=$ **number** $\cdot$ **value**   This is the number-value formula.

$\qquad\qquad\quad\; = (p - 2) \cdot 89$ Substitute $p - 2$ for the number of pounds and 89 for the value (cost per pound) of the apples.

$\qquad\qquad\quad\; = 89(p - 2)$ It is standard practice to write the numerical factor, 89, in front of the quantity $(p - 2)$.

The cost of $(p - 2)$ pounds of apples can be represented by the algebraic expression $89(p - 2)$ cents.

EXAMPLE 6   *Movie Tickets*   Ninety-five people attended a movie matinee. Ticket prices were $6 for adults and $4 for children. Write algebraic expressions that represent the income received from the sale of children's tickets and from the sale of adult tickets.

Strategy In each case, we will multiply the number of tickets sold by their value ($6 for adults and $4 for children) to find the income received.

WHY Since *total value* and *income received* are similar concepts, we can use the number-value formula in this situation.

Solution We will let c = the unknown number of children's tickets sold. If we subtract the number of children's tickets sold from the total number of tickets sold, we obtain an expression for the number of adult tickets sold:

$95 - c$ = number of adult tickets sold

The value of a children's ticket is $4. To find the income from the sale of c children's tickets, we multiply:

Total value = **number** · **value**

$\qquad = \quad c \quad \cdot \quad 4$

$\qquad = 4c$

The income from the sale of the children's tickets is represented by the algebraic expression $4c$ dollars.

The value of an adult ticket is $6. To find the income from the sale of $(95 - c)$ adult tickets we multiply:

Total value = **number** · **value**

$\qquad = (95 - c) \cdot \quad 6$

$\qquad = 6(95 - c)$

The income from the sale of the adult tickets is represented by the algebraic expression $6(95 - c)$ dollars.

These results can be presented in a number-value table, as shown below.

Type of ticket	Number ·	Value ($) =	Total value ($)
Child	c	4	$4c$
Adult	$95 - c$	6	$6(95 - c)$

Multiply to obtain each of these expressions.

Enter this information first.

Self Check 6

FITNESS CLUBS A fitness club has 150 members. Monthly membership fees are $25 for nonseniors and $15 for senior citizens. Find the income the club receives from nonseniors and from seniors each month. Use the table below to present your results. Let m represent the number of nonseniors.

Member's age	Number ·	Fee =	Total income
nonsenior			
senior			

Now Try Problem 29

EXAMPLE 7 *Basketball* On a night when they scored 110 points, a basketball team made only 5 free throws (worth 1 point each). The remainder of their points came from two- and three-point baskets. If the number of two- and three-point baskets totaled 45, how many two-point and how many three-point baskets did they make?

Analyze

- The team scored 110 points. *Given*
- They made 5 free throws (1 point each). *Given*
- They made a total of 45 two- and three-point baskets. *Given*
- How many two-point and three-point baskets were made? *Find*

Self Check 7

FURNISHINGS A restaurant owner purchased $2,720 worth of tables and chairs for the dining area of her cafe. Each table cost $200 and each chair cost $60. If she purchased a total of 36 pieces of furniture, how many tables and how many chairs did she buy?

Now Try Problem 25

Form

The number of two- and three-point baskets totaled 45. If we let $x =$ the number of three-point baskets made, then $45 - x =$ the number of two-point baskets made. We can now organize the data in a table. For each type of basket, multiply the *number* of baskets made by the point *value* to find an expression to represent the total value.

Type of basket	Number	· Value	= Total value
Three-point	x	3	$3x$
Two-point	$45 - x$	2	$2(45 - x)$
Free throw	5	1	5
		Total:	110

Multiply to obtain each of these expressions.

Enter this information first.

Use the information in this column to form an equation.

$3 \cdot$ the number of three-point baskets	plus	$2 \cdot$ the number of two-point baskets	plus	$1 \cdot$ the number of free throws	is	the total points scored.
$3 \cdot \quad x$	$+$	$2 \cdot \quad (45 - x)$	$+$	$1 \cdot \quad 5$	$=$	110

Solve

$$3x + 2(45 - x) + 5 = 110$$ This is the equation to solve.

$$3x + 90 - 2x + 5 = 110$$ Distribute the multiplication by 2.

$$x + 95 = 110$$ Combine like terms: $3x - 2x = x$.

$$x + 95 - 95 = 110 - 95$$ To isolate x, undo the addition of 95 by subtracting 95 from both sides.

$$x = 15$$ Do the subtraction. This is the number of three-point baskets.

$$\begin{array}{r} \overset{10}{\cancel{\overset{\emptyset}{1}}}\overset{10}{\cancel{1}}\emptyset \\ -95 \\ \hline 15 \end{array}$$

We can substitute 15 for x in $45 - x$ to find the number of two-point baskets made.

$$45 - x = 45 - 15$$
$$= 30$$ This is the number of two-point baskets.

State

The basketball team made 15 three-point baskets and 30 two-point baskets.

Check

If we multiply the number of three-point baskets by their value, we get $15 \cdot 3 = 45$ points. If we multiply the number of two-point baskets by their value, we get $30 \cdot 2 = 60$ points. If we add the number of made free throws to these two subtotals, we get $45 + 60 + 5 = 110$ points. The results check.

$$\begin{array}{r} \overset{1}{45} \\ 60 \\ + 5 \\ \hline 110 \end{array}$$

ANSWERS TO SELF CHECKS

1. The student needs to complete 6 more 4-hour shifts of volunteer service.
2. The couple made $570 at the yard sale.
3. His score on the interview was 70 and his score on the written part was 85.
4. The length of the lot is 225 feet and the width of the lot is 175 feet.
5. **a.** $20,000 **b.** $10,000x$ dollars **c.** $10,000(x + 3)$ dollars
6. $m, 25, 25m; 150 - m, 15, 15(150 - m)$
7. She bought 4 tables and 32 chairs.

SECTION 3.6   STUDY SET

VOCABULARY

Fill in the blanks.

1. The five-step problem-solving strategy is:
   - _____ the problem
   - Form an _____
   - _____ the equation
   - State the _____
   - _____ the result

2. Words such as *doubled* and *tripled* indicate the operation of _____.

3. Phrases such as *distributed equally* and *sectioned off uniformly* indicate the operation of _____.

4. Words such as *trimmed*, *removed*, and *melted* indicate the operation of _____.

5. Words such as *extended* and *reclaimed* indicate the operation of _____.

6. A letter (or symbol) that is used to represent a number is called a _____.

Fill in the blanks to complete each formula.

7. _____ · value = total value

8. $P = 2 + 2$

9. **BUSINESS ACCOUNTS** Every month, a salesperson adds five new accounts. Write an algebraic expression that represents the number of new accounts that he will add in *x* months.

10. **ANTIQUE COLLECTING** Every year, a woman purchases four antique spoons to add to her collection. Write an algebraic expression that represents the number of spoons that she will purchase in *x* years.

11. **SERVICE STATIONS** See the illustration below. Write an algebraic expression that represents the number of gallons that the smaller tank holds.

This tank holds *g* gallons.   This tank holds 100 gallons less than the premium tank.

12. **SCHOLARSHIPS** See the illustration below. Write an algebraic expression that represents the number of scholarships that were awarded this year.

Last year, *s* scholarships were awarded.   Six more scholarships were awarded this year than last year.

13. **OCEAN TRAVEL** See the illustration below. Write an algebraic expression that represents the number of miles that the passenger ship traveled.

Port

The freighter traveled *m* miles.   The passenger ship traveled 3 times farther than the freighter.

14. **TAX REFUNDS** See the illustration below. Write an algebraic expression that represents the amount of the tax refund that the husband gets.

A husband and wife received a tax refund of $*d*.   The couple split the refund equally.

15. **GEOMETRY** See the illustration below. The length of a rectangle is twice its width. Write an algebraic expression that represents the length of the rectangle.

w

16. **GEOMETRY** Fill in the blanks to complete the equation that describes the perimeter of the rectangle shown below.

$$2 \cdot + 2 \cdot = 240$$

The perimeter is 240 ft.   *w*

5*w*

17. FOOTWEAR The illustration below shows a rack that contains both dress shoes and athletic shoes.

    a. How many pairs of shoes are stored in the rack?

    b. Suppose there are *d* pairs of dress shoes in the rack. Write an algebraic expression that represents the number of pairs of athletic shoes in the rack.

18. QUIZZES The answers to a Prealgebra quiz are shown below.

    a. How many questions were on the quiz?

    b. Suppose the student answered *c* questions correctly. Write an algebraic expression that represents the number of questions she answered incorrectly.

PREALGEBRA QUIZ CHAPTER 3	
1. 44	6. 250 ft
2. 376	7. 165 mi
3. equal	8. no
4. $9 - x$	9. yes
5. $4x$	10. simplify

In Problems 19 and 20, fill in the blanks to complete each solution.

19. AIRLINE SEATING An 88-seat passenger plane has ten times as many economy seats as first-class seats. Find the number of first-class seats.

Analyze

- There are ▢ seats on the plane.

- There are ▢ times as many economy seats as first-class seats.

- How many _____ seats are there?

Form Since the number of economy seats is related to the number of first-class seats, we let x = the number of _____ seats.

To write an algebraic expression to represent the number of economy seats, we look for a key phrase in the problem.

Key phrase: ten times as many economy seats

Translation: _____ by 10

So ▢ x = the number of economy seats.

Now we translate the words of the problem into an equation.

The number of first-class seats	plus	the number of economy seats	is	88.
x	$+$	▢	$=$	88

Solve

$$x + 10x = ▢$$
$$▢ = 88$$
$$\frac{11x}{▢} = \frac{88}{▢}$$
$$x = ▢$$

State There are ▢ first-class seats

Check If there are 8 first-class seats, there are ▢ $\cdot$ 8 = 80 economy seats. Adding 8 and 80, we get ▢. The result checks.

20. COUPONS A shopper redeemed some 20-cents-off and some 40-cents-off coupons at the supermarket to get $2.60 off her grocery bill. If she used a total of eight coupons, how many 20¢ and how many 40¢ coupons did she redeem?

Analyze

- ▢ ¢ and ▢ ¢ coupons were redeemed.

- She got $2.60, which is ▢ ¢, off her grocery bill.

- She used a total of ▢ coupons.

- How many ▢ ¢ and how many ▢ ¢ coupons did she redeem?

Form The total number of coupons redeemed was 8. If we let x = the number of 20¢ coupons she redeemed, then $8 - x$ = the number of _____ coupons she redeemed.

20 ·	the number of 20¢ coupons redeemed	plus	40 ·	the number of 40¢ coupons redeemed	is 260.
	20▢	$+$		40(▢)	$= 260$

Solve

$$20x + 40(8 - x) = ▢$$
$$20x + ▢ - 40x = 260$$
$$-20x + ▢ = 260$$
$$-20x + 320 - ▢ = 260 - ▢$$
$$-20x = ▢$$
$$\frac{-20x}{▢} = \frac{-60}{▢}$$
$$x = ▢$$

If 3 of the 20¢ coupons were redeemed, then $8 - 3 = \boxed{}$ of the 40¢ coupons were redeemed.

State She redeemed 3 of the 20¢ coupons and $\boxed{}$ of the 40¢ coupons.

Check The value of 3 of the 20¢ coupons is $3 \cdot 20 = \boxed{}$ ¢. The value of 5 of the 40¢ coupons is $5 \cdot 40 = \boxed{}$ ¢. Adding these two subtotals, we get 260¢, which is $2.60. The results check.

GUIDED PRACTICE

Form an equation and solve it to answer the following question.
See Example 1.

21. BUSINESS After beginning a new position with 15 established accounts, a salesman made it his objective to add 5 new accounts every month. His goal was to reach 100 accounts. At this rate, how many months would it take to reach his goal?

Form an equation and solve it to answer the following question.
See Example 2.

22. TAX REFUNDS After receiving their tax refund, a husband and wife split the refunded money equally. The husband then gave $50 of his money to charity, leaving him with $70. What was the amount of the tax refund check?

Form an equation and solve it to answer the following question.
See Example 3.

23. SCHOLARSHIPS Because of increased giving, a college scholarship program awarded six more scholarships this year than last year. If a total of 20 scholarships were awarded over the last two years, how many were awarded last year and how many were awarded this year?

Form an equation and solve it to answer the following question.
See Example 4.

24. GEOMETRY The perimeter of a rectangle is 150 inches. Find the length and the width if the length is four times the width.

In Problems 25 and 26, form an equation and solve it to answer the question. *See Example 7.*

25. PIGGY BANKS When a child emptied her coin bank, she had a collection of pennies, nickels, and dimes. There were a total of 20 pennies, and a combined total of 25 nickels and dimes. If the coins had a total value of 220 cents, how many nickel and dimes were in the bank?

26. WISHING WELLS A city park employee collected 650 cents in nickels, dimes, and quarters at the bottom of a wishing well. There were 20 nickels, and a combined total of 40 dimes and quarters. How many dimes and quarters were at the bottom of the wishing well?

In Problems 27–30, complete the number-value table.
See Examples 5 and 6.

27. COMMISSIONS A shoe salesman receives a commission for every pair of shoes he sells. Complete the table.

Type of shoe	Number sold	·	Commission per pair ($)	=	Total commission ($)
Dress	10		3		
Athletic	12		2		
Child's	x		5		
Sandal	$9 - x$		4		

28. COINS Complete the table.

Type of coin	Number	·	Value (¢)	=	Total value (¢)
Nickel	12				
Dime	d				
Quarter	$q + 2$				

29. TUTORING A tutoring center charges $18 an hour for English tutoring and $20 an hour for mathematics tutoring. One week, forty students are tutored one hour per week in these subjects, and no student took both types of tutoring. Write algebraic expressions that represent the weekly income received by the center from the English tutoring and from the mathematics tutoring. Present your results in the following table.

Type of tutoring	Number of hours	·	Fee ($)	=	Total income ($)
English					
Mathematics					

30. TUXEDOS A formal wear shop rents prom tuxedos for $55 and wedding tuxedos for $75. One weekend, eighty tuxedos were rented. Write algebraic expressions that represent the income received that weekend from the rental of prom tuxedos and from wedding tuxedos. Present your results in the following table.

Type of tuxedo	Number ·	Rental fee ($) =	Total income ($)
Prom			
Wedding			

TRY IT YOURSELF

Form an equation and solve it to answer each question.

31. NUMBER PROBLEMS Eight more than a number is the same as twice the number. What is the number?

32. NUMBER PROBLEMS Twenty more than a number is the same as three times the number. What is the number?

33. NUMBER PROBLEMS Ten less than five times a number is the same as the number increased by six. What is the number?

34. NUMBER PROBLEMS Four less than seven times a number is the same as the number increased by eight. What is the number?

APPLICATIONS

Form an equation and solve it to answer each question.

35. CONSTRUCTION To get a heavy-equipment operator's certificate, 48 hours of on-the-job training are required. If a woman has completed 24 hours, and the training sessions last for 6 hours, how many more sessions must she take to get the certificate?

36. PUBLISHING An editor needs to read a 600-page manuscript. Her goal is to proofread 24 pages each day. If she has already read 96 pages, how many more days will it take her to complete the proofreading?

37. LOANS A student plans to pay back a $600 loan with monthly payments of $30. How many payments has she made if the debt has been reduced to $420?

38. ANTIQUES A woman purchases 4 antique spoons each year. She now owns 56 spoons. In how many years will she have 100 spoons in her collection?

39. CORPORATE DOWNSIZING In an effort to cut costs, a corporation has decided to lay off 5 employees every month until the number of employees totals 465. If 510 people are now employed, how many months will it take to reach the employment goal?

40. BOTTLED WATER DELIVERY A truck driver left the plant carrying 300 bottles of drinking water. His delivery route consisted of office buildings, each of which was to receive 3 bottles of water. The driver returned to the plant at the end of the day with 117 bottles of water on the truck. To how many office buildings did he deliver?

41. OCEAN TRAVEL At noon, a passenger ship and a freighter left a port traveling in opposite directions. By midnight, the passenger ship was 3 times farther from port than the freighter was. How far was the freighter from port if the distance between the ships was 84 miles?

42. RADIO STATIONS The daily listening audience of an AM radio station is four times as large as that of its FM sister station. If 100,000 people listen to these two radio stations, how many listeners does the FM station have?

43. INHERITANCES Five brothers split an inheritance from their father equally. One of the brothers used part of his share to pay off a $5,575 balance on a credit card. That left him with $78,525. Find the total amount of the inheritance that the father left to his sons.

44. APPLIANCES A couple split the rebate check that they received after purchasing a new energy-efficient refrigerator. The wife then spent $25 of her share on a new energy-saving iron. If she was left with $35, what was the amount of the rebate check?

45. RENTALS In renting an apartment with two other friends, Enrique agreed to pay the security deposit of $100 himself. The three of them agreed to contribute equally toward the monthly rent. Enrique's first check to the apartment owner was for $425. What was the monthly rent for the apartment? (*Hint:* First determine how many people are splitting the rent.)

46. LAWYER'S FEES A lawyer and his client split the money that a jury awarded the client in a personal injury lawsuit. From his share, the lawyer paid his two assistants $15,000 each and ended up making $50,000 from the case. What was the amount of the jury award?

47. SERVICE STATIONS At a service station, the underground tank storing regular gas holds 100 gallons less than the tank storing premium gas. If the total storage capacity of the tanks is 700 gallons, how much does the premium gas tank hold?

48. LIBRARIES According to a 2007 survey, the state of New York had the most public libraries of the fifty states. Illinois was in second place with 130 fewer. Together, the two states had a total of 1,376 public libraries. How many public libraries did New York have in 2007? (Source: *The Institute of Museum and Library Service Public Survey,* 2009)

49. COMMERCIALS During a 30-minute television show, a viewer found that the actual program aired a total of 18 minutes more than the time devoted to commercials. How many minutes of commercials were there?

from Campus to Careers
Broadcasting

50. CLASS TIME In a biology course, students spend a total of 250 minutes in lab and lecture each week. The lab time is 50 minutes shorter than the lecture time. How many minutes do the students spend in lecture per week?

51. INTERIOR DECORATING As part of redecorating, crown molding was installed around the ceiling of a room. Sixty feet of molding was needed for the project. Find the width of the room if its length is twice the width.

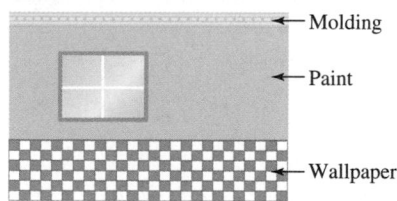

52. SPRINKLER SYSTEMS A landscaper buried a water line around a rectangular lawn to serve as a supply line for a sprinkler system. The length of the lawn is 5 times its width. If 240 feet of pipe was used to do the job, what is the width of the lawn?

53. TENNIS The perimeter of a regulation singles tennis court is 210 feet and the length is 51 feet more than the width. Find the length and width of the court.

54. THE CENTENNIAL STATE The state of Colorado is approximately rectangular-shaped with a perimeter of 1,320 miles. Find the length (east to west) and width (north to south), if the length is 100 miles longer than the width.

Form an equation and then solve it to answer each question. Make a table to organize the data.

55. COMMISSIONS A salesman receives a commission of $3 for every pair of dress shoes he sells. He is paid $2 for every pair of athletic shoes he sells. After selling 9 pairs of shoes in a day, his commission was $24. How many pairs of each kind of shoe did he sell that day?

56. GRADING SCALES For every problem answered correctly on an exam, 3 points are awarded. For every incorrect answer, 4 points are deducted. In a 10-question test, a student scored 16 points. How many correct and incorrect answers did he have on the exam?

57. MOVER'S PAY SCALE A part-time mover's regular pay rate is $60 an hour. If the work involves going up and down stairs, his rate increases to $90 an hour. In one week, he earned $1,380 and worked 20 hours. How many hours did he work at each rate?

58. PRESCHOOL ENROLLMENTS A preschool charges $8 for a child to attend its morning session or $10 to attend the afternoon session. No child can attend both. Thirty children are enrolled in the preschool. If the daily receipts are $264, how many children attend each session?

59. AUTOGRAPHS Martin has collected the autographs of six more movie stars than he has television celebrities. Each movie star autograph is worth $200 and each television celebrity autograph is worth $75. If his collection is valued at $4,500, how many of each type of autograph does he have?

60. RENTALS In an apartment building, seven more 1-bedroom units are rented than 2-bedroom units. The monthly rent for a 1-bedroom is $500 and a 2-bedroom is $700. If the total monthly income from these units is $15,500, how many of each type of unit are there?

WRITING

61. Explain what should be accomplished in each of the five steps of the problem-solving strategy studied in this section.

62. Use an example to explain the difference between the number of quarters a person has and the value of those quarters.

63. Write a problem that could be represented by the following equation.

Age of father	plus	age of son	is	50.
x	$+$	$x - 20$	$=$	50

64. Write a problem that could be represented by the following equation.

2 ·	length of a field	plus	2 ·	width of a field	is	600 ft.
2 ·	$4x$	$+$	2 ·	x	$=$	600

REVIEW

65. What property is illustrated?
$(2 + 9) + 1 = 2 + (9 + 1)$

66. Solve: $4 - x = -8$

67. Evaluate: -10^2

68. List the factors of 18.

69. Fill in the blank: Subtraction of a number is the same as _____ of the opposite of that number.

70. Round 123,808 to the nearest ten thousand.

71. Write this prime factorization using exponents:
$2 \cdot 2 \cdot 2 \cdot 5 \cdot 5$

72. The value of a stock dropped $3 a day for 6 consecutive days. What was the change in the value of the stock over this period?

STUDY SKILLS CHECKLIST

Solving Equations

The first step to solve an equation is often the most difficult for students to determine. Before taking the test on Chapter 3, make sure that you know what to do first when solving the following equations. Put a checkmark in the box if you can answer "yes" to the statement.

☐ I know that the first step to solve
$3(3x + 8) = 51$
is to use the distributive property on the left side.

☐ I know that the first step to solve
$45 = 7a + 8a$
is to combine like terms on the right side.

☐ I know that the first step to solve
$9n + 12 = 6n - 9$
is to subtract 6n from both sides.

☐ I know that the first step to solve
$2(y + 40) - 6y = 3(4y - 80)$
is to use the distributive property on both sides.

☐ I know that the first step to solve
$100 + 25 = 4h - 15 + 3h$
is to combine like terms on both sides.

☐ I know that the first step to solve
$7(r - 3) = r + 4 + r$
is to use the distributive property on the left side and combine like terms on the right side.

CHAPTER 3 SUMMARY AND REVIEW

SECTION 3.1 Algebraic Expressions

DEFINITIONS AND CONCEPTS	EXAMPLES
A **variable** is a letter (or symbol) that stands for a number. Since numbers do not change value, they are called **constants.**	Variables: x, a, and y Constants: 8, -10, $2\frac{3}{5}$, and 3.14
Variables and/or numbers can be combined with the operations of addition, subtraction, multiplication, and division to create **algebraic expressions.** We often refer to *algebraic expressions* as simply **expressions.**	Expressions: $5y + 7$ $\dfrac{12 - x}{5}$ $8a(b - 3)$
Key words and **key phrases** can be translated into algebraic expressions. Review the tables on pages 226 and 227.	*5 more than x* can be expressed as $x + 5$. *25 less than twice y* can be expressed as $2y - 25$. One-half of the cost c can be expressed as $\frac{1}{2}c$.

REVIEW EXERCISES

1. The illustration below shows the distances from two towns to an airport. Which town is closer to the airport? How much closer is it?

   $(x - 250)$ mi x mi

   Brandon   Airport   Mill City

2. See the illustration below. Let h represent the height of the ladder, and write an algebraic expression for the height of the ceiling in feet.

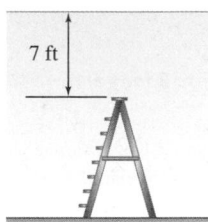

   7 ft

Translate each of the following phrases to an algebraic expression.

3. Five less than n

4. The product of 7 and x

5. The quotient of six and p

6. The sum of s and -15

7. Twice the length l

8. D reduced by 100

9. Two more than r

10. 45 divided by x

11. 100 reduced by twice the cutoff score s

12. The absolute value of the difference of 2 and the square of a

13. Translate the expression $m - 500$ into words.

14. HARDWARE Refer to the illustration below.

    a. Let n represent the length of the nail (in inches). Write an algebraic expression that represents the length of the bolt (in inches).

    b. Let b represent the length of the bolt (in inches). Write an algebraic expression that represents the length of the nail (in inches).

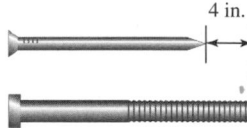

    4 in.

15. CHILD CARE A child care center has six rooms, and the same number of children are in each room. If c children attend the center, write an algebraic expression that represents the number of children in each room.

16. CAR SALES A used car, originally advertised for $1,000, did not sell. The owner decided to drop the price $x. Write an algebraic expression that represents the new price of the car (in dollars).

17. CLOTHES DESIGNERS The legs on a pair of pants are x inches long. The designer then lets the hem down 1 inch. Write an algebraic expression that represents the new length (in inches) of the pants legs.

18. BUTCHERS A roast weighs p pounds. A butcher trimmed the roast into 8 equal-sized servings. Write an algebraic expression that represents the weight (in pounds) of one serving.

19. ROAD TRIPS On a cross-country vacation, a husband drove for twice as many hours as his wife. Choose a variable to represent the hours driven by one of them. Then write an algebraic expression to represent the hours driven by the other.

20. GEOMETRY The length of a rectangle is 3 units more than its width. Choose a variable to represent one of the dimensions. Then write an algebraic expression that represents the other dimension.

21. SPORTS EQUIPMENT An NBA basketball weighs 2 ounces more than twice the weight of a volleyball. Let a variable represent the weight of one of the sports balls. Then write an algebraic expression that represents the weight of the other ball.

22. BEST-SELLING BOOKS *The Lord of the Rings* was first published 6 years before *To Kill a Mockingbird. The Godfather* was first published 9 years after *To Kill a Mockingbird.* Write algebraic expressions to represent the ages of each of those books.

Use a table to help answer Problems 23 and 24.

23. How many eggs are in x dozen?

24. d days is how many weeks?

SECTION 3.2 Evaluating Algebraic Expressions and Formulas

DEFINITIONS AND CONCEPTS	EXAMPLES
To **evaluate algebraic expressions,** we substitute the values of its variables and apply the order of operations rule.	Evaluate $\dfrac{x^2 - y^2}{x + y}$ for $x = 2$ and $y = -3$.

$$\frac{x^2 - y^2}{x + y} = \frac{2^2 - (-3)^2}{2 + (-3)} \qquad \text{Substitute 2 for x and } -3 \text{ for y.}$$

$$= \frac{4 - 9}{-1} \qquad \begin{array}{l}\text{In the numerator, evaluate the exponential} \\ \text{expressions. In the denominator, add.}\end{array}$$

$$= \frac{-5}{-1} \qquad \text{In the numerator, subtract.}$$

$$= 5 \qquad \text{Do the division.}$$

A **formula** is an equation that states a relationship between two or more variables.

Formulas from business:

Sale price = original price − discount

Retail price = cost + markup

Profit = revenue − costs

SMALL BUSINESSES For the month of December, a nail salon's cost of doing business was $6,050. If December revenues totaled $18,295, what was the salon's profit for the month?

$P = r - c$ This is the formula for profit.

$\quad = \mathbf{18{,}295} - \mathbf{6{,}050}$ Substitute 18,295 for the revenue r and 6,050 for the costs c.

$\quad = 12{,}245$ Do the subtraction.

The nail salon made a profit of $12,245 in December.

Formulas from science:

Distance = rate · time

Fahrenheit to Celsius temperature:

$$C = \frac{5(F - 32)}{9}$$

Distance fallen = 16 · (time)^2

WHALES As they migrate from the Bering Sea to Baja California, grey whales swim at an average rate of 3 mph. If they swim for 20 hours a day, find the distance they travel each day.

$d = rt$ This is the formula for distance traveled.

$\quad = \mathbf{3}(\mathbf{20})$ Substitute 3 for the rate r and 20 for the time t.

$\quad = 60$ Do the multiplication.

Grey whales travel a distance of 60 miles each day.

The **mean** (or **average**) of a set of numbers is a value around which the numbers are grouped.

$$\text{Mean} = \frac{\text{sum of values}}{\text{number of values}}$$

GRADES Find the mean of the test scores of 74, 83, 79, 91, and 73.

$$\text{Mean} = \frac{74 + 83 + 79 + 91 + 73}{5}$$

$$= \frac{400}{5}$$

$$= 80$$

The mean test score is 80.

REVIEW EXERCISES

25. RETAINING WALLS The illustration to the right shows the design for a retaining wall. The relationships between the lengths of its important parts are given in words.

    a. Choose a variable to represent one unknown dimension of the wall. Then write algebraic expressions to represent the lengths of the other two parts.

    b. Suppose engineers determine that a 10-foot-high wall is needed. Find the lengths of the upper and lower bases.

The length of the upper base is 5 ft less than the height.

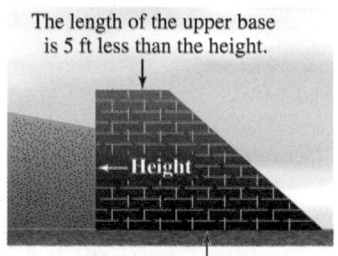

← Height

The length of the lower base is 3 ft less than twice the height

26. SOD FARMS The expression $20,000 - 3s$ gives the number of square feet of sod that are left in a field after s strips have been removed. Suppose a city orders 7,000 strips of sod. Evaluate the expression and explain the result.

Strips of sod, cut and ready to be loaded on a truck for delivery

Evaluate each algebraic expression.

27. $-2x + 6$ for $x = -3$

28. $\dfrac{6 - a}{1 + a}$ for $a = -2$

29. $(-y - 40)^2$ for $y = -50$

30. $|x^3 - 8x^2|$ for $x = 4$

31. $b^2 - 4ac$ for $a = 4, b = 6,$ and $c = -4$

32. $\dfrac{-2k^3}{1 - 2 - 3}$ for $k = -2$

Use the correct formula to solve each problem.

33. SALE PRICE Find the sale price of a trampoline that usually sells for \$315 if a \$37 discount is being offered.

34. RETAIL PRICE Find the retail price of a car if the dealer pays \$14,505 and the markup is \$725.

35. GRAND OPENINGS On its first month of business, a bookstore brought in \$52,895. The costs for the month were \$47,980. Find the profit the store made its first month.

36. DISTANCE TRAVELED Complete the table by finding the distance traveled for a given time at a given rate.

	Rate (mph)	Time (hr)	Distance traveled (mi)
Monorail	65	2	
Subway	38	3	
Train	x	6	
Bus	55	$t + 1$	

37. TEMPERATURE CONVERSION At a summer resort, visitors can relax by taking a dip in a swimming pool or a lake. The pool water is kept at a constant temperature of 77°F. The water in the lake is 23°C. Which water is warmer, and by how many degrees Celsius?

38. DISTANCE FALLEN A steelworker accidentally dropped a wrench while working atop a new high-rise building. How far will the wrench fall in 3 seconds?

39. AVERAGE YEARS OF EXPERIENCE Three generations of Smiths now operate a family-owned real estate office. The two grandparents, who started the business, have been realtors for 40 years. Their son and daughter-in-law joined the company as realtors 18 years ago. Their grandson has worked as a realtor for 4 years. What is the average number of years a member of the Smith family has worked at Smith Realty?

40. SURVEYS Some students were asked to rate their college cafeteria food on a scale from 1 to 5. The responses are shown on the tally sheet. Find the mean rating.

Poor		Fair		Excellent
1	2	3	4	5
	|||	|||	ᵗᴴᴸ	ᵗᴴᴸ ||||

SECTION 3.3   Simplifying Algebraic Expressions and the Distributive Property

DEFINITIONS AND CONCEPTS	EXAMPLES
We often use the *commutative property of multiplication* to reorder factors and the *associative property of multiplication* to regroup factors when **simplifying expressions.**	Simplify: $\begin{aligned} -5 \cdot 3y &= (-5 \cdot 3)y \\ &= -15y \end{aligned}$ Simplify: $\begin{aligned} -45b\left(\dfrac{5}{9}\right) &= \left(-45 \cdot \dfrac{5}{9}\right)b \\ &= -\dfrac{5 \cdot \overset{1}{\cancel{9}} \cdot 5}{\underset{1}{\cancel{9}}}b \\ &= -25b \end{aligned}$

The **distributive property** can be used to remove parentheses: $a(b + c) = ab + ac$ $\quad$ $a(b - c) = ab - ac$	Multiply: $7(x + 3) = 7 \cdot x + 7 \cdot 3$ $\qquad\qquad\qquad = 7x + 21$
The distributive property can be extended to several other useful forms. $a(b + c + d) = ab + ac + ad$ $(b + c)a = ba + ca$ $\quad$ $(b - c)a = ba - ca$	Multiply: $-2(4m - 5n - 7) = -2(4m) - (-2)(5n) - (-2)(7)$ $\qquad\qquad\qquad\qquad = -8m + 10n + 14$ Multiply: $(6y - 10)5 = 6y \cdot 5 - 10 \cdot 5$ $\qquad\qquad\qquad\quad = 30y - 50$
The **opposite of a sum** is the sum of the opposites. $-(a + b) = -a + (-b)$ The result can be obtained very quickly by changing the sign of each term within the parentheses and dropping the parentheses.	Simplify: $-(-3r + 14) = -1(-3r + 14)$ $\qquad$ Replace the − symbol with −1. $\qquad\qquad\quad = (-1)(-3r) + (-1)(14)$ $\quad$ Distribute −1. $\qquad\qquad\quad = 3r - 14$

REVIEW EXERCISES

Simplify each expression.

41. $-2 \cdot 5x$

42. $-7x(-6y)$

43. $4d \cdot 3e \cdot 5$

44. $(4s)8$

45. $-1(-e)(2)$

46. $7x \cdot 7y$

47. $4 \cdot 3k \cdot 7$

48. $(-10t)(-10)$

Multiply.

49. $4(y + 5)$

50. $-5(6t + 9)$

51. $(-3 - 3x)7$

52. $-3(4e - 8x - 1)$

53. $4(6w - 3)2$

54. $-9(x + 1)4$

Simplify.

55. $-(6t - 4)$

56. $-(5 + x)$

57. $-(6t - 3s + 1)$

58. $-(-5a - 3)$

SECTION 3.4   Combining Like Terms

DEFINITIONS AND CONCEPTS	EXAMPLES
A **term** is a product or quotient of numbers and/or variables. A single number or variable is also a term. A term such as 4, that consists of a single number, is called a **constant term.**	Terms: 4, y, $6r$, $-w^3$, $3.7x^5$, $\dfrac{3}{n}$, $-15ab^2$
Addition symbols separate expressions into parts called **terms.** The numerical factor of a term is called the **coefficient** of the term.	Since $6a^2 + a - 5$ can be written as $6a^2 + a + (-5)$, it has three terms. <table><tr><th>Term</th><th>Coefficient</th></tr><tr><td>$6a^2$</td><td>6</td></tr><tr><td>a</td><td>1</td></tr><tr><td>-5</td><td>-5</td></tr></table>
It is important to be able to distinguish between the **terms of an expression** and the **factors of a term.**	$x + 6$ $6x$ x is a term. x is a factor.
Like terms are terms with exactly the same variables raised to exactly the same powers.	$3x$ and $-5x$ are like terms. $-4t^3$ and $3t^2$ are unlike terms because the variable t has different exponents. $0.5xyz$ and $3.7xy$ are unlike terms because they have different variables.
Simplifying the sum or difference of like terms is called **combining like terms.** Like terms can be combined by adding or subtracting the coefficients of the terms and keeping the same variables with the same exponents.	Simplify: $4a + 2a = 6a$ Think: $(4 + 2)a = 6a$. Simplify: $5p^2 + p - p^2 - 9p = 4p^2 - 8p$ Think: $(5 - 1)p^2 = 4p^2$ and $(1 - 9)p = -8p$. Simplify: $2(k - 1) - 3(k + 2) = 2k - 2 - 3k - 6$ $= -k - 8$
The perimeter of a rectangle is given by $$P = 2l + 2w$$ The perimeter of a square is given by $$P = 4s$$	FLAGS Find the perimeter of the flag of Eritrea, a country in east Africa, that is shown to the right. 32 in. 48 in. $P = 2l + 2w$ This is the formula for the perimeter of a rectangle. $= 2(48) + 2(32)$ Substitute 48 for the length l and 32 for the width w. $= 96 + 64$ Do the multiplication. $= 160$ Do the addition. The perimeter of the flag is 160 in.

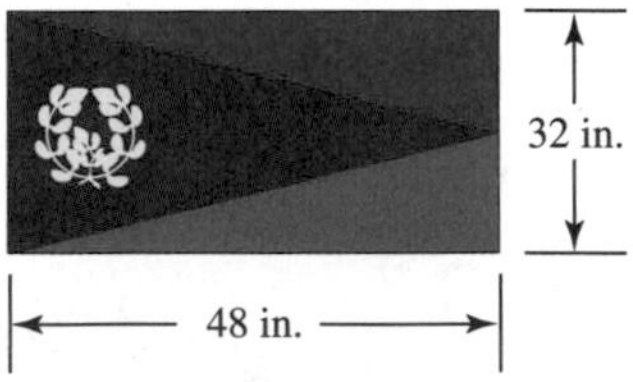

REVIEW EXERCISES

Identify the terms in each expression.

59. $8x^2 - 7x + 9$ **60.** $-15y$

61. $16ab - 6b$ **62.** $4x - 3 + 5x - 7$

Identify the coefficient of each term in the expression.

63. $5x^2 - 4x + 8$ **64.** $7y + 3y + x - y$

65. $t + r - t + 6$ **66.** $-5y^2 + 125$

Determine whether x is used as a factor or a term.

67. $5x - 6y^2$ **68.** $x + 6$

69. $-36 - x + b$ **70.** $6xy$

*Determine whether the following are like terms.
Write yes or no.*

71. $4x, -5x$ **72.** $4x, 4x^2$

73. $3xy, xy$ **74.** $-5b^2c, -5bc^2$

Simplify by combining like terms, if possible.

75. $3x + 4x$ **76.** $-3t^3 - 6t^2$

77. $2z + (-5z)$ **78.** $6x - x$

79. $-6y - 7y - (-y)$ **80.** $5w^2 - 8 - 4w^2 + 3$

81. $-45d - 2a + 4a - d$

82. $5y + 8h - 3 + 7h + 5y + 2$

83. $10a^2 + 6a - 17a + 6$ **84.** $28w + w$

Simplify each expression.

85. $7(y + 6) + 3(2y + 2)$

86. $-4(t - 7) - (t + 6)$

87. $5x - 4 - 2(x - 6)$

88. $6f + (-11) + 7(12 - 8f)$

89. ROBOTS Find an algebraic expression that represents the total length (in feet) of the robotic arm shown below.

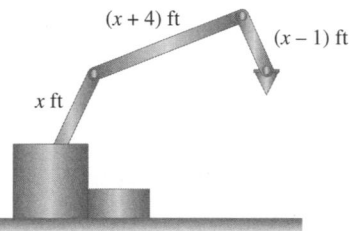

$(x + 4)$ ft

$(x - 1)$ ft

x ft

90. HOLIDAY LIGHTS To decorate a house, lights will be hung around the entire home, as shown. They will also be placed around the two 5-foot-by-5-foot windows in the front. How many feet of lights will be needed?

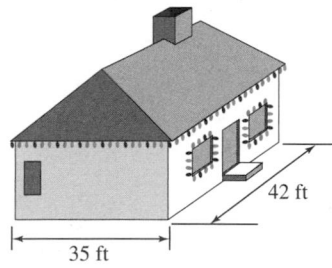

42 ft

35 ft

SECTION 3.5 Simplifying Expressions to Solve Equations

DEFINITIONS AND CONCEPTS	EXAMPLES
A number that makes an equation a true statement when substituted for the variable is called a **solution** of the equation. We say such a number **satisfies** the equation.	Use a check to determine whether -51 is a solution of $3x - 15 = 4x + 36$.

$$3x - 15 = 4x + 36 \qquad \text{The original equation.}$$

$$3(-51) - 15 \stackrel{?}{=} 4(-51) + 36 \qquad \text{Substitute } -51 \text{ for } x.$$

$$-153 - 15 \stackrel{?}{=} -204 + 36 \qquad \text{Do the multiplication.}$$

$$-168 = -168 \qquad \text{True}$$

Since the resulting statement, $-168 = -168$, is true, -51 is a solution of $3x - 15 = 4x + 36$.

When solving equations, we should simplify the expressions that make up the left and right sides before applying any properties of equality.

A strategy for solving equations:

1. *Simplify* each side. Use the distributive property and combine like terms when necessary.
2. *Isolate the variable term.* Use the addition and subtraction properties of equality.
3. *Isolate the variable.* Use the multiplication and division properties of equality.
4. *Check* the result in the original equation.

Solve: $2(y + 2) + 4y = 11 - y$

$2y + 4 + 4y = 11 - y$	Distribute the multiplication by 2.
$6y + 4 = 11 - y$	Combine like terms: $2y + 4y = 6y$.
$6y + 4 + y = 11 - y + y$	To eliminate $-y$ on the right, add y to both sides.
$7y + 4 = 11$	Combine like terms.
$\boxed{7y} + 4 - 4 = 11 - 4$	To isolate the variable term $7y$, subtract 4 from both sides.
$7y = 7$	Simplify each side of the equation.
$\dfrac{7y}{7} = \dfrac{7}{7}$	To isolate y, divide both sides by 7.
$y = 1$	

The solution is 1. Check by substituting it into the original equation.

REVIEW EXERCISES

91. Use a check to determine whether -1 is a solution of $6a + (-7) = 5a - 9$.

92. Use a check to determine whether 4 is a solution of $4(8 - 3t) = 32 - 8(t + 2)$.

Solve each equation. Check the result.

93. $5a - 3a = -36$

94. $3x - 4x = -8$

95. $250 + 350 = x - 10 + x$ **96.** $7x + 1 = 3x - 11$

97. $5(y - 15) = 0$

98. $3a - (2a - 1) = -2$

99. $15 + b = 5b + 1 + 3b$

100. $-6(2x + 3) = -(5x - 10)$

101. $4 + 3(2x + 4) - 4 = -42$

102. $4(9d + 2) = 4(d - 18) - 8d$

SECTION 3.6   Using Equations to Solve Application Problems

DEFINITIONS AND CONCEPTS	EXAMPLES
To solve application problems, use the five-step problem-solving strategy.	SOUND SYSTEMS A 45-foot-long speaker wire is cut into two pieces. One piece is 9 feet longer than the other. Find the length of each piece of wire.
1. Analyze the problem: What information is given? What are you asked to find?	*Analyze*
2. Form an equation: Pick a variable to represent the numerical value to be found. Translate the words of the problem into an equation.	• A 45-foot long wire is cut into two pieces. Given
	• One piece is 9 feet longer than the other. Given
3. Solve the equation.	• What is the length of the shorter piece and the length of the longer piece of wire? Find

4. **State the conclusion clearly:** Be sure to include the units (such as feet, seconds, or pounds) in your answer.

5. **Check the result:** Use the original wording of the problem, not the equation that was formed in step 2 from the words.

The five-step problem-solving strategy can be used to solve application problems to find **two unknowns.**

Form

Since we are told that the length of the longer piece of wire is related to the length of the shorter piece,

Let x = the length of the shorter piece of wire

There is a second unknown quantity. Look for a key phrase to help represent the length of the longer piece of wire using an algebraic expression.

Key Phrase: 9 feet longer   **Translation:** addition

So $x + 9$ = the length of the longer piece of wire.

Now, translate the words of the problem to an equation.

The length of the shorter piece	plus	the length of the longer piece	is	45 feet.
x	$+$	$x + 9$	$=$	45

Solve

$x + x + 9 = 45$	We need to isolate x on the left side.
$2x + 9 = 45$	Combine like terms: x + x = 2x.
$2x + 9 - 9 = 45 - 9$	To isolate 2x, subtract 9 from both sides.
$2x = 36$	Do the subtraction.
$\dfrac{2x}{2} = \dfrac{36}{2}$	To isolate x, undo the multiplication by 2 by dividing both sides by 2.
$x = 18$	Do the division. This is the length of the shorter piece.

To find the second unknown, we substitute 18 for x in the expression that represents the length of the longer piece of wire.

$$x + 9 = \mathbf{18} + 9 = 27$$

State

The length of the shorter piece of wire is 18 feet and the length of the longer piece is 27 feet.

Check

The length of the longer piece of wire, 27 feet, is 9 feet longer than the length of the shorter piece, 18 feet. Adding the two lengths, we get

```
   18
 + 27
   45  ←—This is the original length of the wire,
         before It was cut into two pieces.
```

The results, 18 ft and 27 ft, check.

Be careful to distinguish between the *number* and the *value* of a set of objects.

Total value = number · value

Determine the total value of x \$20 bills.

Total value = **number** · **value**

= x · 20

= $20x$

The total value is $20x$ dollars.

REVIEW EXERCISES

103. CONCERTS The fee to rent a concert hall is $2,250 plus $150 per hour to pay for the support staff. For how many hours can an orchestra rent the hall and stay within a budget of $3,300?

104. COLD STORAGE A meat locker lowers the temperature of a product 7° Fahrenheit every hour. If freshly ground hamburger is placed in the locker, how long would it take to go from a room temperature of 71°F to 29°F?

105. MOVING EXPENSES Tom and his friend split the cost of renting a U-Haul trailer equally. Tom also agreed to pay the $4 to rent a refrigerator dolly. In all, Tom paid $20. What did it cost to rent the trailer?

106. FITNESS The midweek workout for a fitness instructor consists of walking and running. She walks 3 fewer miles than she runs. If her workout covers a total of 15 miles, how many miles does she run and how many miles does she walk?

107. RODEOS Attendance during the first day of a two-day rodeo was low. On the second day, attendance doubled. If a total of 6,600 people attended the show, what was the attendance on the first day and what was the attendance on the second day?

108. PARKING LOTS A rectangular-shaped parking lot is 4 times as long as it is wide. If the perimeter of the parking lot is 250 feet, what is its length and width?

109. Complete the table.

Type of coin	Number	Value (¢)	Total value (¢)
Dime	6		
Quarter	7		
Penny	x		
Nickel	$n + 25$		

110. HEALTH FOOD A fruit juice bar sells two types of drinks: one priced at $3 and the other at $4. One day at lunchtime, business was very brisk. If a total of 50 drinks were sold and the receipts were $185, how many $3 drinks and how many $4 drinks were purchased?

CHAPTER 3   TEST

Fill in the blanks.

1. **a.** _____ are letters (or symbols) that stand for numbers.

    b. To perform the multiplication $3(x + 4)$, we use the _____ property.

    c. Terms such as $7x^2$ and $5x^2$, which have the same variables raised to exactly the same power, are called _____ terms.

    d. To _____ an equation means to find all values of the variable that make the equation true.

    e. The _____ of the term $9y$ is 9.

    f. Variables and/or numbers can be combined with the operations of addition, subtraction, multiplication, and division to create algebraic _____.

    g. To evaluate $y^2 + 9y - 3$ for $y = -5$, we _____ -5 for y and apply the order of operations rule.

    h. An _____ is a statement indicating that two expressions are equal.

    i. When we write $4x + x$ as $5x$, we say we have _____ like terms.

    j. To _____ the solution of an equation, we substitute the value for the variable in the original equation and determine whether the result is a true statement.

2. SALARIES A wife's monthly salary is $1,000 less than twice her husband's monthly salary.

    a. If her husband's monthly salary is h dollars, write an algebraic expression that represents the wife's monthly salary (in dollars).

    b. Suppose the husband's monthly salary is $2,350. Find the wife's monthly salary.

3. REFRESHMENTS How many cups of coffee are left in the coffeemaker shown if c cups have already been poured from it?

4. Let a variable represent the length of one of the fish shown. Then write an expression that represents the length (in inches) of the other fish. Give two possible sets of answers.

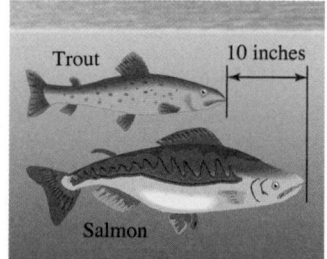

5. Translate each phrase to mathematical symbols.

    a. 2 less than r

    b. The product of $3, x$, and y

    c. x increased by 100

    d. The absolute value of the quotient of x and -9

6. Write an algebraic expression that represents the number of years in d decades.

7. Evaluate each expression.

    a. $x - 16$ for $x = 4$

    b. $2t^2 - 3(t - s)$ for $t = -2$ and $s = 4$

    c. $-a^2 + 10$ for $a = -3$

    d. $\left| \dfrac{-10d + f^3}{-f} \right|$ for $d = -1$ and $f = -5$

8. DISTANCE TRAVELED Find the distance traveled by a motorist who departed from home at 9:00 A.M. and arrived at his destination at noon, traveling at a rate of 55 miles per hour.

9. PROFITS A craft show promoter had revenues and costs as shown. Find the profit.

Revenues	Costs
Ticket sales: $40,000	Supplies: $13,000
Booth rental: $15,000	Facility rental fee: $5,000

10. FALLING OBJECTS If a tennis ball was dropped from the top of a 200-foot-tall building, would it hit the ground after falling for 3 seconds? If not, how far short of the ground would it be?

11. METER READINGS Every hour between 8 A.M. and 5 P.M., a technician noted the value registered by a meter in a power plant and recorded that number on a line graph. Find the mean meter value reading for this period.

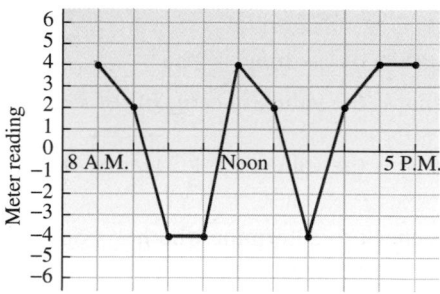

12. LANDMARKS Overlund College is going to construct a gigantic block letter O on a foothill slope near campus. The outline of the letter is to be done using redwood edging. How many feet of edging will be needed?

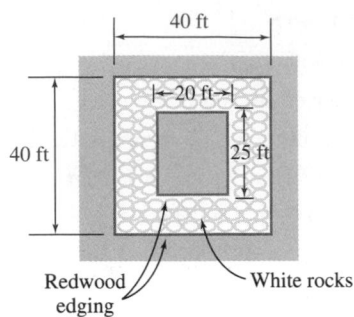

Redwood edging   White rocks

13. AIR CONDITIONING After the air conditioner in a classroom was accidentally left on all night, the room's temperature in the morning was a cool 59°F. What was the temperature in degrees Celsius?

14. Multiply.

  a. $5(5x + 1)$ **b.** $-6(7 - x)$

  c. $-(6y + 4)$ **d.** $3(2a + 3b - 7)$

  e. $(a - 15)8$ **f.** $2(6r + 9)3$

15. Determine whether x is used as a factor or as a term.

  a. $5xy$ **b.** $8y + x + 6$

16. Simplify each expression, if possible.

  a. $7x + 4x$ **b.** $3 \cdot 4e$

  c. $6x^2 - x^2$ **d.** $-5y(-6)$

  e. $0 - 7x$ **f.** $0 + 9y$~

  g. $-8a \cdot 9b$ **h.** $8(-7m)5$

17. a. Identify each term in this algebraic expression: $8x^2 - x - 6$

  b. What is the coefficient of each term?

18. Simplify.

  a. $-20y + 6 - 8y + 4$

  b. $-t - t - t$

  c. $4(y + 3) - 5(2y + 3)$

  d. $-m^4 + 3m^3 + 10m^4 + 20m^3$

19. a. What is the value (in cents of) of k dimes?

  b. What is the value of $p + 2$ twenty-dollar bills?

20. Use a check to determine whether -5 is a solution of $6x - 8 = 12(x - 3)$.

Solve each equation. Check the result.

21. $5x - 3x = -18$ **22.** $6r = r - 12 + r$

23. $-55 + 10 = 3(1 - 4t)$ **24.** $6 - (y - 3) = 19$

25. $8 + 2(3x - 4) = -60$ **26.** $-23 - n - 6n = 13 + 2n$

27. $5(x + 7) - 7x = 9(x - 1)$

28. $-80 + y = -80 - y$

Form an equation and then solve it to answer each question.

29. DRIVING SCHOOLS A driver's training program requires students to attend six equally long classroom sessions. Then the students take a 2-hour final exam at the end of the training. If the entire program requires 20 hours of a student's time, how long is each classroom session?

30. CABLE TELEVISION In order to receive its broadcasting license, a cable television station was required to broadcast locally produced shows in addition to its national programming. During a typical 24-hour period, the national shows aired for 8 hours more than the local shows. How many hours of local shows and how many hours of national shows were broadcast each day?

31. RECREATION A developer donated a large plot of land to a city for a park. Half of the acres will be used for sports fields. From the other half, 4 acres will be used for parking. This will leave 18 acres for a nature habitat. How many acres of land did the developer donate to the city?

32. PICTURE FRAMING A rectangular picture frame is twice as long as it is wide. If 144 inches of framing material were used to make it, what is the width and what is the length of the frame?

33. Do the instructions *simplify* and *solve* mean the same thing? Explain.

34. Explain why we can simplify $5x \cdot 2$ but we cannot simplify $5x + 2$.

1. **GASOLINE** In 2008, the United States produced three billion, two hundred ninety million, fifty-seven thousand barrels of finished motor gasoline. Write this number in standard notation. (Source: U.S Energy Information Administration). [Section 1.1]

2. Round 49,999 to the nearest thousand. [Section 1.1]

Perform each operation.

3. 38,908 [Section 1.2]
   +15,696

4. 9,700 [Section 1.2]
   −5,491

5. 345 [Section 1.3]
   × 67

6. $23\overline{)2,001}$ [Section 1.4]

7. **a.** Explain how to check the following result using addition. [Section 1.2]

   1,142
   − 459
   ─────
     683

   b. Write an expression showing division by 0 and an expression showing division of 0. Which is undefined? [Section 1.4]

8. **VIETNAMESE CALENDAR** An animal represents each Vietnamese lunar year. Recent Years of the Cat are listed below. If the cycle continues, what year will be the next Year of the Cat? [Section 1.2]

   1915   1927   1939   1951   1963   1975   1987   1999

9. Consider the multiplication statement $4 \cdot 5 = 20$.

   Show that multiplication is repeated addition. [Section 1.3]

10. **ROOM DIVIDERS** Four pieces of plywood, each 22 inches wide and 62 inches high, are to be covered with fabric, front and back, to make the room divider shown. How many square inches of fabric will be used? [Section 1.3]

11. **OIL CHANGES** In July of 2009, the 1964 Mercury Comet that Rachel Veitch of Orlando, Florida, drives notched its 558,000 mile. The 90-year-old retired nurse has changed the oil every 3,000 miles since she bought the car new. How many oil changes did the car have to that point? (Source: foxnews.com) [Section 1.4]

12. **a.** Find the factors of 18. [Section 1.5]

    b. Why isn't 27 a prime number? [Section 1.5]

    c. Find the prime factorization of 18. [Section 1.5]

13. Write the first ten prime numbers. [Section 1.5]

14. **a.** Find the LCM of 35 and 45. [Section 1.6]

    b. Find the GCF of 12, 68, and 92. [Section 1.6]

15. Evaluate each expression. [Section 1.7]

    a. $(9 - 2)^2 - 3^3$

    b. $\dfrac{80 - 2[12 - (5 + 4)]}{8 \div 8 \cdot 2}$

16. What property was used to solve the equation shown below? [Section 1.8]

    $$x - 3 = 47$$
    $$x - 3 + 3 = 47 + 3$$
    $$x = 50$$

17. Solve $250 = \dfrac{y}{2}$ and check the result. [Section 1.9]

18. **a.** Simplify: $-(-6)$ [Section 2.1]

    b. Find the absolute value: $|-5|$

    c. Is the statement $-12 > -10$ true or false?

19. Graph the integers greater than -3 but less than 4. [Section 2.1]

20. Translate the following phrase to mathematical symbols: *Negative twenty-one minus negative seventy-three* [Section 2.1]

21. Perform the indicated operations.

    a. $-25 + 5$ [Section 2.2]

    b. $25 - (-5)$ [Section 2.3]

    c. $-25(5)(-1)$ [Section 2.4]

    d. $\dfrac{-25}{-5}$ [Section 2.5]

22. CARD GAMES
Canasta is a card game commonly played by four players in two teams. It is possible for a team to have a negative score for a hand. The canasta scores for two teams are shown to the right. Find the total for each team. [Section 2.2]

	Team 1	Team 2
	305	295
	−75	−120
	600	300
	500	0
	200	100
	100	0
Total:		

23. PLANETS Mercury orbits closer to the sun than any other planet. Temperatures on Mercury can get as high as 810°F and as low as −290°F. What is the temperature range? [Section 2.3]

24. a. Explain how to evaluate -3^2 and $(-3)^2$. [Section 2.4]

   b. What property allows us to rewrite $x \cdot 5$ as $5x$? [Section 2.4]

Evaluate each expression. [Section 2.6]

25. $\dfrac{(-6)^2 - 1^5}{-4 - 3}$

26. $-3 + 3\left(-4 - 4 \cdot 2\right)^2$

Solve each equation and check the result. [Section 2.7]

27. $-4x + 4 = -24$

28. $-y = 10$

29. $\dfrac{m}{-2} - 6 = -9$

30. $-7 + 9 = a + 5(-3)$

31. $-90 = x + (-3)$

32. $16 = -7 - 5 - 7x$

33. Translate each phrase to mathematical symbols. [Section 3.1]

   a. h increased by 12

   b. 4 less than the width w

   c. 1,000 split x equal ways

34. a. TENNIS Write an algebraic expression that represents the length of the handle of the tennis racket in inches. [Section 3.1]

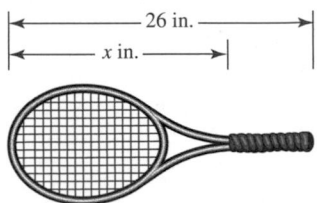

   b. What is the value (in cents) of q quarters? [Section 3.6]

35. Write an algebraic expression that represents the number of inches in f feet. [Section 3.1]

36. Evaluate $x^2 - 2x + 1$ for $x = -5$. [Section 3.2]

37. Complete the table. [Section 3.2]

	Rate (mph)	Time (hr)	Distance traveled (mi)
Truck	55	4	

38. Multiply. [Section 3.3]

   a. $5(2x - 7)$ **b.** $-(5t - 7)$

39. Simplify. [Section 3.3]

   a. $-6(-4t)$ **b.** $4(-3y)(4z)$

40. Complete the table. [Section 3.4]

Term	Coefficient
$4a$	
$-2y^2$	
x	
$-m$	

41. Write an expression in which x is used as a term. Then write an expression in which x is used as a factor. [Section 3.4]

42. Simplify: [Section 3.4]

   a. $5b + 8 - 6b - 7$

   b. $4(x + 5) - 5 - (2x - 4)$

Solve each equation and check the result. [Section 3.5]

43. $8p + 2p - 1 = -11$

44. $7 + 2x = 2 - (4x + 7)$

45. $2b + 15 = -21 - b - 6b$

46. $4(m - 30) - 4m = 6(2m + 6)$

Form an equation and then solve it to answer the following questions. [Section 3.6]

47. CLASS TIME In a chemistry course, students spend a total of 300 minutes in lab and lecture each week. The time spent in lab is 50 minutes less than the time spent in lecture. How many minutes do the students spend in lecture and lab each week?

48. GEOMETRY The perimeter of a rectangle is 120 feet, and the length is five times as long as the width. Find the length and the width.

Fractions and Mixed Numbers

4

© iStockphoto.com/Catherine Yeulet

from *Campus to Careers*

School Guidance Counselor

School guidance counselors plan academic programs and help students choose the best courses to take to achieve their educational goals. Counselors often meet with students to discuss the life skills needed for personal and social growth. To prepare for this career, guidance counselors take classes in an area of mathematics called *statistics,* where they learn how to collect, analyze, explain, and present data.

In **Problem 115** of **Study Set 4.4,** you will see how a counselor must be able to add fractions to better understand a graph that shows students' study habits.

JOB TITLE:
School Guidance Counselor

EDUCATION: A master's degree is usually required to be licensed as a counselor. However, some schools accept a bachelor's degree with the appropriate counseling courses.

JOB OUTLOOK: Excellent.

ANNUAL EARNINGS: The average (median) salary in 2006 was $53,750.

FOR MORE INFORMATION:
www.bls.gov/oco/ocos067.htm

Objectives

1. Identify the numerator and denominator of a fraction.

2. Simplify special fraction forms.

3. Define equivalent fractions.

4. Build equivalent fractions.

5. Simplify fractions.

6. Build and simplify algebraic fractions.

SECTION 4.1

An Introduction to Fractions

Whole numbers are used to count objects, such as CDs, stamps, eggs, and magazines. When we need to describe a part of a whole, such as one-half of a pie, three-quarters of an hour, or a one-third-pound burger, we can use *fractions*.

One-half
of a cherry pie
$\frac{1}{2}$

Three-quarters
of an hour
$\frac{3}{4}$

One-third
pound burger
$\frac{1}{3}$

1 Identify the numerator and denominator of a fraction.

A **fraction** describes the number of equal parts of a whole. For example, consider the figure below with 5 of the 6 equal parts colored red. We say that $\frac{5}{6}$ (five-sixths) of the figure is shaded.

In a fraction, the number above the **fraction bar** is called the **numerator,** and the number below is called the **denominator.**

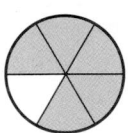

Fraction bar $\longrightarrow \dfrac{5 \longleftarrow \text{numerator}}{6 \longleftarrow \text{denominator}}$

> **The Language of Algebra** The word *fraction* comes from the Latin word *fractio* meaning "breaking in pieces."

Self Check 1

Identify the numerator and denominator of each fraction:

a. $\dfrac{7}{9}$

b. $\dfrac{21}{20}$

Now Try Problem 21

EXAMPLE 1 Identify the numerator and denominator of each fraction:

a. $\dfrac{11}{12}$ b. $\dfrac{8}{3}$

Strategy We will find the number above the fraction bar and the number below it.

WHY The number above the fraction bar is the numerator, and the number below is the denominator.

Solution

a. $\dfrac{11 \longleftarrow \text{numerator}}{12 \longleftarrow \text{denominator}}$ b. $\dfrac{8 \longleftarrow \text{numerator}}{3 \longleftarrow \text{denominator}}$

If the numerator of a fraction is less than its denominator, the fraction is called a **proper fraction.** A proper fraction is less than 1. If the numerator of a fraction is greater than or equal to its denominator, the fraction is called an **improper fraction.** An improper fraction is greater than or equal to 1.

Proper fractions

$\dfrac{1}{4}, \dfrac{2}{3}$, and $\dfrac{98}{99}$

Improper fractions

$\dfrac{7}{2}, \dfrac{98}{97}, \dfrac{16}{16}$, and $\dfrac{5}{1}$

> **The Language of Algebra** The phrase *improper fraction* is somewhat misleading. In algebra and other mathematics courses, we often use such fractions "properly" to solve many types of problems.

EXAMPLE 2 Write fractions that represent the shaded and unshaded portions of the figure below.

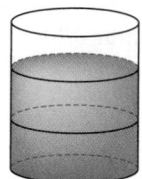

Strategy We will determine the number of equal parts into which the figure is divided. Then we will determine how many of those parts are shaded.

WHY The denominator of a fraction shows the number of equal parts in the whole. The numerator shows how many of those parts are being considered.

Solution

Since the figure is divided into 3 equal parts, the denominator of the fraction is 3. Since 2 of those parts are shaded, the numerator is 2, and we say that

$\dfrac{2}{3}$ of the figure is shaded. Write: $\dfrac{\text{number of parts shaded}}{\text{number of equal parts}}$

Since 1 of the 3 equal parts of the figure is not shaded, the numerator is 1, and we say that

$\dfrac{1}{3}$ of the figure is not shaded. Write: $\dfrac{\text{number of parts not shaded}}{\text{number of equal parts}}$

Self Check 2

Write fractions that represent the portion of the month that has passed and the portion that remains.

DECEMBER

X	X	X	X	X	X	X
X	X	X	X	12	13	14
15	16	17	18	19	20	21
22	23	24	25	26	27	28
29	30	31				

Now Try **Problems 25 and 113**

There are times when a negative fraction is needed to describe a quantity. For example, if an earthquake causes a road to sink seven-eighths of an inch, the amount of downward movement can be represented by $-\dfrac{7}{8}$. Negative fractions can be written in three ways. The negative sign can appear in the numerator, in the denominator, or in front of the fraction.

$$\dfrac{-7}{8} = \dfrac{7}{-8} = -\dfrac{7}{8} \qquad \dfrac{-15}{4} = \dfrac{15}{-4} = -\dfrac{15}{4}$$

Notice that the examples above agree with the rule from Chapter 2 for dividing integers with different (unlike) signs: *the quotient of a negative integer and a positive integer is negative.*

iStockphoto.com/Jamie VanBuskirk

2 Simplify special fraction forms.

Recall from Section 1.4 that a fraction bar indicates division. This fact helps us simplify four special fraction forms.

- **Fractions that have the same numerator and denominator:** In this case, we have a number divided by itself. The result is 1 (provided the numerator and denominator are not 0). We call each of the following fractions a **form of 1.**

$$1 = \frac{1}{1} = \frac{2}{2} = \frac{3}{3} = \frac{4}{4} = \frac{5}{5} = \frac{6}{6} = \frac{7}{7} = \frac{8}{8} = \frac{9}{9} = \cdots$$

- **Fractions that have a denominator of 1:** In this case, we have a number divided by 1. The result is simply the numerator.

$$\frac{5}{1} = 5 \qquad \frac{24}{1} = 24 \qquad \frac{-7}{1} = -7$$

- **Fractions that have a numerator of 0:** In this case, we have division of 0. The result is 0 (provided the denominator is not 0).

$$\frac{0}{8} = 0 \qquad \frac{0}{56} = 0 \qquad \frac{0}{-11} = 0$$

- **Fractions that have a denominator of 0:** In this case, we have division by 0. The division is undefined.

$$\frac{7}{0} \text{ is undefined} \qquad \frac{-18}{0} \text{ is undefined}$$

> **The Language of Algebra**   Perhaps you are wondering about the fraction form $\frac{0}{0}$. It is said to be *undetermined*. This form is important in advanced mathematics courses.

Self Check 3

Simplify, if possible:

a. $\dfrac{4}{4}$ **b.** $\dfrac{51}{1}$ **c.** $\dfrac{45}{0}$ **d.** $\dfrac{0}{6}$

Now Try Problem 33

EXAMPLE 3
Simplify, if possible: **a.** $\dfrac{12}{12}$ **b.** $\dfrac{0}{24}$ **c.** $\dfrac{18}{0}$ **d.** $\dfrac{9}{1}$

Strategy To simplify each fraction, we will divide the numerator by the denominator, if possible.

WHY A fraction bar indicates division.

Solution

a. $\dfrac{12}{12} = 1$ This corresponds to dividing a quantity into 12 equal parts, and then considering all 12 of them. We would get 1 whole quantity.

b. $\dfrac{0}{24} = 0$ This corresponds to dividing a quantity into 24 equal parts, and then considering 0 (none) of them. We would get 0.

c. $\dfrac{18}{0}$ is undefined   This corresponds to dividing a quantity into 0 equal parts, and then considering 18 of them. That is not possible.

d. $\dfrac{9}{1} = 9$ This corresponds to "dividing" a quantity into 1 equal part, and then considering 9 of them. We would get 9 of those quantities.

The Language of Algebra Fractions are often referred to as **rational numbers.** All integers are rational numbers, because every integer can be written as a fraction with a denominator of 1. For example,

$$2 = \frac{2}{1}, \quad -5 = \frac{-5}{1}, \quad \text{and} \quad 0 = \frac{0}{1}$$

3 Define equivalent fractions.

Fractions can look different but still represent the same part of a whole. To illustrate this, consider the identical rectangular regions on the right. The first one is divided into 10 equal parts. Since 6 of those parts are red, $\frac{6}{10}$ of the figure is shaded.

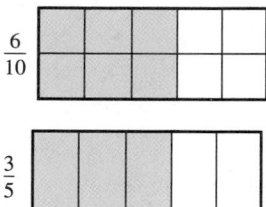

The second figure is divided into 5 equal parts. Since 3 of those parts are red, $\frac{3}{5}$ of the figure is shaded. We can conclude that $\frac{6}{10} = \frac{3}{5}$ because $\frac{6}{10}$ and $\frac{3}{5}$ represent the same shaded portion of the figure. We say that $\frac{6}{10}$ and $\frac{3}{5}$ are *equivalent fractions*.

Equivalent Fractions

Two fractions are **equivalent** if they represent the same number. **Equivalent fractions** represent the same portion of a whole.

4 Build equivalent fractions.

Writing a fraction as an equivalent fraction with a *larger* denominator is called **building** the fraction. To build a fraction, we use a familiar property from Chapter 1 that is also true for fractions:

Multiplication Property of 1

The product of any fraction and 1 is that fraction.

We also use the following rule for multiplying fractions. (It will be discussed in greater detail in the next section.)

Multiplying Fractions

To multiply two fractions, multiply the numerators and multiply the denominators.

To build an equivalent fraction for $\frac{1}{2}$ with a denominator of 8, we first ask, "What number times 2 equals 8?" To answer that question we *divide* 8 by 2 to get 4. Since we need to multiply the denominator of $\frac{1}{2}$ by 4 to obtain a denominator of 8, it follows that $\frac{4}{4}$ should be the form of 1 that is used to build an equivalent fraction for $\frac{1}{2}$.

$$\frac{1}{2} = \frac{1}{2} \cdot \frac{4}{4} \quad \text{Multiply } \frac{1}{2} \text{ by 1 in the form of } \frac{4}{4}. \text{ Note the form of 1 highlighted in red.}$$

$$= \frac{1 \cdot 4}{2 \cdot 4} \quad \text{Use the rule for multiplying two fractions. Multiply the numerators. Multiply the denominators.}$$

$$= \frac{4}{8}$$

We have found that $\frac{4}{8}$ is equivalent to $\frac{1}{2}$. To build an equivalent fraction for $\frac{1}{2}$ with a denominator of 8, we *multiplied by a factor equal to 1* in the form of $\frac{4}{4}$. Multiplying $\frac{1}{2}$ by $\frac{4}{4}$ changes its appearance but does not change its value, because we are multiplying it by 1.

Building Fractions

To build a fraction, *multiply it by a factor equal to 1* in the form of $\frac{2}{2}, \frac{3}{3}, \frac{4}{4}, \frac{5}{5}$, and so on.

> **The Language of Algebra**   Building an equivalent fraction with a larger denominator is also called *expressing a fraction in higher terms.*

Self Check 4

Write $\frac{5}{8}$ as an equivalent fraction with a denominator of 24.

Now Try Problem 37

EXAMPLE 4   Write $\frac{3}{5}$ as an equivalent fraction with a denominator of 35.

Strategy We will compare the given denominator to the required denominator and ask, "What number times 5 equals 35?"

WHY The answer to that question helps us determine the form of 1 to use to build an equivalent fraction.

Solution

To answer the question "What number times 5 equals 35?" we *divide* 35 by 5 to get 7. Since we need to multiply the denominator of $\frac{3}{5}$ by 7 to obtain a denominator of 35, it follows that $\frac{7}{7}$ should be the form of 1 that is used to build an equivalent fraction for $\frac{3}{5}$.

$$\frac{3}{5} = \frac{3}{5} \cdot \frac{\mathbf{7}}{\mathbf{7}} \qquad \text{Multiply } \tfrac{3}{5} \text{ by a form of 1: } \tfrac{7}{7} = 1.$$

$$= \frac{3 \cdot 7}{5 \cdot 7} \qquad \begin{array}{l}\text{Multiply the numerators.}\\ \text{Multiply the denominators.}\end{array}$$

$$= \frac{21}{35}$$

We have found that $\frac{21}{35}$ is equivalent to $\frac{3}{5}$.

> **Success Tip** To build an equivalent fraction in Example 4, we multiplied $\frac{3}{5}$ by 1 in the form of $\frac{7}{7}$. As a result of that step, the numerator and the denominator of $\frac{3}{5}$ were multiplied by 7:
>
> $$\frac{3 \cdot \mathbf{7}}{5 \cdot \mathbf{7}} \quad \begin{array}{l}\longleftarrow \text{ The numerator is multiplied by 7.}\\ \longleftarrow \text{ The denominator is multiplied by 7.}\end{array}$$
>
> This process illustrates the following property of fractions.

The Fundamental Property of Fractions

If the numerator and denominator of a fraction are multiplied by the same nonzero number, the resulting fraction is equivalent to the original fraction.

Since multiplying the numerator and denominator of a fraction by the same nonzero number produces an equivalent fraction, your instructor may allow you to begin your solution to problems like Example 4 as shown in the Success Tip above.

EXAMPLE 5 Write 4 as an equivalent fraction with a denominator of 6.

Strategy We will express 4 as the fraction $\frac{4}{1}$ and build an equivalent fraction by multiplying it by $\frac{6}{6}$.

WHY Since we need to multiply the denominator of $\frac{4}{1}$ by 6 to obtain a denominator of 6, it follows that $\frac{6}{6}$ should be the form of 1 that is used to build an equivalent fraction for $\frac{4}{1}$.

Solution

$$4 = \frac{4}{1}$$ Write 4 as a fraction: $4 = \frac{4}{1}$.

$$= \frac{4}{1} \cdot \frac{6}{6}$$ Build an equivalent fraction by multiplying $\frac{4}{1}$ by a form of 1: $\frac{6}{6} = 1$.

$$= \frac{4 \cdot 6}{1 \cdot 6}$$ Multiply the numerators. Multiply the denominators.

$$= \frac{24}{6}$$

Self Check 5

Write 10 as an equivalent fraction with a denominator of 3.

Now Try Problem 49

5 Simplify fractions.

Every fraction can be written in infinitely many equivalent forms. For example, some equivalent forms of $\frac{10}{15}$ are:

$$\frac{2}{3} = \frac{4}{6} = \frac{6}{9} = \frac{8}{12} = \frac{\mathbf{10}}{\mathbf{15}} = \frac{12}{18} = \frac{14}{21} = \frac{16}{24} = \frac{18}{27} = \frac{20}{30} = \cdots$$

Of all of the equivalent forms in which we can write a fraction, we often need to determine the one that is in *simplest form*.

Simplest Form of a Fraction

A fraction is in **simplest form,** or **lowest terms,** when the numerator and denominator have no common factors other than 1.

EXAMPLE 6 Are the following fractions in simplest form? **a.** $\frac{12}{27}$ **b.** $\frac{5}{8}$

Strategy We will determine whether the numerator and denominator have any common factors other than 1.

WHY If the numerator and denominator have no common factors other than 1, the fraction is in simplest form.

Solution
a. The factors of the numerator, 12, are: **1**, 2, **3**, 4, 6, 12
The factors of the denominator, 27, are: **1**, **3**, 9, 27

Since the numerator and denominator have a common factor of 3, the fraction $\frac{12}{27}$ is *not* in simplest form.

b. The factors of the numerator, 5, are: **1**, 5
The factors of the denominator, 8, are: **1**, 2, 4, 8

Since the only common factor of the numerator and denominator is 1, the fraction $\frac{5}{8}$ is in simplest form.

Self Check 6

Are the following fractions in simplest form?
a. $\frac{4}{21}$
b. $\frac{6}{20}$

Now Try Problem 53

To **simplify a fraction,** we write it in simplest form by *removing a factor equal to 1.* For example, to simplify $\frac{10}{15}$, we note that the greatest factor common to the numerator and denominator is 5 and proceed as follows:

$$\frac{10}{15} = \frac{2 \cdot \mathbf{5}}{3 \cdot \mathbf{5}} \qquad \text{Factor 10 and 15. Note the form of 1 highlighted in red.}$$

$$= \frac{2}{3} \cdot \frac{5}{5} \qquad \text{Use the rule for multiplying fractions in reverse:}$$
$$\text{write } \tfrac{2 \cdot 5}{3 \cdot 5} \text{ as the product of two fractions, } \tfrac{2}{3} \text{ and } \tfrac{5}{5}.$$

$$= \frac{2}{3} \cdot 1 \qquad \text{A number divided by itself is equal to 1: } \tfrac{5}{5} = 1.$$

$$= \frac{2}{3} \qquad \text{Use the multiplication property of 1: the product of any fraction and 1 is that fraction.}$$

We have found that the simplified form of $\frac{10}{15}$ is $\frac{2}{3}$. To simplify $\frac{10}{15}$, we *removed a factor equal to 1* in the form of $\frac{5}{5}$. The result, $\frac{2}{3}$, is equivalent to $\frac{10}{15}$.

To streamline the simplifying process, we can replace pairs of factors common to the numerator and denominator with the equivalent fraction $\frac{1}{1}$.

Self Check 7

Simplify each fraction:

a. $\dfrac{10}{25}$

b. $\dfrac{3}{9}$

Now Try Problems 57 and 61

EXAMPLE 7   Simplify each fraction: **a.** $\dfrac{6}{10}$ **b.** $\dfrac{7}{21}$

Strategy We will factor the numerator and denominator. Then we will look for any factors common to the numerator and denominator and remove them.

WHY We need to make sure that the numerator and denominator have no common factors other than 1. If that is the case, then the fraction is in *simplest form*.

Solution

a. $\dfrac{6}{10} = \dfrac{\mathbf{2} \cdot 3}{\mathbf{2} \cdot 5}$ To prepare to simplify, factor 6 and 10. Note the form of 1 highlighted in red.

$$= \frac{\overset{1}{\cancel{2}} \cdot 3}{\underset{1}{\cancel{2}} \cdot 5} \qquad \begin{array}{l}\text{Simplify by removing the common factor of 2 from the numerator and}\\ \text{denominator. A slash / and the 1's are used to show that } \tfrac{2}{2} \text{ is replaced by}\\ \text{the equivalent fraction } \tfrac{1}{1}. \text{ A factor equal to 1 in the form of } \tfrac{2}{2} \text{ was removed.}\end{array}$$

$$= \frac{3}{5} \qquad \begin{array}{l}\text{Multiply the remaining factors in the numerator: } 1 \cdot 3 = 3. \text{ Multiply the}\\ \text{remaining factors in the denominator: } 1 \cdot 5 = 5.\end{array}$$

Since 3 and 5 have no common factors (other than 1), $\dfrac{3}{5}$ is in simplest form.

b. $\dfrac{7}{21} = \dfrac{7}{3 \cdot 7}$ To prepare to simplify, factor 21.

$$= \frac{\overset{1}{\cancel{7}}}{3 \cdot \underset{1}{\cancel{7}}} \qquad \begin{array}{l}\text{Simplify by removing the common factor of 7 from the numerator}\\ \text{and denominator.}\end{array}$$

$$= \frac{1}{3} \qquad \text{Multiply the remaining factors in the denominator: } 1 \cdot 3 = 3.$$

Caution!   Don't forget to write the 1's when removing common factors of the numerator and the denominator. Failure to do so can lead to the common mistake shown below.

$$\frac{7}{21} = \frac{\cancel{7}}{3 \cdot \cancel{7}} = \frac{0}{3}$$

We can easily identify common factors of the numerator and the denominator of a fraction if we write them in prime-factored form.

EXAMPLE 8   Simplify each fraction, if possible: **a.** $\dfrac{90}{105}$ **b.** $\dfrac{25}{27}$

Simplify each fraction, if possible:

a. $\dfrac{70}{126}$

b. $\dfrac{16}{81}$

Now Try **Problems 65 and 69**

Strategy We begin by prime factoring the numerator, 90, and denominator, 105. Then we look for any factors common to the numerator and denominator and remove them.

WHY When the numerator and/or denominator of a fraction are large numbers, such as 90 and 105, writing their prime factorizations is helpful in identifying any common factors.

Solution

a. $\dfrac{90}{105} = \dfrac{2 \cdot 3 \cdot 3 \cdot 5}{3 \cdot 5 \cdot 7}$ To prepare to simplify, write 90 and 105 in prime-factored form.

$= \dfrac{2 \cdot \overset{1}{\cancel{3}} \cdot 3 \cdot \overset{1}{\cancel{5}}}{\underset{1}{\cancel{3}} \cdot \underset{1}{\cancel{5}} \cdot 7}$ Remove the common factors of 3 and 5 from the numerator and denominator. Slashes and 1's are used to show that $\frac{3}{3}$ and $\frac{5}{5}$ are replaced by the equivalent fraction $\frac{1}{1}$. A factor equal to 1 in the form of $\frac{3 \cdot 5}{3 \cdot 5} = \frac{15}{15}$ was removed.

$= \dfrac{6}{7}$ Multiply the remaining factors in the numerator: $2 \cdot 1 \cdot 3 \cdot 1 = 6$. Multiply the remaining factors in the denominator: $1 \cdot 1 \cdot 7 = 7$.

Since 6 and 7 have no common factors (other than 1), $\dfrac{6}{7}$ is in simplest form.

b. $\dfrac{25}{27} = \dfrac{5 \cdot 5}{3 \cdot 3 \cdot 3}$ Write 25 and 27 in prime-factored form.

Since 25 and 27 have no common factors, other than 1, the fraction $\dfrac{25}{27}$ is in simplest form.

EXAMPLE 9   Simplify: $\dfrac{63}{36}$

Simplify: $\dfrac{162}{72}$

Now Try **Problem 81**

Strategy We will prime factor the numerator and denominator. Then we will look for any factors common to the numerator and denominator and remove them.

WHY We need to make sure that the numerator and denominator have no common factors other than 1. If that is the case, then the fraction is in *simplest form*.

Solution

$\dfrac{63}{36} = \dfrac{3 \cdot 3 \cdot 7}{2 \cdot 2 \cdot 3 \cdot 3}$ To prepare to simplify, write 63 and 36 in prime-factored form.

$\begin{array}{r} 3\,\underline{|\,63} \\ 3\,\underline{|\,21} \\ 7 \end{array}$ $\begin{array}{r} 2\,\underline{|\,36} \\ 2\,\underline{|\,18} \\ 3\,\underline{|\,9} \\ 3 \end{array}$

$= \dfrac{\overset{1}{\cancel{3}} \cdot \overset{1}{\cancel{3}} \cdot 7}{2 \cdot 2 \cdot \underset{1}{\cancel{3}} \cdot \underset{1}{\cancel{3}}}$ Simplify by removing the common factors of 3 from the numerator and denominator.

$= \dfrac{7}{4}$ Multiply the remaining factors in the numerator: $1 \cdot 1 \cdot 7 = 7$. Multiply the remaining factors in the denominator: $2 \cdot 2 \cdot 1 \cdot 1 = 4$.

Success Tip   If you recognized that 63 and 36 have a common factor of 9, you may remove that common factor from the numerator and denominator without writing the prime factorizations. However, make sure that the numerator and denominator of the resulting fraction do not have any common factors. If they do, continue to simplify.

$\dfrac{63}{36} = \dfrac{7 \cdot \overset{1}{\cancel{9}}}{4 \cdot \underset{1}{\cancel{9}}} = \dfrac{7}{4}$ Factor 63 as $7 \cdot 9$ and 36 as $4 \cdot 9$, and then remove the common factor of 9 from the numerator and denominator.

Use the following steps to simplify a fraction.

Simplifying Fractions

To simplify a fraction, *remove factors equal to 1* of the form $\frac{2}{2}, \frac{3}{3}, \frac{4}{4}, \frac{5}{5}$, and so on, using the following procedure:

1. Factor (or prime factor) the numerator and denominator to determine their common factors.

2. Remove factors equal to 1 by replacing each pair of factors common to the numerator and denominator with the equivalent fraction $\frac{1}{1}$.

3. Multiply the remaining factors in the numerator and in the denominator.

Negative fractions are simplified in the same way as positive fractions. Just remember to write a negative sign − in front of each step of the solution. For example, to simplify $-\frac{15}{33}$ we proceed as follows:

$$-\frac{15}{33} = -\frac{\overset{1}{\cancel{3}} \cdot 5}{\underset{1}{\cancel{3}} \cdot 11}$$

$$= -\frac{5}{11}$$

6   Build and simplify algebraic fractions.

Since a variable is a letter that stands for a number, variables can appear in fractions. Fractions that contain a variable (or variables) in the numerator, the denominator, or both are called **algebraic fractions.** Here are some examples of algebraic fractions.

$$\frac{x}{2}, \quad \frac{10}{y}, \quad \frac{m}{25n}, \quad \frac{4a^2b}{6ab^3}, \quad \frac{x+3}{x-5}$$

Algebraic fractions are built up and simplified just like numerical fractions.

Self Check 10

Write $\frac{2}{9}$ as an equivalent fraction

with a denominator of $72y$.

Now Try Problem 91

EXAMPLE 10   Write $\frac{5}{7}$ as an equivalent fraction with a denominator of $42a$.

Strategy We will compare the given denominator to the required denominator and ask, "What expression times 7 equals $42a$?"

WHY The answer to that question helps us determine the form of 1 to use to build an equivalent fraction.

Solution We need to multiply the denominator of $\frac{5}{7}$ by $6a$ to obtain $42a$. It follows that $\frac{6a}{6a}$ should be the form of 1 that is used to build $\frac{5}{7}$.

$$\frac{5}{7} = \frac{5}{7} \cdot \boxed{\frac{6a}{6a}} \qquad \text{Multiply } \tfrac{5}{7} \text{ by a form of 1: } \tfrac{6a}{6a} = 1.$$

$$= \frac{5 \cdot 6a}{7 \cdot 6a} \qquad \begin{array}{l}\text{Multiply the numerators.}\\ \text{Multiply the denominators.}\end{array}$$

$$= \frac{30a}{42a}$$

EXAMPLE 11

Simplify each fraction: **a.** $\dfrac{3}{15x}$ **b.** $\dfrac{9y^3}{10y^2}$ **c.** $\dfrac{24ab^2}{64ab^4}$

Strategy We will factor the numerator and denominator of each algebraic fraction. Then we will look for any factors common to the numerator and denominator and remove them.

WHY We need to make sure that the numerator and denominator have no common factors other than 1. If that is the case, then the fraction is in *simplest form*.

Solution

a. $\dfrac{3}{15x} = \dfrac{3}{3 \cdot 5 \cdot x}$ Recall that 15x means 15 · x. To prepare to simplify, factor 15x as 3 · 5 · x.

$= \dfrac{\overset{1}{\cancel{3}}}{\underset{1}{\cancel{3}} \cdot 5 \cdot x}$ Simplify by removing the common factor of 3 from the numerator and denominator.

$= \dfrac{1}{5x}$ Multiply the remaining factors in the denominator: 1 · 5 · x = 5x.

b. $\dfrac{9y^3}{10y^2} = \dfrac{9 \cdot y \cdot y \cdot y}{10 \cdot y \cdot y}$ To prepare to simplify, use the definition of exponent to write y^3 and y^2 in factored form.

$= \dfrac{9 \cdot \overset{1}{\cancel{y}} \cdot \overset{1}{\cancel{y}} \cdot y}{10 \cdot \underset{1}{\cancel{y}} \cdot \underset{1}{\cancel{y}}}$ Simplify by removing the common factors of y from the numerator and denominator. Slashes and 1's are used o show that $\frac{y}{y}$ is replaced by the equivalent fraction $\frac{1}{1}$. A factor equal to 1 in the form of $\frac{y \cdot y}{y \cdot y} = \frac{y^2}{y^2}$ was removed.

$= \dfrac{9y}{10}$ Multiply the remaining factors in the numerator: 9 · 1 · 1 · y = 9y. Multiply the remaining factors in the denominator: 10 · 1 · 1 = 10.

c. $\dfrac{24ab^2}{64ab^4} = \dfrac{3 \cdot 8 \cdot a \cdot b \cdot b}{8 \cdot 8 \cdot a \cdot b \cdot b \cdot b \cdot b}$ To prepare to simplify, factor 24, b^2, 64, and b^4.

$= \dfrac{3 \cdot \overset{1}{\cancel{8}} \cdot \overset{1}{\cancel{a}} \cdot \overset{1}{\cancel{b}} \cdot \overset{1}{\cancel{b}}}{8 \cdot \underset{1}{\cancel{8}} \cdot \underset{1}{\cancel{a}} \cdot \underset{1}{\cancel{b}} \cdot \underset{1}{\cancel{b}} \cdot b \cdot b}$ Simplify by removing the common factors of 8, a, and b from the numerator and denominator. A factor equal to 1 in the form of $\frac{8 \cdot a \cdot b \cdot b}{8 \cdot a \cdot b \cdot b} = \frac{8ab^2}{8ab^2}$ was removed.

$= \dfrac{3}{8b^2}$ Multiply the remaining factors in the numerator. Multiply the remaining factors in the denominator.

Self Check 11

Simplify each fraction:

a. $\dfrac{2}{16d}$

b. $\dfrac{14x^5}{15x^2}$

c. $\dfrac{35mn^2}{21mn^5}$

Now Try Problems 97, 101, and 105

ANSWERS TO SELF CHECKS

1. a. numerator: 7; denominator: 9 **b.** numerator: 21; denominator: 20   **2.** $\dfrac{11}{31}, \dfrac{20}{31}$

3. a. 1 **b.** 51 **c.** undefined **d.** 0 **4.** $\dfrac{15}{24}$ **5.** $\dfrac{30}{3}$ **6. a.** yes **b.** no **7. a.** $\dfrac{2}{5}$ **b.** $\dfrac{1}{3}$

8. a. $\dfrac{5}{9}$ **b.** in simplest form **9.** $\dfrac{9}{4}$ **10.** $\dfrac{16y}{72y}$ **11. a.** $\dfrac{1}{8d}$ **b.** $\dfrac{14x^3}{15}$ **c.** $\dfrac{5}{3n^3}$

SECTION 4.1 STUDY SET

VOCABULARY

Fill in the blanks.

1. A _____ describes the number of equal parts of a whole.

2. For the fraction $\frac{7}{8}$, the _____ is 7 and the _____ is 8.

3. If the numerator of a fraction is less than its denominator, the fraction is called a _____ fraction. If the numerator of a fraction is greater than or equal to its denominator it is called an _____ fraction.

4. Each of the following fractions is a form of $\boxed{}$.

$$\frac{1}{1} = \frac{2}{2} = \frac{3}{3} = \frac{4}{4} = \frac{5}{5} = \frac{6}{6} = \frac{7}{7} = \frac{8}{8} = \frac{9}{9} = \cdots$$

5. Two fractions are _____ if they represent the same number.

6. Writing a fraction as an equivalent fraction with a larger denominator is called _____ the fraction.

7. A fraction is in _____ form, or lowest terms, when the numerator and denominator have no common factors other than 1.

8. Algebraic fractions, such as $\frac{x}{2}$ and $\frac{4a^2b}{6ab^3}$, are fractions that contain _____ in the numerator, the denominator, or both.

CONCEPTS

9. What concept studied in this section is shown on the right?

10. What concept studied in this section does the following statement illustrate?

$$\frac{1}{2} = \frac{2}{4} = \frac{3}{6} = \frac{4}{8} = \frac{5}{10} = \cdots$$

11. Classify each fraction as a proper fraction or an improper fraction.

   a. $\dfrac{37}{24}$ b. $\dfrac{1}{3}$

   c. $\dfrac{71}{100}$ d. $\dfrac{9}{9}$

12. Remove the common factors of the numerator and denominator to simplify the fraction:

$$\frac{2 \cdot 3 \cdot 3 \cdot 5}{2 \cdot 3 \cdot 5 \cdot 7}$$

13. What common factor (other than 1) do the numerator and the denominator of the fraction $\frac{10}{15}$ have?

Fill in the blank.

14. Multiplication property of 1: The product of any fraction and 1 is that _____.

15. Multiplying fractions: To multiply two fractions, multiply the _____ and multiply the denominators.

16. a. Consider the following solution: $\dfrac{2}{3} = \dfrac{2}{3} \cdot \dfrac{4}{4}$

$$= \frac{8}{12}$$

To build an equivalent fraction for $\frac{2}{3}$ with a denominator of 12, _____ it by a factor equal to 1 in the form of $\boxed{}$.

   b. Consider the following solution: $\dfrac{15}{27} = \dfrac{\overset{1}{\cancel{3}} \cdot 5}{\underset{1}{\cancel{3}} \cdot 9}$

$$= \frac{5}{9}$$

To simplify the fraction $\frac{15}{27}$, _____ a factor equal to 1 of the form $\boxed{}$.

NOTATION

17. Write the fraction $\dfrac{7}{-8}$ in two other ways.

18. Write each integer as a fraction.

   a. 8   b. −25

Complete each solution.

19. Build an equivalent fraction for $\dfrac{1}{6}$ with a denominator of 18.

$$\frac{1}{6} = \frac{1}{6} \cdot \frac{3}{\boxed{}}$$

$$= \frac{\boxed{} \cdot 3}{6 \cdot \boxed{}}$$

$$= \frac{3}{\boxed{}}$$

20. Simplify:

$$\frac{18}{24} = \frac{2 \cdot \boxed{} \cdot 3}{2 \cdot 2 \cdot 2 \cdot \boxed{}}$$

$$= \frac{\overset{1}{\cancel{2}} \cdot 3 \cdot \overset{1}{\cancel{3}}}{\underset{1}{\cancel{2}} \cdot 2 \cdot 2 \cdot \underset{1}{\cancel{3}}}$$

$$= \frac{3}{\boxed{}}$$

▌GUIDED PRACTICE

Identify the numerator and denominator of each fraction.
See Example 1.

21. $\dfrac{4}{5}$ **22.** $\dfrac{7}{8}$

23. $\dfrac{17}{10}$ **24.** $\dfrac{29}{21}$

Write a fraction to describe what part of the figure is shaded.
Write a fraction to describe what part of the figure is not shaded.
See Example 2.

25.    **26.**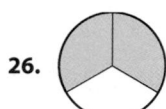

27. 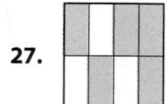   **28.**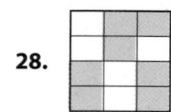

29.    **30.**

31.    **32.**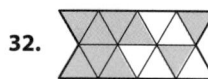

Simplify, if possible. See Example 3.

33. a. $\dfrac{4}{1}$ **b.** $\dfrac{8}{8}$

   c. $\dfrac{0}{12}$ **d.** $\dfrac{1}{0}$

34. a. $\dfrac{25}{1}$ **b.** $\dfrac{14}{14}$

   c. $\dfrac{0}{1}$ **d.** $\dfrac{83}{0}$

35. a. $\dfrac{5}{0}$ **b.** $\dfrac{0}{50}$

   c. $\dfrac{33}{33}$ **d.** $\dfrac{75}{1}$

36. a. $\dfrac{0}{64}$ **b.** $\dfrac{27}{0}$

   c. $\dfrac{125}{125}$ **d.** $\dfrac{98}{1}$

Write each fraction as an equivalent fraction with the indicated denominator. See Example 4.

37. $\dfrac{7}{8}$, denominator 40   **38.** $\dfrac{3}{4}$, denominator 24

39. $\dfrac{4}{9}$, denominator 27   **40.** $\dfrac{5}{7}$, denominator 49

41. $\dfrac{5}{6}$, denominator 54   **42.** $\dfrac{2}{3}$, denominator 27

43. $\dfrac{2}{7}$, denominator 14   **44.** $\dfrac{3}{10}$, denominator 50

Write each whole number as an equivalent fraction with the indicated denominator. See Example 5.

45. 4, denominator 9   **46.** 4, denominator 3

47. 6, denominator 8   **48.** 3, denominator 6

49. 3, denominator 5   **50.** 7, denominator 4

51. 14, denominator 2   **52.** 10, denominator 9

Are the following fractions in simplest form? See Example 6.

53. a. $\dfrac{12}{16}$ **b.** $\dfrac{3}{25}$

54. a. $\dfrac{9}{24}$ **b.** $\dfrac{7}{36}$

55. a. $\dfrac{35}{36}$ **b.** $\dfrac{18}{21}$

56. a. $\dfrac{22}{45}$ **b.** $\dfrac{21}{56}$

Simplify each fraction, if possible. See Example 7.

57. $\dfrac{6}{9}$ **58.** $\dfrac{15}{20}$

59. $\dfrac{16}{20}$ **60.** $\dfrac{25}{35}$

61. $\dfrac{5}{15}$ **62.** $\dfrac{6}{30}$

63. $\dfrac{2}{48}$ **64.** $\dfrac{2}{42}$

Simplify each fraction, if possible. See Example 8.

65. $\dfrac{16}{17}$ 66. $\dfrac{14}{25}$

67. $\dfrac{36}{96}$ 68. $\dfrac{48}{120}$

69. $\dfrac{55}{62}$ 70. $\dfrac{41}{51}$

71. $\dfrac{50}{55}$ 72. $\dfrac{22}{88}$

73. $\dfrac{60}{108}$ 74. $\dfrac{75}{275}$

75. $\dfrac{180}{210}$ 76. $\dfrac{90}{120}$

Simplify each fraction, if possible. See Example 9.

77. $\dfrac{306}{234}$ 78. $\dfrac{208}{117}$

79. $\dfrac{105}{42}$ 80. $\dfrac{120}{80}$

81. $\dfrac{420}{144}$ 82. $\dfrac{216}{189}$

83. $-\dfrac{4}{68}$ 84. $-\dfrac{3}{42}$

85. $-\dfrac{90}{105}$ 86. $-\dfrac{98}{126}$

87. $-\dfrac{16}{26}$ 88. $-\dfrac{81}{132}$

Write each fraction as an equivalent fraction with the indicated denominator. See Example 10.

89. $\dfrac{1}{2}$, denominator $6a$ 90. $\dfrac{1}{3}$, denominator $12b$

91. $\dfrac{9}{10}$, denominator $50c$ 92. $\dfrac{11}{16}$, denominator $32m$

93. $\dfrac{5}{4n}$, denominator $44n$ 94. $\dfrac{9}{7n}$, denominator $63n$

95. $\dfrac{14}{15x}$, denominator $45x$ 96. $\dfrac{12}{13r}$, denominator $39r$

Simplify each fraction, if possible. See Example 11.

97. $\dfrac{7}{14a}$ 98. $\dfrac{5}{25y}$

99. $\dfrac{3x}{12}$ 100. $\dfrac{7x}{35}$

101. $\dfrac{4m^5}{25m^4}$ 102. $\dfrac{2n^4}{13n^2}$

103. $\dfrac{6b^4}{9b}$ 104. $\dfrac{4c^4}{10c^3}$

105. $\dfrac{35a^3b^2}{25a^2b^3}$ 106. $\dfrac{56n^2p^4}{28np^5}$

107. $-\dfrac{16n^5p}{24np}$ 108. $-\dfrac{36cd^6}{54cd^4}$

TRY IT YOURSELF

Tell whether each pair of fractions are equivalent by simplifying each fraction.

109. $\dfrac{2}{14}$ and $\dfrac{6}{36}$ 110. $\dfrac{3}{12}$ and $\dfrac{4}{24}$

111. $\dfrac{22}{34}$ and $\dfrac{33}{51}$ 112. $\dfrac{4}{30}$ and $\dfrac{12}{90}$

APPLICATIONS

113. DENTISTRY Refer to the dental chart.

   a. How many teeth are shown on the chart?

   b. What fraction of this set of teeth have fillings?

114. TIME CLOCKS For each clock, what fraction of the hour has passed? Write your answers in simplified form. (*Hint:* There are 60 minutes in an hour.)

a.    b.

c.    d.

115. RULERS The illustration below shows a ruler.

   a. How many spaces are there between the numbers 0 and 1?

   b. To what fraction is the arrow pointing? Write your answer in simplified form.

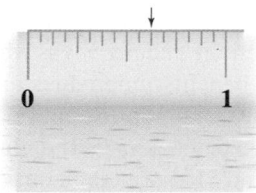

116. SINKHOLES The illustration below shows a side view of a drop in the sidewalk near a sinkhole. Describe the movement of the sidewalk using a signed fraction.

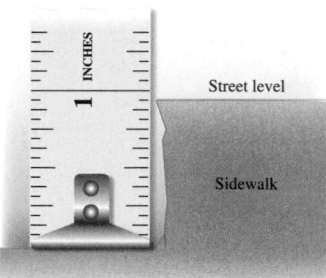

117. POLITICAL PARTIES The graph shows the number of Democrat and Republican governors of the 50 states, as of February 1, 2009.

   a. How many Democrat governors are there? How many Republican governors are there?

   b. What fraction of the governors are Democrats? Write your answer in simplified form.

   c. What fraction of the governors are Republicans? Write your answer in simplified form.

Source: thegreenpapers.com

118. GAS TANKS Write fractions to describe the amount of gas left in the tank and the amount of gas that has been used.

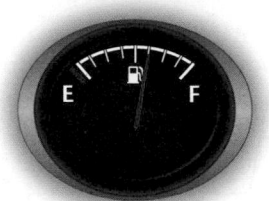

Use unleaded fuel

119. SELLING CONDOS The model below shows a new condominium development. The condos that have been sold are shaded.

   a. How many units are there in the development?

   b. What fraction of the units in the development have been sold? What fraction have not been sold? Write your answers in simplified form.

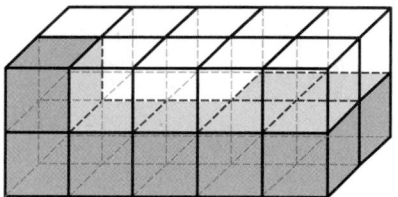

120. MUSIC The illustration shows a side view of the finger position needed to produce a length of string (from the bridge to the fingertip) that gives low C on a violin. To play other notes, fractions of that length are used. Locate these finger positions on the illustration.

   a. $\frac{1}{2}$ of the length gives middle C.

   b. $\frac{3}{4}$ of the length gives F above low C.

   c. $\frac{2}{3}$ of the length gives G.

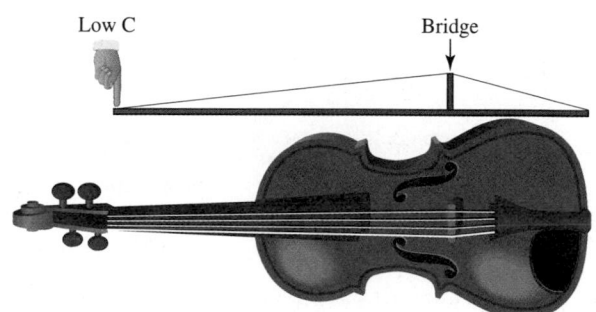

WRITING

121. Explain the concept of equivalent fractions. Give an example.

122. What does it mean for a fraction to be in simplest form? Give an example.

123. Why can't we say that $\frac{2}{5}$ of the figure below is shaded?

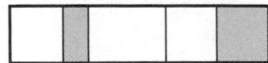

124. Perhaps you have heard the following joke:

> *A pizza parlor waitress asks a customer if he wants the pizza cut into four pieces or six pieces or eight pieces. The customer then declares that he wants either four or six pieces of pizza "because I can't eat eight."*

Explain what is wrong with the customer's thinking.

125. a. What type of problem is shown below? Explain the solution.

$$\frac{1}{2} = \frac{1}{2} \cdot \frac{4}{4} = \frac{4}{8}$$

b. What type of problem is shown below? Explain the solution.

$$\frac{15}{35} = \frac{3 \cdot \overset{1}{\cancel{5}}}{\underset{1}{\cancel{5}} \cdot 7} = \frac{3}{7}$$

126. Explain the difference in the two approaches used to simplify $\frac{20}{28}$. Are the results the same?

$$\frac{\overset{1}{\cancel{4}} \cdot 5}{\underset{1}{\cancel{4}} \cdot 7} \quad \text{and} \quad \frac{\overset{1}{\cancel{2}} \cdot \overset{1}{\cancel{2}} \cdot 5}{\underset{1}{\cancel{2}} \cdot \underset{1}{\cancel{2}} \cdot 7}$$

REVIEW

127. PAYCHECKS *Gross pay* is what a worker makes before deductions and *net pay* is what is left after taxes, health benefits, union dues, and other deductions are taken out. Suppose a worker's monthly gross pay is $3,575. If deductions of $235, $782, $148, and $103 are taken out of his check, what is his monthly net pay?

128. HORSE RACING One day, a man bet on all eight horse races at Santa Anita Racetrack. He won $168 on the first race and he won $105 on the fourth race. He lost his $50-bets on each of the other races. Overall, did he win or lose money betting on the horses? How much?

Objectives

1 Multiply fractions.

2 Simplify answers when multiplying fractions.

3 Multiply algebraic fractions.

4 Evaluate exponential expressions that have fractional bases.

5 Solve application problems by multiplying fractions.

6 Find the area of a triangle.

SECTION 4.2
Multiplying Fractions

In the next three sections, we discuss how to add, subtract, multiply, and divide fractions. We begin with the operation of multiplication.

1 Multiply fractions.

To develop a rule for multiplying fractions, let's consider a real-life application.

Suppose $\frac{3}{5}$ of the last page of a school newspaper is devoted to campus sports coverage. To show this, we can divide the page into fifths, and shade 3 of them red.

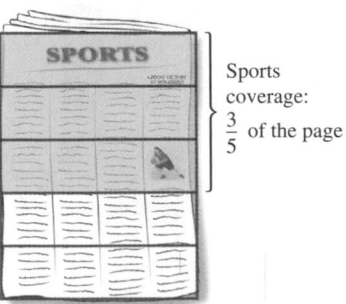

Sports coverage: $\frac{3}{5}$ of the page

Furthermore, suppose that $\frac{1}{2}$ of the sports coverage is about women's teams. We can show that portion of the page by dividing the already colored region into two halves, and shading one of them in purple.

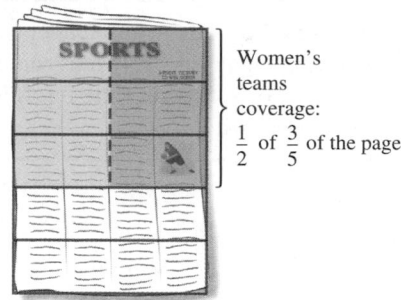

Women's teams coverage: $\frac{1}{2}$ of $\frac{3}{5}$ of the page

To find the fraction represented by the purple shaded region, the page needs to be divided into equal-size parts. If we extend the dashed line downward, we see there are 10 equal-sized parts. The purple shaded parts are 3 out of 10, or $\frac{3}{10}$, of the page. Thus, $\frac{3}{10}$ of the last page of the school newspaper is devoted to women's sports.

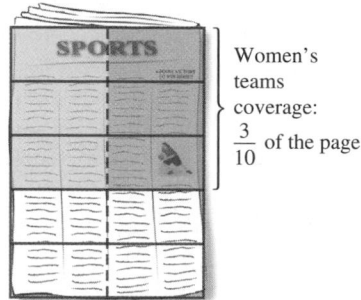

Women's teams coverage: $\frac{3}{10}$ of the page

In this example, we have found that

$$\frac{1}{2} \text{ of } \frac{3}{5} \text{ is } \frac{3}{10}$$
$$\frac{1}{2} \cdot \frac{3}{5} = \frac{3}{10}$$

Since the key word *of* indicates multiplication, and the key word *is* means equals, we can translate this statement to symbols.

Two observations can be made from this result.

- The numerator of the answer is the product of the numerators of the original fractions.

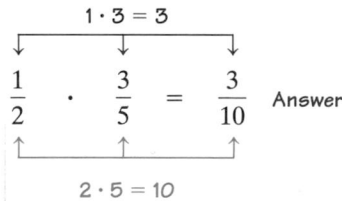

$$\frac{1}{2} \cdot \frac{3}{5} = \frac{3}{10} \quad \text{Answer}$$

$$2 \cdot 5 = 10$$

- The denominator of the answer is the product of the denominators of the original fractions.

These observations illustrate the following rule for multiplying two fractions.

Multiplying Fractions

To multiply two fractions, multiply the numerators and multiply the denominators. Simplify the result, if possible.

> **Success Tip** In the newspaper example, we found a *part of a part* of a page. Multiplying proper fractions can be thought of in this way. When taking a *part of a part* of something, the result is always smaller than the original part that you began with.

Self Check 1

Multiply:

a. $\dfrac{1}{2} \cdot \dfrac{1}{8}$

b. $\dfrac{5}{9} \cdot \dfrac{2}{3}$

Now Try Problems 17 and 21

EXAMPLE 1 Multiply: **a.** $\dfrac{1}{6} \cdot \dfrac{1}{4}$ **b.** $\dfrac{7}{8} \cdot \dfrac{3}{5}$

Strategy We will multiply the numerators and denominators, and make sure that the result is in simplest form.

WHY This is the rule for multiplying two fractions.

a. $\dfrac{1}{6} \cdot \dfrac{1}{4} = \dfrac{1 \cdot 1}{6 \cdot 4}$ Multiply the numerators.
                 Multiply the denominators.

        $= \dfrac{1}{24}$ Since 1 and 24 have no common factors other than 1, the result is in simplest form.

Solution

b. $\dfrac{7}{8} \cdot \dfrac{3}{5} = \dfrac{7 \cdot 3}{8 \cdot 5}$ Multiply the numerators.
                 Multiply the denominators.

        $= \dfrac{21}{40}$ Since 21 and 40 have no common factors other than 1, the result is in simplest form.

The sign rules for multiplying integers also hold for multiplying fractions. When we multiply two fractions with *like* signs, the product is positive. When we multiply two fractions with *unlike* signs, the product is negative.

Self Check 2

Multiply: $\dfrac{5}{6}\left(-\dfrac{1}{3}\right)$

Now Try Problem 25

EXAMPLE 2 Multiply: $-\dfrac{3}{4}\left(\dfrac{1}{8}\right)$

Strategy We will use the rule for multiplying two fractions that have different (unlike) signs.

WHY One fraction is positive and one is negative.

Solution

$-\dfrac{3}{4}\left(\dfrac{1}{8}\right) = -\dfrac{3 \cdot 1}{4 \cdot 8}$ Multiply the numerators.
                     Multiply the denominators.
                     Since the fractions have unlike signs, make the answer negative.

        $= -\dfrac{3}{32}$ Since 3 and 32 have no common factors other than 1, the result is in simplest form.

Self Check 3

Multiply: $\dfrac{1}{3} \cdot 7$

Now Try Problem 29

EXAMPLE 3 Multiply: $\dfrac{1}{2} \cdot 3$

Strategy We will begin by writing the integer 3 as a fraction.

WHY Then we can use the rule for multiplying two fractions to find the product. ▼

Solution

$$\frac{1}{2} \cdot 3 = \frac{1}{2} \cdot \frac{3}{1} \qquad \text{Write 3 as a fraction: } 3 = \frac{3}{1}.$$

$$= \frac{1 \cdot 3}{2 \cdot 1} \qquad \begin{array}{l}\text{Multiply the numerators.}\\\text{Multiply the denominators.}\end{array}$$

$$= \frac{3}{2} \qquad \begin{array}{l}\text{Since 3 and 2 have no common factors other}\\\text{than 1, the result is in simplest form.}\end{array}$$

2 Simplify answers when multiplying fractions.

After multiplying two fractions, we need to simplify the result, if possible. To do that, we can use the procedure discussed in Section 3.1 by removing pairs of common factors of the numerator and denominator.

EXAMPLE 4   Multiply and simplify: $\dfrac{5}{8} \cdot \dfrac{4}{5}$

Strategy We will multiply the numerators and denominators, and make sure that the result is in simplest form.

WHY This is the rule for multiplying two fractions.

Solution

$$\frac{5}{8} \cdot \frac{4}{5} = \frac{5 \cdot 4}{8 \cdot 5} \qquad \begin{array}{l}\text{Multiply the numerators.}\\\text{Multiply the denominators.}\end{array}$$

$$= \frac{5 \cdot 2 \cdot 2}{2 \cdot 2 \cdot 2 \cdot 5} \qquad \begin{array}{l}\text{To prepare to simplify, write 4 and 8 in}\\\text{prime-factored form.}\end{array}$$

$$= \frac{\overset{1}{\cancel{5}} \cdot \overset{1}{\cancel{2}} \cdot \overset{1}{\cancel{2}}}{2 \cdot 2 \cdot \underset{1}{\cancel{2}} \cdot \underset{1}{\cancel{5}}} \qquad \begin{array}{l}\text{To simplify, remove the common factors of 2}\\\text{and 5 from the numerator and denominator.}\end{array}$$

$$= \frac{1}{2} \qquad \begin{array}{l}\text{Multiply the remaining factors in the numerator: } 1 \cdot 1 \cdot 1 = 1.\\\text{Multiply the remaining factors in the denominator: } 1 \cdot 1 \cdot 2 \cdot 1 = 2.\end{array}$$

Success Tip If you recognized that 4 and 8 have a common factor of 4, you may remove that common factor from the numerator and denominator of the product without writing the prime factorizations. However, make sure that the numerator and denominator of the resulting fraction do not have any common factors. If they do, continue to simplify.

$$\frac{5}{8} \cdot \frac{4}{5} = \frac{5 \cdot 4}{8 \cdot 5} = \frac{\overset{1}{\cancel{5}} \cdot \overset{1}{\cancel{4}}}{2 \cdot \underset{1}{\cancel{4}} \cdot \underset{1}{\cancel{5}}} = \frac{1}{2} \qquad \begin{array}{l}\text{Factor 8 as } 2 \cdot 4, \text{ and then remove the}\\\text{common factors of 4 and 5 in the numerator}\\\text{and denominator.}\end{array}$$

The rule for multiplying two fractions can be extended to find the product of three or more fractions.

EXAMPLE 5   Multiply and simplify: $\dfrac{2}{3}\left(-\dfrac{9}{14}\right)\left(-\dfrac{7}{10}\right)$

Strategy We will multiply the numerators and denominators, and make sure that the result is in simplest form.

WHY This is the rule for multiplying three (or more) fractions.

Self Check 4

Multiply and simplify: $\dfrac{11}{25} \cdot \dfrac{10}{11}$

Now Try Problem 33

Self Check 5

Multiply and simplify:

$$\frac{2}{5}\left(-\frac{15}{22}\right)\left(-\frac{11}{26}\right)$$

Now Try Problem 37

Solution   Recall from Section 2.4 that a product is positive when there are an even number of negative factors. Since $\frac{2}{3}\left(-\frac{9}{14}\right)\left(-\frac{7}{10}\right)$ has *two* negative factors, the product is positive.

$$\frac{2}{3}\left(-\frac{9}{14}\right)\left(-\frac{7}{10}\right) = \frac{2}{3}\left(\frac{9}{14}\right)\left(\frac{7}{10}\right)$$

Since the answer is positive, drop both $-$ signs and continue.

$$= \frac{2 \cdot 9 \cdot 7}{3 \cdot 14 \cdot 10}$$

Multiply the numerators.
Multiply the denominators.

$$= \frac{2 \cdot 3 \cdot 3 \cdot 7}{3 \cdot 2 \cdot 7 \cdot 2 \cdot 5}$$

To prepare to simplify, write 9, 14, and 10 in prime-factored form.

$$= \frac{\overset{1}{2} \cdot \overset{1}{3} \cdot 3 \cdot \overset{1}{7}}{\underset{1}{3} \cdot 2 \cdot \underset{1}{7} \cdot \underset{1}{2} \cdot 5}$$

To simplify, remove the common factors of 2, 3, and 7 from the numerator and denominator.

$$= \frac{3}{10}$$

Multiply the remaining factors in the numerator.
Multiply the remaining factors in the denominator. ∎

Caution!   In Example 5, it was very helpful to prime factor and simplify when we did (the third step of the solution). If, instead, you find the product of the numerators and the product of the denominators, the resulting fraction is difficult to simplify because the numerator, 126, and the denominator, 420, are large.

$$\frac{2}{3} \cdot \frac{9}{14} \cdot \frac{7}{10} \quad = \quad \frac{2 \cdot 9 \cdot 7}{3 \cdot 14 \cdot 10} \quad = \quad \cancel{\frac{126}{420}}$$

$\qquad\qquad\qquad\qquad\qquad\uparrow\qquad\qquad\qquad\qquad\quad\uparrow$

Factor and simplify at this stage, before multiplying in the numerator and denominator.

Don't multiply in the numerator and denominator and then try to simplify the result. You will get the same answer, but it takes much more work.

---

3 **Multiply algebraic fractions.**

To multiply two algebraic fractions, we use the same approach as with numerical fractions: *Multiply the numerators and multiply the denominators. Then simplify the result, if possible.*

Self Check 6

Multiply and simplify:

a. $\dfrac{5}{12y} \cdot \dfrac{3y}{8}$

b. $-\dfrac{2a^3}{15b} \cdot \dfrac{35ab}{8a^5}$

Now Try Problems 41 and 45

EXAMPLE 6   Multiply and simplify: **a.** $\dfrac{5b}{2} \cdot \dfrac{4}{7b}$ **b.** $-\dfrac{t^2}{21r} \cdot \dfrac{14r}{t^4}$

Strategy   To find each product, we will use the rule for multiplying fractions. In the process, we must be prepared to factor the numerators and denominators so that any common factors can be removed.

WHY   We want to give each result in simplified form.

Solution

a. $\dfrac{5b}{2} \cdot \dfrac{4}{7b} = \dfrac{5b \cdot 4}{2 \cdot 7b}$ Multiply the numerators.
Multiply the denominators.

It's obvious that the numerator and denominator of $\frac{5b \cdot 4}{2 \cdot 7b}$ have two common factors, 2 and b. These common factors become more apparent when we factor the numerator and denominator completely.

$$= \frac{5 \cdot b \cdot 2 \cdot 2}{2 \cdot 7 \cdot b}$$

To prepare to simplify, factor 4 as $2 \cdot 2$. Write $5b$ as $5 \cdot b$ and $7b$ as $7 \cdot b$.

$$= \frac{5 \cdot \overset{1}{b} \cdot \overset{1}{2} \cdot 2}{\underset{1}{2} \cdot 7 \cdot \underset{1}{b}}$$

To simplify, remove the common factors of 2 and b from the numerator and denominator

$$= \frac{10}{7}$$

Multiply the remaining factors in the numerator: $5 \cdot 1 \cdot 1 \cdot 2 = 10$.
Multiply the remaining factors in the denominator: $1 \cdot 7 \cdot 1 = 7$.

b. $-\dfrac{t^2}{21r} \cdot \dfrac{14r}{t^4} = \dfrac{t^2 \cdot 14r}{\overset{\uparrow}{21r \cdot t^4}}$ Multiply the numerators. Multiply the denominators.

$= -\dfrac{t \cdot t \cdot 7 \cdot 2 \cdot r}{3 \cdot 7 \cdot r \cdot t \cdot t \cdot t \cdot t}$ The product of two fractions with unlike signs is negative.
To prepare to simplify, factor t^2, 14, 21, and t^4.

$= -\dfrac{\overset{1}{\cancel{t}} \cdot \overset{1}{\cancel{t}} \cdot \overset{1}{\cancel{7}} \cdot 2 \cdot \overset{1}{\cancel{r}}}{3 \cdot \underset{1}{\cancel{7}} \cdot \underset{1}{\cancel{r}} \cdot \underset{1}{\cancel{t}} \cdot \underset{1}{\cancel{t}} \cdot t \cdot t}$ To simplify, remove the common factors of 7, r, and t from the numerator and denominator.

$= -\dfrac{2}{3t^2}$ Multiply the remaining factors in the numerator: $1 \cdot 1 \cdot 1 \cdot 2 \cdot 1 = 2$. Multiply the remaining factors in the denominator: $3 \cdot 1 \cdot 1 \cdot 1 \cdot 1 \cdot t \cdot t = 3t^2$. ∎

EXAMPLE 7   Multiply and simplify: **a.** $\dfrac{1}{4}(4y)$ **b.** $9a \cdot \dfrac{1}{3a}$

Self Check 7

Multiply and simplify:

a. $\dfrac{1}{5} \cdot 5m$

b. $12m\left(\dfrac{1}{6m}\right)$

Now Try Problems 49 and 53

Strategy We will write the expressions $4y$ and $9a$ as fractions.

WHY Then we can use the rule for multiplying two fractions to find each product.

Solution

a. $\dfrac{1}{4}(4y) = \dfrac{1}{4} \cdot \left(\dfrac{4y}{1}\right)$ Write $4y$ as a fraction: $4y = \dfrac{4y}{1}$.

$= \dfrac{1 \cdot 4y}{4 \cdot 1}$ Multiply the numerators.
Multiply the denominators.

$= \dfrac{1 \cdot \overset{1}{\cancel{4}} \cdot y}{\underset{1}{\cancel{4}} \cdot 1}$ To simplify, write $4y$ as $4 \cdot y$ and remove the common factor of 4 from the numerator and denominator.

$= \dfrac{y}{1}$ Multiply the remaining factors in the numerator: $1 \cdot 1 \cdot y = y$.
Multiply the remaining factors in the denominator: $1 \cdot 1 = 1$.

$= y$ Any number divided by 1 is equal to that number.

b. $9a \cdot \dfrac{1}{3a} = \dfrac{9a}{1} \cdot \dfrac{1}{3a}$ Write $9a$ as a fraction: $9a = \dfrac{9a}{1}$.

$= \dfrac{9a \cdot 1}{1 \cdot 3a}$ Multiply the numerators.
Multiply the denominators.

$= \dfrac{\overset{1}{\cancel{3}} \cdot 3 \cdot \overset{1}{\cancel{a}} \cdot 1}{1 \cdot \underset{1}{\cancel{3}} \cdot \underset{1}{\cancel{a}}}$ To simplify, factor 9 as $3 \cdot 3$. Then remove the common factors of 3 and a from the numerator and denominator.

$= \dfrac{3}{1}$ Multiply the remaining factors in in the numerator: $1 \cdot 3 \cdot 1 \cdot 1 = 3$.
Multiply the remaining factors in the denominator: $1 \cdot 1 \cdot 1 = 1$.

$= 3$ Any number divided by 1 is equal to that number. ∎

To multiply $\frac{1}{2}$ and x, we can express the product as $\frac{1}{2}x$, or we can use the concept of multiplying fractions to write it in a different form.

$\dfrac{1}{2} \cdot x = \dfrac{1}{2} \cdot \dfrac{x}{1}$ Write x as a fraction: $x = \dfrac{x}{1}$.

$= \dfrac{1 \cdot x}{2 \cdot 1}$ Multiply the numerators.
Multiply the denominators.

$= \dfrac{x}{2}$

The product of $\frac{1}{2}$ and x can be expressed as $\frac{1}{2}x$ or $\frac{x}{2}$. Similarly,

$$\dfrac{3}{4}t = \dfrac{3t}{4} \qquad \text{and} \qquad -\dfrac{5}{16}y = -\dfrac{5y}{16}$$

4 Evaluate exponential expressions that have fractional bases.

We have evaluated exponential expressions that have whole-number bases and integer bases. If the base of an exponential expression is a fraction, the exponent tells us how many times to write that fraction as a factor. For example,

$$\left(\frac{2}{3}\right)^2 = \frac{2}{3} \cdot \frac{2}{3} = \frac{2 \cdot 2}{3 \cdot 3} = \frac{4}{9}$$ Since the exponent is 2, write the base, $\frac{2}{3}$, as a factor 2 times.

Self Check 8

Evaluate each expression.

a. $\left(\frac{2}{5}\right)^3$

b. $\left(-\frac{3}{4}\right)^2$

c. $-\left(\frac{3}{4}\right)^2$

Now Try Problems 57 and 59

EXAMPLE 8

Evaluate each expression: **a.** $\left(\frac{1}{4}\right)^3$ **b.** $\left(-\frac{2}{3}\right)^2$ **c.** $-\left(\frac{2}{3}\right)^2$

Strategy We will write each exponential expression as a product of repeated factors, and then perform the multiplication. This requires that we identify the base and the exponent.

WHY The exponent tells the number of times the base is to be written as a factor.

Solution

Recall that exponents are used to represent repeated multiplication.

a. We read $\left(\frac{1}{4}\right)^3$ as "one-fourth raised to the third power," or as "one-fourth, cubed."

$$\left(\frac{1}{4}\right)^3 = \frac{1}{4} \cdot \frac{1}{4} \cdot \frac{1}{4}$$ Since the exponent is 3, write the base, $\frac{1}{4}$, as a factor 3 times.

$$= \frac{1 \cdot 1 \cdot 1}{4 \cdot 4 \cdot 4}$$ Multiply the numerators. Multiply the denominators.

$$= \frac{1}{64}$$

b. We read $\left(-\frac{2}{3}\right)^2$ as "negative two-thirds raised to the second power," or as "negative two-thirds, squared."

$$\left(-\frac{2}{3}\right)^2 = \left(-\frac{2}{3}\right)\left(-\frac{2}{3}\right)$$ Since the exponent is 2, write the base, $-\frac{2}{3}$, as a factor 2 times.

$$= \frac{2 \cdot 2}{3 \cdot 3}$$ The product of two fractions with like signs is positive: Drop the − signs. Multiply the numerators. Multiply the denominators.

$$= \frac{4}{9}$$

c. We read $-\left(\frac{2}{3}\right)^2$ as "the opposite of two-thirds squared." Recall that if the − symbol is not within the parantheses, it is not part of the base.

$$-\left(\frac{2}{3}\right)^2 = -\frac{2}{3} \cdot \frac{2}{3}$$ Since the exponent is 2, write the base, $\frac{2}{3}$, as a factor 2 times.

$$= -\frac{2 \cdot 2}{3 \cdot 3}$$ Multiply the numerators. Multiply the denominators.

$$= -\frac{4}{9}$$

We can use the rule for multiplying fractions to find powers of algebraic fractions.

Self Check 9

Find the power: $\left(-\frac{3t}{4}\right)^3$

Now Try Problem 61

EXAMPLE 9

Find the power: $\left(-\frac{4x}{5}\right)^2$

Strategy We will write the exponential expression as a product of repeated factors, and then perform the multiplication. This requires that we identify the base and the exponent.

WHY The exponent tells the number of times the base is to be written as a factor. ▼

Solution

$$\left(-\frac{4x}{5}\right)^2 = \left(-\frac{4x}{5}\right)\left(-\frac{4x}{5}\right)$$

Since the exponent is 2, write the base, $-\frac{4x}{5}$, as a factor 2 times.

$$= \frac{4x \cdot 4x}{5 \cdot 5}$$

Since the product of two fractions with like signs is positive, we can drop the − symbols. Multiply the numerators. Multiply the denominators.

$$= \frac{16x^2}{25}$$

Since 16 and 25 have no common factors other than 1, the result is in simplest form.

5 Solve application problems by multiplying fractions.

The key word *of* often appears in application problems involving fractions. When a fraction is followed by the word *of*, such as $\frac{1}{2}$ *of* or $\frac{3}{4}$ *of*, it indicates that we are to find a part of some quantity using multiplication.

EXAMPLE 10 *How a Bill Becomes Law* If the President vetoes (refuses to sign) a bill, it takes $\frac{2}{3}$ of those voting in the House of Representatives (and the Senate) to override the veto for it to become law. If all 435 members of the House cast a vote, how many of their votes does it take to override a presidential veto?

Analyze

- It takes $\frac{2}{3}$ *of* those voting to override a veto. Given
- All 435 members of the House cast a vote. Given
- How many votes does it take to override a presidential veto? Find

Form The key phrase $\frac{2}{3}$ *of* suggests that we are to find a part of the 435 possible votes using multiplication.

We translate the words of the problem to numbers and symbols.

The number of votes needed in the House to override a veto	is equal to	$\frac{2}{3}$	of	the number of House members that vote.

| The number of votes needed in the House to override a veto | = | $\frac{2}{3}$ | · | 435 |

Solve To find the product, we will express 435 as a fraction and then use the rule for multiplying two fractions.

$$\frac{2}{3} \cdot 435 = \frac{2}{3} \cdot \frac{435}{1}$$ Write 435 as a fraction: $435 = \frac{435}{1}$.

$$= \frac{2 \cdot 435}{3 \cdot 1}$$ Multiply the numerators. Multiply the denominators.

$$= \frac{2 \cdot 3 \cdot 5 \cdot 29}{3 \cdot 1}$$ To prepare to simplify, write 435 in prime-factored form: $3 \cdot 5 \cdot 29$.

$$= \frac{2 \cdot \overset{1}{\cancel{3}} \cdot 5 \cdot 29}{\underset{1}{\cancel{3}} \cdot 1}$$ Remove the common factor of 3 from the numerator and denominator.

$$= \frac{290}{1}$$ Multiply the remaining factors in the numerator: $2 \cdot 1 \cdot 5 \cdot 29 = 290$. Multiply the remaining factors in the denominator: $1 \cdot 1 = 1$.

$$= 290$$ Any number divided by 1 is equal to that number.

Self Check 10

HOW A BILL BECOMES LAW If only 96 Senators are present and cast a vote, how many of their votes does it take to override a Presidential veto?

Now Try **Problems 65 and 103**

State It would take 290 votes in the House to override a veto.

Check We can estimate to check the result. We will use 440 to approximate the number of House members voting. Since $\frac{1}{2}$ of 440 is 220, and since $\frac{2}{3}$ is a greater part than $\frac{1}{2}$, we would expect the number of votes needed to be *more than* 220. The result of 290 seems reasonable. ∎

6 Find the area of a triangle.

As the figures below show, a triangle has three sides. The length of the base of the triangle can be represented by the letter b and the height by the letter h. The height of a triangle is always perpendicular (makes a square corner) to the base. This is shown by using the symbol ⌐.

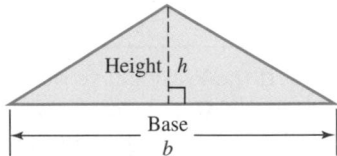 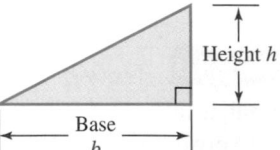

Recall that the area of a figure is the amount of surface that it encloses. The area of a triangle can be found by using the following formula.

Area of a Triangle

The area A of a triangle is one-half the product of its base b and its height h.

$$\text{Area} = \frac{1}{2}(\text{base})(\text{height}) \qquad \text{or} \qquad A = \frac{1}{2} \cdot b \cdot h$$

The Language of Algebra The formula $A = \frac{1}{2} \cdot b \cdot h$ can be written more simply as $A = \frac{1}{2}bh$. The formula for the area of a triangle can also be written as $A = \dfrac{bh}{2}$.

Self Check 11

Find the area of the triangle shown below.

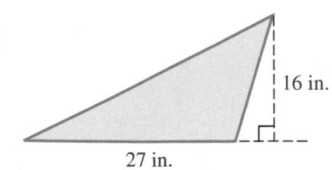

16 in.

27 in.

Now Try Problems 69 and 115

EXAMPLE 11 *Geography* Approximate the area of the state of Virginia (in square miles) using the triangle shown below.

Strategy We will find the product of $\frac{1}{2}$, 405, and 200.

WHY The formula for the area of a triangle is $A = \frac{1}{2}$ (base)(height).

Virginia

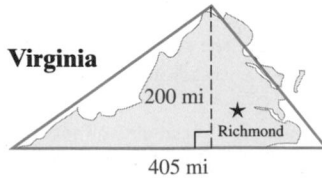

200 mi

★ Richmond

405 mi

Solution

$$A = \frac{1}{2}bh$$ *This is the formula for the area of a triangle.*

$$= \frac{1}{2} \cdot 405 \cdot 200$$ $\frac{1}{2}bh$ *means* $\frac{1}{2} \cdot b \cdot h$. *Substitute 405 for b and 200 for h.*

$$= \frac{1}{2} \cdot \frac{405}{1} \cdot \frac{200}{1}$$ *Write 405 and 200 as fractions.*

$$= \frac{1 \cdot 405 \cdot 200}{2 \cdot 1 \cdot 1}$$ *Multiply the numerators.*
                        Multiply the denominators.

$$= \frac{1 \cdot 405 \cdot \overset{1}{\cancel{2}} \cdot 100}{\underset{1}{\cancel{2}} \cdot 1 \cdot 1}$$ *Factor 200 as 2 · 100. Then remove the common factor of 2 from the numerator and denominator.*

$$= 40{,}500$$ *In the numerator, multiply:* 405 · 100 = 40,500.

The area of the state of Virginia is approximately 40,500 square miles. This can be written as 40,500 mi^2.

> **Caution!** Remember that area is measured in square units, such as $in.^2$, ft^2, and cm^2. Don't forget to write the units in your answer when finding the area of a figure.

ANSWERS TO SELF CHECKS

1. a. $\frac{1}{16}$ **b.** $\frac{10}{27}$ **2.** $-\frac{5}{18}$ **3.** $\frac{7}{3}$ **4.** $\frac{2}{5}$ **5.** $\frac{3}{26}$ **6. a.** $\frac{5}{32}$ **b.** $\frac{7}{12a}$ **7. a.** m **b.** 2

8. a. $\frac{8}{125}$ **b.** $\frac{9}{16}$ **c.** $-\frac{9}{16}$ **9.** $-\frac{27t^3}{64}$ **10.** 64 votes   **11.** 216 $in.^2$

SECTION **4.2**   STUDY SET

▌VOCABULARY

Fill in the blanks.

1. When a fraction is followed by the word *of*, such as $\frac{1}{3}$ *of*, it indicates that we are to find a part of some quantity using _____.

2. The answer to a multiplication is called the _____.

3. To _____ a fraction, we remove common factors of the numerator and denominator.

4. In the expression $\left(\frac{1}{4}\right)^3$, the _____ is $\frac{1}{4}$ and the _____ is 3.

5. The _____ of a triangle is the amount of surface that it encloses.

6. Label the *base* and the *height* of the triangle shown below.

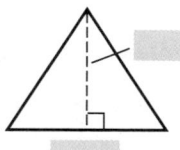

▌CONCEPTS

7. Fill in the blanks: To multiply two fractions, multiply the _____ and multiply the _____. Then _____, if possible.

8. Use the following rectangle to find $\frac{1}{3} \cdot \frac{1}{4}$.

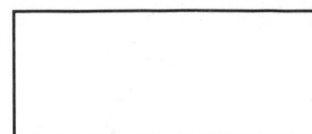

a. Draw three vertical lines that divide the given rectangle into four equal parts and lightly shade one part. What fractional part of the rectangle did you shade?

b. To find $\frac{1}{3}$ of the shaded portion, draw two horizontal lines to divide the given rectangle into three equal parts and lightly shade one part. Into how many equal parts is the rectangle now divided? How many parts have been shaded twice?

c. What is $\frac{1}{3} \cdot \frac{1}{4}$?

9. Determine whether each product is positive or negative. *You do not have to find the answer.*

a. $-\dfrac{1}{8} \cdot \dfrac{3}{5}$

b. $-\dfrac{7}{16}\left(-\dfrac{2}{21}\right)$

c. $-\dfrac{4}{5}\left(\dfrac{1}{3}\right)\left(-\dfrac{1}{8}\right)$

d. $-\dfrac{3}{4}\left(-\dfrac{8}{9}\right)\left(-\dfrac{1}{2}\right)$

10. Translate each phrase to symbols. *You do not have to find the answer.*

a. $\dfrac{7}{10}$ of $\dfrac{4}{9}$

b. $\dfrac{1}{5}$ of 40

11. Fill in the blanks.

a. Area of a triangle $= \frac{1}{2}($ $)($ $)$

or

$A = $

b. Area is measured in _____ units, such as in.^{2} and ft^2.

12. Determine whether each statement is true or false.

a. $\dfrac{1}{2}x = \dfrac{x}{2}$

b. $\dfrac{2t}{3} = \dfrac{2}{3}t$

c. $-\dfrac{3}{8}a = -\dfrac{3}{8a}$

d. $\dfrac{-4e}{7} = -\dfrac{4e}{7}$

NOTATION

13. Write each of the following as a fraction.

a. 4   **b.** -3 **c.** x

14. Fill in the blanks: $\left(\frac{1}{2}\right)^2$ represents the repeated multiplication $\cdot$.

Fill in the blanks to complete each solution.

15. Mutiply and simplify:

$$\frac{5}{8} \cdot \frac{7}{15} = \frac{5 \cdot \boxed{}}{8 \cdot \boxed{}}$$

$$= \frac{5 \cdot 7}{\boxed{} \cdot 2 \cdot 2 \cdot \boxed{} \cdot 5}$$

$$= \frac{\overset{1}{\cancel{5}} \cdot 7}{2 \cdot 2 \cdot 2 \cdot 3 \cdot \underset{1}{\cancel{5}}}$$

$$= \frac{7}{\boxed{}}$$

16. Multiply and simplify:

$$\frac{7a}{12} \cdot \frac{4}{21a} = \frac{7a \cdot 4}{\boxed{} \cdot \boxed{}}$$

$$= \frac{7 \cdot \boxed{} \cdot 4}{\boxed{} \cdot 4 \cdot 3 \cdot \boxed{} \cdot \boxed{}}$$

$$= \frac{\overset{1}{\cancel{7}} \cdot \overset{1}{\cancel{a}} \cdot \overset{1}{\cancel{4}}}{3 \cdot \underset{1}{\cancel{4}} \cdot 3 \cdot \underset{1}{\cancel{7}} \cdot \underset{1}{\cancel{a}}}$$

$$= \frac{\boxed{}}{9}$$

GUIDED PRACTICE

Multiply. See Example 1.

17. $\dfrac{1}{4} \cdot \dfrac{1}{2}$

18. $\dfrac{1}{3} \cdot \dfrac{1}{5}$

19. $\dfrac{1}{9} \cdot \dfrac{1}{5}$

20. $\dfrac{1}{2} \cdot \dfrac{1}{8}$

21. $\dfrac{2}{3} \cdot \dfrac{7}{9}$

22. $\dfrac{3}{4} \cdot \dfrac{5}{7}$

23. $\dfrac{8}{11} \cdot \dfrac{3}{7}$

24. $\dfrac{11}{13} \cdot \dfrac{2}{3}$

Multiply. See Example 2.

25. $-\dfrac{4}{5} \cdot \dfrac{1}{3}$

26. $-\dfrac{7}{9} \cdot \dfrac{1}{4}$

27. $\dfrac{5}{6}\left(-\dfrac{7}{12}\right)$

28. $\dfrac{2}{15}\left(-\dfrac{4}{3}\right)$

Multiply. See Example 3.

29. $\dfrac{1}{8} \cdot 9$

30. $\dfrac{1}{6} \cdot 11$

31. $\dfrac{1}{2} \cdot 5$

32. $\dfrac{1}{2} \cdot 21$

Multiply. Write the product in simplest form. See Example 4.

33. $\dfrac{11}{10} \cdot \dfrac{5}{11}$

34. $\dfrac{5}{4} \cdot \dfrac{2}{5}$

35. $\dfrac{6}{49} \cdot \dfrac{7}{6}$

36. $\dfrac{13}{4} \cdot \dfrac{4}{39}$

Multiply. Write the product in simplest form. See Example 5.

37. $\dfrac{3}{4}\left(-\dfrac{8}{35}\right)\left(-\dfrac{7}{12}\right)$

38. $\dfrac{9}{10}\left(-\dfrac{4}{15}\right)\left(-\dfrac{5}{18}\right)$

39. $-\dfrac{5}{8}\left(\dfrac{16}{27}\right)\left(-\dfrac{9}{25}\right)$

40. $-\dfrac{15}{28}\left(\dfrac{7}{9}\right)\left(-\dfrac{18}{35}\right)$

Multiply. Write the product in simplest form. See Example 6.

41. $\dfrac{3}{4x} \cdot \dfrac{2x}{5}$

42. $\dfrac{4}{15m} \cdot \dfrac{3m}{5}$

43. $\dfrac{6a}{49} \cdot \dfrac{14}{3a}$

44. $\dfrac{9x}{56} \cdot \dfrac{8}{3x}$

45. $-\dfrac{2m^3}{21n} \cdot \dfrac{27mn}{16m^5}$

46. $-\dfrac{2s^2}{27r} \cdot \dfrac{33rs}{16s^5}$

47. $\left(-\dfrac{25xy}{8y}\right)\left(-\dfrac{8x^2}{45x^2y^2}\right)$

48. $\left(-\dfrac{20cd}{21d^2}\right)\left(-\dfrac{14c^2d}{55c^2d^2}\right)$

Multiply. Write the product in simplest form. See Example 7.

49. $\dfrac{1}{8}(8w)$

50. $\dfrac{1}{9}(9e)$

51. $7x \cdot \dfrac{1}{7}$

52. $8r \cdot \dfrac{1}{8}$

53. $12a \cdot \dfrac{1}{6a}$

54. $18b \cdot \dfrac{1}{3b}$

55. $\dfrac{1}{9}(-36y)$

56. $\dfrac{1}{12}(-24d)$

Evaluate each expression. See Example 8.

57. a. $\left(\dfrac{3}{5}\right)^2$ **b.** $\left(-\dfrac{3}{5}\right)^2$

58. a. $\left(\dfrac{4}{9}\right)^2$ **b.** $\left(-\dfrac{4}{9}\right)^2$

59. a. $-\left(-\dfrac{1}{6}\right)^2$ **b.** $\left(-\dfrac{1}{6}\right)^3$

60. a. $-\left(-\dfrac{2}{5}\right)^2$ **b.** $\left(-\dfrac{2}{5}\right)^3$

Find each power. See Example 9.

61. $\left(-\dfrac{6t}{7}\right)^2$

62. $\left(-\dfrac{3s}{8}\right)^2$

63. $\left(-\dfrac{2a}{5}\right)^3$

64. $\left(\dfrac{4x}{5}\right)^3$

Find each product. Write your answer in simplest form. See Example 10.

65. $\dfrac{3}{4}$ of $\dfrac{5}{8}$

66. $\dfrac{4}{5}$ of $\dfrac{3}{7}$

67. $\dfrac{1}{6}$ of 54

68. $\dfrac{1}{9}$ of 36

Find the area of each triangle. See Example 11.

69.
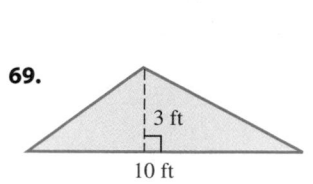

70.
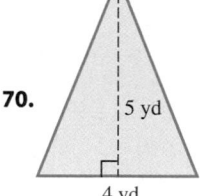

71.
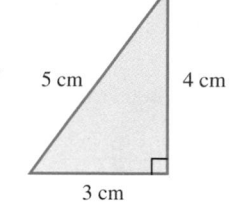

72.
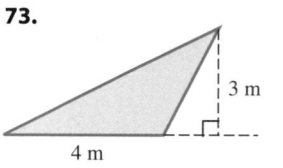

73.
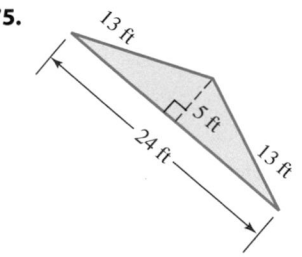

74.
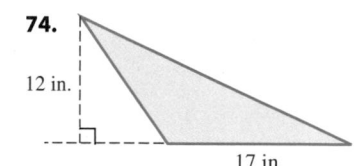

75.
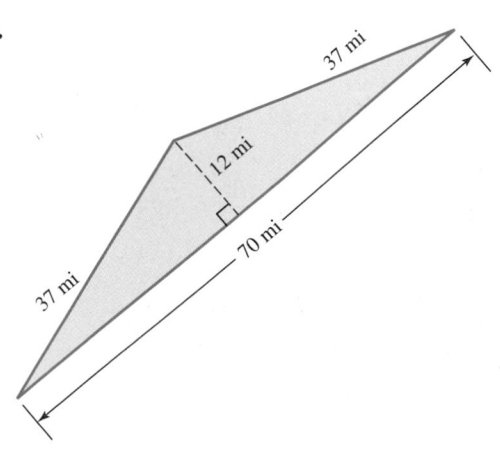

76.

TRY IT YOURSELF

77. Complete the multiplication table of fractions.

$\cdot$	$\dfrac{1}{2}$	$\dfrac{1}{3}$	$\dfrac{1}{4}$	$\dfrac{1}{5}$	$\dfrac{1}{6}$
$\dfrac{1}{2}$					
$\dfrac{1}{3}$					
$\dfrac{1}{4}$					
$\dfrac{1}{5}$					
$\dfrac{1}{6}$					

78. Complete the table by finding the original fraction, given its square.

Original fraction squared	Original fraction
$\dfrac{1}{9}$	
$\dfrac{1}{100}$	
$\dfrac{4}{25}$	
$\dfrac{16}{49}$	
$\dfrac{81}{36}$	
$\dfrac{9}{121}$	

Multiply. Write the product in simplest form.

79. $-\dfrac{15x^3}{24} \cdot \dfrac{8}{25x^2}$

80. $-\dfrac{20}{21x^4} \cdot \dfrac{7x^5}{16}$

81. $\dfrac{3}{8} \cdot \dfrac{7}{16}$

82. $\dfrac{5}{9} \cdot \dfrac{2}{7}$

83. $\left(\dfrac{2}{3}\right)\left(-\dfrac{1}{16}\right)\left(-\dfrac{4}{5}\right)$

84. $\left(\dfrac{3}{8}\right)\left(-\dfrac{2}{3}\right)\left(-\dfrac{12}{27}\right)$

85. $-\dfrac{5}{6} \cdot 18x$

86. $6\left(-\dfrac{2a}{3}\right)$

87. $\left(-\dfrac{3}{4n}\right)^3$

88. $\left(-\dfrac{2}{m}\right)^3$

89. $\dfrac{3a^3}{4} \cdot \dfrac{4}{3a^2}$

90. $\dfrac{4m^2}{5} \cdot \dfrac{5}{4m^3}$

91. $\dfrac{5}{3}\left(-\dfrac{6}{15}\right)(-4)$

92. $\dfrac{5}{6}\left(-\dfrac{2}{3}\right)(-12)$

93. $-\dfrac{11}{12} \cdot \dfrac{18}{55} \cdot 5$

94. $-\dfrac{24}{5} \cdot \dfrac{7}{12} \cdot \dfrac{1}{14}$

95. $\left(-\dfrac{11}{21}\right)\left(-\dfrac{14}{33}\right)$

96. $\left(-\dfrac{16}{35}\right)\left(-\dfrac{25}{48}\right)$

97. $-\left(-\dfrac{5}{9}\right)^2$

98. $-\left(-\dfrac{5}{6}\right)^2$

99. $\dfrac{7x^4}{10xy^3}\left(\dfrac{20y^2}{21x^2}\right)$

100. $\dfrac{7r^4}{6rt^3}\left(\dfrac{9rt^2}{49r^2}\right)$

101. $\dfrac{3}{4}\left(\dfrac{5}{7}\right)\left(\dfrac{2}{3}\right)\left(\dfrac{7}{3}\right)$

102. $-\dfrac{5}{4}\left(\dfrac{8}{15}\right)\left(\dfrac{2}{3}\right)\left(\dfrac{7}{2}\right)$

APPLICATIONS

103. SENATE RULES A *filibuster* is a method U.S. Senators sometimes use to block passage of a bill or appointment by talking endlessly. It takes $\frac{3}{5}$ of those voting in the Senate to break a filibuster. If all 100 Senators cast a vote, how many of their votes does it take to break a filibuster?

104. GENETICS Gregor Mendel (1822–1884), an Augustinian monk, is credited with developing a model that became the foundation of modern genetics. In his experiments, he crossed purple-flowered plants with white-flowered plants and found that $\frac{3}{4}$ of the offspring plants had purple flowers and $\frac{1}{4}$ of them had white flowers. Refer to the illustration below, which shows a group of offspring plants. According to this concept, when the plants begin to flower, how many will have purple flowers?

105. BOUNCING BALLS A tennis ball is dropped from a height of 54 inches. Each time it hits the ground, it rebounds one-third of the previous height that it fell. Find the three missing rebound heights in the illustration.

106. ELECTIONS The final election returns for a city bond measure are shown below.

   a. Find the total number of votes cast.

   b. Find two-thirds of the total number of votes cast.

   c. Did the bond measure pass?

MEASURE 1	
100% of the precincts reporting	
Fire–Police–Paramedics General Obligation Bonds	
(Requires two-thirds vote to pass)	
YES	No
125,599	62,801

107. COOKING Use the recipe below, along with the concept of multiplication of fractions, to find how much sugar and how much molasses are needed to make *one dozen* cookies. (*Hint:* this recipe is for *two dozen* cookies.)

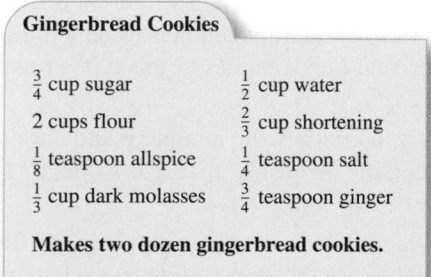

Gingerbread Cookies

$\frac{3}{4}$ cup sugar $\frac{1}{2}$ cup water

2 cups flour $\frac{2}{3}$ cup shortening

$\frac{1}{8}$ teaspoon allspice $\frac{1}{4}$ teaspoon salt

$\frac{1}{3}$ cup dark molasses $\frac{3}{4}$ teaspoon ginger

Makes two dozen gingerbread cookies.

108. THE EARTH'S SURFACE The surface of Earth covers an area of approximately 196,800,000 square miles. About $\frac{3}{4}$ of that area is covered by water. Find the number of square miles of the surface covered by water.

109. BOTANY In an experiment, monthly growth rates of three types of plants doubled when nitrogen was added to the soil. Complete the graph by drawing the improved growth rate bar next to each normal growth rate bar.

Inch   **Growth Rate: June**

1

5/6

2/3

1/2

1/3

1/6

Normal Nitrogen   Normal Nitrogen   Normal Nitrogen
House plants   Tomato plants   Shrubs

110. ICEBERGS About $\frac{9}{10}$ of the volume of an iceberg is below the water line.

   a. What fraction of the volume of an iceberg is *above* the water line?

   b. Suppose an iceberg has a total volume of 18,700 cubic meters. What is the volume of the part of the iceberg that is above the water line?

© Ralph A. Clevenger/Corbis

111. KITCHEN DESIGN Find the area of the *kitchen work triangle* formed by the paths between the refrigerator, the range, and the sink shown below.

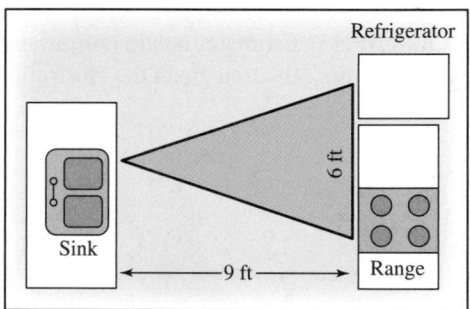

Refrigerator

6 ft

Sink

9 ft   Range

112. STARS AND STRIPES The illustration shows a folded U.S. flag. When it is placed on a table as part of an exhibit, how much area will it occupy?

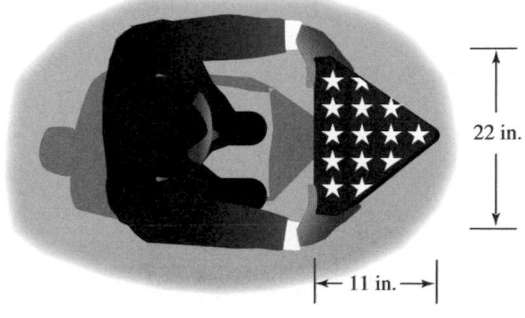

22 in.

11 in.

113. WINDSURFING Estimate the area of the sail on the windsurfing board.

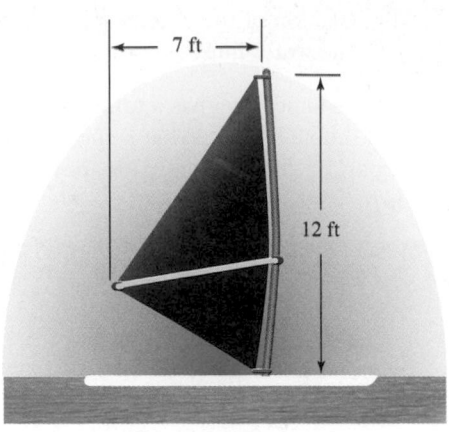

114. TILE DESIGN A design for bathroom tile is shown. Find the amount of area on a tile that is blue.

115. GEOGRAPHY Estimate the area of the state of New Hampshire, using the triangle in the illustration.

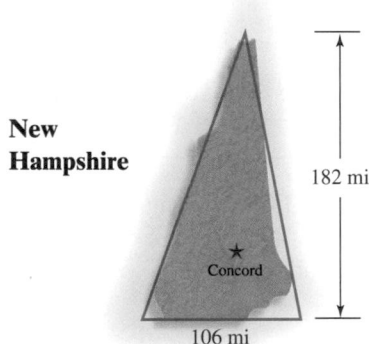

New Hampshire

★
Concord

182 mi

106 mi

116. STAMPS The best designs in a contest to create a wildlife stamp are shown. To save on paper costs, the postal service has decided to choose the stamp that has the smaller area. Which one did the postal service choose? (*Hint:* use the formula for the area of a rectangle.)

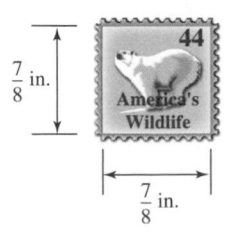

$\frac{7}{8}$ in.

$\frac{7}{8}$ in.

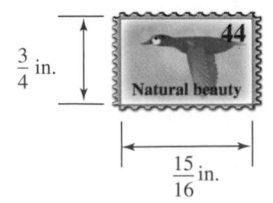

$\frac{3}{4}$ in.

$\frac{15}{16}$ in.

117. VISES Each complete turn of the handle of the bench vise shown below tightens its jaws exactly $\frac{1}{16}$ of an inch. How much tighter will the jaws of the vice get if the handle is turned 12 complete times?

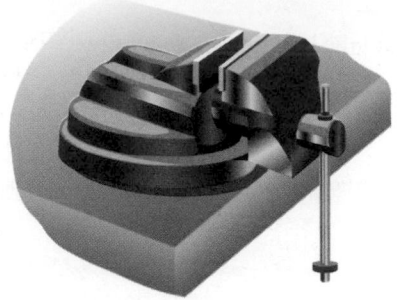

118. WOODWORKING Each time a board is passed through a power sander, the machine removes $\frac{1}{64}$ of an inch of thickness. If a rough pine board is passed through the sander 6 times, by how much will its thickness change?

█ WRITING

119. In a word problem, when a fraction is followed by the word *of,* multiplication is usually indicated. Give three real-life examples of this type of use of the word *of.*

120. Can you multiply the number 5 and another number and obtain an answer that is less than 5? Explain why or why not.

121. In the following solution, what step did the student forget to use that caused him to have to work with such large numbers?

Multiply. Simplify the product, if possible.

$$\frac{44}{63} \cdot \frac{27}{55} = \frac{44 \cdot 27}{63 \cdot 55}$$

$$= \frac{1,188}{3,465}$$

122. Is the product of two proper fractions always smaller than either of those fractions? Explain why or why not.

█ REVIEW

Divide and check each result.

123. $\dfrac{-8}{4}$

124. $21 \div (-3)$

125. $-736 \div (-32)$

126. $\dfrac{-400}{-25}$

SECTION 4.3
Dividing Fractions

We will now discuss how to divide fractions. The fraction multiplication skills that you learned in Section 4.2 will also be useful in this section.

1 Find the reciprocal of a fraction.

Division with fractions involves working with *reciprocals*. To present the concept of reciprocal, we consider the problem $\frac{7}{8} \cdot \frac{8}{7}$.

$$\frac{7}{8} \cdot \frac{8}{7} = \frac{7 \cdot 8}{8 \cdot 7}$$ Multiply the numerators.
Multiply the denominators.

$$= \frac{\overset{1}{7} \cdot \overset{1}{8}}{\underset{1}{8} \cdot \underset{1}{7}}$$ To simplify, remove the common factors of 7 and 8 from the numerator and denominator.

$$= \frac{1}{1}$$ Multiply the remaining factors in the numerator.
Multiply the remaining factors in the denominator.

$$= 1$$ Any whole number divided by 1 is equal to that number.

The product of $\frac{7}{8}$ and $\frac{8}{7}$ is 1.

Whenever the product of two numbers is 1, we say that those numbers are *reciprocals*. Therefore, $\frac{7}{8}$ and $\frac{8}{7}$ are reciprocals. To find the reciprocal of a fraction, *we invert the numerator and the denominator*.

Reciprocals

Two numbers are called **reciprocals** if their product is 1.

Caution! Zero does not have a reciprocal, because the product of 0 and a number can never be 1.

EXAMPLE 1   For each number, find its reciprocal and show that their product is 1:   **a.** $\frac{2}{3}$ **b.** $-\frac{3}{4}$ **c.** 5

Strategy To find each reciprocal, we will invert the numerator and denominator.

WHY This procedure will produce a new fraction that, when multiplied by the original fraction, gives a result of 1.

Solution

a. Fraction   Reciprocal

$$\frac{2}{3} \quad\rightarrow\quad \frac{3}{2}$$
invert

The reciprocal of $\frac{2}{3}$ is $\frac{3}{2}$.

Check: $\frac{2}{3} \cdot \frac{3}{2} = \frac{\overset{1}{2} \cdot \overset{1}{3}}{\underset{1}{3} \cdot \underset{1}{2}} = 1$

Self Check 1

For each number, find its reciprocal and show that their product is 1.

a. $\frac{3}{5}$ **b.** $-\frac{5}{6}$ **c.** 8

Now Try Problem 13

b. Fraction   Reciprocal

$$-\frac{3}{4} \quad\longrightarrow\quad -\frac{4}{3}$$

invert

The reciprocal of $-\dfrac{3}{4}$ is $-\dfrac{4}{3}$.

Check: $-\dfrac{3}{4}\left(-\dfrac{4}{3}\right) = \dfrac{\overset{1}{3} \cdot \overset{1}{4}}{\underset{1}{4} \cdot \underset{1}{3}} = 1$ The product of two fractions with like signs is positive.

c. Since $5 = \dfrac{5}{1}$, the reciprocal of 5 is $\dfrac{1}{5}$.

Check: $5 \cdot \dfrac{1}{5} = \dfrac{5}{1} \cdot \dfrac{1}{5} = \dfrac{\overset{1}{5} \cdot 1}{1 \cdot \underset{1}{5}} = 1$

Caution!   Don't confuse the concepts of the *opposite* of a negative number and the *reciprocal* of a negative number. For example:

The reciprocal of $-\dfrac{9}{16}$ is $-\dfrac{16}{9}$.

The opposite of $-\dfrac{9}{16}$ is $\dfrac{9}{16}$.

---

2 Divide fractions.

To develop a rule for dividing fractions, let's consider a real-life application.

Suppose that the manager of a candy store buys large bars of chocolate and divides each one into four equal parts to sell. How many fourths can be obtained from 5 bars?

We are asking, "How many $\frac{1}{4}$'s are there in 5?" To answer the question, we need to use the operation of division. We can represent this division as $5 \div \frac{1}{4}$.

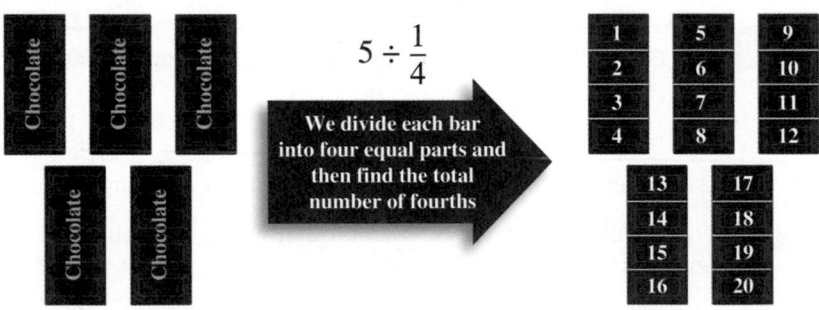

5 bars of chocolate   Total number of fourths = 5 • 4 = 20

There are 20 fourths in the 5 bars of chocolate. Two observations can be made from this result.

- This division problem involves a fraction: $5 \div \frac{1}{4}$.
- Although we were asked to find $5 \div \frac{1}{4}$, we solved the problem using *multiplication* instead of *division*: $5 \cdot 4 = 20$. That is, division by $\frac{1}{4}$ (a fraction) is the same as multiplication by 4 (its reciprocal).

$$5 \div \frac{1}{4} = 5 \cdot 4$$

These observations suggest the following rule for dividing two fractions.

Dividing Fractions

To divide two fractions, multiply the first fraction by the reciprocal of the second fraction. Simplify the result, if possible.

For example, to find $\frac{5}{7} \div \frac{3}{4}$, we multiply $\frac{5}{7}$ by the reciprocal of $\frac{3}{4}$.

Change the division to multiplication.

$$\frac{5}{7} \div \frac{3}{4} = \frac{5}{7} \cdot \frac{4}{3}$$

The reciprocal of $\frac{3}{4}$ is $\frac{4}{3}$.

$$= \frac{5 \cdot 4}{7 \cdot 3} \quad \text{Multiply the numerators.}$$
Multiply the denominators.

$$= \frac{20}{21}$$

Thus, $\frac{5}{7} \div \frac{3}{4} = \frac{20}{21}$. We say that the *quotient* of $\frac{5}{7}$ and $\frac{3}{4}$ is $\frac{20}{21}$.

EXAMPLE 2

Divide: $\frac{1}{3} \div \frac{4}{5}$

Strategy We will multiply the first fraction, $\frac{1}{3}$, by the reciprocal of the second fraction, $\frac{4}{5}$. Then, if possible, we will simplify the result.

WHY This is the rule for dividing two fractions.

Solution

$$\frac{1}{3} \div \frac{4}{5} = \frac{1}{3} \cdot \frac{5}{4} \quad \text{Multiply } \frac{1}{3} \text{ by the reciprocal of } \frac{4}{5}, \text{ which is } \frac{5}{4}.$$

$$= \frac{1 \cdot 5}{3 \cdot 4} \quad \text{Multiply the numerators.}$$
Multiply the denominators.

$$= \frac{5}{12}$$

Since 5 and 12 have no common factors other than 1, the result is in simplest form. ∎

Self Check 2

Divide: $\frac{2}{3} \div \frac{7}{8}$

Now Try Problem 17

EXAMPLE 3

Divide and simplify: $\frac{9}{16} \div \frac{3}{20}$

Strategy We will multiply the first fraction, $\frac{9}{16}$, by the reciprocal of the second fraction, $\frac{3}{20}$. Then, if possible, we will simplify the result.

WHY This is the rule for dividing two fractions.

Self Check 3

Divide and simplify: $\frac{4}{5} \div \frac{8}{25}$

Now Try Problem 21

Solution

$$\frac{9}{16} \div \frac{3}{20} = \frac{9}{16} \cdot \frac{20}{3}$$

Multiply $\frac{9}{16}$ by the reciprocal of $\frac{3}{20}$, which is $\frac{20}{3}$.

$$= \frac{9 \cdot 20}{16 \cdot 3}$$

Multiply the numerators.
Multiply the denominators.

$$= \frac{\overset{1}{\cancel{3}} \cdot 3 \cdot \overset{1}{\cancel{4}} \cdot 5}{\underset{1}{\cancel{4}} \cdot 4 \cdot \underset{1}{\cancel{3}}}$$

To simplify, factor 9 as $3 \cdot 3$, factor 20 as $4 \cdot 5$, and factor 16 as $4 \cdot 4$. Then remove out the common factors of 3 and 4 from the numerator and denominator.

$$= \frac{15}{4}$$

Multiply the remaining factors in the numerator: $1 \cdot 3 \cdot 1 \cdot 5 = 15$
Multiply the remaining factors in the denominator: $1 \cdot 4 \cdot 1 = 4$. ∎

Self Check 4

Divide and simplify:

$$80 \div \frac{20}{11}$$

Now Try Problem 27

EXAMPLE 4   Divide and simplify: $120 \div \frac{10}{7}$

Strategy We will write 120 as a fraction and then multiply the first fraction by the reciprocal of the second fraction.

WHY This is the rule for dividing two fractions.

Solution

$$120 \div \frac{10}{7} = \frac{120}{1} \div \frac{10}{7}$$

Write 120 as a fraction: $120 = \frac{120}{1}$.

$$= \frac{120}{1} \cdot \frac{7}{10}$$

Multiply $\frac{120}{1}$ by the reciprocal of $\frac{10}{7}$, which is $\frac{7}{10}$.

$$= \frac{120 \cdot 7}{1 \cdot 10}$$

Multiply the numerators.
Multiply the denominators.

$$= \frac{\overset{1}{\cancel{10}} \cdot 12 \cdot 7}{1 \cdot \underset{1}{\cancel{10}}}$$

To simplify, factor 120 as $10 \cdot 12$, then remove the common factor of 10 from the numerator and denominator.

$$= \frac{84}{1}$$

Multiply the remaining factors in the numerator: $1 \cdot 12 \cdot 7 = 84$.
Multiply the remaining factors in the denominator: $1 \cdot 1 = 1$.

$$= 84$$

Any whole number divided by 1 is the same number. ∎

Because of the relationship between multiplication and division, the sign rules for *dividing* fractions are the same as those for *multiplying* fractions.

Self Check 5

Divide and simplify:

$$\frac{2}{3} \div \left(-\frac{7}{6}\right)$$

Now Try Problem 29

EXAMPLE 5   Divide and simplify: $\frac{1}{6} \div \left(-\frac{1}{18}\right)$

Strategy We will multiply the first fraction, $\frac{1}{6}$, by the reciprocal of the second fraction, $-\frac{1}{18}$. To determine the sign of the result, we will use the rule for multiplying two fractions that have different (unlike) signs.

WHY One fraction is positive and one is negative.

Solution

$$\frac{1}{6} \div \left(-\frac{1}{18}\right) = \frac{1}{6}\left(-\frac{18}{1}\right) \quad \text{Multiply } \tfrac{1}{6} \text{ by the reciprocal of } -\tfrac{1}{18}, \text{ which is } -\tfrac{18}{1}.$$

$$= -\frac{1 \cdot 18}{6 \cdot 1} \quad \begin{array}{l}\text{Multiply the numerators.}\\ \text{Multiply the denominators.}\\ \text{Since the fractions have unlike signs,}\\ \text{make the answer negative.}\end{array}$$

$$= -\frac{1 \cdot 3 \cdot \overset{1}{\cancel{6}}}{\underset{1}{\cancel{6}} \cdot 1} \quad \begin{array}{l}\text{To simplify, factor 18 as } 3 \cdot 6. \text{ Then remove the common}\\ \text{factor of 6 from the numerator and denominator.}\end{array}$$

$$= -\frac{3}{1} \quad \begin{array}{l}\text{Multiply the remaining factors in the numerator.}\\ \text{Multiply the remaining factors in the denominator.}\end{array}$$

$$= -3$$

EXAMPLE 6

Divide and simplify: $\quad -\dfrac{21}{36} \div (-3)$

Strategy We will multiply the first fraction, $-\frac{21}{36}$, by the reciprocal of -3. To determine the sign of the result, we will use the rule for multiplying two fractions that have the same (like) signs.

WHY Both fractions are negative.

Solution

$$-\frac{21}{36} \div (-3) = -\frac{21}{36}\left(-\frac{1}{3}\right) \quad \text{Multiply } -\tfrac{21}{36} \text{ by the reciprocal of } -3, \text{ which is } -\tfrac{1}{3}.$$

$$= \frac{21}{36}\left(\frac{1}{3}\right) \quad \begin{array}{l}\text{Since the product of two negative fractions is}\\ \text{positive, drop both } - \text{ signs and continue.}\end{array}$$

$$= \frac{21 \cdot 1}{36 \cdot 3} \quad \begin{array}{l}\text{Multiply the numerators.}\\ \text{Multiply the denominators.}\end{array}$$

$$= \frac{\overset{1}{\cancel{3}} \cdot 7 \cdot 1}{36 \cdot \underset{1}{\cancel{3}}} \quad \begin{array}{l}\text{To simplify, factor 21 as } 3 \cdot 7. \text{ Then remove the common}\\ \text{factor of 3 from the numerator and denominator.}\end{array}$$

$$= \frac{7}{36} \quad \begin{array}{l}\text{Multiply the remaining factors in the numerator:}\\ 1 \cdot 7 \cdot 1 = 7.\\ \text{Multiply the remaining factors in the denominator:}\\ 36 \cdot 1 = 36.\end{array}$$

Self Check 6

Divide and simplify:

$$-\frac{35}{16} \div (-7)$$

Now Try Problem 33

3 Divide algebraic fractions.

To work problems involving division of algebraic fractions, we must find the reciprocal of an algebraic fraction. We learned earlier that the reciprocal of a numerical fraction is found by inverting its numerator and denominator. The same is true for algebraic fractions. Here are some examples.

Algebraic fraction   Reciprocal   **Algebraic fraction   Reciprocal**

$$\frac{2a}{3} \quad\longrightarrow\quad \frac{3}{2a} \qquad\qquad\qquad -\frac{m^2}{n} \quad\longrightarrow\quad -\frac{n}{m^2}$$

Invert   Invert

Since $x = \frac{x}{1}$, the reciprocal of x is $\frac{1}{x}$.

To divide algebraic functions, we use the same approach as with numerical fractions.

Divide: $\dfrac{7}{4} \div \dfrac{3}{b}$

Now Try Problem 37

EXAMPLE 7   Divide: $\dfrac{2}{3} \div \dfrac{a}{5}$

Strategy We will multiply the first fraction, $\frac{2}{3}$, by the reciprocal of the second fraction, $\frac{a}{5}$.

WHY This is the rule for dividing two fractions.

Solution

$$\frac{2}{3} \div \frac{a}{5} = \frac{2}{3} \cdot \frac{5}{a} \qquad \text{Multiply } \tfrac{2}{3} \text{ by the reciprocal of } \tfrac{a}{5}, \text{ which is } \tfrac{5}{a}.$$

$$= \frac{2 \cdot 5}{3 \cdot a} \qquad \begin{array}{l}\text{Multiply the numerators.}\\\text{Multiply the denominators.}\end{array}$$

$$= \frac{10}{3a} \qquad \begin{array}{l}\text{Since 10 and 3 have no common factors other than 1,}\\\text{the result is in simplest form.}\end{array}$$

Divide and simplify: $\dfrac{9y^8}{10x} \div \dfrac{18y}{5x}$

Now Try Problem 41

EXAMPLE 8   Divide and simplify: $\dfrac{15x^2}{8y} \div \dfrac{10x}{y^3}$

Strategy We will multiply the first algebraic fraction, $\frac{15x^2}{8y}$, by the reciprocal of the second algebraic fraction, $\frac{10x}{y^3}$.

WHY This is the rule for dividing fractions.

Solution

$$\frac{15x^2}{8y} \div \frac{10x}{y^3} = \frac{15x^2}{8y} \cdot \frac{y^3}{10x} \qquad \text{Multiply } \tfrac{15x^2}{8y} \text{ by the reciprocal of } \tfrac{10x}{y^3}, \text{ which is } \tfrac{y^3}{10x}.$$

$$= \frac{15x^2 \cdot y^3}{8y \cdot 10x} \qquad \begin{array}{l}\text{Multiply the numerators.}\\\text{Multiply the denominators.}\end{array}$$

$$= \frac{3 \cdot 5 \cdot x \cdot x \cdot y \cdot y \cdot y}{8 \cdot y \cdot 2 \cdot 5 \cdot x} \qquad \text{To prepare to simplify, factor 15, } x^2, y^3, \text{ and 10.}$$

$$= \frac{3 \cdot \overset{1}{\cancel{5}} \cdot \overset{1}{\cancel{x}} \cdot x \cdot \overset{1}{\cancel{y}} \cdot y \cdot y}{8 \cdot \underset{1}{\cancel{y}} \cdot 2 \cdot \underset{1}{\cancel{5}} \cdot \underset{1}{\cancel{x}}} \qquad \begin{array}{l}\text{Remove the common factors of 5, } x, \text{ and } y\\\text{from the numerator and denominator.}\end{array}$$

$$= \frac{3xy^2}{16} \qquad \begin{array}{l}\text{Multiply the remaining factors in the}\\\text{numerator: } 3 \cdot 1 \cdot 1 \cdot x \cdot 1 \cdot y \cdot y = 3xy^2.\\\text{Multiply the remaining factors in the}\\\text{denominator: } 8 \cdot 1 \cdot 2 \cdot 1 \cdot 1 = 16.\end{array}$$

---

4 Solve application problems by dividing fractions.

Problems that involve forming equal-sized groups can be solved by division.

Finish:
$\frac{3}{8}$ in. thick

Foam core

EXAMPLE 9   *Surfboard Designs*   Most surfboards are made of a foam core covered with several layers of fiberglass to keep them watertight. How many layers are needed to build up a finish $\frac{3}{8}$ of an inch thick if each layer of fiberglass has a thickness of $\frac{1}{16}$ of an inch?

Analyze

- The surfboard is to have a $\frac{3}{8}$-inch-thick fiberglass finish.   *Given*
- Each layer of fiberglass is $\frac{1}{16}$ of an inch thick.   *Given*
- How many layers of fiberglass need to be applied?   *Find*

Form Think of the $\frac{3}{8}$-inch-thick finish separated into an unknown number of equally thick layers of fiberglass. This indicates division.

We translate the words of the problem to numbers and symbols.

The number of layers of fiberglass that are needed	is equal to	the thickness of the finish	divided by	the thickness of 1 layer of fiberglass.

The number of layers of fiberglass that are needed	=	$\frac{3}{8}$	÷	$\frac{1}{16}$

Solve To find the quotient, we will use the rule for dividing two fractions.

$$\frac{3}{8} \div \frac{1}{16} = \frac{3}{8} \cdot \frac{16}{1}$$
Multiply $\frac{3}{8}$ by the reciprocal of $\frac{1}{16}$, which is $\frac{16}{1}$.

$$= \frac{3 \cdot 16}{8 \cdot 1}$$
Multiply the numerators.
Multiply the denominators.

$$= \frac{3 \cdot 2 \cdot \overset{1}{\cancel{8}}}{\underset{1}{\cancel{8}} \cdot 1}$$
To simplify, factor 16 as 2 · 8. Then remove the common factor of 8 from the numerator and denominator.

$$= \frac{6}{1}$$
Multiply the remaining factors in the numerator: 3 · 2 · 1 = 6.
Multiply the remaining factors in the denominator: 1 · 1 = 1.

$$= 6$$
Any number divided by 1 is the same number.

State The number of layers of fiberglass needed is 6.

Check If 6 layers of fiberglass, each $\frac{1}{16}$ of an inch thick, are used, the finished thickness will be $\frac{6}{16}$ of an inch. If we simplify $\frac{6}{16}$, we see that it is equivalent to the desired finish thickness:

$$\frac{6}{16} = \frac{\overset{1}{\cancel{2}} \cdot 3}{\underset{1}{\cancel{2}} \cdot 8} = \frac{3}{8}$$

The result checks.

Self Check 9

COOKING A recipe calls for 4 cups of sugar, and the only measuring container you have holds $\frac{1}{3}$ cup. How many $\frac{1}{3}$ cups of sugar would you need to add to follow the recipe?

Now Try Problem 89

ANSWERS TO SELF CHECKS

1. a. $\frac{5}{3}$ **b.** $-\frac{6}{5}$ **c.** $\frac{1}{8}$ **2.** $\frac{16}{21}$ **3.** $\frac{5}{2}$ **4.** 44 **5.** $-\frac{4}{7}$ **6.** $\frac{5}{16}$ **7.** $\frac{7b}{12}$ **8.** $\frac{y^7}{4}$ **9.** 12

SECTION 4.3 STUDY SET

VOCABULARY

Fill in the blanks.

1. The _____ of $\frac{5}{12}$ is $\frac{12}{5}$.

2. To find the reciprocal of a fraction, _____ the numerator and denominator.

3. The answer to a division is called the _____.

4. To simplify $\frac{2 \cdot 2 \cdot 3}{2 \cdot 3 \cdot 5 \cdot 7}$, we _____ common factors of the numerator and denominator.

CONCEPTS

5. Fill in the blanks.

   a. To divide two fractions, _____ the first fraction by the _____ of the second fraction.

   b. $\dfrac{1}{2} \div \dfrac{2}{3} = \dfrac{1}{2} \boxed{} \boxed{}$

6. a. What division problem is illustrated below?

   b. What is the answer?

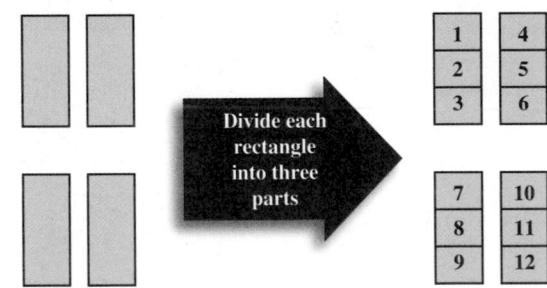

1	4
2	5
3	6

7	10
8	11
9	12

Divide each rectangle into three parts

7. Determine whether each quotient is positive or negative. *You do not have to find the answer.*

   a. $-\dfrac{1}{4} \div \dfrac{3}{4}$ b. $-\dfrac{7}{8} \div \left(-\dfrac{21}{32}\right)$

8. Complete the table.

Number	Opposite	Reciprocal
$\frac{3}{10}$		
$-\frac{7}{11}$		
6		

9. a. Multiply $\frac{4}{5}$ and its reciprocal. What is the result?

   b. Multiply $-\frac{3}{5}$ and its reciprocal. What is the result?

10. a. Find: $15 \div 3$

    b. Rewrite $15 \div 3$ as multiplication by the reciprocal of 3, and find the result.

    c. Complete this statement: Division by 3 is the same as multiplication by $\boxed{}$.

NOTATION

Fill in the blanks to complete each solution.

11. $\dfrac{4}{9} \div \dfrac{8}{27} = \dfrac{4}{9} \cdot \dfrac{\boxed{}}{8}$

$= \dfrac{4 \cdot \boxed{}}{9 \cdot \boxed{}}$

$= \dfrac{4 \cdot 3 \cdot \boxed{}}{9 \cdot \boxed{} \cdot \boxed{}}$

$= \dfrac{\overset{1}{\cancel{4}} \cdot 3 \cdot \overset{1}{\cancel{9}}}{\underset{1}{\cancel{9}} \cdot 2 \cdot \underset{1}{\cancel{4}}}$

$= \dfrac{\boxed{}}{2}$

12. $\dfrac{4a}{9b} \div \dfrac{8a}{27b} = \dfrac{4a}{9b} \cdot \dfrac{\boxed{}}{\boxed{}}$

$= \dfrac{4a \cdot \boxed{}}{9b \cdot \boxed{}}$

$= \dfrac{4 \cdot a \cdot 3 \cdot \boxed{} \cdot b}{9 \cdot b \cdot 2 \cdot \boxed{} \cdot a}$

$= \dfrac{\overset{1}{\cancel{4}} \cdot \overset{1}{\cancel{a}} \cdot 3 \cdot \overset{1}{\cancel{9}} \cdot \overset{1}{\cancel{b}}}{\underset{1}{\cancel{9}} \cdot \underset{1}{\cancel{b}} \cdot 2 \cdot \underset{1}{\cancel{4}} \cdot \underset{1}{\cancel{a}}}$

$= \dfrac{3}{2}$

GUIDED PRACTICE

Find the reciprocal of each number or algebraic fraction. See Example 1 and Objective 3.

13. a. $\dfrac{6}{7}$ b. $-\dfrac{15}{8}$ c. 10

14. a. $\dfrac{2}{9}$ b. $-\dfrac{9}{4}$ c. 7

15. a. $\dfrac{11a}{8}$ b. $-\dfrac{1}{14b}$ c. $-63x$

16. a. $\dfrac{13}{2n}$ b. $-\dfrac{b}{5}$ c. $-21y$

Divide. Simplify each quotient, if possible. See Example 2.

17. $\dfrac{1}{8} \div \dfrac{2}{3}$ 18. $\dfrac{1}{2} \div \dfrac{8}{9}$

19. $\dfrac{2}{23} \div \dfrac{1}{7}$ 20. $\dfrac{4}{21} \div \dfrac{1}{5}$

Divide. Simplify each quotient, if possible. See Example 3.

21. $\dfrac{25}{32} \div \dfrac{5}{28}$

22. $\dfrac{4}{25} \div \dfrac{2}{35}$

23. $\dfrac{27}{32} \div \dfrac{9}{8}$

24. $\dfrac{16}{27} \div \dfrac{20}{21}$

Divide. Simplify each quotient, if possible. See Example 4.

25. $50 \div \dfrac{10}{9}$

26. $60 \div \dfrac{10}{3}$

27. $150 \div \dfrac{15}{32}$

28. $170 \div \dfrac{17}{6}$

Divide. Simplify each quotient, if possible. See Example 5.

29. $\dfrac{1}{8} \div \left(-\dfrac{1}{32}\right)$

30. $\dfrac{1}{9} \div \left(-\dfrac{1}{27}\right)$

31. $\dfrac{2}{5} \div \left(-\dfrac{4}{35}\right)$

32. $\dfrac{4}{9} \div \left(-\dfrac{16}{27}\right)$

Divide. Simplify each quotient, if possible. See Example 6.

33. $-\dfrac{28}{55} \div (-7)$

34. $-\dfrac{32}{45} \div (-8)$

35. $-\dfrac{33}{23} \div (-11)$

36. $-\dfrac{21}{31} \div (-7)$

Divide. See Example 7.

37. $\dfrac{4a}{5} \div \dfrac{3}{7}$

38. $\dfrac{2x}{3} \div \dfrac{3}{2}$

39. $\dfrac{6x}{7} \div \dfrac{5}{11}$

40. $\dfrac{9b}{4} \div \dfrac{4}{3}$

Divide. Simplify each quotient, if possible. See Example 8.

41. $\dfrac{6b^2}{7a} \div \dfrac{9b}{14a^3}$

42. $\dfrac{8c^2}{27d} \div \dfrac{4c^3}{9d^3}$

43. $\dfrac{20x}{33y} \div \dfrac{30x^4}{77y}$

44. $\dfrac{16r^4}{27s^3} \div \dfrac{32r^4}{21s}$

45. $-\dfrac{ab^2}{15} \div \dfrac{a^2b^2}{25}$

46. $-\dfrac{xy^3}{12} \div \dfrac{x^2y^2}{10}$

47. $-\dfrac{48}{m^{11}} \div \left(-\dfrac{16}{m^8}\right)$

48. $-\dfrac{64}{q^{12}} \div \left(-\dfrac{24}{q^6}\right)$

TRY IT YOURSELF

Divide. Simplify each quotient, if possible.

49. $120 \div \dfrac{12}{5x}$

50. $360 \div \dfrac{36}{5y}$

51. $-8x \div \left(-\dfrac{4x^3}{9}\right)$

52. $-12y \div \left(-\dfrac{3y^7}{10}\right)$

53. $\left(-\dfrac{7}{4}\right) \div \left(-\dfrac{21}{8}\right)$

54. $\left(-\dfrac{15}{16}\right) \div \left(-\dfrac{5}{8}\right)$

55. $\dfrac{4}{5} \div \dfrac{4}{5}$

56. $\dfrac{2}{3} \div \dfrac{2}{3}$

57. Divide $-\dfrac{15}{32}$ by $\dfrac{3}{4}$

58. Divide $-\dfrac{7}{10}$ by $\dfrac{4}{5}$

59. $3a \div \dfrac{1}{12a}$

60. $9x \div \dfrac{3}{4x}$

61. $-\dfrac{4}{5} \div (-6)$

62. $-\dfrac{7}{8} \div (-14)$

63. $\dfrac{15}{16} \div 180$

64. $\dfrac{7}{8} \div 210$

65. $-\dfrac{9n^5}{10} \div \dfrac{4n^4}{15}$

66. $-\dfrac{3m^3}{4} \div \dfrac{3m^2}{2}$

67. $\dfrac{9a^3}{10b} \div \left(-\dfrac{3a^4}{25b^3}\right)$

68. $\dfrac{11d^4}{16n} \div \left(-\dfrac{9d^5}{16n^3}\right)$

69. $-\dfrac{x^2}{y^3} \div \dfrac{x}{y}$

70. $-\dfrac{k}{t^8} \div \dfrac{k}{t^2}$

71. $-\dfrac{1}{8} \div 8$

72. $-\dfrac{1}{15} \div 15$

The following problems involve multiplication and division. Perform each operation. Simplify the result, if possible.

73. $-\dfrac{7m^3}{6}\left(-\dfrac{9}{49m^2}\right)$

74. $-\dfrac{7n^4}{10}\left(-\dfrac{20}{21n^3}\right)$

75. $-\dfrac{4}{5} \div \left(-\dfrac{3}{2}\right)$

76. $-\dfrac{2}{3} \div \left(-\dfrac{3}{2}\right)$

77. $\dfrac{13}{16} \div x$

78. $\dfrac{7}{8} \div x$

79. $\left(-\dfrac{11}{21}\right)\left(-\dfrac{14}{33}\right)$

80. $\left(-\dfrac{16}{35}\right)\left(-\dfrac{25}{48}\right)$

81. $-\dfrac{15x}{32y} \div \dfrac{5x}{64y^4}$

82. $-\dfrac{28r}{15s} \div \dfrac{21r^4}{10s}$

83. $11 \cdot \dfrac{1}{6}$

84. $9 \cdot \dfrac{1}{8}$

85. $\dfrac{3x}{4} \cdot \dfrac{5}{7}$

86. $\dfrac{2c}{3} \cdot \dfrac{7}{9}$

87. $\dfrac{25}{7} \div \left(-\dfrac{30}{21}\right)$

88. $\dfrac{39}{25} \div \left(-\dfrac{13}{10}\right)$

APPLICATIONS

89. PATIO FURNITURE A production process applies several layers of a clear plastic coat to outdoor furniture to help protect it from the weather. If each protective coat is $\dfrac{3}{32}$-inch thick, how many applications will be needed to build up $\dfrac{3}{8}$ inch of clear finish?

90. MARATHONS Each lap around a stadium track is $\dfrac{1}{4}$ mile. How many laps would a runner have to complete to get a 26-mile workout?

91 COOKING A recipe calls for $\frac{3}{4}$ cup of flour, and the only measuring container you have holds $\frac{1}{8}$ cup. How many $\frac{1}{8}$ cups of flour would you need to add to follow the recipe?

92. LASERS A technician uses a laser to slice thin pieces of aluminum off the end of a rod that is $\frac{7}{8}$-inch long. How many $\frac{1}{64}$-inch-wide slices can be cut from this rod? (Assume that there is no waste in the process.)

93. UNDERGROUND CABLES Refer to the illustration and table below.

  a. How many days will it take to install underground TV cable from the broadcasting station to the new homes using route 1?

  b. How long is route 2?

  c. How many days will it take to install the cable using route 2?

  d. Which route will require the fewer number of days to install the cable?

Proposal	Amount of cable installed per day	Comments
Route 1	$\frac{2}{5}$ of a mile	Ground very rocky
Route 2	$\frac{3}{5}$ of a mile	Longer than Route 1

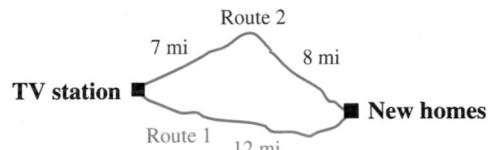

94. PRODUCTION PLANNING The materials used to make a pillow are shown. Examine the inventory list to decide how many pillows can be manufactured in one production run with the materials in stock.

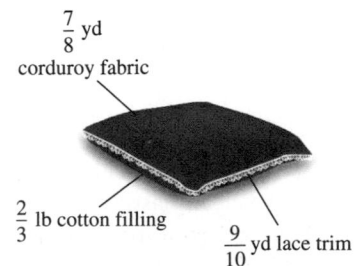

$\frac{7}{8}$ yd corduroy fabric

$\frac{2}{3}$ lb cotton filling

$\frac{9}{10}$ yd lace trim

Factory Inventory List

Materials	Amount in stock
Lace trim	135 yd
Corduroy fabric	154 yd
Cotton filling	98 lb

95. NOTE CARDS Ninety 3 × 5 cards are stacked next to a ruler as shown, below.

  a. Into how many parts is 1 inch divided on the ruler?

  b. How thick is the stack of cards?

  c. How thick is one 3 × 5 card?

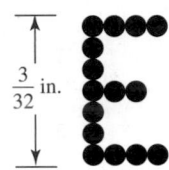

90 note cards

96. COMPUTER PRINTERS The illustration shows how the letter E is formed by a dot matrix printer. What is the height of one dot?

$\frac{3}{32}$ in.

97. FORESTRY A set of forestry maps divides the 6,284 acres of an old-growth forest into $\frac{4}{5}$-acre sections. How many sections do the maps contain?

98. HARDWARE A hardware chain purchases large amounts of nails and packages them in $\frac{9}{16}$-pound bags for sale. How many of these bags of nails can be obtained from 2,871 pounds of nails?

WRITING

99. Explain how to divide two fractions.

100. Why do you need to know how to multiply fractions to be able to divide fractions?

101. Explain why 0 does not have a reciprocal.

102. What number is its own reciprocal? Explain why this is so.

103. Write an application problem that could be solved by finding $10 \div \frac{1}{5}$.

104. Explain why dividing a fraction by 2 is the same as finding $\frac{1}{2}$ of it. Give an example.

REVIEW

Fill in the blanks.

105. The symbol $<$ means ___ ____ ____.

106. The statement $9 \cdot 8 = 8 \cdot 9$ illustrates the
_____ property of multiplication.

107. _____ is neither positive nor negative.

108. The sum of two negative numbers is _____.

109. Graph each of these numbers on a number line:
$-2, 0, |-4|$, and the opposite of 1

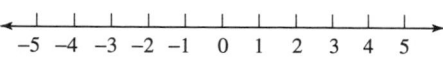

110. Evaluate each expression.

   a. 3^5 **b.** $(-2)^5$

SECTION 4.4
Adding and Subtracting Fractions

In mathematics and everyday life, we can only add (or subtract) objects that are similar. For example, we can add dollars to dollars, but we cannot add dollars to oranges. This concept is important when adding or subtracting fractions.

1   **Add and subtract fractions that have the same denominator.**

Consider the problem $\frac{3}{5} + \frac{1}{5}$. When we write it in words, it is apparent that we are adding similar objects.

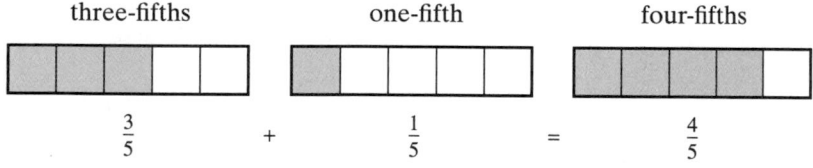

Because the denominators of $\frac{3}{5}$ and $\frac{1}{5}$ are the same, we say that they have a **common denominator.** Since the fractions have a common denominator, we can add them. The following figure explains the addition process.

three-fifths	one-fifth	four-fifths
$\frac{3}{5}$	$\frac{1}{5}$	$\frac{4}{5}$

We can make some observations about the addition shown in the figure.

The sum of the numerators is the numerator of the answer.

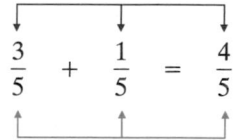

The answer is a fraction that has the same denominator as the two fractions that were added.

These observations illustrate the following rule.

> **Adding and Subtracting Fractions That Have the Same Denominator**
>
> To add (or subtract) fractions that have the same denominator, add (or subtract) their numerators and write the sum (or difference) over the common denominator. Simplify the result, if possible.

Objectives

1 Add and subtract fractions that have the same denominator.

2 Add and subtract fractions that have different denominators.

3 Find the LCD to add and subtract fractions.

4 Add and subtract algebraic fractions.

5 Identify the greater of two fractions.

6 Solve application problems by adding and subtracting fractions.

> **Caution!** We **do not** add fractions by adding the numerators and adding the denominators!
>
> $$\frac{3}{5}+\frac{1}{5}=\frac{3+1}{5+5}=\frac{4}{10}$$
>
> The same caution applies when subtracting fractions.

Self Check 1

Perform each operation and simplify the result, if possible.

a. Add: $\dfrac{5}{12}+\dfrac{1}{12}$

b. Subtract: $\dfrac{8}{9}-\dfrac{1}{9}$

Now Try **Problems 11 and 15**

EXAMPLE 1

Perform each operation and simplify the result, if possible.

a. Add: $\dfrac{1}{8}+\dfrac{5}{8}$ **b.** Subtract: $\dfrac{11}{15}-\dfrac{4}{15}$

Strategy We will use the rule for adding and subtracting fractions that have *the same* denominator.

WHY In part a, the fractions have the same denominator, 8. In part b, the fractions have the same denominator, 15.

Solution

a. $\dfrac{1}{8}+\dfrac{5}{8}=\dfrac{1+5}{8}$ Add the numerators and write the sum over the common denominator 8.

$=\dfrac{6}{8}$ This fraction can be simplified.

$=\dfrac{\overset{1}{2}\cdot 3}{\underset{1}{2}\cdot 4}$ To simplify, factor 6 as 2 · 3 and 8 as 2 · 4. Then remove the common factor of 2 from the numerator and denominator.

$=\dfrac{3}{4}$ Multiply the remaining factors in the numerator: 1 · 3 = 3. Multiply the remaining factors in the denominator: 1· 4 = 4.

b. $\dfrac{11}{15}-\dfrac{4}{15}=\dfrac{11-4}{15}$ Subtract the numerators and write the difference over the common denominator 15.

$=\dfrac{7}{15}$

Since 7 and 15 have no common factors other than 1, the result is in simplest form. ∎

The rule for subtraction from Section 2.3 can be extended to subtraction involving signed fractions:

To subtract two fractions, add the first to the opposite of the fraction to be subtracted.

Self Check 2

Subtract: $-\dfrac{9}{11}-\left(-\dfrac{3}{11}\right)$

Now Try **Problem 19**

EXAMPLE 2

Subtract: $-\dfrac{7}{3}-\left(-\dfrac{2}{3}\right)$

Strategy To find the difference, we will apply the rule for subtraction.

WHY It is easy to make an error when subtracting signed fractions. We will probably be more accurate if we write the subtraction as addition of the opposite. ▼

Solution

We read $-\frac{7}{3} - \left(-\frac{2}{3}\right)$ as "negative seven-thirds *minus* negative two-thirds." Thus, the number to be subtracted is $-\frac{2}{3}$. Subtracting $-\frac{2}{3}$ is the same as adding its opposite, $\frac{2}{3}$.

$$-\frac{7}{3} - \left(-\frac{2}{3}\right) = -\frac{7}{3} + \frac{2}{3} \qquad \text{Add the opposite of } -\frac{2}{3}, \text{ which is } \frac{2}{3}.$$

$$= \frac{-7}{3} + \frac{2}{3} \qquad \text{Write } -\frac{7}{3} \text{ as } \frac{-7}{3}.$$

$$= \frac{-7 + 2}{3} \qquad \text{Add the numerators and write the sum over the common denominator 3.}$$

$$= \frac{-5}{3} \qquad \text{Use the rule for adding two integers with different signs: } -7 + 2 = -5.$$

$$= -\frac{5}{3} \qquad \text{Rewrite the result with the } - \text{ sign in front: } \frac{-5}{3} = -\frac{5}{3}. \text{ This fraction is in simplest form.}$$

EXAMPLE 3

Perform the operations and simplify: $\dfrac{18}{25} - \dfrac{2}{25} - \dfrac{1}{25}$

Strategy We will use the rule for subtracting fractions that have *the same* denominator.

WHY All three fractions have the same denominator, 25.

Solution

$$\frac{18}{25} - \frac{2}{25} - \frac{1}{25} = \frac{18 - 2 - 1}{25} \qquad \text{Subtract the numerators and write the difference over the common denominator 25.}$$

$$= \frac{15}{25} \qquad \text{This fraction can be simplified.}$$

$$= \frac{3 \cdot \overset{1}{\cancel{5}}}{\underset{1}{\cancel{5}} \cdot 5} \qquad \text{To simplify, factor 15 as } 3 \cdot 5 \text{ and 25 as } 5 \cdot 5. \text{ Then remove the common factor of 5 from the numerator and denominator.}$$

$$= \frac{3}{5} \qquad \text{Multiply the remaining factors in the numerator: } 3 \cdot 1 = 3. \text{ Multiply the remaining factors in the denominator: } 1 \cdot 5 = 5.$$

Self Check 3

Perform the operations and simplify:

$$\frac{2}{9} + \frac{2}{9} + \frac{2}{9}$$

Now Try Problem 23

---

2 Add and subtract fractions that have different denominators.

Now we consider the problem $\frac{3}{5} + \frac{1}{3}$. Since the denominators are different, we cannot add these fractions in their present form.

three-**fifths** + one-**third**
└─Not similar objects─┘

To add (or subtract) fractions with different denominators, we express them as equivalent fractions that have a common denominator. The smallest common denominator, called the **least** or **lowest common denominator,** is usually the easiest common denominator to use.

Least Common Denominator

The **least common denominator (LCD)** for a set of fractions is the smallest number each denominator will divide exactly (divide with no remainder).

The denominators of $\frac{3}{5}$ and $\frac{1}{3}$ are 5 and 3. The numbers 5 and 3 divide many numbers exactly (30, 45, and 60, to name a few), but the smallest number that they divide exactly is 15. Thus, 15 is the LCD for $\frac{3}{5}$ and $\frac{1}{3}$.

To find $\frac{3}{5} + \frac{1}{3}$, we *build* equivalent fractions that have denominators of 15. (This procedure was introduced in Section 3.1.) Then we use the rule for adding fractions that have the same denominator.

$$\frac{3}{5} + \frac{1}{3} = \frac{3}{5} \cdot \frac{3}{3} + \frac{1}{3} \cdot \frac{5}{5}$$

We need to multiply this denominator by 5 to obtain 15. It follows that $\frac{5}{5}$ should be the form of 1 used to build $\frac{1}{3}$.

We need to multiply this denominator by 3 to obtain 15. It follows that $\frac{3}{3}$ should be the form of 1 that is used to build $\frac{3}{5}$.

$$= \frac{9}{15} + \frac{5}{15}$$

Multiply the numerators. Multiply the denominators. Note that the denominators are now the same.

$$= \frac{9 + 5}{15}$$

Add the numerators and write the sum over the common denominator 15.

$$= \frac{14}{15}$$

Since 14 and 15 have no common factors other than 1, this fraction is in simplest form.

The figure below shows $\frac{3}{5}$ and $\frac{1}{3}$ expressed as equivalent fractions with a denominator of 15. Once the denominators are the same, the fractions are similar objects and can be added easily.

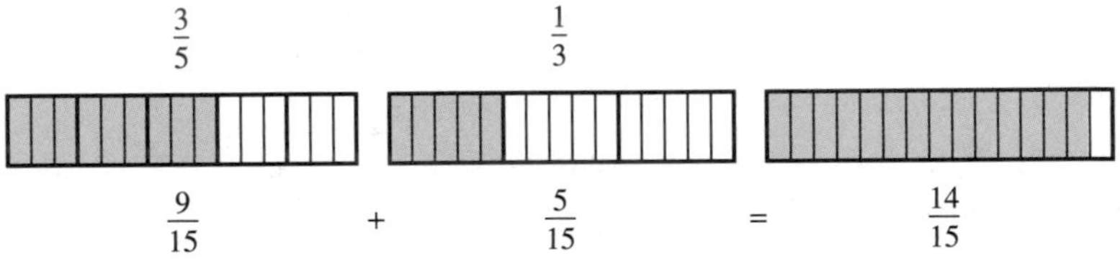

We can use the following steps to add or subtract fractions with different denominators.

Adding and Subtracting Fractions That Have Different Denominators

1. Find the LCD.
2. Rewrite each fraction as an equivalent fraction with the LCD as the denominator. To do so, build each fraction using a form of 1 that involves any factors needed to obtain the LCD.
3. Add or subtract the numerators and write the sum or difference over the LCD.
4. Simplify the result, if possible.

Self Check 4

Add: $\frac{1}{2} + \frac{2}{5}$

Now Try Problem 27

EXAMPLE 4

Add: $\frac{1}{7} + \frac{2}{3}$

Strategy We will express each fraction as an equivalent fraction that has the LCD as its denominator. Then we will use the rule for adding fractions that have the same denominator.

WHY To add (or subtract) fractions, the fractions must have *like* denominators. ▼

Solution

Since the smallest number the denominators 7 and 3 divide exactly is 21, the LCD is 21.

$$\frac{1}{7} + \frac{2}{3} = \frac{1}{7} \cdot \frac{3}{3} + \frac{2}{3} \cdot \frac{7}{7}$$ To build $\frac{1}{7}$ and $\frac{2}{3}$ so that their denominators are 21, multiply each by a form of 1.

$$= \frac{3}{21} + \frac{14}{21}$$ Multiply the numerators. Multiply the denominators. The denominators are now the same.

$$= \frac{3 + 14}{21}$$ Add the numerators and write the sum over the common denominator 21.

$$= \frac{17}{21}$$ Since 17 and 21 have no common factors other than 1, this fraction is in simplest form.

EXAMPLE 5   Subtract: $\dfrac{5}{2} - \dfrac{7}{3}$

Strategy We will express each fraction as an equivalent fraction that has the LCD as its denominator. Then we will use the rule for subtracting fractions that have the same denominator.

WHY To add (or subtract) fractions, the fractions must have *like* denominators.

Solution

Since the smallest number the denominators 2 and 3 divide exactly is 6, the LCD is 6.

$$\frac{5}{2} - \frac{7}{3} = \frac{5}{2} \cdot \frac{3}{3} - \frac{7}{3} \cdot \frac{2}{2}$$ To build $\frac{5}{2}$ and $\frac{7}{3}$ so that their denominators are 6, multiply each by a form of 1.

$$= \frac{15}{6} - \frac{14}{6}$$ Multiply the numerators. Multiply the denominators. The denominators are now the same.

$$= \frac{15 - 14}{6}$$ Subtract the numerators and write the difference over the common denominator 6.

$$= \frac{1}{6}$$ This fraction is in simplest form.

Self Check 5

Subtract: $\dfrac{6}{7} - \dfrac{3}{5}$

Now Try Problem 31

EXAMPLE 6   Subtract: $\dfrac{2}{5} - \dfrac{11}{15}$

Strategy Since the smallest number the denominators 5 and 15 divide exactly is 15, the LCD is 15. We will only need to build an equivalent fraction for $\frac{2}{5}$.

WHY We do not have to build the fraction $\frac{11}{15}$ because it already has a denominator of 15.

Solution

$$\frac{2}{5} - \frac{11}{15} = \frac{2}{5} \cdot \frac{3}{3} - \frac{11}{15}$$ To build $\frac{2}{5}$ so that its denominator is 15, multiply it by a form of 1.

$$= \frac{6}{15} - \frac{11}{15}$$ Multiply the numerators. Multiply the denominators. The denominators are now the same.

$$= \frac{6 - 11}{15}$$ Subtract the numerators and write the difference over the common denominator 15.

$$= -\frac{5}{15}$$ If it is helpful, use the subtraction rule and add the opposite in the numerator: $6 + (-11) = -5$. Write the − sign in front of the fraction.

$$= -\frac{\overset{1}{\cancel{5}}}{3 \cdot \cancel{5}_{1}}$$ To simplify, factor 15 as $3 \cdot 5$. Then remove the common factor of 5 from the numerator and denominator.

$$= -\frac{1}{3}$$ Multiply the remaining factors in the denominator: $3 \cdot 1 = 3$.

Self Check 6

Subtract: $\dfrac{2}{3} - \dfrac{13}{6}$

Now Try Problem 35

> **Success Tip** In Example 6, did you notice that the denominator 5 is a factor of the denominator 15, and that the LCD is 15. In general, when adding (or subtracting) two fractions with different denominators, *if the smaller denominator is a factor of the larger denominator, the larger denominator is the LCD.*

> **Caution!** You might not have to build each fraction when adding or subtracting fractions with different denominators. For instance, the step in blue shown below is unnecessary when solving Example 6.
>
> $$\frac{2}{5} - \frac{11}{15} = \frac{2}{5} \cdot \mathbf{\frac{3}{3}} - \frac{11}{15} \cdot \cancel{\frac{1}{1}}$$

Self Check 7

Add: $-6 + \dfrac{3}{8}$

Now Try Problem 39

EXAMPLE 7   Add: $-5 + \dfrac{3}{4}$

Strategy We will write -5 as the fraction $\frac{-5}{1}$. Then we will follow the steps for adding fractions that have different denominators.

WHY The fractions $\frac{-5}{1}$ and $\frac{3}{4}$ have different denominators.

Solution

Since the smallest number the denominators 1 and 4 divide exactly is 4, the LCD is 4.

$$
\begin{aligned}
-5 + \frac{3}{4} &= \frac{-5}{1} + \frac{3}{4} && \text{Write } -5 \text{ as } \tfrac{-5}{1}. \\[2mm]
&= \frac{-5}{1} \cdot \frac{4}{4} + \frac{3}{4} && \text{To build } \tfrac{-5}{1} \text{ so that its denominator is 4, multiply it by a} \\
& && \text{form of 1.} \\[2mm]
&= \frac{-20}{4} + \frac{3}{4} && \text{Multiply the numerators. Multiply the denominators.} \\
& && \text{The denominators are now the same.} \\[2mm]
&= \frac{-20 + 3}{4} && \text{Add the numerators and write the sum over the} \\
& && \text{common denominator 4.} \\[2mm]
&= \frac{-17}{4} && \text{Use the rule for adding two integers with different signs:} \\
& && -20 + 3 = -17. \\[2mm]
&= -\frac{17}{4} && \text{Write the result with the } - \text{ sign in front: } \tfrac{-17}{4} = -\tfrac{17}{4}. \\
& && \text{This fraction is in simplest form.}
\end{aligned}
$$

3   Find the LCD to add and subtract fractions.

When we add or subtract fractions that have different denominators, the least common denominator is not always obvious. We can use a concept studied earlier to determine the LCD for more difficult problems that involve larger denominators. To illustrate this, let's find the least common denominator of $\frac{3}{8}$ and $\frac{1}{10}$. (Note, the LCD *is not* 80.)

We have learned that both 8 and 10 must divide the LCD exactly. This divisibility requirement should sound familiar. Recall the following fact from Section 1.6.

> **The Least Common Multiple (LCM)**
>
> The **least common multiple (LCM)** of two whole numbers is the smallest whole number that is divisible by both of those numbers.

Thus, the least common denominator of $\frac{3}{8}$ and $\frac{1}{10}$ is simply the *least common multiple* of 8 and 10.

We can find the LCM of 8 and 10 by listing multiples of the larger number, 10, until we find one that is divisible by the smaller number, 8. (This method is explained in Example 2 of Section 1.6.)

Multiples of 10:   10,   20,   30,   **40**,   50,   60,   ...

This is the first multiple of 10 that
is divisible by 8 (no remainder).

Since the LCM of 8 and 10 is 40, it follows that the LCD of $\frac{3}{8}$ and $\frac{1}{10}$ is 40.

We can also find the LCM of 8 and 10 using prime factorization. We begin by prime factoring 8 and 10. (This method is explained in Example 4 of Section 1.6.)

$8 = \boxed{2 \cdot 2 \cdot 2}$
$10 = 2 \cdot \boxed{5}$

The LCM of 8 and 10 is a product of prime factors, where each factor is used the greatest number of times it appears in any one factorization.

- We will use the factor 2 three times, because 2 appears three times in the factorization of 8. Circle $2 \cdot 2 \cdot 2$, as shown above.

- We will use the factor 5 once, because it appears one time in the factorization of 10. Circle 5 as shown above.

Since there are no other prime factors in either prime factorization, we have

Use 2 three times.
Use 5 one time.
$\text{LCM }(8, 10) = 2 \cdot 2 \cdot 2 \cdot 5 = 40$

Finding the LCD

The least common denominator (LCD) of a set of fractions is the least common multiple (LCM) of the denominators of the fractions. Two ways to find the LCM of the denominators are as follows:

- Write the multiples of the largest denominator in increasing order, until one is found that is divisible by the other denominators.

- Prime factor each denominator. The LCM is a product of prime factors, where each factor is used the greatest number of times it appears in any one factorization.

EXAMPLE 8   Add: $\dfrac{7}{15} + \dfrac{3}{10}$

Strategy We begin by expressing each fraction as an equivalent fraction that has the LCD for its denominator. Then we use the rule for adding fractions that have the same denominator.

WHY To add (or subtract) fractions, the fractions must have *like* denominators.

Solution

To find the LCD, we find the prime factorization of both denominators and use each prime factor the *greatest* number of times it appears in any one factorization:

$\left. \begin{array}{l} 15 = \boxed{3} \cdot \boxed{5} \\ 10 = \boxed{2} \cdot 5 \end{array} \right\} \text{LCD} = 2 \cdot 3 \cdot 5 = 30$

2 appears once in the factorization of 10.
3 appears once in the factorization of 15.
5 appears once in the factorizations of 15 and 10.

Self Check 8

Add: $\dfrac{1}{8} + \dfrac{5}{6}$

Now Try Problem 43

The LCD for $\dfrac{7}{15}$ and $\dfrac{3}{10}$ is 30.

$$\dfrac{7}{15} + \dfrac{3}{10} = \dfrac{7}{15} \cdot \dfrac{2}{2} + \dfrac{3}{10} \cdot \dfrac{3}{3}$$

To build $\frac{7}{15}$ and $\frac{3}{10}$ so that their denominators are 30, multiply each by a form of 1.

$$= \dfrac{14}{30} + \dfrac{9}{30}$$

Multiply the numerators. Multiply the denominators. The denominators are now the same.

$$= \dfrac{14 + 9}{30}$$

Add the numerators and write the sum over the common denominator 30.

$$= \dfrac{23}{30}$$

Since 23 and 30 have no common factors other than 1, this fraction is in simplest form.

Self Check 9

Subtract and simplify:

$\dfrac{21}{56} - \dfrac{9}{40}$

Now Try Problem 47

EXAMPLE 9 Subtract and simplify: $\dfrac{13}{28} - \dfrac{1}{21}$

Strategy We begin by expressing each fraction as an equivalent fraction that has the LCD for its denominator. Then we use the rule for subtracting fractions with *like* denominators.

WHY To add (or subtract) fractions, the fractions must have like denominators.

Solution

To find the LCD, we find the prime factorization of both denominators and use each prime factor the *greatest* number of times it appears in any one factorization:

$$\left. \begin{array}{l} 28 = \boxed{2 \cdot 2} \cdot \boxed{7} \\ 21 = \boxed{3} \cdot 7 \end{array} \right\} \text{LCD} = 2 \cdot 2 \cdot 3 \cdot 7 = 84$$

2 appears twice in the factorization of 28.
3 appears once in the factorization of 21.
7 appears once in the factorizations of 28 and 21.

The LCD for $\frac{13}{28}$ and $\frac{1}{21}$ is 84.

We will compare the prime factorizations of 28, 21, and the prime factorization of the LCD, 84, to determine what forms of 1 to use to build equivalent fractions for $\frac{13}{28}$ and $\frac{1}{21}$ with a denominator of 84.

$\text{LCD} = 2 \cdot 2 \cdot 3 \cdot 7$	$\text{LCD} = 2 \cdot 2 \cdot 3 \cdot 7$
Cover the prime factorization of 28. Since 3 is left uncovered, use $\frac{3}{3}$ to build $\frac{13}{28}$.	Cover the prime factorization of 21. Since $2 \cdot 2 = 4$ is left uncovered, use $\frac{4}{4}$ to build $\frac{1}{21}$.

$$\dfrac{13}{28} - \dfrac{1}{21} = \dfrac{13}{28} \cdot \dfrac{3}{3} - \dfrac{1}{21} \cdot \dfrac{4}{4}$$

To build $\frac{13}{28}$ and $\frac{1}{21}$ so that their denominators are 84, multiply each by a form of 1.

$$= \dfrac{39}{84} - \dfrac{4}{84}$$

Multiply the numerators. Multiply the denominators. The denominators are now the same.

$$= \dfrac{39 - 4}{84}$$

Subtract the numerators and write the difference over the common denominator.

$$= \dfrac{35}{84}$$

This fraction is not in simplest form.

$$= \dfrac{5 \cdot \overset{1}{\cancel{7}}}{2 \cdot 2 \cdot 3 \cdot \underset{1}{\cancel{7}}}$$

To simplify, factor 35 and 84. Then remove the common factor of 7 from the numerator and denominator.

Multiply the remaining factors in the numerator: $5 \cdot 1 = 5$. Multiply the remaining factors in the denominator: $2 \cdot 2 \cdot 3 \cdot 1 = 12$.

$$= \dfrac{5}{12}$$

4 Add and subtract algebraic fractions.

To add or subtract algebraic fractions, we use the same approach as with numerical fractions.

EXAMPLE 10 Perform each operation and simplify the result, if possible:

a. $\dfrac{x}{16} + \dfrac{3x}{16}$ **b.** $\dfrac{9}{25m} - \dfrac{2}{25m}$

Strategy We will add (or subtract) the numerators and write the sum (or difference) over the common denominator. Then, if possible, we will simplify the result.

WHY This is the rule for adding (or subtracting) fractions that have like denominators.

Solution

a. $\dfrac{x}{16} + \dfrac{3x}{16} = \dfrac{x + 3x}{16}$ Add the numerators and write the sum over the common denominator, 16.

$= \dfrac{4x}{16}$ Combine like terms in the numerator: x + 3x = 4x. This result can be simplified.

$= \dfrac{\overset{1}{\cancel{4}} \cdot x}{\underset{1}{\cancel{4}} \cdot 4}$ To simplify, factor 16 as 4 · 4 and then remove the common factor of 4 from the numerator and denominator.

$= \dfrac{x}{4}$ Multiply the remaining factors in the numerator: 1 · x = x. Multiply the remaining factors in the denominator: 1 · 4 = 4.

b. $\dfrac{9}{25m} - \dfrac{2}{25m} = \dfrac{9 - 2}{25m}$ Subtract the numerators and write the difference over the common denominator, 25m.

$= \dfrac{7}{25m}$ Since 7 and 25 have no common factors other than 1, the result is in simplified form.

Self Check 10

Perform each operation and simplify the result, if possible:

a. $\dfrac{2x}{15} + \dfrac{4x}{15}$

b. $\dfrac{11}{3n} - \dfrac{7}{3n}$

Now Try Problems 51 and 55

EXAMPLE 11 Perform each operation: **a.** $\dfrac{r}{3} + \dfrac{1}{4}$ **b.** $\dfrac{2}{y} - \dfrac{5}{18}$

Strategy We will use the procedure for adding and subtracting fractions that have unlike denominators. The first step is to determine the LCD.

WHY If we are to add (or subtract) fractions, their denominators must be the same. Since the denominators of these fractions are different, we cannot add (nor subtract) them in their present form.

Solution

a. The denominators of $\frac{r}{3}$ and $\frac{1}{4}$ are 3 and 4. By inspection, we see that the LCD is 3 · 4 = 12.

$\dfrac{r}{3} + \dfrac{1}{4} = \dfrac{r}{3} \cdot \dfrac{4}{4} + \dfrac{1}{4} \cdot \dfrac{3}{3}$ Build each fraction so that each has a denominator of 12.

$= \dfrac{4r}{12} + \dfrac{3}{12}$ Multiply the numerators. Multiply the denominators.

$= \dfrac{4r + 3}{12}$ Add the numerators and write the sum over the common denominator.

Self Check 11

Perform each operation:

a. $\dfrac{y}{5} + \dfrac{7}{9}$

b. $\dfrac{6}{a} - \dfrac{1}{8}$

Now Try Problems 59 and 63

Caution! The result, $\frac{4r + 3}{12}$, does not simplify. Do not make either of the two common mistakes shown below.

$$\frac{4r + 3}{12} = \frac{7r}{12}$$ 4r and 3 are not like terms. Don't add them.

$$\frac{4r + 3}{12} = \frac{4r + \overset{1}{3}}{4 \cdot \underset{1}{3}}$$ 3 is a term of the numerator, not a factor, and therefore cannot be removed.

b. The denominators of $\frac{2}{y}$ and $\frac{5}{18}$ are y and 18. By inspection, we see that the LCD is $y \cdot 18 = 18y$.

$$\frac{2}{y} - \frac{5}{18} = \frac{2}{y} \cdot \frac{\mathbf{18}}{\mathbf{18}} - \frac{5}{18} \cdot \frac{y}{y}$$ Build each fraction so that each has a denominator of 18y.

$$= \frac{36}{18y} - \frac{5y}{18y}$$ Multiply the numerators. Multiply the denominators.

$$= \frac{36 - 5y}{18y}$$ Subtract the numerators and write the difference over the common denominator.

Caution! The result, $\frac{36 - 5y}{18y}$, is in simplest form. Do not make the following common mistake to try to "simplify" it.

$$\frac{36 - 5y}{18y} = \frac{36 - 5\overset{1}{y}}{18\underset{1}{y}} = \frac{31}{18}$$ We cannot remove a common factor of y, because y is not a factor of the entire numerator. It is a factor of one term of the numerator.

5 Identify the greater of two fractions.

If two fractions have the same denominator, the fraction with the greater numerator is the greater fraction.

For example,

$$\frac{7}{8} > \frac{3}{8}$$ because $7 > 3$ $$-\frac{1}{3} > -\frac{2}{3}$$ because $-1 > -2$

If the denominators of two fractions are different, we need to write the fractions with a common denominator (preferably the LCD) before we can make a comparison.

Self Check 12

Which fraction is larger:
$\frac{7}{12}$ or $\frac{3}{5}$?

Now Try Problem 67

EXAMPLE 12   Which fraction is larger: $\frac{5}{6}$ or $\frac{7}{8}$?

Strategy We will express each fraction as an equivalent fraction that has the LCD for its denominator. Then we will compare their numerators.

WHY We cannot compare the fractions as given. They are not similar objects.

five-**sixths**   seven-**eighths**

Solution

Since the smallest number the denominators will divide exactly is 24, the LCD for $\frac{5}{6}$ and $\frac{7}{8}$ is 24.

$$\frac{5}{6} = \frac{5}{6} \cdot \frac{\mathbf{4}}{\mathbf{4}} \qquad \frac{7}{8} = \frac{7}{8} \cdot \frac{\mathbf{3}}{\mathbf{3}}$$ To build $\frac{5}{6}$ and $\frac{7}{8}$ so that their denominators are 24, multiply each by a form of 1.

$$= \frac{20}{24} \qquad\qquad = \frac{21}{24}$$ Multiply the numerators. Multiply the denominators.

Next, we compare the numerators. Since $21 > 20$, it follows that $\frac{21}{24}$ is greater than $\frac{20}{24}$. Thus, $\frac{7}{8} > \frac{5}{6}$.

6 Solve application problems by adding and subtracting fractions.

EXAMPLE 13   *Television Viewing Habits*   Students on a college campus were asked to estimate to the nearest hour how much television they watched each day. The results are given in the **circle graph** below (also called a **pie chart**). For example, the chart tells us that $\frac{1}{4}$ of those responding watched 1 hour per day. What fraction of the student body watches from 0 to 2 hours daily?

Refer to the circle graph for Example 13. Find the fraction of the student body that watches 2 or more hours of television daily.

Now Try Problems 75 and 116

Analyze

- $\frac{1}{6}$ of the student body watches no TV daily.   Given
- $\frac{1}{4}$ of the student body watches 1 hour of TV daily.   Given
- $\frac{7}{15}$ of the student body watches 2 hours of TV daily.   Given
- What fraction of the student body watches 0 to 2 hours of TV daily? Find

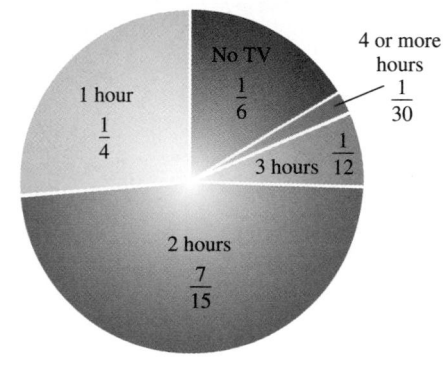

Form We translate the words of the problem to numbers and symbols.

The fraction of the student body that watches from 0 to 2 hours of TV daily	is equal to	the fraction that watches no TV daily	plus	the fraction that watches 1 hour of TV daily	plus	the fraction that watches 2 hours of TV daily.

$$
\begin{array}{c}
\text{The fraction} \\
\text{of the student} \\
\text{body that} \\
\text{watches from} \\
\text{0 to 2 hours} \\
\text{of TV daily}
\end{array}
\quad = \quad \frac{1}{6} \quad + \quad \frac{1}{4} \quad + \quad \frac{7}{15}
$$

Solve We must find the sum of three fractions with different denominators. To find the LCD, we prime factor the denominators and use each prime factor the *greatest* number of times it appears in any one factorization:

$$
\left.\begin{array}{l}
6 = 2 \cdot \textcircled{3} \\
4 = \textcircled{2 \cdot 2} \\
15 = 3 \cdot \textcircled{5}
\end{array}\right\} \text{LCD} = 2 \cdot 2 \cdot 3 \cdot 5 = 60
$$

2 appears twice in the factorization of 4.
3 appears once in the factorization of 6 and 15.
5 appears once in the factorization of 15.

The LCD for $\frac{1}{6}$, $\frac{1}{4}$, and $\frac{7}{15}$ is 60.

$$
\frac{1}{6} + \frac{1}{4} + \frac{7}{15} = \frac{1}{6} \cdot \frac{\mathbf{10}}{\mathbf{10}} + \frac{1}{4} \cdot \frac{\mathbf{15}}{\mathbf{15}} + \frac{7}{15} \cdot \frac{\mathbf{4}}{\mathbf{4}}
$$
Build each fraction so that its denominator is 60.

$$
= \frac{10}{60} + \frac{15}{60} + \frac{28}{60}
$$
Multiply the numerators. Multiply the denominators. The denominators are now the same.

$$
= \frac{10 + 15 + 28}{60}
$$
Add the numerators and write the sum over the common denominator 60.

$$
= \frac{53}{60}
$$
This fraction is in simplest form.

$$
\begin{array}{r}
\frac{1}{}10 \\
15 \\
+\ 28 \\
\hline
53
\end{array}
$$

State The fraction of the student body that watches 0 to 2 hours of TV daily is $\frac{53}{60}$.

Check We can check by estimation. The result, $\frac{53}{60}$, is approximately $\frac{50}{60}$, which simplifies to $\frac{5}{6}$. The red, yellow, and blue shaded areas appear to shade about $\frac{5}{6}$ of the pie chart. The result seems reasonable.

ANSWERS TO SELF CHECKS

1. a. $\frac{1}{2}$ **b.** $\frac{7}{9}$ **2.** $-\frac{6}{11}$ **3.** $\frac{2}{3}$ **4.** $\frac{9}{10}$ **5.** $\frac{9}{35}$ **6.** $-\frac{3}{2}$ **7.** $-\frac{45}{8}$ **8.** $\frac{23}{24}$ **9.** $\frac{3}{20}$

10. a. $\frac{2x}{5}$ **b.** $\frac{4}{3n}$ **11. a.** $\frac{9y+35}{45}$ **b.** $\frac{48-a}{8a}$ **12.** $\frac{3}{5}$ **13.** $\frac{7}{12}$

THINK IT THROUGH   *Budgets*

"Putting together a budget is crucial if you don't want to spend your way into serious problems. You're also developing a habit that can serve you well throughout your life."

Liz Pulliam Weston, MSN Money

The circle graph below shows a suggested budget for new college graduates as recommended by Springboard, a nonprofit consumer credit counseling service. What fraction of net take-home pay should be spent on housing?

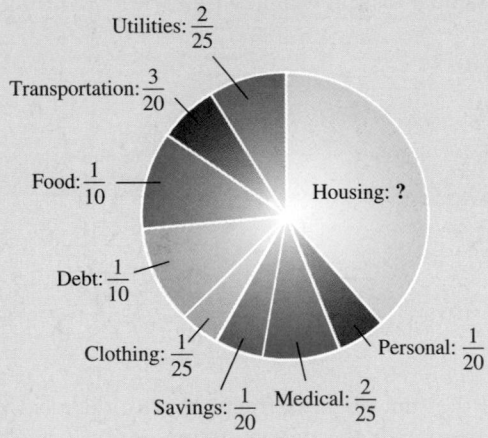

Utilities: $\frac{2}{25}$

Transportation: $\frac{3}{20}$

Food: $\frac{1}{10}$

Debt: $\frac{1}{10}$

Clothing: $\frac{1}{25}$

Savings: $\frac{1}{20}$ Medical: $\frac{2}{25}$

Personal: $\frac{1}{20}$

Housing: **?**

SECTION 4.4   STUDY SET

VOCABULARY

Fill in the blanks.

1. Because the denominators of $\frac{3}{8}$ and $\frac{7}{8}$ are the same number, we say that they have a _____ denominator.

2. Consider the solution below. To _____ an equivalent fraction with a denominator of 18, we multiply $\frac{4}{9}$ by a 1 in the form of ▯.

$$\frac{4}{9} = \frac{4}{9} \cdot \frac{2}{2}$$

$$= \frac{8}{18}$$

CONCEPTS

Fill in the blanks.

3. a. To add (or subtract) fractions that have the same denominator, add (or subtract) their _____ and write the sum (or difference) over the _____ denominator. _____ the result, if possible.

b. To add (or subtract) fractions that have different denominators, we express each fraction as an equivalent fraction that has the _____ for its denominator. Then we use the rule for adding (subtracting) fractions that have the _____ denominator.

4. The least common denominator for a set of fractions is the _____ number each denominator will divide exactly (no remainder).

5. Consider $\frac{3}{4}$. By what form of 1 should we multiply the numerator and denominator to express it as an equivalent fraction with a denominator of 36?

6. The *denominators* of two fractions are given. Find the least common denominator.

a. 2 and 3   **b.** 3 and 5

c. Fill in the blank: When adding (or subtracting) two fractions with different denominators, if the smaller denominator is a factor of the larger denominator, the _____ denominator is the LCD.

7. Consider the following prime factorizations:

$$24 = 2 \cdot 2 \cdot 2 \cdot 3$$
$$90 = 2 \cdot 3 \cdot 3 \cdot 5$$

For any one factorization, what is the greatest number of times

a. a 5 appears?

b. a 3 appears?

c. a 2 appears?

8. a. The *denominators* of two fractions have their prime-factored forms shown below. Fill in the blanks to find the LCD for the fractions.

$$\left.\begin{array}{l}20 = 2 \cdot 2 \cdot 5\\30 = 2 \cdot 3 \cdot 5\end{array}\right\}\text{LCD} = \blacksquare \cdot \blacksquare \cdot \blacksquare \cdot \blacksquare = \blacksquare$$

b. The *denominators* of three fractions have their prime-factored forms shown below. Fill in the blanks to find the LCD for the fractions.

$$\left.\begin{array}{l}20 = 2 \cdot 2 \cdot 5\\30 = 2 \cdot 3 \cdot 5\\90 = 2 \cdot 3 \cdot 3 \cdot 5\end{array}\right\}\text{LCD} = \blacksquare \cdot \blacksquare \cdot \blacksquare \cdot \blacksquare \cdot \blacksquare = \blacksquare$$

NOTATION

Fill in the blanks to complete each solution.

9. $\dfrac{2}{5} + \dfrac{1}{7} = \dfrac{2}{5} \cdot \dfrac{\blacksquare}{\blacksquare} + \dfrac{1}{7} \cdot \dfrac{5}{5}$

$= \dfrac{\blacksquare}{35} + \dfrac{5}{\blacksquare}$

$= \dfrac{\blacksquare + \blacksquare}{35}$

$= \dfrac{\blacksquare}{35}$

10. $\dfrac{7x}{8} - \dfrac{2}{9} = \dfrac{7x}{8} \cdot \dfrac{9}{\blacksquare} - \dfrac{2}{9} \cdot \dfrac{\blacksquare}{8}$

$= \dfrac{63x}{\blacksquare} - \dfrac{16}{\blacksquare}$

$= \dfrac{63x - 16}{\blacksquare}$

GUIDED PRACTICE

Perform each operation and simplify, if possible. See Example 1.

11. $\dfrac{4}{9} + \dfrac{1}{9}$ **12.** $\dfrac{3}{7} + \dfrac{1}{7}$

13. $\dfrac{3}{8} + \dfrac{1}{8}$ **14.** $\dfrac{7}{12} + \dfrac{1}{12}$

15. $\dfrac{11}{15} - \dfrac{7}{15}$ **16.** $\dfrac{10}{21} - \dfrac{5}{21}$

17. $\dfrac{11}{20} - \dfrac{3}{20}$ **18.** $\dfrac{7}{18} - \dfrac{5}{18}$

Subtract and simplify, if possible. See Example 2.

19. $-\dfrac{11}{5} - \left(-\dfrac{8}{5}\right)$ **20.** $-\dfrac{15}{9} - \left(-\dfrac{11}{9}\right)$

21. $-\dfrac{7}{21} - \left(-\dfrac{2}{21}\right)$ **22.** $-\dfrac{21}{25} - \left(-\dfrac{9}{25}\right)$

Perform the operations and simplify, if possible. See Example 3.

23. $\dfrac{19}{40} - \dfrac{3}{40} - \dfrac{1}{40}$ **24.** $\dfrac{11}{24} - \dfrac{1}{24} - \dfrac{7}{24}$

25. $\dfrac{13}{33} + \dfrac{1}{33} + \dfrac{7}{33}$ **26.** $\dfrac{21}{50} + \dfrac{1}{50} + \dfrac{13}{50}$

Add and simplify, if possible. See Example 4.

27. $\dfrac{1}{3} + \dfrac{1}{7}$ **28.** $\dfrac{1}{4} + \dfrac{1}{5}$

29. $\dfrac{2}{5} + \dfrac{1}{9}$ **30.** $\dfrac{2}{7} + \dfrac{1}{2}$

Subtract and simplify, if possible. See Example 5.

31. $\dfrac{4}{5} - \dfrac{3}{4}$ **32.** $\dfrac{2}{3} - \dfrac{3}{5}$

33. $\dfrac{3}{4} - \dfrac{2}{7}$ **34.** $\dfrac{6}{7} - \dfrac{2}{3}$

Subtract and simplify, if possible. See Example 6.

35. $\dfrac{11}{12} - \dfrac{2}{3}$ **36.** $\dfrac{11}{18} - \dfrac{1}{6}$

37. $\dfrac{9}{14} - \dfrac{1}{7}$ **38.** $\dfrac{13}{15} - \dfrac{2}{3}$

Add and simplify, if possible. See Example 7.

39. $-2 + \dfrac{5}{9}$ **40.** $-3 + \dfrac{5}{8}$

41. $-3 + \dfrac{9}{4}$ **42.** $-1 + \dfrac{7}{10}$

Add and simplify, if possible. See Example 8.

43. $\dfrac{1}{6} + \dfrac{5}{8}$ **44.** $\dfrac{7}{12} + \dfrac{3}{8}$

45. $\dfrac{4}{9} + \dfrac{5}{12}$ **46.** $\dfrac{1}{9} + \dfrac{5}{6}$

Subtract and simplify, if possible. See Example 9.

47. $\dfrac{9}{10} - \dfrac{3}{14}$ **48.** $\dfrac{11}{12} - \dfrac{11}{30}$

49. $\dfrac{11}{12} - \dfrac{7}{15}$ **50.** $\dfrac{7}{15} - \dfrac{5}{12}$

Perform each operation and simplify, if possible.
See Example 10.

51. $\dfrac{x}{12} + \dfrac{5x}{12}$ **52.** $\dfrac{d}{16} + \dfrac{7d}{16}$

53. $\dfrac{2c}{7} + \dfrac{3c}{7}$ **54.** $\dfrac{4n}{19} + \dfrac{8n}{19}$

55. $\dfrac{16}{21m} - \dfrac{11}{21m}$ **56.** $\dfrac{9}{17a} - \dfrac{5}{17a}$

57. $\dfrac{16}{15y} - \dfrac{7}{15y}$ **58.** $\dfrac{11}{12r} - \dfrac{1}{12r}$

Perform each operation and simplify, if possible. See Example 11.

59. $\dfrac{a}{5} + \dfrac{2}{3}$ **60.** $\dfrac{d}{4} + \dfrac{2}{7}$

61. $\dfrac{1}{8} + \dfrac{x}{3}$ **62.** $\dfrac{1}{9} + \dfrac{m}{4}$

63. $\dfrac{3}{n} - \dfrac{5}{12}$ **64.** $\dfrac{6}{f} - \dfrac{4}{5}$

65. $\dfrac{8}{9} - \dfrac{11}{d}$ **66.** $\dfrac{12}{13} - \dfrac{6}{r}$

Determine which fraction is larger. See Example 12.

67. $\dfrac{3}{8}$ or $\dfrac{5}{16}$ **68.** $\dfrac{5}{6}$ or $\dfrac{7}{12}$

69. $\dfrac{4}{5}$ or $\dfrac{2}{3}$ **70.** $\dfrac{7}{9}$ or $\dfrac{4}{5}$

71. $\dfrac{7}{9}$ or $\dfrac{11}{12}$ **72.** $\dfrac{3}{8}$ or $\dfrac{5}{12}$

73. $\dfrac{23}{20}$ or $\dfrac{7}{6}$ **74.** $\dfrac{19}{15}$ or $\dfrac{5}{4}$

Add and simplify, if possible. See Example 13.

75. $\dfrac{1}{6} + \dfrac{5}{18} + \dfrac{2}{9}$ **76.** $\dfrac{1}{10} + \dfrac{1}{8} + \dfrac{1}{5}$

77. $\dfrac{4}{15} + \dfrac{2}{3} + \dfrac{1}{6}$ **78.** $\dfrac{1}{2} + \dfrac{3}{5} + \dfrac{3}{20}$

TRY IT YOURSELF

Perform each operation and simplify, if possible.

79. $-\dfrac{1}{12} - \left(-\dfrac{5}{12}\right)$ **80.** $-\dfrac{1}{16} - \left(-\dfrac{15}{16}\right)$

81. $\dfrac{3n}{4} + \dfrac{2}{3}$ **82.** $\dfrac{4b}{5} + \dfrac{2}{3}$

83. $\dfrac{12}{25} - \dfrac{1}{25} - \dfrac{1}{25}$ **84.** $\dfrac{7}{9} + \dfrac{1}{9} + \dfrac{1}{9}$

85. $-\dfrac{7}{20} - \dfrac{1}{5}$ **86.** $-\dfrac{5}{8} - \dfrac{1}{3}$

87. $-\dfrac{7}{16} + \dfrac{1}{4}$ **88.** $-\dfrac{17}{20} + \dfrac{4}{5}$

89. $\dfrac{x}{9} + \dfrac{2x}{9}$ **90.** $\dfrac{a}{8} + \dfrac{5a}{8}$

91. $\dfrac{2}{3} + \dfrac{4}{5} + \dfrac{5}{6}$ **92.** $\dfrac{3}{4} + \dfrac{2}{5} + \dfrac{3}{10}$

93. $\dfrac{9}{20} - \dfrac{1}{30}$ **94.** $\dfrac{5}{6} - \dfrac{3}{10}$

95. $\dfrac{27}{50} + \dfrac{5}{16}$ **96.** $\dfrac{49}{50} - \dfrac{15}{16}$

97. $\dfrac{13}{20} - \dfrac{1}{5}$

98. $\dfrac{71}{100} - \dfrac{1}{10}$

99. $-\dfrac{3}{4} - 5$

100. $-2 - \dfrac{7}{8}$

101. $\dfrac{7}{30} - \dfrac{19}{75}$

102. $\dfrac{73}{75} - \dfrac{31}{30}$

103. Find the difference of $\dfrac{13}{d}$ and $\dfrac{3}{2}$.

104. Find the sum of $\dfrac{25}{x}$ and $\dfrac{4}{3}$.

105. Subtract $\dfrac{5}{12}$ from $\dfrac{2}{15}$.

106. What is the sum of $\dfrac{11}{24}$ and $\dfrac{7}{36}$ increased by $\dfrac{5}{48}$?

APPLICATIONS

107. BOTANY To determine the effects of smog on tree development, a scientist cut down a pine tree and measured the width of the growth rings for the last two years.

   a. What was the growth over this two-year period?

   b. What is the difference in the widths of the two rings?

$\dfrac{5}{32}$ in. $\dfrac{1}{16}$ in.

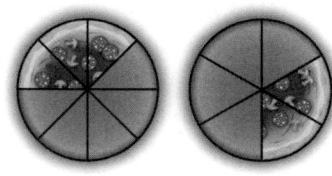

108. GARAGE DOOR OPENERS What is the difference in strength between a $\frac{1}{3}$-hp and a $\frac{1}{2}$-hp garage door opener?

109. MAGAZINE COVERS The page design for the magazine cover shown below includes a blank strip at the top, called a *header*, and a blank strip at the bottom of the page, called a *footer*. How much page length is lost because of the header and footer?

Page length

$\dfrac{3}{8}$ in. header

$\dfrac{5}{16}$ in. footer

110. DELIVERY TRUCKS A truck can safely carry a one-ton load. Should it be used to deliver one-half ton of sand, one-third ton of gravel, and one-fifth ton of cement in one trip to a job site?

111. DINNERS A family bought two large pizzas for dinner. Some pieces of each pizza were not eaten, as shown.

   a. What fraction of the first pizza was not eaten?

   b. What fraction of the second pizza was not eaten?

   c. What fraction of a pizza was left?

   d. Could the family have been fed with just one pizza?

112. GASOLINE BARRELS Three identical-sized barrels are shown below. If the contents of two of the barrels are poured into the empty third barrel, what fraction of the third barrel will be filled?

113. WEIGHTS AND MEASURES A consumer protection agency determines the accuracy of butcher shop scales by placing a known three-quarter-pound weight on the scale and then comparing that to the scale's readout. According to the illustration, by how much is this scale off? Does it result in undercharging or overcharging customers on their meat purchases?

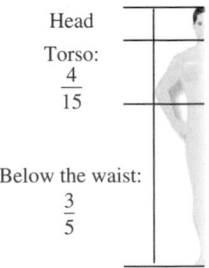

114. FIGURE DRAWING As an aid in drawing the human body, artists divide the body into three parts. Each part is then expressed as a fraction of the total body height. For example, the torso is $\frac{4}{15}$ of the body height. What fraction of body height is the head?

Head

Torso:
$\frac{4}{15}$

Below the waist:
$\frac{3}{5}$

115. Suppose you work as a school guidance counselor at a community college and your department has conducted a survey of the full-time students to learn more about their study habits. As part of a *Power Point* presentation of the survey results to the school board, you show the following circle graph. At that time, you are asked, "What fraction of the full-time students study 2 hours or more daily?" What would you answer?

from Campus to Careers
School Guidance Counselor

© iStockphoto.com/Catherine Yeulet

More than 2 hr
$\frac{3}{10}$

2 hr
$\frac{2}{5}$

Less than 1 hr
$\frac{1}{10}$

$\frac{1}{5}$

1 hr

116. HEALTH STATISTICS The circle graph below shows the leading causes of death in the United States for 2006. For example, $\frac{13}{50}$ of all of the deaths that year were caused by heart disease. What fraction of all the deaths were caused by heart disease, cancer, or stroke, combined?

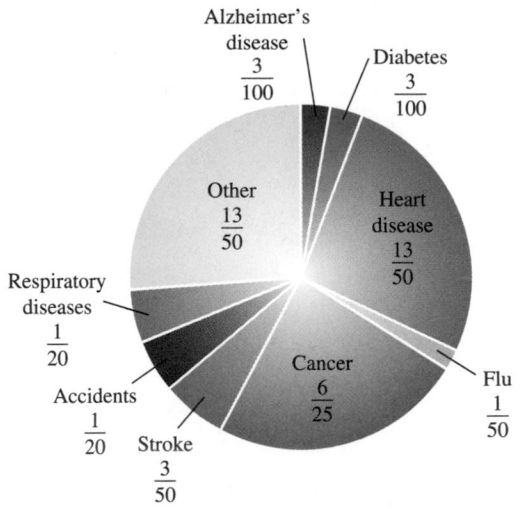

Alzheimer's disease
$\frac{3}{100}$

Diabetes
$\frac{3}{100}$

Other
$\frac{13}{50}$

Heart disease
$\frac{13}{50}$

Respiratory diseases
$\frac{1}{20}$

Accidents
$\frac{1}{20}$

Stroke
$\frac{3}{50}$

Cancer
$\frac{6}{25}$

Flu
$\frac{1}{50}$

Source: National Center for Health Statistics

117. MUSICAL NOTES The notes used in music have fractional values. Their names and the symbols used to represent them are shown in illustration (a). In common time, the values of the notes in each measure must add to 1. Is the measure in illustration (b) complete?

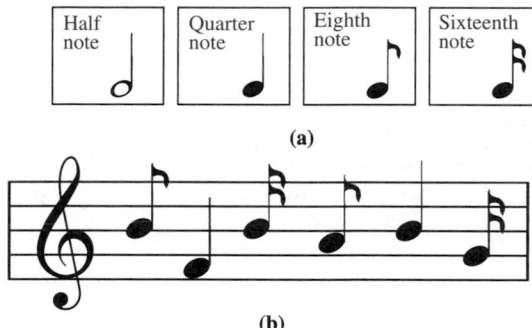

(a)

(b)

118. TOOLS A mechanic likes to hang his wrenches above his tool bench in order of narrowest to widest. What is the proper order of the wrenches in the illustration?

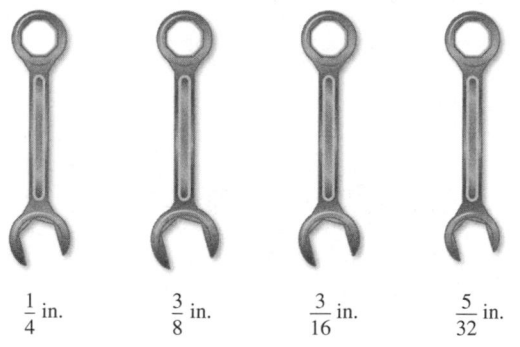

$\frac{1}{4}$ in. $\frac{3}{8}$ in. $\frac{3}{16}$ in. $\frac{5}{32}$ in.

119. TIRE TREAD A mechanic measured the tire tread depth on each of the tires on a car and recorded them on the form shown below. (The letters LF stand for *left front*, RR stands for *right rear*, and so on.)
   a. Which tire has the most tread?
   b. Which tire has the least tread?

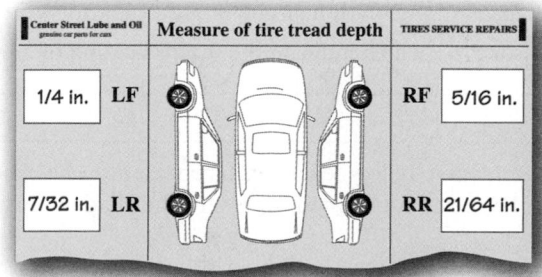

120. HIKING The illustration below shows the length of each part of a three-part hike. Rank the lengths of the parts from longest to shortest.

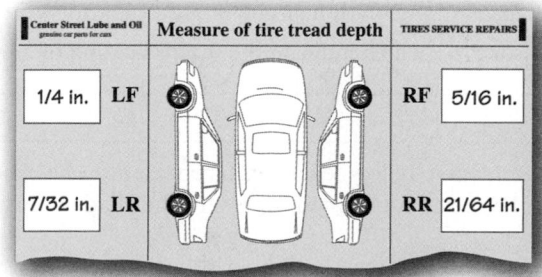

WRITING

121. Explain why we cannot add or subtract the fractions $\frac{2}{9}$ and $\frac{2}{5}$ as they are written.

122. To multiply fractions, must they have the same denominators? Explain why or why not. Give an example.

123. Explain the error in the following work.

$$\frac{5x + 2}{6} = \frac{5x + 2}{2 \cdot 3}$$

$$= \frac{5x + \overset{1}{\cancel{2}}}{\underset{1}{\cancel{2}} \cdot 3}$$

$$= \frac{5x + 1}{3}$$

124. How do we compare the sizes of two fractions with different denominators?

REVIEW

Perform each operation and simplify, if possible.

125. a. $\dfrac{1}{4} + \dfrac{1}{8}$ **b.** $\dfrac{1}{4} - \dfrac{1}{8}$

     c. $\dfrac{1}{4} \cdot \dfrac{1}{8}$ **d.** $\dfrac{1}{4} \div \dfrac{1}{8}$

126. a. $\dfrac{5}{21} + \dfrac{3}{14}$ **b.** $\dfrac{5}{21} - \dfrac{3}{14}$

     c. $\dfrac{5}{21} \cdot \dfrac{3}{14}$ **d.** $\dfrac{5}{21} \div \dfrac{3}{14}$

Objectives

1. Identify the whole-number and fractional parts of a mixed number.
2. Write mixed numbers as improper fractions.
3. Write improper fractions as mixed numbers.
4. Graph fractions and mixed numbers on a number line.
5. Multiply and divide mixed numbers.
6. Solve application problems by multiplying and dividing mixed numbers.

SECTION **4.5**

Multiplying and Dividing Mixed Numbers

In the next two sections, we show how to add, subtract, multiply, and divide *mixed numbers*. These numbers are widely used in daily life.

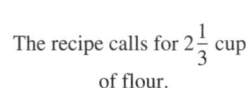

The recipe calls for $2\frac{1}{3}$ cups of flour.

(Read as "two and one-third.")

It took $3\frac{3}{4}$ hours to paint the living room.

(Read as "three and three-fourths.")

The entrance to the park is $1\frac{1}{2}$ miles away.

(Read as "one and one-half.")

1 Identify the whole-number and fractional parts of a mixed number.

A **mixed number** is the *sum* of a whole number and a proper fraction. For example, $3\frac{3}{4}$ is a mixed number.

$$3\frac{3}{4} \qquad = \qquad 3 \qquad + \qquad \frac{3}{4}$$

Mixed number   Whole-number part   Fractional part

Mixed numbers can be represented by shaded regions. In the illustration below, each rectangular region outlined in black represents one whole. To represent $3\frac{3}{4}$, we shade 3 *whole* rectangular regions and 3 out of 4 *parts* of another.

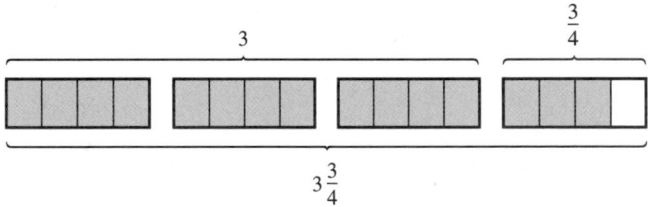

> ***Caution!*** Note that $3\frac{3}{4}$ means $3 + \frac{3}{4}$, even though the + symbol is not written. Do not confuse $3\frac{3}{4}$ with $3 \cdot \frac{3}{4}$ or $3\left(\frac{3}{4}\right)$, which indicate the multiplication of 3 by $\frac{3}{4}$.

Self Check 1

In the illustration below, each oval region represents one whole. Write an improper fraction and a mixed number to represent the shaded portion.

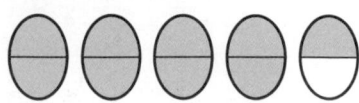

Now Try Problem 19

EXAMPLE 1   In the illustration below, each disk represents one whole. Write an improper fraction and a mixed number to represent the shaded portion.

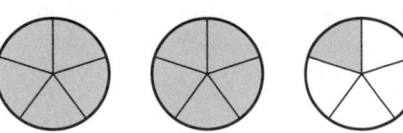

Strategy We will determine the number of equal parts into which a disk is divided. Then we will determine how many of those *parts* are shaded and how many of the *whole* disks are shaded.

WHY To write an improper fraction, we need to find its numerator and its denominator. To write a mixed number, we need to find its whole number part and its fractional part.

Solution

Since each disk is divided into 5 equal parts, the denominator of the improper fraction is 5. Since a total of 11 of those parts are shaded, the numerator is 11, and we say that

$\dfrac{11}{5}$ is shaded.   Write: $\dfrac{\text{total number of parts shaded}}{\text{number of equal parts in one disk}}$

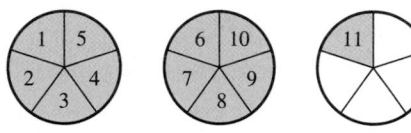

Since 2 whole disks are shaded, the whole number part of the mixed number is 2. Since 1 out of 5 of the parts of the last disk is shaded, the fractional part of the mixed number is $\frac{1}{5}$, and we say that

$2\dfrac{1}{5}$ is shaded.

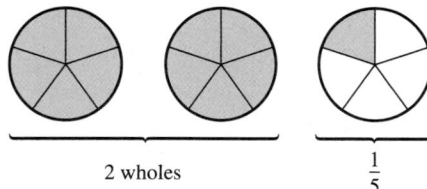

2 wholes $\dfrac{1}{5}$

In this section, we will work with negative as well as positive mixed numbers. For example, the negative mixed number $-3\frac{3}{4}$ could be used to represent $3\frac{3}{4}$ feet below sea level. Think of $-3\frac{3}{4}$ as $-3 - \frac{3}{4}$ or as $-3 + \left(-\frac{3}{4}\right)$.

2 Write mixed numbers as improper fractions.

In Example 1, we saw that the shaded portion of the illustration can be represented by the mixed number $2\frac{1}{5}$ and by the improper fraction $\frac{11}{5}$. To develop a procedure to write any mixed number as an improper fraction, consider the following steps that show how to do this for $2\frac{1}{5}$. The objective is to find how many *fifths* that the mixed number $2\frac{1}{5}$ represents.

$2\dfrac{1}{5} = 2 + \dfrac{1}{5}$ Write the mixed number $2\frac{1}{5}$ as a sum.

$\phantom{2\dfrac{1}{5}} = \dfrac{2}{1} + \dfrac{1}{5}$ Write 2 as a fraction: $2 = \frac{2}{1}$.

$\phantom{2\dfrac{1}{5}} = \dfrac{2}{1} \cdot \dfrac{5}{5} + \dfrac{1}{5}$ To build $\frac{2}{1}$ so that its denominator is 5, multiply it by a form of 1.

$\phantom{2\dfrac{1}{5}} = \dfrac{10}{5} + \dfrac{1}{5}$ Multiply the numerators.
Multiply the denominators.

$\phantom{2\dfrac{1}{5}} = \dfrac{11}{5}$ Add the numerators and write the sum over
the common denominator 5.

Thus, $2\frac{1}{5} = \frac{11}{5}$.

We can obtain the same result with far less work. To change $2\frac{1}{5}$ to an improper fraction, we simply multiply 5 by 2 and add 1 to get the numerator, and keep the denominator of 5.

$$2\frac{1}{5} = \frac{5 \cdot 2 + 1}{5} = \frac{10 + 1}{5} = \frac{11}{5}$$

This example illustrates the following procedure.

Writing a Mixed Number as an Improper Fraction

To write a mixed number as an improper fraction:

1. Multiply the denominator of the fraction by the whole-number part.
2. Add the numerator of the fraction to the result from Step 1.
3. Write the sum from Step 2 over the original denominator.

Self Check 2

Write the mixed number $3\frac{3}{8}$ as an improper fraction.

Now Try Problems 23 and 27

EXAMPLE 2   Write the mixed number $7\frac{5}{6}$ as an improper fraction.

Strategy We will use the 3-step procedure to find the improper fraction.

WHY It's faster than writing $7\frac{5}{6}$ as $7 + \frac{5}{6}$, building to get an LCD, and adding.

Solution
To find the numerator of the improper fraction, multiply 6 by 7, and add 5 to that result. The denominator of the improper fraction is the same as the denominator of the fractional part of the mixed number.

Step 2: add

$$7\frac{5}{6} = \frac{6 \cdot 7 + 5}{6} = \frac{42 + 5}{6} = \frac{47}{6}$$

By the order of operations rule, multiply first, and then add in the numerator.

Step 1: multiply   Step 3: Use the same denominator

To write a *negative mixed number* in fractional form, ignore the − sign and use the method shown in Example 2 on the positive mixed number. Once that procedure is completed, write a − sign in front of the result. For example,

$$-6\frac{1}{4} = -\frac{25}{4} \qquad\qquad -1\frac{9}{10} = -\frac{19}{10} \qquad\qquad -12\frac{3}{8} = -\frac{99}{8}$$

3 Write improper fractions as mixed numbers.

To write an improper fraction as a mixed number, we must find two things: the *whole-number part* and the *fractional part* of the mixed number. To develop a procedure to do this, let's consider the improper fraction $\frac{7}{3}$. To find the number of groups of 3 in 7, we can divide 7 by 3. This will find the whole-number part of the mixed number. The remainder is the numerator of the fractional part of the mixed number.

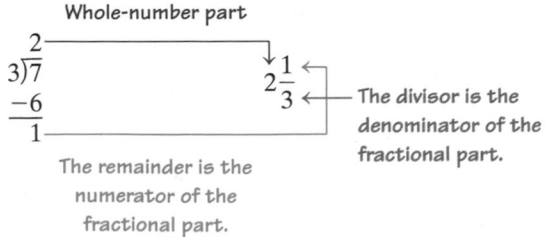

Whole-number part

$$3\overline{)7} \qquad 2\frac{1}{3}$$

The remainder is the numerator of the fractional part.

The divisor is the denominator of the fractional part.

This example suggests the following procedure.

Writing an Improper Fraction as a Mixed Number

To write an improper fraction as a mixed number:

1. Divide the numerator by the denominator to obtain the whole-number part.
2. The remainder over the divisor is the fractional part.

EXAMPLE 3   Write each improper fraction as a mixed number or a whole number: **a.** $\dfrac{29}{6}$ **b.** $\dfrac{40}{16}$ **c.** $\dfrac{84}{3}$ **d.** $-\dfrac{9}{5}$

Strategy We will divide the numerator by the denominator and write the remainder over the divisor.

WHY A fraction bar indicates division.

Solution

a. To write $\dfrac{29}{6}$ as a mixed number, divide 29 by 6:

$$
\begin{array}{r}
4 \leftarrow \text{The whole-number part is 4.}\\
6\overline{)29}\\
-24\\
\hline
5 \leftarrow \text{Write the remainder 5 over the}\\
\text{divisor 6 to get the fractional part.}
\end{array}
$$

Thus, $\dfrac{29}{6} = 4\dfrac{5}{6}$.

b. To write $\dfrac{40}{16}$ as a mixed number, divide 40 by 16:

$$
\begin{array}{r}
2\\
16\overline{)40}\\
-32\\
\hline
8
\end{array}
$$

Thus, $\dfrac{40}{16} = 2\dfrac{8}{16} = 2\dfrac{1}{2}$.   Simplify the fractional part: $\dfrac{8}{16} = \dfrac{\overset{1}{\cancel{8}}}{2 \cdot \underset{1}{\cancel{8}}} = \dfrac{1}{2}$.

c. For $\dfrac{84}{3}$, divide 84 by 3:

$$
\begin{array}{r}
28\\
3\overline{)84}\\
-6\\
\hline
24\\
-24\\
\hline
0 \leftarrow \text{Since the remainder is 0, the improper fraction represents a whole number.}
\end{array}
$$

Thus, $\dfrac{84}{3} = 28$.

d. To write $-\dfrac{9}{5}$ as a mixed number, ignore the – sign, and use the method for the positive improper fraction $\dfrac{9}{5}$. Once that procedure is completed, write a – sign in front of the result.

$$
\begin{array}{r}
1\\
5\overline{)9}\\
-5\\
\hline
4
\end{array}
$$

Thus, $-\dfrac{9}{5} = -1\dfrac{4}{5}$.

Self Check 3

Write each improper fraction as a mixed number or a whole number:

a. $\dfrac{31}{7}$ **b.** $\dfrac{50}{26}$

c. $\dfrac{51}{3}$ **d.** $-\dfrac{10}{3}$

Now Try Problems 31, 35, 39, and 43

4 Graph fractions and mixed numbers on a number line.

In Chapters 1 and 2, we graphed whole numbers and integers on a number line. Fractions and mixed numbers can also be graphed on a number line.

Self Check 4

Graph $-1\frac{7}{8}$, $-\frac{2}{3}$, $\frac{3}{5}$, and $\frac{9}{4}$ on a number line.

Now Try Problem 47

EXAMPLE 4 Graph $-2\frac{3}{4}$, $-1\frac{1}{2}$, $-\frac{1}{8}$, and $\frac{13}{5}$ on a number line.

Strategy We will locate the position of each fraction and mixed number on the number line and draw a bold dot.

WHY To *graph a number* means to make a drawing that represents the number.

Solution

- Since $-2\frac{3}{4} < -2$, the graph of $-2\frac{3}{4}$ is to the left of -2 on the number line.
- The number $-1\frac{1}{2}$ is between -1 and -2.
- The number $-\frac{1}{8}$ is less than 0.
- Expressed as a mixed number, $\frac{13}{5} = 2\frac{3}{5}$.

5 Multiply and divide mixed numbers.

We will use the same procedures for multiplying and dividing mixed numbers as those that were used in Sections 4.2 and 4.3 to multiply and divide fractions. However, we must write the mixed numbers as improper fractions before we actually multiply or divide.

> ### Multiplying and Dividing Mixed Numbers
>
> To multiply or divide mixed numbers, first change the mixed numbers to improper fractions. Then perform the multiplication or division of the fractions. Write the result as a mixed number or a whole number in simplest form.

The sign rules for multiplying and dividing integers also hold for multiplying and dividing mixed numbers.

Self Check 5

Multiply and simplify, if possible.

a. $3\frac{1}{3} \cdot 2\frac{1}{3}$ b. $9\frac{3}{5} \cdot \left(3\frac{3}{4}\right)$

c. $-4\frac{5}{6}(2)$

Now Try Problems 51, 55, and 57

EXAMPLE 5 Multiply and simplify, if possible.

a. $1\frac{3}{4} \cdot 2\frac{1}{3}$ b. $5\frac{1}{5} \cdot \left(1\frac{2}{13}\right)$ c. $-4\frac{1}{9}(3)$

Strategy We will write the mixed numbers and whole numbers as improper fractions.

WHY Then we can use the rule for multiplying two fractions from Section 3.2.

Solution

a.
$$1\frac{3}{4} \cdot 2\frac{1}{3} = \frac{7}{4} \cdot \frac{7}{3}$$ Write $1\frac{3}{4}$ and $2\frac{1}{3}$ as improper fractions.

$$= \frac{7 \cdot 7}{4 \cdot 3}$$ Use the rule for multiplying two fractions. Multiply the numerators and the denominators.

$$= \frac{49}{12}$$ Since there are no common factors to remove, perform the multiplication in the numerator and in the denominator. The result is an improper fraction.

$$= 4\frac{1}{12}$$ Write the improper fraction $\frac{49}{12}$ as a mixed number.

$$\begin{array}{r} 4 \\ 12\overline{)49} \\ -48 \\ \hline 1 \end{array}$$

b. $5\dfrac{1}{5}\left(1\dfrac{2}{13}\right) = \dfrac{26}{5} \cdot \dfrac{15}{13}$ Write $5\dfrac{1}{5}$ and $1\dfrac{2}{13}$ as improper fractions.

$\qquad = \dfrac{26 \cdot 15}{5 \cdot 13}$ Multiply the numerators.
Multiply the denominators.

$\qquad = \dfrac{2 \cdot 13 \cdot 3 \cdot 5}{5 \cdot 13}$ To prepare to simplify, factor 26 as $2 \cdot 13$ and 15 as $3 \cdot 5$.

$\qquad = \dfrac{2 \cdot \overset{1}{\cancel{13}} \cdot 3 \cdot \overset{1}{\cancel{5}}}{\underset{1}{\cancel{5}} \cdot \underset{1}{\cancel{13}}}$ Remove the common factors of 13 and 5 from the numerator and denominator.

$\qquad = \dfrac{6}{1}$ Multiply the remaining factors in the numerator:
$2 \cdot 1 \cdot 3 \cdot 1 = 6$.
Multiply the remaining factors in the denominator: $1 \cdot 1 = 1$.

$\qquad = 6$ Any whole number divided by 1 remains the same.

c. $-4\dfrac{1}{9} \cdot 3 = -\dfrac{37}{9} \cdot \dfrac{3}{1}$ Write $-4\dfrac{1}{9}$ as an improper fraction and write 3 as a fraction.

$\qquad = -\dfrac{37 \cdot 3}{9 \cdot 1}$ Multiply the numerators and multiply the denominators.
Since the fractions have unlike signs, make the answer negative.

$\qquad = -\dfrac{37 \cdot \overset{1}{\cancel{3}}}{\underset{1}{\cancel{3}} \cdot 3 \cdot 1}$ To simplify, factor 9 as $3 \cdot 3$, and then remove the common factor of 3 from the numerator and denominator.

$\qquad = -\dfrac{37}{3}$ Multiply the remaining factors in the numerator and in the denominator.
The result is an improper fraction.

$\qquad = -12\dfrac{1}{3}$ Write the negative improper fraction $-\dfrac{37}{3}$ as a negative mixed number.

$$\begin{array}{r} 12 \\ 3\overline{)37} \\ \underline{-3} \\ 7 \\ \underline{-6} \\ 1 \end{array}$$

Success Tip   We can use rounding to check the results when multiplying mixed numbers. If the fractional part of the mixed number is $\dfrac{1}{2}$ *or greater*, round up by adding 1 to the whole-number part and dropping the fraction. If the fractional part of the mixed number is less than $\dfrac{1}{2}$, round down by dropping the fraction and using only the whole-number part. To check the answer $4\dfrac{1}{12}$ from Example 5, part a, we proceed as follows:

$$1\dfrac{3}{4} \cdot 2\dfrac{1}{3} \approx 2 \cdot 2 = 4 \qquad \begin{array}{l} \text{Since } \dfrac{3}{4} \text{ is greater than } \dfrac{1}{2}, \text{ round } 1\dfrac{3}{4} \text{ up to 2.} \\ \text{Since } \dfrac{1}{3} \text{ is less than } \dfrac{1}{2}, \text{ round } 2\dfrac{1}{3} \text{ down to 2.} \end{array}$$

Since $4\dfrac{1}{12}$ is close to 4, it is a reasonable answer.

EXAMPLE 6   Divide and simplify, if possible:

a. $-3\dfrac{3}{8} \div \left(-2\dfrac{1}{4}\right)$ **b.** $1\dfrac{11}{16} \div \dfrac{3}{4}$

Strategy   We will write the mixed numbers as improper fractions.

WHY   Then we can use the rule for dividing two fractions from Section 4.3.

Solution

a. $-3\dfrac{3}{8} \div \left(-2\dfrac{1}{4}\right) = -\dfrac{27}{8} \div \left(-\dfrac{9}{4}\right)$ Write $-3\dfrac{3}{8}$ and $-2\dfrac{1}{4}$ as improper fractions.

$\qquad\qquad\qquad\qquad = -\dfrac{27}{8}\left(-\dfrac{4}{9}\right)$ Use the rule for dividing two fractions.:
Multiply $-\dfrac{27}{8}$ by the reciprocal of $-\dfrac{9}{4}$, which is $-\dfrac{4}{9}$.

Self Check 6

Divide and simplify, if possible:

a. $-3\dfrac{4}{15} \div \left(-2\dfrac{1}{10}\right)$

b. $5\dfrac{3}{5} \div \dfrac{7}{8}$

Now Try **Problems 59 and 65**

$$= \frac{27}{8}\left(\frac{4}{9}\right)$$ Since the product of two negative fractions is positive, drop both − signs and continue.

$$= \frac{27 \cdot 4}{8 \cdot 9}$$ Multiply the numerators.
Multiply the denominators.

$$= \frac{3 \cdot \overset{1}{\cancel{9}} \cdot \overset{1}{\cancel{4}}}{2 \cdot \underset{1}{\cancel{4}} \cdot \underset{1}{\cancel{9}}}$$ To simplify, factor 27 as 3 · 9 and 8 as 2 · 4. Then remove the common factors of 9 and 4 from the numerator and denominator.

$$= \frac{3}{2}$$ Multiply the remaining factors in the numerator: 3 · 1 · 1 = 3. Multiply the remaining factors in the denominator: 2 · 1 · 1 = 2.

$$= 1\frac{1}{2}$$ Write the improper fraction $\frac{3}{2}$ as a mixed number by dividing 3 by 2.

b. $1\frac{11}{16} \div \frac{3}{4} = \frac{27}{16} \div \frac{3}{4}$ Write $1\frac{11}{16}$ as an improper fraction.

$$= \frac{27}{16} \cdot \frac{4}{3}$$ Multiply $\frac{27}{16}$ by the reciprocal of $\frac{3}{4}$, which is $\frac{4}{3}$.

$$= \frac{27 \cdot 4}{16 \cdot 3}$$ Multiply the numerators.
Multiply the denominators.

$$= \frac{\overset{1}{\cancel{3}} \cdot 9 \cdot \overset{1}{\cancel{4}}}{\underset{1}{\cancel{4}} \cdot 4 \cdot \underset{1}{\cancel{3}}}$$ To simplify, factor 27 as 3 · 9 and 16 as 4 · 4 . Then remove the common factors of 3 and 4 from the numerator and denominator.

$$= \frac{9}{4}$$ Multiply the remaining factors in the numerator and in the denominator. The result is an improper fraction.

$$= 2\frac{1}{4}$$ Write the improper fraction $\frac{9}{4}$ as a mixed number by dividing 9 by 4.

6 Solve application problems by multiplying and dividing mixed numbers.

Self Check 7

BUMPER STICKERS A rectangular-shaped bumper sticker is $8\frac{1}{4}$ inches long by $3\frac{1}{4}$ inches wide. Find its area.

Now Try Problem 99

EXAMPLE 7   *Toys*   The dimensions of the rectangular-shaped screen of an Etch-a-Sketch are shown in the illustration below. Find the area of the screen.

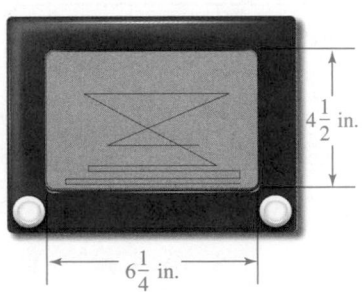

$4\frac{1}{2}$ in.

$6\frac{1}{4}$ in.

Strategy To find the area, we will multiply $6\frac{1}{4}$ by $4\frac{1}{2}$.

WHY The formula for the area of a rectangle is Area = length · width.

Solution

$$A = lw \qquad \text{This is the formula for the area of a rectangle.}$$

$$= 6\frac{1}{4} \cdot 4\frac{1}{2} \qquad \text{Substitute } 6\frac{1}{4} \text{ for } l \text{ and } 4\frac{1}{2} \text{ for } w.$$

$$= \frac{25}{4} \cdot \frac{9}{2} \qquad \text{Write } 6\frac{1}{4} \text{ and } 4\frac{1}{2} \text{ as improper fractions.}$$

$$= \frac{25 \cdot 9}{4 \cdot 2} \qquad \begin{array}{l}\text{Multiply the numerators.}\\\text{Multiply the denominators.}\end{array}$$

$$= \frac{225}{8} \qquad \begin{array}{l}\text{Since there are no common factors to remove,}\\\text{perform the multiplication in the numerator and in}\\\text{the denominator. The result is an improper fraction.}\end{array}$$

$$= 28\frac{1}{8} \qquad \text{Write the improper fraction } \frac{225}{8} \text{ as a mixed number.}$$

$$\begin{array}{r} 28 \\ 8)\overline{225} \\ -16 \\ \hline 65 \\ -64 \\ \hline 1 \end{array}$$

The area of the screen of an Etch-a-Sketch is $28\frac{1}{8}$ in.^{2}.

EXAMPLE 8   *Government Grants*   If $12\frac{1}{2}$ million is to be split equally among five cities to fund recreation programs, how much will each city receive?

Analyze

- There is $12\frac{1}{2}$ million in grant money.   Given
- 5 cities will split the money equally.   Given
- How much grant money will each city receive?   Find

Form   The key phrase *split equally* suggests division. We translate the words of the problem to numbers and symbols.

The amount of money that each city will receive (in millions of dollars)	is equal to	the total amount of grant money (in millions of dollars)	divided by	the number of cities receiving money.
The amount of money that each city will receive (in millions of dollars)	=	$12\frac{1}{2}$	÷	5

Solve   To find the quotient, we will express $12\frac{1}{2}$ and 5 as fractions and then use the rule for dividing two fractions.

$$12\frac{1}{2} \div 5 = \frac{25}{2} \div \frac{5}{1} \qquad \text{Write } 12\frac{1}{2} \text{ as an improper fraction, and write 5 as a fraction.}$$

$$= \frac{25}{2} \cdot \frac{1}{5} \qquad \text{Multiply by the reciprocal of } \frac{5}{1}, \text{ which is } \frac{1}{5}.$$

$$= \frac{25 \cdot 1}{2 \cdot 5} \qquad \begin{array}{l}\text{Multiply the numerators.}\\\text{Multiply the denominators.}\end{array}$$

$$= \frac{\overset{1}{\cancel{5}} \cdot 5 \cdot 1}{2 \cdot \underset{1}{\cancel{5}}} \qquad \begin{array}{l}\text{To simplify, factor 25 as 5 · 5. Then remove the common}\\\text{factor of 5 from the numerator and denominator.}\end{array}$$

$$= \frac{5}{2} \qquad \begin{array}{l}\text{Multiply the remaining factors in the numerator.}\\\text{Multiply the remaining factors in the denominator.}\end{array}$$

$$= 2\frac{1}{2} \qquad \begin{array}{l}\text{Write the improper fraction } \frac{5}{2} \text{ as a mixed number}\\\text{by dividing 5 by 2. The units are in millions of dollars.}\end{array}$$

Self Check 8

TV INTERVIEWS An $18\frac{3}{4}$-minute taped interview with an actor was played in equally long segments over 5 consecutive nights on a celebrity news program. How long was each interview segment?

Now Try Problem 107

State Each city will receive $\$2\frac{1}{2}$ million in grant money.

Check We can estimate to check the result. If there was $10 million in grant money, each city would receive $\frac{\$10\ \text{million}}{5}$, or $2 million. Since there is actually $\$12\frac{1}{2}$ million in grant money, the answer that each city would receive $\$2\frac{1}{2}$ million seems reasonable. ■

ANSWERS TO SELF CHECKS

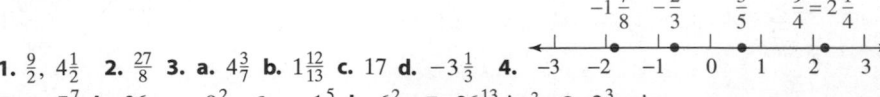

1. $\frac{9}{2}$, $4\frac{1}{2}$ 2. $\frac{27}{8}$ 3. a. $4\frac{3}{7}$ b. $1\frac{12}{13}$ c. 17 d. $-3\frac{1}{3}$ 4.
5. a. $7\frac{7}{9}$ b. 36 c. $-9\frac{2}{3}$ 6. a. $1\frac{5}{9}$ b. $6\frac{2}{5}$ 7. $26\frac{13}{16}$ in.² 8. $3\frac{3}{4}$ min

SECTION 4.5 STUDY SET

VOCABULARY

Fill in the blanks.

1. A _____ number, such as $8\frac{4}{5}$, is the sum of a whole number and a proper fraction.

2. In the mixed number $8\frac{4}{5}$, the _____-number part is 8 and the _____ part is $\frac{4}{5}$.

3. The numerator of an _____ fraction is greater than or equal to its denominator.

4. To _____ a number means to locate its position on the number line and highlight it using a dot.

CONCEPTS

5. What signed mixed number could be used to describe each situation?

   a. A temperature of five and one-third degrees above zero

   b. The depth of a sprinkler pipe that is six and seven-eighths inches below the sidewalk

6. What signed mixed number could be used to describe each situation?

   a. A rain total two and three-tenths of an inch lower than the average

   b. Three and one-half minutes after the liftoff of a rocket

Fill in the blanks.

7. To write a mixed number as an improper fraction:

   1. _____ the denominator of the fraction by the whole-number part.

   2. _____ the numerator of the fraction to the result from Step 1.

   3. Write the sum from Step 2 over the original _____.

8. To write an improper fraction as a mixed number:

   1. _____ the numerator by the denominator to obtain the whole-number part.

   2. The _____ over the divisor is the fractional part.

9. What fractions have been graphed on the number line?

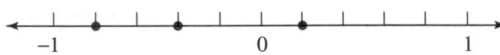

10. What mixed numbers have been graphed on the number line?

11. Fill in the blank: To multiply or divide mixed numbers, first change the mixed numbers to _____ fractions. Then perform the multiplication or division of the fractions as usual.

12. Simplify the fractional part of each mixed number.

    a. $11\frac{2}{4}$

    b. $1\frac{3}{9}$

    c. $7\frac{15}{27}$

13. Use *estimation* to determine whether the following answer seems reasonable:

$$4\frac{1}{5} \cdot 2\frac{5}{7} = 7\frac{2}{35}$$

14. What is the formula for the

    a. area of a rectangle?

    b. area of a triangle?

NOTATION

15. Fill in the blanks.

    a. We read $5\frac{11}{16}$ as "five _____ eleven-_____."

    b. We read $-4\frac{2}{3}$ as "_____ four and _____-thirds."

16. Determine the sign of the result. *You do not have to find the answer.*

    a. $1\frac{1}{9}\left(-7\frac{3}{14}\right)$

    b. $-3\frac{4}{15} \div \left(-1\frac{5}{6}\right)$

Fill in the blanks to complete each solution.

17. Multiply:

$$5\frac{1}{4} \cdot 1\frac{1}{7} = \frac{21}{\boxed{}} \cdot \frac{\boxed{}}{7}$$

$$= \frac{21 \cdot \boxed{}}{\boxed{} \cdot 7}$$

$$= \frac{3 \cdot \overset{1}{\cancel{7}} \cdot 2 \cdot \overset{1}{\cancel{}}}{\underset{1}{\cancel{}} \cdot \underset{1}{\cancel{7}}}$$

$$= \frac{\boxed{}}{1}$$

$$= \boxed{}$$

18. Divide:

$$-5\frac{5}{6} \div 2\frac{1}{12} = -\frac{\boxed{}}{6} \div \frac{25}{\boxed{}}$$

$$= -\frac{\boxed{}}{6} \cdot \frac{12}{\boxed{}}$$

$$= -\frac{35 \cdot 12}{6 \cdot \boxed{}}$$

$$= -\frac{\overset{1}{\cancel{5}} \cdot \boxed{} \cdot 2 \cdot \overset{1}{\cancel{6}}}{\underset{1}{\cancel{6}} \cdot \underset{1}{\cancel{5}} \cdot \boxed{}}$$

$$= -\frac{\boxed{}}{5}$$

$$= -2\frac{\boxed{}}{5}$$

GUIDED PRACTICE

Each region outlined in black represents one whole. Write an improper fraction and a mixed number to represent the shaded portion. **See Example 1.**

19.

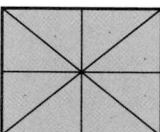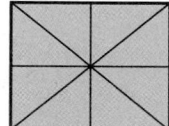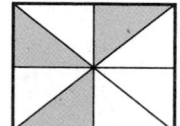

20.

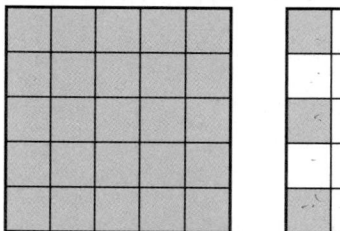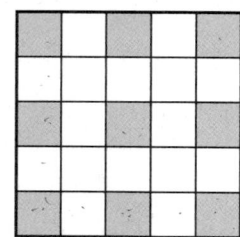

21.

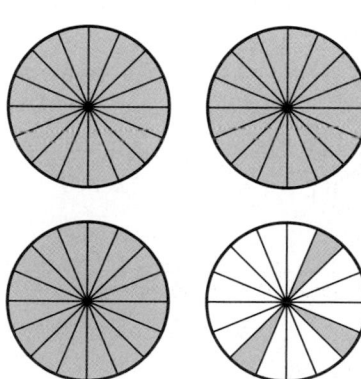

22.

Write each mixed number as an improper fraction.
See Example 2.

23. $6\frac{1}{2}$

24. $8\frac{2}{3}$

25. $20\frac{4}{5}$

26. $15\frac{3}{8}$

27. $-7\frac{5}{9}$

28. $-7\frac{1}{12}$

29. $-8\frac{2}{3}$

30. $-9\frac{3}{4}$

Write each improper fraction as a mixed number or a whole number. Simplify the result, if possible. See Example 3.

31. $\frac{13}{4}$

32. $\frac{41}{6}$

33. $\frac{28}{5}$

34. $\frac{28}{3}$

35. $\frac{42}{9}$

36. $\frac{62}{8}$

37. $\frac{84}{8}$

38. $\frac{93}{9}$

39. $\frac{52}{13}$

40. $\frac{80}{16}$

41. $\frac{34}{17}$

42. $\frac{38}{19}$

43. $-\frac{58}{7}$

44. $-\frac{33}{7}$

45. $-\frac{20}{6}$

46. $-\frac{28}{8}$

Graph the given numbers on a number line. See Example 4.

47. $-2\frac{8}{9}, 1\frac{2}{3}, \frac{16}{5}, -\frac{1}{2}$

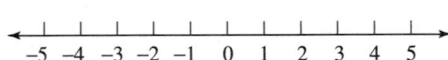

48. $-\frac{3}{4}, -3\frac{1}{4}, \frac{5}{2}, 4\frac{3}{4}$

49. $3\frac{1}{7}, -\frac{98}{99}, -\frac{10}{3}, \frac{3}{2}$

50. $-2\frac{1}{5}, \frac{4}{5}, -\frac{11}{3}, \frac{17}{4}$

Multiply and simplify, if possible. See Example 5.

51. $3\frac{1}{2} \cdot 2\frac{1}{3}$

52. $1\frac{5}{6} \cdot 1\frac{1}{2}$

53. $2\frac{2}{5}\left(3\frac{1}{12}\right)$

54. $\frac{40}{16}\left(\frac{26}{5}\right)$

55. $6\frac{1}{2} \cdot 1\frac{3}{13}$

56. $12\frac{3}{5} \cdot 1\frac{3}{7}$

57. $-2\frac{1}{2}(4)$

58. $-3\frac{3}{4}(8)$

Divide and simplify, if possible. See Example 6.

59. $-1\frac{13}{15} \div \left(-4\frac{1}{5}\right)$

60. $-2\frac{5}{6} \div \left(-8\frac{1}{2}\right)$

61. $15\frac{1}{3} \div 2\frac{2}{9}$

62. $6\frac{1}{4} \div 3\frac{3}{4}$

63. $1\frac{3}{4} \div \frac{3}{4}$

64. $5\frac{3}{5} \div \frac{9}{10}$

65. $1\frac{7}{24} \div \frac{7}{8}$

66. $4\frac{1}{2} \div \frac{3}{17}$

▌ TRY IT YOURSELF

Perform each operation and simplify, if possible.

67. $-6 \cdot 2\frac{7}{24}$

68. $-7 \cdot 1\frac{3}{28}$

69. $-6\frac{3}{5} \div 7\frac{1}{3}$

70. $-4\frac{1}{4} \div 4\frac{1}{2}$

71. $\left(1\frac{2}{3}\right)^2$

72. $\left(3\frac{1}{2}\right)^2$

73. $8 \div 3\frac{1}{5}$

74. $15 \div 3\frac{1}{3}$

75. $-20\frac{1}{4} \div \left(-1\frac{11}{16}\right)$

76. $-2\frac{7}{10} \div \left(-1\frac{1}{14}\right)$

77. $3\frac{1}{16} \cdot 4\frac{4}{7}$

78. $5\frac{3}{5} \cdot 1\frac{11}{14}$

79. Find the quotient of $-4\frac{1}{2}$ and $2\frac{1}{4}$.

80. Find the quotient of 25 and $-10\frac{5}{7}$.

81. $2\frac{1}{2}\left(-3\frac{1}{3}\right)$ **82.** $\left(-3\frac{1}{4}\right)\left(1\frac{1}{5}\right)$

83. $2\frac{5}{8} \cdot \frac{5}{27}$ **84.** $3\frac{1}{9} \cdot \frac{3}{32}$

85. $6\frac{1}{4} \div 20$ **86.** $4\frac{2}{5} \div 11$

87. Find the product of $1\frac{2}{3}$, 6, and $-\frac{1}{8}$.

88. Find the product of $-\frac{5}{6}$, -8, and $-2\frac{1}{10}$.

89. $\left(-1\frac{1}{3}\right)^3$

90. $\left(-1\frac{1}{5}\right)^3$

APPLICATIONS

91. In the illustration below, each barrel represents one whole.

   a. Write a mixed number to represent the shaded portion.

   b. Write an improper fraction to represent the shaded portion.

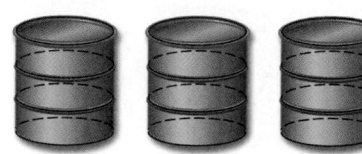

92. Draw $\frac{17}{8}$ pizzas.

93. DIVING Fill in the blank with a mixed number to describe the dive shown below: forward [   ] somersaults

94. PRODUCT LABELING Several mixed numbers appear on the label shown below. Write each mixed number as an improper fraction.

95. READING METERS

   a. Use a mixed number to describe the value to which the arrow is currently pointing.

   b. If the arrow moves twelve tick marks to the left, to what value will it be pointing?

96. READING METERS

   a. Use a mixed number to describe the value to which the arrow is currently pointing.

   b. If the arrow moves up six tick marks, to what value will it be pointing?

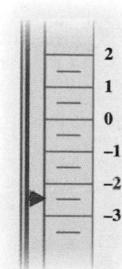

97. ONLINE SHOPPING A mother is ordering a pair of jeans for her daughter from the screen shown below. If the daughter's height is $60\frac{3}{4}$ in. and her waist is $24\frac{1}{2}$ in., on what size and what cut (regular or slim) should the mother point and click?

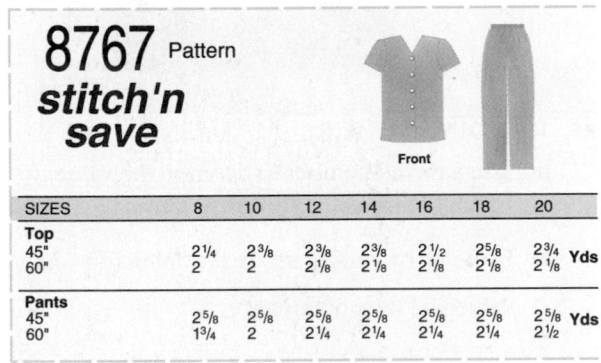

Girl's jeans- regular cut						
Size	7	8	10	12	14	16
Height	50-52	52-54	54-56	56¼-58½	59-61	61-62
Waist	22¼-22¾	22¾-23¼	23¾-24¼	24¾-25¼	25¾-26¼	26¼-28

Girl's jeans- slim cut						
Size	7	8	10	12	14	16
Height	50-52	52-54	54-56	56½-58½	59-61	61-62
Waist	20¾-21¼	21¼-21¾	22¼-22¾	23¼-23¾	24¼-24¾	25-26½

To order:
Point arrow to proper size/cut and click

98. SEWING Use the following table to determine the number of yards of fabric needed . . .

   a. to make a size 16 top if the fabric to be used is 60 inches wide.

   b. to make size 18 pants if the fabric to be used is 45 inches wide.

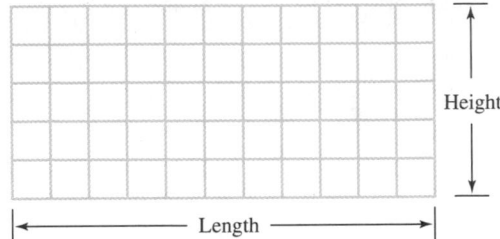

8767 Pattern
stitch'n save

Front

SIZES	8	10	12	14	16	18	20	
Top								
45"		2¼	2⅜	2⅜	2⅜	2½	2⅝	2¾	**Yds**
60"		2	2	2⅛	2⅛	2⅛	2⅛	2⅛	
Pants								
45"		2⅝	2⅝	2⅝	2⅝	2⅝	2⅝	2⅝	**Yds**
60"		1¾	2	2¼	2¼	2¼	2¼	2½	

99. LICENSE PLATES Find the area of the license plate shown below.

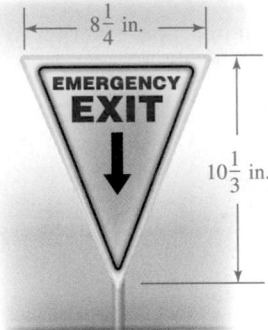

$12\frac{1}{4}$ in.

$6\frac{1}{4}$ in.

WB COUNTY   UTAH   10
123 ABC

100. GRAPH PAPER Mathematicians use specially marked paper, called graph paper, when drawing figures. It is made up of squares that are $\frac{1}{4}$-inch long by $\frac{1}{4}$-inch high.

   a. Find the length of the piece of graph paper shown below.

   b. Find its height.

   c. What is the area of the piece of graph paper?

Height

Length

101. EMERGENCY EXITS The following sign marks the emergency exit on a school bus. Find the area of the sign.

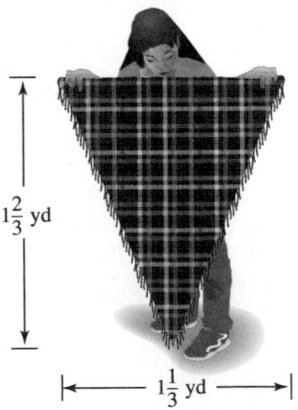

$8\frac{1}{4}$ in.

EMERGENCY EXIT

$10\frac{1}{3}$ in.

102. CLOTHING DESIGN Find the number of square yards of material needed to make the triangular-shaped shawl shown in the illustration.

$1\frac{2}{3}$ yd

$1\frac{1}{3}$ yd

103. CALORIES A company advertises that its mints contain only $3\frac{1}{5}$ calories a piece. What is the calorie intake if you eat an entire package of 20 mints?

104. CEMENT MIXERS A cement mixer can carry $9\frac{1}{2}$ cubic yards of concrete. If it makes 8 trips to a job site, how much concrete will be delivered to the site?

105. SHOPPING In the illustration, what is the cost of buying the fruit in the scale? Give your answer in cents and in dollars.

Oranges
84 cents a pound

106. PICTURE FRAMES How many inches of molding is needed to make the square picture frame below?

$10\frac{1}{8}$ in.

107. BREAKFAST CEREAL A box of cereal contains about $13\frac{3}{4}$ cups. Refer to the nutrition label shown below and determine the recommended size of one serving.

Nutrition Facts
Serving size : ? cups
Servings per container: 11

108. BREAKFAST CEREAL A box of cereal contains about $14\frac{1}{4}$ cups. Refer to the nutrition label shown below. Determine how many servings there are for children under 4 in one box.

Nutrition Facts
Serving size
Children under 4: $\frac{3}{4}$ cup

Servings per Container
Children Under 4: ?

109. CATERING How many people can be served $\frac{1}{3}$-pound hamburgers if a caterer purchases 200 pounds of ground beef?

110. SUBDIVISIONS A developer donated to the county 100 of the 1,000 acres of land she owned. She divided the remaining acreage into $1\frac{1}{3}$-acre lots. How many lots were created?

111. HORSE RACING The race tracks on which thoroughbred horses run are marked off in $\frac{1}{8}$-mile-long segments called *furlongs*. How many furlongs are there in a $1\frac{1}{16}$-mile race?

112. FIRE ESCAPES Part of the fire escape stairway for one story of an office building is shown below. Each riser is $7\frac{1}{2}$ inches high and each story of the building is 105 inches high.

a. How many stairs are there in one story of the fire escape stairway?

b. If the building has 43 stories, how many stairs are there in the entire fire escape stairway?

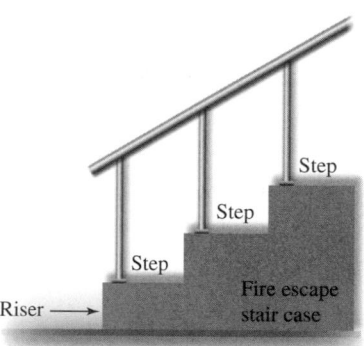

Step

Step

Step

Riser →

Fire escape
stair case

WRITING

113. Explain the difference between $2\frac{3}{4}$ and $2\left(\frac{3}{4}\right)$.

114. Give three examples of how you use mixed numbers in daily life.

REVIEW

Find the LCM of the given numbers.

115. 5, 12, 15   **116.** 8, 12, 16

Find the GCF of the given numbers.

117. 12, 68, 92   **118.** 24, 36, 40

Objectives

1 Add mixed numbers.

2 Add mixed numbers in vertical form.

3 Subtract mixed numbers.

4 Solve application problems by adding and subtracting mixed numbers.

SECTION 4.6
Adding and Subtracting Mixed Numbers

In this section, we discuss several methods for adding and subtracting mixed numbers.

1 Add mixed numbers.

We can add mixed numbers by writing them as improper fractions. To do so, we follow these steps.

Adding Mixed Numbers: Method 1

1. Write each mixed number as an improper fraction.
2. Write each improper fraction as an equivalent fraction with a denominator that is the LCD.
3. Add the fractions.
4. Write the result as a mixed number, if desired.

Method 1 works well when the whole-number parts of the mixed numbers are small.

Self Check 1

Add: $3\frac{2}{3} + 1\frac{1}{5}$

Now Try Problem 13

EXAMPLE 1   Add: $4\frac{1}{6} + 2\frac{3}{4}$

Strategy We will write each mixed number as an improper fraction, and then use the rule for adding two fractions that have different denominators.

WHY We cannot add the mixed numbers as they are; their fractional parts are not similar objects.

$$4\frac{1}{6} + 2\frac{3}{4}$$

Four and one-sixth ⟶   ⟵ Two and three-fourths

Solution

$$4\frac{1}{6} + 2\frac{3}{4} = \frac{25}{6} + \frac{11}{4}$$ Write $4\frac{1}{6}$ and $2\frac{3}{4}$ as improper fractions.

By inspection, we see that the lowest common denominator is 12.

$$= \frac{25\cdot 2}{6\cdot 2} + \frac{11\cdot 3}{4\cdot 3}$$ To build $\frac{25}{6}$ and $\frac{11}{4}$ so that their denominators are 12, multiply each by a form of 1.

$$= \frac{50}{12} + \frac{33}{12}$$ Multiply the numerators. Multiply the denominators.

$$= \frac{83}{12}$$ Add the numerators and write the sum over the common denominator 12. The result is an improper fraction.

$$= 6\frac{11}{12}$$ Write the improper fraction $\frac{83}{12}$ as a mixed number.

$$\begin{array}{r} 6 \\ 12\overline{)83} \\ -72 \\ \hline 11 \end{array}$$

Success Tip We can use rounding to check the results when adding (or subtracting) mixed numbers. To check the answer $6\frac{11}{12}$ from Example 1, we proceed as follows:

$$4\frac{1}{6} + 2\frac{3}{4} \approx 4 + 3 = 7$$

Since $\frac{1}{6}$ is less than $\frac{1}{2}$, round $4\frac{1}{6}$ down to 4.
Since $\frac{3}{4}$ is greater than $\frac{1}{2}$, round $2\frac{3}{4}$ up to 3.

Since $6\frac{11}{12}$ is close to 7, it is a reasonable answer.

EXAMPLE 2

Add: $-3\frac{1}{8} + 1\frac{1}{2}$

Self Check 2

Add: $-4\frac{1}{12} + 2\frac{1}{4}$

Now Try Problem 17

Strategy We will write each mixed number as an improper fraction, and then use the rule for adding two fractions that have different denominators.

WHY We cannot add the mixed numbers as they are; their fractional parts are not similar objects.

$$-3\frac{1}{8} + 1\frac{1}{2}$$

Negative three and one-eighth ⟶   ⟵ One and one-half

Solution

$$-3\frac{1}{8} + 1\frac{1}{2} = -\frac{25}{8} + \frac{3}{2}$$ Write $-3\frac{1}{8}$ and $1\frac{1}{2}$ as improper fractions.

Since the smallest number the denominators 8 and 2 divide exactly is 8, the LCD is 8. We will only need to build an equivalent fraction for $\frac{3}{2}$.

$$= -\frac{25}{8} + \frac{3}{2} \cdot \frac{4}{4}$$ To build $\frac{3}{2}$ so that its denominator is 8, multiply it by a form of 1.

$$= -\frac{25}{8} + \frac{12}{8}$$ Multiply the numerators. Multiply the denominators.

$$= \frac{-25 + 12}{8}$$ Add the numerators and write the sum over the common denominator 8.

$$= \frac{-13}{8}$$ Use the rule for adding integers that have different signs: $-25 + 12 = -13$.

$$= -1\frac{5}{8}$$ Write $\frac{-13}{8}$ as a negative mixed number by dividing 13 by 8.

We can also add mixed numbers by adding their whole-number parts and their fractional parts. To do so, we follow these steps.

Adding Mixed Numbers: Method 2

1. Write each mixed number as the sum of a whole number and a fraction.
2. Use the commutative property of addition to write the whole numbers together and the fractions together.
3. Add the whole numbers and the fractions separately.
4. Write the result as a mixed number, if necessary.

Method 2 works well when the whole number parts of the mixed numbers are large.

Add: $275\dfrac{1}{6} + 81\dfrac{3}{5}$

Now Try Problem 21

EXAMPLE 3 Add: $168\dfrac{3}{7} + 85\dfrac{2}{9}$

Strategy We will write each mixed number as the sum of a whole number and a fraction. Then we will add the whole numbers and the fractions separately.

WHY If we change each mixed number to an improper fraction, build equivalent fractions, and add, the resulting numerators will be very large and difficult to work with.

Solution
We will write the solution in *horizontal* form.

$168\dfrac{3}{7} + 85\dfrac{2}{9} = 168 + \dfrac{3}{7} + 85 + \dfrac{2}{9}$ Write each mixed number as the sum of a whole number and a fraction.

$= 168 + 85 + \dfrac{3}{7} + \dfrac{2}{9}$ Use the commutative property of addition to change the order of the addition so that the whole numbers are together and the fractions are together.

$= 253 + \dfrac{3}{7} + \dfrac{2}{9}$ Add the whole numbers.

$\begin{array}{r}\overset{1\,1}{168}\\+\;\;85\\\hline253\end{array}$

$= 253 + \dfrac{3}{7}\cdot\dfrac{9}{9} + \dfrac{2}{9}\cdot\dfrac{7}{7}$ Prepare to add the fractions. To build $\frac{3}{7}$ and $\frac{2}{9}$ so that their denominators are 63, multiply each by a form of 1.

$= 253 + \dfrac{27}{63} + \dfrac{14}{63}$ Multiply the numerators. Multiply the denominators.

$= 253 + \dfrac{41}{63}$ Add the numerators and write the sum over the common denominator 63.

$\begin{array}{r}\overset{1}{27}\\+\;14\\\hline41\end{array}$

$= 253\dfrac{41}{63}$ Write the sum as a mixed number.

Caution! If we use method 1 to add the mixed numbers in Example 3, the numbers we encounter are very large. As expected, the result is the same: $253\frac{41}{63}$.

$168\dfrac{3}{7} + 85\dfrac{2}{9} = \dfrac{1{,}179}{7} + \dfrac{767}{9}$ Write $168\frac{3}{7}$ and $85\frac{2}{9}$ as improper fractions.

$= \dfrac{1{,}179}{7}\cdot\dfrac{9}{9} + \dfrac{767}{9}\cdot\dfrac{7}{7}$ The LCD is 63.

$= \dfrac{10{,}611}{63} + \dfrac{5{,}369}{63}$ Note how large the numerators are.

$= \dfrac{15{,}980}{63}$ Add the numerators and write the sum over the common denominator 63.

$= 253\dfrac{41}{63}$ To write the improper fraction as a mixed number, divide 15,980 by 63.

Generally speaking, the larger the whole-number parts of the mixed numbers, the more difficult it becomes to add those mixed numbers using method 1.

2 Add mixed numbers in vertical form.

We can add mixed numbers quickly when they are written in **vertical form** by working in columns. The strategy is the same as in Example 2: Add whole numbers to whole numbers and fractions to fractions.

EXAMPLE 4

Add: $25\dfrac{3}{4} + 31\dfrac{1}{5}$

Strategy We will perform the addition in *vertical form* with the fractions in a column and the whole numbers lined up in columns. Then we will add the fractional parts and the whole-number parts separately.

WHY It is often easier to add the fractional parts and the whole-number parts of mixed numbers vertically—especially if the whole-number parts contain two or more digits, such as 25 and 31.

Solution

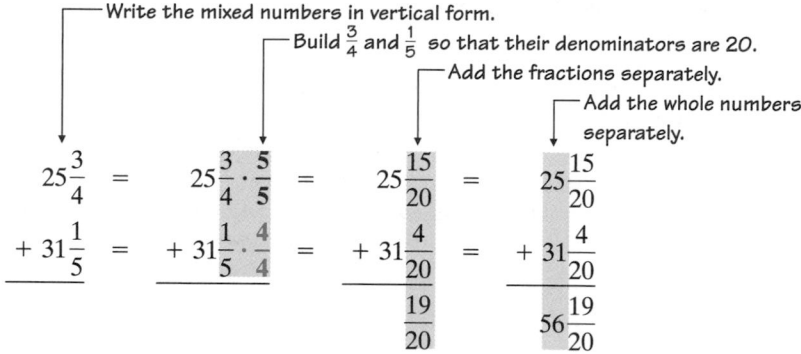

The sum is $56\dfrac{19}{20}$.

Self Check 4

Add: $71\dfrac{5}{8} + 23\dfrac{1}{3}$

Now Try Problem 25

EXAMPLE 5

Add and simplify, if possible: $75\dfrac{1}{12} + 43\dfrac{1}{4} + 54\dfrac{1}{6}$

Strategy We will write the problem in *vertical form*. We will make sure that the fractional part of the answer is in simplest form.

WHY When adding, subtracting, multiplying, or dividing fractions or mixed numbers, the answer should always be written in simplest form.

Solution

The LCD for $\dfrac{1}{12}, \dfrac{1}{4}$, and $\dfrac{1}{6}$ is 12.

Write the mixed numbers in vertical form.
Build $\frac{1}{4}$ and $\frac{1}{6}$ so that their denominators are 12.
Add the fractions separately.
Add the whole numbers separately.

$$75\dfrac{1}{12} \;=\; 75\dfrac{1}{12} \;=\; 75\dfrac{1}{12} \;=\; \overset{11}{75}\dfrac{1}{12}$$

$$43\dfrac{1}{4} \;=\; 43\dfrac{1}{4}\cdot\dfrac{3}{3} \;=\; 43\dfrac{3}{12} \;=\; 43\dfrac{3}{12}$$

$$+\,54\dfrac{1}{6} \;=\; +\,54\dfrac{1}{6}\cdot\dfrac{2}{2} \;=\; +\,54\dfrac{2}{12} \;=\; +\,54\dfrac{2}{12}$$

$$\dfrac{6}{12} \qquad\qquad 172\dfrac{6}{12} = 172\dfrac{1}{2}$$

Simplify: $\dfrac{6}{12} = \dfrac{\cancel{6}}{2\cdot\cancel{6}} = \dfrac{1}{2}$.

The sum is $172\dfrac{1}{2}$.

Self Check 5

Add and simplify, if possible:

$68\dfrac{1}{6} + 37\dfrac{5}{18} + 52\dfrac{1}{9}$

Now Try Problem 29

When we add mixed numbers, sometimes the sum of the fractions is an improper fraction.

Self Check 6

Add: $76\frac{11}{12} + 49\frac{5}{8}$

Now Try Problem 33

EXAMPLE 6   Add: $45\frac{2}{3} + 96\frac{4}{5}$

Strategy We will write the problem in *vertical form*. We will make sure that the fractional part of the answer is in simplest form.

WHY When adding, subtracting, multiplying, or dividing fractions or mixed numbers, the answer should always be written in simplest form.

Solution

The LCD for $\frac{2}{3}$ and $\frac{4}{5}$ is 15.

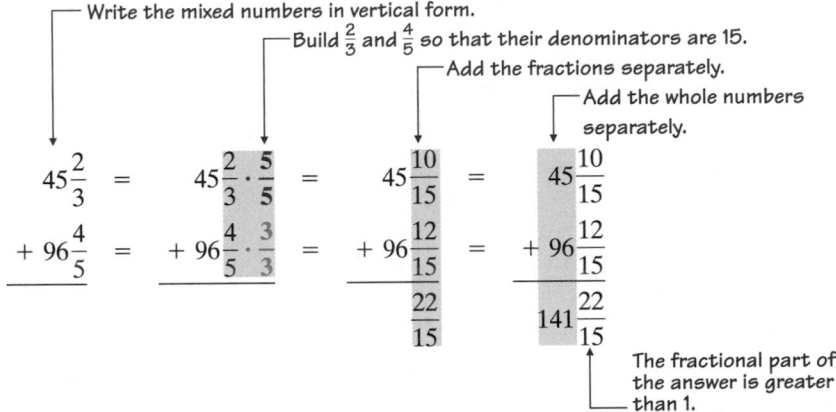

Since we don't want an improper fraction in the answer, we write $\frac{22}{15}$ as a mixed number. Then we *carry* 1 from the fraction column to the whole-number column.

$$141\frac{22}{15} = 141 + \frac{22}{15}$$ Write the mixed number as the sum of a whole number and a fraction.

$$= 141 + 1\frac{7}{15}$$ To write the improper fraction as a mixed number divide 22 by 15.

$$= 142\frac{7}{15}$$ Carry the 1 and add it to 141 to get 142.

$$\begin{array}{r} 1 \\ 15\overline{)22} \\ -15 \\ \hline 7 \end{array}$$

3 Subtract mixed numbers.

Subtracting mixed numbers is similar to adding mixed numbers.

Self Check 7

Subtract and simplify, if possible:

$$12\frac{9}{20} - 8\frac{1}{30}$$

Now Try Problem 37

EXAMPLE 7   Subtract and simplify, if possible: $16\frac{7}{10} - 9\frac{8}{15}$

Strategy We will perform the subtraction in *vertical form* with the fractions in a column and the whole numbers lined up in columns. Then we will subtract the fractional parts and the whole-number parts separately.

WHY It is often easier to subtract the fractional parts and the whole-number parts of mixed numbers vertically.

Solution

The LCD for $\dfrac{7}{10}$ and $\dfrac{8}{15}$ is 30.

Write the mixed numbers in vertical form.
Build $\dfrac{7}{10}$ and $\dfrac{8}{15}$ so that their denominators are 30.
Subtract the fractions separately.
Subtract the whole numbers separately.

$$
\begin{array}{rll}
16\dfrac{7}{10} & = & 16\dfrac{7}{10}\cdot\dfrac{3}{3} = 16\dfrac{21}{30} = 16\dfrac{21}{30} \\[2mm]
- \ 9\dfrac{8}{15} & = & -\ 9\dfrac{8}{15}\cdot\dfrac{2}{2} = -\ 9\dfrac{16}{30} = -\ 9\dfrac{16}{30} \\[2mm]
& & \hspace{5.3cm} \dfrac{5}{30} \qquad 7\dfrac{5}{30} = 7\dfrac{1}{6}
\end{array}
$$

Simplify: $\dfrac{5}{30} = \dfrac{\overset{1}{\cancel{5}}}{\underset{1}{\cancel{5}}\cdot 6} = \dfrac{1}{6}.$

The difference is $7\dfrac{1}{6}$.

Subtraction of mixed numbers (like subtraction of whole numbers) sometimes involves borrowing. When the fraction we are subtracting is greater than the fraction we are subtracting it from, it is necessary to borrow.

EXAMPLE 8   Subtract: $34\dfrac{1}{8} - 11\dfrac{2}{3}$

Strategy We will perform the subtraction in *vertical form* with the fractions in a column and the whole numbers lined up in columns. Then we will subtract the fractional parts and the whole-number parts separately.

WHY It is often easier to subtract the fractional parts and the whole-number parts of mixed numbers vertically.

Solution

The LCD for $\dfrac{1}{8}$ and $\dfrac{2}{3}$ is 24.

Write the mixed number in vertical form.
Build $\dfrac{1}{8}$ and $\dfrac{2}{3}$ so that their denominators are 24.

$$
\begin{array}{rll}
34\dfrac{1}{8} & = & 34\dfrac{1}{8}\cdot\dfrac{3}{3} = 34\dfrac{3}{24} \\[2mm]
- \ 11\dfrac{2}{3} & = & -\ 11\dfrac{2}{3}\cdot\dfrac{8}{8} = -\ 11\dfrac{16}{24}
\end{array}
$$

Note that $\dfrac{16}{24}$ is greater than $\dfrac{3}{24}$.

Since $\dfrac{16}{24}$ is greater than $\dfrac{3}{24}$, borrow 1 (in the form of $\dfrac{24}{24}$) from 34 and add it to $\dfrac{3}{24}$ to get $\dfrac{27}{24}$.
Subtract the fractions separately.
Subtract the whole numbers separately.

$$
\begin{array}{rll}
\overset{33}{34}\dfrac{3}{24} + \dfrac{24}{24} & = & 33\dfrac{27}{24} = 33\dfrac{27}{24} \\[2mm]
- \ 11\dfrac{16}{24} & = & -\ 11\dfrac{16}{24} = -\ 11\dfrac{16}{24} \\[2mm]
& & \hspace{3cm}\dfrac{11}{24} \qquad 22\dfrac{11}{24}
\end{array}
$$

The difference is $22\dfrac{11}{24}$.

Self Check 8

Subtract: $258\dfrac{3}{4} - 175\dfrac{15}{16}$

Now Try Problem 41

Success Tip We can use rounding to check the results when subtracting mixed numbers. To check the answer $22\frac{11}{24}$ from Example 8, we proceed as follows:

$$34\frac{1}{8} - 11\frac{2}{3} \approx 34 - 12 = 22$$

Since $\frac{1}{8}$ is less than $\frac{1}{2}$, round $34\frac{1}{8}$ down to 34.

Since $\frac{2}{3}$ is greater than $\frac{1}{2}$, round $11\frac{2}{3}$ up to 12.

Since $22\frac{11}{24}$ is close to 22, it is a reasonable answer.

Self Check 9

Subtract: $2,300 - 129\frac{31}{32}$

Now Try Problem 45

EXAMPLE 9 Subtract: $419 - 53\frac{11}{16}$

Strategy We will write the numbers in vertical form and borrow 1 (in the form of $\frac{16}{16}$) from 419.

WHY In the fraction column, we need to have a fraction from which to subtract $\frac{11}{16}$.

Solution

Write the mixed number in vertical form.

Borrow 1 (in the form of $\frac{16}{16}$) from 419. Then subtract the fractions separately.

Subtract the whole numbers separately. This also requires borrowing.

$$419 = 418\frac{16}{16} = \overset{311}{\cancel{418}}\frac{16}{16}$$
$$-\ 53\frac{11}{16} = -\ 53\frac{11}{16} = -\ 53\frac{11}{16}$$
$$365\frac{5}{16} \qquad 365\frac{5}{16}$$

The difference is $365\frac{5}{16}$.

4 Solve application problems by adding and subtracting mixed numbers.

Self Check 10

SALADS A three-bean salad calls for one can of green beans ($14\frac{1}{2}$ ounces), one can of garbanzo beans ($10\frac{3}{4}$ ounces), and one can of kidney beans ($15\frac{7}{8}$ ounces). How many ounces of beans are called for in the recipe?

Now Try Problem 89

EXAMPLE 10 *Horse Racing* In order to become the *Triple Crown Champion*, a thoroughbred horse must win three races: the Kentucky Derby ($1\frac{1}{4}$ miles long), the Preakness Stakes ($1\frac{3}{16}$ miles long), and the Belmont Stakes ($1\frac{1}{2}$ miles long). What is the combined length of the three races of the Triple Crown?

Analyze

- The Kentucky Derby is $1\frac{1}{4}$ miles long.
- The Preakness Stakes is $1\frac{3}{16}$ miles long.
- The Belmont Stakes is $1\frac{1}{2}$ miles long.
- What is the combined length of the three races?

Affirmed, in 1978, was the last of only 11 horses in history to win the Triple Crown.

Focus on Sport/Getty Images

Form The key phrase *combined length* indicates addition. We translate the words of the problem to numbers and symbols.

The combined length of the three races	is equal to	the length of the Kentucky Derby	plus	the length of the Preakness Stakes	plus	the length of the Belmont Stakes.
The combined length of the three races	=	$1\frac{1}{4}$	+	$1\frac{3}{16}$	+	$1\frac{1}{2}$

Solve To find the sum, we will write the mixed numbers in vertical form. To add in the fraction column, the LCD for $\frac{1}{4}$, $\frac{3}{16}$, and $\frac{1}{2}$ is 16.

Build $\frac{1}{4}$ and $\frac{1}{2}$ so that their denominators are 16.
Add the fractions separately.
Add the whole numbers separately.

$$
\begin{array}{l}
1\frac{1}{4} = 1\frac{1 \cdot 4}{4 \cdot 4} = 1\frac{4}{16} = 1\frac{4}{16} \\[2mm]
1\frac{3}{16} = 1\frac{3}{16} = 1\frac{3}{16} = 1\frac{3}{16} \\[2mm]
+1\frac{1}{2} = +1\frac{1 \cdot 8}{2 \cdot 8} = +1\frac{8}{16} = +1\frac{8}{16} \\[2mm]
\hline
\frac{15}{16} \qquad\qquad 3\frac{15}{16}
\end{array}
$$

State The combined length of the three races of the Triple Crown is $3\frac{15}{16}$ miles.

Check We can estimate to check the result. If we round $1\frac{1}{4}$ down to 1, round $1\frac{3}{16}$ down to 1, and round $1\frac{1}{2}$ up to 2, the approximate combined length of the three races is $1 + 1 + 2 = 4$ miles. Since $3\frac{15}{16}$ is close to 4, the result seems reasonable. ∎

THINK IT THROUGH

"Americans are not getting the sleep they need which may affect their ability to perform well during the workday."

National Sleep Foundation Report, 2008

The 1,000 people who took part in the 2008 *Sleep in America* poll were asked when they typically wake up, when they go to bed, and how long they sleep on both workdays and non-workdays. The results are shown on the right. Write the average hours slept on a workday and on a non-workday as mixed numbers. How much longer does the average person sleep on a non-workday?

Typical Workday and Non-workday Sleep Schedules

Average workday bedtime
10:53 PM

Average non-workday bedtime
11:24 PM

Average hours slept on workdays
6 hours
40 minutes

Average hours slept on non-workdays
7 hours
25 minutes

5:35 AM
Average workday wake time

7:12 AM
Average non-workday wake time

(Source: National Sleep Foundation, 2008)

Self Check 11

TRUCKING The mixing barrel of a cement truck holds 9 cubic yards of concrete. How much concrete is left in the barrel if $6\frac{3}{4}$ cubic yards have already been unloaded?

Now Try Problem 95

EXAMPLE 11 *Baking* How much butter is left in a 10-pound tub if $2\frac{2}{3}$ pounds are used for a wedding cake?

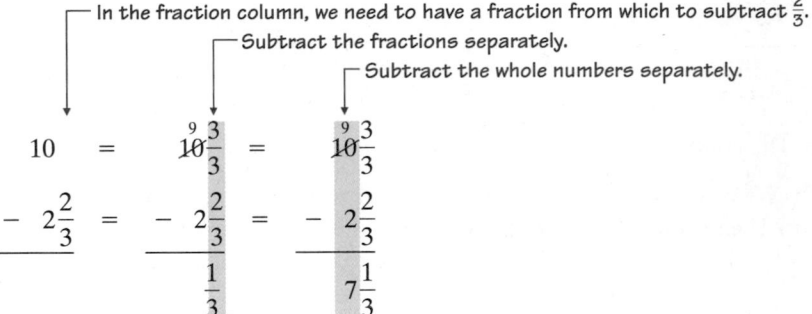

Image copyright Eric Limon, 2009. Used under license from Shutterstock.com

Analyze

- The tub contained 10 pounds of butter.
- $2\frac{2}{3}$ pounds of butter are used for a cake.
- How much butter is left in the tub?

Form The key phrase *how much butter is left* indicates subtraction. We translate the words of the problem to numbers and symbols.

The amount of butter left in the tub	is equal to	the amount of butter in one tub	minus	the amount of butter used for the cake.
The amount of butter left in the tub	=	10	−	$2\frac{2}{3}$

Solve To find the difference, we will write the numbers in vertical form and borrow 1 (in the form of $\frac{3}{3}$) from 10.

In the fraction column, we need to have a fraction from which to subtract $\frac{2}{3}$.
Subtract the fractions separately.
Subtract the whole numbers separately.

$$10 \quad = \quad \overset{9}{\cancel{10}}\frac{3}{3} \quad = \quad \overset{9}{\cancel{10}}\frac{3}{3}$$
$$-\;2\frac{2}{3} \quad = \quad -\;2\frac{2}{3} \quad = \quad -\;2\frac{2}{3}$$
$$\overline{\frac{1}{3}} \qquad \overline{7\frac{1}{3}}$$

State There are $7\frac{1}{3}$ pounds of butter left in the tub.

Check We can check using addition. If $2\frac{2}{3}$ pounds of butter were used and $7\frac{1}{3}$ pounds of butter are left in the tub, then the tub originally contained $2\frac{2}{3} + 7\frac{1}{3} = 9\frac{3}{3} = 10$ pounds of butter. The result checks.

ANSWER TO SELF CHECKS

1. $4\frac{13}{15}$ **2.** $-1\frac{5}{6}$ **3.** $356\frac{23}{30}$ **4.** $94\frac{23}{24}$ **5.** $157\frac{5}{9}$ **6.** $126\frac{13}{24}$ **7.** $4\frac{5}{12}$ **8.** $82\frac{13}{16}$
9. $2,170\frac{1}{32}$ **10.** $41\frac{1}{8}$ oz **11.** $2\frac{1}{4}$ yd³

SECTION 4.6 STUDY SET

▌VOCABULARY

Fill in the blanks.

1. A _____ number, such as $1\frac{7}{8}$, contains a whole-number part and a fractional part.

2. We can add (or subtract) mixed numbers quickly when they are written in _____ form by working in columns.

3. To add (or subtract) mixed numbers written in vertical form, we add (or subtract) the _____ separately and the _____ numbers separately.

4. Fractions such as $\frac{11}{8}$, that are greater than or equal to 1, are called _____ fractions.

5. Consider the following problem:

$$36\frac{5}{7}$$
$$+\ 42\frac{4}{7}$$
$$78\frac{9}{7} = 78 + 1\frac{2}{7} = 79\frac{2}{7}$$

Since we don't want an improper fraction in the answer, we write $\frac{9}{7}$ as $1\frac{2}{7}$, _____ the 1, and add it to 78 to get 79.

6. Consider the following problem:

$$86\frac{1}{3} = \quad 86\frac{\overset{5}{1}}{3} + \frac{3}{3}$$
$$-\ 24\frac{2}{3} = -24\frac{2}{3}$$

To subtract in the fraction column, we _____ 1 from 86 in the form of $\frac{3}{3}$.

CONCEPTS

7. a. For $76\frac{3}{4}$, list the whole-number part and the fractional part.

  b. Write $76\frac{3}{4}$ as a sum.

8. Use the commutative property of addition to rewrite the following expression with the whole numbers together and the fractions together. *You do not have to find the answer.*

$$14 + \frac{5}{8} + 53 + \frac{1}{6}$$

9. The *denominators* of two fractions are given. Find the least common denominator.

  a. 3 and 4   **b.** 5 and 6

  c. 6 and 9   **d.** 8 and 12

10. Simplify.

  a. $9\frac{17}{16}$ **b.** $1{,}288\frac{7}{3}$

  c. $16\frac{12}{8}$ **d.** $45\frac{24}{20}$

NOTATION

Fill in the blanks to complete each solution.

11.
$$6\frac{3}{5} = \quad 6\frac{3}{5}\cdot\frac{7}{7} = \quad 6\frac{\ }{35}$$
$$+\ 3\frac{2}{7} = \quad +3\frac{2}{7}\cdot\frac{\ }{\ } = \quad +3\frac{10}{\ }$$
$$9\frac{\ }{\ }$$

12.
$$67\frac{3}{8} = \quad 67\frac{3}{8}\cdot\frac{\ }{\ } = \quad 67\frac{9}{24} = \quad 67\overset{6}{\ }\frac{9}{24} + \frac{\ }{\ } = \quad 66\frac{\ }{24}$$
$$-\ 23\frac{2}{3} = \quad -23\frac{2}{3}\cdot\frac{8}{8} = \quad -23\frac{16}{24} = \quad -23\frac{16}{24} \quad\quad = -23\frac{16}{24}$$
$$\frac{\ }{24}$$

GUIDED PRACTICE

Add. See Example 1.

13. $1\frac{1}{4} + 2\frac{1}{3}$ **14.** $2\frac{2}{5} + 3\frac{1}{4}$

15. $2\frac{1}{3} + 4\frac{2}{5}$ **16.** $4\frac{1}{3} + 1\frac{1}{7}$

Add. See Example 2.

17. $-4\frac{1}{8} + 1\frac{3}{4}$ **18.** $-3\frac{11}{15} + 2\frac{1}{5}$

19. $-6\frac{5}{6} + 3\frac{2}{3}$ **20.** $-6\frac{3}{14} + 1\frac{2}{7}$

Add. See Example 3.

21. $334\frac{1}{7} + 42\frac{2}{3}$ **22.** $259\frac{3}{8} + 40\frac{1}{3}$

23. $667\frac{1}{5} + 47\frac{3}{4}$ **24.** $568\frac{1}{6} + 52\frac{3}{4}$

Add. See Example 4.

25. $41\frac{2}{9} + 18\frac{2}{5}$ **26.** $60\frac{3}{11} + 24\frac{2}{3}$

27. $89\frac{6}{11} + 43\frac{1}{3}$ **28.** $77\frac{5}{8} + 55\frac{1}{7}$

Add and simplify, if possible. See Example 5.

29. $14\frac{1}{4} + 29\frac{1}{20} + 78\frac{3}{5}$ **30.** $11\frac{1}{12} + 59\frac{1}{4} + 82\frac{1}{6}$

31. $106\frac{5}{18} + 22\frac{1}{2} + 19\frac{1}{9}$ **32.** $75\frac{2}{5} + 43\frac{7}{30} + 54\frac{1}{3}$

Add and simplify, if possible. See Example 6.

33. $39\frac{5}{8} + 62\frac{11}{12}$ **34.** $53\frac{5}{6} + 47\frac{3}{8}$

35. $82\frac{8}{9} + 46\frac{11}{15}$ **36.** $44\frac{2}{9} + 76\frac{20}{21}$

Subtract and simplify, if possible. See Example 7.

37. $19\frac{11}{12} - 9\frac{2}{3}$

38. $32\frac{2}{3} - 7\frac{1}{6}$

39. $21\frac{5}{6} - 8\frac{3}{10}$

40. $41\frac{2}{5} - 6\frac{3}{20}$

Subtract. See Example 8.

41. $47\frac{1}{11} - 15\frac{2}{3}$

42. $58\frac{4}{11} - 15\frac{1}{2}$

43. $84\frac{5}{8} - 12\frac{6}{7}$

44. $95\frac{4}{7} - 23\frac{5}{6}$

Subtract. See Example 9.

45. $674 - 94\frac{11}{15}$

46. $437 - 63\frac{6}{23}$

47. $112 - 49\frac{9}{32}$

48. $221 - 88\frac{35}{64}$

■ **TRY IT YOURSELF**

Add or subtract and simplify, if possible.

49. $140\frac{5}{6} - 129\frac{4}{5}$

50. $291\frac{1}{4} - 289\frac{1}{12}$

51. $4\frac{1}{6} + 1\frac{1}{5}$

52. $2\frac{2}{5} + 3\frac{1}{4}$

53. $5\frac{1}{2} + 3\frac{4}{5}$

54. $6\frac{1}{2} + 2\frac{2}{3}$

55. $2 + 1\frac{7}{8}$

56. $3\frac{3}{4} + 5$

57. $8\frac{7}{9} - 3\frac{1}{9}$

58. $9\frac{9}{10} - 6\frac{3}{10}$

59. $140\frac{3}{16} - 129\frac{3}{4}$

60. $442\frac{1}{8} - 429\frac{2}{3}$

61. $380\frac{1}{6} + 17\frac{1}{4}$

62. $103\frac{1}{2} + 210\frac{2}{5}$

63. $-2\frac{5}{6} + 1\frac{3}{8}$

64. $-4\frac{5}{9} + 2\frac{1}{6}$

65. $3\frac{1}{4} + 4\frac{1}{4}$

66. $2\frac{1}{8} + 3\frac{3}{8}$

67. $-3\frac{3}{4} + \left(-1\frac{1}{2}\right)$

68. $-3\frac{2}{3} + \left(-1\frac{4}{5}\right)$

69. $7 - \frac{2}{3}$

70. $6 - \frac{1}{8}$

71. $12\frac{1}{2} + 5\frac{3}{4} + 35\frac{1}{6}$

72. $31\frac{1}{3} + 20\frac{2}{5} + 10\frac{1}{15}$

73. $16\frac{1}{4} - 13\frac{3}{4}$

74. $40\frac{1}{7} - 19\frac{6}{7}$

75. $-4\frac{5}{8} - 1\frac{1}{4}$

76. $-2\frac{1}{16} - 3\frac{7}{8}$

77. $6\frac{5}{8} - 3$

78. $10\frac{1}{2} - 6$

79. $\frac{7}{3} + 2$

80. $\frac{9}{7} + 3$

81. $58\frac{7}{8} + 340\frac{1}{2} + 61\frac{3}{4}$

82. $191\frac{1}{2} + 233\frac{1}{16} + 16\frac{5}{8}$

83. $9 - 8\frac{3}{4}$

84. $11 - 10\frac{4}{5}$

■ **APPLICATIONS**

85. AIR TRAVEL A businesswoman's flight left Los Angeles and in $3\frac{3}{4}$ hours she landed in Minneapolis. She then boarded a commuter plane in Minneapolis and arrived at her final destination in $1\frac{1}{2}$ hours. Find the total time she spent on the flights.

86. SHIPPING A passenger ship and a cargo ship left San Diego harbor at midnight. During the first hour, the passenger ship traveled south at $16\frac{1}{2}$ miles per hour, while the cargo ship traveled north at a rate of $5\frac{1}{5}$ miles per hour. How far apart were they at 1:00 A.M.?

87. TRAIL MIX How many cups of trail mix will the recipe shown below make?

Trail Mix

A healthy snack–great for camping trips

$2\frac{3}{4}$ cups peanuts $\frac{1}{3}$ cup coconut

$\frac{1}{2}$ cup sunflower seeds $2\frac{2}{3}$ cups oat flakes

$\frac{2}{3}$ cup raisins $\frac{1}{4}$ cup pretzels

88. HARDWARE Refer to the illustration below. How long should the threaded part of the bolt be?

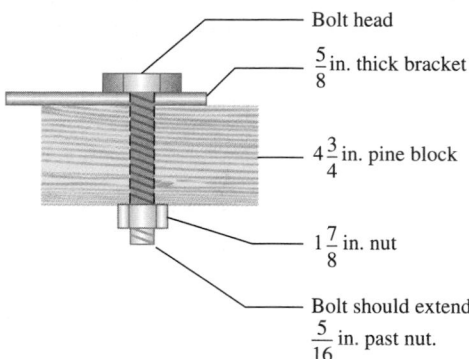

- Bolt head
- $\frac{5}{8}$ in. thick bracket
- $4\frac{3}{4}$ in. pine block
- $1\frac{7}{8}$ in. nut
- Bolt should extend $\frac{5}{16}$ in. past nut.

89. OCTUPLETS On January 26, 2009, at Kaiser Permanente Bellflower Medical Center in California, Nadya Suleman gave birth to eight babies. (The United States' first live octuplets were born in Houston in 1998 to Nkem Chukwu and Iyke Louis Udobi). Find the combined birthweights of the babies from the information shown below. (Source: The Nadya Suleman family website)

No. 1: Noah, male, $2\frac{11}{16}$ pounds

No. 2: Maliah, female, $2\frac{3}{4}$ pounds

No. 3: Isaiah, male, $3\frac{1}{4}$ pounds

No. 4: Nariah, female, $2\frac{1}{2}$ pounds

No. 5: Makai, male, $1\frac{1}{2}$ pounds

No. 6: Josiah, male, $2\frac{3}{4}$ pounds

No. 7: Jeremiah, male, $1\frac{15}{16}$ pounds

No. 8: Jonah, male, $2\frac{11}{16}$ pounds

90. SEPTUPLETS On November 19, 1997, at Iowa Methodist Medical Center, Bobbie McCaughey gave birth to seven babies. Find the combined birthweights of the babies from the following information. (Source: *Los Angeles Times*, Nov. 20, 1997)

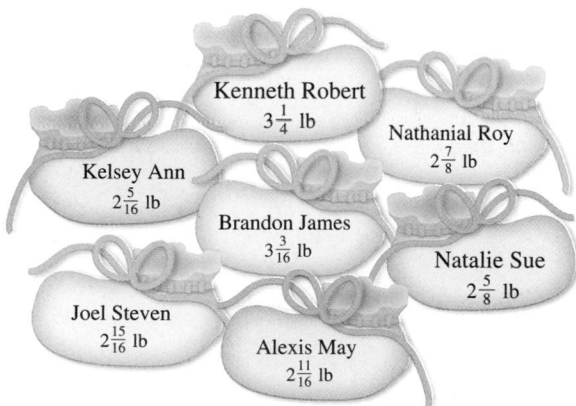

Kenneth Robert $3\frac{1}{4}$ lb

Nathanial Roy $2\frac{7}{8}$ lb

Kelsey Ann $2\frac{5}{16}$ lb

Brandon James $3\frac{3}{16}$ lb

Natalie Sue $2\frac{5}{8}$ lb

Joel Steven $2\frac{15}{16}$ lb

Alexis May $2\frac{11}{16}$ lb

91. HISTORICAL DOCUMENTS The Declaration of Independence on display at the National Archives in Washington, D.C., is $24\frac{1}{2}$ inches wide by $29\frac{3}{4}$ inches high. How many inches of molding would be needed to frame it?

92. STAMP COLLECTING The Pony Express Stamp, shown below, was issued in 1940. It is a favorite of collectors all over the world. A Postal Service document describes its size in an unusual way:

"The dimensions of the stamp are $\frac{84}{100}$ by $1\frac{44}{100}$ inches, arranged horizontally."

To display the stamp, a collector wants to frame it with gold braid. How many inches of braid are needed?

Smithsonian National Postal Museum

93. FREEWAY SIGNS A freeway exit sign is shown. How far apart are the Citrus Ave. and Grand Ave. exits?

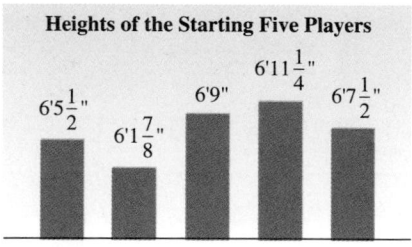

Citrus Ave. $\frac{3}{4}$ mi

Grand Ave. $3\frac{1}{2}$ mi

94. BASKETBALL See the graph below. What is the difference in height between the tallest and the shortest of the starting players?

Heights of the Starting Five Players

6'5$\frac{1}{2}$"   6'1$\frac{7}{8}$"   6'9"   6'11$\frac{1}{4}$"   6'7$\frac{1}{2}$"

95. HOSE REPAIRS To repair a bad connector, a gardener removes $1\frac{1}{2}$ feet from the end of a 50-foot hose. How long is the hose after the repair?

96. HAIRCUTS A mother makes her child get a haircut when his hair measures 3 inches in length. His barber uses clippers with attachment #2 that leaves $\frac{3}{8}$-inch of hair. How many inches does the child's hair grow between haircuts?

97. GASOLINE Use the service station sign below to answer the following questions.

   a. What is the difference in price per gallon between the least and most expensive types of gasoline at the self-serve pump?

   b. For each type of gasoline, how much more is the cost per gallon for full service compared to self-serve?

98. WATER SLIDES An amusement park added a new section to a water slide to create a slide $311\frac{5}{12}$ feet long. How long was the slide before the addition?

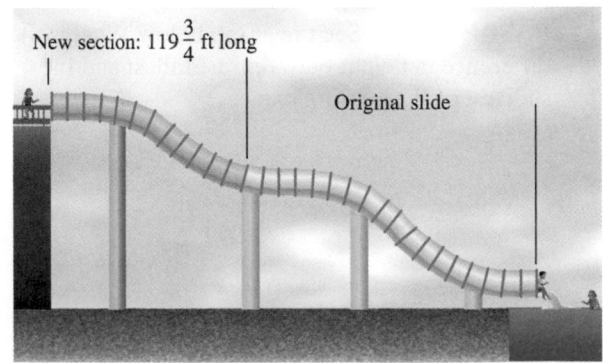

New section: $119\frac{3}{4}$ ft long

Original slide

99. JEWELRY A jeweler cut a 7-inch-long silver wire into three pieces. To do this, he aligned a 6-inch-long ruler directly below the wire and made the proper cuts. Find the length of piece 2 of the wire.

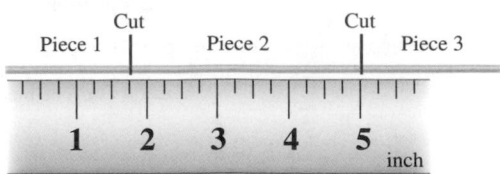

100. SEWING To make some draperies, an interior decorator needs $12\frac{1}{4}$ yards of material for the den and $8\frac{1}{2}$ yards for the living room. If the material comes only in 21-yard bolts, how much will be left over after completing both sets of draperies?

WRITING

101. Of the methods studied to add mixed numbers, which do you like better, and why?

102. LEAP YEAR It actually takes Earth $365\frac{1}{4}$ days, give or take a few minutes, to make one revolution around the sun. Explain why every four years we add a day to the calendar to account for this fact.

103. Explain the process of simplifying $12\frac{7}{5}$.

104. Consider the following problem:

$$108\frac{1}{3}$$
$$-\ 99\frac{2}{3}$$

   a. Explain why borrowing is necessary.

   b. Explain how the borrowing is done.

REVIEW

Perform each operation and simplify, if possible.

105. a. $3\frac{1}{2} + 1\frac{1}{4}$ **b.** $3\frac{1}{2} - 1\frac{1}{4}$

   c. $3\frac{1}{2} \cdot 1\frac{1}{4}$ **d.** $3\frac{1}{2} \div 1\frac{1}{4}$

106. a. $5\frac{1}{10} + \frac{4}{5}$ **b.** $5\frac{1}{10} - \frac{4}{5}$

   c. $5\frac{1}{10} \cdot \frac{4}{5}$ **d.** $5\frac{1}{10} \div \frac{4}{5}$

Order of Operations and Complex Fractions

Objectives

1 Use the order of operations rule.

2 Solve application problems by using the order of operations rule.

3 Evaluate formulas.

4 Simplify complex fractions.

We have seen that the order of operations rule is used to evaluate expressions that contain more than one operation. In Chapter 1, we used it to evaluate expressions involving whole numbers, and in Chapter 2, we used it to evaluate expressions involving integers. We will now use it to evaluate expressions involving fractions and mixed numbers.

1 Use the order of operations rule.

Recall from Section 1.7 that if we don't establish a uniform order of operations, an expression can have more than one value. To avoid this possibility, we must always use the following rule.

Order of Operations

1. Perform all calculations within parentheses and other grouping symbols following the order listed in Steps 2–4 below, working from the innermost pair of grouping symbols to the outermost pair.
2. Evaluate all exponential expressions.
3. Perform all multiplications and divisions as they occur from left to right.
4. Perform all additions and subtractions as they occur from left to right.

When grouping symbols have been removed, repeat Steps 2–4 to complete the calculation.

   If a fraction bar is present, evaluate the expression above the bar (called the **numerator**) and the expression below the bar (called the **denominator**) separately. Then perform the division indicated by the fraction bar, if possible.

EXAMPLE 1

Evaluate: $\dfrac{3}{4} + \dfrac{5}{3}\left(-\dfrac{1}{2}\right)^3$

Strategy We will scan the expression to determine what operations need to be performed. Then we will perform those operations, one at a time, following the order of operations rule.

WHY If we don't follow the correct order of operations, the expression can have more than one value.

Solution

Although the expression contains parentheses, there are no calculations to perform *within* them. We will begin with step 2 of the rule: Evaluate all exponential expressions. We will write the steps of the solution in horizontal form.

$$\frac{3}{4} + \frac{5}{3}\left(-\frac{\mathbf{1}}{\mathbf{2}}\right)^{\mathbf{3}} = \frac{3}{4} + \frac{5}{3}\left(-\frac{\mathbf{1}}{\mathbf{8}}\right) \qquad \text{Evaluate: } \left(-\tfrac{1}{2}\right)^3 = \left(-\tfrac{1}{2}\right)\left(-\tfrac{1}{2}\right)\left(-\tfrac{1}{2}\right) = -\tfrac{1}{8}.$$

$$= \frac{3}{4} + \left(-\frac{5}{24}\right) \qquad \text{Multiply: } \tfrac{5}{3}\left(-\tfrac{1}{8}\right) = -\tfrac{5 \cdot 1}{3 \cdot 8} = -\tfrac{5}{24}.$$

$$= \frac{3}{4} \cdot \frac{6}{6} + \left(-\frac{5}{24}\right) \qquad \begin{array}{l}\text{Prepare to add the fractions: Their LCD}\\ \text{is 24. To build the first fraction so that its}\\ \text{denominator is 24, multiply it by a form of 1.}\end{array}$$

Self Check 1

Evaluate: $\dfrac{7}{8} + \dfrac{3}{2}\left(-\dfrac{1}{4}\right)^2$

Now Try Problem 15

$$= \frac{18}{24} + \left(-\frac{5}{24}\right)$$

Multiply the numerators: $3 \cdot 6 = 18$.
Multiply the denominators: $4 \cdot 6 = 24$.

$$= \frac{13}{24}$$

Add the numerators: $18 + (-5) = 13$. Write the sum over the common denominator 24.

If an expression contains grouping symbols, we perform the operations within the grouping symbols first.

Self Check 2

Evaluate: $\left(\frac{19}{21} - \frac{2}{3}\right) \div \left(-2\frac{1}{7}\right)$

Now Try Problem 19

EXAMPLE 2   Evaluate: $\left(\frac{7}{8} - \frac{1}{4}\right) \div \left(-2\frac{3}{16}\right)$

Strategy We will perform any operations within parentheses first.

WHY This is the first step of the order of operations rule.

Solution

We will begin by performing the subtraction within the first set of parentheses. The second set of parentheses does not contain an operation to perform.

$$\left(\frac{7}{8} - \frac{1}{4}\right) \div \left(-2\frac{3}{16}\right)$$

$$= \left(\frac{7}{8} - \frac{1}{4} \cdot \frac{2}{2}\right) \div \left(-2\frac{3}{16}\right)$$

Within the first set of parentheses, prepare to subtract the fractions: Their LCD is 8. Build $\frac{1}{4}$ so that its denominator is 8.

$$= \left(\frac{7}{8} - \frac{2}{8}\right) \div \left(-2\frac{3}{16}\right)$$

Multiply the numerators: $1 \cdot 2 = 2$.
Multiply the denominators: $4 \cdot 2 = 8$.

$$= \frac{5}{8} \div \left(-2\frac{3}{16}\right)$$

Subtract the numerators: $7 - 2 = 5$.
Write the difference over the common denominator 8.

$$= \frac{5}{8} \div \left(-\frac{35}{16}\right)$$

Write the mixed number as an improper fraction.

$$= \frac{5}{8}\left(-\frac{16}{35}\right)$$

Use the rule for division of fractions:
Multiply the first fraction by the reciprocal of $-\frac{35}{16}$.

$$= -\frac{5 \cdot 16}{8 \cdot 35}$$

Multiply the numerators and multiply the denominators.
The product of two fractions with unlike signs is negative.

$$= -\frac{\overset{1}{\cancel{5}} \cdot 2 \cdot \overset{1}{\cancel{8}}}{8 \cdot \underset{1}{\cancel{5}} \cdot 7}$$

To simplify, factor 16 as $2 \cdot 8$ and factor 35 as $5 \cdot 7$.
Remove the common factors of 5 and 8 from the numerator and denominator.

$$= -\frac{2}{7}$$

Multiply the remaining factors in the numerator.
Multipy the remaining factors in the denominator.

Self Check 3

Add $2\frac{1}{4}$ to the difference of $\frac{7}{8}$ and $\frac{2}{3}$.

Now Try Problem 23

EXAMPLE 3   Add $7\frac{1}{3}$ to the difference of $\frac{5}{6}$ and $\frac{1}{4}$.

Strategy We will translate the words of the problem to numbers and symbols. Then we will use the order of operations rule to evaluate the resulting expression.

WHY Since the expression involves two operations, addition and subtraction, we need to perform them in the proper order.

Solution

The key word *difference* indicates subtraction. Since we are to add $7\frac{1}{3}$ to the difference, the difference should be written first within parentheses, followed by the addition.

Add $7\frac{1}{3}$ to   the difference of $\frac{5}{6}$ and $\frac{1}{4}$.

$\left(\dfrac{5}{6} - \dfrac{1}{4}\right) + 7\dfrac{1}{3}$ Translate from words to numbers and mathematical symbols.

$\left(\dfrac{5}{6} - \dfrac{1}{4}\right) + 7\dfrac{1}{3} = \left(\dfrac{5}{6} \cdot \dfrac{2}{2} - \dfrac{1}{4} \cdot \dfrac{3}{3}\right) + 7\dfrac{1}{3}$ Prepare to subtract the fractions within the parentheses. Build the fractions so that their denominators are the LCD 12.

$= \left(\dfrac{10}{12} - \dfrac{3}{12}\right) + 7\dfrac{1}{3}$ Multiply the numerators. Multiply the denominators.

$= \dfrac{7}{12} + 7\dfrac{1}{3}$ Subtract the numerators: $10 - 3 = 7$. Write the difference over the common denominator 12.

$= \dfrac{7}{12} + 7\dfrac{4}{12}$ Prepare to add the fractions. Build $\frac{1}{3}$ so that its denominator is 12: $\frac{1}{3} \cdot \frac{4}{4} = \frac{4}{12}$.

$= 7\dfrac{11}{12}$ Add the numerators of the fractions: $7 + 4 = 11$. Write the sum over the common denominator 12.

2   Solve application problems by using the order of operations rule.

Sometimes more than one operation is needed to solve a problem.

EXAMPLE 4   *Masonry*   To build a wall, a mason will use blocks that are $5\frac{3}{4}$ inches high, held together with $\frac{3}{8}$-inch-thick layers of mortar. If the plans call for 8 layers, called *courses*, of blocks, what will be the height of the wall when completed?

Blocks $5\frac{3}{4}$ in. high

Mortar $\frac{3}{8}$ in. thick

Self Check 4

MASONRY Find the height of a wall if 8 layers (called *courses*) of $7\frac{3}{8}$-inch-high blocks are held together by $\frac{1}{4}$-inch-thick layers of mortar.

Now Try Problem 77

Analyze

• The blocks are $5\frac{3}{4}$ inches high.   Given
• A layer of mortar is $\frac{3}{8}$ inch thick.   Given
• There are 8 layers (courses) of blocks.   Given
• What is the height of the wall when completed?   Find

Form   To find the height of the wall when it is completed, we could add the heights of 8 blocks and 8 layers of mortar. However, it will be simpler if we find the height of one block and one layer of mortar, and multiply that result by 8.

The height of the wall when completed	is equal to	8	times	the height of one block	plus	the thickness of one layer of mortar.

The height of the wall when completed	=	8		(	$5\frac{3}{4}$	+	$\frac{3}{8}$	)

Solve To evaluate the expression, we use the order of operations rule.

$$8\left(5\frac{3}{4}+\frac{3}{8}\right)=8\left(5\frac{6}{8}+\frac{3}{8}\right)$$

Prepare to add the fractions within the parentheses: Their LCD is 8. Build $\frac{3}{4}$ so that its denominator is 8: $\frac{3}{4}\cdot\frac{2}{2}=\frac{6}{8}$.

$$=8\left(5\frac{9}{8}\right)$$

Add the numerators of the fractions: $6+3=9$. Write the sum over the common denominator 8.

$$=\frac{8}{1}\left(\frac{49}{8}\right)$$

Prepare to multiply the fractions. Write $5\frac{9}{8}$ as an improper fraction.

$$=\frac{\overset{1}{\cancel{8}}\cdot49}{1\cdot\underset{1}{\cancel{8}}}$$

Multiply the numerators and multiply the denominators. To simplify, remove the common factor of 8 from the numerator and denominator.

$$=49$$

Simplify: $\frac{49}{1}=49$.

State The completed wall will be 49 inches high.

Check We can estimate to check the result. Since one block and one layer of mortar is about 6 inches high, eight layers of blocks and mortar would be $8\cdot6$ inches, or 48 inches high. The result of 49 inches seems reasonable.

3 Evaluate formulas.

To evaluate a formula, we replace its letters, called **variables**, with specific numbers and evaluate the right side using the order of operations rule.

Self Check 5

The formula for the area of a triangle is $A=\frac{1}{2}bh$. Find the area of a triangle whose base is $12\frac{1}{2}$ meters long and whose height is $15\frac{1}{3}$ meters.

Now Try Problems 27 and 87

a

h

b

A trapezoid

EXAMPLE 5   The formula for the area of a trapezoid is $A=\frac{1}{2}h(a+b)$, where A is the area, h is the height, and a and b are the lengths of its bases. Find A when $h=1\frac{2}{3}$ in., $a=2\frac{1}{2}$ in., and $b=5\frac{1}{2}$ in.

Strategy In the formula, we will replace the letter h with $1\frac{2}{3}$, the letter a with $2\frac{1}{2}$, and the letter b with $5\frac{1}{2}$.

WHY Then we can use the order of operations rule to find the value of the expression on the right side of the $=$ symbol.

Solution

$$A=\frac{1}{2}h(a+b)$$

This is the formula for the area of a trapezoid.

$$=\frac{1}{2}\left(1\frac{2}{3}\right)\left(2\frac{1}{2}+5\frac{1}{2}\right)$$

Replace h, a, and b with the given values.

$$=\frac{1}{2}\left(1\frac{2}{3}\right)(8)$$

Do the addition within the parentheses: $2\frac{1}{2}+5\frac{1}{2}=8$.

$$=\frac{1}{2}\left(\frac{5}{3}\right)\left(\frac{8}{1}\right)$$

To prepare to multiply fractions, write $1\frac{2}{3}$ as an improper fraction and 8 as $\frac{8}{1}$.

$$=\frac{1\cdot5\cdot8}{2\cdot3\cdot1}$$

Multiply the numerators. Multiply the denominators.

$$=\frac{1\cdot5\cdot\overset{1}{\cancel{2}}\cdot4}{\underset{1}{\cancel{2}}\cdot3\cdot1}$$

To simplify, factor 8 as $2\cdot4$. Then remove the common factor of 2 from the numerator and denominator.

$$=\frac{20}{3}$$

Multiply the remaining factors in the numerator. Multiply the remaining factors in the denominator.

$$=6\frac{2}{3}$$

Write the improper fraction $\frac{20}{3}$ as a mixed number by dividing 20 by 3.

The area of the trapezoid is $6\frac{2}{3}$ in.^{2}.

4 Simplify complex fractions.

Fractions whose numerators and/or denominators contain fractions are called *complex fractions*. Here is an example of a complex fraction:

A fraction in the numerator ⟶ $\dfrac{3}{4}$

$\dfrac{}{}$ ⟵ The main fraction bar

A fraction in the denominator ⟶ $\dfrac{7}{8}$

Complex Fraction

A **complex fraction** is a fraction whose numerator or denominator, or both, contain one or more fractions or mixed numbers.

Here are more examples of complex fractions:

$$\dfrac{-\dfrac{1}{4} - \dfrac{4}{5}}{2\dfrac{4}{5}}$$

⟵ Numerator ⟶

⟵ Main fraction bar ⟶

⟵ Denominator ⟶

$$\dfrac{\dfrac{1}{3} + \dfrac{1}{4}}{\dfrac{1}{3} - \dfrac{1}{4}}$$

To *simplify* a complex fraction means to express it as a fraction in simplified form.

The following method for simplifying complex fractions is based on the fact that the main fraction bar indicates division.

$$\dfrac{\dfrac{1}{4}}{\dfrac{2}{5}}$$

⟵ The main fraction bar means "divide the fraction in the numerator by the fraction in the denominator." ⟶

$$\dfrac{1}{4} \div \dfrac{2}{5}$$

Simplifying a Complex Fraction

To simplify a complex fraction:

1. Add or subtract in the numerator and/or denominator so that the numerator is a single fraction and the denominator is a single fraction.

2. Perform the indicated division by multiplying the numerator of the complex fraction by the reciprocal of the denominator.

3. Simplify the result, if possible.

EXAMPLE 6

Simplify: $\dfrac{\dfrac{1}{4}}{\dfrac{2}{5}}$

Strategy We will perform the division indicated by the main fraction bar using the rule for dividing fractions from Section 4.3.

WHY We can skip step 1 and immediately divide because the numerator and the denominator of the complex fraction are already single fractions.

Self Check 6

Simplify: $\dfrac{\dfrac{1}{6}}{\dfrac{3}{8}}$

Now Try Problem 31

Solution

$$\frac{\dfrac{1}{4}}{\dfrac{2}{5}} = \frac{1}{4} \div \frac{2}{5}$$ Write the division indicated by the main fraction bar using a ÷ symbol.

$$= \frac{1}{4} \cdot \frac{5}{2}$$ Use the rule for dividing fractions: Multiply the first fraction by the reciprocal of $\frac{2}{5}$, which is $\frac{5}{2}$.

$$= \frac{1 \cdot 5}{4 \cdot 2}$$ Multiply the numerators. Multiply the denominators.

$$= \frac{5}{8}$$

Self Check 7

Simplify: $\dfrac{-\dfrac{5}{8} + \dfrac{1}{3}}{\dfrac{3}{4} - \dfrac{1}{3}}$

Now Try Problem 35

EXAMPLE 7

Simplify: $\dfrac{-\dfrac{1}{4} + \dfrac{2}{5}}{\dfrac{1}{2} - \dfrac{4}{5}}$

Strategy Recall that a fraction bar is a type of grouping symbol. We will work above and below the main fraction bar separately to write $-\frac{1}{4} + \frac{2}{5}$ and $\frac{1}{2} - \frac{4}{5}$ as single fractions.

WHY The numerator and the denominator of the complex fraction must be written as single fractions before dividing.

Solution To write the numerator as a single fraction, we build $-\frac{1}{4}$ and $\frac{2}{5}$ to have an LCD of 20, and then add. To write the denominator as a single fraction, we build $\frac{1}{2}$ and $\frac{4}{5}$ to have an LCD of 10, and subtract.

$$\frac{-\dfrac{1}{4} + \dfrac{2}{5}}{\dfrac{1}{2} - \dfrac{4}{5}} = \frac{-\dfrac{1}{4} \cdot \dfrac{5}{5} + \dfrac{2}{5} \cdot \dfrac{4}{4}}{\dfrac{1}{2} \cdot \dfrac{5}{5} - \dfrac{4}{5} \cdot \dfrac{2}{2}}$$

The LCD for the numerator is 20. Build each fraction so that each has a denominator of 20.

The LCD for the denominator is 10. Build each fraction so that each has a denominator of 10.

$$= \frac{-\dfrac{5}{20} + \dfrac{8}{20}}{\dfrac{5}{10} - \dfrac{8}{10}}$$

Multiply in the numerator. Multiply in the denominator.

$$= \frac{\dfrac{3}{20}}{-\dfrac{3}{10}}$$

In the numerator of the complex fraction, add the fractions.

In the denominator, subtract the fractions.

$$= \frac{3}{20} \div \left(-\frac{3}{10}\right)$$

Write the division indicated by the main fraction bar using a ÷ symbol.

$$= \frac{3}{20}\left(-\frac{10}{3}\right)$$

Multiply the first fraction by the reciprocal of $-\frac{3}{10}$, which is $-\frac{10}{3}$.

$$= -\frac{3 \cdot 10}{20 \cdot 3}$$

The product of two fractions with unlike signs is negative. Multiply the numerators. Multiply the denominators.

$$= -\frac{\overset{1}{\cancel{3}} \cdot \overset{1}{\cancel{10}}}{2 \cdot \underset{1}{\cancel{10}} \cdot \underset{1}{\cancel{3}}}$$

To simplify, factor 20 as 2 · 10. Then remove the common factors of 3 and 10 from the numerator and denominator.

$$= -\frac{1}{2}$$

Multiply the remaining factors in the numerator. Multiply the remaining factors in the denominator.

EXAMPLE 8

Simplify: $\dfrac{7 - \dfrac{2}{3}}{4\dfrac{5}{6}}$

Simplify: $\dfrac{5 - \dfrac{3}{4}}{1\dfrac{7}{8}}$

Now Try Problem 39

Strategy Recall that a fraction bar is a type of grouping symbol. We will work above and below the main fraction bar separately to write $7 - \frac{2}{3}$ as a single fraction and $4\frac{5}{6}$ as an improper fraction.

WHY The numerator and the denominator of the complex fraction must be written as single fractions before dividing.

Solution

$\dfrac{7 - \dfrac{2}{3}}{4\dfrac{5}{6}} = \dfrac{\dfrac{7}{1} \cdot \dfrac{3}{3} - \dfrac{2}{3}}{\dfrac{29}{6}}$ In the numerator, write 7 as $\frac{7}{1}$. The LCD for the numerator is 3. Build $\frac{7}{1}$ so that it has a denominator of 3. In the denominator, write $4\frac{5}{6}$ as the improper fraction $\frac{29}{6}$.

$= \dfrac{\dfrac{21}{3} - \dfrac{2}{3}}{\dfrac{29}{6}}$ Multiply in the numerator.

$= \dfrac{\dfrac{19}{3}}{\dfrac{29}{6}}$ In the numerator of the complex fraction, subtract the numerators: $21 - 2 = 19$. Then write the difference over the common denominator 3.

$= \dfrac{19}{3} \div \dfrac{29}{6}$ Write the division indicated by the main fraction bar using a ÷ symbol.

$= \dfrac{19}{3} \cdot \dfrac{6}{29}$ Multiply the first fraction by the reciprocal of $\frac{29}{6}$, which is $\frac{6}{29}$.

$= \dfrac{19 \cdot 6}{3 \cdot 29}$ Multiply the numerators. Multiply the denominators.

$= \dfrac{19 \cdot 2 \cdot \overset{1}{\cancel{3}}}{\underset{1}{\cancel{3}} \cdot 29}$ To simplify, factor 6 as 2 · 3. Then remove the common factor of 3 from the numerator and denominator.

$= \dfrac{38}{29}$ Multiply the remaining factors in the numerator. Multiply the remaining factors in the denominator.

ANSWERS TO SELF CHECKS

1. $\frac{31}{32}$ **2.** $-\frac{1}{9}$ **3.** $2\frac{11}{24}$ **4.** 61 in.   **5.** $95\frac{5}{6}$ m²   **6.** $\frac{4}{9}$ **7.** $-\frac{7}{10}$ **8.** $\frac{34}{15}$

SECTION 4.7   STUDY SET

VOCABULARY

Fill in the blanks.

1. We use the order of _____ rule to evaluate expressions that contain more than one operation.

2. To evaluate a formula such as $A = \frac{1}{2}h(a + b)$, we substitute specific numbers for the letters, called _____, in the formula and find the value of the right side.

3. $\dfrac{\dfrac{1}{2}}{\dfrac{3}{4}}$ and $\dfrac{\dfrac{7}{8}+\dfrac{2}{5}}{\dfrac{1}{2}-\dfrac{1}{3}}$ are examples of _____ fractions.

4. In the complex fraction $\dfrac{\dfrac{2}{5}+\dfrac{1}{4}}{\dfrac{2}{5}-\dfrac{1}{4}}$, the _____

is $\dfrac{2}{5}+\dfrac{1}{4}$ and the _____ is $\dfrac{2}{5}-\dfrac{1}{4}$.

CONCEPTS

5. What operations are involved in this expression?

$$5\left(6\frac{1}{3}\right)+\left(-\frac{1}{4}\right)^3$$

6. a. To evaluate $\frac{7}{8}+\left(\frac{1}{3}\right)\left(\frac{1}{4}\right)$, what operation should be performed first?

  b. To evaluate $\frac{7}{8}+\left(\frac{1}{3}-\frac{1}{4}\right)^2$, what operation should be performed first?

7. Translate the following to numbers and symbols. *You do not have to find the answer.*

Add $1\frac{2}{15}$ to the difference of $\frac{2}{3}$ and $\frac{1}{10}$.

8. Refer to the trapezoid shown below. Label the length of the upper base $3\frac{1}{2}$ inches, the length of the lower base $5\frac{1}{2}$ inches, and the height $2\frac{2}{3}$ inches.

9. What division is represented by this complex fraction?

$$\dfrac{\dfrac{2}{3}}{\dfrac{1}{5}}$$

10. Consider: $\dfrac{\dfrac{2}{3}-\dfrac{1}{5}}{\dfrac{1}{2}+\dfrac{4}{5}}$

  a. What is the LCD for the fractions in the numerator of this complex fraction?

  b. What is the LCD for the fractions in the denominator of this complex fraction?

11. Write the denominator of the following complex fraction as an improper fraction.

$$\dfrac{\dfrac{1}{8}-\dfrac{3}{16}}{5\dfrac{3}{4}}$$

12. When this complex fraction is simplified, will the result be positive or negative?

$$\dfrac{-\dfrac{2}{3}}{\dfrac{3}{4}}$$

NOTATION

Fill in the blanks to complete each solution.

13.
$$\frac{7}{12}-\frac{1}{2}\cdot\frac{1}{3}=\frac{7}{12}-\frac{1\cdot1}{2\cdot\boxed{}}$$
$$=\frac{7}{12}-\frac{1}{\boxed{}}$$
$$=\frac{7}{12}-\frac{1}{6}\cdot\frac{\boxed{}}{\boxed{}}$$
$$=\frac{7}{12}-\frac{\boxed{}}{12}$$
$$=\frac{\boxed{}}{12}$$

14.
$$\dfrac{\dfrac{1}{8}}{\dfrac{3}{4}}=\frac{1}{8}\div\frac{\boxed{}}{\boxed{}}$$
$$=\frac{1}{8}\cdot\frac{\boxed{}}{\boxed{}}$$
$$=\frac{1\cdot\boxed{}}{8\cdot3}$$
$$=\frac{1\cdot\overset{1}{\cancel{}}}{2\cdot\underset{1}{\cancel{}}\cdot3}$$
$$=\frac{1}{\boxed{}}$$

GUIDED PRACTICE

Evaluate each expression. See Example 1.

15. $\dfrac{3}{4}+\dfrac{2}{5}\left(-\dfrac{1}{2}\right)^2$

16. $\dfrac{1}{4}+\dfrac{8}{27}\left(-\dfrac{3}{2}\right)^2$

17. $\dfrac{1}{6}+\dfrac{9}{8}\left(-\dfrac{2}{3}\right)^3$

18. $\dfrac{1}{5}+\dfrac{1}{9}\left(-\dfrac{3}{2}\right)^3$

Evaluate each expression. See Example 2.

19. $\left(\dfrac{3}{4} - \dfrac{1}{6}\right) \div \left(-2\dfrac{1}{6}\right)$

20. $\left(\dfrac{7}{8} - \dfrac{3}{7}\right) \div \left(-1\dfrac{3}{7}\right)$

21. $\left(\dfrac{15}{16} - \dfrac{1}{8}\right) \div \left(-9\dfrac{3}{4}\right)$

22. $\left(\dfrac{19}{36} - \dfrac{1}{6}\right) \div \left(-8\dfrac{2}{3}\right)$

Evaluate each expression. See Example 3.

23. Add $5\dfrac{4}{15}$ to the difference of $\dfrac{5}{6}$ and $\dfrac{2}{3}$.

24. Add $8\dfrac{5}{24}$ to the difference of $\dfrac{3}{4}$ and $\dfrac{1}{6}$.

25. Add $2\dfrac{7}{18}$ to the difference of $\dfrac{7}{9}$ and $\dfrac{1}{2}$.

26. Add $1\dfrac{19}{30}$ to the difference of $\dfrac{4}{5}$ and $\dfrac{1}{2}$.

Evaluate the formula $A = \frac{1}{2}h(a + b)$ for the given values. See Example 5.

27. $a = 2\dfrac{1}{2}, b = 7\dfrac{1}{2}, h = 5\dfrac{1}{4}$

28. $a = 4\dfrac{1}{2}, b = 5\dfrac{1}{2}, h = 2\dfrac{1}{8}$

29. $a = 1\dfrac{1}{4}, b = 6\dfrac{3}{4}, h = 4\dfrac{1}{2}$

30. $a = 1\dfrac{1}{3}, b = 4\dfrac{2}{3}, h = 2\dfrac{2}{5}$

Simplify each complex fraction. See Example 6.

31. $\dfrac{\frac{1}{16}}{\frac{2}{5}}$ **32.** $\dfrac{\frac{2}{11}}{\frac{3}{4}}$

33. $\dfrac{\frac{5}{8}}{\frac{3}{4}}$ **34.** $\dfrac{\frac{1}{5}}{\frac{8}{15}}$

Simplify each complex fraction. See Example 7.

35. $\dfrac{-\frac{1}{4} + \frac{2}{3}}{\frac{5}{6} + \frac{2}{3}}$ **36.** $\dfrac{-\frac{1}{2} + \frac{7}{8}}{\frac{3}{4} - \frac{1}{2}}$

37. $\dfrac{\frac{1}{3} - \frac{3}{4}}{\frac{1}{6} + \frac{2}{3}}$ **38.** $\dfrac{\frac{1}{3} - \frac{3}{4}}{\frac{1}{6} + \frac{1}{3}}$

Simplify each complex fraction. See Example 8.

39. $\dfrac{5 - \frac{5}{6}}{1\frac{1}{12}}$ **40.** $\dfrac{4 - \frac{3}{4}}{1\frac{7}{8}}$

41. $\dfrac{4 - \frac{7}{8}}{3\frac{1}{4}}$ **42.** $\dfrac{6 - \frac{2}{7}}{6\frac{2}{3}}$

TRY IT YOURSELF

Evaluate each expression and simplify each complex fraction.

43. $\dfrac{7}{8} - \left(\dfrac{4}{5} + 1\dfrac{3}{4}\right)$

44. $\left(\dfrac{5}{4}\right)^2 + \left(\dfrac{2}{3} - 2\dfrac{1}{6}\right)$

45. $\dfrac{-\frac{14}{15}}{\frac{7}{10}}$

46. $\dfrac{\frac{5}{27}}{-\frac{5}{9}}$

47. $A = \frac{1}{2}bh$ for $b = 10$ and $h = 7\frac{1}{5}$

48. $V = lwh$ for $l = 12, w = 8\frac{1}{2}$, and $h = 3\frac{1}{3}$

49. $\dfrac{2}{3}\left(-\dfrac{1}{4}\right) + \dfrac{1}{2}$

50. $-\dfrac{7}{8} - \left(\dfrac{1}{8}\right)\left(\dfrac{2}{3}\right)$

51. $\dfrac{4}{5} - \left(-\dfrac{1}{3}\right)^2$

52. $-\dfrac{3}{16} - \left(-\dfrac{1}{2}\right)^3$

53. $\dfrac{\frac{3}{8} + \frac{1}{4}}{\frac{3}{8} - \frac{1}{4}}$

54. $\dfrac{\frac{2}{5} + \frac{1}{4}}{\frac{2}{5} - \frac{1}{4}}$

55. Add $12\dfrac{11}{12}$ to the difference of $5\dfrac{1}{6}$ and $3\dfrac{7}{8}$.

56. Add $18\dfrac{1}{3}$ to the difference of $11\dfrac{3}{5}$ and $9\dfrac{11}{15}$.

57. $\dfrac{5\frac{1}{2}}{-\frac{1}{4}+\frac{3}{4}}$

58. $\dfrac{4\frac{1}{4}}{\frac{2}{3}+\left(-\frac{1}{6}\right)}$

59. $\left|\dfrac{2}{3}-\dfrac{9}{10}\right| \div \left(-\dfrac{1}{5}\right)$

60. $\left|-\dfrac{3}{16} \div 2\frac{1}{4}\right| + \left(-2\frac{1}{8}\right)$

61. $\dfrac{\frac{1}{5}-\left(-\frac{1}{4}\right)}{\frac{1}{4}+\frac{4}{5}}$

62. $\dfrac{\frac{1}{8}-\left(-\frac{1}{2}\right)}{\frac{1}{4}+\frac{3}{8}}$

63. $1\dfrac{3}{5}\left(\dfrac{1}{2}\right)^{2}\left(\dfrac{3}{4}\right)$

64. $2\dfrac{3}{5}\left(-\dfrac{1}{3}\right)^{2}\left(\dfrac{1}{2}\right)$

65. $A = lw$ for $l = 5\frac{5}{6}$ and $w = 7\frac{3}{5}$.

66. $P = 2l + 2w$ for $l = \frac{7}{8}$ and $w = \frac{3}{5}$.

67. $\left(2-\dfrac{1}{2}\right)^{2} + \left(2+\dfrac{1}{2}\right)^{2}$

68. $\left(\dfrac{9}{20} \div 2\dfrac{2}{5}\right) + \left(\dfrac{3}{4}\right)^{2}$

69. $\dfrac{-\frac{5}{6}}{-1\frac{7}{8}}$

70. $\dfrac{-\frac{4}{3}}{-2\frac{5}{6}}$

71. Subtract $9\frac{1}{10}$ from the sum of $7\frac{3}{7}$ and $3\frac{1}{5}$.

72. Subtract $3\frac{2}{3}$ from the sum of $2\frac{5}{12}$ and $1\frac{5}{8}$.

73. $\dfrac{\frac{1}{2}+\frac{1}{4}}{\frac{1}{2}-\frac{1}{4}}$

74. $\dfrac{\frac{1}{3}+\frac{1}{4}}{\frac{1}{3}-\frac{1}{4}}$

75. $\left(\dfrac{8}{5}-1\dfrac{1}{3}\right)-\left(-\dfrac{4}{5}\cdot 10\right)$

76. $\left(1-\dfrac{3}{4}\right)\left(1+\dfrac{3}{4}\right)$

▌APPLICATIONS

77. REMODELING A BATHROOM A handyman installed 20 rows of grout and tile on a bathroom wall using the pattern shown below. How high above floor level does the tile work reach? (*Hint:* There is no grout line above the last row of tiles.)

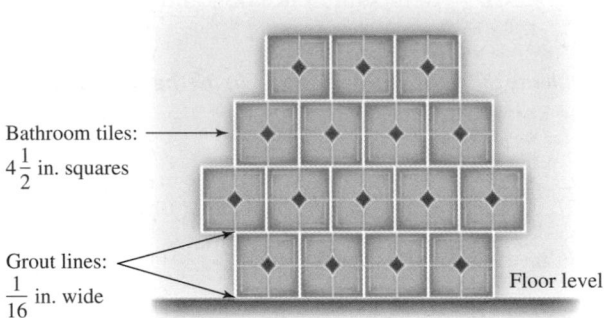

Bathroom tiles: $4\frac{1}{2}$ in. squares

Grout lines: $\frac{1}{16}$ in. wide

Floor level

78. PLYWOOD To manufacture a sheet of plywood, several thin layers of wood are glued together, as shown. Then an exterior finish is attached to the top and the bottom, as shown below. How thick is the final product?

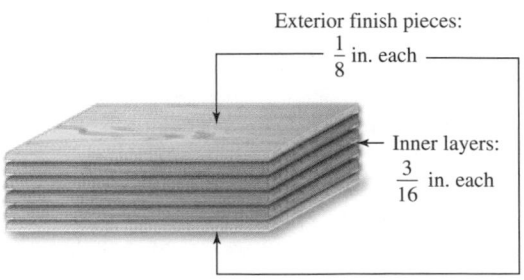

Exterior finish pieces: $\frac{1}{8}$ in. each

Inner layers: $\frac{3}{16}$ in. each

79. POSTAGE RATES Can the advertising package shown below be mailed for the 1-ounce rate?

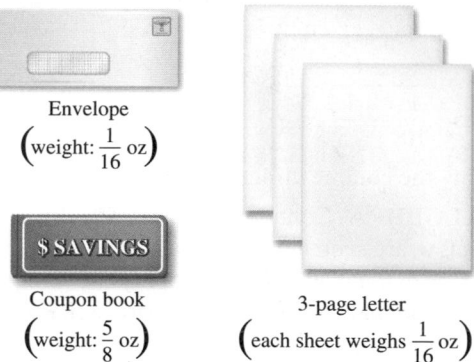

Envelope
$\left(\text{weight: } \frac{1}{16} \text{ oz}\right)$

$ SAVINGS

Coupon book
$\left(\text{weight: } \frac{5}{8} \text{ oz}\right)$

3-page letter
$\left(\text{each sheet weighs } \frac{1}{16} \text{ oz}\right)$

80. PHYSICAL THERAPY After back surgery, a patient followed a walking program shown in the table below to strengthen her muscles. What was the total distance she walked over this three-week period?

Week	Distance per day
#1	$\frac{1}{4}$ mile
#2	$\frac{1}{2}$ mile
#3	$\frac{3}{4}$ mile

81. READING PROGRAMS To improve reading skills, elementary school children read silently at the end of the school day for $\frac{1}{4}$ hour on Mondays and for $\frac{1}{2}$ hour on Fridays. For the month of January, how many total hours did the children read silently in class?

S	M	T	W	T	F	S
	1	2	3	4	5	6
7	8	9	10	11	12	13
14	15	16	17	18	19	20
21	22	23	24	25	26	27
28	29	30	31			

82. PHYSICAL FITNESS Two people begin their workouts from the same point on a bike path and travel in opposite directions, as shown below. How far apart are they in $1\frac{1}{2}$ hours? Use the table to help organize your work.

	Rate (mph)	·	Time (hr)	=	Distance (mi)
Jogger					
Cyclist					

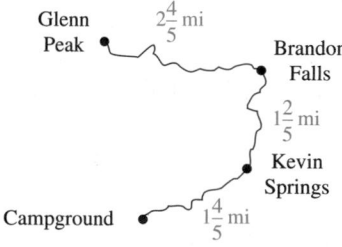

Jogger: $2\frac{1}{2}$ mph   Cyclist: $7\frac{1}{5}$ mph

Start

83. HIKING A scout troop plans to hike from the campground to Glenn Peak, as shown below. Since the terrain is steep, they plan to stop and rest after every $\frac{2}{3}$ mile. With this plan, how many parts will there be to this hike?

Glenn Peak

$2\frac{4}{5}$ mi

Brandon Falls

$1\frac{2}{5}$ mi

Kevin Springs

Campground $1\frac{4}{5}$ mi

84. DELI SHOPS A sandwich shop sells a $\frac{1}{2}$-pound club sandwich made of turkey and ham. The owner buys the turkey in $1\frac{3}{4}$-pound packages and the ham in $2\frac{1}{2}$-pound packages. If he mixes two packages of turkey and one package of ham together, how many sandwiches can he make from the mixture?

85. SKIN CREAMS Using a formula of $\frac{1}{2}$ ounce of sun block, $\frac{2}{3}$ ounce of moisturizing cream, and $\frac{3}{4}$ ounce of lanolin, a beautician mixes her own brand of skin cream. She packages it in $\frac{1}{4}$-ounce tubes. How many full tubes can be produced using this formula? How much skin cream is left over?

86. SLEEP The graph below compares the amount of sleep a 1-month-old baby got to the $15\frac{1}{2}$-hour daily requirement recommended by Children's Hospital of Orange County, California. For the week, how far below the baseline was the baby's daily average?

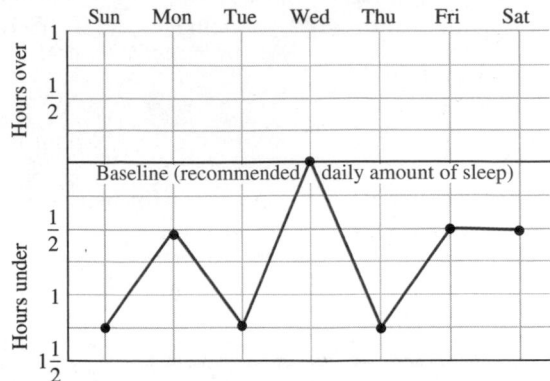

87. CAMPING The four sides of a tent are all the same trapezoid-shape. (See the illustration below.) How many square yards of canvas are used to make one of the sides of the tent?

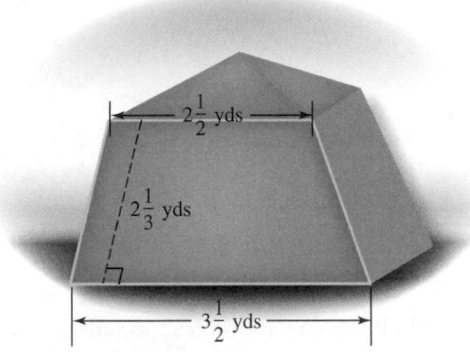

88. SEWING A seamstress begins with a trapezoid-shaped piece of denim to make the back pocket on a pair of jeans. (See the illustration below.) How many square inches of denim are used to make the pocket?

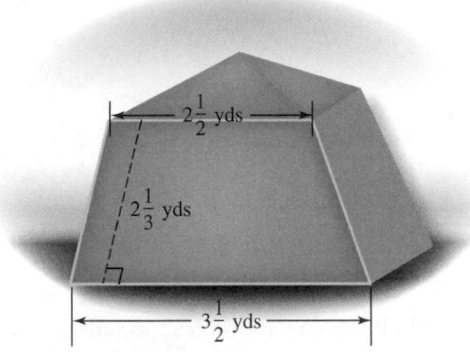

Finished pocket

89. AMUSEMENT PARKS At the end of a ride at an amusement park, a boat splashes into a pool of water. The time (in seconds) that it takes two pipes to refill the pool is given by

$$\frac{1}{\frac{1}{10} + \frac{1}{15}}$$

Simplify the complex fraction to find the time.

90. ALGEBRA Complex fractions, like the one shown below, are seen in an algebra class when the topic of *slope of a line* is studied. Simplify this complex fraction and, as is done in algebra, write the answer as an improper fraction.

$$\frac{\frac{1}{2} - \frac{1}{3}}{\frac{1}{4} - \frac{1}{5}}$$

WRITING

91. Why is an order of operations rule necessary?

92. What does it mean to evaluate a formula?

93. What is a complex fraction?

94. In the complex fraction $\dfrac{\frac{3}{8} + \frac{1}{4}}{\frac{3}{8} - \frac{1}{4}}$, the fraction bar serves as a grouping symbol. Explain why this is so.

REVIEW

95. Find the sum: $8 + 19 + 124 + 2{,}097$

96. Subtract 879 from 1,023.

97. Multiply 879 by 23.

98. Divide 1,665 by 45.

99. List the factors of 24.

100. Find the prime factorization of 24.

Solving Equations That Involve Fractions

In this section, we will discuss how to solve equations that involve fractions and equations whose solutions are fractions. We will make use of several concepts from this chapter, including the reciprocal and the LCD.

Objectives

1. Use the addition and subtraction properties of equality to solve equations that involve fractions.
2. Use reciprocals to solve equations.
3. Clear equations of fractions.
4. Use equations to solve application problems that involve fractions.

1 Use the addition and subtraction properties of equality to solve equations that involve fractions.

Recall that to **solve an equation,** we find all the values of the variable that make the equation true. The properties of equality that we used to solve equations involving whole numbers and integers are also used to solve equations involving fractions.

EXAMPLE 1 Solve: $y - \dfrac{15}{32} = \dfrac{5}{32}$

Strategy We will use the addition property of equality to isolate the variable y on the left side of the equation.

WHY To solve the original equation, we want to find a simpler equivalent equation of the form $y = $ **a number**, whose solution is obvious.

Solution

$$y - \frac{15}{32} = \frac{5}{32}$$ This is the equation to solve.

$$y - \frac{15}{32} + \frac{15}{32} = \frac{5}{32} + \frac{15}{32}$$ To isolate y, undo the subtraction of $\frac{15}{32}$ by adding $\frac{15}{32}$ to both sides.

$$y = \frac{20}{32}$$ On the left side, $-\frac{15}{32} + \frac{15}{32} = 0$. On the right side, $\frac{5}{32} + \frac{15}{32} = \frac{20}{32}$.

$$y = \frac{\overset{1}{\cancel{4}} \cdot 5}{\underset{1}{\cancel{4}} \cdot 8}$$ To simplify the fraction, factor 20 as $4 \cdot 5$ and 32 as $4 \cdot 8$. Then remove the common factor of 4 from the numerator and denominator.

$$y = \frac{5}{8}$$ Multiply the remaining factors in the numerator: $1 \cdot 5 = 5$. Multiply the remaining factors in the denominator: $1 \cdot 8 = 8$.

To check this result, substitute $\frac{5}{8}$ for y in the original equation and simplify.

Check: $y - \dfrac{15}{32} = \dfrac{5}{32}$ This is the original equation.

$$\frac{5}{8} - \frac{15}{32} \overset{?}{=} \frac{5}{32}$$ Substitute $\frac{5}{8}$ for y.

$$\frac{5}{8} \cdot \frac{4}{4} - \frac{15}{32} \overset{?}{=} \frac{5}{32}$$ To prepare to subtract the fractions on the left side, build $\frac{5}{8}$ so that its denominator is 32.

$$\frac{20}{32} - \frac{15}{32} \overset{?}{=} \frac{5}{32}$$ On the left side, multiply the numerators and multiply the denominators.

$$\frac{5}{32} = \frac{5}{32}$$ On the left side, subtract the numerators and write the difference over the common denominator, 32.

Since $\frac{5}{32} = \frac{5}{32}$ is a true statement, $\frac{5}{8}$ is the solution of $y - \frac{15}{32} = \frac{5}{32}$.

Self Check 1

Solve: $a - \dfrac{1}{16} = \dfrac{11}{16}$

Now Try Problem 17

Solve $\frac{2}{3} = y + \frac{1}{5}$ and check the result.

Now Try Problem 21

EXAMPLE 2 Solve: $\frac{3}{4} = x + \frac{1}{6}$

Strategy We will use the subtraction property of equality to isolate the variable x on the right side of the equation.

WHY To solve the original equation, we want to find a simpler equivalent equation of the form **a number** $= x$, whose solution is obvious.

Solution

$$\frac{3}{4} = x + \frac{1}{6}$$ This is the equation to solve.

$$\frac{3}{4} - \mathbf{\frac{1}{6}} = x + \frac{1}{6} - \mathbf{\frac{1}{6}}$$ To isolate x, undo the addition of $\frac{1}{6}$ by subtracting $\frac{1}{6}$ from both sides.

$$\frac{3}{4} - \frac{1}{6} = x$$ On the right side, do the addition: $\frac{1}{6} - \frac{1}{6} = 0$.

$$\frac{3}{4} \cdot \frac{3}{3} - \frac{1}{6} \cdot \frac{2}{2} = x$$ To build $\frac{3}{4}$ and $\frac{1}{6}$ so that their denominators are 12, multiply each by a form of 1.

$$\frac{9}{12} - \frac{2}{12} = x$$ Multiply the numerators. Multiply the denominators.

$$\frac{7}{12} = x$$ On the left side, subtract the numerators and write the difference over the common denominator, 12.

Since 7 and 12 have no common factors other than 1, the result is in simplest form. The solution is $\frac{7}{12}$. Verify this result by substituting it into the original equation. ∎

2 Use reciprocals to solve equations.

Recall that the product of a number and its reciprocal is 1.

$$\frac{2}{3} \cdot \frac{3}{2} = 1 \qquad -\frac{4}{5}\left(-\frac{5}{4}\right) = 1 \qquad 9 \cdot \frac{1}{9} = 1$$

We can use this fact to solve equations such as $\frac{2}{3}x = 6$ and $-\frac{5}{4}x = 3$, where the coefficient of the variable term is a fraction.

$$\frac{2}{3}x = 6 \qquad\qquad -\frac{5}{4}x = 3$$

↑ The coefficient of x is a fraction.　　↑ The coefficient of x is a negative fraction.

Solve and check the result.

a. $\frac{7}{2}b = 21$

b. $-\frac{3}{8}b = 2$

Now Try Problems 25 and 29

EXAMPLE 3 Solve: **a.** $\frac{2}{3}x = 6$ **b.** $-\frac{5}{4}x = 3$

Strategy To isolate the variable x, we will multiply both sides of the equation by the reciprocal of the coefficient of the variable term.

WHY To isolate the variable means that we want its coefficient to be 1. The product of a number and its reciprocal will produce such a coefficient.

Solution

a. Recall that $\frac{2}{3}x = 6$ means $\frac{2}{3} \cdot x = 6$. Since the coefficient of x is $\frac{2}{3}$, we can isolate x by multiplying both sides of the equation by the reciprocal of $\frac{2}{3}$.

$$\frac{2}{3}x = 6 \qquad \text{This is the equation to solve.}$$

$$\frac{3}{2} \cdot \frac{2}{3}x = \frac{3}{2} \cdot 6 \qquad \begin{array}{l} \text{To isolate } x, \text{ undo the multiplication by } \frac{2}{3}, \text{ multiplying both sides} \\ \text{by the reciprocal of } \frac{2}{3}, \text{ which is } \frac{3}{2}. \end{array}$$

$$\left(\frac{3}{2} \cdot \frac{2}{3}\right)x = \frac{3}{2} \cdot \frac{6}{1} \qquad \begin{array}{l} \text{On the left side, use the associative property of multiplication} \\ \text{to group } \frac{3}{2} \text{ and } \frac{2}{3}. \text{ On the right side, write } 6 \text{ as } \frac{6}{1}. \end{array}$$

$$1x = \frac{3 \cdot 6}{2 \cdot 1} \qquad \begin{array}{l} \text{On the left side, the product of a number and its reciprocal is 1:} \\ \frac{3}{2} \cdot \frac{2}{3} = 1. \text{ On the right side, multiply the numerators and multiply} \\ \text{the denominators.} \end{array}$$

$$x = \frac{3 \cdot \overset{1}{\cancel{2}} \cdot 3}{\underset{1}{\cancel{2}} \cdot 1} \qquad \begin{array}{l} \text{On the left side, the coefficient of 1 need not be written since} \\ 1x = x. \text{ To simplify the right side, factor } 6 \text{ as } 2 \cdot 3 \text{ and remove} \\ \text{the common factor of 2.} \end{array}$$

$$x = \frac{9}{1} \qquad \begin{array}{l} \text{On the right side, multiply the remaining factors in the} \\ \text{numerator } (3 \cdot 1 \cdot 3 = 9) \text{ and multiply the remaining factors in} \\ \text{the denominator } (1 \cdot 1 = 1.). \end{array}$$

$$x = 9 \qquad \text{Any number divided by 1 is equal to that number.}$$

To check this result, substitute 9 for x in the original equation and simplify.

Check: $\quad \dfrac{2}{3}x = 6 \qquad$ This is the original equation.

$$\frac{2}{3}(9) \overset{?}{=} 6 \qquad \text{Substitute 9 for } x.$$

$$6 = 6 \qquad \text{On the left side, } \frac{2}{3}(9) = \frac{18}{3} = 6.$$

Since the statement $6 = 6$ is true, 9 is the solution of $\frac{2}{3}x = 6$.

b. Recall that $-\frac{5}{4}x = 3$ means $-\frac{5}{4} \cdot x = 3$. Since the coefficient of x is $-\frac{5}{4}$, we can isolate x by multiplying both sides of the equation by the reciprocal of $-\frac{5}{4}$.

$$-\frac{5}{4}x = 3 \qquad \text{This is the equation to solve.}$$

$$-\frac{4}{5}\left(-\frac{5}{4}x\right) = -\frac{4}{5}(3) \qquad \begin{array}{l} \text{To isolate } x, \text{ undo the multiplication by } -\frac{5}{4} \text{ by multiplying} \\ \text{both sides by the reciprocal of } -\frac{5}{4}, \text{ which is } -\frac{4}{5}. \end{array}$$

$$\left[-\frac{4}{5}\left(-\frac{5}{4}\right)\right]x = -\frac{4}{5}(3) \qquad \begin{array}{l} \text{On the left side, use the associative property of} \\ \text{multiplication to group } -\frac{4}{5} \text{ and } -\frac{5}{4}. \end{array}$$

$$1x = -\frac{4}{5}\left(\frac{3}{1}\right) \qquad \begin{array}{l} \text{On the left side, the product of a number and its reciprocal} \\ \text{is 1: } -\frac{4}{5}\left(-\frac{5}{4}\right) = 1. \text{ On the right side, write } 3 \text{ as } \frac{3}{1}. \end{array}$$

$$x = -\frac{12}{5} \qquad \begin{array}{l} \text{On the left side, the coefficient 1 need not be written since} \\ 1x = x. \text{ On the right side, multiply the numerators and} \\ \text{multiply the denominators.} \\ \text{The product of two numbers with unlike signs is negative.} \end{array}$$

Since 12 and 5 have no common factors other than 1, the result is in simplest form. The solution is $-\frac{12}{5}$. Verify that this is correct by checking.

Caution! In algebra, we usually leave a solution to an equation as an improper fraction (in simplified form) rather than converting it to a mixed number. The one exception is when solving application problems, where presenting the solution in mixed-number form is often more informative.

> ***Success Tip*** Variable terms with fractional coefficients can be written in two ways. For example,
>
> $$\frac{2}{3}x = \frac{2}{3} \cdot \frac{x}{1} = \frac{2x}{3} \qquad \text{Thus,} \frac{2}{3}x = \frac{2x}{3}.$$
>
> Similarly,
>
> $$-\frac{5}{4}x = -\frac{5}{4} \cdot \frac{x}{1} = -\frac{5x}{4} \qquad \text{Thus,} -\frac{5}{4}x = -\frac{5x}{4}.$$

Another method for solving equations such as $\frac{2}{3}x = 6$ uses *two steps* to isolate the variable.

Self Check 4

Solve $\frac{7}{2}b = 21$ using a two-step process.

Now Try Problem 33

EXAMPLE 4   Solve $\frac{2}{3}x = 6$ using a two-step process.

Strategy We will use two properties of equality to isolate the variable x on one side of the equation.

WHY In the expression $\frac{2}{3}x$, we will consider the variable x to be multiplied by 2 and that the product divided by 3.

Solution We will undo the multiplication and division performed on the variable in reverse order.

$$\frac{2}{3}x = 6 \qquad \text{This is the equation to solve.}$$

$$3 \cdot \frac{2}{3}x = 3 \cdot 6 \qquad \begin{array}{l}\text{To isolate 2x on the left side, undo the division by 3}\\\text{by multiplying both sides by 3.}\end{array}$$

$$\left(3 \cdot \frac{2}{3}\right)x = 3 \cdot 6 \qquad \begin{array}{l}\text{On the left side, use the associative property of multiplication to}\\\text{regroup the factors.}\end{array}$$

$$\left(\frac{\overset{1}{\cancel{3}} \cdot 2}{1 \cdot \underset{1}{\cancel{3}}}\right)x = 18 \qquad \begin{array}{l}\text{On the left side, write 3 as } \frac{3}{1}, \text{ multiply the numerators and multiply the}\\\text{demominators, and then remove the common factor of 3. On the right}\\\text{side, do the multiplication.}\end{array}$$

$$2x = 18 \qquad \text{On the left side, simplify the expression within the parentheses.}$$

$$\frac{2x}{2} = \frac{18}{2} \qquad \text{To isolate x, undo the multiplication by 2 by dividing both sides by 2.}$$

$$x = 9 \qquad \text{Do the division.}$$

The solution is 9. (As expected, this is the same as the solution obtained using the reciprocal method in Example 3, part a.)

Self Check 5

Solve $\frac{1}{11}q = \frac{19}{55}$ and check the result.

Now Try Problem 37

EXAMPLE 5   Solve: $\frac{1}{9}r = \frac{17}{27}$

Strategy To isolate the variable r, we will multiply both sides of the equation by the reciprocal of the coefficient of the variable term $\frac{1}{9}r$.

WHY To isolate the variable means that we want the coefficient of r to be 1. The product of $\frac{1}{9}$ and its reciprocal will produce such a coefficient.

Solution Recall that $\frac{1}{9}r = \frac{1}{9} \cdot r$. Since the coefficient of r is $\frac{1}{9}$, we multiply both sides of the equation by the reciprocal of $\frac{1}{9}$.

$$\frac{1}{9}r = \frac{17}{27}$$ This is the equation to solve.

$$\mathbf{9} \cdot \frac{1}{9}r = \mathbf{9} \cdot \frac{17}{27}$$ To isolate r, undo the multiplication by $\frac{1}{9}$ by multiplying both sides by the reciprocal of $\frac{1}{9}$, which is 9.

$$\left(9 \cdot \frac{1}{9} \right)r = \frac{9}{1} \cdot \frac{17}{27}$$ On the left side, use the associative property of multiplication to group 9 and $\frac{1}{9}$. Write 9 as $\frac{9}{1}$.

$$1r = \frac{9 \cdot 17}{1 \cdot 27}$$ On the left side, the product of a number and its reciprocal is 1: $9 \cdot \frac{1}{9} = 1$. On the right side, multiply the numerators and multiply the denominators.

$$r = \left(\frac{\overset{1}{\cancel{9}} \cdot 17}{1 \cdot 3 \cdot \underset{1}{\cancel{9}}} \right)$$ One the left side, the coefficient of 1 need not be written since $1r = r$. To simplify the right side, factor 27 as $3 \cdot 9$ and remove the common factor of 9.

$$r = \frac{17}{3}$$ Multiply the remaining factors in the numerator: $1 \cdot 17 = 17$. Multiply the remaining factors in the denominator: $1 \cdot 3 \cdot 1 = 3$.

To check this result, we substitute $\frac{17}{3}$ for r in the original equation.

Check:

$$\frac{1}{9}r = \frac{17}{27}$$ This is the original equation.

$$\frac{1}{9} \cdot \frac{\mathbf{17}}{\mathbf{3}} \overset{?}{=} \frac{17}{27}$$ Substitute $\frac{17}{3}$ for r.

$$\frac{17}{27} = \frac{17}{27}$$ On the left side, multiply the numerators and multiply the denominators.

Since the statement $\frac{17}{27} = \frac{17}{27}$ is true, $\frac{17}{3}$ is the solution of $\frac{1}{9}r = \frac{17}{27}$.

EXAMPLE 6 Solve: $24z = -\dfrac{11}{3}$

Strategy To isolate the variable z, we will multiply both sides of the equation by the reciprocal of the coefficient of the variable term $24z$.

WHY To isolate z, we can either divide both sides by 24 or multiply both sides by the reciprocal of 24, which is $\frac{1}{24}$. Since it is easier to find $\frac{1}{24}\left(-\frac{11}{3}\right)$ than $\dfrac{-\frac{11}{3}}{24}$, we will use the reciprocal approach.

Solution

$$24z = -\frac{11}{3}$$ This is the equation to solve.

$$\frac{\mathbf{1}}{\mathbf{24}}\left(24z\right) = \frac{\mathbf{1}}{\mathbf{24}}\left(-\frac{11}{3}\right)$$ To isolate z, undo the multiplication by 24 by multiplying both sides by the reciprocal of 24, which is $\frac{1}{24}$.

$$\left(\frac{1}{24} \cdot 24\right)z = -\frac{1 \cdot 11}{24 \cdot 3}$$ On the left side, use the associative property of multiplication to group $\frac{1}{24}$ and 24. On the right side, multiply the numerators and multiply the denominators.
└─ The product of two numbers with unlike signs is negative.

$$1z = -\frac{11}{72}$$ On the left side, the product of a number and its reciprocal is 1: $\frac{1}{24} \cdot 24 = 1$. On the right, do the multiplication in the numerator and the denominator.

$$\begin{array}{r} 24 \\ \times\ 3 \\ \hline 72 \end{array}$$

$$z = -\frac{11}{72}$$ On the left side, the coefficient of 1 need not be written since $1z = z$.

Since 11 and 72 have no common factors other than 1, the result is in simplest form. The solution is $-\frac{11}{72}$. Verify that this is correct by checking.

Self Check 6

Solve $42n = -\dfrac{13}{2}$ and check the result.

Now Try Problem 41

Sometimes several properties of equality must be used to solve an equation.

Solve $\frac{7}{12}a - 6 = -27$ and check the result.

Now Try Problem 45

EXAMPLE 7   Solve: $\frac{5}{8}m - 2 = -12$

Strategy We will use two properties of equality to isolate the variable m on left side of the equation.

WHY To solve the original equation, we want to find a simpler equivalent equation of the form $m = \textbf{a number}$, whose solution is obvious.

Solution We note that the coefficient of m is $\frac{5}{8}$ and proceed as follows.

- To isolate the variable term $\frac{5}{8}m$, we add 2 to both sides to undo the subtraction of 2.

- To isolate the variable m, we multiply both sides by $\frac{8}{5}$ to undo the multiplication by $\frac{5}{8}$.

$$\frac{5}{8}m - 2 = -12 \qquad \text{This is the equation to solve.}$$

$$\frac{5}{8}m - 2 + 2 = -12 + 2 \qquad \text{To isolate the variable term } \tfrac{5}{8}m, \text{ undo the subtraction of 2 by adding 2 to both sides.}$$

$$\frac{5}{8}m = -10 \qquad \text{Do the addition: } -2 + 2 = 0 \text{ and } -12 + 2 = -10.$$

$$\frac{8}{5}\left(\frac{5}{8}m\right) = \frac{8}{5}(-10) \qquad \text{To isolate } m, \text{ undo the multiplication by } \tfrac{5}{8} \text{ by multiplying both sides by the reciprocal of } \tfrac{5}{8}, \text{ which is } \tfrac{8}{5}.$$

$$\left(\frac{8}{5}\cdot\frac{5}{8}\right)m = \frac{8}{5}(-10) \qquad \text{On the left side, use the associative property of multiplication to group } \tfrac{8}{5} \text{ and } \tfrac{5}{8}.$$

$$1m = \frac{8}{5}\left(-\frac{10}{1}\right) \qquad \text{On the left side, the product of a number and its reciprocal is 1: } \tfrac{8}{5}\cdot\tfrac{5}{8} = 1. \text{ On the right side, write } -10 \text{ as } \tfrac{-10}{1}.$$

$$m = -\frac{8\cdot 10}{5\cdot 1} \qquad \text{On the left side, the coefficient of 1 need not be written: } 1m = m. \text{ On the right side, multiply the numerators and multiply the denominators.}$$
The product of two numbers with unlike signs is negative.

$$m = -\frac{8\cdot 2\cdot \overset{1}{5}}{\underset{1}{5}\cdot 1} \qquad \text{To simplify the right side, factor 10 as } 2\cdot 5 \text{ and then remove the common factor of 5.}$$

$$m = -\frac{16}{1} \qquad \text{Multiply the remaining factors in the numerator: } 8\cdot 2\cdot 1 = 16. \text{ Multiply the remaining factors in the denominator: } 1\cdot 1 = 1.$$

$$m = -16 \qquad \text{Any number divided by 1 is the same number.}$$

To check this result, we substitute -16 for m in the original equation and evaluate the left side.

Check: $\quad \dfrac{5}{8}m - 2 = -12 \qquad$ This is the original equation.

$\dfrac{5}{8}(-16) - 2 \stackrel{?}{=} -12 \qquad$ Substitute –16 for m.

$-\dfrac{5 \cdot 16}{8 \cdot 1} - 2 \stackrel{?}{=} -12 \qquad$ On the left side, write 16 as $\frac{16}{1}$. Then multiply the numerators and multiply the denominators. The product of two numbers with unlike signs is negative.

$-\dfrac{5 \cdot 2 \cdot \overset{1}{\cancel{8}}}{\underset{1}{\cancel{8}} \cdot 1} - 2 \stackrel{?}{=} -12 \qquad$ To simplify the fraction, factor 16 as $2 \cdot 8$ and remove the common factor of 8.

$-10 - 2 \stackrel{?}{=} -12 \qquad$ Multiply the remaining factors in the numerator and the denominator. Then simplify: $-\frac{10}{1} = -10$.

$-12 = -12 \qquad$ On the left side, do the subtraction.

Since the statement $-12 = -12$ is true, -16 is the solution of $\frac{5}{8}m - 2 = -12$. ∎

3 Clear equations of fractions.

To solve the equation $\frac{3}{4} = x + \frac{1}{6}$ in Example 2, we had to find an LCD and build equivalent fractions for $\frac{3}{4}$ and $\frac{1}{6}$ to subtract in the third step of the solution. We will now discuss a method in which we *clear* such an equation of fractions.

EXAMPLE 8 Solve $\dfrac{3}{4} = x + \dfrac{1}{6}$ by first clearing the equation of fractions.

Strategy We will use the multiplication property of equality to clear this equation of fractions by multiplying both sides by the LCD.

WHY Equations that involve only integers are usually easier to solve than equations that involve fractions.

Solution Since the denominators of the fractions in the equation are 4 and 6, we multiply both sides of the equation by the LCD, 12.

$\dfrac{3}{4} = x + \dfrac{1}{6} \qquad$ This is the equation to solve.

$12\left(\dfrac{3}{4}\right) = 12\left(x + \dfrac{1}{6}\right) \qquad$ Multiply both sides of the equation by the LCD of $\frac{3}{4}$ and $\frac{1}{6}$, which is 12. Don't forget to write the parentheses on each side.

$12\left(\dfrac{3}{4}\right) = 12(x) + 12\left(\dfrac{1}{6}\right) \qquad$ On the right side, distribute the multiplication by 12.

$\dfrac{12}{1}\left(\dfrac{3}{4}\right) = 12(x) + \dfrac{12}{1}\left(\dfrac{1}{6}\right) \qquad$ Write 12 as $\frac{12}{1}$. This makes the numerators and denominators in the fraction multiplication process clearer.

$\dfrac{3 \cdot \overset{1}{\cancel{4}}}{1}\left(\dfrac{3}{\underset{1}{\cancel{4}}}\right) = 12(x) + \dfrac{2 \cdot \overset{1}{\cancel{6}}}{1}\left(\dfrac{1}{\underset{1}{\cancel{6}}}\right) \qquad$ On the left side, factor 12 as $3 \cdot 4$ and remove the common factor of 4 from the numerator and denominator. On the right side, factor 12 as $2 \cdot 6$ and remove the common factor of 6. Try to do these steps in your head.

Self Check 8

Solve $\dfrac{2}{3} = y + \dfrac{1}{5}$ by first clearing the equation of fractions.

Now Try Problem 49

Success Tip   Here is an alternate way to show how the common factors of the numerator and denominator are removed in the multiplication process:

$$\frac{\overset{3}{\cancel{12}}}{1}\left(\frac{3}{\cancel{4}_{\,1}}\right) = 12(x) + \frac{\overset{2}{\cancel{12}}}{1}\left(\frac{1}{\cancel{6}_{\,1}}\right)$$

$9 = 12x + 2$	Complete each multiplication. Note that the fractions have been cleared from the equation.
$9 - 2 = 12x + 2 - 2$	To isolate the variable term 12x, undo the addition of 2 by subtracting 2 from both sides.
$7 = 12x$	Do the subtraction.
$\dfrac{7}{12} = \dfrac{12x}{12}$	To isolate the variable x, undo the multiplication by 12 by dividing both sides by 12.
$\dfrac{7}{12} = x$	Since the only common factor of 7 and 12 is 1, the fraction is in simplest form.

The solution is $\frac{7}{12}$. Note that this is the same result that we obtained in Example 2. ∎

Self Check 9

Solve $\dfrac{m}{2} - \dfrac{m}{5} = -6$ and check the result.

Now Try Problem 53

EXAMPLE 9   Solve: $\dfrac{n}{3} - \dfrac{n}{5} = -4$

Strategy We will use the multiplication property of equality to clear the equation of fractions by multiplying both sides by the LCD.

WHY Equations that involve only integers are usually easier to solve than equations that involve fractions.

Solution Since the denominators of the fractions in the equation are 3 and 5, we multiply both sides of the equation by the LCD, 15.

$\dfrac{n}{3} - \dfrac{n}{5} = -4$	This is the equation to solve.
$15\left(\dfrac{n}{3} - \dfrac{n}{5}\right) = 15(-4)$	Multiply both sides of the equation by the LCD of $\frac{n}{3}$ and $\frac{n}{5}$, which is 15. Don't forget to write the parentheses on each side.
$15\left(\dfrac{n}{3}\right) - 15\left(\dfrac{n}{5}\right) = 15(-4)$	On the left side, distribute the multiplication by 15.
$\dfrac{15}{1}\left(\dfrac{n}{3}\right) - \dfrac{15}{1}\left(\dfrac{n}{5}\right) = 15(-4)$	Write 15 as $\frac{15}{1}$. This makes the numerators and denominators in the fraction multiplication process clearer.
$\dfrac{\overset{1}{\cancel{3}} \cdot 5}{1}\left(\dfrac{n}{\cancel{3}_{\,1}}\right) - \dfrac{3 \cdot \overset{1}{\cancel{5}}}{1}\left(\dfrac{n}{\cancel{5}_{\,1}}\right) = 15(-4)$	On the left side, factor 15 as $3 \cdot 5$ and remove the common factor of 3 from the numerator and denominator of the first term and the common factor of 5 from the numerator and denominator of the second term. Try to do these steps in your head.

Success Tip   Here is an alternate way to show how the common factors of the numerator and denominator are removed in the multiplication process:

$$\frac{\overset{5}{\cancel{15}}}{1}\left(\frac{n}{\cancel{3}_{\,1}}\right) - \frac{\overset{3}{\cancel{15}}}{1}\left(\frac{n}{\cancel{5}_{\,1}}\right) = 15(-4)$$

$5n - 3n = -60$	Complete each multiplication. Note that the fractions have been cleared from the equation.
$2n = -60$	On the left side, combine like terms: $5n - 3n = 2n$.
$\dfrac{2n}{2} = \dfrac{-60}{2}$	To isolate the variable n, undo the multiplication by 2 by dividing both sides by 2.
$n = -30$	Do the division.

Check: $\dfrac{n}{3} - \dfrac{n}{5} = -4$ This is the original equation.

$\dfrac{-30}{3} - \dfrac{-30}{5} \overset{?}{=} -4$ Substitute -30 for each n.

$-10 - (-6) \overset{?}{=} -4$ On the left side, do each division. Recall that the quotient of two numbers with unlike signs is negative.

$-4 \overset{?}{=} -4$ On the left side, write the subtraction as addition of the opposite: $-10 - (-6) = -10 + 6 = -4$.

Since the statement $-4 = -4$ is true, -30 is the solution of $\frac{n}{3} - \frac{n}{5} = -4$.

EXAMPLE 10

Solve: $\dfrac{3}{4}h - \dfrac{1}{2} = \dfrac{5}{8}h$

Strategy We will use the multiplication property of equality to clear the equation of fractions by multiplying both sides by the LCD.

WHY Equations that involve only integers are usually easier to solve than equations that involve fractions.

Solution Since the denominators of the fractions in the equation are $4, 2,$ and $8,$ we multiply both sides of the equation by the LCD, $8.$

$\dfrac{3}{4}h - \dfrac{1}{2} = \dfrac{5}{8}h$ This is the equation to solve.

$8\left(\dfrac{3}{4}h - \dfrac{1}{2}\right) = 8\left(\dfrac{5}{8}h\right)$ Multiply both sides of the equation by the LCD of $\frac{3}{4}, \frac{1}{2},$ and $\frac{5}{8},$ which is $8.$ Don't forget to write the parentheses on both sides.

$8\left(\dfrac{3}{4}h\right) - 8\left(\dfrac{1}{2}\right) = 8\left(\dfrac{5}{8}h\right)$ On the left side, distribute the multiplication by $8.$

$\dfrac{8}{1}\left(\dfrac{3}{4}h\right) - \dfrac{8}{1}\left(\dfrac{1}{2}\right) = \dfrac{8}{1}\left(\dfrac{5}{8}h\right)$ Write 8 as $\frac{8}{1}.$ This makes the numerators and denominators in the fraction multiplication process clearer.

$\dfrac{2 \cdot \overset{1}{\cancel{4}}}{1}\left(\dfrac{3}{\underset{1}{\cancel{4}}}h\right) - \dfrac{2 \cdot 4}{1}\left(\dfrac{1}{\underset{1}{\cancel{2}}}\right) = \dfrac{\overset{1}{\cancel{8}}}{1}\left(\dfrac{5}{\underset{1}{\cancel{8}}}h\right)$ On the left side, factor 8 as $2 \cdot 4$ and remove the common factor of 4 from the numerator and denominator of the first term and the common factor of 2 from the numerator and denominator of the second term. On the right side, remove the common factor of $8.$ Try to do this step in your head.

Success Tip Here is an alternate way to show how the common factors of the numerator and denominator are removed in the multiplication process.

$\dfrac{\overset{2}{\cancel{8}}}{1}\left(\dfrac{3}{\underset{1}{\cancel{4}}}h\right) - \dfrac{\overset{4}{\cancel{8}}}{1}\left(\dfrac{1}{\underset{1}{\cancel{2}}}\right) = \dfrac{\overset{1}{\cancel{8}}}{1}\left(\dfrac{5}{\underset{1}{\cancel{8}}}h\right)$

$2(3h) - 4(1) = 5h$ Simplify each term. Note that the fractions have been cleared from the equation.

$6h - 4 = 5h$ Complete the multiplication.

$6h - 4 - \mathbf{5h} = 5h - \mathbf{5h}$ To eliminate the term $5h$ from the right side, subtract $5h$ from both sides.

$h - 4 = 0$ Combine like terms: $6h - 5h = h$ and $5h - 5h = 0.$

$h - 4 + \mathbf{4} = 0 + \mathbf{4}$ To isolate the variable h on the left side, undo the subtraction of 4 by adding 4 to both sides.

$h = 4$ Do the addition.

The solution is $4.$ Verify this result by substituting it into the original equation.

Self Check 10

Solve $\dfrac{4}{5}w - \dfrac{1}{2} = \dfrac{3}{4}w$ and check the result.

Now Try Problem 57

> ***Success Tip***   After multiplying both sides by the LCD and simplifying, the equation should not contain any fractions. If it does, check for an algebraic error, or perhaps your LCD is incorrect.

We can now complete the strategy for solving equations discussed in Chapter 3. You won't always have to use all five steps to solve a given equation. If a step doesn't apply, skip it and move to the next step.

Strategy for Solving Equations

1. **Clear the equation of fractions:** Multiply both sides by the LCD to clear fractions.

2. **Simplify each side of the equation:** Use the distributive property to remove parentheses, and then combine like terms on each side.

3. **Isolate the variable term on one side:** Add (or subtract) to get the variable term on one side of the = symbol and a number on the other using the addition (or subtraction) property of equality.

4. **Isolate the variable:** Multiply (or divide) to isolate the variable using the multiplication (or division) property of equality.

5. **Check the result:** Substitute the possible solution for the variable in the *original* equation to see if a true statement results.

4 Use equations to solve application problems that involve fractions.

We can use the concepts of variable and equation to solve application problems involving fractions. Once again, we will follow the strategy of analyze, form, solve, state, and check.

Self Check 11

ANATOMY CLASS A pre-med student has to memorize the name of each bone in the human hand. So far, he has learned 18 of them, which is two-thirds of the total. How many bones are in the human hand?

Now Try Problems 93 and 95

EXAMPLE 11   ***Native Americans***   The U.S. Constitution requires a population count, called a *census*, to be taken every 10 years. In the 2000 census, the population of the Navajo tribe was approximately 298,000. This was about two-fifths of the population of the largest Native American tribe, the Cherokee. What was the population of the Cherokee tribe in 2000?

Analyze

- In 2000, the population of the Navajo tribe was 298,000.   *Given*
- In 2000, the population of the Navajo tribe was about $\frac{2}{5}$ of the population of the Cherokee tribe.   *Given*
- What was the population of the Cherokee tribe in 2000?   *Find*

Form

Let x = the population of the Cherokee tribe in 2000. Next, we look for a key word or phrase in the problem.

Key phrase: *two-fifths of*   **Translation:** multiply by $\frac{2}{5}$

Now we translate the words of the problem into an equation.

In 2000, the population of the Navajo tribe	was	$\frac{2}{5}$	of	the population of the Cherokee tribe.
298,000	$=$	$\frac{2}{5}$	$\cdot$	x

Solve

$$298,000 = \frac{2}{5}x$$

$$\frac{5}{2}(298,000) = \frac{5}{2}\left(\frac{2}{5}x\right)$$

To isolate x, undo the multiplication by $\frac{2}{5}$ by multiplying both sides by the reciprocal of $\frac{2}{5}$, which is $\frac{5}{2}$.

$$\frac{5}{2}\left(\frac{298,000}{1}\right) = \left(\frac{5}{2}\cdot\frac{2}{5}\right)x$$

On the left side, write 298,000 as $\frac{298,000}{1}$. On the right side, use the associative property of multiplication to group $\frac{5}{2}$ and $\frac{2}{5}$.

$$\frac{5\cdot 298,000}{2\cdot 1} = 1x$$

On the left side, multiply the numerators and multiply the denominators. On the right side, the product of a number and its reciprocal is 1: $\frac{5}{2}\cdot\frac{2}{5}=1$.

$$\frac{5\cdot\overset{1}{2}\cdot 149,000}{\underset{1}{2}\cdot 1} = x$$

To simplify the left side, factor 298,000 as $2\cdot 149,000$ and remove the common factor of 2. On the right side, the coefficient of 1 need not be written since 1x = x.

$$745,000 = x$$

On the left side, multiply the remaining factors in the numerator: $5\cdot 1\cdot 149,000 = 745,000$. Then simplify the fraction.

$$\begin{array}{r}\overset{2\,4}{149,000}\\ \times\quad\quad 5\\ \hline 745,000\end{array}$$

State

In 2000, the population of the Cherokee tribe was about 745,000.

Check

If we use a fraction to compare the two populations, we should get $\frac{2}{5}$.

$$\frac{298,000}{745,000} = \frac{298}{745} = \frac{2\cdot\overset{1}{149}}{5\cdot\underset{1}{149}} = \frac{2}{5}$$

Write $\dfrac{\text{the population of the Navajo tribe}}{\text{the population of the Cherokee tribe}}$ and simplify.

The result checks.

EXAMPLE 12 *Filmmaking* A movie director has sketched out a "storyboard" for a film that is in the planning stages. He estimates the amount of time in the film that will be devoted to scenes involving dialogue, action scenes, and scenes that make a transition between the two. From the information on the storyboard shown below, how long will this film be, in minutes?

Storyboard		Film: "Terminating Force"
Dialogue	*Action scenes*	*Transition scenes*
One-half of film	One-third of film	20 minutes

Self Check 12

CASTING A MOVIE Two-thirds of the cast of a movie are male adults, one-fourth are female adults, and there are 6 children in the movie. Find the number of people in the cast.

Now Try Problem 103

Analyze

- $\frac{1}{2}$ of the film is dialogue.   Given
- $\frac{1}{3}$ of the film is action scenes.   Given
- There are 20 minutes of transition scenes.   Given
- How long is the film?   Find

Form

Let x = the length of the film in minutes. To represent the number of minutes for the dialogue and the action scenes, look for a key word or phrase in the storyboard.

Key phrases: *one-half of, one-third of*   **Translation:**   multiply

Thus, $\frac{1}{2}x$ = the number of minutes for dialogue scenes and $\frac{1}{3}x$ = the number of minutes for action scenes. Now translate the words of the problem into an equation.

The time for dialogue scenes	plus	the time for action scenes	plus	the time for transition scenes	is	the total length of the film.
$\frac{1}{2}x$	$+$	$\frac{1}{3}x$	$+$	20	$=$	x

Solve

$$\frac{1}{2}x + \frac{1}{3}x + 20 = x$$

$$6\left(\frac{1}{2}x + \frac{1}{3}x + 20\right) = 6(x)$$

To clear the equation of fractions, multiply both sides by the LCD of $\frac{1}{2}x$ and $\frac{1}{3}x$, which is 6. Don't forget to write the parentheses.

$$6\left(\frac{1}{2}x\right) + 6\left(\frac{1}{3}x\right) + 6(20) = 6(x)$$

On the left side, distribute the multiplication by 6.

$$\frac{6}{1}\left(\frac{1}{2}x\right) + \frac{6}{1}\left(\frac{1}{3}x\right) + 6(20) = 6x$$

Write 6 as $\frac{6}{1}$. This makes the numerators and denominators in the fraction multiplication process clearer.

$$\frac{\overset{1}{2}\cdot 3}{1}\left(\frac{1}{\underset{1}{2}}x\right) + \frac{2\cdot \overset{1}{3}}{1}\left(\frac{1}{\underset{1}{3}}x\right) + 6(20) = 6(x)$$

On the left side, factor 6 as $2 \cdot 3$ and remove the common factors in the numerator and denominator of the first two terms.

$$3x + 2x + 120 = 6x$$

Complete the multiplication on both sides of the equation.

$$5x + 120 = 6x$$

On the left side, combine like terms: $3x + 2x = 5x$.

$$5x + 120 - 5x = 6x - 5x$$

To eliminate the term $5x$ from the left side, subtract $5x$ from both sides.

$$120 = x$$

Combine like terms: $5x - 5x = 0$ and $6x - 5x = x$.

State

The length of the film will be 120 minutes.

Check

If x is **120**, the time for dialogue scenes is $\frac{1}{2}x = \frac{1}{2} \cdot \mathbf{120} = 60$ minutes. The time for action scenes is $\frac{1}{3}x = \frac{1}{3} \cdot \mathbf{120} = 40$ minutes. The time for transition scenes is 20 minutes. Adding the three times, we get $60 + 40 + 20 = 120$ minutes. The result checks. ∎

SECTION 4.8 STUDY SET

VOCABULARY

Fill in the blanks.

1. To _____ an equation, we find all the values of the variable that make the equation true.

2. In the term $\frac{5}{12}x$, the _____ of x is $\frac{5}{12}$.

3. To find the _____ of a fraction, we invert the numerator and the denominator.

4. To _____ the equation $\frac{1}{4}y - \frac{2}{3} = \frac{1}{2}y$ of fractions, we multiply both sides by the LCD of $\frac{1}{4}, \frac{2}{3}$, and $\frac{1}{2}$, which is 12.

CONCEPTS

5. Use a check to determine whether 40 is the solution of $\frac{5}{8}x = 25$.

6. Find the reciprocal of the coefficient of x.

a. $\frac{7}{9}x$ **b.** $-\frac{1}{2}x$

7. What is the result when a fraction is multiplied by its reciprocal?

8. Multiply.

a. $\frac{3}{2}\left(\frac{2}{3}x\right)$ **b.** $-\frac{16}{15}\left(-\frac{15}{16}t\right)$

9. Translate to mathematical symbols.

a. Four-fifths of the population p

b. One-quarter of the time t

10. What property is illustrated by the arrows?

$$12\left(\frac{1}{6}y - \frac{1}{4}\right) = 12\left(\frac{1}{2}\right)$$

11. By what should both sides of the equation be multiplied to clear it of fractions?

a. $\frac{1}{3}x = \frac{5}{2}x - 8$ **b.** $\frac{m}{8} = \frac{m}{3} + \frac{1}{2}$

12. Fill in the blanks.

$$8\left(\frac{1}{4}x\right) = 8\left(\frac{1}{2}x\right) + 8\left(\frac{3}{8}\right)$$

$$\boxed{} = \boxed{} + \boxed{}$$

NOTATION

Complete each solution to solve the equation.

13. $\frac{7}{8}x = 21$

$$\boxed{}\left(\frac{7}{8}x\right) = \boxed{}(21)$$

$$x = \boxed{}$$

The solution is $\boxed{}$.

14. $h + \frac{1}{2} = \frac{2}{3}$

$$\boxed{}\left(h + \frac{1}{2}\right) = \boxed{}\left(\frac{2}{3}\right)$$

$$\boxed{}h + \boxed{}\left(\frac{1}{2}\right) = \boxed{}\left(\frac{2}{3}\right)$$

$$6h + \boxed{} = 4$$

$$6h + 3 - \boxed{} = 4 - \boxed{}$$

$$6h = \boxed{}$$

$$\frac{6h}{\boxed{}} = \frac{1}{\boxed{}}$$

$$h = \frac{1}{6}$$

The solution is $\boxed{}$.

15. Determine whether each statement is true or false.

a. $\frac{1}{2}x = \frac{x}{2}$ **b.** $\frac{1}{8}y = 8y$

c. $-\frac{1}{2}x = \frac{-x}{2} = \frac{x}{-2}$ **d.** $\frac{7p}{8} = \frac{7}{8}p$

16. Write the product of $\frac{4}{7}$ and x in two ways.

GUIDED PRACTICE

Solve each equation and check the result. **See Example 1.**

17. $y - \frac{1}{20} = \frac{13}{20}$ **18.** $a - \frac{1}{24} = \frac{17}{24}$

19. $x - \frac{1}{15} = \frac{11}{15}$ **20.** $y - \frac{1}{16} = \frac{9}{16}$

Solve each equation and check the result. See Example 2.

21. $\dfrac{5}{6} = x + \dfrac{1}{9}$

22. $\dfrac{1}{6} = x + \dfrac{1}{8}$

23. $\dfrac{7}{10} = a + \dfrac{1}{4}$

24. $\dfrac{7}{15} = c + \dfrac{1}{10}$

Solve each equation and check the result. See Example 3.

25. $\dfrac{4}{9}x = 12$

26. $\dfrac{5}{6}x = 10$

27. $\dfrac{3}{7}w = 30$

28. $\dfrac{2}{15}y = 6$

29. $-\dfrac{8}{3}w = 7$

30. $-\dfrac{9}{5}w = 4$

31. $-\dfrac{16}{9}w = 5$

32. $-\dfrac{11}{6}w = 7$

Solve each equation using a two-step process. See Example 4.

33. $\dfrac{4}{9}x = 12$

34. $\dfrac{5}{6}x = 10$

35. $\dfrac{3}{7}w = 30$

36. $\dfrac{2}{15}y = 6$

Solve each equation and check the result. See Example 5.

37. $\dfrac{1}{8}x = \dfrac{25}{72}$

38. $\dfrac{1}{6}x = \dfrac{29}{54}$

39. $\dfrac{1}{5}x = \dfrac{43}{55}$

40. $\dfrac{1}{9}x = \dfrac{20}{27}$

Solve each equation and check the result. See Example 6.

41. $17x = -\dfrac{15}{4}$

42. $21x = -\dfrac{31}{2}$

43. $29x = -\dfrac{13}{3}$

44. $41x = -\dfrac{27}{5}$

Solve each equation and check the result. See Example 7.

45. $\dfrac{5}{6}k - 5 = -15$

46. $\dfrac{2}{5}c - 12 = -32$

47. $\dfrac{7}{16}h + 28 = 21$

48. $\dfrac{5}{8}a + 25 = 15$

Solve each equation by first clearing it of fractions. Check the result. See Example 8.

49. $\dfrac{2}{9} = r - \dfrac{1}{6}$

50. $\dfrac{5}{8} = x - \dfrac{1}{6}$

51. $\dfrac{1}{4} = x - \dfrac{2}{9}$

52. $\dfrac{1}{7} = y - \dfrac{1}{2}$

Solve each equation by first clearing it of fractions. Check the result. See Example 9.

53. $\dfrac{a}{4} - \dfrac{a}{7} = -6$

54. $\dfrac{f}{3} - \dfrac{f}{8} = -10$

55. $\dfrac{k}{7} - \dfrac{k}{3} = -24$

56. $\dfrac{a}{11} - \dfrac{a}{5} = -12$

Solve each equation by first clearing it of fractions. Check the result. See Example 10.

57. $\dfrac{3}{5}d - \dfrac{4}{5} = \dfrac{1}{3}d$

58. $\dfrac{11}{10}x - \dfrac{5}{2} = \dfrac{3}{5}x$

59. $\dfrac{7}{8}w + \dfrac{1}{4} = \dfrac{3}{4}w$

60. $\dfrac{1}{2}d + \dfrac{17}{2} = \dfrac{12}{7}d$

TRY IT YOURSELF

Solve each equation and check the result.

61. $\dfrac{1}{2}x - \dfrac{1}{9} = \dfrac{1}{3}$

62. $\dfrac{1}{4}y - \dfrac{2}{3} = \dfrac{1}{2}$

63. $\dfrac{a}{2} - \dfrac{a}{12} = -\dfrac{5a}{4}$

64. $\dfrac{n}{6} - \dfrac{n}{18} = -\dfrac{4n}{3}$

65. $\dfrac{7}{8}t = -28$

66. $\dfrac{5}{6}c = -25$

67. $x + \dfrac{1}{9} = \dfrac{4}{9}$

68. $x - \dfrac{1}{12} = \dfrac{7}{12}$

69. $\dfrac{5f}{7} = -2$

70. $\dfrac{3h}{5} = -35$

71. $6x = 2x - 11$

72. $5t = t - 7$

73. $2(y - 3) = 7$

74. $3(r + 2) = 10$

75. $\dfrac{1}{3} = 2x - \dfrac{1}{2}$

76. $\dfrac{1}{8} = 3y - \dfrac{2}{5}$

77. $4 = -\dfrac{3}{5}h$

78. $-2 = -\dfrac{5}{6}f$

79. $5 + \dfrac{1}{3}x = \dfrac{1}{2} + \dfrac{1}{6}x$

80. $3 + \dfrac{1}{2}y = \dfrac{3}{5} + \dfrac{1}{10}y$

81. $\dfrac{2}{15}x = \dfrac{4}{5}$

82. $\dfrac{5}{66}y = \dfrac{10}{11}$

83. $\dfrac{2}{5}x + 1 = \dfrac{1}{3} + x$

84. $\dfrac{2}{3}y + 2 = \dfrac{1}{5} + y$

85. $\dfrac{5h}{6} - 8 = 12$

86. $\dfrac{6a}{7} - 1 = 11$

87. $\dfrac{1}{2}y - 2 = \dfrac{1}{5}y + 1$

88. $\dfrac{1}{5}m - 5 = \dfrac{1}{3}m - 3$

In part a of Exercises 89–92, use the methods of this section to solve the equation. In part b, use the methods of Section 4.4 to perform the addition or subtraction.

89. a. $\dfrac{12}{5} + \dfrac{n}{2} = \dfrac{n}{10}$ **b.** $\dfrac{12}{5} + \dfrac{n}{2}$

90. a. $\dfrac{9}{2} + \dfrac{h}{3} = \dfrac{h}{6}$ **b.** $\dfrac{9}{2} + \dfrac{h}{3}$

91. a. $\dfrac{x}{4} - \dfrac{1}{3} = \dfrac{x}{2}$ **b.** $\dfrac{x}{4} - \dfrac{1}{3}$

92. a. $\dfrac{a}{5} - \dfrac{5}{6} = \dfrac{a}{3}$ **b.** $\dfrac{a}{5} - \dfrac{5}{6}$

APPLICATIONS

Complete each solution.

93. CAR REPAIRS One-eighth of the cars that an automobile repair shop serviced last year had transmission problems. If the shop repaired 32 transmissions, how many cars did they service that year?

Analyze

- ☐ of the cars serviced last year had transmission problems.   *Given*

- The shop repaired ☐ transmissions last year.   *Given*

- How many ____ did the shop service last year?   *Find*

Form

Let x = the number of ____ that the repair shop serviced last year.

The key phrase *one-eighth of* suggests _____. We now translate the words of the problem into an equation.

$\dfrac{1}{8}$	of	the number of cars repaired last year	had	transmission problems.

$\dfrac{1}{8}$ ☐ ☐ = ☐

Solve

$$\dfrac{1}{8}x = 32$$

$$\boxed{}\left(\dfrac{1}{8}x\right) = \boxed{}(32)$$

$$x = \boxed{}$$

State   The shop repaired ☐ cars last year.

Check   If we can use a fraction to compare the number of cars with transmission problems to the number of cars serviced, we should get $\frac{1}{8}$.

$$\dfrac{32}{256} = \dfrac{\overset{1}{\cancel{32}}}{8 \cdot \underset{1}{\cancel{32}}} = \dfrac{\boxed{}}{\boxed{}}$$

The result checks.

94. CATTLE RANCHING A rancher is going to fence in a rectangular-shaped grazing area next to a $\frac{3}{4}$-mile-long stretch of shoreline of lake. He has determined that $1\frac{1}{2}$ square miles of land are needed to make sure that overgrazing does not occur. How wide should this grazing area be?

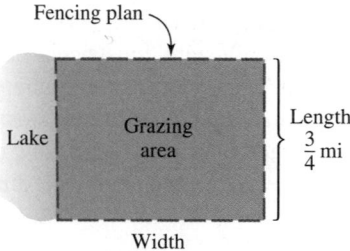
Fencing plan

Analyze

- The rectangular-shaped grazing area is ☐ square miles.   *Given*

- The length of the rectangle is ☐ mile.   *Given*

- How ____ should the grazing area be?   *Find*

Form

Let w = ____ of grazing area.

The key word *area* suggests we multiply the _____ of the rectangle by the width. We now translate the words of the problem into an equation.

The area of the rectangle	is equal to	the length	times	the width.

$\dfrac{3}{2}$ = ☐ $\cdot$ ☐

Solve

$$\dfrac{3}{2} = \dfrac{3}{4}w$$

$$\boxed{}\left(\dfrac{3}{2}\right) = \boxed{}\left(\dfrac{3}{4}w\right)$$

$$\boxed{} = w$$

State

The width of the grazing area must be ☐ miles.

Check If we multiply the length and the width of the rectangular area, we should get $1\frac{1}{2}$ square miles.

$$\frac{3}{4} \cdot 2 = \frac{6}{4} = \boxed{}$$

The result checks.

In Exercises 95–108, let a variable represent the unknown quantity. Then write and solve an equation to answer the question.

95. **TOOTH DEVELOPMENT** During a checkup, a pediatrician found that only four-fifths of a child's baby teeth had emerged. The mother counted 16 teeth in the child's mouth. How many baby teeth will the child eventually have?

96. **GENETICS** Bean plants with inflated pods were cross-bred with bean plants with constricted pods. Of the offspring plants, three-fourths had inflated pods and only one-fourth had constricted pods. If 244 offspring plants had constricted pods, how many offspring plants resulted from the cross-breeding experiment?

   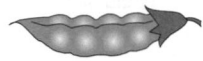

      Inflated pod   Constricted pod

97. **TELEPHONE BOOKS** A telephone book consists of the white pages and the yellow pages. Two-thirds of the book consists of the white pages; the white pages number 300. Find the total number of pages in the telephone book.

98. **BROADWAY MUSICALS** A theater usher at a Broadway musical finds that seven-eighths of the patrons attending a performance, which is 350 people, are in their seats by show time. If the show is always a complete sellout, how many seats does the theater have?

99. **LIGHTING DESIGN** In the warehouse shown below, the distance between light fixtures is two-thirds of the floor-to-ceiling height. What is the height of the warehouse ceiling?

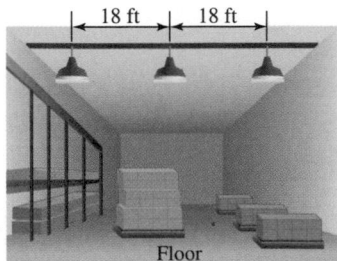

100. **THE VOLUNTEER STATE** The dimensions of the flag of the State of Tennessee were adopted into law in 1905. The official description requires that "the flag is to be a banner whose length is one and two-thirds times its width." If the flag shown below meets this requirement, what is its width?

Length = 30 in.

101. **GRAPHIC ARTS** A design for a yearbook is shown. The page is divided into 12 equal-size parts. Five of the twelve parts (those shaded in red) will contain photographs. If the photographs are to cover an area of 100 square inches, how many square inches are there on the page?

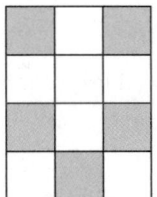

102. **SAFETY REQUIREMENTS** In developing taillights for an automobile, designers must be aware of a safety standard that requires an area of 30 square inches to be visible from behind the vehicle. If the designers want the taillights to be $3\frac{3}{4}$ inches high, how wide must they be to meet safety standards? (*Hint:* Write $3\frac{3}{4}$ as an improper fraction.)

$3\frac{3}{4}$ in.

103. **CPR CLASS** The instructor for a course in CPR (cardiopulmonary resuscitation) has three segments in her lesson plan, as shown below. How many minutes long is the CPR course?

Lecture on subject	Practicing CPR techniques	Legal responsibilities
One-fourth of class	Two-thirds of class	30 min

104. FIREFIGHTING A firefighting crew is composed of three elements, as shown below. How many firefighters are in the crew?

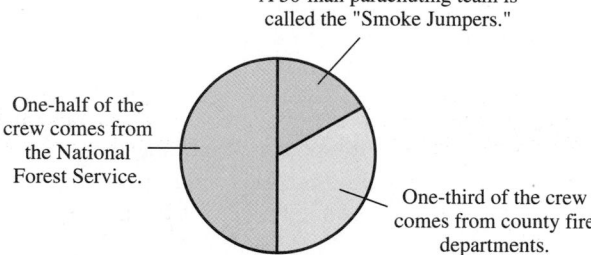

A 50-man parachuting team is called the "Smoke Jumpers."

One-half of the crew comes from the National Forest Service.

One-third of the crew comes from county fire departments.

105. TEAM ROSTERS At the end of the season, major league baseball teams are allowed to add players to the roster. Suppose one-half of the players on a team's expanded roster are pitchers, one-fourth are infielders, three-twentieths are outfielders, and there are 4 catchers. How many players are on the expanded roster?

106. TEAM ROSTERS One-third of the players on a basketball team are forwards and one-fifth are centers. The remaining 7 players are guards. How many players are on the team?

107. HOME SALES In less than a month, three-quarters of the homes in a new subdivision were purchased. This left only 9 homes to be sold. How many homes are there in the subdivision? (*Hint:* First determine what fractional part of the homes in the subdivision were *not* yet sold.)

108. WEDDING GUESTS Of those invited to a wedding, three-tenths were friends of the bride. The friends of the groom numbered 84. How many people were invited to the wedding? (*Hint:* First determine what fractional part of the people invited to the wedding were friends of the groom.)

WRITING

109. What is wrong with the following solution?

Solve:
$$\frac{x}{6} - \frac{3}{5} = 2$$
$$30\left(\frac{x}{6} - \frac{3}{5}\right) = 2$$
$$30\left(\frac{x}{6}\right) - 30\left(\frac{3}{5}\right) = 2$$
$$5x - 18 = 2$$
$$5x = 20$$
$$x = 4$$

110. Explain two ways in which the variable x can be isolated to solve the equation $\frac{2}{3}x = -4$.

111. What does it mean to clear an equation of fractions before solving the equation?

112. Explain how the method used to add $\frac{a}{3} + \frac{1}{5}$ differs from the method used to solve the equation $\frac{a}{3} + \frac{1}{5} = \frac{7}{15}$.

REVIEW

113. Round 12,599,767:
   a. to the nearest million
   b. to the nearest ten thousand
   c. to the nearest hundred

114. Round 1.2599767:
   a. to the nearest millionth
   b. to the nearest ten-thousandth
   c. to the nearest hundredth

STUDY SKILLS CHECKLIST

Working with Fractions

Before taking the test on Chapter 4, make sure that you have a solid understanding of the following methods for simplifying, multiplying, dividing, adding, and subtracting fractions. Put a checkmark in the box if you can answer "yes" to the statement.

☐ I know how to simplify fractions by factoring the numerator and denominator and then removing the common factors.

$$\frac{42}{50} = \frac{2 \cdot 3 \cdot 7}{2 \cdot 5 \cdot 5}$$
$$= \frac{\overset{1}{2} \cdot 3 \cdot 7}{\underset{1}{2} \cdot 5 \cdot 5}$$
$$= \frac{21}{25}$$

☐ When multiplying fractions, I know that it is important to factor and simplify first, before multiplying.

Factor and simplify first

$$\frac{15}{16} \cdot \frac{24}{35} = \frac{15 \cdot 24}{16 \cdot 35}$$
$$= \frac{3 \cdot \overset{1}{5} \cdot 3 \cdot \overset{1}{8}}{2 \cdot \underset{1}{8} \cdot \underset{1}{5} \cdot 7}$$

Don't multiply first

$$\frac{15}{16} \cdot \frac{24}{35} = \frac{15 \cdot 24}{16 \cdot 35}$$
$$= \frac{360}{560}$$

☐ To divide fractions, I know to multiply the first fraction by the reciprocal of the second fraction.

$$\frac{7}{8} \div \frac{23}{24} = \frac{7}{8} \cdot \frac{24}{23}$$

☐ I know that to add or subtract fractions, they must have a common denominator. To multiply or divide fractions, they **do not** need to have a common denominator.

Need an LCD

$$\frac{2}{3} + \frac{1}{5} \qquad \frac{9}{20} - \frac{7}{12}$$

Do not need an LCD

$$\frac{4}{7} \cdot \frac{2}{9} \qquad \frac{11}{40} \div \frac{5}{8}$$

☐ I know how to find the LCD of a set of fractions using one of the following methods.
- Write the multiples of the largest denominator in increasing order, until one is found that is divisible by the other denominators.
- Prime factor each denominator. The LCM is a product of prime factors, where each factor is used the greatest number of times it appears in any one factorization.

☐ I know how to build equivalent fractions by multiplying the given fraction by a form of 1.

$$\frac{2}{3} = \frac{2}{3} \cdot \frac{5}{5}$$
$$= \frac{2 \cdot 5}{3 \cdot 5}$$
$$= \frac{10}{15}$$

CHAPTER 4   SUMMARY AND REVIEW

SECTION 4.1   An Introduction to Fractions

DEFINITIONS AND CONCEPTS	EXAMPLES
A **fraction** describes the number of equal parts of a whole. In a fraction, the number above the **fraction bar** is called the **numerator,** and the number below is called the **denominator.**	Since 3 of 8 equal parts are colored red, $\frac{3}{8}$ (three-eighths) of the figure is shaded. Fraction bar $\longrightarrow \dfrac{3 \leftarrow \text{numerator}}{8 \leftarrow \text{denominator}}$

If the numerator of a fraction is less than its denominator, the fraction is called a **proper fraction.** If the numerator of a fraction is greater than or equal to its denominator, the fraction is called an **improper fraction.**	Proper fractions: $\dfrac{1}{5}, \dfrac{7}{8}$, and $\dfrac{999}{1,000}$ *Proper fractions are less than 1.* Improper fractions: $\dfrac{3}{2}, \dfrac{41}{16}$, and $\dfrac{15}{15}$ *Improper fractions are greater than or equal to 1.*
There are four **special fraction forms** that involve 0 and 1. Each of these fractions is a **form of 1**: $1 = \dfrac{1}{1} = \dfrac{2}{2} = \dfrac{3}{3} = \dfrac{4}{4} = \dfrac{5}{5} = \dfrac{6}{6} = \dfrac{7}{7} = \dfrac{8}{8} = \dfrac{9}{9} = \cdots$	Simplify each fraction: $\dfrac{0}{8} = 0$ $\dfrac{7}{0}$ is undefined $\dfrac{5}{1} = 5$ $\dfrac{20}{20} = 1$
Two fractions are **equivalent** if they represent the same number. **Equivalent fractions** represent the same portion of a whole.	$\dfrac{2}{3}, \dfrac{4}{6}$, and $\dfrac{8}{12}$ are equivalent fractions. They represent the same shaded portion of the figure. $\dfrac{2}{3} \quad = \quad \dfrac{4}{6} \quad = \quad \dfrac{8}{12}$
To **build a fraction,** we multiply it by a factor equal to 1 in the form of $\dfrac{2}{2}, \dfrac{3}{3}, \dfrac{4}{4}, \dfrac{5}{5}$, and so on.	Write $\dfrac{3}{4}$ as an equivalent fraction with a denominator of 36. $\dfrac{3}{4} = \dfrac{3}{4} \cdot$ 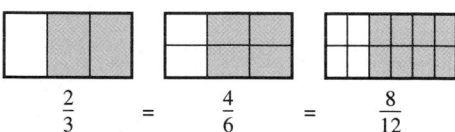 *We must multiply the denominator of $\dfrac{3}{4}$ by 9 to obtain a denominator of 36. It follows that $\dfrac{9}{9}$ should be the form of 1 that is used to build $\dfrac{3}{4}$.* $= \dfrac{3 \cdot 9}{4 \cdot 9}$ *Multiply the numerators.* *Multiply the denominators.* $= \dfrac{27}{36}$ $\dfrac{27}{36}$ is equivalent to $\dfrac{3}{4}$.
A fraction is **in simplest form, or lowest terms,** when the numerator and denominator have no common factors other than 1.	Is $\dfrac{6}{14}$ in simplest form? The factors of the numerator, 6, are: **1, 2,** 3, 6. The factors of the denominator, 14, are: **1, 2,** 7, 14. Since the numerator and denominator have a common factor of 2, the fraction $\dfrac{6}{14}$ is *not* in simplest form.
To **simplify a fraction,** we write it in simplest form by removing a factor equal to 1: 1. Factor (or prime factor) the numerator and denominator to determine their common factors. 2. Remove factors equal to 1 by replacing each pair of factors common to the numerator and denominator with the equivalent fraction $\dfrac{1}{1}$. 3. Multiply the remaining factors in the numerator and in the denominator.	Simplify: $\dfrac{12}{30}$ $\dfrac{12}{30} = \dfrac{2 \cdot 2 \cdot 3}{2 \cdot 3 \cdot 5}$ *Prime factor 12 and 30.* $= \dfrac{\overset{1}{2} \cdot 2 \cdot \overset{1}{3}}{2 \cdot 3 \cdot 5}$ *Remove the common factors of 2 and 3 from the numerator and denominator.* $\quad\;\; \underset{1}{} \quad \underset{1}{}$ $= \dfrac{2}{5}$ *Multiply the remaining factors in the numerator: $1 \cdot 2 \cdot 1 = 2$.* *Multiply the remaining factors in the denominator: $1 \cdot 1 \cdot 5 = 5$.* Since 2 and 5 have no common factors other than 1, we say that $\dfrac{2}{5}$ is in simplest form.

Fractions that contain a variable (or variables) in the numerator, the denominator, or both are called **algebraic fractions**.	Algebraic fractions: $$\frac{x}{5}, \quad \frac{12}{a}, \quad \frac{r}{25s}, \quad \frac{4c^2d}{16cd^3}, \quad \frac{x+1}{x-9}$$
Algebraic fractions are built up just like numerical fractions.	Write $\frac{4}{9}$ as an equivalent fraction with a denominator of $27x$. We need to multiply the denominator of $\frac{4}{9}$ by $3x$ to obtain $27x$. It follows that $\frac{3x}{3x}$ should be the form of 1 that is used to build $\frac{4}{9}$. $$\frac{4}{9} = \frac{4}{9} \cdot \frac{3x}{3x} \qquad \text{Multiply } \tfrac{4}{9} \text{ by a form of 1: } \tfrac{3x}{3x} = 1.$$ $$= \frac{4 \cdot 3x}{9 \cdot 3x} \qquad \begin{array}{l}\text{Multiply the numerators.}\\\text{Multiply the denominators.}\end{array}$$ $$= \frac{12x}{27x}$$
Algebraic fractions are simplified just like numerical fractions.	Simplify: $\dfrac{18x^2y^3}{42x^4y}$ $$\frac{18x^2y^3}{42x^4y} = \frac{2 \cdot 3 \cdot 3 \cdot x \cdot x \cdot y \cdot y \cdot y}{2 \cdot 3 \cdot 7 \cdot x \cdot x \cdot x \cdot x \cdot y} \qquad \begin{array}{l}\text{To prepare to simplify, factor 18,}\\ x^2, y^3, 42, \text{ and } x^4.\end{array}$$ $$= \frac{\overset{1}{2} \cdot \overset{1}{3} \cdot 3 \cdot \overset{1}{\cancel{x}} \cdot \overset{1}{\cancel{x}} \cdot \overset{1}{\cancel{y}} \cdot y \cdot y}{\underset{1}{2} \cdot \underset{1}{3} \cdot 7 \cdot \underset{1}{\cancel{x}} \cdot \underset{1}{\cancel{x}} \cdot x \cdot x \cdot \underset{1}{\cancel{y}}} \qquad \begin{array}{l}\text{Simplify by removing the common}\\\text{factors of 2, 3, }x,\text{ and }y\text{ from the}\\\text{numerator and denominator.}\end{array}$$ $$= \frac{3y^2}{7x^2} \qquad \begin{array}{l}\text{Multiply the remaining factors in the numerator.}\\\text{Multiply the remaining factors in the denominator.}\end{array}$$

REVIEW EXERCISES

1. Identify the numerator and denominator of the fraction $\frac{11}{16}$. Is it a proper or an improper fraction?

2. Write fractions that represent the shaded and unshaded portions of the figure to the right.

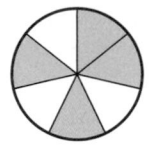

3. In the illustration below, why can't we say that $\frac{3}{4}$ of the figure is shaded?

4. Write the fraction $\frac{2}{-3}$ in two other ways.

5. Simplify, if possible:

   a. $\dfrac{5}{5}$ b. $\dfrac{0}{10}$

   c. $\dfrac{18}{1}$ d. $\dfrac{7}{0}$

6. What concept about fractions is illustrated below?

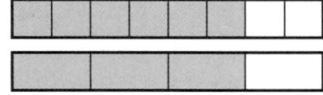

Write each fraction as an equivalent fraction with the indicated denominator.

7. $\dfrac{2}{3}$, denominator 18   8. $\dfrac{3}{8}$, denominator 16

9. $\dfrac{7}{15}$, denominator $45a$ 10. $\dfrac{13}{12x}$, denominator $60x$

11. Write 5 as an equivalent fraction with denominator 9.

12. Are the following fractions in simplest form?

   a. $\dfrac{6}{9}$ b. $\dfrac{10}{81}$

Simplify each fraction, if possible.

13. $\dfrac{15}{45}$

14. $\dfrac{20x^3}{48x^2}$

15. $\dfrac{66}{108}$

16. $\dfrac{117a^2b^6}{208a^5b}$

17. $\dfrac{81}{64}$

18. Tell whether $\frac{8}{12}$ and $\frac{176}{264}$ are equivalent by simplifying each fraction.

19. SLEEP If a woman gets seven hours of sleep each night, write a fraction to describe the part of a whole day that she spends sleeping and another to describe the part of a whole day that she is not sleeping.

20. a. What type of problem is shown below? Explain the solution.

$$\frac{5}{8} = \frac{5}{8} \cdot \frac{\mathbf{2}}{\mathbf{2}} = \frac{10}{16}$$

b. What type of problem is shown below? Explain the solution.

$$\frac{4}{6} = \frac{\overset{1}{\cancel{2}} \cdot 2}{\underset{1}{\cancel{2}} \cdot 3} = \frac{2}{3}$$

SECTION 4.2   Multiplying Fractions

DEFINITIONS AND CONCEPTS	EXAMPLES
To **multiply two fractions,** multiply the numerators and multiply the denominators. Simplify the result, if possible.	Multiply and simplify, if possible: $\dfrac{4}{5} \cdot \dfrac{2}{3}$ $\dfrac{4}{5} \cdot \dfrac{2}{3} = \dfrac{4 \cdot 2}{5 \cdot 3}$ Multiply the numerators. Multiply the denominators. $\qquad = \dfrac{8}{15}$ Since 8 and 15 have no common factors other than 1, the result is in simplest form.

Multiplying signed fractions
 	
 The product of two fractions with the same (like) signs is positive. The product of two fractions with different (unlike) signs is negative.
	Multiply and simplify, if possible: $-\dfrac{3}{4} \cdot \dfrac{2}{27}$
	
	 $-\dfrac{3}{4} \cdot \dfrac{2}{27} = -\dfrac{3 \cdot 2}{4 \cdot 27}$ Multiply the numerators.
	                      Multiply the denominators.
	                      Since the fractions have unlike signs, make the answer negative.
	
	 $\qquad = -\dfrac{\overset{1}{\cancel{3}} \cdot \overset{1}{\cancel{2}}}{2 \cdot \underset{1}{\cancel{2}} \cdot \underset{1}{\cancel{3}} \cdot 3 \cdot 3}$ Prime factor 4 and 27. Then simplify, by removing the common factors of 2 and 3 from the numerator and denominator.
	
	 $\qquad = -\dfrac{1}{18}$ Multiply the remaining factors in the numerator: $1 \cdot 1 = 1$. Multiply the remaining factors in the denominator: $1 \cdot 2 \cdot 1 \cdot 3 \cdot 3 = 18$.

To **multiply two algebraic fractions,** we use the same approach as with numerical fractions.

Multiply and simplify, if possible: $\dfrac{35x^3}{11} \cdot \dfrac{2}{25x}$

$$\frac{35x^3}{11} \cdot \frac{2}{25x} = \frac{35x^3 \cdot 2}{11 \cdot 25x}$$

Multiply the numerators.
Multiply the denominators.

$$= \frac{5 \cdot 7 \cdot x \cdot x \cdot x \cdot 2}{11 \cdot 5 \cdot 5 \cdot x}$$

To prepare to simplify, factor 35, x^3, and 25.

$$= \frac{\overset{1}{\cancel{5}} \cdot 7 \cdot \overset{1}{\cancel{x}} \cdot x \cdot x \cdot 2}{11 \cdot \underset{1}{\cancel{5}} \cdot 5 \cdot \underset{1}{\cancel{x}}}$$

Simplify by removing the common factors of 5 and x from the numerator and denominator.

$$= \frac{14x^2}{55}$$

Multiply the remaining factors in the numerator. Multiply the remaining factors in the denominator.

The base of an **exponential expression** can be a positive or a negative fraction.

The rule for multiplying two fractions can be extended to find the product of three or more fractions.

Evaluate: $\left(\dfrac{2}{3}\right)^3$

$$\left(\frac{2}{3}\right)^3 = \frac{2}{3} \cdot \frac{2}{3} \cdot \frac{2}{3}$$

Write the base, $\frac{2}{3}$, as a factor 3 times.

$$= \frac{2 \cdot 2 \cdot 2}{3 \cdot 3 \cdot 3}$$

Multiply the numerators.
Multiply the denominators.

$$= \frac{8}{27}$$

This fraction is in simplified form.

When a **fraction is followed by the word** *of,* it indicates that we are to find a part of some quantity using multiplication.

To find $\dfrac{2}{5}$ of 35, we multiply:

$$\frac{2}{5} \; of \; 35 = \frac{2}{5} \cdot 35$$

The word *of* indicates multiplication.

$$= \frac{2}{5} \cdot \frac{35}{1}$$

Write 35 as a fraction: $35 = \frac{35}{1}$.

$$= \frac{2 \cdot 35}{5 \cdot 1}$$

Multiply the numerators.
Multiply the denominators.

$$= \frac{2 \cdot \overset{1}{\cancel{5}} \cdot 7}{\underset{1}{\cancel{5}} \cdot 1}$$

Prime factor 35. Then simplify by removing the common factor of 5 from the numerator and denominator.

$$= \frac{14}{1}$$

Multiply the remaining factors in the numerator and in the denominator.

$$= 14$$

Any number divided by 1 is equal to that number.

The formula for the area of a triangle

Area of a triangle = $\dfrac{1}{2}$ (base)(height)

or

$$A = \dfrac{1}{2}bh$$

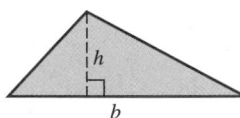

Find the area of the triangle shown on the right.

$A = \dfrac{1}{2}$ **(base)(height)**

$= \dfrac{1}{2}(8)(5)$ Substitute 8 for the base and 5 for the height.

$= \dfrac{1}{2}\left(\dfrac{5}{1}\right)\left(\dfrac{8}{1}\right)$ Write 5 and 8 as fractions.

$= \dfrac{1 \cdot 5 \cdot 8}{2 \cdot 1 \cdot 1}$ Multiply the numerators. Multiply the denominators.

$= \dfrac{1 \cdot 5 \cdot \overset{1}{2} \cdot 2 \cdot 2}{2 \cdot 1 \cdot 1}$ Prime factor 8. Then simplify by removing the common factor of 2 from the numerator and denominator.

$= 20$

The area of the triangle is 20 ft².

REVIEW EXERCISES

21. Fill in the blanks: To multiply two fractions, multiply the _____ and multiply the _____. Then _____, if possible.

22. Translate the following phrase to symbols. *You do not have to find the answer.*

$$\dfrac{5}{6} \text{ of } \dfrac{2}{3}$$

Multiply. Simplify the product, if possible.

23. $\dfrac{1}{2} \cdot \dfrac{1}{3}$

24. $\dfrac{2}{5}\left(-\dfrac{7}{9}\right)$

25. $\dfrac{9c^3}{16d} \cdot \dfrac{20d}{27c}$

26. $-\dfrac{5}{6}\left(-\dfrac{1}{15}\right)\left(-\dfrac{18}{25}\right)$

27. $\dfrac{3}{5} \cdot 7$

28. $-4\left(-\dfrac{9m}{16}\right)$

29. $-3x\left(\dfrac{1}{3}\right)$

30. $-\dfrac{6}{7}\left(-\dfrac{7}{6}\right)$

Evaluate each expression.

31. $-\left(\dfrac{3}{4}\right)^2$

32. $\left(-\dfrac{5a}{2}\right)^3$

33. $\left(-\dfrac{2}{5}\right)^3$

34. $\left(\dfrac{2}{3}\right)^2$

35. DRAG RACING A top-fuel dragster had to make 8 trial runs on a quarter-mile track before it was ready for competition. Find the total distance it covered on the trial runs.

36. GRAVITY Objects on the moon weigh only one-sixth of their weight on Earth. How much will an astronaut weigh on the moon if he weighs 180 pounds on Earth?

37. Find the area of the triangular sign.

38. Find the area of the triangle shown below.

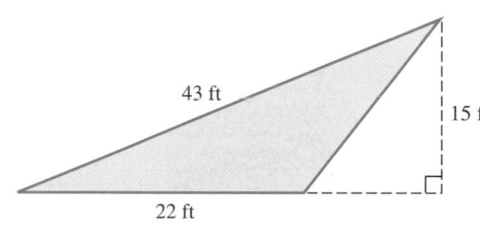

SECTION 4.3   Dividing Fractions

DEFINITIONS AND CONCEPTS	EXAMPLES
One number is the **reciprocal** of another if their product is 1. To find the **reciprocal of a fraction,** invert the numerator and denominator.	The reciprocal of $\frac{4}{5}$ is $\frac{5}{4}$ because $\frac{4}{5} \cdot \frac{5}{4} = 1$. Fraction Reciprocal $\frac{4}{5}$ $\frac{5}{4}$ Invert
To **divide two fractions,** multiply the first fraction by the reciprocal of the second fraction. Simplify the result, if possible.	Divide and simplify, if possible: $\frac{4}{35} \div \frac{2}{21}$ $\frac{4}{35} \div \frac{2}{21} = \frac{4}{35} \cdot \frac{21}{2}$ Multiply $\frac{4}{35}$ by the reciprocal of $\frac{2}{21}$, which is $\frac{21}{2}$. $= \frac{4 \cdot 21}{35 \cdot 2}$ Multiply the numerators. Multiply the denominators. $= \frac{2 \cdot 2 \cdot 3 \cdot 7}{5 \cdot 7 \cdot 2}$ To prepare to simplify, write 4, 21, and 35 in prime-factored form. $= \frac{\overset{1}{2} \cdot 2 \cdot 3 \cdot \overset{1}{7}}{5 \cdot \underset{1}{7} \cdot \underset{1}{2}}$ To simplify, remove the common factors of 2 and 7 from the numerator and denominator. $= \frac{6}{5}$ Multiply the remaining factors in the numerator: $1 \cdot 2 \cdot 3 \cdot 1 = 6$. Multiply the remaining factors in the denominator: $5 \cdot 1 \cdot 1 = 5$.
The **sign rules for dividing fractions** are the same as those for multiplying fractions.	Divide and simplify: $\frac{9}{16} \div (-3)$ $\frac{9}{16} \div (-3) = \frac{9}{16} \cdot \left(-\frac{1}{3}\right)$ Multiply $\frac{9}{16}$ by the reciprocal of -3, which is $-\frac{1}{3}$. $= -\frac{9 \cdot 1}{16 \cdot 3}$ Multiply the numerators. Multiply the denominators. Since the fractions have unlike signs, make the answer negative. $= -\frac{\overset{1}{3} \cdot 3 \cdot 1}{16 \cdot \underset{1}{3}}$ To simplify, factor 9 as $3 \cdot 3$. Then remove the common factor of 3 from the numerator and denominator. $= -\frac{3}{16}$ Multiply the remaining factors in the numerator: $1 \cdot 3 \cdot 1 = 3$. Multiply the remaining factors in the denominator: $16 \cdot 1 = 16$.

To **divide two algebraic fractions,** we use the same approach as with numerical fractions.

The **reciprocal** of an algebraic fraction is found by inverting the numerator and denominator.

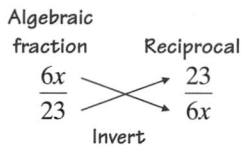

Algebraic
fraction   Reciprocal
$$\frac{6x}{23} \qquad \frac{23}{6x}$$
Invert

Divide and simplify: $-\dfrac{9x}{35y} \div \left(-\dfrac{15x^2}{14y}\right)$

$$-\frac{9x}{35y} \div \left(-\frac{15x^2}{14y}\right) = -\frac{9x}{35y}\left(-\frac{14y}{15x^2}\right)$$

Multiply by the reciprocal of $-\frac{15x^2}{14y}$.

$$= \frac{9x \cdot 14y}{35y \cdot 15x^2}$$

Multiply the numerators. Multiply the denominators. Since the product of two fractions with like signs is positive, drop the negative signs and continue.

$$= \frac{3 \cdot 3 \cdot x \cdot 2 \cdot 7 \cdot y}{5 \cdot 7 \cdot y \cdot 3 \cdot 5 \cdot x \cdot x}$$

To prepare to simplify, factor 9, 14, 35, and $15x^2$.

$$= \frac{3 \cdot \overset{1}{\cancel{3}} \cdot \overset{1}{\cancel{x}} \cdot 2 \cdot \overset{1}{\cancel{7}} \cdot \overset{1}{\cancel{y}}}{5 \cdot \underset{1}{\cancel{7}} \cdot \underset{1}{\cancel{y}} \cdot \underset{1}{\cancel{3}} \cdot 5 \cdot \underset{1}{\cancel{x}} \cdot x}$$

Simplify by removing the common factors of 3, 7, x and y from the numerator and denominator.

$$= \frac{6}{25x}$$

Multiply the remaining factors in the numerator. Multiply the remaining factors in the denominator.

Problems that involve forming **equal-sized groups** can be solved by division.

SEWING How many Halloween costumes, which require $\frac{3}{4}$ yard of material, can be made from 6 yards of material?

Since 6 yards of material is to be separated into an unknown number of equal-sized $\frac{3}{4}$-yard pieces, division is indicated.

$$6 \div \frac{3}{4} = \frac{6}{1} \cdot \frac{4}{3}$$

Write 6 as a fraction: $6 = \frac{6}{1}$.

Multiply $\frac{6}{1}$ by the reciprocal of $\frac{3}{4}$, which is $\frac{4}{3}$.

$$= \frac{6 \cdot 4}{1 \cdot 3}$$

Multiply the numerators. Multiply the denominators.

$$= \frac{2 \cdot \overset{1}{\cancel{3}} \cdot 4}{1 \cdot \underset{1}{\cancel{3}}}$$

To simplify, factor 6 as $2 \cdot 3$. Then remove the common factor of 3 from the numerator and denominator.

$$= \frac{8}{1}$$

Multiply the remaining factors in the numerator. Multiply the remaining factors in the denominator.

$$= 8$$

Any number divided by 1 is the same number.

The number of Halloween costumes that can be made from 6 yards of material is 8.

REVIEW EXERCISES

39. Find the reciprocal of each number or algebraic fraction.

a. $\dfrac{1}{8}$

b. $-\dfrac{11}{12}$

c. 5

d. $\dfrac{8a}{7}$

40. Fill in the blanks: To divide two fractions, _____ the first fraction by the _____ of the second fraction.

Divide. Simplify the quotient, if possible.

41. $\dfrac{1}{6} \div \dfrac{11}{25}$

42. $-\dfrac{7}{32} \div \dfrac{1}{4}$

43. $-\dfrac{39m^4}{25n} \div \left(-\dfrac{13m^3}{10n}\right)$

44. $54d \div \dfrac{63}{5}$

45. $-\dfrac{3}{8} \div \dfrac{1}{4}$

46. $\dfrac{4}{5} \div \dfrac{1}{2}$

47. $\dfrac{2}{3} \div (-120)$

48. $\dfrac{7}{15} \div \dfrac{7}{15}$

49. MAKING JEWELRY How many $\frac{1}{16}$-ounce silver angel pins can be made from a $\frac{3}{4}$-ounce bar of silver?

50. SEWING How many pillow cases, which require $\frac{2}{3}$ yard of material, can be made from 20 yards of cotton cloth?

SECTION 4.4 Adding and Subtracting Fractions

DEFINITIONS AND CONCEPTS	EXAMPLES
To **add (or subtract) fractions that have the same denominator,** add (or subtract) the numerators and write the sum (or difference) over the common denominator. Simplify the result, if possible.	Add: $\dfrac{3}{16} + \dfrac{5}{16}$ $\dfrac{3}{16} + \dfrac{5}{16} = \dfrac{3+5}{16}$ Add the numerators and write the sum over the common denominator 16. $= \dfrac{8}{16}$ The resulting fraction can be simplified. $= \dfrac{\overset{1}{\cancel{8}}}{2 \cdot \underset{1}{\cancel{8}}}$ To simplify, factor 16 as $2 \cdot 8$. Then remove the common factor of 8 from the numerator and denominator. $= \dfrac{1}{2}$ Multiply the remaining factors in the denominator: $2 \cdot 1 = 2$.
Adding and subtracting fractions that have different denominators 1. Find the LCD. 2. Rewrite each fraction as an equivalent fraction with the LCD as the denominator. To do so, build each fraction using a form of 1 that involves any factors needed to obtain the LCD. 3. Add or subtract the numerators and write the sum or difference over the LCD. 4. Simplify the result, if possible.	Subtract: $\dfrac{4}{7} - \dfrac{1}{3}$ Since the smallest number the denominators 7 and 3 divide exactly is 21, the LCD is 21. $\dfrac{4}{7} - \dfrac{1}{3} = \dfrac{4}{7} \cdot \dfrac{3}{3} - \dfrac{1}{3} \cdot \dfrac{7}{7}$ To build $\frac{4}{7}$ and $\frac{1}{3}$ so that their denominators are 21, multiply each by a form of 1. $= \dfrac{12}{21} - \dfrac{7}{21}$ Multiply the numerators. Multiply the denominators. The denominators are now the same. $= \dfrac{12-7}{21}$ Subtract the numerators and write the difference over the common denominator 21. $= \dfrac{5}{21}$ This fraction is in simplest form.

The **least common denominator (LCD)** of a set of fractions is the **least common multiple (LCM)** of the denominators of the fractions. Two ways to find the LCM of the denominators are as follows:

- Write the multiples of the largest denominator in increasing order, until one is found that is divisible by the other denominators.
- Prime factor each denominator. The LCM is a product of prime factors, where each factor is used the greatest number of times it appears in any one factorization.

Add and simplify: $\dfrac{9}{20} + \dfrac{7}{15}$

To find the LCD, find the prime factorization of both denominators and use each prime factor the *greatest* number of times it appears in any one factorization:

$$\left.\begin{array}{l}20 = \boxed{2 \cdot 2}\;\boxed{5}\\[2pt] 15 = \boxed{3} \cdot 5\end{array}\right\} \text{LCD} = 2 \cdot 2 \cdot 3 \cdot 5 = 60$$

$$\dfrac{9}{20} + \dfrac{7}{15} = \dfrac{9}{20} \cdot \dfrac{\mathbf{3}}{\mathbf{3}} + \dfrac{7}{15} \cdot \dfrac{\mathbf{4}}{\mathbf{4}}$$

To build $\frac{9}{20}$ and $\frac{7}{15}$ so that their denominators are 60, multiply each by a form of 1.

$$= \dfrac{27}{60} + \dfrac{28}{60}$$

Multiply the numerators. Multiply the denominators. The denominators are now the same.

$$= \dfrac{27 + 28}{60}$$

Add the numerators and write the sum over the common denominator 60.

$$= \dfrac{55}{60}$$

This fraction is not in simplest form.

$$= \dfrac{\overset{1}{\cancel{5}} \cdot 11}{2 \cdot 2 \cdot 3 \cdot \underset{1}{\cancel{5}}}$$

To simplify, prime factor 55 and 60. Then remove the common factor of 5 from the numerator and denominator.

$$= \dfrac{11}{12}$$

Multiply the remaining factors in the numerator and in the denominator.

To **add or subtract algebraic fractions,** we use the same approach as with numerical fractions.

Add and simplify: $\dfrac{5w}{24} + \dfrac{11w}{24}$

$$\dfrac{5w}{24} + \dfrac{11w}{24} = \dfrac{5w + 11w}{24}$$

Since the fractions have like denominators, add the numerators and write the sum over the common denominator, 24.

$$= \dfrac{16w}{24}$$

Combine like terms in the numerator: $5w + 11w = 16w$.

$$= \dfrac{2 \cdot \overset{1}{\cancel{8}} \cdot w}{3 \cdot \underset{1}{\cancel{8}}}$$

To simplify, factor 16 as $2 \cdot 8$ and 24 as $3 \cdot 8$ and then remove the common factor of 8 from the numerator and denominator.

$$= \dfrac{2w}{3}$$

Multiply the remaining factors in the numerator. Multiply the remaining factors in the denominator.

If we are to **add or subtract algebraic fractions,** their denominators must be the same.

Subtract and simplify, if possible: $\dfrac{6}{m} - \dfrac{3}{4}$

The denominators of $\frac{6}{m}$ and $\frac{3}{4}$ are m and 4. By inspection, we see that the LCD is $m \cdot 4 = 4m$.

$$\frac{6}{m} - \frac{3}{4} = \frac{6}{m} \cdot \frac{\mathbf{4}}{\mathbf{4}} - \frac{3}{4} \cdot \frac{m}{m}$$ Build each fraction so that each has a denominator of 4m.

$$= \frac{24}{4m} - \frac{3m}{4m}$$ Multiply the numerators. Multiply the denominators. The denominators are now the same.

$$= \frac{24 - 3m}{4m}$$ Subtract the numerators and write the difference over the common denominator, 4m.

Caution!   The result, $\frac{24\,-\,3m}{4m}$, is in simplest form. We cannot remove a common factor of m, because m is not a factor of the entire numerator.

$$\frac{24 - 3m}{4m} = \frac{24 - 3\overset{1}{\cancel{m}}}{\underset{1}{\cancel{4m}}} = \frac{21}{4}$$

Comparing fractions

If two fractions have the **same denominator,** the fraction with the greater numerator is the greater fraction.

If two fractions have **different denominators,** express each of them as an equivalent fraction that has the LCD for its denominator. Then compare numerators.

Which fraction is larger: $\dfrac{11}{18}$ or $\dfrac{7}{18}$?

$\dfrac{\mathbf{11}}{\mathbf{18}} > \dfrac{\mathbf{7}}{\mathbf{18}}$ because $11 > 7$

Which fraction is larger: $\dfrac{2}{3}$ or $\dfrac{3}{4}$?

Build each fraction to have a denominator that is the LCD, 12.

$$\frac{\mathbf{2}}{\mathbf{3}} = \frac{2}{3} \cdot \frac{4}{4} = \frac{\mathbf{8}}{\mathbf{12}} \qquad\qquad \frac{\mathbf{3}}{\mathbf{4}} = \frac{3}{4} \cdot \frac{3}{3} = \frac{\mathbf{9}}{\mathbf{12}}$$

Since $9 > 8$, it follows that $\dfrac{9}{12} > \dfrac{8}{12}$ and therefore, $\dfrac{3}{4} > \dfrac{2}{3}$.

REVIEW EXERCISES

Add or subtract and simplify, if possible.

51. $\dfrac{2}{7} + \dfrac{3}{7}$ **52.** $\dfrac{3}{4} - \dfrac{1}{4}$

53. $\dfrac{7x}{8} + \dfrac{3x}{8}$ **54.** $-\dfrac{3}{5} - \dfrac{3}{5}$

55. a. Add the fractions represented by the figures below.

b. Subtract the fractions represented by the figures below.

56. Fill in the blanks. Use the prime factorizations below to find the least common denominator for fractions with denominators of 45 and 30.

$$\left.\begin{array}{l} 45 = 3 \cdot 3 \cdot 5 \\ 30 = 2 \cdot 3 \cdot 5 \end{array}\right\} \text{LCD} = \square \cdot \square \cdot \square \cdot \square = \square$$

Add or subtract and simplify, if possible.

57. $\dfrac{1}{6} + \dfrac{2}{3}$ **58.** $-\dfrac{2}{5} - \dfrac{3}{8}$

59. $\dfrac{5}{24} + \dfrac{3}{16}$ **60.** $3 - \dfrac{1}{7}$

61. $-\dfrac{19}{18} + \dfrac{5}{12}$ **62.** $\dfrac{17}{20} - \dfrac{4}{15}$

63. $-6 + \dfrac{13}{6}$ **64.** $\dfrac{1}{3} + \dfrac{1}{4} + \dfrac{1}{5}$

65. $\dfrac{8}{n} + \dfrac{1}{2}$ **66.** $\dfrac{11}{9} - \dfrac{4}{x}$

67. $\dfrac{a}{3} - \dfrac{4}{11}$ **68.** $\dfrac{7}{8} + \dfrac{r}{7}$

69. MACHINE SHOPS How much must be milled off the $\frac{3}{4}$-inch-thick steel rod below so that the collar will slip over the end of it?

Steel rod

70. POLLS A group of adults were asked to rate the transportation system in their community. The results are shown below in a circle graph. What fraction of the group responded by saying either excellent, good, or fair?

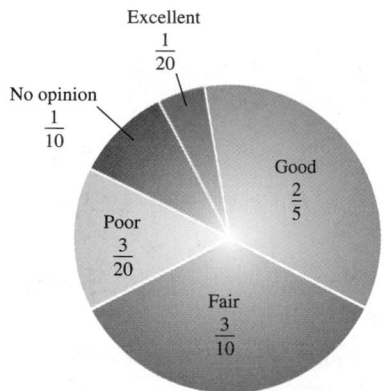

71. TELEMARKETING In the first hour of work, a telemarketer made 2 sales out of 9 telephone calls. In the second hour, she made 3 sales out of 11 calls. During which hour was the rate of sales to calls better?

72. CAMERAS When the shutter of a camera stays open longer than $\frac{1}{125}$ second, any movement of the camera will probably blur the picture. With this in mind, if a photographer is taking a picture of a fast-moving object, should she select a shutter speed of $\frac{1}{60}$ or $\frac{1}{250}$?

SECTION 4.5   Multiplying and Dividing Mixed Numbers

DEFINITIONS AND CONCEPTS	EXAMPLES
A **mixed number** is the sum of a whole number and a proper fraction.	$\underbrace{2\frac{3}{4}}_{\text{Mixed number}} = \underbrace{2}_{\substack{\text{Whole-number} \\ \text{part}}} + \underbrace{\frac{3}{4}}_{\text{Fractional part}}$
There is a relationship between **mixed numbers** and **improper fractions** that can be seen using shaded regions.	Each disk represents one whole. $2\frac{3}{4} = \frac{11}{4}$

To write a mixed number as an improper fraction: 1. Multiply the denominator of the fraction by the whole-number part. 2. Add the numerator of the fraction to the result from Step 1. 3. Write the sum from Step 2 over the original denominator.	Write $3\frac{4}{5}$ as an improper fraction. Step 2: Add $3\frac{4}{5} = \dfrac{5 \cdot 3 + 4}{5} = \dfrac{15 + 4}{5} = \dfrac{19}{5}$ Step 1: Multiply $\quad$ └─Step 3: Use the same denominator Thus, $3\frac{4}{5} = \dfrac{19}{5}$.
To write an improper fraction as a mixed number: 1. Divide the numerator by the denominator to obtain the whole-number part. 2. The remainder over the divisor is the fractional part.	Write $\dfrac{47}{6}$ as a mixed number. $\quad\;\; 7 \leftarrow$ The whole-number part is 7. $6\overline{)47}$ $\;\underline{-42}$ $\quad\;\; 5 \leftarrow$ Write the remainder 5 over the divisor 6 $\qquad\qquad$ to get the fractional part. Thus, $\dfrac{47}{6} = 7\frac{5}{6}$.
Fractions and mixed numbers can be **graphed** on a number line.	Graph $-3\frac{1}{3}, 1\frac{1}{4}, \frac{18}{5}$, and $-\frac{7}{8}$ on a number line. 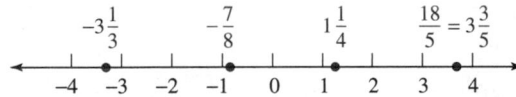
To multiply mixed numbers, first change the mixed numbers to improper fractions. Then perform the multiplication of the fractions. Write the result as a mixed number or whole number in simplest form.	Multiply and simplify: $\;10\frac{1}{2} \cdot 1\frac{1}{6}$ $10\frac{1}{2} \cdot 1\frac{1}{6} = \dfrac{21}{2} \cdot \dfrac{7}{6}$ $\qquad$ Write $10\frac{1}{2}$ and $1\frac{1}{6}$ as improper fractions. $\qquad\qquad = \dfrac{21 \cdot 7}{2 \cdot 6}$ $\qquad$ Use the rule for multiplying two fractions. Multiply the numerators. Multiply the denominators. $\qquad\qquad = \dfrac{\overset{1}{\cancel{3}} \cdot 7 \cdot 7}{2 \cdot 2 \cdot \underset{1}{\cancel{3}}}$ $\qquad$ To simplify, factor 21 as $3 \cdot 7$, and then remove the common factor of 3 from the numerator and denominator. $\qquad\qquad = \dfrac{49}{4}$ $\qquad$ Multiply the remaining factors in the numerator and in the denominator. The result is an improper fraction. $\qquad\qquad = 12\frac{1}{4}$ $\qquad$ Write the improper fraction $\frac{49}{4}$ as a mixed number. $\quad\;\; 12$ $4\overline{)49}$ $\;\underline{-4}$ $\quad\; 09$ $\;\;\underline{-8}$ $\quad\quad 1$

To **divide mixed numbers,** first change the mixed numbers to improper fractions. Then perform the division of the fractions. Write the result as a mixed number or whole number in simplest form.

Divide and simplify: $5\frac{2}{3} \div \left(-3\frac{7}{9}\right)$

$5\frac{2}{3} \div \left(-3\frac{7}{9}\right) = \frac{17}{3} \div \left(-\frac{34}{9}\right)$ Write $5\frac{2}{3}$ and $3\frac{7}{9}$ as improper fractions.

$= \frac{17}{3}\left(-\frac{9}{34}\right)$ Multiply $\frac{17}{3}$ by the reciprocal of $-\frac{34}{9}$, which is $-\frac{9}{34}$.

$= -\frac{17 \cdot 9}{3 \cdot 34}$ Multiply the numerators. Multiply the denominators.

Since the fractions have unlike signs, make the answer negative.

$= -\frac{\overset{1}{\cancel{17}} \cdot \overset{1}{\cancel{3}} \cdot 3}{\cancel{3} \cdot 2 \cdot \cancel{17}}$ To simplify, factor 9 as $3 \cdot 3$ and 34 as $2 \cdot 17$. Then remove the common factors of 3 and 17 from the numerator and denominator.

$= -\frac{3}{2}$ Multiply the remaining factors in the numerator and in the denominator. The result is a negative improper fraction.

$= -1\frac{1}{2}$ Write the negative improper fraction $-\frac{3}{2}$ as a negative mixed number.

REVIEW EXERCISES

73. In the illustration below, each triangular region outlined in black represents one whole. Write a mixed number and an improper fraction to represent what is shaded.

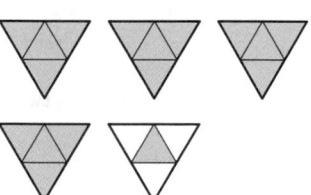

74. Graph $-2\frac{2}{3}, \frac{8}{9}, -\frac{3}{4},$ and $\frac{59}{24}$ on a number line.

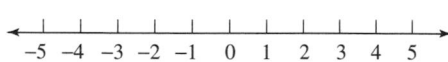

Write each improper fraction as a mixed number or a whole number.

75. $\frac{16}{5}$ **76.** $-\frac{47}{12}$

77. $\frac{51}{3}$ **78.** $\frac{14}{6}$

Write each mixed number as an improper fraction.

79. $9\frac{3}{8}$ **80.** $-2\frac{1}{5}$

81. $3\frac{11}{14}$ **82.** $1\frac{99}{100}$

Multiply or divide and simplify, if possible.

83. $1\frac{2}{5} \cdot 1\frac{1}{2}$ **84.** $-3\frac{1}{2} \div 3\frac{2}{3}$

85. $-6\left(-6\frac{2}{3}\right)$ **86.** $8 \div 3\frac{1}{5}$

87. $-11\frac{1}{5} \div \left(-\frac{7}{10}\right)$ **88.** $5\frac{2}{3}\left(-7\frac{1}{5}\right)$

89. $\left(-2\frac{3}{4}\right)^2$ **90.** $1\frac{5}{16} \cdot 1\frac{7}{9} \cdot 2\frac{2}{3}$

91. PHOTOGRAPHY Each leg of a camera tripod can be extended to become $5\frac{1}{2}$ times its original length. If a leg is originally $8\frac{3}{4}$ inches long, how long will it become when it is completely extended?

92. PET DOORS Find the area of the opening provided by the rectangular-shaped pet door shown below.

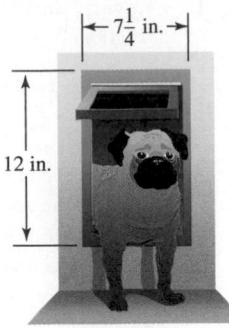

$7\frac{1}{4}$ in.

12 in.

93. PRINTING It takes a color copier $2\frac{1}{4}$ minutes to print a movie poster. How many posters can be printed in 90 minutes?

94. STORM DAMAGE A truck can haul $7\frac{1}{2}$ tons of trash in one load. How many loads would it take to haul away $67\frac{1}{2}$ tons from a hurricane cleanup site?

SECTION 4.6   Adding and Subtracting Mixed Numbers

DEFINITIONS AND CONCEPTS	EXAMPLES
To add (or subtract) mixed numbers, we can change each to an improper fraction and use the method of Section 3.4.	Add: $3\frac{1}{2} + 1\frac{3}{5}$

$$3\frac{1}{2} + 1\frac{3}{5} = \frac{7}{2} + \frac{8}{5}$$

Write $3\frac{1}{2}$ and $1\frac{3}{5}$ as mixed numbers.

$$= \frac{7}{2} \cdot \frac{5}{5} + \frac{8}{5} \cdot \frac{2}{2}$$

To build $\frac{7}{2}$ and $\frac{8}{5}$ so that their denominators are 10, multiply both by a form of 1.

$$= \frac{35}{10} + \frac{16}{10}$$

Multiply the numerators.
Multiply the denominators.

$$= \frac{51}{10}$$

Add the numerators and write the sum over the common denominator 10.

$$= 5\frac{1}{10}$$

To write the improper fraction $\frac{51}{10}$ as a mixed number, divide 51 by 10.

To add (or subtract) mixed numbers, we can also write them in **vertical form** and add (or subtract) the whole-number parts and the fractional parts separately.

Add: $42\frac{1}{3} + 89\frac{6}{7}$

Build to get the LCD, 21.

Add the fractions.

Add the whole numbers.

$$42\frac{1}{3} = \quad 42\frac{1}{3} \cdot \frac{7}{7} = \quad 42\frac{7}{21} = \quad 42\overset{1}{}\frac{7}{21}$$

$$+\ 89\frac{6}{7} = \ +\ 89\frac{6}{7} \cdot \frac{3}{3} = \ +\ 89\frac{18}{21} = \ +\ 89\frac{18}{21}$$

$$\frac{25}{21} \qquad\qquad 131\frac{25}{21}$$

When we add mixed numbers, sometimes the sum of the fractions is an improper fraction. If that is the case, write the improper fraction as a mixed number and **carry** its whole-number part to the whole-number column.

We don't want an improper fraction in the answer.

Write $\frac{25}{21}$ as $1\frac{4}{21}$, carry the 1 to the whole-number column, and add it to 131 to get 132:

$$131\frac{25}{21} = 131 + 1\frac{4}{21} = 132\frac{4}{21}$$

Subtraction of mixed numbers in vertical form sometimes involves **borrowing.** When the fraction we are subtracting is greater than the fraction we are subtracting it from, borrowing is necessary.

Subtract: $23\frac{1}{4} - 17\frac{5}{9}$

Build to get the LCD, 36.

Since $\frac{20}{36}$ is greater than $\frac{9}{36}$, we must borrow from 28.

$$28\frac{1}{4} = 28\frac{1}{4} \cdot \frac{9}{9} = 28\frac{9}{36} = 28\frac{9}{36} + \frac{36}{36} = 28\overset{7}{}\frac{45}{36}$$

$$-17\frac{5}{9} = -17\frac{5}{9} \cdot \frac{4}{4} = -17\frac{20}{36} = -17\frac{20}{36} = -17\frac{20}{36}$$

$$10\frac{25}{36}$$

REVIEW EXERCISES

Add or subtract and simplify, if possible.

95. $1\frac{3}{8} + 2\frac{1}{5}$

96. $3\frac{1}{2} + 2\frac{2}{3}$

97. $157\frac{11}{30} + 98\frac{7}{12}$

98. $6\frac{3}{14} + 17\frac{7}{10}$

99. $33\frac{8}{9} + 49\frac{1}{6}$

100. $98\frac{11}{20} + 14\frac{4}{5}$

101. $23\frac{1}{3} - 2\frac{5}{6}$

102. $39 - 4\frac{5}{8}$

103. PAINTING SUPPLIES In a project to restore a house, painters used $10\frac{3}{4}$ gallons of primer, $21\frac{1}{2}$ gallons of latex paint, and $7\frac{2}{3}$ gallons of enamel. Find the total number of gallons of paint used.

104. PASSPORTS The required dimensions for a passport photograph are shown below. What is the distance from the subject's eyes to the top of the photograph?

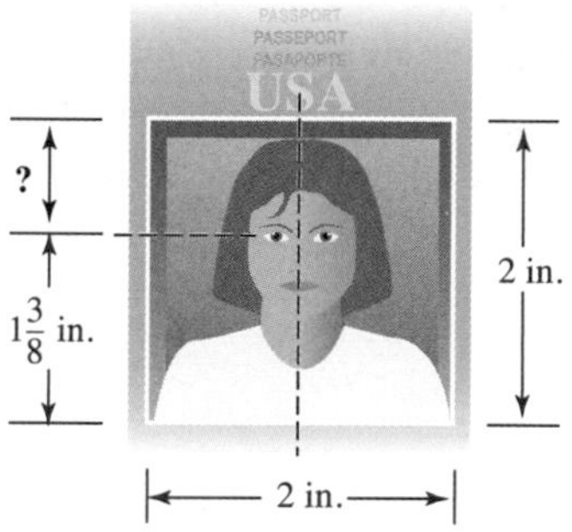

SECTION 4.7   Order of Operations and Complex Fractions

DEFINITIONS AND CONCEPTS	EXAMPLES

Order of Operations

1. Perform all calculations within parentheses and other grouping symbols following the order listed in Steps 2–4 below, working from the innermost pair of grouping symbols to the outermost pair.
2. Evaluate all exponential expressions.
3. Perform all multiplications and divisions as they occur from left to right.
4. Perform all additions and subtractions as they occur from left to right.

When grouping symbols have been removed, repeat Steps 2–4 to complete the calculation.

If a fraction bar is present, evaluate the expression above the bar and the expression below the bar separately. Then perform the division indicated by the fraction bar, if possible.

Evaluate: $\left(\frac{1}{3}\right)^2 \div \left(\frac{3}{4} - \frac{1}{3}\right)$

First, we perform the subtraction within the second set of parentheses. (There is no operation to perform within the first set.)

$\left(\frac{1}{3}\right)^2 \div \left(\frac{3}{4} - \frac{1}{3}\right)$

$= \left(\frac{1}{3}\right)^2 \div \left(\frac{3}{4} \cdot \frac{3}{3} - \frac{1}{3} \cdot \frac{4}{4}\right)$ — Within the parentheses, build each fraction so that its denominator is the LCD 12.

$= \left(\frac{1}{3}\right)^2 \div \left(\frac{9}{12} - \frac{4}{12}\right)$ — Multiply the numerators. Multiply the denominators.

$= \left(\frac{1}{3}\right)^2 \div \frac{5}{12}$ — Subtract the numerators: 9 − 4 = 5. Write the difference over the common denominator 12.

$= \frac{1}{9} \div \frac{5}{12}$ — Evaluate the exponential expression: $\left(\frac{1}{3}\right)^2 = \frac{1}{3} \cdot \frac{1}{3} = \frac{1}{9}$.

$= \frac{1}{9} \cdot \frac{12}{5}$ — Use the rule for dividing fractions: Multiply the first fraction by the reciprocal of $\frac{5}{12}$, which is $\frac{12}{5}$.

$= \frac{1 \cdot 12}{9 \cdot 5}$ — Multiply the numerators. Multiply the denominators.

$= \frac{1 \cdot \overset{1}{\cancel{3}} \cdot 4}{\underset{1}{\cancel{3}} \cdot 3 \cdot 5}$ — To simplify, factor 12 as 3 · 4 and 9 as 3 · 3. Then remove the common factor of 3 from the numerator and denominator.

$= \frac{4}{15}$ — Multiply the remaining factors in the numerator. Multiply the remaining factors in the denominator.

To **evaluate a formula,** we replace its variables (letters) with specific numbers and evaluate the right side using the order of operations rule.

Evaluate: $A = \frac{1}{2}h(a + b)$ for $a = 1\frac{1}{3}, b = 2\frac{2}{3}$, and $h = 2\frac{4}{5}$.

$A = \frac{1}{2}h(a + b)$ — This is the given formula.

$= \frac{1}{2}\left(2\frac{4}{5}\right)\left(1\frac{1}{3} + 2\frac{2}{3}\right)$ — Replace h, a, and b with the given values.

$= \frac{1}{2}\left(2\frac{4}{5}\right)(4)$ — Do the addition within the parentheses.

$= \frac{1}{2}\left(\frac{14}{5}\right)\left(\frac{4}{1}\right)$ — To prepare to multiply fractions, write $2\frac{4}{5}$ as an improper fraction and 4 as $\frac{4}{1}$.

$= \frac{1 \cdot 14 \cdot 4}{2 \cdot 5 \cdot 1}$ — Multiply the numerators. Multiply the denominators.

$= \frac{1 \cdot 14 \cdot \overset{1}{\cancel{2}} \cdot 2}{\underset{1}{\cancel{2}} \cdot 5 \cdot 1}$ — To simplify, factor 4 as 2 · 2. Then remove the common factor of 2 from the numerator and denominator.

$= \frac{28}{5}$ — Multiply the remaining factors in the numerator. Multiply the remaining factors in the denominator.

$= 5\frac{3}{5}$ — Write the improper fraction $\frac{28}{5}$ as a mixed number by dividing 28 by 5.

A **complex fraction** is a fraction whose numerator or denominator, or both, contain one or more fractions or mixed numbers.	Complex fractions: $$\dfrac{\dfrac{9}{10}}{\dfrac{27}{5}} \qquad \dfrac{\dfrac{2}{5}-\dfrac{1}{3}}{\dfrac{3}{7}+\dfrac{1}{5}} \qquad \dfrac{-7\dfrac{1}{4}}{2-\dfrac{1}{9}}$$
The method for **simplifying complex fractions** is based on the fact that the main fraction bar indicates division.	Simplify: $\dfrac{\dfrac{9}{10}}{\dfrac{27}{5}}$ $\dfrac{\dfrac{9}{10}}{\dfrac{27}{5}} = \dfrac{9}{10} \div \dfrac{27}{5}$ Write the division indicated by the main fraction bar using a ÷ symbol. $= \dfrac{9}{10} \cdot \dfrac{5}{27}$ Use the rule for dividing fractions: Multiply the first fraction by the reciprocal of $\frac{27}{5}$, which is $\frac{5}{27}$. $= \dfrac{9 \cdot 5}{10 \cdot 27}$ Multiply the numerators. Multiply the denominators. $= \dfrac{\overset{1}{9} \cdot \overset{1}{5}}{2 \cdot \underset{1}{5} \cdot 3 \cdot \underset{1}{9}}$ To simplify, factor 10 as 2 · 5 and 27 as 3 · 9. Then remove the common factors of 9 and 5 from the numerator and denominator. $= \dfrac{1}{6}$ Multiply the remaining factors in the numerator. Multiply the remaining factors in the denominator.
To **simplify a complex fraction:** 1. Add or subtract in the numerator and/or denominator so that the numerator is a single fraction and the denominator is a single fraction. 2. Perform the indicated division by multiplying the numerator of the complex fraction by the reciprocal of the denominator. 3. Simplify the result, if possible.	Simplify: $\dfrac{\dfrac{2}{5}-\dfrac{1}{3}}{\dfrac{3}{7}+\dfrac{1}{5}}$ $\dfrac{\dfrac{2}{5}-\dfrac{1}{3}}{\dfrac{3}{7}+\dfrac{1}{5}} = \dfrac{\dfrac{2}{5}\cdot\dfrac{3}{3}-\dfrac{1}{3}\cdot\dfrac{5}{5}}{\dfrac{3}{7}\cdot\dfrac{5}{5}+\dfrac{1}{5}\cdot\dfrac{7}{7}}$ In the numerator, build each fraction so that each has a denominator of 15. In the denominator, build each fraction so that each has a denominator of 35. $= \dfrac{\dfrac{6}{15}-\dfrac{5}{15}}{\dfrac{15}{35}+\dfrac{7}{35}}$ Multiply the numerators. Multiply the denominators. $= \dfrac{\dfrac{1}{15}}{\dfrac{22}{35}}$ Subtract the numerators and write the difference over the common denominator 15. Add the numerators and write the sum over the common denominator 35. $= \dfrac{1}{15} \div \dfrac{22}{35}$ Write the division indicated by the main fraction bar using a ÷ symbol. $= \dfrac{1}{15} \cdot \dfrac{35}{22}$ Use the rule for dividing fractions: Multiply the first fraction by the reciprocal of $\frac{22}{35}$, which is $\frac{35}{22}$. $= \dfrac{1 \cdot 35}{15 \cdot 22}$ Multiply the numerators. Multiply the denominators. $= \dfrac{1 \cdot \overset{1}{5} \cdot 7}{3 \cdot \underset{1}{5} \cdot 22}$ To simplify, factor 35 as 5 · 7 and 15 as 3 · 5. Then remove the common factor of 5 from the numerator and denominator. $= \dfrac{7}{66}$ Multiply the remaining factors in the numerator. Multiply the remaining factors in the denominator.

REVIEW EXERCISES

Evaluate each expression.

105. $\dfrac{3}{4} + \left(-\dfrac{1}{3}\right)^2\left(\dfrac{5}{4}\right)$

106. $\left(\dfrac{2}{3} \div \dfrac{16}{9}\right) - \left(1\dfrac{2}{3} \cdot \dfrac{1}{15}\right)$

107. $\left(\dfrac{11}{5} - 1\dfrac{2}{3}\right) - \left(-\dfrac{4}{9} \cdot 18\right)$

108. $\left|-\dfrac{9}{16} \div 2\dfrac{1}{4}\right| + \left(-3\dfrac{7}{8}\right)$

Simplify each complex fraction.

109. $\dfrac{\dfrac{3}{5}}{-\dfrac{17}{20}}$

110. $\dfrac{4 - \dfrac{2}{7}}{4\dfrac{1}{7}}$

111. $\dfrac{\dfrac{2}{3} - \dfrac{1}{6}}{-\dfrac{3}{4} - \dfrac{1}{2}}$

112. $\dfrac{5\dfrac{1}{4}}{\dfrac{7}{4} + \left(-\dfrac{1}{3}\right)}$

113. Subtract $4\dfrac{1}{8}$ from the sum of $5\dfrac{1}{5}$ and $1\dfrac{1}{2}$.

114. Add $12\dfrac{11}{16}$ to the difference of $4\dfrac{5}{8}$ and $3\dfrac{1}{4}$.

115. Evaluate the formula $A = \dfrac{1}{2}h(a + b)$ for $a = 1\dfrac{1}{8}$, $b = 4\dfrac{7}{8}$, and $h = 2\dfrac{7}{9}$.

116. Evaluate the formula $P = 2\ell + 2w$ for $\ell = 2\dfrac{1}{3}$ and $w = 3\dfrac{1}{4}$.

117. DERMATOLOGY A dermatologist mixes $1\dfrac{1}{2}$ ounces of cucumber extract, $2\dfrac{2}{3}$ ounces of aloe vera cream, and $\dfrac{3}{4}$ ounce of vegetable glycerin to make his own brand of anti-wrinkle cream. He packages it in $\dfrac{5}{6}$-ounce tubes. How many full tubes can be produced using this formula? How much cream is left over?

118. GUITAR DESIGN Find the missing dimension on the vintage 1962 Stratocaster body shown below.

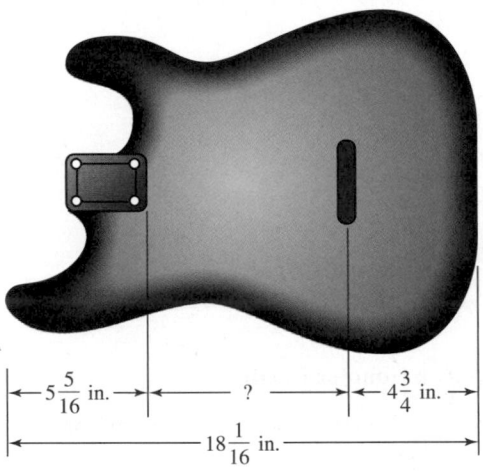

$\leftarrow 5\dfrac{5}{16}$ in. $\rightarrow$ $\leftarrow$? $\rightarrow$ $\leftarrow 4\dfrac{3}{4}$ in. $\rightarrow$

$\longleftarrow 18\dfrac{1}{16}$ in. $\longrightarrow$

SECTION 4.8   Solving Equations That Involve Fractions

DEFINITIONS AND CONCEPTS	EXAMPLES
The properties of equality that we used to solve equations involving whole numbers and integers are also used to **solve equations involving fractions.**	Solve: $\dfrac{2}{21}h = 18$

Recall that the product of a number and its reciprocal is 1. We can use this fact to solve equations like those shown below, where the **coefficient of the variable term** is a fraction.

$$\frac{2}{21}h = 18 \qquad -\frac{7}{3}d = 49$$

The coefficient of h is a fraction.

The coefficient of d is a negative fraction.

Recall that $\frac{2}{21}h = 18$ means $\frac{2}{21} \cdot h = 18$.

$$\frac{2}{21}h = 18 \qquad \text{This is the equation to solve.}$$

$$\frac{21}{2} \cdot \frac{2}{21}h = \frac{21}{2} \cdot 18 \qquad \text{To isolate } h, \text{ undo the multiplication by } \frac{2}{21} \text{ by multiplying both sides by the reciprocal of } \frac{2}{21}, \text{ which is } \frac{21}{2}.$$

$$\left(\frac{21}{2} \cdot \frac{2}{21}\right)h = \frac{21}{2} \cdot \frac{18}{1} \qquad \text{On the left side, use the associative property of multiplication to group } \frac{21}{2} \text{ and } \frac{2}{21}. \text{ On the right side, write 18 as } \frac{18}{1}.$$

$$1h = \frac{21 \cdot 18}{2 \cdot 1} \qquad \text{On the left side, the product of a number and its reciprocal is 1: } \frac{21}{2} \cdot \frac{2}{21} = 1. \text{ On the right side, multiply the numerators and multiply the denominators.}$$

$$h = \frac{21 \cdot \overset{1}{2} \cdot 9}{\underset{1}{2} \cdot 1} \qquad \text{On the left side, the coefficient of 1 need not be written: } 1h = h. \text{ To simplify on the right side, factor 18 as } 2 \cdot 9, \text{ and remove the common factor of 2.}$$

$$h = 189 \qquad \text{Multiply on the right side: } 21 \cdot 9 = 189.$$

The solution is 189. Check this result into the *original* equation.

Equations that involve integers are usually easier to solve than equations that involve fractions. We can **clear an equation of fractions** by multiplying both sides by their LCD.

Strategy for Solving Equations

1. Clear the equation of fractions.
2. Simplify each side of the equation.
3. Isolate the variable term on one side.
4. Isolate the variable.
5. Check the result in the *original* equation.

You won't always have to use all five steps to solve a given equation.

Solve: $\dfrac{6}{5}x = \dfrac{1}{10}x - \dfrac{5}{2}$

$$10\left(\frac{6}{5}x\right) = 10\left(\frac{1}{10}x - \frac{5}{2}\right) \qquad \text{Multiply both sides by the LCD of } \frac{6}{5}, \frac{1}{10}, \text{ and } \frac{5}{2}, \text{ which is 10.}$$

$$10\left(\frac{6}{5}x\right) = 10\left(\frac{1}{10}x\right) - 10\left(\frac{5}{2}\right) \qquad \text{On the right side, distribute the multiplication by 10.}$$

$$\frac{10}{1}\left(\frac{6}{5}x\right) = \frac{10}{1}\left(\frac{1}{10}x\right) - \frac{10}{1}\left(\frac{5}{2}\right) \qquad \text{Write 10 as } \frac{10}{1}. \text{ This makes the numerators and denominators in the fraction multiplication process clearer.}$$

$$\frac{2 \cdot \overset{1}{5}}{1}\left(\frac{6}{\underset{1}{5}}x\right) = \frac{\overset{1}{10}}{1}\left(\frac{1}{\underset{1}{10}}x\right) - \frac{2 \cdot 5}{1}\left(\frac{5}{\underset{1}{2}}\right) \qquad \text{Factor 10 as } 2 \cdot 5. \text{ Remove the common factors of 5, 10, and 2 in the numerator and denominator.}$$

$$12x = x - 25 \qquad \text{Complete each multiplication.}$$

$$12x - x = x - 25 - x \qquad \text{To eliminate } x \text{ on the right side, subtract } x \text{ from both sides.}$$

$$11x = -25 \qquad \text{Combine like terms on each side.}$$

$$\frac{11x}{11} = \frac{-25}{11} \qquad \text{To isolate } x, \text{ undo the multiplication by 11 by dividing both sides by 11.}$$

$$x = -\frac{25}{11}$$

The solution is $-\frac{25}{11}$. Check this result by substituting into the *original* equation.

We can use the **five-step problem-solving strategy** discussed in earlier chapters to solve application problems that involve fractions.

Review Example 11 and Example 12 on pages 408 through 410.

REVIEW EXERCISES

Solve each equation and check the result.

119. $x - \dfrac{1}{40} = \dfrac{11}{40}$

120. $\dfrac{5}{4} = n + \dfrac{2}{3}$

121. $\dfrac{2}{9}x = 22$

122. $\dfrac{4}{25}y = 4$

123. $-\dfrac{7}{3}y = 5$

124. $\dfrac{1}{8}x = \dfrac{27}{32}$

125. $37x = -\dfrac{19}{3}$

126. $\dfrac{7}{8}x + 21 = 7$

127. $\dfrac{d}{4} = \dfrac{d}{7} - 6$

128. $\dfrac{1}{6}x - \dfrac{1}{2} = \dfrac{1}{3}x$

In Exercises 129 and 130, let a variable represent the unknown quantity. Then write and solve an equation to answer the question.

129. TEXTBOOKS In writing a history text, the author decided to devote two-thirds of the book to events prior to World War II. The remainder of the book deals with history after the war. If pre–World War II history is covered in 220 pages, how many pages does the textbook have?

130. SEMINARS A real estate investment seminar has three parts. In the first part, a film is shown that takes one-fourth of the class time. In the second part, the instructor lectures for 35 minutes. In the final part, successful investors take two-fifths of the class time to give their testimonials. How many minutes long is the seminar?

CHAPTER 4   TEST

1. Fill in the blanks.

a. For the fraction $\frac{6}{7}$, the _____ is 6 and the _____ is 7.

b. Two fractions are _____ if they represent the same number.

c. A fraction is in _____ form when the numerator and denominator have no common factors other than 1.

d. To _____ a fraction, we remove common factors of the numerator and denominator.

e. The _____ of $\frac{4}{5}$ is $\frac{5}{4}$.

f. A _____ number, such as $1\frac{9}{16}$, is the sum of a whole number and a proper fraction.

g. $\dfrac{\frac{1}{8}}{\frac{7}{12}}$ and $\dfrac{\frac{3}{4}+\frac{1}{3}}{\frac{5}{12}-\frac{1}{4}}$ are examples of _____ fractions.

2. See the illustration to the right.

a. What fractional part of the plant is above ground?

b. What fractional part of the plant is below ground?

3. Each region outlined in black represents one whole. Write an improper fraction and a mixed number to represent the shaded portion.

4. Graph $2\frac{4}{5}$, $-\frac{2}{5}$, $-1\frac{1}{7}$, and $\frac{7}{6}$ on a number line.

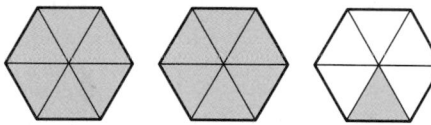

5. Are $\frac{1}{3}$ and $\frac{5}{15}$ equivalent?

6. a. Express $\frac{4}{5}$ as an equivalent fraction with denominator 45.

b. Express $\frac{7}{8}$ as an equivalent fraction with denominator $24x$.

7. Simplify each fraction, if possible.

a. $\dfrac{0}{15}$ **b.** $\dfrac{9}{0}$

8. Simplify each fraction.

a. $\dfrac{27}{36}$ **b.** $\dfrac{72n^3}{180n}$

9. Add and simplify, if possible: $\dfrac{3}{16}+\dfrac{7}{16}$

10. Multiply and simplify, if possible: $-\dfrac{3}{4}\left(\dfrac{1}{5}\right)$

11. Divide and simplify, if possible: $\dfrac{4a}{3b}\div\dfrac{a^2}{9b^5}$

12. Subtract and simplify, if possible: $\dfrac{11}{12}-\dfrac{11}{30}$

13. Add and simplify, if possible: $-\dfrac{3}{7}+2$

14. Multiply and simplify, if possible: $\dfrac{9}{10}\left(-\dfrac{4}{15}\right)\left(-\dfrac{25}{18}\right)$

15. Which fraction is larger: $\dfrac{8}{9}$ or $\dfrac{9}{10}$?

16. Find the reciprocal of:

a. $-\dfrac{17}{53}$ **b.** $\dfrac{7}{3a^2}$

17. Subtract: $\dfrac{x}{6}-\dfrac{4}{5}$

18. COFFEE DRINKERS Two-fifths of 100 adults surveyed said they started their morning with a cup of coffee. Of the 100, how many would this be?

19. THE INTERNET The graph below shows the fraction of the total number of Internet searches that were made using various sites in January 2009. What fraction of the all the searches were done using Google, Yahoo, or Microsoft sites?

Online Search Share January 2009

Google Sites $\frac{16}{25}$

Yahoo Sites $\frac{1}{5}$

Other $\frac{1}{50}$

AOL Sites $\frac{1}{25}$

Microsoft Sites $\frac{1}{10}$

Source: Marketingcharts.com

20. a. Write $\dfrac{55}{6}$ as a mixed number.

b. Write $1\dfrac{18}{21}$ as an improper fraction.

21. Find the sum of $157\dfrac{3}{10}$ and $103\dfrac{13}{15}$. Simplify the result.

22. Subtract and simplify, if possible: $67\dfrac{1}{4} - 29\dfrac{5}{6}$

23. Divide and simplify, if possible: $6\dfrac{1}{4} \div 3\dfrac{3}{4}$

24. BOXING Two of the greatest heavyweight boxers of all time are Muhammad Ali and George Foreman. Refer to the "Tale of the Tape" comparison shown below.

a. Which fighter weighed more? How much more?

b. Which fighter had the larger waist measurement? How much larger?

c. Which fighter had the larger forearm measurement? How much larger?

Tale of the Tape		
Muhammad Ali		George Foreman
6-3	Height	6-4
210½ lb	Weight	250 lb
82 in.	Reach	79 in.
43 in.	Chest (Normal)	48 in.
45½ in.	Chest (Expanded)	50 in.
34 in.	Waist	39½ in.
12½ in.	Fist	13½ in.
15 in.	Forearm	14¾ in.

Source: The International Boxing Hall of Fame

25. Evaluate the formula $P = 2l + 2w$ for $l = \dfrac{1}{3}$ and $w = \dfrac{1}{9}$.

26. SPORTS CONTRACTS A basketball player signed a nine-year contract for $13\dfrac{1}{2}$ million. How much is this per year?

27. SEWING When cutting material for a $10\dfrac{1}{2}$-inch-wide placemat, a seamstress allows $\dfrac{5}{8}$ inch at each end for a hem, as shown below. How wide should the material be cut to make a placemat?

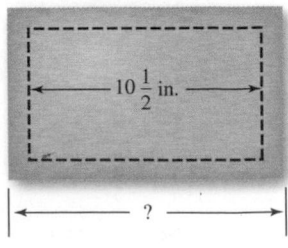

28. Find the perimeter and the area of the triangle shown to the right.

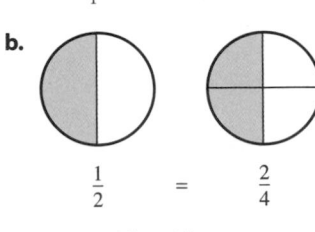

29. NUTRITION A box of Tic Tacs contains 40 of the $1\dfrac{1}{2}$-calorie breath mints. How many calories are there in a box of Tic Tacs?

30. COOKING How many servings are there in an 8-pound roast, if the suggested serving size is $\dfrac{2}{3}$ pound?

31. Evaluate:
$$\left(\dfrac{2}{3} \cdot \dfrac{5}{16}\right) - \left(-1\dfrac{3}{5} \div 4\dfrac{4}{5}\right)$$

32. Evaluate: $\left(\dfrac{1}{2}\right)^3 \div \left(\dfrac{3}{4} - \dfrac{1}{3}\right)$

33. Simplify:
$$\dfrac{\dfrac{5}{6}}{\dfrac{7}{8}}$$

34. Simplify:
$$\dfrac{\dfrac{1}{2} + \dfrac{1}{3}}{-\dfrac{1}{6} - \dfrac{1}{3}}$$

Solve each equation and check the result.

35. $\dfrac{5}{16}y = \dfrac{5}{6}$

36. $-\dfrac{3}{8} = a + \dfrac{1}{2}$

37. $\dfrac{6}{11}n + 35 = -7$

38. $\dfrac{7}{8}x - \dfrac{5}{16} = \dfrac{1}{4}x$

Let a variable represent the unknown quantity. Then write and solve an equation to answer the question.

39. PETS A dog obedience class has four parts. The class begins with a 10-minute demonstration. In the second part, which takes one-half of the class time, an expert explains how to train a dog. Then, a veterinarian speaks about nutrition and health. This takes one-third of the class time. The class ends with a 25-minute question-and-answer session. How many minutes long is the dog obedience class?

40. Explain each mathematical concept that is shown below.

a. $\dfrac{6}{8} = \dfrac{\overset{1}{\cancel{2}} \cdot 3}{\underset{1}{\cancel{2}} \cdot 4} = \dfrac{3}{4}$

b.

$$\dfrac{1}{2} = \dfrac{2}{4}$$

c. $\dfrac{3}{5} = \dfrac{3}{5} \cdot \dfrac{4}{4} = \dfrac{12}{20}$

1. Consider the number 5,896,619. [Section 1.1]

   a. What digit is in the millions column?

   b. What is the place value of the digit 8?

   c. Round to the nearest hundred.

   d. Round to the nearest ten thousand.

2. BANKS In 2008, the world's largest bank, with a net worth of $277,514,000,000, was the Industrial and Commercial Bank of China. In what place-value column is the digit 2? (Source: Skorcareer) [Section 1.1]

3. POPULATION Rank the following counties in order, from greatest to least population. [Section 1.1]

County	2007 Population
Dallas County, TX	2,366,511
Kings County, NY	2,528,050
Miami-Dade County, FL	2,387,170
Orange County, CA	2,997,033
Queens County, NY	2,270,338
San Diego County, CA	2,974,859

   (Source: *The World Almanac and Book of Facts,* 2009)

4. POOL CONSTRUCTION Refer to the rectangular-shaped swimming pool shown below.

   a. Find the perimeter of the pool. [Section 1.2]

   b. Find the area of the pool's surface. [Section 1.3]

   150 ft

   75 ft

5. Add: 7,897 [Section 1.2]
           6,909
           1,812
        + 14,378

6. Subtract 3,456 from 20,000. Check the result. [Section 1.2]

7. SHEETS OF STICKERS There are twenty rows of twelve gold stars on one sheet of stickers. If a packet contains ten sheets, how many stars are there in one packet? [Section 1.3]

8. Multiply: 5,345 [Section 1.3]
             × 56

9. Divide: $35)\overline{34,685}$. Check the result. [Section 1.4]

10. a. List factors of 24, from least to greatest. [Section 1.5]

    b. Find the prime factorization of 450. [Section 1.5]

11. Find the LCM of 16 and 20. [Section 1.6]

12. Find the GCF of 63 and 84. [Section 1.6]

13. Evaluate: $15 + 5[12 - (2^2 + 4)]$ [Section 1.7]

14. REAL ESTATE A homeowner, wishing to sell his house, had it appraised by three different real estate agents. The appraisals were: $158,000, $163,000, and $147,000. He decided to use the mean of the appraisals as the listing price. For what amount was the home listed? [Section 1.7]

Solve each equation and check the result.

15. $18 = n - 47$ [Section 1.8]

16. $23x = 483$ [Section 1.9]

17. a. Write the set of integers. [Section 2.1]

    b. Is the statement $-9 \le -8$ true or false? [Section 2.1]

18. Find the sum of $-20, 6,$ and -1. [Section 2.2]

19. Subtract 453 from 129. [Section 2.3]

20. Subtract: $-50 - (-60)$ [Section 2.3]

21. GOLD MINING An elevator lowers gold miners from the ground level entrance to different depths in the mine. The elevator stops every 25 vertical feet to let off miners. At what depth do the miners work if they get off the elevator at the 8th stop? [Section 2.4]

22. TEMPERATURE DROP During a five-hour period, the temperature steadily dropped 55°F. By how many degrees did the temperature change each hour? [Section 2.5]

Evaluate each expression. [Section 2.6]

23. $6 + (-2)(-5)$

24. $(-2)^3 - 3^3$

25. $-5 + 3|-4 - (-6)|$

26. $\dfrac{2(3^2 - 4^2)}{-2(3) - 1}$

Solve each equation and check the result. [Section 2.9]

27. $-7 - 5 - 7x = 16$ **28.** $-9 = 4 + \dfrac{m}{-4}$

Form an equation and solve it to answer the following questions. [Section 2.9]

29. SHARKS During a research project, a diver inside a shark cage made observations at a depth of 132 feet. For a second set of observations, the cage was raised to a depth of 64 feet. How many feet was the cage raised between observations?

30. PROFITS AND LOSSES In its first year of business, a pet store suffered a loss, ending the year $4,028 in the red. In the second year, it made a sizable profit. If the total profit for the first two years in business was $33,611, how much profit was made the second year?

31. Translate each expression into an algebraic expression involving the variable x. [Section 3.1]

  a. The sum of a number and 15

  b. Eight less than a number

  c. The product of a number and 4

  d. The quotient obtained when a number is divided by 10

32. Evaluate $3x - x^3$ for $x = -4$. [Section 3.2]

33. Simplify each expression. [Section 3.3]

  a. $-3(5x)$ **b.** $-4x(-7y)$

34. Multiply. [Section 3.3]

  a. $2(3x - 4)$ **b.** $-5(3x - 2y + 4)$

Simplify each expression. [Section 3.4]

35. $-3x + 8x$ **36.** $4a^2 - (-3a^2)$

37. $4x - 3y - 5x + 2y$ **38.** $-2(3x - 4) + 2x$

Solve each equation and check the result. [Section 3.5]

39. $6x - 12 = 2x + 4$

40. $3(2y - 8) = -2(y - 4)$

41. $5 - (7 - y) = -5$

42. $14m = m + 5m + 3m - 40$

Form an equation and solve it to answer the following questions. [Section 3.6]

43. OBSERVATION HOURS To get a Master's degree in educational psychology, a student must have 100 hours of observation time at a clinic. If the student has already observed for 37 hours, how many 3-hour shifts must she observe to complete the requirement?

44. GEOMETRY A rectangle is four times as long as it is wide. If its perimeter is 210 feet, find its width and its length.

Simplify each fraction. [Section 4.1]

45. $\dfrac{21}{28}$ **46.** $\dfrac{40x^6y^4}{16x^3y^5}$

Perform each operation.

47. $\dfrac{6}{5}\left(-\dfrac{2}{3}\right)$ **48.** $\dfrac{14p^2}{8} \div \dfrac{7p^3}{2}$

  [Section 4.2]   [Section 4.3]

49. $\dfrac{2}{3} + \dfrac{3}{4}$ **50.** $\dfrac{4}{m} - \dfrac{3}{5}$

  [Section 4.4]   [Section 4.4]

51. FIRE HAZARDS Two terminals in an electrical switch were so close that electricity could jump the gap and start a fire. The illustration below shows a newly designed switch that will keep this from happening. By how much was the distance between the ground terminal and the hot terminal increased? [Section 4.4]

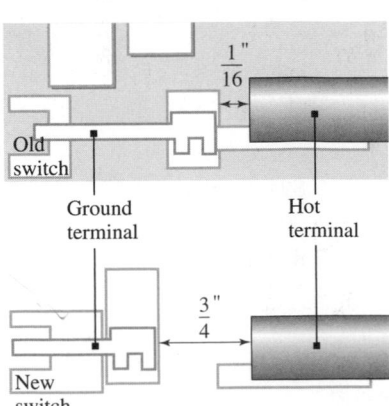

52. a. Write $\dfrac{75}{7}$ as a mixed number. [Section 4.5]

 b. Write $-6\dfrac{5}{8}$ as an improper fraction. [Section 4.5]

Perform each operation. Simplify, if possible.

53. $2\dfrac{2}{5}\left(3\dfrac{1}{12}\right)$ [Section 4.5]

54. $15\dfrac{1}{3} \div 2\dfrac{2}{9}$ [Section 4.5]

55 $4\dfrac{2}{3} + 5\dfrac{1}{4}$ [Section 4.6]

56. $14\dfrac{2}{5} - 8\dfrac{2}{3}$ [Section 4.6]

57. LUMBER As shown below, 2-by-4's from the lumber yard do not really have dimensions of 2 inches by 4 inches. How wide and how high is the stack of 2-by-4's in the illustration? [Section 4.5]

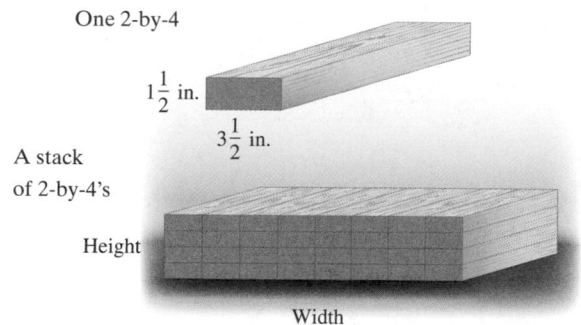

One 2-by-4

$1\dfrac{1}{2}$ in.

$3\dfrac{1}{2}$ in.

A stack of 2-by-4's

Height

Width

58. GAS STATIONS How much gasoline is left in a 500-gallon storage tank if $225\dfrac{3}{4}$ gallons have been pumped out of it? [Section 4.6]

59. Find the perimeter of the triangle shown below. [Section 4.6]

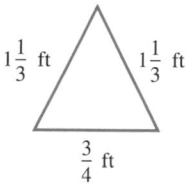

$1\dfrac{1}{3}$ ft $1\dfrac{1}{3}$ ft

$\dfrac{3}{4}$ ft

60. Evaluate: $\left(-\dfrac{3}{4} \cdot \dfrac{9}{16}\right) + \left(\dfrac{1}{2} - \dfrac{1}{8}\right)$ [Section 4.7]

61. Simplify: $\dfrac{\frac{2}{3}}{\frac{4}{5}}$ [Section 4.7]

62. Simplify: $\dfrac{\frac{3}{7} + \left(-\frac{1}{2}\right)}{1\frac{3}{4}}$ [Section 4.7]

Solve each equation and check the result. [Section 4.8]

63. $x + \dfrac{1}{5} = -\dfrac{11}{15}$

64. $\dfrac{3}{4}x = \dfrac{5}{8}x + \dfrac{1}{2}$

65. $\dfrac{2}{3}x = -10$

66. $3y - 8 = 0$

Form an equation and solve it to answer the following questions. [Section 4.8]

67. SHAVING An advertisement for a new, improved model of an electric razor claims that men can shave in just two-thirds of the time it took them with the older model. Using the new model, it took a man 90 seconds to shave. If the advertising claim is true, how long would it have taken him to shave using the older model?

68. GRADING In an economic class, a student's grade is based on the number of points he or she earns on tests, homework, and a term paper. Tests account for three-fourths of the total possible points, homework assigments account for one-fifth of the total possible points, and the term paper is worth 50 points. What is the maximium number of points that a student can earn in the class?

Decimals

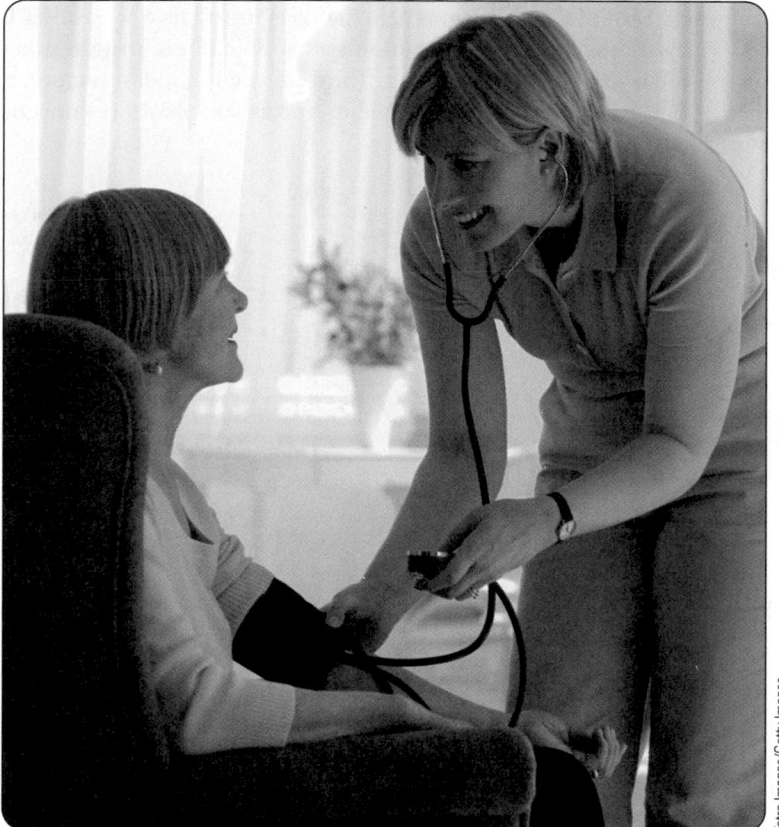

Tetra Images/Getty Images

from **Campus to Careers**

Home Health Aide

Home health aides provide personalized care to the elderly and the disabled in the patient's own home. They help their patients take medicine, eat, dress, and bathe. Home health aides need to have a good number sense. They must accurately take the patient's temperature, pulse, and blood pressure, and monitor the patient's calorie intake and sleeping schedule.

In **Problem 101** of **Study Set 5.2**, you will see how a home health aide uses decimal addition and subtraction to chart a patient's temperature.

JOB TITLE:
Home Health Aide

EDUCATION: Successful completion of a home health aide training program as required by state law or federal regulation.

JOB OUTLOOK: Excellent due to rapid employment growth and high replacement needs.

ANNUAL EARNINGS: The average (median) salary in 2008 was $19,760.

FOR MORE INFORMATION:
www.sbtinsurance.com/filemanager/
download/11410/

Objectives

1 Identify the place value of a digit in a decimal number.

2 Write decimals in expanded form.

3 Read decimals and write them in standard form.

4 Compare decimals using inequality symbols.

5 Graph decimals on a number line.

6 Round decimals.

7 Read tables and graphs involving decimals.

SECTION **5.1**

An Introduction to Decimals

The place value system for whole numbers that was introduced in Section 1.1 can be extended to create the **decimal numeration system.** Numbers written using **decimal notation** are often simply called **decimals.** They are used in measurement, because it is easy to put them in order and compare them. And as you probably know, our money system is based on decimals.

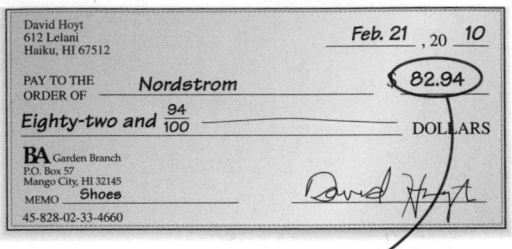

The decimal 1,537.6 on the odometer represents the distance, in miles, that the car has traveled.

The decimal 82.94 repesents the amount of the check, in dollars.

1 Identify the place value of a digit in a decimal number.

Like fraction notation, decimal notation is used to represent part of a whole. However, when writing a number in decimal notation, we don't use a fraction bar, nor is a denominator shown. For example, consider the rectangular region below that has 1 of 10 equal parts colored red. We can use the fraction $\frac{1}{10}$ or the decimal 0.1 to describe the amount of the figure that is shaded. Both are read as "one-tenth," and we can write:

$$\frac{1}{10} = 0.1$$

Fraction: Decimal:
$\frac{1}{10}$ 0.1

The square region on the right has 1 of 100 equal parts colored red. We can use the fraction $\frac{1}{100}$ or the decimal 0.01 to describe the amount of the figure that is shaded. Both are read as "one one-hundredth," and we can write:

$$\frac{1}{100} = 0.01$$

Fraction: $\frac{1}{100}$

Decimal: 0.01

Decimals are written by entering the digits 0, 1, 2, 3, 4, 5, 6, 7, 8, and 9 into place-value columns that are separated by a **decimal point.** The following **place-value chart** shows the names of the place-value columns. Those to the left of the decimal point form the **whole-number part** of the decimal number, and they have the familiar names ones, tens, hundreds, and so on. The columns to the right of the decimal point form the **fractional part.** Their place value names are similar to those in the whole-number part, but they end in "*ths.*" Notice that there is no one*ths* place in the chart.

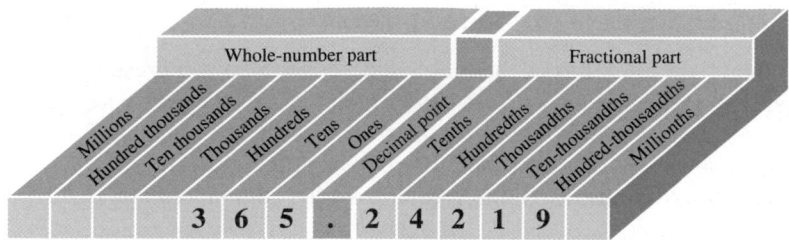

Whole-number part							Fractional part				
Millions	Hundred thousands	Ten thousands	Thousands	Hundreds	Tens	Ones	Decimal point	Tenths	Hundredths	Thousandths	Ten-thousandths	Hundred-thousandths	Millionths

3 6 5 . 2 4 2 1 9

The decimal 365.24219, entered in the place-value chart above, represents the number of days it takes Earth to make one full orbit around the sun. We say that the decimal is written in **standard form** (also called **standard notation**). Each of the 2's in 365.24219 has a different place value because of its position. The place value of the red 2 is two tenths. The place value of the blue 2 is two thousandths.

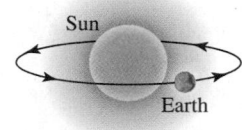

Sun

Earth

EXAMPLE 1   Consider the decimal number: 2,864.709531

a. What is the place value of the digit 5?

b. Which digit tells the number of millionths?

Strategy We will locate the decimal point in 2,864.709531. Then, moving to the right, we will name each column (tenths, hundredths, and so on) until we reach 5.

WHY It's easier to remember the names of the columns if you begin at the decimal point and move to the right.

Solution

a. 2,864.709531   Say "Tenths, hundredths, thousandths, ten-thousandths" as you move from column to column.

5 ten-thousandths is the place value of the digit 5.

b. 2,864.709531   Say "Tenths, hundredths, thousandths, ten-thousandths, hundred thousandths, millionths" as you move from column to column.

The digit 1 is in the millionths column.

Caution! We *do not* separate groups of three digits on the right side of the decimal point with commas as we do on the left side. For example, it would be incorrect to write:

~~2,864.709,531~~

Self Check 1

Consider the decimal number: 56,081.639724
a. What is the place value of the digit 9?
b. Which digit tells the number of hundred-thousandths?

Now Try Problem 17

We can write a whole number in decimal notation by placing a decimal point immediately to its right and then entering a zero, or zeros, to the right of the decimal point. For example,

$$99 \quad = \quad 99.0 \quad = \quad 99.00 \quad \text{Because } 99 = 99\frac{0}{10} = 99\frac{00}{100}.$$

A whole number   Place a decimal point here and enter a zero, or zeros, to the right of it.

When there is no whole-number part of a decimal, we can show that by entering a zero directly to the left of the decimal point. For example,

$$.83 \quad = \quad 0.83 \quad \text{Because } \frac{83}{100} = 0\frac{83}{100}.$$

No whole-number part   Enter a zero here, if desired.

Negative decimals are used to describe many situations that arise in everyday life, such as temperatures below zero and the balance in a checking account that is over-drawn. For example, the coldest natural temperature ever recorded on Earth was −128.6°F at the Russian Vostok Station in Antarctica on July 21, 1983.

© Les Welch/Icon SMI/Corbis

2   Write decimals in expanded form.

The decimal 4.458, entered in the place-value chart below, represents the time (in seconds) that it took women's record holder Melanie Troxel to cover a quarter mile in her top-fuel dragster. Notice that the place values of the columns for the whole-number part are 1, 10, 100, 1,000, and so on. We learned in Section 1.1 that the value of each of those columns is 10 times greater than the column directly to its right.

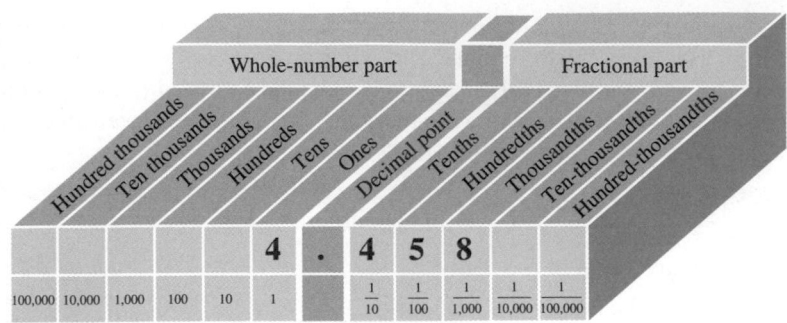

The place values of the columns for the fractional part of a decimal are $\frac{1}{10}$, $\frac{1}{100}$, $\frac{1}{1,000}$, and so on. Each of those columns has a value that is $\frac{1}{10}$ of the value of the place directly to its left. For example,

- The value of the tenths column is $\frac{1}{10}$ of the value of the ones column: $1 \cdot \frac{1}{10} = \frac{1}{10}$.
- The value of the hundredths column is $\frac{1}{10}$ of the value of the tenths column: $\frac{1}{10} \cdot \frac{1}{10} = \frac{1}{100}$.
- The value of the thousandths column is $\frac{1}{10}$ of the value of the hundredths column: $\frac{1}{100} \cdot \frac{1}{10} = \frac{1}{1,000}$.

The meaning of the decimal 4.458 becomes clear when we write it in **expanded form** (also called **expanded notation**).

$$4.458 = 4 \text{ ones} + 4 \text{ tenths} + 5 \text{ hundredths} + 8 \text{ thousandths}$$

which can be written as:

$$4.458 = 4 + \frac{4}{10} + \frac{5}{100} + \frac{8}{1,000}$$

> **The Language of Algebra**   The word *decimal* comes from the Latin word *decima*, meaning a tenth part.

Self Check 2

Write the decimal number 1,277.9465 in expanded form.

Now Try **Problems 23 and 27**

EXAMPLE 2   Write the decimal number 592.8674 in expanded form.

Strategy Working from left to right, we will give the place value of each digit and combine them with + symbols.

WHY The term *expanded form* means to write the number as an addition of the place values of each of its digits.

Solution The expanded form of 592.8674 is:

5 hundreds + **9** tens + **2** ones + **8** tenths + **6** hundredths + **7** thousandths + **4** ten-thousandths

which can be written as

$$500 + 90 + 2 + \frac{8}{10} + \frac{6}{100} + \frac{7}{1,000} + \frac{4}{10,000}$$

3 Read decimals and write them in standard form.

To understand how to read a decimal, we will examine the expanded form of 4.458 in more detail. Recall that

$$4.458 = 4 + \frac{4}{10} + \frac{5}{100} + \frac{8}{1,000}$$

To add the fractions, we need to build $\frac{4}{10}$ and $\frac{5}{100}$ so that each has a denominator that is the LCD, 1,000.

$$4.458 = 4 + \frac{4}{10} \cdot \frac{\mathbf{100}}{\mathbf{100}} + \frac{5}{100} \cdot \frac{\mathbf{10}}{\mathbf{10}} + \frac{8}{1,000}$$

$$= 4 + \frac{400}{1,000} + \frac{50}{1,000} + \frac{8}{1,000}$$

$$= 4 + \frac{458}{1,000}$$

$$= 4\frac{458}{1,000}$$

We have found that 4.458 = $4\frac{458}{1,000}$, with whole-number part and fractional part labeled.

We read 4.458 as "four and four hundred fifty-eight thousandths" because 4.458 is the same as $4\frac{458}{1,000}$. Notice that the last digit in 4.458 is in the thousandths place. This observation suggests the following method for reading decimals.

Reading a Decimal

To read a decimal:

1. Look to the left of the decimal point and say the name of the whole number.
2. The decimal point is read as "and."
3. Say the fractional part of the decimal as a whole number followed by the name of the last place-value column of the digit that is the farthest to the right.

We can use the steps for reading a decimal to write it in words.

EXAMPLE 3
Write each decimal in words and then as a fraction or mixed number. **You do not have to simplify the fraction.**

a. Sputnik, the first satellite launched into space, weighed 184.3 pounds.
b. Usain Bolt of Jamaica holds the men's world record in the 100-meter dash: 9.69 seconds.
c. A one-dollar bill is 0.0043 inch thick.
d. Liquid mercury freezes solid at −37.7°F.

Strategy We will identify the whole number to the left of the decimal point, the fractional part to its right, and the name of the place-value column of the digit the farthest to the right.

WHY We need to know those three pieces of information to read a decimal or write it in words.

Self Check 3
Write each decimal in words and then as a fraction or mixed number. **You do not have to simplify the fraction.**
a. The average normal body temperature is 98.6°F.
b. The planet Venus makes one full orbit around the sun every 224.7007 Earth days.
c. One gram is about 0.035274 ounce.
d. Liquid nitrogen freezes solid at −345.748°F.

Now Try Problems 31, 35, and 39

Solution

a.   **184** . **3**   The whole-number part is 184. The fractional part is 3.
The digit the farthest to the right, 3, is in the tenths place.

One hundred eighty-four and three tenths

Written as a mixed number, 184.3 is $184\frac{3}{10}$.

b.   **9** . **69**   The whole-number part is 9. The fractional part is 69.
The digit the farthest to the right, 9, is in the hundredths place.

Nine and sixty-nine hundredths

Written as a mixed number, 9.69 is $9\frac{69}{100}$.

c.   **0** . **0043**   The whole-number part is 0. The fractional part is 43.
The digit the farthest to the right, 4, is in the ten-thousandths place.

Forty-three ten-thousandths   Since the whole-number part is 0, we need not write it nor the word *and*.

Written as a fraction, 0.0043 is $\frac{43}{10,000}$.

d.   **−37** . **7**   This is a negative decimal.

Negative *thirty-seven and seven tenths.*

Written as a negative mixed number, −37.7 is $-37\frac{7}{10}$.   ■

> **The Language of Mathematics**   Decimals are often read in an informal way.
> For example, we can read 184.3 as "one hundred eighty-four point three" and
> 9.69 as "nine point six nine."

The procedure for reading a decimal can be applied in reverse to convert from written-word form to standard form.

Self Check 4

Write each number in standard form:
a. *Eight hundred six and ninety-two hundredths*
b. *Twelve and sixty-seven ten-thousandths*

Now Try Problems 41, 45, and 47

EXAMPLE 4   Write each number in standard form:

a. *One hundred seventy-two and forty-three hundredths*

b. *Eleven and fifty-one thousandths*

Strategy We will locate the word *and* in the written-word form and translate the phrase that appears before it and the phrase that appears after it separately.

WHY The whole-number part of the decimal is described by the phrase that appears before the word *and*. The fractional part of the decimal is described by the phrase that follows the word *and*.

Solution

a.   *One hundred seventy-two*   *and*   *forty-three hundredths*
↓
172.43
↑___ This is the hundredths place-value column.

b. Sometimes, when changing from written-word form to standard form, we must insert placeholder 0's in the fractional part of a decimal so that that the last digit appears in the proper place-value column.

Eleven   *and*   *fifty-one thousandths*
↓
11.051

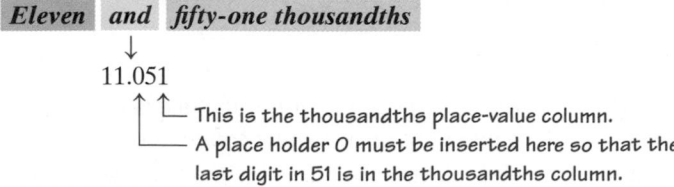
↑ ↑___ This is the thousandths place-value column.
|___ A place holder 0 must be inserted here so that the
last digit in 51 is in the thousandths column.

Caution! If a placeholder 0 is not written in 11.051, an incorrect answer of 11.51 (eleven and fifty-one *hundredths,* not *thousandths*) results.

4 Compare decimals using inequality symbols.

To develop a way to compare decimals, let's consider 0.3 and 0.271. Since 0.271 contains more digits, it may appear that 0.271 is greater than 0.3. However, the opposite is true. To show this, we write 0.3 and 0.271 in fraction form:

$$0.3 = \frac{3}{10} \qquad 0.271 = \frac{271}{1,000}$$

Now we build $\frac{3}{10}$ into an equivalent fraction so that it has a denominator of 1,000, like that of $\frac{271}{1,000}$.

$$0.3 = \frac{3}{10} \cdot \frac{100}{100} = \frac{300}{1,000}$$

Since $\frac{300}{1,000} > \frac{271}{1,000}$, it follows that $0.3 > 0.271$. This observation suggests a quicker method for comparing decimals.

Comparing Decimals

To compare two decimals:

1. Make sure both numbers have the same number of decimal places to the right of the decimal point. Write any additional zeros necessary to achieve this.

2. Compare the digits of each decimal, column by column, working from left to right.

3. *If the decimals are positive:* When two digits differ, the decimal with the greater digit is the greater number. *If the decimals are negative:* When two digits differ, the decimal with the smaller digit is the greater number.

EXAMPLE 5   Place an < or > symbol in the box to make a true statement:

a. 1.2679 ▢ 1.2658   **b.** 54.9 ▢ 54.929   **c.** −10.419 ▢ −10.45

Strategy We will stack the decimals and then, working from left to right, we will scan their place-value columns looking for a difference in their digits.

WHY We need only look in that column to determine which digit is the greater.

Solution

a. Since both decimals have the same number of places to the right of the decimal point, we can immediately compare the digits, column by column.

```
        1.2679
        1.2658
```
Same digit ─┘↑↑↑↑
Same digit ─┘
Same digit ─┘└── These digits are different: Since 7 is greater than 5, it
                 follows that the first decimal is greater than the second.

Thus, 1.2679 is greater than 1.2658 and we can write 1.2679 > 1.2658.

b. We can write two zeros after the 9 in 54.9 so that the decimals have the same number of digits to the right of the decimal point. This makes the comparison easier.

```
        54.900
        54.929
            ↑
```

As we work from left to right, this is the first column in which the digits differ. Since 2 > 0, it follows that 54.929 is greater than 54.9 (or 54.9 is less than 54.929) and we can write 54.9 < 54.929.

Self Check 5

Place an < or > symbol in the box to make a true statement:
a. 3.4308 ▢ 3.4312

b. 678.3409 ▢ 678.34

c. −703.8 ▢ −703.78

Now Try Problems 49, 55, and 59

Success Tip   Writing additional zeros *after the last digit to the right of the decimal point does not change the value of the decimal.* Also, deleting additional zeros after the last digit to the right of the decimal point does not change the value of the decimal. For example,

$$54.9 = 54.90 = 54.900$$
$$\uparrow \qquad \uparrow$$
These additional zeros do not change the value of the decimal.

Because $54\frac{90}{100}$ and $54\frac{900}{1,000}$ in simplest form are equal to $54\frac{9}{10}$.

c. We are comparing two negative decimals. In this case, when two digits differ, the decimal with the smaller digit is the greater number.

$$-10.4\boxed{1}9$$
$$-10.4\boxed{5}0 \qquad \text{Write a zero after 5 to help in the comparison.}$$
$$\uparrow$$

As we work from left to right, this is the first column in which the digits differ. Since $1 < 5$, it follows that -10.419 is greater than -10.45 and we can write $-10.419 > -10.45$.

5 Graph decimals on a number line.

Decimals can be shown by drawing points on a number line.

Self Check 6

Graph -1.1, -1.64, -0.8, and 1.9 on a number line.

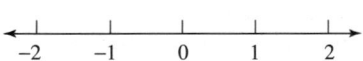

Now Try Problem 61

EXAMPLE 6   Graph -1.8, -1.23, -0.3, and 1.89 on a number line.

Strategy We will locate the position of each decimal on the number line and draw a bold dot.

WHY To *graph a number* means to make a drawing that represents the number.

Solution The graph of each negative decimal is to the left of 0 and the graph of each positive decimal is to the right of 0. Since $-1.8 < -1.23$, the graph of -1.8 is to the left of -1.23.

6 Round decimals.

When we don't need exact results, we can approximate decimal numbers by **rounding.** To round the decimal part of a decimal number, we use a method similar to that used to round whole numbers.

Rounding a Decimal

1. To round a decimal to a certain decimal place value, locate the **rounding digit** in that place.

2. Look at the **test digit** directly to the right of the rounding digit.

3. If the test digit is 5 or greater, round up by adding 1 to the rounding digit and dropping all the digits to its right. If the test digit is less than 5, round down by keeping the rounding digit and dropping all the digits to its right.

EXAMPLE 7   *Chemistry*   A student in a chemistry class uses a digital balance to weigh a compound in grams. Round the reading shown on the balance to the nearest thousandth of a gram.

Strategy   We will identify the digit in the thousandths column and the digit in the ten-thousandths column.

WHY   To round to the nearest thousandth, the digit in the thousandths column is the rounding digit and the digit in the ten-thousandths column is the test digit.

Solution   The rounding digit in the thousandths column is 8. Since the test digit 7 is 5 or greater, we round up.

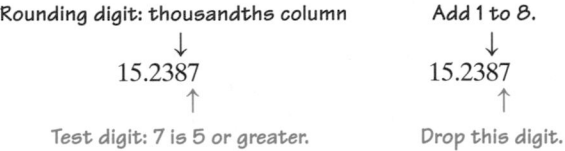

The reading on the balance is approximately 15.239 grams.

EXAMPLE 8   Round each decimal to the indicated place value:
a. −645.1358 to the nearest tenth   **b.** 33.096 to the nearest hundredth

Strategy   In each case, we will first identify the rounding digit. Then we will identify the test digit and determine whether it is less than 5 or greater than or equal to 5.

WHY   If the test digit is less than 5, we round down; if it is greater than or equal to 5, we round up.

Solution
a. Negative decimals are rounded in the same ways as positive decimals. The rounding digit in the tenths column is 1. Since the test digit 3 is less than 5, we round down.

Thus, −645.1358 rounded to the nearest tenth is −645.1.

b. The rounding digit in the hundredths column is 9. Since the test digit 6 is 5 or greater, we round up.

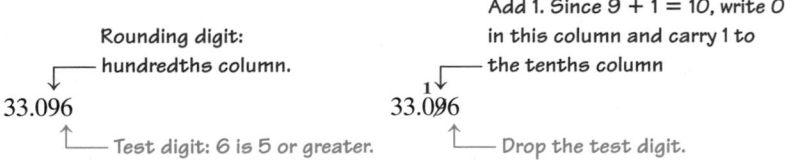

Thus, 33.096 rounded to the nearest hundredth is 33.10.

Caution!   It would be incorrect to drop the 0 in the answer 33.10. If asked to round to a certain place value (in this case, thousandths), that place must have a digit, even if the digit is 0.

Self Check 7
Round 24.41658 to the nearest ten-thousandth.

Now Try   Problems 65 and 69

Self Check 8
Round each decimal to the indicated place value:
a. −708.522 to the nearest tenth
b. 9.1198 to the nearest thousandth

Now Try   Problems 73 and 77

There are many situations in our daily lives that call for rounding amounts of money. For example, a grocery shopper might round the unit cost of an item to the nearest cent or a taxpayer might round his or her income to the nearest dollar when filling out an income tax return.

EXAMPLE 9

a. *Utility Bills*   A utility company calculates a homeowner's monthly electric bill by multiplying the unit cost of $0.06421 by the number of kilowatt hours used that month. Round the unit cost to the nearest cent.

b. *Annual Income*   A secretary earned $36,500.91 dollars in one year. Round her income to the nearest dollar.

Strategy In part a, we will round the decimal to the nearest hundredth. In part b, we will round the decimal to the ones column.

WHY Since there are 100 cents in a dollar, each cent is $\frac{1}{100}$ of a dollar. To round to the *nearest cent* is the same as rounding to the *nearest hundredth* of a dollar. To round to the *nearest dollar* is the same as rounding to the *ones place*.

Solution

a. The rounding digit in the hundredths column is 6. Since the test digit 4 is less than 5, we round down.

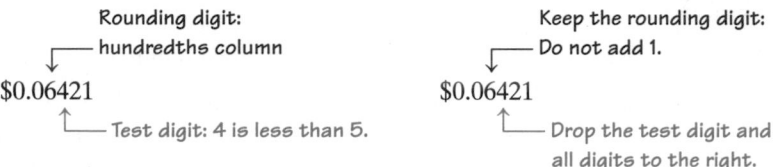

Thus, $0.06421 rounded to the nearest cent is $0.06.

b. The rounding digit in the ones column is 0. Since the test digit 9 is 5 or greater, we round up.

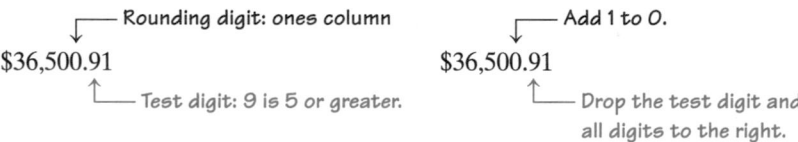

Thus, $36,500.91 rounded to the nearest dollar is $36,501.

7 Read tables and graphs involving decimals.

Year	Pounds
1960	2.68
1970	3.25
1980	3.66
1990	4.50
2000	4.64
2007	4.62

(Source: U.S. Environmental Protection Agency)

The table on the left is an example of the use of decimals. It shows the number of pounds of trash generated daily per person in the United States for selected years from 1960 through 2007.

When the data in the table is presented in the form of a **bar graph,** a trend is apparent. The amount of trash generated daily per person increased steadily until the year 2000. Since then, it appears to have remained about the same.

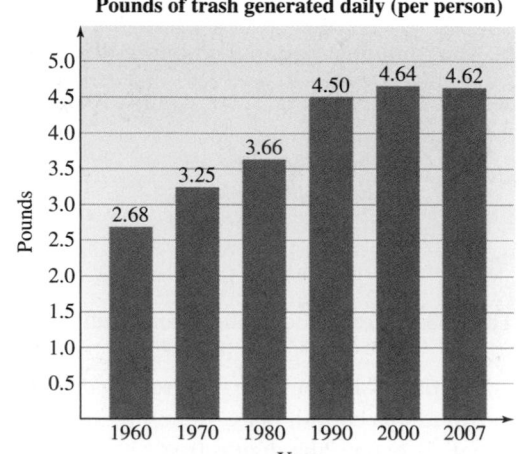

Pounds of trash generated daily (per person)

ANSWERS TO SELF CHECKS

1. a. 9 thousandths   **b.** 2   **2.** $1,000 + 200 + 70 + 7 + \frac{9}{10} + \frac{4}{100} + \frac{6}{1,000} + \frac{5}{10,000}$
3. a. ninety-eight and six tenths, $98\frac{6}{10}$ **b.** two hundred twenty-four and seven thousand
seven ten-thousandths, $224\frac{7,007}{10,000}$ **c.** thirty-five thousand, two hundred seventy-four
millionths, $\frac{35,274}{1,000,000}$ **d.** negative three hundred forty-five and seven hundred forty-eight
thousandths, $-345\frac{748}{1,000}$ **4. a.** 806.92   **b.** 12.0067   **5. a.** <   **b.** >   **c.** <
6.

7. 24.4166   **8. a.** −708.5   **b.** 9.120
9. a. $0.08   **b.** $24,909

SECTION 5.1   STUDY SET

VOCABULARY

Fill in the blanks.

1. Decimals are written by entering the digits 0, 1, 2, 3, 4, 5, 6, 7, 8, and 9 into place-value columns that are separated by a decimal _____.

2. The place-value columns to the left of the decimal point form the whole-number part of a decimal number and the place-value columns to the right of the decimal point form the _____ part.

3. We can show the value represented by each digit of the decimal 98.6213 by using _____ form:

$$98.6213 = 90 + 8 + \frac{6}{10} + \frac{2}{100} + \frac{1}{1,000} + \frac{3}{10,000}$$

4. When we don't need exact results, we can approximate decimal numbers by _____.

CONCEPTS

5. Write the name of each column in the following place-value chart.

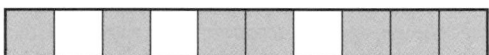

4 , 7   8   9 . 0   2   6   5

6. Write the value of each column in the following place-value chart.

7	2	.	3	1	9	5	8

7. Fill in the blanks.

  a. The value of each place in the whole-number part of a decimal number is ___ times greater than the column directly to its right.

  b. The value of each place in the fractional part of a decimal number is ___ of the value of the place directly to its left.

8. Represent each situation using a signed number.

  a. A checking account overdrawn by $33.45

  b. A river 6.25 feet above flood stage

  c. 3.9 degrees below zero

  d. 17.5 seconds after liftoff

9. a. Represent the shaded part of the rectangular region as a fraction and a decimal.

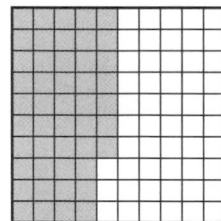

  b. Represent the shaded part of the square region as a fraction and a decimal.

10. Write $400 + 20 + 8 + \frac{9}{10} + \frac{1}{100}$ as a decimal.

11. Fill in the blanks in the following illustration to label the *whole-number part* and the *fractional part*.

$$63.37 \quad = \quad 63\frac{37}{100}$$

12. Fill in the blanks.

  a. To round $0.13506 to the *nearest cent*, the rounding digit is ___ and the test digit is ___.

  b. To round $1,906.47 to the *nearest dollar*, the rounding digit is ___ and the test digit is ___.

NOTATION

Fill in the blanks.

13. The columns to the right of the decimal point in a decimal number form its fractional part. Their place value names are similar to those in the whole-number part, but they end in the letters "____."

14. When reading a decimal, such as 2.37, we can read the decimal point as "____" or as "_____."

15. Write a decimal number that has ...

6 in the ones column,

1 in the tens column,

0 in the tenths column,

8 in the hundreds column,

2 in the hundredths column,

9 in the thousands column,

4 in the thousandths column,

7 in the ten thousands column, and

5 in the ten-thousandths column.

16. Determine whether each statement is true or false.

a. $0.9 = 0.90$

b. $1.260 = 1.206$

c. $-1.2800 = -1.280$

d. $0.001 = .0010$

GUIDED PRACTICE

Answer the following questions about place value. See Example 1.

17. Consider the decimal number: 145.926

a. What is the place value of the digit 9?

b. Which digit tells the number of thousandths?

c. Which digit tells the number of tens?

d. What is the place value of the digit 5?

18. Consider the decimal number: 304.817

a. What is the place value of the digit 1?

b. Which digit tells the number of thousandths?

c. Which digit tells the number of hundreds?

d. What is the place value of the digit 7?

19. Consider the decimal number: 6.204538

a. What is the place value of the digit 8?

b. Which digit tells the number of hundredths?

c. Which digit tells the number of ten-thousandths?

d. What is the place value of the digit 6?

20. Consider the decimal number: 4.390762

a. What is the place value of the digit 6?

b. Which digit tells the number of thousandths?

c. Which digit tells the number of ten-thousandths?

d. What is the place value of the digit 4?

Write each decimal number in expanded form. See Example 2.

21. 37.89

22. 26.93

23. 124.575

24. 231.973

25. 7,498.6468

26. 1,946.7221

27. 6.40941

28. 8.70214

Write each decimal in words and then as a fraction or mixed number. See Example 3.

29. 0.3   **30.** 0.9

31. 50.41   **32.** 60.61

33. 19.529   **34.** 12.841

35. 304.0003   **36.** 405.0007

37. -0.00137   **38.** -0.00613

39. -1,072.499 **40.** -3,076.177

Write each number in standard form. See Example 4.

41. Six and one hundred eighty-seven thousandths

42. Four and three hundred ninety-two thousandths

43. Ten and fifty-six ten-thousandths

44. Eleven and eighty-six ten-thousandths

45. Negative sixteen and thirty-nine hundredths

46. Negative twenty-seven and forty-four hundredths

47. One hundred four and four millionths

48. Two hundred three and three millionths

Place an < or an > symbol in the box to make a true statement. See Example 5.

49. 2.59 ☐ 2.55   **50.** 5.17 ☐ 5.14

51. 45.103 ☐ 45.108   **52.** 13.874 ☐ 13.879

53. 3.28724 ☐ 3.2871   **54.** 8.91335 ☐ 8.9132

55. 379.67 ☐ 379.6088 **56.** 446.166 ☐ 446.2

57. -23.45 ☐ -23.1   **58.** -301.98 ☐ -302.45

59. -0.065 ☐ -0.066   **60.** -3.99 ☐ -3.9888

Graph each number on a number line. See Example 6.

61. 0.8, −0.7, −3.1, 4.5, −3.9

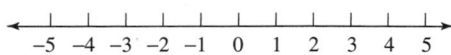

62. 0.6, −0.3, −2.7, 3.5, −2.2

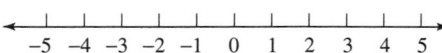

63. −1.21, −3.29, −4.25, 2.75, −1.84

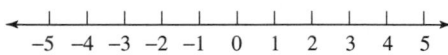

64. −3.19, −0.27, −3.95, 4.15, −1.66

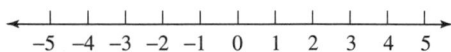

Round each decimal number to the indicated place value. See Example 7.

65. 506.198 nearest tenth

66. 51.451 nearest tenth

67. 33.0832 nearest hundredth

68. 64.0059 nearest hundredth

69. 4.2341 nearest thousandth

70. 8.9114 nearest thousandth

71. 0.36563 nearest ten-thousandth

72. 0.77623 nearest ten-thousandth

Round each decimal number to the indicated place value. See Example 8.

73. −0.137 nearest hundredth

74. −808.0897 nearest hundredth

75. −2.718218 nearest tenth

76. −3,987.8911 nearest tenth

77. 3.14959 nearest thousandth

78. 9.50966 nearest thousandth

79. 1.4142134 nearest millionth

80. 3.9998472 nearest millionth

81. 16.0995 nearest thousandth

82. 67.0998 nearest thousandth

83. 290.303496 nearest hundred-thousandth

84. 970.457297 nearest hundred-thousandth

Round each given dollar amount. See Example 9.

85. $0.284521 nearest cent

86. $0.312906 nearest cent

87. $27,841.52 nearest dollar

88. $44,633.78 nearest dollar

APPLICATIONS

89. READING METERS   To what decimal is the arrow pointing?

90. MEASUREMENT   Estimate a length of 0.3 inch on the 1-inch-long line segment below.

91. CHECKING ACCOUNTS   Complete the check shown by writing in the amount, using a decimal.

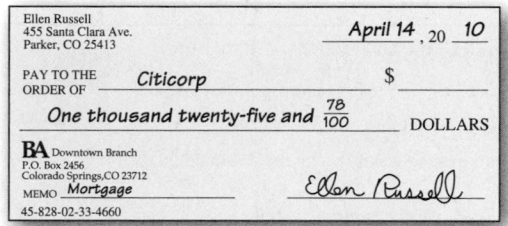

92. MONEY   We use a decimal point when working with dollars, but the decimal point is not necessary when working with cents. For each dollar amount in the table, give the equivalent amount expressed as cents.

Dollars	Cents
$0.50	
$0.05	
$0.55	
$5.00	
$0.01	

93. INJECTIONS   A syringe is shown below. Use an arrow to show to what point the syringe should be filled if a 0.38-cc dose of medication is to be given. ("cc" stands for "cubic centimeters.")

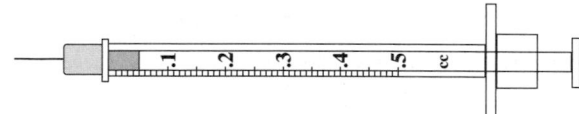

94. LASERS   The laser used in laser vision correction is so precise that each pulse can remove 39 millionths of an inch of tissue in 12 billionths of a second. Write each of these numbers as a decimal.

95. NASCAR The closest finish in NASCAR history took place at the Darlington Raceway on March 16, 2003, when Ricky Craven beat Kurt Busch by a mere 0.002 seconds. Write the decimal in words and then as a fraction in simplest form. (Source: NASCAR)

96. THE METRIC SYSTEM The metric system is widely used in science to measure length (meters), weight (grams), and capacity (liters). Round each decimal to the nearest hundredth.

   a. 1 ft is 0.3048 meter.

   b. 1 mi is 1,609.344 meters.

   c. 1 lb is 453.59237 grams.

   d. 1 gal is 3.785306 liters.

97. UTILITY BILLS A portion of a homeowner's electric bill is shown below. Round each decimal dollar amount to the nearest cent.

Billing Period

From	To	Meter Number		The Gas Company
06/05/10	07/05/10	10694435		

Next Meter Reading Date on or about Aug 03 2010

Summary of Charges

Customer Charge	30 Days	× $0.16438
Baseline	14 Therms	× $1.01857
Over Baseline	11 Therms	× $1.20091
State Regulatory Fee	25 Therms	× $0.00074
Public Purpose Surcharge	25 Therms	× $0.09910

98. INCOME TAX A portion of a W-2 tax form is shown below. Round each dollar amount to the nearest dollar.

Form **W-2** Wage and Tax Statement		**2010**
1 Wages, tips, other comp $35,673.79	**2** Fed inc tax withheld $7,134.28	**3** Social security wages $38,204.16
4 SS tax withheld $2,368.65	**5** Medicare wages & tips $38,204.16	**6** Medicare tax withheld $550.13
7 Social security tips	**8** Allocated tips	**9** Advance EIC payment
10 Depdnt care benefits	**11** Nonqualified plans	**12a**

99. THE DEWEY DECIMAL SYSTEM When stacked on the shelves, the library books shown in the next column are to be in numerical order, least to greatest,

from left to right. How should the titles be rearranged to be in the proper order?

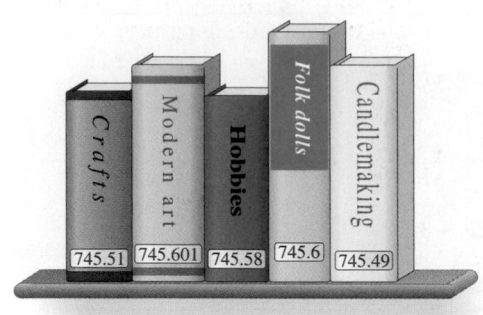

Crafts	Modern art	Hobbies	Folk dolls	Candlemaking
745.51	745.601	745.58	745.6	745.49

100. 2008 OLYMPICS The top six finishers in the women's individual all-around gymnastic competition in the Beijing Olympic Games are shown below in alphabetical order. If the highest score wins, which gymnasts won the gold (1st place), silver (2nd place), and bronze (3rd place) medals?

	Name	Nation	Score
	Yuyuan Jiang	China	60.900
	Shawn Johnson	U.S.A.	62.725
	Nastia Liukin	U.S.A.	63.325
	Steliana Nistor	Romania	61.050
	Ksenia Semenova	Russia	61.925
	Yilin Yang	China	62.650

(Source: SportsIllustrated.cnn.com)

101. TUNEUPS The six spark plugs from the engine of a Nissan Quest were removed, and the spark plug gap was checked. If vehicle specifications call for the gap to be from 0.031 to 0.035 inch, which of the plugs should be replaced?

Spark plug gap

Cylinder 1: 0.035 in.
Cylinder 2: 0.029 in.
Cylinder 3: 0.033 in.
Cylinder 4: 0.039 in.
Cylinder 5: 0.031 in.
Cylinder 6: 0.032 in.

102. GEOLOGY Geologists classify types of soil according to the grain size of the particles that make up the soil. The four major classifications of soil are shown below. Classify each of the samples (A, B, C, and D) in the table as clay, silt, sand, or granule.

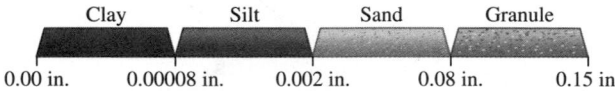

Sample	Location found	Grain size (in.)	Classification
A	Riverbank	0.009	
B	Pond	0.0007	
C	NE corner	0.095	
D	Dry lake	0.00003	

103. MICROSCOPES A microscope used in a lab is capable of viewing structures that range in size from 0.1 to as small as 0.0001 centimeter. Which of the structures listed in the table would be visible through this microscope?

Structure	Size (cm)
Bacterium	0.00011
Plant cell	0.015
Virus	0.000017
Animal cell	0.00093
Asbestos fiber	0.0002

104. FASTEST CARS The graph below shows AutoWeek's list of fastest cars for 2009. Find the time it takes each car to accelerate from 0 to 60 mph.

LIONEL VADAM/
Maxppp/Landov

Time to accelerate from 0 to 60 mph

105. THE STOCK MARKET Refer to the graph below, which shows the earnings (and losses) in the value of one share of Goodyear Tire and Rubber Company stock over twelve quarters. (For accounting purposes, a year is divided into four quarters, each three months long.)

a. In what quarter, of what year, were the earnings per share the greatest? Estimate the gain.

b. In what quarter, of what year, was the loss per share the greatest? Estimate the loss.

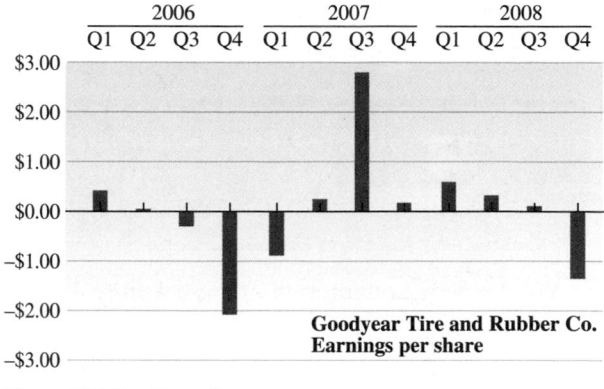

(Source: Wall Street Journal)

106. GASOLINE PRICES Refer to the graph below.

a. In what month, of what year, was the retail price of a gallon of gasoline the lowest? Estimate the price.

b. In what month(s), of what year, was the retail price of a gallon of gasoline the highest? Estimate the price.

c. In what month of 2007 was the price of a gallon of gasoline the greatest? Estimate the price.

U.S. Average Retail Price Regular Unleaded Gasoline*

*Retail price includes state and federal taxes
(Source: EPA Short-Term Energy Outlook, March 2009)

❚ WRITING

107. Explain the difference between ten and one-tenth.

108. "The more digits a number contains, the larger it is." Is this statement true? Explain.

109. Explain why is it wrong to read 2.103 as *"two and one hundred and three thousandths."*

110. SIGNS

    a. A sign in front of a fast food restaurant had the cost of a hamburger listed as .99¢. Explain the error.

    b. The illustration below shows the unusual notation that some service stations use to express the price of a gallon of gasoline. Explain the error.

REGULAR	UNLEADED	UNLEADED +
$2.79\frac{9}{10}$	$2.89\frac{9}{10}$	$2.99\frac{9}{10}$

111. Write a definition for each of these words.

    decade   *decathlon*   *decimal*

112. Show that in the decimal numeration system, each place-value column for the fractional part of a decimal is $\frac{1}{10}$ of the value of the place directly to its left.

❚ REVIEW

113. a. Find the perimeter of the rectangle shown below.

    b. Find the area of the rectangle.

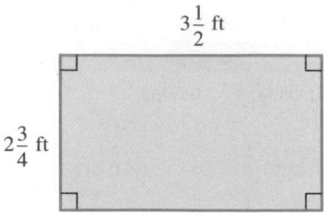

114. a. Find the perimeter of the triangle shown below.

    b. Find the area of the triangle.

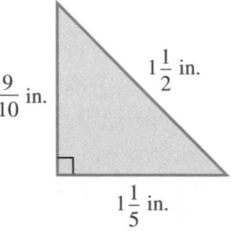

Objectives

1 Add decimals.

2 Subtract decimals.

3 Add and subtract signed decimals.

4 Estimate sums and differences of decimals.

5 Solve application problems by adding and subtracting decimals.

SECTION 5.2

Adding and Subtracting Decimals

To add or subtract objects, they must be similar. The federal income tax form shown below has a vertical line to make sure that dollars are added to dollars and cents added to cents. In this section, we show how decimal numbers are added and subtracted using this type of vertical form.

Form **1040EZ**	Department of the Treasury—Internal Revenue Service **Income Tax Return for Single and Joint Filers With No Dependents** **2010**		
Income **Attach Form(s) W-2 here.** Enclose, but do not attach, any payment.	**1** Wages, salaries, and tips. This should be shown in box 1 of your Form(s) W-2. Attach your Form(s) W-2.	**1**	21,056 | 89
	2 Taxable interest. If the total is over $1,500, you cannot use Form 1040EZ.	**2**	42 | 06
	3 Unemployment compensation and Alaska Permanent Fund dividends (see page 11).	**3**	200 | 00
	4 Add lines 1, 2, and 3. This is your **adjusted gross income.**	**4**	21,298 | 95

1 Add decimals.

Adding decimals is similar to adding whole numbers. We use **vertical form** and stack the decimals with their corresponding place values and decimal points lined up. Then we add the digits in each column, working from right to left, making sure that hundredths are added to hundredths, tenths are added to tenths, ones are added to ones, and so on. We write the decimal point in the **sum** so that it lines up with the decimal points in the **addends.** For example, to find 4.21 + 1.23 + 2.45, we proceed as follows:

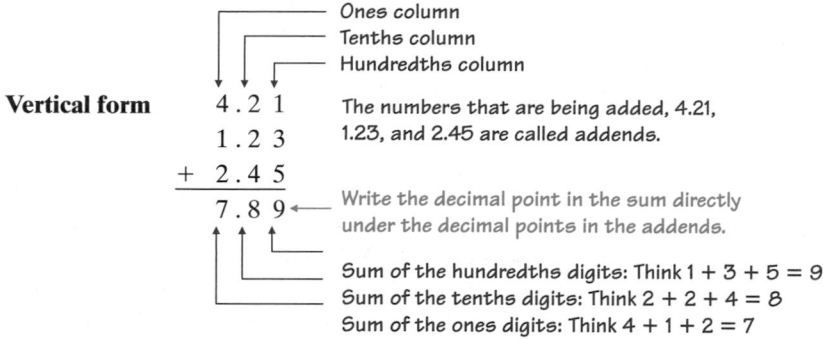

Ones column
Tenths column
Hundredths column

Vertical form

$$\begin{array}{r} 4.21 \\ 1.23 \\ + \ 2.45 \\ \hline 7.89 \end{array}$$

The numbers that are being added, 4.21, 1.23, and 2.45 are called addends.

Write the decimal point in the sum directly under the decimal points in the addends.

Sum of the hundredths digits: Think 1 + 3 + 5 = 9
Sum of the tenths digits: Think 2 + 2 + 4 = 8
Sum of the ones digits: Think 4 + 1 + 2 = 7

The sum is 7.89.

In this example, each addend had two decimal places, tenths and hundredths. If the number of decimal places in the addends are different, we can insert additional zeros so that the number of decimal places match.

Adding Decimals

To add decimal numbers:

1. Write the numbers in vertical form with the decimal points lined up.

2. Add the numbers as you would add whole numbers, from right to left.

3. Write the decimal point in the result from Step 2 directly below the decimal points in the addends.

Like whole number addition, if the sum of the digits in any place-value column is greater than 9, we must **carry.**

EXAMPLE 1   Add: 31.913 + 5.6 + 68 + 16.78

Strategy We will write the addition in vertical form so that the corresponding place values and decimal points of the addends are lined up. Then we will add the digits, column by column, working from right to left.

WHY We can only add digits with the same place value.

Solution To make the column additions easier, we will write two zeros after the 6 in the addend 5.6 and one zero after the 8 in the addend 16.78. Since whole numbers have an "understood" decimal point immediately to the right of their ones digit, we can write the addend 68 as 68.000 to help line up the columns.

$$\begin{array}{r} 31.913 \\ 5.600 \\ 68.000 \\ + \ 16.780 \end{array}$$

Insert two zeros after the 6.
Insert a decimal point and three zeros: 68 = 68.000.
Insert a zero after the 8.

↑ —— Line up the decimal points.

Now we add, right to left, as we would whole numbers, writing the sum from each column below the horizontal bar.

Self Check 1
Add: 41.07 + 35 + 67.888 + 4.1
Now Try Problem 19

$$\overset{2\,2}{31.913}$$ Carry a 2 (shown in blue) to the ones column.

   5.600   Carry a 2 (shown in green) to the tens column.

 68.000

+ 16.780

122.293

⌐⎯⎯⎯ Write the decimal point in the result directly
below the decimal points in the addends.

The sum is 122.293.

Success Tip   In Example 1, the digits in each place-value column were added from *top to bottom*. To check the answer, we can instead add from *bottom to top*. Adding down or adding up should give the same result. If it does not, an error has been made and you should re-add.

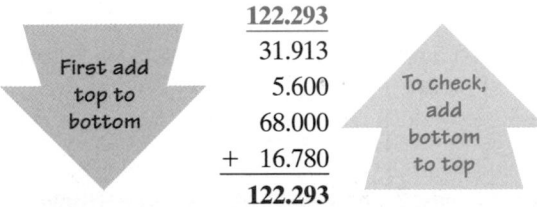

First add
top to
bottom

122.293
31.913
5.600
68.000
+ 16.780
122.293

To check,
add
bottom
to top

Using Your CALCULATOR   **Adding Decimals**

The bar graph on the right shows the number of grams of fiber in a standard serving of each of several foods. It is believed that men can significantly cut their risk of heart attack by eating at least 25 grams of fiber a day. Does this diet meet or exceed the 25-gram requirement?

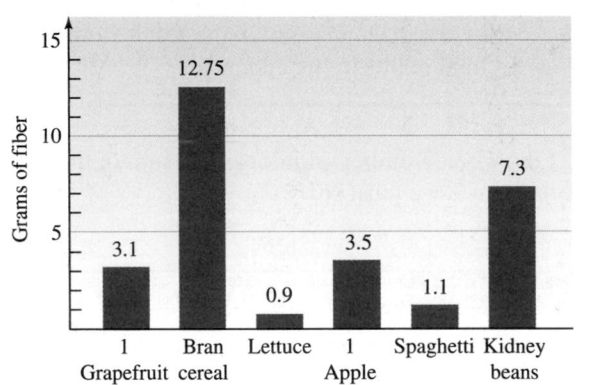

To find the total fiber intake, we add the fiber content of each of the foods. We can use a calculator to add the decimals.

3.1 $\boxed{+}$ 12.75 $\boxed{+}$ .9 $\boxed{+}$ 3.5 $\boxed{+}$ 1.1 $\boxed{+}$ 7.3 $\boxed{=}$ $\boxed{28.65}$

On some calculators, the $\boxed{\text{ENTER}}$ key is pressed to find the sum.

Since 28.65 > 25, this diet exceeds the daily fiber requirement of 25 grams.

---

2 Subtract decimals.

Subtracting decimals is similar to subtracting whole numbers. We use **vertical form** and stack the decimals with their corresponding place values and decimal points lined up so that we subtract similar objects—hundredths from hundredths, tenths from tenths, ones from ones, and so on. We write the decimal point in the **difference** so that it lines up with the decimal points in the **minuend** and **subtrahend.** For example, to find 8.59 − 1.27, we proceed as follows:

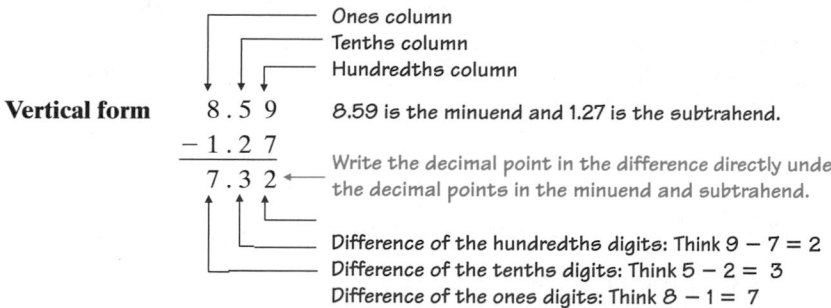

Vertical form

The difference is 7.32.

Subtracting Decimals

To subtract decimal numbers:

1. Write the numbers in vertical form with the decimal points lined up.

2. Subtract the numbers as you would subtract whole numbers from right to left.

3. Write the decimal point in the result from Step 2 directly below the decimal points in the minued and the subtrahend.

As with whole numbers, if the subtraction of the digits in any place-value column requires that we subtract a larger digit from a smaller digit, we must **borrow** or **regroup.**

EXAMPLE 2   Subtract: 279.6 − 138.7

Strategy   As we prepare to subtract in each column, we will compare the digit in the subtrahend (bottom number) to the digit directly above it in the minuend (top number).

WHY   If a digit in the subtrahend is greater than the digit directly above it in the minuend, we must borrow (regroup) to subtract in that column.

Solution   Since 7 in the tenths column of 138.**7** is greater than 6 in the tenths column of 279.**6**, we cannot immediately subtract in that column because 6 − 7 is *not* a whole number. To subtract in the tenths column, we must regroup by borrowing as shown below.

$$
\begin{array}{r}
\overset{\scriptstyle 8\ \ 16}{279.\cancel{6}} \\
-\ 138.7 \\
\hline
140.9
\end{array}
$$

To subtract in the tenths column, borrow 1 one in the form of 10 tenths from the ones column. Add 10 to the 6 in the tenths column to get 16 (shown in blue).

Recall from Section 1.3 that subtraction can be checked by addition. If a subtraction is done correctly, the sum of the difference and the subtrahend will equal the minuend: **Difference + subtrahend = minuend.**

Check:

$$
\begin{array}{r}
\overset{\scriptstyle 1}{140.9} \quad \text{Difference}\\
+\ 138.7 \quad \text{Subtrahend}\\
\hline
279.6 \quad \text{Minuend}
\end{array}
$$

Since the sum of the difference and the subtrahend is the minuend, the subtraction is correct.

Some subtractions require borrowing from two (or more) place-value columns.

Self Check 2

Subtract: 382.5 − 227.1

Now Try   Problem 27

Self Check 3

Subtract 27.122 from 29.7.

Now Try **Problem 31**

EXAMPLE 3   Subtract 13.059 from 15.4.

Strategy We will translate the sentence to mathematical symbols and then perform the subtraction. As we prepare to subtract in each column, we will compare the digit in the subtrahend (bottom number) to the digit directly above it in the minuend (top number).

WHY If a digit in the subtrahend is greater than the digit directly above it in the minuend, we must borrow (regroup) to subtract in that column.

Solution Since 13.059 is the number to be subtracted, it is the subtrahend.

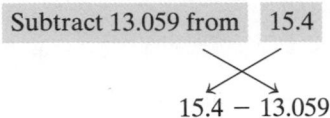

$$15.4 - 13.059$$

To find the difference, we write the subtraction in vertical form. To help with the column subtractions, we write two zeros to the right of 15.4 so that both numbers have three decimal places.

$$
\begin{array}{r}
15.400 \\
-\ 13.059 \\
\end{array}
$$
    Insert two zeros after the 4 so that the decimal places match.

└── Line up the decimal points.

Since 9 in the thousandths column of 13.059 is greater than 0 in the thousandths column of 15.400, we cannot immediately subtract. It is not possible to borrow from the digit **0** in the hundredths column of 15.400. We can, however, borrow from the digit **4** in the tenths column of 15.400.

$$
\begin{array}{r}
\overset{3\,10}{15.4\cancel{0}0} \\
-\ 13.059 \\
\end{array}
$$
    Borrow 1 tenth in the form of 10 hundredths from 4 in the tenths column. Add 10 to 0 in the hundredths column to get 10 (shown in blue).

Now we complete the two-column borrowing process by borrowing from the **10** in the hundredths column. Then we subtract, column-by-column, from the right to the left to find the difference.

$$
\begin{array}{r}
\overset{\ \ \ 9}{\overset{3\,\cancel{10}\,10}{15.\cancel{4}\,\cancel{0}\,\cancel{0}}} \\
-\ 13.0\,5\,9 \\
\hline
2.3\,4\,1 \\
\end{array}
$$
    Borrow 1 hundredth in the form of 10 thousandths from 10 in the hundredths column. Add 10 to 0 in the thousandths column to get 10 (shown in green).

When 13.059 is subtracted from 15.4, the difference is 2.341.

Check:

$$
\begin{array}{r}
\overset{1\,1}{2.341} \\
+\ 13.059 \\
\hline
15.400 \\
\end{array}
$$
    Since the sum of the difference and the subtrahend is the minuend, the subtraction is correct.

Using Your CALCULATOR   **Subtracting Decimals**

A giant weather balloon is made of a flexible rubberized material that has an uninflated thickness of 0.011 inch. When the balloon is inflated with helium, the thickness becomes 0.0018 inch. To find the change in thickness, we need to subtract. We can use a calculator to subtract the decimals.

.011 $\boxed{-}$ .0018 $\boxed{=}$ $\boxed{0.0092}$

On some calculators, the $\boxed{\text{ENTER}}$ key is pressed to find the difference.

After the balloon is inflated, the rubberized material loses 0.0092 inch in thickness.

3 Add and subtract signed decimals.

To add signed decimals, we use the same rules that we used for adding integers.

Adding Two Decimals That Have the Same (Like) Signs

1.   To add two positive decimals, add them as usual. The final answer is positive.

2.   To add two negative decimals, add their absolute values and make the final answer negative.

Adding Two Decimals That Have Different (Unlike) Signs

To add a positive decimal and a negative decimal, subtract the smaller absolute value from the larger.

1.   If the positive decimal has the larger absolute value, the final answer is positive.

2.   If the negative decimal has the larger absolute value, make the final answer negative.

EXAMPLE 4   Add: $-6.1 + (-4.7)$

Strategy We will use the rule for adding two decimals that have the same sign.

WHY Both addends, -6.1 and -4.7, are negative.

Solution Find the absolute values: $|-6.1| = 6.1$ and $|-4.7| = 4.7$.

$$-6.1 + (-4.7) = -10.8$$ Add the absolute values, 6.1 and 4.7, to get 10.8. Then make the final answer negative.

$$\begin{array}{r} 6.1 \\ +\ 4.7 \\ \hline 10.8 \end{array}$$

Self Check 4
Add: $-5.04 + (-2.32)$
Now Try Problem 35

EXAMPLE 5   Add: $5.35 + (-12.9)$

Strategy We will use the rule for adding two integers that have different signs.

WHY One addend is positive and the other is negative.

Solution Find the absolute values: $|5.35| = 5.35$ and $|-12.9| = 12.9$.

$$5.35 + (-12.9) = -7.55$$ Subtract the smaller absolute value from the larger: $12.9 - 5.35 = 7.55$. Since the negative number, -12.9, has the larger absolute value, make the final answer negative.

$$\begin{array}{r} {}^{8\,10} \\ 12.9\cancel{0} \\ -\ 5.3\,5 \\ \hline 7.5\,5 \end{array}$$

Self Check 5
Add: $-21.4 + 16.75$
Now Try Problem 39

The rule for subtraction that was introduced in Section 2.3 can be used with signed decimals: *To subtract two decimals, add the first decimal to the opposite of the decimal to be subtracted.*

EXAMPLE 6   Subtract: $-35.6 - 5.9$

Strategy We will apply the rule for subtraction: Add the first decimal to the opposite of the decimal to be subtracted.

WHY It is easy to make an error when subtracting signed decimals. We will probably be more accurate if we write the subtraction as addition of the opposite.

Self Check 6
Subtract: $-1.18 - 2.88$
Now Try Problem 43

Solution The number to be subtracted is 5.9. Subtracting 5.9 is the same as adding its opposite, −5.9.

Change the subtraction to addition.

$$-35.6 - \mathbf{5.9} = -35.6 + (\mathbf{-5.9}) = -41.5$$

Change the number being subtracted to its opposite.

Use the rule for adding two decimals with the same sign. Make the final answer negative.

$$\begin{array}{r} \overset{1\,1}{35.6} \\ +\ \ 5.9 \\ \hline 41.5 \end{array}$$

Self Check 7
Subtract: −2.56 − (−4.4)
Now Try Problem 47

EXAMPLE 7   Subtract: −8.37 − (−16.2)

Strategy We will apply the rule for subtraction: Add the first decimal to the opposite of the decimal to be subtracted.

WHY It is easy to make an error when subtracting signed decimals. We will probably be more accurate if we write the subtraction as addition of the opposite.

Solution The number to be subtracted is −16.2. Subtracting −16.2 is the same as adding its opposite, 16.2.

Add . . .

$$-8.37 - (\mathbf{-16.2}) = -8.37 + \mathbf{16.2} = 7.83$$

. . . the opposite

Use the rule for adding two decimals with different signs. Since 16.2 has the larger absolute value, the final answer is positive.

$$\begin{array}{r} \overset{\ \ \ \ 11}{\overset{5\ \ \cancel{X}10}{16.2\cancel{0}}} \\ -\ 8.37 \\ \hline 7.83 \end{array}$$

Self Check 8
Evaluate: −4.9 − (−1.2 + 5.6)
Now Try Problem 51

EXAMPLE 8   Evaluate: −12.2 − (−14.5 + 3.8)

Strategy We will perform the operation within the parentheses first.

WHY This is the first step of the order of operations rule.

Solution We perform the addition within the grouping symbols first.

$$\begin{aligned} -12.2 - (\mathbf{-14.5 + 3.8}) &= -12.2 - (\mathbf{-10.7}) \quad &&\text{Perform the addition.} \\ &= -12.2 + 10.7 \quad &&\text{Add the opposite} \\ & &&\text{of −10.7.} \\ &= -1.5 \quad &&\text{Perform the addition.} \end{aligned}$$

$$\begin{array}{r} \overset{3\ 15}{14.\cancel{5}} \\ -\ 3.8 \\ \hline 10.7 \end{array}$$

$$\begin{array}{r} \overset{1\ 12}{12.\cancel{2}} \\ -10.7 \\ \hline 1.5 \end{array}$$

---

4 Estimate sums and differences of decimals.

Estimation can be used to check the reasonableness of an answer to a decimal addition or subtraction. There are several ways to estimate, but the objective is the same: Simplify the numbers in the problem so that the calculations can be made easily and quickly.

Self Check 9
a. Estimate by rounding the addends to the nearest ten: 526.93 + 284.03
b. Estimate using front-end rounding: 512.33 − 36.47
Now Try Problems 55 and 57

EXAMPLE 9

a. Estimate by rounding the addends to the nearest ten: 261.76 + 432.94
b. Estimate using front-end rounding: 381.77 − 57.01

Strategy We will use rounding to approximate each addend, minuend, and subtrahend. Then we will find the sum or difference of the approximations.

WHY Rounding produces numbers that contain many 0's. Such numbers are easier to add or subtract.

Solution

a.
$$\begin{array}{r} 261.76 \to \ \ 260 \\ + \ 432.94 \to + \ 430 \\ \hline 690 \end{array}$$
Round to the nearest ten.
Round to the nearest ten.

The estimate is 690. If we compute 261.76 + 432.94, the sum is 694.7. We can see that the estimate is close; it's just 4.7 less than 694.7.

b. We use front-end rounding. Each number is rounded to its largest place value.

$$\begin{array}{r} 381.77 \to \ \ 400 \\ - \ \ 57.01 \to - \ \ 60 \\ \hline 340 \end{array}$$
Round to the nearest hundred.
Round to the nearest ten.

The estimate is 340. If we compute 381.77 − 57.01, the difference is 324.76. We can see that the estimate is close; it's 15.24 more than 324.76.

5 Solve application problems by adding and subtracting decimals.

To make a profit, a merchant must sell an item for more than she paid for it. The price at which the merchant sells the product, called the **retail price,** is the *sum* of what the item **cost** the merchant plus the **markup.**

Retail price = cost + markup

EXAMPLE 10   *Pricing*   Find the retail price of a Rubik's Cube if a game store owner buys them for $8.95 each and then marks them up $4.25 to sell in her store.

Analyze

- Rubik's Cubes cost the store owner $8.95 each.   Given
- She marks up the price $4.25.   Given
- What is the retail price of a Rubik's Cube?   Find

Andrea Presazzi/Dreamstime.com

Form We translate the words of the problem to numbers and symbols.

The retail price	is equal to	the cost	plus	the markup.
The retail price	=	8.95	+	4.25

Solve Use vertical form to perform decimal addition:

$$\begin{array}{r} \overset{1\ 1}{8.95} \\ + \ \ 4.25 \\ \hline 13.20 \end{array}$$

State The retail price of a Rubik's Cube is $13.20.

Check We can estimate to check the result. If we use $9 to approximate the cost of a Rubik's Cube to the store owner and $4 to be the approximate markup, then the retail price is about $9 + $4 = $13. The result, $13.20, seems reasonable.

Self Check 10

PRICING Find the retail price of a wool coat if a clothing outlet buys them for $109.95 each and then marks them up $99.95 to sell in its stores.

Now Try Problem 91

EXAMPLE 11   *Kitchen Sinks*   One model of kitchen sink is made of 18-gauge stainless steel that is 0.0500 inch thick. Another, less expensive, model is made from 20-gauge stainless steel that is 0.0375 inch thick. How much thicker is the 18-gauge?

Self Check 11

ALUMINUM How much thicker is 16-gauge aluminum that is 0.0508 inch thick than 22-gauge aluminum that is 0.0253 inch thick?

Now Try Problem 97

Analyze

- The 18-gauge stainless steel is 0.0500 inch thick.   *Given*
- The 20-gauge stainless steel is 0.0375 inch thick.   *Given*
- How much thicker is the 18-gauge stainless steel?   *Find*

Image copyright V. J. Matthew, 2009. Used under license from Shutterstock.com

Form Phrases such as *how much older, how much longer,* and, in this case, *how much thicker,* indicate subtraction. We translate the words of the problem to numbers and symbols.

How much thicker	is equal to	the thickness of the 18-gauge stainless steel	minus	the thickness of the 20-gauge stainless steel.
How much thicker	=	0.0500	−	0.0375

Solve Use vertical form to perform subtraction:

$$\begin{array}{r} \overset{9}{}\\ \overset{4}{}\overset{\cancel{10}}{}\overset{10}{}\\ 0.0\cancel{5}\,\cancel{0}\,\cancel{0}\\ -\ 0.0\ 3\ 7\ 5\\ \hline 0.0\ 1\ 2\ 5 \end{array}$$

State The 18-gauge stainless steel is 0.0125 inch thicker than the 20-gauge.

Check We can add to check the subtraction:

$$\begin{array}{r} \overset{11}{}\\ 0.0\overset{}{1}25 \quad \text{Difference}\\ +\ 0.0375 \quad \text{Subtrahend}\\ \hline 0.0500 \quad \text{Minuend} \end{array}$$

The result checks.

Sometimes more than one operation is needed to solve a problem involving decimals.

Self Check 12

WRESTLING A 195.5-pound wrestler had to lose 6.5 pounds to make his weight class. After the weigh-in, he gained back 3.7 pounds. What did he weigh then?

Now Try Problem 103

EXAMPLE 12 *Conditioning Programs* A 350-pound football player lost 15.7 pounds during the first week of practice. During the second week, he gained 4.9 pounds. Find his weight after the first two weeks of practice.

Analyze

- The football player's beginning weight was 350 pounds.   *Given*
- The first week he lost 15.7 pounds.   *Given*
- The second week he gained 4.9 pounds.   *Given*
- What was his weight after two weeks of practice?   *Find*

Form The word *lost* indicates subtraction. The word *gained* indicates addition. We translate the words of the problem to numbers and symbols.

The player's weight after two weeks of practice	is equal to	his beginning weight	minus	the first-week weight loss	plus	the second-week weight gain.
The player's weight after two weeks of practice	=	350	−	15.7	+	4.9

Solve To evaluate $350 - 15.7 + 4.9$, we work from left to right and perform the subtraction first, then the addition.

$$
\begin{array}{r}
\overset{\overset{9}{4\ \cancel{10}\ 10}}{3\ \cancel{5}\ \cancel{0}.\cancel{0}} \\
-\ \ \ \ 1\ 5.7 \\
\hline
3\ 3\ 4.3
\end{array}
$$

Write the whole number 350 as 350.0 and use a two-column borrowing process to subtract in the tenths column.

This is the player's weight after one week of practice.

Next, we add the 4.9-pound gain to the previous result to find the player's weight after two weeks of practice.

$$
\begin{array}{r}
\overset{1}{3}34.3 \\
+\ \ \ \ 4.9 \\
\hline
339.2
\end{array}
$$

State The player's weight was 339.2 pounds after two weeks of practice.

Check We can estimate to check the result. The player lost about 16 pounds the first week and then gained back about 5 pounds the second week, for a net loss of 11 pounds. If we subtract the approximate 11 pound loss from his beginning weight, we get $350 - 11 = 339$ pounds. The result, 339.2 pounds, seems reasonable.

ANSWERS TO SELF CHECKS

1. 148.058   **2.** 155.4   **3.** 2.578   **4.** −7.36   **5.** −4.65   **6.** −4.06   **7.** 1.84   **8.** −9.3
9. a. 810   **b.** 460   **10.** $209.90   **11.** 0.0255 in.   **12.** 192.7 lb

---

SECTION 5.2   STUDY SET

VOCABULARY

Fill in the blanks.

1. In the addition problem shown below, label each *addend* and the *sum*.

$$
\begin{array}{r}
1.72 \leftarrow \rule{2cm}{0.4pt} \\
4.68 \leftarrow \rule{2cm}{0.4pt} \\
+\ 2.02 \leftarrow \rule{2cm}{0.4pt} \\
\hline
8.42 \leftarrow \rule{2cm}{0.4pt}
\end{array}
$$

2. When using the vertical form to add decimals, if the addition of the digits in any one column produces a sum greater than 9, we must _____.

3. In the subtraction problem shown below, label the *minuend, subtrahend,* and the *difference.*

$$
\begin{array}{r}
12.9 \leftarrow \rule{2cm}{0.4pt} \\
-\ \ 4.3 \leftarrow \rule{2cm}{0.4pt} \\
\hline
8.6 \leftarrow \rule{2cm}{0.4pt}
\end{array}
$$

4. If the subtraction of the digits in any place-value column requires that we subtract a larger digit from a smaller digit, we must _____ or *regroup*.

5. To see whether the result of an addition is reasonable, we can round the addends and _____ the sum.

6. In application problems, phrases such as *how much older, how much longer,* and *how much thicker* indicate the operation of _____.

CONCEPTS

7. Check the following result. Use addition to determine if 15.2 is the correct difference.

$$
\begin{array}{r}
28.7 \\
-\ 12.5 \\
\hline
15.2
\end{array}
$$

8. Determine whether the *sign* of each result is positive or negative. *You do not have to find the sum.*
   a. $-7.6 + (-1.8)$
   b. $-24.99 + 29.08$
   c. $133.2 + (-400.43)$

9. Fill in the blank: To subtract signed decimals, add the _____ of the decimal that is being subtracted.

10. Apply the rule for subtraction and fill in the three blanks.

$$
3.6 - (-2.1) = 3.6 \ \boxed{}\ \boxed{} = \boxed{}
$$

11. Fill in the blanks to rewrite each subtraction as addition of the opposite of the number being subtracted.

  a. $6.8 - 1.2 = 6.8 + ($ ⬚ $)$

  b. $29.03 - (-13.55) = 29.03 +$ ⬚

  c. $-5.1 - 7.4 = -5.1 + ($ ⬚ $)$

12. Fill in the blanks to complete the estimation.

$$\begin{array}{r} 567.7 \to \quad \text{Round to the nearest ten.}\\ + 214.3 \to + \quad \text{Round to the nearest ten.}\\ \hline 782.0 \end{array}$$

NOTATION

13. Copy the following addition problem. Insert a decimal point and additional zeros so that the number of decimal places in the addends match.

$$\begin{array}{r} 46.6\\ 11\\ + 15.702\\ \hline \end{array}$$

14. Refer to the subtraction problem below. Fill in the blanks: To subtract in the _____ column, we borrow 1 tenth in the form of 10 hundredths from the 3 in the _____ column.

$$\begin{array}{r} \overset{2\ 11}{29.3\cancel{1}}\\ - 25.16\\ \hline \end{array}$$

GUIDED PRACTICE

Add. See Objective 1.

15. $\begin{array}{r} 32.5\\ + 7.4\\ \hline \end{array}$ **16.** $\begin{array}{r} 16.3\\ + 3.5\\ \hline \end{array}$

17. $\begin{array}{r} 3.04\\ 4.12\\ + 1.43\\ \hline \end{array}$ **18.** $\begin{array}{r} 2.11\\ 5.04\\ + 2.72\\ \hline \end{array}$

Add. See Example 1.

19. $36.821 + 7.3 + 42 + 15.44$

20. $46.228 + 5.6 + 39 + 19.37$

21. $27.471 + 6.4 + 157 + 12.12$

22. $52.763 + 9.1 + 128 + 11.84$

Subtract. See Objective 2.

23. $\begin{array}{r} 6.83\\ - 3.52\\ \hline \end{array}$ **24.** $\begin{array}{r} 9.47\\ - 5.06\\ \hline \end{array}$

25. $\begin{array}{r} 8.97\\ - 6.22\\ \hline \end{array}$ **26.** $\begin{array}{r} 7.56\\ - 2.33\\ \hline \end{array}$

Subtract. See Example 2.

27. $\begin{array}{r} 495.4\\ - 153.7\\ \hline \end{array}$ **28.** $\begin{array}{r} 977.6\\ - 345.8\\ \hline \end{array}$

29. $\begin{array}{r} 878.1\\ - 174.6\\ \hline \end{array}$ **30.** $\begin{array}{r} 767.2\\ - 614.7\\ \hline \end{array}$

Perform the indicated operation. See Example 3.

31. Subtract 11.065 from 18.3.

32. Subtract 15.041 from 17.8.

33. Subtract 23.037 from 66.9.

34. Subtract 31.089 from 75.6.

Add. See Example 4.

35. $-6.3 + (-8.4)$ **36.** $-9.2 + (-6.7)$

37. $-9.5 + (-9.3)$ **38.** $-7.3 + (-5.4)$

Add. See Example 5.

39. $4.12 + (-18.8)$ **40.** $7.24 + (-19.7)$

41. $6.45 + (-12.6)$ **42.** $8.81 + (-14.9)$

Subtract. See Example 6.

43. $-62.8 - 3.9$ **44.** $-56.1 - 8.6$

45. $-42.5 - 2.8$ **46.** $-93.2 - 3.9$

Subtract. See Example 7.

47. $-4.49 - (-11.3)$ **48.** $-5.76 - (-13.6)$

49. $-6.78 - (-24.6)$ **50.** $-8.51 - (-27.4)$

Evaluate each expression. See Example 8.

51. $-11.1 - (-14.4 + 7.8)$

52. $-12.3 - (-13.6 + 7.9)$

53. $-16.4 - (-18.9 + 5.9)$

54. $-15.5 - (-19.8 + 5.7)$

Estimate each sum by rounding the addends to the nearest ten. See Example 9.

55. $510.65 + 279.19$ **56.** $424.08 + 169.04$

Estimate each difference by using front-end rounding. See Example 9.

57. $671.01 - 88.35$ **58.** $447.23 - 36.16$

TRY IT YOURSELF

Perform the indicated operations.

59. $-45.6 + 34.7$ **60.** $-19.04 + 2.4$

61. $-9.5 - 7.1$ **62.** $-7.08 - 14.3$

63. $46.09 + (-7.8)$ **64.** $34.7 + (-30.1)$

65. $\begin{array}{r} 21.88\\ + 33.12\\ \hline \end{array}$ **66.** $\begin{array}{r} 19.05\\ + 31.95\\ \hline \end{array}$

67. $30.03 - (-17.88)$

68. $143.3 - (-64.01)$

69. $645 + 9.90005 + 0.12 + 3.02002$

70. $505.0103 + 23 + 0.989 + 12.0704$

71. Subtract 23.81 from 24.

72. Subtract 5.9 from 7.001.

73. $(3.4 - 6.6) + 7.3$ **74.** $3.4 - (6.6 + 7.3)$

75. $247.9 + 40 + 0.56$

76. $0.0053 + 1.78 + 6$

77.   78.1   **78.**   202.234
    $-$ 7.81 $-$ 19.34

79. $-7.8 + (-6.5)$ **80.** $-5.78 + (-33.1)$

81. $16 - (67.2 + 6.27)$

82. $-43 - (0.032 - 0.045)$

83. Find the sum of *two and forty-three hundredths* and *five and six tenths.*

84. Find the difference of *nineteen hundredths* and *six thousandths.*

85. $|-14.1 + 6.9| + 8$ **86.** $15 - |-2.3 + (-2.4)|$

87. $5 - 0.023$ **88.** $30 - 11.98$

89. $-2.002 - (-4.6)$ **90.** $-0.005 - (-8)$

APPLICATIONS

91. RETAILING Find the retail price of each appliance listed in the following table if a department store purchases them for the given costs and then marks them up as shown.

Appliance	Cost	Markup	Retail price
Refrigerator	$610.80	$205.00	
Washing machine	$389.50	$155.50	
Dryer	$363.99	$167.50	

92. PRICING Find the retail price of a Kenneth Cole two-button suit if a men's clothing outlet buys them for $210.95 each and then marks them up $144.95 to sell in its stores.

93. OFFSHORE DRILLING A company needs to construct a pipeline from an offshore oil well to a refinery located on the coast. Company engineers have come up with two plans for consideration, as shown. Use the information in the illustration to complete the table that is shown in the next column.

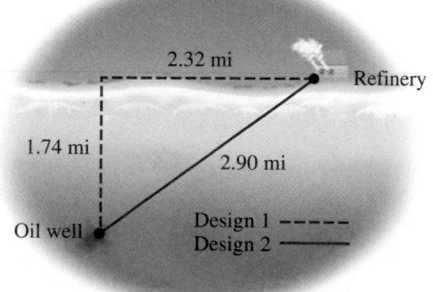

	Pipe underwater (mi)	Pipe underground (mi)	Total pipe (mi)
Design 1			
Design 2			

94. DRIVING DIRECTIONS Find the total distance of the trip using the information in the MapQuest printout shown below.

START	**1:** Start out going EAST on SUNKIST AVE.	**0.0 mi**
	2: Turn LEFT onto MERCED AVE.	**0.4 mi**
	3: Turn Right onto PUENTE AVE.	**0.3 mi**
WEST 10	**4:** Merge onto I-10 W toward LOS ANGELES.	**2.2 mi**
SOUTH 605	**5:** Merge onto I-605 S.	**10.6 mi**
SOUTH 5	**6:** Merge onto I-5 S toward SANTA ANA.	**14.9 mi**
110 A EXIT	**7:** Take the HARBOR BLVD exit, EXIT 110A.	**0.3 mi**
	8: Turn RIGHT onto S HARBOR BLVD.	**0.1 mi**
END	**9:** End at 1313 S Harbor Blvd Anaheim, CA.	

Total Distance: ___?___ miles   **MAPQUEST**®

95. PIPE (PVC) Find the *outside* diameter of the plastic sprinkler pipe shown below if the thickness of the pipe wall is 0.218 inch and the inside diameter is 1.939 inches.

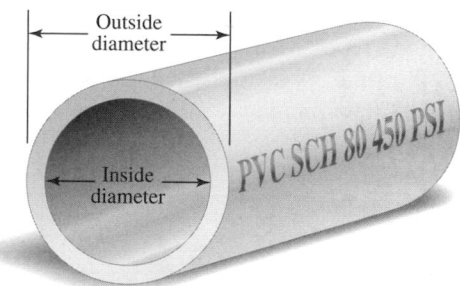

96. pH SCALE The pH scale shown below is used to measure the strength of acids and bases in chemistry. Find the difference in pH readings between

a. bleach and stomach acid.

b. ammonia and coffee.

c. blood and coffee.

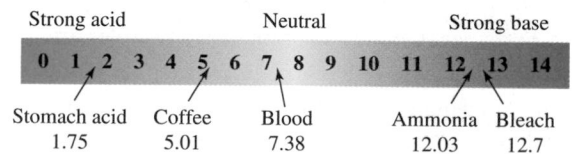

97. RECORD HOLDERS The late Florence Griffith-Joyner of the United States holds the women's world record in the 100-meter sprint: 10.49 seconds. Libby Trickett of Australia holds the women's world record in the 100-meter freestyle swim: 52.88 seconds. How much faster did Griffith-Joyner run the 100 meters than Trickett swam it? (Source: *The World Almanac and Book of Facts*, 2009)

98. WEATHER REPORTS Barometric pressure readings are recorded on the weather map below. In a low-pressure area (L on the map), the weather is often stormy. The weather is usually fair in a high-pressure area (H). What is the difference in readings between the areas of highest and lowest pressure?

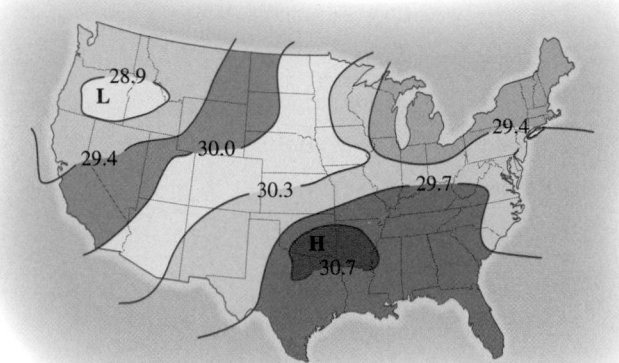

99. BANKING A businesswoman deposited several checks in her company's bank account, as shown on the deposit slip below. Find the *Subtotal* line on the slip by adding the amounts of the checks and total from the reverse side. If the woman wanted to get $25 in cash back from the teller, what should she write as the *Total deposit* on the slip?

Deposit slip		
Cash		
Checks (properly endorsed)	116	10
	47	93
Total from reverse side	359	16
Subtotal		
Less cash	25	00
Total deposit		

100. SPORTS PAGES Decimals are often used in the sports pages of newspapers. Two examples are given below.

   a. "German bobsledders set a world record today with a final run of 53.03 seconds, finishing ahead of the Italian team by only fourteen thousandths of a second." What was the time for the Italian bobsled team?

   b. "The women's figure skating title was decided by only thirty-three hundredths of a point." If the winner's point total was 102.71, what was the second-place finisher's total? (*Hint:* The highest score wins in a figure skating contest.)

101. Suppose certain portions of a patient's morning (A.M.) temperature chart were not filled in. Use the given information to complete the chart below. (*Hint:* 98.6°F is considered normal.)

from Campus to Careers
Home Health Aide

Day of week	Patient's A.M. temperature	Amount above normal
Monday	99.7°	
Tuesday		2.5°
Wednesday	98.6°	
Thursday	100.0°	
Friday		0.9°

102. QUALITY CONTROL An electronics company has strict specifications for the silicon chips it uses in its computers. The company only installs chips that are within 0.05 centimeter of the indicated thickness. The table below gives that specifications for two types of chips. Fill in the blanks to complete the chart.

Chip type	Thickness specification	Acceptable range	
		Low	High
A	0.78 cm		
B	0.643 cm		

103. FLIGHT PATHS Find the added distance a plane must travel to avoid flying through the storm.

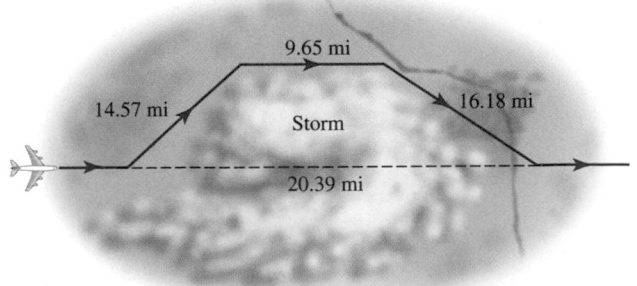

104. TELEVISION The following illustration shows the six most-watched television shows of all time (excluding Super Bowl games and the Olympics).

    a. What was the combined total audience of all six shows?

    b. How many more people watched the last episode of "MASH" than watched the last episode of "Seinfeld"?

    c. How many more people would have had to watch the last "Seinfeld" to move it into a tie for fifth place?

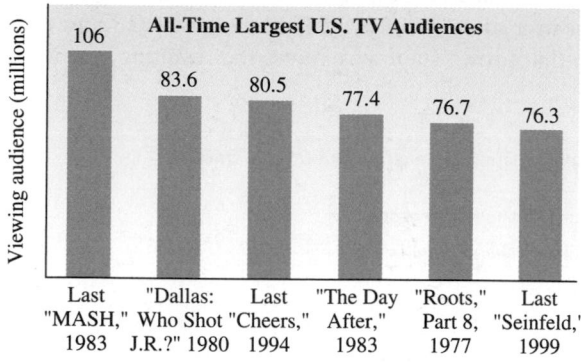

Source: Nielsen Media Research

105. THE HOME SHOPPING NETWORK The illustration shows a description of a cookware set that was sold on television.

    a. Find the difference between the manufacturer's suggested retail price (MSRP) and the sale price.

    b. Including shipping and handling (S & H), how much will the cookware set cost?

Item 229-442	
Continental 9-piece Cookware Set	
Stainless steel	
MSRP	$149.79
HSN Price	$59.85
On Sale	**$47.85**
S & H	$7.95

106. VEHICLE SPECIFICATIONS Certain dimensions of a compact car are shown. Find the wheelbase of the car.

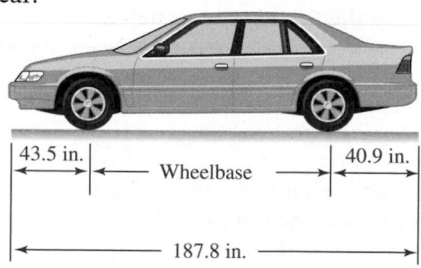

WRITING

107. Explain why we line up the decimal points and corresponding place-value columns when adding decimals.

108. Explain why we can write additional zeros to the right of a decimal such as 7.89 without affecting its value.

109. Explain what is wrong with the work shown below.

$$\begin{array}{r} 203.56 \\ 37 \\ +\ 0.43 \\ \hline 204.36 \end{array}$$

110. Consider the following addition:

$$\begin{array}{r} \overset{2}{23.7} \\ 41.9 \\ +\ 12.8 \\ \hline 78.4 \end{array}$$

Explain the meaning of the small red 2 written above the ones column.

111. Write a set of instructions that explains the two-column borrowing process shown below.

$$\begin{array}{r} \overset{9}{}\overset{4\ \cancel{10}10}{} \\ 2.6\cancel{5}\cancel{0}\,\cancel{0} \\ -\ 1.3246 \\ \hline 1.3254 \end{array}$$

112. Explain why it is easier to add the decimals 0.3 and 0.17 than the fractions $\frac{3}{10}$ and $\frac{17}{100}$.

REVIEW

Perform the indicated operations.

113. **a.** $\dfrac{4}{5} + \dfrac{5}{12}$

    b. $\dfrac{4}{5} - \dfrac{5}{12}$

    c. $\dfrac{4}{5} \cdot \dfrac{5}{12}$

    d. $\dfrac{4}{5} \div \dfrac{5}{12}$

114. **a.** $\dfrac{3}{8} + \dfrac{1}{6}$

    b. $\dfrac{3}{8} - \dfrac{1}{6}$

    c. $\dfrac{3}{8} \cdot \dfrac{1}{6}$

    d. $\dfrac{3}{8} \div \dfrac{1}{6}$

Objectives

1 Multiply decimals.

2 Multiply decimals by powers of 10.

3 Multiply signed decimals.

4 Evaluate exponential expressions that have decimal bases.

5 Use the order of operations rule.

6 Evaluate formulas.

7 Estimate products of decimals.

8 Solve application problems by multiplying decimals.

SECTION 5.3
Multiplying Decimals

Since decimal numbers are *base-ten* numbers, multiplication of decimals is similar to multiplication of whole numbers. However, when multiplying decimals, there is one additional step—we must determine where to write the decimal point in the product.

1 Multiply decimals.

To develop a rule for multiplying decimals, we will consider the multiplication $0.3 \cdot 0.17$ and find the product in a roundabout way. First, we write 0.3 and 0.17 as fractions and multiply them in that form. Then we express the resulting fraction as a decimal.

$$0.3 \cdot 0.17 = \frac{3}{10} \cdot \frac{17}{100} \qquad \text{Express the decimals 0.3 and 0.17 as fractions.}$$

$$= \frac{3 \cdot 17}{10 \cdot 100} \qquad \begin{array}{l}\text{Multiply the numerators.}\\ \text{Multiply the denominators.}\end{array}$$

$$= \frac{51}{1,000}$$

$$= 0.051 \qquad \text{Write the resulting fraction } \tfrac{51}{1,000} \text{ as a decimal.}$$

From this example, we can make observations about multiplying decimals.

- The digits in the answer are found by multiplying 3 and 17.

$$0.3 \quad \cdot \quad 0.17 \quad = \quad 0.051$$
$$3 \cdot 17 = 51$$

- The answer has 3 decimal places. The *sum* of the number of decimal places in the factors 0.3 and 0.17 is also 3.

$$0.3 \quad \cdot \quad 0.17 \quad = \quad 0.051$$
1 decimal   2 decimal   3 decimal
place   places   places

These observations illustrate the following rule for multiplying decimals.

Multiplying Decimals

To multiply two decimals:

1. Multiply the decimals as if they were whole numbers.
2. Find the total number of decimal places in both factors.
3. Insert a decimal point in the result from step 1 so that the answer has the same number of decimal places as the total found in step 2.

Self Check 1

Multiply: $2.7 \cdot 4.3$

Now Try Problem 9

EXAMPLE 1   Multiply: $5.9 \cdot 3.4$

Strategy We will ignore the decimal points and multiply 5.9 and 3.4 as if they were whole numbers. Then we will write a decimal point in that result so that the final answer has two decimal places.

WHY Since the factor 5.9 has 1 decimal place, and the factor 3.4 has 1 decimal place, the product should have $1 + 1 = 2$ decimal places.

Solution We write the multiplication in vertical form and proceed as follows:

Vertical form
$$
\begin{array}{r}
5.9 \leftarrow \text{1 decimal place} \\
\times \quad 3.4 \leftarrow \text{1 decimal place} \\
\hline
236 \\
1770 \\
\hline
20.06
\end{array}
$$

The answer will have $1 + 1 = 2$ decimal places.

Move 2 places from the right to the left and insert a decimal point in the answer.

Thus, $5.9 \cdot 3.4 = 20.06$.

The Language of Algebra   Recall the vocabulary of multiplication.

$$
\begin{array}{r}
5.9 \leftarrow \text{Factor} \\
\times \quad 3.4 \leftarrow \text{Factor} \\
\hline
236 \\
1770 \\
\hline
20.06 \leftarrow \text{Product}
\end{array}
$$

Partial products

Success Tip   When multiplying decimals, we do not need to line up the decimal points, as the next example illustrates.

EXAMPLE 2   Multiply: 1.3(0.005)

Strategy We will ignore the decimal points and multiply 1.3 and 0.005 as if they were whole numbers. Then we will write a decimal point in that result so that the final answer has four decimal places.

WHY Since the factor 1.3 has 1 decimal place, and the factor 0.005 has 3 decimal places, the product should have $1 + 3 = 4$ decimal places.

Solution Since many students find vertical form multiplication of decimals easier if the decimal with the smaller number of nonzero digits is written on the bottom, we will write 0.005 under 1.3.

$$
\begin{array}{r}
1.3 \leftarrow \text{1 decimal place} \\
\times \quad 0.005 \leftarrow \text{3 decimal places} \\
\hline
0.0065
\end{array}
$$

The answer will have $1 + 3 = 4$ decimal places.

Write 2 placeholder zeros in front of 6. Then move 4 places from the right to the left and insert a decimal point in the answer.

Thus, $1.3(0.005) = 0.0065$.

Self Check 2
Multiply: (0.0002)7.2
Now Try Problem 13

EXAMPLE 3   Multiply: 234(5.1)

Strategy We will ignore the decimal point and multiply 234 and 5.1 as if they were whole numbers. Then we will write a decimal point in that result so that the final answer has one decimal place.

WHY Since the factor 234 has 0 decimal places, and the factor 5.1 has 1 decimal place, the product should have $0 + 1 = 1$ decimal place.

Self Check 3
Multiply: 178(4.7)
Now Try Problem 17

Solution We write the multiplication in vertical form, with 5.1 under 234.

$$234 \leftarrow \text{No decimal places}$$
$$\underline{\times \quad 5.1} \leftarrow \text{1 decimal place}$$

} The answer will have
$0 + 1 = 1$ decimal place.

$$23\ 4$$
$$\underline{1170\ 0}$$
$$1193.4$$

Move 1 place from the right to the left and
insert a decimal point in the answer.

Thus, $234(5.1) = 1{,}193.4$.

Using Your CALCULATOR   Multiplying Decimals

When billing a household, a gas company converts the amount of natural gas used to units of heat energy called *therms*. The number of therms used by a household in one month and the cost per therm are shown below.

Customer charge . . . . 39 therms @ $0.72264

To find the total charges for the month, we multiply the number of therms by the cost per therm: $39 \cdot 0.72264$.

$39 \boxed{\times} .72264 \boxed{=} 28.18296$ $\boxed{\text{28.18296}}$

On some calculator models, the $\boxed{\text{ENTER}}$ key is pressed to display the product. Rounding to the nearest cent, we see that the total charge is $28.18.

THINK IT THROUGH   *Overtime*

"Employees covered by the Fair Labor Standards Act must receive overtime pay for hours worked in excess of 40 in a workweek of at least 1.5 times their regular rates of pay."

United States Department of Labor

The map of the United States shown below is divided into nine regions. The average hourly wage for private industry workers in each region is also listed in the legend below the map. Find the average hourly wage for the region where you live. Then calculate the corresponding average hourly overtime wage for that region.

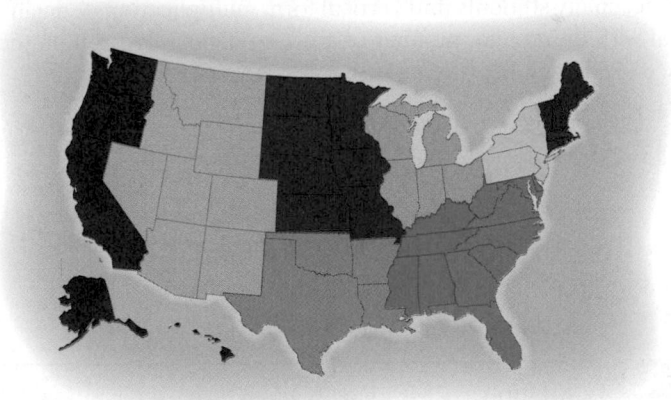

Legend
- West North Central: $17.42
- Mountain: $17.93
- Pacific: $21.68
- East South Central: $16.58
- East North Central: $18.82
- West South Central: $17.17
- New England: $22.38
- Middle Atlantic: $21.31
- South Atlantic: $18.34

(Source: Bureau of Labor Statistics, National Compensation Survey, 2008)

2 Multiply decimals by powers of 10.

The numbers 10, 100, and 1,000 are called **powers of 10,** because they are the results when we evaluate 10^1, 10^2, and 10^3. To develop a rule to find the product when multiplying a decimal by a power of 10, we multiply 8.675 by three different powers of 10.

Multiply: 8.675 · **10**

$$
\begin{array}{r}
8.675 \\
\times \quad 10 \\
\hline
0000 \\
86750 \\
\hline
86.750
\end{array}
$$

Multiply: 8.675 · **100**

$$
\begin{array}{r}
8.675 \\
\times \quad 100 \\
\hline
0000 \\
00000 \\
867500 \\
\hline
867.500
\end{array}
$$

Multiply: 8.675 · **1,000**

$$
\begin{array}{r}
8.675 \\
\times \quad 1000 \\
\hline
0000 \\
00000 \\
000000 \\
8675000 \\
\hline
8675.000
\end{array}
$$

When we inspect the answers, the decimal point in the first factor 8.675 appears to be moved to the right by the multiplication process. The number of decimal places it moves depends on the power of 10 by which 8.675 is multiplied.

One zero in 10

$8.675 \cdot 10 = 86.75$

It moves 1 place
to the right.

Two zeros in 100

$8.675 \cdot 100 = 867.5$

It moves 2 places
to the right.

Three zeros in 1,000

$8.675 \cdot 1,000 = 8675$

It moves 3 places
to the right.

These observations illustrate the following rule.

Multiplying a Decimal by 10, 100, 1,000, and So On

To find the product of a decimal and 10, 100, 1,000, and so on, move the decimal point to the right the same number of places as there are zeros in the power of 10.

EXAMPLE 4   Multiply: **a.** 2.81 · 10   **b.** 0.076(10,000)

Strategy For each multiplication, we will identify the factor that is a power of 10, and count the number of zeros that it has.

WHY To find the product of a decimal and a power of 10 that is greater than 1, we move the decimal point to the right the same number of places as there are zeros in the power of 10.

Solution

a. $2.81 \cdot 10 = 28.1$ Since 10 has one zero, move the decimal point in 2.81 one
place to the right.

b. $0.076(10,000) = 0760.$ Since 10,000 has four zeros, move the decimal point in
0.076 four places to the right. Write a placeholder zero
(shown in blue).

$= 760$

Numbers such as 10, 100, and 1,000 are powers of 10 that are *greater than 1*. There are also powers of 10 that are *less than 1*, such as 0.1, 0.01, and 0.001. To develop a rule to find the product when multiplying a decimal by one tenth, one hundredth, one thousandth, and so on, we will consider three examples:

Multiply: 5.19 · **0.1**

$$
\begin{array}{r}
5.19 \\
\times \quad 0.1 \\
\hline
0.519
\end{array}
$$

Multiply: 5.19 · **0.01**

$$
\begin{array}{r}
5.19 \\
\times \quad 0.01 \\
\hline
0.0519
\end{array}
$$

Multiply: 5.19 · **0.001**

$$
\begin{array}{r}
5.19 \\
\times \quad 0.001 \\
\hline
0.00519
\end{array}
$$

Self Check 4
Multiply:
a. 0.721 · 100
b. 6.08(1,000)

Now Try Problems 21 and 23

When we inspect the answers, the decimal point in the first factor 5.19 appears to be moved to the left by the multiplication process. The number of places that it moves depends on the power of ten by which it is multiplied.

These observations illustrate the following rule.

Multiplying a Decimal by 0.1, 0.01, 0.001, and So On

To find the product of a decimal and 0.1, 0.01, 0.001, and so on, move the decimal point to the left the same number of decimal places as there are in the power of 10.

Self Check 5

Multiply:
a. 0.1(129.9)
b. 0.002 · 0.00001

Now Try Problems 25 and 27

EXAMPLE 5 Multiply: **a.** 145.8 · 0.01 **b.** 9.76(0.0001)

Strategy For each multiplication, we will identify the factor of the form 0.1, 0.01, and 0.001, and count the number of decimal places that it has.

WHY To find the product of a decimal and a power of 10 that is less than 1, we move the decimal point to the left the same number of decimal places as there are in the power of 10.

Solution

a. 145.8 · 0.01 = 1.458   Since 0.01 has *two* decimal places, move the decimal point in 145.8 *two* places to the left.

b. 9.76(0.0001) = 0.000976   Since 0.0001 has *four* decimal places, move the decimal point in 9.76 *four* places to the left. This requires that three placeholder zeros (shown in blue) be inserted in front of the 9.

Quite often, newspapers, websites, and television programs present large numbers in a shorthand notation that involves a decimal in combination with a place-value column name. For example,

- As of December 31, 2008, Sony had sold *21.3 million* Playstation 3 units worldwide. (Source: Sony Computer Entertainment)
- Boston's Big Dig was the most expensive single highway project in U.S. history. It cost about *$14.63 billion.* (Source: Roadtraffic-technology.com)
- The distance that light travels in one year is about *5.878 trillion* miles. (Source: Encyclopaedia Britannica)

We can use the rule for multiplying a decimal by a power of ten to write these large numbers in standard form.

Self Check 6

Write each number in standard notation:
a. 567.1 million
b. 50.82 billion
c. 4.133 trillion

Now Try Problems 29, 31, and 33

EXAMPLE 6 Write each number in standard notation:

a. 21.3 million **b.** 14.63 billion **c.** 5.9 trillion

Strategy We will express each of the large numbers as the product of a decimal and a power of 10.

WHY Then we can use the rule for multiplying a decimal by a power of 10 to find their product. The result will be in the required standard form.

Solution

a. 21.3 million = 21.3 · **1 million**

= 21.3 · **1,000,000**   Write 1 million in standard form.

= 21,300,000   Since 1,000,000 has six zeros, move the decimal point in 21.3 six places to the right.

b. 14.63 billion $= 14.63 \cdot$ **1 billion**

$= 14.63 \cdot 1,000,000,000$ Write 1 billion in standard form.

$= 14,630,000,000$ Since 1,000,000,000 has *nine* zeros, move the decimal point in 14.63 *nine* places to the right.

c. 5.9 trillion $= 5.9 \cdot$ **1 trillion**

$= 5.9 \cdot 1,000,000,000,000$ Write 1 trillion in standard form.

$= 5,900,000,000,000$ Since 1,000,000,000,000 has *twelve* zeros, move the decimal point in 5.9 *twelve* places to the right.

3 Multiply signed decimals.

The rules for multiplying integers also hold for multiplying signed decimals. The product of two decimals with like signs is positive, and the product of two decimals with unlike signs is negative.

EXAMPLE 7   Multiply: **a.** $-1.8(4.5)$ **b.** $(-1,000)(-59.08)$

Strategy In part a, we will use the rule for multiplying signed decimals that have different (unlike) signs. In part b, we will use the rule for multiplying signed decimals that have the same (like) signs.

WHY In part a, one factor is negative and one is positive. In part b, both factors are negative.

Solution

a. Find the absolute values: $|-1.8| = 1.8$ and $|4.5| = 4.5$. Since the decimals have unlike signs, their product is negative.

$-1.8(4.5) = -8.1$ Multiply the absolute values, 1.8 and 4.5, to get 8.1. Then make the final answer negative.

$$\begin{array}{r} 1.8 \\ \times\ 4.5 \\ \hline 90 \\ 720 \\ \hline 8.10 \end{array}$$

b. Find the absolute values: $|-1,000| = 1,000$ and $|-59.08| = 59.08$. Since the decimals have like signs, their product is positive.

$(-1,000)(-59.08) = 1,000(59.08)$

$= 59,080$ Multiply the absolute values, 1,000 and 59.08. Since 1,000 has 3 zeros, move the decimal point in 59.08 3 places to the right. Write a placeholder zero. The answer is positive.

4 Evaluate exponential expressions that have decimal bases.

We have evaluated exponential expressions that have whole number bases, integer bases, and fractional bases. The base of an exponential expression can also be a positive or a negative decimal.

EXAMPLE 8   Evaluate: **a.** $(2.4)^2$ **b.** $(-0.05)^2$

Strategy We will write each exponential expression as a product of repeated factors, and then perform the multiplication. This requires that we identify the base and the exponent.

WHY The exponent tells the number of times the base is to be written as a factor.

Self Check 7

Multiply:
a. $6.6(-5.5)$
b. $-44.968(-100)$

Now Try Problems 37 and 41

Self Check 8

Evaluate:
a. $(-1.3)^2$
b. $(0.09)^2$

Now Try Problems 45 and 47

Solution

a. $(2.4)^2 = 2.4 \cdot 2.4$ The base is 2.4 and the exponent is 2. Write the base as a factor 2 times.

$ = 5.76$ Multiply the decimals.

$$
\begin{array}{r}
2.4 \\
\times\ 2.4 \\
\hline
96 \\
480 \\
\hline
5.76
\end{array}
$$

b. $(-0.05)^2 = (-0.05)(-0.05)$ The base is −0.05 and the exponent is 2. Write the base as a factor 2 times.

$ = 0.0025$ Multiply the decimals. The product of two decimals with like signs is positive.

$$
\begin{array}{r}
0.05 \\
\times\ 0.05 \\
\hline
0.0025
\end{array}
$$

5   Use the order of operations rule.

Recall that the order of operations rule is used to evaluate expressions that involve more than one operation.

Self Check 9
Evaluate:
$-2|-4.4 + 5.6| + (-0.8)^2$
Now Try Problem 49

EXAMPLE 9 Evaluate: $-(0.6)^2 + 5|-3.6 + 1.9|$

Strategy The absolute value bars are grouping symbols. We will perform the addition within them first.

WHY By the order of operations rule, we must perform all calculations within parentheses and other grouping symbols (such as absolute value bars) first.

Solution
$$-(0.6)^2 + 5|-3.6 + 1.9|$$

$$= -(0.6)^2 + 5|-1.7| \quad \begin{array}{l}\text{Do the addition within the absolute value}\\ \text{symbols. Use the rule for adding two}\\ \text{decimals with different signs.}\end{array}$$

$$= -(0.6)^2 + 5(1.7) \quad \text{Simplify: } |-1.7| = 1.7.$$

$$= -0.36 + 5(1.7) \quad \text{Evaluate: } (0.6)^2 = 0.36.$$

$$= -0.36 + 8.5 \quad \text{Do the multiplication: } 5(1.7) = 8.5.$$

$$= 8.14 \quad \begin{array}{l}\text{Use the rule for adding two decimals}\\ \text{with different signs.}\end{array}$$

$$
\begin{array}{r}
\overset{2\ 16}{\cancel{3.6}} \\
-\ 1.9 \\
\hline
1.7
\end{array}
$$

$$
\begin{array}{r}
\overset{3}{1.7} \\
\times\ 5 \\
\hline
8.5
\end{array}
$$

$$
\begin{array}{r}
\overset{4\,10}{8.\cancel{50}} \\
-0.36 \\
\hline
8.14
\end{array}
$$

6   Evaluate formulas.

Recall that to evaluate a formula, we replace the letters (called **variables**) with specific numbers and then use the order of operations rule.

Self Check 10
Evaluate $V = 1.3\pi r^3$ for
$\pi = 3.14$ and $r = 3$.
Now Try Problem 53

EXAMPLE 10 Evaluate the formula $S = 6.28r(h + r)$ for $h = 3.1$ and $r = 6$.

Strategy In the given formula, we will replace the letter r with 6 and h with 3.1.

WHY Then we can use the order of operations rule to find the value of the expression on the right side of the = symbol.

Solution
$$S = 6.28r(h + r) \quad 6.28r(h + r) \text{ means } 6.28 \cdot r \cdot (h + r).$$

$$= 6.28(6)(3.1 + 6) \quad \text{Replace } r \text{ with 6 and } h \text{ with 3.1.}$$

$$= 6.28(6)(9.1) \quad \text{Do the addition within the parentheses.}$$

$$= 37.68(9.1) \quad \text{Do the multiplication: } 6.28(6) = 37.68.$$

$$= 342.888 \quad \text{Do the multiplication.}$$

$$
\begin{array}{r}
37.68 \\
\times\ 9.1 \\
\hline
3768 \\
339120 \\
\hline
342.888
\end{array}
$$

7 Estimate products of decimals.

Estimation can be used to check the reasonableness of an answer to a decimal multiplication. There are several ways to estimate, but the objective is the same: Simplify the numbers in the problem so that the calculations can be made easily and quickly.

EXAMPLE 11

a. Estimate using front-end rounding: $27 \cdot 6.41$

b. Estimate by rounding each factor to the nearest tenth: $13.91 \cdot 5.27$

c. Estimate by rounding: $0.1245(101.4)$

Strategy We will use rounding to approximate the factors. Then we will find the product of the approximations.

WHY Rounding produces factors that contain fewer digits. Such numbers are easier to multiply.

Solution

a. To estimate $27 \cdot 6.41$ by front-end rounding, we begin by rounding both factors to their *largest* place value.

$$
\begin{array}{rcl}
27 & \longrightarrow & 30 \quad \text{Round to the nearest ten.} \\
\times\ 6.41 & \longrightarrow & \times\ \ 6 \quad \text{Round to the nearest one.} \\
\hline
& & 180
\end{array}
$$

The estimate is 180. If we calculate $27 \cdot 6.41$, the product is exactly 173.07. The estimate is close: It's about 7 more than 173.07.

b. To estimate $13.91 \cdot 5.27$, we will round both decimals to the nearest tenth.

$$
\begin{array}{rcl}
13.91 & \longrightarrow & 13.9 \quad \text{Round to the nearest tenth.} \\
\times\ \ 5.27 & \longrightarrow & \times\ \ 5.3 \quad \text{Round to the nearest tenth.} \\
\hline
& & 417 \\
& & 6950 \\
\hline
& & 73.67
\end{array}
$$

The estimate is 73.67. If we calculate $13.91 \cdot 5.27$, the product is exactly 73.3057. The estimate is close: It's just slightly more than 73.3057.

c. Since 101.4 is approximately 100, we can estimate $0.1245(\mathbf{101.4})$ using $0.1245(\mathbf{100})$.

$$0.1245(100) = 12.45 \quad \text{Since 100 has two zeros, move the decimal point in } 0.1245 \text{ two places to the right.}$$

The estimate is 12.45. If we calculate $0.1245(101.4)$, the product is exactly 12.6243. Note that the estimate is close: It's slightly less than 12.6243.

Self Check 11

a. Estimate using front-end rounding: $4.337 \cdot 65$

b. Estimate by rounding the factors to the nearest tenth: $3.092 \cdot 11.642$

c. Estimate by rounding: $0.7899(985.34)$

Now Try **Problems 61 and 63**

8 Solve application problems by multiplying decimals.

Application problems that involve repeated addition are often more easily solved using multiplication.

EXAMPLE 12 *Coins* Banks wrap pennies in rolls of 50 coins. If a penny is 1.55 millimeters thick, how tall is a stack of 50 pennies?

Analyze

- There are 50 pennies in a stack. *Given*
- A penny is 1.55 millimeters thick. *Given*
- How tall is a stack of 50 pennies? *Find*

Cookey/Dreamstime.com

Self Check 12

COINS Banks wrap nickels in rolls of 40 coins. If a nickel is 1.95 millimeters thick, how tall is a stack of 40 nickels?

Now Try **Problem 97**

Form The height (in millimeters) of a stack of 50 pennies, each of which is 1.55 thick, is the sum of fifty 1.55's. This repeated addition can be calculated more simply by multiplication.

The height of a stack of pennies	is equal to	the thickness of one penny	times	the number of pennies in the stack.
The height of stack of pennies	=	1.55	·	50

Solve Use vertical form to perform the multiplication:

$$\begin{array}{r} 1.55 \\ \times \quad 50 \\ \hline 000 \\ 7750 \\ \hline 77.50 \end{array}$$

State A stack of 50 pennies is 77.5 millimeters tall.

Check We can estimate to check the result. If we use 2 millimeters to approximate the thickness of one penny, then the height of a stack of 50 pennies is about 2 · 50 millimeters = 100 millimeters. The result, 77.5 mm, seems reasonable. ∎

Sometimes more than one operation is needed to solve a problem involving decimals.

Self Check 13

WEEKLY EARNINGS A pharmacy assistant's basic workweek is 40 hours. After her daily shift is over, she can work overtime at a rate of 1.5 times her regular rate of $15.90 per hour. How much money will she earn in a week if she works 4 hours of overtime?

Now Try Problem 113

EXAMPLE 13 *Weekly Earnings* A cashier's basic workweek is 40 hours. After his daily shift is over, he can work overtime at a rate 1.5 times his regular rate of $13.10 per hour. How much money will he earn in a week if he works 6 hours of overtime?

Analyze

- A cashier's basic workweek is 40 hours.   Given
- His overtime pay rate is 1.5 times his regular rate of $13.10 per hour. Given
- How much money will he earn in a week if he works his regular shift and 6 hours overtime?   Find

Form To find the cashier's overtime pay rate, we multiply 1.5 times his regular pay rate, $13.10.

$$\begin{array}{r} 13.10 \\ \times \quad 1.5 \\ \hline 6550 \\ 13100 \\ \hline 19.650 \end{array}$$

The cashier's overtime pay rate is $19.65 per hour.
We now translate the words of the problem to numbers and symbols.

The total amount the cashier earns in a week	is equal to	40 hours	times	his regular pay rate	plus	the number of overtime hours	times	his overtime rate.
The total amount the cashier earns in a week	=	40	·	$13.10	+	6	·	$19.65

Solve We will use the rule for the order of operations to evaluate the expression:

$40 \cdot 13.10 + 6 \cdot 19.65 = 524.00 + 117.90$ Do the multiplication first.

$= 641.90$ Do the addition.

$$
\begin{array}{r}
13.10 \\
\times \quad 40 \\
\hline
0000 \\
5240 \\
\hline
524.00
\end{array}
\qquad
\begin{array}{r}
{\scriptstyle 5\,3\,3} \\
19.65 \\
\times \quad 6 \\
\hline
117.90
\end{array}
$$

$$
\begin{array}{r}
{\scriptstyle 1} \\
524.00 \\
+117.90 \\
\hline
641.90
\end{array}
$$

State The cashier will earn a total of $641.90 for the week.

Check We can use estimation to check. The cashier works 40 hours per week for approximately $13 per hour to earn about $40 \cdot \$13 = \$520$. His 6 hours of overtime at approximately $20 per hour earns him about $6 \cdot \$20 = \$120$. His total earnings that week are about $\$520 + \$120 = \$640$. The result, $641.90, seems reasonable. ■

ANSWERS TO SELF CHECKS

1. 11.61 **2.** 0.00144 **3.** 836.6 **4. a.** 72.1 **b.** 6,080 **5. a.** 12.99 **b.** 0.00000002
6. a. 567,100,000 **b.** 50,820,000,000 **c.** 4,133,000,000,000 **7. a.** −36.3 **b.** 4,496.8
8. a. 1.69 **b.** 0.0081 **9.** −1.76 **10.** 110.214 **11. a.** 280 **b.** 35.96 **c.** 789.9
12. 78 mm **13.** $731.40

SECTION 5.3   STUDY SET

VOCABULARY

Fill in the blanks.

1. In the multiplication problem shown below, label each *factor*, the *partial products*, and the *product*.

$$
\begin{array}{r}
3.4 \leftarrow \\
\times\, 2.6 \leftarrow \\
\hline
204 \leftarrow \\
680 \leftarrow \\
\hline
8.84 \leftarrow
\end{array}
$$

2. Numbers such as 10, 100, and 1,000 are called _____ of 10.

CONCEPTS

Fill in the blanks.

3. Insert a decimal point in the correct place for each product shown below. Write placeholder zeros, if necessary.

  a.
$$
\begin{array}{r}
3.8 \\
\times\, 0.6 \\
\hline
228
\end{array}
$$

  b.
$$
\begin{array}{r}
1.79 \\
\times\quad 8.1 \\
\hline
179 \\
14320 \\
\hline
14499
\end{array}
$$

  c.
$$
\begin{array}{r}
2.0 \\
\times\, 7 \\
\hline
140
\end{array}
$$

  d.
$$
\begin{array}{r}
0.013 \\
\times\, 0.02 \\
\hline
0026
\end{array}
$$

4. Fill in the blanks.

  a. To find the product of a decimal and 10, 100, 1,000, and so on, move the decimal point to the _____ the same number of places as there are zeros in the power of 10.

  b. To find the product of a decimal and 0.1, 0.01, 0.001, and so on, move the decimal point to the _____ the same number of places as there are in the power of 10.

5. Determine whether the *sign* of each result is positive or negative. *You do not have to find the product.*

  a. $-7.6(-1.8)$

  b. $-4.09 \cdot 2.274$

6. a. When we move its decimal point to the right, does a decimal number get larger or smaller?

  b. When we move its decimal point to the left, does a decimal number get larger or smaller?

NOTATION

7. a. List the first five powers of 10 that are greater than 1.

  b. List the first five powers of 10 that are less than 1.

8. Write each number in standard notation.

   a. one million

   b. one billion

   c. one trillion

GUIDED PRACTICE

Multiply. See Example 1.

9. $4.8 \cdot 6.2$ **10.** $3.5 \cdot 9.3$

11. $5.6(8.9)$ **12.** $7.2(8.4)$

Multiply. See Example 2.

13. $0.003(2.7)$ **14.** $0.002(2.6)$

15. 5.8 **16.** 8.7
       $\times\ 0.009$ $\times\ 0.004$

Multiply. See Example 3.

17. $179(6.3)$ **18.** $225(4.9)$

19. 316 **20.** 527
      $\times\ 7.4$ $\times\ 3.7$

Multiply. See Example 4.

21. $6.84 \cdot 100$ **22.** $2.09 \cdot 100$

23. $0.041(10,000)$ **24.** $0.034(10,000)$

Multiply. See Example 5.

25. $647.59 \cdot 0.01$ **26.** $317.09 \cdot 0.01$

27. $1.15(0.001)$ **28.** $2.83(0.001)$

Write each number in standard notation. See Example 6.

29. 14.2 million   **30.** 33.9 million

31. 98.2 billion   **32.** 80.4 billion

33. 1.421 trillion   **34.** 3.056 trillion

35. 657.1 billion   **36.** 422.7 billion

Multiply. See Example 7.

37. $-1.9(7.2)$ **38.** $-5.8(3.9)$

39. $-3.3(-1.6)$ **40.** $-4.7(-2.2)$

41. $(-10,000)(-44.83)$ **42.** $(-10,000)(-13.19)$

43. $678.231(-1,000)$ **44.** $491.565(-1,000)$

Evaluate each expression. See Example 8.

45. $(3.4)^2$ **46.** $(5.1)^2$

47. $(-0.03)^2$ **48.** $(-0.06)^2$

Evaluate each expression. See Example 9.

49. $-(0.2)^2 + 4|-2.3 + 1.5|$

50. $-(0.3)^2 + 6|-6.4 + 1.7|$

51. $-(0.8)^2 + 7|-5.1 - 4.8|$

52. $-(0.4)^2 + 6|-6.2 - 3.5|$

Evaluate each formula. See Example 10.

53. $A = P + Prt$ for $P = 85.50, r = 0.08$, and $t = 5$

54. $A = P + Prt$ for $P = 99.95, r = 0.05$, and $t = 10$

55. $A = lw$ for $l = 5.3$ and $w = 7.2$

56. $A = 0.5bh$ for $b = 7.5$ and $h = 6.8$

57. $P = 2l + 2w$ for $l = 3.7$ and $w = 3.6$

58. $P = a + b + c$ for $a = 12.91, b = 19$, and $c = 23.6$

59. $C = 2\pi r$ for $\pi = 3.14$ and $r = 2.5$

60. $A = \pi r^2$ for $\pi = 3.14$ and $r = 4.2$

Estimate each product using front-end rounding. See Example 11.

61. $46 \cdot 5.3$ **62.** $37 \cdot 4.29$

Estimate each product by rounding the factors to the nearest tenth. See Example 11.

63. $17.11 \cdot 3.85$ **64.** $18.33 \cdot 6.46$

TRY IT YOURSELF

Perform the indicated operations.

65. $-0.56 \cdot 0.33$ **66.** $-0.64 \cdot 0.79$

67. $(-1.3)^2$ **68.** $(-2.5)^2$

69. $(-0.7 - 0.5)(2.4 - 3.1)$

70. $(-8.1 - 7.8)(0.3 + 0.7)$

71. 0.008 **72.** 0.003
     $\times\ 0.09$ $\times\ 0.09$

73. $-0.2 \cdot 1,000,000$ **74.** $-1,000,000 \cdot 1.9$

75. $(-5.6)(-2.2)$ **76.** $(-7.1)(-4.1)$

77. $-4.6(23.4 - 19.6)$ **78.** $6.9(9.8 - 8.9)$

79. $(-4.9)(-0.001)$ **80.** $(-0.001)(-7.09)$

81. $(-0.2)^2 + 2(7.1)$ **82.** $(-6.3)(3) - (1.2)^2$

83. 2.13 **84.** 3.06
     $\times\ 4.05$ $\times\ 1.82$

85. $-7(8.1781)$

86. $-5(4.7199)$

87. $-1,000(0.02239)$

88. $-100(0.0897)$

89. $(0.5 + 0.6)^2(-3.2)$

90. $(-5.1)(4.9 - 3.4)^2$

91. $-0.2(306)(-0.4)$

92. $-0.3(417)(-0.5)$

93. $-0.01(|-2.6 - 6.7|)^2$

94. $-0.01(|-8.16 + 9.9|)^2$

Complete each table.

95.

Decimal	Its square
0.1	
0.2	
0.3	
0.4	
0.5	
0.6	
0.7	
0.8	
0.9	

96.

Decimal	Its cube
0.1	
0.2	
0.3	
0.4	
0.5	
0.6	
0.7	
0.8	
0.9	

APPLICATIONS

97. REAMS OF PAPER Find the thickness of a 500-sheet ream of copier paper if each sheet is 0.0038 inch thick.

98. MILEAGE CLAIMS Each month, a salesman is reimbursed by his company for any work-related travel that he does in his own car at the rate of $0.445 per mile. How much will the salesman receive if he traveled a total of 120 miles in his car on business in the month of June?

99. SALARIES Use the following formula to determine the annual salary of a recording engineer who works 38 hours per week at a rate of $37.35 per hour. Round the result to the nearest hundred dollars.

$$\text{Annual salary} = \text{hourly rate} \cdot \text{hours per week} \cdot 52.2\text{ weeks}$$

100. PAYCHECKS If you are paid every other week, your monthly gross income is your gross income from one paycheck times 2.17. Find the monthly gross income of a supermarket clerk who earns $1,095.70 every two weeks. Round the result to the nearest cent.

101. BAKERY SUPPLIES A bakery buys various types of nuts as ingredients for cookies. Complete the table by filling in the cost of each purchase.

Type of nut	Price per pound	Pounds	Cost
Almonds	$5.95	16	
Walnuts	$4.95	25	

102. NEW HOMES Find the cost to build the home shown below if construction costs are $92.55 per square foot.

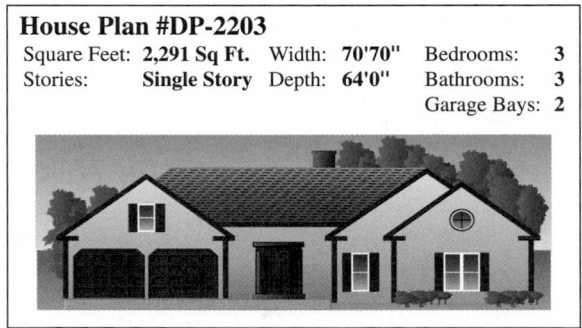

House Plan #DP-2203
Square Feet: **2,291 Sq Ft.** Width: **70'70"** Bedrooms: **3**
Stories: **Single Story** Depth: **64'0"** Bathrooms: **3**
Garage Bays: **2**

103. BIOLOGY Cells contain DNA. In humans, it determines such traits as eye color, hair color, and height. A model of DNA appears below. If 1 Å (angstrom) = 0.000000004 inch, find the dimensions of 34 Å, 3.4 Å, and 10 Å, shown in the illustration.

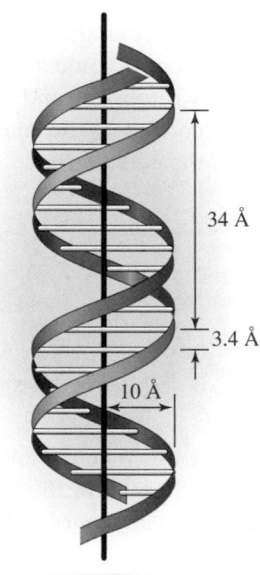

104. TACHOMETERS

a. Estimate the decimal number to which the tachometer needle points in the illustration below.

b. What engine speed (in rpm) does the tachometer indicate?

105. CITY PLANNING The streets shown in blue on the city map below are 0.35 mile apart. Find the distance of each trip between the two given locations.

   a. The airport to the Convention Center

   b. City Hall to the Convention Center

   c. The airport to City Hall

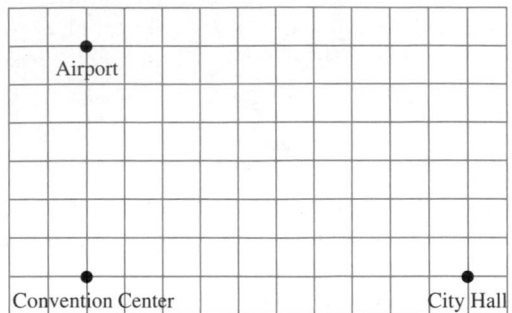

106. RETROFITS The illustration below shows the current widths of the three columns of a freeway overpass. A computer analysis indicated that the width of each column should actually be 1.4 times what it currently is to withstand the stresses of an earthquake. According to the analysis, how wide should each of the columns be?

4.5 ft   3.5 ft   2.5 ft

107. ELECTRIC BILLS When billing a household, a utility company charges for the number of kilowatt-hours used. A kilowatt-hour (kwh) is a standard measure of electricity. If the cost of 1 kwh is $0.14277, what is the electric bill for a household that uses 719 kwh in a month? Round the answer to the nearest cent.

108. UTILITY TAXES Some gas companies are required to tax the number of therms used each month by the customer. What are the taxes collected on a monthly usage of 31 therms if the tax rate is $0.00566 per therm? Round the answer to the nearest cent.

109. Write each highlighted number in standard form.

   a. CONSERVATION The *19.6-million acre* Arctic National Wildlife Refuge is located in the northeast corner of Alaska. (Source: National Wildlife Federation)

   b. POPULATION According to projections by the International Programs Center at the U.S. Census Bureau, at 7:16 P.M. eastern time on Saturday, February 25, 2006, the population of the Earth hit *6.5 billion* people.

   c. DRIVING The U.S. Department of Transportation estimated that Americans drove a total of *3.026 trillion miles* in 2008. (Source: Federal Highway Administration)

110. Write each highlighted number in standard form.

   a. MILEAGE Irv Gordon, of Long Island, New York, has driven a record *2.6 million miles* in his 1966 Volvo P-1800. (Source: autoblog.com)

   b. E-COMMERCE Online spending during the 2008 holiday season (November 1 through December 23) was about *$25.5 billion.* (Source: pcmag.com)

   c. FEDERAL DEBT On March 27, 2009, the U.S. national debt was *$11.073 trillion.* (Source: National Debt Clock)

111. SOCCER A soccer goal is rectangular and measures 24 feet wide by 8 feet high. Major league soccer officials are proposing to increase its width by 1.5 feet and increase its height by 0.75 foot.

   a. What is the area of the goal opening now?

   b. What would the area be if the proposal is adopted?

   c. How much area would be added?

112. SALT INTAKE Studies done by the Centers for Disease Control and Prevention found that the average American eats 3.436 grams of salt each day. The recommended amount is 1.5 grams per day. How many more grams of salt does the average American eat in one week compared with what the Center recommends?

113. CONCERT SEATING Two types of tickets were sold for a concert. Floor seating costs $12.50 a ticket, and balcony seats cost $15.75.

   a. Complete the following table and find the receipts from each type of ticket.

   b. Find the total receipts from the sale of both types of tickets.

Ticket type	Price	Number sold	Receipts
Floor		1,000	
Balcony		100	

114. PLUMBING BILLS A corner of the invoice for plumbing work is torn. What is the labor charge for the 4 hours of work? What is the total charge (standard service charge, parts, labor)?

Carter Plumbing 100 W. Dalton Ave.		Invoice #210
Standard service charge	$	25.75
Parts	$	38.75
Labor: 4 hr @ $40.55/hr	$	
Total charges	$	

115. WEIGHTLIFTING The barbell is evenly loaded with iron plates. How much plate weight is loaded on the barbell?

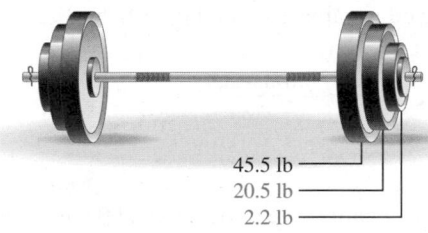

45.5 lb
20.5 lb
2.2 lb

116. SWIMMING POOLS Long bricks, called *coping,* can be used to outline the edge of a swimming pool. How many meters of coping will be needed in the construction of the swimming pool shown?

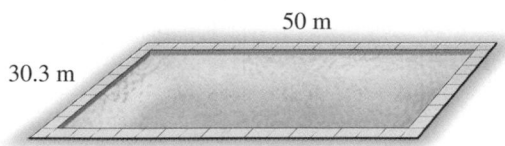

50 m

30.3 m

117. STORM DAMAGE After a rainstorm, the saturated ground under a hilltop house began to give way. A survey team noted that the house dropped 0.57 inch initially. In the next three weeks, the house fell 0.09 inch per week. How far did the house fall during this three-week period?

118. WATER USAGE In May, the water level of a reservoir reached its high mark for the year. During the summer months, as water usage increased, the level dropped. In the months of May and June, it fell 4.3 feet each month. In August, and September, because of high temperatures, it fell another 8.7 feet each month. By the beginning of October, how far below the year's high mark had the water level fallen?

WRITING

119. Explain how to determine where to place the decimal point in the answer when multiplying two decimals.

120. List the similarities and differences between whole-number multiplication and decimal multiplication.

121. Explain how to multiply a decimal by a power of 10 that is greater than 1, and by a power of ten that is less than 1.

122. Is it easier to multiply the decimals 0.4 and 0.16 or the fractions $\frac{4}{10}$ and $\frac{16}{100}$? Explain why.

123. Why do we have to line up the decimal points when adding, but we do not have to when multiplying?

124. Which vertical form for the following multiplication do you like better? Explain why.

$$
\begin{array}{r} 0.000003 \\ \times \quad 2.7 \end{array} \qquad \begin{array}{r} 2.8 \\ \times\ 0.000003 \end{array}
$$

REVIEW

Find the prime factorization of each number. Use exponents in your answer, when helpful.

125. 220   **126.** 400

127. 162   **128.** 735

Objectives

1. Divide a decimal by a whole number.

2. Divide a decimal by a decimal.

3. Round a decimal quotient.

4. Estimate quotients of decimals.

5. Divide decimals by powers of 10.

6. Divide signed decimals.

7. Use the order of operations rule.

8. Evaluate formulas.

9. Solve application problems by dividing decimals.

SECTION 5.4

Dividing Decimals

In Chapter 1, we used a process called long division to divide whole numbers.

Long division form

$$\text{Divisor} \rightarrow 5\overline{)10} \begin{matrix} 2 \leftarrow \text{Quotient} \\ \ \ \leftarrow \text{Dividend} \end{matrix}$$
$$\underline{10}$$
$$0 \leftarrow \text{Remainder}$$

In this section, we consider division problems in which the divisor, the dividend, or both are decimals.

1 Divide a decimal by a whole number.

To develop a rule for decimal division, let's consider the problem $47 \div 10$. If we rewrite the division as $\frac{47}{10}$, we can use the long division method from Chapter 4 for changing an improper fraction to a mixed number to find the answer:

$$10\overline{)47} \quad \begin{matrix} 4\frac{7}{10} \end{matrix}$$
$$\underline{-40}$$
$$7$$

Here the result is written in $\text{quotient} + \dfrac{\text{remainder}}{\text{divisor}}$ form.

To perform this same division using decimals, we write 47 as 47.0 and divide as we would divide whole numbers.

$$10\overline{)47.0} \quad \begin{matrix} 4.7 \end{matrix}$$

Note that the decimal point in the quotient (answer) is placed directly above the decimal point in the dividend.

$$\underline{-40} \downarrow$$
$$70$$

After subtracting 40 from 47, bring down the 0 and continue to divide.

$$\underline{-70}$$
$$0$$

The remainder is 0.

Since $4\frac{7}{10} = 4.7$, either method gives the same answer. This result suggests the following method for dividing a decimal by a whole number.

Dividing a Decimal by a Whole Number

To divide a decimal by a whole number:

1. Write the problem in long division form and place a decimal point in the quotient (answer) directly above the decimal point in the dividend.

2. Divide as if working with whole numbers.

3. If necessary, additional zeros can be written to the right of the last digit of the dividend to continue the division.

Self Check 1

Divide: $20.8 \div 4$. Check the result.

Now Try Problem 15

EXAMPLE 1   Divide: $42.6 \div 6$. Check the result.

Strategy Since the divisor, 6, is a whole number, we will write the problem in long division form and place a decimal point directly above the decimal point in 42.6. Then we will divide as if the problem was $426 \div 6$.

WHY To divide a decimal by a whole number, we divide as if working with whole numbers.

Solution
Step 1

Place a decimal point in the quotient that lines up with the decimal point in the dividend.

$$6 \overline{)42\,.\,6}$$

Step 2 Now divide using the four-step division process: **estimate, multiply, subtract,** and **bring down.**

$$
\begin{array}{r}
7.1 \\
6 \overline{)42.6} \\
-42 \downarrow \\
\hline
06 \\
-\ 6 \\
\hline
0
\end{array}
$$

Ignore the decimal points and divide as if working with whole numbers.

After subtracting 42 from 42, bring down the 6 and continue to divide.

The remainder is 0.

In Section 1.5, we checked whole-number division using multiplication. Decimal division is checked in the same way: *The product of the quotient and the divisor should be the dividend.*

$$
\begin{array}{r}
7.1 \leftarrow \text{Quotient} \\
\times \quad 6 \leftarrow \text{Divisor} \\
\hline
42.6 \leftarrow \text{Dividend}
\end{array}
\qquad
\begin{array}{r}
7.1 \\
6 \overline{)42.6}
\end{array}
$$

The check confirms that $42.6 \div 6 = 7.1$.

EXAMPLE 2   Divide: $71.68 \div 28$

Strategy Since the divisor is a whole number, 28, we will write the problem in long division form and place a decimal point directly above the decimal point in 71.68. Then we will divide as if the problem was $7{,}168 \div 28$.

WHY To divide a decimal by a whole number, we divide as if working with whole numbers.

Solution

Write the decimal point in the quotient (answer) directly above the decimal point in the dividend.

$$
\begin{array}{r}
2.56 \\
28 \overline{)71.68} \\
-56 \downarrow \\
\hline
15\,6 \\
-14\,0 \downarrow \\
\hline
1\,68 \\
-1\,68 \\
\hline
0
\end{array}
$$

Ignore the decimal points and divide as if working with whole numbers.

After subtracting 56 from 71, bring down the 6 and continue to divide.

After subtracting 140 from 156, bring down the 8 and continue to divide.

The remainder is 0.

We can use multiplication to check this result.

$$
\begin{array}{r}
2.56 \\
\times \quad 28 \\
\hline
2048 \\
5120 \\
\hline
71.68
\end{array}
\qquad
\begin{array}{r}
2.56 \\
28 \overline{)71.68}
\end{array}
$$

The check confirms that $71.68 \div 28 = 2.56$.

Self Check 2
Divide: $101.44 \div 32$
Now Try Problem 19

Self Check 3
Divide: 42.8 ÷ 8
Now Try Problem 23

EXAMPLE 3   Divide: 19.2 ÷ 5

Strategy We will write the problem in long division form, place a decimal point directly above the decimal point in 19.2, and divide. If necessary, we will write additional zeros to the right of the 2 in 19.2.

WHY Writing additional zeros to the right of the 2 allows us to continue the division process until we obtain a remainder of 0 or the digits in the quotient repeat in a pattern.

Solution

$$
\begin{array}{r}
3.8 \\
5\overline{)19.2} \\
-15\downarrow \\
\hline
42 \\
-40 \\
\hline
2
\end{array}
$$

After subtracting 15 from 19, bring down the 2 and continue to divide.

All the digits in the dividend have been used, but the remainder is not 0.

We can write a zero to the right of 2 in the dividend and continue the division process. Recall that writing additional zeros to the right of the decimal point does not change the value of the decimal. That is, 19.2 = 19.20.

$$
\begin{array}{r}
3.84 \\
5\overline{)19.20} \\
-15 \\
\hline
42 \\
-40\downarrow \\
\hline
20 \\
-20 \\
\hline
0
\end{array}
$$

Write a zero to the right of the 2 and bring it down.

Continue to divide.

The remainder is 0.

Check:

$$
\begin{array}{r}
3.84 \\
\times\quad 5 \\
\hline
19.20
\end{array}
$$
← Since this is the dividend, the result checks.

2 Divide a decimal by a decimal.

To develop a rule for division involving a decimal divisor, let's consider the problem $0.36\overline{)0.2592}$, where the divisor is the decimal 0.36. First, we express the division in fraction form.

$0.36\overline{)0.2592}$ can be represented by $\dfrac{0.2592}{0.36}$

Divisor

To be able to use the rule for dividing decimals by a *whole number* discussed earlier, we need to move the decimal point in the divisor 0.36 two places to the right. This can be accomplished by multiplying it by 100. However, if the denominator of the fraction is multiplied by 100, the numerator must also be multiplied by 100 so that the fraction maintains the same value. It follows that $\frac{100}{100}$ is the form of 1 that we should use to build $\frac{0.2592}{0.36}$.

$$\frac{0.2592}{0.36} = \frac{0.2592}{0.36} \cdot \frac{100}{100}$$ Multiply by a form of 1.

$$= \frac{0.2592 \cdot 100}{0.36 \cdot 100}$$ Multiply the numerators.
Multiply the denominators.

$$= \frac{25.92}{36}$$ Multiplying both decimals by 100 moves their decimal points two places to the right.

This fraction represents the division problem $36\overline{)25.92}$. From this result, we have the following observations.

- The division problem $0.36\overline{)0.2592}$ is equivalent to $36\overline{)25.92}$; that is, they have the same answer.
- The decimal points in *both* the divisor and the dividend of the first division problem have been moved two decimal places to the right to create the second division problem.

$$0.36\overline{)0.2592} \quad \text{becomes} \quad 36\overline{)25.92}$$

These observations illustrate the following rule for division with a decimal divisor.

Division with a Decimal Divisor

To divide with a decimal divisor:

1. Write the problem in long division form.
2. Move the decimal point of the divisor so that it becomes a whole number.
3. Move the decimal point of the dividend the same number of places to the right.
4. Write the decimal point in the quotient (answer) directly above the decimal point in the dividend. Divide as if working with whole numbers.
5. If necessary, additional zeros can be written to the right of the last digit of the dividend to continue the division.

EXAMPLE 4   Divide: $\dfrac{0.2592}{0.36}$

Strategy We will move the decimal point of the divisor, 0.36, two places to the right and we will move the decimal point of the dividend, 0.2592, the same number of places to the right.

WHY We can then use the rule for dividing a decimal by a *whole number*.

Solution We begin by writing the problem in long division form.

$$0\,36\overline{)0\,25\,.\,92}$$

Move the decimal point two places to the right in the divisor and the dividend. Write the decimal point in the quotient (answer) directly above the decimal point in the dividend.

Since the divisor is now a whole number, we can use the rule for dividing a decimal by a whole number to find the quotient.

```
       0.72
36)25.92     Now divide as with whole numbers.
  -25 2↓
      72
    - 72
       0
```

Check:
```
     0.72
  ×    36
     432
   2160
   25.92   Since this is the dividend, the result checks.
```

Self Check 4

Divide: $\dfrac{0.6045}{0.65}$

Now Try Problem 27

> ***Success Tip*** When dividing decimals, moving the decimal points the same number of places to the right in *both* the divisor and the dividend does not change the answer.

3 Round a decimal quotient.

In Example 4, the division process stopped after we obtained a 0 from the second subtraction. Sometimes when we divide, the subtractions never give a zero remainder, and the division process continues forever. In such cases, we can round the result.

Self Check 5

Divide: 12.82 ÷ 0.9. Round the quotient to the nearest hundredth.

Now Try Problem 33

EXAMPLE 5 Divide: $\dfrac{9.35}{0.7}$. Round the quotient to the nearest hundredth.

Strategy We will use the methods of this section to divide to the thousandths column.

WHY To round to the hundredths column, we need to continue the division process for one more decimal place, which is the thousandths column.

Solution We begin by writing the problem in long division form.

$$0\,7\overline{)93\,.\,5}$$

To write the divisor as a whole number, move the decimal point one place to the right. Do the same for the dividend. Place the decimal point in the quotient (answer) directly above the decimal point in the dividend.

We need to write two zeros to the right of the last digit of the dividend so that we can divide to the thousandths column.

$$7\overline{)93.5\mathbf{00}}$$

After dividing to the thousandths column, we round to the hundredths column.

The rounding digit in the hundredths column is 5.
The test digit in the thousandths column is 7.

```
      13.357
  7)93.500
   - 7↓
     23
   - 21↓
      2 5
    - 2 1↓
       40
     - 35↓
        50
      - 49
         1
```
The division process can stop. We have divided to the thousandths column.

Since the test digit 7 is 5 or greater, we will round 13.357 up to approximate the quotient to the nearest hundredth.

$$\frac{9.35}{0.7} \approx 13.36 \quad \text{Read} \approx \text{as "is approximately equal to."}$$

Check:

```
      13.36  ← The approximation of the quotient
  ×    0.7   ← The original divisor
      9.352  ← Since this is close to the original dividend, 9.35, the result seems reasonable.
```

> **Success Tip**   To round a quotient to a certain decimal place value, continue the division process one more column to its right to find the *test digit*.

Using Your CALCULATOR   Dividing Decimals

The nucleus of a cell contains vital information about the cell in the form of DNA. The nucleus is very small: A typical animal cell has a nucleus that is only 0.00023622 inch across. How many nuclei (plural of *nucleus*) would have to be laid end to end to extend to a length of 1 inch?

To find how many 0.00023622-inch lengths there are in 1 inch, we must use division: $1 \div 0.00023622$.

$$1 \boxed{\div} .00023622 \boxed{=} \qquad \boxed{4233.3418}$$

On some calculators, we press the $\boxed{\text{ENTER}}$ key to display the quotient.

It would take approximately 4,233 nuclei laid end to end to extend to a length of 1 inch.

---

4   Estimate quotients of decimals.

There are many ways to make an error when dividing decimals. Estimation is a helpful tool that can be used to determine whether or not an answer seems reasonable.

To estimate quotients, we use a method that approximates both the dividend and the divisor so that they divide easily. There is one rule of thumb for this method: If possible, round both numbers up or both numbers down.

EXAMPLE 6   Estimate the quotient: $248.687 \div 43.1$

Strategy We will round the dividend and the divisor down and find $240 \div 40$.

WHY The division can be made easier if the dividend and the divisor end with zeros. Also, 40 divides 240 exactly.

Solution

$$248.687 \div 43.1 \qquad 240 \div 40 = 6$$

To divide, drop one zero from 240 and from 40, and find $24 \div 4$.

The estimate is 6.

If we calculate $248.687 \div 43.1$, the quotient is exactly 5.77. Note that the estimate is close: It's just 0.23 more than 5.77.

Self Check 6
Estimate the quotient:
$6{,}229.249 \div 68.9$

Now Try Problems 35 and 39

5   Divide decimals by powers of 10.

To develop a set of rules for division of decimals by a power of 10, we consider the problems $8.13 \div 10$ and $8.13 \div 0.1$.

$$\begin{array}{r} 0.813 \\ 10\overline{)8.130} \\ -8\,0 \\ \hline 13 \\ -10 \\ \hline 30 \\ -30 \\ \hline 0 \end{array}$$ Write a zero to the right of the 3.

$$\begin{array}{r} 81.3 \\ 01\overline{)81.3} \\ -8 \\ \hline 1 \\ -1 \\ \hline 3 \\ -3 \\ \hline 0 \end{array}$$ Move the decimal points in the divisor and dividend one place to the right.

Note that the quotients, 0.813 and 81.3, and the dividend, 8.13, are the same except for the location of the decimal points. The first quotient, 0.813, can be easily obtained by moving the decimal point of the dividend one place to the left. The second quotient, 81.3, is easily obtained by moving the decimal point of the dividend one place to the right. These observations illustrate the following rules for dividing a decimal by a power of 10.

> ### Dividing a Decimal by 10, 100, 1,000, and So On
>
> To find the quotient of a decimal and 10, 100, 1,000, and so on, move the decimal point to the left the same number of places as there are zeros in the power of 10.

> ### Dividing a Decimal by 0.1, 0.01, 0.001, and So On
>
> To find the quotient of a decimal and 0.1, 0.01, 0.001, and so on, move the decimal point to the right the same number of decimal places as there are in the power of 10.

Self Check 7

Find each quotient:
a. $721.3 \div 100$

b. $\dfrac{1.07}{1{,}000}$

c. $19.4407 \div 0.0001$

Now Try **Problems 43 and 49**

EXAMPLE 7 Find each quotient:

a. $16.74 \div 10$ b. $8.6 \div 10{,}000$ c. $\dfrac{290.623}{0.01}$

Strategy We will identify the divisor in each division. If it is a power of 10 greater than 1, we will count the number of zeros that it has. If it is a power of 10 less than 1, we will count the number of decimal places that it has.

WHY Then we will know how many places to the right or left to move the decimal point in the dividend to find the quotient.

Solution

a. $16.74 \div 10 = 1.674$ Since the divisor 10 has one zero, move the decimal point one place to the left.

b. $8.6 \div 10{,}000 = .00086$ Since the divisor 10,000 has four zeros, move the decimal point four places to the left. Write three placeholder zeros (shown in blue).

$= 0.00086$

c. $\dfrac{290.623}{0.01} = 29062.3$ Since the divisor 0.01 has *two decimal places*, move the decimal point in 290.623 two places to the right.

6 Divide signed decimals.

The rules for dividing integers also hold for dividing signed decimals. The quotient of two decimals with *like signs* is positive, and the quotient of two decimals with *unlike signs* is negative.

Self Check 8

Divide:
a. $-100.624 \div 15.2$

b. $\dfrac{-23.9}{-0.1}$

Now Try **Problems 51 and 55**

EXAMPLE 8 Divide: a. $-104.483 \div 16.3$ b. $\dfrac{-38.677}{-0.1}$

Strategy In part a, we will use the rule for dividing signed decimals that have different (unlike) signs. In part b, we will use the rule for dividing signed decimals that have the same (like) signs.

WHY In part a, the divisor is positive and the dividend is negative. In part b, both the dividend and divisor are negative.

Solution

a. First, we find the absolute values: $|-104.483| = 104.483$ and $|16.3| = 16.3$. Then we divide the absolute values, 104.483 by 16.3, using the methods of this section.

$$
\begin{array}{r}
6.41 \\
163\overline{)1044.83} \\
-978 \\
\hline
66\ 8 \\
-65\ 20 \\
\hline
1\ 63 \\
-1\ 63 \\
\hline
0
\end{array}
$$

Move the decimal point in the divisor and the dividend one place to the right.

Write the decimal point in the quotient (answer) directly above the decimal point in the dividend.

Divide as if working with whole numbers.

Since the signs of the original dividend and divisor are unlike, we make the final answer negative. Thus,

$$-104.483 \div 16.3 = -6.41$$

Check the result using multiplication.

b. We can use the rule for dividing a decimal by a power of 10 to find the quotient.

$$\frac{-38.677}{-0.1} = 386.77$$

Since the divisor 0.1 has one decimal place, move the decimal point in 38.677 one place to the right. Since the dividend and divisor have like signs, the quotient is positive.

---

7 Use the order of operations rule.

Recall that the order of operations rule is used to evaluate expressions that involve more than one operation.

EXAMPLE 9

Evaluate: $\dfrac{2(0.351) + 0.5592}{0.2 - 0.6}$

Strategy We will evaluate the expression above and the expression below the fraction bar separately. Then we will do the indicated division, if possible.

WHY Fraction bars are grouping symbols. They group the numerator and denominator.

Solution

$$
\frac{2(0.351) + 0.5592}{0.2 - 0.6}
$$

$$
= \frac{0.702 + 0.5592}{-0.4}
$$

In the numerator, do the multiplication. In the denominator, do the subtraction.

$$
= \frac{1.2612}{-0.4}
$$

In the numerator, do the addition.

$$
= -3.153
$$

Do the division indicated by the fraction bar. The quotient of two numbers with unlike signs is negative.

$$
\begin{array}{r}
\overset{1}{0.351} \\
\times \quad 2 \\
\hline
0.702
\end{array}
\qquad
\begin{array}{r}
\overset{1}{0.7}\overset{1}{0}20 \\
+ 0.5592 \\
\hline
1.2612
\end{array}
$$

$$
\begin{array}{r}
3.153 \\
4\overline{)12.612} \\
-12 \\
\hline
6 \\
-4 \\
\hline
21 \\
-20 \\
\hline
12 \\
-12 \\
\hline
0
\end{array}
$$

Self Check 9

Evaluate: $\dfrac{2.7756 + 3(-0.63)}{0.4 - 1.2}$

Now Try Problem 59

8 Evaluate formulas.

Self Check 10

Evaluate the formula $l = \frac{A}{w}$ for $A = 5.511$ and $w = 1.002$.

Now Try Problem 63

EXAMPLE 10   Evaluate the formula $b = \dfrac{2A}{h}$ for $A = 15.36$ and $h = 6.4$.

Strategy In the given formula, we will replace the letter A with 15.36 and h with 6.4.

WHY Then we can use the order of operations rule to find the value of the expression on the right side of the $=$ symbol.

Solution

$$B = \frac{2A}{h} \qquad \text{This is the given formula.}$$

$$= \frac{2(\mathbf{15.36})}{\mathbf{6.4}} \qquad \text{Replace A with 15.36 and h with 6.4.}$$

$$= \frac{30.72}{6.4} \qquad \text{In the numerator, do the multiplication.}$$

$$= 4.8 \qquad \text{Do the division indicated by the fraction bar.}$$

$$\begin{array}{r} \overset{1}{1}\overset{1}{5}.36 \\ \times\quad 2 \\ \hline 30.72 \end{array}$$

$$\begin{array}{r} 4.8 \\ 64\overline{)307.2} \\ -256 \\ \hline 51\ 2 \\ -51\ 2 \\ \hline 0 \end{array}$$

9 Solve application problems by dividing decimals.

Recall that application problems that involve forming equal-sized groups can be solved by division.

Self Check 11

FRUIT CAKES A 9-inch-long fruit-cake loaf is cut into 0.25-inch-thick slices. How many slices are there in one fruitcake?

Now Try Problem 95

EXAMPLE 11   *French Bread*   A bread slicing machine cuts 25-inch-long loaves of French bread into 0.625-inch-thick slices. How many slices are there in one loaf?

Analyze

- 25-inch-long loaves of French bread are cut into slices.   Given
- Each slice is 0.625-inch thick.   Given
- How many slices are there in one loaf?   Find

Form Cutting a loaf of French bread into equally thick slices indicates division. We translate the words of the problem to numbers and symbols.

The number of slices in a loaf of French bread	is equal to	the length of the loaf of French bread	divided by	the thickness of one slice.
The number of slices in a loaf of French bread	=	25	÷	0.625

Solve When we write $25 \div 0.625$ in long division form, we see that the divisor is a decimal.

$$0.625\,\underset{\curvearrowright}{)}\,25.\underset{\curvearrowright}{000}$$

To write the divisor as a whole number, move the decimal point three places to the right. To move the decimal point three places to the right in the dividend, three placeholder zeros must be inserted (shown in blue).

Now that the divisor is a whole number, we can perform the division.

$$
\begin{array}{r}
40 \\
625\overline{)25000} \\
-2500\downarrow \\
\hline
00 \\
-0 \\
\hline
0
\end{array}
$$

State There are 40 slices in one loaf of French bread.

Check The multiplication below verifies that 40 slices, each 0.625-inch thick, makes a 25-inch-long loaf. The result checks.

$$
\begin{array}{r}
0.625 \leftarrow \text{The thickness of one slice of bread (in inches)} \\
\times\quad 40 \leftarrow \text{The number of slices in one loaf} \\
\hline
0000 \\
25000 \\
\hline
25.000 \leftarrow \text{The length of one loaf of bread (in inches)}
\end{array}
$$

Recall that the **arithmetic mean,** or **average,** of several numbers is a value around which the numbers are grouped. We use addition and division to find the mean (average).

EXAMPLE 12 *Comparison Shopping* An online shopping website, Shopping.com, listed the four best prices for an automobile GPS receiver as shown below. What is the mean (average) price of the GPS?

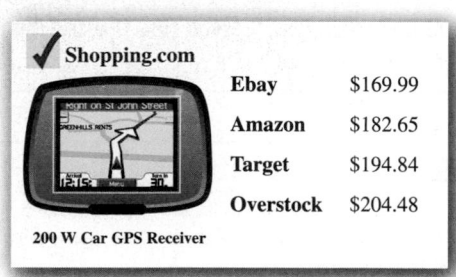

Shopping.com	
Ebay	$169.99
Amazon	$182.65
Target	$194.84
Overstock	$204.48

200 W Car GPS Receiver

Strategy We will add 169.99, 182.65, 194.84, and 204.48 and divide the sum by 4.

WHY To find the mean (average) of a set of values, we divide the sum of the values by the number of values.

Solution

$$
\text{Mean} = \frac{169.99 + 182.65 + 194.84 + 204.48}{4} \quad \text{Since there are 4 prices, divide the sum by 4.}
$$

$$
= \frac{751.96}{4} \quad \text{In the numerator, do the addition.}
$$

$$
= 187.99 \quad \text{Do the indicated division.}
$$

$$
\begin{array}{r}
{}^{2\ 2\ 2\ \ 2} \\
169.99 \\
182.65 \\
194.84 \\
+204.48 \\
\hline
751.96
\end{array}
$$

$$
\begin{array}{r}
187.99 \\
4\overline{)751.96} \\
-4 \\
\hline
35 \\
-32 \\
\hline
31 \\
-28 \\
\hline
39 \\
-36 \\
\hline
36 \\
-36 \\
\hline
0
\end{array}
$$

The mean (average) price of the GPS receiver is $187.99.

Self Check 12

U.S. NATIONAL PARKS Use the following data to determine the average number of visitors per year to the national parks for the years 2004 through 2008. (Source: National Park Service)

Year	Visitors (millions)
2008	2.749
2007	2.756
2006	2.726
2005	2.735
2004	2.769

Now Try Problem 103

THINK IT THROUGH *GPA*

"In considering all of the factors that are important to employers as they recruit students in colleges and universities nationwide, college major, grade point average, and work-related experience usually rise to the top of the list."

Mary D. Feduccia, Ph.D., Career Services Director, Louisiana State University

A grade point average (GPA) is a weighted average based on the grades received and the number of units (credit hours) taken. A GPA for one semester (or term) is defined as

> *the quotient of the sum of the grade points earned for each class and the sum of the number of units taken. The number of grade points earned for a class is the product of the number of units assigned to the class and the value of the grade received in the class.*

1. Use the table of grade values below to compute the GPA for the student whose semester grade report is shown. Round to the nearest hundredth.

Grade	Value
A	4
B	3
C	2
D	1
F	0

Class	Units	Grade
Geology	4	C
Algebra	5	A
Psychology	3	C
Spanish	2	B

2. If you were enrolled in school last semester (or term), list the classes taken, units assigned, and grades received like those shown in the grade report above. Then calculate your GPA.

ANSWERS TO SELF CHECK

1. 5.2   **2.** 3.17   **3.** 5.35   **4.** 0.93   **5.** 14.24   **6.** 6,300 ÷ 70 = 630 ÷ 7 = 90
7. a. 7.213   **b.** 0.00107   **c.** 194,407   **8. a.** −6.62   **b.** 239   **9.** −1.107   **10.** 5.5
11. 36 slices   **12.** 2.747 million visitors

SECTION 5.4   STUDY SET

VOCABULARY

Fill in the blanks.

1. In the division problem shown below, label the *dividend*, the *divisor*, and the *quotient*.

$$3.17 \leftarrow \boxed{}$$
$$\boxed{} \rightarrow 5)\overline{15.85} \nwarrow \boxed{}$$

2. To perform the division $2.7)\overline{9.45}$, we move the decimal point of the divisor so that it becomes the _____ number 27.

CONCEPTS

3. A decimal point is missing in each of the following quotients. Write a decimal point in the proper position.

   a. $\dfrac{526}{4)\overline{21.04}}$ b. $\dfrac{0008}{3)\overline{0.024}}$

4. a. How many places to the right must we move the decimal point in 6.14 so that it becomes a whole number?

   b. When the decimal point in 49.8 is moved three places to the right, what is the resulting number?

5. Move the decimal point in the divisor and the dividend the same number of places so that the divisor becomes a whole number. ***You do not have to find the quotient.***

   a. $1.3 \overline{)10.66}$

   b. $3.71 \overline{)16.695}$

6. Fill in the blanks: To divide with a decimal divisor, write the problem in _____ division form. Move the decimal point of the divisor so that it becomes a _____ number. Then move the decimal point of the dividend the same number of places to the _____. Write the decimal point in the quotient directly _____ the decimal point in the dividend and divide as working with whole _____.

7. To perform the division $7.8 \overline{)14.562}$, the decimal points in the divisor and dividend are moved 1 place to the right. This is equivalent to multiplying $\frac{14.562}{7.8}$ by what form of 1?

8. Use multiplication to check the following division. Is the result correct?

$$\frac{1.917}{0.9} = 2.13$$

9. When rounding a decimal to the hundredths column, to what other column must we look at first?

10. a. When 9.545 is divided by 10, is the answer smaller or larger than 9.545?

    b. When 9.545 is divided by 0.1, is the answer smaller or larger than 9.545?

11. Fill in the blanks.

    a. To find the quotient of a decimal and 10, 100, 1,000, and so on, move the decimal point to the _____ the same number of places as there are zeros in the power of 10.

    b. To find the quotient of a decimal and 0.1, 0.01, 0.001, and so on, move the decimal point to the _____ the same number of decimal places as there are in the power of 10.

12. Determine whether the *sign* of each result is positive or negative. ***You do not have to find the quotient.***

    a. $-15.25 \div (-0.5)$

    b. $\dfrac{-25.92}{3.2}$

NOTATION

13. Explain what the red arrows are illustrating in the division problem below.

$467 \overline{)3208.7}$

14. The division shown below is not finished. Why was the red 0 written after the 7 in the dividend?

$$\begin{array}{r} 2.3 \\ 2\overline{)4.70} \\ -4 \\ \hline 07 \\ -6 \\ \hline 1 \end{array}$$

GUIDED PRACTICE

Divide. Check the result. **See Example 1.**

15. $12.6 \div 6$

16. $40.8 \div 8$

17. $3 \overline{)27.6}$

18. $4 \overline{)28.8}$

Divide. Check the result. **See Example 2.**

19. $98.21 \div 23$

20. $190.96 \div 28$

21. $37 \overline{)320.05}$

22. $32 \overline{)125.12}$

Divide. Check the result. **See Example 3.**

23. $13.4 \div 4$

24. $38.3 \div 5$

25. $5 \overline{)22.8}$

26. $6 \overline{)28.5}$

Divide. Check the result. **See Example 4.**

27. $\dfrac{0.1932}{0.42}$

28. $\dfrac{0.2436}{0.29}$

29. $0.29 \overline{)0.1131}$

30. $0.58 \overline{)0.1566}$

Divide. Round the quotient to the nearest hundredth. Check the result. **See Example 5.**

31. $\dfrac{11.83}{0.6}$

32. $\dfrac{16.43}{0.9}$

33. $\dfrac{17.09}{0.7}$

34. $\dfrac{13.07}{0.6}$

Estimate each quotient. **See Example 6.**

35. $289.842 \div 72.1$

36. $284.254 \div 91.4$

37. $383.76 \div 7.8$

38. $348.84 \div 5.7$

39. $3,883.284 \div 48.12$

40. $5,556.521 \div 67.89$

41. $6.1 \overline{)15,819.74}$

42. $9.2 \overline{)19,460.76}$

Find each quotient. **See Example 7.**

43. $451.78 \div 100$

44. $991.02 \div 100$

45. $\dfrac{30.09}{10,000}$

46. $\dfrac{27.07}{10,000}$

47. $1.25 \div 0.1$

48. $8.62 \div 0.01$

49. $\dfrac{545.2}{0.001}$

50. $\dfrac{67.4}{0.001}$

Divide. See Example 8.

51. $-110.336 \div 12.8$

52. $-121.584 \div 14.9$

53. $-91.304 \div (-22.6)$

54. $-66.126 \div (-32.1)$

55. $\dfrac{-20.3257}{-0.001}$

56. $\dfrac{-48.8933}{-0.001}$

57. $0.003 \div (-100)$

58. $0.008 \div (-100)$

Evaluate each expression. See Example 9.

59. $\dfrac{2(0.614) + 2.3854}{0.2 - 0.9}$

60. $\dfrac{2(1.242) + 0.8932}{0.4 - 0.8}$

61. $\dfrac{5.409 - 3(1.8)}{(0.3)^2}$

62. $\dfrac{1.674 - 5(0.222)}{(0.1)^2}$

Evaluate each formula. See Example 10.

63. $t = \dfrac{d}{r}$ for $d = 211.75$ and $r = 60.5$

64. $h = \dfrac{2A}{b}$ for $A = 9.62$ and $b = 3.7$

65. $r = \dfrac{d}{t}$ for $d = 219.375$ and $t = 3.75$

66. $\pi = \dfrac{C}{d}$ for $C = 14.4513$ and $d = 4.6$ (Round to the nearest hundredth.)

TRY IT YOURSELF

Perform the indicated operations. Round the result to the specified decimal place, when indicated.

67. $4.5\overline{)11.97}$

68. $4.1\overline{)14.637}$

69. $\dfrac{75.04}{10}$

70. $\dfrac{22.32}{100}$

71. $8\overline{)0.036}$

72. $4\overline{)0.073}$

73. $9\overline{)2.889}$

74. $6\overline{)3.378}$

75. $\dfrac{-3(0.2) - 2(3.3)}{30(0.4)^2}$

76. $\dfrac{(-1.3)^2 + 9.2}{-2(0.2) - 0.5}$

77. Divide 1.2202 by -0.01.

78. Divide -0.4531 by -0.001.

79. $-5.714 \div 2.4$ (nearest tenth)

80. $-21.21 \div 3.8$ (nearest tenth)

81. $-39 \div (-4)$

82. $-26 \div (-8)$

83. $7.8915 \div .00001$

84. $23.025 \div 0.0001$

85. $\dfrac{0.0102}{0.017}$

86. $\dfrac{0.0092}{0.023}$

87. $12.243 \div 0.9$ (nearest hundredth)

88. $13.441 \div 0.6$ (nearest hundredth)

89. $1,000\overline{)34.8}$

90. $10,000\overline{)678.9}$

91. $\dfrac{40.7(3 - 8.3)}{0.4 - 0.61}$ (nearest hundredth)

92. $\dfrac{(0.5)^2 - (0.3)^2}{0.005 + 0.1}$ (nearest hundredth)

93. Divide 0.25 by 1.6

94. Divide 1.2 by 0.64

APPLICATIONS

95. BUTCHER SHOPS A meat slicer trims 0.05-inch-thick pieces from a sausage. If the sausage is 14 inches long, how many slices are there in one sausage?

96. ELECTRONICS The volume control on a computer is shown to the right. If the distance between the Low and High settings is 21 cm, how far apart are the equally spaced volume settings?

97. COMPUTERS A computer can do an arithmetic calculation in 0.00003 second. How many of these calculations could it do in 60 seconds?

98. THE LOTTERY In December of 2008, fifteen city employees of Piqua, Ohio, who had played the Mega Millions Lottery as a group, won the jackpot. They were awarded a total of $94.5 million. If the money was split equally, how much did each person receive? (Source: pal-item.com)

99. SPRAY BOTTLES Each squeeze of the trigger of a spray bottle emits 0.017 ounce of liquid. How many squeezes are there in an 8.5-ounce bottle?

100. CAR LOANS See the loan statement below. How many more monthly payments must be made to pay off the loan?

American Finance Company		June
Monthly payment:	Paid to date: $547.30	
$42.10	Loan balance: $631.50	

101. HIKING Refer to the illustration below to determine how long it will take the person shown to complete the hike. Then determine at what time of the day she will complete the hike.

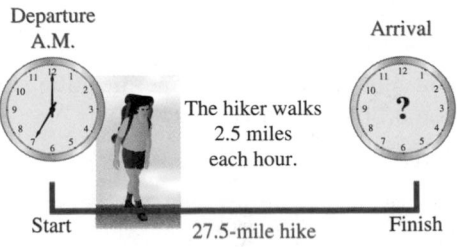

Departure A.M.   Arrival

The hiker walks 2.5 miles each hour.

Start   27.5-mile hike   Finish

102. HOURLY PAY The graph below shows the average hours worked and the average weekly earnings of U.S. production workers in manufacturing for the years 1998 and 2008. What did the average production worker in manufacturing earn per hour

   a. in 1998?   **b.** in 2008?

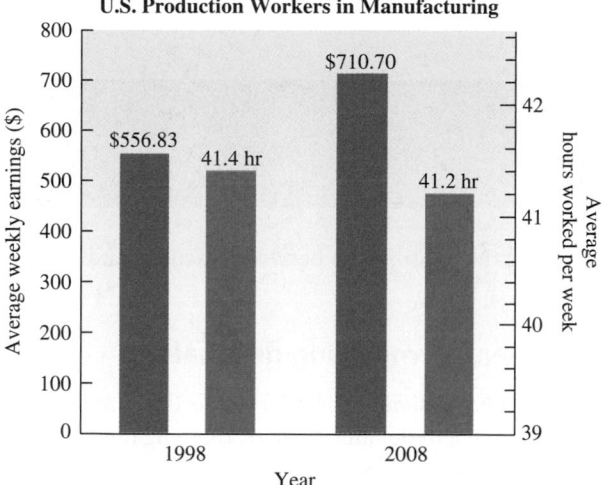

U.S. Production Workers in Manufacturing

Source: *U.S. Department of Labor Statistic*

103. TRAVEL The illustration shows the annual number of person-trips of 50 miles or more (one way) for the years 2002–2007, as estimated by the Travel Industry Association of America. Find the average number of trips per year for this period of time.

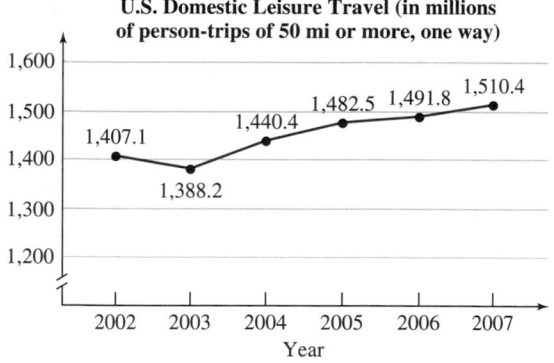

U.S. Domestic Leisure Travel (in millions of person-trips of 50 mi or more, one way)

Source: *U.S. Travel Association*

104. OIL WELLS Geologists have mapped out the types of soil through which engineers must drill to reach an oil deposit. See the illustration below.

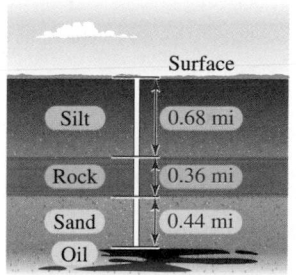

a. How far below the surface is the oil deposit?

b. What is the average depth that must be drilled each week if the drilling is to be a four-week project?

105. REFLEXES An online reaction time test is shown below. When the stop light changes from red to green, the participant is to immediately click on the large green button. The program then displays the participant's reaction time in the table. After the participant takes the test five times, the *average* reaction time is found. Determine the average reaction time for the results shown below.

Test Number	Reaction Time (in seconds)	The stoplight to watch.	The button to click.
1	0.219		
2	0.233		Click here on green light
3	0.204		
4	0.297		
5	0.202		
AVG.	?		

106. INDY 500 Driver Scott Dixon, of New Zealand, had the fastest average qualifying speed for the 2008 Indianapolis 500-mile race. This earned him the *pole position* to begin the race. The speeds for each of his four qualifying laps are shown below. What was his average qualifying speed?

Lap 1: 226.598 mph
Lap 2: 226.505 mph
Lap 3: 226.303 mph
Lap 4: 226.058 mph

(Source: indianapolismotorspeedway.com)

WRITING

107. Explain the process used to divide two numbers when both the divisor and the dividend are decimals. Give an example.

108. Explain why we must sometimes use rounding when we write the answer to a division problem.

109. The division $0.5\overline{)2.005}$ is equivalent to $5\overline{)20.05}$. Explain what equivalent means in this case.

110. In $3\overline{)0.7}$, why can additional zeros be placed to the right of 0.7 without affecting the result?

111. Explain how to estimate the following quotient: $0.75\overline{)2.415}$

112. Explain why multiplying $\frac{4.86}{0.2}$ by the form of 1 shown below moves the decimal points in the dividend, 4.86, and the divisor, 0.2, one place to the right.

$$\frac{4.86}{0.2} = \frac{4.86}{0.2} \cdot \frac{10}{10}$$

▌ REVIEW

113. a. Find the GCF of 10 and 25.

   b. Find the LCM of 10 and 25.

114. a. Find the GCF of 8, 12, and 16.

   b. Find the LCM of 8, 12, and 16.

Objectives

1 Write fractions as equivalent terminating decimals.

2 Write fractions as equivalent repeating decimals.

3 Round repeating decimals.

4 Graph fractions and decimals on a number line.

5 Compare fractions and decimals.

6 Evaluate expressions containing fractions and decimals.

7 Solve application problems involving fractions and decimals.

SECTION 5.5
Fractions and Decimals

In this section, we continue to explore the relationship between fractions and decimals.

1 Write fractions as equivalent terminating decimals.

A fraction and a decimal are said to be **equivalent** if they name the same number. Every fraction can be written in an equivalent decimal form by dividing the numerator by the denominator, as indicated by the fraction bar.

Writing a Fraction as a Decimal

To write a fraction as a decimal, divide the numerator of the fraction by its denominator.

Self Check 1

Write each fraction as a decimal.

a. $\frac{1}{2}$

b. $\frac{3}{16}$

c. $\frac{9}{2}$

Now Try Problems 15, 17, and 21

EXAMPLE 1 Write each fraction as a decimal.

a. $\frac{3}{4}$ **b.** $\frac{5}{8}$ **c.** $\frac{7}{2}$

Strategy We will divide the numerator of each fraction by its denominator. We will continue the division process until we obtain a zero remainder.

WHY We divide the numerator by the denominator because a fraction bar indicates division.

Solution

a. $\frac{3}{4}$ means $3 \div 4$. To find $3 \div 4$, we begin by writing it in long division form as $4\overline{)3}$. To proceed with the division, we must write the dividend 3 with a decimal point and some additional zeros. Then we use the procedure from Section 5.4 for dividing a decimal by a whole number.

$$
\begin{array}{r}
0.75 \\
4\overline{)3.00} \\
-2\,8\!\downarrow \\
\hline
20 \\
-20 \\
\hline
0
\end{array}
$$
Write a decimal point and two additional zeros to the right of 3.

← The remainder is 0.

Thus, $\frac{3}{4} = 0.75$. We say that the **decimal equivalent** of $\frac{3}{4}$ is 0.75.

We can check the result by writing 0.75 as a fraction in simplest form:

$$0.75 = \frac{75}{100} \qquad \textit{0.75 is seventy-five hundredths.}$$

$$= \frac{3 \cdot \overset{1}{\cancel{25}}}{4 \cdot \underset{1}{\cancel{25}}} \qquad \textit{To simplify the fraction, factor 75 as 3 · 25 and 100}$$
$$\textit{as 4 · 25 and remove the common factor of 25.}$$

$$= \frac{3}{4} \qquad \textit{This is the original fraction.}$$

b. $\frac{5}{8}$ means $5 \div 8$.

$$
\begin{array}{r}
0.625 \\
8\overline{)5.000} \\
\end{array}
$$
$\qquad$ Write a decimal point and three additional zeros to the right of 5.

$$
\begin{array}{r}
-48\!\downarrow \\
\hline
20 \\
-16 \\
\hline
40 \\
-40 \\
\hline
0 \\
\end{array}
$$
$\leftarrow$ The remainder is 0.

Thus, $\frac{5}{8} = 0.625$.

c. $\frac{7}{2}$ means $7 \div 2$.

$$
\begin{array}{r}
3.5 \\
2\overline{)7.0} \\
\end{array}
$$
$\qquad$ Write a decimal point and one additional zero to the right of 7.

$$
\begin{array}{r}
-6\!\downarrow \\
\hline
1\,0 \\
-1\,0 \\
\hline
0 \\
\end{array}
$$
$\leftarrow$ The remainder is 0.

Thus, $\frac{7}{2} = 3.5$.

Caution! A common error when finding a decimal equivalent for a fraction is to *incorrectly divide the denominator by the numerator.* An example of this is shown on the right, where the decimal equivalent of $\frac{5}{8}$ (a number less than 1) is incorrectly found to be 1.6 (a number greater than 1).

$$
\begin{array}{r}
1.6 \\
5\overline{)8.0} \\
-5 \\
\hline
3\,0 \\
-3\,0 \\
\hline
0 \\
\end{array}
$$

In parts a, b, and c of Example 1, the division process ended because a remainder of 0 was obtained. When such a division *terminates* with a remainder of 0, we call the resulting decimal a **terminating decimal.** Thus, 0.75, 0.625, and 3.5 are three examples of terminating decimals.

The Language of Algebra To *terminate* means to bring to an end. In the movie *The Terminator,* actor Arnold Schwarzenegger plays a heartless machine sent to Earth to bring an end to his enemies.

---

2 **Write fractions as equivalent repeating decimals.**

Sometimes, when we are finding a decimal equivalent of a fraction, the division process never gives a remainder of 0. In this case, the result is a **repeating decimal.** Examples of repeating decimals are 0.4444 . . . and 1.373737 . . . . The three dots tell us that a block of digits repeats in the pattern shown. Repeating decimals can also be written using a bar over the repeating block of digits. For example, 0.4444 . . . can be written as 0.$\overline{4}$, and 1.373737 . . . can be written as 1.$\overline{37}$.

> **Caution!**   When using an **overbar** to write a repeating decimal, use the least number of digits necessary to show the repeating block of digits.
>
> 0.333 . . . = 0.$\overline{333}$ 6.7454545 . . . = 6.7$\overline{454}$
>
> 0.333 . . . = 0.$\overline{3}$ 6.7454545 . . . = 6.7$\overline{45}$

Some fractions can be written as decimals using an alternate approach. If the denominator of a fraction in simplified form has factors of only 2's or 5's, or a combination of both, it can be written as a decimal by multiplying it by a form of 1. The objective is to write the fraction in an equivalent form with a denominator that is a power of 10, such as 10, 100, 1,000, and so on.

Self Check 2

Write each fraction as a decimal using multiplication by a form of 1:

a. $\dfrac{2}{5}$

b. $\dfrac{8}{25}$

Now Try Problems 27 and 29

EXAMPLE 2   Write each fraction as a decimal using multiplication by a form of 1:   **a.** $\dfrac{4}{5}$ **b.** $\dfrac{11}{40}$

Strategy We will multiply $\frac{4}{5}$ by $\frac{2}{2}$ and we will multiply $\frac{11}{40}$ by $\frac{25}{25}$.

WHY The result of each multiplication will be an equivalent fraction with a denominator that is a power of 10. Such fractions are then easy to write in decimal form.

Solution

a. Since we need to multiply the denominator of $\frac{4}{5}$ by 2 to obtain a denominator of 10, it follows that $\frac{2}{2}$ should be the form of 1 that is used to build $\frac{4}{5}$.

$$\frac{4}{5} = \frac{4}{5} \cdot \frac{2}{2} \quad \text{Multiply } \tfrac{4}{5} \text{ by 1 in the form of } \tfrac{2}{2}.$$
$$= \frac{8}{10} \quad \text{Multiply the numerators. Multiply the denominators.}$$
$$= 0.8 \quad \text{Write the fraction as a decimal.}$$

b. Since we need to multiply the denominator of $\frac{11}{40}$ by 25 to obtain a denominator of 1,000, it follows that $\frac{25}{25}$ should be the form of 1 that is used to build $\frac{11}{40}$.

$$\frac{11}{40} = \frac{11}{40} \cdot \frac{25}{25} \quad \text{Multiply } \tfrac{11}{40} \text{ by 1 in the form of } \tfrac{25}{25}.$$
$$= \frac{275}{1,000} \quad \text{Multiply the numerators. Multiply the denominators.}$$
$$= 0.275 \quad \text{Write the fraction as a decimal.}$$

Mixed numbers can also be written in decimal form.

Self Check 3

Write the mixed number $3\frac{17}{20}$ in decimal form.

Now Try Problem 37

EXAMPLE 3   Write the mixed number $5\frac{7}{16}$ in decimal form.

Strategy We need only find the decimal equivalent for the fractional part of the mixed number.

WHY The whole-number part in the decimal form is the same as the whole-number part in the mixed number form.

Solution To write $\frac{7}{16}$ as a fraction, we find $7 \div 16$.

$$
\begin{array}{r}
0.4375 \\
16\overline{)7.0000} \\
\end{array}
$$
Write a decimal point and four additionl zeros to the right of 7.

$$
\begin{array}{r}
-6\,4\downarrow \\
\hline
60 \\
-48\downarrow \\
\hline
120 \\
-112\downarrow \\
\hline
80 \\
-80 \\
\hline
0 \leftarrow \text{The remainder is 0.}
\end{array}
$$

Since the whole-number part of the decimal must be the same as the whole-number part of the mixed number, we have:

$$5\frac{7}{16} = 5.4375$$

We would have obtained the same result if we changed $5\frac{7}{16}$ to the improper fraction $\frac{87}{16}$ and divided 87 by 16.

EXAMPLE 4   Write $\frac{5}{12}$ as a decimal.

Self Check 4
Write $\frac{1}{12}$ as a decimal.
Now Try Problem 41

Strategy We will divide the numerator of the fraction by its denominator and watch for a repeating pattern of nonzero remainders.

WHY Once we detect a repeating pattern of remainders, the division process can stop.

Solution $\frac{5}{12}$ means $5 \div 12$.

$$
\begin{array}{r}
0.4166 \\
12\overline{)5.0000} \\
\end{array}
$$
Write a decimal point and four additional zeros to the right of 5.

$$
\begin{array}{r}
-4\,8\downarrow \\
\hline
20 \\
-12\downarrow \\
\hline
80 \\
-72\downarrow \\
\hline
80 \\
-72 \\
\hline
8
\end{array}
$$
It is apparent that 8 will continue to reappear as the remainder. Therefore, 6 will continue to reappear in the quotient. Since the repeating pattern is now clear, we can stop the division.

We can use three dots to show that a repeating pattern of 6's appears in the quotient:

$$\frac{5}{12} = 0.416666\ldots$$

Or, we can use an overbar to indicate the repeating part (in this case, only the 6), and write the decimal equivalent in more compact form:

$$\frac{5}{12} = 0.41\overline{6}$$

EXAMPLE 5   Write $-\frac{6}{11}$ as a decimal.

Self Check 5
Write $-\frac{13}{33}$ as a decimal.
Now Try Problem 47

Strategy To find the decimal equivalent for $-\frac{6}{11}$, we will first find the decimal equivalent for $\frac{6}{11}$. To do this, we will divide the numerator of $\frac{6}{11}$ by its denominator and watch for a repeating pattern of nonzero remainders.

WHY Once we detect a repeating pattern of remainders, the division process can stop.

Solution $\frac{6}{11}$ means $6 \div 11$.

```
      0.54545
11)6.00000      Write a decimal point and five additional zeros to the right of 6.
  - 5 5
     50
   - 44
     60
    - 55
      50
    - 44
      60      It is apparent that 6 and 5 will continue to reappear as remainders.
    - 55      Therefore, 5 and 4 will continue to reappear in the quotient. Since the
       5      repeating pattern is now clear, we can stop the division process.
```

We can use three dots to show that a repeating pattern of 5 and 4 appears in the quotient:

$$\frac{6}{11} = 0.545454\ldots \text{ and therefore, } -\frac{6}{11} = -0.545454\ldots$$

Or, we can use an overbar to indicate the repeating part (in this case, 54), and write the decimal equivalent in more compact form:

$$\frac{6}{11} = 0.\overline{54} \text{ and therefore, } -\frac{6}{11} = -0.\overline{54}$$

The repeating part of the decimal equivalent of some fractions is quite long. Here are some examples:

$$\frac{9}{37} = 0.\overline{243} \qquad \text{A block of three digits repeats.}$$

$$\frac{13}{101} = 0.\overline{1287} \qquad \text{A block of four digits repeats.}$$

$$\frac{6}{7} = 0.\overline{857142} \qquad \text{A block of six digits repeats.}$$

Every fraction can be written as either a terminating decimal or a repeating decimal. For this reason, the set of fractions (**rational numbers**) form a subset of the set of decimals called the set of **real numbers.** The set of real numbers corresponds to all points on a number line.

Not all decimals are terminating or repeating decimals. For example,

0.2020020002 . . .

does not terminate, and it has no repeating block of digits. This decimal cannot be written as a fraction with an integer numerator and a nonzero integer denominator. Thus, it is not a rational number. It is an example from the set of **irrational numbers.**

3 Round repeating decimals.

When a fraction is written in decimal form, the result is either a terminating or a repeating decimal. Repeating decimals are often rounded to a specified place value.

EXAMPLE 6 Write $\frac{1}{3}$ as a decimal and round to the nearest hundredth.

Strategy We will use the methods of this section to divide to the thousandths column.

WHY To round to the hundredths column, we need to continue the division process for one more decimal place, which is the thousandths column.

Solution $\frac{1}{3}$ means $1 \div 3$.

$$
\begin{array}{r}
0.333 \\
3\overline{\smash{)}1.000} \\
\underline{-\ 9\downarrow} \\
10 \\
\underline{-\ 9\downarrow} \\
10 \\
\underline{-\ 9} \\
1 \\
\end{array}
$$

Write a decimal point and three additional zeros to the right of 1.

The division process can stop. We have divided to the thousandths column.

After dividing to the thousandths column, we round to the hundredths column.

— The rounding digit in the hundredths column is 3.
— The test digit in the thousandths column is 3.

0.333 . . .

Since 3 is less than 5, we round down, and we have

$$\frac{1}{3} \approx 0.33 \quad \text{Read} \approx \text{as "is approximately equal to."}$$

Self Check 6
Write $\frac{4}{9}$ as a decimal and round to the nearest hundredth.

Now Try Problem 51

EXAMPLE 7 Write $\frac{2}{7}$ as a decimal and round to the nearest thousandth.

Strategy We will use the methods of this section to divide to the ten-thousandths column.

WHY To round to the thousandths column, we need to continue the division process for one more decimal place, which is the ten-thousandths column.

Solution $\frac{2}{7}$ means $2 \div 7$.

$$
\begin{array}{r}
0.2857 \\
7\overline{\smash{)}2.0000} \\
\underline{-\ 1\ 4\downarrow} \\
60 \\
\underline{-\ 56\downarrow} \\
40 \\
\underline{-\ 35\downarrow} \\
50 \\
\underline{-\ 49} \\
1 \\
\end{array}
$$

Write a decimal point and four additional zeros to the right of 2.

The division process can stop.
We have divided to the ten-thousandths column.

After dividing to the ten-thousandths column, we round to the thousandths column.

— The rounding digit in the thousandths column is 5.
— The test digit in the ten-thousandths column is 7.

0.2857

Since 7 is greater than 5, we round up, and $\frac{2}{7} \approx 0.286$.

Self Check 7
Write $\frac{7}{24}$ as a decimal and round to the nearest thousandth.

Now Try Problem 61

Using Your CALCULATOR   The Fixed-Point Key

After performing a calculation, a scientific calculator can round the result to a given decimal place. This is done using the *fixed-point key*. As we did in Example 7, let's find the decimal equivalent of $\frac{2}{7}$ and round to the nearest thousandth. This time, we will use a calculator.

First, we set the calculator to round to the third decimal place (thousandths) by pressing $\boxed{\text{2nd}}$ $\boxed{\text{FIX}}$ 3. Then we press 2 $\boxed{\div}$ 7 $\boxed{=}$ $\boxed{\text{0.286}}$

Thus, $\frac{2}{7} \approx 0.286$. To round to the nearest tenth, we would fix 1; to round to the nearest hundredth, we would fix 2; and so on. After using the FIX feature, don't forget to remove it and return the calculator to the normal mode.

Graphing calculators can also round to a given decimal place. See the owner's manual for the required keystrokes.

---

4 Graph fractions and decimals on a number line.

A number line can be used to show the relationship between fractions and their decimal equivalents. On the number line below, sixteen equally spaced marks are used to scale from 0 to 1. Some commonly used fractions that have terminating decimal equivalents are shown. For example, we see that $\frac{1}{8} = 0.125$ and $\frac{13}{16} = 0.8125$.

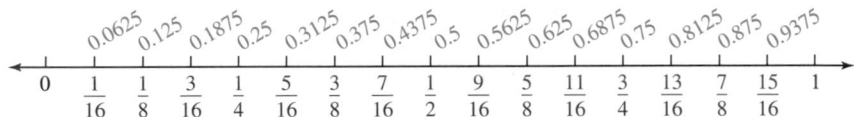

On the next number line, six equally spaced marks are used to scale from 0 to 1. Some commonly used fractions and their repeating decimal equivalents are shown.

5 Compare fractions and decimals.

To compare the size of a fraction and a decimal, it is helpful to write the fraction in its equivalent decimal form.

Self Check 8

Place an $<$, $>$, or an $=$ symbol in the box to make a true statement:

a. $\frac{3}{8}$ ▢ 0.305

b. $0.7\overline{6}$ ▢ $\frac{7}{9}$

c. $\frac{11}{4}$ ▢ 2.75

Now Try Problems 67, 69, and 71

EXAMPLE 8   Place an $<$, $>$, or an $=$ symbol in the box to make a true

statement: **a.** $\frac{4}{5}$ ▢ 0.91   **b.** $0.3\overline{5}$ ▢ $\frac{1}{3}$ **c.** $\frac{9}{4}$ ▢ 2.25

Strategy In each case, we will write the given fraction as a decimal.

WHY Then we can use the procedure for comparing two decimals to determine which number is the larger and which is the smaller.

Solution

a. To write $\frac{4}{5}$ as a decimal, we divide 4 by 5.

$$\begin{array}{r} 0.8 \\ 5\overline{)4.0} \\ -4\,0 \\ \hline 0 \end{array}$$ Write a decimal point and one additional zero to the right of 4.

Thus, $\frac{4}{5} = 0.8$.

To make the comparison of the decimals easier, we can write one zero after 8 so that they have the same number of digits to the right of the decimal point.

0.8̲0   *This is the decimal equivalent for $\frac{4}{5}$.*

0.9̲1
 ↑

As we work from left to right, this is the first column in which the digits differ. Since $8 < 9$, it follows that $0.80 = \frac{4}{5}$ is less than 0.91, and we can write $\frac{4}{5} < 0.91$.

b. In Example 6, we saw that $\frac{1}{3} = 0.3333\ldots$. To make the comparison of these repeating decimals easier, we write them so that they have the same number of digits to the right of the decimal point.

0.3 5̲ 55 . . .   *This is $0.3\overline{5}$.*

0.3 3̲ 33 . . .   *This is the decimal equivalent of $\frac{1}{3}$.*
  ↑

As we work from left to right, this is the first column in which the digits differ. Since $5 > 3$, it follows that $0.3555\ldots = 0.3\overline{5}$ is greater than $0.3333\ldots = \frac{1}{3}$, and we can write $0.3\overline{5} > \frac{1}{3}$.

c. To write $\frac{9}{4}$ as a decimal, we divide 9 by 4.

$$
\begin{array}{r}
2.25 \\
4\overline{)9.00} \\
-8 \\
\hline
1\,0 \\
-8 \\
\hline
20 \\
-20 \\
\hline
0
\end{array}
$$ *Write a decimal point and two additional zeros to the right of 9.*

From the division, we see that $\frac{9}{4} = 2.25$.

EXAMPLE 9   Write the numbers in order from smallest to largest:

$2.168,\ 2\frac{1}{6},\ \frac{20}{9}$

Strategy We will write $2\frac{1}{6}$ and $\frac{20}{9}$ in decimal form.

WHY Then we can do a column-by-column comparison of the numbers to determine the largest and smallest.

Solution From the number line on page 506, we see that $\frac{1}{6} = 0.1\overline{6}$. Thus, $2\frac{1}{6} = 2.1\overline{6}$. To write $\frac{20}{9}$ as a decimal, we divide 20 by 9.

$$
\begin{array}{r}
2.222 \\
9\overline{)20.000} \\
-18 \\
\hline
20 \\
-18 \\
\hline
20 \\
-18 \\
\hline
20 \\
-18 \\
\hline
2
\end{array}
$$ *Write a decimal point and three additional zeros to the right of 20.*

Thus, $\frac{20}{9} = 2.222\ldots$.

Self Check 9

Write the numbers in order from smallest to largest: $1.832,\ \frac{9}{5},\ 1\frac{5}{6}$

Now Try Problem 75

To make the comparison of the three decimals easier, we stack them as shown below.

2.1 6 8 0   This is 2.168 with an additional 0.

2.1 6 6 6...   This is $2\frac{1}{6} = 2.1\overline{6}$.

2.2 2 2 2...   This is $\frac{20}{9}$.

Working from left to right, this is the first column in which the digits differ. Since $2 > 1$, it follows that $2.222... = \frac{20}{9}$ is the largest of the three numbers.

Working from left to right, this is the first column in which the top two numbers differ. Since $8 > 6$, it follows that 2.168 is the next largest number and that $2.1\overline{6} = 2\frac{1}{6}$ is the smallest.

Written in order from smallest to largest, we have :

$$2\frac{1}{6}, \ 2.168, \ \frac{20}{9}$$

6 Evaluate expressions containing fractions and decimals.

Expressions can contain both fractions and decimals. In the following examples, we show two methods that can be used to evaluate expressions of this type. With the first method we find the answer by working in terms of fractions.

Self Check 10

Evaluate by working in terms of fractions: $0.53 + \frac{1}{6}$

Now Try Problem 79

EXAMPLE 10   Evaluate $\frac{1}{3} + 0.27$ by working in terms of fractions.

Strategy We will begin by writing 0.27 as a fraction.

WHY Then we can use the methods of Chapter 3 for adding fractions with unlike denominators to find the sum.

Solution To write 0.27 as a fraction, it is helpful to read it aloud as "twenty-seven hundredths."

$$\frac{1}{3} + 0.27 = \frac{1}{3} + \frac{27}{100}$$ Replace 0.27 with $\frac{27}{100}$.

$$= \frac{1}{3} \cdot \frac{\mathbf{100}}{\mathbf{100}} + \frac{27}{100} \cdot \frac{\mathbf{3}}{\mathbf{3}}$$ The LCD for $\frac{1}{3}$ and $\frac{27}{100}$ is 300. To build each fraction so that its denominator is 300, multiply by a form of 1.

$$= \frac{100}{300} + \frac{81}{300}$$ Multiply the numerators. Multiply the denominators.

$$= \frac{181}{300}$$ Add the numerators and write the sum over the common denominator 300.

Now we will evaluate the expression from Example 10 by working in terms of decimals.

Self Check 11

Estimate the result by working in terms of decimals: $0.53 - \frac{1}{6}$

Now Try Problem 87

EXAMPLE 11   Estimate $\frac{1}{3} + 0.27$ by working in terms of decimals.

Strategy Since 0.27 has two decimal places, we will begin by finding a decimal approximation for $\frac{1}{3}$ to two decimal places.

WHY Then we can use the methods of this chapter for adding decimals to find the sum.

Solution We have seen that the decimal equivalent of $\frac{1}{3}$ is the repeating decimal 0.333 . . . . Rounded to the nearest hundredth: $\frac{1}{3} \approx 0.33$.

$$\frac{1}{3} + 0.27 \approx 0.33 + 0.27 \quad \text{Approximate } \tfrac{1}{3} \text{ with the decimal 0.33.}$$
$$\approx 0.60 \quad \text{Do the addition.}$$

$$\begin{array}{r} \overset{1}{0.33} \\ +0.27 \\ \hline 0.60 \end{array}$$

In Examples 10 and 11, we evaluated $\frac{1}{3} + 0.27$ in different ways. In Example 10, we obtained the exact answer, $\frac{181}{300}$. In Example 11, we obtained an approximation, 0.6. The results seem reasonable when we write $\frac{181}{300}$ in decimal form: $\frac{181}{300} = 0.60333 \ldots$.

EXAMPLE 12 Evaluate: $\left(\dfrac{4}{5}\right)(1.35) + (0.5)^2$

Self Check 12
Evaluate: $(-0.6)^2 + (2.3)\left(\dfrac{1}{8}\right)$
Now Try Problem 99

Strategy We will find the decimal equivalent of $\frac{4}{5}$ and then evaluate the expression in terms of decimals.

WHY Its easier to perform multiplication and addition with the given decimals than it would be converting them to fractions.

Solution We use division to find the decimal equivalent of $\frac{4}{5}$.

$$\begin{array}{r} 0.8 \\ 5\overline{)4.0} \\ -40 \\ \hline 0 \end{array}$$ Write a decimal point and one additional zero to the right of the 4.

Now we use the order of operation rule to evaluate the expression.

$$\left(\dfrac{4}{5}\right)(1.35) + (0.5)^2$$
$$= (\mathbf{0.8})(1.35) + (0.5)^2 \quad \text{Replace } \tfrac{4}{5} \text{ with its decimal equivalent, 0.8.}$$
$$= (0.8)(1.35) + 0.25 \quad \text{Evaluate: } (0.5)^2 = 0.25.$$
$$= 1.08 + 0.25 \quad \text{Do the multiplication: } (0.8)(1.35) = 1.08.$$
$$= 1.33 \quad \text{Do the addition.}$$

$$\begin{array}{r} \overset{2}{0.5} \\ \times\ 0.5 \\ \hline 0.25 \end{array} \qquad \begin{array}{r} \overset{2}{1}\overset{4}{.35} \\ \times\ \ 0.8 \\ \hline 1.080 \end{array} \qquad \begin{array}{r} \overset{1}{1.08} \\ +0.25 \\ \hline 1.33 \end{array}$$

7 Solve application problems involving fractions and decimals.

EXAMPLE 13 *Shopping* A shopper purchased $\frac{3}{4}$ pound of fruit, priced at $0.88 a pound, and $\frac{1}{3}$ pound of fresh-ground coffee, selling for $6.60 a pound. Find the total cost of these items.

Self Check 13
DELICATESSENS A shopper purchased $\frac{2}{3}$ pound of Swiss cheese, priced at $2.19 per pound, and $\frac{3}{4}$ pound of sliced turkey, selling for $6.40 per pound. Find the total cost of these items.
Now Try Problem 111

Analyze

- $\frac{3}{4}$ pound of fruit was purchased at $0.88 per pound.   Given
- $\frac{1}{3}$ pound of coffee was purchased at $6.60 per pound.   Given
- What was the total cost of the items?   Find

Form To find the total cost of each item, multiply the number of pounds purchased by the price per pound.

The total cost of the items	is equal to	the number of pounds of fruit	times	the price per pound	plus	the number of pounds of coffee	times	the price per pound

The total cost of the items	=	$\frac{3}{4}$	$\cdot$	$0.88	+	$\frac{1}{3}$	$\cdot$	$6.60

Solve Because 0.88 is divisible by 4 and 6.60 is divisible by 3, we can work with the decimals and fractions in this form; no conversion is necessary.

$$\frac{3}{4} \cdot 0.88 + \frac{1}{3} \cdot 6.60$$

$$= \frac{3}{4} \cdot \frac{0.88}{1} + \frac{1}{3} \cdot \frac{6.60}{1} \qquad \text{Express 0.88 as } \frac{0.88}{1} \text{ and 6.60 as } \frac{6.60}{1}.$$

$$= \frac{2.64}{4} + \frac{6.60}{3} \qquad \text{Multiply the numerators.}$$
$$\text{Multiply the denominators.}$$

$$= 0.66 + 2.20 \qquad \text{Do each division.}$$

$$= 2.86 \qquad \text{Do the addition.}$$

$$\begin{array}{r} \overset{2}{0.88} \\ \times \quad 3 \\ \hline 2.64 \end{array}$$

$$\begin{array}{r} 0.66 \\ 4\overline{)2.64} \\ -2\,4 \\ \hline 24 \\ -24 \\ \hline 0 \end{array} \qquad \begin{array}{r} 2.20 \\ 3\overline{)6.60} \\ -6 \\ \hline 06 \\ -6 \\ \hline 00 \\ -0 \\ \hline 0 \end{array}$$

$$\begin{array}{r} 0.66 \\ +2.20 \\ \hline 2.86 \end{array}$$

State The total cost of the items is $2.86.

Check If approximately 1 pound of fruit, priced at approximately $1 per pound, was purchased, then about $1 was spent on fruit. If exactly $\frac{1}{3}$ of a pound of coffee, priced at approximately $6 per pound, was purchased, then about $\frac{1}{3} \cdot$ $6, or $2, was spent on coffee. Since the approximate cost of the items $1 + $2 = $3, is close to the result, $2.86, the result seems reasonable.

ANSWERS TO SELF CHECKS

1. a. 0.5   **b.** 0.1875   **c.** 4.5   **2. a.** 0.4   **b.** 0.32   **3.** 3.85   **4.** $0.08\overline{3}$ **5.** $-0.\overline{39}$ **6.** 0.44
7. 0.292   **8. a.** >   **b.** <   **c.** =   **9.** $\frac{9}{5}, 1.832, 1\frac{5}{6}$ **10.** $\frac{209}{300}$ **11.** approximately 0.36
12. 0.6475   **13.** $6.26

SECTION 5.5   STUDY SET

VOCABULARY

Fill in the blanks.

1. A fraction and a decimal are said to be _____ if they name the same number.

2. The _____ equivalent of $\frac{3}{4}$ is 0.75.

3. 0.75, 0.625, and 3.5 are examples of _____ decimals.

4. 0.3333 . . . and 1.666 . . . are examples of _____ decimals.

CONCEPTS

Fill in the blanks.

5. $\frac{7}{8}$ means 7 ▢ 8.

6. To write a fraction as a decimal, divide the _____ of the fraction by its denominator.

7. To perform the division shown below, a decimal point and two additional _____ were written to the right of 3.

$$4\overline{)3.00}$$

8. Sometimes, when finding the decimal equivalent of a fraction, the division process ends because a remainder of 0 is obtained. We call the resulting decimal a _____ decimal.

9. Sometimes, when we are finding the decimal equivalent of a fraction, the division process never gives a remainder of 0. We call the resulting decimal a _____ decimal.

10. If the denominator of a fraction in simplified form has factors of only 2's or 5's, or a combination of both, it can be written as a decimal by multiplying it by a form of ▢ .

11. **a.** Round 0.3777 . . . to the nearest hundredth.

    b. Round 0.212121 . . . to the nearest thousandth.

12. **a.** When evaluating the expression $0.25 + \left(2.3 + \frac{2}{5}\right)^2$, would it be easier to work in terms of fractions or decimals?

    b. What is the first step that should be performed to evaluate the expression?

▎ NOTATION

13. Write each decimal in fraction form.

    a. 0.7   **b.** 0.77

14. Write each repeating decimal in simplest form using an overbar.

    a. 0.888 . . .   **b.** 0.323232 . . .

    c. 0.56333 . . .   **d.** 0.8898989 . . .

▎ GUIDED PRACTICE

Write each fraction as a decimal. **See Example 1.**

15. $\dfrac{1}{2}$ 16. $\dfrac{1}{4}$

17. $\dfrac{7}{8}$ 18. $\dfrac{3}{8}$

19. $\dfrac{11}{20}$ 20. $\dfrac{17}{20}$

21. $\dfrac{13}{5}$ 22. $\dfrac{15}{2}$

23. $\dfrac{9}{16}$ 24. $\dfrac{3}{32}$

25. $-\dfrac{17}{32}$ 26. $-\dfrac{15}{16}$

Write each fraction as a decimal using multiplication by a form of 1. **See Example 2.**

27. $\dfrac{3}{5}$ 28. $\dfrac{13}{25}$

29. $\dfrac{9}{40}$ 30. $\dfrac{7}{40}$

31. $\dfrac{19}{25}$ 32. $\dfrac{21}{50}$

33. $\dfrac{1}{500}$ 34. $\dfrac{1}{250}$

Write each mixed number in decimal form. **See Example 3.**

35. $3\dfrac{3}{4}$ 36. $5\dfrac{4}{5}$

37. $12\dfrac{11}{16}$ 38. $32\dfrac{9}{16}$

Write each fraction as a decimal. Use an overbar in your answer. **See Example 4.**

39. $\dfrac{1}{9}$ 40. $\dfrac{8}{9}$

41. $\dfrac{7}{12}$ 42. $\dfrac{11}{12}$

43. $\dfrac{7}{90}$ 44. $\dfrac{1}{99}$

45. $\dfrac{1}{60}$ 46. $\dfrac{1}{66}$

Write each fraction as a decimal. Use an overbar in your answer. **See Example 5.**

47. $-\dfrac{5}{11}$ 48. $-\dfrac{7}{11}$

49. $-\dfrac{20}{33}$ 50. $-\dfrac{16}{33}$

Write each fraction in decimal form. Round to the nearest hundredth. **See Example 6.**

51. $\dfrac{7}{30}$ 52. $\dfrac{8}{9}$

53. $\dfrac{22}{45}$ 54. $\dfrac{17}{45}$

55. $\dfrac{24}{13}$ 56. $\dfrac{34}{11}$

57. $-\dfrac{13}{12}$ 58. $-\dfrac{25}{12}$

Write each fraction in decimal form. Round to the nearest thousandth. **See Example 7.**

59. $\dfrac{5}{33}$ 60. $\dfrac{5}{24}$

61. $\dfrac{10}{27}$ 62. $\dfrac{17}{21}$

Graph the given numbers on a number line. **See Objective 4.**

63. $1\dfrac{3}{4}$, -0.75, $0.\overline{6}$, $-3.8\overline{3}$

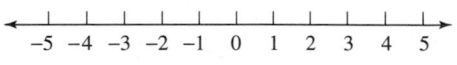

64. $2\frac{7}{8}$, -2.375, $0.\overline{3}$, $4.1\overline{6}$

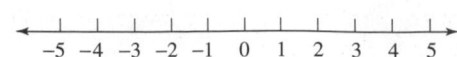

65. 3.875, $-3.\overline{5}$, $0.\overline{2}$, $-1\frac{4}{5}$

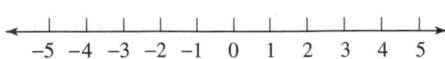

66. 1.375, $-4\frac{1}{7}$, $0.\overline{1}$, $-2.\overline{7}$

Place an $<$, $>$, or an $=$ symbol in the box to make a true statement. See Example 8.

67. $\frac{7}{8}$ ☐ 0.895

68. $\frac{3}{8}$ ☐ 0.381

69. $0.\overline{7}$ ☐ $\frac{17}{22}$

70. $0.\overline{45}$ ☐ $\frac{7}{16}$

71. $\frac{52}{25}$ ☐ 2.08

72. 4.4 ☐ $\frac{22}{5}$

73. $-\frac{11}{20}$ ☐ $-0.\overline{48}$

74. $-0.0\overline{9}$ ☐ $-\frac{1}{11}$

Write the numbers in order from smallest to largest. See Example 9.

75. $6\frac{1}{2}$, 6.25, $\frac{19}{3}$

76. $7\frac{3}{8}$, 7.08, $\frac{43}{6}$

77. $-0.\overline{81}$, $-\frac{8}{9}$, $-\frac{6}{7}$

78. $-0.\overline{19}$, $-\frac{1}{11}$, -0.1

Evaluate each expression. Work in terms of fractions. See Example 10.

79. $\frac{1}{9} + 0.3$

80. $\frac{2}{3} + 0.1$

81. $0.9 - \frac{7}{12}$

82. $0.99 - \frac{5}{6}$

83. $\frac{5}{11}(0.3)$

84. $(0.9)\left(\frac{1}{27}\right)$

85. $\frac{1}{4}(0.25) + \frac{15}{16}$

86. $\frac{2}{5}(0.02) - (0.04)$

Estimate the value of each expression. Work in terms of decimals. See Example 11.

87. $0.24 + \frac{1}{3}$

88. $0.02 + \frac{5}{6}$

89. $5.69 - \frac{5}{12}$

90. $3.19 - \frac{2}{3}$

91. $0.43 - \frac{1}{12}$

92. $0.27 + \frac{5}{12}$

93. $\frac{1}{15} - 0.55$

94. $\frac{7}{30} - 0.84$

Evaluate each expression. Work in terms of decimals. See Example 12.

95. $(3.5 + 6.7)\left(-\frac{1}{4}\right)$

96. $\left(-\frac{5}{8}\right)\left(5.3 - 3\frac{9}{10}\right)$

97. $\left(\frac{1}{5}\right)^2(1.7)$

98. $(2.35)\left(\frac{2}{5}\right)^2$

99. $7.5 - (0.78)\left(\frac{1}{2}\right)^2$

100. $8.1 - \left(\frac{3}{4}\right)^2(0.12)$

101. $\frac{3}{8}(3.2) + \left(4\frac{1}{2}\right)\left(-\frac{1}{4}\right)$

102. $(-0.8)\left(\frac{1}{4}\right) + \left(\frac{1}{5}\right)(0.39)$

APPLICATIONS

103. DRAFTING The architect's scale shown below has several measuring edges. The edge marked 16 divides each inch into 16 equal parts. Find the decimal form for each fractional part of 1 inch that is highlighted with a red arrow.

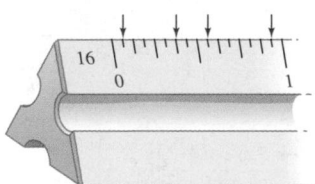

104. MILEAGE SIGNS The freeway sign shown below gives the number of miles to the next three exits. Convert the mileages to decimal notation.

105. GARDENING Two brands of replacement line for a lawn trimmer shown below are labeled in different ways. On one package, the line's thickness is expressed as a decimal; on the other, as a fraction. Which line is thicker?

106. AUTO MECHANICS While doing a tune-up, a mechanic checks the gap on one of the spark plugs of a car to be sure it is firing correctly. The owner's manual states that the gap should be $\frac{2}{125}$ inch. The gauge the mechanic uses to check the gap is in decimal notation; it registers 0.025 inch. Is the spark plug gap too large or too small?

107. HORSE RACING In thoroughbred racing, the time a horse takes to run a given distance is measured using fifths of a second. For example, :23² (read "twenty-three and two") means $23\frac{2}{5}$ seconds. The illustration below lists four split times for a horse named *Speedy Flight* in a $1\frac{1}{16}$-mile race. Express each split time in decimal form.

Speedy Flight	Turfway Park, Ky	3-year–old
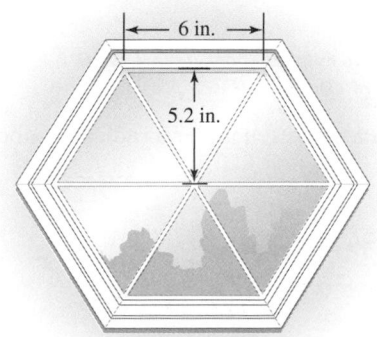	17 May 2010	$1\frac{1}{16}$ mile
Splits	:23² :23⁴ :24¹ :32³	

108. GEOLOGY A geologist weighed a rock sample at the site where it was discovered and found it to weigh $17\frac{7}{8}$ lb. Later, a more accurate digital scale in the laboratory gave the weight as 17.671 lb. What is the difference in the two measurements?

109. WINDOW REPLACEMENTS The amount of sunlight that comes into a room depends on the area of the windows in the room. What is the area of the window shown below? (*Hint:* Use the formula $A = \frac{1}{2}bh$.)

110. FORESTRY A command post asked each of three fire crews to estimate the length of the fire line they were fighting. Their reports came back in different forms, as shown. Find the perimeter of the fire. Round to the nearest tenth.

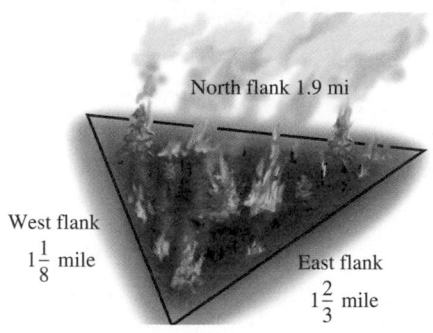

North flank 1.9 mi

West flank $1\frac{1}{8}$ mile

East flank $1\frac{2}{3}$ mile

111. DELICATESSENS A shopper purchased $\frac{2}{3}$ pound of green olives, priced at $4.14 per pound, and $\frac{3}{4}$ pound of smoked ham, selling for $5.68 per pound. Find the total cost of these items.

112. CHOCOLATE A shopper purchased $\frac{3}{4}$ pound of dark chocolate, priced at $8.60 per pound, and $\frac{1}{3}$ pound of milk chocolate, selling for $5.25 per pound. Find the total cost of these items.

WRITING

113. Explain the procedure used to write a fraction in decimal form.

114. How does the terminating decimal 0.5 differ from the repeating decimal $0.\overline{5}$?

115. A student represented the repeating decimal 0.1333 . . . as $0.1\overline{333}$. Is this the best form? Explain why or why not.

116. Is 0.10100100010000 . . . a repeating decimal? Explain why or why not.

117. A student divided 19 by 25 to find the decimal equivalent of $\frac{19}{25}$ to be 0.76. Explain how she can check this result.

118. Explain the error in the following work to find the decimal equivalent for $\frac{5}{6}$.

$$\begin{array}{r} 1.2 \\ 5{\overline{\smash{)}\,6.0}} \\ \underline{-5} \\ 1\,0 \\ \underline{-1\,0} \\ 0 \end{array}$$ Thus, $\frac{5}{6} = 1.2$.

REVIEW

119. Write each set of numbers.
  a. the first ten whole numbers
  b. the first ten prime numbers
  c. the integers

120. Give an example of each property.
  a. the commutative property of addition
  b. the associative property of multiplication
  c. the multiplication property of 1

Objectives

1 Find the square root of a perfect square.

2 Find the square root of fractions and decimals.

3 Evaluate expressions that contain square roots.

4 Evaluate formulas involving square roots.

5 Approximate square roots.

SECTION 5.6
Square Roots

We have discussed the relationships between addition and subtraction and between multiplication and division. In this section, we explore the relationship between raising a number to a power and finding a root. Decimals play an important role in this discussion.

1 Find the square root of a perfect square.

When we raise a number to the second power, we are squaring it, or finding its **square.**

The square of 6 is 36, because $6^2 = 36$.

The square of -6 is 36, because $(-6)^2 = 36$.

The **square root** of a given number is a number whose square is the given number. For example, the square roots of 36 are 6 and -6, because either number, when squared, is 36.

Every positive number has two square roots. The number 0 has only one square root. In fact, it is its own square root, because $0^2 = 0$.

Square Root

A number is a **square root** of a second number if the square of the first number equals the second number.

Self Check 1

Find the two square roots of 64.

Now Try Problem 21

EXAMPLE 1 Find the two square roots of 49.

Strategy We will ask "What positive number and what negative number, when squared, is 49?"

WHY The square root of 49 is a number whose square is 49.

Solution

7 is a square root of 49 because $7^2 = 49$

and

-7 is a square root of 49 because $(-7)^2 = 49$.

In Example 1, we saw that 49 has two square roots—one positive and one negative. The symbol $\sqrt{}$ is called a **radical symbol** and is used to indicate a positive square root of a nonnegative number. When reading this symbol, we usually drop the word *positive* and simply say *square root*. Since 7 is the positive square root of 49, we can write

$\sqrt{49} = 7$ $\sqrt{49}$ *represents the positive number whose square is 49.*
        Read as "the square root of 49 is 7."

When a number, called the **radicand,** is written under a radical symbol, we have a **radical expression.**

Radical symbol
$\sqrt{49}$ ← Radicand

Radical expression

Some other examples of radical expressions are:

$$\sqrt{36} \qquad \sqrt{100} \qquad \sqrt{144} \qquad \sqrt{81}$$

To evaluate (or simplify) a radical expression like those shown above, we need to find the positive square root of the radicand. For example, if we evaluate $\sqrt{36}$ (read as "the square root of 36"), the result is

$$\sqrt{36} = 6$$

because $6^2 = 36$.

Caution!   Remember that the radical symbol asks you to find only the *positive* square root of the radicand. It is incorrect, for example, to say that

$$\sqrt{36} \text{ is } 6 \text{ and } -6$$

The symbol $-\sqrt{}$ is used to indicate the **negative square root** of a positive number. It is the opposite of the positive square root. Since –6 is the negative square root of 36, we can write

$-\sqrt{36} = -6$ *Read as "the negative square root of 36 is −6" or "the opposite of the square root of 36 is −6." $-\sqrt{36}$ represents the negative number whose square is 36.*

If the number under the radical symbol is 0, we have $\sqrt{0} = 0$.

Numbers, such as 36 and 49, that are squares of whole numbers, are called **perfect squares.** To evaluate square root radical expressions, it is helpful to be able to identify perfect square radicands. You need to memorize the following list of perfect squares, shown in red.

Perfect Squares

$0 = 0^2$	$16 = 4^2$	$64 = 8^2$	$144 = 12^2$
$1 = 1^2$	$25 = 5^2$	$81 = 9^2$	$169 = 13^2$
$4 = 2^2$	$36 = 6^2$	$100 = 10^2$	$196 = 14^2$
$9 = 3^2$	$49 = 7^2$	$121 = 11^2$	$225 = 15^2$

A calculator is helpful in finding the square root of a perfect square that is larger than 225.

EXAMPLE 2   Evaluate each square root: **a.** $\sqrt{81}$ **b.** $-\sqrt{100}$

Strategy In each case, we will determine what positive number, when squared, produces the radicand.

WHY The radical symbol $\sqrt{}$ indicates that the positive square root of the number written under it should be found.

Solution

a. $\sqrt{81} = 9$ *Ask: What positive number, when squared, is 81? The answer is 9 because $9^2 = 81$.*

b. $-\sqrt{100}$ is the opposite (or negative) of the square root of 100. Since $\sqrt{100} = 10$, we have

$$-\sqrt{100} = -10$$

Self Check 2

Evaluate each square root:
a. $\sqrt{144}$
b. $-\sqrt{81}$

Now Try Problems 25 and 29

Caution! Radical expressions such as

$$\sqrt{-36} \qquad \sqrt{-100} \qquad \sqrt{-144} \qquad \sqrt{-81}$$

do not represent real numbers, because there are no real numbers that when squared give a negative number.

Be careful to note the difference between expressions such as $-\sqrt{36}$ and $\sqrt{-36}$. We have seen that $-\sqrt{36}$ is a real number: $-\sqrt{36} = -6$. In contrast, $\sqrt{-36}$ is not a real number.

Using Your CALCULATOR **Finding a square root**

We use the $\boxed{\sqrt{}}$ key (square root key) on a scientific calculator to find square roots. For example, to find $\sqrt{729}$, we enter these numbers and press these keys.

729 $\boxed{\sqrt{}}$ $\qquad\qquad\qquad\qquad\qquad\qquad\qquad$ $\boxed{ 27}$

We have found that $\sqrt{729} = 27$. To check this result, we need to square 27. This can be done by entering 27 and pressing the $\boxed{x^2}$ key. We obtain 729. Thus, 27 is the square root of 729.

Some calculator models require keystrokes of $\boxed{\text{2nd}}$ and then $\boxed{\sqrt{}}$ followed by the radicand to find a square root.

---

2 Find the square root of fractions and decimals.

So far, we have found square roots of whole numbers. We can also find square roots of fractions and decimals.

Self Check 3

Evaluate:

a. $\sqrt{\dfrac{16}{49}}$

b. $\sqrt{0.04}$

Now Try Problems 37 and 43

EXAMPLE 3 Evaluate each square root: **a.** $\sqrt{\dfrac{25}{64}}$ **b.** $\sqrt{0.81}$

Strategy In each case, we will determine what positive number, when squared, produces the radicand.

WHY The radical symbol $\sqrt{}$ indicates that the positive square root of the number written under it should be found.

Solution

a. $\sqrt{\dfrac{25}{64}} = \dfrac{5}{8}$ $\quad$ Ask: What positive fraction, when squared, is $\frac{25}{64}$? The answer is $\frac{5}{8}$ because $\left(\frac{5}{8}\right)^2 = \frac{25}{64}$.

b. $\sqrt{0.81} = 0.9$ $\quad$ Ask: What positive decimal, when squared, is 0.81? The answer is 0.9 because $(0.9)^2 = 0.81$.

---

3 Evaluate expressions that contain square roots.

In Chapters 1, 2, 3, and 4, we used the order of operations rule to evaluate expressions that involve more than one operation. If an expression contains any square roots, they are to be evaluated at the same stage in your solution as exponential expressions. (See step 2 in the familiar order of operations rule on the next page.)

Order of Operations

1. Perform all calculations within parentheses and other grouping symbols following the order listed in Steps 2–4 below, working from the innermost pair of grouping symbols to the outermost pair.
2. Evaluate all exponential expressions and **square roots.**
3. Perform all multiplications and divisions as they occur from left to right.
4. Perform all additions and subtractions as they occur from left to right.

EXAMPLE 4   Evaluate: **a.** $\sqrt{64} + \sqrt{9}$ **b.** $-\sqrt{25} - \sqrt{225}$

Strategy We will scan the expression to determine what operations need to be performed. Then we will perform those operations, one-at-a-time, following the order of operations rule.

WHY If we don't follow the correct order of operations, the expression can have more than one value.

Solution Since the expression does not contain any parentheses, we begin with step 2 of the rules for the order of operations: Evaluate all exponential expressions and any square roots.

a. $\sqrt{64} + \sqrt{9} = 8 + 3$ Evaluate each square root first.

$= 11$ Do the addition.

b. $-\sqrt{25} - \sqrt{225} = -5 - 15$ Evaluate each square root first.

$= -20$ Do the subtraction.

Self Check 4
Evaluate:
a. $\sqrt{121} + \sqrt{1}$
b. $-\sqrt{9} - \sqrt{196}$
Now Try Problems 49 and 53

EXAMPLE 5   Evaluate: **a.** $6\sqrt{100}$ **b.** $-5\sqrt{16} + 3\sqrt{9}$

Strategy We will scan the expression to determine what operations need to be performed. Then we will perform those operations, one-at-a-time, following the order of operations rule.

WHY If we don't follow the correct order of operations, the expression can have more than one value.

Solution Since the expression does not contain any parentheses, we begin with step 2 of the rules for the order of operations: Evaluate all exponential expressions and any square roots.

a. We note that $6\sqrt{100}$ means $6 \cdot \sqrt{100}$.

$6\sqrt{100} = 6(10)$ Evaluate the square root first.

$= 60$ Do the multiplication.

b. $-5\sqrt{16} + 3\sqrt{9} = -5(4) + 3(3)$ Evaluate each square root first.

$= -20 + 9$ Do the multiplication.

$= -11$ Do the addition.

Self Check 5
Evaluate:
a. $8\sqrt{121}$
b. $-6\sqrt{25} + 2\sqrt{36}$
Now Try Problems 57 and 61

EXAMPLE 6   Evaluate: $12 + 3[3^2 - (4 - 1)\sqrt{36}]$

Strategy We will work within the parentheses first and then within the brackets. Within each set of grouping symbols, we will follow the order of operations rule.

WHY By the order of operations rule, we must work from the *innermost* pair of grouping symbols to the *outermost.*

Self Check 6
Evaluate:
$10 - 4[2^2 - (3 + 2)\sqrt{4}]$
Now Try Problems 65 and 69

Solution

$$12 + 3\left[3^2 - (\mathbf{4 - 1})\sqrt{36}\right] = 12 + 3\left[3^2 - 3\sqrt{36}\right]$$ Do the subtraction within the parentheses.

$$= 12 + 3[9 - 3(6)]$$ Within the brackets, evaluate the exponential expression and the square root.

$$= 12 + 3[9 - 18]$$ Do the multiplication within the brackets.

$$= 12 + 3[-9]$$ Do the subtraction within the brackets.

$$= 12 + (-27)$$ Do the multiplication.

$$= -15$$ Do the addition. ∎

4 Evaluate formulas involving square roots.

To evaluate formulas that involve square roots, we replace the letters with specific numbers and the then use the order of operations rule.

Self Check 7

Evaluate $a = \sqrt{c^2 - b^2}$ for $c = 17$ and $b = 15$.

Now Try Problem 81

EXAMPLE 7 Evaluate $c = \sqrt{a^2 + b^2}$ for $a = 3$ and $b = 4$.

Strategy In the given formula, we will replace the letter a with 3 and b with 4. Then we will use the order of operations rule to find the value of the radicand.

WHY We need to know the value of the radicand before we can find its square root.

Solution

$$c = \sqrt{a^2 + b^2}$$ This is the formula to evaluate.

$$= \sqrt{3^2 + 4^2}$$ Replace a with 3 and b with 4.

$$= \sqrt{9 + 16}$$ Evaluate the exponential expressions.

$$= \sqrt{25}$$ Do the addition.

$$= 5$$ Evaluate the square root. ∎

5 Approximate square roots.

In Examples 2–7, we have found square roots of perfect squares. If a number is not a perfect square, we can use the $\boxed{\sqrt{}}$ key on a calculator or a table of square roots to find its *approximate* square root. For example, to find $\sqrt{17}$ using a scientific calculator, we enter 17 and press the square root key:

17 $\boxed{\sqrt{}}$

The display reads

4.123105626

This result is an approximation, because the exact value of $\sqrt{17}$ is a **nonterminating decimal** that never repeats. If we round to the nearest thousandth, we have

$$\sqrt{17} \approx 4.123$$ Read $\approx$ as "is approximately equal to."

To check this approximation, we square 4.123.

$$(4.123)^2 = 16.999129$$

Since the result is close to 17, we know that $\sqrt{17} \approx 4.123$.

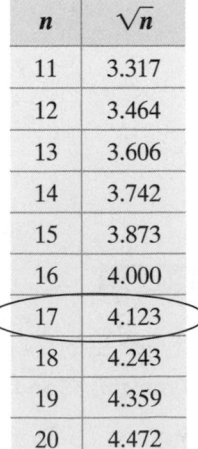

n	$\sqrt{n}$
11	3.317
12	3.464
13	3.606
14	3.742
15	3.873
16	4.000
17	4.123
18	4.243
19	4.359
20	4.472

A portion of the table of square roots from Appendix I on page A-9 is shown in the margin on the previous page. The table gives decimal approximations of square roots of whole numbers that are not perfect squares. To find an approximation of $\sqrt{17}$ to the nearest thousandth, we locate 17 in the n-column of the table and scan directly right, to the $\sqrt{n}$-column, to find that $\sqrt{17} \approx 4.123$.

EXAMPLE 8 Use a calculator to approximate each square root. Round to the nearest hundredth. **a.** $\sqrt{373}$ **b.** $\sqrt{56.2}$ **c.** $\sqrt{0.0045}$

Strategy We will identify the radicand and find the square root using the $\sqrt{\ }$ key. Then we will identify the digit in the thousandths column of the display.

WHY To round to the hundredths column, we must determine whether the digit in the thousandths column is less than 5, or greater than or equal to 5.

Solution
a. From the calculator, we get $\sqrt{373} \approx 19.31320792$. Rounded to the nearest hundredth, $\sqrt{373} \approx 19.31$.

b. From the calculator, we get $\sqrt{56.2} \approx 7.496665926$. Rounded to the nearest hundredth, $\sqrt{56.2} \approx 7.50$.

c. From the calculator, we get $\sqrt{0.0045} \approx 0.067082039$. Rounded to the nearest hundredth, $\sqrt{0.0045} \approx 0.07$.

Self Check 8
Use a scientific calculator to approximate each square root. Round to the nearest hundredth.
a. $\sqrt{153}$
b. $\sqrt{607.8}$
c. $\sqrt{0.076}$
Now Try Problems 87 and 91

ANSWERS TO SELF CHECKS
1. 8 and -8 **2. a.** 12 **b.** -9 **3. a.** $\frac{4}{7}$ **b.** 0.2 **4. a.** 12 **b.** -17 **5. a.** 88 **b.** -18
6. 34 **7.** 8 **8. a.** 12.37 **b.** 24.65 **c.** 0.28

SECTION 5.6 STUDY SET

VOCABULARY
Fill in the blanks.

1. When we raise a number to the second power, we are squaring it, or finding its _____.
2. The square _____ of a given number is a number whose square is the given number.
3. The symbol $\sqrt{\ }$ is called a _____ symbol.
4. Label the *radicand,* the *radical expression,* and the *radical symbol* in the illustration below.

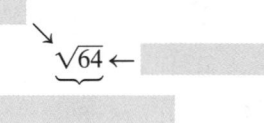

5. Whole numbers such as 36 and 49, that are squares of whole numbers, are called _____ squares.
6. The exact value of $\sqrt{17}$ is a _____ decimal that never repeats.

CONCEPTS
Fill in the blanks.

7. **a.** The square of 5 is ___, because $5^2 =$ ___.
   b. The square of $\frac{1}{4}$ is ___, because $\left(\frac{1}{4}\right)^2 =$ ___.
8. Complete the list of perfect squares: 1, 4, ___, 16, ___, 36, 49, 64, ___, 100, ___, 144, ___, 196, ___.
9. **a.** $\sqrt{49} = 7$, because ___$^2 = 49$.
   b. $\sqrt{4} = 2$, because ___$^2 = 4$.
10. **a.** $\sqrt{\frac{9}{16}} =$ ___, because $\left(\frac{3}{4}\right)^2 = \frac{9}{16}$.
    b. $\sqrt{0.16} =$ ___, because $(0.4)^2 = 0.16$.
11. Evaluate each square root.
    a. $\sqrt{1}$ **b.** $\sqrt{0}$
12. Evaluate each square root.
    a. $\sqrt{121}$ **b.** $\sqrt{144}$ **c.** $\sqrt{169}$
    d. $\sqrt{196}$ **e.** $\sqrt{225}$

13. In what step of the order of operations rule are square roots to be evaluated?

14. Graph $\sqrt{9}$ and $-\sqrt{4}$ on a number line.

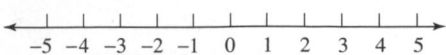

15. Graph $-\sqrt{3}$ and $\sqrt{7}$ on a number line. (*Hint:* Use a calculator or square root table to approximate each square root first.)

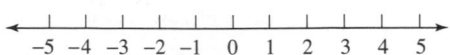

16. a. Between what two whole numbers would $\sqrt{19}$ be located when graphed on a number line?

   b. Between what two whole numbers would $\sqrt{50}$ be located when graphed on a number line?

NOTATION

Fill in the blanks.

17. a. The symbol $\sqrt{}$ is used to indicate a positive _____ _____.

   b. The symbol $-\sqrt{}$ is used to indicate the _____ square root of a positive number.

18. $4\sqrt{9}$ means $4 \boxed{} \sqrt{9}$.

Complete each solution to evaluate the expression.

19. $-\sqrt{49} + \sqrt{64} = \boxed{} + \boxed{}$
$$= 1$$

20. $2\sqrt{100} - 5\sqrt{25} = 2(\boxed{}) - 5(\boxed{})$
$$= \boxed{} - 25$$
$$= -5$$

GUIDED PRACTICE

Find the two square roots of each number. See Example 1.

21. 25

22. 1

23. 16

24. 144

Evaluate each square root without using a calculator. See Example 2.

25. $\sqrt{16}$

26. $\sqrt{64}$

27. $\sqrt{9}$

28. $\sqrt{16}$

29. $-\sqrt{144}$

30. $-\sqrt{121}$

31. $-\sqrt{49}$

32. $-\sqrt{81}$

Use a calculator to evaluate each square root. See Objective 1, Using Your Calculator.

33. $\sqrt{961}$

34. $\sqrt{841}$

35. $\sqrt{3,969}$

36. $\sqrt{5,625}$

Evaluate each square root without using a calculator. See Example 3.

37. $\sqrt{\dfrac{4}{25}}$

38. $\sqrt{\dfrac{36}{121}}$

39. $-\sqrt{\dfrac{16}{9}}$

40. $-\sqrt{\dfrac{64}{25}}$

41. $-\sqrt{\dfrac{1}{81}}$

42. $-\sqrt{\dfrac{1}{4}}$

43. $\sqrt{0.64}$

44. $\sqrt{0.36}$

45. $-\sqrt{0.81}$

46. $-\sqrt{0.49}$

47. $\sqrt{0.09}$

48. $\sqrt{0.01}$

Evaluate each expression without using a calculator. See Example 4.

49. $\sqrt{36} + \sqrt{1}$

50. $\sqrt{100} + \sqrt{16}$

51. $\sqrt{81} + \sqrt{49}$

52. $\sqrt{4} + \sqrt{36}$

53. $-\sqrt{144} - \sqrt{16}$

54. $-\sqrt{1} - \sqrt{196}$

55. $-\sqrt{225} + \sqrt{144}$

56. $-\sqrt{169} + \sqrt{16}$

Evaluate each expression without using a calculator. See Example 5.

57. $4\sqrt{25}$

58. $2\sqrt{81}$

59. $-10\sqrt{196}$

60. $-40\sqrt{4}$

61. $-4\sqrt{169} + 2\sqrt{4}$

62. $-6\sqrt{81} + 5\sqrt{1}$

63. $-8\sqrt{16} + 5\sqrt{225}$

64. $-3\sqrt{169} + 2\sqrt{225}$

Evaluate each expression without using a calculator. See Example 6.

65. $15 + 4\left[5^2 - (6 - 1)\sqrt{4}\right]$

66. $18 + 2\left[4^2 - (7 - 3)\sqrt{9}\right]$

67. $50 - \left[(6^2 - 24) + 9\sqrt{25}\right]$

68. $40 - \left[(7^2 - 40) + 7\sqrt{64}\right]$

69. $\sqrt{196} + 3\left(5^2 - 2\sqrt{225}\right)$

70. $\sqrt{169} + 2\left(7^2 - 3\sqrt{144}\right)$

71. $\dfrac{\sqrt{16} - 6(2^2)}{\sqrt{4}}$

72. $\dfrac{\sqrt{49} - 3(1^6)}{\sqrt{16} - \sqrt{64}}$

73. $\sqrt{\dfrac{1}{16}} - \sqrt{\dfrac{9}{25}}$

74. $\sqrt{\dfrac{25}{9}} - \sqrt{\dfrac{64}{81}}$

75. $5\left(-\sqrt{49}\right)(-2)^2$

76. $\left(-\sqrt{64}\right)(-2)(3)^3$

77. $(6^2)\sqrt{0.04} + 2.36$

78. $(5^2)\sqrt{0.25} + 4.7$

79. $-\left(-3\sqrt{1.44} + 5\right)$

80. $-\left(-2\sqrt{1.21} - 6\right)$

Evaluate each formula without using a calculator. See Example 7.

81. Evaluate $c = \sqrt{a^2 + b^2}$ for $a = 9$ and $b = 12$.

82. Evaluate $c = \sqrt{a^2 + b^2}$ for $a = 6$ and $b = 8$.

83. Evaluate $a = \sqrt{c^2 - b^2}$ for $c = 25$ and $b = 24$.

84. Evaluate $b = \sqrt{c^2 - a^2}$ for $c = 17$ and $a = 8$.

Use a calculator (or the square root table in Appendix II) to complete each square root table. Round to the nearest thousandth when an answer is not exact. **See Example 8.**

85.

Number	Square Root
1	
2	
3	
4	
5	
6	
7	
8	
9	
10	

86.

Number	Square Root
10	
20	
30	
40	
50	
60	
70	
80	
90	
100	

Use a calculator (or a square root table) to approximate each of the following to the nearest hundredth. **See Example 8.**

87. $\sqrt{15}$ **88.** $\sqrt{51}$

89. $\sqrt{66}$ **90.** $\sqrt{204}$

Use a calculator to approximate each of the following to the nearest thousandth. **See Example 8.**

91. $\sqrt{24.05}$ **92.** $\sqrt{70.69}$

93. $-\sqrt{11.1}$ **94.** $\sqrt{0.145}$

APPLICATIONS

In the following problems, some lengths are expressed as square roots. Solve each problem by evaluating any square roots. You may need to use a calculator. If so, round to the nearest tenth when an answer is not exact.

95. CARPENTRY Find the length of the slanted side of each roof truss shown below.

a.

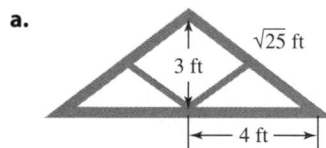

b.

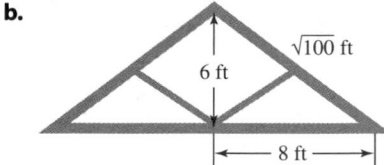

96. RADIO ANTENNAS Refer to the illustration below. How far from the base of the antenna is each guy wire anchored to the ground? (The measurements are in feet.)

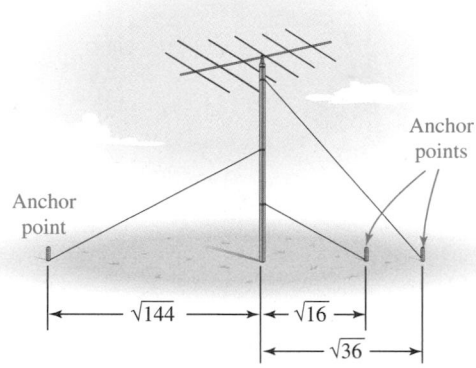

97. BASEBALL The illustration below shows some dimensions of a major league baseball field. How far is it from home plate to second base?

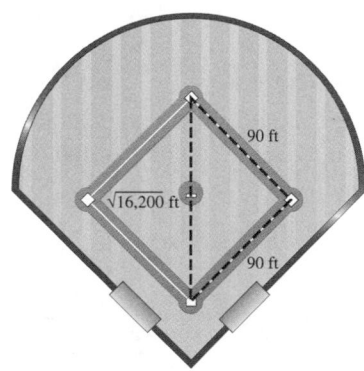

98. SURVEYING Refer to the illustration below. Use the imaginary triangles set up by a surveyor to find the length of each lake. (The measurements are in meters.)

a.

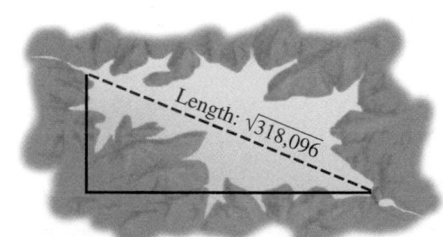

b.

99. FLATSCREEN TELEVISIONS The picture screen on a television set is measured diagonally. What size screen is shown below?

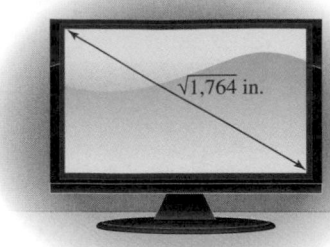

$\sqrt{1{,}764}$ in.

100. LADDERS A painter's ladder is shown below. How long are the legs of the ladder?

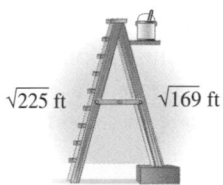

$\sqrt{225}$ ft $\sqrt{169}$ ft

■ WRITING

101. When asked to find $\sqrt{16}$, a student answered 8. Explain his misunderstanding of the concept of square root.

102. Explain the difference between the *square* and the *square root* of a number. Give an example.

103. What is a *nonterminating* decimal? Use an example in your explanation.

104. a. How would you check whether $\sqrt{389} = 17$?

   b. How would you check whether $\sqrt{7} \approx 2.65$?

105. Explain why $\sqrt{-4}$ does not represent a real number.

106. Is there a difference between $-\sqrt{25}$ and $\sqrt{-25}$? Explain.

107. $\sqrt{6} \approx 2.449$. Explain why an $\approx$ symbol is used and not an $=$ symbol.

108. Without evaluating the following square roots, determine which is the largest and which is the smallest. Explain how you decided.

$$\sqrt{23}, \ \sqrt{27}, \ \sqrt{11}, \ \sqrt{6}, \ \sqrt{20}$$

■ REVIEW

109. Multiply: $6.75 \cdot 12.2$

110. Divide: $5.7)\overline{18.525}$

111. Evaluate: $(3.4)^3$

112. Add: $23.45 + 76 + 0.009 + 3.8$

Objectives

1 Simplify products.

2 Use the distributive property.

3 Simplify expressions by combining like terms.

4 Use one property of equality to solve equations that involve decimals.

5 Use more than one property of equality to solve equations that involve decimals.

6 Use equations to solve application problems that involve decimals.

SECTION **5.7**

Solving Equations That Involve Decimals

In this section, we revisit the topic of solving equations. The equations that we will solve involve decimals, and some of the solutions are decimals as well. But first, we need to practice simplifying algebraic expressions that involve decimals.

1 Simplify products.

In algebra, we often replace one algebraic expression with another that is equivalent and simpler in form. That process is called *simplifying an expression.*

EXAMPLE 1 Simplify:

a. $5.4 \cdot 6x$ **b.** $9.91y(-8)$ **c.** $-1.2(-7.1t)$ **d.** $2(6.78s)4$

Strategy We will use the commutative and associative properties of multiplication to reorder and regroup the factors in each expression.

WHY We want to group all of the numerical factors of an expression together so that we can find their product.

Solution

a. $5.4 \cdot 6x = (5.4 \cdot 6)x$ Use the associative property of
multiplication to regroup the factors.

$= 32.4x$ Do the multiplication within the parentheses.

$$\overset{2}{5.4}$$
$$\underline{\times\ \ 6}$$
$$32.4$$

b. $9.91y(-8) = 9.91(-8)y$ Use the commutative property of
multiplication to reorder the factors.

$= -79.28y$ Do the multiplication, working from
left to right. Since the signs of 9.91 and -8
are unlike, make the final answer negative.

$$\overset{7}{9.91}$$
$$\underline{\times\ \ \ 8}$$
$$79.28$$

c. $-1.2(-7.1t) = [-1.2(-7.1)]t$ Use the associative property of
multiplication to regroup the factors.

$= 8.52t$ Do the multiplication within the brackets.
Since the signs are like, make the
final answer positive.

$$1.2$$
$$\underline{\times\ 7.1}$$
$$12$$
$$\underline{840}$$
$$8.52$$

d. $2(6.78s)4 = (2 \cdot 6.78 \cdot 4)s$ Use the commutative and associative
properties of multiplication to reorder
and regroup the factors.

$= (8 \cdot 6.78)s$ Within the parentheses, use the
commutative property of multiplication:
multiply 2 and 4 to get 8.

$= 54.24s$ Complete the multiplication within
the parentheses.

$$\overset{6\ \ 6}{6.78}$$
$$\underline{\times\ \ \ 8}$$
$$54.24$$

Self Check 1

Simplify:
a. $1.6 \cdot 9a$
b. $4.23d(-3)$
c. $-8.9(-1.5x)$
d. $3(7.7g)2$

Now Try Problems 13, 15, and 19

2 Use the distributive property.

Another property that is often used to simplify algebraic expressions is the *distributive property*.

EXAMPLE 2   Multiply:

a. $3.2(3b + 4)$ **b.** $-6(5.9y - 8.2)$ **c.** $(t - 9.1)0.07$

Strategy In each case, we will distribute the multiplication by the factor outside the parentheses.

WHY In each case, we cannot simplify the expression within the parentheses. To multiply, we must use the distributive property.

Solution

a. $\mathbf{3.2}(3b + 4) = \mathbf{3.2}(3b) + \mathbf{3.2}(4)$ Distribute the multiplication
by 3.2.

$= (3.2 \cdot 3)b + 3.2(4)$ In the first term, use the
associative property of
multiplication to regroup
the factors.

$= 9.6b + 12.8$ Do the multiplication. Try to
go directly to this step.

$$3.2$$
$$\underline{\times\ 3}$$
$$9.6$$

$$3.2$$
$$\underline{\times\ 4}$$
$$12.8$$

Self Check 2

Multiply:
a. $4.7(2x + 8)$
b. $-9(5.3y - 3.9)$
c. $(c - 2.2)0.08$

Now Try Problems 21, 23, and 25

b. $-6(5.9y - 8.2) = -6(5.9y) - (-6)(8.2)$ Distribute the multiplication by −6.

$= (-6 \cdot 5.9)y - (-6)(8.2)$ In the first term, use the associative property of multiplication to regroup the factors.

$= -35.4y - (-49.2)$ Do the multiplication.

$= -35.4y + 49.2$ Write the result in simpler form. Add the opposite of −49.2. Try to go directly to this step.

$$\begin{array}{r} \overset{5}{5.9} \\ \times\ 6 \\ \hline 35.4 \end{array}$$

$$\begin{array}{r} \overset{1}{8.2} \\ \times\ 6 \\ \hline 49.2 \end{array}$$

c. $(t - 9.1)\,\mathbf{0.07} = (t)0.07 - (9.1)0.07$ Distribute the multiplication by 0.07.

$= 0.07t - 0.637$ Do the multiplication. Try to go directly to this step.

$$\begin{array}{r} 9.1 \\ \times 0.07 \\ \hline 0.637 \end{array}$$

3 Simplify expressions by combining like terms.

Recall that **like terms** are terms containing exactly the same variables raised to exactly the same powers. Here are several examples.

Like terms	*Unlike terms*	
$4.5x$ and $7.3x$	$4.5x$ and $7.3y$	The variables are not the same.
$-1.4p^2$ and $2.8p^2$	$-1.4p$ and $2.8p^2$	Same variable, but different powers.
$0.005c^3d$ and $1.22c^3d$	$0.005c^3d$ and $1.22a^3b$	The variables are not the same.

Like terms are combined by adding or subtracting the coefficients of the terms and keeping the same variables with the same exponents.

Self Check 3

Simplify by combining like terms:

a. $5.2y + 1.6y$

b. $-3.9s + (-8.5s) + 2.8s$

c. $0.55n^2 - 0.93n^2$

d. $6.72x + 7 + 3.04x - 2.81$

Now Try Problems 29, 33, and 35

EXAMPLE 3   Simplify by combining like terms: **a.** $2.6x + 9.3x$

b. $-8.4p + (-6.7p) + 4.5p$ **c.** $0.38s^2 - 0.52s^2$ **d.** $4.16w + 9 + 3.85w - 6.32$

Strategy We will use the distributive property in reverse to add (or subtract) the coefficients of the like terms. We will keep the same variables raised to the same powers.

WHY To combine like terms means to add or subtract the like terms in an expression.

Solution

a. Since $2.6x$ and $9.3x$ are like terms with the common variable x, we can combine them.

$2.6x + 9.3x = 11.9x$ Think: $(2.6 + 9.3)x = 11.9x$.

$$\begin{array}{r} 2.6 \\ +\ 9.3 \\ \hline 11.9 \end{array}$$

b. Since $-8.4p$, $-6.7p$, and $4.5p$ are like terms with the common variable p, we can combine them.

$-8.4p + (-6.7p) + 4.5p$

$= -15.1p + 4.5p$ Work left to right.
Think: $[-8.4 + (-6.7)]p = -15.1p$.

$= -10.6p$ Think: $(-15.1 + 4.5)p = -10.6p$.

$$\begin{array}{r} \overset{1}{8.4} \\ +\ 6.7 \\ \hline 15.1 \end{array}$$

$$\begin{array}{r} \overset{4\ 11}{1\cancel{5}.\cancel{1}} \\ -\ 4.5 \\ \hline 10.6 \end{array}$$

c. Since $0.38s^2$ and $0.52s^2$ are like terms with the same variable s raised to the same power, 2, we can combine them.

$$0.38s^2 - 0.52s^2 = -0.14s^2 \quad \text{Think: } (0.38 - 0.52)s^2 = -0.14s^2.$$

$$\begin{array}{r} \overset{4\,12}{0.\cancel{5}\cancel{2}} \\ -\,0.38 \\ \hline 0.14 \end{array}$$

d. We can combine the like terms that involve w. Since the constant terms in the expression, 9 and -6.32, are like terms, we can combine them, as well.

Like terms
Think: $(4.16 + 3.85)w = 8.01w.$

$$4.16w + 9 + 3.85w - 6.32 = 8.01w + 2.68$$

Like terms
Think: $9 - 6.32 = 2.68.$

$$\begin{array}{r} \overset{1\;1}{4.16} \\ +\,3.85 \\ \hline 8.01 \end{array}$$

$$\begin{array}{r} \overset{9}{8}\;\overset{10}{\cancel{1}0} \\ 9.\cancel{0}\cancel{0} \\ -\,6.32 \\ \hline 2.68 \end{array}$$

EXAMPLE 4   Simplify: $4.9(x + 5) - 3.7 - (2.5x - 6.3)$

Strategy First, we will use the distributive property to remove the parentheses. Then we will identify any like terms and combine them.

WHY Any multiplication should be performed before the addition or subtraction.

Solution

$4.9(x + 5) - 3.7 - (2.5x - 6.3)$

$= 4.9(x + 5) - 3.7 - 1(2.5x - 6.3)$ Replace the − symbol in front of (2.5x − 6.3) with −1.

$= 4.9x + 24.5 - 3.7 - 2.5x + 6.3$ Distribute the multiplication by 4.9 and −1.

$= 2.4x + 20.8 + 6.3$ Combine like terms.
  Think: $(4.9 - 2.5)x = 2.4x.$
  Think: $24.5 - 3.7 = 20.8.$

$= 2.4x + 27.1$ Combine like terms.
  Think: $20.8 + 6.3 = 27.1.$

$$\begin{array}{r} \overset{4}{4}.9 \\ \times\;\;5 \\ \hline 24.5 \end{array}$$

$$\begin{array}{r} 4.9 \\ -\,2.5 \\ \hline 2.4 \end{array}$$

$$\begin{array}{r} \overset{3\;15}{24.\cancel{5}} \\ -\,3.7 \\ \hline 20.8 \end{array}$$

$$\begin{array}{r} \overset{1}{20.8} \\ +\,6.3 \\ \hline 27.1 \end{array}$$

Self Check 4

Simplify:
$6.2(3y - 4) + 2.3 - (-5.8y + 4.7)$

Now Try Problem 37

4 Use one property of equality to solve equations that involve decimals.

Recall that to **solve an equation,** we find all the values of the variable that make the equation true. The properties of equality that we used to solve equations involving whole numbers, integers, and fractions are also used to solve equations that involve decimals.

EXAMPLE 5   Solve each equation and check the result:

a. $x + 3.5 = 7.8$ **b.** $x - 1.23 = -4.52$

Strategy We will use a property of equality to isolate the variable on one side of the equation.

WHY To solve the original equation, we want to find a simpler equivalent equation of the form $x = $ **a number**, whose solution is obvious.

Self Check 5

Solve each equation and check the result:

a. $4.6 + x = 15.7$

b. $-1.24 = r - 0.04$

Now Try Problems 45 and 49

Solution

a. We will use the subtraction property of equality to isolate x on the left side of the equation. We can undo the addition of 3.5 by subtracting 3.5 from both sides.

$$x + 3.5 = 7.8 \qquad \text{This is the equation to solve.}$$

$$x + 3.5 - \mathbf{3.5} = 7.8 - \mathbf{3.5} \qquad \text{Subtract 3.5 from both sides.}$$

$$x = 4.3 \qquad \begin{array}{l}\text{On the left side, subtract: } 3.5 - 3.5 = 0.\\ \text{On the right side, subtract:}\\ 7.8 - 3.5 = 4.3.\end{array}$$

$$\begin{array}{r} 7.8 \\ - 3.5 \\ \hline 4.3 \end{array}$$

To check, we substitute 4.3 for x in the original equation.

$$x + 3.5 = 7.8 \qquad \text{This is the original equation.}$$

$$\mathbf{4.3} + 3.5 \overset{?}{=} 7.8 \qquad \text{Substitute 4.3 for } x.$$

$$7.8 = 7.8 \qquad \text{On the left side, do the addition.}$$

$$\begin{array}{r} 4.3 \\ + 3.5 \\ \hline 7.8 \end{array}$$

Since the resulting statement $7.8 = 7.8$ is true, 4.3 is the solution.

b. We will use the addition property of equality to isolate x on the left side. We can undo the subtraction of 1.23 by adding 1.23 to both sides.

$$x - 1.23 = -4.52 \qquad \text{This is the equation to solve.}$$

$$x - 1.23 + \mathbf{1.23} = -4.52 + \mathbf{1.23} \qquad \text{Add 1.23 to both sides.}$$

$$x = -3.29 \qquad \begin{array}{l}\text{On the left side, } -1.23 +\\ 1.23 = 0. \text{ On the right side,}\\ \text{do the addition: } -4.52 +\\ 1.23 = -3.29.\end{array}$$

$$\begin{array}{r} {\scriptstyle 4\,1\,2} \\ 4.\cancel{5}\cancel{2} \\ - 1.23 \\ \hline 3.29 \end{array}$$

Check:

$$x - 1.23 = -4.52 \qquad \text{This is the original equation.}$$

$$\mathbf{-3.29} - 1.23 \overset{?}{=} -4.52 \qquad \text{Substitute } -3.29 \text{ for } x.$$

$$-4.52 = -4.52 \qquad \begin{array}{l}\text{On the left side, do the subtraction:}\\ -3.29 - 1.23 = -3.29 + (-1.23) = -4.52.\end{array}$$

$$\begin{array}{r} {\scriptstyle 1} \\ 3.29 \\ + 1.23 \\ \hline 4.52 \end{array}$$

Since the resulting statement $-4.52 = -4.52$ is true, -3.29 is the solution. ∎

Self Check 6

Solve: $\dfrac{y}{3} = -13.11$

Now Try Problem 53

EXAMPLE 6   Solve: $\dfrac{m}{2} = -24.8$

Strategy We will use a property of equality to isolate the variable on one side of the equation.

WHY To solve the original equation, we want to find a simpler equivalent equation of the form $m =$ **a number**, whose solution is obvious.

Solution We will use the multiplication property of equality to isolate m on the left side. We can undo the division by 2 by multiplying both sides by 2.

$$\frac{m}{2} = -24.8 \qquad \text{This is the equation to solve.}$$

$$2\left(\frac{m}{2}\right) = 2(-24.8) \qquad \text{Multiply both sides by 2.}$$

$$m = -49.6 \qquad \text{Do the multiplication.}$$

$$\begin{array}{r} {\scriptstyle 1} \\ 24.8 \\ \times \quad 2 \\ \hline 49.6 \end{array}$$

Check the result to verify that -49.6 is the solution. ∎

EXAMPLE 7   Solve: $-9.66 = -4.6x$

Self Check 7
Solve: $-22.32 = -3.1m$
Now Try Problem 57

Strategy We will use a property of equality to isolate the variable on one side of the equation.

WHY To solve the original equation, we need to find a simpler equivalent equation of the form **a number = x**, whose solution is obvious.

Solution We will use the division property of equality to isolate x on the right side. We can undo multiplication by -4.6 by dividing both sides by -4.6.

$-9.66 = -4.6x$ This is the equation to solve.

$\dfrac{-9.66}{-4.6} = \dfrac{-4.6x}{-4.6}$ Divide both sides by -4.6.

$2.1 = x$ Do the division.

$$\begin{array}{r} 2.1 \\ 4.6\overline{)9.66} \\ \underline{-92} \\ 46 \\ \underline{-46} \\ 0 \end{array}$$

Check the result to verify that 2.1 is the solution.

5 Use more than one property of equality to solve equations that involve decimals.

Sometimes, more than one property must be used to solve an equation that involves decimals.

EXAMPLE 8   Solve: $8.1y - 6.04 = -13.33$

Self Check 8
Solve: $-4.2h + 3.14 = 1.88$
Now Try Problem 61

Strategy First we will use a property of equality to isolate the *variable term* on one side of the equation. Then we will use a second property of equality to isolate the *variable* itself.

WHY To solve the original equation, we want to find a simpler equivalent equation of the form $y = $ **a number**, whose solution is obvious.

Solution On the left side of the equation, y is multiplied by 8.1 and then 6.04 is subtracted from that product. To solve the equation, we undo the operations in the opposite order.

- To isolate the variable term $8.1y$, we add 6.04 to both sides to undo the subtraction of 6.04.

- To isolate the variable y, we divide both sides by 8.1 to undo the multiplication by 8.1.

$8.1y - 6.04 = -13.33$ This is the equation to solve.

$8.1y - 6.04 + \mathbf{6.04} = -13.33 + \mathbf{6.04}$ Use the addition property of equality: Add 6.04 to both sides to isolate 8.1y.

$8.1y = -7.29$ Do the addition: $-6.04 + 6.04 = 0$ and $-13.33 + 6.04 = -7.29$. Now isolate y.

$$\begin{array}{r} ^{213} \\ 13.\overset{\cdot\cdot}{3}\overset{\cdot}{3} \\ \underline{-\ 6.04} \\ 7.29 \end{array}$$

$\dfrac{8.1y}{\mathbf{8.1}} = \dfrac{-7.29}{\mathbf{8.1}}$ Use the division property of equality: Divide both sides by 8.1 to isolate y.

$$\begin{array}{r} 0.9 \\ 8.1\overline{)7.29} \\ \underline{-7\ 29} \\ 0 \end{array}$$

$y = -0.9$ Do the division.

Check:

$$8.1y - 6.04 = -13.33 \quad \text{This is the original equation.}$$

$$8.1(-0.9) - 6.04 \overset{?}{=} -13.33 \quad \text{Substitute } -0.9 \text{ for } y.$$

$$-7.29 - 6.04 \overset{?}{=} -13.33 \quad \text{Do the multiplication:}$$
$$8.1(-0.9) = -7.29.$$

$$-13.33 = -13.33 \quad \text{Do the subtraction by adding the}$$
$$\text{opposite: } -7.29 - 6.04 =$$
$$-7.29 + (-6.04) = -13.33.$$

$$\begin{array}{r} 8.1 \\ \times\, 0.9 \\ \hline 7.29 \end{array}$$

$$\begin{array}{r} \overset{1}{7}.29 \\ +\, 6.04 \\ \hline 13.33 \end{array}$$

Since the resulting statement $-13.33 = -13.33$ is true, -0.9 is the solution. ∎

Self Check 9

Solve the equation and check the result:
$6.1b - 5.5 = 5.2b + 5.3$

Now Try Problem 65

EXAMPLE 9   Solve: $0.2s - 3 = 0.7s + 1.5$

Strategy There are variable terms ($0.2s$ and $0.7s$) on both sides of the equation. We will eliminate $0.2s$ on the left side by subtracting $0.2s$ from both sides.

WHY To solve for s, all the terms containing s must be on the same side of the equation.

Solution   We isolate the variable terms on the right side and isolate the constant terms on the left side of the equation.

$$0.2s - 3 = 0.7s + 1.5 \quad \text{This is the equation to solve.}$$

$$0.2s - 3 - \mathbf{0.2s} = 0.7s + 1.5 - \mathbf{0.2s} \quad \begin{array}{l} \text{Subtract } 0.2s \text{ from} \\ \text{both sides to isolate the} \\ \text{variable term on the right.} \end{array}$$

$$-3 = 0.5s + 1.5 \quad \begin{array}{l} \text{Combine like terms:} \\ 0.2s - 0.2s = 0 \text{ and} \\ 0.7s - 0.2s = 0.5s. \end{array}$$

$$-3 - \mathbf{1.5} = \boxed{0.5s} + 1.5 - \mathbf{1.5} \quad \begin{array}{l} \text{To isolate the variable term} \\ 0.5s, \text{ undo the addition of } 1.5 \\ \text{by subtracting } 1.5 \text{ from both} \\ \text{sides.} \end{array}$$

$$-3 + (-1.5) = 0.5s \quad \begin{array}{l} \text{On the left, write the subtraction as addition} \\ \text{of the opposite. On the right, subtract.} \end{array}$$

$$-4.5 = 0.5s \quad \text{Now isolate } s. \text{ Do the addition: } -3 + (-1.5) = -4.5.$$

$$\frac{-4.5}{\mathbf{0.5}} = \frac{0.5s}{\mathbf{0.5}} \quad \begin{array}{l} \text{To isolate } s, \text{ undo the multiplication} \\ \text{by } 0.5 \text{ by dividing both sides by } 0.5. \end{array}$$

$$-9 = s \quad \text{Do the division.}$$

$$\begin{array}{r} 9 \\ 0.5\overline{)4.5} \\ -4\,5 \\ \hline 0 \end{array}$$

Check the result in the original equation to verify that -9 is the solution. ∎

Success Tip   In Example 9, we could have eliminated $0.7s$ from the right side of the equation by subtracting $0.7s$ from both sides.

$$0.2s - 3 - \mathbf{0.7s} = 0.7s + 1.5 - \mathbf{0.7s}$$

$$-0.5s - 3 = 1.5$$

However, it is usually easier to isolate the variable term on the side that will result in a *positive* coefficient. With this approach, the resulting coefficient of s is negative.

EXAMPLE 10   Solve: $5(x + 1.3) + 1.1 = -8.8$

Strategy We will use the distributive property along with the process of combining like terms to simplify the left side of the equation.

WHY It is best to simplify each side of the equation before using a property of equality.

Solution   First, we remove the parentheses by applying the distributive property.

$5(x + 1.3) + 1.1 = -8.8$	This is the equation to solve.	$\begin{array}{r} 6.5 \\ + 1.1 \\ \hline 7.6 \end{array}$
$5x + 6.5 + 1.1 = -8.8$	Distribute the multiplication by 5.	
$5x + 7.6 = -8.8$	Combine like terms: $6.5 + 1.1 = 7.6$.	$\begin{array}{r} \overset{1}{8}.8 \\ + 7.6 \\ \hline 16.4 \end{array}$
$5x + 7.6 - \mathbf{7.6} = -8.8 - \mathbf{7.6}$	To isolate the variable term 5x, undo the addition of 7.6 by subtracting 7.6 from both sides.	
$5x = -8.8 + (-7.6)$	On the right side, add the opposite of 7.6.	$\begin{array}{r} 3.28 \\ 5\overline{)16.40} \\ -15 \\ \hline 1\,4 \\ -1\,0 \\ \hline 40 \\ -40 \\ \hline 0 \end{array}$
$5x = -16.4$	Do the addition: $-8.8 + (-7.6) = -16.4$. Now isolate x.	
$\dfrac{5x}{5} = \dfrac{-16.4}{5}$	To isolate the variable x, undo the multiplication by 5 by dividing both sides by 5.	
$x = -3.28$	Do the division.	

Check:

$5(x + 1.3) + 1.1 = -8.8$	This is the original equation.	$\begin{array}{r} \overset{2\ 12}{\cancel{3}.\cancel{2}8} \\ -1.30 \\ \hline 1.98 \end{array}$
$5(\mathbf{-3.28} + 1.3) + 1.1 \overset{?}{=} -8.8$	Substitute −3.28 for x.	
$5(-1.98) \overset{?}{=} -8.8$	Do the addition within the parentheses.	$\begin{array}{r} \overset{4\ 4}{1.98} \\ \times\ \ 5 \\ \hline 9.90 \end{array}$
$-9.9 + 1.1 \overset{?}{=} -8.8$	Do the multiplication: $5(-1.98) = -8.8$.	$\begin{array}{r} 9.9 \\ -1.1 \\ \hline 8.8 \end{array}$
$-8.8 = -8.8$	Do the addition: $-9.9 + 1.1 = -8.8$.	

Since the resulting statement $-8.8 = -8.8$ is true, -3.28 is the solution.

6 Use equations to solve application problems that involve decimals.

We can use the concepts of variable and equation to solve application problems involving decimals. Once again, we will follow the five-step problem-solving strategy of analyze, form, solve, state, and check.

EXAMPLE 11   ***Business Expenses***   A business decides to rent a copy machine instead of buying one. Under the rental agreement, the company is charged a basic rental fee of $65 per month plus 2¢ for every copy made. If the business has budgeted $125 for copier expenses each month, how many copies can be made before exceeding the budget?

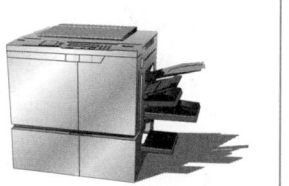

Analyze

- The basic rental charge is $65 a month.   Given
- There is a 2¢ charge for each copy made.   Given
- $125 is budgeted for copier expenses each month.   Given
- What is the maximum number of copies that can
  be made each month?   Find

Form

We will let x = the maximum number of copies that can be made each month. To begin to translate the words of the problem to numbers and symbols, we first write:

The basic fee	plus	the total cost of the copies	must equal	the amount budgeted each month.

We can find the total cost of the copies by multiplying the cost per copy by the maximum number of copies that can be made. Notice that the monthly budgeted cost is $125 and the cost per copy is 2¢. We need to work in terms of one unit, so we write 2¢ as $0.02 and work in terms of dollars.

The basic fee	plus	the cost per copy	times	the maximum number of copies made	must equal	the amount budgeted each month.
65	+	0.02	·	x	=	125

Caution! Don't forget to write the 2¢ per copy cost in terms of dollars: $0.02 per copy. If you incorrectly form the equation as $65 + 2x = 125$, you are saying that the copies cost $2 each!

Solve

$$65 + 0.02x = 125$$

$$65 + \underline{0.02x} - \mathbf{65} = 125 - \mathbf{65}$$ To isolate the variable term 0.02x, undo the addition of 65 by subtracting 65 from both sides.

$$0.02x = 60$$ Do the subtraction. Now isolate x.

$$\frac{0.02x}{\mathbf{0.02}} = \frac{60}{\mathbf{0.02}}$$ To isolate x, undo the multiplication by 0.02 by dividing both sides by 0.02.

$$x = 3{,}000$$ Do the division.

$$\begin{array}{r} {\scriptstyle 012} \\ \cancel{1}2\cancel{5} \\ -\ 65 \\ \hline 60 \end{array}$$

$$\begin{array}{r} 3000 \\ .02\overline{)60.00} \end{array}$$

State

The business can make up to 3,000 copies each month without exceeding its budget.

Check

If we multiply the cost per copy and the maximum number of copies, we get $0.02 · 3,000 = $60. Then we add the $65 monthly fee: $60 + $65 = $125. The result checks.

$$\begin{array}{r} 3{,}000 \\ \times\ 0.02 \\ \hline 60.00 \end{array} \qquad \begin{array}{r} 60 \\ +\ 65 \\ \hline 125 \end{array}$$

SECTION 5.7 STUDY SET

VOCABULARY

Fill in the blanks.

1. In algebra, we often replace one algebraic expression with another that is equivalent and simpler in form. That process is called _____ an expression.

2. 4.1(x + 3) = 4.1x + 4.1(3) is an example of the use of the _____ property.

3. When we write 6.2x + 2.7x as 8.9x, we say we have _____ like terms.

4. To _____ an equation means to find all the values of the variable that make the equation true.

CONCEPTS

5. **a.** Fill in the blanks to simplify the expression.

   $4(3.2t) = ($ ▢ · ▢ $)t = $ ▢ t

   b. What property did you use in part a?

   c. Fill in the blanks to simplify the expression.

   $-6.1y \cdot 2 = $ ▢ · ▢ · $y = $ ▢ y

   d. What property did you use in part c?

6. Fill in the blanks.
   a. 2.9(x + 4) = 2.9x ▢ 11.6
   b. 2.9(x − 4) = 2.9x ▢ 11.6
   c. −2.9(x + 4) = −2.9x ▢ 11.6
   d. −2.9(−x − 4) = 2.9x ▢ 11.6

7. Fill in the blanks to combine like terms.
   a. 4.2m + 6.3m = (▢ + ▢)m = ▢ m
   b. 3.6n^2 − 5.8n^2 = (▢ − ▢)n^2 = ▢ n^2
   c. 1.2 + 3.2d + 1.5 = 3.2d + ▢
   d. Like terms can be combined by adding or subtracting the _____ of the terms and keeping the same variables with the same exponents.

8. Simplify each expression, if possible.
   a. 5(2x) and 5 + 2x
   b. 6(−7x) and 6 − 7x
   c. 2(3x)(3) and 2 + 3x + 3

9. Fill in the blanks.

   It takes two steps to solve the equation 2.1m − 9.1 = 5.6.
   - To isolate the variable term 2.1m, we undo the subtraction of 9.1 by _____ 9.1 to both sides.
   - To isolate the variable m, we undo the multiplication by 2.1 by _____ both sides by 2.1.

10. Write each amount of money as a dollar amount.
    a. 25 cents **c.** 1 penny
    b. 250 cents **d.** 99 cents

NOTATION

Complete each solution to solve the equation.

11. $0.6a - 2.3 = -1.82$

    $0.6a - 2.3 + $ ▢ $ = -1.82 + $ ▢

    ▢ $ = 0.48$

    $\dfrac{0.6a}{▢} = \dfrac{0.48}{▢}$

    $a = $ ▢

    Check: $0.6a - 2.3 = -1.82$

    $0.6($ ▢ $) - 2.3 \overset{?}{=} -1.82$

    $0.48 - 2.3 \overset{?}{=}$

    ▢ $ = -1.82$ True

    The solution is ▢ .

12. $\dfrac{x}{2} + 1 = -5.2$

$\dfrac{x}{2} + 1 - \boxed{} = -5.2 - \boxed{}$

$\boxed{} = -6.2$

$\boxed{}\left(\dfrac{x}{2}\right) = \boxed{}(-6.2)$

$x = \boxed{}$

Check: $\dfrac{x}{2} + 1 = -5.2$

$\dfrac{\boxed{}}{2} + 1 \overset{?}{=} -5.2$

$\boxed{} + 1 \overset{?}{=} \boxed{}$

$\boxed{} = -5.2$ True

The solution is $\boxed{}$.

GUIDED PRACTICE

Simplify each expression. See Example 1.

13. $3.2 \cdot 4t$

14. $9.7 \cdot 3s$

15. $8.06m(-7)$

16. $7.16n(-4)$

17. $-5(-6.7t)$

18. $-8(-1.3a)$

19. $2(4.4c)(3)$

20. $3(2.7h)(2)$

Multiply. See Example 2.

21. $3.7(4x + 3)$

22. $5.9(7a + 6)$

23. $-6(1.9m - 2.8)$

24. $-8(2.8r - 1.4)$

25. $(y - 9.4)0.06$

26. $(h - 3.9)0.07$

27. $-2.5(-1.2t + 11)$

28. $-2.5(-5.2y + 13)$

Simplify each expression by combining like terms. See Example 3.

29. $3.2x + 4.7x$

30. $1.8m + 8.1m$

31. $-2.4v + (-9.8v) + 3.4v$

32. $-9.5t + (-4.6t) + 6.4t$

33. $0.32b^2 - 0.59b^2$

34. $0.26d^2 - 0.89d^2$

35. $7.55a + 3 + 1.12a - 1.56$

36. $4.05y + 5 + 4.39y - 3.67$

Simplify each expression. See Example 4.

37. $5.2(m + 3) - 1.2 - (3.7m - 4.1)$

38. $6.3(n + 4) - 4.3 - (1.5n - 2.9)$

39. $3.8(d + 2) - 4.5 - (1.8d - 3.7)$

40. $7.6(s + 2) - 8.1 - (5.2s - 6.8)$

41. $7(3.4y - 1.1) - (6.7y + 4.9) + y$

42. $4(6.2r - 2.2) - (4.9r + 7.4) + r$

43. $9(2.7b - 4.2) - (14.2b + 37.8) - b$

44. $3(8.4y - 6.3) - (9.1y + 18.9) - y$

Solve each equation and check the result. See Example 5.

45. $x + 8.1 = 9.8$

46. $x + 4.3 = 8.9$

47. $6.75 + y = 8.99$

48. $2.61 + y = 9.93$

49. $m - 1.4 = -5.8$

50. $h - 6.3 = -10.8$

51. $7.08 = c - 0.03$

52. $14.1 = k - 13.1$

Solve each equation and check the result. See Example 6.

53. $\dfrac{x}{6} = -4.7$

54. $\dfrac{x}{8} = -8.4$

55. $\dfrac{y}{7} = 0.06$

56. $\dfrac{r}{3} = 0.23$

Solve each equation and check the result. See Example 7.

57. $-3.51 = -2.7x$

58. $-10.15 = -3.5m$

59. $-1.95 = 0.5f$

60. $-4.92 = 0.6m$

Solve each equation and check the result. See Example 8.

61. $3.2x - 3.01 = -5.25$

62. $6.4a - 1.29 = -6.41$

63. $-1.5b + 2.7 = 1.2$

64. $-2.1x - 3.1 = 5.3$

Solve each equation and check the result. See Example 9.

65. $0.4y - 1 = 0.8y + 3.4$

66. $0.6w - 2 = 0.9w + 1.6$

67. $53.7r + 2.6 = 46.3r + 17.4$

68. $37.1w + 12.2 = 16.8w + 93.4$

Solve each equation and check the result. See Example 10.

69. $3(y - 1.1) + 0.5 = -0.4$

70. $4(b - 2.7) + 1.6 = -6.4$

71. $8(m - 1.9) + 17.1 = -14.5$

72. $2(w - 6.2) + 5.4 = -13.1$

TRY IT YOURSELF

Simplify each expression or solve each equation.

73. $2x = -8.72$

74. $3y = -12.63$

75. $2.3 = 3(a - 1.1) + 3.2$

76. $-6.2 = 2(m - 4.3) + 1.2$

77. $5.6x - 8.3 - 6.1x + 12.2$

78. $-17.3y - 8.01 + 12.2y - 4.4$

79. $1.2x - 1.3 = 2.4x + 0.02$

80. $-4.4y - 1.3 = -5.1y - 5.08$

81. $-6.7 = \dfrac{x}{2.04} - 2.7$

82. $-7.5 = \dfrac{y}{2.22} - 1.5$

83. $0.06x + 0.09(100 - x) = 8.85$

84. $0.08(1,000 - x) + 0.6x = 72.72$

85. $3.1r - 5.5r - 1.3r$ **86.** $3.8x - 6.5x - 2.4x$

87. $\frac{1}{3}x = -7.06$ **88.** $\frac{1}{5}x = -3.02$

89. $2(x + 3.9) = 3.4$ **90.** $3(x - 0.4) = -4.8$

91. $7.75 = n - (-7.85)$ **92.** $3.33 = y - (-5.55)$

93. $-5.6 - h = -17.1$ **94.** $-0.05 - x = -1.25$

95. $\frac{x}{100} = 0.004$ **96.** $\frac{y}{1,000} = 0.0606$

APPLICATIONS

Complete each solution.

97. PETITIONS On weekends, a college student works for a political organization collecting signatures for a petition drive. Her pay is $48 a day plus 95¢ for each signature she obtains. How many signatures does she have to collect to make $200 a day?

Analyze
- Her base pay is $ ☐ a day.   *Given*
- She makes ☐ ¢ for each signature.   *Given*
- She wants to make $ ☐ a day.   *Given*
- How many _____ does she need to collect?   *Find*

Form Let $x =$ the number of _____ she needs to collect.

We need to work in terms of the same units, so we write 95¢ as $ ☐ . We now translate the words of the problem into an equation.

Her base pay	plus	the amount per signature	times	the number of signatures	should equal	$200.
☐	+	0.95	·	x	=	☐

Solve

$48 + ☐ = ☐$

$48 + 0.95x - ☐ = 200 - ☐$

$0.95x = ☐$

$\frac{0.95x}{☐} = \frac{152}{☐}$

$x = ☐$

State

She needs to collect ☐ signatures to make $200 a day.

Check

If she collects ☐ signatures, she will make $0.95 · ☐ = $152 from signatures. If we add this to $48, we get $ ☐ . The result checks.

98. CONSTRUCTION A 12.78-mile-long highway is in its third and final year of construction. In the first year, 2.31 miles of the highway were completed. In the second year, 4.93 miles were finished. How many more miles of the highway need to be completed?

Analyze
- The planned highway is ☐ miles long.   *Given*
- The first year, ☐ miles were completed.   *Given*
- The second year, ☐ miles were completed.   *Given*
- How many more _____ of highway need to be completed?   *Find*

Form We will let $x =$ the number of _____ of highway that need to be completed. We now translate the words of the problem into an equation.

The miles completed in the first year	plus	the miles completed in the 2nd year	plus	the miles that need to be completed	is	the length of the highway.
☐	+	4.93	+	☐	=	12.78

Solve

$2.31 + 4.93 + ☐ = 12.78$

$☐ + x = 12.78$

$7.24 + x - ☐ = 12.78 - ☐$

$x = ☐$

State

The number of miles of highway that need to be completed is ☐ .

Check

Add:

$$\begin{array}{r} 2.31 \\ 4.93 \\ + ☐ \\ \hline 12.78 \end{array}$$

The result checks.

In Exercises 99–110, let a variable represent the unknown quantity. Then write and solve an equation to answer the question.

99. DISASTER RELIEF After hurricane damage estimated at $27.9 million, a county looked to three sources for relief. Local agencies contributed $6.8 million toward the cleanup. A state emergency fund offered another $12.5 million. When applying for federal government help, how much money should the county ask for?

100. TELETHONS Midway through a telethon, the donations had reached $16.7 million. How much more was donated in the second half of the program if the total pledged during the program was $30 million?

101. GRADE POINT AVERAGES After receiving her grades for the fall semester, a college student noticed that her overall GPA had dropped by 0.18. If her new GPA was 3.09, what was her GPA at the beginning of the fall semester?

102. MONTHLY PAYMENTS A food dehydrator offered on a home shopping channel can be purchased by making 3 equal monthly payments. If the price is $113.25, how much is each monthly payment?

103. AWARDS A city honors its citizen of the year with a framed certificate. A calligrapher charges $20 for the frame and then 15 cents a word for writing out the proclamation. If the city charter prohibits gifts in excess of $50, what is the maximum number of words that can be printed on the award?

104. GIVE-AWAYS A promotions company sells key chains with a personalized message printed on them. The key chains cost 78¢ each and there is a $32 shipping and handling fee on every order. If a car repair shop plans to spend $500 for this type of advertising, how many key chains can they order?

105. HELIUM BALLOONS The organizer of a jog-a-thon wants an archway of balloons constructed at the finish line of the race. A company charges a $100 setup fee and then 8 cents for every balloon. How many balloons will be used if $300 is spent for the decoration?

106. UTILITY BILLS An electric company charges customers a $17.50 basic monthly service fee plus 18¢ for every kilowatt of energy used. One resident's bill was $43.96. How many kilowatt hours did the resident use that month?

107. TUTORING Elementary school students enrolling in a reading improvement program must first take a placement test (cost $60) before receiving tutoring (cost $9.75 per hour). If a family has set aside $450 to get their child extra help, how many hours of tutoring can they afford?

108. DATA CONVERSION The *Books2Bytes* service converts old print books to Microsoft Word electronic files for $40 per book plus $2.15 per page. If it cost $900 to convert a novel, how many pages did the novel have?

109. VIDEO CASSETTES A family has a large number of VHS cassettes that they want to transfer to DVDs. The conversion costs $14.50 per tape and there is a $31 shipping and handling charge on any order. If they have $350 to spend on the project, how many VHS cassettes can they have converted to DVDs?

110. LETTERMAN JACKETS At one high school, athletes pay $185 for a varsity letterman jacket. It also costs them $4.25 per letter if they want their last name embroidered on the back. If a soccer player spent $223.25 on his letterman jacket, how many letters are in his last name?

WRITING

111. In the following case, why is it rather easy to apply the distributive property?
$$100(0.07x + 5.16)$$

112. Explain why the expression $5.6A + 3.4a$ cannot be simplified.

REVIEW

113. a. Add: $\frac{3}{4} + \frac{2}{3}$

 b. Add: $\frac{1}{x} + \frac{2}{3}$

114. a. Multiply: $\frac{3}{4} \cdot \frac{2}{3}$

 b. Multiply: $\frac{3}{4x} \cdot \frac{2x}{3}$

115. a. Subtract: $\frac{4}{5} - \frac{1}{7}$

 b. Subtract: $\frac{4}{5} - \frac{n}{7}$

116. a. Divide: $\frac{5}{8} \div \frac{15}{16}$

 b. Divide: $\frac{5b}{8} \div \frac{15b}{16}$

STUDY SKILLS CHECKLIST

Do You Know the Basics?

The key to mastering the material in Chapter 5 is to know the basics. Put a checkmark in the box if you can answer "yes" to the statement.

☐ I have memorized the *place-value chart* on page 3.

☐ I know the rules for rounding a decimal to a certain decimal place value by identifying the *rounding digit* and the *test digit*.

☐ I know how to add decimals using *carrying* and how to subtract decimals using *borrowing*.

$$\begin{array}{r} \overset{1\ \ 1}{7.18} \\ 154.20 \\ +\ 46.03 \\ \hline 207.41 \end{array} \qquad \begin{array}{r} \overset{6\ \ 10\,14}{537.0\,4} \\ -\ 23.98 \\ \hline 513.0\,6 \end{array}$$

☐ I have memorized the list of *perfect squares* on page 515 and can find their *square roots*.

$$\sqrt{16} = 4 \qquad \sqrt{121} = 11$$

☐ I know how to *multiply* and *divide* decimals and locate the decimal point in the answer.

$$\begin{array}{r} 1.84 \\ \times\ 7.6 \\ \hline 1104 \\ 12880 \\ \hline 13.984 \end{array} \qquad \begin{array}{r} 2.8 \\ 3.4\overline{)9.52} \\ -68\downarrow \\ \hline 272 \\ -272 \\ \hline 0 \end{array}$$

☐ I know how to use division to write a *fraction as a decimal*.

$$\frac{3}{5} = 0.6 \qquad \begin{array}{r} 0.6 \\ 5\overline{)3.0} \\ -30 \\ \hline 0 \end{array}$$

CHAPTER 5   SUMMARY AND REVIEW

SECTION 5.1   An Introduction to Decimals

DEFINITIONS AND CONCEPTS

EXAMPLES

The place-value system for whole numbers can be extended to create the **decimal numeration system.**

The place-value columns to the left of the decimal point form the **whole-number part** of the decimal number. The value of each of those columns is 10 times greater than the column directly to its right.

The columns to the right of the decimal point form the **fractional part.** Each of those columns has a value that is $\frac{1}{10}$ of the value of the place directly to its left.

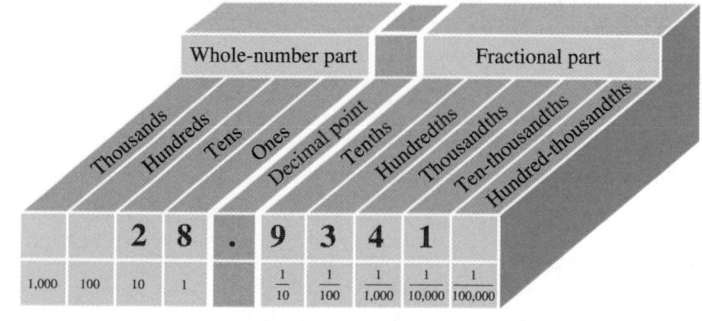

The place value of the digit 3 is *3 hundredths.*
The digit that tells the number of *ten-thousandths* is 1.

To write a decimal number in **expanded form** (**expanded notation**) means to write it as an addition of the place values of each of its digits.	Write 28. 9341 in expanded notation: $$28.9341 = 20 + 8 + \frac{9}{10} + \frac{3}{100} + \frac{4}{1{,}000} + \frac{1}{10{,}000}$$

To read a decimal:

1. Look to the left of the decimal point and say the name of the whole number.

2. The decimal point is read as "and."

3. Say the fractional part of the decimal as a whole number followed by the name of the last place-value column of the digit that is the farthest to the right.

We can use the steps for reading a decimal to **write it in words.**

Write the decimal in words and then as a fraction or mixed number:

28.9341   The whole-number part is 28. The fractional part is 9341. The digit the farthest to the right, 1, is in the ten-thousandths place.

Twenty-eight and nine thousand three hundred forty-one ten-thousandths

Written as a mixed number, 28.9341 is $28\frac{9{,}341}{10{,}000}$.

Write the decimal in words and then as a fraction or mixed number:

0.079   The whole-number part is 0. The fractional part is 79. The digit the farthest to the right, 9, is in the thousandths place.

Seventy-nine thousandths

Written as a fraction, 0.079 is $\frac{79}{1{,}000}$.

The procedure for **reading a decimal** can be applied in reverse to convert from written-word form to standard form.	Write the decimal number in standard form: 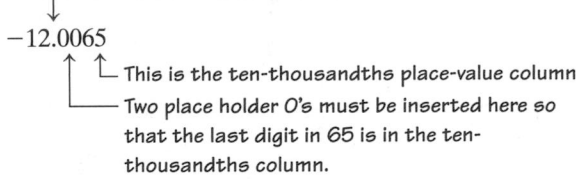 **Negative twelve and sixty-five ten-thousandths** -12.0065 ⟶ This is the ten-thousandths place-value column. Two place holder 0's must be inserted here so that the last digit in 65 is in the ten-thousandths column.

To compare two decimals:

1. Make sure both numbers have the same number of decimal places to the right of the decimal point. Write any additional zeros necessary to achieve this.

2. Compare the digits of each decimal, column by column, working from left to right.

3. If the decimals are *positive*: When two digits differ, the decimal with the greater digit is the greater number.

   If the decimals are *negative*: When two digits differ, the decimal with the smaller digit is the greater number.

Compare 47.31572 and 47.31569.

47.315 7 2
47.315 6 9

⟶ As we work from left to right, this is the first column in which the digits differ. Since 7 > 6, it follows that 47.31572 is greater than 47.31569.

Thus, 47.31572 > 47.31569.

Compare -6.418 and -6.41.

-6.41 8   These decimals are negative.
-6.41 0   Write a zero after 1 to help in the comparison.

⟶ As we work from left to right, this is the first column in which the digits differ. Since 0 < 8, it follows that -6.410 is greater than -6.418.

Thus, $-6.41 > -6.418$.

To **graph a decimal number** means to make a drawing that represents the number.	Graph -2.17, 0.6, -2.89, 3.99, and -0.5 on a number line. 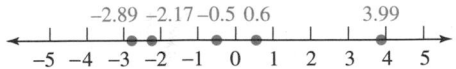

1. To **round a decimal** to a certain decimal place value, locate the **rounding digit** in that place. 2. Look at the **test digit** directly to the right of the rounding digit. 3. If the test digit is 5 or greater, round up by adding 1 to the rounding digit and dropping all the digits to its right. If the test digit is less than 5, round down by keeping the rounding digit and dropping all the digits to its right.	Round 33.41632 to the nearest thousandth. Rounding digit: thousandths column ↓ 33.41632 ↑ Test digit: 3 is less than 5. Keep the rounding digit: Do not add 1. → 33.41632 ← Drop the test digit and all digits to its right. Thus, 33.41632 rounded to the nearest thousandth is 33.416. Round 2.798 to the nearest hundredth. Rounding digit: hundredths column ↓ 2.798 ↑ Test digit: 8 is 5 or greater. Add 1. Since 9 + 1 = 10, write 0 in this column and carry 1 to the tenths column. → 2.798 ← Drop the test digit. Thus, 2.798 rounded to the nearest hundredth is 2.80.
There are many situations in our daily lives that call for **rounding amounts of money.**	Rounded to the *nearest cent,* \$0.14672 is \$0.15. Rounded to the *nearest dollar,* \$142.39 is \$142.

REVIEW EXERCISES

1. a. Represent the amount of the square region that is shaded, using a decimal and a fraction.

   b. Shade 0.8 of the region shown below.

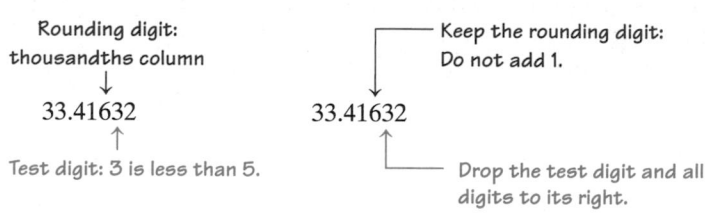

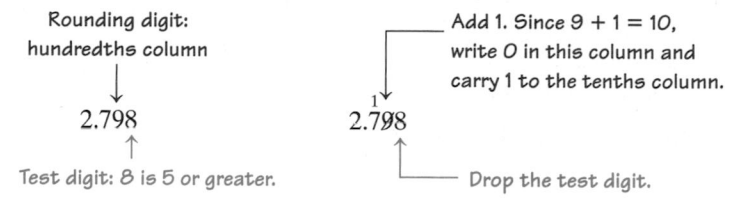

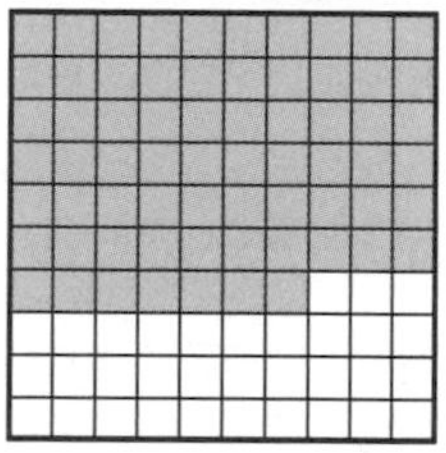

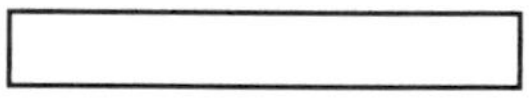

2. Consider the decimal number 2,809.6735.
   a. What is the place value of the digit 7?
   b. Which digit tells the number of thousandths?
   c. Which digit tells the number of hundreds?
   d. What is the place value of the digit 5?

3. Write 16.4523 in expanded notation.

Write each decimal in words and then as a fraction or mixed number.

4. 2.3
5. –615.59
6. 0.0601
7. 0.00001

Write each number in standard form.

8. One hundred and sixty-one hundredths
9. Eleven and nine hundred ninety-seven thousandths
10. Three hundred one and sixteen millionths

Place an < or an > symbol in the box to make a true statement.

11. 5.68 ☐ 5.75
12. 106.8199 ☐ 106.82
13. −78.23 ☐ −78.303
14. −555.098 ☐ −555.0991
15. Graph: 1.55, −0.8, −2.1, and −2.7.

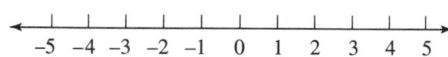

16. Determine whether each statement is true or false.
   a. 78 = 78.0
   b. 6.910 = 6.901
   c. −3.4700 = −3.470
   d. 0.008 = .00800

Round each decimal to the indicated place value.

17. 3,706.0815 nearest thousandth
18. −0.0614 nearest tenth
19. 11.314964 nearest ten-thousandth
20. 0.2222282 nearest millionth

Round each given dollar amount.

21. $0.671456 to the nearest cent

22. $12.82 to the nearest dollar

23. VALEDICTORIANS At the end of the school year, the five students listed below were in the running to be class valedictorian (the student with the highest grade point average). Rank the students in order by GPA, beginning with the valedictorian.

Name	GPA
Diaz, Cielo	3.9809
Chou, Wendy	3.9808
Washington, Shelly	3.9865
Gerbac, Lance	3.899
Singh, Amani	3.9713

24. ALLERGY FORECAST The graph below shows a four-day forecast of pollen levels for Las Vegas, Nevada. Determine the decimal-number forecast for each day.

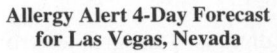
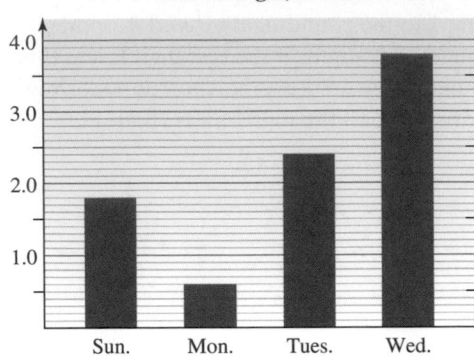

Allergy Alert 4-Day Forecast
for Las Vegas, Nevada

SECTION 5.2   Adding and Subtracting Decimals

DEFINITIONS AND CONCEPTS

To **add or subtract decimals:**

1. Write the numbers in **vertical form** with the decimal points lined up.

2. Add (or subtract) as you would whole numbers.

3. Write the decimal point in the result from Step 2 below the decimal points in the problem.

If the number of decimal places in the problem are different, insert additional zeros so that the number of decimal places match.

If the sum of the digits in any place-value column is greater than 9, we must **carry.**

If the subtraction of the digits in any place-value column requires that we subtract a larger digit from a smaller digit, we must **borrow** or **regroup.**

EXAMPLES

Add: $15.82 + 19 + 32.995$

Write the problem in vertical form and add, column-by-column, working right to left.

$$
\begin{array}{r}
\overset{1\,1\ \ 1}{15.820} \quad \text{Insert an extra zero.}\\
19.000 \quad \text{Insert a decimal point and extra zeros.}\\
+\,32.995\\
\hline
67.815\\
\end{array}
$$

↳ Line up the decimal points.

To **check** the result, add *bottom to top.*

Subtract: $8.4 - 3.029$

Write the problem in **vertical form** and subtract, column-by-column, working right to left.

$$
\begin{array}{r}
\overset{9}{}\,\overset{3\ \cancel{10}10}{8.\cancel{4}\cancel{0}\cancel{0}} \quad \text{Insert extra zeros.}\\
-\,3.029 \quad \text{First, borrow from the tenths column: then}\\
\hline
5.371 \quad \text{borrow from the hundredths column.}\\
\end{array}
$$

To **check**: The sum of the difference and the subtrahend should equal the minuend.

$$
\begin{array}{r}
\overset{1\,1}{5.371} \leftarrow \text{Difference}\\
+\,3.029 \leftarrow \text{Subtrahend}\\
\hline
8.400 \leftarrow \text{Minuend}\\
\end{array}
$$

To **add signed decimals,** we use the same rules that are used for adding integers.

With like signs: Add their absolute values and attach their common sign to the sum.

With unlike signs: Subtract their absolute values (the smaller from the larger). If the positive decimal has the larger absolute value, the final answer is positive. If the negative decimal has the larger absolute value, make the final answer negative.

Add: $-21.35 + (-64.52)$

Find the absolute values: $|-21.35| = 21.35$ and $|-64.52| = 64.52$

$$-21.35 + (-64.52) = -85.87$$

Add the absolute values, 21.35 and 64.52, to get 85.87. Since both decimals are negative, make the final result negative.

Add: $-7.4 + 9.8$

Find the absolute values: $|-7.4| = 7.4$ and $|9.8| = 9.8$

$$-7.4 + 9.8 = 2.4$$

Subtract the smaller absolute value from the larger: $9.8 - 7.4 = 2.4$. Since the positive number, 9.8, has the larger absolute value, the final answer is positive.

To **subtract two signed decimals,** add the first decimal to the opposite of the decimal to be subtracted.

Subtract: $-8.62 - (-1.4)$

The number to be subtracted is -1.4. Subtracting -1.4 is the same as adding its opposite, 1.4.

Add . . .

$$-8.62 - (-\mathbf{1.4}) = -8.62 + \mathbf{1.4} = -7.22$$

. . . the opposite

Use the rule for adding two decimals with different signs.

Estimation can be used to check the reasonableness of an answer to a decimal addition or subtraction.

Estimate the sum by rounding the addends to the nearest ten: $328.99 + 459.02$

$$
\begin{array}{rcl}
328.99 & \longrightarrow & 330 \\
+\,459.02 & \longrightarrow & +460 \\
\hline
788.01 & & 790 \\
\end{array}
$$

Round to the nearest ten.
Round to the nearest ten.
This is the estimate.

Estimate the difference by using **front-end rounding:** $302.47 - 36.9$

Each number is rounded to its largest place value.

$$
\begin{array}{rcl}
302.47 & \longrightarrow & 300 \\
-\;\;36.9 & \longrightarrow & -\;40 \\
\hline
265.57 & & 260 \\
\end{array}
$$

Round to the nearest hundred.
Round to the nearest ten.
This is the estimate.

We can use the five-step **problem-solving strategy** to solve application problems that involve decimals.

See Examples 10–12 that begin on page 465 to review how to solve application problems by adding and subtracting decimals.

REVIEW EXERCISES

Perform each indicated operation.

25. $19.5 + 34.4 + 12.8$

26. $45.8 - 17.372$

27. $9{,}000.09 - 7{,}067.445$

28.
$$
\begin{array}{r}
8.61 \\
5.97 \\
+\,9.72 \\
\hline
\end{array}
$$

29. $-16.1 + 8.4$

30. $-4.8 - (-7.9)$

31. $-3.55 + (-1.25)$

32. $-15.1 - 13.99$

Evaluate each expression.

33. $-8.8 + (-7.3 - 9.5)$

34. $(5 - 0.096) - (-0.035)$

35. **a.** Estimate the sum by rounding the addends to the nearest ten: $612.05 + 145.006$

  b. Estimate the difference by using front-end rounding: $289.43 - 21.86$

36. COINS The thicknesses of a penny, nickel, dime, quarter, half-dollar, and presidential $1 coin are 1.55 millimeters, 1.95 millimeters, 1.35 millimeters, 1.75 millimeters, 2.15 millimeters, and 2.00 millimeters, in that order. Find the height of a stack made from one of each type of coin.

37. SALE PRICES A calculator normally sells for $52.20. If it is being discounted $3.99, what is the sale price?

38. MICROWAVE OVENS A microwave oven is shown below. How tall is the window?

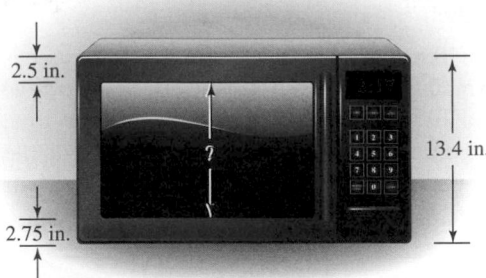

2.5 in.

?

2.75 in.

13.4 in.

SECTION 5.3   Multiplying Decimals

DEFINITIONS AND CONCEPTS	EXAMPLES
To multiply two decimals: **1.** Multiply the decimals as if they were whole numbers. **2.** Find the total number of decimal places in both factors. **3.** Insert a decimal point in the result from step 1 so that the answer has the same number of decimal places as the total found in step 2. When multiplying decimals, we *do not* need to line up the decimal points.	Multiply: $2.76 \cdot 4.3$ Write the problem in vertical form and multiply 2.76 and 4.3 as if they were whole numbers. $\begin{array}{r} 2.76 \\ \times\ 4.3 \\ \hline 828 \\ 11040 \\ \hline 11.868 \end{array}$ 2 decimal places / 1 decimal place $\}$ The answer will have $2+1=3$ decimal places. 11.868 Move 3 places from right to left and insert a decimal point in the answer. Thus, $2.76 \cdot 4.3 = 11.868$.
Multiplying a decimal by 10, 100, 1,000, and so on To find the product of a decimal and 10, 100, 1,000, and so on, move the decimal point to the right the same number of places as there are zeros in the power of 10. **Multiplying a decimal by 0.1, 0.01, 0.001, and so on** To find the product of a decimal and 0.1, 0.01, 0.001, and so on, move the decimal point to the left the same number of places as there are in the power of 10.	Multiply: $84.561 \cdot 10,000 = 845,610$ Since 10,000 has *four zeros*, move the decimal point in 84.561 four places to the right. Write a placeholder zero (shown in blue). Multiply: $32.67 \cdot 0.01 = 0.3267$ Since 0.01 has *two decimal places*, move the decimal point in 32.67 two places to the left.

The rules for multiplying integers also hold for **multiplying signed decimals:** The product of two decimals with **like signs** is positive, and the product of two decimals with **unlike signs** is negative.	Multiply: $(-0.03)(-4.1)$ Find the absolute values: $\|-0.03\| = 0.03$ and $\|4.1\| = 4.1$ Since the decimals have like signs, the product is positive. $\quad (-0.03)(-4.1) = 0.123$ Multiply the absolute values, $\qquad\qquad\qquad\qquad\qquad$ 0.03 and 4.1, to get 0.123. Multiply: $-5.7(0.4)$ Find the absolute values: $\|-5.7\| = 5.7$ and $\|0.4\| = 0.4$ Since the decimals have unlike signs, the product is negative. $\quad -5.7(0.4) = -2.28$ Multiply the absolute values, 5.7 $\qquad\qquad\qquad\uparrow\qquad$ and 0.4, to get 2.28. $\qquad\qquad\qquad\underline{\quad\quad}$ Make the final answer negative.
We can use the rule for multiplying a decimal by a power of ten to **write large numbers in standard form.**	Write *4.16 billion* in standard notation: 4.16 billion $= 4.16 \cdot$ **1 billion** $\qquad\qquad\quad = 4.16 \cdot$ **1,000,000,000** Write 1 billion in standard form. $\qquad\qquad\quad = 4,160,000,000$ Since 1,000,000,000 has *nine* $\qquad\qquad\qquad\underwave{\qquad\qquad}$ zeros, move the decimal point in $\qquad\qquad\qquad\qquad\qquad\qquad$ 4.16 *nine* places to the right.
The base of an **exponential expression** can be a positive or a negative decimal.	Evaluate: $(1.5)^2$ $(1.5)^2 = 1.5 \cdot 1.5$ The base is 1.5 and the exponent is 2. Write the base $\qquad\qquad\qquad\qquad$ as a factor 2 times. $\qquad\; = 2.25$ Multiply the decimals. Evaluate: $(-0.02)^2$ $(-0.02)^2 = (-0.02)(-0.02)$ The base is -0.02 and the exponent is 2. $\qquad\qquad\qquad\qquad\qquad\qquad$ Write the base as a factor 2 times. $\qquad\qquad = 0.0004$ Multiply the decimals. The product of two $\qquad\qquad\qquad\qquad\qquad$ decimals with like signs is positive.
To **evaluate a formula,** we replace the letters with specific numbers and then use the order of operations rule.	Evaluate $P = 2l + 2w$ for $l = 4.9$ and $w = 3.4$. $P = 2l + 2w$ $\quad = 2(\mathbf{4.9}) + 2(\mathbf{3.4})$ Replace *l* with 4.9 and *w* with 3.4. $\quad = 9.8 + 6.8$ Do the multiplication. $\quad = 16.6$ Do the addition.
Estimation can be used to check the reasonableness of an answer to a decimal multiplication.	Estimate $37 \cdot 8.49$ by **front-end rounding.** $\quad 37 \longrightarrow \quad 40$ Round to the nearest ten. $\underline{\times 8.49} \longrightarrow \underline{\times\; 8}$ Round to the nearest one. $\qquad\qquad\qquad\; 320$ This is the estimate. The estimate is 320. If we calculate $37 \cdot 8.49$, the product is exactly 314.13.
We can use the five-step **problem-solving strategy** to solve application problems that involve decimals.	See Examples 12 and 13 that begin on page 479 to review how to solve application problems by multiplying decimals.

REVIEW EXERCISES

Multiply.

39. $2.3 \cdot 6.9$

40.
$$\begin{array}{r} 1.7 \\ \times 0.004 \\ \hline \end{array}$$

41. $15.5(-9.8)$

42. $(-0.003)(-0.02)$

43. $1,000(90.1452)$

44. $0.001(2.897)$

Evaluate each expression.

45. $(0.2)^2$

46 $(0.6 + 0.7)^2 - (-3)(-4.1)$

47. $(-3.3)^2(0.00001)$

48. $(0.1)^3 + 2|-45.63 - 12.24|$

Write each number in standard notation.

49. a. GEOGRAPHY China is the third largest country in land area with territory that extends over *9.6 million* square kilometers. (Source: china.org)

b. PLANTING TREES In 2008, the Chinese people planted *2.31 billion* trees in mountains, city parks, and along highways to increase the number of forests in their country. (Source: xinhuanet.com)

50. a. Estimate the product using front-end rounding: $193.28 \cdot 7.63$

b. Estimate the product by rounding the factors to the nearest tenth: $12.42 \cdot 7.38$

51. Evaluate the formula $A = P + Prt$ for $P = 70.05$, $r = 0.08$, and $t = 5$.

52. SHOPPING If crab meat sells for $12.95 per pound, what would 1.5 pounds of crab meat cost? Round to the nearest cent.

53. AUTO PAINTING A manufacturer uses a three-part process to finish the exterior of the cars it produces.

Step 1: A 0.03-inch-thick rust-prevention undercoat is applied.

Step 2: Three layers of color coat, each 0.015 inch thick, are sprayed on.

Step 3: The finish is then buffed down, losing 0.005 inch of its thickness.

What is the resulting thickness of the automobile's finish?

54. WORD PROCESSORS The Page Setup screen for a word processor is shown. Find the area that can be filled with text on an 8.5 in. × 11 in. piece of paper if the margins are set as shown.

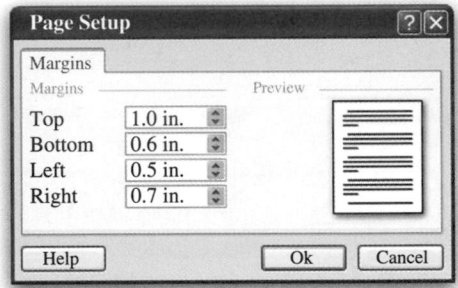

SECTION 5.4   Dividing Decimals

DEFINITIONS AND CONCEPTS

To **divide a decimal by a whole number:**

1. Write the problem in long division form and place a decimal point in the quotient (answer) directly above the decimal point in the dividend.

2. Divide as if working with whole numbers.

3. If necessary, additional zeros can be written to the right of the last digit of the dividend to continue the division.

EXAMPLES

Divide: $6.2 \div 4$

Place a decimal point in the quotient that lines up with the decimal point in the dividend.

$$\begin{array}{r} 1.55 \\ 4\overline{)6.20} \\ -4\downarrow \\ \hline 2\,2 \\ -2\,0\downarrow \\ \hline 20 \\ -20 \\ \hline 0 \end{array}$$

Ignore the decimal points and divide as if working with whole numbers.

Write a zero to the right of the 2 and bring it down. Continue to divide.

The remainder is 0.

To **check** the result, we multiply the divisor by the quotient. The result should be the dividend.	*Check:* $\quad$ 1.55 ← Quotient $\quad \underline{\times \quad 4}$ ← Divisor $\quad$ 6.20 ← Dividend The check confirms that 6.2 ÷ 4 = 1.55.
To divide with a decimal divisor: **1.** Write the problem in long division form. **2.** Move the decimal point of the divisor so that it becomes a whole number. **3.** Move the decimal point of the dividend the same number of places to the right. **4.** Write a decimal point in the quotient (answer) directly above the decimal point in the dividend. Divide as if working with whole numbers. **5.** If necessary, additional zeros can be written to the right of the last digit of the dividend to continue the division.	Divide: $\dfrac{1.462}{3.4}$ 3.4)1.462 Write the problem in long division form. Move the decimal point of the divisor, 3.4, one place to the right to make it a whole number. Move the decimal point of the dividend, 1.462, the same number of places to the right. Now use the rule for dividing a decimal by a *whole number*. $\quad$ 0.4 3 Write a decimal point in the quotient (answer) directly 34)14.6 2 above the decimal point in the dividend. $\underline{-13\ 6↓}$ $\quad$ 1 0 2 Divide as with whole numbers. $\quad \underline{-1\ 0\ 2}$ $\qquad$ 0
Sometimes when we divide decimals, the subtractions never give a zero remainder, and the division process continues forever. In such cases, we can **round the result.**	Divide: 0.77 ÷ 6. Round the quotient to the nearest hundredth. To round to the hundredths column, we need to continue the division process for one more decimal place, which is the thousandths column. 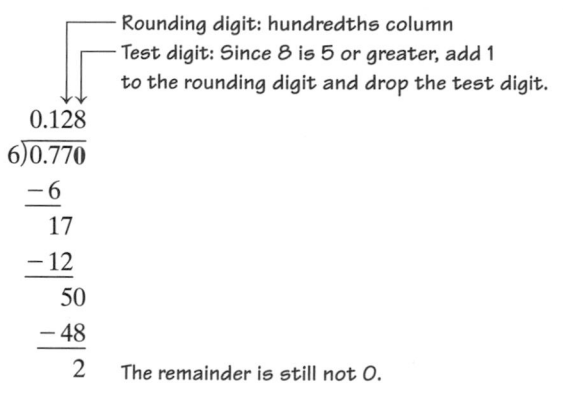 Thus, 0.77 ÷ 6 ≈ 0.13.
To **estimate quotients,** we use a method that approximates both the dividend and the divisor so that they divide easily. There is one rule of thumb for this method: If possible, round both numbers up or both numbers down.	Estimate the quotient: 337.96 ÷ 23.8 $\qquad$ ⌐ The dividend is ⌐ $\qquad\quad$ approximately 337.96 ÷ 23.8$\qquad$ 320 ÷ 20 = 16$\qquad$ To divide, drop one zero from $\qquad\qquad\qquad\qquad\qquad\qquad\qquad\qquad$ 320 and one zero from 20, $\qquad$ └ The divisor is ┘$\qquad\qquad\qquad\quad$ and then find 32 ÷ 2. $\qquad\quad$ approximately The estimate is 16. (The exact answer is 14.2.)
Dividing a decimal by 10, 100, 1,000, and so on To find the quotient of a decimal and 10, 100, 1,000, and so on, move the decimal point to the left the same number of places as there are zeros in the power of 10.	Divide: 79.36 ÷ 10,000 $\quad$ 79.36 ÷ 10,000 = 0.007936$\qquad$ Since the divisor 10,000 has four zeros, move the decimal point four places to the left. Insert two placeholder zeros (shown in blue).

Dividing a decimal by 0.1, 0.01, 0.001, and so on	Divide: $\dfrac{1.6402}{0.001}$
To find the quotient of a decimal and 0.1, 0.01, 0.001, and so on, move the decimal point to the right the same number of decimal places as there are in the power of 10.	$\dfrac{1.6402}{0.001} = 1{,}640.2$ Since the divisor 0.001 has *three* decimal places, move the decimal point in 1.6402 *three* places to the right.
The rules for dividing integers also hold for **dividing signed decimals**. The quotient of two decimals with *like signs* is positive, and the quotient of two decimals with *unlike signs* is negative.	Divide: $-1.53 \div 0.3 = -5.1$ Since the signs of the dividend and divisor are unlike, the final answer is negative. Divide: $\dfrac{-0.84}{-4.2} = 0.2$ Since the dividend and divisor have like signs, the quotient is positive.
We use the order of operations rule to **evaluate expressions** and **formulas**.	Evaluate: $\dfrac{37.8 - (1.2)^2}{0.1 + 0.3}$ $\dfrac{37.8 - (1.2)^2}{0.1 + 0.3} = \dfrac{37.8 - 1.44}{0.4}$ In the numerator, evaluate $(1.2)^2$. In the denominator, do the addition. $= \dfrac{36.36}{0.4}$ In the numerator, do the subtraction. $= 90.9$ Do the division indicated by the fraction bar.
We can use the five-step **problem-solving strategy** to solve application problems that involve decimals.	See Examples 10 and 11 that begin on page 494 to review how to solve application problems by dividing decimals.

REVIEW EXERCISES

Divide. Check the result.

55. $3\overline{)27.9}$

56. $\dfrac{-29.67}{-23}$

57. $-80.625 \div 12.9$

58. $\dfrac{0.0742}{1.4}$

59. $\dfrac{15.75}{0.25}$

60. $\dfrac{-0.003726}{-0.0046}$

61. $89.76 \div 1{,}000$

62. $\dfrac{0.0112}{-10}$

63. Divide -0.8765 by -0.001.

64. $77.021 \div 0.0001$

Estimate each quotient:

65. $4{,}983.01 \div 41.33$

66. $8.8\overline{)25{,}904.39}$

Divide and round each result to the specified decimal place.

67. $78.98 \div 6.1$ (nearest tenth)

68. $\dfrac{-5.438}{0.007}$ (nearest hundredth)

69. Evaluate: $\dfrac{(1.4)^2 - 2(-4.6)}{0.5 + 0.3}$

70. Evaluate the formula $C = \dfrac{5}{9}(F - 32)$ for $F = 68.9$.

71. THANKSGIVING DINNER The cost of purchasing the food for a Thanksgiving dinner for a family of 5 was $41.70. What was the cost of the dinner per person?

72. DRINKING WATER Water samples from five wells were taken and tested for PCBs (polychlorinated biphenyls). The number of parts per billion (ppb) found in each sample is given below. Find the average number of parts per billion for these samples.

Sample #1: 0.44 ppb
Sample #2: 0.50 ppb
Sample #3: 0.46 ppb
Sample #4: 0.52 ppb
Sample #5: 0.63 ppb

73. SERVING SIZE The illustration below shows the package labeling on a box of children's cereal. Use the information given to find the number of servings.

Nutrition Facts	
Serving size	1.1 ounce
Servings per container	?
Package weight	**15.4 ounces**

74. TELESCOPES To change the position of a focusing mirror on a telescope, an adjustment knob is used. The mirror moves 0.025 inch with each revolution of the knob. The mirror needs to be moved 0.2375 inch to improve the sharpness of the image. How many revolutions of the adjustment knob does this require?

SECTION 5.5   Fractions and Decimals

DEFINITIONS AND CONCEPTS	EXAMPLES
A fraction and a decimal are said to be **equivalent** if they name the same number. To **write a fraction as a decimal,** divide the numerator of the fraction by its denominator. Sometimes, when finding the decimal equivalent of a fraction, the division process ends because a remainder of 0 is obtained. We call the resulting decimal a **terminating decimal.**	Write $\frac{3}{5}$ as a decimal. We divide the numerator by the denominator because a fraction bar indicates division: $\frac{3}{5}$ means $3 \div 5$. $$\begin{array}{r} 0.6 \\ 5\overline{)3.0} \\ \downarrow \\ -30 \\ \hline 0 \end{array}$$ Write a decimal point and one additional zero to the right of 3. $0 \leftarrow$ Since a zero remainder is obtained, the result is a terminating decimal. Thus, $\frac{3}{5} = 0.6$. We say that 0.6 is the **decimal equivalent** of $\frac{3}{5}$.
If the denominator of a fraction in simplified form has factors of only 2's or 5's, or a combination of both, it can be written as a decimal by **multiplying it by a form of 1.** The objective is to write the fraction in an equivalent form with a denominator that is a power of 10, such as 10, 100, 1,000, and so on.	Write $\frac{3}{25}$ as a decimal. Since we need to multiply the denominator of $\frac{3}{25}$ by 4 to obtain a denominator of 100, it follows that $\frac{4}{4}$ should be the form of 1 that is used to build $\frac{3}{25}$. $\frac{3}{25} = \frac{3}{25} \cdot \frac{4}{4}$ Multiply $\frac{3}{25}$ by 1 in the form of $\frac{4}{4}$. $= \frac{12}{100}$ Multiply the numerators. Multiply the denominators. $= 0.12$ Write the fraction as a decimal.
Sometimes, when we are finding the decimal equivalent of a fraction, the division process never gives a remainder of 0. We call the resulting decimal a **repeating decimal.** An **overbar** can be used instead of the three dots ... to represent the repeating pattern in a **repeating decimal.**	Write $\frac{5}{6}$ as a decimal. $$\begin{array}{r} 0.833 \\ 6\overline{)5.000} \\ -4\,8\downarrow \\ \hline 20 \\ -18\downarrow \\ \hline 20 \\ -18 \\ \hline 2 \end{array}$$ Write a decimal point and three additional zeros to the right of 5. It is apparent that 2 will continue to reappear as the remainder. Therefore, 3 will continue to reappear in the quotient. Since the repeating pattern is now clear, we can $\leftarrow$ stop the division. Thus, $\frac{5}{6} = 0.8333\ldots$, or, using an overbar, we have $\frac{5}{6} = 0.8\overline{3}$.
When a fraction is written in decimal form, the result is either a terminating or repeating decimal. Repeating decimals are often **rounded** to a specified place value.	The decimal equivalent for $\frac{5}{11}$ is $0.454545\ldots$. Round it to the nearest hundredth. Rounding digit: hundredths column. Test digit: Since 4 is less than 5, round down. $\frac{5}{11} = 0.454545\ldots$ Thus, $\frac{5}{11} \approx 0.45$.

To write a mixed number in decimal form, we need only find the decimal equivalent for the fractional part of the mixed number. The whole-number part in the decimal form is the same as the whole-number part in the mixed-number form.	Whole-number part $$4\frac{7}{8} = 4.875$$ Write the fraction as a decimal.
A number line can be used to show the relationship between fractions and decimals.	Graph 3.125, $-4\frac{5}{7}$, $0.\overline{6}$, $-1.\overline{09}$ on a number line. $-4\frac{5}{7}$ $-1.\overline{09}$ $0.\overline{6}$ 3.125 $-5\ -4\ -3\ -2\ -1\ 0\ 1\ 2\ 3\ 4\ 5$
To compare the size of a fraction and a decimal, it is helpful to write the fraction in its equivalent decimal form.	Place an $<$, $>$, or an $=$ symbol in the box to make a true statement: $$\frac{3}{50}\ \boxed{}\ 0.07$$ To write $\frac{3}{50}$ as a decimal, divide 50 by 3: $\frac{3}{50} = 0.06$. Since 0.06 is less than 0.07, we have: $\frac{3}{50} < 0.07$.
To evaluate expressions that can contain both fractions and decimals, we can work in terms of decimals or in terms of fractions.	Evaluate: $\frac{1}{6} + 0.31$ If we work in terms of fractions, we have: $\frac{1}{6} + 0.31 = \frac{1}{6} + \frac{31}{100}$ Write 0.31 in fraction form. $= \frac{1}{6}\cdot\frac{50}{50} + \frac{31}{100}\cdot\frac{3}{3}$ The LCD is 300. Build each fraction by multiplying by a form of 1. $= \frac{50}{300} + \frac{93}{300}$ Multiply the numerators. Multiply the denominators. $= \frac{143}{300}$ Add the numerators and write the sum over the common denominator 300. If we work in terms of decimals, we have: $\frac{1}{6} + 0.31 \approx \mathbf{0.17} + 0.31$ Approximate $\frac{1}{6}$ with the decimal 0.17. ≈ 0.48 Do the addition.
We can use the five-step problem-solving strategy to solve application problems that involve fractions and decimals.	See Example 13 on page 509 to review how to solve application problems involving fractions and decimals.

REVIEW EXERCISES

Write each fraction or mixed number as a decimal. Use an overbar when necessary.

75. $\frac{7}{8}$ **76.** $-\frac{2}{5}$

77. $\frac{9}{16}$ **78.** $\frac{3}{50}$

79. $\frac{6}{11}$ **80.** $-\frac{4}{3}$

81. $3\frac{7}{125}$ **82.** $\frac{26}{45}$

Write each fraction as a decimal. Round to the nearest hundredth.

83. $\frac{19}{33}$ **84.** $\frac{31}{30}$

Place an $<$, $>$, or an $=$ symbol in the box to make a true statement.

85. $\frac{13}{25}\ \boxed{}\ 0.499$ **86.** $-\frac{4}{15}\ \boxed{}\ -0.2\overline{6}$

87. Write the numbers in order from smallest to largest: $\frac{10}{33}$, $0.\overline{3}$, 0.3

88. Graph 1.125, $-3.\overline{3}$, $2\frac{3}{4}$, and $-\frac{9}{10}$ on a number line.

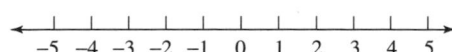

Evaluate each expression. Work in terms of fractions.

89. $\frac{1}{3} + 0.4$ **90.** $\frac{5}{6} + 0.19$

Evaluate each expression. Work in terms of decimals.

91. $\frac{1}{2}(9.7 + 8.9)(10)$ **92.** $7.5 - (0.78)\left(\frac{1}{2}\right)^2$

93. ROADSIDE EMERGENCY What is the area of the reflector shown below?

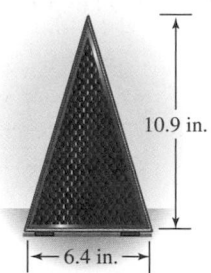

10.9 in.

6.4 in.

94. SEAFOOD A shopper purchased $\frac{3}{4}$ pound of crab meat, priced at $13.80 per pound, and $\frac{1}{3}$ pound of lobster meat, selling for $35.70 per pound. Find the total cost of these items.

SECTION 5.6   Square Roots

DEFINITIONS AND CONCEPTS	EXAMPLES
The **square root** of a given number is a number whose square is the given number. Every positive number has two square roots. The number 0 has only one square root.	Find the two square roots of 81. 9 is a square root of 81 because $9^2 = 81$ and -9 is a square root of 81 because $(-9)^2 = 81$.
A **radical symbol** $\sqrt{}$ is used to indicate a positive square root. To **evaluate a radical expression** such as $\sqrt{4}$, find the positive square root of the radicand. Radical symbol $\underbrace{\sqrt{4}}$ ← Radicand Read as "the square root of 4." Radical expression Numbers such as 4, 64, and 225, that are squares of whole numbers, are called **perfect squares.** To evaluate square root radical expressions, it is helpful to be able to identify **perfect square radicands.** Review the list of perfect squares on page 00.	Evaluate each square root: $\sqrt{4} = 2$ Ask: What positive number, when squared, is 4? The answer is 2 because $2^2 = 4$. $\sqrt{64} = 8$ Ask: What positive number, when squared, is 64? The answer is 8 because $8^2 = 64$. $\sqrt{225} = 15$ Ask: What positive number, when squared, is 225? The answer is 15 because $15^2 = 225$.
The symbol $-\sqrt{}$ is used to indicate the **negative square root** of a positive number. It is the opposite of the positive square root.	Evaluate: $-\sqrt{36}$ $-\sqrt{36}$ is the opposite (or negative) of the positive square root of 36. Since $\sqrt{36} = 6$, we have: $-\sqrt{36} = -6$
We can find the square root of fractions and decimals.	Evaluate each square root: $\sqrt{\dfrac{49}{100}}$ Ask: What positive fraction, when squared, is $\frac{49}{100}$? The answer is $\frac{7}{10}$ because $\left(\frac{7}{10}\right)^2 = \frac{49}{100}$. $\sqrt{0.25}$ Ask: What positive decimal, when squared, is 0.25? The answer is 0.5 because $(0.5)^2 = 0.25$.

When **evaluating an expression containing square roots,** evaluate square roots at the same stage in your solution as exponential expressions.

Evaluate: $20 + 6\left(2^3 - 4\sqrt{9}\right)$

Perform the operations within the parentheses first.

$$20 + 6\left(2^3 - 4\sqrt{9}\right) = 20 + 6(8 - 4 \cdot 3) \quad \begin{array}{l}\text{Within the parentheses,}\\ \text{evaluate the exponential}\\ \text{expression and the square}\\ \text{root.}\end{array}$$

$$= 20 + 6(8 - 12) \quad \begin{array}{l}\text{Within the parentheses, do}\\ \text{the multiplication.}\end{array}$$

$$= 20 + 6(-4) \quad \begin{array}{l}\text{Within the parentheses, do}\\ \text{the subtraction.}\end{array}$$

$$= 20 + (-24) \quad \text{Do the multiplication.}$$

$$= -4 \quad \text{Do the addition.}$$

To **evaluate formulas that involve square roots,** we replace the letters with specific numbers and then use the order of operations rule.

Evaluate $a = \sqrt{c^2 - b^2}$ for $c = 25$ and $b = 20$.

$$a = \sqrt{c^2 - b^2} \quad \text{This is the formula to evaluate.}$$

$$= \sqrt{25^2 - 20^2} \quad \text{Replace } c \text{ with 25 and } b \text{ with 20.}$$

$$= \sqrt{625 - 400} \quad \text{Evaluate the exponential expressions.}$$

$$= \sqrt{225} \quad \text{Do the subtraction.}$$

$$= 15 \quad \text{Evaluate the square root.}$$

If a number is not a perfect square, we can use the square root key $\boxed{\sqrt{}}$ on a calculator (or a table of square roots) to find its **approximate square root.**

Approximate $\sqrt{149}$. Round to the nearest hundredth.

From a scientific calculator we get $\sqrt{149} \approx 12.20655562$. Rounded to the nearest hundredth,

$$\sqrt{149} \approx 12.21$$

REVIEW EXERCISES

95. Find the two square roots of 25.

96. Fill in the blanks: $\sqrt{49} = \boxed{}$ because $\boxed{}^2 = 49$.

Evaluate each square root without using a calculator.

97. $\sqrt{49}$

98. $-\sqrt{16}$

99. $\sqrt{\dfrac{64}{169}}$

100. $\sqrt{0.81}$

101. Graph each square root: $\sqrt{9}, -\sqrt{2}, \sqrt{3}, -\sqrt{16}$ (*Hint:* Use a calculator or square root table to approximate any square roots, when necessary.)

$$\begin{array}{cccccccccccc} & & & & & & & & & & & \\ \hline -5 & -4 & -3 & -2 & -1 & 0 & 1 & 2 & 3 & 4 & 5 \end{array}$$

102. Use a calculator to approximate each square root to the nearest hundredth.

    a. $\sqrt{19}$ **b.** $\sqrt{598}$ **c.** $\sqrt{12.75}$

Evaluate each expression without using a calculator.

103. $-3\sqrt{49} - \sqrt{36}$

104. $\sqrt{\dfrac{100}{9}} + \sqrt{225}$

105. $40 + 6[5^2 - (7 - 2)\sqrt{16}]$

106. $1 - 7[6^2 + (1 + 2)\sqrt{81}]$

107. Evaluate $b = \sqrt{c^2 - a^2}$ for $c = 17$ and $a = 15$.

108. SHEET METAL Find the length of the side of the range hood shown in the illustration below.

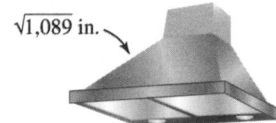

$\sqrt{1,089}$ in.

109. Between what two whole numbers would $\sqrt{83}$ be located when graphed on a number line?

110. $\sqrt{7} \approx 2.646$. Explain why an $\approx$ symbol is used and not an $=$ symbol.

SECTION 5.7   Solving Equations That Involve Decimals

DEFINITIONS AND CONCEPTS	EXAMPLES
We often use the *commutative property of multiplication* to reorder factors and the *associative property of multiplication* to regroup factors when **simplifying expressions.**	Simplify: $5(3.1y) = (5 \cdot 3.1)y$ Use the associative property of multiplication to regroup the factors. $= 15.5y$ Do the multiplication within the parentheses. Simplify: $6.2m(-4) = 6.2(-4)m$ Use the commutative property of multiplication to reorder the factors. $= -24.8m$ Do the multiplication, working from left to right.
The **distributive property** can be used to remove parentheses	Multiply: $7.7(2x + 3) = 7.7(2x) + 7.7(3)$ Distribute the multiplication by 7.7. $= 15.4x + 23.1$ Do the multiplication. Try to go directly to this step.
Like terms are terms with exactly the same variables raised to exactly the same powers.	$3.5x$ and $5.8x$ are like terms. $-4.2t^3$ and $3.9t^2$ are unlike terms because the variable t has different exponents.
Simplifying the sum or difference of like terms is called **combining like terms.** Like terms can be combined by adding or subtracting the coefficients of the terms and keeping the same variables with the same exponents.	Simplify: $4.3a + 3.6a = 7.9a$ Think: $(4.3 + 3.6)a = 7.9a$. Simplify: $5.4p + 17.8 - 4.2p = 1.2p + 17.8$ Think: $(5.4 - 4.2)p = 1.2p$. $\underset{\text{Like terms}}{\uparrow \qquad \qquad \uparrow}$ Simplify: $5.2(2y - 1) - (3.1y - 9.6)$ $= 5.2(2y - 1) - 1(3.1y - 9.6)$ Replace the $-$ symbol in front of $(3.1y - 9.6)$ with -1. $= 10.4y - 5.2 - 3.1y + 9.6$ Distribute the multiplication by 5.2 and -1. $= 7.3y + 4.4$ Combine like terms. Think: $(10.4 - 3.1)y = 7.3y$. Think: $-5.2 + 9.6 = 4.4$.
To **solve an equation** means to find all the values of the variable that make the equations true. To isolate the variable on one side of the equation, we use the: **1.** Addition property of equality **2.** Subtraction property of equality **3.** Multiplication property of equality **4.** Division property of equality	Solve: $x + 15.6 = 39.8$ We can use the subtraction property of equality to isolate x on the left side of the equation. $x + 15.6 - 15.6 = 39.8 - 15.6$ To undo the addition of 15.6, subtract 15.6 from both sides. $x = 24.2$ Do the subtraction. ***Check:*** $x + 15.6 = 39.8$ This is the original equation. $24.2 + 15.6 \stackrel{?}{=} 39.8$ Substitute 24.2 for x. $39.8 = 39.8$ Do the addition. Since the resulting statement $39.8 = 39.8$ is true, 24.2 is the solution.

Sometimes we must use two (or more) properties of equality to solve more complicated equations.

Solve: $9.3x - 2.6 = 6.2x + 9.8$

$9.3x - 2.6 - 6.2x = 6.2x + 9.8 - 6.2x$ To eliminate the term 6.2x on the right side, subtract 6.2x from both sides.

$3.1x - 2.6 = 9.8$ Combine like terms on both sides.

$3.1x - 2.6 + 2.6 = 9.8 + 2.6$ To isolate 3.1x, undo the subtraction of 2.6 by adding 2.6 to both sides.

$3.1x = 12.4$ Do the addition. Now isolate x.

$\dfrac{3.1x}{3.1} = \dfrac{12.4}{3.1}$ To isolate x, undo the multiplication by 3.1 by dividing both sides by 3.1.

$x = 4$ Do the division.

Check the result in the original equation to verify that 4 is the solution.

We can use the concepts of *variable* and *equation* to solve application problems involving decimals.

ADVERTISING A promotions company sells pens with a personalized marketing message printed on them. They charge a $12 setup fee and the pens cost 85¢ each. If a business has budgeted $250 for this type of promotion, how many pens can they order?

Analyze

- There is a setup fee of $12.   Given
- The pens cost 85¢ each. (This is $0.85 each.)   Given
- The company has budgeted $250 for the promotion.   Given
- How many pens can they order?   Find

Form Let x = the number of pens that they can order.

The setup fee	plus	the cost per pen	times	the number of pens ordered	must equal	the amount budgeted.
12	+	0.85	·	x	=	250

Solve $12 + 0.85x = 250$

$12 + 0.85x - 12 = 250 - 12$ To isolate the variable term 0.85x, subtract 12 from both sides to undo the addition of 12.

$0.85x = 238$ Do the subtraction. Now isolate x.

$\dfrac{0.85x}{0.85} = \dfrac{238}{0.85}$ To isolate x, divide both sides by 0.85 to undo the multiplication by 0.85.

$x = 280$ Do the division.

State The business can order 280 pens.

Check If we multiply the number of pens (280) by the cost per pen ($0.85), and add the setup fee ($12) to that product, we should get the amount budgeted ($250).

$$\begin{array}{r} 280 \\ \times\ 0.85 \\ \hline 14\ 00 \\ 224\ 00 \\ \hline 238.00 \end{array}$$
$$\begin{array}{r} \overset{1}{238} \leftarrow \text{Cost of pens} \\ +\ \ 12 \leftarrow \text{Setup fee} \\ \hline 250 \leftarrow \text{Budget} \end{array}$$

The result checks.

REVIEW EXERCISES

Simplify each expression.

111. $4.6(7w)$

112. $3.8(-2t)(4)$

Multiply.

113. $5.3(2y + 3)$

114. $-7(2.9x - 8.4)$

Simplify each expression by combining like terms.

115. $8.2p + 5.9p$

116. $0.57m^2 - 0.69m^2$

117. $5.7a - 12.4 - 2.9a$

118. $2(0.3t - 0.4) + 3(0.8t - 0.2)$

Solve each equation and check the result.

119. $y + 12.4 = -6.01$

120. $x - 0.23 = 5$

121. $\dfrac{x}{1.78} = -3$

122. $-1.61b = -27.37$

123. $1.7y + 1.24 = -1.4y - 0.62$

124. $0.05(1{,}000 - x) + 0.9x = 60.2$

In Problems 125 and 126, let a variable represent the unknown quantity. Then write and solve an equation to answer the question.

125. U.S. GASOLINE PRICES The average price of regular-grade gasoline in 2008 was $3.26 per gallon. This was $0.45 more than the average price in 2007. What was the price per gallon of regular gasoline in 2007? (Source: Energy Information Administration)

126. BOWLING If it costs $4.00 to rent shoes and $4.50 a game to use a lane, how many games can be bowled for $40?

1. Fill in the blanks.
   a. Copy the following addition. Label each *addend* and the *sum*.

   $$2.67 \leftarrow \underline{\hspace{2cm}}$$
   $$+6.01 \leftarrow \underline{\hspace{2cm}}$$
   $$8.68 \leftarrow \underline{\hspace{2cm}}$$

   b. Copy the following subtraction. Label the *minuend*, the *subtrahend*, and the *difference*.

   $$9.6 \leftarrow \underline{\hspace{2cm}}$$
   $$-6.2 \leftarrow \underline{\hspace{2cm}}$$
   $$3.4 \leftarrow \underline{\hspace{2cm}}$$

   c. Copy the following multiplication. Label the *factors* and the *product*.

   $$1.3 \leftarrow \underline{\hspace{2cm}}$$
   $$\times 7 \leftarrow \underline{\hspace{2cm}}$$
   $$9.1 \leftarrow \underline{\hspace{2cm}}$$

   d. Copy the following division. Label the *dividend*, the *divisor*, and the *quotient*.

   $$3.4 \leftarrow \underline{\hspace{2cm}}$$
   $$\underline{\hspace{1.5cm}} \rightarrow 2\overline{)6.8} \leftarrow \underline{\hspace{2cm}}$$

   e. 0.6666 . . . and 0.8333 . . . are examples of _____ decimals.

   f. The $\sqrt{}$ symbol is called a _____ symbol.

   g. When we write $4.1x + 2.7x$ as $6.8x$, we say we have _____ like terms.

   h. To _____ an equation means to find all the values of the variable that make the equations true.

2. Express the amount of the square region that is shaded using a fraction and a decimal.

   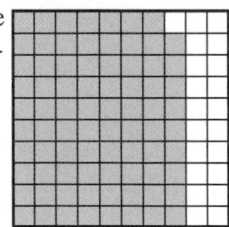

3. Consider the decimal number: 629.471
   a. What is the place value of the digit 1?
   b. Which digit tells the number of tenths?
   c. Which digit tells the number of hundreds?
   d. What is the place value of the digit 2?

4. WATER PURITY A county health department sampled the pollution content of tap water in five cities, with the results shown. Rank the cities in order, from dirtiest tap water to cleanest.

City	Pollution, parts per million
Monroe	0.0909
Covington	0.0899
Paston	0.0901
Cadia	0.0890
Selway	0.1001

5. Write *four thousand five hundred nineteen and twenty-seven ten-thousandths* in standard form.

6. Write each decimal in:
   • expanded form
   • words
   • as a fraction or mixed number. (You do not have to simplify the fraction.)

   a. SKATEBOARDING Gary Hardwick of Carlsbad, California, set the skateboard speed record of 62.55 mph in 1998. (Source: skateboardballbearings.com)

   b. MONEY A dime weighs 0.08013 ounce.

7. Round each decimal number to the indicated place value.
   a. 461.728, nearest tenth
   b. 2,733.0495, nearest thousandth
   c. −1.9833732, nearest millionth

8. Round $0.648209 to the nearest cent.

Perform each operation.

9. $4.56 + 2 + 0.896 + 3.3$

10. Subtract 39.079 from 45.2

11. $(0.32)^2$

12. $\dfrac{0.1368}{0.24}$

13. $-6.7(-2.1)$

14. $$\begin{array}{r} 8.7 \\ \times\ 0.004 \\ \hline \end{array}$$

15. $11\overline{)13}$

16. $-2.4 - (-1.6)$

17. Divide. Round the quotient to the nearest hundredth:
    $$\dfrac{12.146}{-5.3}$$

18. a. Estimate the product using front-end rounding: $34 \cdot 6.83$

    b. Estimate the quotient: $3,907.2 \div 19.3$

19. Perform each operation in your head.

  a. $567.909 \div 1{,}000$

  b. $0.00458 \cdot 100$

20. Write 61.4 billion in standard notation.

21. NEW YORK CITY Refer to the illustration on the right. Central Park, which lies in the middle of Manhattan, is the city's best-known park. If it is 2.5 miles long and 0.5 mile wide, what is its area?

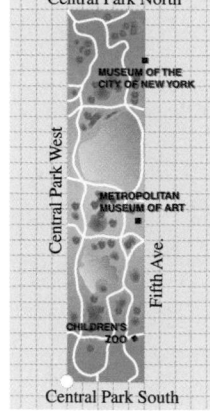

22. TELEPHONE BOOKS To print a telephone book, 565 sheets of paper were used. If the book is 2.26 inches thick, what is the thickness of each sheet of paper?

23. ACCOUNTING At an ice-skating complex, receipts on Friday were $130.25 for indoor skating and $162.25 for outdoor skating. On Saturday, the corresponding amounts were $140.50 and $175.75. On which day, Friday or Saturday, were the receipts higher? How much higher?

24. WEIGHT OF WATER One gallon of water weighs 8.33 pounds. How much does the water in a $2\frac{1}{2}$-gallon jug weigh?

25. Evaluate the formula $C = 2\pi r$ for $\pi = 3.14$ and $r = 1.7$.

26. Write each fraction as a decimal.

  a. $\dfrac{17}{50}$

  b. $\dfrac{5}{12}$

Evaluate each expression.

27. $4.1 - (3.2)(0.4)^2$

28. $\left(\dfrac{2}{5}\right)^2 + 6\left|-6.2 - 3\dfrac{1}{4}\right|$

29. $8 - 2\left(2^4 - 60 + 6\sqrt{81}\right)$

30. $\dfrac{2}{3} + 0.7$ (Work in terms of fractions.)

31. a. Graph $\frac{3}{8}$, $\frac{2}{3}$, and $-\frac{4}{5}$ on the number line. Label each point using the decimal equivalent of the fraction.

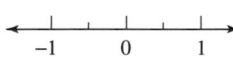

  b. Graph $\sqrt{16}$, $\sqrt{2}$, $-\sqrt{9}$, and $-\sqrt{5}$ on the number line below. (*Hint:* When necessary, use a calculator or square root table to approximate a square root.)

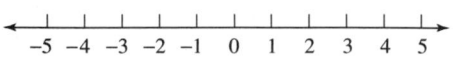

32. SALADS A shopper purchased $\frac{3}{4}$ pound of potato salad, priced at $5.60 per pound, and $\frac{1}{3}$ pound of coleslaw, selling for $4.35 per pound. Find the total cost of these items.

33. Use a calculator to evaluate $c = \sqrt{a^2 + b^2}$ for $a = 12$ and $b = 35$.

34. Write each number as a decimal.

  a. $-\dfrac{27}{25}$

  b. $2\dfrac{9}{16}$

35. Fill in the blank: $\sqrt{144} = \boxed{}$ because $\boxed{}^2 = 144$.

36. Place an $<$, $>$, or an $=$ symbol in the box to make a true statement.

  a. $-6.78 \ \boxed{} \ -6.79$

  b. $0.3 \ \boxed{} \ \dfrac{3}{8}$

  c. $\sqrt{\dfrac{16}{81}} \ \boxed{} \ 0.\overline{4}$

  d. $0.45 \ \boxed{} \ 0.\overline{45}$

Evaluate each expression without using a calculator.

37. $-2\sqrt{25} + 3\sqrt{49}$

38. $\sqrt{\dfrac{1}{36}} - \sqrt{\dfrac{1}{25}}$

39. Evaluate each square root without using a calculator.

  a. $-\sqrt{0.04}$

  b. $\sqrt{1.69}$

  c. $\sqrt{225}$

  d. $-\sqrt{121}$

40. Although the decimal 3.2999 contains more digits than 3.3, it is smaller than 3.3. Explain why this is so.

Simplify each expression.

41. a. $9(1.6t)$

  b. $-3(4.7a)(2)$

42. a. $6.18s - 1.22s$

  b. $2.1(x - 3) + 3.1(x - 4)$

Solve each equation and check the result.

43. $-2.4d = 16.8$

44. $\dfrac{x}{2.04} + 2.2 = 6.7$

45. $2(y + 4.3) - 0.3 = 7.1$

46. $0.3x + 6.9 = 1.6x - 8.7$

In Problems 47 and 48, let a variable represent the unknown quantity. Then write and solve an equation to answer the question.

47. CHEMISTRY In a lab experiment, a chemist mixed three compounds together to form a mixture weighing 4.37 grams. Later, she discovered that she had forgotten to record the weight of compound C in her notes. Find the weight of compound C used in the experiment.

	Weight
Compound A	1.86 grams
Compound B	2.09 grams
Compound C	?
Mixture total	4.37 grams

48. WEDDING COSTS A printer charges a setup fee of $24 and then 95 cents for each wedding announcement printed (tax included). If a couple has budgeted $100 for printing costs, how many announcements can they have printed?

CHAPTERS 1–5 CUMULATIVE REVIEW

1. Write 154,302

   a. in words [Section 1.1]

   b. in expanded form [Section 1.1]

2. Use 3, 4, and 5 to express the associative property of addition. [Section 1.2]

3. Add: $9,339 + 471 + 6,883$ [Section 1.2]

4. Subtract 199 from 301. [Section 1.2]

5. SUDOKU The world's largest Sudoku puzzle was carved into a hillside near Bristol, England. It measured 275 ft by 275 ft. Find the area covered by the square-shaped puzzle. (Source: joe-ks.com) [Section 1.3]

Tim Anderson Photography Ltd/Sky 1

6. Divide: $43\overline{)1,203}$ [Section 1.4]

7. List the factors of 20, from smallest to largest. [Section 1.5]

8. Find the prime factorization of 220. [Section 1.5]

9. Find the LCM and the GCF of 100 and 120. [Section 1.6]

10. Find the mean (average) of 7, 1, 8, 2, and 2. [Section 1.7]

11. Solve each equation and check the result.

   a. $x + 19 = 285$ [Section 1.8]

   b. $19x = 285$ [Section 1.9]

12. Place an $<$ or an $>$ symbol in the box to make a true statement: $|-50|$ $-(-40)$ [Section 2.1]

13. Add: $-8 + (-5)$ [Section 2.2]

14. Fill in the blank: Subtraction is the same as _____ the opposite. [Section 2.3]

15. WEATHER Marsha flew from her Minneapolis home to Hawaii for a week of vacation. She left blizzard conditions and a temperature of $-11°F$, and stepped off the airplane into $72°F$ weather. What temperature change did she experience? [Section 2.3]

16. Multiply: $-3(-5)(2)(-9)$ [Section 2.4]

17. Evaluate: $(-1)^5$ [Section 2.4]

18. SUBMARINES As part of a training exercise, the captain of a submarine ordered it to descend 350 feet, level off for 10 minutes, and then repeat the process several times. If the sub was on the ocean's surface at the beginning of the exercise, find its depth after the 6th dive. [Section 2.4]

19. Consider the division statement $\frac{-15}{-5} = 3$. What is its related multiplication statement? [Section 2.5]

20. Divide: $-420,000 \div (-7,000)$ [Section 2.5]

21. Evaluate: $(-6)^2 - 2(5 - 4 \cdot 2)$ [Section 2.6]

22. Solve $2y + 8 = -6$ and check the result. [Section 2.7]

Write an equation and solve it to answer the following question. [Section 2.7]

23. MERCURY The freezing point of mercury is $-38°F$. By how many degrees must it be heated to reach its boiling point, which is $674°F$?

24. Translate to mathematical symbols: The cost c split five equal ways. [Section 3.1]

25. Write an algebraic expression that represents the number of minutes in h hours. [Section 3.1]

26. BOWLING Find the average for a bowler who rolled scores of 233, 218, and 206. [Section 3.2]

27. Simplify: $-5(-6x)$ [Section 3.3]

28. Combine like terms: $5x + 5x + 5x$ [Section 3.4]

29. Solve $9a + 3(a - 7) = 18 - a$ and check the result. [Section 3.5]

Write an equation and solve it to answer the following question. [Section 3.6]

30. AIRLINE SEATING A 132-seat passenger plane has ten times as many economy seats as first-class seats. Find the number of first-class seats.

31. FLAGS What fraction of the stripes on a U.S. flag are white? [Section 4.1]

32. Simplify: $\dfrac{90}{126}$ [Section 4.1]

Perform the operations. Simplify the result.

33. $\dfrac{3}{8} \cdot \dfrac{7}{16}$ [Section 4.2]

34. $-\dfrac{15}{8a^4} \div \dfrac{10}{a^5}$ [Section 4.3]

35. $\dfrac{a}{9} + \dfrac{5}{7}$ [Section 4.4]

36. $-4\dfrac{1}{4}\left(-4\dfrac{1}{2}\right)$ [Section 4.5]

37. $76\dfrac{1}{6} - 49\dfrac{7}{8}$ [Section 4.6]

38. $\dfrac{\dfrac{5}{27}}{-\dfrac{5}{9}}$ [Section 4.7]

39. What is $\frac{1}{4}$ of $\frac{7}{16}$? [Section 4.2]

40. TAPE MEASURES Use the information shown in the illustration below to determine the inside length of the drawer. [Section 4.6]

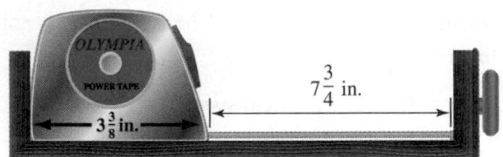

$7\dfrac{3}{4}$ in.

$3\dfrac{3}{8}$ in.

41. Evaluate: $\left(\dfrac{9}{20} \div 2\dfrac{2}{5}\right) + \left(\dfrac{3}{4}\right)^2$ [Section 4.7]

42. Solve $\dfrac{3}{4}m = -27$ and check the result. [Section 4.8]

43. GLASS Some electronic and medical equipment uses glass that is only 0.00098 inch thick. Round this number to the nearest thousandth. [Section 5.1]

44. Graph $-3\frac{1}{4}, 0.75, -1.5, -\frac{9}{8}, 3.8$, and $\sqrt{4}$ on a number line. [Section 5.1]

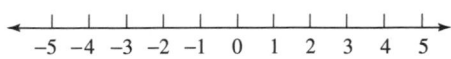

Perform the operations.

45. $7.001 - 5.9$ [Section 5.2]

46. $-1.8(4.52)$ [Section 5.3]

47. $56.012(0.001)$ [Section 5.3]

48. $\dfrac{-21.28}{-3.8}$ [Section 5.4]

49. KITES Find the area of the front of the kite shown below. [Section 5.3]

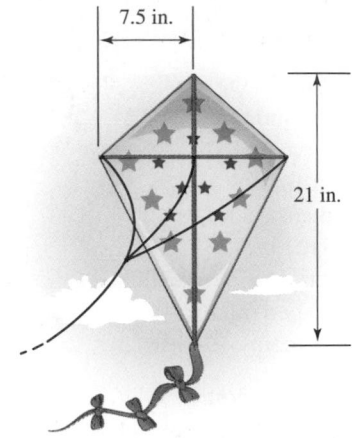

7.5 in.

21 in.

50. Evaluate the formula $C = \frac{5}{9}(F - 32)$ for $F = 451$. Round to the nearest tenth. [Section 5.4]

51. Write the fraction $\frac{5}{12}$ as a decimal. [Section 5.5]

52. Evaluate: $\dfrac{3}{8}(-3.2) + \left(4\dfrac{1}{2}\right)\left(-\dfrac{1}{4}\right)$ [Section 5.5]

53. Evaluate: $-4\sqrt{36} + 2\sqrt{81}$ [Section 5.6]

54. Solve $-5.6 - 2h = -17.1 - h$ and check the result. [Section 5.7]

Ratio, Proportion, and Measurement

Nick White/Getty Images

from *Campus to Careers*

Chef

Chefs prepare and cook a wide range of foods—from soups, snacks, and salads to main dishes, side dishes, and desserts. They work in a variety of restaurants and food service kitchens. They measure, mix, and cook ingredients according to recipes, using a variety of equipment and tools. They are also responsible for directing the tasks of other kitchen workers, estimating food requirements, and ordering food supplies.

In **Problem 90** of **Study Set 6.2**, you will see how a chef can use proportions to determine the correct amounts of each ingredient needed to make a large batch of brownies.

JOB TITLE: Chef

EDUCATION: Training programs are available through culinary schools, 2- or 4-year college degree programs, and the armed forces.

JOB OUTLOOK: Job openings are expected to be plentiful through 2016.

ANNUAL EARNINGS: The average (median) salary in 2008 was $55,976.

FOR MORE INFORMATION: www.searchbydegree.com/chef-cook-career.html

Objectives

1 Write ratios as fractions.

2 Simplify ratios involving decimals and mixed numbers.

3 Convert units to write ratios.

4 Write rates as fractions.

5 Find unit rates.

6 Find the best buy based on unit price.

SECTION 6.1

Ratios

Ratios are often used to describe important relationships between two quantities. Here are three examples:

To prepare fuel for an outboard marine engine, gasoline must be mixed with oil in the ratio of 50 to 1.

To make 14-karat jewelry, gold is combined with other metals in the ratio of 14 to 10.

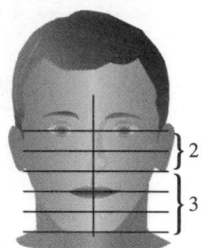

In this drawing, the eyes-to-nose distance and the nose-to-chin distance are drawn using a ratio of 2 to 3.

1 Write ratios as fractions.

Ratios give us a way to compare two numbers or two quantities measured in the same units.

Ratios

A **ratio** is the quotient of two numbers or the quotient of two quantities that have the same units.

There are three ways to write a ratio. The most comon way is as a fraction. Ratios can also be written as two numbers separated by the word *to,* or as two numbers separated by a colon. For example, the ratios described in the illustrations above can be expressed as:

$$\frac{50}{1}, \qquad 14 \text{ to } 10, \qquad \text{and} \qquad 2\colon 3$$

- The fraction $\frac{50}{1}$ is read as "the ratio of 50 to 1."   *A fraction bar separates the numbers being compared.*

- 14 **to** 10 is read as "the ratio of 14 to 10."   *The word "to" separates the numbers being compared.*

- 2:3 is read as "the ratio of 2 to 3."   *A colon separates the numbers being compared.*

Writing a Ratio as a Fraction

To **write a ratio as a fraction,** write the first number (or quantity) mentioned as the numerator and the second number (or quantity) mentioned as the denominator. Then simplify the fraction, if possible.

EXAMPLE 1   Write each ratio as a fraction: **a.** 3 to 7   **b.** 10:11

Strategy We will identify the numbers before and after the word *to* and the numbers before and after the colon.

WHY The word *to* and the colon separate the numbers to be compared in a ratio.

Solution

To write the ratio as a fraction, the first number mentioned is the numerator and the second number mentioned is the denominator.

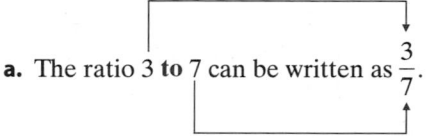

a. The ratio 3 **to** 7 can be written as $\frac{3}{7}$.   The fraction $\frac{3}{7}$ is in simplest form.

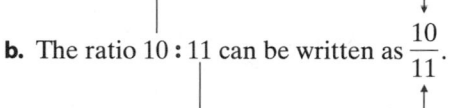

b. The ratio 10 **:** 11 can be written as $\frac{10}{11}$.   The fraction $\frac{10}{11}$ is in simplest form.

Caution! When a ratio is written as a fraction, the fraction should be in simplest form. (Recall from Chapter 4 that a fraction is in **simplest form,** or **lowest terms,** when the numerator and denominator have no common factors other than 1.)

Self Check 1

Write each ratio as a fraction:

a. 4 to 9   **b.** 8:15

Now Try Problem 13

EXAMPLE 2   Write the ratio 35 to 10 as a fraction in simplest form.

Strategy We will translate the ratio from its given form in words to fractional form. Then we will look for any factors common to the numerator and denominator and remove them.

WHY We need to make sure that the numerator and denominator have no common factors other than 1. If that is the case, the ratio will be in *simplest form.*

Solution

The ratio 35 **to** 10 can be written as $\frac{35}{10}$.   The fraction $\frac{35}{10}$ is not in simplest form.

Now, we simplify the fraction using the method discussed in Section 4.1.

$$\frac{35}{10} = \frac{\overset{1}{\cancel{5}} \cdot 7}{2 \cdot \underset{1}{\cancel{5}}}$$ Factor 35 as 5 · 7 and 10 as 2 · 5. Then remove the common factor of 5 in the numerator and denominator.

$$= \frac{7}{2}$$

The ratio 35 to 10 can be written as the fraction $\frac{35}{10}$, which simplifies to $\frac{7}{2}$ (read as "7 to 2"). Because the fractions $\frac{35}{10}$ and $\frac{7}{2}$ represent equal numbers, they are called **equal ratios.**

Self Check 2

Write the ratio 12 to 9 as a fraction in simplest form.

Now Try Problems 17 and 23

> **Caution!** Since ratios are comparisons of two numbers, it would be *incorrect* in Example 2 to write the ratio $\frac{7}{2}$ as the mixed number $3\frac{1}{2}$. Ratios written as improper fractions are perfectly acceptable—just make sure the numerator and denominator have no common factors other than 1.

To write a ratio in simplest form, we remove any common factors of the numerator and denominator as well as any common units.

Self Check 3

CARRY-ON LUGGAGE

a. Write the ratio of the height to the length of the carry-on space shown in the illustration in Example 3 as a fraction in simplest form.

b. Write the ratio of the length of the carry-on space to its height in simplest form.

Now Try Problem 27

EXAMPLE 3   *Carry-on Luggage*   An airline allows its passengers to carry a piece of luggage onto an airplane only if it will fit in the space shown below.

a. Write the ratio of the width of the space to its length as a fraction in simplest form.

b. Write the ratio of the length of the space to its width as a fraction in simplest form.

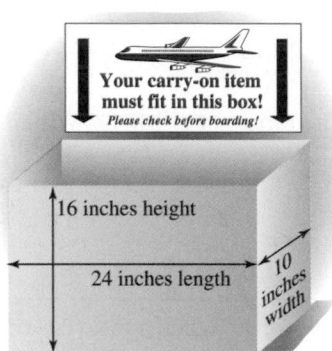

Strategy To write each ratio as a fraction, we will identify the quantity before the word *to* and the quantity after it.

WHY The first quantity mentioned is the numerator of the fraction and the second quantity mentioned is the denominator.

Solution

a. The ratio of the width of the space to its length is $\dfrac{10 \text{ inches}}{24 \text{ inches}}$.

To write a ratio in simplest form, we remove the common factors *and* the common units of the numerator and denominator.

$$\frac{10 \text{ inches}}{24 \text{ inches}} = \frac{\overset{1}{2} \cdot 5 \text{ inches}}{\underset{1}{2} \cdot 12 \text{ inches}}$$

Factor 10 as $2 \cdot 5$ and 24 as $2 \cdot 12$. Then remove the common factor of 2 and the common units of inches from the numerator and denominator.

$$= \frac{5}{12}$$

The width-to-length ratio of the carry-on space is $\dfrac{5}{12}$ (read as "5 to 12").

b. The ratio of the length of the space to its width is $\dfrac{24 \text{ inches}}{10 \text{ inches}}$.

$$\frac{24 \text{ inches}}{10 \text{ inches}} = \frac{\overset{1}{2} \cdot 12 \text{ inches}}{\underset{1}{2} \cdot 5 \text{ inches}}$$

Factor 24 and 10. Then remove the common factor of 2 and the common units of inches from the numerator and denominator.

$$= \frac{12}{5}$$

The length-to-width ratio of the carry-on space is $\dfrac{12}{5}$ (read as "12 to 5").

> **Caution!**   Example 3 shows that order is important when writing a ratio. The width-to-length ratio is $\frac{5}{12}$ while the length-to-width ratio is $\frac{12}{5}$.

2   Simplify ratios involving decimals and mixed numbers.

EXAMPLE 4   Write the ratio 0.3 to 1.8 as a fraction in simplest form.

Strategy   After writing the ratio as a fraction, we will multiply it by a form of 1 to obtain an equivalent ratio of whole numbers.

WHY   A ratio of whole numbers is easier to understand than a ratio of decimals.

Solution

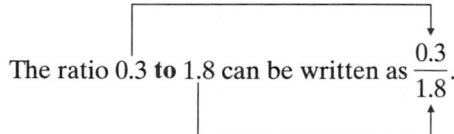

The ratio 0.3 **to** 1.8 can be written as $\dfrac{0.3}{1.8}$.

To write this as a ratio of *whole numbers*, we need to move the decimal points in the numerator and denominator one place to the right. Recall that to find the product of a decimal and 10, we simply move the decimal point one place to the right. Therefore, it follows that $\frac{10}{10}$ is the form of 1 that we should use to build $\frac{0.3}{1.8}$ into an equivalent ratio.

$$\frac{0.3}{1.8} = \frac{0.3}{1.8} \cdot \frac{\mathbf{10}}{\mathbf{10}}$$ Multiply the ratio by a form of 1.

$$\frac{0.3}{1.8} = \frac{0.3 \cdot \mathbf{10}}{1.8 \cdot \mathbf{10}}$$ Multiply the numerators. Multiply the denominators.

$$= \frac{3}{18}$$ Do the multiplications by moving each decimal point one place to the right. $0.3 \cdot 10 = 3$ and $1.8 \cdot 10 = 18$.

$$= \frac{1}{6}$$ Simplify the fraction: $\frac{3}{18} = \frac{\overset{1}{\cancel{3}}}{\underset{1}{\cancel{3}} \cdot 6} = \frac{1}{6}$.

Self Check 4

Write the ratio 0.8 to 2.4 as a fraction in simplest form.

Now Try Problems 29 and 33

THINK IT THROUGH   *Student-to-Instructor Ratio*

"A more personal classroom atmosphere can sometimes be an easier adjustment for college freshmen. They are less likely to feel like a number, a feeling that can sometimes impact students' first semester grades."
From *The Importance of Class Size* by Stephen Pemberton

The data below come from a nationwide study of mathematics programs at two-year colleges. Determine which course has the lowest student-to-instructor ratio. (Assume that there is one instructor per section.)

	Basic Mathematics	Elementary Algebra	Intermediate Algebra
Students enrolled	101,200	367,920	318,750
Number of sections	4,400	15,330	12,750

Source: Conference Board of the Mathematical Science, 2005 CBMS Survey of Undergraduate Programs (The data has been rounded to yield ratios involving whole numbers.)

Self Check 5
Write the ratio $3\frac{1}{3}$ to $1\frac{1}{9}$ as a fraction in simplest form.

Now Try Problem 37

EXAMPLE 5   Write the ratio $4\frac{2}{3}$ to $1\frac{1}{6}$ as a fraction in simplest form.

Strategy After writing the ratio as a fraction, we will use the method for simplifying a complex fraction from Section 4.7 to obtain an equivalent ratio of whole numbers.

WHY A ratio of whole numbers is easier to understand than a ratio of mixed numbers.

Solution

The ratio of $4\frac{2}{3}$ **to** $1\frac{1}{6}$ can be written as $\dfrac{4\frac{2}{3}}{1\frac{1}{6}}$.

The resulting ratio is a complex fraction. To write the ratio in simplest form, we perform the division indicated by the main fraction bar (shown in red).

$$\dfrac{4\frac{2}{3}}{1\frac{1}{6}} = \dfrac{\frac{14}{3}}{\frac{7}{6}}$$ Write $4\frac{2}{3}$ and $1\frac{1}{6}$ as improper fractions.

$$= \frac{14}{3} \div \frac{7}{6}$$ Write the division indicated by the main fraction bar using a $\div$ symbol.

$$= \frac{14}{3} \cdot \frac{6}{7}$$ Use the rule for dividing fractions: Multiply the first fraction by the reciprocal of $\frac{7}{6}$, which is $\frac{6}{7}$.

$$= \frac{14 \cdot 6}{3 \cdot 7}$$ Multiply the numerators.
Multiply the denominators.

$$= \frac{2 \cdot \overset{1}{\cancel{7}} \cdot 2 \cdot \overset{1}{\cancel{3}}}{\underset{1}{\cancel{3}} \cdot \underset{1}{\cancel{7}}}$$ To simplify the fraction, factor 14 as $2 \cdot 7$ and 6 as $2 \cdot 3$. Then remove the common factors 3 and 7.

$$= \frac{4}{1}$$ Multiply the remaining factors in the numerator.
Multiply the remaining factors in the denominator.

We would normally simplify the result $\frac{4}{1}$ and write it as 4. But since a ratio compares two numbers, we leave the result in fractional form.

3 Convert units to write ratios.

When a ratio compares 2 quantities, both quantities must be measured in the same units. For example, inches must be compared to inches, pounds to pounds, and seconds to seconds.

Self Check 6
Write the ratio *6 feet to 3 yards* as a fraction in simplest form. (*Hint:* 3 feet = 1 yard.)

Now Try Problem 41

EXAMPLE 6   Write the ratio *12 ounces to 2 pounds* as a fraction in simplest form.

Strategy We will convert 2 pounds to ounces and write a ratio that compares ounces to ounces. Then we will simplify the ratio.

WHY A ratio compares two quantities that have the *same* units. When the units are different, it's usually easier to write the ratio using the smaller unit of measurement. Since ounces are smaller than pounds, we will compare in ounces.

Solution

To express 2 pounds in ounces, we use the fact that there are 16 ounces in one pound.

$$2 \cdot 16 \text{ ounces} = 32 \text{ ounces}$$

We can now express the ratio *12 ounces to 2 pounds* using the same units:

12 **ounces** to 32 **ounces**

Next, we write the ratio in fraction form and simplify.

$$\frac{12 \text{ ounces}}{32 \text{ ounces}} = \frac{3 \cdot \overset{1}{\cancel{4}} \text{ \cancel{ounces}}}{\underset{1}{\cancel{4}} \cdot 8 \text{ \cancel{ounces}}}$$ To simplify, factor 12 as 3 · 4 and 32 as 4 · 8. Then remove the common factor of 4 and the common units of ounces from the numerator and denominator.

$$= \frac{3}{8}$$

The ratio in simplest form is $\frac{3}{8}$.

---

4 Write rates as fractions.

When we compare two quantities that have different units (and neither unit can be converted to the other), we call the comparison a **rate,** and we can write it as a fraction. For example, on the label of the can of paint shown on the right, we see that 1 quart of paint is needed for every 200 square feet to be painted. Writing this as a rate in fractional form, we have

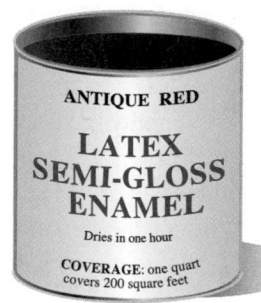

$$\frac{1 \text{ quart}}{200 \text{ square feet}}$$ Read as "1 quart per 200 square feet."

> ***The Language of Algebra*** The word *per* is associated with the operation of division, and it means "for each" or "for every." For example, when we say 1 quart of paint *per* 200 square feet, we mean 1 quart of paint *for every* 200 square feet.

Rates

A **rate** is a quotient of two quantities that have different units.

When writing a rate, always include the units. Some other examples of rates are:

- 16 computers **for** 75 students
- 1,550 feet **in** 4.5 seconds
- 88 tomatoes **from** 3 plants
- 250 miles **on** 2 gallons of gasoline

> ***The Language of Algebra*** As seen above, words such as *per, for, in, from,* and *on* are used to separate the two quantities that are compared in a rate.

> ### Writing a Rate as a Fraction
>
> To **write a rate as a fraction,** write the first quantity mentioned as the numerator and the second quantity mentioned as the denominator, and then simplify, if possible. Write the units as part of the fraction.

Self Check 7

GROWTH RATES The fastest-growing flowering plant on record grew 12 feet in 14 days. Write the rate of growth as a fraction in simplest form.

Now Try **Problems 49 and 53**

EXAMPLE 7   *Snowfall* According to the *Guinness Book of World Records,* a total of 78 inches of snow fell at Mile 47 Camp, Cooper River Division, Arkansas, in a 24-hour period in 1963. Write the rate of snowfall as a fraction in simplest form.

Strategy We will use a fraction to compare the amount of snow that fell (in inches) to the amount of time in which it fell (in hours). Then we will simplify it.

WHY A rate is a quotient of two quantities with different units.

Solution

78 inches **in** 24 hours can be written as $\dfrac{78 \text{ inches}}{24 \text{ hours}}$.

Now, we simplify the fraction.

$$\frac{78 \text{ inches}}{24 \text{ hours}} = \frac{\overset{1}{\cancel{6}} \cdot 13 \text{ inches}}{4 \cdot \underset{1}{\cancel{6}} \text{ hours}}$$

To simplify, factor 78 as 6 · 13 and 24 as 4 · 6. Then remove the common factor of 6 from the numerator and denominator.

$$= \frac{13 \text{ inches}}{4 \text{ hours}}$$

Since the units are different, they cannot be removed.

The snow fell at a rate of 13 inches per 4 hours.

5 Find unit rates.

> ### Unit Rate
>
> A **unit rate** is a rate in which the denominator is 1.

To illustrate the concept of a unit rate, suppose a driver makes the 354-mile trip from Pittsburgh to Indianapolis in 6 hours. Then the motorist's rate (or more specifically, rate of speed) is given by

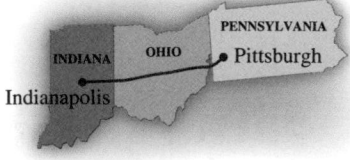

$$\frac{354 \text{ miles}}{6 \text{ hours}} = \frac{\overset{1}{\cancel{6}} \cdot 59 \text{ miles}}{\underset{1}{\cancel{6}} \cdot \text{ hours}}$$

Factor 354 as 6 · 59 and remove the common factor of 6 from the numerator and denominator.

$$= \frac{59 \text{ miles}}{1 \text{ hour}}$$

Since the units are different, they cannot be removed. Note that the denominator is 1.

We can also find the unit rate by dividing 354 by 6.

Rate: **Unit rate:**

$$\frac{354 \text{ miles}}{6 \text{ hours}}$$

$$\begin{array}{r} 59 \\ 6\overline{)354} \\ -30 \\ \hline 54 \\ -54 \\ \hline 0 \end{array}$$ ——This quotient is the numerical part of the unit rate, written as a fraction. ⟶ $\dfrac{59 \text{ miles}}{1 \text{ hour}}$

The numerical part of the ⟶ denominator is always 1.

The unit rate $\frac{59 \text{ miles}}{1 \text{ hour}}$ can be expressed in any of the following forms:

$$59 \,\frac{\text{miles}}{\text{hour}}, \quad 59 \text{ miles per hour}, \quad 59 \text{ miles/hour}, \quad \text{or} \quad 59 \text{ mph}$$

> **The Language of Algebra**   A slash mark / is often used to write a unit rate. In such cases, we read the slash mark as "per." For example, 33 pounds/gallon is read as 33 pounds *per* gallon.

Writing a Rate as a Unit Rate

To **write a rate as a unit rate,** divide the numerator of the rate by the denominator.

EXAMPLE 8   *Coffee*   There are 384 calories in a 16-ounce cup of caramel Frappuccino blended coffee with whip cream. Write this rate as a unit rate. (*Hint:* Find the number of calories in 1 ounce.)

Strategy   We will translate the rate from its given form in words to fractional form. Then we will perform the indicated division.

WHY   To write a rate as a unit rate, we divide the numerator of the rate by the denominator.

Solution

384 calories **in** 16 ounces can be written as $\dfrac{384 \text{ calories}}{16 \text{ ounces}}$.

To find the number of calories in 1 ounce of the coffee (the unit rate), we perform the division as indicated by the fraction bar:

$$\begin{array}{r} 24 \\ 16\overline{)384} \\ -32 \\ \hline 64 \\ -64 \\ \hline 0 \end{array}$$ Divide the numerator of the rate by the denominator.

For the caramel Frappuccino blended coffee with whip cream, the unit rate is $\frac{24 \text{ calories}}{1 \text{ ounce}}$, which can be written as 24 calories per ounce or 24 calories /ounce.

Self Check 8

NUTRITION   There are 204 calories in a 12-ounce can of cranberry juice. Write this rate as a unit rate. (*Hint:* Find the number of calories in 1 ounce.)

Now Try Problem 57

Self Check 9

FULL-TIME JOBS Joan earns $436 per 40-hour week managing a dress shop. Write this rate as a unit rate. (*Hint:* Find her hourly rate of pay.)

Now Try **Problem 61**

EXAMPLE 9   *Part-time Jobs*   A student earns $74 for working 8 hours in a bookstore. Write this rate as a unit rate. (*Hint:* Find his hourly rate of pay.)

Strategy We will translate the rate from its given form in words to fractional form. Then we will perform the indicated division.

WHY To write a rate as a unit rate, we divide the numerator of the rate by the denominator.

Solution

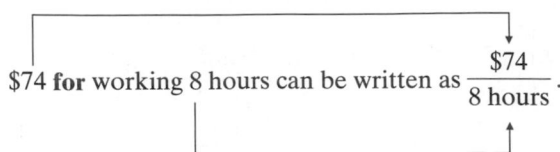

$74 **for** working 8 hours can be written as $\dfrac{\$74}{8 \text{ hours}}$.

To find the rate of pay for 1 hour of work (the unit rate), we divide 74 by 8.

```
    9.25
8)74.00     Write a decimal point and two additional zeros to the right of 4.
 -72
   2 0
  -1 6
    40
   -40
     0
```

The unit rate of pay is $\dfrac{\$9.25}{1 \text{ hour}}$, which can be written as $9.25 per hour or $9.25/hr.

6 Find the best buy based on unit price.

If a grocery store sells a 5-pound package of hamburger for $18.75, a consumer might want to know what the hamburger costs per pound. When we find the cost of 1 pound of the hamburger, we are finding a **unit price.** To find the unit price of an item, we begin by comparing its price to the number of units.

$$\dfrac{\$18.75 \; \leftarrow \text{Price}}{5 \text{ pounds} \; \leftarrow \text{Number of units}}$$

Then we divide the price by the number of units.

```
    3.75
5)18.75
```

The unit price of the hamburger is $3.75 per pound.
   Other examples of unit prices are:

- $8.15 per ounce
- $200 per day
- $0.75 per foot

Unit Price

A **unit price** is a rate that tells how much is paid for *one* unit (or *one* item). It is the quotient of price to the number of units.

$$\text{Unit price} = \dfrac{\text{price}}{\text{number of units}}$$

When shopping, it is often difficult to determine the best buys because the items that we purchase come in so many different sizes and brands. Comparison shopping can be made easier by finding unit prices. *The best buy is the item that has the lowest unit price.*

EXAMPLE 10 *Comparison Shopping*

Olives come packaged in a 10-ounce jar, which sells for $2.49, or in a 6-ounce jar, which sells for $1.53. Which is the better buy?

Strategy We will find the unit price for each jar of olives. Then we will identify which jar has the lower unit price.

WHY The better buy is the jar of olives that has the lower unit price.

Solution

To find the unit price of each jar of olives, we write the quotient of its price and its weight, and then perform the indicated division. Before dividing, we convert each price from dollars to cents so that the unit price can be expressed in cents per ounce.

The 10-ounce jar:

$$\frac{\$2.49}{10 \text{ oz}} = \frac{249¢}{10 \text{ oz}}$$ Write the rate: $\frac{\text{price}}{\text{number of units}}$.
                                              Then change $2.49 to 249 cents.

$$= 24.9¢ \text{ per oz}$$ Divide 249 by 10 by moving the
                               decimal point 1 place to the left.

The 6-ounce jar:

$$\frac{\$1.53}{6 \text{ oz}} = \frac{153¢}{6 \text{ oz}}$$ Write the rate: $\frac{\text{price}}{\text{number of units}}$.
                                            Then change $1.53 to 153 cents.

$$= 25.5¢ \text{ per oz}$$ Do the division.

```
        25.5
  6)153.0
   -12
    33
   -30
     3 0
    -3 0
       0
```

One ounce for 24.9¢ is a better buy than one ounce for 25.5¢. The unit price is less when olives are packaged in 10-ounce jars, so that is the better buy.

Self Check 10

COMPARISON SHOPPING A fast-food restaurant sells a 12-ounce cola for 72¢ and a 16-ounce cola for 99¢. Which is the better buy?

Now Try Problems 65 and 101

ANSWERS TO SELF CHECKS

1. a. $\frac{4}{9}$ **b.** $\frac{8}{15}$ **2.** $\frac{4}{3}$ **3. a.** $\frac{2}{3}$ **b.** $\frac{3}{2}$ **4.** $\frac{1}{3}$ **5.** $\frac{3}{1}$ **6.** $\frac{2}{3}$ **7.** $\frac{6 \text{ feet}}{7 \text{ days}}$

8. 17 calories/oz **9.** $10.90 per hour **10.** the 12-oz cola

---

SECTION 6.1   STUDY SET

VOCABULARY

Fill in the blanks.

1. A _____ is the quotient of two numbers or the quotient of two quantities that have the same units.

2. A _____ is the quotient of two quantities that have different units.

3. A _____ rate is a rate in which the denominator is 1.

4. A unit _____ is a rate that tells how much is paid for one unit or one item.

CONCEPTS

5. To write the ratio $\frac{15}{24}$ in lowest terms, we remove any common factors of the numerator and denominator. What common factor do they have?

6. Complete the solution. Write the ratio $\frac{14}{21}$ in lowest terms.

$$\frac{14}{21} = \frac{2 \cdot 7}{\boxed{} \cdot \boxed{}} = \frac{2 \cdot \overset{1}{7}}{\boxed{} \cdot \underset{1}{7}} = \frac{\boxed{}}{\boxed{}}$$

7. Consider the ratio $\frac{0.5}{0.6}$. By what number should we multiply numerator and denominator to make this a ratio of whole numbers?

8. What should be done to write the ratio $\frac{15 \text{ inches}}{22 \text{ inches}}$ in simplest form?

9. Write $\frac{11 \text{ minutes}}{1 \text{ hour}}$ so that it compares the same units and then simplify.

10. a. Consider the rate $\frac{\$248}{16 \text{ hours}}$. What division should be performed to find the unit rate in dollars per hour?

   b. Suppose 3 pairs of socks sell for $7.95: $\frac{\$7.95}{3 \text{ pairs}}$. What division should be performed to find the unit price of one pair of socks?

NOTATION

11. Write the ratio of the flag's length to its width using a fraction, using the word *to*, and using a colon.

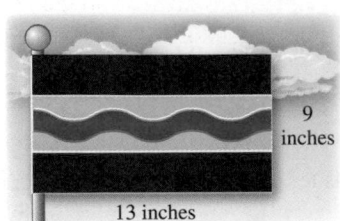

13 inches / 9 inches

12. The rate $\frac{55 \text{ miles}}{1 \text{ hour}}$ can be expressed as

   • 55 _____ _____ _____ (in three words)

   • 55 _____ / _____ (in two words with a slash)

   • 55 ___ ___ ___ (in three letters)

GUIDED PRACTICE

Write each ratio as a fraction. See Example 1.

13. 5 to 8   **14.** 3 to 23

15. 11:16   **16.** 9:25

Write each ratio as a fraction in simplest form. See Example 2.

17. 25 to 15   **18.** 45 to 35

19. 63:36   **20.** 54:24

21. 22:33   **22.** 14:21

23. 17 to 34   **24.** 19 to 38

Write each ratio as a fraction in simplest form. See Example 3.

25. 4 ounces to 12 ounces   **26.** 3 inches to 15 inches

27. 24 miles to 32 miles   **28.** 56 yards to 64 yards

Write each ratio as a fraction in simplest form. See Example 4.

29. 0.3 to 0.9   **30.** 0.2 to 0.6

31. 0.65 to 0.15   **32.** 2.4 to 1.5

33. 3.8:7.8   **34.** 4.2:8.2

35. 7:24.5   **36.** 5:22.5

Write each ratio as a fraction in simplest form. See Example 5.

37. $2\frac{1}{3}$ to $4\frac{2}{3}$ **38.** $1\frac{1}{4}$ to $1\frac{1}{2}$

39. $10\frac{1}{2}$ to $1\frac{3}{4}$ **40.** $12\frac{3}{4}$ to $2\frac{1}{8}$

Write each ratio as a fraction in simplest form. See Example 6.

41. 12 minutes to 1 hour   **42.** 8 ounces to 1 pound

43. 3 days to 1 week   **44.** 4 inches to 1 yard

45. 18 months to 2 years   **46.** 8 feet to 4 yards

47. 21 inches to 3 feet   **48.** 32 seconds to 2 minutes

Write each rate as a fraction in simplest form. See Example 7.

49. 64 feet in 6 seconds

50. 45 applications for 18 openings

51. 75 days on 20 gallons of water

52. 3,000 students over a 16-year career

53. 84 made out of 100 attempts

54. 16 right compared to 34 wrong

55. 18 beats every 12 measures

56. 10 inches as a result of 30 turns

Write each rate as a unit rate. See Example 8.

57. 60 revolutions in 5 minutes

58. 14 trips every 2 months

59. $50,000 paid over 10 years

60. 245 presents for 35 children

Write each rate as a unit rate. **See Example 9.**

61. 12 errors in 8 hours

62. 114 times in a 12-month period

63. 4,007,500 people living in 12,500 square miles

64. 117.6 pounds of pressure on 8 square inches

Find the unit price of each item. **See Example 10.**

65. They charged $48 for 12 minutes.

66. 150 barrels cost $4,950.

67. Four sold for $272.

68. 7,020 pesos will buy six tickets.

69. 65 ounces sell for 78 cents.

70. For 7 dozen, you will pay $10.15.

71. $3.50 for 50 feet

72. $4 billion over a 5-month span

APPLICATIONS

73. GEAR RATIOS Refer to the illustration below.

  a. Write the ratio of the number of teeth of the smaller gear to the number of teeth of the larger gear in simplest form.

  b. Write the ratio of the number of teeth of the larger gear to the number of teeth of the smaller gear in simplest form.

74. CARDS The suit of hearts from a deck of playing cards is shown below. What is the ratio of the number of face cards to the total number of cards in the suit? (*Hint:* A face card is a Jack, Queen, or King.)

75. SKIN Refer to the cross-section of human skin shown below. Write the ratio of the thickness of the stratum corneum to the thickness of the dermis in simplest form. (*Source:* Philips Research Laboratories)

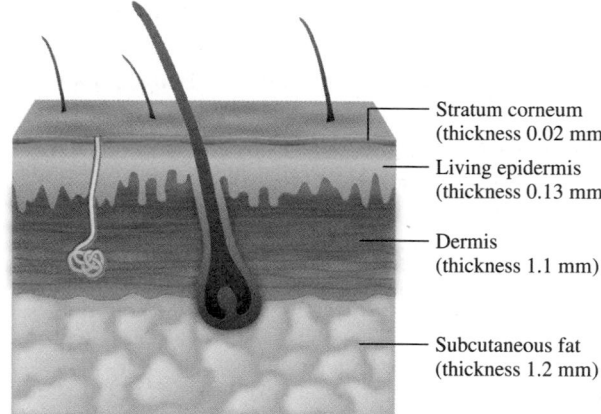

Stratum corneum (thickness 0.02 mm)

Living epidermis (thickness 0.13 mm)

Dermis (thickness 1.1 mm)

Subcutaneous fat (thickness 1.2 mm)

76. PAINTING A 9.5-mil thick coat of fireproof paint is applied with a roller to a wall. (A *mil* is a unit of measure equal to 1/1,000 of an inch.) The coating dries to a thickness of 5.7 mils. Write the ratio of the thickness of the coating when wet to the thickness when dry in simplest form.

77. BAKING A recipe for sourdough bread calls for $5\frac{1}{4}$ cups of all-purpose flour and $1\frac{3}{4}$ cups of water. Write the ratio of flour to water in simplest form.

78. DESSERTS Refer to the recipe card shown below. Write the ratio of milk to sugar in simplest form.

Frozen Chocolate Slush
(Serves 8)

Once frozen, this chocolate can be cut into cubes and stored in sealed plastic bags for a spur-of-the-moment dessert.

$\frac{1}{2}$ cup Dutch cocoa powder, sifted

$1\frac{1}{2}$ cups sugar

$3\frac{1}{2}$ cups skim milk

79. BUDGETS Refer to the circle graph below that shows a monthly budget for a family. Write each ratio in simplest form.

   a. Find the total amount for the monthly budget.

   b. Write the ratio of the amount budgeted for rent to the total budget.

   c. Write the ratio of the amount budgeted for food to the total budget.

   d. Write the ratio of the amount budgeted for the phone to the total budget.

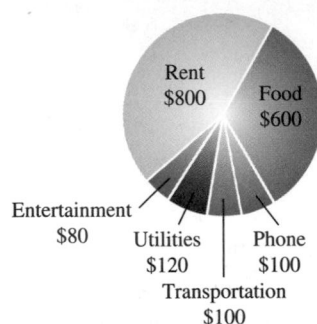

80. TAXES Refer to the list of tax deductions shown below. Write each ratio in simplest form.

   a. Write the ratio of the real estate tax deduction to the total deductions.

   b. Write the ratio of the charitable contributions to the total deductions.

   c. Write the ratio of the mortgage interest deduction to the union dues deduction.

Item	Amount
Medical expenses	$875
Real estate taxes	$1,250
Charitable contributions	$1,750
Mortgage interest	$4,375
Union dues	$500
Total deductions	$8,750

81. ART HISTORY Leonardo da Vinci drew the human figure shown within a square. Write the ratio of the length of the man's outstretched arms to his height. (*Hint:* All four sides of a square are the same length.)

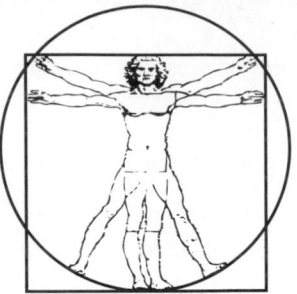

82. FLAGS The checkered flag is composed of 24 equal-sized squares. What is the ratio of the width of the flag to its length? (*Hint:* All four sides of a square are the same length.)

83. BANKRUPTCY After declaring bankruptcy, a company could pay its creditors only 5¢ on the dollar. Write this as a ratio in simplest form.

84. EGGS An average-sized ostrich egg weighs 3 pounds and an average-sized chicken egg weighs 2 ounces. Write the ratio of the weight of an ostrich egg to the weight of a chicken egg in simplest form.

85. CPR A paramedic performed 125 compressions to 50 breaths on an adult with no pulse. What compressions-to-breaths rate did the paramedic use?

86. FACULTY–STUDENT RATIOS At a college, there are 125 faculty members and 2,000 students. Find the rate of faculty to students. (This is often referred to as the faculty–student *ratio*, even though the units are different.)

87. AIRLINE COMPLAINTS An airline had 3.29 complaints for every 1,000 passengers. Write this as a rate of whole numbers.

88. FINGERNAILS On average, fingernails grow 0.02 inch per week. Write this rate using whole numbers.

89. INTERNET SALES A website determined that it had 112,500 hits in one month. Of those visiting the site, 4,500 made purchases.

   a. Those that visited the site, but did not make a purchase, are called *browsers*. How many browsers visited the website that month?

   b. What was the browsers-to-buyers unit rate for the website that month?

90. TYPING A secretary typed a document containing 330 words in 5 minutes. Write this rate as a unit rate.

91. UNIT PRICES A 12-ounce can of cola sells for 84¢. Find the unit price in cents per ounce.

92. DAYCARE A daycare center charges $32 for 8 hours of supervised care. Find the unit price in dollars per hour for the daycare.

93. PARKING A parking meter requires 25¢ for 20 minutes of parking. Find the unit price to park.

94. GASOLINE COST A driver pumped 17 gallons of gasoline into the tank of his pickup truck at a cost of $32.13. Find the unit price of the gasoline.

95. LANDSCAPING A 50-pound bag of grass seed sells for $222.50. Find the unit price of grass seed.

96. UNIT COSTS A 24-ounce package of green beans sells for $1.29. Find the unit price in cents per ounce.

97. DRAINING TANKS An 11,880-gallon tank of water can be emptied in 27 minutes. Find the unit rate of flow of water out of the tank.

98. PAY RATE Ricardo worked for 27 hours to help insulate a hockey arena. For his work, he received $337.50. Find his hourly rate of pay.

99. AUTO TRAVEL A car's odometer reads 34,746 at the beginning of a trip. Five hours later, it reads 35,071.

   a. How far did the car travel?

   b. What was its rate of speed?

100. RATES OF SPEED An airplane travels from Chicago to San Francisco, a distance of 1,883 miles, in 3.5 hours. Find the rate of speed of the plane.

101. COMPARISON SHOPPING A 6-ounce can of orange juice sells for 89¢, and an 8-ounce can sells for $1.19. Which is the better buy?

102. COMPARISON SHOPPING A 30-pound bag of planting mix costs $12.25, and an 80-pound bag costs $30.25. Which is the better buy?

103. COMPARISON SHOPPING A certain brand of cold and sinus medication is sold in 20-tablet boxes for $4.29 and in 50-tablet boxes for $9.59. Which is the better buy?

104. COMPARISON SHOPPING Which tire shown is the better buy?

ECONOMY	PREMIUM
$30.99	$37.50
35,000-mile warranty	40,000-mile warranty

105. COMPARING SPEEDS A car travels 345 miles in 6 hours, and a truck travels 376 miles in 6.2 hours. Which vehicle is going faster?

106. READING One seventh-grader read a 54-page book in 40 minutes. Another read an 80-page book in 62 minutes. If the books were equally difficult, which student read faster?

107. GAS MILEAGE One car went 1,235 miles on 51.3 gallons of gasoline, and another went 1,456 miles on 55.78 gallons. Which car got the better gas mileage?

108. ELECTRICITY RATES In one community, a bill for 575 kilowatt-hours of electricity is $38.81. In a second community, a bill for 831 kwh is $58.10. In which community is electricity cheaper?

WRITING

109. Are the ratios 3 to 1 and 1 to 3 the same? Explain why or why not.

110. Give three examples of ratios (or rates) that you have encountered in the past week.

111. How will the topics studied in this section make you a better shopper?

112. What is a unit rate? Give some examples.

REVIEW

Use front-end rounding to estimate each result.

113. $12,897 + 29,431 + 2,595$

114. $6,302 - 788$

115. $410 \cdot 21$

116. $63,467 \div 3,103$

Objectives

1 Write proportions.

2 Determine whether proportions are true or false.

3 Solve a proportion to find an unknown term.

4 Write proportions to solve application problems.

Proportions

One of the most useful concepts in mathematics is the *equation*. Recall that an **equation** is a statement indicating that two expressions are equal. All equations contain an = symbol. Some examples of equations are:

$$4 + 4 = 8, \qquad 15.6 - 4.3 = 11.3, \qquad \frac{1}{2} \cdot 10 = 5, \qquad \text{and} \qquad -16 \div 8 = -2$$

Each of the equations shown above is true. Equations can also be false. For example,

$$3 + 2 = 6 \quad \text{and} \quad -40 \div (-5) = -8$$

are false equations.

In this section, we will work with equations that state that two ratios (or rates) are equal.

1 Write proportions.

Like any tool, a ladder can be dangerous if used improperly. When setting up an extension ladder, users should follow the *4-to-1 rule:* For every 4 feet of ladder height, position the legs of the ladder 1 foot away from the base of the wall. The 4-to-1 rule for ladders can be expressed using a ratio.

$$\frac{4 \text{ feet}}{1 \text{ foot}} = \frac{4 \text{ \cancel{feet}}}{1 \text{ \cancel{foot}}} = \frac{4}{1} \qquad \textit{Remove the common units of feet.}$$

The figure on the right shows how the 4-to-1 rule was used to properly position the legs of a ladder 3 feet from the base of a 12-foot-high wall. We can write a ratio comparing the ladder's height to its distance from the wall.

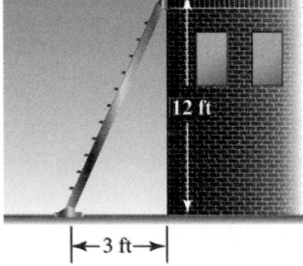

$$\frac{12 \text{ feet}}{3 \text{ feet}} = \frac{12 \text{ \cancel{feet}}}{3 \text{ \cancel{feet}}} = \frac{12}{3} \qquad \textit{Remove the common units of feet.}$$

Since this ratio satisfies the 4-to-1 rule, the two ratios $\frac{4}{1}$ and $\frac{12}{3}$ must be equal. Therefore, we have

$$\frac{4}{1} = \frac{12}{3}$$

Equations like this, which show that two ratios are equal, are called *proportions*.

> **Proportion**
>
> A **proportion** is a statement that two ratios (or rates) are equal.

Some examples of proportions are

- $\dfrac{1}{2} = \dfrac{3}{6}$ *Read as "1 is to 2 as 3 is to 6."*

- $\dfrac{3 \text{ waiters}}{7 \text{ tables}} = \dfrac{9 \text{ waiters}}{21 \text{ tables}}$ *Read as "3 waiters are to 7 tables as 9 waiters are to 21 tables."*

EXAMPLE 1   Write each statement as a proportion.

a. 22 is to 6 as 11 is to 3.

b. 1,000 administrators is to 8,000 teachers as 1 administrator is to 8 teachers.

Strategy We will locate the word *as* in each statement and identify the ratios (or rates) before and after it.

WHY The word *as* translates to the = symbol that is needed to write the statement as a proportion (equation).

Solution

a. This proportion states that two ratios are equal.

$$\underbrace{22 \text{ is } \textbf{to } 6}_{\frac{22}{6}} \quad \overset{\textbf{as}}{=} \quad \underbrace{11 \text{ is } \textbf{to } 3}_{\frac{11}{3}}.$$ Recall that the word *"to"* is used to separate the numbers being compared.

b. This proportion states that two rates are equal.

$$\underbrace{1{,}000 \text{ administrators is } \textbf{to } 8{,}000 \text{ teachers}}_{\frac{1{,}000 \text{ administrators}}{8{,}000 \text{ teachers}}} \quad \overset{\textbf{as}}{=} \quad \underbrace{1 \text{ administrator is } \textbf{to } 8 \text{ teachers}}_{\frac{1 \text{ administrator}}{8 \text{ teachers}}}$$

When proportions involve rates, the units are often written outside of the proportion, as shown below:

$$\text{Administrators} \longrightarrow \frac{1{,}000}{8{,}000} = \frac{1}{8} \longleftarrow \text{Administrators}$$
$$\text{Teachers} \longrightarrow \qquad\qquad\quad \longleftarrow \text{Teachers}$$

Self Check 1

Write each statement as a proportion.

a. 16 is to 28 as 4 is to 7.

b. 300 children is to 500 adults as 3 children is to 5 adults.

Now Try Problems 17 and 19

2 Determine whether proportions are true or false.

Since a proportion is an equation, a proportion can be true or false. A proportion is true if its ratios (or rates) are equal and false if it its ratios (or rates) are not equal. One way to determine whether a proportion is true is to use the fraction simplifying skills of Chapter 3.

EXAMPLE 2   Determine whether each proportion is true or false by simplifying.

a. $\dfrac{3}{8} = \dfrac{21}{56}$ **b.** $\dfrac{30}{4} = \dfrac{45}{12}$

Strategy We will simplify any ratios in the proportion that are not in simplest form. Then we will compare them to determine whether they are equal.

WHY If the ratios are equal, the proportion is true. If they are not equal, the proportion is false.

Solution

a. On the left side of the proportion $\frac{3}{8} = \frac{21}{56}$, the ratio $\frac{3}{8}$ is in simplest form. On the right side, the ratio $\frac{21}{56}$ can be simplified.

$$\frac{21}{56} = \frac{3 \cdot \overset{1}{\cancel{7}}}{\underset{1}{\cancel{7}} \cdot 8} = \frac{3}{8}$$ Factor 21 and 56 and then remove the common factor of 7 in the numerator and denominator.

Since the ratios on the left and right sides of the proportion are equal, the proportion is true.

Self Check 2

Determine whether each proportion is true or false by simplifying.

a. $\dfrac{4}{5} = \dfrac{16}{20}$ **b.** $\dfrac{30}{24} = \dfrac{28}{16}$

Now Try Problem 23

b. Neither ratio in the proportion $\frac{30}{4} = \frac{45}{12}$ is in simplest form. To simplify each ratio, we proceed as follows:

$$\frac{30}{4} = \frac{\overset{1}{2} \cdot 15}{\underset{1}{2} \cdot 2} = \frac{15}{2} \qquad \frac{45}{12} = \frac{\overset{1}{3} \cdot 15}{\underset{1}{3} \cdot 4} = \frac{15}{4}$$

Since the ratios on the left and right sides of the proportion are not equal $\left(\frac{15}{2} \neq \frac{15}{4}\right)$, the proportion is false. ∎

There is another way to determine whether a proportion is true or false. Before we can discuss it, we need to introduce some more vocabulary of proportions.

Each of the four numbers in a proportion is called a **term.** The first and fourth terms are called the **extremes,** and the second and third terms are called the **means.**

First term (extreme) ⟶ $\dfrac{1}{2} = \dfrac{3}{6}$ ⟵ Third term (mean)
Second term (mean) ⟶　　　⟵ Fourth term (extreme)

In the proportion shown above, the *product of the extremes is equal to the product of the means.*

$$1 \cdot 6 = 6 \qquad \text{and} \qquad 2 \cdot 3 = 6$$

These products can be found by multiplying diagonally in the proportion in such a way that one numerator is multiplied by the other denominator. We call $1 \cdot 6$ and $2 \cdot 3$ **cross products.**

⎰— Cross products —⎱

$$1 \cdot 6 = 6 \qquad\qquad 2 \cdot 3 = 6$$

$$\frac{1}{2} \underset{\times}{=} \frac{3}{6}$$

Multiplication along one diagonal is shown in red.
Multiplication along the other diagonal is shown in blue.

Note that the cross products are equal. To see why this is true in general, consider the proportion $\frac{a}{b} = \frac{c}{d}$. If the ratios $\frac{a}{b}$ and $\frac{c}{d}$ are equal, we can show that the cross products are equal by multiplying both sides of the proportion by bd.

$$\frac{a}{b} = \frac{c}{d}$$
Assume that this proportion is true and that neither denominator is 0.

$$bd \cdot \frac{a}{b} = bd \cdot \frac{c}{d}$$
To clear the equation of fractions, multiply both sides by the LCD of $\frac{a}{b}$ and $\frac{c}{d}$, which is bd.

$$\frac{bd}{1} \cdot \frac{a}{b} = \frac{bd}{1} \cdot \frac{c}{d}$$
Write bd as a fraction: $bd = \frac{bd}{1}$.

$$\frac{abd}{b} = \frac{bcd}{d}$$
Multiply the numerators. Write the factors in alphabetical order. Multiply the denominators.

$$\frac{a\overset{1}{\cancel{b}}d}{\underset{1}{\cancel{b}}} = \frac{\overset{1}{\cancel{d}}cd}{\underset{1}{\cancel{d}}}$$
Simplify each fraction. On the left, remove the common factor of b in the numerator and denominator. On the right, remove the common factor of d.

$$ad = bc$$
The cross products are equal.

If neither b nor d is 0, then the steps shown above are reversible, and it is also true that if $ad = bc$, then $\frac{a}{b} = \frac{c}{d}$. This observation leads to the following property of proportions.

Cross-Products Property (Means-Extremes Property)

To determine whether a proportion is true or false, first multiply along one diagonal, and then multiply along the other diagonal.

- If the cross products are *equal,* the proportion is true.
- If the cross products are *not equal,* the proportion is false.

(If the product of the extremes is *equal* to the product of the means, the proportion is true. If the product of the extremes is *not equal* to the product of the means, the proportion is false.)

EXAMPLE 3   Determine whether each proportion is true or false.

a. $\dfrac{3}{7} = \dfrac{9}{21}$ **b.** $\dfrac{8}{3} = \dfrac{13}{5}$

Strategy We will check to see whether the cross products are equal (the product of the extremes is equal to the product of the means).

WHY If the cross products are equal, the proportion is true. If the cross products are not equal, the proportion is false.

Solution

a. $3 \cdot 21 = 63$ $7 \cdot 9 = 63$

$$\dfrac{3}{7} \bowtie \dfrac{9}{21}$$ Each cross product is 63.

Since the cross products are equal, the proportion is true.

b. $8 \cdot 5 = 40$ $3 \cdot 13 = 39$

$$\dfrac{8}{3} \bowtie \dfrac{13}{5}$$ One cross product is 40 and the other is 39.

Since the cross products are not equal, the proportion is false.

Caution! We cannot remove common factors "across" an = symbol. When this is done, the true proportion from Example 3 part a, $\frac{3}{7} = \frac{9}{21}$, is changed into the false proportion $\frac{1}{7} = \frac{9}{7}$.

$$\dfrac{\overset{1}{\cancel{3}}}{7} \bowtie \dfrac{9}{\underset{7}{\cancel{21}}}$$

EXAMPLE 4   Determine whether each proportion is true or false.

a. $\dfrac{0.9}{0.6} = \dfrac{2.4}{1.5}$ **b.** $\dfrac{2\frac{1}{3}}{3\frac{1}{2}} = \dfrac{4\frac{2}{3}}{7}$

Strategy We will check to see whether the cross products are equal (the product of the extremes is equal to the product of the means).

WHY If the cross products are equal, the proportion is true. If the cross products are not equal, the proportion is false.

Self Check 3

Determine whether the proportion

$$\dfrac{6}{13} = \dfrac{18}{39}$$

is true or false.

Now Try Problem 25

Self Check 4

Determine whether each proportion is true or false.

a. $\dfrac{9.9}{13.2} = \dfrac{1.125}{1.5}$

b. $\dfrac{3\frac{3}{16}}{2\frac{1}{2}} = \dfrac{4\frac{1}{4}}{3\frac{1}{3}}$

Now Try Problems 31 and 35

Solution

a.
$$1.5 \times 0.9 = 1.35 \qquad 2.4 \times 0.6 = 1.44$$

$$\frac{0.9}{0.6} = \frac{2.4}{1.5}$$ *One cross product is 1.35 and the other is 1.44.*

Since the cross products are not equal, the proportion is not true.

b.
$$2\frac{1}{3} \cdot 7 = \frac{7}{3} \cdot \frac{7}{1} \qquad\qquad 3\frac{1}{2} \cdot 4\frac{2}{3} = \frac{7}{2} \cdot \frac{14}{3}$$

$$= \frac{49}{3} \qquad\qquad\qquad = \frac{7 \cdot 2 \cdot \overset{1}{7}}{2 \cdot 3}$$

$$= \frac{49}{3}$$

$$\frac{2\frac{1}{3}}{3\frac{1}{2}} = \frac{4\frac{2}{3}}{7}$$ *Each cross product is $\frac{49}{3}$.*

Since the cross products are equal, the proportion is true.

When two pairs of numbers such as 2, 3 and 8, 12 form a true proportion, we say that they are **proportional.** To show that 2, 3 and 8, 12 are proportional, we check to see whether the equation

$$\frac{2}{3} = \frac{8}{12}$$

is a true proportion. To do so, we find the cross products.

$$2 \cdot 12 = 24 \qquad\qquad 3 \cdot 8 = 24$$

Since the cross products are equal, the proportion is true, and the numbers are proportional.

Self Check 5

Determine whether 6, 11 and 54, 99 are proportional.

Now Try Problem 37

EXAMPLE 5   Determine whether 3, 7 and 36, 91 are proportional.

Strategy We will use the given pairs of numbers to write two ratios and form a proportion. Then we will find the cross products.

WHY If the cross products are equal, the proportion is true, and the numbers are proportional. If the cross products are not equal, the proportion is false, and the numbers are not proportional.

Solution
The pair of numbers 3 and 7 form one ratio and the pair of numbers 36 and 91 form a second ratio. To write a proportion, we set the ratios equal. Then we find the cross products.

$$3 \cdot 91 = 273 \qquad\qquad 7 \cdot 36 = 252$$

$$\frac{3}{7} = \frac{36}{91}$$ *One cross product is 273 and the other is 252.*

Since the cross products are not equal, the numbers are not proportional.

3 Solve a proportion to find an unknown term.

Suppose that we know three of the four terms in the following proportion.

$$\frac{?}{5} = \frac{24}{20}$$

If we let the variable x represent the unknown term, we can write:

$$\frac{x}{5} = \frac{24}{20}$$

If the proportion is to be true, the cross products must be equal.

$x \cdot 20 = 5 \cdot 24$ Find the cross products for $\frac{x}{5} = \frac{24}{20}$ and set them equal.

$20x = 120$ On the left side, rewrite $x \cdot 20$ as $20x$. On the right side, do the multiplication: $5 \cdot 24 = 120$.

$$\begin{array}{r} \overset{2}{24} \\ \times\ 5 \\ \hline 120 \end{array}$$

On the left side of the equation, the unknown number x is multiplied by 20. To undo the multiplication by 20 and isolate x, we divide both sides of the equation by 20.

$\dfrac{20x}{20} = \dfrac{120}{20}$ Use the division property of equality.

$x = 6$ Do the division: $120 \div 20 = 6$.

$$\begin{array}{r} 6 \\ 20\overline{)120} \\ -\ 120 \\ \hline 0 \end{array}$$

Thus, x is 6. We have found that the unknown term in the proportion is 6 and we can write:

$$\frac{6}{5} = \frac{24}{20}$$

To check this result, we find the cross products.

Check:

$$\frac{6}{5} \overset{?}{=} \frac{24}{20} \qquad \begin{array}{l} 6 \cdot 20 = \mathbf{120} \\ 5 \cdot 24 = \mathbf{120} \end{array}$$

Since the cross products are equal, the result, 6, checks.

In the previous example, when we find the value of the variable x that makes the given proportion true, we say that we have *solved the proportion* to find the unknown term.

> ***The Language of Algebra*** We solve proportions by writing a series of steps that result in an equation of the form $x = \textbf{a number}$ or $\textbf{a number} = x$. We say that the variable x is *isolated* on one side of the equation. *Isolated* means alone or by itself.

Solving a Proportion to Find an Unknown Term

1. Set the cross products equal to each other to form an equation.
2. Isolate the variable on one side of the equation by dividing both sides by the number that is multiplied by that variable.
3. Check by substituting the result into the original proportion and finding the cross products.

EXAMPLE 6

Solve the proportion: $\dfrac{12}{20} = \dfrac{3}{x}$

Strategy We will set the cross products equal to each other to form an equation.

WHY Then we can isolate the variable x on one side of the equation to find the unknown term in the proportion that it represents.

Self Check 6

Solve the proportion: $\dfrac{15}{x} = \dfrac{20}{32}$

Now Try Problem 41

Solution

$$\frac{12}{20} = \frac{3}{x}$$ This is the proportion to solve.

$$12 \cdot x = 20 \cdot 3$$ Set the cross products equal to each other to form an equation.

$$12x = 60$$ On the right side, do the multiplication: $20 \cdot 3 = 60$.

$$\frac{12x}{12} = \frac{60}{12}$$ To isolate x, undo the multiplication by 12 by dividing both sides by 12.

$$x = 5$$ Do the division: $60 \div 12 = 5$.

$$\begin{array}{r} 5 \\ 12\overline{)60} \\ -60 \\ \hline 0 \end{array}$$

Thus, x is 5. To check this result, we substitute 5 for x in the original proportion.

Check:

$$\frac{12}{20} \overset{?}{=} \frac{3}{5} \qquad 12 \cdot 5 = 60$$
$$\qquad\qquad 20 \cdot 3 = 60$$

Since the cross products are equal, the result, 5, checks.

Self Check 7

Solve the proportion:
$$\frac{6.7}{x} = \frac{33.5}{38}$$

Now Try Problem 45

EXAMPLE 7

Solve the proportion: $\dfrac{3.5}{7.2} = \dfrac{x}{15.84}$

Strategy We will set the cross products equal to each other to form an equation.

WHY Then we can isolate the variable x on one side of the equation to find the unknown term in the proportion that it represents.

Solution

$$\frac{3.5}{7.2} = \frac{x}{15.84}$$ This is the proportion to solve.

$$3.5 \cdot 15.84 = 7.2 \cdot x$$ Set the cross products equal to each other to form an equation.

$$55.44 = 7.2x$$ On the left side, do the multiplication: $3.5 \cdot 15.84 = 55.44$.

$$\frac{55.44}{7.2} = \frac{7.2x}{7.2}$$ To isolate x, undo the multiplication by 7.2 by dividing both sides by 7.2

$$7.7 = x$$ Do the division: $55.44 \div 7.2 = 7.7$.

$$\begin{array}{r} 15.84 \\ \times\,3.5 \\ \hline 7920 \\ 47520 \\ \hline 55.440 \end{array}$$

$$\begin{array}{r} 7.7 \\ 7.2\overline{)\,55.44} \\ -50\,4 \\ \hline 5\,04 \\ -5\,04 \\ \hline 0 \end{array}$$

Thus, x is 7.7. Check the result in the original proportion.

Self Check 8

Solve the proportion:

$$\frac{x}{2\frac{1}{3}} = \frac{2\frac{1}{4}}{1\frac{1}{2}}$$

Write the result as a mixed number.

Now Try Problem 49

EXAMPLE 8

Solve the proportion $\dfrac{x}{4\frac{1}{5}} = \dfrac{5\frac{1}{2}}{16\frac{1}{2}}$. Write the result as a mixed number.

Strategy We will set the cross products equal to each other to form an equation.

WHY Then we can isolate the variable x on one side of the equation to find the unknown term in the proportion that it represents.

Solution

$$\frac{x}{4\frac{1}{5}} = \frac{5\frac{1}{2}}{16\frac{1}{2}}$$ This is the proportion to solve.

$$x \cdot 16\frac{1}{2} = 4\frac{1}{5} \cdot 5\frac{1}{2}$$ Set the cross products equal to each other to form an equation.

$$x \cdot \frac{33}{2} = \frac{21}{5} \cdot \frac{11}{2}$$ Write each mixed number as an improper fraction.

$$\frac{33}{2}x = \frac{21}{5} \cdot \frac{11}{2}$$ On the left side, write $x \cdot \frac{33}{2}$ as $\frac{33}{2}x$.

$$\frac{2}{33} \cdot \frac{33}{2}x = \frac{2}{33} \cdot \frac{21}{5} \cdot \frac{11}{2}$$ To isolate x, multiply both sides by the reciprocal of the coefficient of the variable term, $\frac{33}{2}x$. The reciprocal of $\frac{33}{2}$ is $\frac{2}{33}$.

$$1x = \frac{2 \cdot 21 \cdot 11}{33 \cdot 5 \cdot 2}$$ On the left side, the product of a number and its reciprocal is 1: $\frac{2}{33} \cdot \frac{33}{2} = 1$. On the right side, multiply the numerators and multiply the denominators.

$$x = \frac{\overset{1}{2} \cdot \overset{1}{3} \cdot 7 \cdot \overset{1}{11}}{\underset{1}{3} \cdot \underset{1}{11} \cdot 5 \cdot \underset{1}{2}}$$ On the left side, the coefficient of 1 need not be written since 1x = x. To simplify the right side, factor 21 and 33. Then remove the common factors 2, 3, and 11.

$$x = \frac{7}{5}$$ Multiply the remaining factors in the numerator: $1 \cdot 1 \cdot 7 \cdot 1 = 7$.
Multiply the remaining factors in the denominator: $1 \cdot 1 \cdot 5 \cdot 1 = 5$.

$$\begin{array}{r} 1 \\ 5\overline{)7} \\ -5 \\ \hline 2 \end{array}$$

$$x = 1\frac{2}{5}$$ Write the improper fraction as a mixed number.

Thus, x is $1\frac{2}{5}$. Check this result in the original proportion.   ■

Using Your CALCULATOR   Solving Proportions with a Calculator

To solve the proportion in Example 7, we set the cross products equal and divided both sides by 7.2 to isolate the variable x.

$$\frac{3.5 \cdot 15.84}{7.2} = x$$

We can find x by entering these numbers and pressing these keys on a calculator.

3.5 $\boxed{\times}$ 15.84 $\boxed{\div}$ 7.2 $\boxed{=}$ $\boxed{7.7}$

Thus, x is 7.7.

---

4 **Write proportions to solve application problems.**

Proportions can be used to solve application problems from a wide variety of fields such as medicine, accounting, construction, and business. It is easy to spot problems that can be solved using a proportion. You will be given a ratio (or rate) and asked to find the missing part of another ratio (or rate). It is helpful to follow the five-step problem-solving strategy seen earlier in the text to solve proportion problems.

EXAMPLE 9   *Shopping*   If 5 apples cost $1.15, find the cost of 16 apples.

Analyze

- We can express the fact that 5 apples cost $1.15 using the rate: $\dfrac{5 \text{ apples}}{\$1.15}$.
- What is the cost of 16 apples?

Form We will let the variable c represent the unknown cost of 16 apples. If we compare the number of apples to their cost, we know that the two rates must be equal and we can write a proportion.

5 apples is **to** $1.15   16 apples is **to** $c.

$$\underset{\text{Cost of 5 apples} \longrightarrow}{5 \text{ apples} \longrightarrow} \frac{5}{1.15} = \frac{16}{c} \underset{\longleftarrow \text{ Cost of 16 apples}}{\longleftarrow 16 \text{ apples}}$$

The units can be written outside of the proportion.

Self Check 9

CONCERT TICKETS If 9 tickets to a concert cost $112.50, find the cost of 15 tickets.

Now Try Problem 73

Solve To find the cost of 16 apples, we solve the proportion for c.

$5 \cdot c = 1.15 \cdot 16$ *Set the cross products equal to each other to form an equation.*

$5c = 18.4$ *On the right side, do the multiplication: 1.15(16) = 18.4.*

$\dfrac{5c}{5} = \dfrac{18.4}{5}$ *To isolate c, undo the multiplication by 5 by dividing both sides by 5.*

$c = 3.68$ *Do the division: 18.4 ÷ 5 = 3.68.*

$$
\begin{array}{r}
3.68 \\
5\overline{)18.40} \\
-15 \\
\hline
3\,4 \\
-3\,0 \\
\hline
40 \\
-40 \\
\hline
0
\end{array}
$$

State Sixteen apples will cost $3.68.

Check If 5 apples cost $1.15, then 15 apples would cost 3 times as much: $3 \cdot \$1.15 = \$3.45$. It seems reasonable that 16 apples would cost $3.68.

We could have compared the cost of the apples to the number of apples as shown below. If we solve that proportion for c, we obtain the same result: 3.68.

Cost of 5 apples ⟶ $\dfrac{1.15}{5}$ = $\dfrac{c}{16}$ ⟵ Cost of 16 apples
5 apples ⟶   16 ⟵ 16 apples

Caution! When solving problems using proportions, make sure that the units of the numerators are the same and the units of the denominators are the same. For Example 9, it would be incorrect to write

Cost of 5 apples ⟶ $\dfrac{1.15}{5}$ = $\dfrac{16}{c}$ ⟵ 16 apples
5 apples ⟶   ⟵ Cost of 16 apples

Self Check 10

SCALE MODELS In a scale model of a city, a 300-foot-tall building is 4 inches high. An observation tower in the model is 9 inches high. How tall is the actual tower?

Now Try Problem 83

EXAMPLE 10 *Scale Drawings* A **scale** is a ratio (or rate) that compares the size of a model, drawing, or map to the size of an actual object. The airplane shown below is drawn using a scale of 1 inch: 6 feet. This means that 1 inch on the drawing is actually 6 feet on the plane. The distance from wing tip to wing tip (the wingspan) on the drawing is 4.5 inches. What is the actual wingspan of the plane?

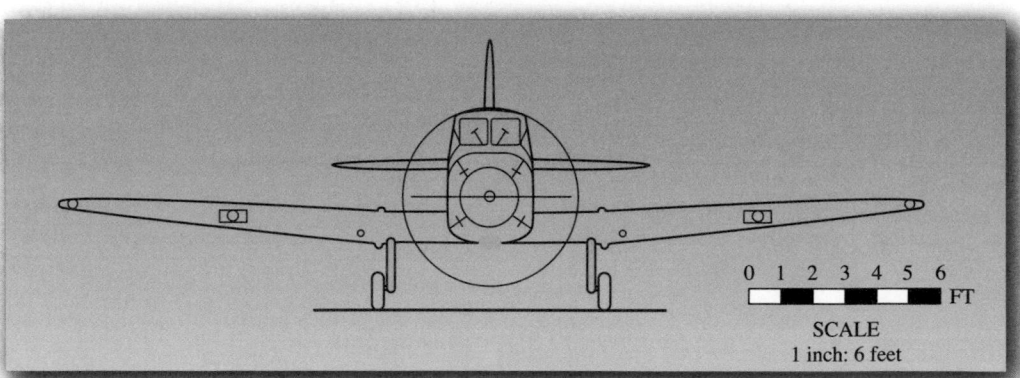

0 1 2 3 4 5 6 FT
SCALE
1 inch: 6 feet

Analyze

- The airplane is drawn using a scale of 1 inch: 6 feet, which can be written as a rate in fraction form as: $\dfrac{1 \text{ inch}}{6 \text{ feet}}$.
- The wingspan of the airplane on the drawing is 4.5 inches.
- What is the actual wingspan of the plane?

Form We will let w represent the unknown actual wingspan of the plane. If we compare the measurements on the drawing to their actual measurement of the plane, we know that those two rates must be equal and we can write a proportion.

1 inch corresponds **to** 6 feet **as** 4.5 inches corresponds **to** w feet.

Measure on the drawing → $\dfrac{1}{6}$ = $\dfrac{4.5}{w}$ ← Measure on the drawing
Measure on the plane →   ← Measure on the plane

Solve To find the actual wingspan of the airplane, we solve the proportion for w.

$1 \cdot w = 6 \cdot 4.5$ *Set the cross products equal to form an equation.*

$w = 27$ *Do the multiplication: $6 \cdot 4.5 = 27$.*

$$\begin{array}{r} \overset{3}{4.5} \\ \times\ 6 \\ \hline 27.0 \end{array}$$

State The actual wingspan of the plane is 27 feet.

Check Every 1 inch on the scale drawing corresponds to an actual length of 6 feet on the plane. Therefore, a 5-inch measurement corresponds to an actual wingspan of $5 \cdot 6$ feet, or 30 feet. It seems reasonable that a 4.5-inch measurement corresponds to an actual wingspan of 27 feet.

EXAMPLE 11   *Baking*   A recipe for chocolate cake calls for $1\frac{1}{2}$ cups of sugar for every $2\frac{1}{4}$ cups of flour. If a baker has only $\frac{1}{2}$ cup of sugar on hand, how much flour should he add to it to make chocolate cake batter?

Analyze

- The sugar-to-flour rate can be expressed as: $\dfrac{1\frac{1}{2} \text{ cups sugar}}{2\frac{1}{4} \text{ cups flour}}$

- How much flour should be added to $\frac{3}{4}$ cups of sugar?

Form We will let the variable f represent the unknown cups of flour. If we compare the cups of sugar to the cups of flour, we know that the two rates must be equal and we can write a proportion.

$1\frac{1}{2}$ cups of sugar is **to** $2\frac{1}{4}$ cups of flour **as** $\frac{1}{2}$ cup of sugar is **to** f cups of flour

Cups of sugar → $\dfrac{1\frac{1}{2}}{2\frac{1}{4}}$ = $\dfrac{\frac{1}{2}}{f}$ ← Cup of sugar
Cups of flour →   ← Cups of flour

Solve To find the amount of flour that is needed, we solve the proportion for f.

$\dfrac{1\frac{1}{2}}{2\frac{1}{4}} = \dfrac{\frac{1}{2}}{f}$ *This is the proportion to solve.*

$1\frac{1}{2} \cdot f = 2\frac{1}{4} \cdot \frac{1}{2}$ *Set the cross products equal to each other to form an equation.*

$\dfrac{3}{2} \cdot f = \dfrac{9}{4} \cdot \dfrac{1}{2}$ *Write each mixed number as an improper fraction.*

$\dfrac{2}{3} \cdot \dfrac{3}{2}f = \dfrac{2}{3} \cdot \dfrac{9}{4} \cdot \dfrac{1}{2}$ *To isolate f, multiply both sides by the reciprocal of the coefficient of the variable term $\frac{3}{2}f$. The reciprocal of $\frac{3}{2}$ is $\frac{2}{3}$.*

Self Check 11

BAKING See Example 11. How many cups of flour will be needed to make several chocolate cakes that will require a total of $12\frac{1}{2}$ cups of sugar?

Now Try Problem 89

$$1f = \frac{2 \cdot 9 \cdot 1}{3 \cdot 4 \cdot 2}$$

On the left side, the product of a number and its reciprocal is 1: $\frac{2}{3} \cdot \frac{3}{2} = 1$. On the right side, multiply the numerators and multiply the denominators.

$$f = \frac{\overset{1}{2} \cdot \overset{1}{\cancel{3}} \cdot 3 \cdot 1}{\cancel{3} \cdot 4 \cdot \cancel{2}}$$

On the left side, the coefficient of 1 need not be written since $1f = f$. To simplify the right side, factor 9. Then remove the common factors 2 and 3.

$$f = \frac{3}{4}$$

Multiply the remaining factors in the numerator: $1 \cdot 1 \cdot 3 \cdot 1 = 3$. Multiply the remaining factors in the denominator: $1 \cdot 4 \cdot 1 = 4$.

State   The baker should use $\frac{3}{4}$ cups of flour.

Check   The rate of $1\frac{1}{2}$ cups of sugar for every $2\frac{1}{4}$ cups of flour is about 1 to 2. The rate of $\frac{1}{2}$ cup of sugar to $\frac{3}{4}$ cup flour is also about 1 to 2. The result, $\frac{3}{4}$, seems reasonable.

Success Tip   In Example 11, an alternate approach would be to write each term of the proportion in its equivalent decimal form and then solve for f.

Fractions and mixed numbers   **Decimals**

$$\frac{1\frac{1}{2}}{2\frac{1}{4}} = \frac{\frac{1}{2}}{f} \longrightarrow \frac{1.5}{2.25} = \frac{0.5}{f}$$

ANSWERS TO SELF CHECKS

1. a. $\frac{16}{28} = \frac{4}{7}$ **b.** $\frac{300 \text{ children}}{500 \text{ adults}} = \frac{3 \text{ children}}{5 \text{ adults}}$ **2. a.** true **b.** false   **3.** true   **4. a.** true **b.** true
5. yes   **6.** 24   **7.** 7.6   **8.** $3\frac{1}{2}$ **9.** \$187.50   **10.** 675 ft   **11.** $18\frac{3}{4}$ cups

---

SECTION 6.2   STUDY SET

VOCABULARY

Fill in the blanks.

1. A _____ is a statement that two ratios (or rates) are equal.

2. In $\frac{1}{2} = \frac{5}{10}$, the terms 1 and 10 are called the _____ of the proportion and the terms 2 and 5 are called the _____ of the proportion.

3. The _____ products for the proportion $\frac{4}{7} = \frac{36}{x}$ are $4 \cdot x$ and $7 \cdot 36$.

4. When two pairs of numbers form a proportion, we say that the numbers are _____.

5. A letter that is used to represent an unknown number is called a _____.

6. When we find the value of x that makes the proportion $\frac{3}{8} = \frac{x}{16}$ true, we say that we have _____ the proportion.

7. We solve proportions by writing a series of steps that result in an equation of the form $x =$ a number or a number $= x$. We say that the variable x is _____ on one side of the equation.

8. A _____ is a ratio (or rate) that compares the size of a model, drawing, or map to the size of an actual object.

CONCEPTS

Fill in the blanks.

9. If the cross products of a proportion are equal, the proportion is _____. If the cross products are *not equal,* the proportion is _____.

10. The proportion $\frac{2}{5} = \frac{4}{10}$ will be true if the product $\blacksquare \cdot 10$ is equal to the product $\blacksquare \cdot 4$.

11. Complete the cross products.

$$\blacksquare \cdot 10 = \blacksquare \qquad 2 \cdot \blacksquare = \blacksquare$$
$$\frac{9}{2} \times \frac{45}{10}$$

12. In the equation $6 \cdot x = 2 \cdot 12$, to undo the multiplication by 6 and isolate x, _____ both sides of the equation by 6.

13. Label the missing units in the proportion.

$$\text{Teacher's aides} \longrightarrow \frac{12}{100} = \frac{3}{25} \longleftarrow \text{Children}$$

14. Consider the following problem: *For every 15 feet of chain link fencing, 4 support posts are used. How many support posts will be needed for 300 feet of chain link fencing?* Which of the proportions below could be used to solve this problem?

i. $\frac{15}{4} = \frac{300}{x}$ **ii.** $\frac{15}{4} = \frac{x}{300}$

iii. $\frac{4}{15} = \frac{300}{x}$ **iv.** $\frac{4}{15} = \frac{x}{300}$

NOTATION

Complete each solution.

15. Solve the proportion: $\frac{2}{3} = \frac{x}{9}$

$$2 \cdot 9 = \blacksquare$$
$$\blacksquare = 3x$$
$$\frac{18}{\blacksquare} = \frac{3x}{\blacksquare}$$
$$\blacksquare = x$$

The solution is $\blacksquare$.

16. Solve the proportion: $\frac{14}{x} = \frac{49}{17.5}$

$$14 \cdot \blacksquare = x \cdot 49$$
$$\blacksquare = x \cdot 49$$
$$\frac{245}{\blacksquare} = \frac{x \cdot 49}{\blacksquare}$$
$$\blacksquare = x$$

The solution is $\blacksquare$.

GUIDED PRACTICE

Write each statement as a proportion. **See Example 1.**

17. 20 is to 30 as 2 is to 3.

18. 9 is to 36 as 1 is to 4.

19. 400 sheets is to 100 beds as 4 sheets is to 1 bed.

20. 50 shovels is to 125 laborers as 2 shovels is to 5 laborers.

Determine whether each proportion is true or false by simplifying. **See Example 2.**

21. $\frac{7}{9} = \frac{70}{81}$ **22.** $\frac{2}{5} = \frac{8}{20}$

23. $\frac{21}{14} = \frac{18}{12}$ **24.** $\frac{42}{38} = \frac{95}{60}$

Determine whether each proportion is true or false by finding cross products. **See Example 3.**

25. $\frac{4}{32} = \frac{2}{16}$ **26.** $\frac{6}{27} = \frac{4}{18}$

27. $\frac{9}{19} = \frac{38}{80}$ **28.** $\frac{40}{29} = \frac{29}{22}$

Determine whether each proportion is true or false by finding cross products. **See Example 4.**

29. $\frac{0.5}{0.8} = \frac{1.1}{1.3}$ **30.** $\frac{0.6}{1.4} = \frac{0.9}{2.1}$

31. $\frac{1.2}{3.6} = \frac{1.8}{5.4}$ **32.** $\frac{3.2}{4.5} = \frac{1.6}{2.7}$

33. $\frac{1\frac{4}{5}}{3\frac{3}{7}} = \frac{2\frac{3}{16}}{4\frac{1}{6}}$ **34.** $\frac{2\frac{1}{2}}{1\frac{1}{5}} = \frac{3\frac{3}{4}}{2\frac{9}{10}}$

35. $\frac{\frac{1}{5}}{1\frac{1}{6}} = \frac{1\frac{1}{7}}{11\frac{2}{3}}$ **36.** $\frac{11\frac{1}{4}}{2\frac{1}{2}} = \frac{\frac{3}{4}}{\frac{1}{6}}$

Determine whether the numbers are proportional. **See Example 5.**

37. 18, 54 and 3, 9   **38.** 4, 3 and 12, 9

39. 8, 6 and 21, 16   **40.** 15, 7 and 13, 6

Solve each proportion. Check each result. **See Example 6.**

41. $\dfrac{5}{10} = \dfrac{3}{c}$

42. $\dfrac{7}{14} = \dfrac{2}{x}$

43. $\dfrac{2}{3} = \dfrac{x}{6}$

44. $\dfrac{3}{6} = \dfrac{x}{8}$

Solve each proportion. Check each result. **See Example 7.**

45. $\dfrac{0.6}{9.6} = \dfrac{x}{4.8}$

46. $\dfrac{0.4}{3.4} = \dfrac{x}{13.6}$

47. $\dfrac{2.75}{x} = \dfrac{1.5}{1.2}$

48. $\dfrac{9.8}{x} = \dfrac{2.8}{5.4}$

Solve each proportion. Check each result. Write each result as a fraction or mixed number. **See Example 8.**

49. $\dfrac{x}{1\frac{1}{2}} = \dfrac{10\frac{1}{2}}{4\frac{1}{2}}$

50. $\dfrac{x}{3\frac{1}{3}} = \dfrac{1\frac{1}{2}}{1\frac{9}{11}}$

51. $\dfrac{x}{1\frac{1}{6}} = \dfrac{2\frac{5}{8}}{3\frac{1}{2}}$

52. $\dfrac{x}{2\frac{2}{3}} = \dfrac{1\frac{1}{20}}{3\frac{1}{2}}$

TRY IT YOURSELF

Solve each proportion.

53. $\dfrac{4{,}000}{x} = \dfrac{3.2}{2.8}$

54. $\dfrac{0.4}{1.6} = \dfrac{96.7}{x}$

55. $\dfrac{12}{6} = \dfrac{x}{\frac{1}{4}}$

56. $\dfrac{15}{10} = \dfrac{x}{\frac{1}{3}}$

57. $\dfrac{x}{800} = \dfrac{900}{200}$

58. $\dfrac{x}{200} = \dfrac{1{,}800}{600}$

59. $\dfrac{x}{2.5} = \dfrac{3.7}{9.25}$

60. $\dfrac{8.5}{x} = \dfrac{4.25}{1.7}$

61. $\dfrac{0.8}{2} = \dfrac{x}{5}$

62. $\dfrac{0.9}{0.3} = \dfrac{6}{x}$

63. $\dfrac{x}{4\frac{1}{10}} = \dfrac{3\frac{3}{4}}{1\frac{7}{8}}$

64. $\dfrac{x}{2\frac{1}{4}} = \dfrac{\frac{1}{2}}{\frac{1}{5}}$

65. $\dfrac{340}{51} = \dfrac{x}{27}$

66. $\dfrac{480}{36} = \dfrac{x}{15}$

67. $\dfrac{0.4}{1.2} = \dfrac{6}{x}$

68. $\dfrac{5}{x} = \dfrac{2}{4.4}$

69. $\dfrac{4.65}{7.8} = \dfrac{x}{5.2}$

70. $\dfrac{8.6}{2.4} = \dfrac{x}{6}$

71. $\dfrac{\frac{3}{4}}{\frac{1}{2}} = \dfrac{0.25}{x}$

72. $\dfrac{\frac{7}{8}}{\frac{1}{2}} = \dfrac{0.25}{x}$

APPLICATIONS

To solve each problem, write and then solve a proportion.

73. SCHOOL LUNCHES A manager of a school cafeteria orders 750 pudding cups. What will the order cost if she purchases them wholesale, 6 cups for $1.75?

74. CLOTHES SHOPPING As part of a spring clearance, a men's store put dress shirts on sale, 2 for $25.98. How much will a businessman pay if he buys five shirts?

75. ANNIVERSARY GIFTS A florist sells a dozen long-stemmed red roses for $57.99. In honor of their 16th wedding anniversary, a man wants to buy 16 roses for his wife. What will the roses cost? (*Hint:* How many roses are in one dozen?)

76. COOKING A recipe for spaghetti sauce requires four 16-ounce bottles of ketchup to make 2 gallons of sauce. How many bottles of ketchup are needed to make 10 gallons of sauce? (*Hint:* Read the problem very carefully.)

77. BUSINESS PERFORMANCE The following bar graph shows the yearly costs and the revenue received by a business. Are the ratios of costs to revenue for 2009 and 2010 equal?

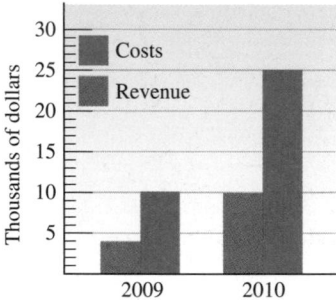

78. RAMPS Write a ratio of the rise to the run for each ramp shown. Set the ratios equal.
    a. Is the resulting proportion true?
    b. Is one ramp steeper than the other?

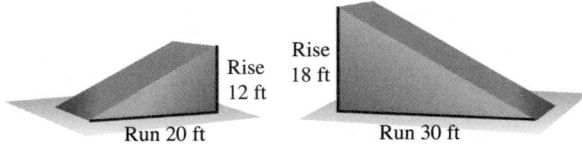

79. MIXING PERFUMES A perfume is to be mixed in the ratio of 3 drops of pure essence to 7 drops of alcohol. How many drops of pure essence should be mixed with 56 drops of alcohol?

80. MAKING COLOGNE A cologne can be made by mixing 2 drops of pure essence with 5 drops of distilled water. How much water should be used with 15 drops of pure essence?

81. LAB WORK In a red blood cell count, a drop of the patient's diluted blood is placed on a grid like that shown below. Instead of counting each and every red blood cell in the 25-square grid, a technician counts only the number of cells in the five highlighted squares. Then he or she uses a proportion to estimate the total red blood cell count. If there are 195 red blood cells in the blue squares, about how many red blood cells are in the entire grid?

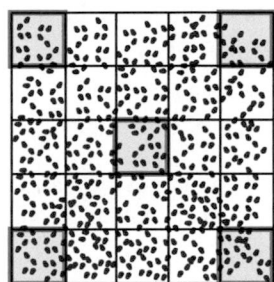

82. DOSAGES The proper dosage of a certain medication for a 30-pound child is shown. At this rate, what would be the dosage for a 45-pound child?

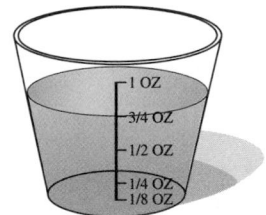

83. DRAFTING In a scale drawing, a 280-foot antenna tower is drawn 7 inches high. The building next to it is drawn 2 inches high. How tall is the actual building?

84. BLUEPRINTS The scale for the drawing in the blueprint tells the reader that a $\frac{1}{4}$-inch length $\left(\frac{1}{4}''\right)$ on the drawing corresponds to an actual size of 1 foot $(1'0'')$. Suppose the length of the kitchen is $2\frac{1}{2}$ inches on the blueprint. How long is the actual kitchen?

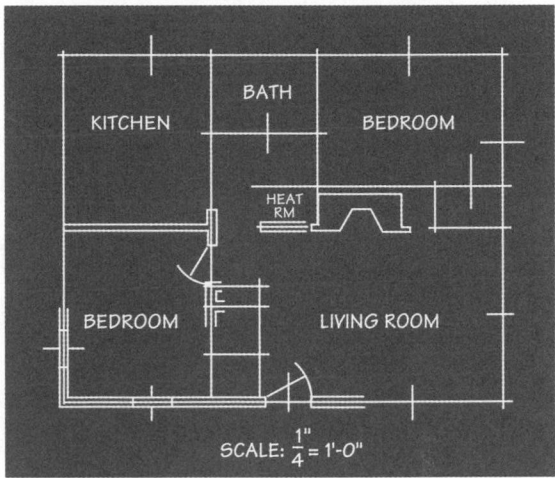

85. MODEL RAILROADS An HO-scale model railroad engine is 9 inches long. If HO scale is 87 feet to 1 foot, how long is a real engine? (*Hint:* Compare feet to inches. How many inches are in one foot?)

86. MODEL RAILROADS An N-scale model railroad caboose is 4 inches long. If N scale is 169 feet to 1 foot, how long is a real caboose? (*Hint:* Compare feet to inches. How many inches are in one foot?)

87. CAROUSELS The ratio in the illustration below indicates that 1 inch on the model carousel is equivalent to 160 inches on the actual carousel. How wide should the model be if the actual carousel is 35 feet wide? (*Hint:* Convert 35 feet to inches.)

Carousel ratio
1:160

88. MIXING FUELS The instructions on a can of oil intended to be added to lawn mower gasoline read as shown. Are these instructions correct? (*Hint*: There are 128 ounces in 1 gallon.)

Recommended	Gasoline	Oil
50 to 1	6 gal	16 oz

89. MAKING COOKIES A recipe for chocolate chip cookies calls for $1\frac{1}{4}$ cups of flour and 1 cup of sugar. The recipe will make $3\frac{1}{2}$ dozen cookies. How many cups of flour will be needed to make 12 dozen cookies?

90. MAKING BROWNIES
A recipe for brownies calls for 4 eggs and $1\frac{1}{2}$ cups of flour. If the recipe makes 15 brownies, how many cups of flour will be needed to make 130 brownies?

from Campus to Careers
Chef

Nick White/Getty Images

91. COMPUTER SPEED Using the *Mathematica 3.0* program, a Dell Dimension XPS R350 (Pentium II) computer can perform a set of 15 calculations in 2.85 seconds. How long will it take the computer to perform 100 such calculations?

92. QUALITY CONTROL Out of a sample of 500 men's shirts, 17 were rejected because of crooked collars. How many crooked collars would you expect to find in a run of 15,000 shirts?

93. DOGS Refer to the illustration below. A Saint Bernard website lists the "ideal proportions for the *height at the withers* to *body length* as 5:6." What is the ideal height at the withers for a Saint Bernard whose body length is $37\frac{1}{2}$ inches?

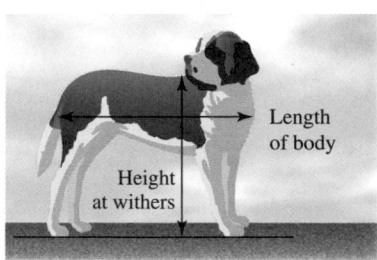

Length of body

Height at withers

94. MILEAGE Under normal conditions, a Hummer can travel 325 miles on a full tank (25 gallons) of diesel. How far can it travel on its auxiliary tank, which holds 17 gallons of diesel?

95. PAYCHECKS Billie earns $412 for a 40-hour week. If she missed 10 hours of work last week, how much did she get paid?

96. STAFFING A school board has determined that there should be 3 teachers for every 50 students. Complete the table by filling in the number of teachers needed at each school.

	Glenwood High	Goddard Junior High	Sellers Elementary
Enrollment	2,700	1,900	850
Teachers			

WRITING

97. Explain the difference between a ratio and a proportion.

98. The following paragraph is from a book about dollhouses. What concept from this section is mentioned?

Today, the internationally recognized scale for dollhouses and miniatures is 1 in. = 1 ft. This is small enough to be defined as a miniature, yet not too small for all details of decoration and furniture to be seen clearly.

99. Write a problem that could be solved using the following proportion.

Ounces of cashews $\longrightarrow$ $\dfrac{4}{639} = \dfrac{10}{x}$ $\longleftarrow$ Ounces of cashews
Calories $\longrightarrow$ $\qquad\quad$ $\longleftarrow$ Calories

100. Write a problem about a situation you encounter in your daily life that could be solved by using a proportion.

REVIEW

Perform each operation.

101. $7.4 + 6.78 + 35 + 0.008$

102. $29.5 + 34.4 + 12.8$

103. $48.8 - 17.372$

104. $78.47 - 53.3$

105. $-3.8 - (-7.9)$

106. $-17.1 + 8.4$

107. $-35.1 - 13.99$

108. $-5.55 + (-1.25)$

SECTION 6.3
American Units of Measurement

Two common systems of measurement are the **American** (or **English**) **system** and the **metric system.** We will discuss American units of measurement in this section and metric units in the next. Some common American units are *inches, feet, miles, ounces, pounds, tons, cups, pints, quarts,* and *gallons.* These units are used when measuring length, weight, and capacity.

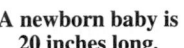
A newborn baby is 20 inches long.

First-class postage for a letter that weighs less than 1 ounce is 44¢.

Milk is sold in gallon containers.

1 Use a ruler to measure lengths in inches.

A ruler is one of the most common tools used for measuring distances or lengths. The figure below shows part of a ruler. Most rulers are 12 inches (1 foot) long. Since 12 inches = 1 foot, a ruler is divided into 12 equal lengths of 1 inch. Each inch is divided into halves of an inch, quarters of an inch, eighths of an inch, and sixteenths of an inch.

The left end of a ruler can be (but sometimes isn't) labeled with a 0. Each point on a ruler, like each point on a number line, has a number associated with it. That number is the distance between the point and 0. Several lengths on the ruler are shown below.

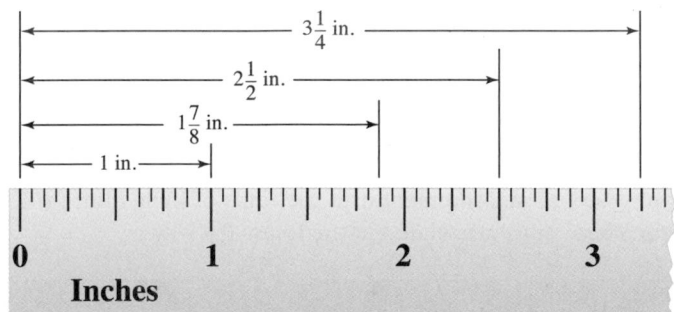

Actual size

EXAMPLE 1 Find the length of the paper clip shown here.

Strategy We will place a ruler below the paper clip, with the left end of the ruler (which could be thought of as 0) directly underneath one end of the paper clip.

WHY Then we can find the length of the paper clip by identifying where its other end lines up on the tick marks printed in black on the ruler.

Self Check 1

Self Check 1

Find the length of the jumbo paper clip.

Now Try **Problem 27**

Solution

Since the tick marks between 0 and 1 on the ruler create eight equal spaces, the ruler is scaled in eighths of an inch. The paper clip is $1\frac{3}{8}$ inches long.

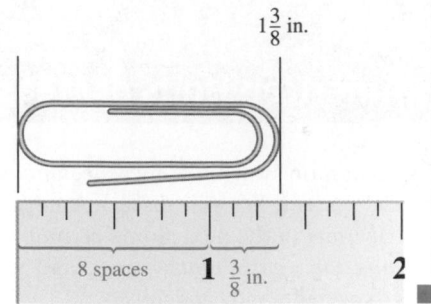

Self Check 2

Find the width of the circle.

Now Try **Problem 29**

EXAMPLE 2   Find the length of the nail shown below.

Strategy We will place a ruler below the nail, with the left end of the ruler (which could be thought of as 0) directly underneath the head of the nail.

WHY Then we can find the length of the nail by identifying where its pointed end lines up on the tick marks printed in black on the ruler.

Solution

Since the tick marks between 0 and 1 on the ruler create sixteen equal spaces, the ruler is scaled in sixteenths of an inch.

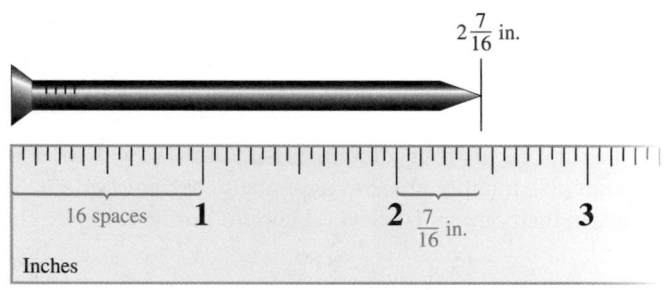

The nail is $2\frac{7}{16}$ inches long.

2 Define American units of length.

The American system of measurement uses the units of **inch, foot, yard,** and **mile** to measure length. These units are related in the following ways.

American Units of Length	
1 foot (ft) = 12 inches (in.)	1 yard (yd) = 36 inches
1 yard = 3 feet	1 mile (mi) = 5,280 feet

The abbreviation for each unit is written within parentheses.

> **The Language of Algebra** According to some sources, the inch was originally defined as the length from the tip of the thumb to the first knuckle. In some languages the word for *inch* is similar to or the same as *thumb*. For example, in Spanish, *pulgada* is inch and *pulgar* is thumb. In Swedish, *tum* is inch and *tumme* is thumb. In Italian, *pollice* is both inch and thumb.

3 Convert from one American unit of length to another.

To convert from one unit of length to another, we use *unit conversion factors*. To find the unit conversion factor between yards and feet, we begin with this fact:

$$3 \text{ ft} = 1 \text{ yd}$$

If we divide both sides of this equation by 1 yard, we get

$$\frac{3 \text{ ft}}{\textbf{1 yd}} = \frac{1 \text{ yd}}{\textbf{1 yd}}$$

$$\frac{3 \text{ ft}}{1 \text{ yd}} = 1 \qquad \text{Simplify the right side of the equation. A number divided by itself is 1: } \frac{1 \text{ yd}}{1 \text{ yd}} = 1.$$

The fraction $\frac{3 \text{ ft}}{1 \text{ yd}}$ is called a **unit conversion factor,** because its value is 1. It can be read as "3 feet per yard." Since this fraction is equal to 1, multiplying a length by this fraction does not change its measure; it changes only the *units* of measure.

To convert units of length in the American system of measurement, we use the following unit conversion factors. Each conversion factor shown below is a form of 1.

To convert from	Use the unit conversion factor	To convert from	Use the unit conversion factor
feet to inches	$\frac{12 \text{ in.}}{1 \text{ ft}}$	inches to feet	$\frac{1 \text{ ft}}{12 \text{ in.}}$
yards to feet	$\frac{3 \text{ ft}}{1 \text{ yd}}$	feet to yards	$\frac{1 \text{ yd}}{3 \text{ ft}}$
yards to inches	$\frac{36 \text{ in.}}{1 \text{ yd}}$	inches to yards	$\frac{1 \text{ yd}}{36 \text{ in.}}$
miles to feet	$\frac{5{,}280 \text{ ft}}{1 \text{ mi}}$	feet to miles	$\frac{1 \text{ mi}}{5{,}280 \text{ ft}}$

EXAMPLE 3   Convert 8 yards to feet.

Strategy We will multiply 8 yards by a carefully chosen unit conversion factor.

WHY If we multiply by the proper unit conversion factor, we can eliminate the unwanted units of yards and convert to feet.

Solution
To convert from yards to feet, we must use a unit conversion factor that relates feet to yards. Since there are 3 feet per yard, we multiply 8 yards by the unit conversion factor $\frac{3 \text{ ft}}{1 \text{ yd}}$.

$$8 \text{ yd} = \frac{8 \text{ yd}}{1} \cdot \frac{\textbf{3 ft}}{\textbf{1 yd}} \qquad \text{Write 8 yd as a fraction: } 8 \text{ yd} = \frac{8 \text{ yd}}{1}. \text{ Then multiply by a form of 1: } \frac{3 \text{ ft}}{1 \text{ yd}}.$$

$$= \frac{8 \text{ y\!\!\!/d}}{1} \cdot \frac{3 \text{ ft}}{1 \text{ y\!\!\!/d}} \qquad \text{Remove the common units of yards from the numerator and denominator. Notice that the units of feet remain.}$$

$$= 8 \cdot 3 \text{ ft} \qquad \text{Simplify.}$$

$$= 24 \text{ ft} \qquad \text{Multiply: } 8 \cdot 3 = 24.$$

8 yards is equal to 24 feet.

Self Check 3

Convert 9 yards to feet.

Now Try Problem 35

Success Tip   Notice that in Example 3, we eliminated the units of yards and introduced the units of feet by multiplying by the appropriate unit conversion factor. In general, a unit conversion factor is a fraction with the following form:

$$\frac{\text{Unit we want to introduce} \leftarrow \text{Numerator}}{\text{Unit we want to eliminate} \leftarrow \text{Denominator}}$$

Self Check 4

Convert $1\frac{1}{2}$ feet to inches.

Now Try Problem 39

EXAMPLE 4   Convert $1\frac{3}{4}$ feet to inches.

Strategy We will multiply $1\frac{3}{4}$ feet by a carefully chosen unit conversion factor.

WHY If we multiply by the proper unit conversion factor, we can eliminate the unwanted units of feet and convert to inches.

Solution

To convert from feet to inches, we must choose a unit conversion factor whose numerator contains the units we want to introduce (inches), and whose denominator contains the units we want to eliminate (feet). Since there are 12 inches per foot, we will use

$$\frac{12 \text{ in.}}{1 \text{ ft}}$$ ← This is the unit we want to introduce.
← This is the unit we want to eliminate (the original unit).

To perform the conversion, we multiply.

$$1\frac{3}{4} \text{ ft} = \frac{7}{4} \text{ ft} \cdot \frac{12 \text{ in.}}{1 \text{ ft}}$$ Write $1\frac{3}{4}$ as an improper fraction: $1\frac{3}{4} = \frac{7}{4}$. Then multiply by a form of 1: $\frac{12 \text{ in.}}{1 \text{ ft}}$.

$$= \frac{7}{4} \text{ ft} \cdot \frac{12 \text{ in.}}{1 \text{ ft}}$$ Remove the common units of feet from the numerator and denominator. Notice that the units of inches remain.

$$= \frac{7 \cdot 12}{4 \cdot 1} \text{ in.}$$ Multiply the fractions.

$$= \frac{7 \cdot 3 \cdot \overset{1}{\cancel{4}}}{\underset{1}{\cancel{4}} \cdot 1} \text{ in.}$$ To simplify the fraction, factor 12. Then remove the common factor of 4 from the numerator and denominator.

$$= 21 \text{ in.}$$ Simplify.

$1\frac{3}{4}$ feet is equal to 21 inches.

> **Caution!** When converting lengths, if no common units appear in the numerator and denominator to remove, you have chosen the wrong conversion factor.

Sometimes we must use two (or more) unit conversion factors to eliminate the given units while introducing the desired units. The following example illustrates this concept.

Self Check 5

MARATHONS The *marathon* is a long-distance race with an official distance of 26 miles 385 yards. Convert 385 yards to miles. Give the exact answer and a decimal approximation, rounded to the nearest hundredth of a mile.

Now Try Problem 43

EXAMPLE 5   *Football* A football field (including both end zones) is 120 yards long. Convert this length to miles. Give the exact answer and a decimal approximation, rounded to the nearest hundredth of a mile.

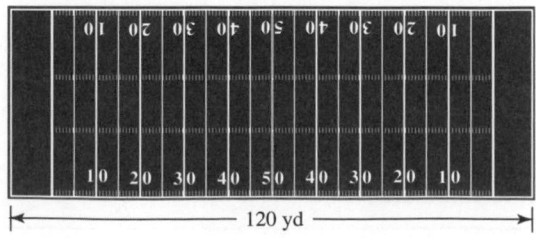

120 yd

Strategy We will use a two-part multiplication process that converts 120 yards to feet and then converts that result to miles.

WHY We must use a two-part process because the table on page 589 does not contain a single unit conversion factor that converts from yards to miles.

Solution

Since there are 3 feet per yard, we can convert 120 yards to feet by multiplying by the unit conversion factor $\frac{3ft}{1yd}$. Since there is 1 mile for every 5,280 feet, we can convert that result to miles by multiplying by the unit conversion factor $\frac{1\,mi}{5,280\,ft}$.

$$120 \text{ yd} = \frac{120 \text{ yd}}{1} \cdot \frac{3 \text{ ft}}{1 \text{ yd}} \cdot \frac{1 \text{ mi}}{5,280 \text{ ft}}$$

Write 120 yd as a fraction: 120 yd = $\frac{120 \text{ yd}}{1}$. Then multiply by two unit conversion factors: $\frac{3\,ft}{1\,yd} = 1$ and $\frac{1\,mi}{5,280\,ft} = 1$.

$$= \frac{120 \text{ y\!d}}{1} \cdot \frac{3 \text{ ft}}{1 \text{ y\!d}} \cdot \frac{1 \text{ mi}}{5,280 \text{ f\!t}}$$

Remove the common units of yards and feet in the numerator and denominator. Notice that all the units are removed except for miles.

$$= \frac{120 \cdot 3}{5,280} \text{ mi}$$

Multiply the fractions.

$$= \frac{\overset{1}{2} \cdot \overset{1}{2} \cdot \overset{1}{2} \cdot \overset{1}{3} \cdot \overset{1}{5} \cdot 3}{\underset{1}{2} \cdot \underset{1}{2} \cdot \underset{1}{2} \cdot 2 \cdot 2 \cdot \underset{1}{3} \cdot \underset{1}{5} \cdot 11} \text{ mi}$$

To simplify the fraction, prime factor 120 and 5,280, and remove the common factors 2, 3, and 5.

$$= \frac{3}{44} \text{ mi}$$

Multiply the remaining factors in the numerator. Multiply the remaining factors in the denominator.

```
         0.068
44)3.000
       − 0
       3 00
     − 2 64
         360
       − 352
           8
```

A football field (including the end zones) is *exactly* $\frac{3}{44}$ miles long.

We can also present this conversion as a decimal. If we divide 3 by 44 (as shown on the right), and round the result to the nearest hundredth, we see that a football field (including the end zones) is *approximately* 0.07 mile long.

---

4 Define American units of weight.

The American system of measurement uses the units of **ounce, pound,** and **ton** to measure weight. These units are related in the following ways.

> **American Units of Weight**
>
> 1 pound (lb) = 16 ounces (oz)   1 ton (T) = 2,000 pounds
>
> The abbreviation for each unit is written within parentheses.

---

5 Convert from one American unit of weight to another.

To convert units of weight in the American system of measurement, we use the following unit conversion factors. Each conversion factor shown below is a form of 1.

To convert from	Use the unit conversion factor	To convert from	Use the unit conversion factor
pounds to ounces	$\frac{16 \text{ oz}}{1 \text{ lb}}$	ounces to pounds	$\frac{1 \text{ lb}}{16 \text{ oz}}$
tons to pounds	$\frac{2,000 \text{ lb}}{1 \text{ ton}}$	pounds to tons	$\frac{1 \text{ ton}}{2,000 \text{ lb}}$

EXAMPLE 6   Convert 40 ounces to pounds.

Strategy We will multiply 40 ounces by a carefully chosen unit conversion factor.

WHY If we multiply by the proper unit conversion factor, we can eliminate the unwanted units of ounces and convert to pounds.

Solution

To convert from ounces to pounds, we must chose a unit conversion factor whose numerator contains the units we want to introduce (pounds), and whose denominator contains the units we want to eliminate (ounces). Since there is 1 pound for every 16 ounces, we will use

$$\frac{1\ \text{lb}}{16\ \text{oz}}$$ ← This is the unit we want to introduce.
       ← This is the unit we want to eliminate (the original unit).

To perform the conversion, we multiply.

$$40\ \text{oz} = \frac{40\ \text{oz}}{1} \cdot \frac{\mathbf{1\ lb}}{\mathbf{16\ oz}}$$ Write 40 oz as a fraction: $40\ \text{oz} = \frac{40\ \text{oz}}{1}$. Then multiply by a form of 1: $\frac{1\ \text{lb}}{16\ \text{oz}}$.

$$= \frac{40\ \cancel{\text{oz}}}{1} \cdot \frac{1\ \text{lb}}{16\ \cancel{\text{oz}}}$$ Remove the common units of ounces from the numerator and denominator. Notice that the units of pounds remain.

$$= \frac{40}{16}\ \text{lb}$$ Multiply the fractions.

There are two ways to complete the solution. First, we can remove any common factors of the numerator and denominator to simplify the fraction. Then we can write the result as a mixed number.

$$\frac{40}{16}\ \text{lb} = \frac{5 \cdot \overset{1}{\cancel{8}}}{2 \cdot \underset{1}{\cancel{8}}}\ \text{lb} = \frac{5}{2}\ \text{lb} = 2\frac{1}{2}\ \text{lb}$$

A second approach is to divide the numerator by the denominator and express the result as a decimal.

$$\frac{40}{16}\ \text{lb} = 2.5\ \text{lb}$$ Perform the division: $40 \div 16$.

40 ounces is equal to $2\frac{1}{2}$ lb (or 2.5 lb). ∎

```
       2.5
 16)40.0
    -32
      8 0
     -8 0
        0
```

EXAMPLE 7   Convert 25 pounds to ounces.

Strategy We will multiply 25 pounds by a carefully chosen unit conversion factor.

WHY If we multiply by the proper unit conversion factor, we can eliminate the unwanted units of pounds and convert to ounces.

Solution

To convert from pounds to ounces, we must chose a unit conversion factor whose numerator contains the units we want to introduce (ounces), and whose denominator contains the units we want to eliminate (pounds). Since there are 16 ounces per pound, we will use

$$\frac{16\ \text{oz}}{1\ \text{lb}}$$ ← This is the unit we want to introduce.
       ← This is the unit we want to eliminate (the original unit).

To perform the conversion, we multiply.

$$25 \text{ lb} = \frac{25 \text{ lb}}{1} \cdot \frac{\mathbf{16 \text{ oz}}}{\mathbf{1 \text{ lb}}}$$
Write 25 lb as a fraction: $25 \text{ lb} = \frac{25 \text{ lb}}{1}$. Then multiply by a form of 1: $\frac{16 \text{ oz}}{1 \text{ lb}}$.

$$= \frac{25 \text{ l\!b}}{1} \cdot \frac{16 \text{ oz}}{1 \text{ l\!b}}$$
Remove the common units of pounds from the numerator and denominator. Notice that the units of ounces remain.

$$= 25 \cdot 16 \text{ oz}$$ Simplify.

$$= 400 \text{ oz}$$ Multiply: $25 \cdot 16 = 400$.

25 pounds is equal to 400 ounces.

$$\begin{array}{r} 25 \\ \times\, 16 \\ \hline 150 \\ 250 \\ \hline 400 \end{array} \blacksquare$$

6 Define American units of capacity.

The American system of measurement uses the units of **ounce, cup, pint, quart,** and **gallon** to measure capacity. These units are related as follows.

> **The Language of Algebra** The word *capacity* means the amount that can be contained. For example, a gas tank might have a *capacity* of 12 gallons.

American Units of Capacity

1 cup (c) = 8 fluid ounces (fl oz)	1 pint (pt) = 2 cups
1 quart (qt) = 2 pints	1 gallon (gal) = 4 quarts

The abbreviation for each unit is written within parentheses.

7 Convert from one American unit of capacity to another.

To convert units of capacity in the American system of measurement, we use the following unit conversion factors. Each conversion factor shown below is a form of 1.

To convert from	Use the unit conversion factor	To convert from	Use the unit conversion factor
cups to ounces	$\frac{8 \text{ fl oz}}{1 \text{ c}}$	ounces to cups	$\frac{1 \text{ c}}{8 \text{ fl oz}}$
pints to cups	$\frac{2 \text{ c}}{1 \text{ pt}}$	cups to pints	$\frac{1 \text{ pt}}{2 \text{ c}}$
quarts to pints	$\frac{2 \text{ pt}}{1 \text{ qt}}$	pints to quarts	$\frac{1 \text{ qt}}{2 \text{ pt}}$
gallons to quarts	$\frac{4 \text{ qt}}{1 \text{ gal}}$	quarts to gallons	$\frac{1 \text{ gal}}{4 \text{ qt}}$

Self Check 8

Convert 2.5 pints to fluid ounces.

Now Try Problem 55

© Felix Wirth/Corbis

EXAMPLE 8 *Cooking* If a recipe calls for 3 pints of milk, how many fluid ounces of milk should be used?

Strategy We will use a two-part multiplication process that converts 3 pints to cups and then converts that result to fluid ounces.

WHY We must use a two-part process because the table on page 593 does not contain a single unit conversion factor that converts from pints to fluid ounces.

Solution

Since there are 2 cups per pint, we can convert 3 pints to cups by multiplying by the unit conversion factor $\frac{2\,c}{1\,pt}$. Since there are 8 fluid ounces per cup, we can convert that result to fluid ounces by multiplying by the unit conversion factor $\frac{8\,fl\,oz}{1\,c}$.

$$3\text{ pt} = \frac{3\text{ pt}}{1} \cdot \frac{2\text{ c}}{1\text{ pt}} \cdot \frac{8\text{ fl oz}}{1\text{ c}}$$

Write 3 pt as a fraction: $3\text{ pt} = \frac{3\text{ pt}}{1}$.
Multiply by two unit conversion factors: $\frac{2\,c}{1\,pt} = 1$ and $\frac{8\,fl\,oz}{1\,c} = 1$.

$$= \frac{3\cancel{\text{ pt}}}{1} \cdot \frac{2\cancel{\text{ c}}}{1\cancel{\text{ pt}}} \cdot \frac{8\text{ fl oz}}{1\cancel{\text{ c}}}$$

Remove the common units of pints and cups in the numerator and denominator. Notice that all the units are removed except for fluid ounces.

$$= 3 \cdot 2 \cdot 8 \text{ fl oz} \qquad \text{Simplify.}$$

$$= 48 \text{ fl oz} \qquad \text{Multiply.}$$

Since 3 pints is equal to 48 fluid ounces, 48 fluid ounces of milk should be used. ∎

8 Define units of time.

The American system of measurement (and the metric system) use the units of **second, minute, hour,** and **day** to measure time. These units are related as follows.

Units of Time

1 minute (min) = 60 seconds (sec)   1 hour (hr) = 60 minutes

1 day = 24 hours

The abbreviation for each unit is written within parentheses.

To convert units of time, we use the following unit conversion factors. Each conversion factor shown below is a form of 1.

To convert from	Use the unit conversion factor	To convert from	Use the unit conversion factor
minutes to seconds	$\frac{60\text{ sec}}{1\text{ min}}$	seconds to minutes	$\frac{1\text{ min}}{60\text{ sec}}$
hours to minutes	$\frac{60\text{ min}}{1\text{ hr}}$	minutes to hours	$\frac{1\text{ hr}}{60\text{ min}}$
days to hours	$\frac{24\text{ hr}}{1\text{ day}}$	hours to days	$\frac{1\text{ day}}{24\text{ hr}}$

9 Convert from one unit of time to another.

EXAMPLE 9   *Astronomy*   A lunar eclipse occurs when the Earth is between the sun and the moon in such a way that Earth's shadow darkens the moon. (See the figure below, which is not to scale.) A total lunar eclipse can last as long as 105 minutes. Express this time in hours.

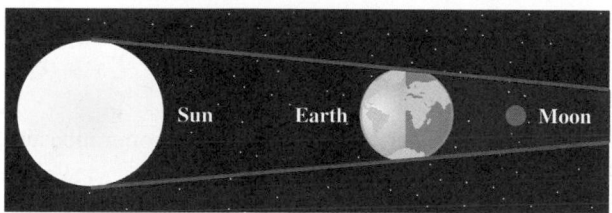

Strategy We will multiply 105 minutes by a carefully chosen unit conversion factor.

WHY If we multiply by the proper unit conversion factor, we can eliminate the unwanted units of minutes and convert to hours.

Solution

To convert from minutes to hours, we must chose a unit conversion factor whose numerator contains the units we want to introduce (hours), and whose denominator contains the units we want to eliminate (minutes). Since there is 1 hour for every 60 minutes, we will use

$$\dfrac{1 \ \text{hr}}{60 \ \text{min}}$$ ←— This is the unit we want to introduce.
←— This is the unit we want to eliminate (the original unit).

To perform the conversion, we multiply.

$$105 \ \text{min} = \dfrac{105 \ \text{min}}{1} \cdot \dfrac{1 \ \text{hr}}{60 \ \text{min}}$$ Write 105 min as a fraction: $105 = \frac{105 \ \text{min}}{1}$. Then multiply by a form of 1: $\frac{1 \ \text{hr}}{60 \ \text{min}}$.

$$= \dfrac{105 \ \cancel{\text{min}}}{1} \cdot \dfrac{1 \ \text{hr}}{60 \ \cancel{\text{min}}}$$ Remove the common units of minutes in the numerator and denominator. Notice that the units of hours remain.

$$= \dfrac{105}{60} \ \text{hr}$$ Multiply the fractions.

$$= \dfrac{\overset{1}{\cancel{3}} \cdot \overset{1}{\cancel{5}} \cdot 7}{2 \cdot 2 \cdot \underset{1}{\cancel{3}} \cdot \underset{1}{\cancel{5}}} \ \text{hr}$$ To simplify the fraction, prime factor 105 and 60. Then remove the common factors 3 and 5 in the numerator and denominator.

$$= \dfrac{7}{4} \ \text{hr}$$ Multiply the remaining factors in the numerator. Multiply the remaining factors in the denominator.

$$= 1\dfrac{3}{4} \ \text{hr}$$ Write $\frac{7}{4}$ as a mixed number.

A total lunar eclipse can last as long as $1\frac{3}{4}$ hours.

SECTION **6.3** **STUDY SET**

VOCABULARY

Fill in the blanks.

1. A ruler is used for measuring _____.

2. Inches, feet, and miles are examples of American units of _____.

3. $\frac{3\text{ ft}}{1\text{ yd}}$, $\frac{1\text{ ton}}{2,000\text{ lb}}$, and $\frac{4\text{ qt}}{1\text{ gal}}$ are examples of _____ conversion factors.

4. Ounces, pounds, and tons are examples of American units of _____.

5. Some examples of American units of _____ are cups, pints, quarts, and gallons.

6. Some units of _____ are seconds, minutes, hours, and days.

CONCEPTS

Fill in the blanks.

7. **a.** 12 inches = ▢ foot

   b. ▢ feet = 1 yard

   c. 1 yard = ▢ inches

   d. 1 mile = ▢ feet

8. **a.** ▢ ounces = 1 pound

   b. ▢ pounds = 1 ton

9. **a.** 1 cup = ▢ fluid ounces

   b. 1 pint = ▢ cups

   c. 2 pints = ▢ quart

   d. 4 quarts = ▢ gallon

10. **a.** 1 day = ▢ hours

    b. 2 hours = ▢ minutes

11. The value of any unit conversion factor is ▢.

12. In general, a unit conversion factor is a fraction with the following form:

$$\frac{\text{Unit that we want to ▢}}{\text{Unit that we want to ▢}} \begin{array}{l} \leftarrow \text{Numerator} \\ \leftarrow \text{Denominator} \end{array}$$

13. Consider the work shown below.

$$\frac{48\text{ oz}}{1} \cdot \frac{1\text{ lb}}{16\text{ oz}}$$

   a. What units can be removed?

   b. What units remain?

14. Consider the work shown below.

$$\frac{600\text{ yd}}{1} \cdot \frac{3\text{ ft}}{1\text{ yd}} \cdot \frac{1\text{ mi}}{5,280\text{ ft}}$$

   a. What units can be removed?

   b. What units remain?

15. Write a unit conversion factor to convert

    a. pounds to tons

    b. quarts to pints

16. Write the two unit conversion factors used to convert

    a. inches to yards

    b. days to minutes

17. Match each item with its proper measurement.

    a. Length of the U.S. coastline

    b. Height of a Barbie doll

    c. Span of the Golden Gate Bridge

    d. Width of a football field

    i. $11\frac{1}{2}$ in.

    ii. 4,200 ft

    iii. 53.5 yd

    iv. 12,383 mi

18. Match each item with its proper measurement.

    a. Weight of the men's shot put used in track and field

    b. Weight of an African elephant

    c. Amount of gold that is worth $500

    i. $1\frac{1}{2}$ oz

    ii. 16 lb

    iii. 7.2 tons

19. Match each item with its proper measurement.

    a. Amount of blood in an adult

    b. Size of the Exxon Valdez oil spill in 1989

    c. Amount of nail polish in a bottle

    d. Amount of flour to make 3 dozen cookies

    i. $\frac{1}{2}$ fluid oz

    ii. 2 cups

    iii. 5 qt

    iv. 10,080,000 gal

20. Match each item with its proper measurement.

    a. Length of first U.S. manned space flight

    b. A leap year

    c. Time difference between New York and Fairbanks, Alaska

    d. Length of Wright Brothers' first flight

    i. 12 sec

    ii. 15 min

    iii. 4 hr

    iv. 366 days

NOTATION

21. What unit does each abbreviation represent?

    a. lb   **b.** oz

    c. fl oz

22. What unit does each abbreviation represent?

    a. qt   **b.** c

    c. pt

Complete each solution.

23. Convert 2 yards to inches.

$$2 \text{ yd} = \frac{2 \text{ yd}}{1} \cdot \frac{\boxed{} \text{ in.}}{1 \text{ yd}}$$

$$= 2 \cdot 36 \boxed{}$$

$$= \boxed{} \text{ in.}$$

24. Convert 24 pints to quarts.

$$24 \text{ pt} = \frac{24 \text{ pt}}{1} \cdot \frac{1 \text{ qt}}{\boxed{} \text{ pt}}$$

$$= \frac{24}{1} \cdot \frac{1}{2} \boxed{}$$

$$= \boxed{} \text{ qt}$$

25. Convert 1 ton to ounces.

$$1 \text{ ton} = \frac{1 \text{ ton}}{1} \cdot \frac{\boxed{} \text{ lb}}{1 \text{ ton}} \cdot \frac{\boxed{} \text{ oz}}{1 \text{ lb}}$$

$$= 1 \cdot 2{,}000 \cdot 16 \boxed{}$$

$$= \boxed{} \text{ oz}$$

26. Convert 37,440 minutes to days.

$$37{,}440 \text{ min} = 37{,}440 \text{ min} \cdot \frac{1 \text{ hr}}{\boxed{} \text{ min}} \cdot \frac{1 \text{ day}}{\boxed{} \text{ hr}}$$

$$= \frac{37{,}440}{60 \cdot 24} \boxed{}$$

$$= \boxed{} \text{ days}$$

GUIDED PRACTICE

Refer to the given ruler to answer each question. See Example 1.

27. a. Each inch is divided into how many equal parts?

    b. Determine which measurements the arrows point to on the ruler.

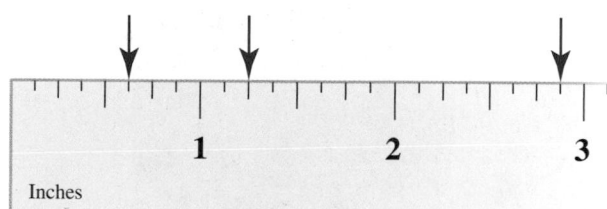

28. Find the length of the needle.

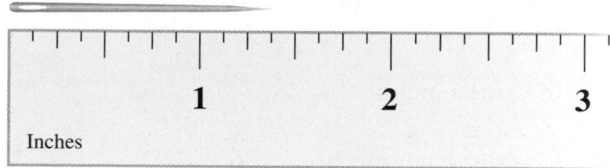

Refer to the given ruler to answer each question. See Example 2.

29. a. Each inch is divided into how many equal parts?

    b. Determine which measurements the arrows point to on the ruler.

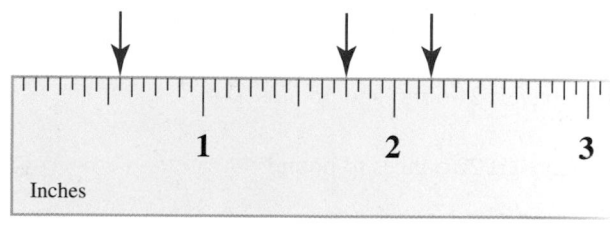

30. Find the length of the bolt.

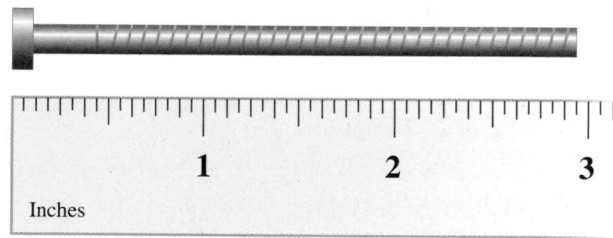

Use a ruler scaled in sixteenths of an inch to measure each object. See Example 2.

31. The width of a dollar bill

32. The length of a dollar bill

33. The length (top to bottom) of this page

34. The length of the word as printed here: supercalifragilisticexpialidocious

Perform each conversion. See Example 3.

35. 4 yards to feet   **36.** 6 yards to feet

37. 35 yards to feet   **38.** 33 yards to feet

Perform each conversion. See Example 4.

39. $3\frac{1}{2}$ feet to inches   **40.** $2\frac{2}{3}$ feet to inches

41. $5\frac{1}{4}$ feet to inches   **42.** $6\frac{1}{2}$ feet to inches

Use two unit conversion factors to perform each conversion. Give the exact answer and a decimal approximation, rounded to the nearest hundredth, when necessary. See Example 5.

43. 105 yards to miles

44. 198 yards to miles

45. 1,540 yards to miles

46. 1,512 yards to miles

Perform each conversion. See Example 6.

47. Convert 44 ounces to pounds.

48. Convert 24 ounces to pounds.

49. Convert 72 ounces to pounds.

50. Convert 76 ounces to pounds.

Perform each conversion. See Example 7.

51. 50 pounds to ounces
52. 30 pounds to ounces
53. 87 pounds to ounces
54. 79 pounds to ounces

Perform each conversion. See Example 8.

55. 8 pints to fluid ounces
56. 5 pints to fluid ounces
57. 21 pints to fluid ounces
58. 30 pints to fluid ounces

Perform each conversion. See Example 9.

59. 165 minutes to hours

60. 195 minutes to hours

61. 330 minutes to hours

62. 80 minutes to hours

▌ TRY IT YOURSELF

Perform each conversion.

63. 3 quarts to pints   **64.** 20 quarts to gallons

65. 7,200 minutes to days   **66.** 691,200 seconds to days

67. 56 inches to feet   **68.** 44 inches to feet

69. 4 feet to inches   **70.** 7 feet to inches

71. 16 pints to gallons   **72.** 3 gallons to fluid ounces

73. 80 ounces to pounds   **74.** 8 pounds to ounces

75. 240 minutes to hours   **76.** 2,400 seconds to hours

77. 8 yards to inches   **78.** 324 inches to yards

79. 90 inches to yards   **80.** 12 yards to inches

81. 5 yards to feet   **82.** 21 feet to yards

83. 12.4 tons to pounds   **84.** 48,000 ounces to tons

85. 7 feet to yards   **86.** $4\frac{2}{3}$ yards to feet

87. 15,840 feet to miles   **88.** 2 miles to feet

89. $\frac{1}{2}$ mile to feet   **90.** 1,320 feet to miles

91. 7,000 pounds to tons   **92.** 2.5 tons to ounces

93. 32 fluid ounces to pints   **94.** 2 quarts to fluid ounces

▌ APPLICATIONS

95. THE GREAT PYRAMID The Great Pyramid in Egypt is about 450 feet high. Express this distance in yards.

96. THE WRIGHT BROTHERS In 1903, Orville Wright made the world's first sustained flight. It lasted 12 seconds, and the plane traveled 120 feet. Express the length of the flight in yards.

Hulton Archive/Getty Images

97. THE GREAT SPHINX The Great Sphinx of Egypt is 240 feet long. Express this in inches.

98. HOOVER DAM The Hoover Dam in Nevada is 726 feet high. Express this distance in inches.

99. THE SEARS TOWER The Sears Tower in Chicago has 110 stories and is 1,454 feet tall. To the nearest hundredth, express this height in miles.

100. NFL RECORDS Emmit Smith, the former Dallas Cowboys and Arizona Cardinals running back, holds the National Football League record for yards rushing in a career: 18,355. How many miles is this? Round to the nearest tenth of a mile.

101. NFL RECORDS When Dan Marino of the Miami Dolphins retired, it was noted that Marino's career passing total was nearly 35 miles! How many yards is this?

102. LEWIS AND CLARK The trail traveled by the Lewis and Clark expedition is shown below. When the expedition reached the Pacific Ocean, Clark estimated that they had traveled 4,162 miles. (It was later determined that his guess was within 40 miles of the actual distance.) Express Clark's estimate of the distance in feet.

103. WEIGHT OF WATER One gallon of water weighs about 8 pounds. Express this weight in ounces.

104. WEIGHT OF A BABY A newborn baby boy weighed 136 ounces. Express this weight in pounds.

105. HIPPOS An adult hippopotamus can weigh as much as 9,900 pounds. Express this weight in tons.

106. ELEPHANTS An adult elephant can consume as much as 495 pounds of grass and leaves in one day. How many ounces is this?

107. BUYING PAINT A painter estimates that he will need 17 gallons of paint for a job. To take advantage of a closeout sale on quart cans, he decides to buy the paint in quarts. How many cans will he need to buy?

108. CATERING How many cups of apple cider are there in a 10-gallon container of cider?

109. SCHOOL LUNCHES Each student attending Eagle River Elementary School receives 1 pint of milk for lunch each day. If 575 students attend the school, how many gallons of milk are used each day?

110. RADIATORS The radiator capacity of a piece of earth-moving equipment is 39 quarts. If the radiator is drained and new coolant put in, how many gallons of new coolant will be used?

111. CAMPING How many ounces of camping stove fuel will fit in the container shown?

FUEL $2\frac{1}{2}$ gal

112. HIKING A college student walks 11 miles in 155 minutes. To the nearest tenth, how many hours does he walk?

113. SPACE TRAVEL The astronauts of the Apollo 8 mission, which was launched on December 21, 1968, were in space for 147 hours. How many days did the mission take?

114. AMELIA EARHART In 1935, Amelia Earhart became the first woman to fly across the Atlantic Ocean alone, establishing a new record for the crossing: 13 hours and 30 minutes. How many minutes is this?

WRITING

115. a. Explain how to find the unit conversion factor that will convert feet to inches.

    b. Explain how to find the unit conversion factor that will convert pints to gallons.

116. Explain why the unit conversion factor $\frac{1\ lb}{16\ oz}$ is a form of 1.

REVIEW

117. Round 3,673.263 to the

    a. nearest hundred

    b. nearest ten

    c. nearest hundredth

    d. nearest tenth

118. Round 0.100602 to the

    a. nearest thousandth

    b. nearest hundredth

    c. nearest tenth

    d. nearest one

Objectives

1. Define metric units of length.

2. Use a metric ruler to measure lengths.

3. Use unit conversion factors to convert metric units of length.

4. Use a conversion chart to convert metric units of length.

5. Define metric units of mass.

6. Convert from one metric unit of mass to another.

7. Define metric units of capacity.

8. Convert from one metric unit of capacity to another.

9. Define a cubic centimeter.

SECTION 6.4

Metric Units of Measurement

The metric system is the system of measurement used by most countries in the world. All countries, including the United States, use it for scientific purposes. The metric system, like our decimal numeration system, is based on the number 10. For this reason, converting from one metric unit to another is easier than with the American system.

1 Define metric units of length.

The basic metric unit of length is the **meter** (m). One meter is approximately 39 inches, which is slightly more than 1 yard. The figure below compares the length of a yardstick to a meterstick.

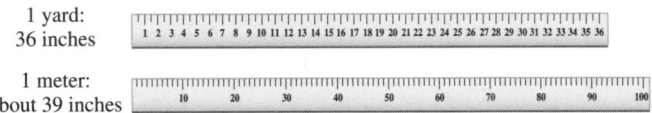

1 yard: 36 inches

1 meter: about 39 inches

Longer and shorter metric units of length are created by adding **prefixes** to the front of the basic unit, *meter*.

kilo means thousands	*deci* means tenths
hecto means hundreds	*centi* means hundredths
deka means tens	*milli* means thousandths

Metric Units of Length

Prefix	kilo-meter	hecto-meter	deka-meter	meter	deci-meter	centi-meter	milli-meter
Meaning	1,000 meters	100 meters	10 meters	1 meter	$\frac{1}{10}$ or 0.1 of a meter	$\frac{1}{100}$ or 0.01 of a meter	$\frac{1}{1,000}$ or 0.001 of a meter
Abbreviation	km	hm	dam	m	dm	cm	mm

> **The Language of Algebra** It is helpful to memorize the prefixes listed above because they are also used with metric units of weight and capacity.

The most often used metric units of length are kilometers, meters, centimeters, and millimeters. It is important that you gain a practical understanding of metric lengths just as you have for the length of an inch, a foot, and a mile. Some examples of metric lengths are shown below.

1 kilometer is about the length of 60 train cars.

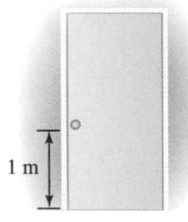

1 m

1 meter is about the distance from a doorknob to the floor.

1 cm

1 centimeter is about as wide as the nail on your little finger.

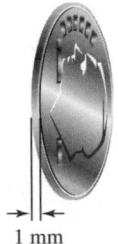

1 mm

1 millimeter is about the thickness of a dime.

2 Use a metric ruler to measure lengths.

Parts of a metric ruler, scaled in centimeters, and a ruler scaled in inches are shown below. Several lengths on the metric ruler are highlighted.

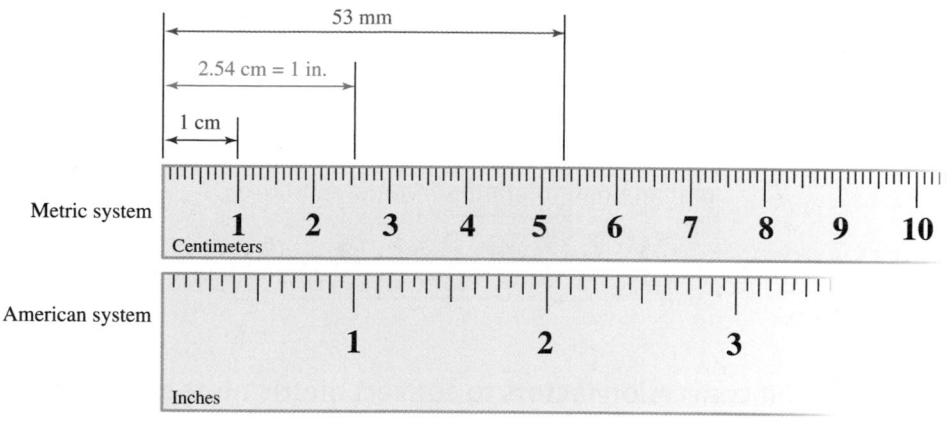

(Actual size)

EXAMPLE 1   Find the length of the nail shown below.

Strategy We will place a metric ruler below the nail, with the left end of the ruler (which could be thought of as 0) directly underneath the head of the nail.

WHY Then we can find the length of the nail by identifying where its pointed end lines up on the tick marks printed in black on the ruler.

Solution
The longest tick marks on the ruler (those labeled with numbers) mark lengths in centimeters. Since the pointed end of the nail lines up on 6, the nail is 6 centimeters long.

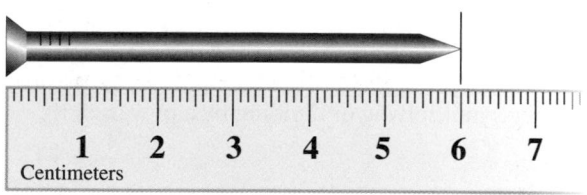

Self Check 1

To the nearest centimeter, find the width of the circle.

Now Try Problem 23

EXAMPLE 2   Find the length of the paper clip shown below.

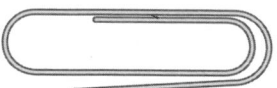

Strategy We will place a metric ruler below the paper clip, with the left end of the ruler (which could be thought of as 0) directly underneath one end of the paper clip.

WHY Then we can find the length of the paper clip by identifying where its other end lines up on the tick marks printed in black on the ruler.

Self Check 2

Find the length of the jumbo paper clip.

Now Try Problem 25

Solution
On the ruler, the shorter tick marks divide each centimeter into 10 millimeters, as shown. If we begin at the left end of the ruler and count by tens as we move right to 3, and then add an additional 6 millimeters to that result, we find that the length of the paper clip is $30 + 6 = 36$ millimeters.

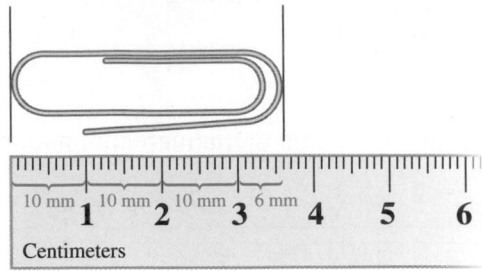

3 Use unit conversion factors to convert metric units of length.

Metric units of length are related as shown in the following table.

Metric Units of Length	
1 kilometer (km) = 1,000 meters	1 meter = 10 decimeters (dm)
1 hectometer (hm) = 100 meters	1 meter = 100 centimeters (cm)
1 dekameter (dam) = 10 meters	1 meter = 1,000 millimeters (mm)
The abbreviation for each unit is written within parentheses.	

We can use the information in the table to write unit conversion factors that can be used to convert metric units of length. For example, in the table we see that

1 meter = 100 centimeters

From this fact, we can write two unit conversion factors.

$$\frac{1\ m}{100\ cm} = 1 \quad \text{and} \quad \frac{100\ cm}{1\ m} = 1$$

To obtain the first unit conversion factor, divide both sides of the equation 1 m = 100 cm by 100 cm. To obtain the second unit conversion factor, divide both sides by 1 m.

One advantage of the metric system is that multiplying or dividing by a unit conversion factor involves multiplying or dividing by a power of 10.

Self Check 3

Convert 860 centimeters to meters.

Now Try **Problem 31**

EXAMPLE 3   Convert 350 centimeters to meters.

Strategy We will multiply 350 centimeters by a carefully chosen unit conversion factor.

WHY If we multiply by the proper unit conversion factor, we can eliminate the unwanted units of centimeters and convert to meters.

Solution
To convert from centimeters to meters, we must choose a unit conversion factor whose numerator contains the units we want to introduce (meters), and whose denominator contains the units we want to eliminate (centimeters). Since there is 1 meter for every 100 centimeters, we will use

$$\frac{1\ m}{100\ cm}$$

⟵ This is the unit we want to introduce.

⟵ This is the unit we want to eliminate (the original unit).

To perform the conversion, we multiply 350 centimeters by the unit conversion factor $\frac{1\text{ m}}{100\text{ cm}}$.

$$350 \text{ cm} = \frac{350 \text{ cm}}{1} \cdot \frac{\mathbf{1 \text{ m}}}{\mathbf{100 \text{ cm}}}$$

Write 350 cm as a fraction: $350 \text{ cm} = \frac{350 \text{ cm}}{1}$.
Multiply by a form of 1: $\frac{1 \text{ m}}{100 \text{ cm}}$.

$$= \frac{350 \ \cancel{\text{cm}}}{1} \cdot \frac{1 \text{ m}}{100 \ \cancel{\text{cm}}}$$

Remove the common units of centimeters from the numerator and denominator. Notice that the units of meter remain.

$$= \frac{350}{100} \text{ m}$$

Multiply the fractions.

$$= \frac{350.0}{100} \text{ m}$$

Write the whole number 350 as a decimal by placing a decimal point immediately to its right and entering a zero: $350 = 350.0$

$$= 3.5 \text{ m}$$

Divide 350.0 by 100 by moving the decimal point 2 places to the left: 3.500.

Thus, 350 centimeters = 3.5 meters.

4   Use a conversion chart to convert metric units of length.

In Example 3, we converted 350 centimeters to meters using a unit conversion factor. We can also make this conversion by recognizing that all units of length in the metric system are powers of 10 of a meter.

To see this, review the table of metric units of length on page 600. Note that each unit has a value that is $\frac{1}{10}$ of the value of the unit immediately to its left and 10 times the value of the unit immediately to its right. Converting from one unit to another is as easy as multiplying (or dividing) by the correct power of 10 or, simply moving a decimal point the correct number of places to the right (or left). For example, in the **conversion chart** below, we see that to convert from centimeters to meters, we move 2 places to the left.

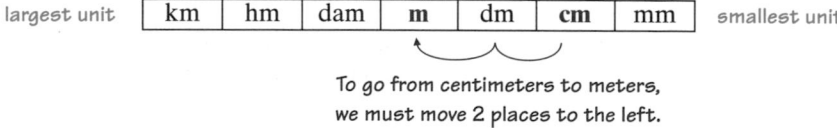

largest unit | km | hm | dam | **m** | dm | **cm** | mm | smallest unit

To go from centimeters to meters,
we must move 2 places to the left.

If we write 350 centimeters as 350.0 centimeters, we can convert to meters by moving the decimal point 2 places to the left.

350.0 centimeters = 3.500 meters = 3.5 meters

Move 2 places to the left.

With the unit conversion factor method or the conversion chart method, we get 350 cm = 3.5 m.

> ***Caution!*** When using a chart to help make a metric conversion, be sure to list the units from *largest to smallest* when reading from left to right.

EXAMPLE 4   Convert 2.4 meters to millimeters.

Strategy On a conversion chart, we will count the places and note the direction as we move from the original units of meters to the conversion units of millimeters.

WHY The decimal point in 2.4 must be moved the same number of places and in that same direction to find the conversion to millimeters.

Self Check 4

Convert 5.3 meters to millimeters.

Now Try Problem 35

Solution

To construct a conversion chart, we list the metric units of length from largest (kilometers) to smallest (millimeters), working from left to right. Then we locate the original units of meters and move to the conversion units of millimeters, as shown below.

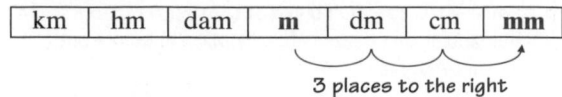

3 places to the right

We see that the decimal point in 2.4 should be moved 3 places to the right to convert from meters to millimeters.

$$2.4 \text{ meters} = 2\,400. \text{ millimeters} = 2,400 \text{ millimeters}$$

Move 3 places to the right.

We can use the unit conversion factor method to confirm this result. Since there are 1,000 millimeters per meter, we multiply 2.4 meters by the unit conversion factor $\frac{1,000 \text{ mm}}{1 \text{ m}}$.

$$2.4 \text{ m} = \frac{2.4 \text{ m}}{1} \cdot \frac{\textbf{1,000 mm}}{\textbf{1 m}}$$ Write 2.4 m as a fraction: 2.4 m = $\frac{2.4 \text{ m}}{1}$. Multiply by a form 1: $\frac{1,000 \text{ mm}}{1 \text{ m}}$.

$$= \frac{2.4 \text{ m\!\!\!\!/}}{1} \cdot \frac{1,000 \text{ mm}}{1 \text{ m\!\!\!\!/}}$$ Remove the common units of meters from the numerator and denominator. Notice that the units of millimeters remain.

$$= 2.4 \cdot 1,000 \text{ mm}$$ Multiply the fractions and simplify.

$$= 2,400 \text{ mm}$$ Multiply 2.4 by 1,000 by moving the decimal point 3 places to the right: 2 400.

Self Check 5

Convert 5.15 centimeters to kilometers.

Now Try Problem 39

EXAMPLE 5   Convert 3.2 centimeters to kilometers.

Strategy On a conversion chart, we will count the places and note the direction as we move from the original units of centimeters to the conversion units of kilometers.

WHY The decimal point in 3.2 must be moved the same number of places and in that same direction to find the conversion to kilometers.

Solution

We locate the original units of centimeters on a conversion chart, and then move to the conversion units of kilometers, as shown below.

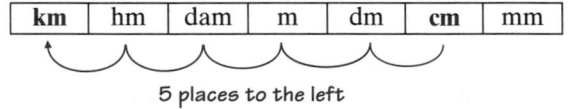

5 places to the left

We see that the decimal point in 3.2 should be moved 5 places to the left to convert centimeters to kilometers.

$$3.2 \text{ centimeters} = 0.000032 \text{ kilometers} = 0.000032 \text{ kilometers}$$

Move 5 places to the left.

We can use the unit conversion factor method to confirm this result. To convert to kilometers, we must use two unit conversion factors so that the units of centimeters drop out and the units of kilometers remain. Since there is 1 meter for every 100 centimeters and 1 kilometer for every 1,000 meters, we multiply by $\frac{1\text{ m}}{100\text{ cm}}$ and $\frac{1\text{ km}}{1,000\text{ m}}$.

$$3.2\text{ cm} = \frac{3.2\ \cancel{\text{cm}}}{1} \cdot \frac{1\ \cancel{\text{m}}}{100\ \cancel{\text{cm}}} \cdot \frac{1\text{ km}}{1,000\ \cancel{\text{m}}}$$

Remove the common units of centimeters and meters. The units of km remain.

$$= \frac{3.2}{100 \cdot 1,000}\text{ km}$$

Multiply the fractions.

$$= 0.000032\text{ km}$$

Divide 3.2 by 1,000 and 100 by moving the decimal point 5 places to the left.

5 Define metric units of mass.

The **mass** of an object is a measure of the amount of material in the object. When an object is moved about in space, its mass does not change. One basic unit of mass in the metric system is the **gram** (g). A gram is defined to be the mass of water contained in a cube having sides 1 centimeter long. (See the figure below.)

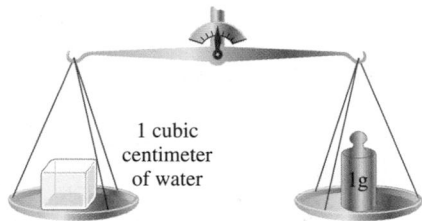

1 cubic centimeter of water

1 g

Other units of mass are created by adding prefixes to the front of the basic unit, *gram.*

Metric Units of Mass							
Prefix	kilo- gram	hecto- gram	deka- gram	**gram**	deci- gram	centi- gram	milli- gram
Meaning	1,000 grams	100 grams	10 grams	1 gram	$\frac{1}{10}$ or 0.1 of a gram	$\frac{1}{100}$ or 0.01 of a gram	$\frac{1}{1,000}$ or 0.001 of a gram
Abbreviation	kg	hg	dag	g	dg	cg	mg

The most often used metric units of mass are kilograms, grams, and milligrams. Some examples are shown below.

An average bowling ball weighs about 6 kilograms.

A raisin weighs about 1 gram.

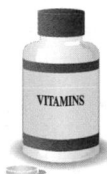

A certain vitamin tablet contains 450 milligrams of calcium.

The **weight** of an object is determined by the Earth's gravitational pull on the object. Since gravitational pull on an object decreases as the object gets farther from Earth, the object weighs less as it gets farther from Earth's surface. This is why astronauts experience weightlessness in space. However, since most of us remain near Earth's surface, we will use the words *mass* and *weight* interchangeably. Thus, a mass of 30 grams is said to weigh 30 grams.

Metric units of mass are related as shown in the following table.

Metric Units of Mass

1 kilogram (kg) = 1,000 grams   1 gram = 10 decigrams (dg)

1 hectogram (hg) = 100 grams   1 gram = 100 centigrams (cg)

1 dekagram (dag) = 10 grams   1 gram = 1,000 milligrams (mg)

The abbreviation for each unit is written within parentheses.

We can use the information in the table to write unit conversion factors that can be used to convert metric units of mass. For example, in the table we see that

$$1 \text{ kilogram} = 1,000 \text{ grams}$$

From this fact, we can write two unit conversion factors.

$$\frac{1 \text{ kg}}{1,000 \text{ g}} = 1 \quad \text{and} \quad \frac{1,000 \text{ g}}{1 \text{ kg}} = 1$$

To obtain the first unit conversion factor, divide both sides of the equation 1 kg = 1,000 g by 1,000 g. To obtain the second unit conversion factor, divide both sides by 1 kg.

6 Convert from one metric unit of mass to another.

Self Check 6

Convert 5.83 kilograms to grams.

Now Try Problem 43

EXAMPLE 6   Convert 7.86 kilograms to grams.

Strategy On a conversion chart, we will count the places and note the direction as we move from the original units of kilograms to the conversion units of grams.

WHY The decimal point in 7.86 must be moved the same number of places and in that same direction to find the conversion to grams.

Solution

To construct a conversion chart, we list the metric units of mass from largest (kilograms) to smallest (milligrams), working from left to right. Then we locate the original units of kilograms and move to the conversion units of grams, as shown below.

largest unit | kg | hg | dag | g | dg | cg | mg | smallest unit

3 places to the right

We see that the decimal point in 7.86 should be moved 3 places to the right to change kilograms to grams.

7.86 kilograms = 7 860. grams = 7,860 grams

Move 3 places to the right.

We can use the unit conversion factor method to confirm this result. To convert to grams, we must chose a unit conversion factor such that the units of kilograms drop out and the units of grams remain. Since there are 1,000 grams per 1 kilogram, we multiply 7.86 kilograms by $\frac{1,000\,g}{1\,kg}$.

$$7.86\ kg = \frac{7.86\ kg}{1} \cdot \frac{\mathbf{1,000\ g}}{\mathbf{1\ kg}} \qquad \text{Remove the common units of kilograms in the numerator and denominator. The units of g remain.}$$

$$= 7.86 \cdot 1,000\ g \qquad \text{Simplify.}$$

$$= 7,860\ g \qquad \text{Multiply 7.86 by 1,000 by moving the decimal point 3 places to the right.}$$

EXAMPLE 7   *Medications*   A bottle of Verapamil, a drug taken for high blood pressure, contains 30 tablets. If each tablet has 180 mg of active ingredient, how many grams of active ingredient are in the bottle?

Strategy We will multiply the number of tablets in one bottle by the number of milligrams of active ingredient in each tablet.

WHY We need to know the total number of milligrams of active ingredient in one bottle before we can convert that number to grams.

Solution

Since there are 30 tablets, and each one contains 180 mg of active ingredient, there are

$$30 \cdot 180\ mg = 5,400\ mg = 5400.0\ mg$$

$$\begin{array}{r} 180 \\ \times\ 30 \\ \hline 000 \\ 5400 \\ \hline 5,400 \end{array}$$

of active ingredient in the bottle. To use a conversion chart to solve this problem, we locate the original units of milligrams and then move to the conversion units of grams, as shown below.

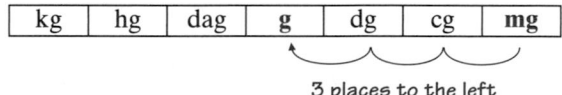

3 places to the left

We see that the decimal point in 5,400.0 should be moved 3 places to the left to convert from milligrams to grams.

$$5,400\ \text{milligrams} = 5.400\ \text{grams}$$

Move 3 places to the left.

There are 5.4 grams of active ingredient in the bottle.

We can use the unit conversion factor method to confirm this result. To convert milligrams to grams, we multiply 5,400 milligrams by $\frac{1\,g}{1,000\,mg}$.

$$5,400\ mg = \frac{5,400\ mg}{1} \cdot \frac{1\ g}{1,000\ mg} \qquad \text{Remove the common units of milligrams from the numerator and denominator. The units of g remain.}$$

$$= \frac{5,400}{1,000}\ g \qquad \text{Multiply the fractions.}$$

$$= 5.4\ g \qquad \text{Divide 5,400 by 1,000 by moving the understood decimal point in 5,400 three places to the left.}$$

7 Define metric units of capacity.

In the metric system, one basic unit of capacity is the **liter** (L), which is defined to be the capacity of a cube with sides 10 centimeters long. Other units of capacity are created by adding prefixes to the front of the basic unit, liter.

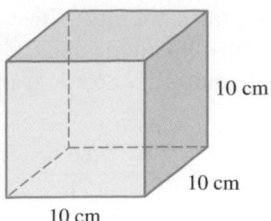

10 cm

10 cm

10 cm

Metric Units of Capacity							
Prefix	kilo- liter	hecto- liter	deka- liter	**liter**	deci- liter	centi- liter	milli- liter
Meaning	1,000 liters	100 liters	10 liters	1 liter	$\frac{1}{10}$ or 0.1 of a liter	$\frac{1}{100}$ or 0.01 of a liter	$\frac{1}{1,000}$ or 0.001 of a liter
Abbreviation	kL	hL	daL	L	dL	cL	mL

The most often used metric units of capacity are liters and milliliters. Here are some examples.

Soft drinks are sold in 2-liter plastic bottles.

The fuel tank of a minivan can hold about 75 liters of gasoline.

A teaspoon holds about 5 milliliters.

Metric units of capacity are related as shown in the following table.

Metric Units of Capacity	
1 kiloliter (kL) = 1,000 liters	1 liter = 10 deciliters (dL)
1 hectoliter (hL) = 100 liters	1 liter = 100 centiliters (cL)
1 dekaliter (daL) = 10 liters	1 liter = 1,000 milliliters (mL)
The abbreviation for each unit is written within parentheses.	

We can use the information in the table to write unit conversion factors that can be used to convert metric units of capacity. For example, in the table we see that

1 liter = 1,000 milliliters

From this fact, we can write two unit conversion factors.

$$\frac{1\ \text{L}}{1,000\ \text{mL}} = 1 \qquad \text{and} \qquad \frac{1,000\ \text{mL}}{1\ \text{L}} = 1$$

8   Convert from one metric unit of capacity to another.

EXAMPLE 8   *Soft Drinks*   How many milliliters are in *three* 2-liter bottles of cola?

Strategy We will multiply the number of bottles of cola by the number of liters of cola in each bottle.

WHY We need to know the total number of liters of cola before we can convert that number to milliliters.

Solution
Since there are three bottles, and each contains 2 liters of cola, there are

$$3 \cdot 2 \text{ L} = 6 \text{ L} = 6.0 \text{ L}$$

of cola in the bottles. To construct a conversion chart, we list the metric units of capacity from largest (kiloliters) to smallest (milliliters), working from left to right. Then we locate the original units of liters and move to the conversion units of milliliters, as shown below.

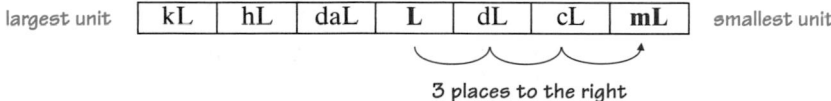

largest unit | kL | hL | daL | **L** | dL | cL | **mL** | smallest unit

3 places to the right

We see that the decimal point in 6.0 should be moved 3 places to the right to convert from liters to milliliters.

6 liters = 6 000. milliliters = 6,000 milliliters

   Move 3 places to the right.

Thus, there are 6,000 milliliters in *three* 2-liter bottles of cola.
   We can use the unit conversion factor method to confirm this result. To convert to milliliters, we must chose a unit conversion factor such that liters drop out and the units of milliliters remain. Since there are 1,000 milliliters per 1 liter, we multiply 6 liters by the unit conversion factor $\frac{1,000 \text{ mL}}{1 \text{ L}}$.

$$6 \text{ L} = \frac{6 \cancel{\text{L}}}{1} \cdot \frac{\mathbf{1,000 \text{ mL}}}{\mathbf{1 \cancel{\text{L}}}}$$ Remove the common units of liters in the numerator and denominator. The units of mL remain.

$$= 6 \cdot 1,000 \text{ mL}$$ Simplify.

$$= 6,000 \text{ mL}$$ Multiply 6 by 1,000 by moving the understood decimal point in 6 three places to the right.

Self Check 8

SOFT DRINKS How many milliliters are in a case of *twelve* 2-liter bottles of cola?

Now Try Problems 51 and 97

9   Define a cubic centimeter.

Another metric unit of capacity is the **cubic centimeter,** which is represented by the notation cm³ or, more simply, cc. One milliliter and one cubic centimeter represent the same capacity.

$$1 \text{ mL} = 1 \text{ cm}^3 = 1 \text{ cc}$$

The units of cubic centimeters are used frequently in medicine. For example, when a nurse administers an injection containing 5 cc of medication, the dosage can also be expressed using milliliters.

$$5 \text{ cc} = 5 \text{ mL}$$

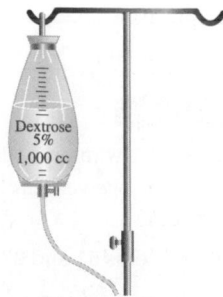

When a doctor orders that a patient be put on 1,000 cc of dextrose solution, the request can be expressed in different ways.

1,000 cc = 1,000 mL = 1 liter

ANSWERS TO SELF CHECKS

1. 3 cm   **2.** 47 mm   **3.** 8.6 m   **4.** 5,300 mm   **5.** 0.0000515 km   **6.** 5,830 g   **7.** 1.8 g
8. 24,000 mL

SECTION 6.4   STUDY SET

VOCABULARY

Fill in the blanks.

1. The meter, the gram, and the liter are basic units of measurement in the _____ system.

2. a. The basic unit of length in the metric system is the _____.

   b. The basic unit of mass in the metric system is the _____.

   c. The basic unit of capacity in the metric system is the _____.

3. a. *Deka* means _____.

   b. *Hecto* means _____.

   c. *Kilo* means _____.

4. a. *Deci* means _____.

   b. *Centi* means _____.

   c. *Milli* means _____.

5. We can convert from one unit to another in the metric system using _____ conversion factors or a conversion _____ like that shown below.

km	hm	dam	m	dm	cm	mm

6. The _____ of an object is a measure of the amount of material in the object.

7. The _____ of an object is determined by the Earth's gravitational pull on the object.

8. Another metric unit of capacity is the cubic _____, which is represented by the notation cm^3, or, more simply, cc.

CONCEPTS

Fill in the blanks.

9. a. 1 kilometer = _____ meters

   b. _____ centimeters = 1 meter

   c. _____ millimeters = 1 meter

10. a. 1 gram = _____ milligrams

    b. 1 kilogram = _____ grams

11. a. _____ milliliters = 1 liter

    b. 1 dekaliter = _____ liters

12. a. 1 milliliter = _____ cubic centimeter

    b. 1 liter = _____ cubic centimeters

13. Write a unit conversion factor to convert

    a. meters to kilometers

    b. grams to centigrams

    c. liters to milliliters

14. Use the chart to determine how many decimal places and in which direction to move the decimal point when converting the following.

    a. Kilometers to centimeters

km	hm	dam	m	dm	cm	mm

    b. Milligrams to grams

kg	hg	dag	g	dg	cg	mg

    c. Hectoliters to centiliters

kL	hL	daL	L	dL	cL	mL

15. Match each item with its proper measurement.

    a. Thickness of a phone book   i. 6,275 km

    b. Length of the Amazon River   ii. 2 m

    c. Height of a soccer goal   iii. 6 cm

16. Match each item with its proper measurement.

    a. Weight of a giraffe   i. 800 kg

    b. Weight of a paper clip   ii. 1 g

    c. Active ingredient in an aspirin tablet   iii. 325 mg

17. Match each item with its proper measurement.

   a. Amount of blood in an adult   **i.** 290,000 kL

   b. Cola in an aluminum can   **ii.** 6 L

                             iii. 355 mL

   c. Kuwait's daily production of crude oil

18. Of the objects shown below, which can be used to measure the following?

   a. Millimeters

   b. Milligrams

   c. Milliliters

Balance

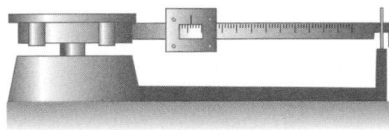

Beaker

Micrometer

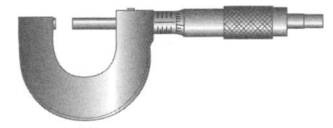

▌NOTATION

Complete each solution.

19. Convert 20 centimeters to meters.

$$20 \text{ cm} = \frac{20 \text{ cm}}{1} \cdot \frac{\boxed{} \text{ m}}{100 \text{ cm}}$$

$$= \frac{20}{\boxed{}} \text{ m}$$

$$= \boxed{} \text{ m}$$

20. Convert 3,000 milligrams to grams.

$$3,000 \text{ mg} = \frac{3,000 \text{ mg}}{1} \cdot \frac{1 \text{ g}}{1,000 \boxed{}}$$

$$= \frac{3,000}{1,000} \boxed{}$$

$$= \boxed{} \text{ g}$$

21. Convert 0.2 kilograms to milligrams.

$$0.2 \text{ kg} = \frac{0.2 \text{ kg}}{1} \cdot \frac{\boxed{} \text{ g}}{1 \text{ kg}} \cdot \frac{1,000 \text{ mg}}{\boxed{} \text{ g}}$$

$$= 0.2 \cdot 1,000 \cdot 1,000 \boxed{}$$

$$= \boxed{} \text{ mg}$$

22. Convert 400 milliliters to kiloliters.

$$400 \text{ mL} = \frac{400 \text{ mL}}{1} \cdot \frac{1 \text{ L}}{\boxed{} \text{ mL}} \cdot \frac{1}{1,000 \text{ L}}$$

$$= \frac{\boxed{}}{1,000 \cdot 1,000} \text{ kL}$$

$$= 0.0004 \text{ kL}$$

▌GUIDED PRACTICE

Refer to the given ruler to answer each question. See Example 1.

23. Determine which measurements the arrows point to on the metric ruler.

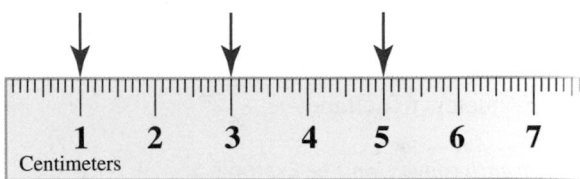

24. Find the length of the birthday candle (including the wick).

Refer to the given ruler to answer each question. See Example 2.

25. a. Refer to the metric ruler below. Each centimeter is divided into how many equal parts? What is the length of one of those parts?

    b. Determine which measurements the arrows point to on the ruler.

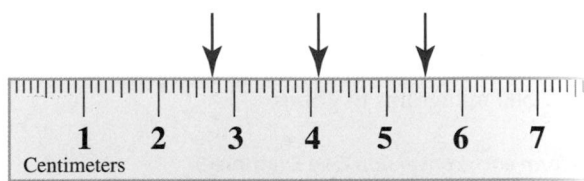

26. Find the length of the stick of gum.

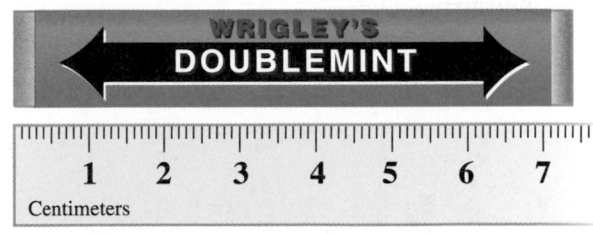

Use a metric ruler scaled in millimeters to measure each object.
See Example 2.

27. The length of a dollar bill
28. The width of a dollar bill
29. The length (top to bottom) of this page
30. The length of the word antidisestablishmentarianism as printed here.

Perform each conversion. See Example 3.

31. 380 centimeters to meters
32. 590 centimeters to meters
33. 120 centimeters to meters
34. 640 centimeters to meters

Perform each conversion. See Example 4.

35. 8.7 meters to millimeters
36. 1.3 meters to millimeters
37. 2.89 meters to millimeters
38. 4.06 meters to millimeters

Perform each conversion. See Example 5.

39. 4.5 centimeters to kilometers
40. 6.2 centimeters to kilometers
41. 0.3 centimeters to kilometers
42. 0.4 centimeters to kilometers

Perform each conversion. See Example 6.

43. 1.93 kilograms to grams
44. 8.99 kilograms to grams
45. 4.531 kilograms to grams
46. 6.077 kilograms to grams

Perform each conversion. See Example 7.

47. 6,000 milligrams to grams
48. 9,000 milligrams to grams
49. 3,500 milligrams to grams
50. 7,500 milligrams to grams

Perform each conversion. See Example 8.

51. 3 liters to milliliters
52. 4 liters to milliliters
53. 26.3 liters to milliliters
54. 35.2 liters to milliliters

TRY IT YOURSELF

Perform each conversion.

55. 0.31 decimeters to centimeters
56. 73.2 meters to decimeters
57. 500 milliliters to liters
58. 500 centiliters to milliliters
59. 2 kilograms to grams
60. 4,000 grams to kilograms
61. 0.074 centimeters to millimeters
62. 0.125 meters to millimeters
63. 1,000 kilograms to grams
64. 2 kilograms to centigrams
65. 658.23 liters to kiloliters
66. 0.0068 hectoliters to kiloliters
67. 4.72 cm to dm
68. 0.593 cm to dam
69. 10 mL = ___ cc
70. 2,000 cc = __ L
71. 500 mg to g
72. 500 mg to cg
73. 5,689 g to kg
74. 0.0579 km to mm
75. 453.2 cm to m
76. 675.3 cm to m
77. 0.325 dL to L
78. 0.0034 mL to L
79. 675 dam = _____ cm
80. 76.8 hm = _____ mm
81. 0.00777 cm = _____ dam
82. 400 liters to hL
83. 134 m to hm
84. 6.77 mm to cm
85. 65.78 km to dam
86. 5 g to cg

APPLICATIONS

87. SPEED SKATING American Eric Heiden won an unprecedented five gold medals by capturing the men's 500-m, 1,000-m, 1,500-m, 5,000-m, and 10,000-m races at the 1980 Winter Olympic Games in Lake Placid, New York. Convert each race length to kilometers.

88. THE SUEZ CANAL The 163-km-long Suez Canal connects the Mediterranean Sea with the Red Sea. It provides a shortcut for ships operating between European and American ports. Convert the length of the Suez Canal to meters.

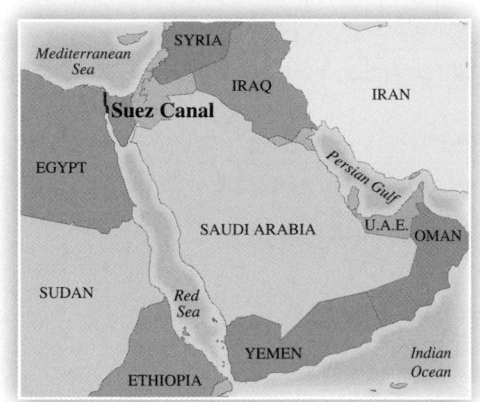

89. SKYSCRAPERS The John Hancock Center in Chicago has 100 stories and is 343 meters high. Give this height in hectometers.

90. WEIGHT OF A BABY A baby weighs 4 kilograms. Give this weight in centigrams.

91. HEALTH CARE Blood pressure is measured by a sphygmomanometer *(see at right)*. The measurement is read at two points and is expressed, for example, as 120/80. This indicates a *systolic* pressure of 120 millimeters of mercury and a *diastolic* pressure of 80 millimeters of mercury. Convert each measurement to centimeters of mercury.

92. JEWELRY A gold chain weighs 1,500 milligrams. Give this weight in grams.

93. EYE DROPPERS One drop from an eye dropper is 0.05 mL. Convert the capacity of one drop to liters.

94. BOTTLING How many liters of wine are in a 750-mL bottle?

95. MEDICINE A bottle of hydrochlorothiazine contains 60 tablets. If each tablet contains 50 milligrams of active ingredient, how many grams of active ingredient are in the bottle?

96. IBUPROFEN What is the total weight, in grams, of all the tablets in the box shown at right?

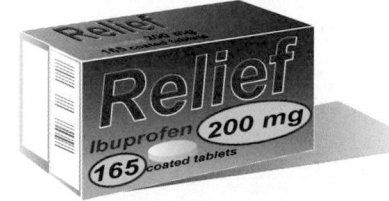

97. SIX PACKS Some stores sell Fanta orange soda in 0.5 liter bottles. How many milliliters are there in a six pack of this size bottle?

98. CONTAINERS How many deciliters of root beer are in *two* 2-liter bottles?

99. OLIVES The net weight of a bottle of olives is 284 grams. Find the smallest number of bottles that must be purchased to have at least 1 kilogram of olives.

100. COFFEE A can of Cafe Vienna has a net weight of 133 grams. Find the smallest number of cans that must be packaged to have at least 1 metric ton of coffee. (*Hint:* 1 metric ton = 1,000 kg.)

101. INJECTIONS The illustration below shows a 3cc syringe. Express its capacity using units of milliliters.

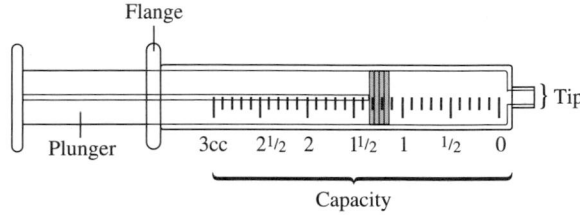

102. MEDICAL SUPPLIES A doctor ordered 2,000 cc of a saline (salt) solution from a pharmacy. How many liters of saline solution is this?

WRITING

103. To change 3.452 kilometers to meters, we can move the decimal point in 3.452 three places to the right to get 3,452 meters. Explain why.

104. To change 7,532 grams to kilograms, we can move the decimal point in 7,532 three places to the left to get 7.532 kilograms. Explain why.

105. A *centimeter* is one hundredth of a meter. Make a list of five other words that begin with the prefix *centi* or *cent* and write a definition for each.

106. List the advantages of the metric system of measurement as compared to the American system. There have been several attempts to bring the metric system into general use in the United States. Why do you think these efforts have been unsuccessful?

REVIEW

Write each fraction as a decimal. Use an overbar in your answer.

107. $\dfrac{8}{9}$

108. $\dfrac{11}{12}$

109. $\dfrac{7}{90}$

110. $\dfrac{1}{66}$

Objectives

1. Use unit conversion factors to convert between American and metric units.
2. Convert between Fahrenheit and Celsius temperatures.

SECTION 6.5
Converting between American and Metric Units

It is often necessary to convert between American units and metric units. For example, we must convert units to answer the following questions.

- Which is higher: Pikes Peak (elevation 14,110 feet) or the Matterhorn (elevation 4,478 meters)?
- Does a 2-pound tub of butter weigh more than a 1-kilogram tub?
- Is a quart of soda pop more or less than a liter of soda pop?

In this section, we discuss how to answer such questions.

1 Use unit conversion factors to convert between American and metric units.

The following table shows some conversions between American and metric units of length. In all but one case, the conversions are rounded approximations. An ≈ symbol is used to show this. The one exact conversion in the table is 1 inch = 2.54 centimeters.

1 foot

1 yard

1 meter

Equivalent Lengths	
American to metric	**Metric to American**
1 in. = 2.54 cm	1 cm ≈ 0.39 in.
1 ft ≈ 0.30 m	1 m ≈ 3.28 ft
1 yd ≈ 0.91 m	1 m ≈ 1.09 yd
1 mi ≈ 1.61 km	1 km ≈ 0.62 mi

Unit conversion factors can be formed from the facts in the table to make specific conversions between American and metric units of length.

Self Check 1

CLOTHING LABELS Refer to the figure in Example 1. What is the inseam length, to the nearest inch?

Now Try Problem 13

EXAMPLE 1   *Clothing Labels*   The figure shows a label sewn into some pants made in Mexico that are for sale in the United States. Express the waist size to the nearest inch.

WAIST: 82 cm
INSEAM: 76 cm
RN-80811
SEE REVERSE FOR CARE

MADE IN MEXICO

Strategy   We will multiply 82 centimeters by a carefully chosen unit conversion factor.

WHY   If we multiply by the proper unit conversion factor, we can eliminate the unwanted units of centimeters and convert to inches.

Solution

To convert from centimeters to inches, we must choose a unit conversion factor whose numerator contains the units we want to introduce (inches), and whose denominator contains the units we want to eliminate (centimeters). From the first row of the *Metric to American* column of the table, we see that there is approximately 0.39 inch per centimeter. Thus, we will use the unit conversion factor:

$$\frac{0.39 \text{ in.}}{1 \text{ cm}}$$

0.39 in. ⟵ This is the unit we want to introduce.

1 cm ⟵ This is the unit we want to eliminate (the original unit).

To perform the conversion, we multiply.

$$82 \text{ cm} \approx \frac{82 \text{ cm}}{1} \cdot \frac{\textbf{0.39 in.}}{\textbf{1 cm}}$$ Write 82 cm as a fraction: 82 cm = $\frac{82 \text{ cm}}{1}$.
                                               Multiply by a form of 1: $\frac{0.39 \text{ in.}}{1 \text{ cm}}$.

$$\approx \frac{82 \text{ c\hspace{-0.4em}/\hspace{-0.2em}m}}{1} \cdot \frac{0.39 \text{ in.}}{1 \text{ c\hspace{-0.4em}/\hspace{-0.2em}m}}$$ Remove the common units of centimeters from the numerator and denominator. The units of inches remain.

$$\approx 82 \cdot 0.39 \text{ in.}$$ Simplify.

$$\approx 31.98 \text{ in.}$$ Do the multiplication.

$$\approx 32 \text{ in.}$$ Round to the nearest inch (ones column).

$$\begin{array}{r} 0.39 \\ \times\ 82 \\ \hline 78 \\ 3120 \\ \hline 31.98 \end{array}$$

To the nearest inch, the waist size is 32 inches.

EXAMPLE 2   *Mountain Elevations*   Pikes Peak, one of the most famous peaks in the Rocky Mountains, has an elevation of 14,110 feet. The Matterhorn, in the Swiss Alps, rises to an elevation of 4,478 meters. Which mountain is higher?

Strategy We will convert the elevation of Pikes Peak, which given in feet, to meters.

WHY Then we can compare the mountain's elevations in the same units, meters.

Solution

To convert Pikes Peak elevation from feet to meters we must choose a unit conversion factor whose numerator contains the units we want to introduce (meters) and whose denominator contains the units we want to eliminate (feet). From the second row of the *American to metric* column of the table, we see that there is approximately 0.30 meter per foot. Thus, we will use the unit conversion factor:

$\dfrac{0.30 \text{ m}}{1 \text{ ft}}$ ←— This is the unit we want to introduce.
       ←— This is the unit we want to eliminate (the original unit).

To perform the conversion, we multiply.

$$14,110 \text{ ft} \approx \frac{14,110 \text{ ft}}{1} \cdot \frac{\textbf{0.30 m}}{\textbf{1 ft}}$$ Write 14,110 ft as a fraction: 14,110 ft = $\frac{14,110 \text{ ft}}{1}$.
        Multiply by a form of 1: $\frac{0.30 \text{ m}}{1 \text{ ft}}$.

$$\approx \frac{14,110 \text{ f\hspace{-0.4em}/\hspace{-0.2em}t}}{1} \cdot \frac{0.30 \text{ m}}{1 \text{ f\hspace{-0.4em}/\hspace{-0.2em}t}}$$ Remove the common units of feet from the numerator and denominator. The units of meters remain.

$$\approx 14,110 \cdot 0.30 \text{ m}$$ Simplify.

$$\approx 4,233 \text{ m}$$ Do the multiplication.

$$\begin{array}{r} \overset{1}{14,110} \\ \times\ 0.30 \\ \hline 000\ 00 \\ 4233\ 00 \\ \hline 4233.00 \end{array}$$

Since the elevation of Pikes Peak is about 4,233 meters, we can conclude that the Matterhorn, with an elevation of 4,478 meters, is higher.

We can convert between American units of weight and metric units of mass using the rounded approximations in the following table.

Equivalent Weights and Masses	
American to metric	**Metric to American**
1 oz ≈ 28.35 g	1 g ≈ 0.035 oz
1 lb ≈ 0.45 kg	1 kg ≈ 2.20 lb

Self Check 2

TRACK AND FIELD Which is longer: a 500-meter race or a 550-yard race?

Now Try Problem 17

1 pound

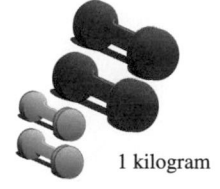

1 kilogram

Self Check 3

Convert 68 pounds to grams. Round to the nearest gram.

Now Try Problem 21

EXAMPLE 3 Convert 50 pounds to grams.

Strategy We will use a two-part multiplication process that converts 50 pounds to ounces, and then converts that result to grams.

WHY We must use a two-part process because the conversion table on page 615 does not contain a single unit conversion factor that converts from pounds to grams.

Solution

Since there are 16 ounces per pound, we can convert 50 pounds to ounces by multiplying by the unit conversion factor $\frac{16 \text{ oz}}{1 \text{ lb}}$. Since there are approximately 28.35 g per ounce, we can convert that result to grams by multiplying by the unit conversion factor $\frac{28.35 \text{ g}}{1 \text{ oz}}$.

$$50 \text{ lb} \approx \frac{50 \text{ lb}}{1} \cdot \frac{\textbf{16 oz}}{\textbf{1 lb}} \cdot \frac{\textbf{28.35 g}}{\textbf{1 oz}}$$

Write 50 lb as a fraction: $50 \text{ lb} = \frac{50 \text{ lb}}{1}$. Multiply by two forms of 1: $\frac{16 \text{ oz}}{1 \text{ lb}}$ and $\frac{28.35 \text{ g}}{1 \text{ oz}}$.

$$\approx \frac{50 \cancel{\text{lb}}}{1} \cdot \frac{16 \cancel{\text{oz}}}{1 \cancel{\text{lb}}} \cdot \frac{28.35 \text{ g}}{1 \cancel{\text{oz}}}$$

Remove the common units of pounds and ounces from the numerator and denominator. The units of grams remain.

$$\approx 50 \cdot 16 \cdot 28.35 \text{ g} \qquad \text{Simplify.}$$

$$\approx 800 \cdot 28.35 \text{ g} \qquad \text{Multiply: } 50 \cdot 16 = 800.$$

$$\approx 22{,}680 \text{ g} \qquad \text{Do the multiplication.}$$

$$
\begin{array}{r}
\overset{3}{16} \\
\times\, 50 \\
\hline
800
\end{array}
\qquad
\begin{array}{r}
\overset{6\,2\ 4}{28.35} \\
\times\, 800 \\
\hline
22680.00
\end{array}
$$

Thus, 50 pounds $\approx$ 22,680 grams.

Self Check 4

BODY WEIGHT Who weighs more, a person who weighs 165 pounds or one who weighs 76 kilograms?

Now Try Problem 25

EXAMPLE 4 *Packaging* Does a 2.5 pound tub of butter weigh more than a 1.5-kilogram tub?

Strategy We will convert the weight of the 1.5-kilogram tub of butter to pounds.

WHY Then we can compare the weights of the tubs of butter in the same units, pounds.

Solution

To convert 1.5 kilograms to pounds we must choose a unit conversion factor whose numerator contains the units we want to introduce (pounds), and whose denominator contains the units we want to eliminate (kilograms). From the second row of the *Metric to American* column of the table, we see that there are approximately 2.20 pounds per kilogram. Thus, we will use the unit conversion factor:

$$\frac{2.20 \text{ lb}}{1 \text{ kg}}$$
$\leftarrow$ This is the unit we want to introduce.
$\leftarrow$ This is the unit we want to eliminate (the original unit).

To perform the conversion, we multiply.

$$1.5 \text{ kg} \approx \frac{1.5 \text{ kg}}{1} \cdot \frac{\textbf{2.20 lb}}{\textbf{1 kg}}$$

Write 1.5 kg as a fraction: $1.5 \text{ kg} = \frac{1.5 \text{ kg}}{1}$. Multiply by a form of 1: $\frac{2.20 \text{ lb}}{1 \text{ kg}}$.

$$\approx \frac{1.5 \cancel{\text{kg}}}{1} \cdot \frac{2.20 \text{ lb}}{1 \cancel{\text{kg}}}$$

Remove the common units of kilograms from the numerator and denominator. The units of pounds remain.

$$\approx 1.5 \cdot 2.20 \text{ lb} \qquad \text{Simplify.}$$

$$\approx 3.3 \text{ lb} \qquad \text{Do the multiplication.}$$

$$
\begin{array}{r}
2.20 \\
\times\, 1.5 \\
\hline
1100 \\
2200 \\
\hline
3.300
\end{array}
$$

Since a 1.5-kilogram tub of butter weighs about 3.3 pounds, the 1.5-kilogram tub weighs more.

We can convert between American and metric units of capacity using the rounded approximations in the following table.

Equivalent Capacities	
American to metric	**Metric to American**
1 fl oz ≈ 29.57 mL	1 L ≈ 33.81 fl oz
1 pt ≈ 0.47 L	1 L ≈ 2.11 pt
1 qt ≈ 0.95 L	1 L ≈ 1.06 qt
1 gal ≈ 3.79 L	1 L ≈ 0.264 gal

1 liter   1 quart

THINK IT THROUGH   *Studying in Other Countries*

"Over the past decade, the number of U.S. students studying abroad has more than doubled."

From The Open Doors 2008 Report

In 2006/2007, a record number of 241,791 college students received credit for study abroad. Since students traveling to other countries are almost certain to come into contact with the metric system of measurement, they need to have a basic understanding of metric units.

Suppose a student studying overseas needs to purchase the following school supplies. For each item in red, choose the appropriate metric units.

1. $8\frac{1}{2}$ in. × 11 in. notebook paper:

   216 meters × 279 meters   216 centimeters × 279 centimeters

   216 millimeters × 279 millimeters

2. A backpack that can hold 20 pounds of books:

   9 kilograms   9 grams   9 milligrams

3. $\frac{3}{4}$ fluid ounce bottle of Liquid Paper correction fluid:

   22.5 hectoliters   2.5 liters   22.2 milliliters

EXAMPLE 5   *Cleaning Supplies*   A bottle of window cleaner contains 750 milliliters of solution. Convert this measure to quarts. Round to the nearest tenth.

Strategy We will use a two-part multiplication process that converts 750 milliliters to liters, and then converts that result to quarts.

WHY We must use a two-part process because the conversion table at the top of this page does not contain a single unit conversion factor that converts from milliliters to quarts.

Solution
Since there is 1 liter for every 1,000 mL, we can convert 750 milliliters to liters by multiplying by the unit conversion factor $\frac{1\,L}{1,000\,mL}$. Since there are approximately

---

Self Check 5

DRINKING WATER   A student bought a 360-mL bottle of water. Convert this measure to quarts. Round to the nearest tenth.

Now Try Problem 29

1.06 qt per liter, we can convert that result to quarts by multiplying by the unit conversion factor $\frac{1.06 \text{ qt}}{1 \text{ L}}$.

$$750 \text{ mL} \approx \frac{750 \text{ mL}}{1} \cdot \frac{1 \text{ L}}{1,000 \text{ mL}} \cdot \frac{1.06 \text{ qt}}{1 \text{ L}}$$

Write 750 mL as a fraction:
750 mL = $\frac{750 \text{ mL}}{1}$. Multiply by
two forms of 1: $\frac{1 \text{ L}}{1,000 \text{ mL}}$ and $\frac{1.06 \text{ qt}}{1 \text{ L}}$.

$$\approx \frac{750 \text{ mL}}{1} \cdot \frac{1 \text{ L}}{1,000 \text{ mL}} \cdot \frac{1.06 \text{ qt}}{1 \text{ L}}$$

Remove the common units of milliliters and liters from the numerator and denominator. The units of quarts remain.

$$\approx \frac{750 \cdot 1.06}{1,000} \text{ qt}$$

Multiply the fractions.

$$\approx \frac{795}{1,000} \text{ qt}$$

Multiply: 750 · 1.06 = 795.

$$
\begin{array}{r}
750 \\
\times\, 1.06 \\
\hline
4500 \\
0000 \\
75000 \\
\hline
795.00
\end{array}
$$

$$\approx 0.795 \text{ qt}$$

Divide 795 by 1,000 by moving the decimal point 3 places to the left.

$$\approx 0.8 \text{ qt}$$

Round to the nearest tenth.

The bottle contains approximately 0.8 qt of cleaning solution.

2 Convert between Fahrenheit and Celsius temperatures.

In the American system, we measure temperature using **degrees Fahrenheit** (°F). In the metric system, we measure temperature using **degrees Celsius** (°C). These two scales are shown on the thermometers on the right. From the figures, we can see that

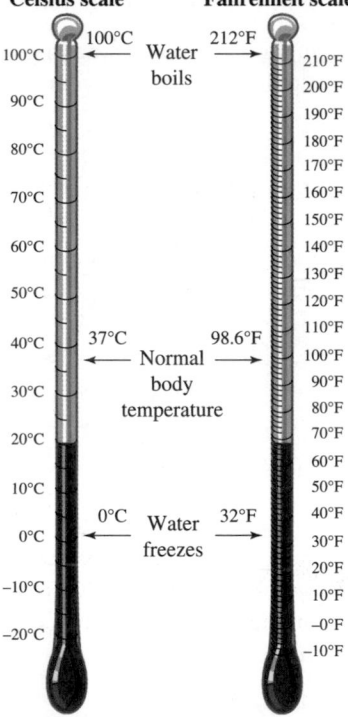

- 212°F ≈ 100°C   Water boils
- 32°F ≈ 0°C   Water freezes
- 5°F ≈ −15°C   A cold winter day
- 95°F ≈ 35°C   A hot summer day

There are formulas that enable us to convert from degrees Fahrenheit to degrees Celsius and from degrees Celsius to degrees Fahrenheit.

Conversion Formulas for Temperature

If F is the temperature in degrees Fahrenheit and C is the corresponding temperature in degrees Celsius, then

$$C = \frac{5}{9}(F - 32) \quad \text{and} \quad F = \frac{9}{5}C + 32$$

EXAMPLE 6 *Bathing* Warm bath water is 90°F. Express this temperature in degrees Celsius. Round to the nearest tenth of a degree.

Strategy We will substitute 90 for F in the formula $C = \frac{5}{9}(F - 32)$.

WHY Then we can use the rule for the order of operations to evaluate the right side of the equation and find the value of C, the temperature in degrees Celsius of the bath water.

Solution

$$C = \frac{5}{9}(F - 32) \quad \text{This is the formula to find degrees Celsius.}$$

$$= \frac{5}{9}(90 - 32) \quad \text{Substitute 90 for } F.$$

$$= \frac{5}{9}(58) \quad \begin{array}{l}\text{Do the subtraction within the}\\\text{parentheses first: } 90 - 32 = 58.\end{array}$$

$$= \frac{5}{9}\left(\frac{58}{1}\right) \quad \text{Write 58 as a fraction: } 58 = \frac{58}{1}.$$

$$= \frac{290}{9} \quad \begin{array}{l}\text{Multiply the numerators.}\\\text{Multiply the denominators.}\end{array}$$

$$= 32.222\ldots \quad \text{Do the division.}$$

$$\approx 32.2 \quad \text{Round to the nearest tenth.}$$

$$\begin{array}{r} \overset{4}{58} \\ \times\, 5 \\ \hline 290 \end{array}$$

$$\begin{array}{r} 32.22 \\ 9\overline{)290.00} \\ -27 \\ \hline 20 \\ -18 \\ \hline 20 \\ -18 \\ \hline 20 \\ -18 \\ \hline 2 \end{array}$$

To the nearest tenth of a degree, the temperature of the bath water is 32.2°C. ◼

Self Check 6

COFFEE Hot coffee is 110°F. Express this temperature in degrees Celsius. Round to the nearest tenth of a degree.

Now Try Problem 33

EXAMPLE 7 *Dishwashers* A dishwasher manufacturer recommends that dishes be rinsed in hot water with a temperature of 60°C. Express this temperature in degrees Fahrenheit.

Strategy We will substitute 60 for C in the formula $F = \frac{9}{5}C + 32$.

WHY Then we can use the rule for the order of operations to evaluate the right side of the equation and find the value of F, the temperature in degrees Fahrenheit of the water.

Solution

$$F = \frac{9}{5}C + 32 \quad \text{This is the formula to find degrees Fahrenheit.}$$

$$= \frac{9}{5}(60) + 32 \quad \text{Substitute 60 for } C.$$

$$= \frac{540}{5} + 32 \quad \text{Multiply: } \frac{9}{5}(60) = \frac{9}{5}\left(\frac{60}{1}\right) = \frac{540}{5}.$$

$$= 108 + 32 \quad \text{Do the division.}$$

$$= 140 \quad \text{Do the addition.}$$

$$\begin{array}{r} 60 \\ \times\, 9 \\ \hline 540 \end{array}$$

$$\begin{array}{r} 108 \\ 5\overline{)540} \\ -5 \\ \hline 4 \\ -0 \\ \hline 40 \\ -40 \\ \hline 0 \end{array}$$

The manufacturer recommends that dishes be rinsed in 140°F water. ◼

Self Check 7

FEVERS To determine whether a baby has a fever, her mother takes her temperature with a Celsius thermometer. If the reading is 38.8°C, does the baby have a fever? (*Hint:* Normal body temperature is 98.6°F.)

Now Try Problem 37

ANSWERS TO SELF CHECKS

1. 30 in. **2.** the 550-yard race **3.** 30,845 g **4.** the person who weighs 76 kg
5. 0.4 qt **6.** 43.3°C **7.** yes

SECTION 6.5 STUDY SET

VOCABULARY

Fill in the blanks.

1. In the American system, temperatures are measured in degrees _____. In the metric system, temperatures are measured in degrees _____.

2. **a.** Inches and centimeters are units used to measure _____.

   b. Pounds and grams are used to measure _____ (weight).

   c. Gallons and liters are units used to measure _____.

CONCEPTS

3. Which is longer:

   a. A yard or a meter?

   b. A foot or a meter?

   c. An inch or a centimeter?

   d. A mile or a kilometer?

4. Which is heavier:

   a. An ounce or a gram?

   b. A pound or a kilogram?

5. Which is the greater unit of capacity:

   a. A pint or a liter?

   b. A quart or a liter?

   c. A gallon or a liter?

6. **a.** What formula is used for changing degrees Celsius to degrees Fahrenheit?

   b. What formula is used for changing degrees Fahrenheit to degrees Celsius?

7. Write a unit conversion factor to convert

   a. feet to meters

   b. pounds to kilograms

   c. gallons to liters

8. Write a unit conversion factor to convert

   a. centimeters to inches

   b. grams to ounces

   c. liters to fluid ounces

NOTATION

Complete each solution.

9. Convert 4,500 feet to meters.

$$4{,}500 \text{ ft} \approx \frac{4{,}500\text{ft}}{1} \cdot \frac{\boxed{}}{1\text{ft}}$$

$$\approx 1{,}350 \;\boxed{}$$

10. Convert 8 liters to gallons.

$$8 \text{ L} \approx \frac{8 \text{ L}}{1} \cdot \frac{\boxed{} \text{ gal}}{1 \text{ L}}$$

$$\approx 2.112 \;\boxed{}$$

11. Convert 3 kilograms to ounces.

$$3 \text{ kg} \approx \frac{3 \text{ kg}}{1} \cdot \frac{1{,}000 \text{ g}}{1 \text{ kg}} \cdot \frac{\boxed{} \text{ oz}}{1 \text{ g}}$$

$$\approx 3 \cdot \boxed{} \cdot 0.035 \text{ oz}$$

$$\approx 105 \;\boxed{}$$

12. Convert 70°C to degrees Fahrenheit.

$$F = \frac{9}{5}C + 32$$

$$= \frac{9}{5}(\boxed{}) + 32$$

$$= \boxed{} + 32$$

$$= 158$$

Thus, 70°C = 158 $\boxed{}$

GUIDED PRACTICE

Perform each conversion. Round to the nearest inch. See Example 1.

13. 25 centimeters to inches

14. 35 centimeters to inches

15. 88 centimeters to inches

16. 91 centimeters to inches

Perform each conversion. See Example 2.

17. 8,400 feet to meters

18. 7,300 feet to meters

19. 25,115 feet to meters

20. 36,242 feet to meters

Perform each conversion. See Example 3.

21. 20 pounds to grams

22. 30 pounds to grams

23. 75 pounds to grams

24. 95 pounds to grams

Perform each conversion. See Example 4.

25. 6.5 kilograms to pounds

26. 7.5 kilograms to pounds

27. 300 kilograms to pounds

28. 800 kilograms to pounds

Perform each conversion. Round to the nearest tenth.
See Example 5.

29. 650 milliliters to quarts

30. 450 milliliters to quarts

31. 1,200 milliliters to quarts

32. 1,500 milliliters to quarts

Express each temperature in degrees Celsius. Round to the nearest tenth of a degree. See Example 6.

33. 120°F   **34.** 110°F

35. 35°F   **36.** 45°F

Express each temperature in degrees Fahrenheit. See Example 7.

37. 75°C   **38.** 85°C

39. 10°C   **40.** 20°C

TRY IT YOURSELF

Perform each conversion. If necessary, round answers to the nearest tenth. Since most conversions are approximate, answers will vary slightly depending on the method used.

41. 25 pounds to grams

42. 7.5 ounces to grams

43. 50°C to degrees Fahrenheit

44. 36.2°C to degrees Fahrenheit

45. 0.75 quarts to milliliters

46. 3 pints to milliliters

47. 0.5 kilograms to ounces

48. 35 grams to pounds

49. 3.75 meters to inches

50. 2.4 kilometers to miles

51. 3 fluid ounces to liters

52. 2.5 pints to liters

53. 12 kilometers to feet

54. 3,212 centimeters to feet

55. 37 ounces to kilograms

56. 10 pounds to kilograms

57. −10°C to degrees Fahrenheit

58. −22.5°C to degrees Fahrenheit

59. 17 grams to ounces

60. 100 kilograms to pounds

61. 7.2 liters to fluid ounces

62. 5 liters to quarts

63. 3 feet to centimeters

64. 7.5 yards to meters

65. 500 milliliters to quarts

66. 2,000 milliliters to gallons

67. 50°F to degrees Celsius

68. 67.7°F to degrees Celsius

69. 5,000 inches to meters

70. 25 miles to kilometers

71. − 5°F to degrees Celsius

72. − 10°F to degrees Celsius

APPLICATIONS

Since most conversions are approximate, answers will vary slightly depending on the method used.

73. THE MIDDLE EAST The distance between Jerusalem and Bethlehem is 8 kilometers. To the nearest mile, give this distance in miles.

74. THE DEAD SEA The Dead Sea is 80 kilometers long. To the nearest mile, give this distance in miles.

75. CHEETAHS A cheetah can run 112 kilometers per hour. Express this speed in mph. Round to the nearest mile.

76. LIONS A lion can run 50 mph. Express this speed in kilometers per hour.

77. MOUNT WASHINGTON The highest peak of the White Mountains of New Hampshire is Mount Washington, at 6,288 feet. Give this height in kilometers. Round to the nearest tenth.

78. TRACK AND FIELD Track meets are held on an oval track. One lap around the track is usually 400 meters. However, some older tracks in the United States are 440-yard ovals. Are these two types of tracks the same length? If not, which is longer?

79. HAIR GROWTH When hair is short, its rate of growth averages about $\frac{3}{4}$ inch per month. How many centimeters is this a month? Round to the nearest tenth of a centimeter.

80. WHALES An adult male killer whale can weigh as much as 12,000 pounds and be as long as 25 feet. Change these measurements to kilograms and meters.

81. WEIGHTLIFTING The table lists the personal best bench press records for two of the world's best powerlifters. Change each metric weight to pounds. Round to the nearest pound.

Name	Hometown	Bench press
Liz Willet	Ferndale, Washington	187 kg
Brian Siders	Charleston, W. Virginia	350 kg

82. WORDS OF WISDOM Refer to the wall hanging. Convert the first metric weight to ounces and the second to pounds. What famous saying results?

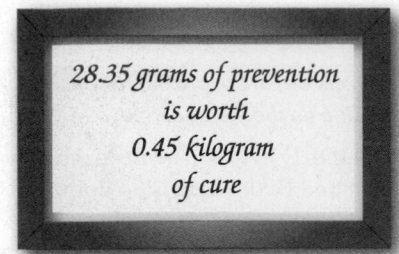

28.35 grams of prevention is worth 0.45 kilogram of cure

83. OUNCES AND FLUID OUNCES

a. There are 310 calories in 8 ounces of broiled chicken. Convert 8 ounces to grams.

b. There are 112 calories in a glass of fresh Valencia orange juice that holds 8 fluid ounces. Convert 8 fluid ounces to liters. Round to the nearest hundredth.

84. TRACK AND FIELD A shot-put weighs 7.264 kilograms. Convert this weight to pounds. Round to the nearest pound.

85. POSTAL REGULATIONS You can mail a package weighing up to 70 pounds via priority mail. Can you mail a package that weighs 32 kilograms by priority mail?

86. NUTRITION Refer to the nutrition label shown below for a packet of oatmeal. Change each circled weight to ounces.

Nutrition Facts
Serving Size: 1 Packet (46g)
Servings Per Container: 10

Amount Per Serving
Calories 170   Calories from Fat 20

	% Daily Value
Total fat 2g	3%
Saturated fat 0.5g	2%
Polyunsaturated Fat 0.5g	
Monounsaturated Fat 1g	
Cholesterol 0mg	0%
Sodium 250mg	10%
Total carbohydrate 35g	12%
Dietary fiber 3g	12%
Soluble Fiber 1g	
Sugars 16g	
Protein 4g	

87. HOT SPRINGS The thermal springs in Hot Springs National Park in central Arkansas emit water as warm as 143°F. Change this temperature to degrees Celsius.

88. COOKING MEAT Meats must be cooked at temperatures high enough to kill harmful bacteria. According to the USDA and the FDA, the internal temperature for cooked roasts and steaks should be at least 145°F, and whole poultry should be 180°F. Convert these temperatures to degrees Celsius. Round up to the next degree.

89. TAKING A SHOWER When you take a shower, which water temperature would you choose: 15°C, 28°C, or 50°C?

90. DRINKING WATER To get a cold drink of water, which temperature would you choose: −2°C, 10°C, or 25°C?

91. SNOWY WEATHER At which temperatures might it snow: −5°C, 0°C, or 10°C?

92. AIR CONDITIONING At which outside temperature would you be likely to run the air conditioner: 15°, 20°C, or 30°C?

93. COMPARISON SHOPPING Which is the better buy: 3 quarts of root beer for $4.50 or 2 liters of root beer for $3.60?

94. COMPARISON SHOPPING Which is the better buy: 3 gallons of antifreeze for $10.35 or 12 liters of antifreeze for $10.50?

WRITING

95. Explain how to change kilometers to miles.

96. Explain how to change 50°C to degrees Fahrenheit.

97. The United States is the only industrialized country in the world that does not officially use the metric system. Some people claim this is costing American businesses money. Do you think so? Why?

98. What is meant by the phrase *a table of equivalent measures*?

REVIEW

Perform each operation.

99. $\dfrac{3}{5} + \dfrac{4}{3}$

100. $\dfrac{3}{5} - \dfrac{4}{3}$

101. $\dfrac{3}{5} \cdot \dfrac{4}{3}$

102. $\dfrac{3}{5} \div \dfrac{4}{3}$

103. $3.25 + 4.8$

104. $3.25 - 4.8$

105. $3.25 \cdot 4.8$

106. $4.8\overline{)15.6}$

STUDY SKILLS CHECKLIST

Proportions and Unit Conversion Factors

Before taking the test on Chapter 6, make sure that you have a solid understanding of how to write proportions and how to choose unit conversion factors. Put a checkmark in the box if you can answer "yes" to the statement.

☐ When writing a proportion, I know that the units of the numerators must be the same and the units of the denominators must be the same.

This proportion is correctly written:

$$\text{Ounces} \longrightarrow \frac{150}{x} = \frac{3}{2.75} \longleftarrow \text{Ounces}$$
$$\text{Cost} \longrightarrow \qquad \qquad \longleftarrow \text{Cost}$$

This proportion is incorrectly written:

$$\text{Ounces} \longrightarrow \frac{50}{x} = \frac{2.75}{3} \longleftarrow \text{Cost}$$
$$\text{Cost} \longrightarrow \qquad \qquad \longleftarrow \text{Ounces}$$

☐ When converting from one unit to another, I know that I must choose a unit conversion factor with the following form:

$$\frac{\text{Unit I want to introduce}}{\text{Unit I want to eliminate}}$$

For example, in the following conversion of 15 pints to cups, the units of pints are eliminated and the units of cups are introduced by choosing the unit conversion factor $\frac{2\text{ c}}{1\text{ pt}}$.

$$15\text{ pt} = \frac{15\text{ pt}}{1} \cdot \frac{2\text{ c}}{1\text{ pt}} = 30\text{ c}$$

CHAPTER 6   SUMMARY AND REVIEW

SECTION 6.1   Ratios and Rates

DEFINITIONS AND CONCEPTS	EXAMPLES
Ratios are often used to describe important relationships between two quantities. A **ratio** is the quotient of two numbers or the quotient of two quantities that have the same units. Ratios are written in three ways: as fractions, in words separated by the word *to*, and using a colon.	The ratio 4 **to** 5 can be written as $\frac{4}{5}$. The ratio 5 **:** 12 can be written as $\frac{5}{12}$.
To **write a ratio as a fraction,** write the first number (or quantity) mentioned as the numerator and the second number (or quantity) mentioned as the denominator. Then simplify the fraction, if possible.	Write the ratio 30 to 36 as a fraction in simplest form. The word *to* separates the numbers to be compared. $\frac{30}{36} = \frac{5 \cdot \overset{1}{\cancel{6}}}{\underset{1}{\cancel{6}} \cdot 6}$ To simplify, factor 30 and 36. Then remove the common factor of 6 from the numerator and denominator. $= \frac{5}{6}$

To write a **ratio in simplest form,** remove any common factors of the numerator and denominator as well as any common units.	Write the ratio *14 feet: 2 feet* as a fraction in simplest form. A colon separates the quantities to be compared. $$\frac{14 \text{ feet}}{2 \text{ feet}} = \frac{\overset{1}{\cancel{2}} \cdot 7 \ \cancel{\text{feet}}}{\underset{1}{\cancel{2} \ \cancel{\text{feet}}}}$$ To simplify, factor 14. Then remove the common factor of 2 and the common units of feet from the numerator and denominator. $$= \frac{7}{1}$$ Since a ratio compares two numbers, we leave the result in fractional form. Do not simplify further.
To **simplify ratios involving decimals,** multiply the ratio by a form of 1 so that the numerator and denominator become whole numbers. Then simplify, if possible.	Write the ratio 0.23 to 0.71 as a fraction in simplest form. To write this as a ratio of *whole numbers,* we need to move the decimal points in the numerator and denominator two places to the right. This will occur if they are both multiplied by 100. $$\frac{0.23}{0.71} = \frac{0.23}{0.71} \cdot \frac{\mathbf{100}}{\mathbf{100}}$$ Multiply the ratio by a form of 1. $$= \frac{0.23 \cdot \mathbf{100}}{0.71 \cdot \mathbf{100}}$$ Multiply the numerators. Multiply the denominators. $$= \frac{23}{71}$$ To find the product of each decimal and 100, simply move the decimal point two places to the right. The resulting fraction is in simplest form.
To **simplify ratios involving mixed numbers,** use the method for simplifying complex fractions from Section 4.7. Perform the division indicated by the main fraction bar.	Write the ratio $3\frac{1}{3}$ to $4\frac{1}{6}$ as a fraction in simplest form. $$\frac{3\frac{1}{3}}{4\frac{1}{6}} = \frac{\frac{10}{3}}{\frac{25}{6}}$$ Write $3\frac{1}{3}$ and $4\frac{1}{6}$ and as improper fractions. $$= \frac{10}{3} \div \frac{25}{6}$$ Write the division indicated by the main fraction bar using a ÷ symbol. $$= \frac{10}{3} \cdot \frac{6}{25}$$ Use the rule for dividing fractions: Multiply the first fraction by the reciprocal of $\frac{25}{6}$, which is $\frac{6}{25}$. $$= \frac{10 \cdot 6}{3 \cdot 25}$$ Multiply the numerators. Multiply the denominators. $$= \frac{2 \cdot \overset{1}{\cancel{5}} \cdot 2 \cdot \overset{1}{\cancel{3}}}{\underset{1}{\cancel{3}} \cdot \underset{1}{\cancel{5}} \cdot 5}$$ To simplify the fraction, factor 10, 6, and 25. Then remove the common factors 3 and 5. $$= \frac{4}{5}$$ Multiply the remaining factors in the numerator. Multiply the remaining factors in the denominator.
When a ratio compares two quantities, both quantities must be measured in the **same units.** When the units are different, it's usually easier to write the ratio using the smaller unit of measurement.	Write the ratio *5 inches to 2 feet* as a fraction in simplest form. Since inches are smaller than feet, compare in inches: **5 inches** to 24 **inches** Because 2 feet = 24 inches. Next, write the ratio in fraction form and simplify. $$\frac{5 \ \cancel{\text{inches}}}{24 \ \cancel{\text{inches}}} = \frac{5}{24}$$ Remove the common units of inches.

When we compare two quantities that have different units (and neither unit can be converted to the other), we call the comparison a **rate.**

To **write a rate as a fraction,** write the first quantity mentioned as the numerator and the second quantity mentioned as the denominator, and then simplify, if possible. Write the units as part of the fraction.

Words such as *per, for, in, from,* and *on* are used to separate the two quantities that are compared in a rate.

Write the rate *33 miles in 6 hours* as a fraction in simplest form.

33 miles **in** 6 hours can be written as $\dfrac{33 \text{ miles}}{6 \text{ hours}}$

$\dfrac{33 \text{ miles}}{6 \text{ hours}} = \dfrac{\overset{1}{\cancel{3}} \cdot 11 \text{ miles}}{2 \cdot \underset{1}{\cancel{3}} \text{ hours}}$

To simplify, factor 33 and 6. Then remove the common factor of 3 from the numerator and denominator.

$= \dfrac{11 \text{ miles}}{2 \text{ hours}}$ Write the units as part of the rate.

The rate can be written as 11 miles per 2 hours.

A **unit rate** is a rate in which the denominator is 1.

To **write a rate as a unit rate,** divide the numerator of the rate by the denominator.

A **slash mark** / is often used to write a unit rate.

Write as a unit rate: 2,490 apples from 6 trees.

To find the unit rate, divide 2,490 by 6.

$\dfrac{415}{6)\overline{2{,}490}}$

The unit rate is $\frac{415 \text{ apples}}{1 \text{ tree}}$. This rate can also be expressed as: $415 \frac{\text{apples}}{\text{tree}}$, 415 apples per tree, or 415 apples/tree.

A **unit price** is a rate that tells how much is paid for *one* unit (or *one* item). It is the quotient of price to the number of units.

$$\text{Unit price} = \frac{\text{price}}{\text{number of units}}$$

Comparison shopping can be made easier by finding **unit prices.** The best buy is the item that has the lowest unit price.

Which is the better buy for shampoo?

    12 ounces for $3.84   or   16 ounces for $4.64

To find the unit price of a bottle of shampoo, write the quotient of its price and its weight, and then perform the indicated division. Before dividing, convert each price from dollars to cents so that the unit price can be expressed in cents per ounce.

$\dfrac{\$3.84}{12 \text{ oz}} = \dfrac{384¢}{12 \text{ oz}}$ $\dfrac{\$4.64}{16 \text{ oz}} = \dfrac{464¢}{16 \text{ oz}}$

$= 32¢ \text{ per oz}$ $= 29¢ \text{ per oz}$

One ounce of shampoo for 29¢ is better than one ounce for 32¢. Thus, the 16-ounce bottle is the better buy.

REVIEW EXERCISES

Write each ratio as a fraction in simplest form.

1. 7 to 25

2. 15:16

3. 24 to 36

4. 21:14

5. 4 inches to 12 inches

6. 63 meters to 72 meters

7. 0.28 to 0.35

8. 5.1:1.7

9. $2\frac{1}{3}$ to $2\frac{2}{3}$

10. $4\frac{1}{6}:3\frac{1}{3}$

11. 15 minutes : 3 hours

12. 8 ounces to 2 pounds

Write each rate as a fraction in simplest form.

13. 64 centimeters in 12 years

14. $15 for 25 minutes

Write each rate as a unit rate.

15. 600 tickets in 20 minutes

16. 45 inches every 3 turns

17. 195 feet in 6 rolls

18. 48 calories in 15 pieces

Find the unit price of each item.

19. 5 pairs cost $11.45.

20. $3 billion in a 12-month span

21. AIRCRAFT Specifications for a Boeing B-52 Stratofortress are shown below. What is the ratio of the airplane's wingspan to its length?

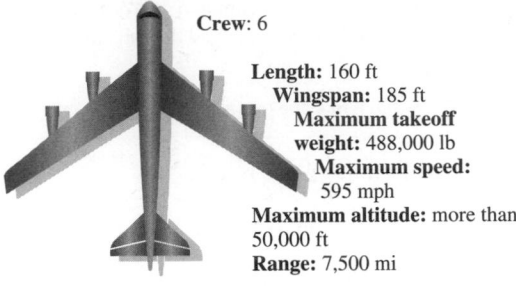

Crew: 6

Length: 160 ft
Wingspan: 185 ft
Maximum takeoff weight: 488,000 lb
Maximum speed: 595 mph
Maximum altitude: more than 50,000 ft
Range: 7,500 mi

22. PAY RATES Find the hourly rate of pay for a student who earned $333.25 for working 43 hours.

23. CROWD CONTROL After a concert is over, it takes 48 minutes for a crowd of 54,000 people to exit a stadium. Find the unit rate of people exiting the stadium.

24. COMPARISON SHOPPING Mixed nuts come packaged in a 12-ounce can, which sells for $4.95, or an 8-ounce can, which sells for $3.25. Which is the better buy?

SECTION 6.2   Proportions

DEFINITIONS AND CONCEPTS	EXAMPLES
A **proportion** is a statement that two ratios or two rates are equal.	Write each statement as a proportion. $\underbrace{6 \text{ is to } 10}$ **as** $\underbrace{3 \text{ is to } 5}$ The word *"to"* is used to separate the numbers $$\frac{6}{10} = \frac{3}{5}$$ to be compared in a ratio (or rate). $\underbrace{\$300 \text{ is to } 500 \text{ minutes}}$ **as** $\underbrace{\$3 \text{ is to } 5 \text{ minutes}}$ $$\frac{\$300}{500 \text{ minutes}} = \frac{\$3}{5 \text{ minutes}}$$
Each of the four numbers in a proportion is called a **term.** The first and fourth terms are called the **extremes,** and the second and third terms are called the **means.**	First term (extreme) Third term (mean) $$\frac{1}{2} = \frac{3}{6}$$ Second term (mean) Fourth term (extreme)
Since a proportion is an equation, **a proportion can be true or false**. A proportion is true if its ratios (or rates) are equivalent and false if its ratios (or rates) are not equivalent. One way to determine whether a proportion is true or false is to use the fraction simplifying skills of Chapter 3. The two products found by multiplying diagonally in a proportion are called **cross products.** Another way to determine whether a proportion is true or false involves the cross products. If the **cross products are equal,** the proportion is true. If the cross products are *not equal,* the proportion is false.	Determine whether the proportion $\dfrac{3}{5} = \dfrac{15}{27}$ is true or false. ***Method 1*** Simplify any ratios in the proportion that are not in simplest form. Then compare them to determine whether they are equal. $$\frac{15}{27} = \frac{\overset{1}{\cancel{3}} \cdot 5}{\underset{1}{\cancel{3}} \cdot 9} = \frac{5}{9}$$ Simplify the ratio on the right side. Since the ratios on the left and right sides of the proportion are not equal, the proportion is false. ***Method 2*** Check to see whether the cross products are equal. Cross products $$3 \cdot 27 = 81 \qquad\qquad 5 \cdot 15 = 75$$ $$\frac{3}{5} = \frac{15}{27}$$ Since the cross products are not equal, the proportion is not true.

When two pairs of numbers form a proportion, we say that they are **proportional.**	Determine whether 0.7, 0.3 and 2.1, 0.9 are proportional. Write two ratios and form a proportion. Then find the cross products. $$\frac{0.7}{0.3} = \frac{2.1}{0.9} \qquad 0.7 \cdot 0.9 = \mathbf{0.63} \qquad 0.3 \cdot 2.1 = \mathbf{0.63}$$ Since the cross products are equal, the numbers are proportional.

Solving a proportion to find an unknown term:

1. Set the cross products equal to each other to form an equation.

2. Isolate the variable on one side of the equation by dividing both sides by the number that is multiplied by that variable.

3. Check by substituting the result into the original proportion and finding the cross products.

Solve the proportion: $\dfrac{5}{37.5} = \dfrac{2}{x}$

$\dfrac{5}{37.5} = \dfrac{2}{x}$	This is the proportion to solve.
$5 \cdot x = 37.5 \cdot 2$	Set the cross products equal to each other to form an equation.
$5x = 75$	On right side, do the multiplication: $37.5 \cdot 2 = 75$.
$\dfrac{5x}{\mathbf{5}} = \dfrac{75}{\mathbf{5}}$	To isolate x, undo the multiplication by 5 by dividing both sides by 5.
$x = 15$	Do the division: $75 \div 5 = 15$.

Thus, x is 15. Check this result in the original proportion by finding the cross products.

Proportions can be used to **solve application problems.** It is easy to spot problems that can be solved using a proportion. You will be given a ratio (or rate) and asked to find the missing part of another ratio (or rate).

It is helpful to follow the **five-step problem-solving strategy** seen earlier in the text to solve proportion problems.

PEANUT BUTTER It takes 360 peanuts to make 8 ounces of peanut butter. How many peanuts does it take to make 12 ounces? (Source: National Peanut Board)

Analyze

- We can express the fact that it takes 360 peanuts to make 8 ounces of peanut butter as a rate: $\frac{360 \text{ peanuts}}{8 \text{ ounces}}$.
- How many peanuts does it take to make 12 ounces?

Form We will let the variable p represent the unknown number of peanuts.

360 peanuts is **to** 8 ounces **as** p peanuts is **to** 12 ounces.

Number of peanuts ⟶ $\dfrac{360}{8} = \dfrac{p}{12}$ ⟵ Number of peanuts
Ounces of peanuts ⟶   ⟵ Ounces of peanuts

Solve To find the number of peanuts needed, solve the proportion for p.

$360 \cdot 12 = 8 \cdot p$	Set the cross products equal to each other to form an equation.
$4{,}320 = 8p$	On the left side, do the multiplication: $360 \cdot 12 = 4{,}320$.
$\dfrac{4{,}320}{\mathbf{8}} = \dfrac{8p}{\mathbf{8}}$	To isolate p, undo the multiplication by 8 by dividing both sides by 8.
$540 = p$	Do the division: $4{,}320 \div 8 = 540$.

State It takes 540 peanuts to make 12 ounces of peanut butter.

Check 16 ounces of peanut butter would require twice as many peanuts as 8 ounces: $2 \cdot 360$ peanuts = 720 peanuts. It seems reasonable that 12 ounces would require 540 peanuts.

REVIEW EXERCISES

25. Write each statement as a proportion.

   a. 20 is to 30 as 2 is to 3.

   b. 6 buses replace 100 cars as 36 buses replace 600 cars.

26. Complete the cross products.

$$\blacksquare \cdot 27 = \blacksquare \qquad 9 \cdot \blacksquare = \blacksquare$$

$$\frac{2}{9} \diagdown\!\!\!\!\diagup \frac{6}{27}$$

Determine whether each proportion is true or false by simplifying.

27. $\dfrac{8}{12} = \dfrac{3}{7}$ **28.** $\dfrac{4}{18} = \dfrac{10}{45}$

Determine whether each proportion is true or false by finding cross products.

29. $\dfrac{9}{27} = \dfrac{2}{6}$ **30.** $\dfrac{17}{7} = \dfrac{51}{21}$

31. $\dfrac{3.5}{9.3} = \dfrac{1.2}{3}$ **32.** $\dfrac{1\frac{1}{2}}{3\frac{1}{3}} = \dfrac{\frac{1}{4}}{1\frac{1}{7}}$

Determine whether the numbers are proportional.

33. 5, 9 and 20, 36   **34.** 7, 13 and 29, 54

Solve each proportion.

35. $\dfrac{12}{18} = \dfrac{3}{x}$ **36.** $\dfrac{4}{x} = \dfrac{2}{8}$

37. $\dfrac{4.8}{6.6} = \dfrac{x}{9.9}$ **38.** $\dfrac{0.08}{x} = \dfrac{0.04}{0.06}$

39. $\dfrac{1\frac{9}{11}}{x} = \dfrac{3\frac{1}{3}}{2\frac{3}{4}}$ **40.** $\dfrac{\frac{4}{5}}{1\frac{1}{20}} = \dfrac{2\frac{2}{3}}{x}$

41. $\dfrac{\frac{2}{3}}{\frac{1}{2}} = \dfrac{x}{0.25}$ **42.** $\dfrac{x}{300} = \dfrac{5,000}{1,500}$

43. TRUCKS A Dodge Ram pickup truck can go 35 miles on 2 gallons of gas. How far can it go on 11 gallons?

44. QUALITY CONTROL In a manufacturing process, 12 parts out of 66 were found to be defective. How many defective parts will be expected in a run of 1,650 parts?

45. SCALE DRAWINGS The illustration below shows an architect's drawing of a kitchen using a scale of $\frac{1}{8}$ inch to 1 foot $\left(\frac{1}{8}'' : 1'0''\right)$. On the drawing, the length of the kitchen is $1\frac{1}{2}$ inches. How long is the actual kitchen? (The symbol ″ means inch and ′ means foot.)

ELEVATION B-B
SCALE: $\frac{1''}{8}$ to 1'0"

46. DOGS The American Kennel Club website gives the ideal *length to height proportions* for a German Shepherd as $10 : 8\frac{1}{2}$. What is the ideal length of a German Shepherd that is $25\frac{1}{2}$ inches high at the shoulder?

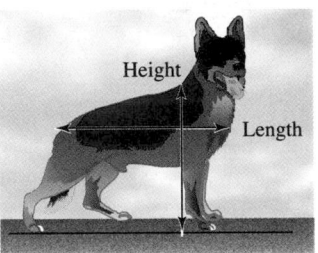

Height

Length

SECTION **6.3**   **American Units of Measurement**

DEFINITIONS AND CONCEPTS	EXAMPLES
The **American system of measurement** uses the units of **inch, foot, yard,** and **mile** to measure **length.** A **ruler** is one of the most common tools for measuring lengths. Most rulers are 12 inches long. Each inch is divided into halves of an inch, quarters of an inch, eighths of an inch, and sixteenths of an inch.	$1 \text{ ft} = 12 \text{ in.}$ $\qquad\qquad 1 \text{ yd} = 3 \text{ ft}$ $1 \text{ yd} = 36 \text{ in.}$ $\qquad\quad 1 \text{ mi} = 5,280 \text{ ft}$ Since the black tick marks between 0 and 1 on the ruler create sixteen equal spaces, the ruler is scaled in sixteenths. 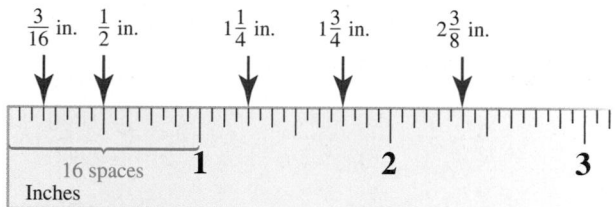

To convert from one unit of length to another, we use **unit conversion factors.** They are called unit conversion factors because their value is 1.

Multiplying a measurement by a unit conversion factor does not change the measure; it only changes the units of the measure.

A list of unit conversion factors for American units of length is given on page 588.

Convert 4 yards to inches.

To convert from yards to inches, we select a unit conversion factor that introduces the units of inches and eliminates the units of yards. Since there are 36 inches per yard, we will use:

$\dfrac{36 \text{ in.}}{1 \text{ yd}}$ ← This is the unit we want to introduce.
← This is the unit we want to eliminate (the original unit).

To perform the conversion, we multiply.

$4 \text{ yd} = \dfrac{4 \text{ yd}}{1} \cdot \dfrac{36 \text{ in.}}{1 \text{ yd}}$ Write 4 yd as a fraction. Then multiply by a form of 1: $\frac{36 \text{ in.}}{1 \text{ yd}}$.

$= \dfrac{4\ \cancel{\text{yd}}}{1} \cdot \dfrac{36 \text{ in.}}{1\ \cancel{\text{yd}}}$ Remove the common units of yards from the numerator and denominator. The units of inches remain.

$= 4 \cdot 36 \text{ in.}$ Simplify.

$= 144 \text{ in.}$ Do the multiplication.

Thus, 4 yards = 144 inches.

The American system of measurement uses the units of **ounce, pound,** and **ton** to measure **weight.**

$1 \text{ lb} = 16 \text{ oz} \qquad\qquad 1 \text{ ton} = 2,000 \text{ lb}$

A list of unit conversion factors for American units of weight is given on page 591.

Convert 9,000 pounds to tons.

To convert from pounds to tons, we select a unit conversion factor that introduces the units of tons and eliminates the units of pounds. Since there is 1 ton for every 2,000 pounds, we will use:

$\dfrac{1 \text{ ton}}{2,000 \text{ lb}}$ ← This is the unit we want to introduce.
← This is the unit we want to eliminate (the original unit).

To perform the conversion, we multiply.

$$9,000 \text{ lb} = \frac{9,000 \text{ lb}}{1} \cdot \frac{1 \text{ ton}}{2,000 \text{ lb}}$$

Write 9,000 lb as a fraction. Then multiply by a form of 1: $\frac{1 \text{ ton}}{2,000 \text{ lb}}$.

$$= \frac{9,000 \overset{1}{\cancel{\text{lb}}}}{1} \cdot \frac{1 \text{ ton}}{2,000 \underset{1}{\cancel{\text{lb}}}}$$

Remove the common units of pounds from the numerator and denominator. The units of tons remains.

$$= \frac{9,000}{2,000} \text{ ton}$$

Multiply the fractions.

There are two ways to complete the solution. First, we can remove any common factors of the numerator and denominator to simplify the fraction. Then we can write the result as a mixed number.

$$\frac{9,000}{2,000} \text{ tons} = \frac{9 \cdot \overset{1}{\cancel{1,000}}}{2 \cdot \underset{1}{\cancel{1,000}}} \text{ tons} = \frac{9}{2} \text{ tons} = 4\frac{1}{2} \text{ tons}$$

A second approach is to divide the numerator by the denominator and express the result as a decimal.

$$\frac{9,000}{2,000} \text{ tons} = 4.5 \text{ tons}$$

Thus, 9,000 pounds is equal to $4\frac{1}{2}$ tons (or 4.5 tons).

The American system of measurement uses the units of **ounce, cup, pint, quart,** and **gallon** to measure **capacity.**	1 c = 8 fl oz 1 pt = 2 c 1 qt = 2 pt 1 gal = 4 qt

A list of unit conversion factors for American units of capacity is given on page 593.

Some conversions require the use of **two** (or more) **unit conversion factors.**

Convert 5 gallons to pints.

There is not a single unit conversion factor that converts from gallons to pints. We must use two unit conversion factors.

Since there are 4 quarts per gallon, we can convert 5 gallons to quarts by multiplying by the unit conversion factor $\frac{4 \text{ qt}}{1 \text{ gal}}$. Since there are 2 pints per quart, we can convert that result to pints by multiplying by the unit conversion factor $\frac{2 \text{ pt}}{1 \text{ qt}}$.

$$5 \text{ gal} = \frac{5 \text{ gal}}{1} \cdot \frac{4 \text{ qt}}{1 \text{ gal}} \cdot \frac{2 \text{ pt}}{1 \text{ qt}}$$

$$= \frac{5 \cancel{\text{ gal}}}{1} \cdot \frac{4 \cancel{\text{ qt}}}{1 \cancel{\text{ gal}}} \cdot \frac{2 \text{ pt}}{1 \cancel{\text{ qt}}}$$

Remove the common units of gallons and quarts in the numerator and denominator. The units of pints remain.

$$= 40 \text{ pt}$$

Do the multiplication: $5 \cdot 4 \cdot 2 = 40$.

Thus, 5 gallons = 40 pints.

The American (and metric) system of measurement use the units of **seconds, minutes, hours,** and **days** to measure time.	1 min = 60 sec 1 hr = 60 min 1 day = 24 hr

A list of unit conversion factors for units of time is given on page 594.

Convert 240 minutes to hours.

To convert from minutes to hours, we select a unit conversion factor that introduces the units of hours and eliminates the units of minutes. Since there is 1 hour for every 60 minutes, we will use:

$$\frac{1\ hr}{60\ min}$$
← This is the unit we want to introduce.
← This is the unit we want to eliminate (the original unit).

To perform the conversion, we multiply.

$$240\ min = \frac{240\ min}{1} \cdot \frac{1\ hr}{60\ min}$$

Write 240 min as a fraction. Then multiply by a form of 1: $\frac{1\ hr}{60\ min}$.

$$= \frac{240\ \cancel{min}}{1} \cdot \frac{1\ hr}{60\ \cancel{min}}$$

Remove the common units of minutes from the numerator and denominator. The units of hours remain.

$$= \frac{240}{60}\ hr$$

Multiply the fractions.

$$= 4\ hr$$

Do the division.

Thus, 240 minutes = 4 hours.

REVIEW EXERCISES

47. a. Refer to the ruler below. Each inch is divided into how many equal parts?

  b. Determine which measurements the arrows point to on the ruler.

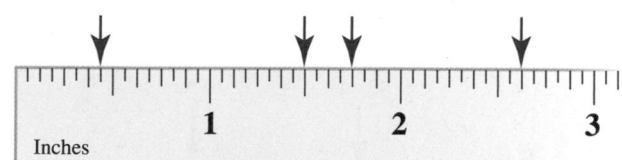

48. Use a ruler to measure the length of the computer mouse.

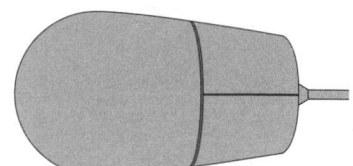

49. Write two unit conversion factors using the fact that 1 mile = 5,280 ft.

50. Consider the work shown below.

$$\frac{100\ min}{1} \cdot \frac{60\ sec}{1\ min}$$

  a. What units can be removed?

  b. What units remain?

Perform each conversion.

51. 5 yards to feet

52. 6 yards to inches

53. 66 inches to feet

54. 9,240 feet to miles

55. $4\frac{1}{2}$ feet to inches

56. 1 mile to yards

57. 32 ounces to pounds

58. 17.2 pounds to ounces

59. 3 tons to ounces

60. 4,500 pounds to tons

61. 5 pints to fluid ounces

62. 8 cups to gallons

63. 17 quarts to cups

64. 176 fluid ounces to quarts

65. 5 gallons to pints

66. 3.5 gallons to cups

67. 20 minutes to seconds

68. 900 seconds to minutes

69. 200 hours to days

70. 6 hours to minutes

71. 4.5 days to hours

72. 1 day to seconds

73. Convert 210 yards to miles. Give the exact answer and a decimal approximation, rounded to the nearest hundredth.

74. TRUCKING Large concrete trucks can carry roughly 40,500 pounds of concrete. Express this weight in tons.

Image copyright Elemental Imaging 2009. Used under license from Shutterstock.com

75. SKYSCRAPERS The Sears Tower in Chicago is 1,454 feet high. Express this height in yards.

76. BOTTLING A magnum is a 2-quart bottle of wine. How many magnums will be needed to hold 50 gallons of wine?

SECTION 6.4 Metric Units of Measurement

DEFINITIONS AND CONCEPTS	EXAMPLES
The basic metric unit of measurement is the **meter,** which is abbreviated **m.** Longer and shorter metric units are created by adding **prefixes** to the front of the basic unit, meter.	*kilo* means thousands *deci* means tenths *hecto* means hundreds *centi* means hundredths *deka* means tens *milli* means thousandths
Common metric units of length are the **kilometer, hectometer, dekameter, decimeter, centimeter,** and **millimeter.** Abbreviations are often used when writing these units. See the table on page 600.	1 km = 1,000 m 1 m = 10 dm 1 hm = 100 m 1 m = 100 cm 1 dam = 10 m 1 m = 1,000 mm
A **metric ruler** can be used for measuring lengths. On most metric rulers, each centimeter is divided into 10 millimeters.	2 cm 43 mm 65 mm 10 mm **1** **2** **3** **4** **5** **6** **7** Centimeters
To convert from one metric unit of length to another, we use **unit conversion factors.**	Convert 4 meters to centimeters. To convert from meters to centimeters, we select a unit conversion factor that introduces the units of centimeters and eliminates the units of meters. Since there are 100 centimeters per meter, we will use: $\dfrac{100 \text{ cm}}{1 \text{ m}}$ ← This is the unit we want to introduce. ← This is the unit we want to eliminate (the original unit). To perform the conversion, we multiply. $4 \text{ m} = \dfrac{4 \text{ m}}{1} \cdot \dfrac{\mathbf{100 \text{ cm}}}{\mathbf{1 \text{ m}}}$ Write 4 m as a fraction. Then multiply by a form of 1: $\frac{100 \text{ cm}}{1 \text{ m}}$. $= \dfrac{4 \text{ m}}{1} \cdot \dfrac{100 \text{ cm}}{1 \text{ m}}$ Remove the common units of meters from the numerator and denominator. The units of cm remain. $= 400 \text{ cm}$ Multiply the fractions and simplify. Thus, 4 meters = 400 centimeters.

The **mass** of an object is a measure of the amount of material in the object. Common metric units of mass are the **kilogram, hectogram, dekagram, decigram, centigram** and **milligram.** Abbreviations are often used when writing these units. See the table on page 605.	1 kg = 1,000 g 1 g = 10 dg 1 hg = 100 g 1 g = 100 cg 1 dag = 10 g 1 g = 1,000 mg

Converting from one metric unit to another can be done using **unit conversion factors or** a **conversion chart.**

In a conversion chart, the units are listed from largest to smallest, reading left to right. We **count the places** and note the **direction** as we move from the original units to the conversion units.

Convert 820 grams to kilograms.

To use a conversion chart, locate the original units of grams and move to the conversion units of kilograms.

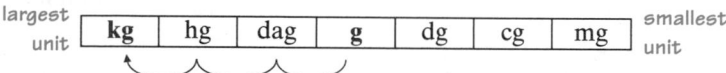

To go from grams to kilograms, we must move 3 places to the left.

If we write 820 grams as 820.0 grams, we can convert to kilograms by moving the decimal point 3 places to the left.

820.0 grams = 0.820 0 kilograms = 0.82 kilograms

The unit conversion factor method gives the same result:

$$820\ g = \frac{820\ g}{1} \cdot \frac{1\ kg}{1{,}000\ g}$$
$$= \frac{820}{1{,}000}kg$$
$$= 0.82\ kg$$

Thus, 820 grams = 0.82 kilograms.

Common metric units of capacity are the **kiloliter, hectoliter, dekaliter, deciliter, centiliter** and **milliliter.** Abbreviations are often used when writing these units. See the table on page 608.	1 kL = 1,000 L 1 L = 10 dL 1 hL = 100 L 1 L = 100 cL 1 daL = 10 L 1 L = 1,000 mL

Converting from one metric unit to another can be done using **unit conversion factors** or a **conversion chart.**

Convert 0.7 kiloliters to milliliters.

To use a conversion chart, locate the original units of kiloliters and move to the conversion units of milliliters.

To go from kiloliters to milliliters, we must move 6 places to the right.

We can convert to milliliters by moving the decimal point 6 places to the right.

0.7 kiloliters = 0 700000. milliliters = 700,000 milliliters

Another metric unit of capacity is the **cubic centimeter,** written cm³, or, more simply, cc.

The unit conversion factor method gives the same result:

$$0.7 \text{ kL} = \frac{0.7 \text{ k\cancel{L}}}{1} \cdot \frac{\mathbf{1,000} \text{ \cancel{L}}}{\mathbf{1} \text{ k\cancel{L}}} \cdot \frac{\mathbf{1,000} \text{ mL}}{1 \text{ \cancel{L}}}$$

$$= 0.7 \cdot 1,000 \cdot 1,000 \text{ mL}$$

$$= 700,000 \text{ mL}$$

Thus, 0.7 kiloliters = 700,000 milliliters.

The units of cubic centimeters are used frequently in medicine.	1 milliliter = 1 cm³ = 1 cc 5 milliliters = 5 cm³ = 5 cc 0.6 milliliters = 0.6 cm³ = 0.6 cc

REVIEW EXERCISES

77. a. Refer to the metric ruler below. Each centimeter is divided into how many equal parts? What is the length of one of those parts?

   b. Determine which measurements the arrows point to on the ruler.

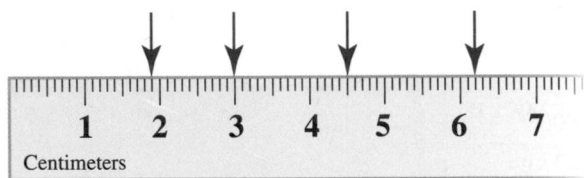

Centimeters

78. Use a metric ruler to measure the length of the computer mouse to the nearest centimeter.

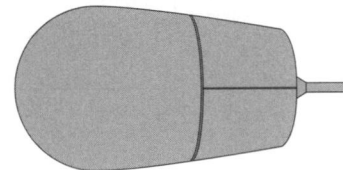

79. Write two unit conversion factors using the given fact.

   a. 1 km = 1,000 m

   b. 1 g = 100 cg

80. Use the chart to determine how many decimal places and in which direction to move the decimal point when converting from centimeters to kilometers.

km	hm	dam	m	dm	cm	mm

Perform each conversion.

81. 475 centimeters to meters

82. 8 meters to millimeters

83. 165.7 kilometers to meters

84. 6,789 centimeters to decimeters

85. 5,000 centigrams to kilograms

86. 800 centigrams to grams

87. 5,425 grams to kilograms

88. 5,425 grams to milligrams

89. 150 centiliters to liters

90. 3,250 liters to kiloliters

91. 400 milliliters to centiliters

92. 1 hectoliter to deciliters

93. THE BRAIN The adult human brain weighs about 1,350 g. Convert the weight to kilograms.

94. TEST TUBES A rack holds *one dozen* 20-mL test tubes. Find the total capacity of the test tubes in the rack in liters.

95. TYLENOL A bottle of Extra-Strength Tylenol contains 100 caplets of 500 milligrams each. How many grams of Tylenol are in the bottle?

96. SURGERY A dextrose solution is being administered to a patient intravenously as shown to the right. How many milliliters of solution does the IV bag hold?

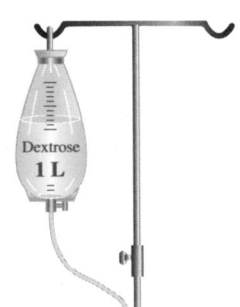

Dextrose
1 L

SECTION 6.5 Converting between American and Metric Units

DEFINITIONS AND CONCEPTS	EXAMPLES
We **convert between American and metric units** of length using the facts on the right. In all but one case, the conversions are rounded approximations.	**American to metric** **Metric to American** 1 in. = 2.54 cm 1 cm ≈ 0.39 in. 1 ft ≈ 0.30 m 1 m ≈ 3.28 ft 1 yd ≈ 0.91 m 1 m ≈ 1.09 yd 1 mi ≈ 1.61 km 1 km ≈ 0.62 mi

Unit conversion factors can be formed from the facts in the tables on the right to make specific conversions between American and metric units of length.	Convert 15 inches to centimeters. To convert from inches to centimeters, we select a unit conversion factor that introduces the units of centimeters and eliminates the units of inches. Since there are 2.54 centimeters for every inch, we will use: $$\frac{2.54 \text{ cm}}{1 \text{ in.}}$$ ← This is the unit we want to introduce. ← This is the unit we want to eliminate (the original unit). To perform the conversion, we multiply. $15 \text{ in.} = \dfrac{15 \text{ in.}}{1} \cdot \dfrac{\mathbf{2.54 \text{ cm}}}{\mathbf{1 \text{ in.}}}$ Write 15 in. as a fraction. Then multiply by a form of 1: $\frac{2.54 \text{ cm}}{1 \text{ in.}}$. $= \dfrac{15 \text{ in.}}{1} \cdot \dfrac{2.54 \text{ cm}}{1 \text{ in.}}$ Remove the common units of inches from the numerator and denominator. The units of cm remain. $= 15 \cdot 2.54 \text{ cm}$ Simplify. $= 38.1 \text{ cm}$ Do the multiplication. Thus, 15 inches = 38.1 centimeters.

We **convert between American and metric units** of mass (weight) using the facts on the right. The conversions are rounded approximations.	**American to metric** **Metric to American** 1 oz ≈ 28.35 g 1 g ≈ 0.035 oz 1 lb ≈ 0.45 kg 1 kg ≈ 2.20 lb

Unit conversion factors can be formed from the facts in the tables on the right to make specific conversions between American and metric units of mass (weight).	Convert 6 kilograms to ounces. There is not a single unit conversion factor that converts from kilograms to ounces. We must use **two unit conversion factors.** One to convert kilograms to grams, and another to convert that result to ounces. $6 \text{ kg} \approx \dfrac{6 \text{ kg}}{1} \cdot \dfrac{\mathbf{1{,}000 \text{ g}}}{\mathbf{1 \text{ kg}}} \cdot \dfrac{\mathbf{0.035 \text{ oz}}}{\mathbf{1 \text{ g}}}$ $\approx \dfrac{6 \text{ kg}}{1} \cdot \dfrac{1{,}000 \text{ g}}{1 \text{ kg}} \cdot \dfrac{0.035 \text{ oz}}{1 \text{ g}}$ Remove the common units of kilograms and grams in the numerator and denominator. The units of oz remain. $\approx 6 \cdot 1{,}000 \cdot 0.035 \text{ oz}$ Simplify. $\approx 6 \cdot 35 \text{ oz}$ Multiply the last two factors: $1{,}000 \cdot 0.035 = 35$. $\approx 210 \text{ oz}$ Do the multiplication. Thus, 6 kilograms ≈ 210 ounces.

We **convert between American and metric units** of capacity using the facts on the right. The conversions are rounded approximations.

American to metric	Metric to American
1 fl oz ≈ 29.57 mL	1 L ≈ 33.81 fl oz
1 pt ≈ 0.47 L	1 L ≈ 2.11 pt
1 qt ≈ 0.95 L	1 L ≈ 1.06 qt
1 gal ≈ 3.79 L	1 L ≈ 0.264 gal

Unit conversion factors can be formed from the facts in the tables on the right to make specific conversions between American and metric units of capacity.

Convert 5 fluid ounces to milliliters. Round to the nearest tenth.

To convert from fluid ounces to milliliters, we select a unit conversion factor that introduces the units of milliliters and eliminates the units of fluid ounces. Since there are 29.57 milliliters for every fluid ounce, we will use:

$$\frac{29.57 \text{ mL}}{1 \text{ fl oz}}$$

 ← This is the unit we want to introduce.
 ← This is the unit we want to eliminate (the original unit).

To perform the conversion, we multiply.

$$5 \text{ fl oz} \approx \frac{5 \text{ fl oz}}{1} \cdot \frac{29.57 \text{ mL}}{1 \text{ fl oz}}$$ Write 5 fl oz as a fraction. Then multiply by a form of 1: $\frac{29.57 \text{ mL}}{1 \text{ fl oz}}$.

$$\approx \frac{5 \cancel{\text{ fl oz}}}{1} \cdot \frac{29.57 \text{ mL}}{1 \cancel{\text{ fl oz}}}$$ Remove the common units of fluid ounces from the numerator and denominator. The units of mL remain.

$$\approx 5 \cdot 29.57 \text{ mL}$$ Simplify.

$$\approx 147.85 \text{ mL}$$ Do the multiplication.

$$\approx 147.9 \text{ mL}$$ Round to the nearest tenth.

Thus, 5 fluid ounces ≈ 147.9 milliliters.

In the American system, we measure temperature using **degrees Fahrenheit** (°F). In the metric system, we measure temperature using **degrees Celsius** (°C).

If F is the temperature in degrees Fahrenheit and C is the corresponding temperature in degrees Celsius, then

$$C = \frac{5}{9}(F - 32) \quad \text{and} \quad F = \frac{9}{5}C + 32$$

Convert 92°F to degrees Celsius. Round to the nearest tenth of a degree.

$$C = \frac{5}{9}(F - 32)$$ This is the formula to find degrees Celsius.

$$= \frac{5}{9}(92 - 32)$$ Substitute 92 for F.

$$= \frac{5}{9}(60)$$ Do the subtraction within the parentheses first.

$$= \frac{5}{9}\left(\frac{60}{1}\right)$$ Write 60 as a fraction.

$$= \frac{300}{9}$$ Multiply the numerators: 5 · 60 = 300. Multiply the denominators.

$$= 33.333\ldots$$ Do the division.

$$\approx 33.3$$ Round to the nearest tenth.

Thus, 92°F ≈ 33.3°C.

REVIEW EXERCISES

97. SWIMMING Olympic-size swimming pools are 50 meters long. Express this distance in feet.

98. HIGH-RISE BUILDINGS The Sears Tower is 443 meters high, and the Empire State Building is 1,250 feet high. Which building is taller?

99. WESTERN SETTLERS The Oregon Trail was an overland route pioneers used from the 1840s through the 1870s to reach the Oregon Territory. It stretched 1,930 miles from Independence, Missouri, to Oregon City, Oregon. Find this distance to the nearest kilometer.

100. AIR JORDAN Michael Jordan is 78 inches tall (6 feet, 6 inches). Express his height in centimeters. Round to the nearest centimeter.

Perform each conversion. Since most conversions are approximate, answers will vary slightly depending on the method used.

101. 30 ounces to grams

102. 15 kilograms to pounds

103. 50 pounds to grams

104. 2,000 pounds to kilograms

105. POLAR BEARS At birth, polar bear cubs weigh less than human babies—about 910 grams. Convert this to pounds.

106. BOTTLED WATER LaCroix bottled water can be purchased in bottles containing 17 fluid ounces. Mountain Valley water can be purchased in half-liter bottles. Which bottle contains more water?

107. CRUDE OIL There are 42 gallons in a barrel of crude oil. How many liters of crude oil is that?

108. Convert 105°C to degrees Fahrenheit.

109. Convert 77°F to degrees Celsius.

110. RECREATION Which water temperature is appropriate for swimming: 10°C, 30°C, 50°C, or 70°C?

CHAPTER 6   TEST

1. Fill in the blanks.

   a. A _____ is the quotient of two numbers or the quotient of two quantities that have the same units.

   b. A _____ is the quotient of two quantities that have different units.

   c. A _____ is a statement that two ratios (or rates) are equal.

   d. The _____ products for the proportion $\frac{3}{8} = \frac{6}{16}$ are $3 \cdot 16$ and $8 \cdot 6$.

   e. *Deci* means _____, *centi* means _____, and *milli* means _____.

   f. The meter, the gram, and the liter are basic units of measurement in the _____ system.

   g. In the American system, temperatures are measured in degrees _____. In the metric system, temperatures are measured in degrees _____.

2. PIANOS A piano keyboard is made up of a total of eighty-eight keys, as shown below. What is the ratio of the number of black keys to white keys?

Middle C

Write each ratio as a fraction in simplest form.

3. 6 feet to 8 feet

4. 8 ounces to 3 pounds

5. 0.26 : 0.65

6. $3\frac{1}{3}$ to $3\frac{8}{9}$

7. Write the rate 54 feet in 36 seconds as a fraction in simplest form.

8. COMPARISON SHOPPING A 2-pound can of coffee sells for $3.38, and a 5-pound can of the same brand of coffee sells for $8.50. Which is the better buy?

9. UTILITY COSTS A household used 675 kilowatt-hours of electricity during a 30-day month. Find the rate of electric usage in kilowatt-hours per day.

10. Write the following statement as a proportion: 15 billboards to 50 miles as 3 billboards to 10 miles.

11. Determine whether each proportion is true.

    a. $\frac{25}{33} = \frac{2}{3}$ b. $\frac{2.2}{3.5} = \frac{1.76}{2.8}$

12. Are the numbers 7, 15 and 35, 75 proportional?

Solve each proportion.

13. $\frac{x}{3} = \frac{35}{7}$ 14. $\frac{15.3}{x} = \frac{3}{12.4}$

15. $\frac{2\frac{2}{9}}{\frac{4}{3}} = \frac{x}{1\frac{1}{2}}$ 16. $\frac{25}{\frac{1}{10}} = \frac{50}{x}$

17. SHOPPING If 13 ounces of tea costs $2.79, how much would you expect to pay for 16 ounces of tea?

18. BAKING A recipe calls for $1\frac{2}{3}$ cup of sugar and 5 cups of flour. How much sugar should be used with 6 cups of flour?

19. a. Refer to the ruler below. Each inch is divided into how many equal parts?

    b. Determine which measurements the arrows point to on the ruler.

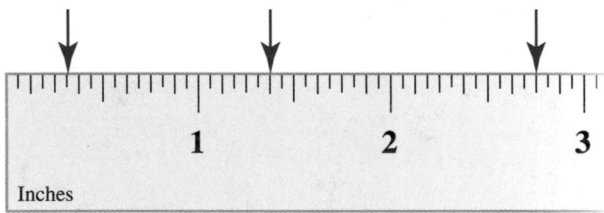

20. Fill in the blanks. In general, a unit conversion factor is a fraction with the following form:

Unit that we want to _____	← Numerator
Unit that we want to _____	← Denominator

21. Convert 180 inches to feet.

22. TOOLS If a 25-foot tape measure is completely extended, how many yards does it stretch? Write your answer as a mixed number.

23. Convert $10\frac{3}{4}$ pounds to ounces.

24. AUTOMOBILES A car weighs 1.6 tons. Find its weight in pounds.

25. CONTAINERS How many fluid ounces are in a 1-gallon carton of milk?

26. LITERATURE An excellent work of early science fiction is the book *Around the World in 80 Days* by Jules Verne (1828–1905). Convert 80 days to minutes.

27. a. A quart and a liter of fruit punch are shown below. Which is the 1-liter carton: The one on the left side or the right side?

b. The figures below show the relative lengths of a yardstick and a meterstick. Which one represents the meterstick: the longer one or the shorter one?

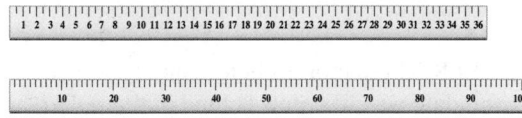

c. One ounce and one gram weights are placed on a balance, as shown below. On which side is the gram: the left side or the right side?

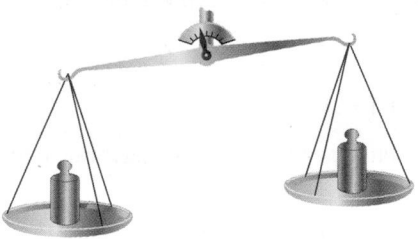

28. Determine which measurements the arrows point to on the metric ruler shown below.

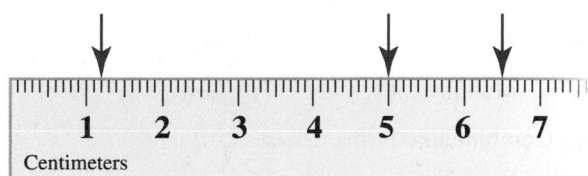

29. SPEED SKATING American Bonnie Blair won gold medals in the women's 500-meter speed skating competitions at the 1988, 1992, and 1994 Winter Olympic Games. Convert the race length to kilometers.

30. How many centimeters are in 5 meters?

31. Convert 8,000 centigrams to kilograms.

32. Convert 70 liters to milliliters.

33. PRESCRIPTIONS A bottle contains 50 tablets, each containing 150 mg of medicine. How many grams of medicine does the bottle contain?

34. TRACK Which is the longer distance: a 100-yard race or an 80-meter race?

35. BODY WEIGHT Which person is heavier: Jim, who weighs 160 pounds, or Ricardo, who weighs 71 kilograms?

36. Convert 810 milliliters to quarts. Round to the nearest tenth.

37. Convert 16.5 inches to centimeters. Round to the nearest centimeter.

38. COOKING MEAT The USDA recommends that turkey be cooked to a temperature of 83°C. Change this to degrees Fahrenheit. To be safe, *round up* to the next degree. (*Hint:* $F = \frac{9}{5}C + 32$.)

39. What is a scale drawing? Give an example.

40. Explain the benefits of the metric system of measurement as compared to the American system.

CHAPTERS 1–6   CUMULATIVE REVIEW

1. Write 5,764,502:

   a. in words

   b. in expanded notation [Section 1.1]

2. BASKETBALL RECORDS On December 13, 1983, the Detroit Pistons and the Denver Nuggets played in the highest-scoring game in NBA history. See the game summary below. [Section 1.2]

   a. What was the final score?

   b. Which team won?

   c. What was the total number of points scored in the game?

	Quarter				Overtime			
	1	**2**	**3**	**4**	**1**	**2**	**3**	**Total**
Detroit	38	36	34	37	14	12	15	
Denver	34	40	39	32	14	12	13	

(*Source*: ESPN.com)

3. Subtract: $70,006 - 348$ [Section 1.2]

4. Multiply: $504 \cdot 729$ [Section 1.3]

5. Divide: $37\overline{)743}$ [Section 1.4]

6. List the factors of 30, from smallest to largest. [Section 1.5]

7. Find the prime factorization of 360. [Section 1.5]

8. Find the LCM and the GCF of 20 and 28. [Section 1.6]

9. Evaluate: $81 + 9\left[7^2 - 7(11 - 4)\right]$ [Section 1.7]

10. Solve each equation and check the result.

    a. $m + 158 = 203$ [Section 1.8]

    b. $\dfrac{n}{57} = 300$ [Section 1.9]

11. Place an $<$ or an $>$ symbol in the box to make a true statement: $-(-10)$ ☐ $|-11|$ [Section 2.1]

12. Evaluate: $(-12 + 6) + (-6 + 8)$ [Section 2.2]

13. GOLF Tiger Woods won the 100th U.S. Open in June of 2000 by the largest margin in the history of that tournament. If he shot 12 under par (-12) and the second-place finisher, Miguel Angel Jimenez, shot 3 over par $(+3)$, what was Tiger's margin of victory? [Section 2.3]

14. Evaluate: -3^2 and $(-3)^2$ [Section 2.4]

15. Evaluate each expression, if possible. [Sections 2.2–2.5]

    a. $0 + (-8)$ **b.** $\dfrac{-8}{0}$

    c. $0 - |-8|$ **d.** $\dfrac{0}{-8}$

    e. $0 - (-8)$ **f.** $0(-8)$

16. Evaluate: $\dfrac{3 + 3\left[5(-6) - (1 - 10)\right]}{-1 + (-1)}$ [Section 2.6]

17. Estimate the value of the following expression by rounding each number to the nearest hundred. [Section 2.6]

    $$-3,887 + (-5,806) + 4,701$$

Write and then solve an equation to answer the following question. [Section 2.7]

18. POLLS Three months before an election, a political candidate was 18 points behind in the polls. The election was held and the candidate won by 3 points. How much support did the candidate gain over the last three months?

Solve each equation and check the result. [Section 2.7]

19. $t - 4 = -8 - (-2)$

20. $-3a + (-2) = 16$

21. Translate each phrase to an algebraic expression. [Section 3.1]

    a. Twenty-nine less than the weight w

    b. The quotient of 10 and the cube of m

22. Evaluate $\dfrac{-x + 3(1 + x)}{x^2 - 3}$ for $x = -2$ [Section 3.2]

23. **a.** Simplify: $-6(15a)$ [Section 3.3]

    b. Multiply: $5(4x - 2y + 7)$ [Section 3.3]

24. Combine like terms. [Section 3.4]

    a. $5x - 11x$

    b. $-4(x - 3y) + 5x - 2y$

Solve each equation and check the result. [Section 3.5]

25. $z + 12 + z = 8z - 4 - 2z$

26. $2y + 7 = 2 - (4y + 7)$

Write and then solve an equation to answer the following question. [Section 3.6]

27. SPRINKLERS A landscaper buried a water line around a rectangular lawn to serve as a supply line for a sprinkler system. The length of the lawn is 3 times its width. If 144 feet of pipe was used to do the job, what is the width and the length of the lawn?

28. Simplify: $-\dfrac{16}{20}$ [Section 4.1]

29. Express $\dfrac{9}{10}$ as an equivalent fraction with a denominator of $60a$. [Section 4.1]

30. GEOGRAPHY Earth has a surface area of about 197,000,000 square miles. Use the information in the circle graph below to determine the number of square miles of Earth's surface covered by land. (*Source:* scienceclarified.com) [Section 4.2]

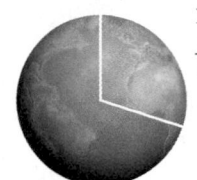

Land covers about $\dfrac{3}{10}$ of the Earth's surface

Water covers about $\dfrac{7}{10}$ of the Earth's surface

31. What is the formula for the area of a triangle? [Section 4.2]

32. Divide: $\dfrac{5}{10b^3} \div \dfrac{1}{2b^2}$ [Section 4.3]

33. Subtract: $\dfrac{11}{12} - \dfrac{7}{15}$ [Section 4.4]

34. Determine which fraction is larger: $\dfrac{19}{15}$ or $\dfrac{5}{4}$

[Section 4.4]

35. Add: $\dfrac{5}{m} + \dfrac{7}{8}$ [Section 4.4]

36. HARDWARE Find the length of the wood screw shown below. [Section 4.4]

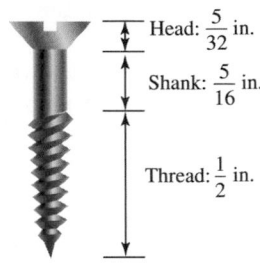

Head: $\dfrac{5}{32}$ in.

Shank: $\dfrac{5}{16}$ in.

Thread: $\dfrac{1}{2}$ in.

37. Multiply: $-15\dfrac{1}{3}\left(-\dfrac{9}{20}\right)$ [Section 4.5]

38. PAPER SHREDDERS A paper shredder cuts paper into $\frac{1}{4}$-inch-wide strips. If an $8\frac{1}{2}$ in. by 11 in. piece of notebook paper is fed into the shredder lengthwise, as shown, into how many strips will it be shredded? [Section 4.5]

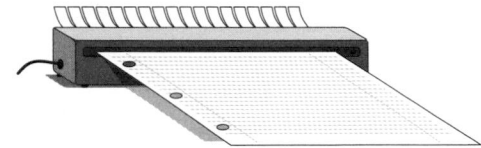

39. MOTORS What is the difference in horsepower (hp) between the two motors shown? [Section 4.6]

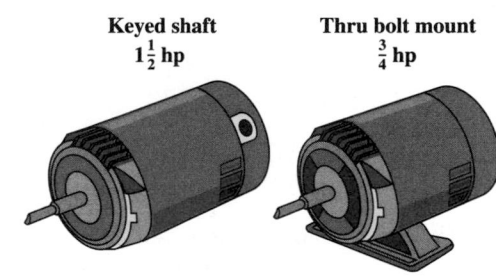

Keyed shaft $1\frac{1}{2}$ hp

Thru bolt mount $\frac{3}{4}$ hp

40. Simplify: $\dfrac{4 - \dfrac{3}{4}}{-1\dfrac{7}{8}}$ [Section 4.7]

Solve each equation and check the result. [Section 4.8]

41. $\dfrac{7}{8}t = -28$

42. $\dfrac{4}{5}x = \dfrac{3}{4}x + \dfrac{1}{2}$

Write and then solve an equation to answer the following question. [Section 4.8]

43. CONCERT TICKETS Seven-eighths of the total number of tickets sold for a concert were ordered by mail. The remaining 200 tickets were purchased at the concert hall box office. How many tickets were sold for the concert?

44. Place an $<$ or $>$ symbol in the box to make a true statement. [Section 5.1]

$$-64.22 \quad\rule{1cm}{0.4pt}\quad -64.238$$

45. Graph $-1\frac{3}{4}, 2.25, -0.5, \frac{11}{8}, -3.2,$ and $\sqrt{9}$ on a number line. [Section 5.1]

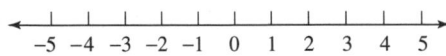

46. Add: $-20.04 + 2.4$ [Section 5.2]

47. Subtract: $-8.08 - 15.3$ [Section 5.2]

48. Multiply: $2.5 \cdot 100$ [Section 5.3]

49. AQUARIUMS One gallon of water weighs 8.33 pounds. What is the weight of the water in an aquarium that holds 55 gallons of water? [Section 5.3]

50. Divide: $2.5 \div 100$ [Section 5.4]

51. Evaluate the formula $t = \frac{d}{r}$ for $d = 107.95$ and $r = 8.5$. [Section 5.4]

52. Write $\frac{1}{12}$ as a decimal. [Section 5.5]

53. LUNCH MEATS A shopper purchased $\frac{3}{4}$ pound of barbequed beef, priced at $8.60 per pound, and $\frac{2}{3}$ pound of ham, selling for $5.25 per pound. Find the total cost of these items. [Section 5.5]

54. Evaluate: $3\sqrt{25} + 4\sqrt{4}$ [Section 5.6]

Solve each equation and check the result. [Section 5.7]

55. $3.2x = 74.46 - 1.9x$

56. $-5.2x = 108 - 6.1x$

57. $-2(x - 2.1) = -2.4$

58. $\frac{1}{5}x - 2.5 = -17.2$

Write and then solve an equation to answer the following question. [Section 5.7]

59. PETITION DRIVES A worker for a political organization is to collect signatures for a petition drive. Her pay is $20 plus 5¢ per signature. How many signatures must she get to earn $60?

60. Express the phrase "3 inches to 15 inches" as a ratio in simplest form. [Section 6.1]

61. BUILDING MATERIALS Which is the better buy: a 94-pound bag of cement for $4.48 or a 100-pound bag of cement for $4.80? [Section 6.1]

62. Determine whether the proportion $\frac{25}{33} = \frac{12}{17}$ is true or false. [Section 6.2]

63. CAFFEINE There are 55 milligrams of caffeine in 12 ounces of Mountain Dew. How many milligrams of caffeine are there in a super-size 44-ounce cup of Mountain Dew? Round to the nearest milligram. [Section 6.2]

64. Solve the proportion: $\dfrac{x}{3} = \dfrac{35}{7}$ [Section 6.2]

65. SURVIVAL GUIDE [Section 6.3]

   a. A person can go without food for about 40 days. How many hours is this?

   b. A person can go without water for about 3 days. How many minutes is that?

   c. A person can go without breathing oxygen for about 8 minutes. How many seconds is that?

66. Convert 40 ounces to pounds. [Section 6.3]

67. Convert 2.4 meters to millimeters. [Section 6.4]

68. Convert 320 grams to kilograms. [Section 6.4]

69. **a.** Which holds more: a 2-liter bottle or a 1-gallon bottle? [Section 6.5]

   b. Which is longer: a meterstick or a yardstick?

70. BELTS A leather belt made in Mexico is 92 centimeters long. Express the length of the belt to the nearest inch. [Section 6.5]

Percent

Ariel Skelley/Getty Images

from **Campus to Careers**

Loan Officer

Loan officers help people apply for loans. *Commercial loan officers* work with businesses, *mortgage loan officers* work with people who want to buy a house or other real estate, and *consumer loan officers* work with people who want to buy a boat, a car, or need a loan for college. Loan officers analyze the applicant's financial history and often use banking formulas to determine the possibility of granting a loan.

In **Problem 43** of **Study Set 7.5,** you will see how a credit union loan officer calculates the interest to be charged on a loan.

JOB TITLE:
Loan Officer

EDUCATION: Most have a degree in finance, economics, or a similar field. Mathematics and computer classes are good preparation for this job.

JOB OUTLOOK: Employment of loan officers is expected to grow about as fast as the average for all jobs through 2016.

ANNUAL EARNINGS: In 2009, the average salary for a consumer loan officer was about $39,000 and the average salary for a commercial loan officer was about $64,000.

FOR MORE INFORMATION:
www.bls.gov/k12/money03.htm

Objectives

1 Explain the meaning of percent.

2 Write percents as fractions.

3 Write percents as decimals.

4 Write decimals as percents.

5 Write fractions as percents.

SECTION 7.1
Percents, Decimals, and Fractions

We see percents everywhere, everyday. Stores use them to advertise discounts, manufacturers use them to describe the contents of their products, and banks use them to list interest rates for loans and savings accounts. Newspapers are full of information presented in percent form. In this section, we introduce percents and show how fractions, decimals, and percents are related.

1 Explain the meaning of percent.

A percent tells us the number of parts per one hundred. You can think of a percent as the *numerator* of a fraction (or ratio) that has a denominator of 100.

Percent

Percent means parts per one hundred.

> ***The Language of Algebra***   The word *percent* is formed from the prefix *per*, which means ratio, and the suffix *cent*, which comes from the Latin word *centum*, meaning 100.
>
> $$\text{per} \bullet \text{cent}$$
> ratio ⟶   ⟵ 100

In the figure below, there are 100 equal-sized square regions, and 93 of them are shaded. Thus, $\frac{93}{100}$ or 93 percent of the figure is shaded. The word *percent* can be written using the symbol %, so we say that 93% of the figure is shaded.

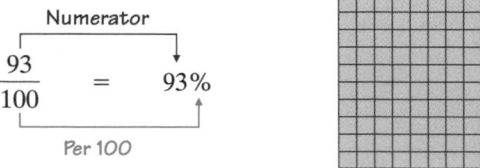

If the entire figure had been shaded, we would say that 100 out of the 100 square regions, or 100%, was shaded. Using this fact, we can determine what percent of the figure is *not* shaded by subtracting the percent of the figure that is shaded from 100%.

$$100\% - 93\% = 7\%$$

So 7% of the figure is *not* shaded.

To illustrate a percent greater than 100%, say 121%, we would shade one entire figure and 21 of the 100 square regions in a second, equal-sized grid.

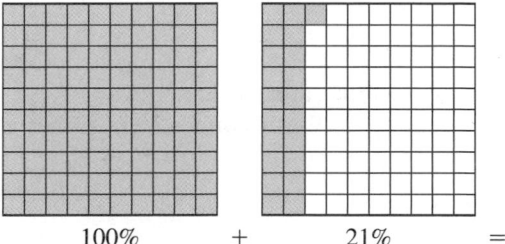

100%   +   21%   = 121%

EXAMPLE 1   *Tossing a Coin*   A coin was tossed 100 times and it landed heads up 51 times.

a. What percent of the time did the coin land heads up?

b. What percent of the time did it land tails up?

Strategy We will write a fraction that compares the number of times that the coin landed heads up (or tails up) to the total number of tosses.

WHY Since the denominator in each case will be 100, the numerator of the fraction will give the percent.

Solution

a. If a coin landed heads up 51 times after being tossed 100 times, then

$$\frac{51}{100} = \mathbf{51\%}$$

of the time it landed heads up.

b. The number of times the coin landed tails up is $100 - 51 = 49$ times. If a coin landed tails up 49 times after being tossed 100 times, then

$$\frac{49}{100} = \mathbf{49\%}$$

of the time it landed tails up.

Self Check 1

BOARD GAMES A standard Scrabble game contains 100 tiles. There are 42 vowel tiles, 2 blank tiles, and the rest are consonant tiles.

a. What percent of the tiles are vowels?

b. What percent of the letter tiles are consonants?

Now Try Problem 13

2 Write percents as fractions.

We can use the definition of percent to write any percent in an equivalent fraction form.

Writing Percents as Fractions

To write a percent as a fraction, drop the % symbol and write the given number over 100. Then simplify the fraction, if possible.

EXAMPLE 2   *Earth*   The chemical makeup of Earth's atmosphere is 78% nitrogen, 21% oxygen, and 1% other gases. Write each percent as a fraction in simplest form.

Strategy We will drop the % symbol and write the given number over 100. Then we will simplify the resulting fraction, if possible.

WHY *Percent* means parts per one hundred, and the word *per* indicates a ratio (fraction).

Self Check 2

WATERMELONS An average watermelon is 92% water. Write this percent as a fraction in simplest form.

Now Try Problems 17 and 23

Solution We begin with nitrogen.

$$78\% = \frac{78}{100} \qquad \text{Drop the \% symbol and write 78 over 100.}$$

$$= \frac{\overset{1}{\cancel{2}} \cdot 39}{\underset{1}{\cancel{2}} \cdot 50} \qquad \begin{array}{l}\text{To simplify the fraction, factor 78 as } 2 \cdot 39 \text{ and 100 as } 2 \cdot 50. \text{ Then} \\ \text{remove the common factor of 2 from the numerator and denominator.}\end{array}$$

$$= \frac{39}{50}$$

Nitrogen makes up $\frac{78}{100}$, or $\frac{39}{50}$, of Earth's atmosphere.

Oxygen makes up 21%, or $\frac{21}{100}$, of Earth's atmosphere. Other gases make up 1%, or $\frac{1}{100}$, of the atmosphere. ■

Self Check 3

UNIONS In 2002, 13.3% of the U.S. labor force belonged to a union. Write this percent as a fraction in simplest form.

Now Try Problems 27 and 31

EXAMPLE 3   ***Unions***   In 2007, 12.1% of the U.S. labor force belonged to a union. Write this percent as a fraction in simplest form. (*Source:* Bureau of Labor Statistics)

Strategy We will drop the % symbol and write the given number over 100. Then we will multiply the resulting fraction by a form of 1 and simplify, if possible.

WHY When writing a percent as a fraction, the numerator and denominator of the fraction should be whole numbers that have no common factors (other than 1).

Solution

$$12.1\% = \frac{12.1}{100} \qquad \text{Drop the \% symbol and write 12.1 over 100.}$$

To write this as an equivalent fraction of *whole numbers*, we need to move the decimal point in the numerator one place to the right. (Recall that to find the product of a decimal and 10, we simply move the decimal point one place to the right.) Therefore, it follows that $\frac{10}{10}$ is the form of 1 that we should use to build $\frac{12.1}{100}$.

$$\frac{12.1}{100} = \frac{12.1}{100} \cdot \frac{\mathbf{10}}{\mathbf{10}} \qquad \text{Multiply the fraction by a form of 1.}$$

$$= \frac{12.1 \cdot \mathbf{10}}{100 \cdot \mathbf{10}} \qquad \begin{array}{l}\text{Multiply the numerators.} \\ \text{Multiply the denominators.}\end{array}$$

$$= \frac{121}{1,000} \qquad \begin{array}{l}\text{Since 121 and 1,000 do not have any common factors} \\ \text{(other than 1), the fraction is in simplest form.}\end{array}$$

Thus, $12.1\% = \frac{121}{1,000}$. This means that 121 out of every 1,000 workers in the U.S. labor force belonged to a union in 2007. ■

Self Check 4

Write $83\frac{1}{3}$% as a fraction in simplest form.

Now Try Problem 35

EXAMPLE 4   Write $66\frac{2}{3}$% as a fraction in simplest form.

Strategy We will drop the % symbol and write the given number over 100. Then we will perform the division indicated by the fraction bar and simplify, if possible.

WHY When writing a percent as a fraction, the numerator and denominator of the fraction should be whole numbers that have no common factors (other than 1).

Solution

$$66\frac{2}{3}\% = \frac{66\frac{2}{3}}{100} \qquad \text{Drop the \% symbol and write } 66\frac{2}{3} \text{ over 100.}$$

To write this as a fraction of whole numbers, we will perform the division indicated by the fraction bar.

$$\frac{66\frac{2}{3}}{100} = 66\frac{2}{3} \div 100 \qquad \text{The fraction bar indicates division.}$$

$$= \frac{200}{3} \cdot \frac{1}{100} \qquad \begin{array}{l}\text{Write } 66\frac{2}{3} \text{ as a mixed number and then multiply} \\ \text{by the reciprocal of 100.}\end{array}$$

$$= \frac{200 \cdot 1}{3 \cdot 100} \qquad \begin{array}{l}\text{Multiply the numerators.} \\ \text{Multiply the denominators.}\end{array}$$

$$= \frac{2 \cdot \overset{1}{\cancel{100}} \cdot 1}{3 \cdot \underset{1}{\cancel{100}}} \qquad \begin{array}{l}\text{To simplify the fraction, factor 200 as } 2 \cdot 100. \text{ Then remove the} \\ \text{common factor of 100 from the numerator and denominator.}\end{array}$$

$$= \frac{2}{3}$$

EXAMPLE 5   **a.** Write 175% as a fraction in simplest form.

b. Write 0.22% as a fraction in simplest form.

Strategy We will drop the % symbol and write each given number over 100. Then we will simplify the resulting fraction, if possible.

WHY *Percent* means parts per one hundred and the word *per* indicates a ratio (fraction).

Solution

a. $175\% = \dfrac{175}{100}$ Drop the % symbol and write 175 over 100.

$$= \frac{\overset{1}{\cancel{5}} \cdot \overset{1}{\cancel{5}} \cdot 7}{2 \cdot 2 \cdot \underset{1}{\cancel{5}} \cdot \underset{1}{\cancel{5}}} \qquad \begin{array}{l}\text{To simplify the fraction, prime factor 175} \\ \text{and 100. Remove the common factors of} \\ 5 \text{ from the numerator and denominator.}\end{array} \qquad \begin{array}{r}5\,\vert\,\underline{175} \\ 5\,\vert\,\underline{35} \\ 7\end{array} \quad \begin{array}{r}2\,\vert\,\underline{100} \\ 2\,\vert\,\underline{50} \\ 5\,\vert\,\underline{25} \\ 5\end{array}$$

$$= \frac{7}{4}$$

Thus, $175\% = \dfrac{7}{4}$.

b. $0.22\% = \dfrac{0.22}{100}$ Drop the % symbol and write 175 over 100.

To write this as an equivalent fraction of *whole numbers,* we need to move the decimal point in the numerator two places to the right. (Recall that to find the product of a decimal and 100, we simply move the decimal point two places to the right.) Therefore, it follows that $\frac{100}{100}$ is the form of 1 that we should use to build $\frac{0.22}{100}$.

$$\frac{0.22}{100} = \frac{0.22}{100} \cdot \boxed{\frac{\mathbf{100}}{\mathbf{100}}} \qquad \text{Multiply the fraction by a form of 1.}$$

$$= \frac{0.22 \cdot 100}{100 \cdot 100} \qquad \begin{array}{l}\text{Multiply the numerators.} \\ \text{Multiply the denominators.}\end{array}$$

$$= \frac{22}{10{,}000}$$

$$= \frac{\overset{1}{\cancel{2}} \cdot 11}{\underset{1}{\cancel{2}} \cdot 5{,}000} \qquad \begin{array}{l}\text{To simplify the fraction, factor 22 and 10,000.} \\ \text{Remove the common factor of 2 from the} \\ \text{numerator and denominator.}\end{array}$$

$$= \frac{11}{5{,}000}$$

Thus, $0.22\% = \dfrac{11}{5{,}000}$.

Self Check 5

a. Write 210% as a fraction in simplest form.

b. Write 0.54% as a fraction in simplest form.

Now Try **Problems 39 and 43**

> **Success Tip**   When percents that are greater than 100% are written as fractions, the fractions are greater than 1. When percents that are less than 1% are written as fractions, the fractions are less than $\frac{1}{100}$.

3 Write percents as decimals.

To write a percent as a decimal, recall that a percent can be written as a fraction with denominator 100 and that a denominator of 100 indicates division by 100.

For example, consider 14%, which means 14 parts per 100.

$$14\% = \frac{14}{100} \qquad \text{Use the definition of percent: write 14 over 100.}$$

$$= 14 \div 100 \qquad \text{The fraction bar indicates division.}$$

$$= 14.0 \div 100 \qquad \text{Write the whole number 14 in decimal notation by placing a decimal point immediately to its right and entering a zero to the right of the decimal point.}$$

$$= .140 \qquad \text{Since the divisor 100 has two zeros, move the decimal point 2 places to the left.}$$

$$= 0.14 \qquad \text{Write a zero to the left of the decimal point.}$$

We have found that 14% = 0.14. This example suggests the following procedure.

Writing Percents as Decimals

To write a percent as a decimal, drop the % symbol and divide the given number by 100 by moving the decimal point 2 places to the left.

Self Check 6

a. Write 16.43% as a decimal.

b. Write 2.06% as a decimal.

Now Try Problems 51 and 57

EXAMPLE 6   *TV Websites*   The graph below shows the percent of market share for the top 5 network TV show websites.

a. Write the percent of market share for the *American Idol* website as a decimal.

b. Write the percent of market share for the *Deal or No Deal* website as a decimal.

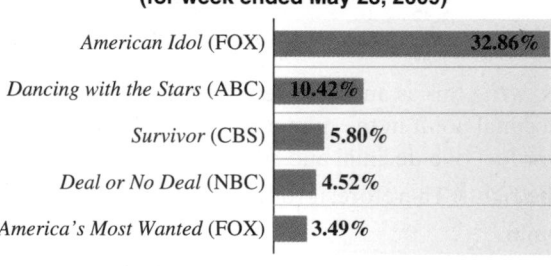

Top Five Network TV Show Websites
by Market Share of Visits (%)
(for week ended May 23, 2009)

American Idol (FOX) 32.86%
Dancing with the Stars (ABC) 10.42%
Survivor (CBS) 5.80%
Deal or No Deal (NBC) 4.52%
America's Most Wanted (FOX) 3.49%

(*Source:* marketingcharts.com)

Strategy   We will drop the % symbol and divide each given number by 100 by moving the decimal point 2 places to the left.

WHY   Recall from Section 5.4 that to find the quotient of a decimal and 10, 100, 1,000, and so on, move the decimal point to the left the same number of places as there are zeros in the power of 10.

Solution

a. From the graph, we see that the percent market share for the *American Idol* website is 32.86%. To write this percent as a decimal, we proceed as follows.

$$32.86\% = .32\,86 \qquad \text{Drop the \% symbol and divide 32.86 by 100 by moving the decimal point 2 places to the left.}$$

$$= 0.3286 \qquad \text{Write a zero to the left of the decimal point.}$$

32.86%, written as a decimal, is 0.3286.

b. From the graph, we see that the percent market share for the *Deal or No Deal* website is 4.52%. To write this percent as a decimal, we proceed as follows.

$$4.52\% = {\underset{\curvearrowleft}{.}}0452 \quad$$ Drop the % symbol and divide 4.52 by 100 by moving the decimal point 2 places to the left. This requires that a placeholder zero (shown in blue) be inserted in front of the 4.

$$= 0.0452 \quad$$ Write a zero to the left of the decimal point.

4.52%, written as a decimal, is 0.0452.

EXAMPLE 7   *Population*

The population of the state of Oregon is approximately $1\frac{1}{4}\%$ of the population of the United States. Write this percent as a decimal. (*Source:* U.S. Census Bureau)

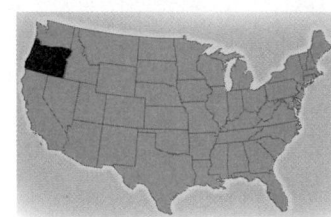

Strategy We will write the mixed number $1\frac{1}{4}$ in decimal notation.

WHY With $1\frac{1}{4}$ in mixed-number form, we cannot apply the rule for writing a percent as a decimal; there is no decimal point to move 2 places to the left.

Solution To change a percent to a decimal, we drop the percent symbol and divide by 100 by moving the decimal point 2 places to the left. In this case, however, there is no decimal point to move in $1\frac{1}{4}\%$. Since $1\frac{1}{4} = 1 + \frac{1}{4}$, and since the decimal equivalent of $\frac{1}{4}$ is 0.25, we can write $1\frac{1}{4}\%$ in an equivalent form as 1.25%.

$$1\frac{1}{4}\% = 1.25\% \quad$$ Write $1\frac{1}{4}$ as 1.25.

$$= {\underset{\curvearrowleft}{.}}0125 \quad$$ Drop the % symbol and divide 1.25 by 100 by moving the decimal point 2 places to the left. This requires that a placeholder zero (shown in blue) be inserted in front of the 1.

$$= 0.0125 \quad$$ Write a zero to the left of the decimal point.

$1\frac{1}{4}\%$, written as a decimal, is 0.0125.

Self Check 7

POPULATION The population of the state of Ohio is approximately $3\frac{3}{4}\%$ of the population of the United States. Write this percent as a decimal. (*Source:* U.S. Census Bureau)

Now Try Problem 59

EXAMPLE 8

a. Write 310% as a decimal.   **b.** Write 0.9% as a decimal.

Strategy We will drop the % symbol and divide each given number by 100 by moving the decimal point two places to the left.

WHY Recall that to find the quotient of a decimal and 100, we move the decimal point to the left the same number of places as there are zeros in 100.

Solution
a. $310\% = 310.0\% \quad$ Write the whole number 310 in decimal notation: 310 = 310.0.

$$= 3{\underset{\curvearrowleft}{.}}10\,0 \quad$$ Drop the % symbol and divide 310 by 100 by moving the decimal point 2 places to the left.

$$= 3.1 \quad$$ Drop the unnecessary zeros to the right of the 1.

310%, written as a decimal, is 3.1.

b. $0.9\% = {\underset{\curvearrowleft}{.}}009 \quad$ Drop the % symbol and divide 0.9 by 100 by moving the decimal point 2 places to the left. This requires that a placeholder zero (shown in blue) be inserted in front of the 0.

$$= 0.009 \quad$$ Write a zero to the left of the decimal point.

0.9%, written as a decimal, is 0.009.

Self Check 8

a. Write 600% as a decimal.

b. Write 0.8% as a decimal.

Now Try Problems 63 and 67

> **Success Tip** When percents that are greater than 100% are written as decimals, the decimals are greater than 1.0. When percents that are less than 1% are written as decimals, the decimals are less than 0.01.

4 Write decimals as percents.

To write a percent as a decimal, we drop the % symbol and move the decimal point 2 places to the left. To write a decimal as a percent, we do the opposite: we move the decimal point 2 places to the right and insert a % symbol.

Writing Decimals as Percents

To write a decimal as a percent, multiply the decimal by 100 by moving the decimal point 2 places to the right, and then insert a % symbol.

Self Check 9

Write 0.5343 as a percent.

Now Try Problems 71 and 75

EXAMPLE 9 *Geography* Land areas make up 0.291 of Earth's surface. Write this decimal as a percent.

Strategy We will multiply the decimal by 100 by moving the decimal point 2 places to the right, and then insert a % symbol.

WHY To write a *decimal as a percent,* we reverse the steps used to write a *percent as a decimal.*

Solution

$$0.291 = 029.1\%$$ Multiply 0.291 by 100 by moving the decimal point 2 places to the right, and then insert a % symbol.

$$= 29.1\%$$

0.291, written as a percent, is 29.1%

5 Write fractions as percents.

We use a two-step process to write a fraction as a percent. First, we write the fraction as a decimal. Then we write that decimal as a percent.

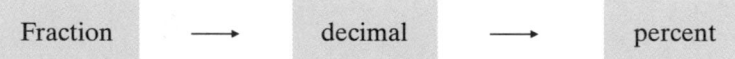

$$\boxed{\text{Fraction}} \longrightarrow \boxed{\text{decimal}} \longrightarrow \boxed{\text{percent}}$$

Writing Fractions as Percents

To write a fraction as a percent:

1. Write the fraction as a decimal by dividing its numerator by its denominator.
2. Multiply the decimal by 100 by moving the decimal point 2 places to the right, and then insert a % symbol.

Self Check 10

Write 7 out of 8 as a percent.

Now Try Problem 79

EXAMPLE 10 *Television* The highest-rated television show of all time was a special episode of *M*A*S*H* that aired February 28, 1983. Surveys found that three out of every five American households watched this show. Express the rating as a percent.

Strategy First, we will translate the phrase *three out of every five* to fraction form and write that fraction as a decimal. Then we will write that decimal as a percent.

WHY A fraction-to-decimal-to-percent approach must be used to write a fraction as a percent.

Solution

Step 1 The phrase *three out of every five* can be expressed as $\frac{3}{5}$. To write this fraction as a decimal, we divide the numerator, 3, by the denominator, 5.

$$
\begin{array}{r}
0.6 \\
5\overline{)3.0} \\
-3\,0 \\
\hline
0
\end{array}
$$
Write a decimal point and one additional zero to the right of 3.

$0 \leftarrow$ The remainder is 0.

The result is a terminating decimal.

Step 2 To write 0.6 as a percent, we proceed as follows.

$$\frac{3}{5} = 0.6$$

$0.6 = 060.\%$ Write a placeholder 0 to the right of the 6 (shown in blue). Multiply 0.60 by 100 by moving the decimal point 2 places to the right, and then insert a % symbol.

$= 60\%$

60% of American households watched the special episode of *M*A*S*H*.

EXAMPLE 11   Write $\frac{13}{4}$ as a percent.

Self Check 11
Write $\frac{5}{2}$ as a percent.
Now Try Problem 85

Strategy We will write the fraction $\frac{13}{4}$ as a decimal. Then we will write that decimal as a percent.

WHY A fraction-to-decimal-to-percent approach must be used to write a fraction as a percent.

Solution

Step 1 To write $\frac{13}{4}$ as a decimal, we divide the numerator, 13, by the denominator, 4.

$$
\begin{array}{r}
3.25 \\
4\overline{)13.00} \\
-12 \\
\hline
1\,0 \\
-8 \\
\hline
20 \\
-20 \\
\hline
0
\end{array}
$$
Write a decimal point and two additional zeros to the right of 3.

$0 \leftarrow$ The remainder is 0.

The result is a terminating decimal.

Step 2 To write 3.25 as a percent, we proceed as follows.

$3.25 = 325.\%$ Multiply 3.25 by 100 by moving the decimal point 2 places to the right, and then insert a % symbol.

$= 325\%$

The fraction $\frac{13}{4}$, written as a percent, is 325%.

Success Tip When fractions that are greater than 1 are written as percents, the percents are greater than 100%.

In Examples 10 and 11, the result of the division was a terminating decimal. Sometimes when we write a fraction as a decimal, the result of the division is a repeating decimal.

Self Check 12

Write $\frac{2}{3}$ as a percent. Give the exact answer and an approximation to the nearest tenth of one percent.

Now Try Problem 91

EXAMPLE 12   Write $\frac{5}{6}$ as a percent. Give the exact answer and an approximation to the nearest tenth of one percent.

Strategy We will write the fraction $\frac{5}{6}$ as a decimal. Then we will write that decimal as a percent.

WHY A fraction-to-decimal-to-percent approach must be used to write a fraction as a percent.

Solution

Step 1 To write $\frac{5}{6}$ as a decimal, we divide the numerator, 5, by the denominator, 6.

```
      0.8333
   6)5.0000     Write a decimal point and several zeros to the right of 5.
    -4 8
      20
     -18
      20
     -18
      20
     -18
       2  ← The repeating pattern is now clear. We can stop the division.
```

The result is a repeating decimal.

Step 2 To write the decimal as a percent, we proceed as follows.

$$\frac{5}{6} = 0.8333\ldots$$

$$0.833\ldots = 0\,83.33\ldots\% $$
$$= 83.33\ldots\%$$

Multiply 0.8333... by 100 by moving the decimal point 2 places to the right, and then insert a % symbol.

We must now decide whether we want an exact answer or an approximation. For an exact answer, we can represent *the repeating part of the decimal using an equivalent fraction.* For an approximation, we can round 83.333...% to a specific place value.

Exact answer:

$$\frac{5}{6} = 83.\underline{3333}\ldots\%$$
$$= 83\frac{1}{3}\%$$ Use the fraction $\frac{1}{3}$ to represent .3333 ....

Thus,

$$\frac{5}{6} = 83\frac{1}{3}\%$$

Approximation:

$$\frac{5}{6} = 83.33\ldots\%$$
$$\approx 83.3\%$$ Round to the nearest tenth.

Thus,

$$\frac{5}{6} \approx 83.3\%$$

Some percents occur so frequently that it is useful to memorize their fractional and decimal equivalents.

Percent	Decimal	Fraction	Percent	Decimal	Fraction
1%	0.01	$\frac{1}{100}$	$33\frac{1}{3}\%$	0.3333...	$\frac{1}{3}$
10%	0.1	$\frac{1}{10}$	50%	0.5	$\frac{1}{2}$
$16\frac{2}{3}\%$	0.1666...	$\frac{1}{6}$	$66\frac{2}{3}\%$	0.6666...	$\frac{2}{3}$
20%	0.2	$\frac{1}{5}$	$83\frac{1}{3}\%$	0.8333...	$\frac{5}{6}$
25%	0.25	$\frac{1}{4}$	75%	0.75	$\frac{3}{4}$

SECTION **7.1**   STUDY SET

VOCABULARY

Fill in the blanks.

1. _____ means parts per one hundred.

2. The word *percent* is formed from the prefix *per*, which means _____, and the suffix *cent*, which comes from the Latin word *centum*, meaning _____.

CONCEPTS

Fill in the blanks.

3. To write a percent as a fraction, drop the % symbol and write the given number over _____. Then _____ the fraction, if possible.

4. To write a percent as a decimal, drop the % symbol and divide the given number by 100 by moving the decimal point 2 places to the _____.

5. To write a decimal as a percent, multiply the decimal by 100 by moving the decimal point 2 places to the _____, and then insert a % symbol.

6. To write a fraction as a percent, first write the fraction as a _____. Then multiply the decimal by 100 by moving the decimal point 2 places to the right, and then insert a ____ symbol.

NOTATION

7. What does the symbol % mean?

8. Write the whole number 45 as a decimal.

GUIDED PRACTICE

What percent of the figure is shaded? What percent of the figure is not shaded? See Objective 1.

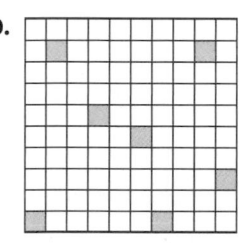

9.   10.

In the following illustrations, each set of 100 square regions represents 100%. What percent is shaded?

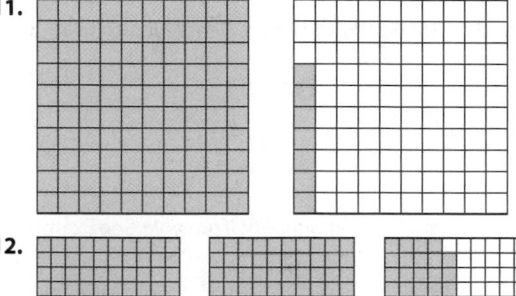

11.

12.

For Problems 13–16, see Example 1.

13. THE INTERNET The following sentence appeared on a technology blog: "Ask Internet users what they want from their service and 99 times out of 100 the answer will be the same: more speed." According to the blog, what percent of the time do Internet users give that answer?

14. BASKETBALL RECORDS In 1962, Wilt Chamberlain of the Philadelphia Warriors scored a total of 100 points in an NBA game. If twenty-eight of his points came from made free throws, what percent of his point total came from free throws?

15. QUILTS A quilt is made from 100 squares of colored cloth.

    a. If fifteen of the squares are blue, what percent of the squares in the quilt are blue?

    b. What percent of the squares are not blue?

16. DIVISIBILITY Of the natural numbers from 1 through 100, only fourteen of them are divisible by 7.

    a. What percent of the numbers are divisible by 7?

    b. What percent of the numbers are not divisible by 7?

Write each percent as a fraction. Simplify, if possible. See Example 2.

17. 17%   **18.** 31%

19. 91%   **20.** 89%

21. 4%   **22.** 5%

23. 60%   **24.** 40%

Write each percent as a fraction. Simplify, if possible. See Example 3.

25. 1.9%   **26.** 2.3%

27. 54.7%   **28.** 97.1%

29. 12.5%   **30.** 62.5%

31. 6.8%   **32.** 4.2%

Write each percent as a fraction. Simplify, if possible. See Example 4.

33. $1\frac{1}{3}\%$ **34.** $3\frac{1}{3}\%$

35. $14\frac{1}{6}\%$ **36.** $10\frac{5}{6}\%$

Write each percent as a fraction. Simplify, if possible. See Example 5.

37. 130%   **38.** 160%

39. 220%   **40.** 240%

41. 0.35%   **42.** 0.45%

43. 0.25%   **44.** 0.75%

Write each percent as a decimal. See Objective 3.

45. 16%   **46.** 11%

47. 81%   **48.** 93%

Write each percent as a decimal. See Example 6.

49. 34.12%   **50.** 27.21%

51. 50.033%   **52.** 40.083%

53. 6.99%   **54.** 4.77%

55. 1.3%   **56.** 8.6%

Write each percent as a decimal. See Example 7.

57. $7\frac{1}{4}\%$ **58.** $9\frac{3}{4}\%$

59. $18\frac{1}{2}\%$ **60.** $25\frac{1}{2}\%$

Write each percent as a decimal. See Example 8.

61. 460%   **62.** 230%

63. 316%   **64.** 178%

65. 0.5%   **66.** 0.9%

67. 0.03%   **68.** 0.06%

Write each decimal or whole number as a percent. See Example 9.

69. 0.362   **70.** 0.245

71. 0.98   **72.** 0.57

73. 1.71   **74.** 4.33

75. 4   **76.** 9

Write each fraction as a percent. See Example 10.

77. $\frac{2}{5}$ **78.** $\frac{1}{5}$

79. $\frac{4}{25}$ **80.** $\frac{9}{25}$

81. $\frac{5}{8}$ **82.** $\frac{3}{8}$

83. $\frac{7}{16}$ **84.** $\frac{9}{16}$

Write each fraction as a percent. See Example 11.

85. $\frac{9}{4}$ **86.** $\frac{11}{4}$

87. $\frac{21}{20}$ **88.** $\frac{33}{20}$

Write each fraction as a percent. Give the exact answer and an approximation to the nearest tenth of one percent. See Example 12.

89. $\frac{1}{6}$ **90.** $\frac{2}{9}$

91. $\frac{5}{3}$ **92.** $\frac{4}{3}$

TRY IT YOURSELF

Complete the table. Give an exact answer and an approximation to the nearest tenth of one percent when necessary. Round decimals to the nearest hundredth when necessary.

	Fraction	Decimal	Percent
93.		0.0314	
94.		0.0021	
95.			40.8%
96.			34.2%
97.			$5\frac{1}{4}\%$
98.			$6\frac{3}{4}\%$
99.	$\frac{7}{3}$		
100.	$\frac{7}{9}$		

APPLICATIONS

101. THE RED CROSS A fact sheet released by the American Red Cross in 2008 stated, "An average of 91 cents of every dollar donated to the Red Cross is spent on services and programs." What percent of the money donated to the Red Cross went to services and programs?

102. SAVING MONEY According to an article on the CNN website, in 1970 Americans saved 14 cents out of every dollar earned. (*Source:* CNN.com/living, May 21, 2009)

   a. Express the amount saved for every dollar earned as a fraction in simplest form.

   b. Write your answer to part a as a percent.

103. REGIONS OF THE COUNTRY The continental United States is divided into seven regions as shown below.

   a. What percent of the 50 states are in the Rocky Mountain region?

   b. What percent of the 50 states are in the Midwestern region?

   c. What percent of the 50 states are not located in any of the seven regions shown here?

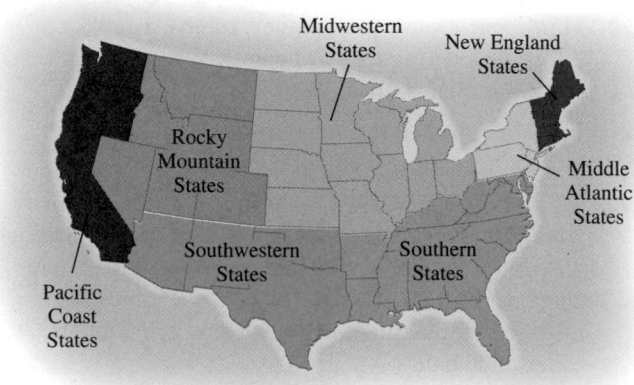

104. ROAD SIGNS Sometimes, signs like that shown below are posted to warn truckers when they are approaching a steep grade on the highway.

   a. Write the grade shown on the sign as a fraction.

   b. Write the grade shown on the sign as a decimal.

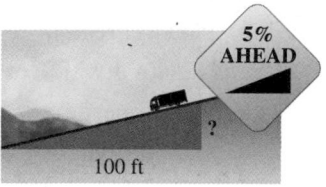

105. INTEREST RATES Write each interest rate for the following accounts as a decimal.

   a. Home loan: 7.75%

   b. Savings account: 5%

   c. Credit card: 14.25%

106. DRUNK DRIVING In most states, it is illegal to drive with a blood alcohol concentration of 0.08% or higher.

   a. Write this percent as a fraction. Do not simplify.

   b. Use your answer to part a to fill in the blanks: A blood alcohol concentration of 0.08% means ___ parts alcohol to ___ parts blood.

107. HUMAN SKIN The illustration below shows what percent of the total skin area that each section of the body covers. Find the missing percent for the torso, and then complete the bar graph. (*Source:* Burn Center at Sherman Oaks Hospital, American Medical Assn. Encyclopedia of Medicine)

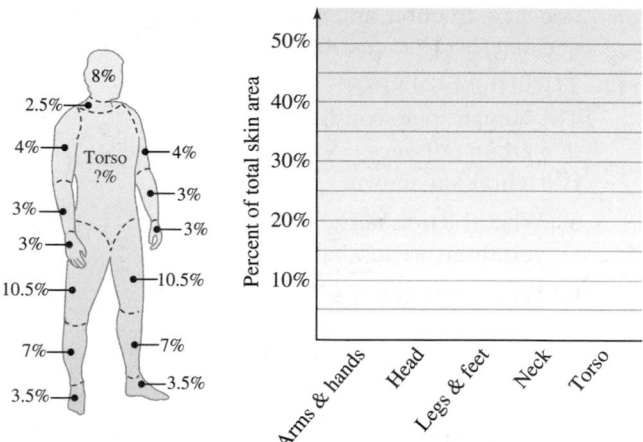

108. RAP MUSIC The table below shows what percent rap/hip-hop music sales were of total U.S. dollar sales of recorded music for the years 2001–2007. Use the data to construct a line graph.

2001	2002	2003	2004	2005	2006	2007
11.4%	13.8%	13.3%	12.1%	13.3%	11.4%	10.8%

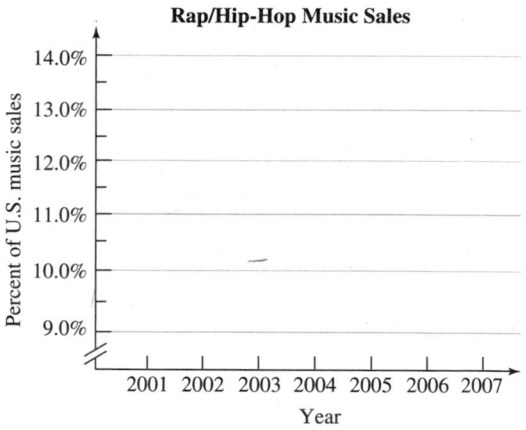

Source: Recording Industry Association of America

109. THE U.N. SECURITY COUNCIL The United Nations has 192 members. The United States, Russia, the United Kingdom, France, and China, along with ten other nations, make up the Security Council. (*Source: The World Almanac and Book of Facts,* 2009)

   a. What fraction of the members of the United Nations belong to the Security Council? Write your answer in simplest form.

   b. Write your answer to part a as a decimal. (*Hint:* Divide to six decimal places. The result is a terminating decimal.)

   c. Write your answer to part b as a percent.

110. SOAP Ivory soap claims to be $99\frac{44}{100}\%$ pure. Write this percent as a decimal.

111. LOGOS In the illustration, what part of the company's logo is shaded red? Express your answer as a percent (exact), a fraction, and a decimal (using an overbar).

Recycling Industries Inc.

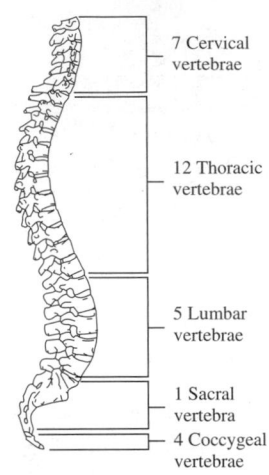

112. THE HUMAN SPINE The human spine consists of a group of bones (vertebrae) as shown.

   a. What fraction of the vertebrae are lumbar?

   b. What percent of the vertebrae are lumbar? (Round to the nearest one percent.)

   c. What percent of the vertebrae are cervical? (Round to the nearest one percent.)

7 Cervical vertebrae

12 Thoracic vertebrae

5 Lumbar vertebrae

1 Sacral vertebra

4 Coccygeal vertebrae

113. BOXING Oscar De La Hoya won 39 out of 45 professional fights.

   a. What fraction of his fights did he win?

   b. What percent of his fights did he win? Give the exact answer and an approximation to the nearest tenth of one percent.

114. MAJOR LEAGUE BASEBALL In 2008, the Milwaukee Brewers won 90 games and lost 72 during the regular season.

   a. What was the total number of regular season games that the Brewers played in 2008?

   b. What percent of the games played did the Brewers win in 2008? Give the exact answer and an approximation to the nearest tenth of one percent.

115. ECONOMIC FORECASTS One economic indicator of the national economy is the number of orders placed by manufacturers. One month, the number of orders rose *one-fourth of 1 percent.*

   a. Write this using a % symbol.

   b. Express it as a fraction.

   c. Express it as a decimal.

116. TAXES In August of 2008, Springfield, Missouri, voters approved a *one-eighth of one percent* sales tax to fund transportation projects in the city.

   a. Write the percent as a decimal.

   b. Write the percent as a fraction.

117. BIRTHDAYS If the day of your birthday represents $\frac{1}{365}$ of a year, what percent of the year is it? Round to the nearest hundredth of a percent.

118. POPULATION As a fraction, each resident of the United States represents approximately $\frac{1}{305,000,000}$ of the U.S. population. Express this as a percent. Round to one nonzero digit.

WRITING

119. If you were writing advertising, which form do you think would attract more customers: "25% off" or "$\frac{1}{4}$ off"? Explain your reasoning.

120. Many coaches ask their players to give a 110% effort during practices and games. What do you think this means? Is it possible?

121. Explain how an amusement park could have an attendance that is 103% of capacity.

122. WON-LOST RECORDS In sports, when a team wins as many games as it loses, it is said to be playing "500 ball." Suppose in its first 40 games, a team wins 20 games and loses 20 games. Use the concepts in this section to explain why such a record could be called "500 ball."

REVIEW

123. The width of a rectangle is 6.5 centimeters and its length is 10.5 centimeters.

   a. Find its perimeter.

   b. Find its area.

124. The length of a side of a square is 9.8 meters.

   a. Find its perimeter.

   b. Find its area.

SECTION 7.2
Solving Percent Problems Using Percent Equations and Proportions

The articles on the front page of the newspaper on the right illustrate three types of percent problems.

Type 1 In the labor article, if we want to know how many union members voted to accept the new offer, we would ask:

> What number is 84% of 500?

Type 2 In the article on drinking water, if we want to know what percent of the wells are safe, we would ask:

> 38 is what percent of 40?

Type 3 In the article on new appointees, if we want to know how many members are on the State Board of Examiners, we would ask:

> 6 is 75% of what number?

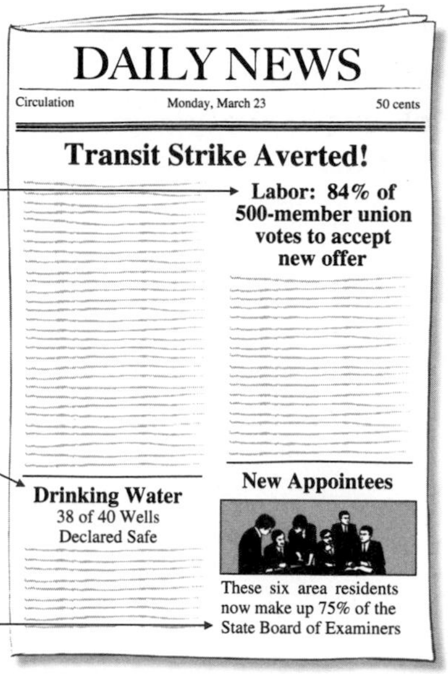

DAILY NEWS

Circulation   Monday, March 23   50 cents

Transit Strike Averted!

→ Labor: 84% of 500-member union votes to accept new offer

Drinking Water
38 of 40 Wells Declared Safe

New Appointees

These six area residents now make up 75% of the State Board of Examiners

This section introduces two methods that can be used to solve the percent problems shown above. The first method involves writing and solving *percent equations*. The second method involves writing and solving *percent proportions*. If your instructor only requires you to learn the proportion method, then turn to page 664 and begin reading Objective 1.

METHOD 1: PERCENT EQUATIONS

1 Translate percent sentences to percent equations.

The **percent sentences** highlighted in blue in the introduction above have three things in common.

- Each contains the word *is*. Here, *is* can be translated as an = symbol.
- Each contains the word *of*. In this case, *of* means multiply.
- Each contains a phrase such as *what number* or *what percent*. In other words, there is an unknown number that can be represented by a variable.

These observations suggest that each percent sentence contains key words that can be translated to form an equation. The equation, called a **percent equation,** will contain three numbers (two known and one unknown represented by a variable), the operation of multiplication, and, of course, an = symbol.

> **The Language of Algebra**   The key words in a percent sentence translate as follows:
>
> - *is* translates to an equal symbol = .
> - *of* translates to multiplication that is shown with a raised dot ·
> - *what number* or *what percent* translates to an unknown number that is represented by a variable.

Translate each percent sentence to a percent equation.

a. What number is 33% of 80?

b. What percent of 55 is 6?

c. 172% of what number is 4?

Now Try Problem 17

EXAMPLE 1   Translate each percent sentence to a percent equation.

a. What number is 12% of 64?

b. What percent of 88 is 11?

c. 165% of what number is 366?

Strategy We will look for the key words *is, of,* and *what number* (or *what percent*) in each percent sentence.

WHY These key words translate to mathematical symbols that form the percent equation.

Solution In each case, we will let the variable x represent the unknown number. However, any letter can be used.

a.   What number   is   12%   of   64?   This is the given percent sentence.
           x =   12%   ·   64   This is the percent equation.

b.   What percent   of   88   is   11?   This is the given percent sentence.
           x ·   88   =   11   This is the percent equation.

c.   165%   of   what number   is   366?   This is the given percent sentence.
         165%   · x =   366   This is the percent equation.

2   Solve percent equations to find the amount.

To solve the labor union percent problem (Type 1 from the newspaper), we translate the percent sentence into a percent equation and then find the unknown number.

What number is 36% of 400?

Now Try Problems 19 and 71

EXAMPLE 2   What number is 84% of 500?

Strategy We will look for the key words *is, of,* and *what number* in the percent sentence and translate them to mathematical symbols to form a percent equation.

WHY Then it will be clear what operation should be performed to find the unknown number.

Solution   First, we translate.

What number   is   84%   of   500?
   x =   84%   ·   500   Translate to a percent equation.

Now we perform the multiplication on the right side of the equation.

$x = 0.84 \cdot 500$ Write 84% as a decimal: 84% = 0.84.

$x = 420$ Do the multiplication.

We have found that 420 is 84% of 500. That is, 420 union members mentioned in the newspaper article voted to accept the new offer.

> ***The Language of Algebra***   When we find the value of the variable that makes a percent equation true, we say that we have **solved the equation.** In Example 2, we solved $x = 84\% \cdot 500$ to find that x is 420.

> ***Caution!***   When solving percent equations, always write the percent as a decimal (or a fraction) before performing any calculations. In Example 2, we wrote 84% as 0.84 before multiplying by 500.

Percent sentences involve a comparison of numbers. In the statement "420 is 84% of 500," the number 420 is called the **amount,** 84% is the **percent,** and 500 is called the **base.** Think of the base as the standard of comparison—it represents the **whole** of some quantity. The amount is a **part** of the base, but it can exceed the base when the percent is more than 100%. The percent, of course, has the % symbol.

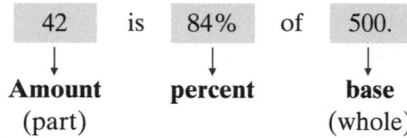

In any percent problem, the relationship between the amount, the percent, and the base is as follows: *Amount is percent of base.* This relationship is shown below as the **percent equation** (also called the **percent formula**).

Percent Equation (Formula)

Any percent sentence can be translated to a percent equation that has the form:

$$\text{Amount} = \text{percent} \cdot \text{base} \qquad \text{or} \qquad \text{Part} = \text{percent} \cdot \text{whole}$$

EXAMPLE 3   What number is 160% of 15.8?

Strategy We will look for the key words *is, of,* and *what number* in the percent sentence and translate them to mathematical symbols to form a percent equation.

WHY Then it will be clear what operation needs to be performed to find the unknown number.

Solution First, we translate.

What number	is	160%	of	15.8?
↓	↓	↓	↓	↓
x	$=$	160%	·	15.8

x is the amount, 160% is the percent, and 15.8 is the base.

Now we solve the equation by performing the multiplication on the right side.

$x = 1.6 \cdot 15.8$ Write 160% as a decimal: 160% = 1.6.

$x = 25.28$ Do the multiplication.

$$\begin{array}{r} 15.8 \\ \times\ 1.6 \\ \hline 948 \\ 1580 \\ \hline 25.28 \end{array}$$

Thus, 25.28 is 160% of 15.8. In this case, the amount exceeds the base because the percent is more than 100%.

Self Check 3

What number is 240% of 80.3?

Now Try **Problem 23**

3   Solve percent equations to find the percent.

In the drinking water problem (Type 2 from the newspaper), we must find the percent. Once again, we translate the words of the problem into a percent equation and solve it.

> **The Language of Algebra** We solve percent equations by writing a series of steps that result in an equation of the form $x =$ **a number** or **a number** $= x$. We say that the variable x is *isolated* on one side of the equation. *Isolated* means alone or by itself.

Self Check 4

4 is what percent of 80?

Now Try Problems 27 and 79

EXAMPLE 4   38 is what percent of 40?

Strategy We will look for the key words *is, of,* and *what percent* in the percent sentence and translate them to mathematical symbols to form a percent equation.

WHY Then we can solve the equation to find the unknown percent.

Solution First, we translate.

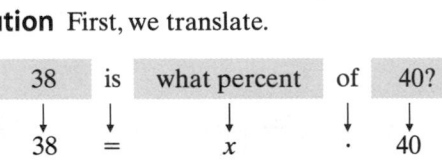

38	is	what percent	of	40?
↓	↓	↓	↓	↓
38	=	x	·	40

38 is the amount, x is the percent, and 40 is the base.

$38 = 40x$ Use the commutative property of multiplication to write $x \cdot 40$ as $40x$.

$\dfrac{38}{40} = \dfrac{40x}{40}$ To isolate x, undo the multiplication by 40 by dividing both sides by 40.

$0.95 = x$ Do the division: $38 \div 40 = 0.95$.

$$
\begin{array}{r}
0.95 \\
40\overline{)38.00} \\
-36\,0 \\
\hline
2\,00 \\
-2\,00 \\
\hline
0
\end{array}
$$

Since we want to find the percent, we need to write the decimal 0.95 as a percent.

$0\,95\% = x$ To write 0.95 as a percent, multiply it by 100 by moving the decimal point two places to the right, and then insert a % symbol.

$95\% = x$

We have found that 38 is 95% of 40. That is, 95% of the wells mentioned in the newspaper article were declared safe.

Self Check 5

9 is what percent of 16?

Now Try Problem 31

EXAMPLE 5   14 is what percent of 32?

Strategy We will look for the key words *is, of,* and *what percent* in the percent sentence and translate them to mathematical symbols to form a percent equation.

WHY Then we can solve the equation to find the unknown percent.

Solution First, we translate.

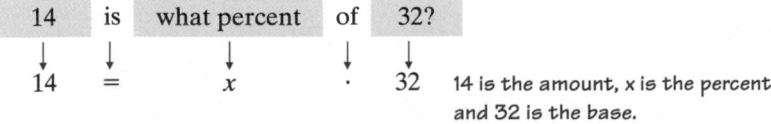

14	is	what percent	of	32?
↓	↓	↓	↓	↓
14	=	x	·	32

14 is the amount, x is the percent, and 32 is the base.

$14 = 32x$ Use the commutative property of multiplication to
write x · 32 as 32x.

$\dfrac{14}{32} = \dfrac{32x}{32}$ To isolate x, undo the multiplication by 32 by
dividing both sides by 32.

$0.4375 = x$ Do the division: 14 ÷ 32 = 0.4375.

$0\underset{\curvearrowright}{.}43.75\% = x$ To write the decimal 0.4375 as a percent, multiply
it by 100 by moving the decimal point two places
to the right, and then insert a % symbol.

$43.75\% = x$

```
        0.4375
32) 14.0000
  − 12 8
    1 20
    − 96
     240
   − 224
     160
   − 160
       0
```

Thus, 14 is 43.75% of 32.   ◼

Using Your CALCULATOR   Cost of an Air Bag

An air bag is estimated to add an additional $500 to the cost of a car. What percent of the $16,295 sticker price is the cost of the air bag?

First, we translate the words of the problem into a percent equation.

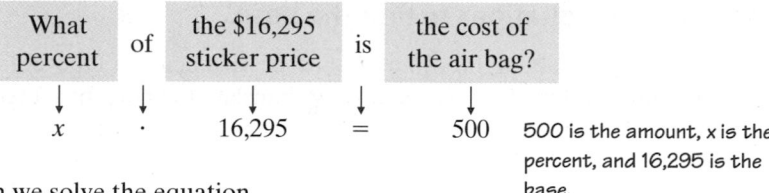

| What percent | of | the $16,295 sticker price | is | the cost of the air bag? |

x ·   16,295   =   500   500 is the amount, x is the percent, and 16,295 is the base.

Then we solve the equation.

$16{,}295x = 500$ Write x · 16,295 as 16,295x.

$\dfrac{16{,}295x}{16{,}295} = \dfrac{500}{16{,}295}$ To undo the multiplication by 16,295 and isolate x on the left side, divide both sides of the equation by 16,295.

$x = \dfrac{500}{16{,}295}$

To perform the division on the right side using a scientific calculator, enter the following:

500 ÷ 16295 = $\boxed{0.03068\overline{4}259}$

This display gives the answer in decimal form. To change it to a percent, we multiply the result by 100. This moves the decimal point 2 places to the right. (See the display.) Then we insert a % symbol. If we round to the nearest tenth of a percent, the cost of the air bag is about 3.1% of the sticker price.

$\boxed{3.068425898}$

EXAMPLE 6   What percent of 6 is 7.5?

Strategy We will look for the key words *is, of,* and *what percent* in the percent sentence and translate them to mathematical symbols to form a percent equation.

WHY Then we can solve the equation to find the unknown percent.

Self Check 6

What percent of 5 is 8.5?

Now Try Problem 35

Solution First, we translate.

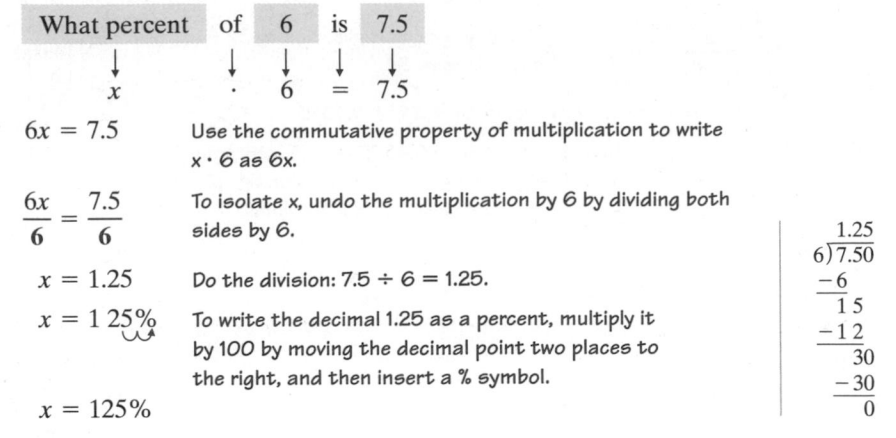

What percent	of	6	is	7.5	
↓		↓	↓	↓	
x	·	6	=	7.5	

$6x = 7.5$ Use the commutative property of multiplication to write $x \cdot 6$ as $6x$.

$\dfrac{6x}{6} = \dfrac{7.5}{6}$ To isolate x, undo the multiplication by 6 by dividing both sides by 6.

$x = 1.25$ Do the division: $7.5 ÷ 6 = 1.25$.

$x = 1\underset{\smile}{25}\%$ To write the decimal 1.25 as a percent, multiply it by 100 by moving the decimal point two places to the right, and then insert a % symbol.

$x = 125\%$

$$
\begin{array}{r}
1.25 \\
6\overline{)7.50} \\
-6 \\
\hline
1\,5 \\
-1\,2 \\
\hline
30 \\
-30 \\
\hline
0
\end{array}
$$

Thus, 7.5 is 125% of 6.

---

4 Solve percent equations to find the base.

In the percent problem about the State Board of Examiners (Type 3 from the newspaper), we must find the base. As before, we translate the percent sentence into a percent equation and then find the unknown number.

Self Check 7

3 is 5% of what number?

Now Try Problem 39

EXAMPLE 7   6 is 75% of what number?

Strategy We will look for the key words *is, of,* and *what number* in the percent sentence and translate them to mathematical symbols to form a percent equation.

WHY Then we can solve the equation to find the unknown number.

Solution First, we translate.

6	is	75%	of	what number?
↓	↓	↓	↓	↓
6	=	75%	·	x

6 is the amount, 75% is the percent, and x is the base.

Now we solve the equation.

$6 = 0.75x$ Write 75% as a decimal: 75% = 0.75. It is not necessary to write the multiplication raised dot.

$\dfrac{6}{0.75} = \dfrac{0.75x}{0.75}$ To isolate x, undo the multiplication by 0.75 by dividing both sides by 0.75.

$8 = x$ Do the division: $6 ÷ 0.75 = 8$.

$$
\begin{array}{r}
8 \\
75\overline{)600} \\
-600 \\
\hline
0
\end{array}
$$

Thus, 6 is 75% of 8. That is, there are 8 members on the State Board of Examiners mentioned in the newspaper article.

> **Success Tip** Sometimes the calculations to solve a percent problem are made easier if we write the percent as a fraction instead of a decimal. This is the case with percents that have *repeating* decimal equivalents such as $33\frac{1}{3}\%$, $66\frac{2}{3}\%$, and $16\frac{2}{3}\%$. You may want to review the table of percents and their fractional equivalents on page 652.

EXAMPLE 8 31.5 is $33\frac{1}{3}\%$ of what number?

Strategy We will look for the key words *is*, *of*, and *what number* in the percent sentence and translate them to mathematical symbols to form a percent equation.

WHY Then we can solve the equation to find the unknown number.

Solution First, we translate.

31.5	is	$33\frac{1}{3}\%$	of	what number?
↓	↓	↓	↓	↓
31.5	=	$33\frac{1}{3}\%$	·	x

31.5 is the amount, $33\frac{1}{3}\%$ is the percent, and x is the base.

In this case, the calculations can be made easier by writing $33\frac{1}{3}\%$ as a fraction instead of as a repeating decimal.

$31.5 = \dfrac{1}{3}x$ Recall from Section 7.1 that $33\frac{1}{3}\% = \frac{1}{3}$. It is not necessary to write the multiplication raised dot.

$3(31.5) = 3\left(\dfrac{1}{3}x\right)$ The coefficient of x is the fraction $\frac{1}{3}$. To isolate x, multiply both sides by the reciprocal of $\frac{1}{3}$, which is 3.

$94.5 = \left(3 \cdot \dfrac{1}{3}\right)x$ On the left side, do the multiplication: $3(31.5) = 94.5$. On the right side, use the associative property of multiplication to group 3 and $\frac{1}{3}$.

$$\begin{array}{r} \overset{1}{31.5} \\ \times \ 3 \\ \hline 94.5 \end{array}$$

$94.5 = 1x$ On the right side, the product of a number and its reciprocal is 1.

$94.5 = x$ On the right side, the coefficient of 1 need not be written since $1x = x$.

Thus, 31.5 is $33\frac{1}{3}\%$ of 94.5.

To solve percent application problems, we often have to rewrite the facts of the problem in percent sentence form before we can translate to an equation.

EXAMPLE 9 *Rentals* In an apartment complex, 198 of the units are currently occupied. If this represents an 88% occupancy rate, how many units are in the complex?

Strategy We will carefully read the problem and use the given facts to write them in the form of a percent sentence.

Self Check 8

150 is $66\frac{2}{3}\%$ of what number?

Now Try **Problems 43 and 83**

Self Check 9

CAPACITY OF A GYM A total of 784 people attended a graduation in a high school gymnasium. If this was 98% of capacity, what is the total capacity of the gym?

Now Try Problem 81

WHY Then we can translate the sentence into a percent equation and solve it to find the unknown number of units in the complex.

Solution An occupancy rate of 88% means that 88% of the units are occupied. Thus, the 198 units that are currently occupied are 88% of some unknown number of units in the complex, and we can write:

198	is	88%	of	what number?
↓	↓	↓	↓	↓
198	=	88%	·	x

198 is the amount, 88% is the percent, and x is the base.

Now we solve the equation.

$198 = 0.88x$ Write 88% as a decimal: 88% = 0.88. It is not necessary to write the multiplication raised dot.

$\dfrac{198}{0.88} = \dfrac{0.88x}{0.88}$ To isolate x, undo the multiplication by 0.88 by dividing both sides by 0.88.

$225 = x$ Do the division: 198 ÷ 0.88 = 225.

$$\begin{array}{r} 225 \\ 88\overline{)19800} \\ -176 \\ \hline 220 \\ -176 \\ \hline 440 \\ -440 \\ \hline 0 \end{array}$$

The apartment complex has 225 units, of which 198, or 88%, are occupied.

If you are only learning the percent equation method for solving percent problems, turn to page 671 and pick up your reading at Objective 5.

METHOD 2: PERCENT PROPORTIONS

1 Write percent proportions.

Another method to solve percent problems involves writing and then solving a proportion. To introduce this method, consider the figure on the right. The vertical line down its middle divides the figure into two equal-sized parts. Since 1 of the 2 parts is shaded red, the shaded portion of the figure can be described by the ratio $\frac{1}{2}$. We call this an **amount-to-base** (or **part-to-whole**) **ratio.**

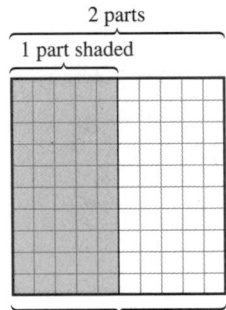

2 parts

1 part shaded

50 of the 100 parts shaded: 50% shaded

Now consider the 100 equal-sized square regions within the figure. Since 50 of them are shaded red, we say that $\frac{50}{100}$, or 50% of the figure is shaded. The ratio $\frac{50}{100}$ is called a **percent ratio.**

Since the amount-to-base ratio, $\frac{1}{2}$, and the percent ratio, $\frac{50}{100}$, represent the same shaded portion of the figure, they must be equal, and we can write

The amount-to-base ratio ⟹ $\dfrac{1}{2} = \dfrac{50}{100}$ ⟸ The percent ratio

Recall from Section 6.2 that statements of this type stating that two ratios are equal are called *proportions*. We call $\frac{1}{2} = \frac{50}{100}$ a **percent proportion.** The four terms of a percent proportion are shown on the following page.

Percent Proportion

To translate a percent sentence to a **percent proportion,** use the following form:

Amount is to base as percent is to 100.   *Part is to whole as percent is to 100.*

$$\frac{\text{amount}}{\text{base}} = \frac{\text{percent}}{100} \qquad \text{or} \qquad \frac{\text{part}}{\text{whole}} = \frac{\text{percent}}{100}$$

This is always 100 because percent
means parts per one hundred.

To write a percent proportion, you must identify 3 of the terms as you read the problem. (Remember, the fourth term of the proportion is always 100.) Here are some ways to identify those terms.

- The **percent** is easy to find. Look for the % symbol or the words *what percent.*
- The **base** (or **whole**) usually follows the word *of.*
- The **amount** (or **part**) is compared to the base (or whole).

EXAMPLE 1   Translate each percent sentence to a percent proportion.

a. What number is 12% of 64?

b. What percent of 88 is 11?

c. 165% of what number is 366?

Strategy A percent proportion has the form $\frac{\text{amount}}{\text{base}} = \frac{\text{percent}}{100}$. Since one of the terms of the percent proportion is always 100, we only need to identify three terms to write the proportion. We will begin by identifying the percent and the base in the given sentence.

WHY The remaining number (or unknown) must be the amount.

Solution

a. We will identify the terms in this order:

- *First:* the percent (next to the % symbol)
- *Second:* the base (usually after the word *of*)
- *Last:* the amount (the number that remains)

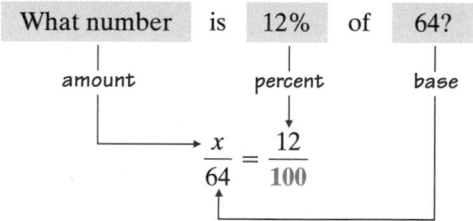

b.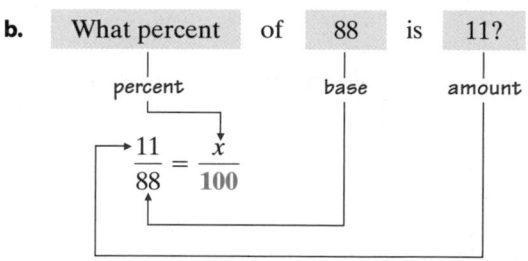

Self Check 1

Translate each percent sentence to a percent proportion.

a. What number is 33% of 80?

b. What percent of 55 is 6?

c. 172% of what number is 4?

Now Try **Problem 17**

c. 165% of what number is 366?

percent   base   amount

$$\frac{366}{x} = \frac{165}{100}$$

2 Solve percent proportions to find the amount.

Recall the labor union problem from the newspaper example in the introduction to this section. We can write and solve a percent proportion to find the unknown amount.

Self Check 2

What number is 36% of 400?

Now Try Problems 19 and 71

EXAMPLE 2   What number is 84% of 500?

Strategy We will identify the percent, the base, and the amount and write a percent proportion of the form $\frac{amount}{base} = \frac{percent}{100}$.

WHY Then we can solve the proportion to find the unknown number.

Solution First, we write the percent proportion.

What number   is   84%   of   500?

amount   percent   base

$$\frac{x}{500} = \frac{84}{100}$$ This is the proportion to solve.

To make the calculations easier, it is helpful to simplify the ratio $\frac{84}{100}$ at this time.

$$\frac{x}{500} = \frac{21}{25}$$ On the right side, simplify: $\frac{84}{100} = \frac{\overset{1}{\cancel{4}} \cdot 21}{\cancel{4} \cdot 25} = \frac{21}{25}$.

Recall from Section 6.2 that to solve a proportion we use the cross products.

$x \cdot 25 = 500 \cdot 21$ Find the cross products: $\frac{x}{500} = \frac{21}{25}$.
    Then set them equal.

$25x = 10,500$ On the left side, use the commutative property of multiplication to write $x \cdot 25$ as $25x$. On the right side, do the multiplication: $500 \cdot 21 = 10,500$.

$$\frac{25x}{25} = \frac{10,500}{25}$$ To isolate x, undo the multiplication by 25 by dividing both sides by 25.

$x = 420$ Do the division: $10,500 \div 25 = 420$.

$$\begin{array}{r} 500 \\ \times\ 21 \\ \hline 500 \\ 10\ 000 \\ \hline 10,500 \end{array}$$

$$\begin{array}{r} 420 \\ 25\overline{)10,500} \\ -10\ 0 \\ \hline 50 \\ -50 \\ \hline 00 \\ -0 \\ \hline 0 \end{array}$$

We have found that 420 is 84% of 500. That is, 420 union members mentioned in the newspaper article voted to accept the new offer.

The Language of Algebra   When we find the value of the variable that makes a percent proportion true, we say that we have **solved the proportion.** In Example 2, we solved $\frac{x}{500} = \frac{84}{100}$ to find that x is 420.

EXAMPLE 3   What number is 160% of 15.8?

Strategy We will identify the percent, the base, and the amount and write a percent proportion of the form $\frac{amount}{base} = \frac{percent}{100}$.

WHY Then we can solve the proportion to find the unknown number.

Solution First, we write the percent proportion.

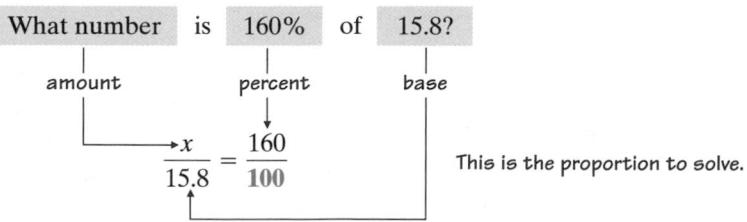

This is the proportion to solve.

To make the calculations easier, it is helpful to simplify the ratio $\frac{160}{100}$ at this time.

$\dfrac{x}{15.8} = \dfrac{8}{5}$ On the right side, simplify: $\dfrac{160}{100} = \dfrac{8 \cdot \overset{1}{\cancel{20}}}{5 \cdot \underset{1}{\cancel{20}}} = \dfrac{8}{5}$.

$x \cdot 5 = 15.8 \cdot 8$ Find the cross products: $\dfrac{x}{15.8} \bcancel{=} \dfrac{8}{5}$.
Then set them equal.

$5x = 126.4$ On the left side, use the commutative property of multiplication to write $x \cdot 5$ as $5x$. On the right side, do the multiplication: $15.8 \cdot 8 = 126.4$.

$\dfrac{5x}{5} = \dfrac{126.4}{5}$ To isolate x, undo the multiplication by 5 by dividing both sides by 5.

$x = 25.28$ Do the division: $126.4 \div 5 = 25.28$.

$$\begin{array}{r} \overset{4\ 6}{15.8} \\ \times\ 8 \\ \hline 126.4 \end{array}$$

$$\begin{array}{r} 25.28 \\ 5\overline{)126.40} \\ -10\ \ \ \ \ \\ \hline 26\ \ \ \ \ \\ -25\ \ \ \ \ \\ \hline 1\ 4\ \ \ \\ -1\ 0\ \ \ \\ \hline 40\ \\ -40\ \\ \hline 0 \end{array}$$

Thus, 25.28 is 160% of 15.8.

Self Check 3

What number is 240% of 80.3?

Now Try Problem 23

3 Solve percent proportions to find the percent.

Recall the drinking water problem from the newspaper example in the introduction to this section. We can write and solve a percent proportion to find the unknown percent.

EXAMPLE 4   38 is what percent of 40?

Strategy We will identify the percent, the base, and the amount and write a percent proportion of the form $\frac{amount}{base} = \frac{percent}{100}$.

WHY Then we can solve the proportion to find the unknown percent.

Solution First, we write the percent proportion.

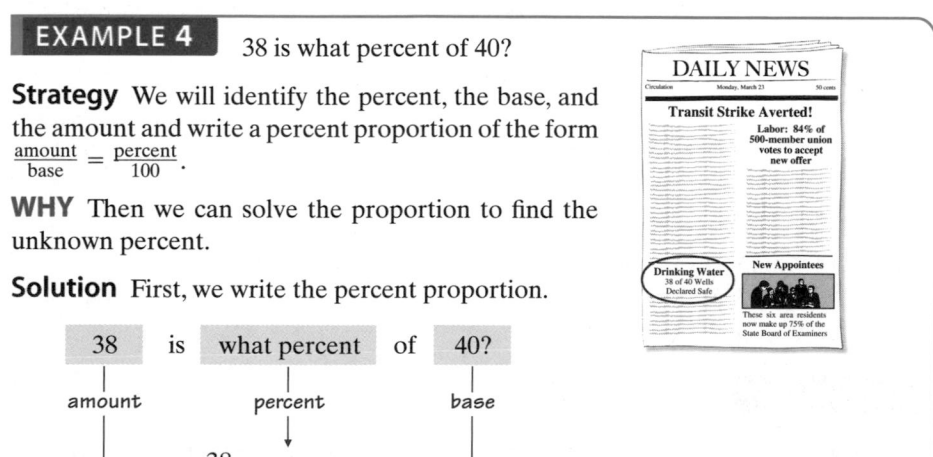

This is the proportion to solve.

Self Check 4

4 is what percent of 80?

Now Try Problems 27 and 79

To make the calculations easier, it is helpful to simplify the ratio $\frac{38}{40}$ at this time.

$$\frac{19}{20} = \frac{x}{100}$$ On the left side, simplify: $\frac{38}{40} = \frac{\overset{1}{2} \cdot 19}{2 \cdot 20} = \frac{19}{20}$.

$19 \cdot 100 = 20 \cdot x$ To solve the proportion, find the cross products: $\frac{19}{20} \bowtie \frac{x}{100}$.
Then set them equal.

$1{,}900 = 20x$ On the left side, do the multiplication: $19 \cdot 100 = 1{,}900$.
On the right side, write $20 \cdot x$ as 20x.

$\dfrac{1{,}900}{20} = \dfrac{20x}{20}$ To isolate x, undo the multiplication by 20 by
dividing both sides by 20.

$95 = x$ Do the division: $1{,}900 \div 20 = 95$.

$$\begin{array}{r} 95 \\ 20\overline{)1{,}900} \\ -1\,80 \\ \hline 100 \\ -100 \\ \hline 0 \end{array}$$

We have found that 38 is 95% of 40. That is, 95% of the wells mentioned in the newspaper article were declared safe. ∎

Self Check 5
9 is what percent of 16?
Now Try Problem 31

EXAMPLE 5   14 is what percent of 32?

Strategy We will identify the percent, the base, and the amount and write a percent proportion of the form $\frac{\text{amount}}{\text{base}} = \frac{\text{percent}}{100}$.

WHY Then we can solve the proportion to find the unknown percent.

Solution First, we write the percent proportion.

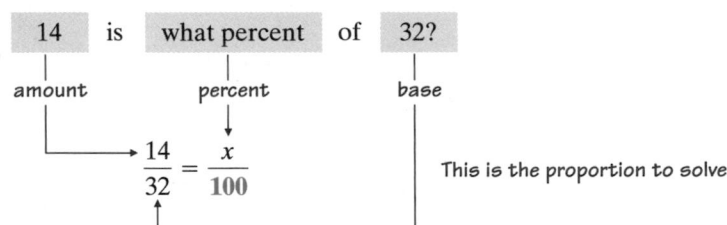

$$\begin{array}{ccc} \boxed{14} & \text{is} & \boxed{\text{what percent}} & \text{of} & \boxed{32?} \\ | & & | & & | \\ \text{amount} & & \text{percent} & & \text{base} \end{array}$$

$$\frac{14}{32} = \frac{x}{100}$$ This is the proportion to solve.

To make the calculations easier, it is helpful to simplify the ratio $\frac{14}{32}$ at this time.

$$\frac{7}{16} = \frac{x}{100}$$ On the left side, simplify: $\frac{14}{32} = \frac{\overset{1}{2} \cdot 7}{2 \cdot 16} = \frac{7}{16}$.

$7 \cdot 100 = 16 \cdot x$ To solve the proportion, find the cross products: $\frac{7}{16} \bowtie \frac{x}{100}$.
Then set them equal.

$700 = 16x$ On the left side, do the multiplication:
$7 \cdot 100 = 700$. On the right side, write
$16 \cdot x$ as 16x.

$\dfrac{700}{16} = \dfrac{16x}{16}$ To isolate x, undo the multiplication by 16 by
dividing both sides by 16.

$43.75 = x$ Do the division: $700 \div 16 = 43.75$.

$$\begin{array}{r} 43.75 \\ 16\overline{)700.00} \\ -64 \\ \hline 60 \\ -48 \\ \hline 12\,0 \\ -11\,2 \\ \hline 80 \\ -80 \\ \hline 0 \end{array}$$

Thus, 14 is 43.75% of 32. ∎

Self Check 6
What percent of 5 is 8.5?
Now Try Problem 35

EXAMPLE 6   What percent of 6 is 7.5?

Strategy We will identify the percent, the base, and the amount and write a percent proportion of the form $\frac{\text{amount}}{\text{base}} = \frac{\text{percent}}{100}$.

WHY Then we can solve the proportion to find the unknown percent.

Solution First, we write the percent proportion.

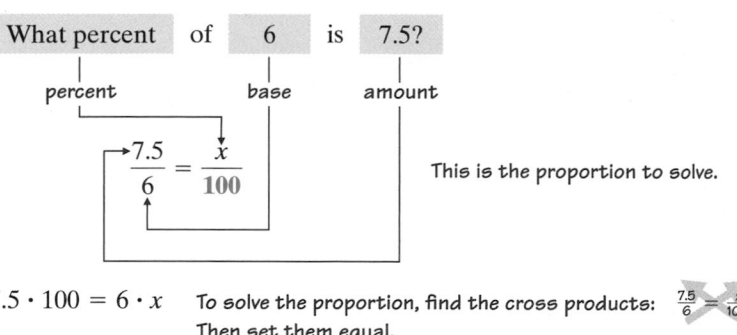

$7.5 \cdot 100 = 6 \cdot x$ To solve the proportion, find the cross products: $\frac{7.5}{6} \diagup\!\!\!\!\! = \diagup\!\!\!\!\! \frac{x}{100}$.
Then set them equal.

$750 = 6x$ On the left side, do the multiplication:
7.5 · 100 = 750. On the right side, write
6 · x as 6x.

$\dfrac{750}{6} = \dfrac{6x}{6}$ To isolate x, undo the multiplication by 6 by
dividing both sides by 6.

$125 = x$ Do the division: 750 ÷ 6 = 125.

$$
\begin{array}{r}
125 \\
6\overline{)750} \\
\underline{-6} \\
15 \\
\underline{-12} \\
30 \\
\underline{-30} \\
0
\end{array}
$$

Thus, 7.5 is 125% of 6.

---

4 **Solve percent proportions to find the base.**

Recall the State Board of Examiners problem from the newspaper example in the introduction to this section. We can write and solve a percent proportion to find the unknown base.

EXAMPLE 7   6 is 75% of what number?

Strategy We will identify the percent, the base, and the amount and write a percent proportion of the form $\frac{\text{amount}}{\text{base}} = \frac{\text{percent}}{100}$.

WHY Then we can solve the proportion to find the unknown number.

Solution First, we write the percent proportion.

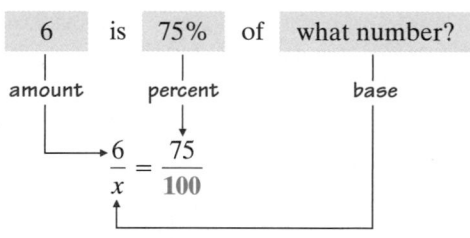

Self Check 7
3 is 5% of what number?
Now Try Problem 39

To make the calculations easier, it is helpful to simplify the ratio $\frac{75}{100}$ at this time.

$\dfrac{6}{x} = \dfrac{3}{4}$ Simplify: $\dfrac{75}{100} = \dfrac{3 \cdot \overset{1}{\cancel{25}}}{4 \cdot \underset{1}{\cancel{25}}} = \dfrac{3}{4}$.

$6 \cdot 4 = x \cdot 3$ To solve the proportion, find the cross products: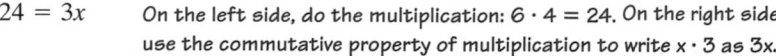
Then set them equal.

$24 = 3x$ On the left side, do the multiplication: 6 · 4 = 24. On the right side,
use the commutative property of multiplication to write x · 3 as 3x.

$$\frac{24}{3} = \frac{3x}{3}$$ To isolate x, undo the multiplication by 3 by dividing both sides by 3.

$$8 = x$$ Do the division: 24 ÷ 3 = 8.

Thus, 6 is 75% of 8. That is, there are 8 members on the State Board of Examiners mentioned in the newspaper article.

Self Check 8

150 is $66\frac{2}{3}$% of what number?

Now Try Problems 43 and 83

EXAMPLE 8   31.5 is $33\frac{1}{3}$% of what number?

Strategy We will identify the percent, the base, and the amount and write a percent proportion of the form $\frac{\text{amount}}{\text{base}} = \frac{\text{percent}}{100}$.

WHY Then we can solve the proportion to find the unknown number.

Solution First, we write the percent proportion.

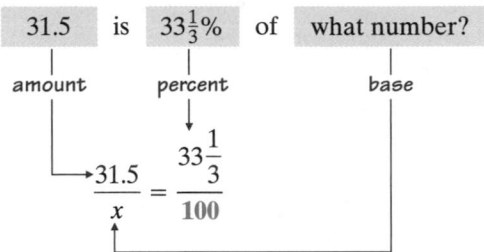

To make the calculations easier, it is helpful to write the mixed number $33\frac{1}{3}$ as the improper fraction $\frac{100}{3}$.

$$\frac{31.5}{x} = \frac{\frac{100}{3}}{100}$$ Write $33\frac{1}{3}$ as $\frac{100}{3}$.

$$31.5 \cdot 100 = x \cdot \frac{100}{3}$$ To solve the proportion, find the cross products: $\frac{31.5}{x} = \frac{\frac{100}{3}}{100}$. Then set them equal.

$$3{,}150 = \frac{100}{3}x$$ On the left side, do the multiplication: 31.5 · 100 = 3,150. On the right side, use the commutative property of multiplication to write $x \cdot \frac{100}{3}$ as $\frac{100}{3}x$.

$$\frac{3}{100}(3{,}150) = \frac{3}{100}\left(\frac{100}{3}x\right)$$ The coefficient of x is the fraction $\frac{100}{3}$. To isolate x, multiply both sides by the reciprocal of $\frac{100}{3}$, which is $\frac{3}{100}$.

$$\frac{3}{100}\left(\frac{3{,}150}{1}\right) = \left(\frac{3}{100} \cdot \frac{100}{3}\right)x$$ On the left side, write 3,150 as a fraction: 3,150 = $\frac{3{,}150}{1}$. On the right, use the associative property of multiplication to group $\frac{3}{100}$ and $\frac{100}{3}$.

$$\frac{3 \cdot 3{,}150}{100 \cdot 1} = 1x$$ On the left side, multiply the numerators and multiply the denominators. On the right side, the product of a number and its reciprocal is 1.

$$\frac{9{,}450}{100} = x$$ On the left side, 3 · 3,150 = 9,450. On the right side, the coefficient of 1 need not be written since 1x = x.

$$\begin{array}{r}\overset{1}{3{,}150}\\ \underline{\times\ \ 3}\\ 9{,}450\end{array}$$

$$94.50 = x$$ Divide 9,450 by 100 by moving the understood decimal point in 9,450 two places to the left.

Thus, 31.5 is $33\frac{1}{3}$% of 94.5.

To solve percent application problems, we often have to rewrite the facts of the problem in percent sentence form before we can translate to an equation.

EXAMPLE 9   *Rentals*   In an apartment complex, 198 of the units are currently occupied. If this represents an 88% occupancy rate, how many units are in the complex?

Strategy   We will carefully read the problem and use the given facts to write them in the form of a percent sentence.

WHY   Then we can write and solve a percent proportion to find the unknown number of units in the complex.

Solution   An occupancy rate of 88% means that 88% of the units are occupied. Thus, the 198 units that are currently occupied are 88% of some unknown number of units in the complex, and we can write:

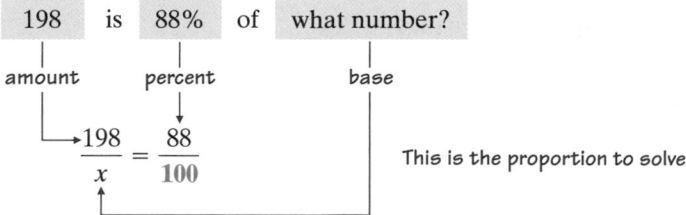

To make the calculations easier, it is helpful to simplify the ratio $\frac{88}{100}$ at this time.

$$\frac{198}{x} = \frac{22}{25} \qquad \text{On the right side, simplify:} \quad \frac{88}{100} = \frac{\overset{1}{\cancel{4}} \cdot 22}{\underset{1}{\cancel{4}} \cdot 25} = \frac{22}{25}.$$

$$198 \cdot 25 = x \cdot 22 \qquad \text{Find the cross products. Then set them equal.}$$

$$4{,}950 = 22x \qquad \begin{array}{l}\text{On the left side, do the multiplication:} \\ 198 \cdot 25 = 4{,}950. \text{ On the right side, use the} \\ \text{commutative property of multiplication to write} \\ x \cdot 22 \text{ as } 22x.\end{array}$$

$$\frac{4{,}950}{22} = \frac{22x}{22} \qquad \begin{array}{l}\text{To isolate } x, \text{ undo the multiplication by 22 by} \\ \text{dividing both sides by 22.}\end{array}$$

$$225 = x \qquad \begin{array}{l}\text{On the left side, do the division:} \\ 4{,}950 \div 22 = 225.\end{array}$$

$$\begin{array}{r}198 \\ \times 25 \\ \hline 990 \\ 3960 \\ \hline 4{,}950\end{array}$$

$$\begin{array}{r}225 \\ 22\overline{)4{,}950} \\ -44 \\ \hline 55 \\ -44 \\ \hline 110 \\ -110 \\ \hline 0\end{array}$$

The apartment complex has 225 units, of which 198, or 88%, are occupied. ∎

Self Check 9

CAPACITY OF A GYM   A total of 784 people attended a graduation in a high school gymnasium. If this was 98% of capacity, what is the total capacity of the gym?

Now Try Problem 81

5   Read circle graphs.

Percents are used with **circle graphs,** or **pie charts,** as a way of presenting data for comparison. In the figure below, the entire circle represents the total amount of electricity generated in the United States in 2008. The pie-shaped pieces of the graph show the relative sizes of the energy sources used to generate the electricity. For example, we see that the greatest amount of electricity (50%) was generated from coal. Note that if we add the percents from all categories (50% + 3% + 7% + 18% + 20% + 2%), the sum is 100%.

The 100 tick marks equally spaced around the circle serve as a visual aid when constructing a circle graph. For example, to represent hydropower as 7%, a line was drawn from the center of the circle to a tick mark. Then we counted off 7 ticks and drew a second line from the center to that tick to complete the pie-shaped wedge.

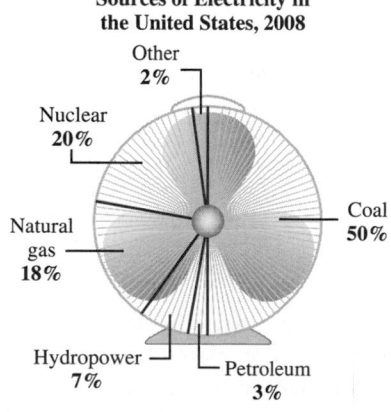

Sources of Electricity in the United States, 2008

Source: Energy Information Administration

Self Check 10

PRESIDENTIAL ELECTIONS Results from the 2004 U.S. presidential election are shown in the circle graph below. Find the number of states won by President Bush.

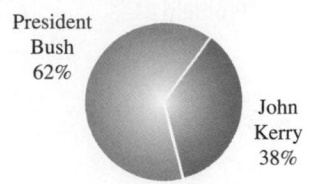

2004 Presidential Election
States won by each candidate

Now Try Problem 85

EXAMPLE 10   *Presidential Elections*

Results from the 2008 U.S. presidential election are shown in the circle graph to the right. Find the number of states won by Barack Obama.

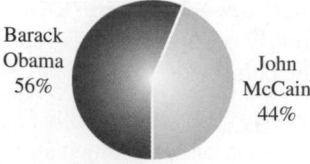

2008 Presidential Election
States won by each candidate

Strategy We will rewrite the facts of the problem in percent sentence form.

WHY Then we can translate the sentence to a percent equation (or percent proportion) to find the number of states won by Barack Obama.

Solution The circle graph shows that Barack Obama won 56% of the 50 states. Thus, the percent is 56% and the base is 50. One way to find the unknown amount is to write and then solve a percent equation.

What number	is	56%	of	50?
↓	↓	↓	↓	↓
x	$=$	56%	$\cdot$	50

Translate to a percent equation.

Now we perform the multiplication on the right side of the equation.

$x = 0.56 \cdot 50$ Write 56% as a decimal: 56% = 0.56.

$x = 28$ Do the multiplication.

$$\begin{array}{r} 50 \\ \times 0.56 \\ \hline 3\ 00 \\ 25\ 00 \\ \hline 28.00 \end{array}$$

Thus, Barack Obama won 28 of the 50 states in the 2008 U.S. presidential election.
    Another way to find the unknown amount is to write and then solve a percent proportion.

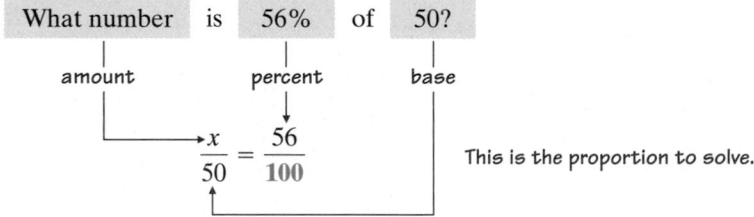

This is the proportion to solve.

To make the calculations easier, it is helpful to simplify the ratio $\frac{56}{100}$ at this time.

$\dfrac{x}{50} = \dfrac{14}{25}$ On the right side, simplify: $\dfrac{56}{100} = \dfrac{\overset{1}{\cancel{4}} \cdot 14}{\cancel{4} \cdot 25} = \dfrac{14}{25}$.

$x \cdot 25 = 50 \cdot 14$ Find the cross products: $\frac{x}{50} \bowtie \frac{14}{25}$. Then set them equal.

$25x = 700$ On the left side, use the commutative property of multiplication to write $x \cdot 25$ as $25x$. On the right side, do the multiplication: $50 \cdot 14 = 700$.

$\dfrac{25x}{25} = \dfrac{700}{25}$ To isolate x, undo the multiplication by 25 by dividing both sides by 25.

$x = 28$ On the right side, do the division: $700 \div 25 = 28$.

$$\begin{array}{r} 50 \\ \times 14 \\ \hline 200 \\ 500 \\ \hline 700 \end{array}$$

$$\begin{array}{r} 28 \\ 25\overline{)700} \\ -50 \\ \hline 200 \\ -200 \\ \hline 0 \end{array}$$

As we would expect, the percent proportion method gives the same answer as the percent equation method. Barack Obama won 28 of the 50 states in the 2008 U.S. presidential election.

THINK IT THROUGH   *Community College Students*

"When the history of American higher education is updated years from now, the story of our current times will highlight the pivotal role community colleges played in developing human capital and bolstering the nation's educational system."

Community College Survey of Student Engagement, 2007

More than 310,000 students responded to the 2007 Community College Survey of Student Engagement. Some results are shown below. Study each circle graph and then complete its legend.

Enrollment in Community Colleges	**Community College Students Who Work More Than 20 Hours per Week**	**Community College Students Who Discussed Their Grades or Assignments with an Instructor**
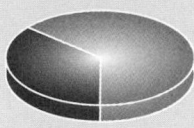	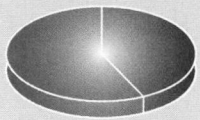	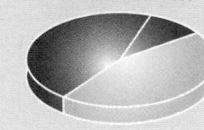

■ 64% are enrolled in college part time.

■ ?

■ 57% of the students work more than 20 hours per week.

■ ?

■ 45% often or very often

■ 45% sometimes

■ ?

ANSWERS TO SELF CHECKS

1. a. $x = 33\% \cdot 80$ or $\frac{x}{80} = \frac{33}{100}$ **b.** $x \cdot 55 = 6$ or $\frac{6}{55} = \frac{x}{100}$ **c.** $172\% \cdot x = 4$ or $\frac{4}{x} = \frac{172}{100}$
2. 144 **3.** 192.72 **4.** 5% **5.** 56.25% **6.** 170% **7.** 60 **8.** 225 **9.** 800 people
10. 31 states

SECTION 7.2   STUDY SET

▌VOCABULARY

Fill in the blanks.

1. We call "What number is 15% of 25?" a percent _____. It translates to the percent _____ $x = 15\% \cdot 25$.

2. The key words in a percent sentence translate as follows:

- ___ translates to an equal symbol =
- ___ translates to multiplication that is shown with a raised dot ·
- _____ *number* or _____ *percent* translates to an unknown number that is represented by a variable.

3. When we find the value of the variable that makes a percent equation true, we say that we have _____ the equation.

4. In the percent sentence "45 is 90% of 50," 45 is the _____, 90% is the percent, and 50 is the _____.

5. The amount is _____ of the base. The base is the standard of comparison—it represents the _____ of some quantity.

6. a. *Amount is to base as percent is to 100:*

$$\frac{}{\text{base}} = \frac{\text{percent}}{}$$

b. *Part is to whole as percent is to 100:*

$$\frac{\text{part}}{} = \frac{}{100}$$

7. The _____ products for the proportion $\frac{24}{x} = \frac{36}{100}$ are $24 \cdot 100$ and $x \cdot 36$.

8. In a _____ graph, pie-shaped wedges are used to show the division of a whole quantity into its component parts.

CONCEPTS

9. Fill in the blanks to complete the percent equation (formula):

$$\boxed{} = \text{percent} \cdot \boxed{}$$

or

$$\text{Part} = \boxed{} \cdot \boxed{}$$

10. a. Without doing the calculation, tell whether 12% of 55 is more than 55 or less than 55.

    b. Without doing the calculation, tell whether 120% of 55 is more than 55 or less than 55.

11. CANDY SALES The circle graph shows the percent of the total candy sales for each of four holiday seasons in 2008. What is the sum of all the percents?

Percent of Total Candy Sales, 2008

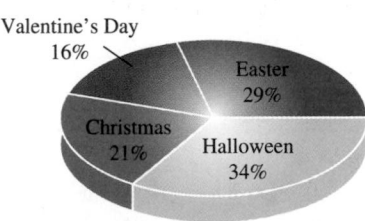

Valentine's Day 16%
Easter 29%
Christmas 21%
Halloween 34%

Source: National Confectioners Association, Annual Industry Review, 2009

12. SMARTPHONES The circle graph shows the percent U.S. market share for the leading smartphone companies. What is the sum of all the percents?

U.S. Smartphone Marketshare

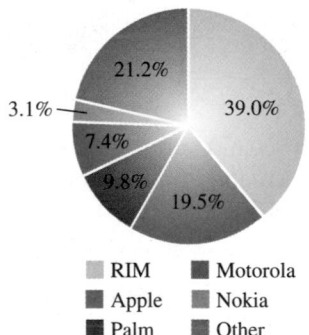

21.2%
3.1%
7.4%
9.8%
19.5%
39.0%

◻ RIM   ◼ Motorola
◼ Apple   ◼ Nokia
◼ Palm   ◼ Other

NOTATION

13. When computing with percents, we must change the percent to a decimal or a fraction. Change each percent to a decimal.

    a. 12%

    b. 5.6%

    c. 125%

    d. $\frac{1}{4}\%$

14. When computing with percents, we must change the percent to a decimal or a fraction. Change each percent to a fraction.

    a. $33\frac{1}{3}\%$ **b.** $66\frac{2}{3}\%$

    c. $16\frac{2}{3}\%$ **d.** $83\frac{1}{3}\%$

GUIDED PRACTICE

Translate each percent sentence to a percent equation or percent proportion. Do not solve. **See Example 1.**

15. a. What number is 7% of 16?

    b. 125 is what percent of 800?

    c. 1 is 94% of what number?

16. a. What number is 28% of 372?

    b. 9 is what percent of 21?

    c. 4 is 17% of what number?

17. a. 5.4% of 99 is what number?

    b. 75.1% of what number is 15?

    c. What percent of 33.8 is 3.8?

18. a. 1.5% of 3 is what number?

    b. 49.2% of what number is 100?

    c. What percent of 100.4 is 50.2?

Translate to a percent equation or percent proportion and then solve to find the unknown number. **See Example 2.**

19. What is 34% of 200?

20. What is 48% of 600?

21. What is 88% of 150?

22. What number is 52% of 350?

Translate to a percent equation or percent proportion and then solve to find the unknown number. **See Example 3.**

23. What number is 224% of 7.9?

24. What number is 197% of 6.3?

25. What number is 105% of 23.2?

26. What number is 228% of 34.5?

Translate to a percent equation or percent proportion and then solve to find the unknown number. **See Example 4.**

27. 8 is what percent of 32?

28. 9 is what percent of 18?

29. 51 is what percent of 60?

30. 52 is what percent of 80?

Translate to a percent equation or percent proportion and then solve to find the unknown number. **See Example 5.**

31. 5 is what percent of 8?

32. 7 is what percent of 8?

33. 7 is what percent of 16?

34. 11 is what percent of 16?

Translate to a percent equation or percent proportion and then solve to find the unknown number. **See Example 6.**

35. What percent of 60 is 66?

36. What percent of 50 is 56?

37. What percent of 24 is 84?

38. What percent of 14 is 63?

Translate to a percent equation or percent proportion and then solve to find the unknown number. **See Example 7.**

39. 9 is 30% of what number?

40. 8 is 40% of what number?

41. 36 is 24% of what number?

42. 24 is 16% of what number?

Translate to a percent equation or percent proportion and then solve to find the unknown number. **See Example 8.**

43. 19.2 is $33\frac{1}{3}$% of what number?

44. 32.8 is $33\frac{1}{3}$% of what number?

45. 48.4 is $66\frac{2}{3}$% of what number?

46. 56.2 is $16\frac{2}{3}$% of what number?

TRY IT YOURSELF

Translate to a percent equation or percent proportion and then solve to find the unknown number.

47. What percent of 40 is 0.5?

48. What percent of 15 is 0.3?

49. 7.8 is 12% of what number?

50. 39.6 is 44% of what number?

51. $33\frac{1}{3}$% of what number is 33?

52. $66\frac{2}{3}$% of what number is 28?

53. What number is 36% of 250?

54. What number is 82% of 300?

55. 16 is what percent of 20?

56. 13 is what percent of 25?

57. What number is 0.8% of 12?

58. What number is 5.6% of 40?

59. 3.3 is 7.5% of what number?

60. 8.4 is 20% of what number?

61. What percent of 0.05 is 1.25?

62. What percent of 0.06 is 2.46?

63. 102% of 105 is what number?

64. 210% of 66 is what number?

65. $9\frac{1}{2}$% of what number is 5.7?

66. $\frac{1}{2}$% of what number is 5,000?

67. What percent of 8,000 is 2,500?

68. What percent of 3,200 is 1,400?

69. Find $7\frac{1}{4}$% of 600.

70. Find $1\frac{3}{4}$% of 800.

APPLICATIONS

71. DOWNLOADING The message on the computer monitor screen shown below indicates that 24% of the 50K bytes of information that the user has decided to view have been downloaded to her computer at that time. Find the number of bytes of information that have been downloaded. (50K stands for 50,000.)

72. LUMBER The rate of tree growth for walnut trees is about 3% per year. If a walnut tree has 400 board feet of lumber that can be cut from it, how many more board feet will it produce in a year? (*Source:* Iowa Department of Natural Resources)

73. REBATES A telephone company offered its customers a rebate of 20% of the cost of all long-distance calls made in the month of July. One customer's long-distance calls for July are shown below.

    a. Find the total amount of the customer's long-distance charges for July.

    b. How much will this customer receive in the form of a rebate for these calls?

Date	Time	Place called	Min.	Amount
Jul 4	3:48 P.M.	Denver	47	$3.80
Jul 9	12:00 P.M.	Detroit	68	$7.50
Jul 20	8:59 A.M.	San Diego	70	$9.45
July Totals			185	?

74. PRICE GUARANTEES To assure its customers of low prices, the Home Club offers a "10% Plus" guarantee. If the customer finds the same item selling for less somewhere else, he or she receives the difference in price, plus 10% of the difference. A woman bought miniblinds at the Home Club for $120 but later saw the same blinds on sale for $98 at another store.

    a. What is the difference in the prices of the miniblinds?

    b. What is 10% of the difference in price?

    c. How much money can the woman expect to receive if she takes advantage of the "10% Plus" guarantee from the Home Club?

75. ENLARGEMENTS The enlarge feature on a copier is set at 180%, and a 1.5-inch wide picture is to be copied. What will be the width of the enlarged picture?

76. COPY MACHINES The reduce feature on a copier is set at 98%, and a 2-inch wide picture is to be copied. What will be the width of the reduced picture?

77. DRIVER'S LICENSE On the written part of his driving test, a man answered 28 out of 40 questions correctly. If 70% correct is passing, did he pass the test?

78. HOUSING A general budget rule of thumb is that your rent or mortgage payment should be less than 30% of your income. Together, a couple earns $4,500 per month and they pay $1,260 in rent. Are they following the budget rule of thumb for housing?

79. INSURANCE The cost to repair a car after a collision was $4,000. The automobile insurance policy paid the entire bill except for a $200 deductible, which the driver paid. What percent of the cost did he pay?

80. FLOOR SPACE A house has 1,200 square feet on the first floor and 800 square feet on the second floor.

    a. What is the total square footage of the house?

    b. What percent of the square footage of the house is on the first floor?

81. CHILD CARE After the first day of registration, 84 children had been enrolled in a new day care center. That represented 70% of the available slots. What was the maximum number of children the center could enroll?

82. RACING PROGRAMS One month before a stock car race, the sale of ads for the official race program was slow. Only 12 pages, or 60% of the available pages, had been sold. What was the total number of pages devoted to advertising in the program?

83. WATER POLLUTION A 2007 study found that about 4,500 kilometers, or $33\frac{1}{3}\%$ of China's Yellow River and its tributaries were not fit for any use. What is the combined length of the river and its tributaries? (*Source:* Discovermagazine.com)

84. FINANCIAL AID The National Postsecondary Student Aid Study found that in 2008 about 14 million, or $66\frac{2}{3}\%$, of the nation's undergraduate students received some type of financial aid. How many undergraduate students were there in 2008?

85. GOVERNMENT SPENDING The circle graph below shows the breakdown of federal spending for fiscal year 2007. If the total spending was approximately $2,700 billion, how many dollars were spent on Social Security, Medicare, and other retirement programs?

Source: 2008 Federal Income Tax Form 1040

86. WASTE The circle graph below shows the types of trash U.S. residents, businesses, and institutions generated in 2007. If the total amount of trash produced that year was about 254 million tons, how many million tons of yard trimmings was there?

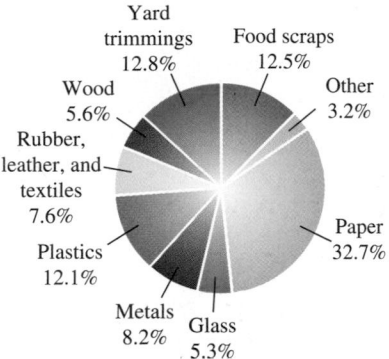

**U.S. Trash Generation by Material
Before Recycling, 2007
(254 Million Tons)**

Yard trimmings 12.8%
Food scraps 12.5%
Wood 5.6%
Other 3.2%
Rubber, leather, and textiles 7.6%
Paper 32.7%
Plastics 12.1%
Metals 8.2%
Glass 5.3%

Source: Environmental Protection Agency

87. PRODUCT PROMOTION To promote sales, a free 6-ounce bottle of shampoo is packaged with every large bottle. Use the information on the package to find how many ounces of shampoo the large bottle contains.

SHAMPOO
25% MORE–FREE!

88. NUTRITION FACTS The nutrition label on a package of corn chips is shown.

   a. How many milligrams of sodium are in one serving of chips?

   b. According to the label, what percent of the daily value of sodium is this?

   c. What daily value of sodium intake is considered healthy?

Nutrition Facts
Serving Size: 1 oz. (28g/About 29 chips)
Servings Per Container: About 11

Amount Per Serving	
Calories 160	Calories from Fat 90
	% Daily Value
Total fat 10g	**15%**
Saturated fat 1.5 g	**7%**
Cholesterol 0mg	**0%**
Sodium 240mg	**12%**
Total carbohydrate 15g	**5%**
Dietary fiber 1g	**4%**
Sugars less than 1g	
Protein 2g	

89. MIXTURES Complete the table to find the number of gallons of sulfuric acid in each of two storage tanks.

	Gallons of solution in tank	% Sulfuric acid	Gallons of sulfuric acid in tank
Tank 1	60	50%	
Tank 2	40	30%	

90. THE ALPHABET What percent of the English alphabet do the vowels a, e, i, o, and u make up? (Round to the nearest 1 percent.)

91. TIPS In August of 2006, a customer left Applebee's employee Cindy Kienow of Hutchinson, Kansas, a $10,000 tip for a bill that was approximately $25. What percent tip is this? (*Source:* cbsnews.com)

92. ELECTIONS In Los Angeles City Council races, if no candidate receives more than 50% of the vote, a runoff election is held between the first- and second-place finishers.

   a. How many total votes were cast?

   b. Determine whether there must be a runoff election for District 10.

City council	District 10
Nate Holden	8,501
Madison T. Shockley	3,614
Scott Suh	2,630
Marsha Brown	2,432

Use a circle graph to illustrate the given data. A circle divided into 100 sections is provided to help in the graphing process.

93. ENERGY Draw a circle graph to show what percent of the total U.S. energy produced in 2007 was provided by each source.

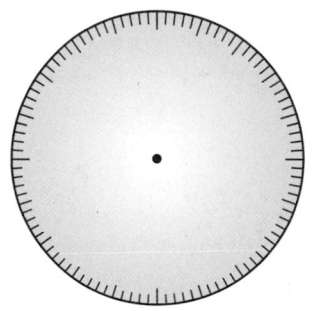

Renewable	10%
Nuclear	12%
Coal	32%
Natural gas	32%
Petroleum	14%

Source: Energy Information Administration

94. GREENHOUSE GASSES Draw a circle graph to show what percent of the total U.S. greenhouse gas emissions in 2007 came from each economic sector.

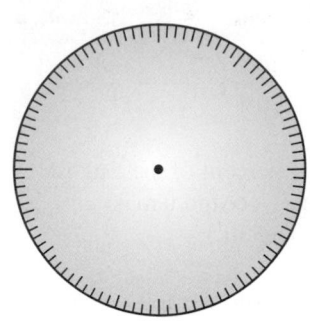

Electric power	34%
Transportation	28%
Industry	20%
Agriculture	7%
Commercial	6%
Residential	5%

Source: Environmental Protection Agency, *Time Magazine,* June 8, 2009

95. GOVERNMENT INCOME Complete the following table by finding what percent of total federal government income in 2007 each source provided. Then draw a circle graph for the data.

Total Income, Fiscal Year 2007: $2,600 Billion

Source of income	Amount	Percent of total
Social Security, Medicare, unemployment taxes	$832 billion	
Personal income taxes	$1,118 billion	
Corporate income taxes	$338 billion	
Excise, estate, customs taxes	$156 billion	
Borrowing to cover deficit	$156 billion	

Source: 2008 Federal Income Tax Form

2007 Federal Income Sources

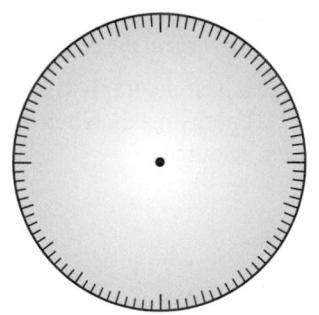

96. WATER USAGE The per-person indoor water use in the typical single family home is about 70 gallons per day. Complete the following table. Then draw a circle graph for the data.

Use	Gallons per person per day	Percent of total daily use
Showers	11.9	
Clothes washer	15.4	
Dishwasher	0.7	
Toilets	18.9	
Baths	1.4	
Leaks	9.8	
Faucets	10.5	
Other	1.4	

Source: American Water Works Association

Daily Water Use per Person

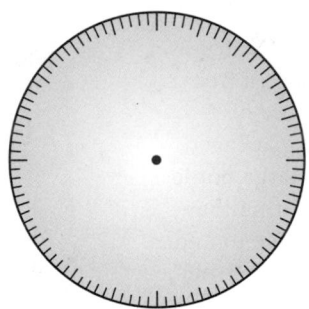

WRITING

97. Write a real-life situation that can be described by "9 is what percent of 20?"

98. Write a real-life situation that can be translated to $15 = 25\% \cdot x$.

99. Explain why 150% of a number is more than the number.

100. Explain why each of the following problems is easy to solve.

    a. What is 9% of 100?

    b. 16 is 100% of what number?

    c. 27 is what percent of 27?

101. When solving percent problems, when is it best to write a given percent as a fraction instead of as a decimal?

102. Explain how to identify the amount, the percent, and the base in a percent problem.

REVIEW

103. Add: $2.78 + 6 + 9.09 + 0.3$

104. Evaluate: $\sqrt{64} + 3\sqrt{9}$

105. On the number line, which is closer to 5: the number 4.9 or the number 5.001?

106. Multiply: $34.5464 \cdot 1{,}000$

107. Evaluate: $(0.2)^3$

108. Evaluate the formula $d = 4t$ for $t = 25$.

SECTION 7.3
Applications of Percent

Objectives

1 Calculate sales taxes, total cost, and tax rates.

2 Calculate commissions and commission rates.

3 Find the percent of increase or decrease.

4 Calculate the amount of discount, the sale price and the discount rate.

In this section, we discuss applications of percent. Three of them (taxes, commissions, and discounts) are directly related to purchasing. A solid understanding of these concepts will make you a better shopper and consumer. The fourth uses percent to describe increases or decreases of such things as population and unemployment.

1 Calculate sales taxes, total cost, and tax rates.

The department store sales receipt shown below gives a detailed account of what items were purchased, how many of each were purchased, and the price of each item.

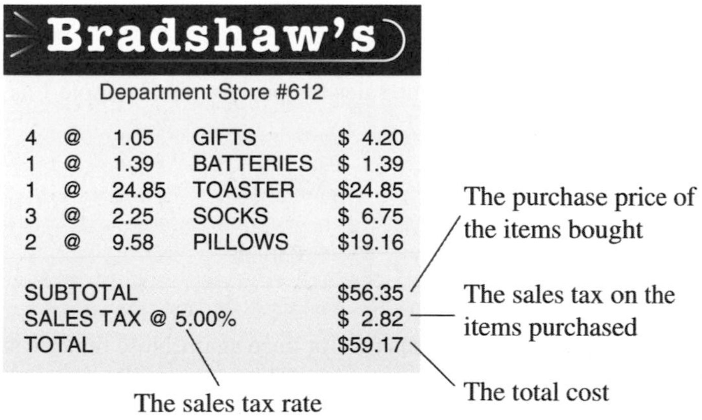

> **Bradshaw's**
>
> Department Store #612
>
> | 4 | @ | 1.05 | GIFTS | $ 4.20 |
> |---|---|------|-------|--------|
> | 1 | @ | 1.39 | BATTERIES | $ 1.39 |
> | 1 | @ | 24.85 | TOASTER | $24.85 |
> | 3 | @ | 2.25 | SOCKS | $ 6.75 |
> | 2 | @ | 9.58 | PILLOWS | $19.16 |
>
> | SUBTOTAL | | | | $56.35 |
> | SALES TAX @ 5.00% | | | | $ 2.82 |
> | TOTAL | | | | $59.17 |

The purchase price of the items bought

The sales tax on the items purchased

The sales tax rate

The total cost

    The receipt shows that the $56.35 purchase price (labeled *subtotal*) was taxed at a rate of 5%. Sales tax of $2.82 was charged.

    This example illustrates the following sales tax formula. Notice that the formula is based on the percent equation discussed in Section 7.2.

Finding the Sales Tax

The sales tax on an item is a percent of the purchase price of the item.

Sales tax = sales tax rate · purchase price

amount = percent · base

Sales tax rates are usually expressed as a percent and, when necessary, sales tax dollar amounts are rounded to the nearest cent.

Self Check 1

SALES TAX What would the sales tax be if the $56.35 purchase were made in a state that has a 6.25% state sales tax?

Now Try Problem 13

EXAMPLE 1 **Sales Tax** Find the sales tax on a purchase of $56.35 if the sales tax rate is 5%. (This is the purchase on the sales receipt shown on the previous page.)

Strategy We will identify the sales tax rate and the purchase price.

WHY Then we can use the sales tax formula to find the unknown sales tax.

Solution The sales tax rate is 5% and the purchase price is $56.35.

Sales tax = **sales tax rate · purchase price**   This is the sales tax formula.

= 5% · $56.35   Substitute 5% for the sales tax rate and $56.35 for the purchase price.

= 0.05 · $56.35   Write 5% as a decimal: 5% = 0.05.

$$\begin{array}{r} {}^{3\,1\,2}\\ 56.35 \\ \times\,0.05 \\ \hline 2.8175 \end{array}$$

= $2.8175   Do the multiplication.

The rounding digit in the hundredths column is 1.

= $2.8175   Prepare to round the sales tax to the nearest cent (hundredth) by identifying the rounding digit and test digit.

The test digit is 7.

≈ $2.82   Since the test digit is 5 or greater, round up.

The sales tax on the $56.35 purchase is $2.82. The sales receipt shown on the previous page is correct.

Success Tip It is helpful to see the sales tax problem in Example 1 as a type of percent problem from Section 7.2.

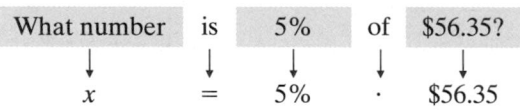

What number is 5% of $56.35?

x = 5% · $56.35

Look at the department store sales receipt once again. Note that the sales tax was added to the purchase price to get the total cost. This example illustrates the following formula for total cost.

Finding the Total Cost

The total cost of an item is the sum of its purchase price and the sales tax on the item.

Total cost = purchase price + sales tax

EXAMPLE 2   *Total Cost*   Find the total cost of the child's car seat shown on the right if the sales tax rate is 7.2%.

Saftey-T First
Child's
Car
Seat
$249.50
Buy today!
Ships next business day

Strategy First, we will find the sales tax on the child's car seat.

WHY Then we can add the purchase price and the sales tax to find the total of the car seat.

Solution The sales tax rate is 7.2% and the purchase price is $249.50.

Sales tax = **sales tax rate · purchase price**   This is the sales tax formula.

$\quad\quad\quad = \quad$ **7.2**% $\quad\cdot\quad$ **$249.50**   Substitute 7.2% for the sales tax rate and $249.50 for the purchase price.

$\quad\quad\quad = 0.072 \cdot \$249.50$ Write 7.2% as a decimal: 7.2% = 0.072.

$\quad\quad\quad = \$17.964$ Do the multiplication.

$$\begin{array}{r} 249.50 \\ \times\ 0.072 \\ \hline 49900 \\ 1746500 \\ \hline 17.96400 \end{array}$$

The rounding digit in the hundredths column is 6.

$\quad\quad\quad = \$17.964$ Prepare to round the sales tax to the nearest cent (hundredth) by identifying the rounding digit and test digit.

The test digit is 4.

$\quad\quad\quad \approx \$17.96$ Since the test digit is less than 5, round down.

Thus, the sales tax on the $249.50 purchase is $17.96. The total cost of the car seat is the sum of its purchase price and the sales tax.

Total cost = **purchase price** + **sales tax**   This is the formula for the total cost.

$\quad\quad\quad = \$249.50 + \$17.96$ Substitute $249.50 for the purchase price and $17.96 for the sales tax.

$$\begin{array}{r} \overset{1}{249.50} \\ +\ 17.96 \\ \hline 267.46 \end{array}$$

$\quad\quad\quad = \$267.46$ Do the addition. ∎

In addition to sales tax, we pay many other taxes in our daily lives. Income tax, gasoline tax, and Social Security tax are just a few. To find such tax rates, we can use an approach like that discussed in Section 7.2.

EXAMPLE 3   *Withholding Tax*   A waitress found that $11.04 was deducted from her weekly gross earnings of $240 for federal income tax. What withholding tax rate was used?

Strategy We will carefully read the problem and use the given facts to write them in the form of a percent sentence.

WHY Then we can translate the sentence into a percent equation (or percent proportion) and solve it to find the unknown withholding tax rate.

Solution There are two methods that can be used to solve this problem.

The percent equation method: Since the withholding tax of $11.04 is some unknown percent of her weekly gross earnings of $240, the percent sentence is:

$$\boxed{\$11.04}\quad \text{is}\quad \boxed{\text{what percent}}\quad \text{of}\quad \boxed{\$240?}$$

$$11.04 \quad = \quad\quad x \quad\quad \cdot \quad 240 \quad\quad \text{This is the percent equation to solve.}$$

Self Check 2

TOTAL COST Find the total cost of a $179.95 baby stroller if the sales tax rate on the purchase is 3.2%.

Now Try Problem 17

Self Check 3

INHERITANCE TAX A tax of $5,250 was paid on an inheritance of $15,000. What was the inheritance tax rate?

Now Try Problem 21

$11.04 = 240x$ On the right side, use the commutative property of multiplication to write $x \cdot 240$ as $240x$.

$$\frac{11.04}{240} = \frac{240x}{240}$$ To isolate x, undo the multiplication by 240 by dividing both sides by 240.

```
        0.046
240)11.0400
     - 0
      11 04
    - 9 60
      1 440
    - 1 440
          0
```

$0.046 = x$ Do the division: $11.04 \div 240 = 0.046$.

$0\,04.6\% = x$ To write the decimal 0.046 as a percent, multiply it by 100 by moving the decimal point two places to the right, and then insert a % symbol.

$4.6\% = x$

The withholding tax rate was 4.6%.

The percent proportion method: Since the withholding tax of $11.04 is some unknown percent of her weekly gross earnings of $240, the percent sentence is:

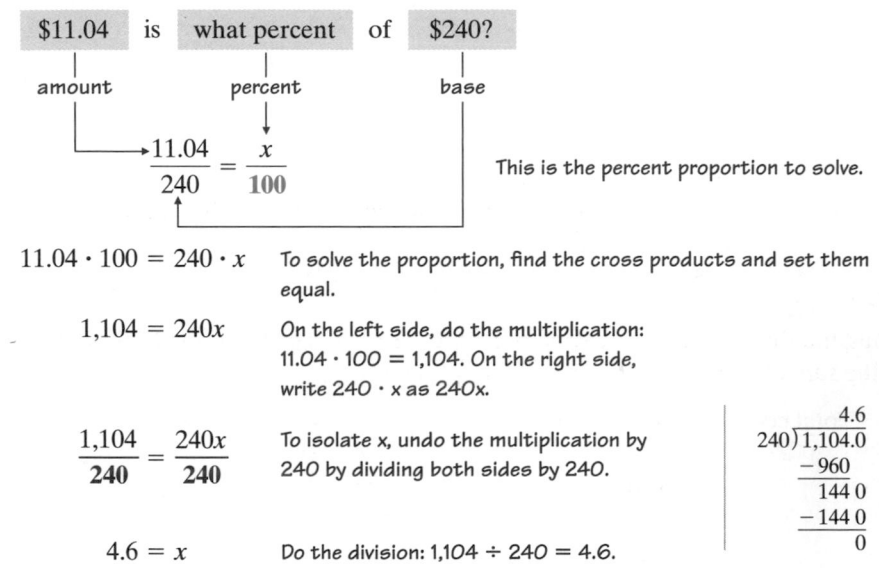

$11.04 \cdot 100 = 240 \cdot x$ To solve the proportion, find the cross products and set them equal.

$1{,}104 = 240x$ On the left side, do the multiplication: $11.04 \cdot 100 = 1{,}104$. On the right side, write $240 \cdot x$ as $240x$.

$$\frac{1{,}104}{240} = \frac{240x}{240}$$ To isolate x, undo the multiplication by 240 by dividing both sides by 240.

```
        4.6
240)1,104.0
    - 960
      144 0
    - 144 0
          0
```

$4.6 = x$ Do the division: $1{,}104 \div 240 = 4.6$.

The withholding tax rate was 4.6%.

2 Calculate commissions and commission rates.

Instead of working for a salary or getting paid at an hourly rate, many salespeople are paid on **commission.** They earn a certain percent of the total dollar amount of the goods or services that they sell. The following formula to calculate a commission is based on the percent equation discussed in Section 7.2.

Finding the Commission

The amount of commission paid is a percent of the total dollar sales of goods or services.

Commission = commission rate · sales

amount   =   percent   ·   base

EXAMPLE 4 *Appliance Sales* The commission rate for a salesperson at an appliance store is 16.5%. Find his commission from the sale of a refrigerator that costs $500.

Strategy We will identify the commission rate and the dollar amount of the sale.

WHY Then we can use the commission formula to find the unknown amount of the commission.

Solution The commission rate is 16.5% and the dollar amount of the sale is $500.

Commission = **commission rate · sales** This is the commission formula.

$= \quad 16.5\% \quad · \$500$ Substitute 16.5% for the commission rate and $500 for the sales.

$= 0.165 · \$500$ Write 16.5% as a decimal: 16.5% = 0.165.

$= \$82.50$ Do the multiplication.

$$\begin{array}{r} {\scriptstyle 3\,2} \\ 0.165 \\ \times\ 500 \\ \hline 82.500 \end{array}$$

The commission earned on the sale of the $500 refrigerator is $82.50.

Self Check 4

SELLING INSURANCE An insurance salesperson receives a 4.1% commission on each $120 premium paid by a client. What is the amount of the commission on this premium?

Now Try Problem 25

EXAMPLE 5 *Jewelry Sales* A jewelry salesperson earned a commission of $448 for selling a diamond ring priced at $5,600. Find the commission rate.

Strategy We will identify the commission and the dollar amount of the sale.

WHY Then we can use the commission formula to find the unknown commission rate.

Solution The commission is $448 and the dollar amount of the sale is $5,600.

Commission = commission rate · **sales** This is the commission formula.

$\mathbf{\$448} \quad = \quad x \quad · \mathbf{\$5{,}600}$ Substitute $448 for the commission and $5,600 for the sales. Let x represent the unknown commission rate.

$448 = 5{,}600x$ We can drop the dollar signs. On the right side, use the commutative property of multiplication to write x · 5,600 as 5,600x.

$\dfrac{448}{5{,}600} = \dfrac{5{,}600x}{5{,}600}$ To isolate x, undo the multiplication by 5,600 by dividing both sides by 5,600.

$0.08 = x$ Do the division: 448 ÷ 5,600 = 0.08.

$$\begin{array}{r} 0.08 \\ 5{,}600\overline{)448.00} \\ -448\ 00 \\ \hline 0 \end{array}$$

$008\%\, = x$ To write the decimal 0.08 as a percent, multiply it by 100 by moving the decimal point two places to the right, and then insert a % symbol.

$8\% = x$

The commission rate paid the salesperson on the sale of the diamond ring was 8%.

Self Check 5

SELLING ELECTRONICS If the commission on a $430 digital camcorder is $21.50, what is the commission rate?

Now Try Problem 29

3 **Find the percent of increase or decrease.**

Percents can be used to describe how a quantity has changed. For example, consider the table on the right, which shows the number of television channels that the average U.S. home received in 2000 and 2007.

Year	Number of television channels that the average U.S. home received
2000	61
2007	119

Source: The Nielsen Company

From the table, we see that the number of television channels received increased considerably from 2000 to 2007. To describe this increase using a percent, we first subtract to find the **amount of increase.**

$$119 - 61 = 58 \quad \text{Subtract the number of TV channels received in 2000}$$
$$\text{from the number received in 2007.}$$

Thus, the number of channels received increased by 58 from 2000 to 2007.

Next, we find what percent of the *original* 61 channels received in 2000 that the 58 channel increase represents. To do this, we translate the problem into a percent equation (or percent proportion) and solve it.

The percent equation method:

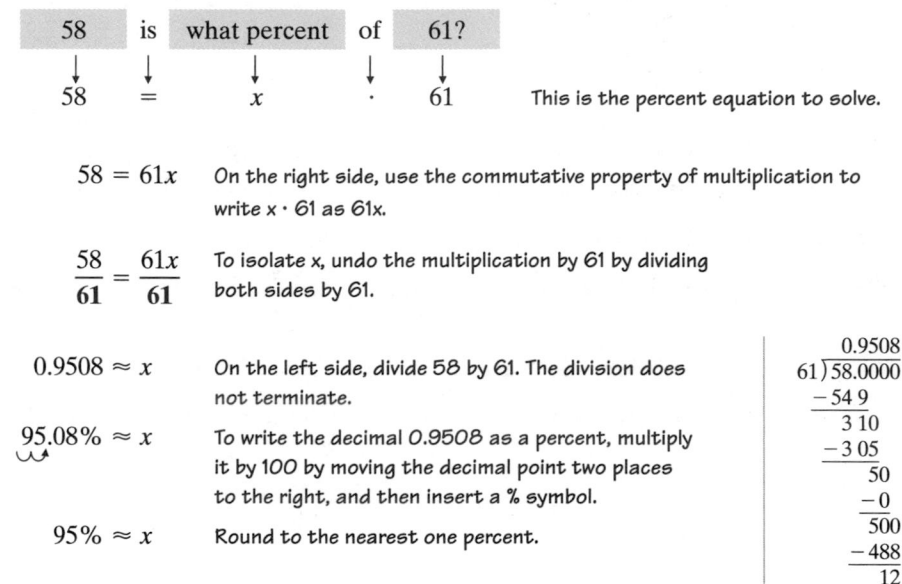

$$58 = 61x \quad \text{On the right side, use the commutative property of multiplication to}$$
$$\text{write } x \cdot 61 \text{ as } 61x.$$

$$\frac{58}{61} = \frac{61x}{61} \quad \text{To isolate } x, \text{ undo the multiplication by 61 by dividing}$$
$$\text{both sides by 61.}$$

$$0.9508 \approx x \quad \text{On the left side, divide 58 by 61. The division does}$$
$$\text{not terminate.}$$

$$95.08\% \approx x \quad \text{To write the decimal 0.9508 as a percent, multiply}$$
$$\text{it by 100 by moving the decimal point two places}$$
$$\text{to the right, and then insert a \% symbol.}$$

$$95\% \approx x \quad \text{Round to the nearest one percent.}$$

```
        0.9508
61) 58.0000
    -54 9
      3 10
     -3 05
        50
       - 0
       500
      -488
        12
```

The percent proportion method:

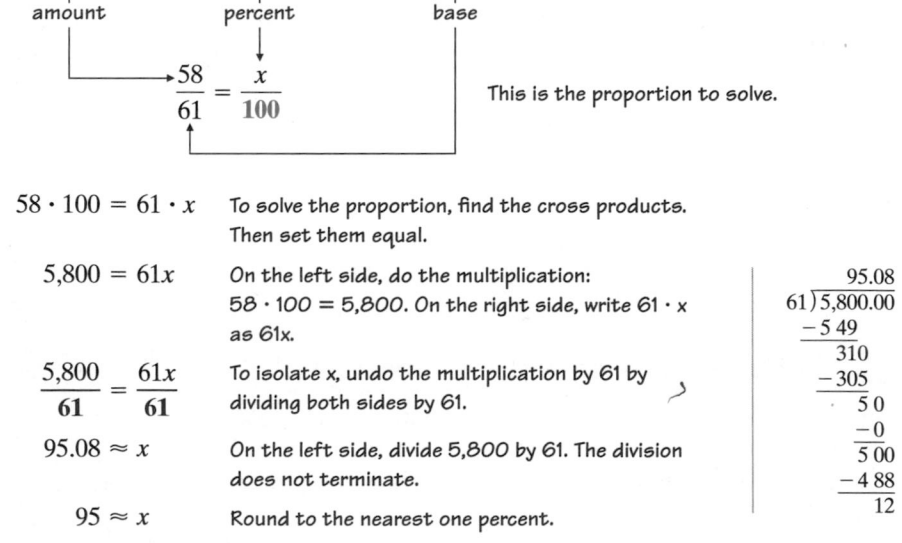

$$58 \cdot 100 = 61 \cdot x \quad \text{To solve the proportion, find the cross products.}$$
$$\text{Then set them equal.}$$

$$5{,}800 = 61x \quad \text{On the left side, do the multiplication:}$$
$$58 \cdot 100 = 5{,}800. \text{ On the right side, write } 61 \cdot x$$
$$\text{as } 61x.$$

$$\frac{5{,}800}{61} = \frac{61x}{61} \quad \text{To isolate } x, \text{ undo the multiplication by 61 by}$$
$$\text{dividing both sides by 61.}$$

$$95.08 \approx x \quad \text{On the left side, divide 5,800 by 61. The division}$$
$$\text{does not terminate.}$$

$$95 \approx x \quad \text{Round to the nearest one percent.}$$

```
        95.08
61) 5,800.00
   -5 49
     310
    -305
      5 0
     - 0
     5 00
    -4 88
      12
```

With either method, we see that there was a 95% increase in the number of television channels received by the average American home from 2000 to 2007.

EXAMPLE 6   *JFK*   A 1996 auction included an oak rocking chair used by President John F. Kennedy in the Oval Office. The chair, originally valued at $5,000, sold for $453,500. Find the percent of increase in the value of the rocking chair.

Paul Schutzer/Time & Life Pictures/Getty Images

Strategy We will begin by finding the amount of increase in the value of the rocking chair.

WHY Then we can calculate what percent of the original $5,000 value of the chair that the increase represents.

Solution First, we find the amount of increase in the value of the rocking chair.

$$453,500 - 5,000 = 448,500$$ Subtract the original value from the price paid at auction.

The rocking chair increased in value by $448,500. Next, we find what percent of the original $5,000 value of the rocking chair the $448,500 increase represents by translating the problem into a percent equation (or percent proportion) and solving it.

The percent equation method:

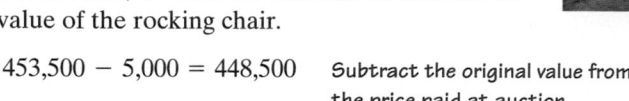

| $448,500 | is | what percent | of | $5,000? |

$$448,500 \quad = \quad x \quad \cdot \quad 5,000$$ This is the percent equation to solve.

$$448,500 = 5,000x$$ On the right side, use the commutative property of multiplication to write x · 5,000 as 5,000x.

$$\frac{448,500}{5,000} = \frac{5,000x}{5,000}$$ To isolate x, undo the multiplication by 5,000 by dividing both sides by 5,000.

On the left side, recall that there is a shortcut for dividing a dividend by a divisor when both end with zeros. Remove two of the ending zeros in the divisor 5,000 and remove the same number of ending zeros in the dividend 448,500.

$$\frac{4,485}{50} = x$$

$$89.7 = x$$ Do the division: 4,485 ÷ 50 = 89.7.

$$89\underset{\curvearrowright}{.7}0\% = x$$ To write the decimal 89.7 as a percent, multiply it by 100 by moving the decimal point two places to the right, and then insert a % symbol.

$$8,970\% = x$$

```
          89.7
   50) 4,485.0
      - 4 00
        485
      - 450
         35 0
       - 35 0
            0
```

The percent proportion method:

| $448,500 | is | what percent | of | $5,000? |

amount   percent   base

$$\frac{448,500}{5,000} = \frac{x}{100}$$ This is the proportion to solve.

$$448,500 \cdot 100 = 5,000 \cdot x$$ To solve the proportion, find the cross products. Then set them equal.

$$44,850,000 = 5,000x$$ On the left side, do the multiplication: 448,500 · 100 = 44,850,000. On the right side, write 5,000 · x as 5,000x.

Self Check 6

HOME SCHOOLING In one school district, the number of home-schooled children increased from 15 to 150 in 4 years. Find the percent of increase.

Now Try Problem 33

$$\frac{44{,}850{,}000}{5{,}000} = \frac{5{,}000x}{5{,}000}$$ To isolate x, undo the multiplication by 5,000 by dividing both sides by 5,000.

On the left side, recall that there is a shortcut for dividing a dividend by a divisor when both end with zeros. Remove the three ending zeros in the divisor 5,000 and remove the same number of ending zeros in the dividend 44,850,000.

$$\frac{44{,}850}{5} = x$$

$$8{,}970 = x$$ Divide 44,850 by 5.

```
        8970
   5) 44,850
      - 40
        4 8
       -4 5
          35
         -35
           0
          -0
           0
```

With either method, we see that there was an amazing 8,970% increase in the value of the Kennedy rocking chair.

> **Caution!** The percent of increase (or decrease) is a percent of the *original number,* that is, the number before the change occurred. Thus, in Example 6, it would be incorrect to write a percent sentence that compares the increase to the *new value* of the Kennedy rocking chair.
>
> $448,500   is   what percent   of $453,500?

Finding the Percent of Increase or Decrease

To find the percent of increase or decrease:

1. Subtract the smaller number from the larger to find the amount of increase or decrease.

2. Find what percent the amount of increase or decrease is of the original amount.

Self Check 7

REDUCING FAT INTAKE One serving of the original *Jif* peanut butter has 16 grams of fat per serving. The new *Jif Reduced Fat* product contains 12 grams of fat per serving. What is the percent decrease in the number of grams of fat per serving?

Now Try Problem 37

EXAMPLE 7 *Commercials* Jared Fogle credits his tremendous weight loss to exercise and a diet of low-fat Subway sandwiches. His maximum weight (reached in March of 1998) was 425 pounds. His current weight is about 187 pounds. Find the percent of decrease in his weight.

Zack Seckler/Getty Images

Strategy We will begin by finding the amount of decrease in Jared Fogle's weight.

WHY Then we can calculate what percent of his original 425-pound weight that the decrease represents.

Solution First, we find the amount of decrease in his weight.

$$425 - 187 = 238$$ Subtract his new weight from his weight before going on the weight-loss program.

His weight decreased by 238 pounds.

Next, we find what percent of his original 425 weight the 238-pound decrease represents by translating the problem into a percent equation (or percent proportion) and solving it.

The percent equation method:

238	is	what percent	of	425?
↓	↓	↓	↓	↓
238	=	x	·	425

$238 = x \cdot 425$ This is the percent equation to solve.

$238 = 425x$ On the right side, use the commutative property of multiplication to write x · 425 as 425x.

$\dfrac{238}{425} = \dfrac{425x}{425}$ To isolate x, undo the multiplication by 425 by dividing both sides by 425.

$0.56 = x$ Do the division: 238 ÷ 425 = 0.56.

$0\underset{\curvearrowright}{.}56\,\% = x$ To write the decimal 0.56 as a percent, multiply it by 100 by moving the decimal point two places to the right, and then insert a % symbol.

$56\% = x$

$$
\begin{array}{r}
0.56 \\
425\overline{)238.00} \\
-212\ 5 \\
\hline
25\ 50 \\
-25\ 50 \\
\hline
0
\end{array}
$$

The percent proportion method:

238	is	what percent	of	425?
amount		percent		base

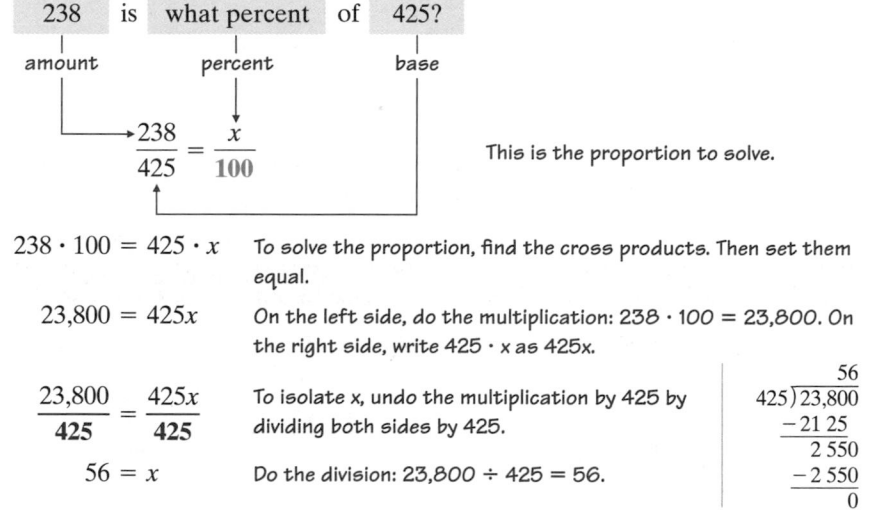

$\dfrac{238}{425} = \dfrac{x}{100}$ This is the proportion to solve.

$238 \cdot 100 = 425 \cdot x$ To solve the proportion, find the cross products. Then set them equal.

$23{,}800 = 425x$ On the left side, do the multiplication: 238 · 100 = 23,800. On the right side, write 425 · x as 425x.

$\dfrac{23{,}800}{425} = \dfrac{425x}{425}$ To isolate x, undo the multiplication by 425 by dividing both sides by 425.

$56 = x$ Do the division: 23,800 ÷ 425 = 56.

$$
\begin{array}{r}
56 \\
425\overline{)23{,}800} \\
-21\ 25 \\
\hline
2\ 550 \\
-2\ 550 \\
\hline
0
\end{array}
$$

With either method, we see that there was a 56% decrease in Jared Fogle's weight. ∎

THINK IT THROUGH   *Studying Mathematics*

"All students, regardless of their personal characteristics, backgrounds, or physical challenges, must have opportunities to study—and support to learn—mathematics."
National Council of Teachers of Mathematics

The table below shows the number of students enrolled in Basic Mathematics classes at two-year colleges.

Year	1970	1975	1980	1985	1990	1995	2000	2005
Enrollment	57,000	100,000	146,000	142,000	147,000	134,000	122,000	104,000

Source: 2005 CBMS Survey of Undergraduate Programs

1. Over what five-year span was there the greatest percent increase in enrollment in Basic Mathematics classes? What was the percent increase?

2. Over what five-year span was there the greatest percent decrease in enrollment in Basic Mathematics classes? What was the percent increase?

4 Calculate the amount of discount, the sale price, and the discount rate.

While shopping, you have probably noticed that many stores display signs advertising sales. Store managers have found that offering discounts attracts more customers. To be a smart shopper, it is important to know the vocabulary of discount sales.

The difference between the **original price** and the **sale price** of an item is called the **amount of discount,** or simply the **discount.** If the discount is expressed as a percent of the selling price, it is called the **discount rate.**

If we know the original price and the sale price of an item, we can use the following formula to find the amount of discount.

Finding the Discount

The amount of discount is the difference between the original price and the sale price.

Amount of discount = original price − sale price

If we know the original price of an item and the discount rate, we can use the following formula to find the amount of discount. Like several other formulas in this section, it is based on the percent equation discussed in Section 7.2.

Finding the Discount

The amount of discount is a percent of the original price.

Amount of discount = discount rate · original price

    ↑   ↑   ↑

  amount   =   percent   ·   base

We can use the following formula to find the sale price of an item that is being discounted.

Finding the Sale Price

To find the sale price of an item, subtract the discount from the original price.

Sale price = original price − discount

EXAMPLE 8   *Shoe Sales*   Use the information in the advertisement shown on the previous page to find the amount of the discount on the pair of men's basketball shoes. Then find the sale price.

Strategy We will identify the discount rate and the original price of the shoes and use a formula to find the amount of the discount.

WHY Then we can subtract the discount from the original price to find the sale price of the shoes.

Solution From the advertisement, we see that the discount rate on the men's shoes is 25% and the original price is $89.80.

Amount of discount = **discount rate · original price**   This is the amount of discount formula.

= **25%** · **$89.80**   Substitute 25% for the discount rate and $89.80 for the original price.

= 0.25 · $89.80   Write 25% as a decimal: 25% = 0.25.

= $22.45   Do the multiplication.

$$89.80$$
$$\times 0.25$$
$$\overline{44900}$$
$$179600$$
$$\overline{22.4500}$$

The discount on the men's shoes is $22.45. To find the sale price, we use subtraction.

Sale price = **original price − discount**   This is the sale price formula.

= **$89.80** − **$22.45**   Substitute $89.80 for the original price and $22.45 for the discount.

= $67.35   Do the subtraction.

$$\overset{710}{89.8\cancel{0}}$$
$$-22.45$$
$$\overline{67.35}$$

The sale price of the men's basketball shoes is $67.35.

Self Check 8

SUNGLASSES SALES Sunglasses, regularly selling for $15.40, are discounted 15%. Find the amount of the discount. Then find the sale price.

Now Try Problem 41

EXAMPLE 9   *Discounts*   Find the discount rate on the ladies' cross trainer shoes shown in the advertisement on the previous page. Round to the nearest one percent.

Strategy We will think of this as a percent-of-decrease problem.

WHY We want to find what percent of the $59.99 original price the amount of discount represents.

Solution From the advertisement, we see that the original price of the women's shoes is $59.99 and the sale price is $33.99. The discount (decrease in price) is found using subtraction.

$59.99 − $33.99 = $26   Use the formula:
Amount of discount = original price − sale price.

The shoes are discounted $26. Now we find what percent of the original price the $26 discount represents.

Amount of discount = discount rate · original price   This is the amount of discount formula.

26 = *x* · $59.99   Substitute 26 for the amount of discount and $59.99 for the original price. Let *x* represent the unkown discount rate.

26 = 59.99*x*   We can drop the dollar signs. On the right side, use the commutative property of multiplication to write *x* · 59.99 as 59.99x.

Self Check 9

DINING OUT An early-bird special at a restaurant offers a $10.99 prime rib dinner for only $7.95 if it is ordered before 6 P.M. Find the rate of discount. Round to the nearest one percent.

Now Try Problem 45

$$\frac{26}{59.99} = \frac{59.99x}{59.99}$$

To isolate x, undo the multiplication by 59.99 by dividing both sides by 59.99.

$0.433 \approx x$

On the left side, divide 26 by 59.99. The division does not terminate.

$$\begin{array}{r} 0.433 \\ 59.99\overline{)26\,00.000} \\ \underline{-23\,99\,6} \\ 2\,00\,40 \\ \underline{-1\,79\,97} \\ 20\,430 \\ \underline{-17\,997} \\ 2\,433 \end{array}$$

$043.3\% \approx x$

To write the decimal 0.433 as a percent, multiply it by 100 by moving the decimal point two places to the right, and then insert a % symbol.

$43\% \approx x$

Round to the nearest one percent.

To the nearest one percent, the discount rate on the women's shoes is 43%.

ANSWERS TO SELF CHECKS

1. $3.52 **2.** $185.71 **3.** 35% **4.** $4.92 **5.** 5% **6.** 900% **7.** 25%
8. $2.31, $13.09 **9.** 28%

SECTION 7.3 STUDY SET

VOCABULARY

Fill in the blanks.

1. Instead of working for a salary or getting paid at an hourly rate, some salespeople are paid on _____. They earn a certain percent of the total dollar amount of the goods or services they sell.

2. Sales tax _____ are usually expressed as a percent.

3. **a.** When we use percent to describe how a quantity has increased compared to its original value, we are finding the percent of _____.

   b. When we use percent to describe how a quantity has decreased compared to its _____ value, we are finding the percent of decrease.

4. Refer to the advertisement below for a ceiling fan on sale.

   a. The _____ price of the ceiling fan was $199.99.

   b. The amount of the _____ is $40.00.

   c. The discount _____ is 20%.

   d. The _____ price of the ceiling fan is $159.00.

Ceiling Fan

Hampton Bay
52 in.
Quick install
Antique Brass

20% OFF

Was: $199.99
−40.00
Now: $159.00

CONCEPTS

Fill in the blanks in each of the following formulas.

5. Sales tax = sales tax rate · [____]

6. Total cost = [____] + sales tax

7. Commission = commission rate · [____]

8. **a.** Amount of discount = original price − [____]

   b. Amount of discount = [____] · original price

   c. Sale price = [____] − discount

9. **a.** The sales tax on an item priced at $59.32 is $4.75. What is the total cost of the item?

   b. The original price of an item is $150.99. The amout of discount is $15.99. What is the sale price of the item?

10. Round each dollar amount to the nearest cent.

    a. $168.257

    b. $57.234

    c. $3.396

11. Fill in the blanks: To find the percent decrease, _____ the smaller number from the larger number to find the amount of decrease. Then find what percent that difference is of the _____ amount.

12. NEWSPAPERS The table below shows how the circulations of two daily newspapers changed from 2003 to 2007.

Daily Circulation

	Miami Herald	*USA Today*
2003	315,850	2,154,539
2007	255,844	2,293,137

Source: The World Almanac, 2009

a. What was the *amount of decrease* of the *Miami Herald*'s circulation?

b. What was the *amount of increase* of *USA Today*'s circulation?

▌ GUIDED PRACTICE

Solve each problem to find the sales tax. See Example 1.

13. Find the sales tax on a purchase of $92.70 if the sales tax rate is 4%.

14. Find the sales tax on a purchase of $33.60 if the sales tax rate is 8%.

15. Find the sales tax on a purchase of $83.90 if the sales tax rate is 5%.

16. Find the sales tax on a purchase of $234.80 if the sales tax rate is 2%.

Solve each problem to find the total cost. See Example 2.

17. Find the total cost of a $68.24 purchase if the sales tax rate is 3.8%.

18. Find the total cost of a $86.56 purchase if the sales tax rate is 4.3%.

19. Find the total cost of a $60.18 purchase if the sales tax rate is 6.4%.

20. Find the total cost of a $70.73 purchase if the sales tax rate is 5.9%.

Solve each problem to find the tax rate. See Example 3.

21. SALES TAX The purchase price for a blender is $140. If the sales tax is $7.28, what is the sales tax rate?

22. SALES TAX The purchase price for a camping tent is $180. If the sales tax is $8.64, what is the sales tax rate?

23. SELF-EMPLOYED TAXES A business owner paid self-employment taxes of $4,590 on a taxable income of $30,000. What is the self-employment tax rate?

24. CAPITAL GAINS TAXES A couple paid $3,000 in capital gains tax on a profit of $20,000 made from the sale of some shares of stock. What is the capital gains tax rate?

Solve each problem to find the commission. See Example 4.

25. SELLING SHOES A shoe salesperson earns a 12% commission on all sales. Find her commission if she sells a pair of dress shoes for $95.

© iStockphoto.com/Cameron Pashak

26. SELLING CARS A used car salesperson earns an 11% commission on all sales. Find his commission if he sells a 2001 Chevy Malibu for $4,800.

27. EMPLOYMENT AGENCIES An employment counselor receives a 35% commission on the first week's salary of anyone that she places in a new job. Find her commission if one of her clients is hired as a secretary at $480 per week.

28. PHARMACEUTICAL SALES A medical sales representative is paid an 18% commission on all sales. Find her commission if she sells $75,000 of Coumadin, a blood-thinning drug, to a pharmacy chain.

Solve each problem to find the commission rate. See Example 5.

29. AUCTIONS An auctioneer earned a $15 commission on the sale of an antique chair for $750. What is the commission rate?

30. SELLING TIRES A tire salesman was paid a $28 commission after one of his customers purchased a set of new tires for $560. What is the commission rate?

31. SELLING ELECTRONICS If the commission on a $500 laptop computer is $20, what is the commission rate?

32. SELLING CLOCKS If the commission on a $600 grandfather clock is $54, what is the commission rate?

Solve each problem to find the percent of increase. See Example 6.

33. CLUBS The number of members of a service club increased from 80 to 88. What was the percent of increase in club membership?

34. SAVINGS ACCOUNTS The amount of money in a savings account increased from $2,500 to $3,000. What was the percent of increase in the amount of money saved?

35. RAISES After receiving a raise, the salary of a secretary increased from $300 to $345 dollars per week. What was the percent of increase in her salary?

36. TUITION The tuition at a community college increased from $2,500 to $2,650 per semester. What was the percent of increase in the tuition?

Solve each problem to find the percent of decrease. **See Example 7.**

37. TRAVEL TIME After a new freeway was completed, a commuter's travel time to work decreased from 30 minutes to 24 minutes. What was the percent of decrease in travel time?

38. LAYOFFS A printing company reduced the number of employees from 300 to 246. What was the percent of decrease in the number of employees?

39. ENROLLMENT Thirty-six of the 40 students originally enrolled in an algebra class completed the course. What was the percent of decrease in the number of students in the class?

40. DECLINING SALES One year, a pumpkin patch sold 1,200 pumpkins. The next year, they only sold 900 pumpkins. What was the percent of decrease in the number of pumpkins sold?

Image Copyright Eye for Africa, 2009. Used under license from Shutterstock.com

Solve each problem to find the amount of the discount and the sale price. **See Example 8.**

41. DINNERWARE SALES Find the amount of the discount on a six-place dinnerware set if it regularly sells for $90, but is on sale for 33% off. Then find the sale price of the dinnerware set.

42. BEDDING SALES Find the amount of the discount on a $130 bedspread that is now selling for 20% off. Then find the sale price of the bedspread.

43. MEN'S CLOTHING SALES 501 Levi jeans that regularly sell for $58 are now discounted 15%. Find the amount of the discount. Then find the sale price of the jeans.

44. BOOK SALES At a bookstore, the list price of $23.50 for the *Merriam-Webster's Collegiate Dictionary* is crossed out, and a 30% discount sticker is pasted on the cover. Find the amount of the discount. Then find the sale price of the dictionary.

Solve each problem to find the discount rate. **See Example 9.**

45. LADDER SALES Find the discount rate on an aluminum ladder regularly priced at $79.95 that is on sale for $64.95. Round to the nearest one percent.

46. OFFICE SUPPLIES SALES Find the discount rate on an electric pencil sharpener regularly priced at $49.99 that is on sale for $45.99. Round to the nearest one percent.

47. DISCOUNT TICKETS The price of a one-way airline ticket from Atlanta to New York City was reduced from $209 to $179. Find the discount rate. Round to the nearest one percent.

48. DISCOUNT HOTELS The cost of a one-night stay at a hotel was reduced from $245 to $200. Find the discount rate. Round to the nearest one percent.

▮ **APPLICATIONS**

49. SALES TAX The Utah state sales tax rate is 5.95%. Find the sales tax on a dining room set that sells for $900.

50. SALES TAX Find the sales tax on a pair of jeans costing $40 if they are purchased in Missouri, which has a state sales tax rate of 4.225%.

51. SALES RECEIPTS Complete the sales receipt below by finding the subtotal, the sales tax, and the total cost of the purchase.

NURSERY CENTER
Your one-stop garden supply

3 @ 2.99	PLANTING MIX	$ 8.97
1 @ 9.87	GROUND COVER	$ 9.87
2 @ 14.25	SHRUBS	$28.50
SUBTOTAL		$
SALES TAX @ 6.00%		$
TOTAL		$

52. SALES RECEIPTS Complete the sales receipt below by finding all three prices, the subtotal, the sales tax, and the total cost of the purchase.

McCOY'S FURNITURE

1 @ 450.00	SOFA	$
2 @ 90.00	END TABLES	$
1 @ 350.00	LOVE SEAT	$
SUBTOTAL		$
SALES TAX @ 4.20%		$
TOTAL		$

53. ROOM TAX After checking out of a hotel, a man noticed that the hotel bill included an additional charge labeled *room tax*. If the price of the room was $129 plus a room tax of $10.32, find the room tax rate.

54. EXCISE TAX While examining her monthly telephone bill, a woman noticed an additional charge of $1.24 labeled *federal excise tax*. If the basic service charges for that billing period were $42, what is the federal excise tax rate? Round to the nearest one percent.

55. GAMBLING For state authorized wagers (bets) placed with legal bookmakers and lottery operators, there is a federal excise tax on the wager. What is the excise tax rate if there is an excise tax of $5 on a $2,000 bet?

56. BUYING FISHING EQUIPMENT There are federal exercise taxes on the retail price when purchasing fishing equipment. The taxes are intended to help pay for parks and conservation. What is the federal excise tax rate if there is an excise tax of $17.50 on a fishing rod and reel that has a retail price of $175?

57. TAX HIKES In order to raise more revenue, some states raise the sales tax rate. How much additional money will be collected on the sale of a $15,000 car if the sales tax rate is raised 1%?

58. FOREIGN TRAVEL Value-added tax (VAT) is a consumer tax on goods and services. Currently, VAT systems are in place all around the world. (The United States is one of the few nations not using a value-added tax system.) Complete the table by determining the VAT a traveler would pay in each country on a dinner that cost $25. Round to the nearest cent.

Country	VAT rate	Tax on a $25 dinner
Mexico	15%	
Germany	19%	
Ireland	21.5%	
Sweden	25%	

Source: www.worldwide-tax.com

59. PAYCHECKS Use the information on the paycheck stub to find the tax rate for the federal withholding, worker's compensation, Medicare, and Social Security taxes that were deducted from the gross pay.

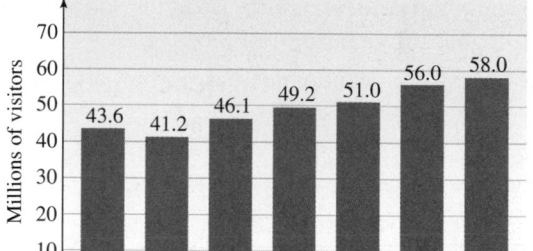

6286244	
Issue date: 03-27-10	
GROSS PAY	$360.00
TAXES	
FED. TAX	$ 28.80
WORK. COMP.	$ 13.50
MEDICARE	$ 4.32
SOCIAL SECURITY	$ 22.32
NET PAY	$291.06

60. GASOLINE TAX In one state, a gallon of unleaded gasoline sells for $3.05. This price includes federal and state taxes that total approximately $0.64. Therefore, the price of a gallon of gasoline, before taxes, is $2.41. What is the tax rate on gasoline? Round to the nearest one percent.

61. POLICE FORCE A police department plans to increase its 80-person force to 84 persons. Find the percent increase in the size of the police force.

62. COST-OF-LIVING INCREASES A woman making $32,000 a year receives a cost-of-living increase that raises her salary to $32,768 per year. Find the percent of increase in her yearly salary.

63. LAKE SHORELINES Because of a heavy spring runoff, the shoreline of a lake increased from 5.8 miles to 7.6 miles. What was the percent of increase in the length of the shoreline? Round to the nearest one percent.

© iStockphoto.com

64. CROP DAMAGE After flooding damaged much of the crop, the cost of a head of lettuce jumped from $0.99 to $2.20. What percent of increase is this? Round to the nearest one percent.

65. OVERTIME From May to June, the number of overtime hours for employees at a printing company increased from 42 to 106. What is the percent of increase in the number of overtime hours? Round to the nearest percent.

66. TOURISM The graph below shows the number of international visitors (travelers) to the United States each year from 2002 to 2008.

a. The greatest percent of increase in the number of travelers was between 2003 and 2004. Find the percent increase. Round to the nearest one percent.

b. The only decrease in the number of travelers was between 2002 and 2003. Find the percent decrease. Round to the nearest one percent.

International Travel to the U.S.

Year	Millions of visitors
2002	43.6
2003	41.2
2004	46.1
2005	49.2
2006	51.0
2007	56.0
2008	58.0

Source: U.S. Department of Commerce

67. REDUCED CALORIES A company advertised its new, improved chips as having 96 calories per serving. The original style contained 150 calories. What percent of decrease in the number of calories per serving is this?

68. CAR INSURANCE A student paid a car insurance premium of $420 every three months. Then the premium dropped to $370, because she qualified for a good-student discount. What was the percent of decrease in the premium? Round to the nearest percent.

69. BUS PASSES To increase the number of riders, a bus company reduced the price of a monthly pass from $112 to $98. What was the percent of decrease in the cost of a bus pass?

70. BASEBALL The illustration below shows the path of a baseball hit 110 mph, with a launch angle of 35 degrees, at sea level and at Coors Field, home of the Colorado Rockies. What is the percent of increase in the distance the ball travels at Coors Field?

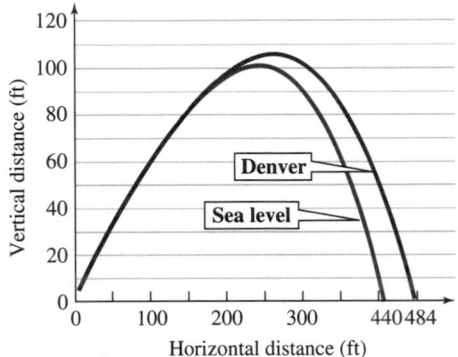

Source: Los Angeles Times, September 16, 1996

71. EARTH MOVING The illustration below shows the typical soil volume change during earth moving. (One cubic yard of soil fits in a cube that is 1 yard long, 1 yard wide, and 1 yard high.)

  a. Find the percent of increase in the soil volume as it goes through step 1 of the process.

  b. Find the percent of decrease in the soil volume as it goes through step 2 of the process.

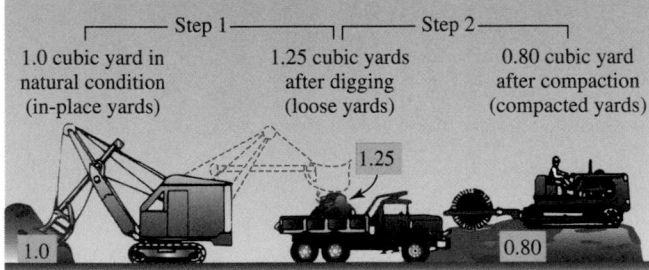

Source: U.S. Department of the Army

72. PARKING The management of a mall has decided to increase the parking area. The plans are shown in the next column. What will be the percent of increase in the parking area when the project is completed?

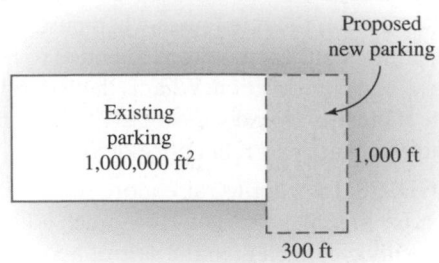

73. REAL ESTATE After selling a house for $98,500, a real estate agent split the 6% commission with another agent. How much did each person receive?

74. COMMISSIONS A salesperson for a medical supplies company is paid a commission of 9% for orders less than $8,000. For orders exceeding $8,000, she receives an additional 2% in commission on the total amount. What is her commission on a sale of $14,600?

75. SPORTS AGENTS A sports agent charges her clients a fee to represent them during contract negotiations. The fee is based on a percent of the contract amount. If the agent earned $37,500 when her client signed a $2,500,000 professional football contract, what rate did she charge for her services?

76. ART GALLERIES An art gallery displays paintings for artists and receives a commission from the artist when a painting is sold. What is the commission rate if a gallery received $135.30 when a painting was sold for $820?

77. WHOLE LIFE INSURANCE For the first 12 months, insurance agents earn a very large commission on the monthly premium of any whole life policy that they sell. After that, the commission rate is lowered significantly. Suppose on a new policy with monthly premiums of $160, an agent is paid monthly commissions of $144. Find the commission rate.

78. TERM INSURANCE For the first 12 months, insurance agents earn a large commission on the monthly premium of any term life policy that they sell. After that, the commission rate is lowered significantly. Suppose on a new policy with monthly premiums of $180, an agent is paid monthly commissions of $81. Find the commission rate.

79. CONCERT PARKING A concert promoter gets a commission of $33\frac{1}{3}\%$ of the revenue an arena receives from parking the night of the performance. How much can the promoter make if 6,000 cars are expected and parking costs $6 a car?

80. PARTIES A homemaker invited her neighbors to a kitchenware party to show off cookware and utensils. As party hostess, she received 12% of the total sales. How much was purchased if she received $41.76 for hosting the party?

81. WATCH SALE Refer to the advertisement below.

    a. Find the amount of the discount on the watch.

    b. Find the sale price of the watch.

82. SCOOTER SALE Refer to the advertisement below.

    a. Find the amount of the discount on the scooter.

    b. Find the sale price of the scooter.

83. SEGWAYS Find the discount rate on a Segway PT shown in the advertisement. Round to the nearest one percent.

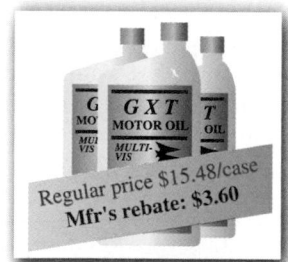

84. FAX MACHINES An HP 3180 fax machine, regularly priced at $160, is on sale for $116. What is the discount rate?

85. DISC PLAYERS What are the sale price and the discount rate for a Blu-ray disc player that regularly sells for $399.97 and is being discounted $50? Round to the nearest one percent.

86. CAMCORDER SALE What are the sale price and the discount rate for a camcorder that regularly sells for $559.97 and is being discounted $80? Round to the nearest one percent.

87. REBATES Find the discount rate and the new price for a case of motor oil if a shopper receives the manufacturer's rebate mentioned in the advertisement. Round to the nearest one percent.

88. DOUBLE COUPONS Find the discount, the discount rate, and the reduced price for a box of cereal that normally sells for $3.29 if a shopper presents the coupon at a store that doubles the value of the coupon.

89. TV SHOPPING Determine the Home Shopping Network (HSN) price of the ring described in the illustration if it sells it for 55% off of the retail price. Ignore shipping and handling costs.

Item 169-117
2.75 lb ctw
10K
Blue Topaz
Ring
6, 7, 8, 9, 10
Retail value $170
HSN Price
$??.??
S&H $5.95

90. INFOMERCIALS The host of a TV infomercial says that the suggested retail price of a rotisserie grill is $249.95 and that it is now offered "for just 4 easy payments of only $39.95." What is the discount, and what is the discount rate?

91. RING SALE What does a ring regularly sell for if it has been discounted 20% and is on sale for $149.99? (*Hint:* The ring is selling for 80% of its regular price.)

92. BLINDS SALE What do vinyl blinds regularly sell for if they have been discounted 55% and are on sale for $49.50? (*Hint:* The blinds are selling for 45% of their regular price.)

▌ WRITING

93. Explain the difference between a sales tax and a sales tax rate.

94. List the pros and cons of working on commission.

95. Suppose the price of an item increases $25 from $75 to $100. Explain why the following percent sentence *cannot* be used to find the percent of increase in the price of the item.

    25   is   what percent   of   100?

96. Explain how to find the sale price of an item if you know the regular price and the discount rate.

▌ REVIEW

97. Multiply: $-5(-5)(-2)$

98. Divide: $\dfrac{-320}{40}$

99. Subtract: $-4 - (-7)$

100. Add: $-17 + 6 + (-12)$

101. Evaluate: $|-5 - 8|$

102. Evaluate: $\sqrt{25} - \sqrt{16}$

Objectives

1 Estimate answers to percent problems involving 1% and 10%.

2 Estimate answers to percent problems involving 50%, 25%, 5%, and 15%.

3 Estimate answers to percent problems involving 200%.

4 Use estimation to solve percent application problems.

SECTION 7.4

Estimation with Percent

Estimation can be used to find approximations when exact answers aren't necessary. For example, when dining at a restaurant, it's helpful to be able to estimate the amount of the tip. When shopping, the ability to estimate a discount or the sale price of an item also comes in handy. In this section, we will discuss some estimation methods that can be used to make quick calculations involving percents.

1 Estimate answers to percent problems involving 1% and 10%.

There is an easy way to find 1% of a number that does not require any calculations. First, recall that $1\% = \frac{1}{100} = 0.01$. Thus, to find 1% of a number, we multiply it by 0.01, and a quick way to multiply the number by 0.01 is to move its decimal point *two places to the left.*

> #### Finding 1% of a Number
>
> To find 1% of a number, move the decimal point in the number two places to the left.

Self Check 1

What is 1% of 519.3? Find the exact answer and an estimate using front-end rounding.

Now Try Problem 11

EXAMPLE 1   What is 1% of 423.1? Find the exact answer and an estimate using front-end rounding.

Strategy To find the exact answer, we will move the decimal point in 423.1 two places to the left. To find an estimate, we will move the decimal point in an approximation of 423.1 two places to the left.

WHY We move the decimal point *two* places to the left because 1% of a number means 0.01 of (times) the number.

Solution
Exact answer:

$$1\% \text{ of } 423.1 = 4.231 \quad \text{Move the decimal point in 423.1 two places to the left.}$$

Estimate: Recall from Chapter 1 that with **front-end rounding,** a number is rounded to its largest place value so that all but its first digit is zero. To estimate 1% of 423.1, we can front-end round 423.1 to 400 and find 1% of 400. If we move the understood decimal point in 400 two places to the left, we get 4. Thus,

$$1\% \text{ of } 423.1 \approx 4 \quad \text{Because 1\% of 400 = 4.}$$

> *Success Tip*   To quickly find 2% of a number, find 1% of the number by moving the decimal point two places to the left, and then double (multiply by 2) the result. In Example 1, we found that 1% of 423.1 is **4.231.** Thus, 2% of 423.1 is $2 \cdot$ **4.231** $= 8.462$. A similar approach can be used to find 3% of a number, 4% of a number, and so on.

There is also an easy way to find 10% of a number that doesn't require any calculations. First, recall that $10\% = \frac{10}{100} = \frac{1}{10}$. Thus, to find 10% of a number, we multiply the number by 0.1, and a quick way to multiply the number by 0.1 is to move its decimal point *one place to the left.*

Finding 10% of a Number

To find 10% of a number, move the decimal point in the number one place to the left.

EXAMPLE 2   What is 10% of 6,872 feet? Find the exact answer and an estimate using front-end rounding.

Strategy To find the exact answer, we will move the decimal point in 6,872 one place to the left. To find an estimate, we will move the decimal point in an approximation of 6,872 one place to the left.

WHY We move the decimal point *one* place to the left because 10% of a number means 0.10 of (times) the number.

Solution
Exact answer:

$$10\% \text{ of } 6{,}872 \text{ feet} = 687.2 \text{ feet}$$ Move the understood decimal point in 6,872 one place to the left.

Estimate: To estimate 10% of 6,872 feet, we can front-end round 6,872 to 7,000 and find 10% of 7,000 feet . If we move the understood decimal point in 7,000 one place to the left, we get 700. Thus,

$$10\% \text{ of } 6{,}872 \text{ feet} \approx 700 \text{ feet}$$ Because 10% of 7,000 = 700.

> **Caution!** In Examples 1 and 2, *front-end rounding* was used to find estimates of answers to percent problems. Since there are other ways to approximate (round) the numbers involved in a percent problem, the answers to estimation problems may vary.

The rule for finding 10% of a number can be extended to help us quickly find multiples of 10% of a number.

Finding 20%, 30%, 40%, . . . of a Number

To find 20% of a number, find 10% of the number by moving the decimal point one place to the left, and then double (multiply by 2) the result. A similar approach can be used to find 30% of a number, 40% of a number, and so on.

EXAMPLE 3   Estimate the answer: What is 20% of 416?

Strategy We will estimate 10% of 416, and double (multiply by 2) the result.

WHY 20% of a number is twice as much as 10% of a number.

Solution Since 10% of 416 is 41.6 (or about **42**), it follows that 20% of 416 is about $2 \cdot \mathbf{42}$, which is 84. Thus,

$$20\% \text{ of } 416 \approx 84$$ Because 10% of 416 = 41.6 ≈ 42 and 2 · 42 = 84.

Self Check 2
What is 10% of 3,536 pounds? Find the exact answer and an estimate using front-end rounding.
Now Try Problem 15

Self Check 3
Estimate the answer: What is 20% of 129?
Now Try Problem 19

2   Estimate answers to percent problems involving 50%, 25%, 5%, and 15%.

There is an easy way to find 50% of a number. First, recall that $50\% = \frac{50}{100} = \frac{1}{2}$. Thus, to find 50% of a number means to find $\frac{1}{2}$ of that number, and to find $\frac{1}{2}$ of a number we simply divide it by 2.

> ### Finding 50% of a Number
>
> To find 50% of a number, divide the number by 2.

Self Check 4

Estimate the answer: What is 50% of 14,272,549?

Now Try   Problem 23

EXAMPLE 4   Estimate the answer: What is 50% of 2,595,603?

Strategy We will divide an approximation of 2,595,603 by 2.

WHY To find 50% of a number, we divide the number by 2.

Solution To estimate 50% of 2,595,603, we will find 50% of 2,600,000. We use 2,600,000 as an approximation because it is close to 2,595,603, because it is even, and, therefore, divisible by 2, and because it ends with many zeros.

$\qquad$ 50% of 2,595,603 $\approx$ 1,300,000   *Because 50% of 2,600,000 = $\frac{2,600,000}{2}$ = 1,300,000*

There is also an easy way to find 25% of a number. First, find 50% of the number by dividing the number by 2. Then, since 25% is one-half of 50%, divide that result by 2. Or, to save time, simply divide the original number by 4.

> ### Finding 25% of a Number
>
> To find 25% of a number, divide the number by 4.

Self Check 5

Estimate the answer: What is 25% of 27.16?

Now Try   Problem 27

EXAMPLE 5   Estimate the answer: What is 25% of 43.02?

Strategy We will divide an approximation of 43.02 by 4.

WHY To find 25% of a number, divide the number by 4.

Solution To estimate 25% of 43.02, we will find 25% of 44. We use 44 as an approximation because it is close to 43.02 and because it is divisible by 4.

$\qquad$ 25% of 43.02 $\approx$ 11   *Because 25% of 44 = $\frac{44}{4}$ = 11.*

There is a quick way to find 5% of a number. First, find 10% of the number by moving the decimal point in the number one place to the left. Then, since 5% is one-half of 10%, divide that result by 2.

> ### Finding 5% of a Number
>
> To find 5% of a number, find 10% of the number by moving the decimal point in the number one place to the left. Then, divide that result by 2.

EXAMPLE 6   *Electricity Usage*   The average U.S. household uses 10,656 kilowatt-hours of electricity each year. Several energy conservation groups would like each household to take steps to reduce its electricity usage by 5%. Estimate 5% of 10,656 kilowatt-hours. (*Source:* U.S. Department of Energy)

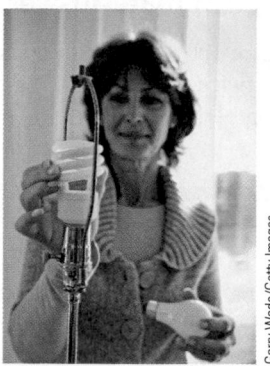

Garry Wade/Getty Images

Self Check 6

Estimate the answer: What is 5% of 24,198?

Now Try Problems 31

Strategy We will find 10% of 10,656. Then, we will divide an approximation of that result by 2.

WHY 5% of a number is one-half of 10% of a number.

Solution First, we find 10% of 10,656.

$$10\% \text{ of } 10{,}656 = 1{,}065.6$$ Move the understood decimal point in 10,656 one place to the left.

We will use 1,066 as an approximation of this result because it is close to 1,065.6 and because it is even, and, therefore, divisible by 2. Next, we divide the approximation by 2 to estimate 5% of 10,656.

$$\frac{1{,}066}{2} = 533$$ Divide the approximation of 10% of 10,656 by 2.

Thus, 5% of 10,656 ≈ 533. A 5% reduction in electricity usage by the average U.S. household is about 533 kilowatt-hours.

We can use the shortcuts for finding 10% and 5% of a number to find 15% of a number.

Finding 15% of a Number

To find 15% of a number, find the sum of 10% of the number and 5% of the number.

EXAMPLE 7   *Tipping*   As a general rule, if the service in a restaurant is acceptable, a tip of 15% of the total bill should be left for the server. Estimate the 15% tip on a $77.55 dinner bill.

Self Check 7

TIPPING Estimate the 15% tip on a $29.55 breakfast bill.

Now Try Problems 35 and 75

tetra images/First Light

Strategy We will find 10% and 5% of an approximation of $77.55. Then we will add those results.

WHY To find 15% of a number, find the sum of 10% of the number and 5% of the number.

Solution To simplify the calculations, we will estimate the cost of the $77.55 dinner to be $80. Then, to estimate the tip, we find 10% of $80 and 5% of $80, and add.

10% of $80 is $8 ————————→ $8
5% of $80 (half as much as 10% of $80) ——→ + $4
                                                  $12   Add to get the estimated tip.

The tip should be $12.

3 Estimate answers to percent problems involving 200%.

Since 100% of a number is the number itself, it follows that 200% of a number would be twice the number. We can extend this rule to quickly find multiples of 100% of a number.

> ### Finding 200%, 300%, 400%, ... of a Number
>
> To find 200% of a number, multiply the number by 2. A similar approach can be used to find 300% of a number, 400% of a number, and so on.

Self Check 8

Estimate the answer: What is 200% of 12.437?

Now Try Problem 43

EXAMPLE 8   Estimate the answer: What is 200% of 5.673?

Strategy We will multiply an approximation of 5.673 by 2.

WHY To find 200% of a number, multiply the number by 2.

Solution To estimate 200% of 5.673, we will find 200% of 6. We use 6 as an approximation because it is close to 5.673 and it makes the multiplication by 2 easy.

$$200\% \text{ of } 5.673 \approx 12 \quad \textit{Because 200\% of } 6 = 2 \cdot 6 = 12.$$

4 Use estimation to solve percent application problems.

In the previous examples of this section, we were given the percent (1%, 10%, 50%, 25%, 5%, 15%, or 200%), we approximated the base, and then we estimated the amount. Sometimes we must approximate the percent, as well, to estimate an answer.

Self Check 9

STUDENT DRIVERS Of the 1,550 students attending a high school, 26% of them drive to school. Estimate the number of students that drive to school.

Now Try Problem 85

EXAMPLE 9   *Music Education*   Of the 350 children attending an elementary school, 24% of them are enrolled in the instrumental music program. Estimate the number of children taking instrumental music.

Strategy We will use the rule from this section for finding 25% of a number.

WHY 24% is approximately 25%, and there is a quick way to find 25% of a number.

Solution 24% of the 350 children in the school are taking instrumental music. To estimate 24% of 350, we will find 25% of 360. We use 360 as an approximation because it is close to 350 and it is divisible by 4.

$$24\% \text{ of } 350 \approx 90 \quad \textit{Because 25\% of } 360 = \frac{360}{4} = 90.$$

There are approximately 90 children in the school taking instrumental music.

> **ANSWERS TO SELF CHECKS**
> **1.** 5.193, 5   **2.** 353.6 lb, 400 lb   **3.** 26   **4.** 7,000,000   **5.** 7   **6.** 1,210   **7.** $4.50
> **8.** 24   **9.** 400 students

SECTION 7.4   STUDY SET

VOCABULARY

Fill in the blanks.

1. _____ can be used to find approximations when exact answers aren't necessary.

2. With _____-end rounding, a number is rounded to its largest place value so that all but its first digit is zero.

CONCEPTS

Fill in the blanks.

3. To find 1% of a number, move the decimal point in the number _____ places to the left.

4. To find 10% of a number, move the decimal point in the number _____ place to the left.

5. To find 20% of a number, find 10% of the number by moving the decimal point one place to the left, and then double (multiply by ___) the result.

6. To find 50% of a number, divide the number by ___.

7. To find 25% of a number, divide the number by ___.

8. To find 5% of a number, find 10% of the number by moving the decimal point in the number one place to the left. Then, divide that result by ___.

9. To find 15% of a number, find the sum of ___% of the number and ___% of the number.

10. To find 200% of a number, multiply the number by ___.

GUIDED PRACTICE

What is 1% of the given number? Find the exact answer and an estimate using front-end rounding. See Example 1.

11. 275.1   12. 460.9

13. 12.67   14. 92.11

What is 10% of the given number? Find the exact answer and an estimate using front-end rounding. See Example 2.

15. 4,059 pounds

16. 7,435 hours

17. 691.4 minutes

18. 881.2 kilometers

Estimate each answer. (Answers may vary.) See Example 3.

19. What is 20% of 346?

20. What is 20% of 409?

21. What is 20% of 67?

22. What is 20% of 32?

Estimate each answer. (Answers may vary.) See Example 4.

23. What is 50% of 4,195,898?

24. What is 50% of 6,802,117?

25. What is 50% of 397,020?

26. What is 50% of 793,288?

Estimate each answer. (Answers may vary.) See Example 5.

27. What is 25% of 15.49?

28. What is 25% of 7.02?

29. What is 25% of 49.33?

30. What is 25% of 39.74?

Estimate each answer. (Answers may vary because of the approximation used.) See Example 6.

31. What is 5% of 16,359?

32. What is 5% of 44,191?

33. What is 5% of 394.182?

34. What is 5% of 176.001?

Estimate a 15% tip on each dollar amount. (Answers may vary.) See Example 7.

35. $58.99   36. $38.60

37. $27.16   38. $49.05

39. $115.75   40. $135.88

41. $9.74   42. $11.75

Estimate each answer. (Answers may vary.) See Example 8.

43. What is 200% of 4.212?

44. What is 200% of 5.189?

45. What is 200% of 35.77?

46. What is 200% of 80.32?

TRY IT YOURSELF

Find the exact answer using methods from this section.

47. What is 2% of 600?

48. What is 3% of 700?

49. What is 30% of 18?

50. What is 40% of 45?

Estimate each answer. (Answers may vary.)

51. What is 300% of 59.2?

52. What is 400% of 203.77?

53. What is 5% of 4,605?

54. What is 5% of 8,401?

55. What is 1% of 628.21?

56. What is 1% of 12,847.9?

57. What is 15% of 119?

58. What is 15% of 237?

59. What is 10% of 67.0056?

60. What is 10% of 94.2424?

61. What is 25% of 275?

62. What is 25% of 313?

63. What is 50% of 23,898?

64. What is 25% of 56,716?

65. What is 200% of 0.9123?

66. What is 200% of 0.4189?

Find the exact answer.

67. What is 1% of 50% of 98?

68. What is 10% of 25% of 20?

69. What is 15% of 20% of 400?

70. What is 5% of 10% of 30?

■ APPLICATIONS

Estimate each answer unless stated otherwise. (Answers may vary.)

71. COLLEGE COURSES 20% of the 815 students attending a small college were enrolled in a science course. How many students is this?

72. SPECIAL OFFERS In the grocery store, a 65-ounce bottle of window cleaner was marked "25% free." How many ounces are free?

73. DISCOUNTS By how much is the price of a coat discounted if the regular price of $196.88 is reduced by 30%?

74. SIGNS The nation's largest electronic billboard is at the south intersection of Times Square in New York City. It has 12,000,000 LED lights. If just 1% of these lights burnt out, how many lights would have to be replaced? Give the exact answer.

75. TIPPING A restaurant tip is normally 15% of the cost of the meal. Find the tip on a dinner costing $38.64.

76. VISA RECEIPTS Refer to the receipt to the right. Estimate the 15% gratuity (tip) and then find the total.

CLARK'S SEAFOOD
OKLAHOMA CITY, OK

Date:
Card Type: VISA
Acct Num: ★★★★★★★★★★★★0241
Exp Date: ★★/★★
Customer: WONG/TOM
Server: 209 Colleen

Amount: $58.47
Gratuity: ?
Total: ?

77. DINING OUT A couple went out to eat at a restaurant. The food they ordered cost $28.55 and the drinks they ordered cost $19.75. Estimate a 15% tip on the total bill.

78. SPLITTING THE TIP The total bill for three businessmen who went out to eat at a Chinese restaurant was $121.10. If they split the tip equally, estimate each person's share.

79. FIRE DAMAGE An insurance company paid 25% of the $118,000 it cost to rebuild a home that was destroyed by fire. How much did the insurance company pay?

80. SAFETY INSPECTIONS Of the 2,513 vehicles inspected at a safety checkpoint, 10% had code violations. How many cars had code violations?

81. WEIGHTLIFTING A 158-pound weightlifter can bench press 200% of his body weight. How many pounds can he bench press?

82. TESTING On a 60-question true/false test, 5% of a student's answers were wrong. How many questions did she miss?

83. TRAFFIC STUDIES According to an electronic traffic monitor, 30% of the 690 motorists who passed it were speeding. How many of these motorists were speeding?

84. SELLING A HOME A homeowner has been told she will get back 50% of her $6,125 investment if she paints her home before selling it. How much will she get back if she paints her home?

Approximate the percent and then estimate each answer. (Answers may vary.)

85. NO-SHOWS The attendance at a seminar was only 24% of what the organizers had anticipated. If 875 people were expected, how many actually attended the seminar?

86. HONOR ROLL Of the 900 students in a school, 16% were on the principal's honor roll. How many students were on the honor roll?

87. INTERNET SURVEYS The illustration shows an online survey question. How many people voted yes?

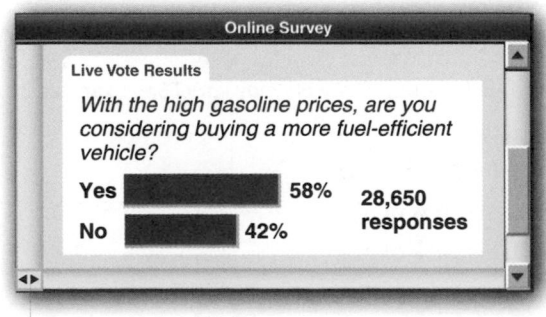

88. SALES TAX The state sales tax rate in Kansas is 5.3%. Estimate the sales tax on a purchase of $596.

89. VOTING On election day, 48% of the 6,200 workers at the polls were volunteers. How many volunteers helped with the election?

90. BUDGETS Each department at a college was asked to cut its budget by 21%. By how much money should the mathematics department budget be reduced if it is currently $4,715?

WRITING

91. Explain why 200% of a number is twice the number.

92. If you know 10% of a number, explain how you can find 30% of the same number.

93. If you know 10% of a number, explain how you can find 5% of the same number.

94. Explain why 25% of a number is the same as $\frac{1}{4}$ of the number.

REVIEW

Perform each operation and simplify, if possible.

95. a. $\frac{5}{6} + \frac{1}{2}$ **b.** $\frac{5}{6} - \frac{1}{2}$

 c. $\frac{5}{6} \cdot \frac{1}{2}$ **d.** $\frac{5}{6} \div \frac{1}{2}$

96. a. $\frac{7}{15} + \frac{7}{18}$ **b.** $\frac{7}{15} - \frac{7}{18}$

 c. $\frac{7}{15} \cdot \frac{7}{18}$ **d.** $\frac{7}{15} \div \frac{7}{18}$

SECTION 7.5

Interest

Objectives

1 Calculate simple interest.

2 Calculate compound interest.

When money is borrowed, the lender expects to be paid back the amount of the loan plus an additional charge for the use of the money. The additional charge is called **interest.** When money is deposited in a bank, the depositor is paid for the use of the money. The money the deposit earns is also called interest. In general, *interest is money that is paid for the use of money.*

1 Calculate simple interest.

Interest is calculated in one of two ways: either as **simple interest** or as **compound interest.** We begin by discussing simple interest. First, we need to introduce some key terms associated with borrowing or lending money.

- **Principal:** the amount of money that is invested, deposited, loaned, or borrowed.

- **Interest rate:** a percent that is used to calculate the amount of interest to be paid. The interest rate is assumed to be per year (annual interest) unless otherwise stated.

- **Time:** the length of time that the money is invested, deposited, or borrowed.

The amount of interest to be paid depends on the principal, the rate, and the time. That is why all three are usually mentioned in advertisements for bank accounts, investments, and loans, as shown below.

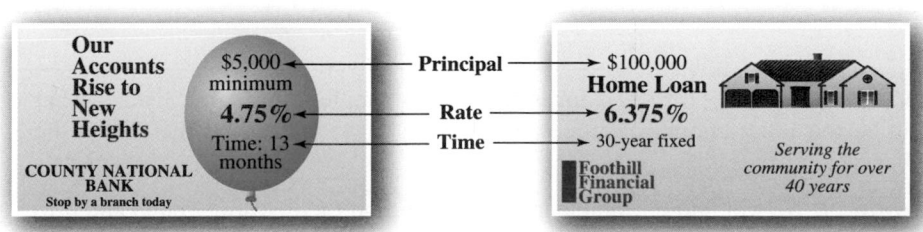

Simple interest is interest earned only on the original principal. It is found using the following formula.

Simple Interest Formula

Interest = principal · rate · time   or $I = P \cdot r \cdot t$

where the rate r is expressed as an annual (yearly) rate and the time t is expressed in years. This formula can be written more simply without the multiplication raised dots as

$I = Prt$

Self Check 1

If $4,200 is invested for 2 years at a rate of 4%, how much simple interest is earned?

Now Try Problem 17

EXAMPLE 1   If $3,000 is invested for 1 year at a rate of 5%, how much simple interest is earned?

Strategy We will identify the principal, rate, and time for the investment.

WHY Then we can use the formula $I = Prt$ to find the unknown amount of simple interest earned.

Solution The principal is $3,000, the interest rate is 5%, and the time is 1 year.

$P = \$3,000$ $r = 5\% = 0.05$ $t = 1$

$I = Prt$	This is the simple interest formula.
$I = \$3,000 \cdot 0.05 \cdot 1$	Substitute the values for P, r, and t. Remember to write the rate r as a decimal.
$I = \$3,000 \cdot 0.05$	Multiply: $0.05 \cdot 1 = 0.05$.
$I = \$150$	Complete the multiplication.

$$\begin{array}{r} 3,000 \\ \times 0.05 \\ \hline 150.00 \end{array}$$

The simple interest earned in 1 year is $150.

The information given in this problem and the result can be presented in a table.

Principal	Rate	Time	Interest earned
$3,000	5%	1 year	$150

If no money is withdrawn from an investment, the investor receives the principal *and* the interest at the end of the time period. Similarly, a borrower must repay the principal *and* the interest when taking out a loan. In each case, the **total amount** of money involved is given by the following formula.

Finding the Total Amount

The total amount in an investment account or the total amount to be repaid on a loan is the sum of the principal and the interest.

Total amount = principal + interest

Self Check 2

If $600 is invested at 2.5% simple interest for 4 years, what will be the total amount of money in the investment account at the end of the 4 years?

EXAMPLE 2   If $800 is invested at 4.5% simple interest for 3 years, what will be the total amount of money in the investment account at the end of the 3 years?

Strategy We will find the simple interest earned on the investment and add it to the principal.

WHY At the end of 3 years, the total amount of money in the account is the sum of the principal and the interest earned.

Now Try **Problem 21**

Solution The principal is $800, the interest rate is 4.5%, and the time is 3 years. To find the interest the investment earns, we use multiplication.

$$P = \$800 \qquad r = 4.5\% = 0.045 \qquad t = 3$$

$I = Prt$	This is the simple interest formula.
$I = \$800 \cdot 0.045 \cdot 3$	Substitute the values for P, r, and t. Remember to write the rate r as a decimal.
$I = \$36 \cdot 3$	Multiply: $\$800 \cdot 0.045 = \$36$.
$I = \$108$	Complete the multiplication.

$$\begin{array}{r} \overset{4}{0.045} \\ \times\ 800 \\ \hline 36.000 \end{array} \qquad \begin{array}{r} \overset{1}{36} \\ \times 3 \\ \hline 108 \end{array}$$

The simple interest earned in 3 years is $108. To find the total amount of money in the account, we add.

Total amount =	**principal** + **interest**	This is the total amount formula.
=	**$800** + **$108**	Substitute $800 for the principal and $108 for the interest.
=	$908	Do the addition.

At the end of 3 years, the total amount of money in the account will be $908.

> **Caution!** When we use the formula $I = Prt$, the time must be expressed in years. If the time is given in days or months, we rewrite it as a fractional part of a year. For example, a 30-day investment lasts $\frac{30}{365}$ of a year, since there are 365 days in a year. For a 6-month loan, we express the time as $\frac{6}{12}$ or $\frac{1}{2}$ of a year, since there are 12 months in a year.

EXAMPLE 3 *Education Costs* A student borrowed $920 at 3% for 9 months to pay some college tuition expenses. Find the simple interest that must be paid on the loan.

Self Check 3

SHORT-TERM LOANS Find the simple interest on a loan of $810 at 9% for 8 months.

Now Try **Problem 25**

Strategy We will rewrite 9 months as a fractional part of a year, and then we will use the formula $I = Prt$ to find the unknown amount of simple interest to be paid on the loan.

WHY To use the formula $I = Prt$, the time must be expressed in years, or as a fractional part of a year.

Solution Since there are 12 months in a year, we have

$$9 \text{ months} = \frac{9}{12} \text{ year} = \frac{\overset{1}{\cancel{3}} \cdot 3}{\underset{1}{\cancel{3}} \cdot 4} \text{ year} = \frac{3}{4} \text{ year}$$

Simplify the fraction $\frac{9}{12}$ by removing a common factor of 3 from the numerator and denominator.

The time of the loan is $\frac{3}{4}$ year. To find the amount of interest, we multiply.

$$P = \$920 \qquad r = 3\% = 0.03 \qquad t = \frac{3}{4}$$

$I = Prt$	This is the simple interest formula.
$I = \$920 \cdot 0.03 \cdot \dfrac{3}{4}$	Substitute the values for P, r, and t. Remember to write the rate r as a decimal.
$I = \dfrac{\$920}{1} \cdot \dfrac{0.03}{1} \cdot \dfrac{3}{4}$	Write $920 and 0.03 as fractions.
$I = \dfrac{\$82.80}{4}$	Multiply the numerators. Multiply the denominators.
$I = \$20.70$	Do the division: $82.80 \div 4 = 20.70$.

$$\begin{array}{r} 920 \\ \times 0.03 \\ \hline 27.60 \end{array} \qquad \begin{array}{r} \overset{2\,1}{27.60} \\ \times\ \ 3 \\ \hline 82.80 \end{array}$$

$$\begin{array}{r} 20.70 \\ 4\overline{)82.80} \\ -8 \\ \hline 02 \\ -0 \\ \hline 28 \\ -28 \\ \hline 00 \\ -0 \\ \hline 00 \end{array}$$

The simple interest to be paid on the loan is $20.70.

Self Check 4

ACCOUNTING To cover payroll expenses, a small business owner borrowed $3,200 at a simple interest rate of 15%. Find the total amount he must repay at the end of 120 days.

Now Try Problem 29

EXAMPLE 4 *Short-term Business Loans* To start a business, a couple borrowed $5,500 for 90 days to purchase equipment and supplies. If the loan has a 14% simple interest rate, find the total amount they must repay at the end of the 90-day period.

Strategy We will rewrite 90 days as a fractional part of a year, and then we will use the formula $I = Prt$ to find the unknown amount of simple interest to be paid on the loan.

WHY To use the formula $I = Prt$, the time must be expressed in years, or as a fractional part of a year.

Solution Since there are 365 days in a year, we have

$$90 \text{ days} = \frac{90}{365} \text{ year} = \frac{\overset{1}{\cancel{5}} \cdot 18}{\cancel{5} \cdot 73} \text{ year} = \frac{18}{73} \text{ year}$$

Simplify the fraction $\frac{90}{365}$ by removing a common factor of 5 from the numerator and denominator.

The time of the loan is $\frac{18}{73}$ year. To find the amount of interest, we multiply.

$$P = \$5,500 \qquad r = 14\% = 0.14 \qquad t = \frac{90}{365} = \frac{18}{73}$$

$$I = Prt \qquad \text{This is the simple interest formula.}$$

$$I = \$5,500 \cdot 0.14 \cdot \frac{18}{73} \qquad \text{Substitute the values for } P, r, \text{ and } t.$$

$$I = \frac{\$5,500}{1} \cdot \frac{0.14}{1} \cdot \frac{18}{73} \qquad \text{Write } \$5,500 \text{ and } 0.14 \text{ as fractions.}$$

$$I = \frac{\$13,860}{73} \qquad \text{Multiply the numerators. Multiply the denominators.}$$

$$I \approx \$189.86 \qquad \text{Divide 13,860 by 73. The division does not terminate. Round to the nearest cent.}$$

5,500	770
×0.14	×18
22000	6160
55000	7700
770.00	13,860

The interest on the loan is $189.86. To find how much they must pay back, we add.

$$\text{Total amount} = \textbf{principal} + \textbf{interest} \qquad \text{This is the total amount formula.}$$

$$= \$5,500 + \$189.86 \qquad \text{Substitute \$5,500 for the principal and \$189.86 for the interest.}$$

$$= \$5,689.86 \qquad \text{Do the addition.}$$

The couple must pay back $5,689.86 at the end of 90 days.

2 Calculate compound interest.

Most savings accounts and investments pay *compound interest* rather than simple interest. We have seen that simple interest is paid only on the original principal. **Compound interest** is paid on the principal and *previously earned interest*. To illustrate this concept, suppose that $2,000 is deposited in a savings account at a rate of 5% for 1 year. We can use the formula $I = Prt$ to calculate the interest earned at the end of 1 year.

$$I = Prt \qquad \text{This is the simple interest formula.}$$

$$I = \$2,000 \cdot 0.05 \cdot 1 \qquad \text{Substitute for } P, r, \text{ and } t.$$

$$I = \$100 \qquad \text{Do the multiplication.}$$

Interest of $100 was earned. At the end of the first year, the account contains the interest ($100) plus the original principal ($2,000), for a balance of $2,100.

Suppose that the money remains in the savings account for another year at the same interest rate. For the second year, interest will be paid on a principal of $2,100.

That is, during the second year, we earn *interest on the interest* as well as on the original \$2,000 principal. Using $I = Prt$, we can find the interest earned in the second year.

$I = \textbf{\textit{Prt}}$	This is the simple interest formula.
$I = \textbf{\$2,100} \cdot 0.05 \cdot 1$	Substitute for *P*, *r*, and *t*.
$I = \$105$	Do the multiplication.

In the second year, \$105 of interest is earned. The account now contains that interest plus the \$2,100 principal, for a total of \$2,205.

As the figure below shows, we calculated the simple interest two times to find the compound interest.

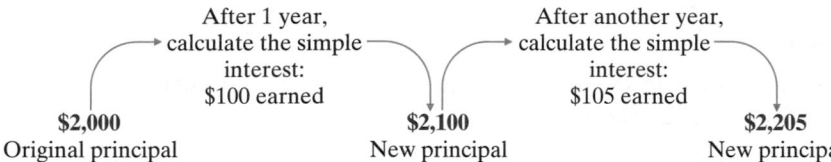

After 1 year,
calculate the simple
interest:
\$100 earned

After another year,
calculate the simple
interest:
\$105 earned

\$2,000
Original principal

\$2,100
New principal

\$2,205
New principal

If we compute only the *simple interest* on \$2,000, at 5% for 2 years, the interest earned is $I = \$2,000 \cdot 0.05 \cdot 2 = \$200$. Thus, the account balance would be \$2,200. Comparing the balances, we find that the account earning compound interest will contain \$5 more than the account earning simple interest.

In the previous example, the interest was calculated at the end of each year, or **annually.** When compounding, we can compute the interest in other time spans, such as **semiannually** (twice a year), **quarterly** (four times a year), or even **daily.**

EXAMPLE 5   *Compound Interest*   As a special gift for her newborn granddaughter, a grandmother opens a \$1,000 savings account in the baby's name. The interest rate is 4.2%, compounded quarterly. Find the amount of money the child will have in the bank on her first birthday.

Strategy We will use the simple interest formula $I = Prt$ four times in a series of steps to find the amount of money in the account after 1 year. Each time, the time t is $\frac{1}{4}$.

WHY The interest is compounded *quarterly*.

Solution If the interest is compounded quarterly, the interest will be computed four times in one year. To find the amount of interest \$1,000 will earn in the first quarter of the year, we use the simple interest formula, where t is $\frac{1}{4}$ of a year.

Interest earned in the first quarter:

$$P_{\text{1st Qtr}} = \$1,000 \qquad r = 4.2\% = 0.042 \qquad t = \frac{1}{4}$$

$I = \textbf{\textit{Prt}}$	This is the simple interest formula.
$I = \textbf{\$1,000} \cdot 0.042 \cdot \dfrac{1}{4}$	Substitute for *P*, *r*, and *t*.
$I = \$42 \cdot \dfrac{1}{4}$	Multiply: \$1,000 · 0.042 = \$42.
$I = \dfrac{\$42}{4}$	Do the multiplication. Think of 42 as $\frac{42}{1}$.
$I = \$10.50$	Do the division: 42 ÷ 4 = 10.5.

$$
\begin{array}{r}
10.5 \\
4\overline{)42.0} \\
-4 \\
\hline
02 \\
-0 \\
\hline
2\,0 \\
-2\,0 \\
\hline
0
\end{array}
$$

The interest earned in the first quarter is \$10.50. This now becomes part of the principal for the second quarter.

$P_{\text{2nd Qtr}} = \$1,000 + \$10.50 = \$1,010.50$ Add the original principal and the interest that it earned to find the second-quarter principal.

Self Check 5

COMPOUND INTEREST Suppose \$8,000 is deposited in an account that earns 2.3% compounded quarterly. Find the amount of money in an account at the end of the first year.

Now Try **Problem 33**

To find the amount of interest $1,010.50 will earn in the second quarter of the year, we use the simple interest formula, where t is again $\frac{1}{4}$ of a year.

Interest earned in the second quarter:

$P_{\text{2nd Qtr}} = \$1,010.50 \qquad r = 0.042 \qquad t = \dfrac{1}{4}$

$I = Prt$ — This is the simple interest formula.

$I = \$1,010.50 \cdot 0.042 \cdot \dfrac{1}{4}$ — Substitute for P, r, and t.

$I = \dfrac{\$1,010.50 \cdot 0.042 \cdot 1}{4}$ — Multiply.

$I \approx \$10.61$ — Use a calculator. Round to the nearest cent (hundredth).

The interest earned in the second quarter is $10.61. This becomes part of the principal for the third quarter.

$P_{\text{3rd Qtr}} = \$1,010.50 + \$10.61 = \$1,021.11$ — Add the second-quarter principal and the interest that it earned to find the third-quarter principal.

To find the interest $1,021.11 will earn in the third quarter of the year, we proceed as follows.

Interest earned in the third quarter:

$P_{\text{3rd Qtr}} = \$1,021.11 \qquad r = 0.042 \qquad t = \dfrac{1}{4}$

$I = Prt$ — This is the simple interest formula.

$I = \$1,021.11 \cdot 0.042 \cdot \dfrac{1}{4}$ — Substitute for P, r, and t.

$I = \dfrac{\$1,021.11 \cdot 0.042 \cdot 1}{4}$ — Multiply.

$I \approx \$10.72$ — Use a calculator. Round to the nearest cent (hundredth).

The interest earned in the third quarter is $10.72. This now becomes part of the principal for the fourth quarter.

$P_{\text{4th Qtr}} = \$1,021.11 + \$10.72 = \$1,031.83$ — Add the third-quarter principal and the interest that it earned to find the fourth-quarter principal.

To find the interest $1,031.83 will earn in the fourth quarter, we again use the simple interest formula.

Interest earned in the fourth quarter:

$P_{\text{4th Qtr}} = \$1,031.83 \qquad r = 0.042 \qquad t = \dfrac{1}{4}$

$I = Prt$ — This is the simple interest formula.

$I = \$1,031.83 \cdot 0.042 \cdot \dfrac{1}{4}$ — Substitute for P, r, and t.

$I = \dfrac{\$1,031.83 \cdot 0.042 \cdot 1}{4}$ — Multiply.

$I \approx \$10.83$ — Use a calculator. Round to the nearest cent (hundredth).

The interest earned in the fourth quarter is $10.83. Adding this to the existing principal, we get

Total amount $= \$1,031.83 + \$10.83 = \$1,042.66$ — Add the fourth-quarter principal and the interest that it earned.

The total amount in the account after four quarters, or 1 year, is $1,042.66.

Calculating compound interest by hand can take a long time. The **compound interest formula** can be used to find the total amount of money that an account will contain at the end of the term quickly.

Compound Interest Formula

The total amount A in an account can be found using the formula

$$A = P\left(1 + \frac{r}{n}\right)^{nt}$$

where P is the principal, r is the annual interest rate expressed as a decimal, t is the length of time in years, and n is the number of compoundings in one year.

A calculator is very helpful in performing the operations on the right side of the compound interest formula.

Using Your CALCULATOR   Compound Interest

A businessperson invests $9,250 at 7.6% interest, to be compounded monthly. To find what the investment will be worth in 3 years, we use the compound interest formula with the following values.

$P = \$9,250 \quad r = 7.6\% = 0.076 \quad t = 3$ years $\quad n = 12$ times a year (monthly)

$$A = P\left(1 + \frac{r}{n}\right)^{nt} \qquad \text{This is the compound interest formula.}$$

$$A = 9,250\left(1 + \frac{0.076}{12}\right)^{12(3)} \qquad \begin{array}{l}\text{Substitute the values of } P, r, t, \text{ and } n.\\ \text{In the exponent, } nt \text{ means } n \cdot t.\end{array}$$

$$A = 9,250\left(1 + \frac{0.076}{12}\right)^{36} \qquad \text{Evaluate the exponent: } 12(3) = 36.$$

To evaluate the expression on the right-hand side of the equation using a calculator, we enter these numbers and press these keys.

9250 $\times$ (1 + .076 $\div$ 12) y^x 36 $=$ ⬚ 11610.43875 ⬚

On some calculator models, the $\boxed{\wedge}$ key is used in place of the $\boxed{y^x}$ key. Also, the $\boxed{\text{ENTER}}$ key is pressed instead of the $\boxed{=}$ key for the result to be displayed.

Rounded to the nearest cent, the amount in the account after 3 years will be $11,610.44.

If your calculator does not have parenthesis keys, calculate the sum within the parentheses first. Then find the power. Finally, multiply by 9,250.

EXAMPLE 6   *Compounding Daily*   An investor deposited $50,000 in a long-term account at 6.8% interest, compounded daily. How much money will he be able to withdraw in 7 years if the principal is to remain in the bank?

Strategy We will use the compound interest formula to find the *total amount* in the account after 7 years. Then we will subtract the original principal from that result.

WHY When the investor withdraws money, he does not want to touch the original $50,000 principal in the account.

Self Check 6

COMPOUNDING DAILY Find the amount of interest $25,000 will earn in 10 years if it is deposited in an account at 5.99% interest, compounded daily.

Now Try Problem 37

Solution "Compounded daily" means that compounding will be done 365 times in a year for 7 years.

$$P = \$50{,}000 \qquad r = 6.8\% = 0.068 \qquad t = 7 \qquad n = 365$$

$$A = P\left(1 + \frac{r}{n}\right)^{nt} \qquad \text{This is the compound interest formula.}$$

$$A = 50{,}000\left(1 + \frac{0.068}{365}\right)^{365(7)} \qquad \begin{array}{l}\text{Substitute the values of } P, r, t, \text{ and } n.\\ \text{In the exponent, } nt \text{ means } n \cdot t.\end{array}$$

$$A = 50{,}000\left(1 + \frac{0.068}{365}\right)^{2{,}555} \qquad \text{Evaluate the exponent: } 365 \cdot 7 = 2{,}555.$$

$$A \approx 80{,}477.58 \qquad \text{Use a calculator. Round to the nearest cent.}$$

$$\begin{array}{r} \overset{4\,3}{365} \\ \times\ \ 7 \\ \hline 2{,}555 \end{array}$$

The account will contain $80,477.58 at the end of 7 years. To find how much money the man can withdraw, we must subtract the original principal of $50,000 from the total amount in the account.

$$80{,}477.58 - 50{,}000 = 30{,}477.58$$

The man can withdraw $30,477.58 without having to touch the $50,000 principal. ∎

ANSWERS TO SELF CHECKS

1. $336   **2.** $660   **3.** $48.60   **4.** $3,357.81   **5.** $8,185.59   **6.** $20,505.20

SECTION 7.5   STUDY SET

▌VOCABULARY

Fill in the blanks.

1. In general, _____ is money that is paid for the use of money.

2. In banking, the original amount of money invested, deposited, loaned, or borrowed is known as the _____.

3. The percent that is used to calculate the amount of interest to be paid is called the interest ____.

4. _____ interest is interest earned only on the original principal.

5. The _____ amount in an investment account is the sum of the principal and the interest.

6. _____ interest is interest paid on the principal and previously earned interest.

▌CONCEPTS

7. Refer to the home loan advertisement below.

a. What is the principal?

b. What is the interest rate?

c. What is the time?

8. Refer to the investment advertisement below.

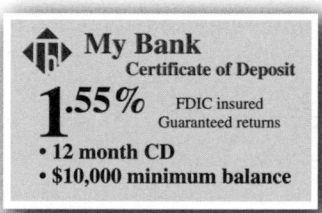

a. What is the principal?

b. What is the interest rate?

c. What is the time?

9. When making calculations involving percents, they must be written as decimals or fractions. Change each percent to a decimal.

a. 7%   **b.** 9.8%   **c.** $6\frac{1}{4}\%$

10. Express each of the following as a fraction of a year. Simplify the fraction.

a. 6 months   **b.** 90 days

c. 120 days   **d.** 1 month

11. Complete the table by finding the simple interest earned.

Principal	Rate	Time	Interest earned
$10,000	6%	3 years	

12. Determine how many times a year the interest on a savings account is calculated if the interest is compounded
    a. annually   b. semiannually
    c. quarterly   d. daily
    e. monthly

13. a. What concept studied in this section is illustrated by the diagram below?
    b. What was the original principal?
    c. How many times was the interest found?
    d. How much interest was earned on the first compounding?
    e. For how long was the money invested?

    ┌─ 1st qtr ─┐┌─2nd qtr─┐┌─3rd qtr─┐┌─4th qtr─┐
    $1,000 $1,050 $1,102.50 $1,157.63 $1,215.51

14. $3,000 is deposited in a savings account that earns 10% interest compounded annually. Complete the series of calculations in the illustration below to find how much money will be in the account at the end of 2 years.

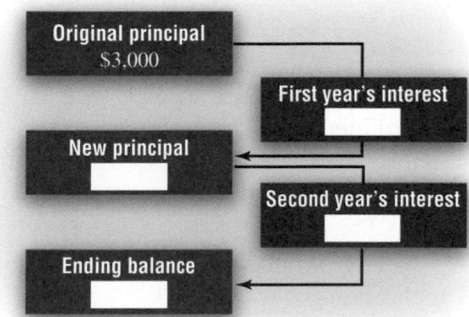

NOTATION

15. Write the simple interest formula $I = P \cdot r \cdot t$ without the multiplication raised dots.

16. In the formula $A = P\left(1 + \dfrac{r}{n}\right)^{nt}$, how many operations must be performed to find A?

GUIDED PRACTICE

Calculate the simple interest earned. See Example 1.

17. If $2,000 is invested for 1 year at a rate of 5%, how much simple interest is earned?

18. If $6,000 is invested for 1 year at a rate of 7%, how much simple interest is earned?

19. If $700 is invested for 4 years at a rate of 9%, how much simple interest is earned?

20. If $800 is invested for 5 years at a rate of 8%, how much simple interest is earned?

Calculate the total amount in each account. See Example 2.

21. If $500 is invested at 2.5% simple interest for 2 years, what will be the total amount of money in the investment account at the end of the 2 years?

22. If $400 is invested at 6.5% simple interest for 6 years, what will be the total amount of money in the investment account at the end of the 6 years?

23. If $1,500 is invested at 1.2% simple interest for 5 years, what will be the total amount of money in the investment account at the end of the 5 years?

24. If $2,500 is invested at 4.5% simple interest for 8 years, what will be the total amount of money in the investment account at the end of the 8 years?

Calculate the simple interest. See Example 3.

25. Find the simple interest on a loan of $550 borrowed at 4% for 9 months.

26. Find the simple interest on a loan of $460 borrowed at 9% for 9 months.

27. Find the simple interest on a loan of $1,320 borrowed at 7% for 4 months.

28. Find the simple interest on a loan of $1,250 borrowed at 10% for 3 months.

Calculate the total amount that must be repaid at the end of each short-term loan. See Example 4.

29. $12,600 is loaned at a simple interest rate of 18% for 90 days. Find the total amount that must be repaid at the end of the 90-day period.

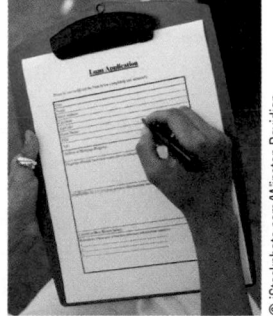

30. $45,000 is loaned at a simple interest rate of 12% for 90 days. Find the total amount that must be repaid at the end of the 90-day period.

31. $40,000 is loaned at 10% simple interest for 45 days. Find the total amount that must be repaid at the end of the 45-day period.

32. $30,000 is loaned at 20% simple interest for 60 days. Find the total amount that must be repaid at the end of the 60-day period.

Calculate the total amount in each account. See Example 5.

33. Suppose $2,000 is deposited in a savings account that pays 3% interest, compounded quarterly. How much money will be in the account in one year?

34. Suppose $3,000 is deposited in a savings account that pays 2% interest, compounded quarterly. How much money will be in the account in one year?

35. If $5,400 earns 4% interest, compounded quarterly, how much money will be in the account at the end of one year?

36. If $10,500 earns 8% interest, compounded quarterly, how much money will be in the account at the end of one year?

Use a calculator to solve the following problems. See Example 6.

37. A deposit of $30,000 is placed in a savings account that pays 4.8% interest, compounded daily. How much money can be withdrawn at the end of 6 years if the principal is to remain in the bank?

38. A deposit of $12,000 is placed in a savings account that pays 5.6% interest, compounded daily. How much money can be withdrawn at the end of 8 years if the principal is to remain in the bank?

39. If 8.55% interest, compounded daily, is paid on a deposit of $55,250, how much money will be in the account at the end of 4 years?

40. If 4.09% interest, compounded daily, is paid on a deposit of $39,500, how much money will be in the account at the end of 9 years?

APPLICATIONS

41. RETIREMENT INCOME A retiree invests $5,000 in a savings plan that pays a simple interest rate of 6%. What will the account balance be at the end of the first year?

42. INVESTMENTS A developer promised a return of 8% simple interest on an investment of $15,000 in her company. How much could an investor expect to make in the first year?

43. A member of a credit union was loaned $1,200 to pay for car repairs . The loan was made for 3 years at a simple interest rate of 5.5%. Find the interest due on the loan.

from Campus to Careers
Loan Officer

Ariel Skelley/Getty Images

44. REMODELING A homeowner borrows $8,000 to pay for a kitchen remodeling project. The terms of the loan are 9.2% simple interest and repayment in 2 years. How much interest will be paid on the loan?

45. SMOKE DAMAGE The owner of a café borrowed $4,500 for 2 years at 12% simple interest to pay for the cleanup after a kitchen fire. Find the total amount due on the loan.

46. ALTERNATIVE FUELS To finance the purchase of a fleet of natural-gas–powered vehicles, a city borrowed $200,000 for 4 years at a simple interest rate of 3.5%. Find the total amount due on the loan.

47. SHORT-TERM LOANS A loan of $1,500 at 12.5% simple interest is paid off in 3 months. What is the interest charged?

48. FARM LOANS An apple orchard owner borrowed $7,000 from a farmer's co-op bank. The money was loaned at 8.8% simple interest for 18 months. How much money did the co-op charge him for the use of the money?

49. MEETING PAYROLLS In order to meet end-of-the-month payroll obligations, a small business had to borrow $4,200 for 30 days. How much did the business have to repay if the simple interest rate was 18%?

50. CAR LOANS To purchase a car, a man takes out a loan for $2,000 for 120 days. If the simple interest rate is 9% per year, how much interest will he have to pay at the end of the 120-day loan period?

51. SAVINGS ACCOUNTS Find the interest earned on $10,000 at $7\frac{1}{4}$% for 2 years. Use the table to organize your work.

P	*r*	*t*	*I*

52. TUITION A student borrows $300 from an educational fund to pay for books for spring semester. If the loan is for 45 days at $3\frac{1}{2}$% annual interest, what will the student owe at the end of the loan period?

53. LOAN APPLICATIONS Complete the following loan application.

Loan Application Worksheet

1. Amount of loan (principal) ___$1,200.00___

2. Length of loan (time) ___2 YEARS___

3. Annual percentage rate ___8%___
   (simple interest)

4. Interest charged _____

5. Total amount to be repaid _____

6. Check method of repayment:
   ☐ 1 lump sum   ☑ monthly payments

   Borrower agrees to pay ___24___ equal payments of _____ to repay loan.

54. LOAN APPLICATIONS Complete the following loan application.

> **Loan Application Worksheet**
>
> 1. Amount of loan (principal)   *$810.00*
>
> 2. Length of loan (time)   *9 mos.*
>
> 3. Annual percentage rate   *12%*
>   (simple interest)
>
> 4. Interest charged _____
>
> 5. Total amount to be repaid _____
>
> 6. Check method of repayment:
>   ☐ 1 lump sum   ☑ monthly payments
>
>   Borrower agrees to pay   *9*   equal
>   payments of _____ to repay loan.

55. LOW-INTEREST LOANS An underdeveloped country receives a low-interest loan from a bank to finance the construction of a water treatment plant. What must the country pay back at the end of $3\frac{1}{2}$ years if the loan is for $18 million at 2.3% simple interest?

56. REDEVELOPMENT A city is awarded a low-interest loan to help renovate the downtown business district. The $40-million loan, at 1.75% simple interest, must be repaid in $2\frac{1}{2}$ years. How much interest will the city have to pay?

A calculator will be helpful in solving the following problems.

57. COMPOUNDING ANNUALLY If $600 is invested in an account that earns 8%, compounded annually, what will the account balance be after 3 years?

58. COMPOUNDING SEMIANNUALLY If $600 is invested in an account that earns annual interest of 8%, compounded semiannually, what will the account balance be at the end of 3 years?

59. COLLEGE FUNDS A ninth-grade student opens a savings account that locks her money in for 4 years at an annual rate of 6%, compounded daily. If the initial deposit is $1,000, how much money will be in the account when she begins college in 4 years?

60. CERTIFICATE OF DEPOSITS A 3-year certificate of deposit pays an annual rate of 5%, compounded daily. The maximum allowable deposit is $90,000. What is the most interest a depositor can earn from the CD?

61. TAX REFUNDS A couple deposits an income tax refund check of $545 in an account paying an annual rate of 4.6%, compounded daily. What will the size of the account be at the end of 1 year?

62. INHERITANCES After receiving an inheritance of $11,000, a man deposits the money in an account paying an annual rate of 7.2%, compounded daily. How much money will be in the account at the end of 1 year?

63. LOTTERIES Suppose you won $500,000 in the lottery and deposited the money in a savings account that paid an annual rate of 6% interest, compounded daily. How much interest would you earn each year?

64. CASH GIFTS After receiving a $250,000 cash gift, a university decides to deposit the money in an account paying an annual rate of 5.88%, compounded quarterly. How much money will the account contain in 5 years?

65. WITHDRAWING ONLY INTEREST A financial advisor invested $90,000 in a long-term account at 5.1% interest, compounded daily. How much money will she be able to withdraw in 20 years if the principal is to remain in the account?

66. LIVING ON THE INTEREST A couple sold their home and invested the profit of $490,000 in an account at 6.3% interest, compounded daily. How much money will they be able to withdraw in 2 years if they don't want to touch the principal?

WRITING

67. What is the difference between simple and compound interest?

68. Explain this statement: *Interest is the amount of money paid for the use of money.*

69. On some accounts, banks charge a penalty if the depositor withdraws the money before the end of the term. Why would a bank do this?

70. Explain why it is better for a depositor to open a savings account that pays 5% interest, compounded daily, than one that pays 5% interest, compounded monthly.

REVIEW

71. Evaluate: $\sqrt{\dfrac{1}{4}}$ **72.** Evaluate: $\left(\dfrac{1}{4}\right)^2$

73. Add: $\dfrac{3}{7} + \dfrac{2}{5}$ **74.** Subtract: $\dfrac{3}{7} - \dfrac{2}{5}$

75. Multiply: $2\dfrac{1}{2} \cdot 3\dfrac{1}{3}$ **76.** Divide: $-12\dfrac{1}{2} \div 5$

77. Evaluate: -6^2 **78.** Evaluate: $(0.2)^2 - (0.3)^2$

STUDY SKILLS CHECKLIST

Percents, Decimals, and Fractions

Before taking the test on Chapter 7, read the following checklist. These skills are sometimes misunderstood by students. Put a checkmark in the box if you can answer "yes" to the statement.

☐ I know that to write a decimal as a percent, the decimal point is moved two places *to the right* and a % symbol is inserted.

Decimal	Percent
0.23	23%
0.768	76.8%
1.50	150%
0.9	90%

☐ I know that to write a percent as a decimal, the % symbol is dropped and the decimal point is moved two places *to the left*.

Percent	Decimal
44%	0.44
98.7%	0.987
0.5%	0.005
178.3%	1.783

☐ I know that to write a fraction as a percent, *a two-step process* is used:

Fraction	→	decimal	→	percent
		Divide the numerator by the denominator		Move the decimal point two places to the right

$$\frac{3}{4} \longrightarrow \begin{array}{r} 0.75 \\ 4\overline{)3.00} \\ -2\,8 \\ \hline 20 \\ -20 \\ \hline 0 \end{array} \longrightarrow 75\%$$

☐ I know that to find the percent increase (or decrease), we find what percent the amount of increase (or decrease) is of the *original amount*.

The number of phone calls increased from 10 to 18 per day.

Original amount   Amount of increase: 18 − 10 = 8

CHAPTER 7   SUMMARY AND REVIEW

SECTION 7.1   Percents, Decimals, and Fractions

DEFINITIONS AND CONCEPTS	EXAMPLES
Percent means parts per one hundred. The word *percent* can be written using the symbol %.	In the figure below, there are 100 equal-sized square regions, and 37 of them are shaded. We say that $\frac{37}{100}$, or 37%, of the figure is shaded. Numerator $\frac{37}{100} = 37\%$ Per 100 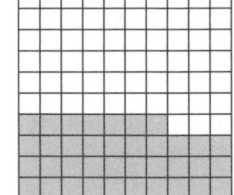

To **write a percent as a fraction,** drop the % symbol and write the given number over 100. Then simplify the fraction, if possible.	Write 22% as a fraction.
	$22\% = \dfrac{22}{100}$ Drop the % symbol and write 22 over 100.
	$= \dfrac{\overset{1}{\cancel{2}} \cdot 11}{\underset{1}{\cancel{2}} \cdot 50}$ To simplify the fraction, factor 22 and 100. Then remove the common factor of 2 from the numerator and denominator.
	Thus, $22\% = \frac{11}{50}$.
Percents such as 9.1% and 36.23% can be written as fractions of whole numbers by multiplying the numerator and denominator by a power of 10.	Write 9.1% as a fraction.
	$9.1\% = \dfrac{9.1}{100}$ Drop the % symbol and write 9.1 over 100.
	$= \dfrac{9.1}{100} \cdot \dfrac{\mathbf{10}}{\mathbf{10}}$ To obtain an equivalent fraction of *whole numbers*, we need to move the decimal point in the numerator one place to the right. Choose $\frac{10}{10}$ as the form of 1 to build the fraction.
	$= \dfrac{91}{1{,}000}$ Multiply the numerators. Multiply the denominators.
	Thus, $9.1\% = \frac{91}{1{,}000}$.
Mixed number percents, such as $2\frac{1}{3}\%$ and $23\frac{5}{6}\%$, can be written as fractions of whole numbers by performing the indicated division.	Write $2\frac{1}{3}\%$ as a fraction.
	$2\dfrac{1}{3}\% = \dfrac{2\frac{1}{3}}{100}$ Drop the % symbol and write $2\frac{1}{3}$ over 100.
	$= 2\dfrac{1}{3} \div 100$ The fraction bar indicates division.
	$= \dfrac{7}{3} \cdot \dfrac{1}{100}$ Write $2\frac{1}{3}$ as an improper fraction and then multiply by the reciprocal of 100.
	$= \dfrac{7}{300}$ Multiply the numerators. Multiply the denominators.
	Thus, $2\frac{1}{3}\% = \frac{7}{300}$.
When **percents that are greater than 100%** are written as fractions, the fractions are greater than 1.	Write 170% as a fraction.
	$170\% = \dfrac{170}{100}$ Drop the % symbol and write 170 over 100.
	$= \dfrac{\overset{1}{\cancel{10}} \cdot 17}{\underset{1}{\cancel{10}} \cdot 10}$ To simplify the fraction, factor 170 and 100. Then remove the common factor of 10 from the numerator and denominator.
	Thus, $170\% = \frac{17}{10}$.
When **percents that are less than 1%** are written as fractions, the fractions are less than $\frac{1}{100}$.	Write 0.03% as a fraction.
	$0.03\% = \dfrac{0.03}{100}$ Drop the % symbol and write 0.03 over 100.
	$= \dfrac{0.03}{100} \cdot \dfrac{\mathbf{100}}{\mathbf{100}}$ To obtain an equivalent fraction of *whole numbers*, we need to move the decimal point in the numerator two places to the right. Choose $\frac{100}{100}$ as the form of 1 to build the fraction.
	$= \dfrac{3}{10{,}000}$ Multiply the numerators and multiply the denominators. Since the numerator and denominator do not have any common factors (other than 1), the fraction is in simplified form.
	Thus, $0.03\% = \frac{3}{10{,}000}$.

To **write a percent as a decimal,** drop the % symbol and divide the given number by 100 by moving the decimal point 2 places to the left.	Write each percent as a decimal. $14\% = 14.0\% = 0.14$ Write a decimal point and 0 to the right of the 4 in 14%. $9.35\% = 0.0935$ Write a placeholder 0 (shown in blue) to the left of the 9. $198\% = 198.0\% = 1.98$ Write a decimal point and 0 to the right of the 8 in 198%. $0.75\% = 0.0075$
Mixed number percents, such as $1\frac{3}{4}\%$ and $10\frac{1}{2}\%$, can be written as decimals by writing the fractional part of the mixed number in its equivalent decimal form.	Write $1\frac{3}{4}\%$ as a decimal. There is no decimal point to move in $1\frac{3}{4}\%$. Since $1\frac{3}{4} = 1 + \frac{3}{4}$ and since the decimal equivalent of $\frac{3}{4}$ is 0.75, we can write $1\frac{3}{4}\%$ as 1.75% $1\frac{3}{4}\% = 1.75\% = 0.0175$ Write a placeholder 0 (shown in blue) to the left of the 1.
To **write a decimal as a percent,** multiply the decimal by 100 by moving the decimal point 2 places to the right, and then insert a % symbol.	Write each decimal as a percent. $0.501 = 50.1\%$ $3.66 = 366\%$ $0.002 = 000.2\% = 0.2\%$
To **write a fraction as a percent,** 1. Write the fraction as a decimal by dividing its numerator by its denominator. 2. Multiply the decimal by 100 by moving the decimal point 2 places to the right, and then insert a % symbol. Fraction ⟶ decimal ⟶ percent	Write $\frac{3}{4}$ as a percent. **Step 1** Divide the numerator by the denominator. $\begin{array}{r} 0.75 \\ 4\overline{)3.00} \\ -2\,8 \\ \hline 20 \\ -20 \\ \hline 0 \end{array}$ Write a decimal point and some additional zeros to the right of 3. The remainder is 0. **Step 2** Write the decimal 0.75 as a percent. $\frac{3}{4} = 0.75 = 75\%$
Sometimes, when we want to write a fraction as a percent, the result of the division is a **repeating decimal.** In such cases, we can give an **exact answer** or an **approximate answer.**	Write $\frac{2}{3}$ as a percent. **Step 1** Divide the numerator by the denominator. $\begin{array}{r} 0.666 \\ 3\overline{)2.000} \\ -1\,8 \\ \hline 20 \\ -18 \\ \hline 20 \\ -18 \\ \hline 2 \end{array}$ Write a decimal point and some additional zeros to the right of 2. The repeating pattern is now clear. We can stop the division. **Step 2** Write the decimal 0.6666... as a percent. $0.6666 = 66.66...\%$ **Exact Answer:** Use $\frac{2}{3}$ to represent 0.666.... $\frac{2}{3} = 66.66...\% = 66\frac{2}{3}\%$ **Approximation:** Round to the nearest tenth. $\frac{2}{3} = 66.66...\% \approx 66.7\%$

REVIEW EXERCISES

Express the amount of each figure that is shaded as a percent, as a decimal, and as a fraction. Each set of squares represents 100%.

1.

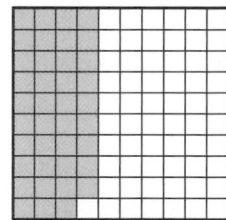

2.

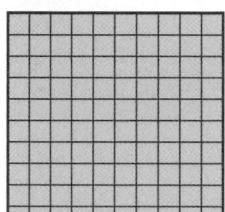

3. In Problem 1, what percent of the figure is not shaded?

4. THE INTERNET The following sentence appeared on a technology blog: "54 out of the top 100 websites failed Yahoo's performance test."

   a. What percent of the websites failed the test?

   b. What percent of the websites passed the test?

Write each percent as a fraction.

5. 15%   **6.** 120%   **7.** $9\frac{1}{4}\%$ **8.** 0.2%

Write each percent as a decimal.

9. 27%   **10.** 8%   **11.** 655%   **12.** $1\frac{4}{5}\%$

13. 0.75%   **14.** 0.23%

Write each decimal or whole number as a percent.

15. 0.83   **16.** 1.625   **17.** 0.051   **18.** 6

Write each fraction as a percent.

19. $\frac{1}{2}$ **20.** $\frac{4}{5}$ **21.** $\frac{7}{8}$ **22.** $\frac{1}{16}$

Write each fraction as a percent. Give the exact answer and an approximation to the nearest tenth of a percent.

23. $\frac{1}{3}$ **24.** $\frac{5}{6}$ **25.** $\frac{11}{12}$ **26.** $\frac{15}{9}$

27. WATER DISTRIBUTION The oceans contain 97.2% of all of the water on Earth. (*Source:* National Ground Water Association)

   a. Write this percent as a decimal.

   b. Write this percent as a fraction in simplest form.

28. BILL OF RIGHTS There are 27 amendments to the Constitution of the United States. The first ten are known as the Bill of Rights. What percent of the amendments were adopted after the Bill of Rights? (Round to the nearest one percent.)

29. TAXES The city of Grand Prairie, Texas, has a *one-fourth of one percent* sales tax to help fund park improvements.

   a. Write this percent as a decimal.

   b. Write this percent as a fraction.

30. SOCIAL SECURITY If your retirement age is 66, your Social Security benefits are reduced by $\frac{1}{15}$ if you retire at age 65. Write this fraction as a percent. Give the exact answer and an approximation to the nearest tenth of a percent. (*Source:* Social Security Administration)

SECTION 7.2   **Solving Percent Problems Using Percent Equations and Proportions**

DEFINITIONS AND CONCEPTS	EXAMPLES

The key words in a **percent sentence** can be translated to a percent equation.
 	
 • Each *is* translates to an equal symbol =
 	
 • *of* translates to multiplication that is shown with a raised dot ·
 	
 • *what* number or *what* percent translates to an unknown number that is represented by a variable.
	Translate the percent sentence to a percent equation.
	
	 What number is 26% of 180?
	   ↓ ↓ ↓ ↓ ↓
	   x = 26% · 180 This is the percent equation.

Percent sentences involve a comparison of numbers. The relationship between the **base** (the standard of comparison, the whole), the **amount** (a part of the base), and the **percent** is:

Amount = percent · base

or

Part = percent · whole

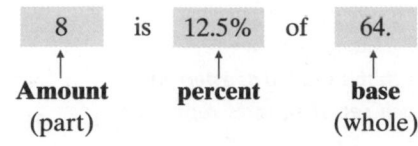

The percent equation method
We can translate percent sentences to percent equations and solve to **find the amount.**

Caution! When solving percent equations, always write the percent as a decimal (or fraction) before performing any calculations.

What number	is	45%	of	120?
↓	↓	↓	↓	↓
x	$=$	45%	$\cdot$	120

Translate.

Now, solve the percent equation.

$x = 0.45 \cdot 120$ Write 45% as a decimal.

$x = 54$ Do the multiplication.

Thus, 54 is 45% of 120.

We can translate percent sentences to percent equations and solve to **find the percent.**

12	is	what percent	of	192?
↓	↓	↓	↓	↓
12	$=$	x	$\cdot$	192

Translate.

Now, solve the percent equation.

$12 = 192x$ Write x · 192 as 192x.

$\dfrac{12}{192} = \dfrac{192x}{192}$ To isolate x, undo the multiplication by 192 by dividing both sides by 192.

$0.0625 = x$ Do the division: 12 ÷ 192 = 0.0625.

$06.25\% = x$ To write 0.0625 as a percent, multiply it by 100 by moving the decimal point two places to the right, and then insert a % symbol.

Thus, 12 is 6.25% of 192.

We can translate percent sentences to percent equations and solve to **find the base.**

Caution! Sometimes the calculations to solve a percent problem are made easier if we **write the percent as a fraction instead of a decimal.** This is the case with percents that have *repeating* decimal equivalents such as $33\frac{1}{3}\%$, $66\frac{2}{3}\%$, and $16\frac{2}{3}\%$.

8.2	is	$33\frac{1}{3}\%$	of	what number?
↓	↓	↓		↓
8.2	$=$	$33\frac{1}{3}\%$	$\cdot$	x

Translate.

Now, solve the percent equation.

$8.2 = \dfrac{1}{3}x$ Write the percent as a fraction: $33\frac{1}{3}\% = \frac{1}{3}$.

$3(8.2) = 3\left(\dfrac{1}{3}x\right)$ To isolate x, multiply both sides by the reciprocal of $\frac{1}{3}$, which is 3.

$24.6 = 1x$ On the left side, do the multiplication: 3(8.2) = 24.6. On the right side, the product of a number and its reciprocal is 1: $3\left(\frac{1}{3}\right) = 1$.

$24.6 = x$ On the right side, the coefficient of 1 need not be written: 1x = x.

Thus, 8.2 is $33\frac{1}{3}\%$ of 24.6.

We can translate percent sentences to **percent proportions** and solve to **find the amount.**

To translate a percent sentence to a **percent proportion,** use the following form:

Amount is to base as percent is to 100:

$$\frac{\text{amount}}{\text{base}} = \frac{\text{percent}}{100}$$

or

Part is to whole as percent is to 100:

$$\frac{\text{part}}{\text{whole}} = \frac{\text{percent}}{100}$$

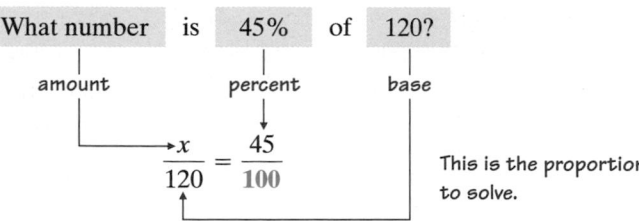

This is the proportion to solve.

To make the calculations easier, simplify the ratio $\frac{45}{100}$.

$$\frac{x}{120} = \frac{9}{20}$$ Simplify: $\frac{45}{100} = \frac{\overset{1}{\cancel{5}} \cdot 9}{\underset{1}{\cancel{5}} \cdot 20} = \frac{9}{20}$.

To solve the proportion we use the cross products.

$x \cdot 20 = 120 \cdot 9$ Find the cross products and set them equal.

$20x = 1{,}080$ On the left side, write x · 20 as 20x. On the right side, do the multiplication: 120 · 9 = 1,080.

$$\frac{20x}{20} = \frac{1{,}080}{20}$$ To isolate x, undo the multiplication by 20 by dividing both sides by 20.

$x = 54$ Do the division: 1,080 ÷ 20 = 54.

Thus, 54 is 45% of 120.

We can translate percent sentences to **percent proportions** and solve to **find the percent.**

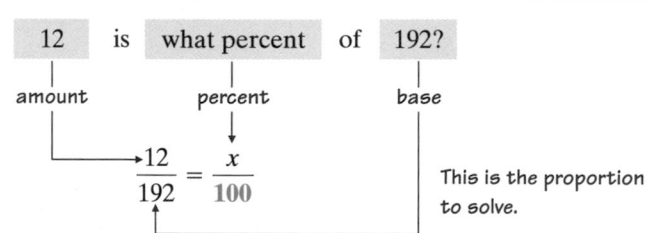

This is the proportion to solve.

To make the calculations easier, simplify the ratio $\frac{12}{192}$ first.

$$\frac{1}{16} = \frac{x}{100}$$ Simplify: $\frac{12}{192} = \frac{\overset{1}{\cancel{2}} \cdot \overset{1}{\cancel{2}} \cdot \overset{1}{\cancel{3}}}{\underset{1}{\cancel{2}} \cdot \underset{1}{\cancel{2}} \cdot 2 \cdot 2 \cdot 2 \cdot 2 \cdot \underset{1}{\cancel{3}}} = \frac{1}{16}$.

$1 \cdot 100 = 16 \cdot x$ Find the cross products and set them equal.

$100 = 16x$ On the left side, do the multiplication: 1 · 100 = 100. On the right side, write 16 · x as 16x.

$$\frac{100}{16} = \frac{16x}{16}$$ To isolate x, undo the multiplication by 16 by dividing both sides by 16.

$6.25 = x$ Do the division: 100 ÷ 16 = 6.25.

Thus, 12 is 6.25% of 192.

We can translate percent sentences to **percent proportions** and solve to **find the base.**

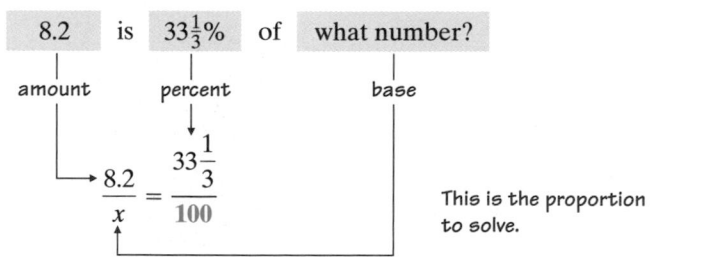

This is the proportion to solve.

To make the calculations easier, write the mixed number $33\frac{1}{3}$ as the improper fraction $\frac{100}{3}$.

$$\frac{8.2}{x} = \frac{\frac{100}{3}}{100}$$ Write $33\frac{1}{3}$ as $\frac{100}{3}$.

$$8.2 \cdot 100 = x \cdot \frac{100}{3}$$ To solve the proportion, find the cross products and set them equal.

$$820 = \frac{100}{3}x$$ On the left side, do the multiplication: $8.2 \cdot 100 = 820$. On the right side, write $x \cdot \frac{100}{3}$ as $\frac{100}{3}x$.

$$\frac{3}{100}(820) = \frac{3}{100}\left(\frac{100}{3}x\right)$$ To isolate x, multiply both sides by the reciprocal of $\frac{100}{3}$, which is $\frac{3}{100}$.

$$\frac{3}{100}\left(\frac{820}{1}\right) = 1x$$ On the left side, write 820 as a fraction: $820 = \frac{820}{1}$. On the right side, the product of a number and its reciprocal is 1: $\frac{3}{100}\left(\frac{100}{3}\right) = 1$.

$$\frac{2,460}{100} = x$$ On the left side, multiply the numerators and multiply the denominators. On the right side, the coefficient of 1 is not needed: $1x = x$.

$$24.6 = x$$ Divide 2,460 by 100 by moving the understood decimal point in 2,460 two places to the left.

Thus, 8.2 is $33\frac{1}{3}\%$ of 24.6.

A **circle graph** is a way of presenting data for comparison. The pie-shaped pieces of the graph show the relative sizes of each category.

The 100 tick marks equally spaced around the circle serve as a visual aid when constructing a circle graph.

FACEBOOK As of April 2009, Facebook had approximately 195 million users worldwide. Use the information in the circle graph to the right to find how many of them were male.

The circle graph shows that 46% of the 195 million users of Facebook were male.

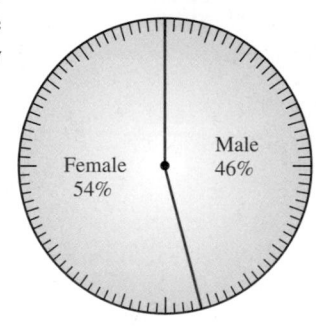

Facebook Users Worldwide 195 Million

Female 54%   Male 46%

(*Source:* O'Reilly Radar)

Method 1: To find the unknown amount write and then solve a **percent equation.**

What number	is	46%	of	195 million?
↓	↓	↓	↓	↓
x	=	46%	·	195 Translate.

To solve percent application problems, we often have to **rewrite the facts** of the problem in percent sentence form before we can translate to an equation.

Now, solve the percent equation.

$$x = 0.46 \cdot 195$$ Write 46% as a decimal: $46\% = 0.46$.

$$x = 89.7$$ Do the multiplication. The answer is in millions.

In April of 2009, there were approximately 89.7 million male users of Facebook worldwide.

Method 2: To find the unknown amount write and then solve a **percent proportion.**

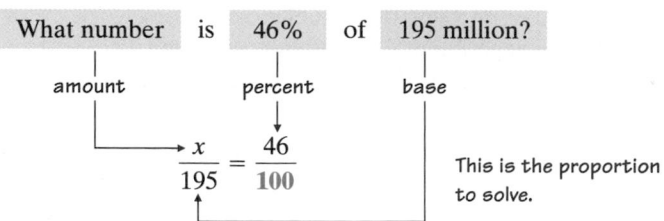

$$\frac{x}{195} = \frac{23}{50}$$ Simplify the ratio: $\frac{46}{100} = \frac{\overset{1}{\cancel{2}} \cdot 23}{\cancel{2} \cdot 50} = \frac{23}{50}.$

$x \cdot 50 = 195 \cdot 23$ Find the cross products and set them equal.

$50x = 4{,}485$ On the left side, write $x \cdot 50$ as $50x$. On the right side, do the multiplication.

$\dfrac{50x}{50} = \dfrac{4{,}485}{50}$ To isolate x, undo the multiplication by 50 by dividing both sides by 50.

$x = 89.7$ Do the division: $4{,}485 \div 50 = 89.7$. The answer is in millions.

In April of 2009, there were approximately 89.7 million male users of Facebook worldwide.

REVIEW EXERCISES

31. a. Identify the amount, the base, and the percent in the statement "15 is $33\frac{1}{3}\%$ of 45."

  b. Fill in the blanks to complete the percent equation (formula):

$$\boxed{} = \text{percent} \cdot \boxed{}$$

or

$$\text{Part} = \boxed{} \cdot \text{whole}$$

32. When computing with percents, we must change the percent to a decimal or a fraction. Change each percent to a decimal.

  a. 13%   **b.** 7.1%

  c. 195%   **d.** $\frac{1}{4}\%$

When computing with percents, we must change the percent to a decimal or a fraction. Change each percent to a fraction.

  e. $33\frac{1}{3}\%$

  f. $66\frac{2}{3}\%$

  g. $16\frac{2}{3}\%$

33. Translate each percent sentence into a *percent equation*. **Do not solve.**

  a. What number is 32% of 96?

  b. 64 is what percent of 135?

  c. 9 is 47.2% of what number?

34. Translate each percent sentence into a *percent proportion*. **Do not solve.**

  a. What number is 32% of 96?

  b. 64 is what percent of 135?

  c. 9 is 47.2% of what number?

Translate to a percent equation or percent proportion and then solve to find the unknown number.

35. What number is 40% of 500?

36. 16% of what number is 20?

37. 1.4 is what percent of 80?

38. $66\frac{2}{3}\%$ of 3,150 is what number?

39. Find 220% of 55.

40. What is 0.05% of 60,000?

41. 43.5 is $7\frac{1}{4}\%$ of what number?

42. What percent of 0.08 is 4.24?

43. RACING The nitro–methane fuel mixture used to power some experimental cars is 96% nitro and 4% methane. How many gallons of methane are needed to fill a 15-gallon fuel tank?

44. HOME SALES After the first day on the market, 51 homes in a new subdivision had already sold. This was 75% of the total number of homes available. How many homes were originally for sale?

45. HURRICANE DAMAGE In a mobile home park, 96 of the 110 trailers were either damaged or destroyed by hurricane winds. What percent is this? (Round to the nearest 1 percent.)

46. TIPPING The cost of dinner for a family of five at a restaurant was $36.20. Find the amount of the tip if it should be 15% of the cost of dinner.

47. COLLEGE EXPENSES In 2008, Survey.com asked 500 college students and parents of students who needed a loan, where they turned first to pay for college costs. The results of the survey are shown below in the table. Draw a circle graph for the data.

College	57%
Family/Friends	5%
Local bank	18%
Internet	15%
Other	5%

48. EARTH'S SURFACE The surface of Earth is approximately 196,800,000 square miles. Use the information in the circle graph to determine the number of square miles of Earth's surface that are covered with water.

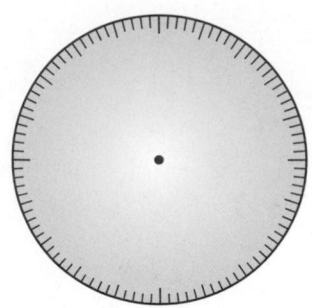

Water 70.9%   Land 29.1%

SECTION 7.3   Applications of Percent

DEFINITIONS AND CONCEPTS	EXAMPLES
The **sales tax** on an item is a percent of the purchase price of the item. Sales tax = sales tax rate · purchase price Amount = percent · base Notice that the formula is based on the percent equation discussed in Section 7.2. **Sales tax dollar amounts** are rounded to the nearest cent (hundredth). The **total cost** of an item is the sum of its purchase price and the sales tax on the item. Total cost = purchase price + sales tax	SHOPPING Find the sales tax and total cost of a $50.95 purchase if the sales tax rate is 8%. Sales tax = **sales tax rate · purchase price** = **8%** · **$50.95** = 0.08 · $50.95 Write 8% as a decimal: 8% = 0.08. = $4.076 Do the multiplication. ≈ $4.08 Round the sales tax to the nearest cent (hundredth). Thus, the sales tax is $4.08. The total cost is the sum of its purchase price and the sales tax. Total cost = **purchase price** + **sales tax rate** = **$50.95** + **$4.08** = $55.03 Do the addition. The total cost of the purchase is $55.03.
Sales tax rates are usually expressed as a percent.	APPLIANCES The purchase price of a toaster is $82. If the sales tax is $5.33, what is the sales tax rate? The sales tax of $5.33 is some unknown percent of the purchase price of $82. There are two methods that can be used to solve this problem.

There are two methods that can be used to find the unknown sales tax rate:

- The percent equation method
- The percent proportion method

The percent equation method:

| $5.33 | is | what percent | of | 82? |

$$5.33 = x \cdot 82 \qquad \text{Translate.}$$

Now, solve the percent equation.

$$5.33 = 82x \qquad \text{Write } x \cdot 82 \text{ as } 82x.$$

$$\frac{5.33}{82} = \frac{x \cdot 82}{82} \qquad \begin{array}{l}\text{To isolate } x, \text{ undo the multiplication by 82 by}\\ \text{dividing both sides by 82.}\end{array}$$

$$0.065 = x \qquad \text{Do the division: } 5.33 \div 82 = 0.065.$$

$$006.5\% = x \qquad \text{Write the decimal 0.065 as a percent.}$$

$$6.5\% = x$$

The sales tax rate is 6.5%.

The percent proportion method:

| 5.33 | is | what percent | of | 82? |

amount — percent — base

$$\frac{5.33}{82} = \frac{x}{100} \qquad \begin{array}{l}\text{This is the percent}\\ \text{proportion to solve.}\end{array}$$

$$5.33 \cdot 100 = 82 \cdot x \qquad \begin{array}{l}\text{To solve the proportion, find the cross products}\\ \text{and set them equal.}\end{array}$$

$$533 = 82x \qquad \begin{array}{l}\text{On the left side, do the multiplication:}\\ 5.33 \cdot 100 = 533. \text{ On the right side, write}\\ 82 \cdot x \text{ as } 82x.\end{array}$$

$$\frac{533}{82} = \frac{82x}{82} \qquad \begin{array}{l}\text{To isolate } x, \text{ undo the multiplication by 82 by}\\ \text{dividing both sides by 82.}\end{array}$$

$$6.5 = x \qquad \text{Do the division: } 533 \div 82 = 6.5.$$

The sales tax rate is 6.5%.

Instead of working for a salary or getting paid at an hourly rate, many salespeople are paid on **commission.**

The **amount of commission** paid is a percent of the total dollar sales of goods or services.

Commission = commission rate · sales

COMMISSIONS A salesperson earns an 11% commission on all appliances that she sells. If she sells a $450 dishwasher, what is her commission?

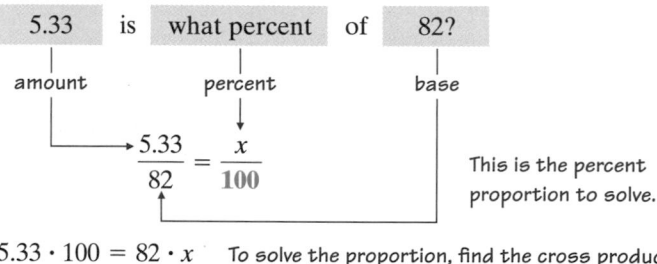

$$\begin{aligned}\text{Commission} &= \textbf{commission rate} \cdot \textbf{sales}\\ &= \quad 11\% \quad \cdot \ \$450\\ &= 0.11 \cdot \$450 \qquad \text{Write 11\% as a decimal.}\\ &= \$49.50 \qquad \text{Do the multiplication.}\end{aligned}$$

The commission earned on the sale of the $450 dishwasher is $49.50.

The **commission rate** is usually expressed as a percent.

TELEMARKETING A telemarketer made a commission of $600 in one week on sales of $4,000. What is his commission rate?

$$\begin{aligned}\textbf{Commission} &= \text{commission rate} \cdot \textbf{sales}\\ \$600 \quad &= \qquad x \qquad \cdot \ \$4{,}000 \qquad \begin{array}{l}\text{Let } x \text{ represent the}\\ \text{unknown commission}\\ \text{rate.}\end{array}\end{aligned}$$

$$600 = 4{,}000x \qquad \text{Drop the dollar signs. Write } x \cdot 4{,}000 \text{ as } 4{,}000x.$$

$$\frac{600}{4{,}000} = \frac{4{,}000x}{4{,}000} \qquad \begin{array}{l}\text{To isolate } x, \text{ undo the multiplication by 4,000 by}\\ \text{dividing both sides by 4,000.}\end{array}$$

$$0.15 = x \qquad \textit{Do the division: } 600 \div 4,000 = 0.15.$$
$$0\underset{\curvearrowright}{1}5\% = x \qquad \textit{Write the decimal 0.15 as a percent.}$$

The commission rate is 15%.

To find percent of increase or decrease:	WATCHING TELEVISION According to the Nielsen Company, the average American watched 145 hours of TV a month in 2007. That increased to 151 hours per month in 2008. Find the percent of increase. Round to the nearest one percent.

To find percent of increase or decrease:

1. Subtract the smaller number from the larger to find the amount of increase or decrease.
2. Find what percent the amount of increase or decrease is of the original amount.

There are two methods that can be used to find the unknown percent of increase (or decrease):

- The percent equation method
- The percent proportion method

Caution! The percent of increase (or decrease) is a percent of the *original number,* that is, the number before the change occurred.

WATCHING TELEVISION According to the Nielsen Company, the average American watched 145 hours of TV a month in 2007. That increased to 151 hours per month in 2008. Find the percent of increase. Round to the nearest one percent.

First, subtract to find the amount of increase.

$$151 - 145 = 6 \qquad \textit{Subtract the smaller number from the larger number.}$$

The number of hours watched per month increased by 6.

Next, find what percent of the *original* 145 hours the 6 hour increase represents.

The percent equation method:

$$\boxed{6} \quad \text{is} \quad \boxed{\text{what percent}} \quad \text{of} \quad \boxed{145?}$$
$$6 \quad = \quad x \quad \cdot \quad 145 \qquad \textit{Translate.}$$

Now, solve the percent equation.

$$6 = 145x \qquad \textit{Write } x \cdot 145 \textit{ as } 145x.$$
$$\frac{6}{145} = \frac{145x}{145} \qquad \textit{To isolate x, undo the multiplication by 145 by dividing both sides by 145.}$$
$$0.041 \approx x \qquad \textit{On the left side, divide 6 by 145. The division does not terminate.}$$
$$00\underset{\curvearrowright}{4}.1\% \approx x \qquad \textit{Write the decimal 0.041 as a percent.}$$
$$4\% \approx x \qquad \textit{Round to the nearest one percent.}$$

Between 2007 and 2008, the number of hours of television watched by the average American each month increased by 4%.

If the **percent proportion method** is used, solve the following proportion for x to find the percent of increase.

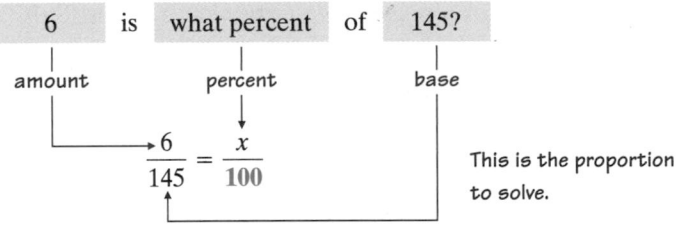

The **amount of discount** is a percent of the original price.	TOOL SALES Find the amount of the discount on a tool kit if it is normally priced at \$89.95, but is currently on sale for 35% off. Then find the sale price.

The **amount of discount** is a percent of the original price.

$$\underset{\substack{\uparrow \\ \text{amount}}}{\substack{\text{Amount of} \\ \text{discount}}} = \underset{\substack{\uparrow \\ \text{percent}}}{\substack{\text{discount} \\ \text{rate}}} \cdot \underset{\substack{\uparrow \\ \text{base}}}{\substack{\text{original} \\ \text{price}}}$$

Notice that the formula is based on the percent equation discussed in Section 7.2.

TOOL SALES Find the amount of the discount on a tool kit if it is normally priced at \$89.95, but is currently on sale for 35% off. Then find the sale price.

$$\text{Amount of discount} = \textbf{discount rate} \cdot \textbf{original price}$$
$$= \quad \textbf{35\%} \quad \cdot \quad \textbf{\$89.95}$$
$$= 0.35 \cdot \$89.95 \qquad \textit{Write 35\% as a decimal.}$$
$$= \$31.4825 \qquad \textit{Do the multiplication.}$$
$$\approx \$31.48 \qquad \textit{Round to the neaerst cent (hundredth).}$$

To find the **sale price** of an item, subtract the discount from the original price. Sale price = original price − discount	The discount on the tool kit is $31.48. To find the sale price, we use subtraction. Sale price = **original price** − **discount** 　　　　= 　**$89.95**　 − 　**$31.48** 　　　　= $58.47　　　　*Do the subtraction.* The sale price of the tool kit is $58.47.
The difference between the original price and the sale price is the **amount of discount.** Amount of discount = original price − sale price	FURNITURE SALES Find the discount rate on a living room set regularly priced at $2,500 that is on sale for $1,870. Round to the nearest one percent. We will think of this as a *percent-of-decrease problem*. The discount (decrease in price) is found using subtraction. $2,500 − $1,870 = $630　*Discount = original price − sale price* The living room set is discounted $630. Now we find what percent of the original price the $630 discount represents. **Amount of discount** = discount rate · **original price** 　　**$630**　　　　= 　x　·　**$2,500** $630 = 2,500x　*Drop the dollar signs. Write x · 2,500 as 2,500x.* $\dfrac{630}{2,500} = \dfrac{2,500x}{2,500}$　*To isolate x, undo the multiplication by 2,500 by dividing both sides by 2,500.* $0.252 = x$　*Do the division: 630 ÷ 2,500 = 0.252.* $025.2\% = x$　*Write the decimal 0.252 as a percent.* $25\% \approx x$　*Round to the nearest one percent.* To the nearest one percent, the discount rate on the living room set is 25%.

REVIEW EXERCISES

49. SALES RECEIPTS Complete the sales receipt shown below by finding the sales tax and total cost of the camera.

CAMERA CENTER

35mm Canon Camera	$59.99
SUBTOTAL	$59.99
SALES TAX @ 5.5%	?
TOTAL	?

50. SALES TAX RATES Find the sales tax rate if the sales tax is $492 on the purchase of an automobile priced at $12,300.

51. COMMISSIONS If the commission rate is 6%, find the commission earned by an appliance salesperson who sells a washing machine for $369.97 and a dryer for $299.97.

52. SELLING MEDICAL SUPPLIES A salesperson made a commission of $646 on a $15,200 order of antibiotics. What is her commission rate?

53. T-SHIRT SALES A stadium owner earns a commission of $33\frac{1}{3}\%$ of the T-shirt sales from any concert or sporting event. How much can the owner make if 12,000 T-shirts are sold for $25 each at a soccer match?

54. Fill in the blank: The percent of increase (or decrease) is a percent of the _____ number, that is, the number before the change occurred.

55. THE UNITED NATIONS In 2008, the U.N. Security Council voted to increase the size of a peacekeeping force from 17,000 to 20,000 troops. Find the percent of increase in the number of troops. Round to the nearest one percent. (*Source:* Reuters)

56. GAS MILEAGE A woman found that the gas mileage fell from 18.8 to 17.0 miles per gallon when she experimented with a new brand of gasoline in her truck. Find the percent of decrease in her mileage. Round to the nearest tenth of one percent.

57. Fill in the blanks.

   a. Sales tax = sales tax rate · ◻

   b. Total cost = purchase price + ◻

   c. Commission = ◻ · sales

58. Fill in the blanks.

   a. Amount of discount = original price − ◻

   b. Amount of discount = discount rate · ◻

   c. Sale price = original price − ◻

59. TOOL CHESTS Use the information in the advertisement below to find the discount, the original price, and the discount rate on the tool chest.

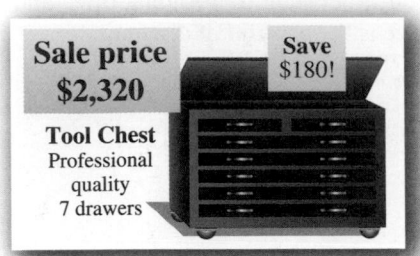

60. RENTS Find the discount rate if the monthly rent for an apartment is reduced from $980 to $931 per month.

SECTION 7.4   Estimation with Percent

DEFINITIONS AND CONCEPTS	EXAMPLES

Estimation can be used to find approximations when exact answers aren't necessary.
 	
To find **1% of a number,** move the decimal point in the number two places to the left.
	What is 1% of 291.4? Find the exact answer and an estimate using front-end rounding.
	
	Exact answer:
	
	   $1\% \text{ of } 291.4 = 2.914$ Move the decimal point two places to the left.
	
	Estimate: 291.4 front-end rounds to 300. If we move the understood decimal point in 300 two places to the left, we get 3. Thus
	
	   $1\% \text{ of } 291.4 \approx 3$ Because 1% of 300 = 3.

To find **10% of a number,** move the decimal point in the number one place to the left.	What is 10% of 40,735 pounds? Find the exact answer and an estimate using front-end rounding.
	
	Exact answer:
	
	   $10\% \text{ of } 40,735 = 4,073.5$ Move the decimal point one place to the left.
	
	Estimate: 40,735 front-end rounds to 40,000. If we move the understood decimal point in 40,000 one place to the left, we get 4,000. Thus
	
	   $1\% \text{ of } 40,735 \approx 4,000$ Because 10% of 40,000 = 4,000.

To find **20% of a number,** find 10% of the number by moving the decimal point one place to the left, and then double (multiply by 2) the result. A similar approach can be used to find 30% of a number, 40% of a number, and so on.	Estimate the answer: What is 20% of 809? Since 10% of 809 is 80.9 (or about **81**), it follows that 20% of 809 is about $2 \cdot 81$, which is 162. Thus, $20\% \text{ of } 809 \approx 162$ Because 10% of 809 ≈ 81.

To find **50% of a number,** divide the number by 2.	Estimate the answer: What is 50% of 1,442,957? We use 1,400,000 as an approximation of 1,442,957 because it is even, divisible by 2, and ends with many zeros. $$50\% \text{ of } 1{,}442{,}957 \approx 700{,}000 \quad \text{Because 50\% of 1,400,000} = \frac{1{,}400{,}000}{2} = 700{,}000.$$
To find **25% of a number,** divide the number by 4.	Estimate the answer: What is 25% of 21.004? We use 20 as an approximation because it is close to 21.004 and because it is divisible by 4. $$25\% \text{ of } 21.004 \approx 5 \quad \text{Because 25\% of 20} = \frac{20}{4} = 5.$$
To find **5% of a number,** find 10% of the number by moving the decimal point in the number one place to the left. Then, divide that result by 2.	Estimate the answer: What is 5% of 36,150? First, we find 10% of 36,150: $$10\% \text{ of } 36{,}150 = 3{,}615$$ We use 3,600 as an approximation of this result because it is close to 3,615 and because it is even, and therefore divisible by 2. Next, we divide the approximation by 2 to estimate 5% of 36,150. $$\frac{3{,}600}{2} = 1{,}800$$ Thus, 5% of 36,150 $\approx$ 1,800.
To find **15% of a number,** find the sum of 10% of the number and 5% of the number.	TIPPING Estimate the 15% tip on a dinner costing $88.55. To simplify the calculations, we will estimate the cost of the $88.55 dinner to be $90. Then, to estimate the tip, we find 10% of $90 and 5% of $90, and add. 10% of $90 is $9 $\longrightarrow$ $9 5% of $90 (half as much as 10% of $90) $\longrightarrow$ + $4.50 $13.50 The tip should be $13.50.
To find **200% of a number,** multiply the number by 2. A similar approach can be used to find 300% of a number, 400% of a number, and so on.	Estimate the answer: What is 200% of 3.509? To estimate 200% of 3.509, we will find 200% of 4. We use 4 as an approximation because it is close to 3.509 and it makes the multiplication by 2 easy. $$200\% \text{ of } 3.509 \approx 8 \quad \text{Because 200\% of 4} = 2 \cdot 4 = 8.$$
Sometimes we must **approximate the percent,** to estimate an answer.	QUALITY CONTROL In a production run of 145,350 ceramic tiles, 3% were found to be defective. Estimate the number of defective tiles. To estimate 3% of 145,350, we will find 1% of 150,000, and multiply the result by 3. We use 150,000 as the approximation because it is close to 145,350 and it ends with several zeros. $$3\% \text{ of } 145{,}350 \approx 4{,}500 \quad \text{Because 1\% of 150,000} = 1{,}500 \text{ and } 3 \cdot 1{,}500 = 4{,}500.$$ There were about 4,500 defective tiles in the production run.

REVIEW EXERCISES

What is 1% of the given number? Find the exact answer and an estimate using front-end rounding.

61. 342.03

62. 8,687

What is 10% of the given number? Find the exact answer and an estimate using front-end rounding.

63. 43.4 seconds

64. 10,900 liters

Estimate each answer. (Answers may vary.)

65. What is 20% of 63?

66. What is 20% of 612?

67. What is 50% of 279,985?

68. What is 50% of 327?

69. What is 25% of 13.02?

70. What is 25% of 39.9?

71. What is 5% of 7,150?

72. What is 5% of 19,359?

73. What is 200% of 29.78?

74. What is 200% of 1.125?

Estimate a 15% tip on each dollar amount. (Answers may vary.)

75. $243.55

76. $46.99

Estimate each answer. (Answers may vary.)

77. SPECIAL OFFERS A home improvement store sells a 50-fluid ounce pail of asphalt driveway sealant that is labeled "25% free." How many ounces are free?

78. JOB TRAINING 15% of the 785 people attending a job training program had a college degree. How many people is this?

Approximate the percent and then estimate each answer. (Answers may vary.)

79. SEAT BELTS A state trooper survey on an interstate highway found that of the 3,850 cars that passed the inspection point, 6% of the drivers were not wearing a seat belt. Estimate the number not wearing a seat belt.

80. DOWN PAYMENTS Estimate the amount of an 11% down payment on a house that is selling for $279,950.

SECTION 7.5   Interest

DEFINITIONS AND CONCEPTS	EXAMPLES
Interest is money that is paid for the use of money.	If $4,000 is invested for 3 years at a rate of 7.2%, how much simple interest is earned?
Simple interest is interest earned on the original principal and is found using the formula $$I = Prt$$ where P is the principal, r is the annual (yearly) interest rate, and t is the length of time in years.	$P = \$4,000 \qquad r = 7.2\% = 0.072 \qquad t = 3$

$$I = Prt \qquad \text{This is the simple interest formula.}$$
$$I = \$4,000 \cdot 0.072 \cdot 3 \qquad \text{Substitute the values for } P, r, \text{ and } t.$$
$$\text{Remember to write the rate } r \text{ as a decimal.}$$
$$I = \$288 \cdot 3 \qquad \text{Multiply: } \$4,000 \cdot 0.072 = \$288.$$
$$I = \$864 \qquad \text{Complete the multiplication.}$$

The simple interest earned in 3 years is $864.

The **total amount** in an investment account or the total amount to be repaid on a loan is the sum of the principal and the interest.

$$\text{Total amount} = \text{principal} + \text{interest}$$

HOME REPAIRS A homeowner borrowed $5,600 for 2 years at 10% simple interest to pay for a new concrete driveway. Find the total amount due on the loan.

$$P = \$5,600 \qquad r = 10\% = 0.10 \qquad t = 2$$

$$I = Prt \qquad \text{This is the simple interest formula.}$$
$$I = \$5,600 \cdot 0.10 \cdot 2 \qquad \text{Write the rate } r \text{ as a decimal.}$$
$$I = \$560 \cdot 2 \qquad \text{Multiply: } \$5,600 \cdot 0.10 = \$560.$$
$$I = \$1,120 \qquad \text{Complete the multiplication.}$$

The interest due in 2 years is $1,120. To find the total amount of money due on the loan, we add.

$$\begin{aligned} \text{Total amount} &= \textbf{principal} + \textbf{interest} \\ &= \$5,600 + \$1,120 \\ &= \$6,720 \qquad \text{Do the addition.} \end{aligned}$$

At the end of 2 years, the total amount of money due on the loan is $6,720.

When using the formula $I = Prt$, the **time must be expressed in years.** If the time is given in days or months, rewrite it as a fractional part of a year.

Here are two examples:

- Since there are 365 days in a year,

$$60 \text{ days} = \frac{60}{365} \text{ year} = \frac{\overset{1}{\cancel{5}} \cdot 12}{\cancel{5} \cdot 73} \text{ year} = \frac{12}{73} \text{ year}$$

- Since there are 12 months in a year,

$$4 \text{ months} = \frac{4}{12} \text{ year} = \frac{\overset{1}{\cancel{4}}}{3 \cdot \cancel{4}} \text{ year} = \frac{1}{3} \text{ year}$$

FINES A man borrowed \$300 at 15% for 45 days to get his car out of an impound parking garage. Find the simple interest that must be paid on the loan.

Since there are 365 days in a year, we have

$$45 \text{ days} = \frac{45}{365} \text{ year} = \frac{\overset{1}{\cancel{5}} \cdot 9}{\cancel{5} \cdot 73} \text{ year} = \frac{9}{73} \text{ year} \quad \text{Simplify the fraction.}$$

The time of the loan is $\frac{9}{73}$ year. To find the amount of interest, we multiply.

$$P = \$300 \qquad r = 15\% = 0.15 \qquad t = \frac{9}{73}$$

$I = Prt$ This is the simple interest formula.

$I = \$300 \cdot 0.15 \cdot \dfrac{9}{73}$ Write the rate r as a decimal.

$I = \dfrac{\$300}{1} \cdot \dfrac{0.15}{1} \cdot \dfrac{9}{73}$ Write \$300 and 0.15 as fractions.

$I = \dfrac{\$405}{73}$ Multiply the numerators. Multiply the denominators.

$I \approx \$5.55$ Do the division. Round to the nearest cent.

The simple interest that must be paid on the loan is \$5.55.

Compound interest is interest earned on the original principal and previously earned interest.

When compounding, we can calculate interest:

- **annually:** once a year
- **semiannually:** twice a year
- **quarterly:** four times a year
- **daily:** 365 times a year

COMPOUND INTEREST Suppose \$10,000 is deposited in an account that earns 6.5% compounded semiannually. Find the amount of money in an account at the end of the first year.

The word *semiannually* means that the interest will be compounded two times in one year. To find the amount of interest \$10,000 will earn in the first half of the year, use the simple interest formula, where t is $\frac{1}{2}$ of a year.

Interest earned in the first half of the year:

$$P = \$10,000 \qquad r = 6.5\% = 0.065 \qquad t = \frac{1}{2}$$

$I = Prt$ This is the simple interest formula.

$I = \$10,000 \cdot 0.065 \cdot \dfrac{1}{2}$ Write the rate r as a decimal.

$I = \dfrac{\$10,000}{1} \cdot \dfrac{0.065}{1} \cdot \dfrac{1}{2}$ Write \$10,000 and 0.065 as fractions.

$I = \dfrac{\$650}{2}$ Multiply the numerators. Multiply the denominators.

$I = \$325$ Do the division.

The interest earned in the first half of the year is \$325. The original principal and this interest now become the principal for the second half of the year.

$$\$10,000 + \$325 = \$10,325$$

To find the amount of interest \$10,325 will earn in the second half of the year, use the simple interest formula, where t is again $\frac{1}{2}$ of a year.

Interest earned in the second half of the year:

$$P = \$10{,}325 \qquad r = 6.5\% = 0.065 \qquad t = \frac{1}{2}$$

$I = Prt$ — This is the simple interest formula.

$I = \$10{,}325 \cdot 0.065 \cdot \dfrac{1}{2}$ — Write the rate r as a decimal.

$I = \dfrac{\$10{,}325}{1} \cdot \dfrac{0.065}{1} \cdot \dfrac{1}{2}$ — Write $10,325 and 0.065 as fractions.

$I = \dfrac{\$671.125}{2}$ — Multiply the numerators. Multiply the denominators.

$I \approx \$335.56$ — Do the division. Round to the nearest cent.

The interest earned in the second half of the year is $335.56. Adding this to the principal for the second half of the year, we get

$$\$10{,}325 + \$335.56 = \$10{,}660.56$$

The total amount in the account after one year is $10,660.56

Computing compound interest by hand can take a long time. The **compound interest formula** can be used to find the amount of money that an account will contain at the end of the term.

$$A = P\left(1 + \frac{r}{n}\right)^{nt}$$

where A is the amount in the account, P is the principal, r is the annual interest rate, n is the number of compoundings in one year, and t is the length of time in years.

A **calculator** is helpful in performing the operations on the right side of the compound interest formula.

COMPOUNDING DAILY A mini-mall developer promises investors in his company $3\frac{1}{4}\%$ interest, compounded daily. If a businessman decides to invest $80,000 with the developer, how much money will be in his account in 8 years?

Compounding daily means the compounding will be done 365 times a year.

$$P = \$80{,}000 \qquad r = 3\frac{1}{4}\% = 0.0325 \qquad t = 8 \qquad n = 365$$

$A = P\left(1 + \dfrac{r}{n}\right)^{nt}$ — This is the compound interest formula.

$A = 80{,}000\left(1 + \dfrac{0.0325}{365}\right)^{365(8)}$ — Substitute for P, r, n, and t.

$A = 80{,}000\left(1 + \dfrac{0.0325}{365}\right)^{2{,}920}$ — Evaluate the exponent: $365 \cdot 8 = 2{,}920$.

$A \approx 103{,}753.21$ — Use a calculator. Round to the nearest cent.

There will be $103,753.21 in the account in 8 years.

REVIEW EXERCISES

81. INVESTMENTS Find the simple interest earned on $6,000 invested at 8% for 2 years. Use the following table to organize your work.

P	r	t	I

82. INVESTMENT ACCOUNTS If $24,000 is invested at a simple interest rate of 4.5% for 3 years, what will be the total amount of money in the investment account at the end of the term?

83. EMERGENCY LOANS A teacher's credit union loaned a client $2,750 at a simple interest rate of 11% so that he could pay an overdue medical bill. How much interest does the client pay if the loan must be paid back in 3 months?

84. CODE VIOLATIONS A business was ordered to correct safety code violations in a production plant. To pay for the needed corrections, the company borrowed $10,000 at 12.5% simple interest for 90 days. Find the total amount that had to be paid after 90 days.

85. MONTHLY PAYMENTS A couple borrows $1,500 for 1 year at a simple interest rate of $7\frac{3}{4}$%.

a. How much interest will they pay on the loan?

b. What is the total amount they must repay on the loan?

c. If the couple decides to repay the loan by making 12 equal monthly payments, how much will each monthly payment be?

86. SAVINGS ACCOUNTS Find the amount of money that will be in a savings account at the end of 1 year if $2,000 is the initial deposit and the interest rate of 7% is compounded semi-annually. (*Hint:* Find the simple interest twice.)

87. SAVINGS ACCOUNTS Find the amount that will be in a savings account at the end of 3 years if a deposit of $5,000 earns interest at a rate of $6\frac{1}{2}$%, compounded daily.

88. CASH GRANTS Each year a cash grant is given to a deserving college student. The grant consists of the interest earned that year on a $500,000 savings account. What is the cash award for the year if the money is invested at a rate of 8.3%, compounded daily?

1. Fill in the blanks.

    a. _____ means parts per one hundred.

    b. The key words in a percent sentence translate as follows:

        • ___ translates to an equal symbol =

        • ___ translates to multiplication that is shown with a raised dot ·

        • _____ *number* or _____ *percent* translates to an unknown number that is represented by a variable.

    c. In the percent sentence "5 is 25% of 20," 5 is the _____, 25% is the percent, and 20 is the _____.

    d. When we use percent to describe how a quantity has increased compared to its original value, we are finding the percent of _____.

    e. _____ interest is interest earned only on the original principal. _____ interest is interest paid on the principal and previously earned interest.

2. a. Express the amount of the figure that is shaded as a percent, as a fraction, and as a decimal.

    b. What percent of the figure is not shaded?

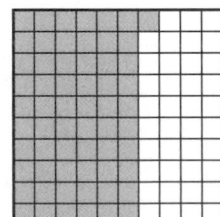

3. In the illustration below, each set of 100 square regions represents 100%. Express as a percent the amount of the figure that is shaded. Then express that percent as a fraction and as a decimal.

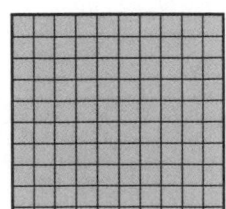 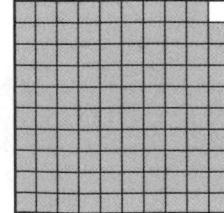

4. Write each percent as a decimal.

    a. 67%   b. 12.3%   c. $9\frac{3}{4}\%$

5. Write each percent as a decimal.

    a. 0.06%   b. 210%   c. 55.375%

6. Write each fraction as a percent.

    a. $\frac{1}{4}$ b. $\frac{5}{8}$ c. $\frac{28}{25}$

7. Write each decimal as a percent.

    a. 0.19   b. 3.47   c. 0.005

8. Write each decimal or whole number as a percent.

    a. 0.667   b. 2   c. 0.9

9. Write each percent as a fraction. Simplify, if possible.

    a. 55%   b. 0.01%   c. 125%

10. Write each percent as a fraction. Simplify, if possible.

    a. $6\frac{2}{3}\%$ b. 37.5%   c. 8%

11. Write each fraction as a percent. Give the exact answer and an approximation to the nearest tenth of a percent.

    a. $\frac{1}{30}$ b. $\frac{16}{9}$

12. 65 is what percent of 1,000?

13. What percent of 14 is 35?

14. FUGITIVES As of November 29, 2008, exactly 460 of the 491 fugitives who have appeared on the FBI's Ten Most Wanted list have been captured or located. What percent is this? Round to the nearest tenth of one percent. (*Source:* www.fbi.gov/wanted)

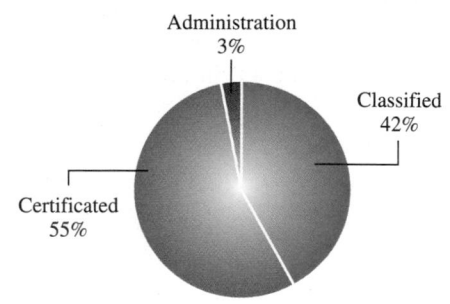

15. SWIMMING WORKOUTS A swimmer was able to complete 18 laps before a shoulder injury forced him to stop. This was only 20% of a typical workout. How many laps does he normally complete during a workout?

16. COLLEGE EMPLOYEES The 700 employees at a community college fall into three major categories, as shown in the circle graph. How many employees are in administration?

Administration
3%

Classified
42%

Certificated
55%

17. What number is 224% of 60?

18. 2.6 is $33\frac{1}{3}$% of what number?

19. SHRINKAGE See the following label from a new pair of jeans. The measurements are in inches. (*Inseam* is a measure of the length of the jeans.)

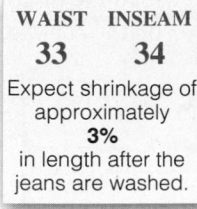

WAIST	INSEAM
33	34

Expect shrinkage of approximately **3%** in length after the jeans are washed.

    a. How much length will be lost due to shrinkage?

    b. What will be the length of the jeans after being washed?

20. TOTAL COST Find the total cost of a $25.50 purchase if the sales tax rate is 2.9%.

21. SALES TAX The purchase price for a watch is $90. If the sales tax is $2.70, what is the sales tax rate?

22. POPULATION INCREASES After a new freeway was completed, the population of a city it passed through increased from 2,800 to 3,444 in two years. Find the percent of increase.

23. INSURANCE An automobile insurance salesperson receives a 4% commission on the annual premium of any policy she sells. Find her commission on a policy if the annual premium is $898.

24. TELEMARKETING A telemarketer earned a commission of $528 on $4,800 worth of new business that she obtained over the telephone. Find her rate of commission.

25. COST-OF-LIVING A teacher earning $40,000 just received a cost-of-living increase of 3.6%. What is the teacher's new salary?

26. AUTO CARE Refer to the advertisement below. Find the discount, the sale price, and the discount rate on the car waxing kit.

SAVE! SAVE! SAVE! SAVE!
CAR WAX KIT
$9 OFF
CLEAN & SHINE COMPLETE
Regularly $75.00

27. TOWEL SALES Find the amount of the discount on a beach towel if it regularly sells for $20, but is on sale for 33% off. Then find the sale price of the towel.

28. Fill in the blanks.

   a. To find 1% of a number, move the decimal point in the number _____ places to the _____.

   b. To find 10% of a number, move the decimal point in the number _____ place to the _____.

29. Estimate each answer. (Answers may vary.)

   a. What is 20% of 396?

   b. What is 50% of 6,189,034?

   c. What is 200% of 21.2?

30. BRAKE INSPECTIONS Of the 1,920 trucks inspected at a safety checkpoint, 5% had problems with their brakes. Estimate the number of trucks that had brake problems?

31. TIPPING Estimate the amount of a 15% tip on a lunch costing $28.40.

32. CAR SHOWS 24% of 63,400 people that attended a five-day car show were female. Estimate the number of females that attended the car show.

33. INTEREST CHARGES Find the simple interest on a loan of $3,000 at 5% per year for 1 year.

34. INVESTMENTS If $23,000 is invested at $4\frac{1}{2}\%$ simple interest for 5 years, what will be the total amount of money in the investment account at the end of the 5 years?

35. SHORT-TERM LOANS Find the simple interest on a loan of $2,000 borrowed at 8% for 90 days.

36. Use the formula $A = P\left(1 + \dfrac{r}{n}\right)^{nt}$ to find the amount of interest earned on an investment of $24,000 paying an annual rate of 6.4% interest, compounded daily for 3 years.

1. Write 6,054,346 [Section 1.1]

   a. in words

   b. in expanded notation

2. WEATHER The tables below shows the average number of cloudy days in Anchorage, Alaska, each month. Find the total number of cloudy days in a year. (*Source:* Western Regional Climate Center) [Section 1.2]

Jan	Feb	Mar	Apr	May	June
19	18	18	18	20	20

July	Aug	Sept	Oct	Nov	Dec
22	21	21	21	20	21

3. Subtract: $50,055 - 7,899$ [Section 1.2]

4. Multiply: $308 \cdot 75$ [Section 1.3]

5. PAINTING A square tarp has sides 8 feet long. When it is laid out on a floor, how much area will it cover? [Section 1.3]

6. Divide: $37\overline{)561}$ [Section 1.4]

7. **a.** List the factors of 40, from smallest to largest. [Section 1.5]

   b. Find the prime factorization of 294. [Section 1.5]

8. Find the LCM and the GCF of 24 and 30. [Section 1.6]

9. Evaluate: $\dfrac{39 + 3[4^3 - 2(2^2 - 3)]}{4 \cdot 2^2 - 1}$ [Section 1.7]

10. AUTO INSURANCE See the premium comparison below. What is the mean six-month insurance premium for the companies listed?

Allstate	$2,672	Geico	$1,370
Auto Club	$1,680	State Farm	$2,737
Farmers	$2,485	20th Century	$1,692

11. Solve each equation and check the result.

    a. $23 + a = 207$ [Section 1.8]

    b. $23a = 207$ [Section 1.9]

12. Place an $<$ or an $>$ symbol in the box to make a true statement: $|-8|$ ☐ $-(-5)$ [Section 2.1]

13. Evaluate: $(-20 + 9) + (-13 + 24)$ [Section 2.2]

14. OVERDRAFT PROTECTION A student forgot that she had only $55 in her bank account and wrote a check for $75, used an ATM to get $60 cash, and used her debit card to buy $25 worth of groceries. On each of the three transactions, the bank charged her a $10 overdraft protection fee. Find the new account balance. [Section 2.3]

15. Evaluate: -6^2 and $(-6)^2$ [Section 2.4]

16. Evaluate each expression, if possible. [Sections 2.2–2.5]

    a. $\dfrac{-14}{0}$ **b.** $\dfrac{0}{-12}$

    c. $-3(-4)(-5)(0)$ **d.** $0 - (-14)$

17. Evaluate: $24 \div 2(3) - 3^3$ [Section 2.6]

18. Estimate the following sum by rounding each number to the nearest hundred. [Section 2.6]

    $$-5,684 + (-2,270) + 3,404 + 2,689$$

19. Solve $6 = 2 - 2x$ and check the result. [Section 2.7]

Write and solve an equation to answer the following question. [Section 2.7]

20. MARKET SHARE After its first year of business, the manufacturer of an energy drink found its market share 51 points behind the industry leader. Five years later, it trailed the leader by only 15 points. How many points of market share did the company gain over this five-year span?

21. Translate into mathematical symbols: 16 less than twice the total t. [Section 3.1]

22. FRUIT STORAGE Use the formula $C = \frac{5(F - 32)}{9}$ to complete the label on the box of bananas shown below. [Section 3.2]

PREMIUM
BANANAS
Keep at 59°F or ?°C
Imported by Pacific Fruit, Inc.

23. a. Simplify: $4(8m)$ [Section 3.3]

    b. Multiply: $(4d + 8)2$ [Section 3.3]

24. Combine like terms. [Section 3.4]

    a. $5x - 4x$

    b. $l + w + l + w$

25. Solve $-(3x - 3) = 6(2x - 7)$ and check the result. [Section 3.5]

Write and then solve an equation to answer the following question. [Section 3.6]

26. BUSINESS After beginning a new position with 34 established accounts, a salesperson made it her objective to add 6 new accounts every month. At this rate, how many months will it take her to reach the goal of 100 accounts?

27. SPELLING What fraction of the letters in the word *Mississippi* are vowels? [Section 4.1]

28. Simplify: $\dfrac{10y}{15y}$ [Section 4.1]

29. Express $\dfrac{4}{5}$ as an equivalent fraction with a denominator of 45. [Section 4.1]

30. What is $\dfrac{1}{4}$ of -240? [Section 4.2]

31. KITES Find the number of square inches of nylon cloth used to make the kite shown below. (*Hint:* Find the area.) [Section 4.2]

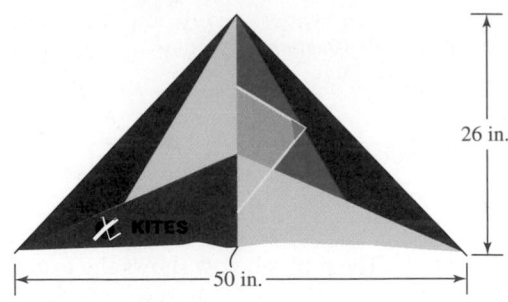

26 in.

KITES

50 in.

32. Multiply: $-\dfrac{16a}{35} \cdot \dfrac{25}{48a^2}$ [Section 4.2]

33. Divide: $\dfrac{4}{9} \div \left(-\dfrac{16}{27}\right)$ [Section 4.3]

34. Subtract: $\dfrac{9}{10} - \dfrac{3}{14}$ [Section 4.4]

35. Add: $\dfrac{4}{m} + \dfrac{2}{7}$ [Section 4.4]

36. Determine which fraction is larger: $\dfrac{23}{20}$ or $\dfrac{7}{6}$ [Section 4.4]

37. HAMBURGERS What is the difference in weight between a $\frac{1}{4}$-pound and a $\frac{1}{3}$-pound hamburger? [Section 4.4]

38. Multiply: $-3\dfrac{3}{4}(8)$ [Section 4.5]

39. BELTS Refer to the belt shown below. What is the maximum waist size that the belt will fit if it is fastened using the last hole? [Section 4.6]

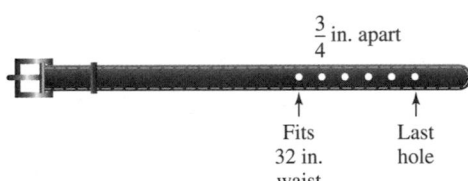

$\frac{3}{4}$ in. apart

Fits
32 in.
waist

Last
hole

40. Subtract: $34\dfrac{1}{9} - 13\dfrac{5}{6}$ [Section 4.6]

41. Simplify: $\dfrac{\dfrac{1}{3} - \dfrac{3}{4}}{\dfrac{1}{6} + \dfrac{1}{3}}$ [Section 4.7]

Write and then solve an equation to answer the following question. [Section 4.8]

42. TELEPHONE BOOKS A telephone book consists of white pages and yellow pages. Find the total number of pages if the 350 white pages make up two-thirds of the telephone book.

Solve each equation and check the result. [Section 4.8]

43. $\frac{5}{6}y = -25$ **44.** $\frac{y}{6} = \frac{y}{12} + \frac{2}{3}$

45. Round each decimal. [Section 5.1]

   a. Round 452.0298 to the nearest hundredth.

   b. Round 452.0298 to the nearest thousandth.

46. Evaluate: $3.4 - (6.6 + 7.3) + 5$ [Section 5.2]

47. WEEKLY EARNINGS A welder's basic work week is 40 hours. After his daily shift is over, he can work overtime at a rate of 1.5 times his regular rate of $15.90 per hour. How much money will he earn in a week if he works 4 hours of overtime? [Section 5.3]

48. Divide: $0.58\overline{)0.1566}$ [Section 5.4]

49. Write $\frac{11}{15}$ as a decimal. Use an overbar. [Section 5.5]

50. Evaluate: $3\sqrt{81} - 8\sqrt{49}$ [Section 5.6]

Solve each equation and check the result. [Section 5.7]

51. $37.1n + 12.2 = 12.4n + 93.4 + 4.4n$

52. $2(a - 4.3) + 1.2 = -6.2$

Write and then solve an equation to answer the following question. [Section 5.7]

53. LABOR COSTS On the repair bill shown below, one line cannot be read. How many hours of labor did it take to repair the car?

Brian Wood Auto Repair

Parts..	$175.00
Total labor (at $35 an hour)........................	
Total..	$297.50

54. Write the ratio $1\frac{1}{4}$ to $1\frac{1}{2}$ as a fraction in simplest form. [Section 6.1]

55. Solve the proportion: $\frac{7}{14} = \frac{2}{x}$ [Section 6.2]

56. TYPING A secretary typed a document containing 385 words in 7 minutes. How long will it take her to type a document containing 495 words? [Section 6.2]

57. How many days are in 960 hours? [Section 6.3]

58. Convert 2,400 millimeters to meters. [Section 6.4]

59. Convert 6.5 kilograms to pounds. [Section 6.5]

60. Complete the table. [Section 7.1]

Percent	Decimal	Fraction
	0.29	
47.3%		
		$\frac{7}{8}$

61. 16% of what number is 20? [Section 7.2]

62. GENEALOGY Through an extensive computer search, a genealogist determined that worldwide, 180 out of every 10 million people had his last name. What percent is this? [Section 7.2]

63. HEALTH CLUBS The number of members of a health club increased from 300 to 534. What was the percent of increase in club membership? [Section 7.3]

64. GUITAR SALE What are the regular price, the sale price, the discount, and discount rate for the guitar shown in the advertisement below? [Section 7.3]

Save on the Standard Strat

Fender

Now Only

$321⁰⁰

Save $107

65. TIPPING Refer to the sales receipt below. [Section 7.4]

    a. Estimate the 15% tip.

    b. Find the total.

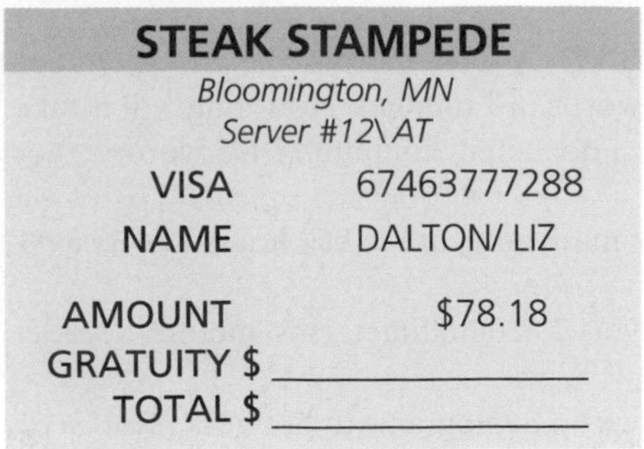

STEAK STAMPEDE
Bloomington, MN
Server #12\ AT

VISA	67463777288
NAME	DALTON/ LIZ
AMOUNT	$78.18
GRATUITY $	_____
TOTAL $	_____

66. INVESTMENTS Find the simple interest earned on $10,000 invested for 2 years at 7.25%. [Section 7.5]

Graphs and Statistics

Kim Steele/Photodisc/Getty Images

from **Campus to Careers**

Postal Service Mail Carrier

Mail carriers follow schedules as they collect and deliver mail to homes and businesses. They must have the ability to quickly and accurately compare similarities and differences among sets of letters, numbers, objects, pictures, and patterns. They also need to have strong problem-solving skills to redirect mislabeled letters and packages. Mail carriers weigh items on postal scales and make calculations with money as they read postage rate tables.

In **Problem 19** of **Study Set 8.1**, you will see how a mail carrier must be able to read a postal rate table and know American units of weight to determine the cost to send a package using priority mail.

JOB TITLE:
Postal Service Mail Carrier

EDUCATION: A high school diploma (or equivalent) and a passing score on a written exam are required.

JOB OUTLOOK: Competition for jobs is high since positions usually come open only upon retirement of current mail carriers.

ANNUAL EARNINGS: Average (mean) salary $46,970

FOR MORE INFORMATION:
http://stats.bls.gov/oco/ocos141.HTM

Objectives

1 Read tables.

2 Read bar graphs.

3 Read pictographs.

4 Read circle graphs.

5 Read line graphs.

6 Read histograms and frequency polygons.

SECTION 8.1

Reading Graphs and Tables

We live in an information age. Never before have so many facts and figures been right at our fingertips. Since information is often presented in the form of tables or graphs, we need to be able to read and make sense of data displayed in that way.

The following **table, bar graph,** and **circle graph** (or **pie chart**) show the results of a shopper survey. A large sample of adults were asked how far in advance they typically shop for a gift. In the bar graph, the length of a bar represents the percent of responses for a given shopping method. In the circle graph, the size of a colored region represents the percent of responses for a given shopping method.

Shopper Survey
How far in advance gift givers typically shop

A Table
Survey responses

Time in advance	Percent
A month or longer	8%
Within a month	12%
Within 3 weeks	12%
Within 2 weeks	23%
Within a week	41%
The same day as giving it	4%

A Bar Graph
Survey responses

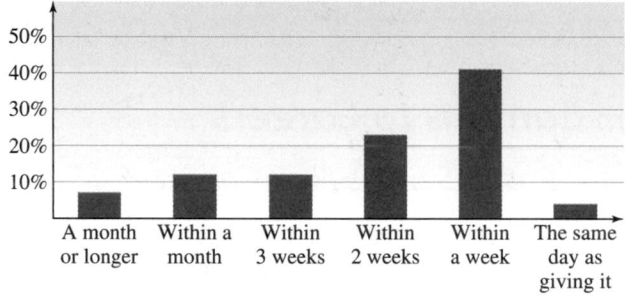

A Circle Graph
Survey responses

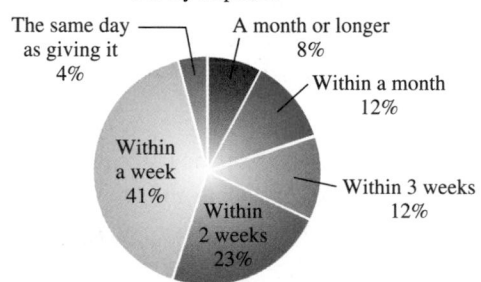

(*Source:* Harris interactive online study via QuickQuery for Gifts.com)

It is often said that a picture is worth a thousand words. That is the case here, where the graphs display the results of the survey more clearly than the table. It's easy to see from the graphs that most people shop within a week of when they need to purchase a gift. It is also apparent that same-day shopping for a gift was the least popular response. That information also appears in the table, but it is just not as obvious.

1 Read tables.

Data are often presented in tables, with information organized in **rows** and **columns.** To read a table, we must find the *intersection* of the row and column that contains the desired information.

EXAMPLE 1   *Postal Rates*   Refer to the table of priority mail postal rates (from 2009) below. Find the cost of mailing an $8\frac{1}{2}$-pound package by priority mail to postal zone 4.

Self Check 1

POSTAL RATES Refer to the table of priority mail postal rates. Find the cost of mailing a 3.75-pound package by priority mail to postal zone 8.

Now Try Problem 17

Postage Rate for Priority Mail 2009							
Weight Not Over (pounds)	**Zones**						
	Local, 1 & 2	**3**	**4**	**5**	**6**	**7**	**8**
1	$4.95	$4.95	$4.95	$4.95	$4.95	$4.95	$4.95
2	4.95	5.20	5.75	7.10	7.60	8.10	8.70
3	5.50	6.25	7.10	9.05	9.90	10.60	11.95
4	6.10	7.10	8.15	10.80	11.95	12.95	14.70
5	6.85	8.15	9.45	12.70	13.75	15.20	17.15
6	7.55	9.25	10.75	14.65	15.50	17.50	19.60
7	8.30	10.30	12.05	16.55	17.30	19.75	22.05
8	8.80	10.70	13.10	17.95	18.80	21.70	24.75
9	9.25	11.45	**13.95**	19.15	20.30	23.60	27.55
10	9.90	12.35	15.15	20.75	22.50	25.90	29.95
11	10.55	13.30	16.40	22.40	24.75	28.20	32.40
12	11.20	14.20	17.60	24.00	26.95	30.50	34.80

Strategy We will read the number at the intersection of the 9th row and the column labeled Zone 4.

WHY Since $8\frac{1}{2}$ pounds is more than 8 pounds, we cannot use the 8th row. Since $8\frac{1}{2}$ pounds does not exceed 9 pounds, we use the 9th row of the table.

Solution

The number at the intersection of the 9th row (in red) and the column labeled Zone 4 (in blue) is 13.95 (in purple). This means it would cost $13.95 to mail the $8\frac{1}{2}$-pound package by priority mail.

2 Read bar graphs.

Another popular way to display data is to use a **bar graph** with bars drawn vertically or horizontally. The relative heights (or lengths) of the bars make for easy comparisons of values. A horizontal or vertical line used for reference in a bar graph is called an **axis.** The **horizontal axis** and the **vertical axis** of a bar graph serve to frame the graph, and they are scaled in units such as years, dollars, minutes, pounds, and percent.

SPEED OF ANIMALS Refer to the bar graph of Example 2.
a. What is the maximum speed of a giraffe?
b. How much greater is the maximum speed of a coyote compared to that of a reindeer?
c. Which animals listed in the graph have a maximum speed that is slower than that of a domestic cat?

Now Try Problem 21

EXAMPLE 2   *Speed of Animals*   The following bar graph shows the maximum speeds for several animals over a given distance.

a. What animal in the graph has the fastest maximum speed?
b. What animal in the graph has the slowest maximum speed?
c. How much greater is the maximum speed of a lion compared to that of a coyote?

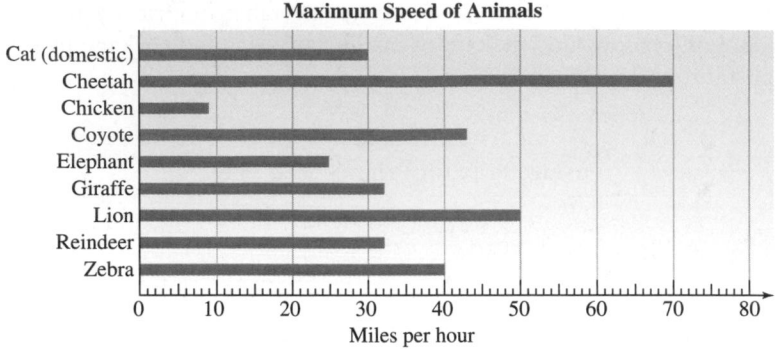

Maximum Speed of Animals

Source: Infoplease.com

Strategy We will locate the name of each desired animal on the vertical axis and move right to the end of its corresponding bar.

WHY Then we can extend downward and read the animal's maximum speed on the horizontal axis scale.

Solution

a. The longest bar in the graph has a length of 70 units and corresponds to a cheetah. Of all the animals listed in the graph, the cheetah has the fastest maximum speed at 70 mph.

b. The shortest bar in the graph has a length of approximately 9 units and corresponds to a chicken. Of all the animals listed in the graph, the chicken has the slowest maximum speed at 9 mph.

c. The length of the bar that represents a lion's maximum speed is 50 units long and the length of the bar that represents a coyote's maximum speed appears to be 43 units long. To find how much greater is the maximum speed of a lion compared to that of a coyote, we subtract

50 mph – 43 mph = 7 mph   Subtract the coyote's maximum speed from the lion's maximum speed.

The maximum speed of a lion is about 7 mph faster than the maximum speed of a coyote.

To compare sets of related data, groups of two (or three) bars can be shown. For **double-bar** or **triple-bar graphs**, a **key** is used to explain the meaning of each type of bar in a group.

EXAMPLE 3   *The U.S. Economy*   The following bar graph shows the total income generated by three sectors of the U.S. economy in each of three years.

a. What income was generated by retail sales in 2000?
b. Which sector of the economy consistently generated the most income?
c. By what amount did the income from the wholesale sector increase from 1990 to 2007?

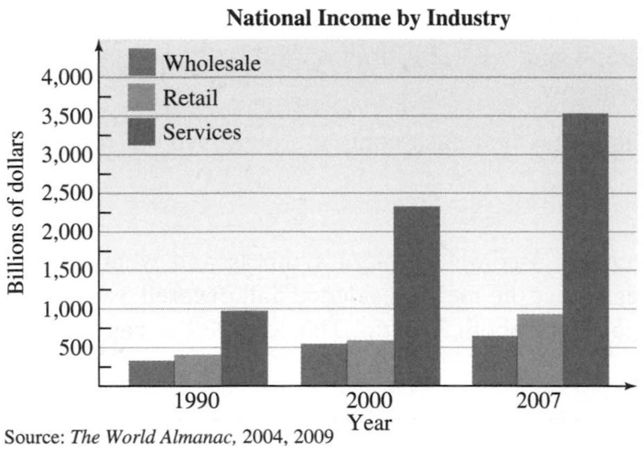

National Income by Industry

Source: *The World Almanac*, 2004, 2009

Self Check 3

THE U.S. ECONOMY Refer to the bar graph of Example 3.
a. What income was generated by retail sales in 1990?
b. What income was generated by the wholesale sector in 2007?
c. In 2000, by what amount did the income from the services sector exceed the income from the retail sector?

Now Try Problems 25 and 31

Strategy To answer questions about years, we will locate the correct colored bar and look at the *horizontal axis* of the graph. To answer questions about the income, we will locate the correct colored bar and extend to the left to look at the *vertical axis* of the graph.

WHY The years appear on the horizontal axis. The height of each bar, representing income in billions of dollars, is measured on the scale on the vertical axis.

Solution

a. The second group of bars indicates income in the year 2000. According to the color key, the blue bar of that group shows the retail sales. Since the vertical axis is scaled in units of $250 billion, the height of that bar is approximately 500 plus one-half of 250, or 125. Thus, the height of the blue bar is approximately 500 + 125 = 625, which represents $625 billion in retail sales in 2000.

b. In each group, the green bar is the tallest. That bar, according to the color key, represents the income from the services sector of the economy. Thus, services consistently generated the most income.

c. According to the color key, the orange bar in each group shows income from the wholesale sector. That sector generated about $260 billion of income in 1990 and $700 billion in income in 2007. The amount of increase is the difference of these two quantities.

$700 billion − $260 billion = $440 billion   Subtract the 1990 wholesale income from the 2007 wholesale income.

Wholesale income increased by about $440 billion between 1990 and 2007. ∎

3 Read pictographs.

A **pictograph** is like a bar graph, but the bars are made from pictures or symbols. A **key** tells the meaning (or value) of each symbol.

EXAMPLE 4 *Pizza Deliveries*

The pictograph on the right shows the number of pizzas delivered to the three residence halls on a college campus during final exam week. In the graph, what information does the top row of pizzas give?

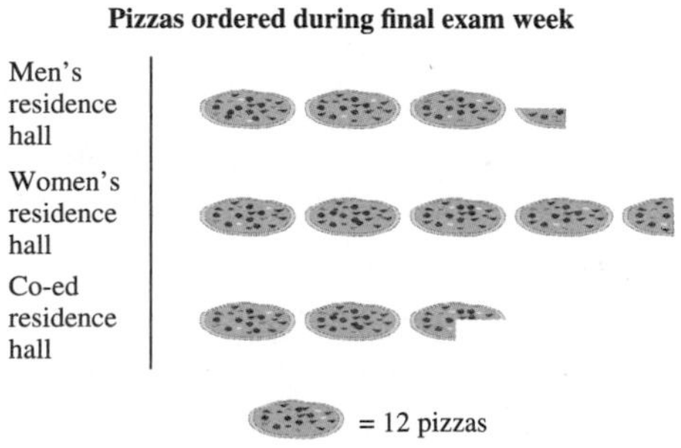

Pizzas ordered during final exam week

= 12 pizzas

Self Check 4

PIZZA DELIVERIES In the pictograph of Example 4, what information does the last row of pizzas give?

Now Try Problems 33 and 35

Strategy We will count the number of complete pizza symbols that appear in the top row of the graph, and we will estimate what fractional part of a pizza symbol also appears in that row.

WHY The key indicates that each complete pizza symbol represents one dozen (12) pizzas.

Solution

The top row contains 3 complete pizza symbols and what appears to be $\frac{1}{4}$ of another. This means that the men's residence hall ordered $3 \cdot 12$, or 36 pizzas, plus approximately $\frac{1}{4}$ of 12, or about 3 pizzas. This totals 39 pizzas.

> **Caution!** One drawback of a pictograph is that it can be difficult to determine what fractional amount is represented by a portion of a symbol. For example, if the CD shown to the right represents 1,000 units sold, we can only estimate that the partial CD symbol represents about 600 units sold.
>
>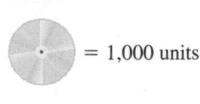
> = 1,000 units
>
> ≈ 600 units

---

4 Read circle graphs.

In **a circle graph**, regions called **sectors** are used to show what part of the whole each quantity represents.

> **The Language of Algebra** A *sector* has the shape of a slice of pizza or a slice of pie. Thus, circle graphs are also called **pie charts.**

GOLD PRODUCTION Refer to the circle graph of Example 5. To the nearest tenth of a million, how many ounces of gold did Russia produce in 2008?

Now Try Problems 37, 41, and 43

EXAMPLE 5 *Gold Production* The circle graph to the right gives information about world gold production. The entire circle represents the world's total production of 78 million troy ounces in 2008. Use the graph to answer the following questions.

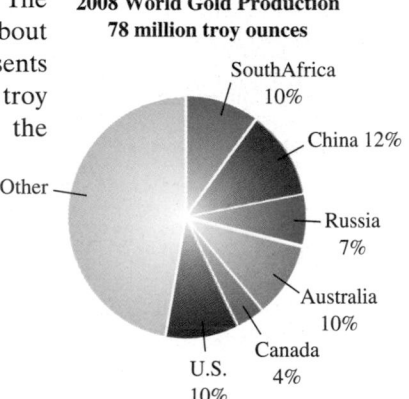

2008 World Gold Production
78 million troy ounces

Source: Goldsheet Mining Directory

a. What percent of the total was the combined production of the United States and Canada?

b. What percent of the total production came from sources other than those listed?

c. To the nearest tenth of a million, how many ounces of gold did China produce in 2008?

Strategy We will look for the key words in each problem.

WHY Key words tell us what operation (addition, subtraction, multiplication, or division) must be performed to answer each question.

Solution

a. The key word *combined* indicates addition. According to the graph, the United States produced 10% and Canada produced 4% of the total amount of gold in 2008. Together, they produced 10% + 4%, or 14% of the total.

b. The phrase *from sources other than those listed* indicates subtraction. To find the percent of gold produced by countries that are not listed, we add the contributions of all the listed sources and subtract that total from 100%.

$$100\% - (10\% + 12\% + 7\% + 10\% + 4\% + 10\%) = 100\% - 53\% = 47\%$$

Countries that are not listed in the graph produced 47% of the world's total production of gold in 2008.

c. From the graph we see that China produced 12% of the world's gold in 2008. To find the number of ounces produced by China (the amount), we use the method for solving percent problems from Section 6.2.

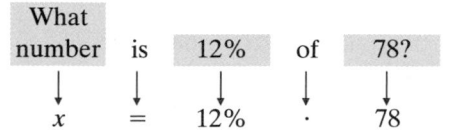

What number is 12% of 78?   This is the percent sentence. The units are millions of ounces.

$$x = 12\% \cdot 78 \qquad \text{Translate to a percent equation.}$$

Now we perform the multiplication on the right side of the equation.

$x = 0.12 \cdot 78$ Write 12% as a decimal: 12% = 0.12.

$x = 9.36$ Do the multiplication.

$$\begin{array}{r} 78 \\ \times\, 0.12 \\ \hline 156 \\ 780 \\ \hline 9.36 \end{array}$$

Rounded to the nearest tenth of a million, China produced 9.4 million ounces of gold in 2008. ■

5 Read line graphs.

Another type of graph, called a **line graph,** is used to show how quantities change with time. From such a graph, we can determine when a quantity is increasing and when it is decreasing.

> **The Language of Algebra** The symbol ⸲ is often used when graphing to show a break in the scale on an axis. Such a break enables us to omit large portions of empty space on a graph.

EXAMPLE 6   **ATMs** The line graph below shows the number of automated teller machines (ATMs) in the United States for the years 2000 through 2007. Use the graph to answer the following questions.

a. How many ATMs were there in the United States in 2001?

b. Between which two years was there the greatest increase in the number of ATMs?

c. When did the number of ATMs decrease?

d. Between which two years did the number of ATMs remain about the same?

Self Check 6

ATMs Refer to the line graph of Example 6.
a. Find the increase in the number of ATMs between 2002 and 2003.
b. How many more ATMs were there in the United States in 2007 as compared to 2000?

Now Try Problems 45, 47, and 51

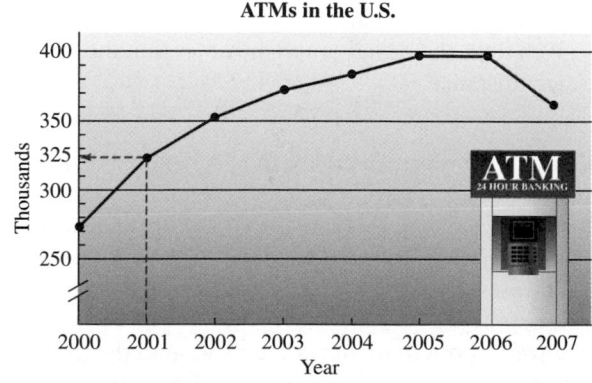

ATMs in the U.S.

Thousands (vertical axis: 250, 300, 325, 350, 400)

Year (horizontal axis: 2000 2001 2002 2003 2004 2005 2006 2007)

Source: The Federal Reserve and *ATM & Debit News*

Strategy We will determine whether the graph is rising, falling, or is horizontal.

WHY When the graph rises as we read from left to right, the number of ATMs is increasing. When the graph falls as we read from left to right, the number of ATMs is decreasing. If the graph is horizontal, there is no change in the number of ATMs.

Solution

a. To find the number of ATMs in 2001, we follow the dashed blue line from the label 2001 on the horizontal axis straight up to the line graph. Then we extend directly over to the scale on the vertical axis, where the arrowhead points to approximately 325. Since the vertical scale is in thousands of ATMs, there were about 325,000 ATMs in 2001 in the United States.

b. This line graph is composed of seven line segments that connect pairs of consecutive years. The steepest of those seven segments represents the greatest increase in the number of ATMs. Since that segment is between the 2000 and 2001, the greatest increase in the number of ATMs occurred between 2000 and 2001.

c. The only line segment of the graph that falls as we read from left to right is the segment connecting the data points for the years 2006 and 2007. Thus, the number of ATMs decreased from 2006 to 2007.

d. The line segment connecting the data points for the years 2005 and 2006 appears to be horizontal. Since there is little or no change in the number of ATMS for those years, the number of ATMs remained about the same from 2005 to 2006. ∎

Two quantities that are changing with time can be compared by drawing both lines on the same graph.

Self Check 7

TRAINS In the graph for Exercise 7, what is train 1 doing at time D?

Now Try Problems 53, 55, and 59

EXAMPLE 7   *Trains*   The line graph below shows the movements of two trains. The horizontal axis represents time, and the vertical axis represents the distance that the trains have traveled.

a. How are the trains moving at time A?

b. At what time (A, B, C, D, or E) are both trains stopped?

c. At what times have both trains traveled the same distance?

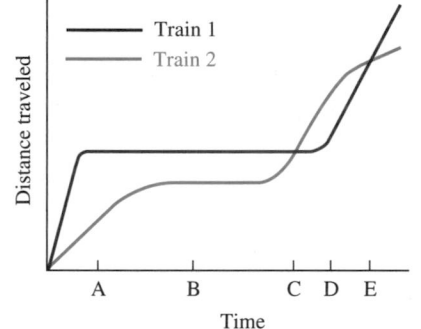

Strategy We will determine whether the graphs are rising or are horizontal. We will also consider the relative positions of the graphs for a given time.

WHY A rising graph indicates the train is moving and a horizontal graph means it is stopped. For any given time, the higher graph indicates that the train it represents has traveled the greater distance.

Solution

The movement of train 1 is represented by the red line, and that of train 2 is represented by the blue line.

a. At time A, the blue line is rising. This shows that the distance traveled by train 2 is increasing. Thus, at time A, train 2 is moving. At time A, the red line is horizontal. This indicates that the distance traveled by train 1 is not changing: At time A, train 1 is stopped.

b. To find the time at which both trains are stopped, we find the time at which both the red and the blue lines are horizontal. At time B, both trains are stopped.

c. At any time, the height of a line gives the distance a train has traveled. Both trains have traveled the same distance whenever the two lines are the same height—that is, at any time when the lines intersect. This occurs at times C and E. ■

6 Read histograms and frequency polygons.

A company that makes vitamins is sponsoring a program on a cable TV channel. The marketing department must choose from three advertisements to show during the program.

1. Children talking about a chewable vitamin that the company makes.

2. A college student talking about an active-life vitamin that the company makes.

3. A grandmother talking about a multivitamin that the company makes.

A survey of the viewing audience records the age of each viewer, counting the number in the 6-to-15-year-old age group, the 16-to-25-year-old age group, and so on. The graph of the data is displayed in a special type of bar graph called a **histogram,** as shown on the right. The vertical axis, labeled **Frequency**, indicates the number of viewers in each age group. For example, the histogram shows that 105 viewers are in the 36-to-45-year-old age group.

A histogram is a bar graph with three important features.

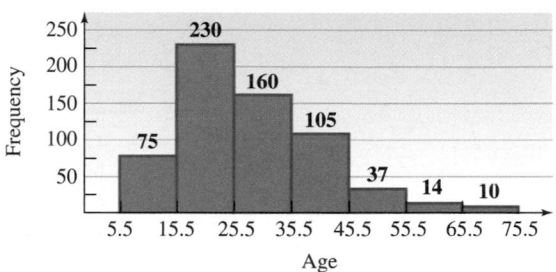

Age of Viewers of a Cable TV Channel

1. The bars of a histogram touch.

2. Data values never fall at the edge of a bar.

3. The widths of each bar are equal and represent a range of values.

The width of each bar of a histogram represents a range of numbers called a **class interval.** The histogram above has 7 class intervals, each representing an age span of 10 years. Since most viewers are in the 16-to-25-year-old age group, the marketing department decides to advertise the active-life vitamins in commercials that appeal to young adults.

EXAMPLE 8   *Carry-on Luggage*   An airline weighed the carry-on luggage of 2,260 passengers. The data is displayed in the histogram below.
a. How many passengers carried luggage in the 8-to-11-pound range?
b. How many carried luggage in the 12-to-19-pound range?

Strategy We will examine the scale on the horizontal axis of the histogram and identify the interval that contains the given range of weight for the carry-on luggage.

WHY Then we can read the height of the corresponding bar to answer the question.

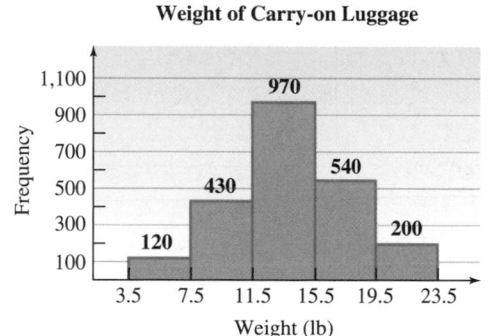

Weight of Carry-on Luggage

Self Check 8

CARRY-ON LUGGAGE Refer to the histogram of Example 8. How many passengers carried luggage in the 20-to-23-pound range?

Now Try Problem 61

Solution

a. The second bar, with edges at 7.5 and 11.5 pounds, corresponds to the 8-to-11-pound range. Use the height of the bar (or the number written there) to determine that 430 passengers carried such luggage.

b. The 12-to-19-pound range is covered by two bars. The total number of passengers with luggage in this range is 970 + 540, or 1,510. ■

A special line graph, called a **frequency polygon,** can be constructed from the carry-on luggage histogram by joining the center points at the top of each bar. (See the graphs below.) On the horizontal axis, we write the coordinate of the middle value of each bar. After erasing the bars, we get the frequency polygon shown on the right below.

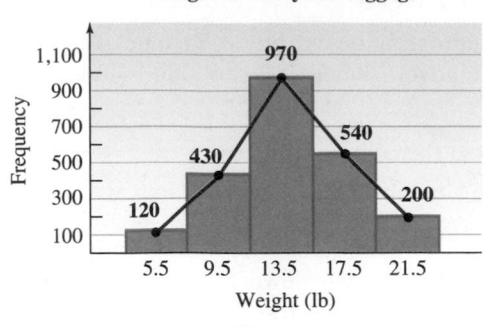

Weight of Carry-on Luggage

Histogram

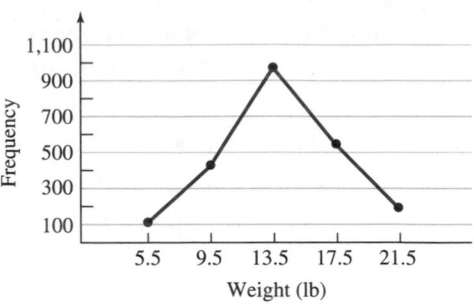

Weight of Carry-on Luggage

Frequency polygon

ANSWERS TO SELF CHECKS

1. $14.70 **2. a.** 32 mph **b.** 11 mph **c.** a chicken and an elephant **3. a.** about $400 billion **b.** about $700 billion **c.** about $170 billion **4.** 33 pizzas were delivered to the co-ed residence hall. **5.** 5.5 million ounces **6. a.** about 20,000 **b.** about 90,000 **7.** Train 1, which had been stopped, is beginning to move. **8.** 200

SECTION 8.1   STUDY SET

VOCABULARY

For problems 1-6, refer to graphs a through f below. Fill in the blanks with the correct letter.

1. Graph _____ is a bar graph.

2. Graph _____ is a circle graph.

3. Graph _____ is a pictograph.

4. Graph _____ is a line graph.

5. Graph _____ is a histogram.

6. Graph _____ is a frequency polygon.

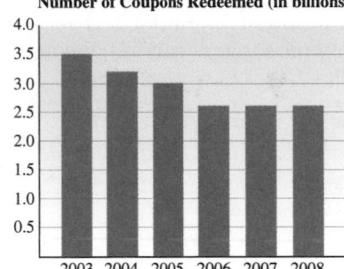

Number of Coupons Redeemed (in billions)

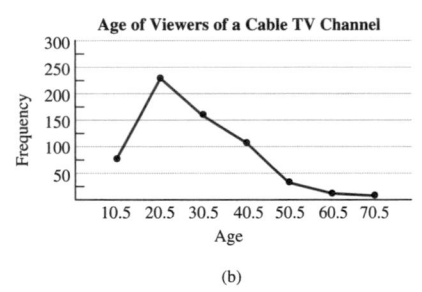

(b)

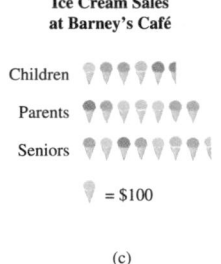

(c)

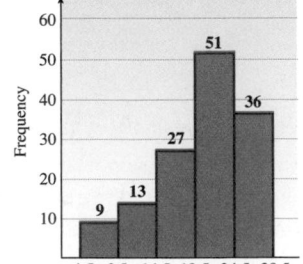

(d)

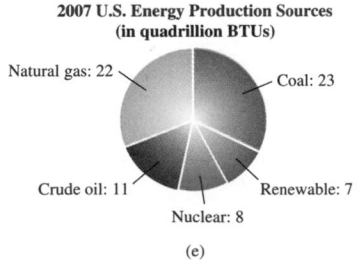

(e)

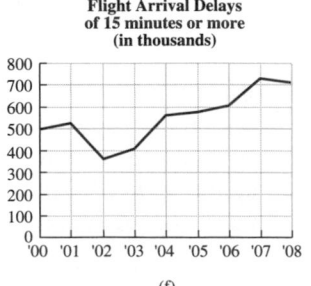

(f)

7. A horizontal or vertical line used for reference in a bar graph is called an _____.

8. In a circle graph, slice-of-pie–shaped figures called _____ are used to show what part of the whole each quantity represents.

CONCEPTS

Fill in the blanks.

9. To read a table, we must find the _____ of the row and column that contains the desired information.

10. The _____ axis and the vertical axis of a bar graph serve to frame the graph, and they are scaled in units such as years, dollars, minutes, pounds, and percent.

11. A pictograph is like a bar graph, but the bars are made from _____ or symbols.

12. Line graphs are often used to show how a quantity changes with _____. On such graphs, we can easily see when a quantity is increasing and when it is _____.

13. A histogram is a bar graph with three important features.
- The _____ of a histogram touch.
- Data values never fall at the _____ of a bar.
- The widths of the bars of a histogram are _____ and represent a range of values.

14. A frequency polygon can be constructed from a histogram by joining the _____ points at the top of each bar.

NOTATION

15. If the symbol 🚌 =1,000 buses, estimate what the symbol 🚌 represents.

16. Fill in the blank: The symbol $\frac{\,}{\,}$ is used when graphing to show a _____ in the scale on an axis.

GUIDED PRACTICE

Refer to the postal rate table on page 741 to answer the following questions. See Example 1.

17. Find the cost of using priority mail to send a package weighing $7\frac{1}{4}$ pounds to zone 3.

18. Find the cost of sending a package weighing $2\frac{1}{4}$ pounds to zone 5 by priority mail.

19. A woman wants to send a birthday gift and an anniversary gift to her brother, who lives in zone 6, using priority mail. One package weighs 2 pounds 9 ounces, and the other weighs 3 pounds 8 ounces. Suppose you are the woman's mail carrier and she asks you how much money will be saved by sending both gifts as one package instead of two. Make the necessary calculations to answer her question. (Hint: 16 ounces = 1 pound.)

from Campus to Careers
Postal Service Mail Carrier

Kim Steele/Photodisc/Getty Images

20. Juan wants to send a package weighing 6 pounds 1 ounce to a friend living in zone 2. Standard postage would be $3.25. How much could he save by sending the package standard postage instead of priority mail?

Refer to the bar graph below to answer the following questions. See Example 2.

21. List the top three most commonly owned pets in the United States.

22. There are four types of pets that are owned in approximately equal numbers. What are they?

23. Together, are there more pet dogs and cats than pet fish?

24. How many more pet cats are there than pet dogs?

Total Number of Pets Owned in the United States, 2009

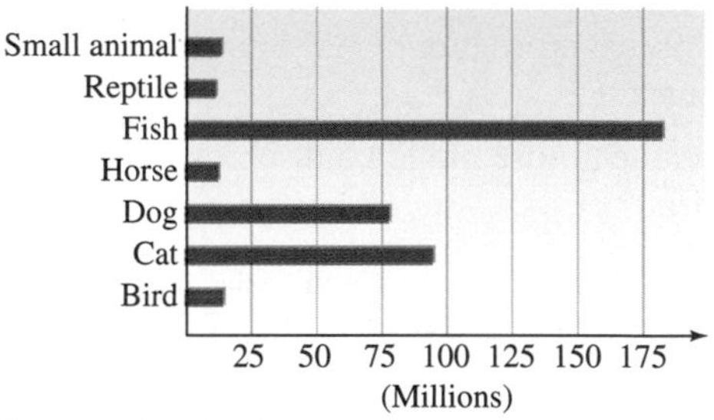

Source: National Pet Owners Survey, AAPA

Refer to the bar graph on the next page to answer the following questions. See Example 3.

25. For the years shown in the graph, has the production of zinc always exceeded the production of lead?

26. Estimate how many times greater the amount of zinc produced in 2000 was compared to the amount of lead produced that year?

27. What is the sum of the amounts of lead produced in 1990, 2000, and 2007?

28. For which metal, lead or zinc, has the production remained about the same over the years?

29. In what years was the amount of zinc produced at least twice that of lead?

30. Find the difference in the amount of zinc produced in 2007 and the amount produced in 2000.

31. By how many metric tons did the amount of zinc produced increase between 1990 and 2007?

32. Between which two years did the production of lead decrease?

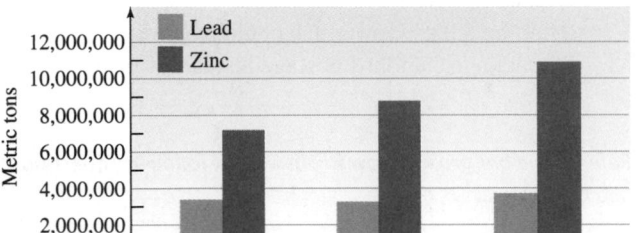

World Lead and Zinc Production

Source: U.S. Geological Survey

Refer to the pictograph below to answer the following questions. See Example 4.

33. Which group (children, parents, or seniors) spent the most money on ice cream at Barney's Café?

34. How much money did parents spend on ice cream?

35. How much more money did seniors spend than parents?

36. How much more money did seniors spend than children?

Ice Cream Sales at Barney's Café

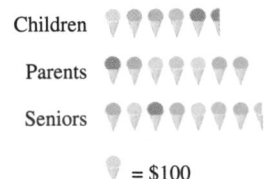

Children
Parents
Seniors

= $100

Refer to the circle graph in the next column to answer the following questions. See Example 5.

37. Of the languages in the graph, which is spoken by the greatest number of people?

38. Do more people speak Spanish or French?

39. Together, do more people speak English, French, Spanish, Russian, and German combined than Chinese?

40. Three pairs of languages shown in the graph are spoken by groups of the same size. Which pairs of languages are they?

41. What percent of the world's population speak a language other than the eight shown in the graph?

42. What percent of the world's population speak Russian or English?

43. To the nearest one million, how many people in the world speak Chinese?

44. To the nearest one million, how many people in the world speak Arabic?

World Languages
and the percents of the world population that speak them

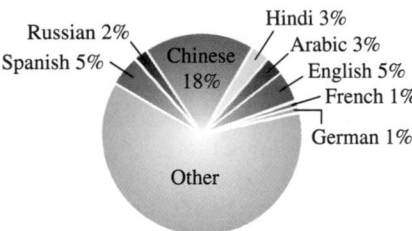

Estimated world population (2009): 6,771,000,000

Source: *The World Almanac*, 2009

Refer to the line graph on the next page to answer the following questions. See Example 6.

45. How many U.S. ski resorts were in operation in 2004?

46. How many U.S. ski resorts were in operation in 2008?

47. Between which two years was there a decrease in the number of ski resorts in operation? (*Hint:* there is more than one answer.)

48. Between which two years was there an increase in the number of ski resorts in operation? (*Hint:* there is more than one answer.)

49. For which two years were the number of ski resorts in operation the same?

50. Find the difference in the number of ski resorts in operation in 2001 and 2008.

51. Between which two years was there the greatest decrease in the number of ski resorts in operation? What was the decrease?

52. Between which two years was there the greatest increase in the number of ski resorts in operation? What was the increase?

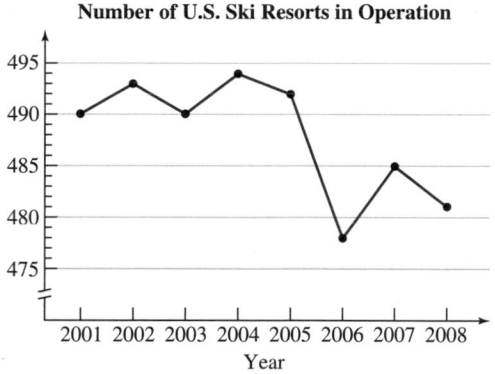

Number of U.S. Ski Resorts in Operation

Source: National Ski Area Assn.

Refer to the line graph below to answer the following questions. See Example 7.

53. Which runner ran faster at the start of the race?

54. At time A, which runner was ahead in the race?

55. At what time during the race were the runners tied for the lead?

56. Which runner stopped to rest first?

57. Which runner dropped his watch and had to go back to get it?

58. At which of these times (A, B, C, D, E) was runner 1 stopped and runner 2 running?

59. Describe what was happening at time E. Who was running? Who was stopped?

60. Which runner won the race?

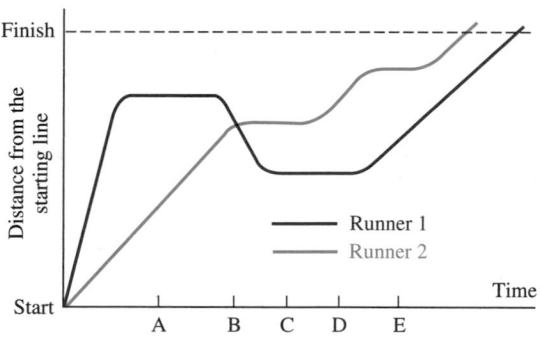

Refer to the histogram and frequency polygon below to answer the following questions. See Example 8.

61. COMMUTING MILES An insurance company collected data on the number of miles its employees drive to and from work. The data are presented in the histogram below.

   a. How many employees have a commute that is in the range of 15 to 19 miles per week?

   b. How many employees commute 14 miles or less per week?

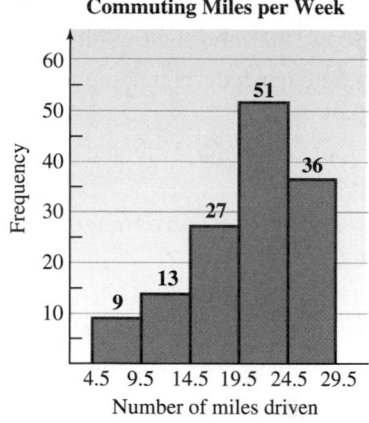

Commuting Miles per Week

62. NIGHT SHIFT STAFFING A hospital administrator surveyed the medical staff to determine the number of room calls during the night. She constructed the frequency polygon below.

   a. On how many nights were there about 30 room calls?

   b. On how many nights were there about 60 room calls?

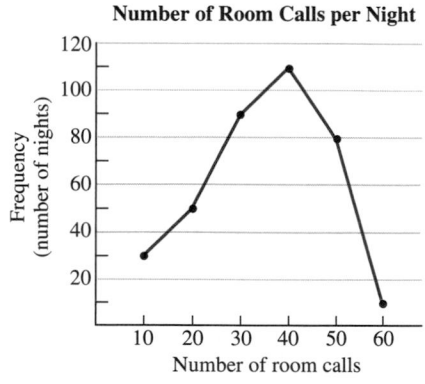

Number of Room Calls per Night

TRY IT YOURSELF

Refer to the 2008 federal income tax table below.

63. FILING A SINGLE RETURN Herb is single and has an adjusted income of $79,250. Compute his federal income tax.

64. FILING A JOINT RETURN Raul and his wife have a combined adjusted income of $57,100. Compute their federal income tax if they file jointly.

65. TAX-SAVING STRATEGY Angelina is single and has an adjusted income of $53,000. If she gets married, she will gain other deductions that will reduce her income by $2,000, and she can file a joint return.

   a. Compute her federal income tax if she remains single.

   b. Compute her federal income tax if she gets married.

   c. How much will she save in federal income tax by getting married?

66. THE MARRIAGE PENALTY A single man with an adjusted income of $80,000 is dating a single woman with an adjusted income of $75,000.

   a. Find the amount of federal income tax each person would pay on their adjusted income.

   b. Add the results from part a.

   c. If they get married and file a joint return, how much federal income tax will they have to pay on their combined adjusted incomes?

   d. Would they have saved on their federal income taxes if they did not get married and paid as two single persons? Find the amount of the "marriage penalty."

Refer to the following bar graph.

67. In which year was the largest percent of flights cancelled? Estimate the percent.

68. In which year was the smallest percent of flights cancelled? Estimate the percent.

69. Did the percent of cancelled flights increase or decrease between 2006 and 2007? By how much?

70. Did the percent of cancelled flights increase or decrease between 2007 and 2008? By how much?

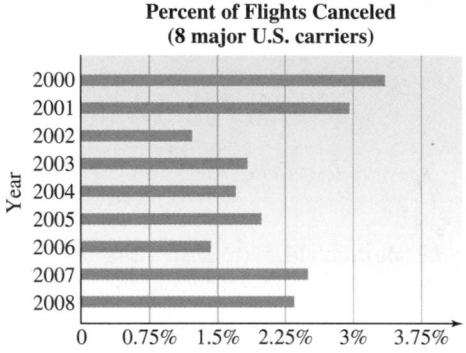

Percent of Flights Canceled
(8 major U.S. carriers)

Source: Bureau of Transportation Statistics

Revised 2008 Tax Rate Schedules

	If TAXABLE INCOME		The TAX is		
	THEN				
	Is Over	But Not Over	This Amount	Plus This %	Of the Amount Over
SCHEDULE X —					
Single	$0	$8,025	$0.00	10%	$0.00
	$8,025	$32,550	$802.50	15%	$8,025
	$32,550	$78,850	$4,481.25	25%	$32,550
	$78,850	$164,550	$16,056.25	28%	$78,850
	$164,550	$357,700	$40,052.25	33%	$164,550
	$357,700	—	$103,791.75	35%	$357,700
SCHEDULE Y-1 —					
Married Filing	$0	$16,050	$0.00	10%	$0.00
Jointly or	$16,050	$65,100	$1,605.00	15%	$16,050
Qualifying	$65,100	$131,450	$8,962.50	25%	$65,100
Widow(er)	$131,450	$200,300	$25,550.00	28%	$131,450
	$200,300	$357,700	$44,828.00	33%	$200,300
	$357,700	—	$96,770.00	35%	$357,700

Refer to the following line graph, which shows the altitude of a small private airplane.

71. How did the plane's altitude change between times B and C?

72. At what time did the pilot first level off the airplane?

73. When did the pilot first begin his descent to land the airplane?

74. How did the plane's altitude change between times D and E?

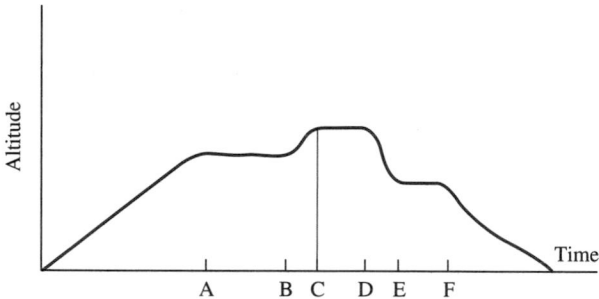

Refer to the following double-bar graph.

75. In which categories of moving violations have violations decreased since last month?

76. Last month, which violation occurred most often?

77. This month, which violation occurred least often?

78. Which violation has shown the greatest decrease in number since last month?

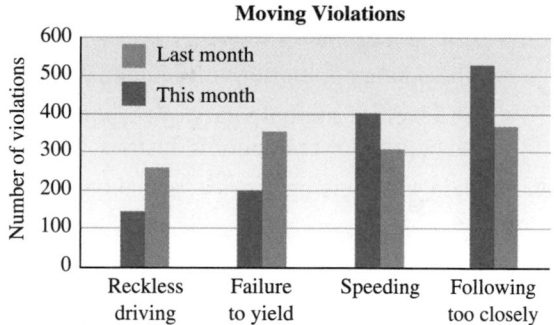

Refer to the following line graph.

79. What were the average weekly earnings in mining for the year 1980?

80. What were the average weekly earnings in construction for the year 1980?

81. Were the average weekly earnings in mining and construction ever the same?

82. What was the difference in a miner's and a construction worker's weekly earnings in 1995?

83. In the period between 2005 and 2008, which occupation's weekly earnings were increasing more rapidly, the miner's or the construction worker's?

84. Did the weekly earnings of a miner or a construction worker ever decrease over a five-year span?

85. In the period from 1980 to 2008, which workers received the greatest increase in weekly earnings?

86. In what five-year span was the miner's increase in weekly earnings the smallest?

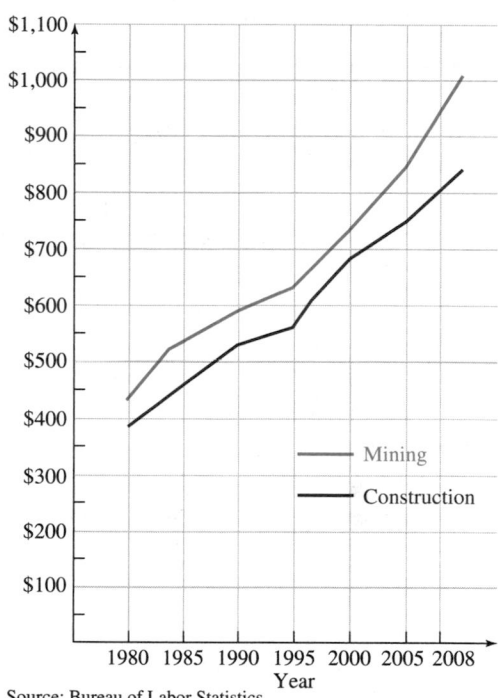

Refer to the following pictograph.

87. What is the daily parking rate for Midtown New York?

88. What is the daily parking rate for Boston?

89. How much more would it cost to park a car for five days in Boston compared to five days in San Francisco?

90. How much more would it cost to park a car for five days in Midtown New York compared to five days in Boston?

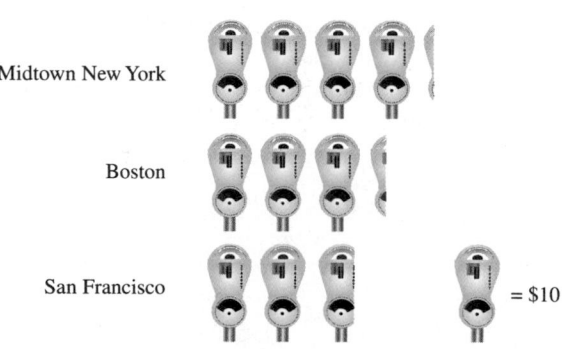

Daily Parking Rates

Source: Colliers International

Refer to the following circle graph.

91. What percent of U.S. energy production comes from nuclear energy? Round to the nearest percent.

92. What percent of U.S. energy production comes from natural gas? Round to the nearest percent.

93. What percent of the total energy production comes from renewable and nuclear combined?

94. By what percent does energy produced from coal exceed that produced from crude oil?

2007 U.S. Energy Production Sources
(in quadrillion BTUs)

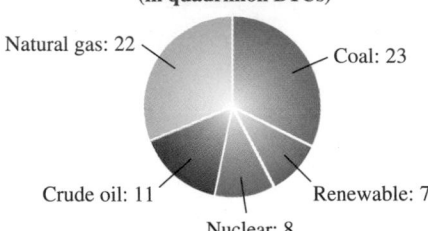

Natural gas: 22   Coal: 23

Crude oil: 11   Renewable: 7

Nuclear: 8

Total production: 71 quadrillion BTUs

Source: Energy Information Administration

95. NUMBER OF U.S. FARMS Use the data in the table below to make a bar graph showing the number of U.S. farms for selected years from 1950 through 2007.

96. SIZE OF U.S. FARMS Use the data in the table below to make a line graph showing the average acreage of U.S. farms for selected years from 1950 through 2007.

Year	Number of U.S. farms (in millions)	Average size of U.S. farms (acres)
1950	5.6	213
1960	4.0	297
1970	2.9	374
1980	2.4	426
1990	2.1	460
2000	2.2	436
2007	2.1	449

Source: U.S. Dept. of Agriculture

97. COUPONS Each coupon value shown in the table below provides savings for shoppers. Make a line graph that relates the original price (in dollars, on the horizontal axis) to the sale price (in dollars, on the vertical axis).

Coupon value: amount saved	Original price of the item
$10	$100, but less than $250
$25	$250, but less than $500
$50	$500 or more

98. DENTISTRY To study the effect of fluoride in preventing tooth decay, researchers counted the number of fillings in the teeth of 28 patients and recorded these results:

3, 7, 11, 21, 16, 22, 18, 8, 12, 3, 7, 2, 8, 19, 12, 19, 12, 10, 13, 10, 14, 15, 14, 14, 9, 10, 12, 13

Tally the results by completing the table. Then make a histogram. The first bar extends from 0.5 to 5.5, the second bar from 5.5 to 10.5, and so on.

Number of fillings	Frequency
1–5	
6–10	
11–15	
16–20	
21–25	

WRITING

99. What kind of presentation (table, bar graph, line graph, circle graph, pictograph, or histogram) is most appropriate for displaying each type of information? Explain your choices.

- The percent of students at a college, classified by major
- The percent of biology majors at a college each year since 1970
- The number of hours a group of students spent studying for final exams
- The ethnic populations of the ten largest cities
- The average annual salary of corporate executives for ten major industries

100. Explain why a histogram is a special type of bar graph.

REVIEW

101. Write the prime numbers between 10 and 30.

102. Write the first ten composite numbers.

103. Write the even whole numbers less than 6 that are not prime.

104. Write the odd whole numbers less than 20 that are not prime.

Objectives

1 Find the mean (average) of a set of values.

2 Find the weighted mean of a set of values.

3 Find the median of a set of values.

4 Find the mode of a set of values.

5 Use the mean, median, mode, and range to describe a set of values.

SECTION 8.2
Mean, Median, and Mode

Graphs are not the only way of describing sets of numbers in a compact form. Another way to describe a set of numbers is to find *one* value around which the numbers in the set are grouped. We call such a value a **measure of central tendency.** In Section 1.7, we studied the most popular measure of central tendency, the *mean* or *average*. In this section we will examine two other measures of central tendency, called the *median* and the *mode*.

1 Find the mean (average) of a set of values.

Recall that the *mean* or *average* of a set of values gives an indication of the "center" of the set of values. To review this concept, let's consider the case of a student who has taken five tests this semester in a history class scoring 87, 73, 89, 92, and 84. To find out how well she is doing, she calculates the mean, or the average, of these scores, by finding their sum and then dividing it by 5.

$$\text{Mean} = \frac{87 + 73 + 89 + 92 + 84}{5} \quad \leftarrow \text{The sum of the test scores}$$
$$\leftarrow \text{The number of test scores}$$

$$= \frac{425}{5} \quad \text{In the numerator, do the addition.}$$

$$= 85 \quad \text{Do the division.}$$

$$
\begin{array}{r}
\overset{2}{8}7 \\
73 \\
89 \\
92 \\
+\,84 \\
\hline
425
\end{array}
\qquad
\begin{array}{r}
85 \\
5\overline{)425} \\
-40 \\
\hline
25 \\
-25 \\
\hline
0
\end{array}
$$

The mean is 85. Some scores were better and some were worse, but 85 is a good indication of her performance in the class.

> **Success Tip** The mean (average) is a single value that is "typical" of a set of values. It can be, but is not necessarily, one of the values in the set. In the previous example, note that the student's mean score was 85; however, she did not score 85 on any of the tests.

Finding the Mean (Arithmetic Average)

The **mean,** or the **average,** of a set of values is given by the formula:

$$\text{Mean (average)} = \frac{\text{the sum of the values}}{\text{the number of values}}$$

> **The Language of Algebra** The *mean* (*average*) of a set of values is more formally called the **arithmetic mean** (pronounced air-rith-MET-tick).

EXAMPLE 1   *Store Sales*   One week's sales in men's, women's, and children's departments of the Clothes Shoppe are given in the table on the next page. Find the mean of the daily sales in the women's department for the week.

Strategy We will add $3,135, $2,310, $3,206, $2,115, $1,570, and $2,100 and divide the sum by 6.

Self Check 1

STORE SALES Find the mean of the daily sales in the men's department of the Clothes Shoppe for the week.

Now Try Problems 9 and 41

© iStockphoto.com/craftvision

Total Daily Sales Per Department—Clothes Shoppe			
Day	**Men's department**	**Women's department**	**Children's department**
Monday	$2,315	$3,135	$1,110
Tuesday	2,020	2,310	890
Wednesday	1,100	3,206	1,020
Thursday	2,000	2,115	880
Friday	955	1,570	1,010
Saturday	850	2,100	1,000

WHY To find the mean (average) of a set of values, we divide the sum of the values by the number of values. In this case, there are 6 days of sales (Monday through Saturday).

Solution

Since there are 6 days of sales, divide the sum by 6.

$$\text{Mean} = \frac{\$3,135 + \$2,310 + \$3,206 + \$2,115 + \$1,570 + \$2,100}{6}$$

$$= \frac{\$14,436}{6} \qquad \text{In the numerator, do the addition.}$$

$$= \$2,406 \qquad \text{Do the division.}$$

```
          1 11          2406
        3,135      6)14,436
        2,310      − 12
        3,206        2 4
        2,115      − 2 4
        1,570        0 3
      + 2,100      − 0
       14,436         36
                    − 36
                       0
```

The mean of the week's daily sales in the women's department is $2,406.

Using Your CALCULATOR Finding the Mean

Most scientific calculators do statistical calculations and can easily find the mean of a set of numbers. To use a scientific calculator in statistical mode to find the mean in Example 1, try these keystrokes:

- Set the calculator to statistical mode.
- Reset the calculator to clear the *statistical registers*.
- Enter each number, followed by the $\boxed{\Sigma+}$ key instead of the $\boxed{+}$ key. That is, enter 3,135, press $\boxed{\Sigma+}$, enter 2,310, press $\boxed{\Sigma+}$, and so on.
- When all data are entered, find the mean by pressing the $\boxed{\bar{x}}$ key. You may need to press $\boxed{2^{nd}}$ first. The mean is 2,406.

Because keystrokes vary among calculator brands, you might have to check the owner's manual if these instructions don't work.

Self Check 2

TRUCKING If a trucker drove 3,360 miles in February, how many miles did he drive per day, on average? (Assume it is not a leap year.)

Now Try Problem 43

EXAMPLE 2 *Driving* In the month of January, a trucker drove a total of 4,805 miles. On the average, how many miles did he drive per day?

Strategy We will divide 4,805 by 31 (the number of days in the month of January).

January						
S	M	T	W	T	F	S
				1	2	3	4
5	6	7	8	9	10	11
12	13	14	15	16	17	18
19	20	21	22	23	24	25
26	27	28	29	30	31	

WHY We do not have to find the sum of the miles driven each day in January. That total is given in the problem as 4,805 miles.

Solution

$$\text{Average number of miles driven per day} = \frac{\text{the total miles driven}}{\text{the number of days}}$$

$$= \frac{4,805}{31} \xleftarrow{\text{This is given.}} \xleftarrow{\text{January has 31 days.}}$$

$$= 155 \quad \text{Do the division.}$$

On average, the trucker drove 155 miles per day.

```
      155
31)4,805
    -3 1
     1 70
    -1 55
       155
      -155
         0
```

2 Find the weighted mean of a set of values.

When a value in a set appears more than once, that value has a greater "influence" on the mean than another value that only occurs a single time. To simplify the process of finding a mean, any value that appears more than once can be "weighted" by multiplying it by the number of times it occurs. A mean that is found in this way is called a **weighted mean**.

EXAMPLE 3 *Hotel Reservations*

A hotel electronically recorded the number of times the reservation desk telephone rang before it was answered by a receptionist. The results of the week-long survey are shown in the table on the right. Find the average number of times the phone rang before a receptionist answered.

Number of rings	Number of calls
1	11
2	46
3	45
4	28
5	20

Strategy First, we will determine the total number of times the reservation desk telephone rang during the week before it was answered. Then we will divide that result by the total number of calls received.

WHY To find the average of a set of values, we divide the sum of the values by the number of values.

Solution To find the total number of times the reservation desk telephone rang during the week before it was answered, we multiply each number of rings (1, 2, 3, 4, and 5) by the number of times it occurred and add those results to get 450. The calculations are shown in blue in the "Weighted number of rings" column.

Number of rings	Number of calls	Weighted number of rings	
1	11	$1 \cdot 11 \rightarrow$	11
2	46	$2 \cdot 46 \rightarrow$	92
3	45	$3 \cdot 45 \rightarrow$	135
4	28	$4 \cdot 28 \rightarrow$	112
5	+ 20	$5 \cdot 20 \rightarrow$	+ 100
Totals	**150**		**450**

QUIZ RESULTS The class results on a five-question true-or-false Spanish quiz are shown in the table below. Find the average number of incorrect answers on the quiz.

Total number of incorrect answers on the quiz	Number of students
0	8
1	8
2	5
3	15
4	3
5	1

Now Try Problem 45

To find the total number of calls received, we add the values in the "Number of calls" column of the table and get 150, as shown in red. To find the average, we divide.

$$\text{Average} = \frac{450}{150} \begin{matrix} \leftarrow \text{ The total number of rings} \\ \leftarrow \text{ The total number of calls} \end{matrix}$$

$$= 3 \qquad \text{Do the division.}$$

$$\begin{array}{r} 3 \\ 150\overline{)450} \\ -450 \\ \hline 0 \end{array}$$

The average number of times the phone rang before it was answered was 3. ∎

Finding the Weighted Mean

To find the weighted mean of a set of values:

1. Multiply each value by the number of times it occurs.
2. Find the sum of products from step 1.
3. Divide the sum from step 2 by the total number of individual values.

Another example of a weighted mean is a **grade point average (GPA).** To find a GPA, we divide:

$$\text{GPA} = \frac{\text{total number of grade points}}{\text{total number of credit hours}}$$

The Language of Algebra Some schools assign a certain number of **credit hours (credits)** to a course while others assign a certain number of **units.** For example, at San Antonio College, the Basic Mathematics course is 3 credit hours while the same course at Los Angeles City College is 3 units.

Self Check 4

FINDING GPAs Find the semester grade point average for a student that received the following grades.

Course	Grade	Credits
MATH 130	A	4
ENG 101	D	3
PHY 080	B	4
SWIM 100	C	1

Now Try Problem 51

EXAMPLE 4 *Finding GPAs* Find the semester grade point average for a student that received the following grades. Round to the nearest hundredth.

Course	Grade	Credits
Speech	C	2
Basic Mathematics	A	4
French	B	4
Business Law	D	3
Study Skills	A	1

Strategy First, we will determine the total number of grade points earned by the student. Then we will divide that result by the total number of credits.

WHY To find the mean of a set of values, we divide the sum of the values by the number of values.

Solution

The point values of grades that are used at most colleges and universities are:

A: 4 pts **B:** 3 pts **C:** 2 pts **D:** 1 pt **F:** 0 pt

To find the total number of grade points that the student earned, we multiply the number of credits for each course by the point value of the grade received. Then we add those results to find that the total number of grade points is 39. The calculations are shown in blue in the "Weighted grade points" column on the next page.

To find the total number of credits, we add the values in that column (shown in red), to get 14.

Course	Grade	Credits	Weighted grade points	
Speech	C	2	$2 \cdot 2 \rightarrow$	4
Basic Mathematics	A	4	$4 \cdot 4 \rightarrow$	16
French	B	4	$3 \cdot 4 \rightarrow$	12
Business Law	D	3	$1 \cdot 3 \rightarrow$	3
Study Skills	A	+ 1	$4 \cdot 1 \rightarrow$	+ 4
Totals		**14**		**39**

To find the GPA, we divide.

$$\text{GPA} = \frac{39}{14} \begin{matrix} \leftarrow \text{The total number of grade points} \\ \leftarrow \text{The total number of credits} \end{matrix}$$

$$\approx 2.785 \quad \text{Do the division.}$$

$$= 2.79 \quad \text{Round 2.785 to the nearest hundredth.}$$

The student's semester GPA is 2.79.

```
        2.785
  14)39.000
    − 28
      11 0
     − 9 8
       1 20
      − 1 12
          80
        − 70
          10
```

3 Find the median of a set of values.

The mean is not always the best measure of central tendency. It can be effected by very high or very low values. For example, suppose the weekly earnings of four workers in a small company are $280, $300, $380, and $240, and the owner pays himself $5,000 a week. At that company, the mean salary per week is

$$\text{Mean} = \frac{\$280 + \$300 + \$380 + \$240 + \$5,000}{5} \quad \begin{matrix}\text{There are 4 employees plus} \\ \text{the owner: } 4 + 1 = 5.\end{matrix}$$

$$= \frac{\$6,200}{5} \quad \text{In the numerator, do the addition.}$$

$$= \$1,240 \quad \text{Do the division.}$$

The owner could say, "Our employees earn an average of $1,240 per week." Clearly, the mean does not fairly represent the typical worker's salary there.

A better measure of the company's typical salary is the *median:* the salary in the middle when all of them are arranged by size.

Smallest $240 $280   **$300** $380 $5,000   Largest

Two salaries   The middle salary   Two salaries

The typical worker earns $300 per week, far less than the mean salary.

The Median

The **median** of a set of values is the middle value. To find the median:

1. Arrange the values in increasing order.
2. If there is an odd number of values, the median is the middle value.
3. If there is an even number of values, the median is the mean (average) of the middle two values.

Self Check 5

Find the median of the following set of values:

$$1\frac{7}{8} \quad 2\frac{1}{2} \quad 3\frac{3}{5} \quad \frac{1}{2} \quad 2\frac{3}{4}$$

Now Try Problems 17 and 21

EXAMPLE 5   Find the median of the following set of values:

| 7.5 | 20.9 | 9.9 | 4.4 | 9.8 | 5.3 | 6.2 | 7.5 | 4.9 |

Strategy We will arrange the nine values in increasing order.

WHY It is easier to find the middle value when they are written in that way.

Solution

Since there is an odd number of values, the median is the middle value.

Smallest   4.4   4.9   5.3   6.2   **7.5**   7.5   9.8   9.9   20.9   Largest

$\underbrace{\qquad\qquad}_{\text{Four values}}$ ↑ $\underbrace{\qquad\qquad}_{\text{Four values}}$

The middle value

The median is 7.5

If there is an even number of values in a set, there is no middle value. In that case, the median is the mean (average) of the two values closest to the middle.

Self Check 6

GRADE DISTRIBUTIONS On a mathematics exam, there were four scores of 68, five scores of 83, and scores of 72, 78, and 90. Find the median score.

Now Try Problems 25 and 29

EXAMPLE 6   *Grade Distributions*   On an exam, there were three scores of 59, four scores of 77, and scores of 43, 47, 53, 60, 68, 82, and 97. Find the median score.

Strategy We will arrange the fourteen exam scores in increasing order.

WHY It is easier to find the two middle scores when they are written in that way.

Solution

Since there is an even number of exam scores, we need to identify the two middle scores.

Smallest   43   47   53   59   59   59   **60   68**   77   77   77   77   82   97   Largest

$\underbrace{\qquad\qquad}_{\text{Six scores}}$ ↑ $\underbrace{\qquad\qquad}_{\text{Six scores}}$

Two middle scores

Since there is an even number of scores, the median is the average (mean) of the two scores closest to the middle: the 60 and the 68.

$$\text{Median} = \frac{60 + 68}{2} = \frac{128}{2} = 64$$

The median is 64.

Success Tip The median is a single value that is "typical" of a set of values. It can be, but is not necessarily, one of the values in the set. In Example 5, the median, 7.5, was one of the given values. In Example 6, the median exam score, 64, was not in the given set of exam scores.

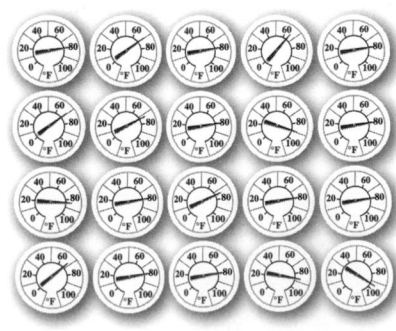

4 Find the mode of a set of values.

The mean and the median are not always the best measure of central tendency. For example, suppose a hardware store displays 20 outdoor thermometers. Ten of them read 80°, and the other ten all have different readings.

To choose an accurate thermometer, should we choose one with a reading that is closest to the *mean* of all 20, or to their *median*? Neither. Instead, we should choose one of the 10 that all read the same, figuring that any of those that agree will likely be correct.

By choosing that temperature that appears most often, we have chosen the *mode* of the 20 values.

The Mode

The **mode** of a set of values is the single value that occurs most often. The mode of several values is also called the **modal value.**

EXAMPLE 7   Find the mode of these values:

3   6   5   7   3   7   2   4   3   5   3   7   8   7   3   7   6   3   4

Strategy We will determine how many times each of the values, 2, 3, 4, 5, 6, 7, and 8 occurs.

WHY We need to know which values occur most often.

Solution

It is not necessary to list the values in increasing order. Instead, we can make a chart and use **tally marks** to keep track of the number of times that the values 2, 3, 4, 5, 6, 7, and 8 occur.

2	3	4	5	6	7	8	← These values appear in the list.
/	### /	//	//	//	###	/	← Tally marks

Because 3 occurs more times than any other value, it is the mode.   ∎

The Language of Algebra   In Example 7, the given set of values has one mode. If a set of values has two modes (exactly two values that occur an equal number of times and more often than any other value) it is said to be **bimodal.** If no value in a set occurs more often than another, then there is *no mode*.

Self Check 7

Find the mode of these values:
2   3   4   6   2   4
3   4   3   4   2   5

Now Try Problems 33 and 37

---

5 **Use the mean, median, mode, and range to describe a set of values.**

Another measure that is used to describe a set of values is the **range.** It indicates the spread of the values.

The Range

The **range** of a set of values is the difference between the largest value and the smallest value.

EXAMPLE 8   *Machinist's Tools*   The diameters (distances across) of eight stainless steel bearings were found using the calipers shown below. Find **a.** the mean, **b.** the median, **c.** the mode, and **d.** the range of the set of measurements listed below.

3.43 cm   3.25 cm   3.48 cm   3.39 cm   3.54 cm   3.48 cm   3.23 cm   3.24 cm

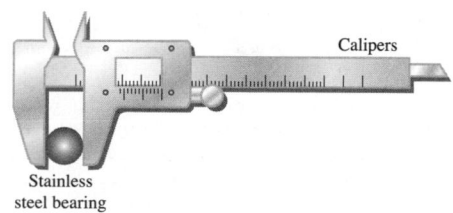

Calipers

Stainless
steel bearing

Strategy We will determine the sum of the measurements, the number of measurements, the middle measurement(s), the most often occurring measurement, and the difference between the largest and smallest measurement.

WHY We need to know that information to find the mean, median, mode, and range.

Self Check 8

MOBILE PHONES The weights of eight different makes of mobile phones are: 4.37 oz, 5.98 oz, 4.36 oz, 4.95 oz, 5.05 oz, 5.95 oz, 4.95 oz, and 5.27 oz. Find the mean, median, and mode weight. Then find the range of the weights.

Now Try Problem 47

Solution

a. To find the mean, we add the measurements and divide by the number of values, which is 8.

$$\text{Mean} = \frac{3.43 + 3.25 + 3.48 + 3.39 + 3.54 + 3.48 + 3.23 + 3.24}{8}$$

$$= \frac{27.04}{8} \qquad \text{In the numerator, do the addition.}$$

$$= 3.38 \qquad \text{Do the division.}$$

The mean is 3.38 cm.

$$
\begin{array}{r}
{\scriptstyle 3\ 4} \\
3.43 \\
3.25 \\
3.48 \\
3.39 \\
3.54 \\
3.48 \\
3.23 \\
+\ 3.24 \\
\hline
27.04
\end{array}
\qquad
\begin{array}{r}
3.38 \\
8\overline{)27.04} \\
-24 \\
\hline
3\ 0 \\
-2\ 4 \\
\hline
64 \\
-64 \\
\hline
0
\end{array}
$$

b. To find the median, we first arrange the eight measurements in increasing order.

Smallest   3.23   3.24   3.25   **3.39   3.43**   3.48   3.48   3.54   Largest

↑

Two middle measurements

Because there is an even number of measurements, the median is the average of the two middle values.

$$\text{Median} = \frac{\mathbf{3.39 + 3.43}}{2} = \frac{6.82}{2} = 3.41 \text{ cm}$$

c. Since the measurement 3.48 cm occurs most often (twice), it is the mode.

d. In part b, we see that the smallest value is 3.23 and the largest value is 3.54. To find the range, we subtract the smallest value from the largest value.

Range = 3.54 − 3.23 = 0.31 cm

$$
\begin{array}{r}
3.54 \\
-3.23 \\
\hline
0.31
\end{array}
$$

THINK IT THROUGH   *The Value of an Education*

"Additional education makes workers more productive and enables them to increase their earnings."

Virginia Governor, Mark R. Warner, 2004

As college costs increase, some people wonder if it is worth it to spend years working toward a degree when that same time could be spent earning money. The following median income data makes it clear that, over time, additional education is well worth the investment. Use the given facts to complete the bar graph.

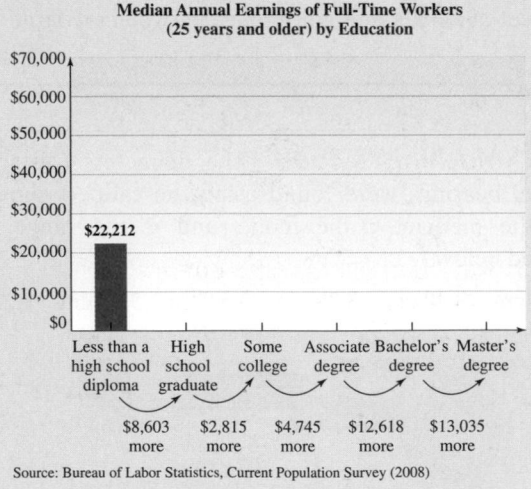

Median Annual Earnings of Full-Time Workers (25 years and older) by Education

Source: Bureau of Labor Statistics, Current Population Survey (2008)

ANSWERS TO SELF CHECKS

1. $1,540   **2.** 120 miles per day   **3.** 2 incorrect answers   **4.** 2.75   **5.** $2\frac{1}{2}$ **6.** 80.5
7. 4   **8.** mean: 5.11 oz; median: 5.00 oz; mode: 4.95 oz; range: 1.62 oz

SECTION 8.2   STUDY SET

VOCABULARY

Fill in the blanks.

1. The _____ (average) of a set of values is the sum of the values divided by the number of values in the set.

2. The _____ of a set of values written in increasing order is the middle value.

3. The _____ of a set of values is the single value that occurs most often.

4. The _____ of a set of values is the difference between the largest value and the smallest value.

CONCEPTS

5. Fill in the blank. The mean of a set of values is given by the formula

   $$\text{Mean} = \frac{\text{the sum of the values}}{\rule{3cm}{0.4pt}}$$

6. Consider the following set of values written in increasing order:

   $$3 \quad 6 \quad 8 \quad 10 \quad 11 \quad 15 \quad 16$$

   a. Is there an even or an odd number of values?

   b. What is the middle number of the list?

   c. What is the median of the set of values?

   d. What is the largest value? What is the smallest value? What is the range of the values?

7. Consider the following set of values written in increasing order:

   $$4 \quad 5 \quad 5 \quad 6 \quad 8 \quad 9 \quad 9 \quad 15$$

   a. Is there an even or odd number of values?

   b. What are the middle numbers of the set of values?

   c. Fill in the blanks:

   $$\text{Median} = \frac{\boxed{} + \boxed{}}{2} = \frac{\boxed{}}{2} = \boxed{}$$

   d. What is the largest value? What is the smallest value? What is the range of the values?

8. Consider the following set of values:

   $$1 \quad 6 \quad 8 \quad 6 \quad 10 \quad 9 \quad 10 \quad 2 \quad 6$$

   a. What value occurs the most often? How many times does it occur?

   b. What is the mode of the set of values?

   c. What is the largest value? What is the smallest value? What is the range of the values?

GUIDED PRACTICE

Find the mean of each set of values. See Example 1.

9. 3   4   7   7   8   11   16

10. 13   15   17   17   15   13

11. 5   9   12   35   37   45   60   77

12. 0   0   3   4   7   9   12

13. 15   7   12   19   27   17   19   35   20

14. 45   67   42   35   86   52   91   102

15. 4.2   3.6   7.1   5.9   8.2

16. 19.1   12.8   16.5   20.0

Find the median of each set of values. See Example 5.

17. 29   5   1   9   11   17   2

18. 20   4   3   2   9   8   1

19. 7   5   4   7   3   6   7   4   1

20. 0   0   3   4   0   0   3   4   5

21. 15.1   44.9   19.7   13.6   17.2

22. 22.4   22.1   50.5   22.3   22.2

23. $\dfrac{1}{100} \quad \dfrac{999}{1{,}000} \quad \dfrac{16}{15} \quad \dfrac{1}{3} \quad \dfrac{5}{8}$

24. $\dfrac{1}{30} \quad \dfrac{17}{30} \quad \dfrac{7}{30} \quad \dfrac{29}{30} \quad \dfrac{11}{30}$

Find the median of each set of values. See Example 6.

25. 8   10   16   63   6   7

26. 7   2   11   5   4   17

27. 39   1   50   41   51   47

28. 47   18   35   29   27   16

29. 1.8   1.7   2.0   9.0   2.1   2.3   2.1   2.0

30. 5.0   1.3   5.0   2.3   4.3   5.6   3.2   4.5

31. $\dfrac{1}{5} \quad \dfrac{11}{5} \quad \dfrac{13}{5} \quad \dfrac{2}{5} \quad \dfrac{3}{5} \quad \dfrac{7}{5}$

32. $\dfrac{1}{9} \quad \dfrac{2}{9} \quad \dfrac{7}{9} \quad \dfrac{11}{9} \quad \dfrac{13}{9} \quad \dfrac{29}{9}$

Find the mode (if any) of each set of values. See Example 7.

33. 3 5 7 3 5 4 6 7 2 3 1 4

34. 12 12 17 17 12 13 17 12

35. −6 −7 −6 −4 −3 −6 −7

36. 0 3 0 2 7 0 6 0 3 4 2 0

37. 23.1 22.7 23.5 22.7 34.2 22.7

38. 21.6 19.3 1.3 19.3 1.6 9.3 2.6

39. $\dfrac{1}{2}$ $\dfrac{1}{3}$ $\dfrac{1}{3}$ 2 $\dfrac{1}{2}$ 2 $\dfrac{1}{5}$ $\dfrac{1}{2}$ 5 $\dfrac{1}{3}$

40. 5 9 12 35 37 45 60

APPLICATIONS

41. SEMESTER GRADES Frank's algebra grade is based on the average of four exams, which count equally. His grades are 75, 80, 90, and 85.

   a. Find his average exam score.

   b. If Frank's professor decided to count the fourth exam double, what would Frank's average be?

42. HURRICANES The table lists the number of major hurricanes to strike the mainland of the United States by decade. Find the average number per decade. Round to the nearest one.

Decade	Number	Decade	Number
1901–1910	4	1951–1960	8
1911–1920	7	1961–1970	6
1921–1930	5	1971–1980	4
1931–1940	8	1981–1990	5
1941–1950	10	1991–2000	5

Source: National Hurricane Center

43. FLEET MILEAGE An insurance company's sales force uses 37 cars. Last June, those cars logged a total of 98,790 miles.

   a. On average, how many miles did each car travel that month?

   b. Find the average number of miles driven daily for each car.

44. BUDGETS The Hinrichs family spent $519 on groceries last April.

   a. On average, how much did they spend on groceries each day?

   b. The Hinrichs family has five members. What is the average spent for groceries for one family member for one day?

45. CASH AWARDS A contest is to be part of a promotional kickoff for a new children's cereal. The prizes to be awarded are shown.

   a. How much money will be awarded in the promotion?

   b. How many cash prizes will be awarded?

   c. What is the average cash prize?

> ### Coloring Contest
> **Grand prize:** Disney World vacation plus $2,500
> Four 1st place prizes of $500
> Thirty-five 2nd place prizes of $150
> Eighty-five 3rd place prizes of $25

46. SURVEYS Some students were asked to rate their college cafeteria food on a scale from 1 to 5. The responses are shown on the tally sheet. Find the average rating.

Poor		Fair		Excellent
1	2	3	4	5
	III	III	⊪	⊪ IIII

47. CANDY BARS The prices (in cents) of the different types of candy bars sold in a drug store are: 50, 60, 50, 50, 70, 75, 50, 45, 50, 50, 60, 75, 60, 75, 100, 50, 80, 75, 100, 75.

   a. Find the mean price of a candy bar.

   b. Find the median price for a candy bar.

   c. Find the mode of the prices of the candy bars.

   d. Find the range of the candy bar prices.

48. COMPUTER SUPPLIES Several computer stores reported differing prices for toner cartridges for a laser printer (in dollars): 51, 55, 73, 75, 72, 70, 53, 59, 75.

   a. Find the mean price of a toner cartridge.

   b. Find the median price for a toner cartridge.

   c. Find the mode of the prices for a toner cartridge.

   d. Find the range of the toner cartridge prices.

49. TEMPERATURE CHANGES Temperatures were recorded at hourly intervals and listed in the table below. Find the average temperature of the period from midnight to 11:00 A.M.

Time	Temperature	Time	Temperature
12:00 A.M.	53	12:00 noon	71
1:00	54	1:00 P.M.	73
2:00	57	2:00	76
3:00	58	3:00	77
4:00	59	4:00	78
5:00	59	5:00	71
6:00	61	6:00	70
7:00	62	7:00	64
8:00	64	8:00	61
9:00	66	9:00	59
10:00	68	10:00	53
11:00	71	11:00	51

50. AVERAGE TEMPERATURES Find the average temperature for the 24-hour period shown in the table in Exercise 49.

For Exercises 51–54, find the semester grade point average for a student that received the following grades. Round to the nearest hundredth, when necessary.

51.

Course	Grade	Credits
MATH 210	C	5
ACCOUNTING 175	A	3
HEALTH 090	B	1
JAPANESE 010	D	4

52.

Course	Grade	Credits
NURSING 101	D	3
READING 150	B	4
PAINTING 175	A	2
LATINO STUDIES 090	C	3

53.

Course	Grade	Credits
PHOTOGRAPHY	D	3
MATH 020	B	4
CERAMICS 175	A	1
ELECTRONICS 090	C	3
SPANISH 130	B	5

54.

Course	Grade	Credits
ANTHROPOLOGY 050	D	3
STATISTICS 100	A	4
ASTRONOMY 100	C	1
FORESTRY 130	B	5
CHOIR 130	C	1

55. EXAM AVERAGES Roberto received the same score on each of five exams, and his mean score is 85. Find his median score and the mode of his scores.

56. EXAM SCORES The scores on the first exam of the students in a history class were 57, 59, 61, 63, 63, 63, 87, 89, 95, 99, and 100. Kia got a score of 70 and claims that "70 is better than average." Which of the three measures of central tendency is she better than: the mean, the median, or the mode?

57. COMPARING GRADES A student received scores of 37, 53, and 78 on three quizzes. His sister received scores of 53, 57, and 58. Who had the better average? Whose grades were more consistent?

58. What is the average of all of the integers from -100 to 100, inclusive?

59. OCTUPLETS In December 1998, Nkem Chukwu gave birth to eight babies in Texas Children's Hospital. Find the mean, the median, and the range of their birth weights listed below.

Ebuka (girl)	24 oz	Odera (girl)	11.2 oz
Chidi (girl)	27 oz	Ikem (boy)	17.5 oz
Echerem (girl)	28 oz	Jioke (boy)	28.5 oz
Chima (girl)	26 oz	Gorom (girl)	18 oz

60. COMPARISON SHOPPING A survey of grocery stores found the price of a 15-ounce box of Cheerios cereal ranging from $3.89 to $4.39, as shown below. What are the mean, median, mode, and range of the prices listed?

$4.29 $3.89 $4.29 $4.09 $4.24 $3.99
$3.98 $4.19 $4.19 $4.39 $3.97 $4.29

61. EARTHQUAKES The magnitudes of 2008's major earthquakes are listed below. Find the mean (round to the nearest tenth), the median, and the range.

Date	Location	Magnitude
Jan. 5	Queen Charlotte Islands Region	6.6
Jan. 10	Off the coast of Oregon	6.4
Feb. 20	Simeulue, Indonesia	7.4
Feb. 24	Nevada	6.0
Feb. 25	Kepulauan Mentawai Region, Indonesia	7.0
March 21	Xinjiang-Xizang Border Region	7.2
April 9	Loyalty Islands	7.3
May 12	China	7.9
June 13	Eastern Honshu, Japan	6.9
July 19	Honshu, Japan	7.0
Oct. 6	Kyrgyzstan	6.6
Oct. 11	Russia	6.3
Oct. 29	Pakistan	6.4
Nov. 16	Indonesia	7.3
Dec. 20	Japan	6.3

Source: Incorporated Research Institutions for Seismology

62. FUEL EFFICIENCY The ten most fuel-efficient cars in 2009, based on manufacturer's estimated city and highway average miles per gallon (mpg), are shown in the table below.

a. Find the mean, median, and mode of the city mileage.

b. Find the mean, median, and mode of the highway mileage.

Model	mpg city/hwy
Toyota Prius	50/49
Honda Civic Hybrid	40/45
Honda Insight	40/43
Ford Fusion Hybrid	41/36
Mercury Milan Hybrid	41/36
VW Jetta TDI	30/41
Nissan Altima Hybrid	35/33
Toyota Camry Hybrid	33/34
Toyota Yaris	29/36
Toyota Corolla	26/35

Source: edmonds.com

63. SPORT FISHING The report shown below lists the fishing conditions at Pyramid Lake for a Saturday in January. Find the median, the mode, and the range of the weights of the striped bass caught at the lake.

Pyramid Lake—Some striped bass are biting but are on the small side. Striking jigs and plastic worms. Water is cold: 38°. Weights of fish caught (lb): 6, 9, 4, 7, 4, 3, 3, 5, 6, 9, 4, 5, 8, 13, 4, 5, 4, 6, 9

64. NUTRITION Refer to the table below.

a. Find the mean number of calories in one serving of the meats shown.

b. Find the median.

c. Find the mode.

d. Find the range of the number of calories.

NUTRITIONAL COMPARISONS
Per 3.5 oz. serving of cooked meat

Species	Calories
Bison	143
Beef (Choice)	283
Beef (Select)	201
Pork	212
Chicken (Skinless)	190
Sockeye Salmon	216

Source: The National Bison Association

WRITING

65. Explain how to find the mean, the median, the mode, and the range of a set of values.

66. The mean, median, and mode are used to measure the central tendency of a set of values. What is meant by central tendency?

67. Which measure of central tendency, mean, median, or mode, do you think is the best for describing the salaries at a large company? Explain your reasoning.

68. When is the mode a better measure of central tendency than the mean or the median? Give an example and explain why.

REVIEW

Translate to a percent equation (or percent proportion) and then solve to find the unknown number.

69. 52 is what percent of 80?

70. What percent of 50 is 56?

71. $66\frac{2}{3}\%$ of what number is 28?

72. 56.2 is $16\frac{1}{3}\%$ of what number?

73. 5 is what percent of 8?

74. What number is 52% of 350?

75. Find $7\frac{1}{4}\%$ of 600.

76. $\frac{1}{2}\%$ of what number is 5,000?

SECTION 8.3
Equations in Two Variables; The Rectangular Coordinate System

We have seen that information is often presented in the form of tables or graphs. In algebra, we also present information that way. For example, the following table and graph are related to the equation $d = 4t$. This formula gives the distance d (in miles) that a hiker can walk in a time t (in hours) at a rate of 4 miles per hour.

To find the distance the hiker can walk in 3 hours, we substitute 3 for t in the formula and evaluate the right side.

$d = 4t$ This is the given formula.

$ = 4(3)$ Substitute 3 for t, the time.

$ = 12$ Do the multiplication.

In 3 hours, the hiker can walk 12 miles. This result and others are shown in the table and graph below.

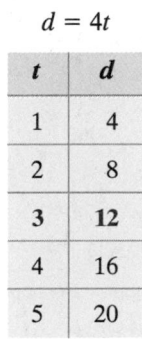

$d = 4t$

t	d
1	4
2	8
3	12
4	16
5	20

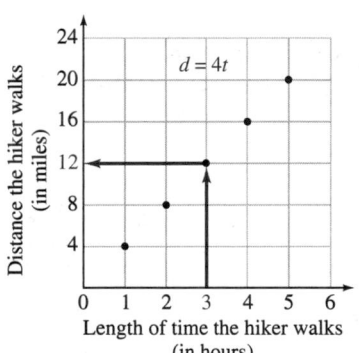

Both the table and the graph show the time-distance relationship for the hiker as paired data.

- In 1 hour, the hiker can walk a distance of 4 miles.
- In 2 hours, the hiker can walk a distance of 8 miles.
- In 3 hours, the hiker can walk a distance of 12 miles.
- In 4 hours, the hiker can walk a distance of 16 miles.
- In 5 hours, the hiker can walk a distance of 20 miles.

In the next two sections, we will discuss how to construct tables and graphs like those shown above.

1 Determine whether an ordered pair is a solution of an equation.

So far, we have worked with **equations in one variable.** For example, $x + 3 = 9$ is an equation in x. If we subtract 3 from both sides, we see that 6 is the solution. To check, we replace x with 6 and note that the result is a true statement: $9 = 9$.

We will now extend our equation-solving skills to find solutions of **equations in two variables.** To begin, let's consider $2x + y = 12$, an equation in x and y.

A solution of $2x + y = 12$ is a pair of values, one for x and one for y, that make the equation true. To illustrate, suppose x is 3 and y is 6. Then we have:

$2x + y = 12$ This is the given equation.

$2(3) + 6 \stackrel{?}{=} 12$ Substitute 3 for x and 6 for y.

$6 + 6 \stackrel{?}{=} 12$ Do the multiplication: $2(3) = 6$.

$12 = 12$ Do the addition: $6 + 6 = 12$.

Since the result $12 = 12$ is a true statement, $x = 3$ and $y = 6$ is a solution of $2x + y = 12$. We write the solution as the **ordered pair** $(3, 6)$, with the value of x listed first. We say that $(3, 6)$ *satisfies* the equation.

An ordered pair: $(3, 6)$
$$x \uparrow \quad \uparrow y$$

In general, a **solution of an equation in two variables** is an ordered pair of numbers that makes the equation a true statement.

> **Caution!** Don't be confused by this new use of parentheses. $(3, 6)$ represents an ordered pair, whereas $3(6)$ indicates multiplication.

Self Check 1

Is $(-4, 2)$ a solution of
$2x - y = -10$?

Now Try Problem 27

EXAMPLE 1   Is $(-2, 4)$ a solution of $3x - 4y = -22$?

Strategy We will substitute -2 for x and 4 for y and see whether the resulting equation is true.

WHY An ordered pair is a *solution* of $3x - 4y = -22$ if replacing the variables with the values of the ordered pair results in a true statement.

Solution

$$
\begin{aligned}
3x - 4y &= -22 & &\text{This is the given equation.} \\
3(-2) - 4(4) &\overset{?}{=} -22 & &\text{Substitute } -2 \text{ for } x \text{ and } 4 \text{ for } y. \\
-6 - 16 &\overset{?}{=} -22 & &\text{On the left side, do the multiplication.} \\
-6 + (-16) &\overset{?}{=} -22 & &\text{Write the subtraction as addition of the opposite.} \\
-22 &= -22 & &\text{Do the addition: } -6 + (-16) = -22.
\end{aligned}
$$

Since $-22 = -22$ is a true statement, $(-2, 4)$ is a solution of $3x - 4y = -22$.

Self Check 2

Is $(8, 9)$ a solution of
$y = x - 1$?

Now Try Problem 35

EXAMPLE 2   Is $(-1, -3)$ a solution of $y = x - 1$?

Strategy We will substitute -1 for x and -3 for y in $y = x - 1$ and see whether the resulting equation is true.

WHY An ordered pair is a solution of $y = x - 1$ if replacing the variables with the values of the ordered pair results in a true statement.

Solution

$$
\begin{aligned}
&= x - 1 & &\text{This is the given equation.} \\
&\overset{?}{=} -1 - 1 & &\text{Substitute } -1 \text{ for } x \text{ and } -3 \text{ for } y. \\
-3 &\overset{?}{=} -1 + (-1) & &\text{Write the subtraction as addition of the opposite.} \\
-3 &= -2 & &\text{Do the addition.}
\end{aligned}
$$

Since $-3 = -2$ is false, $(-1, -3)$ is not a solution of $y = x - 1$.

2 Complete ordered-pair solutions of equations.

If only one of the values of an ordered-pair solution is known, we can substitute it into the equation to find the other value.

EXAMPLE 3   Complete the following ordered pairs so that each one is a solution of the equation $4x + 2y = 2$.

a. $(0, \boxed{})$ **b.** $(\boxed{}, 2)$

Strategy In each case, we will substitute the known value of the solution into $4x + 2y = 2$.

WHY Then we can solve the resulting equation in one variable to find the unknown value of the solution.

Solution

a. For $(0, \boxed{})$, we are given the x-value of the solution is 0. To find the corresponding y-value, we substitute 0 for x in $4x + 2y = 2$ and solve for y.

$$4x + 2y = 2 \quad \text{This is the given equation.}$$
$$4(\mathbf{0}) + 2y = 2 \quad \text{Substitute 0 for x.}$$
$$0 + 2y = 2 \quad \text{On the left side, do the multiplication: 4(0) = 0.}$$
$$2y = 2 \quad \text{On the left side, do the addition: 0 + 2y = 2y.}$$
$$\frac{2y}{2} = \frac{2}{2} \quad \text{To isolate y, undo the multiplication by 2 by dividing both sides by 2.}$$
$$y = 1 \quad \text{Do the division. This is the missing y-value of the solution.}$$

When $x = 0$, $y = 1$. The completed ordered pair is $(0, 1)$.

b. For $(\boxed{}, 2)$, we are given the y-value of the solution is 2. To find the corresponding x-value, we substitute 2 for y in $4x + 2y = 2$ and solve for x.

$$4x + 2y = 2 \quad \text{This is the given equation.}$$
$$4x + 2(\mathbf{2}) = 2 \quad \text{Substitute 2 for y.}$$
$$4x + 4 = 2 \quad \text{On the left side, do the multiplication: 2(2) = 4.}$$
$$4x + 4 - \mathbf{4} = 2 - \mathbf{4} \quad \text{To isolate the variable term 4x, undo the addition of 4 by subtracting 4 from both sides.}$$
$$4x = -2 \quad \text{Do the subtraction: 2 - 4 = 2 + (-4) = -2.}$$
$$\frac{4x}{4} = \frac{-2}{4} \quad \text{To isolate x, undo the multiplication by 4 by dividing both sides by 4.}$$
$$x = -\frac{2}{4} \quad \text{Write the - sign in front of the fraction.}$$
$$x = -\frac{1}{2} \quad \text{Simplify the fraction. This is the missing x-value of the solution.}$$

When $y = 2$, $x = -\frac{1}{2}$. The completed ordered pair is $\left(-\frac{1}{2}, 2\right)$.

Self Check 3

Complete the following ordered pairs so that each one is a solution of the equation $2x - 7y = 14$.

a. $(7, \boxed{})$ **b.** $(\boxed{}, 3)$

Now Try Problems 43 and 49

Solutions of equations in two variables are often listed in a **table of solutions** (or a **table of values**). The solutions of $4x + 2y = 2$ that we found in Example 3 are shown in the table below.

x	y	(x, y)
0	1	$(0, 1)$
$-\frac{1}{2}$	2	$\left(-\frac{1}{2}, 2\right)$

A table of solutions

Complete the table of solutions for $3x + 2y = 5$.

x	y	(x, y)
▓	−2	(▓, −2)
5	▓	(5, ▓)

Now Try Problem 51

EXAMPLE 4 Complete the table of solutions for $3x + 2y = 5$.

x	y	(x, y)
7	▓	(7, ▓)
▓	4	(▓, 4)

Strategy In each case we will substitute the known value of the solution into the equation $3x + 2y = 5$.

WHY Then we can solve the resulting equation in one variable to find the unknown value of the solution.

Solution In the first row of the table, we are given an x-value of 7. To find the corresponding y-value, we substitute 7 for x and solve for y.

x	y	(x, y)
7	−8	(7, −8)

$$3x + 2y = 5 \qquad \text{This is the given equation.}$$
$$3(\mathbf{7}) + 2y = 5 \qquad \text{Substitute 7 for x.}$$
$$21 + 2y = 5 \qquad \text{On the left side, do the multiplication: 3(7) = 21.}$$
$$21 + 2y - \mathbf{21} = 5 - \mathbf{21} \qquad \text{To isolate the variable term 2y, subtract 21 from both sides.}$$
$$2y = -16 \qquad \text{Do the subtraction: 5 − 21 = 5 + (−21) = −16.}$$
$$\frac{2y}{2} = \frac{-16}{2} \qquad \text{To isolate y, undo the multiplication by 2 by dividing both sides by 2.}$$
$$y = -8 \qquad \text{Do the division. This is the missing y-value of the solution.}$$

When $x = 7$, $y = -8$. The completed ordered pair $(7, -8)$ is entered in the first row in the table on the left.

In the second row of the table, we are given a y-value of 4. To find the corresponding x-value, we substitute 4 for y and solve for x.

x	y	(x, y)
7	−8	(7, −8)
−1	4	(−1, 4)

$$3x + 2y = 5 \qquad \text{This is the given equation.}$$
$$3x + 2(\mathbf{4}) = 5 \qquad \text{Substitute 4 for y.}$$
$$3x + 8 = 5 \qquad \text{On the left side, do the multiplication: 2(4) = 8.}$$
$$3x + 8 - \mathbf{8} = 5 - \mathbf{8} \qquad \text{To isolate the variable term 3x, subtract 8 from both sides.}$$
$$3x = -3 \qquad \text{Do the subtraction: 5 − 8 = 5 + (−8) = −3.}$$
$$\frac{3x}{3} = \frac{-3}{3} \qquad \text{To isolate x, undo the multiplication by 3 by dividing both sides by 3.}$$
$$x = -1 \qquad \text{Do the division. This is the missing x-value of the solution.}$$

When $y = 4$, $x = -1$. The completed ordered pair $(-1, 4)$ is entered in the second row of the table on the left.

We have seen that solutions of an equation containing two variables are ordered pairs and that the ordered pairs can be listed in a table. We will now introduce a way to represent ordered pairs as points on a graph.

3 Construct a rectangular coordinate system.

Ordered pairs of numbers can be displayed on a grid called a **rectangular coordinate system.** This system is also called the *Cartesian coordinate system* after its developer, René Descartes, a 17th-century French mathematician.

> ***The Language of Algebra*** A rectangular coordinate system is a *grid*—a network of uniformly spaced lines. At times, some large U.S. cities have such horrible traffic congestion that vehicles can barely move, if at all. The condition is called *gridlock*.

A rectangular coordinate system is formed by two perpendicular number lines. The horizontal number line is usually called the **x-axis,** and the vertical number line is usually called the *y*-axis. On the *x*-axis, the positive direction is to the right. On the *y*-axis, the positive direction is upward.

> **The Language of Algebra**   The word *axis* is used in mathematics and science. For example, Earth rotates on its *axis* once every 24 hours. The plural of *axis* is **axes.**

The point where the axes intersect is called the **origin.** This is the zero point on each axis. The axes form a **coordinate plane,** and they divide it into four regions called **quadrants,** which are numbered counterclockwise using the Roman numerals I, II, III, and IV.

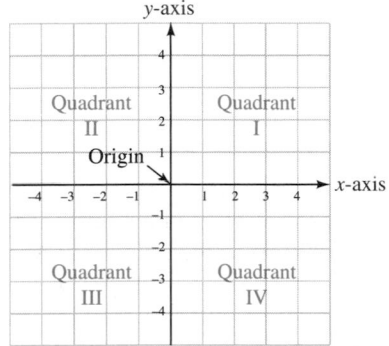

4 Plot ordered pairs and determine the coordinates of a point.

Each point in a coordinate plane can be identified by an **ordered pair** of numbers x and y written in the form (x, y). The first number in the pair is called the **x-coordinate** and the second number is called the **y-coordinate.** Some examples of such pairs are $(3, 2), (-4, 4), (-2, -3),$ and $(4, -4)$.

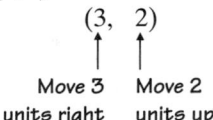

The process of locating a point on a rectangular coordinate system is called **graphing** or **plotting** the point. Here are four examples:

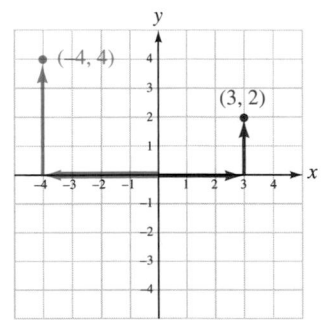

- Red arrows are used to show how to graph (plot) the point $(3, 2)$ on a rectangular coordinate system. Since the x-coordinate is positive, we start at the origin and move 3 units to the *right* along the x-axis. Since the y-coordinate is positive, we then move *up* 2 units and draw a dot. This locates the point $(3, 2)$.

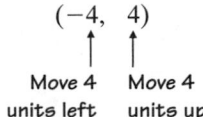

- Blue arrows are used to show how to graph (plot) the point $(-4, 4)$. Since the x-coordinate is negative, we start at the origin and move 4 units to the *left* along the x-axis. Since the y-coordinate is positive, we then move *up* 4 units and draw a dot. This locates the point $(-4, 4)$.

$$(-4, \quad 4)$$

Move 4   Move 4
units left   units up

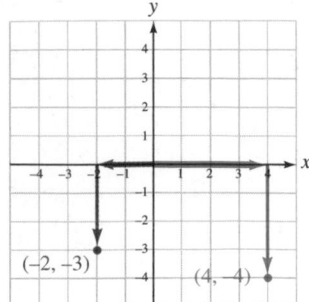

- Purple arrows are used to show how to graph (plot) the point $(-2, -3)$ on a rectangular coordinate system. Since the x-coordinate is negative, we start at the origin and move 2 units to the *left* along the x-axis. Since the y-coordinate is negative, we then move *down* 3 units and draw a dot. This locates the point $(-2, -3)$.

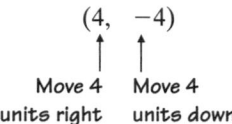

$$(-2, \quad -3)$$

Move 2   Move 3
units left   units down

- Green arrows are used to show how to graph (plot) the point $(4, -4)$. Since the x-coordinate is positive, we start at the origin and move 4 units to the *right* along the x-axis. Since the y-coordinate is negative, we then move *down* 4 units and draw a dot. This locates the point $(4, -4)$.

$$(4, \quad -4)$$

Move 4   Move 4
units right   units down

> **Caution!** The order of the coordinates of a point is important. The point with coordinates $(-4, 4)$ is not the same as the point with coordinates $(4, -4)$.

> **Success Tip** Points with an x-coordinate that is 0 lie on the y-axis. Points with a y-coordinate that is 0 lie on the x-axis. Points that lie on an axis are not considered to be in any quadrant.

Self Check 5

Graph (plot) the points $(2, -2)$, $(-4, 0)$, $\left(1.5, \frac{5}{2}\right)$, and $(0, 5)$.

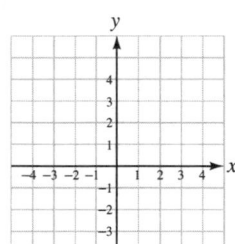

Now Try Problem 55

EXAMPLE 5   Graph (plot) each point. Then state the quadrant in which it lies or the axis on which it lies. **a.** $(4, 4)$ **b.** $(2, -3)$ **c.** $(0, 2.5)$ **d.** $(-3, 0)$

e. $(0, 0)$ **f.** $\left(-1, -\frac{7}{2}\right)$

Strategy After identifying the x- and y-coordinates of the ordered pair, we will move the corresponding number of units left, right, up, or down to locate the point.

WHY The coordinates of a point determine its location on the coordinate plane.

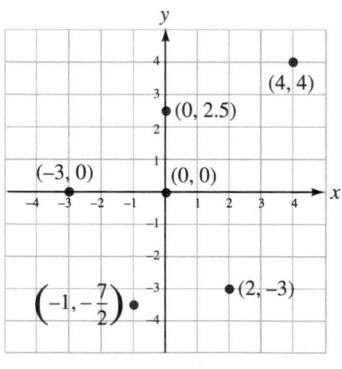

Solution

a. Since the x-coordinate, 4, is positive, we start at the origin and move 4 units to the *right* along the x-axis. Since the y-coordinate, 4, is positive, we then move *up* 4 units and draw a dot. This locates the point $(4, 4)$. The point lies in quadrant I.

b. To plot $(2, -3)$, we begin at the origin and move 2 units to the *right,* because the x-coordinate is 2. Then, since the y-coordinate is negative, we move *down* 3 units. The point lies in quadrant IV.

c. To plot $(0, 2.5)$, we begin at the origin and do not move right or left, because the x-coordinate is 0. Since the y-coordinate is positive, we move 2.5 units *up*. The point lies on the y-axis.

d. To plot $(-3, 0)$, we begin at the origin and move 3 units to the *left*, because the x-coordinate is -3. Since the y-coordinate is 0, we do not move up or down. The point lies on the x-axis.

e. To plot $(0, 0)$, we begin at the origin, and we remain there because both coordinates are 0. The point with coordinates $(0, 0)$ is the origin.

f. To plot $\left(-1, -\frac{7}{2}\right)$, we begin at the origin and move 1 unit to the *left*, because the x-coordinate is -1. The y-coordinate of the given point is negative. To better understand how many units to move *down*, we note that $-\frac{7}{2} = -3\frac{1}{2}$. After moving down $3\frac{1}{2}$ units, we draw a dot. The point lies in quadrant III.

$$\begin{array}{r} 3 \\ 2\overline{)7} \\ \underline{-6} \\ 1 \end{array}$$

> **Success Tip** To graph the point $\left(-1, -\frac{7}{2}\right)$ in Example 5, part f, we expressed the y-coordinate as $-3\frac{1}{2}$. When graphing, if the x- or y-coordinate of a point is an improper fraction, it is helpful to express such a coordinate in equivalent mixed-number or decimal form. Thus, $\left(-1, -\frac{7}{2}\right)$, $\left(-1, -3\frac{1}{2}\right)$, and $(-1, -\textbf{3.5})$ all name the same point.

To find the coordinates of a point on the rectangular coordinate system, we use the numbering (scaling) on the x- and y-axes.

> **The Language of Algebra** Points are often labeled with capital letters. For example, the notation $A(2, 3)$ indicates that point A has coordinates $(2, 3)$.

EXAMPLE 6 Find the coordinates of points $A, B, C, D, E,$ and F plotted in figure (a) below.

Self Check 6
Find the coordinates of each point in figure (b) of Example 6.

Now Try Problem 59

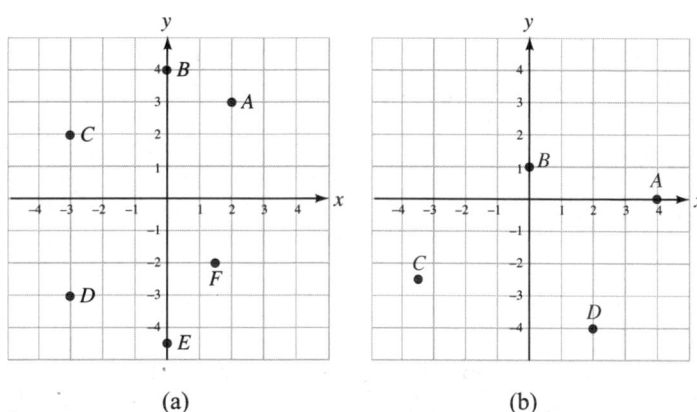

(a)   (b)

Strategy We will start at the origin and count to the left or right on the x-axis, and then up or down to reach each point.

WHY The movement left or right gives the x-coordinate of the ordered pair and the movement up or down gives the y-coordinate.

Solution To locate point A, we start at the origin, move 2 units to the right on the x-axis, and then 3 units up. Its coordinates are $(2, 3)$. The coordinates of the other points are found in the same manner.

$B(0, 4),\quad C(-3, 2),\quad D(-3, -3),\quad E(0, -4.5)\quad$ or $\quad E\left(0, -4\frac{1}{2}\right),\quad F(1.5, -2)\quad$ or $\quad F\left(1\frac{1}{2}, -2\right)$

THINK IT THROUGH   *Population Shift*

"Since 1950, the median center of the U.S. population has moved south and west at every census."

U.S. Census Bureau

In the illustration below, data from the 2000 census were used to draw a north–south line and an east–west line, so equal numbers of the nation's population lived in each "quadrant." The point of intersection of the lines occurs in northeast Daviess County, Indiana. It could be thought of as the "center" of the U.S. population in 2000. If the 2000 census recorded the population of the United States to be 285,230,516, how many people lived in each quadrant created by the lines in the illustration?

Median U. S. population center, Daviess County, Indiana

Source: U.S. Census Bureau

ANSWERS TO SELF CHECKS

1. yes   **2.** no   **3. a.** $(7, 0)$ **b.** $\left(\dfrac{35}{2}, 3\right)$ **4.** $3, 3, -5, -5$

5. 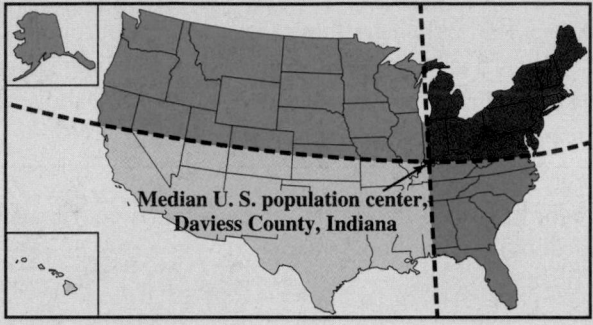   **6.** $A(4, 0), B(0, 1), C(-3.5, -2.5)$ or $C\left(-3\dfrac{1}{2}, -2\dfrac{1}{2}\right), D(2, -4)$

SECTION 8.3   STUDY SET

VOCABULARY

Fill in the blanks.

1. $x + y = 4$ is an equation in two _____.

2. A _____ of an equation in two variables is an ordered pair of numbers that makes the equation a true statement.

3. When we substitute 1 for x and 3 for y in the equation $x + y = 4$, the result is the true statement $4 = 4$. We say that $(1, 3)$ _____ the equation.

4. $x = 2$ and $y = 3$ is a solution of the equation $x + y = 5$. We can write the solution as the _____ pair $(2, 3)$.

5. Solutions of equations in two variables can be listed in a _____ of solutions like the one shown below.

x	y	(x, y)
1	3	(1, 3)
4	7	(4, 7)

6. A rectangular coordinate _____ is shown below. Label the x-axis and the y-axis.

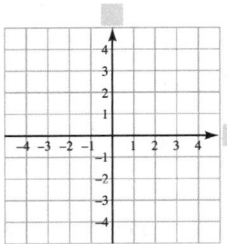

7. The point with coordinates $(0, 0)$ is called the _____.

8. The x- and y-axes divide the rectangular coordinate system into four regions called _____.

9. In the ordered pair $(-2, 4)$, -2 is the ___-coordinate and 4 is the ___-coordinate.

10. The process of locating a point on a rectangular coordinate system is called graphing or _____ the point.

CONCEPTS

11. BURNING CALORIES The table below shows the number of calories a 140-pound woman would burn doing light activities such as office work, cleaning house, or playing golf. Create a graph of the paired data.

Minutes of activity	Calories burned
1	4
2	8
3	12
4	16

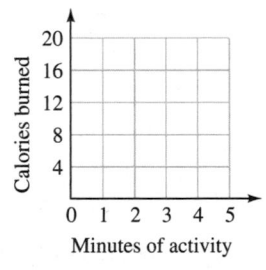

12. CONCRETE The graph in the next column shows the number of parts of sand that should be used for a given number of parts of cement when mixing concrete for a walkway. Create a table of the paired data.

Parts cement	Parts sand

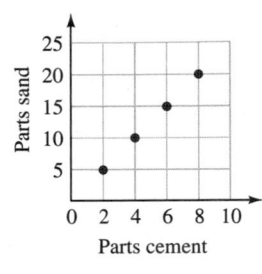

Fill in the blanks.

13. a. $2x + 5 = 10$ is an equation in _____ variable.

b. $2x + 5y = 10$ is an equation in _____ variables.

14. If only one of the values of an ordered-pair solution is known, we can _____ it into the equation to find the other value.

15. a. On the x-axis, the positive direction is to the _____.

b. On the y-axis, the positive direction is _____.

16. To plot the point $\left(\frac{9}{2}, 6\right)$, it is helpful to write the x-coordinate as the mixed number ▢ or as the decimal ▢.

17. To plot the point $(3, -4)$, we start at the _____ and move 3 units to the _____ on the x-axis and then move 4 units _____.

18. To plot the point $(-2, 3)$, we start at the _____ and move 2 units to the _____ on the x-axis and then move 3 units _____.

19. Refer to the graph below. In which quadrant does each point lie? If a point does not lie in a quadrant, tell on what axis (or axes) it lies.

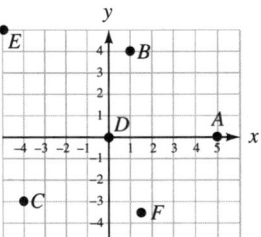

20. In which quadrant does each point lie?

a. $(-1, 2.5)$ **b.** $\left(6, -\frac{5}{2}\right)$

c. $(8, 10)$ **d.** $(-2, -3)$

NOTATION

21. Label each coordinate with the correct letter, x or y.

$$(3, 6)$$
□ ↰ ↱ □

22. Label the top row of the table of solutions shown below with the correct letters.

□	□	(,)
2	5	(2, 5)
6	1	(6, 1)

23. Does the ordered pair $\left(1\frac{1}{3}, -\frac{5}{2}\right)$ name the same point as $\left(\frac{4}{3}, -2.5\right)$?

24. List the Roman numerals from 1 to 4. What are they used to label?

Complete each solution.

25. For the equation $4x + 3y = 14$, find the value of y when $x = 2$.

$$4x + 3y = 14$$
$$4(\square) + 3y = 14$$
$$\square + 3y = 14$$
$$8 + 3y - \square = 14 - \square$$
$$3y = \square$$
$$\frac{3y}{\square} = \frac{6}{\square}$$
$$y = \square$$

26. For the equation $2x - 5y = 20$, find the value of x when $y = 2$.

$$2x - 5y = 20$$
$$2x - 5(\square) = 20$$
$$2x - \square = 20$$
$$2x - 10 + \square = 20 + \square$$
$$2x = \square$$
$$\frac{2x}{\square} = \frac{30}{\square}$$
$$x = \square$$

GUIDED PRACTICE

Determine whether the given ordered pair is a solution of the equation. See Example 1.

27. Is $(2, 3)$ a solution of $2x + 3y = 13$?

28. Is $(4, 1)$ a solution of $3x - 2y = 10$?

29. Is $(9, -3)$ a solution of $x + 5y = -7$?

30. Is $(3, -1)$ a solution of $x + 6y = -1$?

31. Is $(-9, -9)$ a solution of $-3x + 4y = -11$?

32. Is $(-6, -4)$ a solution of $-11x + 6y = 40$?

33. Is $(0, 0)$ a solution of $10x - y = 0$?

34. Is $(0, 0)$ a solution of $6x + 7y = 0$?

Determine whether the given ordered pair is a solution of the equation. See Example 2.

35. Is $(2, 1)$ a solution of $y = 5x - 4$?

36. Is $(5, 8)$ a solution of $y = 2x - 4$?

37. Is $(-2, 6)$ a solution of $y = -x + 4$?

38. Is $(-3, 7)$ a solution of $y = -x + 4$?

39. Is $(-4, 25)$ a solution of $y = -3x + 13$?

40. Is $(-6, 20)$ a solution of $y = -3x + 2$?

41. Is $(0, -9)$ a solution of $y = 7x - 9$?

42. Is $(9, 0)$ a solution of $y = 5x - 9$?

Complete the following ordered pairs so that each one is a solution of the given equation. See Example 3.

43. $2x + y = 8$
   a. $(0, \square)$ **b.** $(\square, 2)$

44. $3x + y = 14$
   a. $(0, \square)$ **b.** $(\square, 2)$

45. $x - 3y = 5$
   a. $(2, \square)$ **b.** $(\square, 3)$

46. $x - 7y = 10$
   a. $(3, \square)$ **b.** $(\square, 4)$

47. $5x + 3y = 15$
   a. $(-6, \square)$ **b.** $\left(\square, 4\right)$

48. $9x + 4y = 36$
   a. $(-4, \square)$ **b.** $\left(\square, 8\right)$

49. $8x - 15y = -4$
   a. $\left(1, \square\right)$ **b.** $(\square, -4)$

50. $7x - 4y = 9$
   a. $\left(1, \square\right)$ **b.** $(\square, 3)$

Complete each table of solutions for the given equation. See Example 4.

51. $4x + 3y = 24$

x	y	(x, y)
0	□	$(0, \square)$
□	0	$(\square, 0)$
−3	□	$(-3, \square)$

52. $3x + y = 12$

x	y	(x, y)
0		(0,)
	0	(, 0)
−6		(−6,)

53. $5x - 4y = 20$

x	y	(x, y)
0		(0,)
	0	(, 0)
−4		(−4,)

54. $7x - 3y = 21$

x	y	(x, y)
0		(0,)
	0	(, 0)
−3		(−3,)

Graph (plot) each point on a rectangular coordinate system.
See Example 5.

55. $(1, 3), (-2, 4), (4, 0), (0, 1), (-3, -2), (0, -5),$
$(3, -2), (0, 0)$

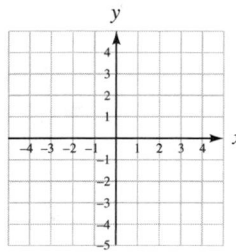

56. $\left(-\frac{3}{2}, 2\right), \left(3, -\frac{5}{2}\right), (0, 0), (0, 2), (0, -3), (3, 0),$
$(-4, -1), (4.5, 4)$

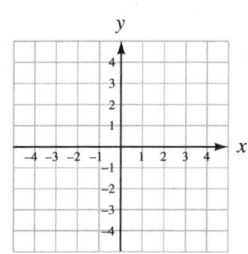

57. $(-4, -3), (1.5, 1.5), \left(\frac{7}{2}, 4\right), (0, -1), (-3.5, 0), (0, 3.5),$
$\left(0, -\frac{9}{2}\right), (5, -2)$

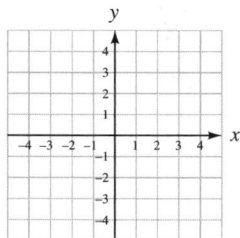

58. $(0, 0), \left(-\frac{1}{2}, \frac{5}{2}\right), (3.5, 0), \left(-\frac{7}{2}, -5\right), (5, -5), (-5, 5),$
$\left(5, -\frac{3}{2}\right), (2, 4)$

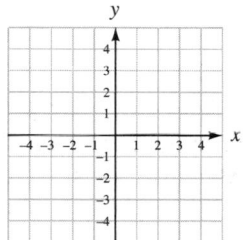

Find the coordinates of each point shown in the graph.
See Example 6.

59.

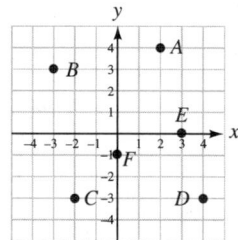

60.

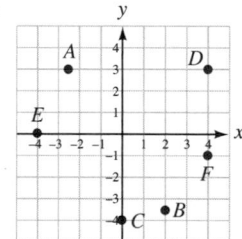

61.

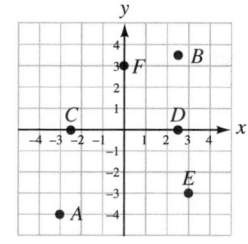

62.

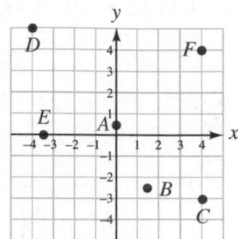

TRY IT YOURSELF

Determine whether the given ordered pair is a solution of the equation.

63. $3x - 6y = 12; (-3.6, -3.8)$

64. $8x + 4y = 10; (-0.5, 3.5)$

65. $y - 6x = 12; \left(\frac{5}{6}, 7\right)$

66. $y + 8x = 4; \left(\frac{3}{4}, 2\right)$

APPLICATIONS

67. MAPS Road maps usually have a coordinate system to help locate cities. Use the following map to locate Rockford, Forreston, Harvard, and the intersection of State Highway 251 and U.S. Highway 30. Express each answer in the form (number, letter).

68. GAMES In the game Battleship, coordinates are used to locate ships. What are the coordinates of the ship shown below? Express each answer in the form (letter, number).

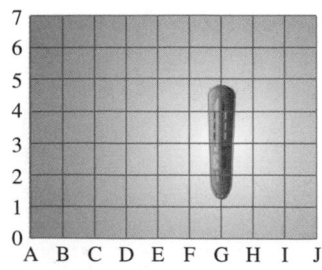

69. EARTHQUAKES On thc map below, the circular area that is shaded blue shows where damage was caused by an earthquake. Important roads and freeways are also labeled.

   a. Find the coordinates of the epicenter (the source of the quake).

   b. Was damage done at the point $(4, 5)$?

   c. Was damage done at the point $(-1, -4)$?

   d. Did Highway 220 suffer any damage?

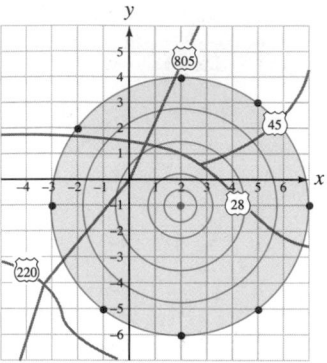

70. AUTOMATION A robot can be programmed to make welds on a car frame. To do this, an imaginary coordinate system is superimposed on the side of the car. Using the commands Up, Down, Left, and Right, write a set of instructions for the robot arm to move from its beginning position to weld the points A, B, C, and D, in that order.

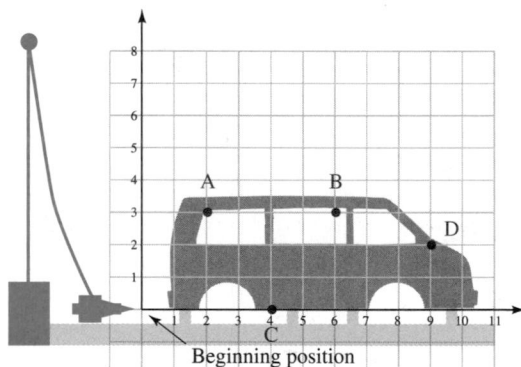

71. THE GLOBE A coordinate system that is used to locate places on the surface of Earth uses a series of curved lines running north and south and east and west, as shown on the next page. List the cities in order, beginning with the one that is farthest east on this map.

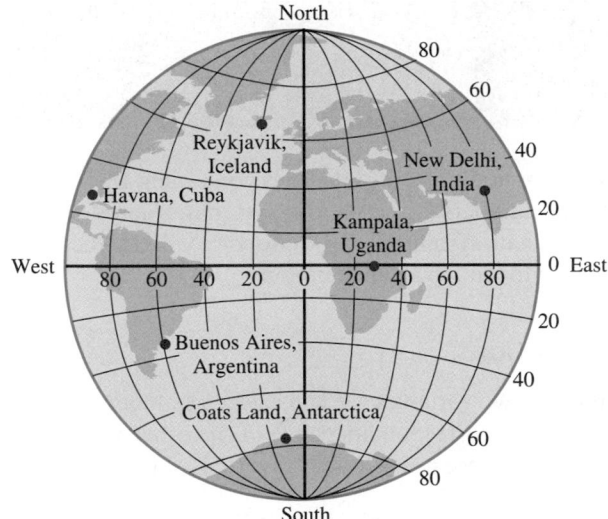

72. BLOOD TRANSFUSIONS The red shaded boxes in the illustration below show which pairs of the major blood groups (AB, A, B, and O) can be mixed without clumping occurring. List all of the ordered pairs of blood types that do not clump when combined. Express your answers in the form (donor blood type, recipient blood type).

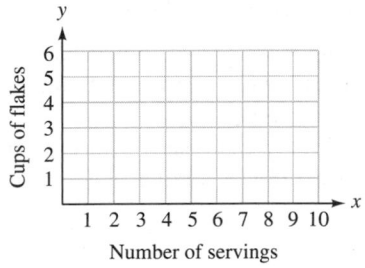

73. COOKING Use the information from the table below to complete the graph for 2, 4, 6, 8, and 10 servings of instant mashed potatoes.

Number of servings	2	4	6	8	10
Flakes (cups)	$\frac{2}{3}$	$1\frac{1}{3}$	2	$2\frac{2}{3}$	$3\frac{1}{3}$

74. DICE The red point in figure (a) represents one of the 36 possible outcomes when two fair dice are rolled a single time. Draw the correct number of dots on the top face of each die in figure (b) to illustrate this outcome.

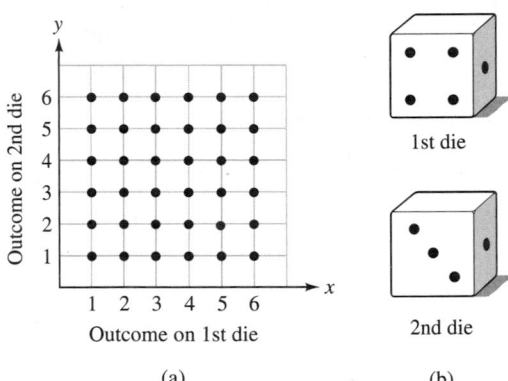

(a)   (b)

WRITING

75. Explain why the point with coordinates $(-4, 4)$ is not the same as the point with coordinates $(4, -4)$.

76. Explain the difference between $2(4)$ and $(2, 4)$.

77. Explain how to plot the point with coordinates of $(1, -7)$.

78. Explain how to plot the point with coordinates of $\left(-\frac{9}{2}, 5\right)$.

79. Explain why the coordinates of the origin are $(0, 0)$.

80. Explain the diagram shown below.

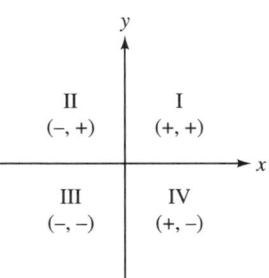

REVIEW

Evaluate each expression.

81. $(-8 - 5) - 3$

82. $-1 - [5 - (-3)]$

83. $-4^2 - 3^2$

84. $-5 - \dfrac{24}{6} - 8(-3)$

Solve each equation and check the result.

85. $\dfrac{x}{3} + 3 = 10$

86. $-3x - 4 = 8$

87. $5 - (7 - x) = -5$

88. $2(y + 6) - 4 = 2$

Objectives

1 Construct a table of solutions.

2 Graph linear equations that are solved for y.

3 Graph linear equations by finding intercepts.

4 Graph equations of the form $y = b$ and $x = a$.

SECTION **8.4**

Graphing Linear Equations

In the previous section, we saw that solutions of equations containing the variables x and y were ordered pairs of real numbers (x, y). We also saw that ordered pairs can be graphed on a rectangular coordinate system. In this section, we will use these skills to see how plotting points can give the graph of an equation.

1 Construct a table of solutions.

To find a solution of an equation in two variables, we can select a number, substitute it for one of the variables, and find the corresponding value of the other variable. For example, to find a solution of $y = x - 1$, we can select a value for x, say, -4, substitute -4 for x in the equation, and find y.

x	y	(x, y)
-4	-5	$(-4, -5)$

$y = x - 1$	This is the given equation.
$y = -4 - 1$	Substitute -4 for x.
$y = -5$	Do the subtraction: $-4 - 1 = -4 + (-1) = -5$.

The ordered pair $(-4, -5)$ is a solution of $y = x - 1$. We list it in the table on the left.

To find another solution of $y = x - 1$, we select another value for x, say, -2, and find the corresponding y-value.

x	y	(x, y)
-4	-5	$(-4, -5)$
-2	-3	$(-2, -3)$

$y = x - 1$	This is the given equation.
$y = -2 - 1$	Substitute -2 for x.
$y = -3$	Do the subtraction: $-2 - 1 = -2 + (-1) = -3$.

A second solution is $(-2, -3)$, and we list it in the table of solutions.

If we let $x = 0$, we can find a third ordered pair that satisfies $y = x - 1$.

x	y	(x, y)
-4	-5	$(-4, -5)$
-2	-3	$(-2, -3)$
0	-1	$(0, -1)$

$y = x - 1$	This is the given equation.
$y = 0 - 1$	Substitute 0 for x.
$y = -1$	Do the subtraction: $0 - 1 = 0 + (-1) = -1$.

A third solution is $(0, -1)$, which we also add to the table of solutions.

We can find a fourth solution by letting $x = 2$, and a fifth solution by letting $x = 4$.

x	y	(x, y)
-4	-5	$(-4, -5)$
-2	-3	$(-2, -3)$
0	-1	$(0, -1)$
2	1	$(2, 1)$
4	3	$(4, 3)$

$y = x - 1$		$y = x - 1$	
$y = 2 - 1$	Substitute 2 for x.	$y = 4 - 1$	Substitute 4 for x.
$y = 1$	Subtract.	$y = 3$	Subtract.

A fourth solution is $(2, 1)$ and a fifth solution is $(4, 3)$. We add them to the table.

Since we can choose any number for x, and since any choice of x will give a corresponding value of y, it is apparent that the equation $y = x - 1$ has *infinitely many solutions*. We have found five of them: $(-4, -5)$, $(-2, -3)$, $(0, -1)$, $(2, 1)$, and $(4, 3)$.

2 Graph linear equations that are solved for y.

It is impossible to list the infinitely many solutions of the equation $y = x - 1$ in a table. However, to show all of its solutions, we can draw a mathematical "picture" of them. We call this picture the *graph of the equation*.

To graph $y = x - 1$, we plot the ordered pairs shown in the table above on a rectangular coordinate system. Then we draw a straight line through the points, because the graph of any solution of $y = x - 1$ will lie on this line. Furthermore, every point on this line represents a solution. We call the line the **graph of the equation.** It represents all of the solutions of $y = x - 1$. The graph below only shows a part of the line. The arrowheads indicate that it extends indefinitely in both directions.

$y = x - 1$

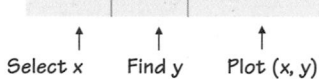

x	y	(x, y)
-4	-5	(-4, -5)
-2	-3	(-2, -3)
0	-1	(0, -1)
2	1	(2, 1)
4	3	(4, 3)

↑ Select x   ↑ Find y   ↑ Plot (x, y)

Construct a table of solutions.

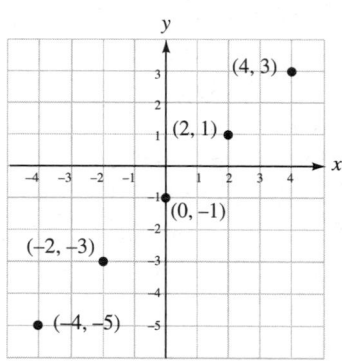

Plot the ordered pairs.

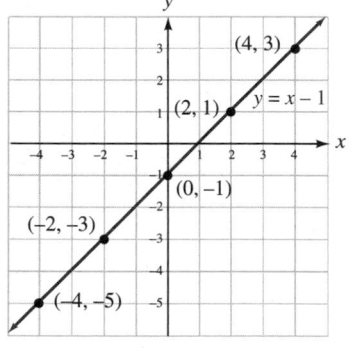

Draw a straight line through the points. This is the *graph of the equation.*

The equation $y = x - 1$ is said to be *linear* and its graph is a straight line. Some more examples of linear equations are

$$y = 2x + 4, \quad x - 3y = 6, \quad \text{and} \quad 40x + 3y = -120$$

When we graphed $y = x - 1$, we did more work than necessary. Since *two points determine a line,* only two points are needed to graph the line. However, it is always a good idea to plot a third point as a check. If the three points do not lie on a straight line, then at least one of them is incorrect.

Linear equations can be graphed in several ways. Generally, the form in which an equation is written determines the method that we use to graph it. To graph linear equations solved for y, such as $y = x - 1$ and $y = 2x + 4$, we can use the following method.

Graphing Linear Equations Solved for y by Plotting Points

1. Find three ordered pairs that are solutions of the equation by selecting three values for x and calculating the corresponding values of y.
2. Plot the solutions on a rectangular coordinate system.
3. Draw a straight line passing through the points. If the points do not lie on a line, check your calculations.

EXAMPLE 1 Graph: $y = 2x + 4$

Strategy We will find three solutions of the given equation, plot them on a rectangular coordinate system, and then draw a straight line passing through the points.

WHY To *graph* a linear equation in two variables means to make a drawing that represents all of its solutions.

Self Check 1

Graph: $y = 2x - 2$

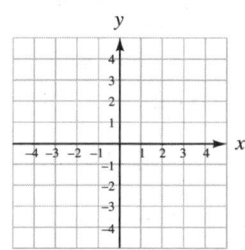

Now Try Problem 23

Solution To find three solutions of $y = 2x + 4$, we select three values for x that will make the calculations easy. Then we find each corresponding value of y.

If $x = -2$:	If $x = 0$:	If $x = 2$:
$y = 2x + 4$	$y = 2x + 4$	$y = 2x + 4$
$y = 2(-2) + 4$	$y = 2(0) + 4$	$y = 2(2) + 4$
$y = -4 + 4$	$y = 0 + 4$	$y = 4 + 4$
$y = 0$	$y = 4$	$y = 8$
$(-2, 0)$ is a solution.	$(0, 4)$ is a solution.	$(2, 8)$ is a solution.

We enter these results in a table of solutions and plot the points. Then we draw a straight line through the points and label it $y = 2x + 4$.

$y = 2x + 4$

x	y	(x, y)
-2	0	$(-2, 0)$
0	4	$(0, 4)$
2	8	$(2, 8)$

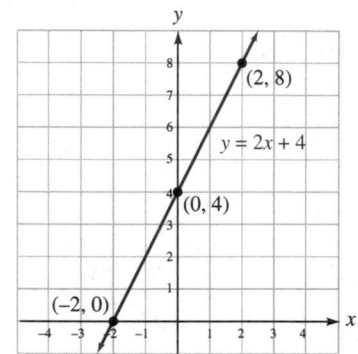

As a check, we can pick two points that the line appears to pass through, such as $(1, 6)$ and $(-1, 2)$. When we substitute their coordinates into the given equation, the two true statements that result indicate that $(1, 6)$ and $(-1, 2)$ are solutions and that the graph of the line is correctly drawn.

Check (1, 6):
$y = 2x + 4$
$6 \stackrel{?}{=} 2(1) + 4$
$6 \stackrel{?}{=} 2 + 4$
$6 = 6$ True

Check (−1, 2):
$y = 2x + 4$
$2 \stackrel{?}{=} 2(-1) + 4$
$2 \stackrel{?}{=} -2 + 4$
$2 = 2$ True

> **Success Tip** When selecting x-values for a table of solutions, a rule of thumb is to choose a negative number, a positive number, and 0. When $x = 0$, the calculations to find y are usually quite simple.

EXAMPLE 2   Graph: $y = -\dfrac{1}{3}x - 2$

Strategy We will find three solutions of the given equation, plot them on a rectangular coordinate system, and then draw a straight line passing through the points.

WHY To *graph* a linear equation in two variables means to make a drawing that represents all of its solutions.

Solution To find three solutions of $y = -\frac{1}{3}x - 2$, each value of x must be multiplied by $-\frac{1}{3}$. This calculation is made easier if we select x-values that are *multiples of the denominator 3,* such as $-3, 0,$ and 3.

If $x = -3$:

$y = -\frac{1}{3}x - 2$

$y = -\frac{1}{3}(-3) - 2$

$y = -\frac{1}{3}\left(-\frac{3}{1}\right) - 2$

$y = \frac{3}{3} - 2$

$y = 1 - 2$

$y = -1$

$(-3, -1)$ is a solution.

If $x = 0$:

$y = -\frac{1}{3}x - 2$

$y = -\frac{1}{3}(0) - 2$

$y = 0 - 2$

$y = -2$

$(0, -2)$ is a solution.

If $x = 3$:

$y = -\frac{1}{3}x - 2$

$y = -\frac{1}{3}(3) - 2$

$y = -\frac{1}{3}\left(\frac{3}{1}\right) - 2$

$y = -\frac{3}{3} - 2$

$y = -1 - 2$

$y = -3$

$(3, -3)$ is a solution.

We enter these results in a table of solutions and plot the points. Then we draw a straight line through the points and label it $y = -\frac{1}{3}x - 2$.

$$y = -\frac{1}{3}x - 2$$

x	y	(x, y)
-3	-1	$(-3, -1)$
0	-2	$(0, -2)$
3	-3	$(3, -3)$

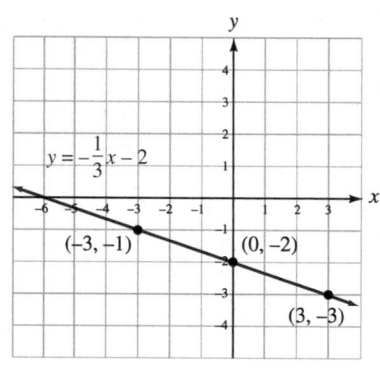

Self Check 2

Graph: $y = -\frac{1}{4}x + 3$

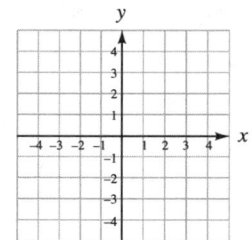

Now Try Problem 27

Success Tip In Example 2, when we select x-values that are multiples of the denominator 3, the corresponding y-values are integers, and not difficult-to-plot fractions.

EXAMPLE 3 Graph: $y = 20x$

Strategy We will find three solutions of the given equation, plot them on a rectangular coordinate system, and then draw a straight line passing through the points.

WHY To *graph* a linear equation in two variables means to make a drawing that represents all of its solutions.

Solution We begin by selecting three values for x: $-2, 0,$ and 2. If $x = -2$, we can calculate the corresponding value of y by substituting -2 for x in $y = 20x$.

$y = 20x$ This is the equation to graph.

$y = 20(-2)$ Substitute -2 for x.

$y = -40$ Do the multiplication.

Self Check 3

Graph: $y = 25x$

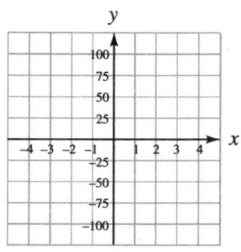

Now Try Problem 31

We see that $x = -2$ and $y = -40$ is a solution of $y = 20x$. In a similar manner, we find the corresponding values for y when x is 0 and 2 and enter them in the table below.

Because of the sizes of the y-coordinates of the points $(-2, -40)$ and $(2, 40)$, we must adjust the scale on the y-axis. (If we used grid lines 1 unit apart, the graph would be very large.)

One way to make these points fit is to scale the y-axis in units of 5, 10, or 20. If we choose divisions of 20 units, plot the three solutions from the table, and draw a line through them, we get the graph shown below.

$y = 20x$

x	y	(x, y)
-2	-40	$(-2, -40)$
0	0	$(0, 0)$
2	40	$(2, 40)$

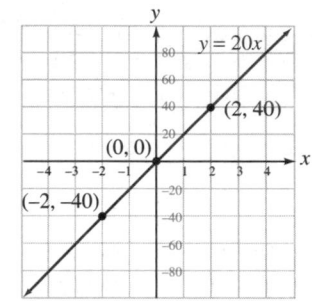

3 Graph linear equations by finding intercepts.

The graph of $y = 2x + 4$ from Example 1 is shown below. We see that the graph crosses the x-axis at $(-2, 0)$; this point is called the **x-intercept** of the graph. The graph crosses the y-axis at the point $(0, 4)$; this point is called the **y-intercept** of the graph.

> **The Language of Algebra**   The point where a line intersects the x- or y-axis is called an *intercept*.

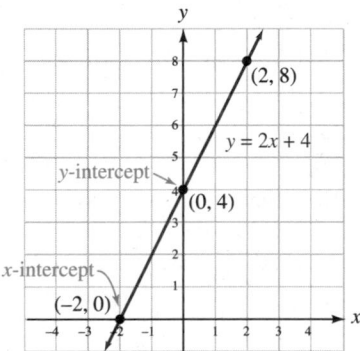

We see that the x-intercept has a y-coordinate of 0, and the y-intercept has an x-coordinate of 0. These observations suggest the following procedures for finding the intercepts of a graph from its equation.

Finding Intercepts

To find the y-intercept, substitute 0 for x in the given equation and solve for y.
To find the x-intercept, substitute 0 for y in the given equation and solve for x.

Plotting the x- and y-intercepts of a graph and drawing a line through them is called the **intercept method of graphing a line.** This method is useful when graphing linear equations that have x- and y-terms on one side and a constant on the other side, such as $x - 3y = 6$ and $40x + 3y = -120$.

EXAMPLE 4 Graph $x - 3y = 6$ by finding the x- and y-intercepts.

Strategy We will let $y = 0$ to find the x-intercept. We will then let $x = 0$ to find the y-intercept of the graph.

WHY Since two points determine a line, the x-intercept and y-intercept are enough information to graph this linear equation.

Solution

x-intercept: let y = 0

$$x - 3y = 6$$
$$x - 3(0) = 6 \quad \text{Substitute 0 for y.}$$
$$x - 0 = 6$$
$$x = 6$$

The x-intercept is $(6, 0)$.

y-intercept: let x = 0

$$x - 3y = 6$$
$$0 - 3y = 6 \quad \text{Substitute 0 for x.}$$
$$-3y = 6$$
$$y = -2 \quad \text{To isolate y, divide both sides by } -3.$$

The y-intercept is $(0, -2)$.

Since each intercept of the graph is a solution of the equation, we enter the intercepts in the table of solutions below.

As a check, we find one more point on the line. We select a convenient value for x, say, 3, and find the corresponding value of y. The check point should lie on the same line as the x- and y-intercepts. If it does not, check your work to find the incorrect coordinate or coordinates.

$$x - 3y = 6$$
$$3 - 3y = 6 \quad \text{Substitute 3 for x.}$$
$$-3y = 3 \quad \text{To isolate the variable term, } -3y, \text{ subtract 3 from both sides.}$$
$$y = -1 \quad \text{To isolate y, divide both sides by } -3.$$

Therefore, $(3, -1)$ is a solution. It is also entered in the table.

We plot the intercepts and the check point, draw a straight line through them, and label the line as $x - 3y = 6$.

$x - 3y = 6$

x	y	(x, y)	
6	0	$(6, 0)$	← x-intercept
0	−2	$(0, -2)$	← y-intercept
3	−1	$(3, -1)$	← Check point

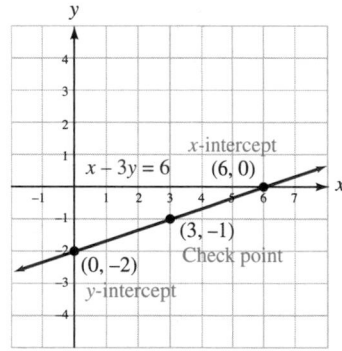

The calculations for finding intercepts can be simplified if we realize what occurs when we substitute 0 for y or 0 for x in a linear equation.

EXAMPLE 5 Graph $40x + 3y = -120$ by finding the x- and y-intercepts.

Strategy We will let $y = 0$ to find the x-intercept. We will then let $x = 0$ to find the y-intercept of the graph.

WHY Since two points determine a line, the x-intercept and y-intercept are enough information to graph this linear equation.

Self Check 4

Graph $x - 2y = 2$ by finding the x- and y-intercepts.

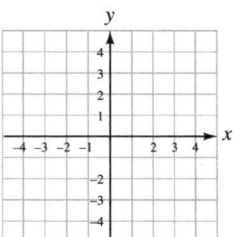

Now Try **Problem 35**

Graph $32x + 5y = -160$ by finding the x- and y-intercepts.

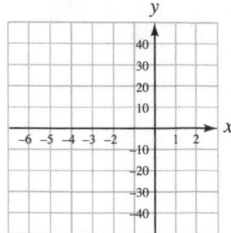

Now Try Problem 39

Solution When we substitute 0 for y, it follows that the term $3y$ will be equal to 0. Therefore, to find the x-intercept, we can cover the $3y$ and solve the remaining equation for x.

$$40x \;\boxed{+\; 3y}\; = -120 \qquad \text{If } y = 0, \text{ then } 3y = 3(0) = 0. \text{ Cover the } 3y \text{ term.}$$
$$x = -3 \qquad \text{To solve } 40x = -120, \text{ divide both sides by 40.}$$

The x-intercept is $(-3, 0)$.

When we substitute 0 for x, it follows that the term $40x$ will be equal to 0. Therefore, to find the y-intercept, we can cover the $40x$ and solve the remaining equation for y.

$$\boxed{40x}\; + 3y = -120 \qquad \text{If } x = 0, \text{ then } 40x = 40(0) = 0. \text{ Cover the } 40x \text{ term.}$$
$$y = -40 \qquad \text{To solve } 3y = -120, \text{ divide both sides by 3.}$$

The y-intercept is $(0, -40)$.

> **Caution!** When using the cover-over method to find the y-intercept, be careful not to cover the sign in front of the y-term.

We can find a third solution by selecting a convenient value for x and finding the corresponding value for y. If we choose $x = -6$, we find that $y = 40$. The solution $(-6, 40)$ is entered in the table, and the equation is graphed as shown.

> **Success Tip** To fit y-values of 40 and -40 on the graph, the y-axis was scaled in units of 10.

$$40x + 3y = -120$$

x	y	(x, y)	
-3	0	$(-3, 0)$	← x-intercept
0	-40	$(0, -40)$	← y-intercept
-6	40	$(-6, 40)$	← Check point

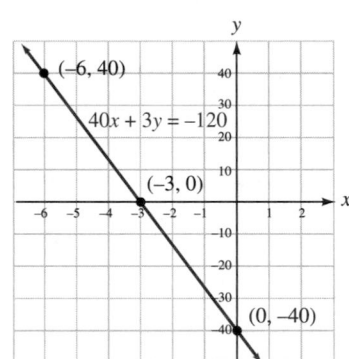

> **The Language of Algebra** The method to find the intercepts of the graph of a linear equation shown in Example 5 is commonly referred to as the *cover-over method*.

4 Graph equations of the form $y = b$ and $x = a$.

When a linear equation contains only one variable, such as $y = 4$ or $x = -2$, its graph is either a horizontal or a vertical line.

EXAMPLE 6   Graph: $y = 4$

Strategy To find three ordered-pair solutions of this equation to plot, we will select three values for x and use 4 for y each time.

WHY The given equation requires that $y = 4$.

Solution We can write the equation in the form $0x + y = 4$. Since the coefficient of x is 0, the numbers chosen for x have no effect on y. The value of y is always 4. For example, if $x = 2$, we have

$$0x + y = 4$$
$$0(2) + y = 4 \qquad \text{Substitute 2 for } x.$$
$$y = 4 \qquad \text{Simplify the left side.}$$

One solution is $(2, 4)$. To find two more solutions, we select $x = 0$ and $x = -3$. For any x-value, the y-value is always 4, so we enter $(0, 4)$ and $(-3, 4)$ in the table. If we plot the ordered pairs and draw a straight line through the points, the result is a horizontal line. The y-intercept is $(0, 4)$ and there is no x-intercept.

Self Check 6

Graph: $y = -2$

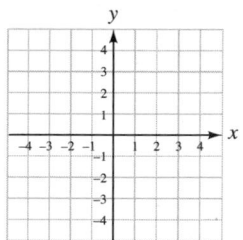

Now Try Problem 43

$y = 4$

x	y	(x, y)
2	4	$(2, 4)$
0	4	$(0, 4)$
-3	4	$(-3, 4)$

↑   ↑
Select any   Each value of y
number for x.   must be 4.

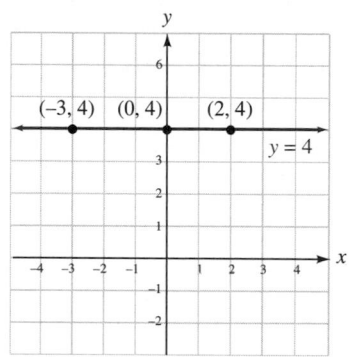

EXAMPLE 7   Graph: $x = -3$

Strategy To find three ordered-pair solutions of this equation to plot, we must select -3 for x each time.

WHY The given equation requires that $x = -3$.

Solution We can write the equation in the form $x + 0y = -3$. Since the coefficient of y is 0, the numbers chosen for y have no effect on x. The value of x is always -3. For example, if $y = -2$, we have

$$x + 0y = -3$$
$$x + 0(-2) = -3 \qquad \text{Substitute } -2 \text{ for } y.$$
$$x = -3 \qquad \text{Simplify the left side.}$$

One solution is $(-3, -2)$. To find two more solutions, we select $y = 0$ and $y = 3$. For any y-value, the x-value is always -3, so we enter $(-3, 0)$ and $(-3, 3)$ in the table. If we plot the ordered pairs and draw a straight line through the points, the result is a vertical line. The x-intercept is $(-3, 0)$ and there is no y-intercept.

Self Check 7

Graph: $x = 4$

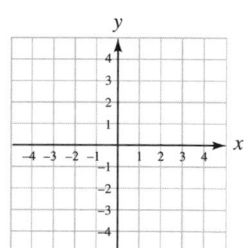

Now Try Problem 47

$x = -3$

x	y	(x, y)
-3	-2	$(-3, -2)$
-3	0	$(-3, 0)$
-3	3	$(-3, 3)$

↑   ↑
Each value of x   Select any
must be -3.   number for y.

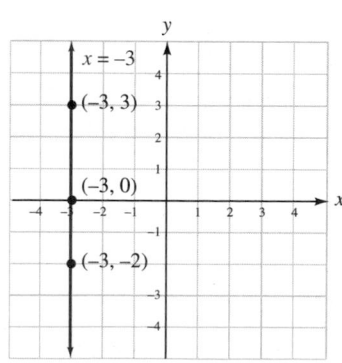

From the results of Examples 6 and 7, we have the following facts.

Equations of Horizontal and Vertical Lines

The equation $y = b$ represents the horizontal line that intersects the y-axis at $(0, b)$.

The equation $x = a$ represents the vertical line that intersects the x-axis at $(a, 0)$.

The graph of the equation $y = 0$ has special importance; it is the x-axis. Similarly, the graph of the equation $x = 0$ is the y-axis.

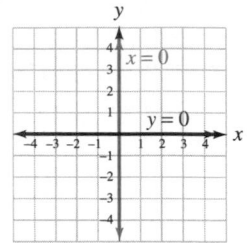

ANSWERS TO SELF CHECKS

1.

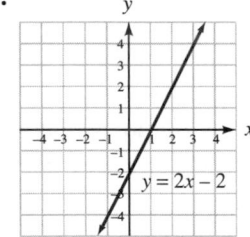

$y = 2x - 2$

2.

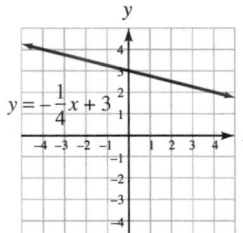

$y = -\dfrac{1}{4}x + 3$

3.

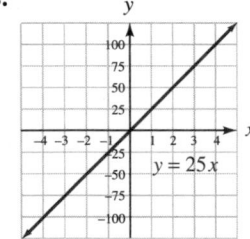

$y = 25x$

4.

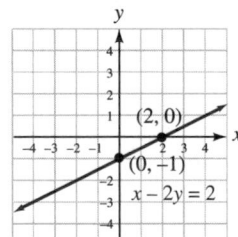

$(2, 0)$
$(0, -1)$
$x - 2y = 2$

5.

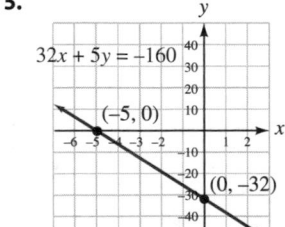

$32x + 5y = -160$
$(-5, 0)$
$(0, -32)$

6.

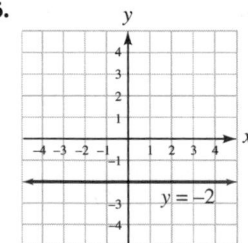

$y = -2$

7.

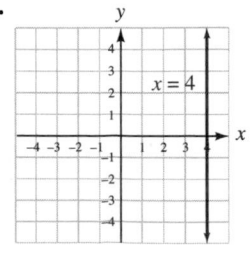

$x = 4$

SECTION 8.4   STUDY SET

VOCABULARY

Fill in the blanks.

1. $y = 2x + 3$ is an equation in _____ variables, x and y.

2. A _____ of an equation in two variables is an ordered pair of numbers that makes the equation a true statement.

3. The _____ of a linear equation is a mathematical "picture" of all of its solutions.

4. The graph of a linear equation is a straight _____.

5. The point where the graph of a linear equation crosses the x-axis is called the _____.

6. The y-intercept of the graph of a linear equation is the point where it crosses the ___-axis.

CONCEPTS

Fill in the blanks.

7. To graph $y = 2x + 3$, we can use the following steps:

   Step 1. Find three ordered pairs that are solutions of $y = 2x + 3$ by selecting three values for ___ and calculating the corresponding values of ___.

   Step 2. _____ the solutions on a rectangular coordinate system.

   Step 3. Draw a straight _____ passing through the points. If the points do not lie on a line, _____ your calculations.

8. The graph of a linear equation is shown here. What three points were plotted to obtain the graph? Enter them in the table of solutions.

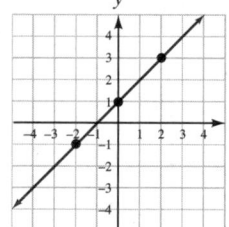

9. **a.** Find the y-intercept of the line graphed on the right.

   b. What is its x-intercept?

   c. Does the line pass through the point $(4, 3)$?

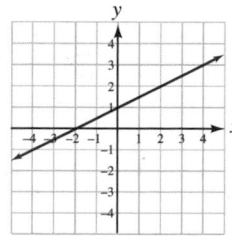

10. **a.** Find the y-intercept of the line graphed on the right.

    b. What is its x-intercept?

    c. Does the line pass through the point $(1, -1)$?

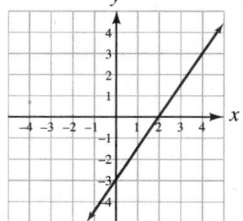

11. Fill in the blanks. To graph $3x + 4y = 12$, we can use the following two-step process:

    Step 1. To find the x-intercept, substitute ▨ for y in $3x + 4y = 12$ and solve for x.

    Step 2. To find the y-intercept, substitute 0 for ___ in $3x + 4y = 12$ and solve for y.

12. Fill in the blanks in the following table of solutions.

$3x + 2y = 6$

x	y	(x, y)	
2	0	$(2, 0)$	← ▨-intercept
0	3	$(0, 3)$	← ▨-intercept

13. **a.** Fill in the blanks. The graph of the equation $y = 3$ is a _____ line.

    b. The graph of the equation $x = -2$ is a _____ line.

14. **a.** What is the equation of the x-axis?

    b. What is the equation of the y-axis?

15. **a.** Name three points on the line graphed in figure (a) below.

    b. Name three points on the line graphed in figure (b) below.

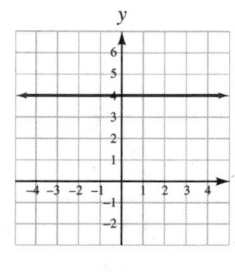 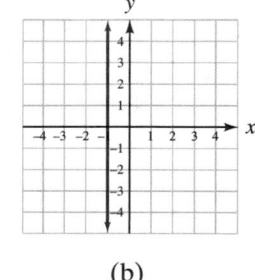

(a)   (b)

16. The graph of a linear equation is shown below. Name six ordered-pair solutions of the equation from the graph.

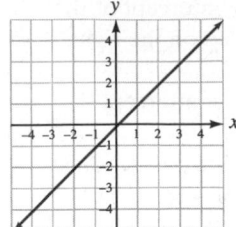

NOTATION

Complete each solution

17. Find the *x*-intercept of the graph of $2x - 4y = 8$.

$$2x - 4y = 8$$
$$2x - 4(\ ) = 8$$
$$2x - \ = 8$$
$$2x = \ $$
$$\frac{2x}{\ } = \frac{8}{\ }$$
$$x = \ $$

The *x*-intercept of the graph is (,0).

18. Find the *y*-intercept of the graph of $2x - 4y = 8$.

$$2x - 4y = 8$$
$$2(\ ) - 4y = 8$$
$$\ - 4y = 8$$
$$-4y = \ $$
$$\frac{-4y}{\ } = \frac{8}{\ }$$
$$y = \ $$

The *y*-intercept of the graph is (0,).

GUIDED PRACTICE

Complete each table of solutions. See Objective 1.

19. $y = 3x + 3$

x	y	(x, y)
-2		(,)
0		(,)
2		(,)

20. $y = 4x + 2$

x	y	(x, y)
-2		(,)
0		(,)
2		(,)

21. $y = -2x - 3$

x	y	(x, y)
-2		(,)
0		(,)
2		(,)

22. $y = -3x + 4$

x	y	(x, y)
-2		(,)
0		(,)
2		(,)

Graph each equation. **See Example 1.**

23. $y = 2x - 5$

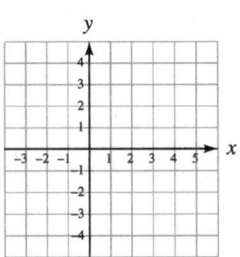

24. $y = 3x + 1$

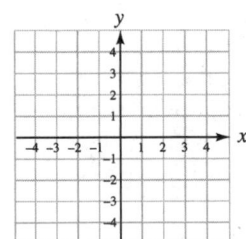

25. $y = -3x + 2$

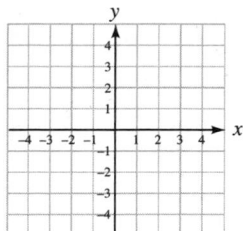

26. $y = -4x - 2$

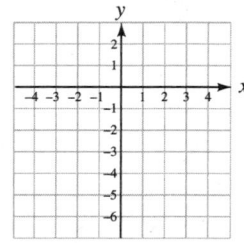

Graph each equation. See Example 2.

27. $y = -\dfrac{1}{3}x + 1$

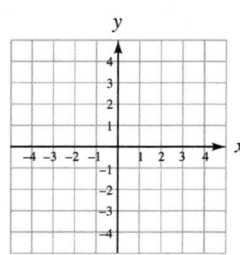

28. $y = -\dfrac{1}{3}x - 1$

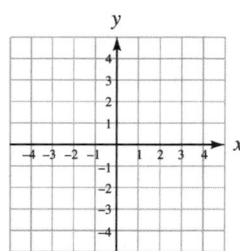

29. $y = -\dfrac{1}{2}x + 1$

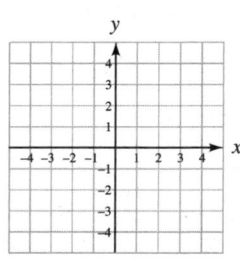

30. $y = -\dfrac{1}{2}x - 2$

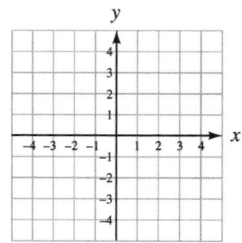

Graph each equation. See Example 3.

31. $y = 100x$

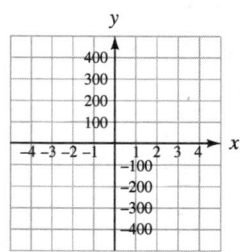

32. $y = 50x$

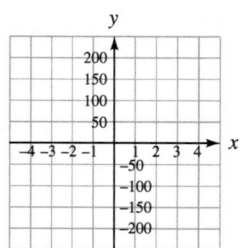

33. $y = -30x$

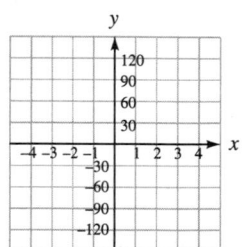

34. $y = -20x$

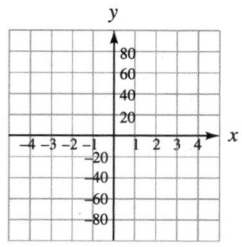

Complete the table and graph the equation using the intercept method. **See Example 4.**

35. $x - 2y = -4$

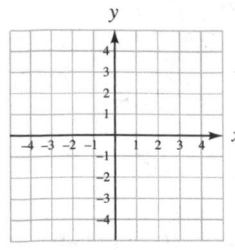

36. $3x + y = -3$

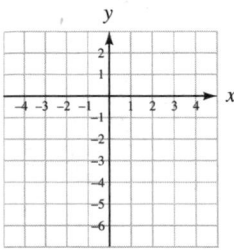

37. $4x + 5y = 20$

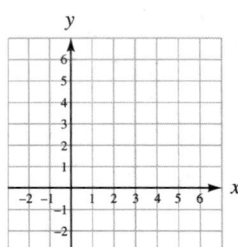

38. $3x - 5y = 15$

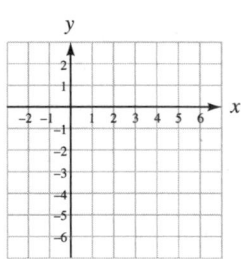

Use the intercept method to graph each equation.
See Example 5.

39. $30x + y = -30$ **40.** $20x - y = -20$

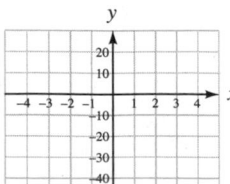 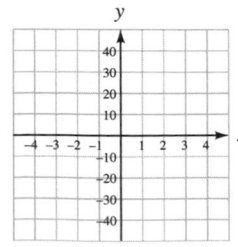

41. $4x - 20y = 60$ **42.** $6x - 30y = 30$

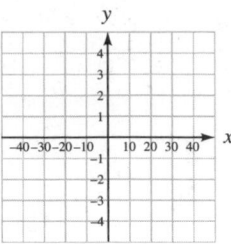 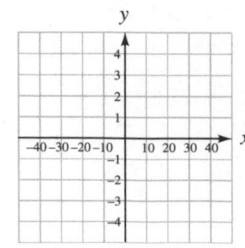

Graph each equation. **See Example 6.**

43. $y = 5$ **44.** $y = 1$

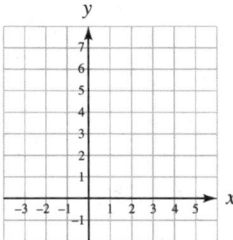 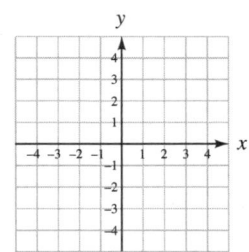

45. $y = -4$ **46.** $y = -3$

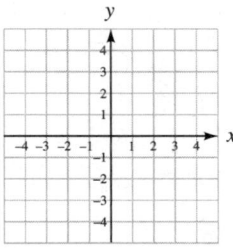 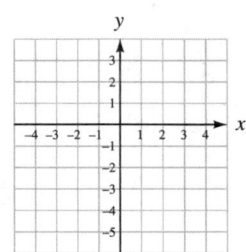

Graph each equation. **See Example 7.**

47. $x = 4$ **48.** $x = 5$

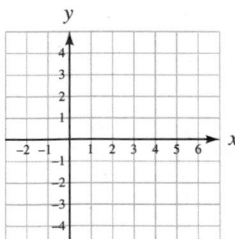 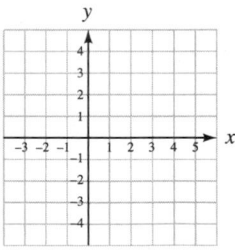

49. $x = -2$ **50.** $x = -1$

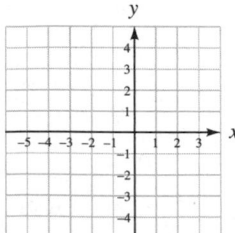 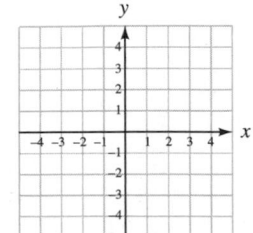

TRY IT YOURSELF

Find the coordinates of the x- and y-intercepts of the graph of each equation. You do not have to graph the equation.

51. $x + y = 8$

52. $x - y = 9$

53. $4x + 5y = 100$

54. $3x - 5y = 75$

Graph each equation.

55. $y = \dfrac{2}{3}x - 2$

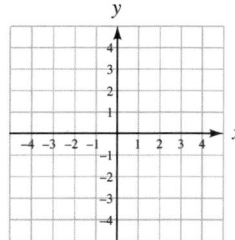

56. $y = \dfrac{5}{6}x - 5$

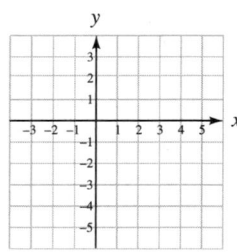

57. $x + y = 5$

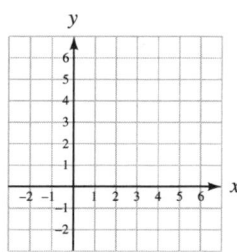

58. $x - y = 2$

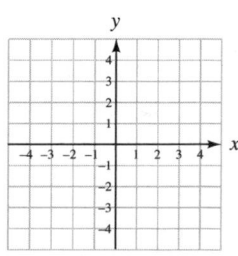

59. $y = x$

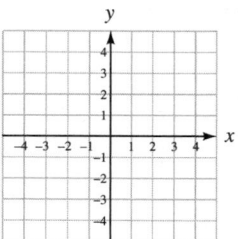

60. $y = -2x$

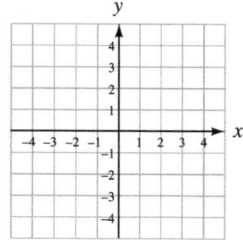

61. $x = 0$

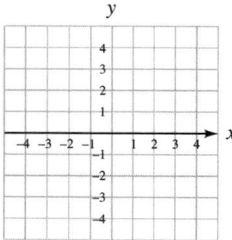

62. $y = 0$

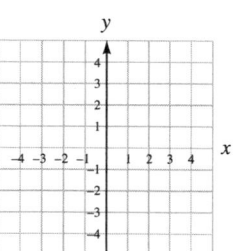

63. $3x + 5y = -150$

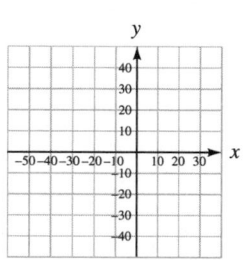

64. $x + 5y = 50$

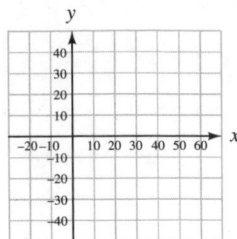

65. $y = \dfrac{x}{3}$

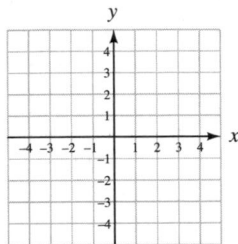

66. $y = \dfrac{3}{4}x$

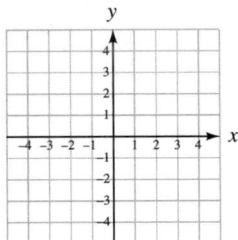

67. $y = -50x - 25$

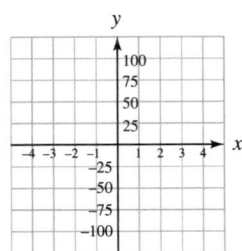

68. $y = 200x - 400$

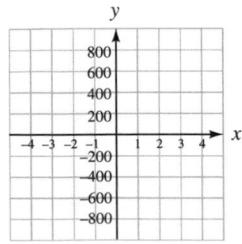

69. $4x - 3y = 12$

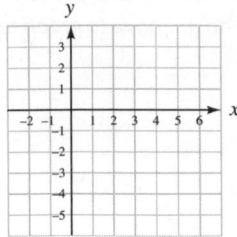

70. $5x - 10y = 20$

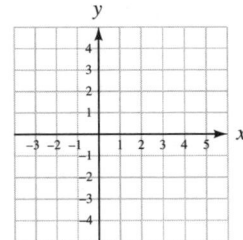

71. $y = \dfrac{5}{2}$

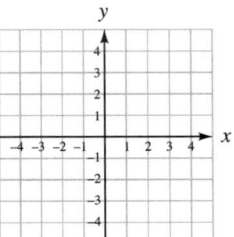

72. $x = \dfrac{4}{3}$

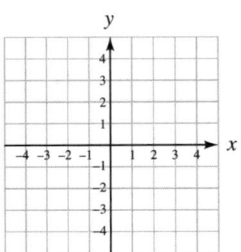

APPLICATIONS

73. HOURLY WAGES The following table gives the amount *y* (in dollars) that a student can earn by working *x* hours. Plot the ordered pairs in the table and draw a straight line through the points. Then estimate how much the student will earn in 3 hours.

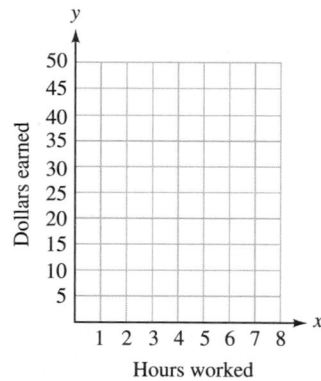

x	y
2	15
4	30
6	45

74. VALUE OF A CAR The following table shows the value *y* (in thousands of dollars) of a car that is *x* years old. Plot the ordered pairs in the table and draw a straight line through the points. Then estimate the value of the car when it is 7 years old.

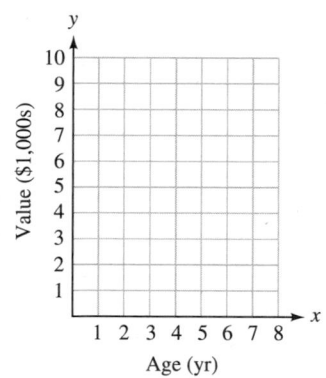

x	y
3	7
4	5.5
5	4

75. DISTANCE, RATE, AND TIME The formula $d = 2t$ gives the distance *d* (in miles) that a child can walk in a time *t* (in hours) at the rate of 2 mph. Complete the table of solutions in the next column, and then graph the equation to get a picture of the relationship between distance and time. (*Hint:* Plot *t* on the horizontal axis and *d* on the vertical axis.)

$d = 2t$

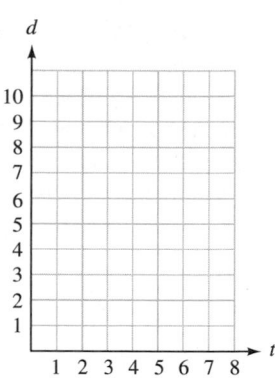

t	d
1	
2	
3	
4	
5	

76. INVESTMENTS If \$100 is invested in a savings account paying 6% per year simple interest, the amount *A* in the account over a period of time *t* is given by the formula $A = 6t + 100$. Complete the table of solutions, and then graph this equation to get a picture of how the account grows over a period of time. (*Hint:* Plot *t* on the horizontal axis and *A* on the vertical axis.)

$A = 6t + 100$

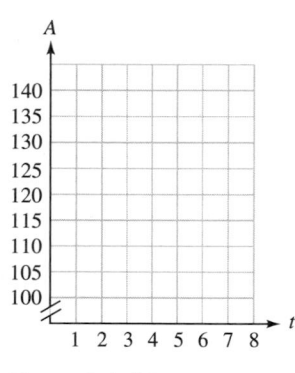

t	A
1	
2	
3	
4	
5	
6	

The symbol ⌇ is used to indicate a break in the labeling of the vertical axis.

77. BILLIARDS Refer to the billiard table shown below. The path traveled by the black 8-ball is described by the equations $y = 2x - 4$ and $y = -2x + 12$. Construct a table of solutions for the equation $y = 2x - 4$ using the *x*-values 1, 2, and 4. Do the same for the equation $y = -2x + 12$, using the *x*-values 4, 6, and 8. Then graph the path of the 8-ball.

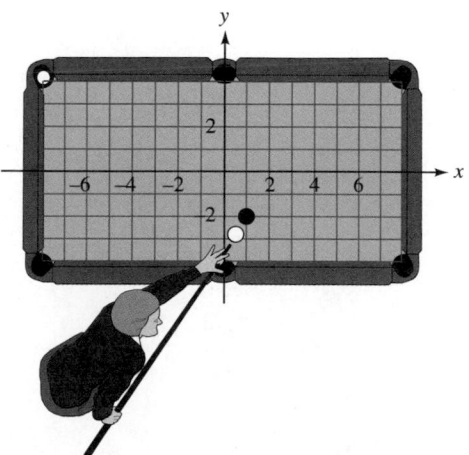

78. AIR TRAFFIC CONTROL The equations describing the paths of two airplanes are $y = -\frac{1}{2}x + 3$ and $2x - 3y = -2$. Each equation is graphed on the radar screen shown below. If the planes are flying at the same altitude, is there a possibility of a midair collision? If so, where?

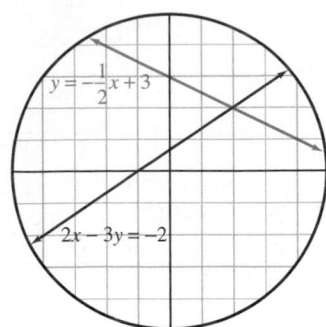

WRITING

79. When we say that $(-2, -1)$ is a solution of $5x - 6y = -4$, what do we mean?

80. What does it mean when we say that the equation $y = 2x + 4$ has *infinitely many* solutions?

81. On a quiz, students were asked to graph $y = 3x - 1$. One student made the table of solutions on the left below. Another student made the table of solutions on the right. Which table is incorrect? Or could they both be correct? Explain.

x	y	(x, y)
0	−1	(0, −1)
2	5	(2, 5)
3	8	(3, 8)

x	y	(x, y)
−2	−7	(−2, −7)
−1	−4	(−1, −4)
1	2	(1, 2)

82. Explain how the *cover-over method* can be used to quickly find the x- and y-intercepts of the graph of $3x - 2y = 12$.

83. To graph $y = -x + 1$, a student constructed a table of solutions and plotted the ordered pairs as shown. Instead of drawing a curve through the points, what should he have done?

$$y = -x + 1$$

x	y	(x, y)
−3	−2	(−3, −2)
0	1	(0, 1)
2	−1	(2, −1)

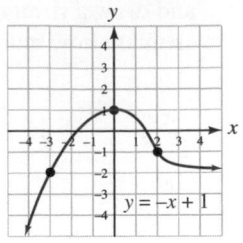

84. What is wrong with the graph of $x - y = 3$ shown below?

x	y	(x, y)
0	−3	(0, −3)
3	0	(3, 0)
1	−2	(1, −2)

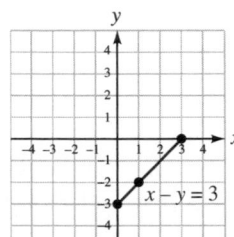

REVIEW

Find the prime factorization of each number.

85. 180   **86.** 270

Evaluate each expression for a = −2 and b = 3.

87. $\dfrac{3(b - a)}{5a + 7}$ **88.** $\dfrac{-2b^2 - b}{b}$

89. LIGHTNING The average flash of lightning lasts 0.25 second. Write the decimal as a fraction in simplest form.

90. Simplify: $4(a + 1) - 5(6 - a)$

STUDY SKILLS CHECKLIST

Know the Definitions

Before taking the test on Chapter 8, make sure that you have memorized the definitions of *mean*, *median*, *mode*, and *range*. Put a checkmark in the box if you can answer "yes" to the statement.

☐ I know that the *mean* of a set of values is often referred to as the *average*.

☐ I know that the *mean* of a set of values is given by the formula:

$$\text{Mean} = \frac{\text{sum of the values}}{\text{number of values}}$$

☐ I know that the *median* of a set of values is the middle value when they are arranged in increasing order.

☐ I know that the *range* of a set of values is the difference of the largest value and the smallest value.

☐ I know how to find the *median* of a set of values if there is an odd number of values.

$$2 \quad 4 \quad 5 \quad \underset{\uparrow}{8} \quad 10 \quad 13 \quad 14 \qquad 7 \text{ values}$$

Median = Middle value

☐ I know how to find the *median* of a set of values if there is an even number of values.

$$2 \quad 4 \quad 5 \quad \underline{8 \quad 10} \quad 13 \quad 14 \quad 16 \qquad 8 \text{ values}$$

$$\text{Median} = \frac{8 + 10}{2} = 9$$

☐ I know that the *mode* of a set of values is the value that occurs most often.

☐ I know that a set of values may have one *mode*, or more than one *mode*.

$$2 \quad 8 \quad 5 \quad 8 \quad 10 \quad 8 \quad 14 \qquad \text{mode: } 8$$

$$2 \quad 8 \quad 5 \quad 8 \quad 2 \quad 8 \quad 2 \qquad \text{two modes: } 2, 8$$

CHAPTER 8   SUMMARY AND REVIEW

SECTION 8.1   Reading Graphs and Tables

DEFINITIONS AND CONCEPTS	EXAMPLES
To read a **table** and locate a specific fact in it, we find the *intersection* of the correct row and column that contains the desired information.	SALARY SCHEDULES Find the annual salary for a teacher with a master's degree plus 15 additional units of study who is beginning her 4th year of teaching.

Teacher Salary Schedule

Step	BA	BA+15	BA+30	BA+45	MA	MA+15	MA+30
1	37,295	38,362	39,416	40,480	41,556	42,612	43,669
2	38,504	39,581	40,652	41,728	42,812	43,879	44,952
3	39,716	40,802	41,885	42,973	44,066	45,147	46,234
4	40,926	42,021	43,120	44,220	45,321	46,417	47,514
5	42,135	43,240	44,356	45,465	46,577	47,682	48,795
6	44,458	45,567	46,683	47,782	48,897	50,010	51,113
7	46,780	47,891	49,003	50,115	51,226	52,330	53,438

The annual salary is $46,417. It can be found by looking on the fourth row (labeled Step 4) in the 6th column (labeled MA + 15).

A **bar graph** presents data using vertical or horizontal bars. A **horizontal axis** and vertical axis serve to frame the graph and they are scaled in units such as years, dollars, minutes, pounds, and percent.

CANCER DEATHS Refer to the bar graph below. How many more deaths were caused by lung cancer than by colon cancer in the United States in 2007?

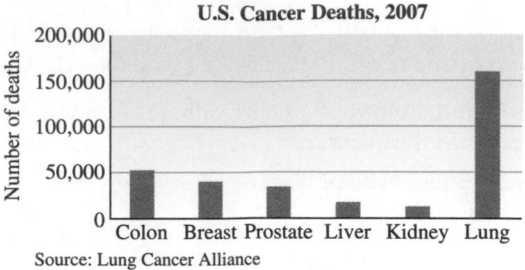

U.S. Cancer Deaths, 2007

Source: Lung Cancer Alliance

From the graph, we see that there were about 160,000 deaths caused by lung cancer and about 50,000 deaths from colon cancer. To find the difference, we subtract:

$$160,000 - 50,000 = 110,000$$

There were about 110,000 more deaths caused by lung cancer than deaths caused by colon cancer in the United States in 2007.

To compare sets of related data, groups of two (or three) bars can be shown. For **double-bar** or **triple-bar graphs,** a key is used to explain the meaning of each type of bar in a group.

SEAT BELTS Refer to the double-bar graph below. How did the percent of male high school students that rarely or never wore seat belts change from 2001 to 2007?

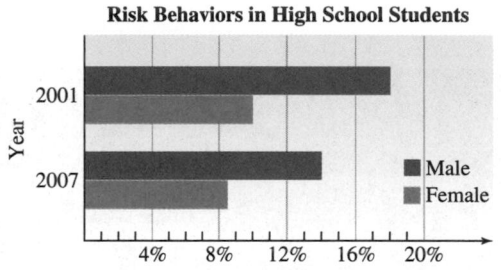

Risk Behaviors in High School Students

Percent that rarely or never wear seat belts

Source: *The World Almanac,* 2003, 2009

From the graph, we see that in 2001 about 18% of male high school students rarely or never wore seat belts. By 2007, the percent was about 14%, a decrease of 18% − 14%, or 4%.

A **pictograph** is like a bar graph, but the bars are made from pictures or symbols. A **key** tells the meaning (or value) of each symbol.

MEDICAL SCHOOLS Refer to the pictograph below. In 2008, how many students were enrolled in California medical schools?

Total Medical School Enrollment by State, 2008

California

Missouri

Virginia

= 1,000 medical students

The California row contains 4 complete symbols and almost all of another. This means that there were 4 · 1,000, or 4,000 medical students, plus approximately 900 more. In 2008, about 4,900 students were enrolled in California medical schools.

In a **circle graph,** regions called *sectors* (they look like slices of pizza) are used to show what part of the whole each quantity represents.

CHECKING E-MAIL The circle graph to the right shows the results of a survey of adults who were asked how many personal e-mail addresses they regularly check. What percent of the adults surveyed check 4 or more e-mail addresses regularly?

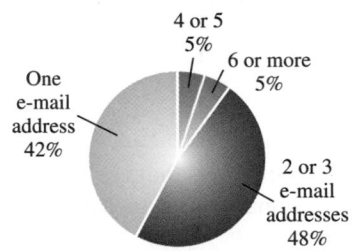

Source: Ipsos for Habeas

We add the percent of the responses for 4 or 5 e-mail addresses and the percent of the responses for 6 or more e-mail addresses:

$$5\% + 5\% = 10\%$$

Thus, 10% of the adults surveyed check 4 or more e-mail addresses regularly.

Use the survey results to predict the number of adults in a group of 5,000 that would check only one e-mail address regularly.

In the survey, 42% said they check only one e-mail address. We need to find:

What number	is	42%	of	5,000?
↓	↓	↓	↓	↓

$$x \;=\; 42\% \;\cdot\; 5{,}000 \qquad \text{Translate.}$$
$$x = 0.42 \cdot 5{,}000 \qquad \text{Write 42\% as a decimal.}$$
$$x = 2{,}100 \qquad \text{Do the multiplication.}$$

According to the survey, about 2,100 of the 5,000 adults would check only one e-mail address regularly.

A **line graph** is used to show how quantities change with time. From such a graph, we can determine when a quantity is increasing and when it is decreasing.

SNOWBOARDING The line graph below shows the number of people who participated in snowboarding in the United States for the years 2000–2007.

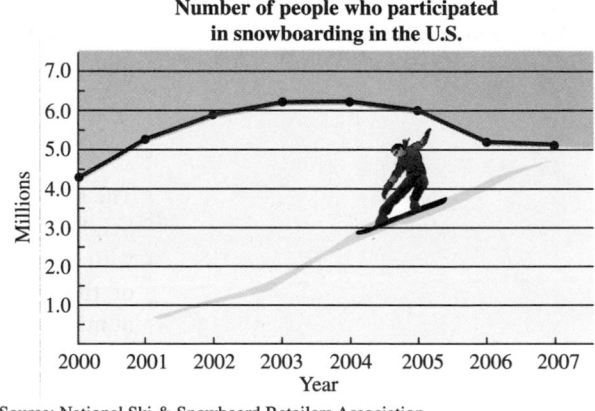

Number of people who participated in snowboarding in the U.S.

Source: National Ski & Snowboard Retailers Association

When did the popularity of snowboarding seem to peak?

The years with the highest participation were 2003 and 2004.

Between which two years was there the greatest decrease in the number of snowboarding participants?

The line segment with the greatest "fall" as we read left to right is the segment connecting the data points for the years 2005 and 2006. Thus, the greatest decrease in the number of snowboarding participants occurred between 2005 and 2006.

Two quantities that are changing with time can be compared by **drawing both lines on the same graph.**	SKATEBOARDING Refer to the line graphs below that show the results of a skateboarding race.

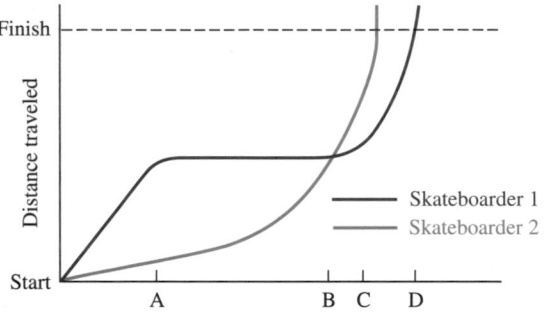

Observations:

- Since the red graph is well above the blue graph at time A, skateboarder 1 was well ahead of skateboarder 2 at that stage of the race.
- Since the red graph is horizontal from time A to time B, skateboarder 1 had stopped.
- Since the blue graph crosses the red graph at time B, at that instant, the skateboarders are tied for the lead.
- Since the blue graph crosses the dashed finish line at time C, which is sooner than time D, skateboarder 2 won the race.

A **histogram** is a bar graph with these features: **1.** The bars of the histogram touch. **2.** Data values never fall at the edge of a bar. **3.** The widths of the bars are equal and represent a range of values.	SLEEP A group of parents of junior high students were surveyed and asked to estimate the number of hours that their children slept each night. The results are displayed in the histogram to the right. How many children sleep 6 to 9 hours a night?

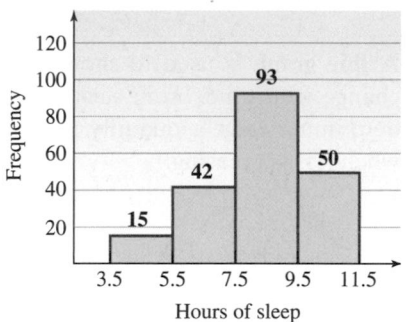

The bar with edges 5.5 and 7.5 corresponds to the 6 to 7 hour range. The height of that bar indicates that 42 children sleep 6 to 7 hours. The bar with edges 7.5 and 9.5 corresponds to the 8 to 9 hour range. The height of that bar indicates that 93 children sleep 8 to 9 hours. The total number of children sleeping 6 to 9 hours is found using addition:

$$42 + 93 = 135$$

135 of the junior high children sleep 6 to 9 hours a night.

A **frequency polygon** is a special line graph formed from a histogram by joining the center points at the top of each bar. On the horizontal axis, we write the coordinate of the middle value of each **class interval**. Then we erase the bars.

Frequency polygon

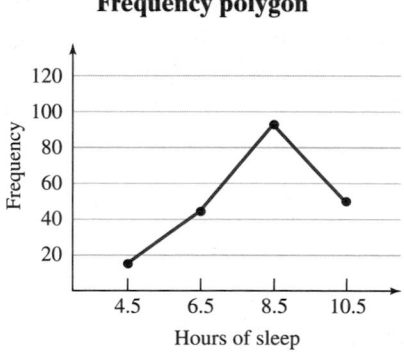

REVIEW EXERCISES

Refer to the table below to answer the following questions.

1. WINDCHILL TEMPERATURES

    a. Find the windchill temperature on a 10°F day when a 15-mph wind is blowing.

    b. Find the windchill temperature on a –15°F day when a 30-mph wind is blowing.

2. WIND SPEEDS

    a. The windchill temperature is −25°F, and the actual outdoor temperature is 15°F. How fast is the wind blowing?

    b. The windchill temperature is −38°F, and the actual outdoor temperature is –5°F. How fast is the wind blowing?

As of 2008, the United States had the most nuclear power plants in operation worldwide, with 104. The following bar graph shows the remainder of the top ten countries and the number of nuclear power plants they have in operation.

3. How many nuclear power plants does Korea have in operation?

4. How many nuclear power plants does France have in operation?

5. Which countries have the same number of nuclear power plants in operation? How many?

6. How many more nuclear power plants in operation does Japan have than Canada?

Number of Nuclear Power Plants in Operation

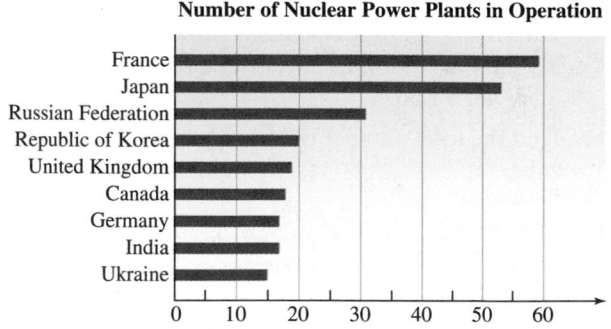

Source: International Atomic Energy Agency

Determining the Windchill Temperature

Wind speed	Actual temperature							
	20°F	15°F	10°F	5°F	0°F	−5°F	−10°F	−15°F
5 mph	16°	12°	7°	0°	−5°	−10°	−15°	−21°
10 mph	3°	−3°	−9°	−15°	−22°	−27°	−34°	−40°
15 mph	−5°	−11°	−18°	−25°	−31°	−38°	−45°	−51°
20 mph	−10°	−17°	−24°	−31°	−39°	−46°	−53°	−60°
25 mph	−15°	−22°	−29°	−36°	−44°	−51°	−59°	−66°
30 mph	−18°	−25°	−33°	−41°	−49°	−56°	−64°	−71°
35 mph	−20°	−27°	−35°	−43°	−52°	−58°	−67°	−74°

In a workplace survey, employed adults were asked if they would date a co-worker. The results of the survey are shown below. Use the double-bar graph to answer the following questions.

7. What percent of the women said they would not date a co-worker?

8. Did more men or women say that they would date a co-worker? What percent more?

9. When asked, were more men or more women unsure if they would date a co-worker?

10. Which of the three responses to the survey was given by approximately the same percent of men and women?

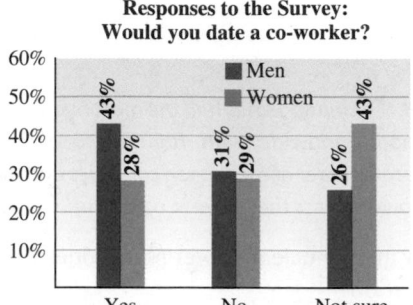

**Responses to the Survey:
Would you date a co-worker?**

Source: Spherion Workplace Survey

Refer to the pictograph below to answer the following questions.

11. How many animals are there at the San Diego Zoo?

12. Which of the zoos listed has the most animals? How many?

13. How many animals would have to be added to the Phoenix Zoo for it to have the same number as the San Diego Zoo?

14. Find the total number of animals in all three zoos.

**America's Best Zoos
Number of Animals**

San Diego Zoo

Columbus Zoo, Ohio

Phoenix Zoo

= 1,000 animals

Source: USA Travel Guide

Refer to the circle graph below to answer the following questions.

15. What element makes up the largest percent of the body weight of a human?

16. Elements *other than oxygen, carbon, hydrogen, and nitrogen* account for what percent of the weight of a human body?

17. Hydrogen accounts for how much of the body weight of a 135-pound woman?

18. Oxygen and carbon account for how much of the body weight of a 200-pound man?

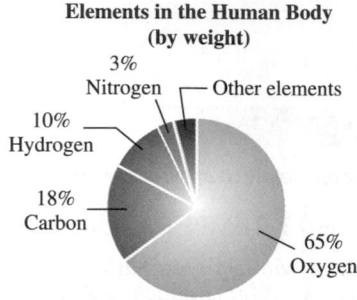

**Elements in the Human Body
(by weight)**

3% Nitrogen — Other elements

10% Hydrogen

18% Carbon

65% Oxygen

Source: General Chemistry Online

Refer to the line graph on the next page to answer the following questions.

19. How many eggs were produced in Nebraska in 2001?

20. How many eggs were produced in North Carolina in 2008?

21. In what year was the egg production of Nebraska equal to that of North Carolina? How many eggs?

22. What was the total egg production of Nebraska and North Carolina in 2005?

23. Between what two years did the egg production in North Carolina increase dramatically?

24. Between what two years did the egg production in Nebraska decrease dramatically?

25. How many more eggs did North Carolina produce in 2008 compared to Nebraska?

26. How many more eggs did Nebraska produce in 2000 compared to North Carolina?

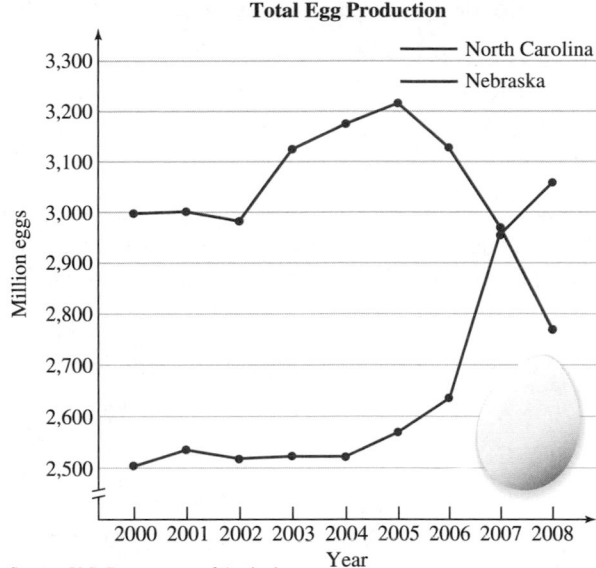

Total Egg Production

Source: U.S. Department of Agriculture

A survey of the weekly television viewing habits of 320 households produced the histogram in the next column. Use the graph to answer the following questions.

27. How many households watch between 1 and 5 hours of TV each week?

28. How many households watch between 6 and 15 hours of TV each week?

29. How many households watch 11 hours or more each week?

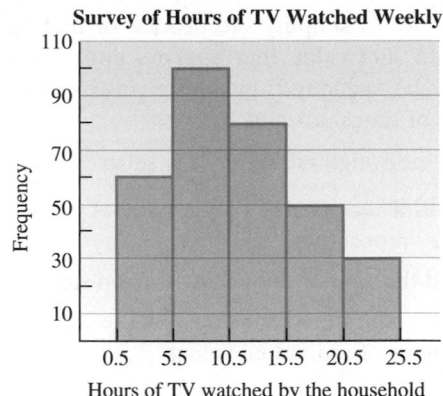

Survey of Hours of TV Watched Weekly

30. Create a frequency polygon from the histogram shown above.

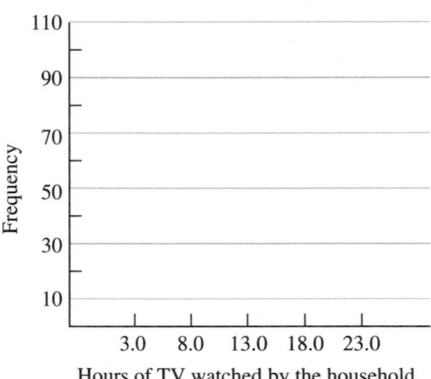

DEFINITIONS AND CONCEPTS	EXAMPLES
It is often beneficial to use one number to represent the "center" of all the numbers in a set of data. There are three measures of **central tendency:** mean, median, mode.	Find the mean of the following set of values:
	6 8 3 5 9 8 10 7 8 5
The **mean** of a set of values is given by the formula	To find the mean, we divide the sum of the values by the number of values, which is 10.
$$\text{Mean} = \frac{\text{sum of the values}}{\text{number of values}}$$	$$\frac{6 + 8 + 3 + 5 + 9 + 8 + 10 + 7 + 8 + 5}{10} = \frac{69}{10}$$ $$= 6.9$$
	Thus, 6.9 is the mean.

When a value in a set appears more than once, that value has a greater "influence" on the mean than another value that only occurs a single time. To simplify the process of finding the mean, any value that appears more than once can be "weighted" by multiplying it by the number of times it occurs.

To find the **weighted mean** of a set of values:

1. Multiply each value by the number of times it occurs.

2. Find the sum of the products from step 1.

3. Divide the sum from step 2 by the total number of individual values.

A student's **grade point average (GPA)** can be found using a weighted mean.

Some schools assign a certain number of **credit hours** to a course while others assign a certain number of **units.**

GPAs Find the semester grade point average for a student that received the following grades. (The point values are A = 4, B = 3, C = 2, D = 1, and F = 0.)

Course	Grade	Credits
Algebra	A	5
History	C	3
Art	D	4

Multiply the number of credits for each course by the point value of the grade received. Add the results (as shown in blue) to get the total number of grade points. To find the total number of credits, add as shown in red.

Course	Grade	Credits	Weighted grade points
Algebra	A	5	$4 \cdot 5 \rightarrow$ 20
History	C	3	$2 \cdot 3 \rightarrow$ 6
Art	D	+ 4	$1 \cdot 4 \rightarrow$ + 4
Totals		12	30

To find the GPA, we divide.

$$\text{GPA} = \frac{30}{12} \quad \begin{matrix} \leftarrow \text{The total number of grade points} \\ \leftarrow \text{The total number of credits} \end{matrix}$$

$$= 2.5 \quad \textit{Do the division.}$$

The student's semester GPA is 2.5.

To find the **median** of a set of values:

1. Arrange the values in increasing order.

2. If there is an odd number of values, the median is the middle value.

3. If there is an even number of values, the median is the mean (average) of the middle two values.

To find the median of

6   8   3   5   9   8   10   7   8   5

arrange them in increasing order:

Smallest   Largest

3   5   5   6   **7   8**   8   8   9   10   There are 10 values.

↑
Middle two values

Since there are an even number of values, the median is the mean (average) of the two middle values:

$$\frac{7 + 8}{2} = \frac{15}{2} = 7.5$$

Thus, 7.5 is the median.

The **range** of a set of values is the difference of the largest value and the smallest value.

The range of the data listed above is:

$$\text{range} = 10 - 3 = 7 \quad \textit{Subtract the smallest value, 3, from the largest value, 10.}$$

The **mode** of a set of values is the single value that occurs most often.	To find the mode of
	6 8 3 5 9 8 10 7 8 5
	we find the value that occurs most often.
	6 **8** 3 5 9 **8** 10 7 **8** 5
	3 times
	Since 8 occurs the most times, it is the mode.
When a collection of values has two modes, it is called **bimodal.**	The collection of values
	1 2 **3 3** 4 5 **6 6** 7 8
	has two modes: 3 and 6.

REVIEW EXERCISES

31. GRADES Jose worked hard this semester, earning grades of 87, 92, 97, 100, 100, 98, 90, and 98. If he needs a 95 average to earn an A in the class, did he make it?

32. GRADE SUMMARIES The students in a mathematics class had final averages of 43, 83, 40, 100, 40, 36, 75, 39, and 100. When asked how well her students did, their teacher answered, "43 was typical." What measure was the teacher using: mean, median, or mode?

33. PRETZEL PACKAGING Samples of SnacPak pretzels were weighed to find out whether the package claim "Net weight 1.2 ounces" was accurate. The tally appears in the table. Find the mode of the weights.

Weights of SnacPak Pretzels	
Ounces	Number
0.9	1
1.0	6
1.1	18
1.2	23
1.3	2
1.4	0

34. Find the mean weight and the range of the weights of the samples in Exercise 33.

35. BLOOD SAMPLES A medical laboratory technician examined a blood sample under a microscope and measured the sizes (in microns) of the white blood cells. The data are listed below. Find the mean, median, mode, and range.

7.8   6.9   7.9   6.7   6.8   8.0   7.2   6.9   7.5

36. SUMMER READING A paperback version of the classic *Gone With the Wind* is 960 pages long. If a student wants to read the entire book during the month of June, how many pages must she average per day?

37. WALK-A-THONS Use the data in the table to find the mean (average) donation to a charity walk-a-thon.

Donation amount	$5	$10	$20	$50	$100
Number received	20	65	25	5	10

38. GPAs Find the semester grade point average for a student that received the grades shown below. Round to the nearest hundredth. (Assume the following standard point values for the letter grades: A = 4, B = 3, C = 2, D = 1, and F = 0.)

Course	Grade	Credits
Chemistry	A	5
Sociology	C	3
Economics	D	4
Archery	A	1

DEFINITIONS AND CONCEPTS	EXAMPLES
A **solution of an equation in two variables** is an **ordered pair** of numbers that makes the equation a true statement when the numbers are substituted for the variables. An ordered pair: $(2, -3)$ $x\!\uparrow$ $\uparrow y$	Is $(2, -3)$ a solution of $2x - y = 7$? $2x - y = 7$ This is the given equation. $2(2) - (-3) \stackrel{?}{=} 7$ Substitute 2 for x and −3 for y. $4 + 3 \stackrel{?}{=} 7$ Evaluate the left side. $7 = 7$ Do the addition. Since $7 = 7$ is a true statement, $(2, -3)$ is a solution of $2x - y = 7$.
If only one coordinate of an ordered-pair solution is known: 1. Substitute it into the equation for the proper variable. 2. Solve the resulting equation to find the unknown coordinate.	To complete the solution $(\ \ , 8)$ for $3x + y = -1$, we substitute 8 for y and solve the resulting equation for x. $3x + y = -1$ This is the given equation. $3x + 8 = -1$ Substitute 8 for y. $3x + 8 - 8 = -1 - 8$ To isolate the variable term 3x, subtract 8 from both sides. $3x = -9$ Do the subtraction. $\dfrac{3x}{3} = \dfrac{-9}{3}$ To isolate x, undo the multiplication by 3 by dividing both sides by 3. $x = -3$ Do the division. This is the missing x-value of the solution. When $y = 8, x = -3$. The completed ordered pair is $(-3, 8)$.
Solutions of an equation in two variables can be listed in a **table of solutions.**	Two solutions of $y = x + 4$ are shown below. 	x	y	(x, y)	 |---|---|---| 	0	4	$(0, 4)$	 	6	10	$(6, 10)$	
A **rectangular coordinate system** is composed of a horizontal number line, called the **x-axis,** and a vertical number line, called the **y-axis.** The two axes intersect at the **origin.** To **plot** or **graph** ordered pairs means to locate their position on a rectangular coordinate system. The x- and y-axes divide the coordinate plane into four regions called **quadrants.**	Plot the points: $(2, 3), (-4, 2), (-3, -1), (0, -2.5),$ and $(4, -2)$ To graph each point, start at the origin and move the given number of units to the right or left on the x-axis and then the given number of units up or down. Then draw a dot and label it. 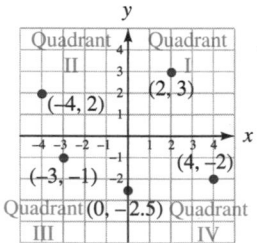

REVIEW EXERCISES

39. Is $(2, -3)$ a solution of $2x + 5y = -11$?

40. Is $(-3, 2)$ a solution of $y = 8x + 15$?

41. Complete the solutions of the equation
$3x - 4y = 12$: $(0, \quad)$ and $(\quad, 0)$

42. Complete the table of solutions for $y = -3x - 2$.

x	y	(x, y)
1		$(1, \quad)$
	-11	$(\quad, -11)$
-2		$(-2, \quad)$

43. Graph the points $(2, 3), (-3, 4), (-5, 0), (0, -4),$
$(-1.5, -3),$ and $\left(\frac{7}{2}, -1\right)$.

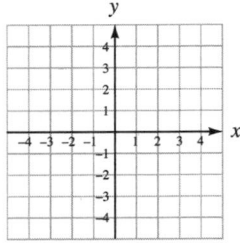

44. Give the coordinates of each point graphed below.

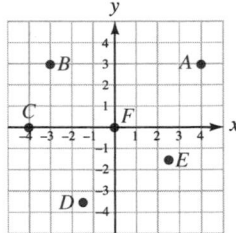

45. In what quadrant does the point $(-3, -4)$ lie?

46. THEATER SEATING Your ticket at the theater is
for seat B-10. Locate your seat on the diagram.

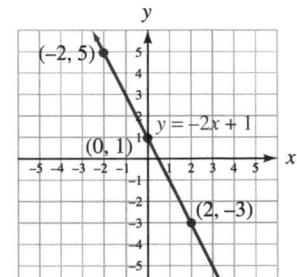

SECTION 8.4   Graphing Linear Equations

DEFINITIONS AND CONCEPTS	EXAMPLES

The **graph of an equation** in two variables is a mathematical "picture" of all of its solutions.

To **graph a linear equation** solved for y:

1. Find three solutions by selecting three values of x and finding the corresponding values of y.

2. Plot each ordered-pair solution.

3. Draw a straight line through the points.

Graph: $y = -2x + 1$

We construct a table of solutions, plot the points, and draw the line.

$y = -2x + 1$

x	y	(x, y)
-2	5	$(-2, 5)$
0	1	$(0, 1)$
2	-3	$(2, -3)$

↑ Select x   ↑ Find y   ↑ Plot (x, y)

The point where a line intersects the x-axis is called the **x-intercept.** The point where a line intersects the y-axis is called the **y-intercept.**

To **find the x-intercept,** substitute 0 for y and solve for x. To **find the y-intercept,** substitute 0 for x in the given equation and solve for y.

Plotting the x- and y-intercepts of a graph and drawing a line through them is called the **intercept method for graphing a line.**

Use the x- and y-intercepts to graph $3x + 2y = -6$.

x-intercept: $y = 0$	y-intercept: $x = 0$
$3x + 2y = -6$	$3x + 2y = -6$
$3x + 2(0) = -6$	$3(0) + 2y = -6$
$3x = -6$	$2y = -6$
$x = -2$	$y = -3$

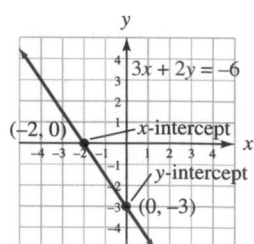

The x-intercept is $(-2, 0)$ and the y-intercept is $(0, -3)$.

As a check, a third point on the line can be found by selecting a convenient value for x and finding the corresponding value for y.

$$3x + 2y = -6$$

x	y	(x, y)	
-2	0	$(-2, 0)$	← x-intercept
0	-3	$(0, -3)$	← y-intercept
-4	3	$(-4, 3)$	← Check point

The equation $y = b$ represents the **horizontal line** that intersects the y-axis at $(0, b)$.

The equation $x = a$ represents the **vertical line** that intersects the x-axis at $(a, 0)$.

Graph: $y = 3$

x	y	(x, y)
-2	3	$(-2, 3)$
0	3	$(0, 3)$
2	3	$(2, 3)$

↑
Each value of y must be 3.

Graph: $x = -3$

x	y	(x, y)
-3	-2	$(-3, -2)$
-3	0	$(-3, 0)$
-3	2	$(-3, 2)$

↑
Each value of x must be -3.

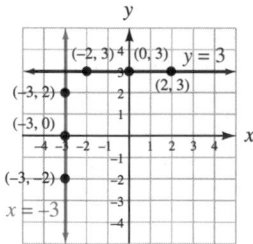

REVIEW EXERCISES

Graph each equation.

47. $y = 2x - 3$

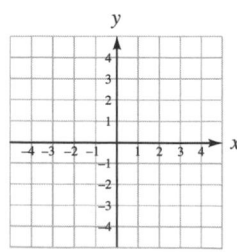

48. $y = \dfrac{1}{2}x - 1$

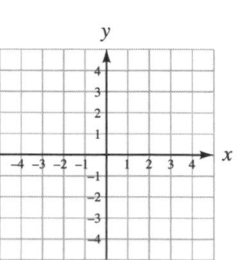

49. $y = -3x + 2$

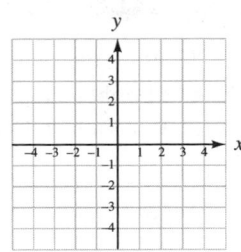

50. $y = 4x$

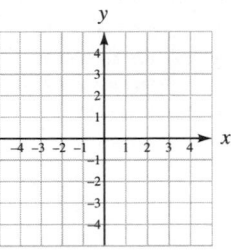

51. The graph of a linear equation is shown on the right. Name six ordered-pair solutions of the equations from the graph.

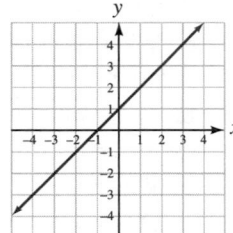

52. Identify the x- and y-intercepts of the graph shown on the right.

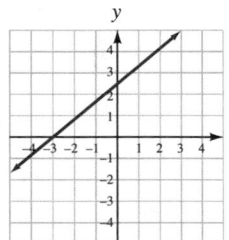

Use the intercept method to graph each equation.

53. $8x + 4y = -24$

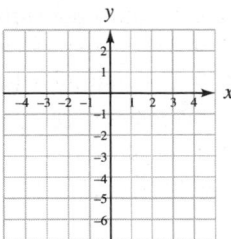

54. $30x - y = 30$

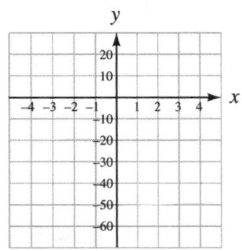

Graph each equation.

55. $y = 2$

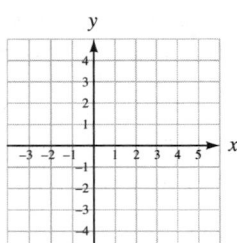

56. $x = 1$

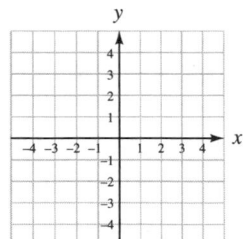

Fill in the blanks.

1. **a.** A horizontal or vertical line used for reference in a bar graph is called an _____.

    b. The _____ (average) of a set of values is the sum of the values divided by the number of values in the set.

    c. The _____ of a set of values written in increasing order is the middle value.

    d. The _____ of a set of values is the single value that occurs most often.

    e. The mean, median, and mode are three measures of _____ tendency.

2. **WORKOUTS** Refer to the table below to answer the following questions.

Number of Calories Burned While Running for One Hour

Running speed (mph)	Body Weight		
	130 lb	**155 lb**	**190 lb**
5	472	563	690
6	590	704	863
7	679	809	992
8	797	950	1,165
9	885	1,056	1,294

Source: nutristrategy.com

    a. How many calories will a 155-pound person burn if she runs for one hour at a rate of 5 mph?

    b. In one hour, how many more calories will a 190-pound person burn if he runs at a rate of 7 mph instead of 6 mph?

    c. At what rate does a 130-pound person have to run for one hour to burn approximately 800 calories?

3. **MOVING** Refer to the bar graph in the next column to answer the following questions.

    a. Which piece of furniture shown in the graph requires the greatest number of feet of bubble wrap? How much?

    b. How many more feet of bubble wrap is needed to wrap a desk than a coffee table?

    c. How many feet of bubble wrap is needed to cover a bedroom set that has a headboard, a dresser, and two end tables?

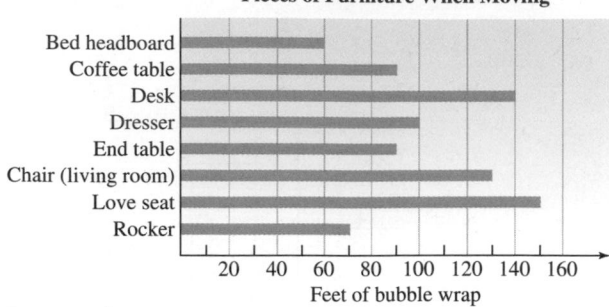

Source: transitsystems.com

4. **CANCER SURVIVAL RATES** Refer to the graph below to answer the following questions.

    a. What was the survival rate (in percent) from breast cancer in 1976?

    b. By how many percent did the cancer survival rate for breast cancer increase by 2006?

    c. Which type of cancer shown in the graph has the lowest survival rate?

    d. Which type of cancer has had the greatest increase in survival rate from 1976 to 2006? How much of an increase?

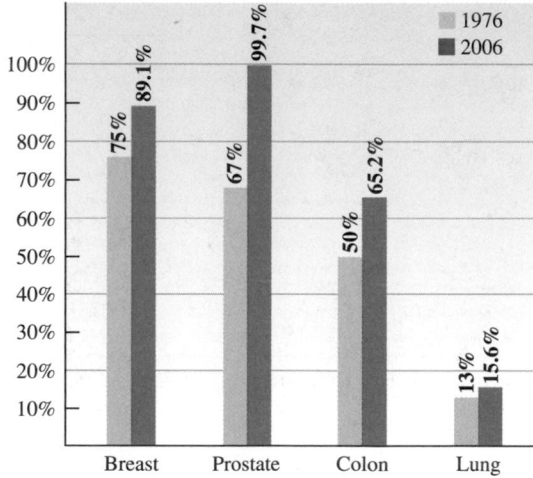

Source: SEER Cancer Statistics Review

5. ENERGY DRINKS Refer to the pictograph below to answer the following questions.

Sugar Content in Energy Drinks and Coffee
(12-ounce serving)

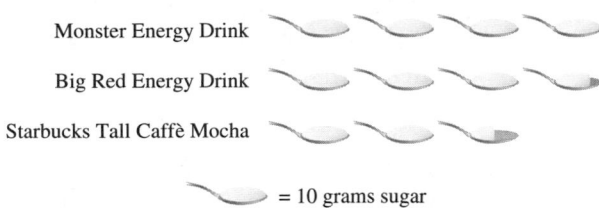

Monster Energy Drink

Big Red Energy Drink

Starbucks Tall Caffè Mocha

 = 10 grams sugar

Source: energyfiend.com

a. How many grams of sugar are there in 12 ounces of Big Red?

b. For a 12-ounce serving, how many more grams of sugar are there in Monster Energy Drink than in Starbucks Tall Caffè Mocha?

6. FIRES Refer to the graph below to answer the following questions.

a. In 2007, what percent of the fires in the United States were vehicle fires?

b. In 2007, there were a total of 1,557,500 fires in the United States. How many were structure fires?

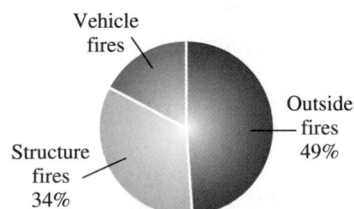

Where Fires Occurred, 2007

Vehicle fires

Outside fires
49%

Structure fires
34%

Source: U.S. Fire Administration

7. NYPD Refer to the graph in the next column to answer the following questions.

a. How many uniformed police officers did the NYPD have in 1987?

b. When was the number of uniformed police officers the least? How many officers were there at that time?

c. When was the number of uniformed police officers the greatest? How many officers were there at that time?

d. Find the decrease in the number of uniformed police officers from 2000 to 2003.

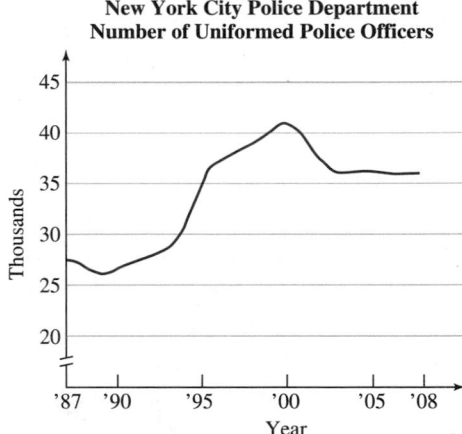

New York City Police Department
Number of Uniformed Police Officers

Thousands

'87 '90   '95   '00   '05 '08
Year

Source: *New York Times*, July 17, 2009

8. BICYCLE RACES Refer to the graph below to answer each of following questions about a two-man bicycle race.

a. Which bicyclist had traveled farther at time A?

b. Explain what was happening in the race at time B.

c. When was the first time that bicyclist 2 stopped to rest?

d. Did bicyclist 2 ever lead the race? If so, at what time?

e. Which bicyclist won the race?

Ten-Mile Bicycle Race

Finish

Distance traveled

Bicyclist 1
Bicyclist 2

Start
A   B C   D   Time

9. COMMUTING TIME A school district collected data on the number of minutes it took its employees to drive to work in the morning. The results are presented in the histogram below.

   a. How many employees have a commute time that is in the 7-to-10-minute range?

   b. How many employees have a commute time that is less than 10 minutes?

   c. How many employees have a commute that takes 15 minutes or more each day?

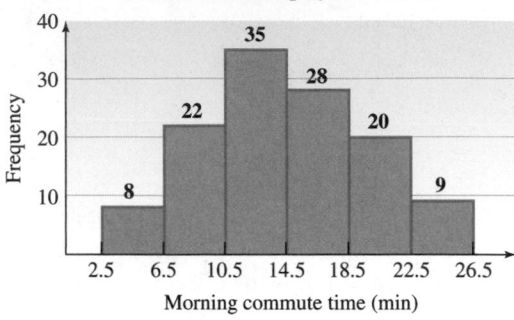

School District Employees' Commute

10. VOLUNTEER SERVICE The number of hours served last month by each of the volunteers at a homeless shelter are listed below:

    4 6 8 2 8 10 11 9 5 12 5 18 7 5 1 9

    a. Find the mean (average) of the hours of volunteer service.

    b. Find the median of the hours of volunteer service.

    c. Find the mode of the hours of volunteer service.

    d. Find the range of the number of hours of volunteer service.

11. RATING MOVIES Netflicks, a popular online DVD rental system, allows members to rate movies using a 5-star system. The table below shows a tally of the ratings that a group of college students gave a movie. Find the mean (average) rating of the movie.

Number of Stars	Comments	Tally
★ ★ ★ ★ ★	Loved it	III
★ ★ ★ ★	Really liked it	IIII
★ ★ ★	Liked it	⊞
★ ★	Didn't like it	⊞ I
★	Hated it	II

12. GPAs Find the semester grade point average for a student who received the following grades. Round to the nearest hundredth.

Course	Grade	Credits
WEIGHT TRAINING	C	1
TRIGONOMETRY	A	3
GOVERNMENT	B	2
PHYSICS	A	4
PHYSICS LAB	D	1

13. RATINGS The seven top-rated cable television programs for the week of March 30–April 5, 2009, are given below. What are the mean, median, mode, and range of the viewer data?

Show/day/time/network	Millions of viewers
WCW Raw, Mon. 10 P.M., USA	5.39
WCW Raw, Mon. 9 P.M., USA	4.99
NCIS, Tue. 7 P.M., USA	4.25
NCIS, Wed. 7 P.M., USA	4.25
NCIS, Mon. 7 P.M., USA	4.04
Penguins of Madagascar, Sun. 10 A.M., Nickelodeon	4.02
The O'Reilly Factor, Wed. 8 P.M., Fox	3.93

Source: Bay Ledger News Zone

14. REAL ESTATE In May of 2009, the median sales price of an existing single-family home in the United States was $172,900. Explain what is meant by the median sales price. (Source: National Association of Realtors)

15. Is $(-1, 2)$ a solution of $4x + 5y = 6$?

16. Is $(3, -2)$ a solution of $y = 12x - 34$?

17. Complete the ordered pairs so that each one is a solution of the equation $x - 2y = 4$.

    $(0, \boxed{})$, $(\boxed{}, 0)$ and $(2, \boxed{})$

18. Complete the table of solutions for $x - 3y = -3$.

x	y	(x, y)
0		$(0,)$
	2	$(, 2)$
-3		$(-3,)$

19. PANTS SALE In the illustration below, an X indicates the sizes of jeans that a store has in stock. List the jeans sizes that are not available as ordered pairs of the form (waist, length).

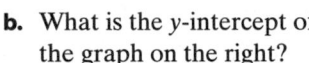

These pants in your size or they're free. Guaranteed!*

Stonewash jeans, $31.99-$39.99

*Our size guarantee is good only for the following sizes:

Waist	30	31	32	33	34	36	38
Length 30	X	X	X	X	X	X	
32		X	X	X	X	X	X
34			X	X	X	X	X

20. Graph the points: $(0, -4.5)$, $\left(\frac{3}{2}, 1\right)$, $(4, 2)$, $(-1, 3)$, $(-2, 0)$, and $(4, -3)$

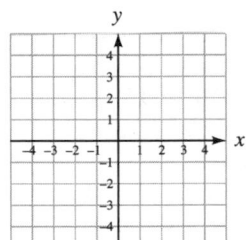

21. Give the coordinates of each point on the graph.

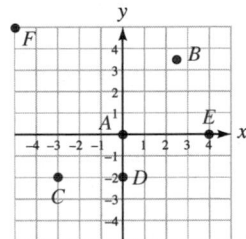

22. In what quadrant does the point $(-7, -1)$ lie?

Graph each equation.

23. $y = 4x - 2$

24. $y = -\frac{3}{2}x - 1$

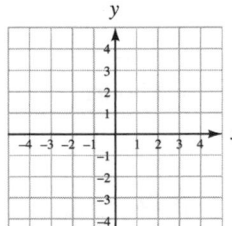
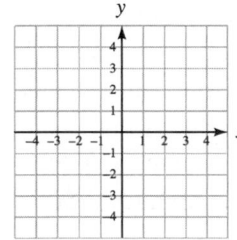

25. a. What is the x-intercept of the graph on the right?

b. What is the y-intercept of the graph on the right?

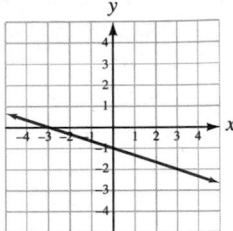

26. The graph of a linear equation is shown on the right. Name three ordered-pair solutions of the equation from the graph.

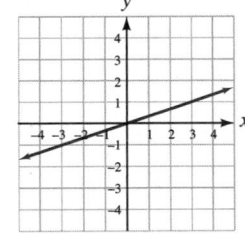

Use the intercept method to graph each equation.

27. $3x + 4y = 12$

28. $20x + y = -20$

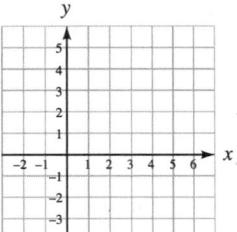
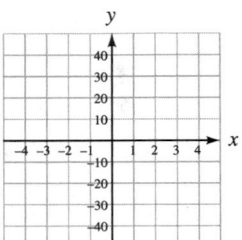

Graph each equation.

29. $y = -2$

30. $x = 3$

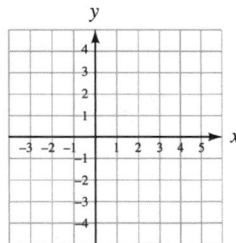
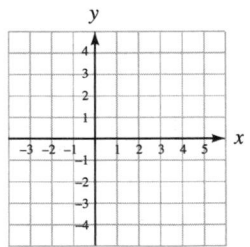

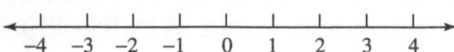

1. AUTOMOBILES In 2008, a total of 52,940,559 cars were produced in the world. Write this number in words and in expanded notation. (Source: Worldometers) [Section 1.1]

2. Round 59,999 to the nearest hundred. [Section 1.1]

Perform each operation.

3. $\begin{array}{r} 48,908 \\ + \ 5,696 \end{array}$ [Section 1.2]

4. $\begin{array}{r} 8,700 \\ - \ 5,491 \end{array}$ [Section 1.2]

5. $\begin{array}{r} 408 \\ \times \ 67 \end{array}$ [Section 1.3]

6. $87\overline{)2,001}$ [Section 1.4]

7. Explain how to check the following result using addition. [Section 1.2]

$$\begin{array}{r} 2,142 \\ - \ 459 \\ \hline 1,683 \end{array}$$

8. GEOMETRY Find the perimeter and the area of the rectangle shown below. [Section 1.3]

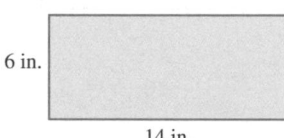

6 in.

14 in.

9. **a.** Find the factors of 36. [Section 1.5]

   b. Write the first ten prime numbers. [Section 1.5]

   c. Find the prime factorization of 36. [Section 1.5]

10. **a.** Find the LCM of 8 and 12. [Section 1.6]

    b. Find the GCF of 8 and 12.

Evaluate each expression. [Section 1.7]

11. $15 + 5\left[12 - (2^2 + 4)\right]$

12. $\dfrac{12 + 5 \cdot 3}{3^2 - 2 \cdot 3}$

Solve each equation and check the result.

13. $96 = m - 83$ [Section 1.8]

14. $73n = 219$ [Section 1.9]

15. Graph the integers greater than -3 but less than 4. [Section 2.1]

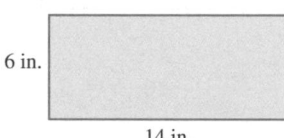

16. **a.** Simplify: $-(-6)$ [Section 2.1]

    b. Find the absolute value: $|-5|$

    c. Is the statement $-12 > -10$ true or false?

17. *Perform each operation.*

    a. $-35 + 5$ [Section 2.2]

    b. $-35 - (-5)$ [Section 2.3]

    c. $-35(5)$ [Section 2.4]

    d. $\dfrac{-35}{-5}$ [Section 2.5]

18. PLANETS Mercury orbits closer to the sun than does any other planet. Temperatures on Mercury can get as high as 810°F and as low as -290°F. What is the temperature range? [Section 2.3]

Evaluate each expression. [Section 2.6]

19. $\dfrac{(-6)^2 - 1^5}{-4 - 3}$

20. $-3 + 3(-4 - 4 \cdot 2)^2$

21. $-\left|\dfrac{45}{-9} - (-9)\right|$

22. $-10^2 - (-10)^2$

23. Solve $\dfrac{x}{-4} + 4 = -5$ and check the result. [Section 2.7]

Write and then solve an equation to answer the following question.

24. WEATHER FORECASTS The weather forecast for Barrow, Alaska, warned listeners that the daytime high temperature of 3° below zero would drop to a nighttime low of 31° below. By how many degrees did the temperature fall overnight? [Section 2.8]

25. Translate to mathematical symbols: 4 less than the square of x. [Section 3.1]

26. ROAD TRIPS Find the distance covered by a car traveling 45 miles per hour for 5 hours. [Section 3.2]

27. **a.** Simplify: $(-9t)(-8)$ [Section 3.3]

    b. Multiply: $-5(3x - 2y - 10)$ [Section 3.3]

28. Combine like terms. [Section 3.4]

    a. $x + x + x + x$

    b. $5(3 - 2x) + 4(2 - 3x) + 19x$

29. Solve $8 + 4(2a - 2) = -16 + 4a$ and check the result. [Section 3.5]

Write and then solve an equation to answer the following question. [Section 3.6]

30. CIVIL SERVICE A candidate for a position with the Department of Homeland Security scored 4 points higher on the written part of the civil service exam than he did on his interview. If his combined score was 98, what were his scores on the interview and on the written part of the exam?

31. Simplify each fraction. [Section 4.1]

    a. $\dfrac{60}{108}$ **b.** $\dfrac{24a^3}{16a}$

32. Simplify, if possible. [Section 4.1]

    a. $\dfrac{0}{64}$ **b.** $\dfrac{27}{0}$

Perform each operation. Simplify, if possible.

33. $\dfrac{4}{5} \cdot \dfrac{2}{7}$ [Section 4.2]

34. $\dfrac{8m^2}{63n^5} \div \dfrac{2m^2}{n^4}$ [Section 4.3]

35. Subtract $\dfrac{2}{3}$ from $\dfrac{1}{2}$. [Section 4.4]

36. $\dfrac{11}{12} + \dfrac{1}{30}$ [Section 4.4]

37. $\dfrac{x}{9} - \dfrac{1}{8}$ [Section 4.4]

38. CLASS TIME In a chemistry course, students spend a total of 300 minutes in lab and lecture each week. If $\frac{7}{15}$ of the time is spent in lab each week, how many minutes are spent in *lecture* each week? [Section 4.3]

39. Divide: $2\dfrac{4}{5} \div \left(-2\dfrac{2}{3}\right)$ [Section 4.5]

40. TENNIS Find the length of the handle on the tennis racquet shown below. [Section 4.6]

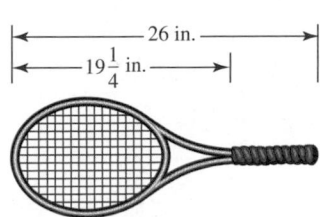

41. Evaluate the formula $A = \frac{1}{2}h(a + b)$ for $a = 4\frac{1}{2}$, $b = 5\frac{1}{2}$, and $h = 2\frac{1}{8}$. [Section 4.7]

42. Simplify the complex fraction: $\dfrac{-\dfrac{1}{5}}{\dfrac{8}{15}}$ [Section 4.7]

Solve each equation and check the result. [Section 4.8]

43. $-\dfrac{5}{6}y = -2$

44. $\dfrac{a}{5} - \dfrac{5}{6} = \dfrac{a}{3}$

Write and then solve an equation to answer the following question. [Section 4.8]

45. TEAM ROSTERS One-third of the players on a women's basketball team are forwards and one-fifth are centers. The remaining 7 players are guards. How many players are on the team?

46. Write $400 + 20 + 8 + \frac{9}{10} + \frac{1}{100}$ as a decimal. [Section 5.1]

47. CHECKBOOKS Find the total dollar amount of checks written in the register shown below. [Section 5.2]

DATE	CHECK NUMBER	TRANSACTION DESCRIPTION	✓ T	(•) AMOUNT OF PAYMENT OR DEBIT
3 17	703	TO: *Albertsons*		$ 213 16
		FOR: *Groceries*		
3 19	704	TO: *Brian Auto*		$1,504 80
		FOR: *Car Repair*		
3 19	705	TO: *Nordstrom*		$ 89 73
		FOR: *Sweater*		
3 21	706	TO: *Girl Scouts*		$ 7 50
		FOR: *Cookies*		

48. Perform each operation in your head. [Section 5.3]

    a. Multiply: $3.45 \cdot 100$

    b. Divide: $3.45 \div 10,000$

Perform each operation.

49. Subtract: $\begin{array}{r} 760.2 \\ -\ 614.7 \end{array}$ [Section 5.2]

50. Multiply: $(-0.31)(2.4)$ [Section 5.3]

51. Divide: $0.72\overline{)536.4}$ [Section 5.4]

52. Divide: $4\overline{)0.073}$ [Section 5.4]

53. Write $\dfrac{8}{11}$ as a decimal. [Section 5.5]

54. Evaluate: $15 + \sqrt{16}\left[5^2 - \left(\sqrt{9} + 2\right)\sqrt{4}\right]$
[Section 5.6]

55. Solve $2(3.6t - 4.1) + 0.9t = 16.1$ and check the result. [Section 5.7]

Write and then solve an equation to answer the following question. [Section 5.7]

56. BUSINESS EXPENSES A business decides to rent a copy machine instead of buying one. Under the rental agreement, the company is charged a basic rental fee of $35 per month plus 3¢ for every copy made. If the business has budgeted $110 for copier expenses each month, how many copies can be made before exceeding the budget?

57. Express the phrase "8 feet to 4 yards" as a ratio in simplest form. [Section 6.1]

58. CLOTHES SHOPPING As part of a summer clearance, a women's store put turtleneck sweaters on sale, 3 for $35.97. How much will five turtleneck sweaters cost? [Section 6.2]

59. Solve the proportion: $\dfrac{\frac{7}{8}}{\frac{1}{2}} = \dfrac{\frac{1}{4}}{x}$ [Section 6.2]

60. Convert 8 pints to fluid ounces. [Section 6.3]

61. Convert 640 centimeters to meters. [Section 6.4]

62. Convert 67.7°F to degrees Celsius. Round to the nearest tenth. [Section 6.5]

63. Complete the table below. [Section 7.1]

Fraction	Decimal	Percent
		3%
$\frac{9}{4}$		
	0.041	

64. 90 is what percent of 525? Round to the nearest one percent. [Section 7.2]

65. What number is 105% of 23.2? [Section 7.2]

66. 19.2 is $33\frac{1}{3}\%$ of what number? [Section 7.2]

67. SALES TAX Find the sales tax on a purchase of $98.95 if the sales tax rate is 8%. [Section 7.3]

68. SELLING ELECTRONICS If the commission on a $1,500 laptop computer is $240, what is the commission rate? [Section 7.3]

69. TIPPING Estimate the 15% tip on a $77.55 dinner bill. [Section 7.4]

70. REMODELING A homeowner borrows $18,000 to pay for a kitchen remodeling project. The terms of the loan are 9.2% annual simple interest and repayment in 2 years. How much interest will be paid on the loan? [Section 7.5]

71. LOANS $12,600 is loaned at an annual simple interest rate of 18%. Find the total amount that must be repaid at the end of a 90-day period. [Section 7.5]

72. SPINAL CORD INJURIES Refer to the circle graph below. [Section 8.1]

   a. What percent of spinal cord injuries are caused by sports accidents?

   b. If there are approximately 12,000 new cases of spinal cord injury each year, according to the graph, how many of them were caused by motor vehicle crashes?

Causes of Spinal Cord Injury in the United States

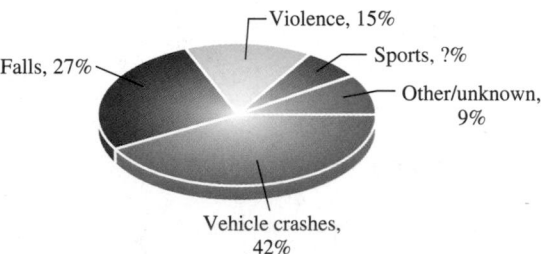

Source: National Spinal Cord Injury Statistical Center

73. AVALANCHES The bar graph shows the number of deaths from avalanches in the United States for the winter seasons ending in the years 2000 to 2009. Use the graph to answer the following questions. [Section 8.1]

   a. In which year were there the most deaths from avalanches? How many deaths were there?

   b. Between what two years was there the greatest increase in the number of deaths from avalanches? What was the increase?

   c. Between what two years was there the greatest decrease in the number of deaths from avalanches? What was the decrease?

U.S. Annual Avalanche Deaths

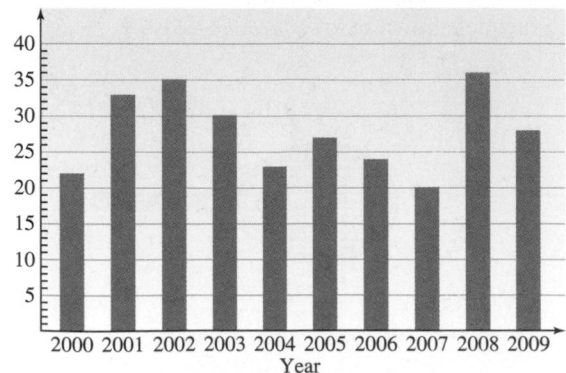

Source: Northwest Weather and Avalanche Center

74. TEAM GPA The grade point averages of the players on a badminton team are listed below. Find the mean, median, mode, and range of the team's GPAs. [Section 8.2]

  3.04   4.00   2.75   3.23   3.87   2.21
  3.02   2.25   2.98   2.56   3.58   2.75

75. Is $(-2, 3)$ a solution of $4x - 5y = -23$? [Section 8.3]

76. Complete the solution (, 4) for the equation $4x - 5y = -4$. [Section 8.3]

77. Graph the points: $(-1, 3), (0, 1.5), (-4, -4), \left(2, \frac{7}{2}\right)$, and $(4, 0)$ [Section 8.3]

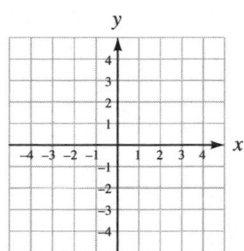

78. Graph: $x = 4$ [Section 8.4]

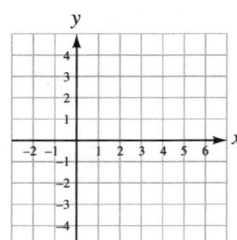

79. Graph: $y = -x - 1$ [Section 8.4]

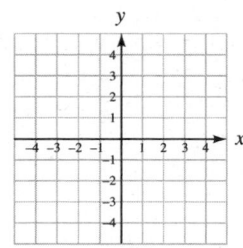

80. Graph: $3x - 3y = 9$ [Section 8.4]

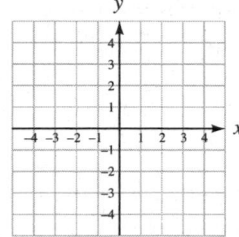

An Introduction to Geometry

© iStockphoto.com/Lukaz Laska

from *Campus to Careers*

Surveyor

Surveyors measure distances, directions, elevations (heights), contours (curves), and angles between lines on Earth's surface. Surveys are also done in the air and underground. Surveyors often work in teams. They use a variety of instruments and electronics, including the Global Positioning System (GPS). In general, people who like surveying also like math—primarily geometry and trigonometry. The field attracts people with geology, forestry, history, engineering, computer science, and astronomy backgrounds, too.

In **Problem 83** of **Study Set 9.5,** you will see how a surveyor, using geometry, can stay on dry land and yet measure the width of a river.

JOB TITLE: Surveyor

EDUCATION: Courses in algebra, geometry, trigonometry, and computer science are required.

JOB OUTLOOK: Job growth is expected to be 21% through 2016—much faster than the average for all occupations.

ANNUAL EARNINGS: In 2008, the annual median income was $53,120.

FOR MORE INFORMATION: http://www.bls.gov/k12/math03.htm

Objectives

1 Identify and name points, lines, and planes.

2 Identify and name line segments and rays.

3 Identify and name angles.

4 Use a protractor to measure angles.

5 Solve problems involving adjacent angles.

6 Use the property of vertical angles to solve problems.

7 Solve problems involving complementary and supplementary angles.

© INTERFOTO/Alamy

SECTION 9.1
Basic Geometric Figures; Angles

Geometry is a branch of mathematics that studies the properties of two- and three-dimensional figures such as triangles, circles, cylinders, and spheres. More than 5,000 years ago, Egyptian surveyors used geometry to measure areas of land in the flooded plains of the Nile River after heavy spring rains. Even today, engineers marvel at the Egyptians' use of geometry in the design and construction of the pyramids. History records many other practical applications of geometry made by Babylonians, Chinese, Indians, and Romans.

> **The Language of Algebra**   The word *geometry* comes from the Greek words *geo* (meaning earth) and *metron* (meaning measure).

Many scholars consider **Euclid** (330?–275? BCE) to be the greatest of the Greek mathematicians. His book *The Elements* is an impressive study of geometry and number theory. It presents geometry in a highly structured form that begins with several simple assumptions and then expands on them using logical reasoning. For more than 2,000 years, *The Elements* was the textbook that students all over the world used to learn geometry.

1 Identify and name points, lines, and planes.

Geometry is based on three undefined words: *point, line,* and *plane.* Although we will make no attempt to define these words formally, we can think of a **point** as a geometric figure that has position but no length, width, or depth. Points can be represented on paper by drawing small dots, and they are labeled with capital letters. For example, point A is shown in figure (a) below.

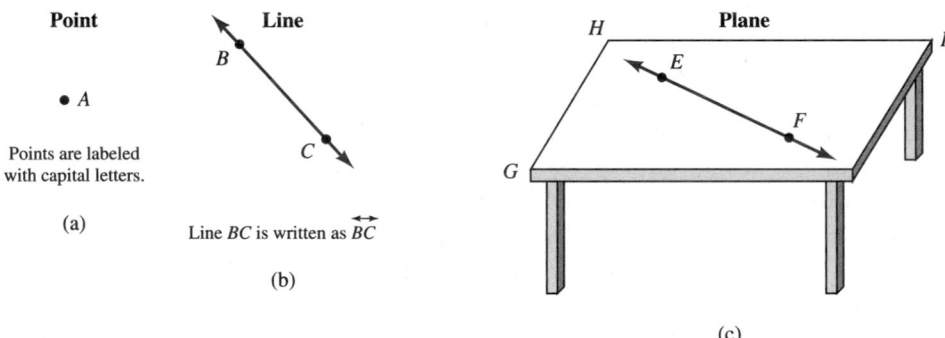

Lines are made up of points. A line extends infinitely far in both directions, but has no width or depth. Lines can be represented on paper by drawing a straight line with arrowheads at either end. We can name a line using any two points on the line. In figure (b) above, the line that passes through points B and C is written as $\overleftrightarrow{BC}$.

Planes are also made up of points. A plane is a flat surface, extending infinitely far in every direction, that has length and width but no depth. The top of a table, a floor, or a wall is part of a plane. We can name a plane using any three points that lie in the plane. In figure (c) above, $\overleftrightarrow{EF}$ lies in plane GHI.

As figure (b) illustrates, points B and C determine exactly one line, the line $\overleftrightarrow{BC}$. In figure (c), the points E and F determine exactly one line, the line $\overleftrightarrow{EF}$. In general, any two different points determine exactly one line.

As figure (c) illustrates, points *G, H,* and *I* determine exactly one plane. In general, any three different points determine exactly one plane.

Other geometric figures can be created by using parts or combinations of points, lines, and planes.

2 Identify and name line segments and rays.

> ### Line Segment
>
> The **line segment** *AB,* written as $\overline{AB}$, is the part of a line that consists of points *A* and *B* and all points in between (see the figure below). Points *A* and *B* are the **endpoints** of the segment.

Line segment

Line segment *AB* is written as $\overline{AB}$.

Every line segment has a **midpoint,** which divides the segment into two parts of equal length. In the figure below, *M* is the midpoint of segment *AB,* because the measure of $\overline{AM}$, which is written as m$(\overline{AM})$, is equal to the measure of $\overline{MB}$ which is written as m$(\overline{MB})$.

$$\text{m}(\overline{AM}) = 4 - 1$$
$$= 3$$

and

$$\text{m}(\overline{MB}) = 7 - 4$$
$$= 3$$

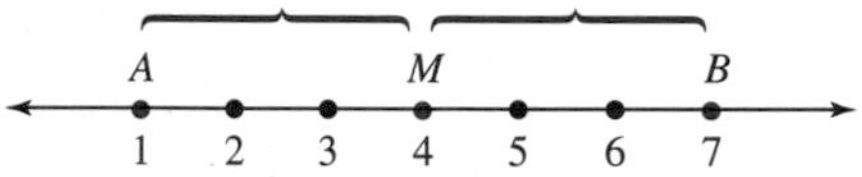

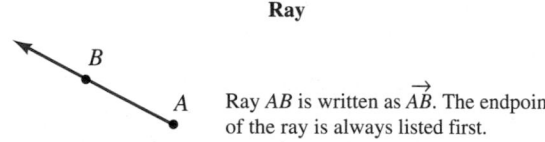

Since the measure of both segments is 3 units, we can write m$(\overline{AM})$ = m$(\overline{MB})$.

When two line segments have the same measure, we say that they are **congruent.** Since m$(\overline{AM})$ = m$(\overline{MB})$, we can write

$$\overline{AM} \cong \overline{MB} \quad \text{Read the symbol} \cong \text{as "is congruent to."}$$

Another geometric figure is the ray, as shown below.

> ### Ray
>
> A **ray** is the part of a line that begins at some point (say, *A*) and continues forever in one direction. Point *A* is the **endpoint** of the ray.

Ray

Ray *AB* is written as $\overrightarrow{AB}$. The endpoint of the ray is always listed first.

To name a ray, we list its endpoint and then one other point on the ray. Sometimes it is possible to name a ray in more than one way. For example, in the figure on the right, $\overrightarrow{DE}$ and $\overrightarrow{DF}$ name the same ray. This is because both have point *D* as their endpoint and extend forever in the same direction. In contrast, $\overrightarrow{DE}$ and $\overrightarrow{ED}$ are not the same ray. They have different endpoints and point in opposite directions.

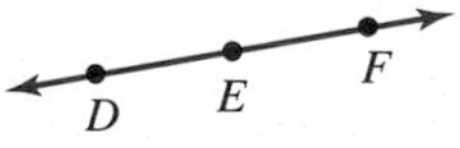

3 Identify and name angles.

> **Angle**
>
> An **angle** is a figure formed by two rays with a common endpoint. The common endpoint is called the **vertex,** and the rays are called **sides.**

The angle shown below can be written as $\angle BAC$, $\angle CAB$, $\angle A$, or $\angle 1$. The symbol $\angle$ means angle.

Angle

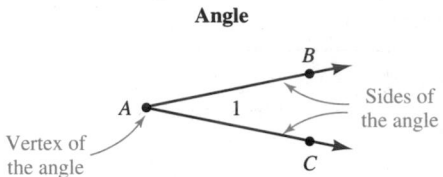

Vertex of the angle

Sides of the angle

> **Caution!** When using three letters to name an angle, be sure the letter name of the vertex is the middle letter. Furthermore, we can only name an angle using a single vertex letter when there is no possibility of confusion. For example, in the figure on the right, we cannot refer to any of the angles as simply $\angle X$, because we would not know if that meant $\angle WXY$, $\angle WXZ$, or $\angle YXZ$.

4 Use a protractor to measure angles.

One unit of measurement of an angle is the **degree.** The symbol for degree is a small raised circle, °. An angle measure of 1° (read as "one degree") means that one side of an angle is rotated $\frac{1}{360}$ of a complete revolution about the vertex from the other side of the angle. The measure of $\angle ABC$, shown below, is 1°. We can write this in symbols as m($\angle ABC$) = 1°.

This side of the angle is rotated $\frac{1}{360}$ of a complete revolution from the other side of the angle.

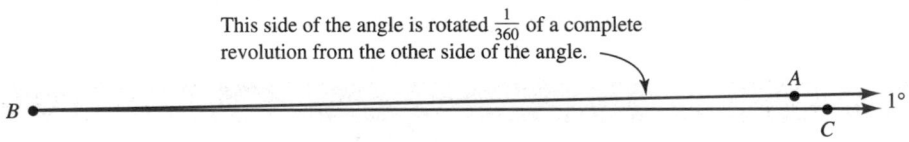

The following figures show the measures of several other angles. An angle measure of 90° is equivalent to $\frac{90}{360} = \frac{1}{4}$ of a complete revolution. An angle measure of 180° is equivalent to $\frac{180}{360} = \frac{1}{2}$ of a complete revolution, and an angle measure of 270° is equivalent to $\frac{270}{360} = \frac{3}{4}$ of a complete revolution.

m($\angle FED$) = 90°   m($\angle IHG$) = 180°   m($\angle JKL$) = 270°

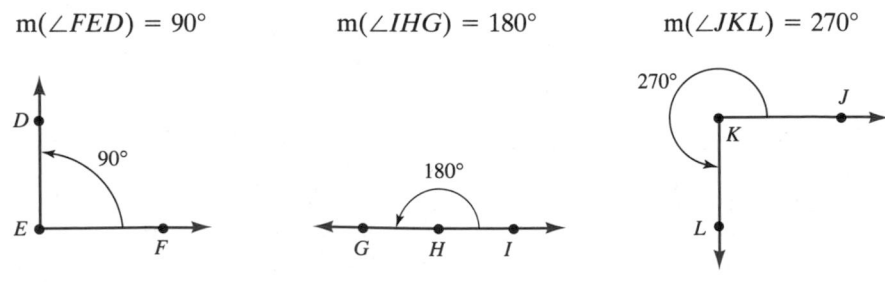

We can use a **protractor** to measure angles. To begin, we place the center of the protractor at the vertex of the angle, with the edge of the protractor aligned with one side of the angle, as shown below. The angle measure is found by determining where the other side of the angle crosses the scale. Be careful to use the appropriate scale, inner or outer, when reading an angle measure.

If we read the protractor from right to left, using the outer scale, we see that m($\angle ABC$) = 30°. If we read the protractor from left to right, using the inner scale, we can see that m($\angle GBF$) = 30°.

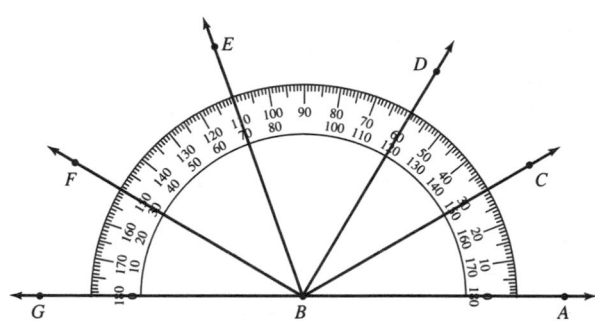

Angle	Measure in degrees
$\angle ABC$	30°
$\angle ABD$	60°
$\angle ABE$	110°
$\angle ABF$	150°
$\angle ABG$	180°
$\angle GBF$	30°
$\angle GBC$	150°

When two angles have the same measure, we say that they are **congruent.** Since m($\angle ABC$) = 30° and m($\angle GBF$) = 30°, we can write

$\angle ABC \cong \angle GBF$ Read the symbol $\cong$ as "is congruent to."

We classify angles according to their measure.

Classifying Angles

Acute angles: Angles whose measures are greater than 0° but less than 90°.

Right angles: Angles whose measures are 90°.

Obtuse angles: Angles whose measures are greater than 90° but less than 180°.

Straight angles: Angles whose measures are 180°.

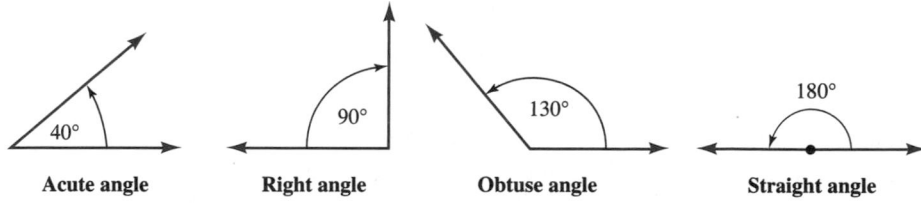

| Acute angle | Right angle | Obtuse angle | Straight angle |

The Language of Algebra   A ⌐ symbol is often used to label a right angle. For example, in the figure on the right, the ⌐ symbol drawn near the vertex of $\angle ABC$ indicates that m($\angle ABC$) = 90°.

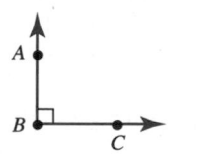

Self Check 1

Classify $\angle EFG$, $\angle DEF$, $\angle 1$, and $\angle GED$ in the figure as an acute angle, a right angle, an obtuse angle, or a straight angle.

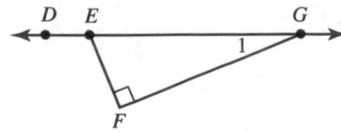

Now Try Problems 57, 59, and 61

EXAMPLE 1 Classify each angle in the figure as an acute angle, a right angle, an obtuse angle, or a straight angle.

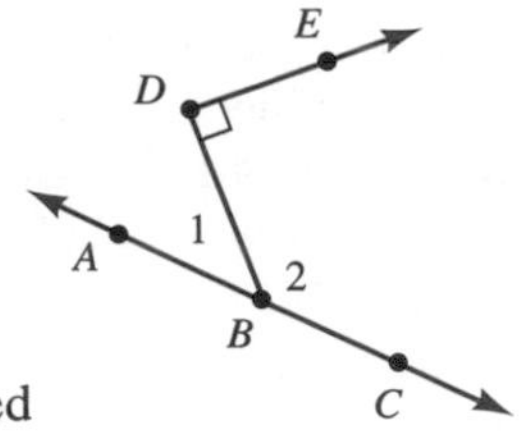

Strategy We will determine how each angle's measure compares to 90° or to 180°.

WHY Acute, right, obtuse, and straight angles are defined with respect to 90° and 180° angle measures.

Solution

Since m($\angle 1$) < 90°, it is an acute angle.

Since m($\angle 2$) > 90° but less than 180°, it is an obtuse angle.

Since m($\angle BDE$) = 90°, it is a right angle.

Since m($\angle ABC$) = 180°, it is a straight angle.

---

5 Solve problems involving adjacent angles.

Two angles that have a common vertex and a common side are called **adjacent angles** if they are side-by-side and their interiors do not overlap.

> **Success Tip** We can use the algebra concepts of variable and equation that were introduced in Chapter 3 to solve many types of geometry problems.

Self Check 2

Use the information in the figure to find x.

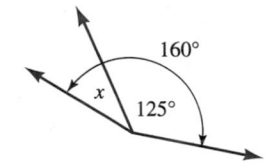

Now Try Problem 65

EXAMPLE 2 Two angles with degree measures of x and 35° are adjacent angles, as shown. Use the information in the figure to find x.

Strategy We will write an equation involving x that mathematically models the situation.

WHY We can then solve the equation to find the unknown angle measure.

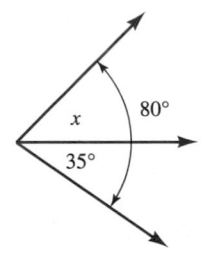

Adjacent angles

Solution

Since the sum of the measures of the two adjacent angles is 80°, we have

$$x + 35° = 80° \qquad \text{The word sum indicates addition.}$$
$$x + 35° - 35° = 80° - 35° \qquad \text{To isolate x, undo the addition of 35° by subtracting 35° from both sides.}$$
$$x = 45° \qquad \text{Do the subtractions: } 35° - 35° = 0° \text{ and } 80° - 35° = 45°.$$

$$\begin{array}{r} 7\,10 \\ 8\cancel{0} \\ -\ 35 \\ \hline 45 \end{array}$$

Thus, x is 45°. As a check, we see that 45° + 35° = 80°.

> **Caution!** In the figure for Example 2, we used the variable x to represent an unknown angle measure. In such cases, we will assume that the variable "carries" with it the associated units of degrees. That means we do not have to write a ° symbol next to the variable. Furthermore, if x represents an unknown number of degrees, then expressions such as $3x$, $x + 15°$, and $4x - 20°$ also have units of degrees.

---

6 Use the property of vertical angles to solve problems.

When two lines intersect, pairs of nonadjacent angles are called **vertical angles.** In the following figure, $\angle 1$ and $\angle 3$ are vertical angles and $\angle 2$ and $\angle 4$ are vertical angles.

Vertical angles

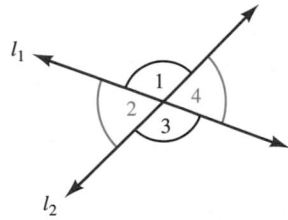

- ∠1 and ∠3
- ∠2 and ∠4

The Language of Algebra   When we work with two (or more) lines at one time, we can use **subscripts** to name the lines. The prefix *sub* means below or beneath, as in *sub*marine or *sub*way. To name the first line in the figure above, we use l_1, which is read as "*l* sub one." To name the second line, we use l_2, which is read as "*l* sub two."

To illustrate that vertical angles always have the same measure, refer to the figure below, with angles having measures of x, y, and 30°. Since the measure of any straight angle is 180°, we have

$$\mathbf{30°} + x = 180° \qquad \text{and} \qquad \mathbf{30°} + y = 180°$$
$$x = 150° \qquad\qquad\qquad\quad y = 150°$$

To undo the addition of 30°, subtract 30° from both sides.

Since x and y are both 150°, we conclude that $x = y$.

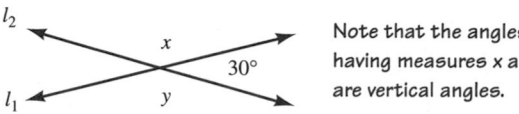

Note that the angles having measures x and y are vertical angles.

The previous example illustrates that vertical angles have the same measure. Recall that when two angles have the same measure, we say that they are *congruent*. Therefore, we have the following important fact.

Property of Vertical Angles

Vertical angles are congruent (have the same measure).

EXAMPLE 3   Refer to the figure. Find:

a. m(∠1)   **b.** m(∠ABF)

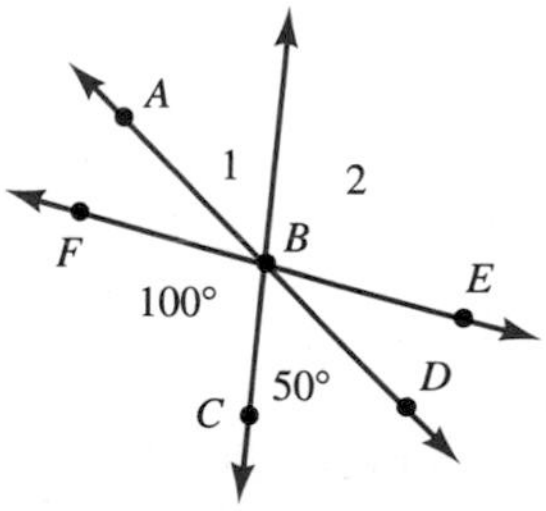

Strategy To answer part a, we will use the property of vertical angles. To answer part b, we will write an equation involving m(∠ABF) that mathematically models the situation.

WHY For part a, we note that $\overleftrightarrow{AD}$ and $\overleftrightarrow{BC}$ intersect to form vertical angles. For part b, we can solve the equation to find the unknown, m(∠ABF).

Solution

a. If we ignore $\overleftrightarrow{FE}$ for the moment, we see that $\overleftrightarrow{AD}$ and $\overleftrightarrow{BC}$ intersect to form the pair of vertical angles ∠CBD and ∠1. By the property of vertical angles,

$$\angle CBD \cong \angle 1 \qquad \text{Read as "angle CBD is congruent to angle one."}$$

Self Check 3

Refer to the figure for Example 3. Find:

a. m(∠2)

b. m(∠DBE)

Now Try Problems 69 and 71

Since congruent angles have the same measure,

$$m(\angle CBD) = m(\angle 1)$$

In the figure, we are given $m(\angle CBD) = 50°$. Thus, $m(\angle 1)$ is also $50°$, and we can write $m(\angle 1) = 50°$.

b. Since $\angle ABD$ is a straight angle, the *sum* of the measures of $\angle ABF$, the $100°$ angle, and the $50°$ angle is $180°$. If we let $x = m(\angle ABF)$, we have

$$x + 100° + 50° = 180°$$ The word *sum* indicates addition.

$$x + 150° = 180°$$ On the left side, combine like terms: $100° + 50° = 150°$.

$$x = 30°$$ To isolate x, undo the addition of $150°$ by subtracting $150°$ from both sides: $180° - 150° = 30°$.

Thus, $m(\angle ABF) = 30°$

Self Check 4

In the figure below, find:

a. y

b. $m(\angle XYZ)$

c. $m(\angle MYX)$

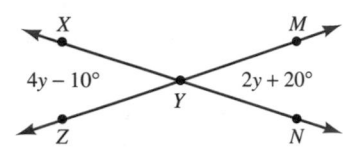

Now Try Problem 75

EXAMPLE 4   In the figure on the right, find:

a. x **b.** $m(\angle ABC)$ **c.** $m(\angle CBE)$

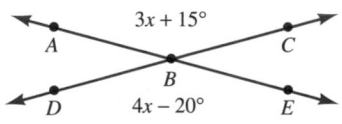

Strategy We will use the property of vertical angles to write an equation that mathematically models the situation.

WHY $\overleftrightarrow{AE}$ and $\overleftrightarrow{DC}$ intersect to form two pairs of vertical angles.

Solution

a. In the figure, two vertical angles have degree measures that are represented by the algebraic expressions $4x - 20°$ and $3x + 15°$. Since the angles are vertical angles, they have equal measures.

$$4x - 20° = 3x + 15°$$ Set the algebraic expressions equal.

$$4x - 20° - 3x = 3x + 15° - 3x$$ To eliminate $3x$ from the right side, subtract $3x$ from both sides.

$$x - 20° = 15°$$ Combine like terms: $4x - 3x = x$ and $3x - 3x = 0$.

$$x = 35°$$ To isolate x, undo the subtraction of $20°$ by adding $20°$ to both sides.

Thus, x is $35°$.

b. To find $m(\angle ABC)$, we evaluate the expression $3x + 15°$ for $x = 35°$.

$$3x + 15° = 3(35°) + 15°$$ Substitute $35°$ for x.

$$= 105° + 15°$$ Do the multiplication.

$$= 120°$$ Do the addition.

$$\begin{array}{r} 1 \\ 35 \\ \times 3 \\ \hline 105 \end{array}$$

Thus, $m(\angle ABC) = 120°$.

c. $\angle ABE$ is a straight angle. Since the measure of a straight angle is $180°$ and $m(\angle ABC) = 120°$, $m(\angle CBE)$ must be $180° - 120°$, or $60°$.

7 **Solve problems involving complementary and supplementary angles.**

Complementary and Supplementary Angles

Two angles are **complementary angles** when the sum of their measures is $90°$.

Two angles are **supplementary angles** when the sum of their measures is $180°$.

In figure (a) below, ∠*ABC* and ∠*CBD* are complementary angles because the sum of their measures is 90°. Each angle is said to be the **complement** of the other. In figure (b) below, ∠*X* and ∠*Y* are also complementary angles, because m(∠*X*) + m(∠*Y*) = 90°. Figure (b) illustrates an important fact: Complementary angles need not be adjacent angles.

Complementary angles

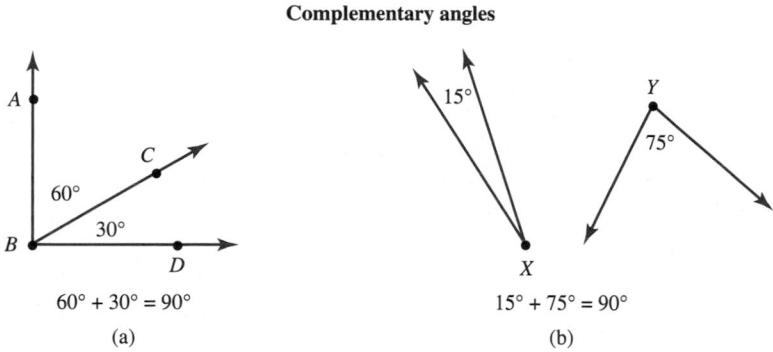

60° + 30° = 90°   15° + 75° = 90°

(a)   (b)

In figure (a) below, ∠*MNO* and ∠*ONP* are supplementary angles because the sum of their measures is 180°. Each angle is said to be the **supplement** of the other. Supplementary angles need not be adjacent angles. For example, in figure (b) below, ∠*G* and ∠*H* are supplementary angles, because m(∠*G*) + m(∠*H*) = 180°.

Supplementary angles

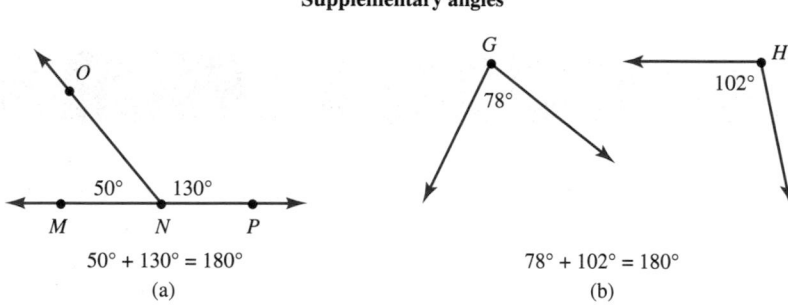

50° + 130° = 180°   78° + 102° = 180°

(a)   (b)

Caution! The definition of supplementary angles requires that the sum of *two* angles be 180°. Three angles of 40°, 60°, and 80° are not supplementary even though their sum is 180°.

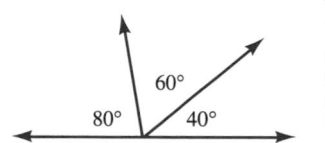

EXAMPLE 5

a. Find the complement of a 35° angle.

b. Find the supplement of a 105° angle.

Strategy We will use the definitions of complementary and supplementary angles to write equations that mathematically model each situation.

WHY We can then solve each equation to find the unknown angle measure.

Self Check 5

a. Find the complement of a 50° angle.

b. Find the supplement of a 50° angle.

Now Try **Problems 77 and 79**

Solution

a. It is helpful to draw a figure, as shown to the right. Let x represent the measure of the complement of the 35° angle. Since the angles are complementary, we have

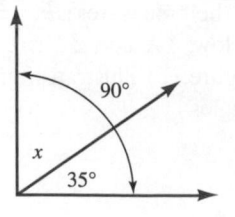

$x + 35° = 90°$ The sum of the angles' measures must be 90°.

$x = 55°$ To isolate x, undo the addition of 35° by subtracting 35° from both sides: 90° − 35° = 55°.

The complement of a 35° angle has measure 55°.

b. It is helpful to draw a figure, as shown on the right. Let y represent the measure of the supplement of the 105° angle. Since the angles are supplementary, we have

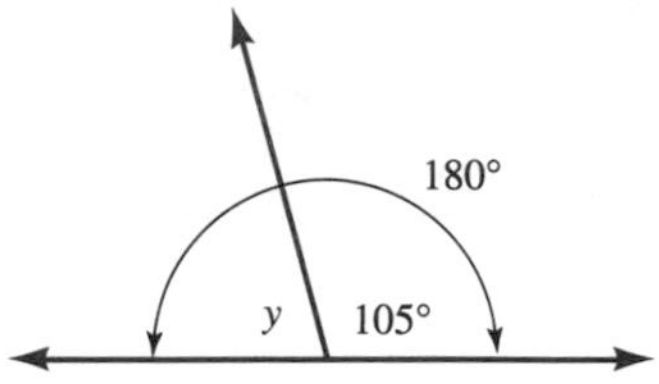

$y + 105° = 180°$ The sum of the angles' measures must be 180°.

$y = 75°$ To isolate y, undo the addition of 105° by subtracting 105° from both sides: 180° − 105° = 75°.

The supplement of a 105° angle has measure 75°.

ANSWERS TO SELF CHECKS

1. right angle, obtuse angle, acute angle, straight angle   **2.** 35°   **3. a.** 100°   **b.** 30°
4. a. 15°   **b.** 50°   **c.** 130°   **5. a.** 40°   **b.** 130°

SECTION 9.1   STUDY SET

VOCABULARY

Fill in the blanks.

1. Three undefined words in geometry are _____, _____, and _____.

2. A line _____ has two endpoints.

3. A _____ divides a line segment into two parts of equal length.

4. A ____ is the part of a line that begins at some point and continues forever in one direction.

5. An _____ is formed by two rays with a common endpoint.

6. An angle is measured in _____.

7. A _____ is used to measure angles.

8. The measure of an _____ angle is less than 90°.

9. The measure of a _____ angle is 90°.

10. The measure of an _____ angle is greater than 90° but less than 180°.

11. The measure of a straight angle is _____.

12. When two segments have the same length, we say that they are _____.

13. _____ angles have the same vertex, are side-by-side, and their interiors do not overlap.

14. When two lines intersect, pairs of nonadjacent angles are called _____ angles.

15. When two angles have the same measure, we say that they are _____.

16. The word *sum* indicates the operation of _____.

17. The sum of two complementary angles is ____.

18. The sum of two _____ angles is 180°.

CONCEPTS

19. **a.** Given two points (say, M and N), how many different lines pass through these two points?

    b. Fill in the blank: In general, two different points determine exactly one ____.

20. Refer to the figure.

    a. Name $\overrightarrow{NM}$ in another way.

    b. Do $\overrightarrow{MN}$ and $\overrightarrow{NM}$ name the same ray?

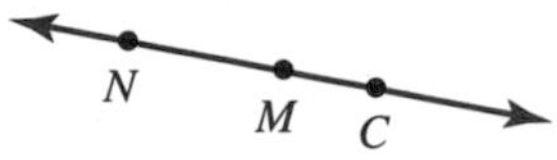

21. Consider the acute angle shown below.

  a. What two rays are the sides of the angle?

  b. What point is the vertex of the angle?

  c. Name the angle in four ways.

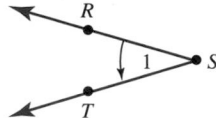

22. Estimate the measure of each angle. Do not use a protractor.

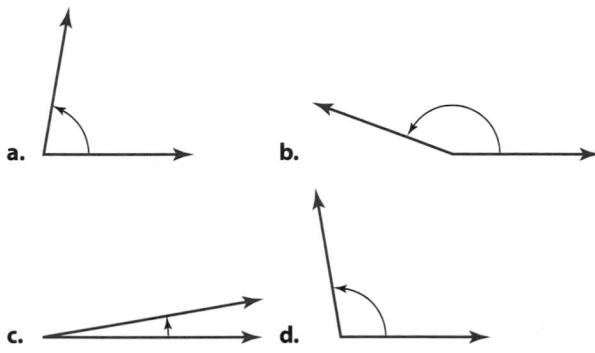

23. Draw an example of each type of angle.

  a. an acute angle   **b.** an obtuse angle

  c. a right angle   **d.** a straight angle

24. Fill in the blanks with the correct symbol.

  a. If m($\overline{AB}$) = m($\overline{CD}$), then $\overline{AB}$ ▯ $\overline{CD}$.

  b. If ∠ABC ≅ ∠DEF, then m(∠ABC) ▯ m(∠DEF).

25. a. Draw a pair of adjacent angles. Label them ∠ABC and ∠CBD.

  b. Draw two intersecting lines. Label them lines l_1 and l_2. Label one pair of vertical angles that are formed as ∠1 and ∠2.

  c. Draw two adjacent complementary angles.

  d. Draw two adjacent supplementary angles.

26. Fill in the blank:

  If ∠MNO ≅ ∠BFG, then m(∠MNO) ▯ m(∠BFG).

27. Fill in the blank:

  The vertical angle property: Vertical angles are _____.

28. Refer to the figure below. Fill in the blanks.

  a. ∠XYZ and ∠_____ are vertical angles.

  b. ∠XYZ and ∠ZYW are _____ angles.

  c. ∠ZYW and ∠XYV are _____ angles.

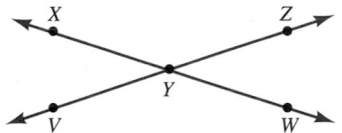

29. Refer to the figure below and tell whether each statement is true.

  a. ∠AGF and ∠BGC are vertical angles.

  b. ∠FGE and ∠BGA are adjacent angles.

  c. m(∠AGB) = m(∠BGC).

  d. ∠AGC ≅ ∠DGF.

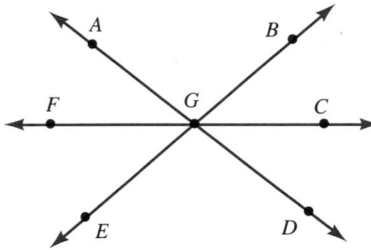

30. Refer to the figure below and tell whether the angles are congruent.

  a. ∠1 and ∠2   **b.** ∠FGB and ∠CGE

  c. ∠AGF and ∠FGE   **d.** ∠CGD and ∠CGB

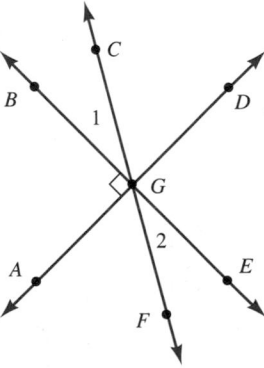

Refer to the figure above and tell whether each statement is true.

31. ∠1 and ∠CGD are adjacent angles.

32. ∠FGA and ∠AGC are supplementary.

33. ∠AGB and ∠BGC are complementary.

34. ∠AGF and ∠2 are complementary.

NOTATION

Fill in the blanks.

35. The symbol $\overleftrightarrow{AB}$ is read as "_____ AB."

36. The symbol $\overline{AB}$ is read as "_____ AB."

37. The symbol $\overrightarrow{AB}$ is read as "_____ AB."

38. We read $m(\overline{AB})$ as "the _____ of segment AB."

39. We read $\angle ABC$ as "_____ ABC."

40. We read $m(\angle ABC)$ as "the _____ of angle ABC."

41. The symbol for _____ is a small raised circle, °.

42. The symbol ⌐ indicates a _____ angle.

43. The symbol ≅ is read as "is _____ to."

44. The symbol l_1 can be used to name a line. It is read as "line l _____ one."

GUIDED PRACTICE

45. Draw each geometric figure and label it completely. **See Objective 1.**

    a. Point T

    b. $\overleftrightarrow{JK}$

    c. Plane ABC

46. Draw each geometric figure and label it completely. **See Objectives 2 and 3.**

    a. $\overline{RS}$

    b. $\overrightarrow{PQ}$

    c. $\angle XYZ$

    d. $\angle L$

47. Refer to the figure and find the length of each segment. **See Objective 2.**

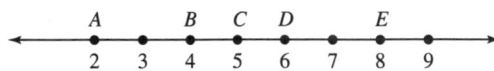

    a. $\overline{AB}$ **b.** $\overline{CE}$

    c. $\overline{DC}$ **d.** $\overline{EA}$

48. Refer to the figure above and find each midpoint. **See Objective 2.**

    a. Find the midpoint of $\overline{AD}$.

    b. Find the midpoint of $\overline{BE}$.

    c. Find the midpoint of $\overline{EA}$.

Use the protractor to find each angle measure listed below. **See Objective 4.**

49. $m(\angle GDE)$ **50.** $m(\angle ADE)$

51. $m(\angle EDS)$ **52.** $m(\angle EDR)$

53. $m(\angle CDR)$ **54.** $m(\angle CDA)$

55. $m(\angle CDG)$ **56.** $m(\angle CDS)$

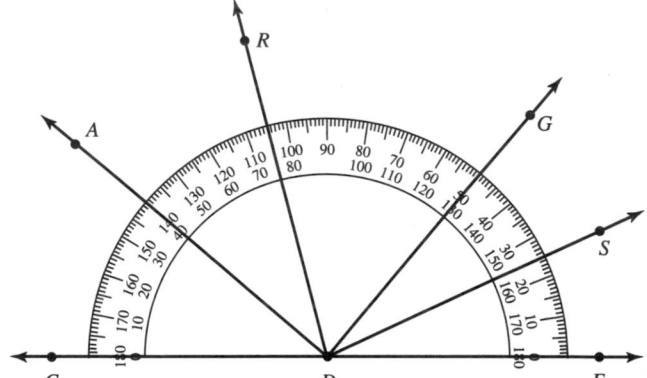

Classify the following angles in the figure as an acute angle, a right angle, an obtuse angle, or a straight angle. **See Example 1.**

57. $\angle MNO$ **58.** $\angle OPN$

59. $\angle NOP$ **60.** $\angle POS$

61. $\angle MPQ$ **62.** $\angle PNO$

63. $\angle QPO$ **64.** $\angle MNQ$

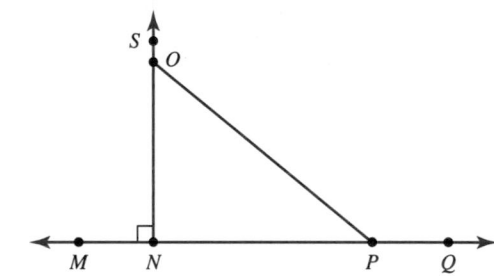

Find x. **See Example 2.**

65.

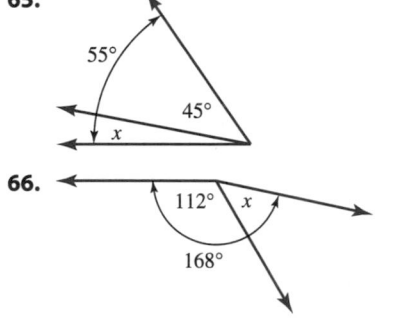

66.

67.   **68.**

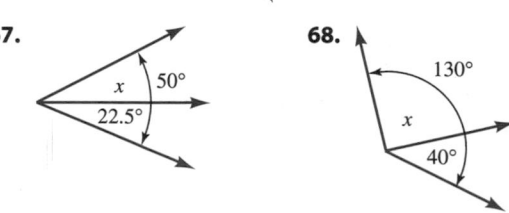

Refer to the figure below. Find the measure of each angle.
See Example 3.

69. $\angle 1$ **70.** $\angle MYX$

71. $\angle NYZ$ **72.** $\angle 2$

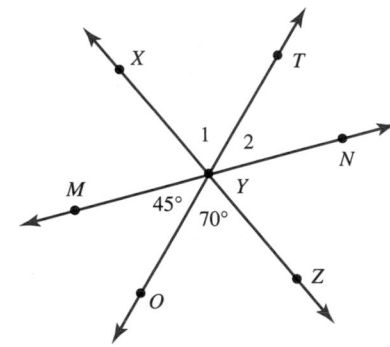

First find x. Then find m($\angle ABD$) *and* m($\angle DBE$). *See Example 4.*

73.   **74.**

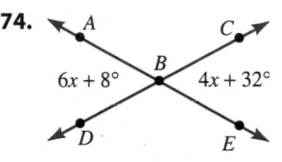

First find x. Then find m($\angle ZYQ$) *and* m($\angle PYQ$). *See Example 4.*

75.   **76.**

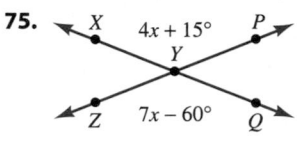

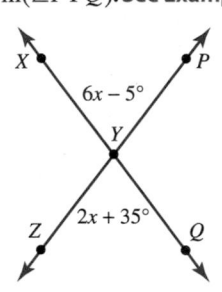

Let x represent the unknown angle measure. Write an equation and solve it to find x. **See Example 5.**

77. Find the complement of a 30° angle.

78. Find the supplement of a 30° angle.

79. Find the supplement of a 105° angle.

80. Find the complement of a 75° angle.

TRY IT YOURSELF

81. Refer to the figure in the next column and tell whether each statement is true. If a statement is false, explain why.

   a. $\overrightarrow{GF}$ has point G as its endpoint.

   b. $\overline{AG}$ has no endpoints.

   c. $\overleftrightarrow{CD}$ has three endpoints.

   d. Point D is the vertex of $\angle DGB$.

   e. m($\angle AGC$) = m($\angle BGD$)

   f. $\angle AGF \cong \angle BGE$

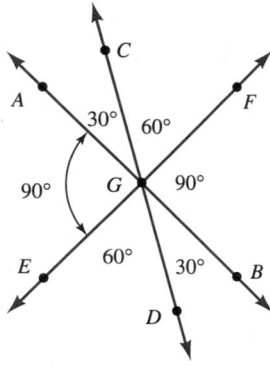

82. Refer to the figure for Problem 81 and tell whether each angle is an acute angle, a right angle, an obtuse angle, or a straight angle.

   a. $\angle AGC$ **b.** $\angle EGA$

   c. $\angle FGD$ **d.** $\angle BGA$

Use a protractor to measure each angle.

83.   **84.**

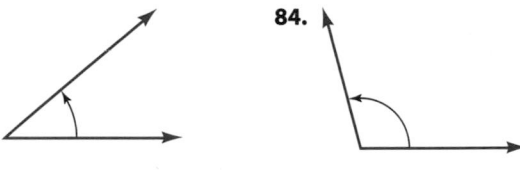

85.   **86.**

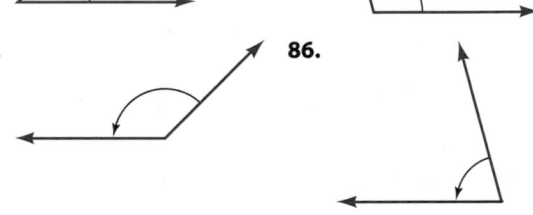

87. Refer to the figure below, in which m($\angle 1$) = 50°. Find the measure of each angle or sum of angles.

   a. $\angle 3$

   b. $\angle 4$

   c. m($\angle 1$) + m($\angle 2$) + m($\angle 3$)

   d. m($\angle 2$) + m($\angle 4$)

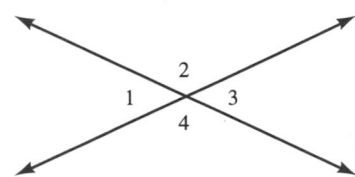

88. Refer to the figure below, in which m($\angle 1$) + m($\angle 3$) + m($\angle 4$) = 180°, $\angle 3 \cong \angle 4$, and $\angle 4 \cong \angle 5$. Find the measure of each angle.

   a. $\angle 1$ **b.** $\angle 2$

   c. $\angle 3$ **d.** $\angle 6$

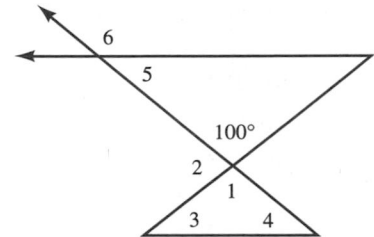

89. Refer to the figure below where $\angle 1 \cong \angle ACD$, $\angle 1 \cong \angle 2$, and $\angle BAC \cong \angle 2$.

  a. What is the complement of $\angle BAC$?

  b. What is the supplement of $\angle BAC$?

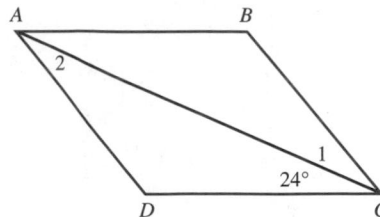

90. Refer to the figure below where $\angle EBS \cong \angle BES$.

  a. What is the measure of $\angle AEF$?

  b. What is the supplement of $\angle AET$?

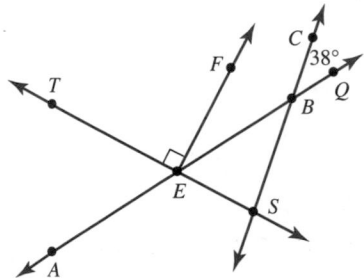

91. Find the supplement of the complement of a 51° angle.

92. Find the complement of the supplement of a 173° angle.

93. Find the complement of the complement of a 1° angle.

94. Find the supplement of the supplement of a 6° angle.

APPLICATIONS

95. MUSICAL INSTRUMENTS Suppose that you are a beginning band teacher describing the correct posture needed to play various instruments. Using the diagrams shown below, approximate the angle measure (in degrees) at which each instrument should be held in relation to the student's body.

  a. flute   **b.** clarinet   **c.** trumpet

96. PLANETS The figures below show the direction of rotation of several planets in our solar system. They also show the angle of tilt of each planet.

  a. Which planets have an angle of tilt that is an acute angle?

  b. Which planets have an angle of tilt that is an obtuse angle?

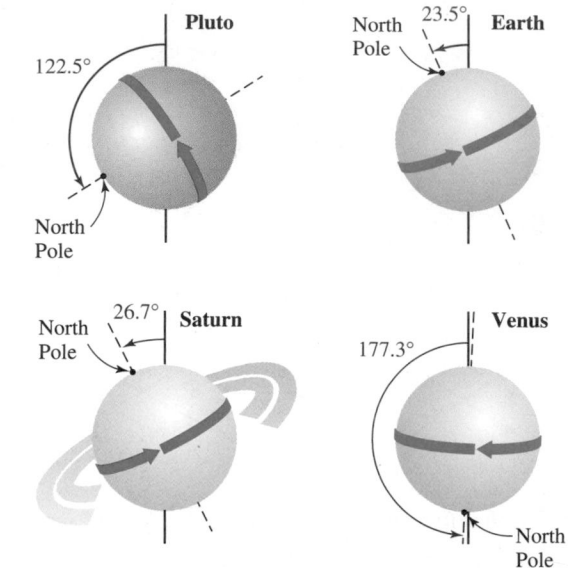

97. a. AVIATION How many degrees from the horizontal position are the wings of the airplane?

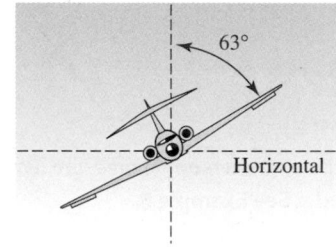

  b. GARDENING What angle does the handle of the lawn mower make with the ground?

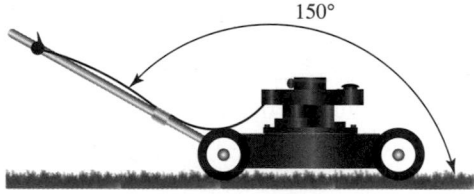

98. SYNTHESIZER Find x and y.

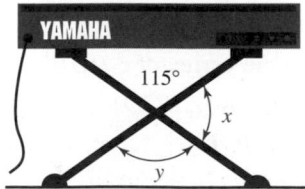

WRITING

99. PHRASES Explain what you think each of these phrases means. How is geometry involved?

    a. The president did a complete 180-degree flip on the subject of a tax cut.

    b. The rollerblader did a "360" as she jumped off the ramp.

100. In the statements below, the ° symbol is used in two different ways. Explain the difference.

$$m(\angle A) = 85° \quad \text{and} \quad 85°F$$

101. Can two angles that are complementary be equal? Explain.

102. Explain why the angles highlighted below are not vertical angles.

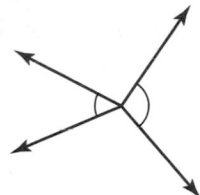

REVIEW

103. Add: $\dfrac{1}{2} + \dfrac{2}{3} + \dfrac{3}{4}$

104. Subtract: $\dfrac{3}{4} - \dfrac{1}{8} - \dfrac{1}{2}$

105. Multiply: $\dfrac{5}{8} \cdot \dfrac{2}{15} \cdot \dfrac{6}{5}$

106. Divide: $\dfrac{12}{17} \div \dfrac{4}{34}$

SECTION 9.2
Parallel and Perpendicular Lines

Objectives

1 Identify and define parallel and perpendicular lines.

2 Identify corresponding angles, interior angles, and alternate interior angles.

3 Use properties of parallel lines cut by a transversal to find unknown angle measures.

In this section, we will consider *parallel* and *perpendicular* lines. Since parallel lines are always the same distance apart, the railroad tracks shown in figure (a) illustrate one application of parallel lines. Figure (b) shows one of the events of men's gymnastics, the parallel bars. Since perpendicular lines meet and form right angles, the monument and the ground shown in figure (c) illustrate one application of perpendicular lines.

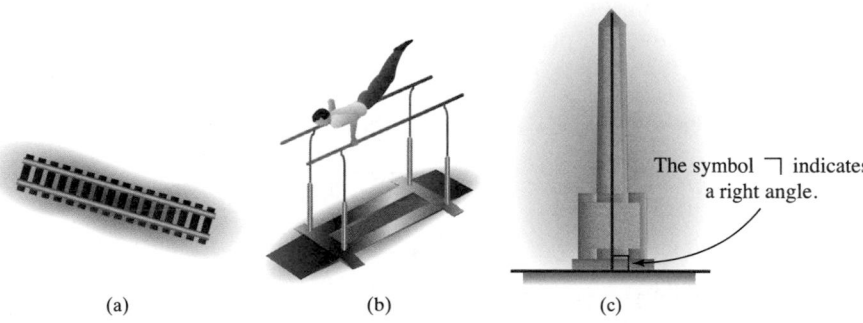

The symbol ⌐ indicates a right angle.

(a)   (b)   (c)

1 Identify and define parallel and perpendicular lines.

If two lines lie in the same plane, they are called **coplanar.** Two coplanar lines that do not intersect are called **parallel lines.** See figure (a) on the next page. If two lines do not lie in the same plane, they are called noncoplanar. Two noncoplanar lines that do not intersect are called **skew lines.**

Parallel lines   **Perpendicular lines**

l_1 l_2 l_1 l_2

(a)   (b)

Parallel Lines

Parallel lines are coplanar lines that do not intersect.

Some lines that intersect are perpendicular. See figure (b) above.

Perpendicular Lines

Perpendicular lines are lines that intersect and form right angles.

> ***The Language of Algebra***   If lines l_1 (read as "l sub 1") and l_2 (read as "l sub 2") are parallel, we can write $l_1 \parallel l_2$, where the symbol $\parallel$ is read as "is parallel to."
>
> If lines l_1 and l_2 are perpendicular, we can write $l_1 \perp l_2$, where the symbol $\perp$ is read as "is perpendicular to."

2 Identify corresponding angles, interior angles, and alternate interior angles.

A line that intersects two coplanar lines in two distinct (different) points is called a **transversal.** For example, line l_1 in the figure to the right is a transversal intersecting lines l_2 and l_3.

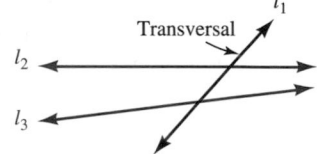

When two lines are cut by a transversal, all eight angles that are formed are important in the study of parallel lines. Descriptive names are given to several pairs of these angles.

In the figure below, four pairs of **corresponding angles** are formed.

Corresponding angles

- $\angle 1$ and $\angle 5$
- $\angle 3$ and $\angle 7$
- $\angle 2$ and $\angle 6$
- $\angle 4$ and $\angle 8$

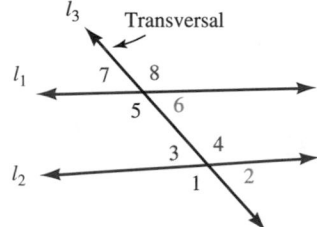

Corresponding Angles

If two lines are cut by a transversal, then the angles on the same side of the transversal and in corresponding positions with respect to the lines are called corresponding angles.

In the figure below, four **interior angles** are formed.

Interior angles

• ∠3, ∠4, ∠5, and ∠6

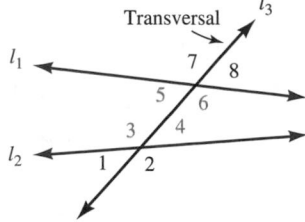

In the figure below, two pairs of **alternate interior angles** are formed.

Alternate interior angles

• ∠4 and ∠5
• ∠3 and ∠6

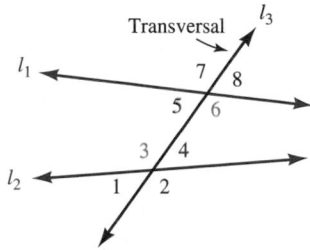

Alternate Interior Angles

If two lines are cut by a transversal, then the nonadjacent angles on opposite sides of the transversal and on the interior of the two lines are called alternate interior angles.

Success Tip Alternate interior angles are easily spotted because they form a Z-shape or a backward Z-shape, as shown below.

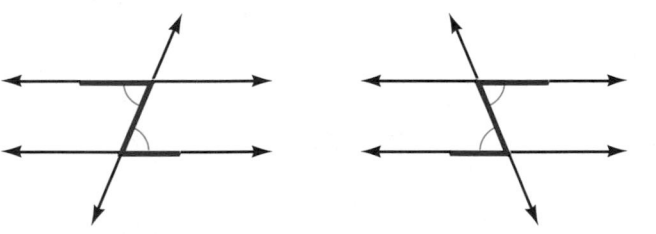

EXAMPLE 1 Refer to the figure. Identify:

a. all pairs of corresponding angles

b. all interior angles

c. all pairs of alternate interior angles

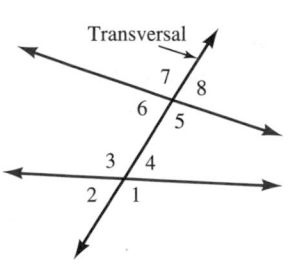

Strategy When two lines are cut by a transversal, eight angles are formed. We will consider the relative position of the angles with respect to the two lines and the transversal.

WHY There are four pairs of corresponding angles, four interior angles, and two pairs of alternate interior angles.

Self Check 1

Refer to the figure below. Identify:
a. all pairs of corresponding angles

b. all interior angles

c. all pairs of alternate interior angles

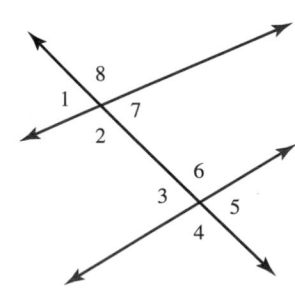

Now Try Problem 21

Solution

a. To identify corresponding angles, we examine the angles to the right of the transversal and the angles to the left of the transversal. The pairs of corresponding angles in the figure are

- $\angle 1$ and $\angle 5$
- $\angle 4$ and $\angle 8$
- $\angle 2$ and $\angle 6$
- $\angle 3$ and $\angle 7$

b. To identify the interior angles, we determine the angles inside the two lines cut by the transversal. The interior angles in the figure are

$\angle 3, \angle 4, \angle 5,$ and $\angle 6$

c. Alternate interior angles are nonadjacent angles on opposite sides of the transversal inside the two lines. Thus, the pairs of alternate interior angles in the figure are

- $\angle 3$ and $\angle 5$
- $\angle 4$ and $\angle 6$

3 Use properties of parallel lines cut by a transversal to find unknown angle measures.

Lines that are cut by a transversal may or may not be parallel. When a pair of parallel lines are cut by a transversal, we can make several important observations about the angles that are formed.

1. **Corresponding angles property:** If two parallel lines are cut by a transversal, each pair of corresponding angles are congruent. In the figure below, if $l_1 \parallel l_2$, then $\angle 1 \cong \angle 5$, $\angle 3 \cong \angle 7$, $\angle 2 \cong \angle 6$, and $\angle 4 \cong \angle 8$.

2. **Alternate interior angles property:** If two parallel lines are cut by a transversal, alternate interior angles are congruent. In the figure below, if $l_1 \parallel l_2$, then $\angle 3 \cong \angle 6$ and $\angle 4 \cong \angle 5$.

3. **Interior angles property:** If two parallel lines are cut by a transversal, interior angles on the same side of the transversal are supplementary. In the figure below, if $l_1 \parallel l_2$, then $\angle 3$ is supplementary to $\angle 5$ and $\angle 4$ is supplementary to $\angle 6$.

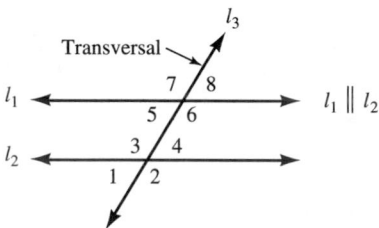

4. If a transversal is perpendicular to one of two parallel lines, it is also perpendicular to the other line. In figure (a) below, if $l_1 \parallel l_2$ and $l_3 \perp l_1$, then $l_3 \perp l_2$.

5. If two lines are parallel to a third line, they are parallel to each other. In figure (b) below, if $l_1 \parallel l_2$ and $l_1 \parallel l_3$, then $l_2 \parallel l_3$.

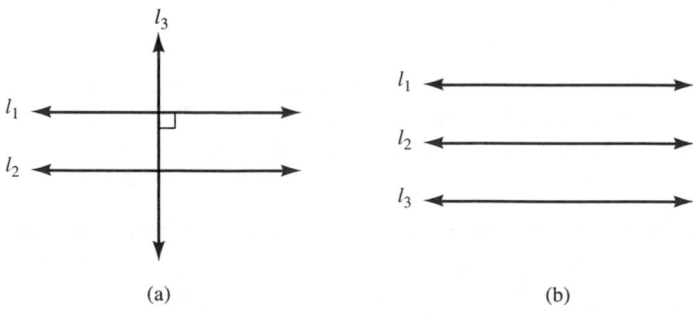

(a) (b)

EXAMPLE 2 Refer to the figure. If $l_1 \parallel l_2$ and m($\angle 3$) = 120°, find the measures of the other seven angles that are labeled.

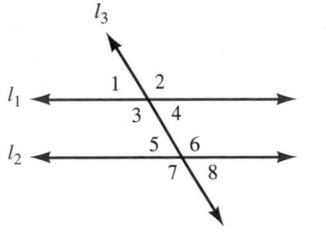

Strategy We will look for vertical angles, supplementary angles, and alternate interior angles in the figure.

WHY The facts that we have studied about vertical angles, supplementary angles, and alternate interior angles enable us to use known angle measures to find unknown angle measures.

Solution

m($\angle 1$) = 60° $\angle 3$ and $\angle 1$ are supplementary: m($\angle 3$) + m($\angle 1$) = 180°.

m($\angle 2$) = 120°   Vertical angles are congruent: m($\angle 2$) = m($\angle 3$).

m($\angle 4$) = 60°   Vertical angles are congruent: m($\angle 4$) = m($\angle 1$).

m($\angle 5$) = 60°   If two parallel lines are cut by a transversal, alternate interior angles are congruent: m($\angle 5$) = m($\angle 4$).

m($\angle 6$) = 120°   If two parallel lines are cut by a transversal, alternate interior angles are congruent: m($\angle 6$) = m($\angle 3$).

m($\angle 7$) = 120°   Vertical angles are congruent: m($\angle 7$) = m($\angle 6$).

m($\angle 8$) = 60°   Vertical angles are congruent: m($\angle 8$) = m($\angle 5$).

Self Check 2
Refer to the figure for Example 2. If $l_1 \parallel l_2$ and m($\angle 8$) = 50°, find the measures of the other seven angles that are labeled.

Now Try Problem 23

Some geometric figures contain two transversals.

EXAMPLE 3 Refer to the figure. If $\overline{AB} \parallel \overline{DE}$, which pairs of angles are congruent?

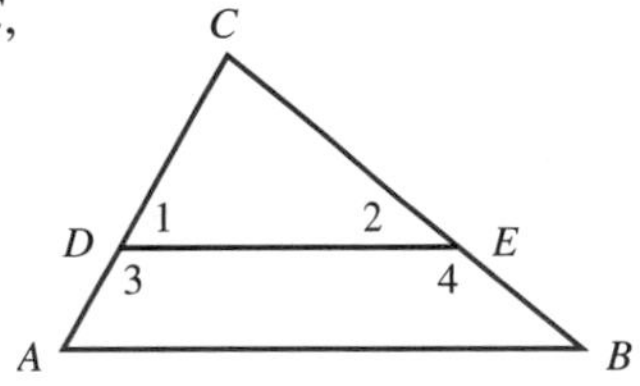

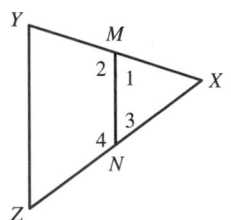

Strategy We will use the corresponding angles property twice to find two pairs of congruent angles.

WHY Both $\overleftrightarrow{AC}$ and $\overleftrightarrow{BC}$ are transversals cutting the parallel line segments $\overline{AB}$ and $\overline{DE}$.

Solution Since $\overline{AB} \parallel \overline{DE}$, and $\overleftrightarrow{AC}$ is a transversal cutting them, corresponding angles are congruent. So we have

$\angle A \cong \angle 1$

Since $\overline{AB} \parallel \overline{DE}$ and $\overleftrightarrow{BC}$ is a transversal cutting them, corresponding angles must be congruent. So we have

$\angle B \cong \angle 2$

Self Check 3
See the figure below. If $\overline{YZ} \parallel \overline{MN}$, which pairs of angles are congruent?

Now Try Problem 25

EXAMPLE 4 In the figure, $l_1 \parallel l_2$. Find x.

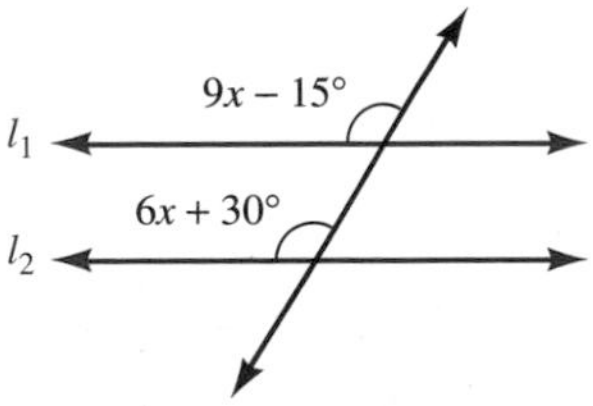

Strategy We will use the corresponding angles property to write an equation that mathematically models the situation.

WHY We can then solve the equation to find x.

Solution In the figure, two corresponding angles have degree measures that are represented by the algebraic expressions $9x - 15°$ and $6x + 30°$. Since $l_1 \parallel l_2$, this pair of corresponding angles are congruent.

Self Check 4
In the figure below, $l_1 \parallel l_2$. Find y.

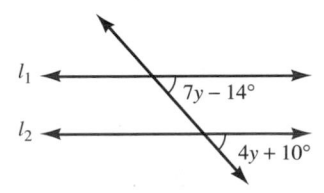

Now Try Problem 27

$$9x - 15° = 6x + 30°$$ Since the angles are congruent, their measures are equal.

$$3x - 15° = 30°$$ To eliminate 6x from the right side, subtract 6x from both sides.

$$3x = 45°$$ To isolate the variable term 3x, undo the subtraction of 15° by adding 15° to both sides: 30° + 15° = 45°.

$$x = 15°$$ To isolate x, undo the multiplication by 3 by dividing both sides by 3.

Thus, x is 15°.

Self Check 5

In the figure below, $l_1 \parallel l_2$.

a. Find x.

b. Find the measures of both angles labeled in the figure.

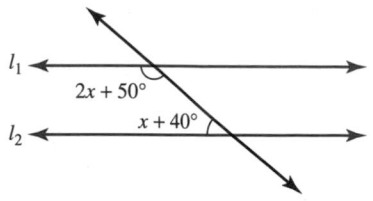

Now Try Problem 29

EXAMPLE 5   In the figure, $l_1 \parallel l_2$.

a. Find x.

b. Find the measures of both angles labeled in the figure.

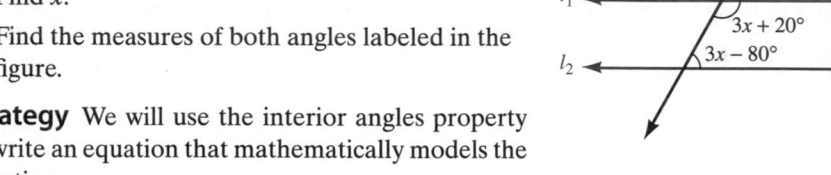

Strategy We will use the interior angles property to write an equation that mathematically models the situation.

WHY We can then solve the equation to find x.

Solution

a. Because the angles are interior angles on the same side of the transversal, they are supplementary.

$$3x - 80° + 3x + 20° = 180°$$ The sum of the measures of two supplementary angles is 180°.

$$6x - 60° = 180°$$ Combine like terms: 3x + 3x = 6x.

$$6x = 240°$$ To undo the subtraction of 60°, add 60° to both sides: 180° + 60° = 240°.

$$x = 40°$$ To isolate x, undo the multiplication by 6 by dividing both sides by 6.

Thus, x is 40°.

This problem may be solved using a different approach. In the figure below, we see that $\angle 1$ and the angle with measure $3x - 80°$ are corresponding angles.

Because l_1 and l_2 are parallel, all pairs of corresponding angles are congruent. Therefore,

$$m(\angle 1) = 3x - 80°$$

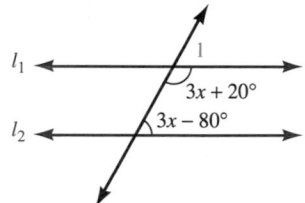

In the figure, we also see that $\angle 1$ and the angle with measure $3x + 20°$ are supplementary. That means that the sum of their measures must be 180°. We have

$$m(\angle 1) + 3x + 20° = 180°$$

$$3x - 80° + 3x + 20° = 180°$$ Replace m(∠1) with 3x − 80°.

This is the same equation that we obtained in the previous solution. When it is solved, we find that x is 40°.

b. To find the measures of the angles in the figure, we evaluate the expressions $3x + 20°$ and $3x - 80°$ for $x = 40°$.

$$3x + 20° = 3(\mathbf{40°}) + 20° \qquad 3x - 80° = 3(\mathbf{40°}) - 80°$$
$$= 120° + 20° \qquad\qquad\quad = 120° - 80°$$
$$= 140° \qquad\qquad\qquad\quad = 40°$$

The measures of the angles labeled in the figure are 140° and 40°.

ANSWERS TO SELF CHECKS

1. a. $\angle 1$ and $\angle 3$, $\angle 2$ and $\angle 4$, $\angle 8$ and $\angle 6$, $\angle 7$ and $\angle 5$ **b.** $\angle 2, \angle 7, \angle 3$, and $\angle 6$ **c.** $\angle 2$ and $\angle 6$, $\angle 7$ and $\angle 3$ **2.** $m(\angle 5) = 50°, m(\angle 7) = 130°, m(\angle 6) = 130°, m(\angle 3) = 130°,$ $m(\angle 4) = 50°, m(\angle 1) = 50°$, and $m(\angle 2) = 130°$ **3.** $\angle 1 \cong \angle Y, \angle 3 \cong \angle Z$ **4.** 8°
5. a. 30° **b.** 110°, 70°

SECTION 9.2   STUDY SET

VOCABULARY

Fill in the blanks.

1. Two lines that lie in the same plane are called _____. Two lines that lie in different planes are called _____.

2. Two coplanar lines that do not intersect are called _____ lines. Two noncoplanar lines that do not intersect are called _____ lines.

3. _____ lines are lines that intersect and form right angles.

4. A line that intersects two coplanar lines in two distinct (different) points is called a _____.

5. In the figure below, $\angle 4$ and $\angle 6$ are _____ interior angles.

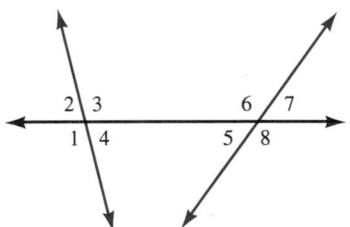

6. In the figure above, $\angle 2$ and $\angle 6$ are _____ angles.

CONCEPTS

7. **a.** Draw two parallel lines. Label them l_1 and l_2.

   b. Draw two lines that are not parallel. Label them l_1 and l_2.

8. **a.** Draw two perpendicular lines. Label them l_1 and l_2.

   b. Draw two lines that are not perpendicular. Label them l_1 and l_2.

9. **a.** Draw two parallel lines cut by a transversal. Label the lines l_1 and l_2 and label the transversal l_3.

   b. Draw two lines that are not parallel and cut by a transversal. Label the lines l_1 and l_2 and label the transversal l_3.

10. Draw three parallel lines. Label them $l_1, l_2,$ and l_3.

In Problems 11–14, two parallel lines are cut by a transversal. Fill in the blanks.

11. In the figure below, on the left, $\angle ABC \cong \angle BEF$. When two parallel lines are cut by a transversal, _____ angles are congruent.

12. In the figure below, on the right, $\angle 1 \cong \angle 2$. When two parallel lines are cut by a transversal, _____ _____ angles are congruent.

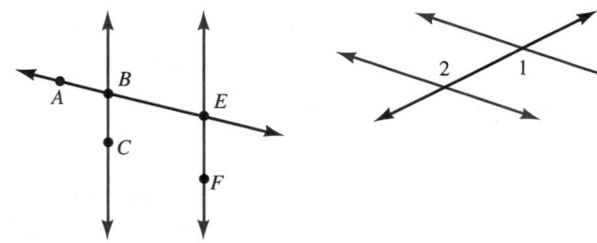

13. In the figure below, on the left, m($\angle ABC$) + m($\angle BCD$) = 180°. When two parallel lines are cut by a transversal, _____ angles on the same side of the transversal are supplementary.

14. In the figure below, on the right, $\angle 8 \cong \angle 6$. When two parallel lines are cut by a transversal, _____ angles are congruent.

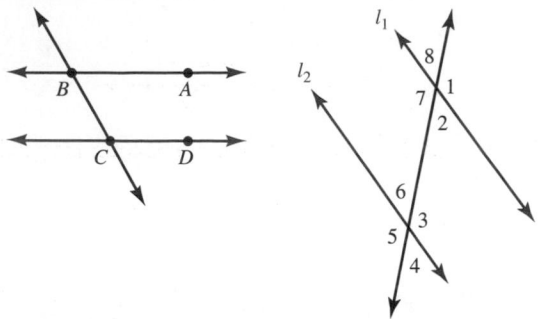

15. In the figure below, on the left, $l_1 \parallel l_2$. What can you conclude about l_1 and l_3?

16. In the figure below, on the right, $l_1 \parallel l_2$ and $l_2 \parallel l_3$. What can you conclude about l_1 and l_3?

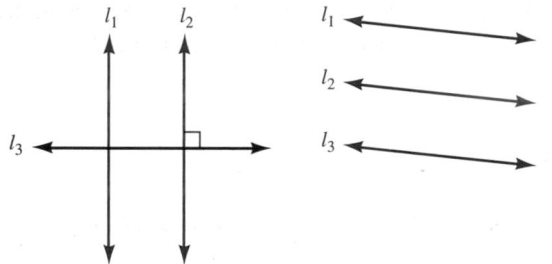

NOTATION

Fill in the blanks.

17. The symbol ⌐ indicates a _____ angle.

18. The symbol $\parallel$ is read as "is _____ to."

19. The symbol $\perp$ is read as "is _____ to."

20. The symbol l_1 is read as "line l _____ one."

GUIDED PRACTICE

21. Refer to the figure below and identify each of the following. **See Example 1.**

   a. corresponding angles

   b. interior angles

   c. alternate interior angles

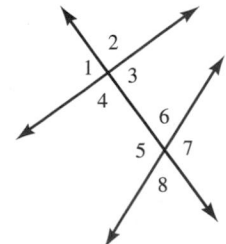

22. Refer to the figure below and identify each of the following. **See Example 1.**

   a. corresponding angles

   b. interior angles

   c. alternate interior angles

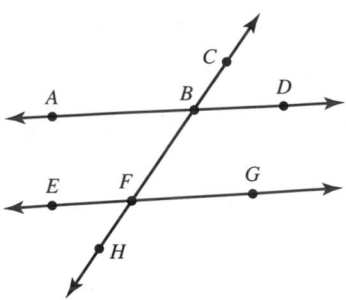

23. In the figure below, $l_1 \parallel l_2$ and m($\angle 4$) = 130°. Find the measures of the other seven angles that are labeled. **See Example 2.**

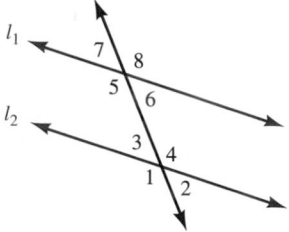

24. In the figure below, $l_1 \parallel l_2$ and m($\angle 2$) = 40°. Find the measures of the other angles. **See Example 2.**

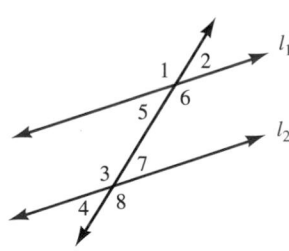

25. In the figure below, $\overline{YM} \parallel \overline{XN}$. Which pairs of angles are congruent? **See Example 3.**

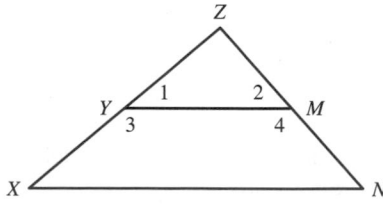

26. In the figure below, $\overline{AE} \parallel \overline{BD}$. Which pairs of angles are congruent? **See Example 3.**

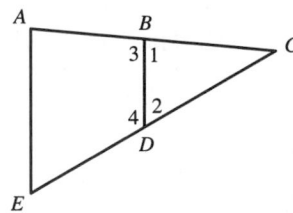

In Problems 27 and 28, $l_1 \parallel l_2$. First find x. Then determine the measure of each angle that is labeled in the figure.
See Example 4.

27.

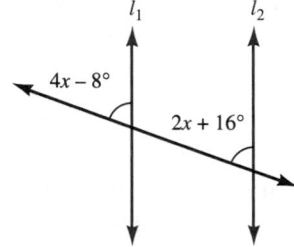

28.

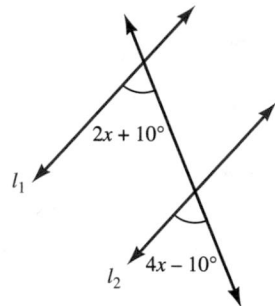

In Problems 29 and 30, $l_1 \parallel l_2$. First find x. Then determine the measure of each angle that is labeled in the figure.
See Example 5.

29.

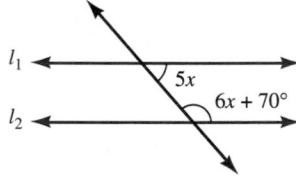

30.

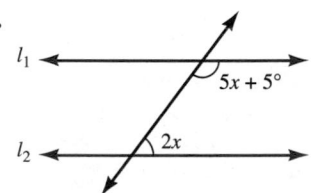

TRY IT YOURSELF

31. In the figure below, $l_1 \parallel AB$. Find:
  a. $m(\angle 1), m(\angle 2), m(\angle 3)$, and $m(\angle 4)$
  b. $m(\angle 3) + m(\angle 4) + m(\angle ACD)$
  c. $m(\angle 1) + m(\angle ABC) + m(\angle 4)$

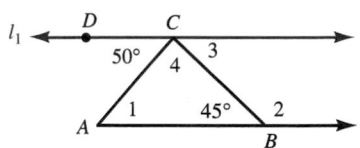

32. In the figure below, $\overline{AB} \parallel \overline{DE}$. Find $m(\angle B), m(\angle E)$, and $m(\angle 1)$.

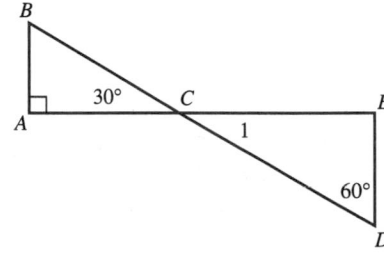

33. In the figure below, $\overline{AB} \parallel \overline{DE}$. What pairs of angles are congruent? Explain your reasoning.

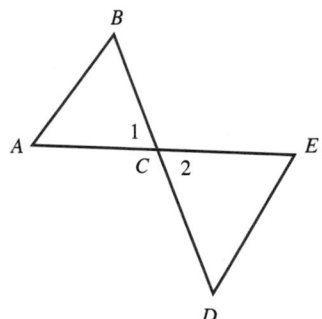

34. In the figure below, $l_1 \parallel l_2$. Find x.

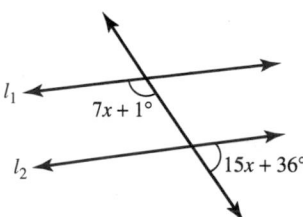

In Problems 35–38, first find x. Then determine the measure of each angle that is labeled in the figure.

35. $l_1 \parallel \overline{CA}$

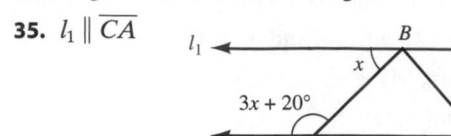

36. $\overline{AB} \parallel \overline{DE}$

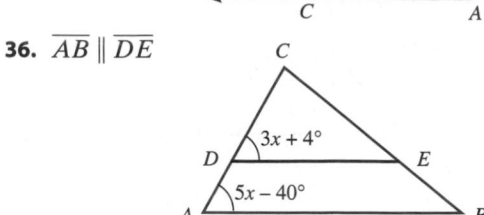

37. $\overline{AB} \parallel \overline{DE}$

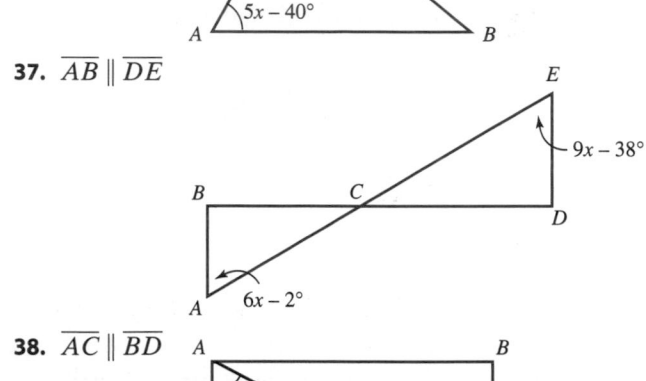

38. $\overline{AC} \parallel \overline{BD}$

APPLICATIONS

39. CONSTRUCTING PYRAMIDS The Egyptians used a device called a **plummet** to tell whether stones were properly leveled. A plummet (shown below) is made up of an A-frame and a plumb bob suspended from the peak of the frame. How could a builder use a plummet to tell that the two stones on the left are not level and that the three stones on the right are level?

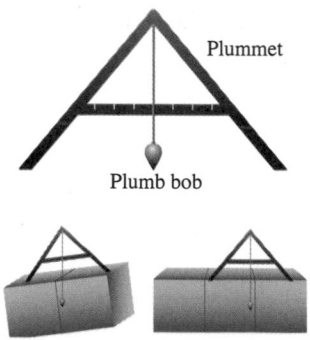

40. DIAGRAMMING SENTENCES English instructors have their students diagram sentences to help teach proper sentence structure. A diagram of the sentence *The cave was rather dark and damp* is shown below. Point out pairs of parallel and perpendicular lines used in the diagram.

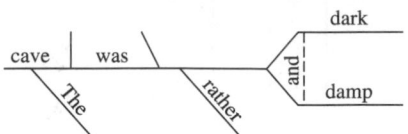

41. BEAUTY TIPS The figure to the right shows how one website illustrated the "geometry" of the ideal eyebrow. If $l_1 \parallel l_2$ and $m(\angle DCF) = 130°$, find $m(\angle ABE)$.

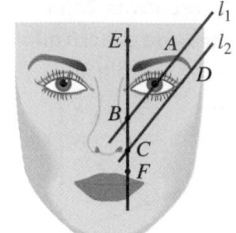

42. PAINTING SIGNS For many sign painters, the most difficult letter to paint is a capital E because of all the right angles involved. How many right angles are there?

43. HANGING WALLPAPER Explain why the concepts of *perpendicular* and *parallel* are both important when hanging wallpaper.

44. TOOLS What geometric concepts are seen in the design of the rake shown here?

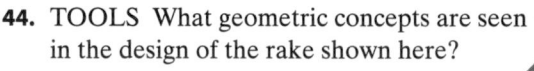

45. SEISMOLOGY The figure shows how an earthquake fault occurs when two blocks of earth move apart and one part drops down. Determine the measures of $\angle 1$, $\angle 2$, and $\angle 3$.

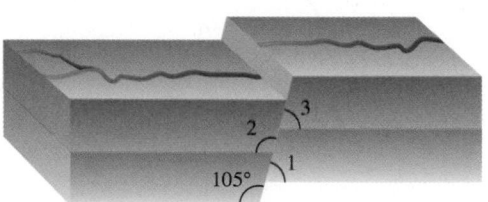

46. CARPENTRY A carpenter cross braced three
2 × 4's as shown below and then used a tool to
measure the three highlighted angles in red. Are the
2 × 4's parallel? Explain your answer.

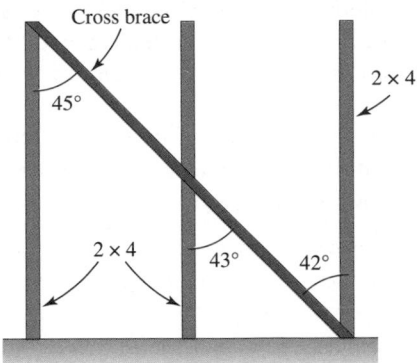

WRITING

47. PARKING DESIGN Using terms from this section,
write a paragraph describing the parking layout
shown below.

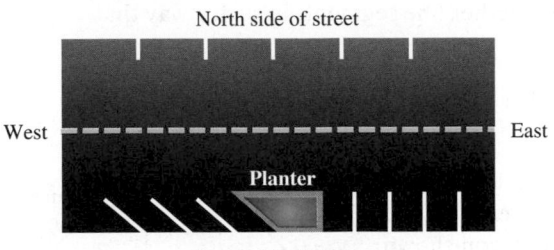

48. In the figure below, $l_1 \parallel l_2$. Explain why
m($\angle BDE$) = 91°.

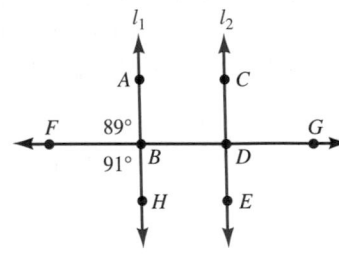

49. In the figure below, $l_1 \parallel l_2$. Explain why
m($\angle FEH$) = 100°.

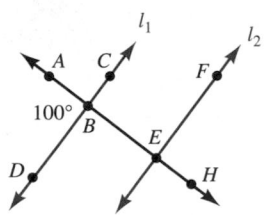

50. In the figure below, $l_1 \parallel l_2$. Explain why the figure must
be mislabeled.

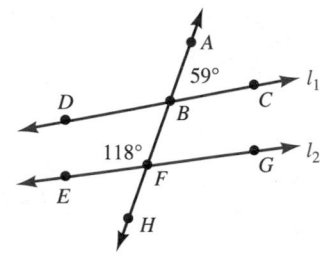

51. Are pairs of alternate interior angles always
congruent? Explain.

52. Are pairs of interior angles on the same side of a
transversal always supplementary? Explain.

REVIEW

53. Find 60% of 120.

54. 80% of what number is 400?

55. What percent of 500 is 225?

56. Simplify: $3.45 + 7.37 \cdot 2.98$

57. Is every whole number an integer?

58. Multiply: $2\frac{1}{5} \cdot 4\frac{3}{7}$

59. Express the phrase as a ratio in lowest terms:
4 ounces to 12 ounces

60. Convert 5,400 milligrams to kilograms.

Objectives

1 Classify polygons.

2 Classify triangles.

3 Identify isosceles triangles.

4 Find unknown angle measures of triangles.

The House of the Seven Gables, Salem, Massachusetts

© William Owens/Alamy

SECTION 9.3

Triangles

We will now discuss geometric figures called *polygons*. We see these shapes every day. For example, the walls of most buildings are rectangular in shape. Some tile and vinyl floor patterns use the shape of a pentagon or a hexagon. Stop signs are in the shape of an octagon.

In this section, we will focus on one specific type of polygon called a *triangle*. Triangular shapes are especially important because triangles contribute strength and stability to walls and towers. The gable roofs of houses are triangular, as are the sides of many ramps.

1 Classify polygons.

Polygon

A **polygon** is a closed geometric figure with at least three line segments for its sides.

Polygons are formed by fitting together line segments in such a way that

* no two of the segments intersect, except at their endpoints, and
* no two line segments with a common endpoint lie on the same line.

The line segments that form a polygon are called its **sides.** The point where two sides of a polygon intersect is called a **vertex** of the polygon (plural **vertices**). The polygon shown to the right has 5 sides and 5 vertices.

Polygons are classified according to the number of sides that they have. For example, in the figure below, we see that a polygon with four sides is called a *quadrilateral,* and a polygon with eight sides is called an *octagon.* If a polygon has sides that are all the same length and angles that are the same measure, we call it a **regular polygon.**

	Triangle 3 sides	Quadrilateral 4 sides	Pentagon 5 sides	Hexagon 6 sides	Heptagon 7 sides	Octagon 8 sides	Nonagon 9 sides	Decagon 10 sides	Dodecagon 12 sides
Polygons									
Regular polygons									

Self Check 1

Give the number of vertices of:

a. a quadrilateral

b. a pentagon

Now Try Problems 25 and 27

EXAMPLE 1 Give the number of vertices of:

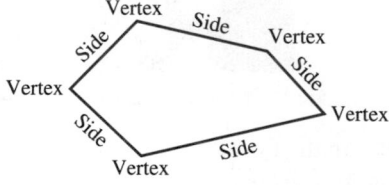

a. a triangle   **b.** a hexagon

Strategy We will determine the number of angles that each polygon has.

WHY The number of its vertices is equal to the number of its angles.

Solution

a. From the figure on the previous page, we see that a triangle has three angles and therefore three vertices.

b. From the figure on the previous page, we see that a hexagon has six angles and therefore six vertices. ■

> **Success Tip**   From the results of Example 1, we see that *the number of vertices of a polygon is equal to the number of its sides.*

2 Classify triangles.

A **triangle** is a polygon with three sides (and three vertices). Recall that in geometry points are labeled with capital letters. We can use the capital letters that denote the vertices of a triangle to name the triangle. For example, when referring to the triangle in the box below, with vertices *A*, *B*, and *C*, we can use the notation △*ABC* (read as "triangle *ABC*").

> **The Language of Algebra**   When naming a triangle, we may begin with any vertex. Then we move around the figure in a clockwise (or counterclockwise) direction as we list the remaining vertices. Other ways of naming the triangle shown here are △*ACB*, △*BCA*, △*BAC*, △*CAB*, and △*CBA*.
>
>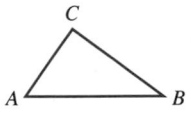

> **The Language of Algebra**   The figures below show how triangles can be classified according to the lengths of their sides. The single **tick marks** drawn on each side of the equilateral triangle indicate that the sides are of equal length. The double tick marks drawn on two of the sides of the isosceles triangle indicate that they have the same length. Each side of the scalene triangle has a different number of tick marks to indicate that the sides have different lengths.

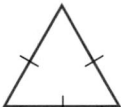 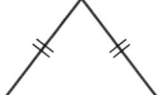 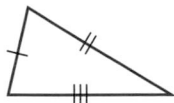

Equilateral triangle   **Isosceles triangle**   **Scalene triangle**
(all sides equal length)   (at least two sides of   (no sides of equal length)
                                    equal length)

> **The Language of Algebra**   Since every angle of an equilateral triangle has the same measure, an equilateral triangle is also equiangular.

> **The Language of Algebra**   Since equilateral triangles have at least two sides of equal length, they are also isosceles. However, isosceles triangles are not necessarily equilateral.

Triangles may also be classified by their angles, as shown below.

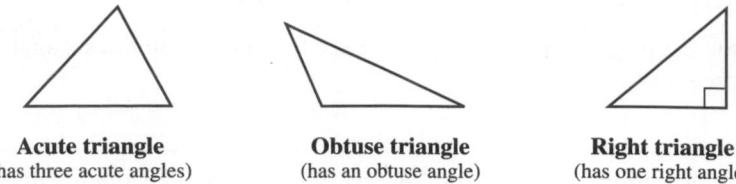

Acute triangle
(has three acute angles)

Obtuse triangle
(has an obtuse angle)

Right triangle
(has one right angle)

Right triangles have many real-life applications. For example, in figure (a) below, we see that a right triangle is formed when a ladder leans against the wall of a building.

The longest side of a right triangle is called the **hypotenuse,** and the other two sides are called **legs.** The hypotenuse of a right triangle is always opposite the 90° (right) angle. The legs of a right triangle are adjacent to (next to) the right angle, as shown in figure (b).

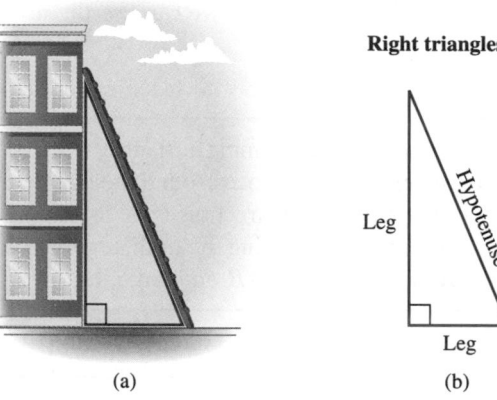

(a)   (b)

3 Identify isosceles triangles.

In an isosceles triangle, the angles opposite the sides of equal length are called **base angles,** the sides of equal length form the **vertex angle,** and the third side is called the **base.** Two examples of isosceles triangles are shown below.

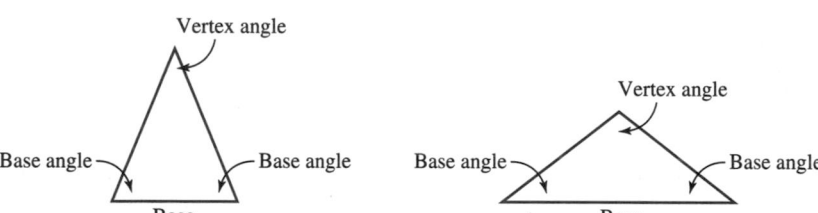

We have seen that isosceles triangles have two sides of equal length. The **isosceles triangle theorem** states that such triangles have one other important characteristic: Their base angles are congruent.

Isosceles Triangle Theorem

If two sides of a triangle are congruent, then the angles opposite those sides are congruent.

The Language of Algebra **Tick marks** can be used to denote the sides of a triangle that have the same length. They can also be used to indicate the angles of a triangle with the same measure. For example, we can show that the base angles of the isosceles triangle below are congruent by using single tick marks.

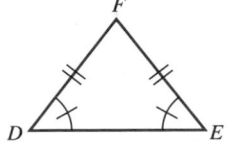

∠D is opposite $\overline{FE}$, and ∠E is opposite $\overline{FD}$. By the isosceles triangle theorem, if m($\overline{FD}$) = m($\overline{FE}$), then m(∠D) = m(∠E).

If a mathematical statement is written in the form *if p . . . , then q . . .* , we call the statement *if q . . . , then p . . .* its **converse.** The converses of some statements are true, while the converses of other statements are false. It is interesting to note that the converse of the isosceles triangle theorem is true.

Converse of the Isosceles Triangle Theorem

If two angles of a triangle are congruent, then the sides opposite the angles have the same length, and the triangle is isosceles.

EXAMPLE 2   Is the triangle shown here an isosceles triangle?

Strategy We will consider the measures of the angles of the triangle.

WHY If two angles of a triangle are congruent, then the sides opposite the angles have the same length, and the triangle is isosceles.

Solution ∠A and ∠B have the same measure, 50°. By the converse of the isosceles triangle theorem, if m(∠A) = m(∠B), we know that m($\overline{BC}$) = m($\overline{AC}$) and that △ABC is isosceles.

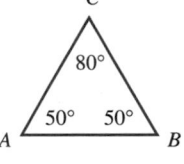

Self Check 2

Is the triangle shown below an isosceles triangle?

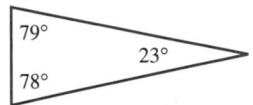

Now Try Problems 33 and 35

4 Find unknown angle measures of triangles.

If you draw several triangles and carefully measure each angle with a protractor, you will find that the sum of the angle measures of each triangle is 180°. Two examples are shown below.

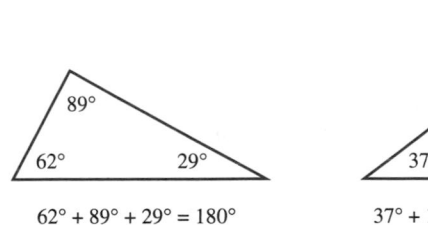

62° + 89° + 29° = 180°   37° + 110° + 33° = 180°

Another way to show this important fact about the sum of the angle measures of a triangle is discussed in Problem 82 of the Study Set at the end of this section.

Angles of a Triangle

The sum of the angle measures of any triangle is 180°.

Self Check 3

In the figure, find y.

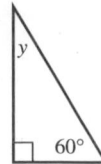

Now Try Problem 37

EXAMPLE 3   In the figure, find x.

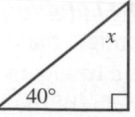

Strategy We will use the fact that the sum of the angle measures of any triangle is 180° to write an equation that models the situation.

WHY We can then solve the equation to find the unknown angle measure, x.

Solution Since the sum of the angle measures of any triangle is 180°, we have

$x + 40° + 90° = 180°$ The ⌐ symbol indicates that the measure of the angle is 90°.

$$\begin{array}{r} 90 \\ +\,40 \\ \hline 130 \end{array}$$

$x + 130° = 180°$ Do the addition.

$x = 50°$ To isolate x, undo the addition of 130° by subtracting 130° from both sides.

Thus, x is 50°.

Self Check 4

In $\triangle DEF$, the measure of $\angle D$ exceeds the measure of $\angle E$ by 5°, and the measure of $\angle F$ is three times the measure of $\angle E$. Find the measure of each angle of $\triangle DEF$.

Now Try Problem 41

EXAMPLE 4   In the figure, find the measure of each angle of $\triangle ABC$.

Strategy We will use the fact that the sum of the angle measures of any triangle is 180° to write an equation that models the situation.

WHY We can then solve the equation to find the unknown angle measure x, and use it to evaluate the expressions $2x$ and $x + 32°$.

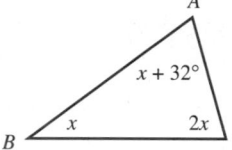

Solution

$x + 32° + x + 2x = 180°$ The sum of the angle measures of any triangle is 180°.

$$\begin{array}{r} {}^{7\,10} \\ 18\cancel{0} \\ -\,32 \\ \hline 148 \end{array}$$

$4x + 32° = 180°$ Combine like terms: $x + x + 2x = 4x$.

$4x + 32° - \mathbf{32}° = 180° - \mathbf{32}°$ To isolate the variable term, $4x$, subtract 32° from both sides.

$$\begin{array}{r} 37 \\ 4\overline{)148} \\ -\,12 \\ \hline 28 \\ -\,28 \\ \hline 0 \end{array}$$

$4x = 148°$ Do the subtractions.

$\dfrac{4x}{4} = \dfrac{148°}{4}$ To isolate x, divide both sides by 4.

$x = 37°$ Do the divisions. This is the measure of $\angle B$.

To find the measures of $\angle A$ and $\angle C$, we evaluate the expressions $x + 32°$ and $2x$ for $x = 37°$.

$x + 32° = \mathbf{37}° + 32°$ Substitute 37 for x. | $2x = 2(\mathbf{37}°)$ Substitute 37 for x.

$= 69°$ | $= 74°$

The measure of $\angle B$ is 37°, the measure of $\angle A$ is 69°, and the measure of $\angle C$ is 74°. ∎

Self Check 5

If one base angle of an isosceles triangle measures 33°, what is the measure of the vertex angle?

Now Try Problem 45

EXAMPLE 5   If one base angle of an isosceles triangle measures 70°, what is the measure of the vertex angle?

Strategy We will use the isosceles triangle theorem and the fact that the sum of the angle measures of any triangle is 180° to write an equation that models the situation.

WHY We can then solve the equation to find the unknown angle measure.

Solution By the isosceles triangle theorem, if one of the base angles measures 70°, so does the other. (See the figure on the right.) If we let x represent the measure of the vertex angle, we have

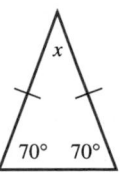

$x + 70° + 70° = 180°$ The sum of the measures of the angles of a triangle is 180°.

$x + 140° = 180°$ Combine like terms: 70° + 70° = 140°.

$x = 40°$ To isolate x, undo the addition of 140° by subtracting 140° from both sides.

The vertex angle measures 40°.

EXAMPLE 6   If the vertex angle of an isosceles triangle measures 99°, what are the measures of the base angles?

Strategy We will use the fact that the base angles of an isosceles triangle have the same measure and the sum of the angle measures of any triangle is 180° to write an equation that mathematically models the situation.

WHY We can then solve the equation to find the unknown angle measures.

Solution The base angles of an isosceles triangle have the same measure. If we let x represent the measure of one base angle, the measure of the other base angle is also x. (See the figure to the right.) Since the sum of the measures of the angles of any triangle is 180°, the sum of the measures of the base angles and of the vertex angle is 180°. We can use this fact to form an equation.

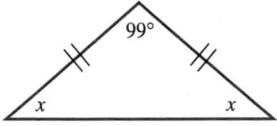

$x + x + 99° = 180°$

$2x + 99° = 180°$ Combine like terms: x + x = 2x.

$2x = 81°$ To isolate the variable term, 2x, undo the addition of 99° by subtracting 99° from both sides.

$\dfrac{2x}{2} = \dfrac{81°}{2}$ To isolate x, undo the multiplication by 2 by dividing both sides by 2.

$x = 40.5°$

$$\begin{array}{r} 40.5 \\ 2\overline{)81.0} \\ -8 \\ \hline 01 \\ -0 \\ \hline 1\,0 \\ -1\,0 \\ \hline 0 \end{array}$$

The measure of each base angle is 40.5°.

Self Check 6

If the vertex angle of an isosceles triangle measures 57°, what are the measures of the base angles?

Now Try Problem 49

ANSWERS TO SELF CHECKS

1. a. 4 **b.** 5   **2.** no   **3.** 30°   **4.** m($\angle D$) = 40°, m($\angle E$) = 35°, m($\angle F$) = 105°
5. 114°   **6.** 61.5°

SECTION 9.3   STUDY SET

VOCABULARY

Fill in the blanks.

1. A _____ is a closed geometric figure with at least three line segments for its sides.

2. The polygon shown to the right has seven _____ and seven vertices.

3. A point where two sides of a polygon intersect is called a _____ of the polygon.

4. A _____ polygon has sides that are all the same length and angles that all have the same measure.

5. A triangle with three sides of equal length is called an _____ triangle. An _____ triangle has at least two sides of equal length. A _____ triangle has no sides of equal length.

6. An _____ triangle has three acute angles. An _____ triangle has one obtuse angle. A _____ triangle has one right angle.

7. The longest side of a right triangle is called the _____. The other two sides of a right triangle are called _____.

8. The _____ angles of an isosceles triangle have the same measure. The sides of equal length of an isosceles triangle form the _____ angle.

9. In this section, we discussed the sum of the measures of the angles of a triangle. The word *sum* indicates the operation of _____.

10. Complete the table.

Number of Sides	Name of Polygon
3	
4	
5	
6	
7	
8	
9	
10	
12	

CONCEPTS

11. Draw an example of each type of regular polygon.

   a. hexagon   **b.** octagon

   c. quadrilateral   **d.** triangle

   e. pentagon   **f.** decagon

12. Refer to the triangle below.

   a. What are the names of the vertices of the triangle?

   b. How many sides does the triangle have? Name them.

   c. Use the vertices to name this triangle in three ways.

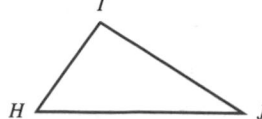

13. Draw an example of each type of triangle.

   a. isosceles   **b.** equilateral

   c. scalene   **d.** obtuse

   e. right   **f.** acute

14. Classify each triangle as an acute, an obtuse, or a right triangle.

   a. 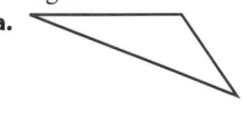   **b.**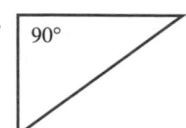

   c. 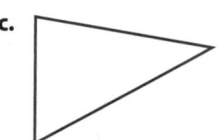   **d.**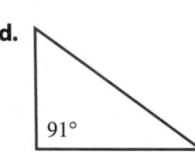

15. Refer to the triangle shown below.

   a. What is the measure of ∠B?

   b. What type of triangle is it?

   c. What two line segments form the legs?

   d. What line segment is the hypotenuse?

   e. Which side of the triangle is the longest?

   f. Which side is opposite ∠B?

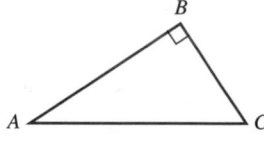

16. Fill in the blanks.

   a. The sides of a right triangle that are adjacent to the right angle are called the _____.

   b. The hypotenuse of a right triangle is the side _____ the right angle.

17. Fill in the blanks.

   a. The _____ triangle theorem states that if two sides of a triangle are congruent, then the angles opposite those sides are congruent.

   b. The _____ of the isosceles triangle theorem states that if two angles of a triangle are congruent, then the sides opposite the angles have the same length, and the triangle is isosceles.

18. Refer to the given triangle.

   a. What two sides are of equal length?

   b. What type of triangle is △XYZ?

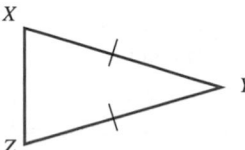

c. Name the base angles.

d. Which side is opposite $\angle X$?

e. What is the vertex angle?

f. Which angle is opposite side $\overline{XY}$?

g. Which two angles are congruent?

19. Refer to the triangle below.

a. What do we know about $\overline{EF}$ and $\overline{GF}$?

b. What type of triangle is $\triangle EFG$?

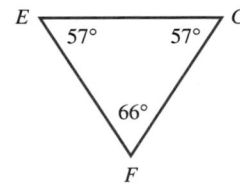

20. a. Find the sum of the measures of the angles of $\triangle JKL$, shown in figure (a).

b. Find the sum of the measures of the angles of $\triangle CDE$, shown in figure (b).

c. What is the sum of the measures of the angles of *any* triangle?

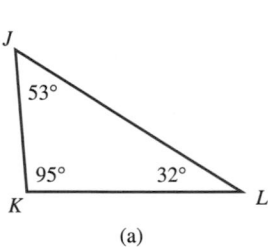 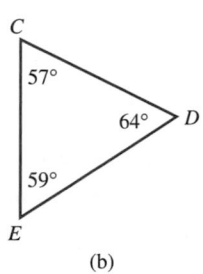

(a)   (b)

NOTATION

Fill in the blanks.

21. The symbol $\triangle$ means _____.

22. The symbol m($\angle A$) means the _____ of angle A.

Refer to the triangle below.

23. What fact about the sides of $\triangle ABC$ do the tick marks indicate?

24. What fact about the angles of $\triangle ABC$ do the tick marks indicate?

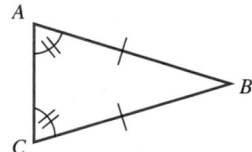

For each polygon, give the number of sides it has, give its name, and then give the number of vertices that it has. See Example 1.

25. a. 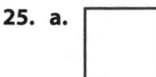   **b.**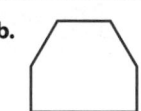

26. a. 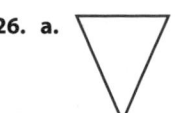   **b.**

27. a. 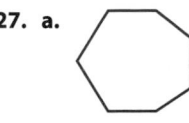   **b.**

28. a. 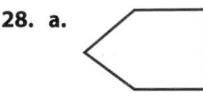   **b.**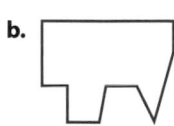

Classify each triangle as an equilateral triangle, an isosceles triangle, or a scalene triangle. See Objective 2.

29. a. 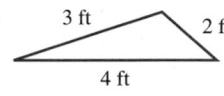   **b.**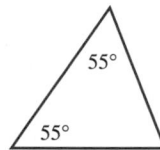

30. a. 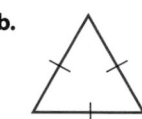   **b.**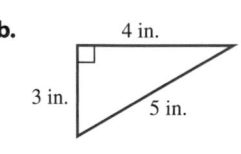

31. a. 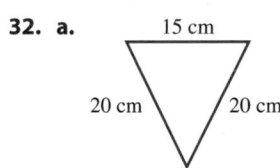   **b.**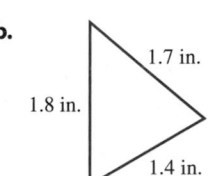

32. a. 15 cm, 20 cm, 20 cm   **b.** 1.7 in., 1.8 in., 1.4 in.

State whether each of the triangles is an isosceles triangle.
See Example 2.

33.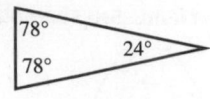

34.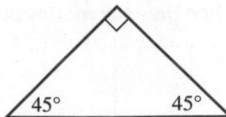

35.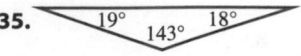

36.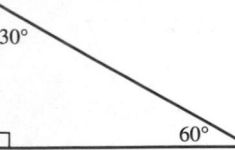

Find y. **See Example 3.**

37.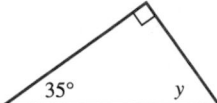

38.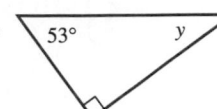

39.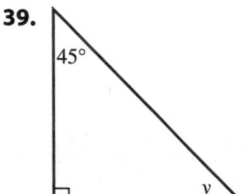

40.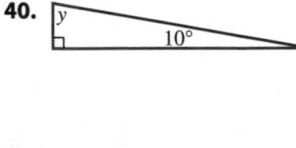

The degree measures of the angles of a triangle are represented by algebraic expressions. First find x. Then determine the measure of each angle of the triangle. **See Example 4.**

41.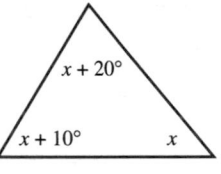

42.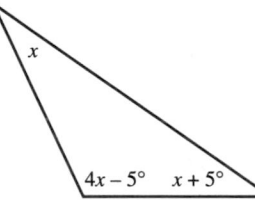

43.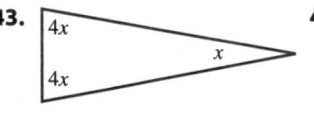

44.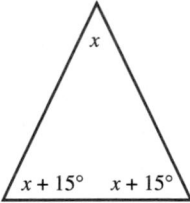

Find the measure of the vertex angle of each isosceles triangle given the following information. **See Example 5.**

45. The measure of one base angle is 56°.

46. The measure of one base angle is 68°.

47. The measure of one base angle is 85.5°.

48. The measure of one base angle is 4.75°.

Find the measure of one base angle of each isosceles triangle given the following information. **See Example 6.**

49. The measure of the vertex angle is 102°.

50. The measure of the vertex angle is 164°.

51. The measure of the vertex angle is 90.5°.

52. The measure of the vertex angle is 2.5°.

Find the measure of each vertex angle.

53.

54.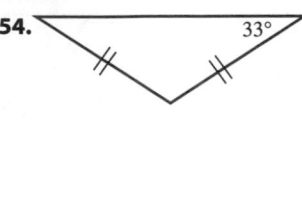

55.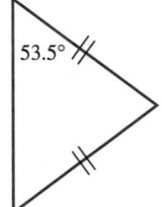

56.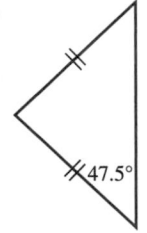

The measures of two angles of $\triangle ABC$ are given. Find the measure of the third angle.

57. m($\angle A$) = 30° and m($\angle B$) = 60°; find m($\angle C$).

58. m($\angle A$) = 45° and m($\angle C$) = 105°; find m($\angle B$).

59. m($\angle B$) = 100° and m($\angle A$) = 35°; find m($\angle C$).

60. m($\angle B$) = 33° and m($\angle C$) = 77°; find m($\angle A$).

61. m($\angle A$) = 25.5° and m($\angle B$) = 63.8°; find m($\angle C$).

62. m($\angle B$) = 67.25° and m($\angle C$) = 72.5°; find m($\angle A$).

63. m($\angle A$) = 29° and m($\angle C$) = 89.5°; find m($\angle B$).

64. m($\angle A$) = 4.5° and m($\angle B$) = 128°; find m($\angle C$).

In Problems 65–68, find x.

65.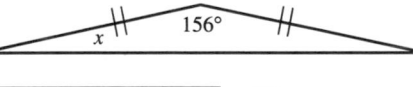

66.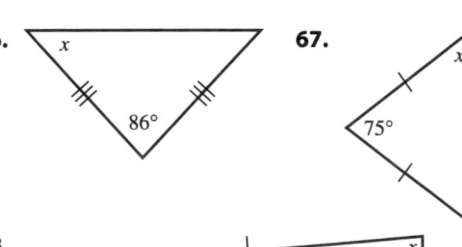

67.

68.

69. One angle of an isosceles triangle has a measure of 39°. What are the possible measures of the other angles?

70. One angle of an isosceles triangle has a measure of 2°. What are the possible measures of the other angles?

71. Find m($\angle C$).

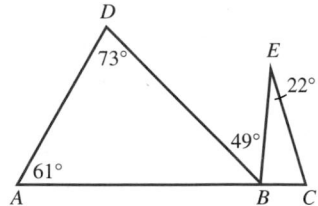

72. Find:
  a. m($\angle MXZ$)
  b. m($\angle MYN$)

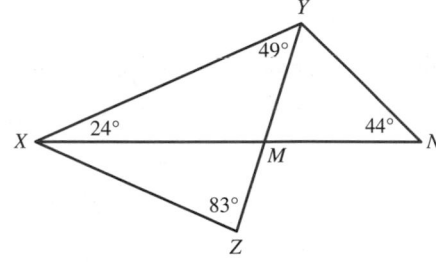

73. Find m($\angle NOQ$).

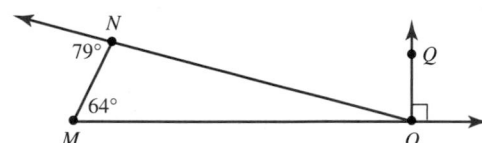

74. Find m($\angle S$).

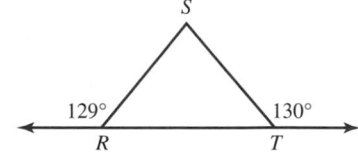

APPLICATIONS

75. POLYGONS IN NATURE As seen below, a starfish fits the shape of a pentagon. What polygon shape do you see in each of the other objects?
  a. lemon
  b. chili pepper
  c. apple

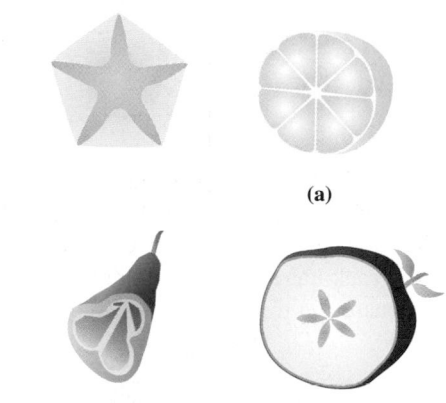

(a)

(b)   (c)

76. CHEMISTRY Polygons are used to represent the chemical structure of compounds. In the figure below, what types of polygons are used to represent methylprednisolone, the active ingredient in an anti-inflammatory medication?

Methylprednisolone

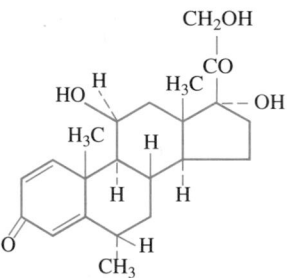

77. AUTOMOBILE JACK Refer to the figure below. No matter how high the jack is raised, it always forms two isosceles triangles. Explain why.

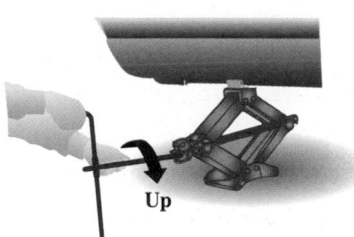

78. EASELS Refer to the figure below. What type of triangle studied in this section is used in the design of the legs of the easel?

79. POOL The rack shown below is used to set up the billiard balls when beginning a game of pool. Although it does not meet the strict definition of a polygon, the rack has a shape much like a type of triangle discussed in this section. Which type of triangle?

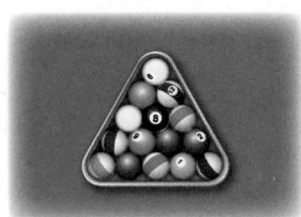

80. DRAFTING Among the tools used in drafting are the two clear plastic triangles shown below. Classify each according to the lengths of its sides and then according to its angle measures.

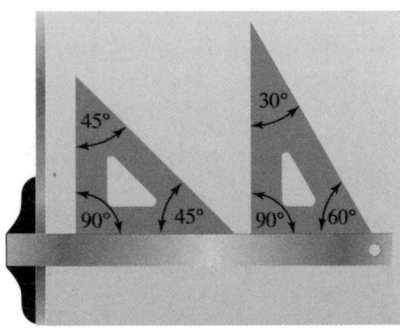

WRITING

81. In this section, we discussed the definition of a pentagon. What is *the* Pentagon? Why is it named that?

82. A student cut a triangular shape out of blue construction paper and labeled the angles ∠1, ∠2, and ∠3, as shown in figure (a) below. Then she tore off each of the three corners and arranged them as shown in figure (b). Explain what important geometric concept this model illustrates.

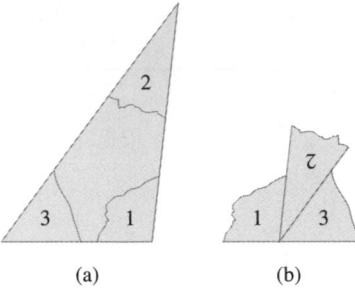

(a)   (b)

83. Explain why a triangle cannot have two right angles.

84. Explain why a triangle cannot have two obtuse angles.

REVIEW

85. Find 20% of 110.

86. Find 15% of 50.

87. What percent of 200 is 80?

88. 20% of what number is 500?

89. Evaluate: $0.85 \div 2(0.25)$

90. FIRST AID When checking an accident victim's pulse, a paramedic counted 13 beats during a 15-second span. How many beats would be expected in 60 seconds?

SECTION 9.4
The Pythagorean Theorem

A **theorem** is a mathematical statement that can be proven. In this section, we will discuss one of the most widely used theorems of geometry—the Pythagorean theorem. It is named after Pythagoras, a Greek mathematician who lived about 2,500 years ago. He is thought to have been the first to develop a proof of it. The Pythagorean theorem expresses the relationship between the lengths of the sides of any right triangle.

1 Use the Pythagorean theorem to find the exact length of a side of a right triangle.

Recall that a right triangle is a triangle that has a right angle (an angle with measure 90°). In a right triangle, the longest side is called the **hypotenuse.** It is the side opposite the right angle. The other two sides are called **legs.** It is common practice to let the variable c represent the length of the hypotenuse and the variables a and b represent the lengths of the legs, as shown on the right.

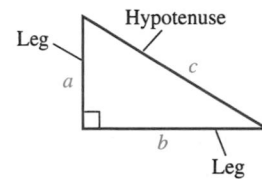

Pythagoras
© SEF/Art Resource, NY

If we know the lengths of any two sides of a right triangle, we can find the length of the third side using the **Pythagorean theorem.**

Pythagorean Theorem

If a and b are the lengths of two legs of a right triangle and c is the length of the hypotenuse, then

$$a^2 + b^2 = c^2$$

In words, the Pythagorean theorem is expressed as follows:

In a right triangle, the sum of the squares of the lengths of the two legs is equal to the square of the length of the hypotenuse.

Caution! When using the **Pythagorean equation** $a^2 + b^2 = c^2$, we can let a represent the length of either leg of the right triangle. We then let b represent the length of the other leg. The variable c must always represent the length of the hypotenuse.

EXAMPLE 1   Find the length of the hypotenuse of the right triangle shown here.

Strategy We will use the Pythagorean theorem to find the length of the hypotenuse.

WHY If we know the lengths of any two sides of a right triangle, we can find the length of the third side using the Pythagorean theorem.

3 in.
4 in.

Self Check 1

Find the length of the hypotenuse of the right triangle shown below.

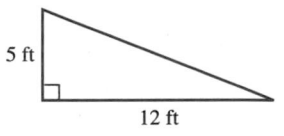

5 ft
12 ft

Now Try Problem 15

Solution We will let $a = 3$ and $b = 4$, and substitute into the Pythagorean equation to find c.

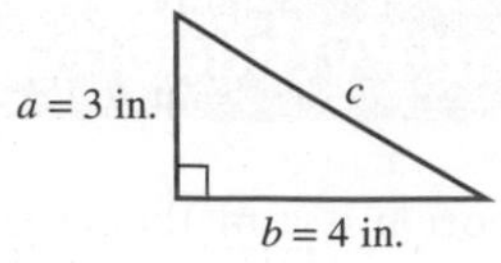

$a^2 + b^2 = c^2$ This is the Pythagorean equation.

$3^2 + 4^2 = c^2$ Substitute 3 for a and 4 for b.

$9 + 16 = c^2$ Evaluate each exponential expression.

$25 = c^2$ Do the addition.

$c^2 = 25$ Reverse the sides of the equation so that c^2 is on the left.

[figure: right triangle with $a = 3$ in. (vertical), $b = 4$ in. (horizontal), hypotenuse c]

To find c, we must find a number that, when squared, is 25. There are two such numbers, one positive and one negative; they are the square roots of 25. Since c represents the length of a side of a triangle, c cannot be negative. For this reason, we need only find the positive square root of 25 to get c.

$c = \sqrt{25}$ The symbol $\sqrt{}$ is used to indicate the positive square root of a number.

$c = 5$ $\sqrt{25} = 5$ because $5^2 = 25$.

The length of the hypotenuse is 5 in.

Success Tip The Pythagorean theorem is used to find the lengths of sides of right triangles. A calculator with a square root key $\boxed{\sqrt{}}$ is often helpful in the final step of the solution process when we must find the positive square root of a number.

Self Check 2

In Example 2, can the crews communicate by radio if the distance from point B to point C remains the same but the distance from point A to point C increases to 2,520 yards?

Now Try Problems 19 and 43

EXAMPLE 2   *Firefighting* To fight a forest fire, the forestry department plans to clear a rectangular fire break around the fire, as shown in the following figure. Crews are equipped with mobile communications that have a 3,000-yard range. Can crews at points A and B remain in radio contact?

Strategy We will use the Pythagorean theorem to find the distance between points A and B.

WHY If the distance is less than 3,000 yards, the crews can communicate by radio. If it is greater than 3,000 yards, they cannot.

Solution The line segments connecting points A, B, and C form a right triangle. To find the distance c from point A to point B, we can use the Pythagorean equation, substituting 2,400 for a and 1,000 for b and solving for c.

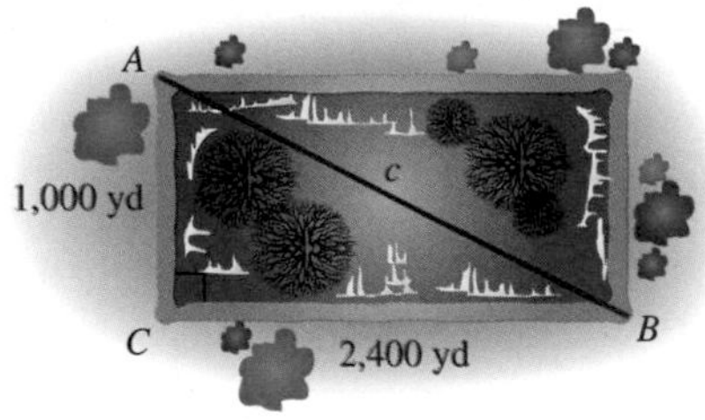

[figure: rectangular fire break with points A (top left), B (bottom right), C (bottom left); left side labeled 1,000 yd; bottom labeled 2,400 yd; diagonal labeled c]

$a^2 + b^2 = c^2$ This is the Pythagorean equation.

$2,400^2 + 1,000^2 = c^2$ Substitute for a and b.

$5,760,000 + 1,000,000 = c^2$ Evaluate each exponential expression.

$6,760,000 = c^2$ Do the addition.

$c^2 = 6,760,000$ Reverse the sides of the equation so that c^2 is on the left.

$c = \sqrt{6,760,000}$ If $c^2 = 6,760,000$, then c must be a square root of 6,760,000. Because c represents a length, it must be the positive square root of 6,760,000.

$c = 2,600$ Use a calculator to find the square root.

The two crews are 2,600 yards apart. Because this distance is less than the 3,000-yard range of the radios, they can communicate by radio.

EXAMPLE 3   The lengths of two sides of a right triangle are given in the figure. Find the missing side length.

Strategy We will use the Pythagorean theorem to find the missing side length.

WHY If we know the lengths of any two sides of a right triangle, we can find the length of the third side using the Pythagorean theorem.

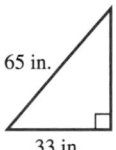

61 ft

11 ft

Solution We may substitute 11 for either a or b, but 61 must be substituted for the length c of the hypotenuse. If we choose to substitute 11 for b, we can find the unknown side length a as follows.

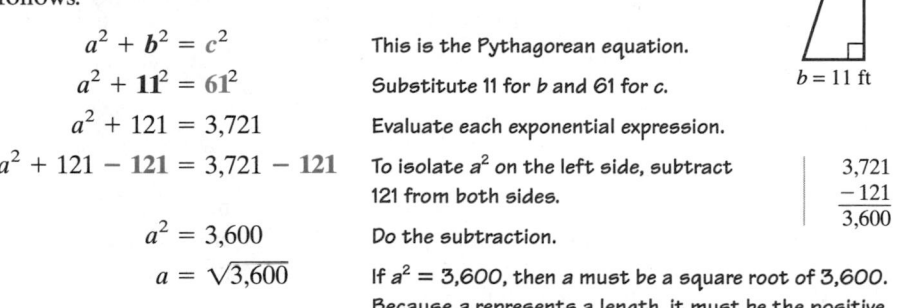

$c = 61$ ft a

$b = 11$ ft

$a^2 + b^2 = c^2$	This is the Pythagorean equation.
$a^2 + 11^2 = 61^2$	Substitute 11 for b and 61 for c.
$a^2 + 121 = 3{,}721$	Evaluate each exponential expression.
$a^2 + 121 - 121 = 3{,}721 - 121$	To isolate a^2 on the left side, subtract 121 from both sides.
$a^2 = 3{,}600$	Do the subtraction.
$a = \sqrt{3{,}600}$	If $a^2 = 3{,}600$, then a must be a square root of 3,600. Because a represents a length, it must be the positive square root of 3,600.
$a = 60$	Use a calculator, if necessary, to find the square root.

$$\begin{array}{r} 3{,}721 \\ -\,121 \\ \hline 3{,}600 \end{array}$$

The missing side length is 60 ft.

Self Check 3

The lengths of two sides of a right triangle are given. Find the missing side length.

65 in.

33 in.

Now Try Problem 23

---

2   Use the Pythagorean theorem to approximate the length of a side of a right triangle.

When we use the Pythagorean theorem to find the length of a side of a right triangle, the solution is sometimes the square root of a number that is not a perfect square. In that case, we can use a calculator to *approximate* the square root.

EXAMPLE 4   Refer to the right triangle shown here. Find the missing side length. Give the exact answer and an approximation to the nearest hundredth.

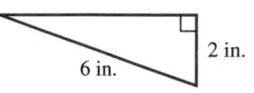

6 in.   2 in.

Strategy We will use the Pythagorean theorem to find the missing side length.

WHY If we know the lengths of any two sides of a right triangle, we can find the length of the third side using the Pythagorean theorem.

Solution We may substitute 2 for either a or b, but 6 must be substituted for the length c of the hypotenuse. If we choose to substitute 2 for a, we can find the unknown side length b as follows.

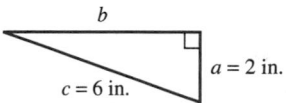

b

$c = 6$ in. $a = 2$ in.

$a^2 + b^2 = c^2$	This is the Pythagorean equation.
$2^2 + b^2 = 6^2$	Substitute 2 for a and 6 for c.
$4 + b^2 = 36$	Evaluate each exponential expression.
$4 + b^2 - 4 = 36 - 4$	To isolate b^2 on the left side, undo the addition of 4 by subtracting 4 from both sides.
$b^2 = 32$	Do the subtraction.

Self Check 4

Refer to the triangle below. Find the missing side length. Give the exact answer and an approximation to the nearest hundredth.

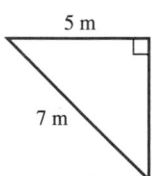

5 m

7 m

Now Try Problem 35

We must find a number that, when squared, is 32. Since b represents the length of a side of a triangle, we consider only the positive square root.

$b = \sqrt{32}$ *This is the exact length.*

The missing side length is exactly $\sqrt{32}$ inches long. Since 32 is not a perfect square, its square root is not a whole number. We can use a calculator to *approximate* $\sqrt{32}$. To the nearest hundredth, the missing side length is 5.66 inches.

$\sqrt{32}$ in. ≈ 5.66 in.

Using Your CALCULATOR   Finding the Width of a TV Screen

The size of a television screen is the diagonal measure of its rectangular screen. To find the length of a 27-inch screen that is 17 inches high, we use the Pythagorean theorem with $c = 27$ and $b = 17$.

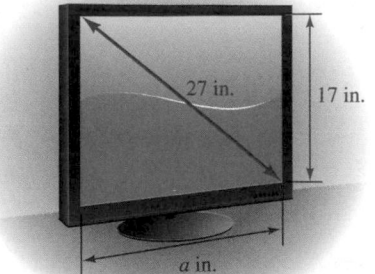

$$c^2 = a^2 + b^2$$
$$27^2 = a^2 + 17^2$$
$$27^2 - 17^2 = a^2$$

Since the variable a represents the length of the television screen, it must be positive. To find a, we find the positive square root of the result when 17^2 is subtracted from 27^2.

Using a radical symbol to indicate this, we have

$$\sqrt{27^2 - 17^2} = a$$

We can evaluate the expression on the left side by entering:

$(\;27\;\boxed{x^2}\;\boxed{-}\;17\;\boxed{x^2}\;\boxed{)}\;\boxed{\sqrt{}}$ $\boxed{20.97617696}$

To the nearest inch, the length of the television screen is 21 inches.

---

3 Use the converse of the Pythagorean theorem.

If a mathematical statement is written in the form *if p . . . , then q . . .* , we call the statement *if q . . . , then p . . .* its **converse.** The converses of some statements are true, while the converses of other statements are false. It is interesting to note that the converse of the Pythagorean theorem is true.

Converse of the Pythagorean Theorem

If a triangle has three sides of lengths a, b, and c, such that $a^2 + b^2 = c^2$, then the triangle is a right triangle.

Self Check 5

Is the triangle below a right triangle?

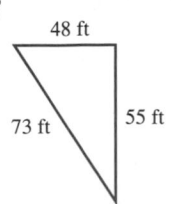

EXAMPLE 5   Is the triangle shown here a right triangle?

Strategy We will substitute the side lengths, 6, 8, and 11, into the Pythagorean equation $a^2 + b^2 = c^2$.

WHY By the converse of the Pythagorean theorem, the triangle is a right triangle if a true statement results. The triangle is not a right triangle if a false statement results.

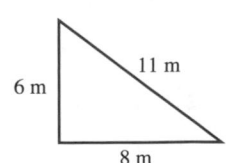

Solution We must substitute the longest side length, 11, for c, because it is the possible hypotenuse. The lengths of 6 and 8 may be substituted for either a or b.

Now Try **Problem 39**

$a^2 + b^2 = c^2$ This is the Pythagorean equation.

$6^2 + 8^2 \overset{?}{=} 11^2$ Substitute 6 for a, 8 for b, and 11 for c.

$36 + 64 \overset{?}{=} 121$ Evaluate each exponential expression.

$100 = 121$ This is a false statement.

$$\begin{array}{r} \overset{1}{3}6 \\ +64 \\ \hline 100 \end{array}$$

Since $100 \neq 121$, the triangle is not a right triangle.

ANSWERS TO SELF CHECKS

1. 13 ft   **2.** no   **3.** 56 in.   **4.** $\sqrt{24}$ m ≈ 4.90 m   **5.** yes

---

SECTION 9.4   STUDY SET

VOCABULARY

Fill in the blanks.

1. In a right triangle, the side opposite the 90° angle is called the _____. The other two sides are called ____.

2. The Pythagorean theorem is named after the Greek mathematician, _____, who is thought to have been the first to prove it.

3. The _____ theorem states that in any right triangle, the square of the length of the hypotenuse is equal to the sum of the squares of the lengths of the two legs.

4. $a^2 + b^2 = c^2$ is called the Pythagorean _____.

CONCEPTS

Fill in the blanks.

5. If a and b are the lengths of two legs of a right triangle and c is the length of the hypotenuse, then ▢ + ▢ = ▢ .

6. The two solutions of $c^2 = 36$ are $c = $ ▢ or $c = $ ▢ . If c represents the length of the hypotenuse of a right triangle, then we can discard the solution ▢ .

7. The converse of the Pythagorean theorem: If a triangle has three sides of lengths a, b, and c, such that $a^2 + b^2 = c^2$, then the triangle is a _____ triangle.

8. Use a protractor to draw an example of a right triangle.

9. Refer to the triangle on the right.
   a. What side is the hypotenuse?
   b. What side is the longer leg?
   c. What side is the shorter leg?

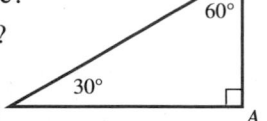

10. What is the first step when solving the equation $25 + b^2 = 81$ for b?

NOTATION

Complete the solution to solve the equation, where $a > 0$ and $c > 0$.

11. $8^2 + 6^2 = c^2$

   ▢ $+ 36 = c^2$

   ▢ $= c^2$

   $\sqrt{} = c$

   $10 = c$

12. $a^2 + 15^2 = 17^2$

   $a^2 + $ ▢ $ = $ ▢

   $a^2 + 225 - $ ▢ $ = 289 - $ ▢

   $a^2 = $ ▢

   $a = \sqrt{}$

   $a = $ ▢

GUIDED PRACTICE

Find the length of the hypotenuse of the right triangle shown below if it has the given side lengths. **See Examples 1 and 2.**

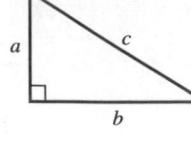

13. $a = 6$ ft and $b = 8$ ft

14. $a = 12$ mm and $b = 9$ mm

15. $a = 5$ m and $b = 12$ m

16. $a = 16$ in. and $b = 12$ in.

17. $a = 48$ mi and $b = 55$ mi

18. $a = 80$ ft and $b = 39$ ft

19. $a = 88$ cm and $b = 105$ cm

20. $a = 132$ mm and $b = 85$ mm

Refer to the right triangle below. **See Example 3.**

21. Find b if $a = 10$ cm and $c = 26$ cm.

22. Find b if $a = 14$ in. and $c = 50$ in.

23. Find a if $b = 18$ m and $c = 82$ m.

24. Find a if $b = 9$ yd and $c = 41$ yd.

25. Find a if $b = 21$ m and $c = 29$ m.

26. Find a if $b = 16$ yd and $c = 34$ yd.

27. Find b if $a = 180$ m and $c = 181$ m.

28. Find b if $a = 630$ ft and $c = 650$ ft.

The lengths of two sides of a right triangle are given. Find the missing side length. Give the exact answer and an approximation to the nearest hundredth. **See Example 4.**

29. $a = 5$ cm and $c = 6$ cm

30. $a = 4$ in. and $c = 8$ in.

31. $a = 12$ m and $b = 8$ m

32. $a = 10$ ft and $b = 4$ ft

33. $a = 9$ in. and $b = 3$ in.

34. $a = 5$ mi and $b = 7$ mi

35. $b = 4$ in. and $c = 6$ in.

36. $b = 9$ mm and $c = 12$ mm

Is a triangle with the following side lengths a right triangle? **See Example 5.**

37. 12, 14, 15

38. 15, 16, 22

39. 33, 56, 65

40. 20, 21, 29

APPLICATIONS

41. ADJUSTING LADDERS A 20-foot ladder reaches a window 16 feet above the ground. How far from the wall is the base of the ladder?

42. LENGTH OF GUY WIRES A 30-foot tower is to be fastened by three guy wires attached to the top of the tower and to the ground at positions 20 feet from its base. How much wire is needed? Round to the nearest tenth.

43. PICTURE FRAMES After gluing and nailing two pieces of picture frame molding together, a frame maker checks her work by making a diagonal measurement. If the sides of the frame form a right angle, what measurement should the frame maker read on the yardstick?

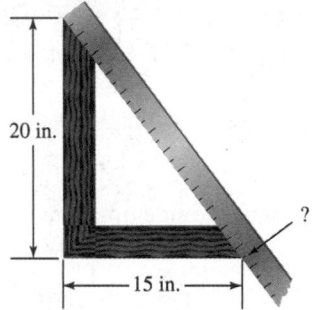

20 in.

15 in.

44. CARPENTRY The gable end of the roof shown is divided in half by a vertical brace, 8 feet in height. Find the length of the roof line.

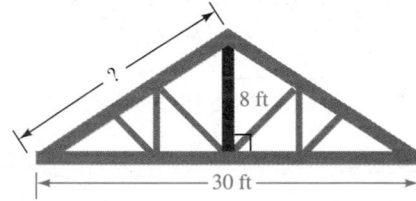

8 ft

30 ft

45. BASEBALL A baseball diamond is a square with each side 90 feet long. How far is it from home plate to second base? Round to the nearest hundredth.

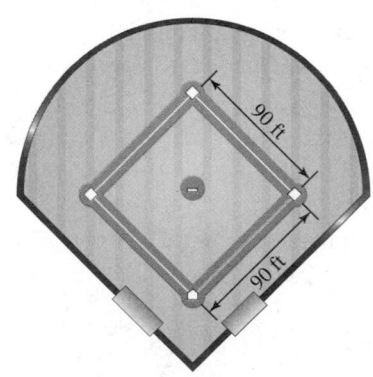

90 ft

90 ft

46. PAPER AIRPLANE The figure below gives the directions for making a paper airplane from a square piece of paper with sides 8 inches long. Find the length of the plane when it is completed in Step 3. Round to the nearest hundredth.

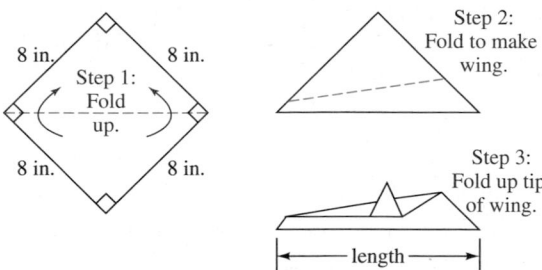

47. FIREFIGHTING The base of the 37-foot ladder shown in the figure below is 9 feet from the wall. Will the top reach a window ledge that is 35 feet above the ground? Explain how you arrived at your answer.

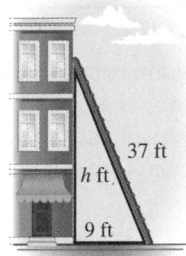

48. WIND DAMAGE A tree was blown over in a wind storm. Find the height of the tree when it was standing vertically upright.

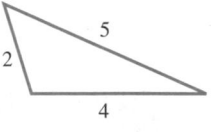

WRITING

49. State the Pythagorean theorem in your own words.

50. When the lengths of the sides of the triangle shown below are substituted into the equation $a^2 + b^2 = c^2$, the result is a false statement. Explain why.

$$a^2 + b^2 = c^2$$
$$2^2 + 4^2 = 5^2$$
$$4 + 16 = 25$$
$$20 = 25$$

51. In the figure below, equal-sized squares have been drawn on the sides of right triangle $\triangle ABC$. Explain how this figure demonstrates that $3^2 + 4^2 = 5^2$.

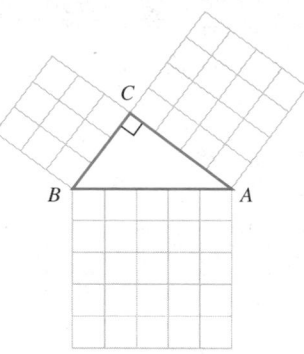

52. In the movie *The Wizard of Oz,* the scarecrow was in search of a brain. To prove that he had found one, he recited the following:

"The sum of the square roots of any two sides of an isosceles triangle is equal to the square root of the remaining side."

Unfortunately, this statement is not true. Correct it so that it states the Pythagorean theorem.

REVIEW

Use a check to determine whether the given number is a solution of the equation.

53. $2b + 3 = -15, -8$

54. $5t - 4 = -16, -2$

55. $0.5x = 2.9, 5$

56. $1.2 + x = 4.7, 3.5$

57. $33 - \dfrac{x}{2} = 30, -6$

58. $\dfrac{x}{4} + 98 = 100, -8$

59. $3x - 2 = 4x - 5, 12$

60. $5y + 8 = 3y - 2, 5$

Objectives

1 Identify corresponding parts of congruent triangles.

2 Use congruence properties to prove that two triangles are congruent.

3 Determine whether two triangles are similar.

4 Use similar triangles to find unknown lengths in application problems.

© iStockphoto.com/Lucinda Deitman

SECTION **9.5**

Congruent Triangles and Similar Triangles

In our everyday lives, we see many types of triangles. Triangular-shaped kites, sails, roofs, tortilla chips, and ramps are just a few examples. In this section, we will discuss how to compare the size and shape of two given triangles. From this comparison, we can make observations about their side lengths and angle measures.

1   Identify corresponding parts of congruent triangles.

Simply put, two geometric figures are congruent if they have the same shape and size. For example, if $\triangle ABC$ and $\triangle DEF$ shown below are congruent, we can write

$\triangle ABC \cong \triangle DEF$ Read as "Triangle ABC is congruent to triangle DEF."

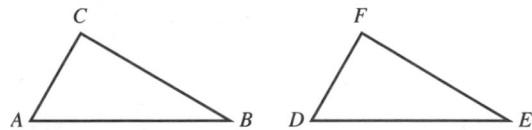

One way to determine whether two triangles are congruent is to see if one triangle can be moved onto the other triangle in such a way that it fits exactly. When we write $\triangle ABC \cong \triangle DEF$, we are showing how the vertices of one triangle are matched to the vertices of the other triangle to obtain a "perfect fit." We call this matching of points a **correspondence.**

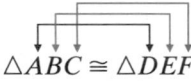

$$\triangle ABC \cong \triangle DEF$$

$A \leftrightarrow D$ Read as "Point A corresponds to point D."

$B \leftrightarrow E$ Read as "Point B corresponds to point E."

$C \leftrightarrow F$ Read as "Point C corresponds to point F."

When we establish a correspondence between the vertices of two congruent triangles, we also establish a correspondence between the angles and the sides of the triangles. Corresponding angles and corresponding sides of congruent triangles are called **corresponding parts.** *Corresponding parts of congruent triangles are always congruent.* That is, corresponding parts of congruent triangles always have the same measure. For the congruent triangles shown above, we have

$$m(\angle A) = m(\angle D) \qquad m(\angle B) = m(\angle E) \qquad m(\angle C) = m(\angle F)$$
$$m(\overline{BC}) = m(\overline{EF}) \qquad m(\overline{AC}) = m(\overline{DF}) \qquad m(\overline{AB}) = m(\overline{DE})$$

Congruent Triangles

Two triangles are congruent if and only if their vertices can be matched so that the corresponding sides and the corresponding angles are congruent.

EXAMPLE 1   Refer to the figure below, where $\triangle XYZ \cong \triangle PQR$.

a. Name the six congruent corresponding parts of the triangles.

b. Find $m(\angle P)$.

c. Find $m(\overline{XZ})$.

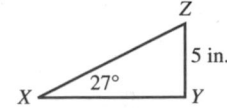

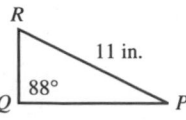

Strategy We will establish the correspondence between the vertices of △*XYZ* and the vertices of △*PQR*.

WHY This will, in turn, establish a correspondence between the congruent corresponding angles and sides of the triangles.

Solution

a. The correspondence between the vertices is

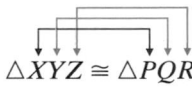

$$\triangle XYZ \cong \triangle PQR$$

$$X \leftrightarrow P \qquad Y \leftrightarrow Q \qquad Z \leftrightarrow R$$

Corresponding parts of congruent triangles are congruent. Therefore, the congruent corresponding angles are

$$\angle X \cong \angle P \qquad \angle Y \cong \angle Q \qquad \angle Z \cong \angle R$$

The congruent corresponding sides are

$$\overline{YZ} \cong \overline{QR} \qquad \overline{XZ} \cong \overline{PR} \qquad \overline{XY} \cong \overline{PQ}$$

b. From the figure, we see that $m(\angle X) = 27°$. Since $\angle X \cong \angle P$, it follows that $m(\angle P) = 27°$.

c. From the figure, we see that $m(\overline{PR}) = 11$ inches. Since $\overline{XZ} \cong \overline{PR}$, it follows that $m(\overline{XZ}) = 11$ inches. ∎

2 Use congruence properties to prove that two triangles are congruent.

Sometimes it is possible to conclude that two triangles are congruent without having to show that three pairs of corresponding angles are congruent and three pairs of corresponding sides are congruent. To do so, we apply one of the following properties.

SSS Property

If three sides of one triangle are congruent to three sides of a second triangle, the triangles are congruent.

We can show that the triangles shown below are congruent by the SSS property:

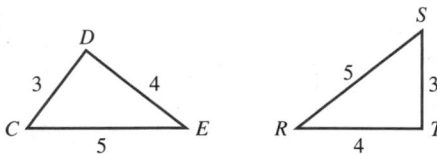

$\overline{CD} \cong \overline{ST}$ Since $m(\overline{CD}) = 3$ and $m(\overline{ST}) = 3$, the segments are congruent.

$\overline{DE} \cong \overline{TR}$ Since $m(\overline{DE}) = 4$ and $m(\overline{TR}) = 4$, the segments are congruent.

$\overline{EC} \cong \overline{RS}$ Since $m(\overline{EC}) = 5$ and $m(\overline{RS}) = 5$, the segments are congruent.

Therefore, $\triangle CDE \cong \triangle STR$.

SAS Property

If two sides and the angle between them in one triangle are congruent, respectively, to two sides and the angle between them in a second triangle, the triangles are congruent.

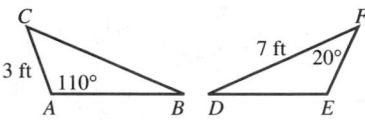

We can show that the triangles shown below are congruent by the SAS property:

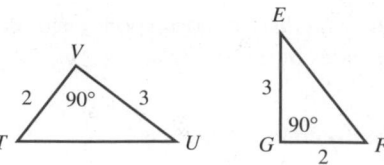

$\overline{TV} \cong \overline{FG}$ Since m($\overline{TV}$) = 2 and m($\overline{FG}$) = 2, the segments are congruent.

$\angle V \cong \angle G$ Since m($\angle V$) = 90° and m($\angle G$) = 90°, the angles are congruent.

$\overline{UV} \cong \overline{EG}$ Since m($\overline{UV}$) = 3 and m($\overline{EG}$) = 3, the segments are congruent.

Therefore, $\triangle TVU \cong \triangle FGE$.

ASA Property

If two angles and the side between them in one triangle are congruent, respectively, to two angles and the side between them in a second triangle, the triangles are congruent.

We can show that the triangles shown below are congruent by the ASA property:

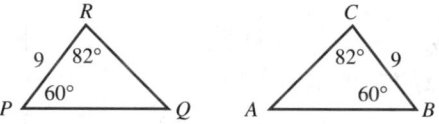

$\angle P \cong \angle B$ Since m($\angle P$) = 60° and m($\angle B$) = 60°, the angles are congruent.

$\overline{PR} \cong \overline{BC}$ Since m($\overline{PR}$) = 9 and m($\overline{BC}$) = 9, the segments are congruent.

$\angle R \cong \angle C$ Since m($\angle R$) = 82° and m($\angle C$) = 82°, the angles are congruent.

Therefore, $\triangle PQR \cong \triangle BAC$.

Caution!   There is no SSA property. To illustrate this, consider the triangles shown below. Two sides and an angle of $\triangle ABC$ are congruent to two sides and an angle of $\triangle DEF$. But the congruent angle is not between the congruent sides.

We refer to this situation as SSA. Obviously, the triangles are not congruent because they are not the same shape and size.

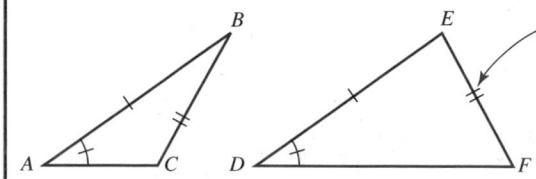

The tick marks indicate congruent parts. That is, the sides with one tick mark are the same length, the sides with two tick marks are the same length, and the angles with one tick mark have the same measure.

EXAMPLE 2   Explain why the triangles in the figure on the following page are congruent.

Strategy We will show that two sides and the angle between them in one triangle are congruent, respectively, to two sides and the angle between them in a second triangle.

WHY Then we know that the two triangles are congruent by the SAS property.

Solution Since vertical angles are congruent,

$$\angle 1 \cong \angle 2$$

From the figure, we see that

$$\overline{AC} \cong \overline{EC} \quad \text{and} \quad \overline{BC} \cong \overline{DC}$$

Since two sides and the angle between them in one triangle are congruent, respectively, to two sides and the angle between them in a second triangle, $\triangle ABC \cong \triangle EDC$ by the SAS property.

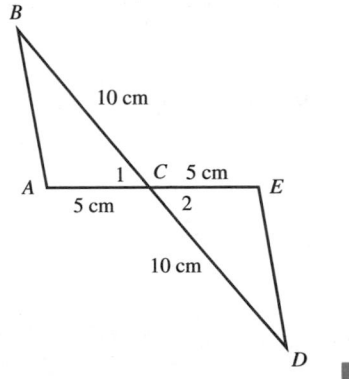

Self Check 2

Are the triangles in the figure below congruent? Explain why or why not.

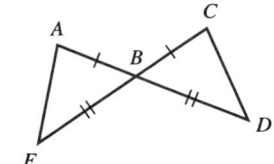

Now Try Problem 35

EXAMPLE 3 Are $\triangle RST$ and $\triangle RUT$ in the figure on the right congruent?

Strategy We will show that two angles and the side between them in one triangle are congruent, respectively, to two angles and the side between them in a second triangle.

WHY Then we know that the two triangles are congruent by the ASA property.

Solution From the markings on the figure, we know that two pairs of angles are congruent.

$\angle SRT \cong \angle URT$ These angles are marked with 1 tick mark, which indicates that they have the same measure.

$\angle STR \cong \angle UTR$ These angles are marked with 2 tick marks, which indicates that they have the same measure.

From the figure, we see that the triangles have side $\overline{RT}$ in common. Furthermore, $\overline{RT}$ is between each pair of congruent angles listed above. Since every segment is congruent to itself, we also have

$$\overline{RT} \cong \overline{RT}$$

Knowing that two angles and the side between them in $\triangle RST$ are congruent, respectively, to two angles and the side between them in $\triangle RUT$, we can conclude that $\triangle RST \cong \triangle RUT$ by the ASA property.

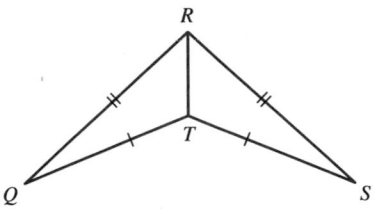

Self Check 3

Are the triangles in the following figure congruent? Explain why or why not.

Now Try Problem 37

---

3 Determine whether two triangles are similar.

We have seen that congruent triangles have the same shape and size. **Similar triangles** have the same shape, but not necessarily the same size. That is, one triangle is an exact scale model of the other triangle. If the triangles in the figure below are similar, we can write $\triangle ABC \sim \triangle DEF$ (read the symbol $\sim$ as "is similar to").

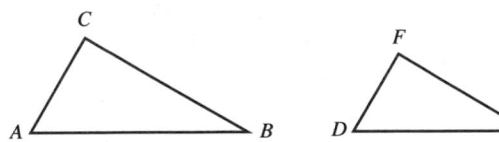

> **Success Tip** Note that congruent triangles are always similar, but similar triangles are not always congruent.

The formal definition of similar triangles requires that we establish a correspondence between the vertices of the triangles. The definition also involves the word *proportional*.

Recall that a **proportion** is a mathematical statement that two ratios (fractions) are equal. An example of a proportion is

$$\frac{1}{2} = \frac{4}{8}$$

In this case, we say that $\frac{1}{2}$ and $\frac{4}{8}$ are *proportional*.

> ## Similar Triangles
>
> Two triangles are similar if and only if their vertices can be matched so that corresponding angles are congruent and the lengths of corresponding sides are proportional.

Self Check 4

If $\triangle GEF \sim \triangle IJH$, name the congruent angles and the sides that are proportional.

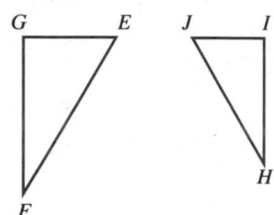

Now Try Problem 39

EXAMPLE 4   Refer to the figure below. If $\triangle PQR \sim \triangle CDE$, name the congruent angles and the sides that are proportional.

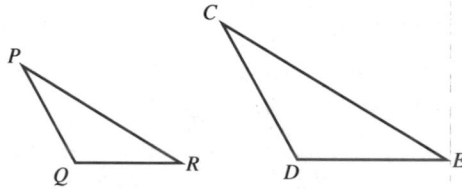

Strategy We will establish the correspondence between the vertices of $\triangle PQR$ and the vertices of $\triangle CDE$.

WHY This will, in turn, establish a correspondence between the congruent corresponding angles and proportional sides of the triangles.

Solution When we write $\triangle PQR \sim \triangle CDE$, a correspondence between the vertices of the triangles is established.

$$\triangle PQR \sim \triangle CDE$$

Since the triangles are similar, corresponding angles are congruent:

$$\angle P \cong \angle C \qquad \angle Q \cong \angle D \qquad \angle R \cong \angle E$$

The lengths of the corresponding sides are proportional. To simplify the notation, we will now let $PQ = \text{m}(\overline{PQ})$, $CD = \text{m}(\overline{CD})$, $QR = \text{m}(\overline{QR})$, and so on.

$$\frac{PQ}{CD} = \frac{QR}{DE} \qquad \frac{QR}{DE} = \frac{PR}{CE} \qquad \frac{PQ}{CD} = \frac{PR}{CE}$$

Written in a more compact way, we have

$$\frac{PQ}{CD} = \frac{QR}{DE} = \frac{PR}{CE}$$

> ## Property of Similar Triangles
>
> If two triangles are similar, all pairs of corresponding sides are in proportion.

It is possible to conclude that two triangles are similar without having to show that all three pairs of corresponding angles are congruent and that the lengths of all three pairs of corresponding sides are proportional.

AAA Similarity Theorem

If the angles of one triangle are congruent to corresponding angles of another triangle, the triangles are similar.

EXAMPLE 5   In the figure on the right, $\overline{PR} \parallel \overline{MN}$. Are $\triangle PQR$ and $\triangle NQM$ similar triangles?

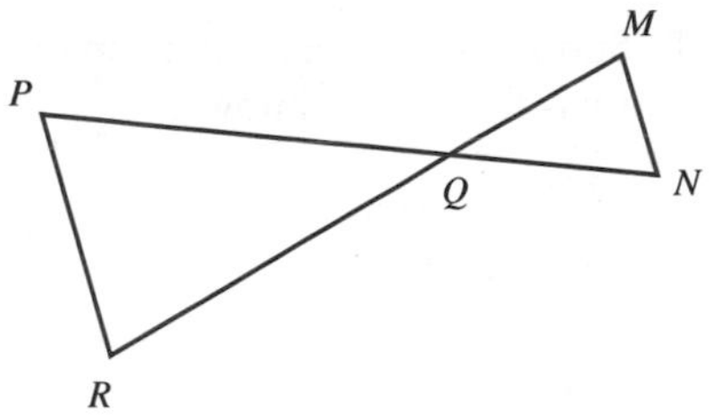

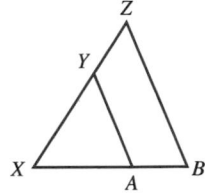

Strategy We will show that the angles of one triangle are congruent to corresponding angles of another triangle.

WHY Then we know that the two triangles are similar by the AAA property.

Solution Since vertical angles are congruent,

$\angle PQR \cong \angle NQM$ *This is one pair of congruent corresponding angles.*

In the figure, we can view $\overleftrightarrow{PN}$ as a transversal cutting parallel line segments $\overline{PR}$ and $\overline{MN}$. Since alternate interior angles are then congruent, we have:

$\angle RPQ \cong \angle MNQ$ *This is a second pair of congruent corresponding angles.*

Furthermore, we can view $\overleftrightarrow{RM}$ as a transversal cutting parallel line segments $\overline{PR}$ and $\overline{MN}$. Since alternate interior angles are then congruent, we have:

$\angle QRP \cong \angle QMN$ *This is a third pair of congruent corresponding angles.*

These observations are summarized in the figure on the right. We see that corresponding angles of $\triangle PQR$ are congruent to corresponding angles of $\triangle NQM$. By the AAA similarity theorem, we can conclude that

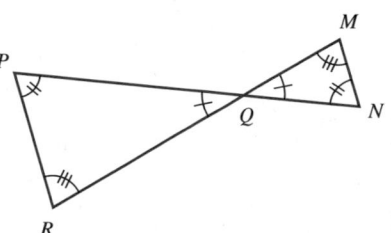

$\triangle PQR \sim \triangle NQM$

Self Check 5

In the figure below, $\overline{YA} \parallel \overline{ZB}$. Are $\triangle XYA$ and $\triangle XZB$ similar triangles?

Now Try **Problems 41 and 43**

EXAMPLE 6   In the figure below, $\triangle RST \sim \triangle JKL$. Find: **a.** x **b.** y

Strategy To find x, we will write a proportion of corresponding sides so that x is the only unknown. Then we will solve the proportion for x. We will use a similar method to find y.

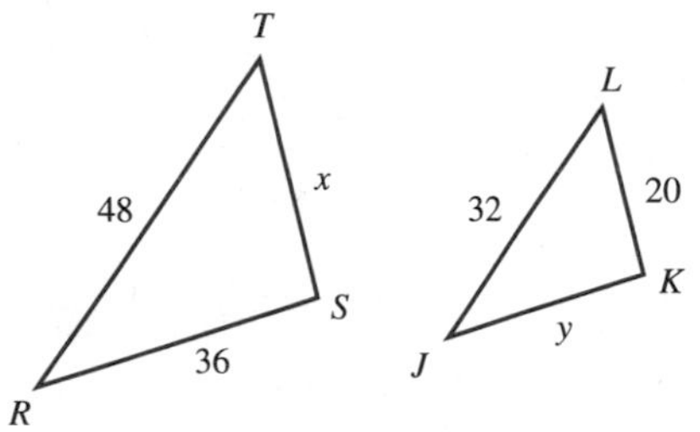

WHY Since $\triangle RST \sim \triangle JKL$, we know that the lengths of corresponding sides of $\triangle RST$ and $\triangle JKL$ are proportional.

Solution

a. When we write $\triangle RST \sim \triangle JKL$, a correspondence between the vertices of the two triangles is established.

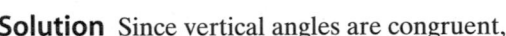

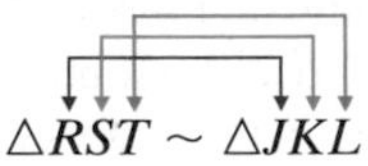

$\triangle RST \sim \triangle JKL$

Self Check 6

In the figure below, $\triangle DEF \sim \triangle GHI$. Find:

a. x **b.** y

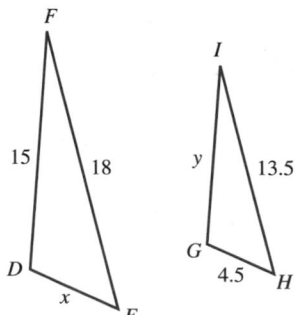

Now Try **Problem 53**

The lengths of corresponding sides of these similar triangles are proportional.

$$\frac{RT}{JL} = \frac{ST}{KL}$$ Each fraction is a ratio of a side length of $\triangle RST$ to its corresponding side length of $\triangle JKL$.

$$\frac{48}{32} = \frac{x}{20}$$ Substitute: $RT = 48$, $JL = 32$, $ST = x$, and $KL = 20$.

$$48(20) = 32x$$ Find each cross product and set them equal.

$$960 = 32x$$ Do the multiplication.

$$30 = x$$ To isolate x, undo the multiplication by 32 by dividing both sides by 32.

Thus, x is 30.

$$
\begin{array}{r}
48 \\
\times\,20 \\
\hline
960
\end{array}
$$

$$
\begin{array}{r}
30 \\
32\overline{)960} \\
-96 \\
\hline
00 \\
-00 \\
\hline
0
\end{array}
$$

b. To find y, we write a proportion of corresponding side lengths in such a way that y is the only unknown.

$$\frac{RT}{JL} = \frac{RS}{JK}$$

$$\frac{48}{32} = \frac{36}{y}$$ Substitute: $RT = 48$, $JL = 32$, $RS = 36$, and $JK = y$.

$$48y = 32(36)$$ Find each cross product and set them equal.

$$48y = 1,152$$ Do the multiplication.

$$y = 24$$ To isolate y, undo the multiplication by 48 by dividing both sides by 48.

Thus, y is 24.

$$
\begin{array}{r}
36 \\
\times\,32 \\
\hline
72 \\
1080 \\
\hline
1152
\end{array}
$$

$$
\begin{array}{r}
24 \\
48\overline{)1,152} \\
-96 \\
\hline
192 \\
-192 \\
\hline
0
\end{array}
$$

4 Use similar triangles to find unknown lengths in application problems.

Similar triangles and proportions can be used to find lengths that would normally be difficult to measure. For example, we can use the reflective properties of a mirror to calculate the height of a flagpole while standing safely on the ground.

Self Check 7

In the figure below, $\triangle ABC \sim \triangle EDC$. Find h.

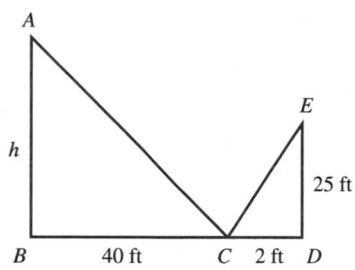

Now Try Problem 85

EXAMPLE 7   To determine the height of a flagpole, a woman walks to a point 20 feet from its base, as shown below. Then she takes a mirror from her purse, places it on the ground, and walks 2 feet farther away, where she can see the top of the pole reflected in the mirror. Find the height of the pole.

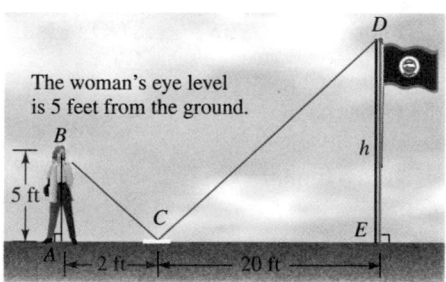

Strategy We will show that $\triangle ABC \sim \triangle EDC$.

WHY Then we can write a proportion of corresponding sides so that h is the only unknown and we can solve the proportion for h.

Solution To show that $\triangle ABC \sim \triangle EDC$, we begin by applying an important fact about mirrors. When a beam of light strikes a mirror, it is reflected at the same angle as it hits the mirror. Therefore, $\angle BCA \cong \angle DCE$. Furthermore, $\angle A \cong \angle E$ because the woman and the flagpole are perpendicular to the ground. Finally, if two pairs of corresponding angles are congruent, it follows that the third pair of corresponding angles are also congruent: $\angle B \cong \angle D$. By the AAA similarity theorem, we conclude that $\triangle ABC \sim \triangle EDC$.

Since the triangles are similar, the lengths of their corresponding sides are in proportion. If we let h represent the height of the flagpole, we can find h by solving the following proportion.

Height of the flagpole → $\dfrac{h}{5} = \dfrac{20}{2}$ ← Distance from flagpole to mirror

Height of the woman → ← Distance from woman to mirror

$$2h = 5(20) \qquad \text{Find each cross product and set them equal.}$$
$$2h = 100 \qquad \text{Do the multiplication.}$$
$$h = 50 \qquad \text{To isolate } h, \text{ divide both sides by 2.}$$

The flagpole is 50 feet tall.

ANSWERS TO SELF CHECKS

1. a. $\angle A \cong \angle E, \angle B \cong \angle D, \angle C \cong \angle F, \overline{AB} \cong \overline{ED}, \overline{BC} \cong \overline{DF}, \overline{CA} \cong \overline{FE}$ **b.** 20°   **c.** 3 ft
2. yes, by the SAS property   **3.** yes, by the SSS property   **4.** $\angle G \cong \angle I, \angle E \cong \angle J,$
$\angle F \cong \angle H; \dfrac{EG}{JI} = \dfrac{GF}{IH}, \dfrac{GF}{IH} = \dfrac{FE}{HJ}, \dfrac{EG}{JI} = \dfrac{FE}{HJ}$ **5.** yes, by the AAA similarity theorem:
$\angle X \cong \angle X, \angle XYA \cong \angle XZB, \angle XAY \cong \angle XBZ$ **6. a.** 6   **b.** 11.25   **7.** 500 ft

---

SECTION 9.5   STUDY SET

VOCABULARY

Fill in the blanks.

1. _____ triangles are the same size and the same shape.

2. When we match the vertices of $\triangle ABC$ with the vertices of $\triangle DEF$, as shown below, we call this matching of points a _____.

$$A \leftrightarrow D \qquad B \leftrightarrow E \qquad C \leftrightarrow F$$

3. Two angles or two line segments with the same measure are said to be _____.

4. Corresponding _____ of congruent triangles are congruent.

5. If two triangles are _____, they have the same shape but not necessarily the same size.

6. A mathematical statement that two ratios (fractions) are equal, such as $\frac{x}{18} = \frac{4}{9}$, is called a _____.

CONCEPTS

7. Refer to the triangles below.

a. Do these triangles appear to be congruent? Explain why or why not.

b. Do these triangles appear to be similar? Explain why or why not.

8. a. Draw a triangle that is congruent to $\triangle CDE$ shown below. Label it $\triangle ABC$.

b. Draw a triangle that is similar to, but not congruent to, $\triangle CDE$. Label it $\triangle MNO$.

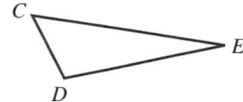

Fill in the blanks.

9. $\triangle XYZ \cong \triangle$ _____

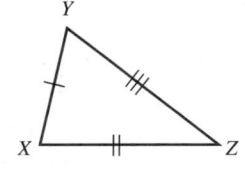 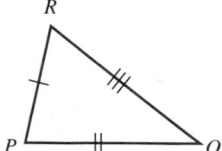

10. △ _____ ≅ △DEF

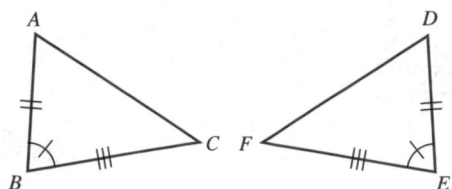

11. △RST ~ △_____

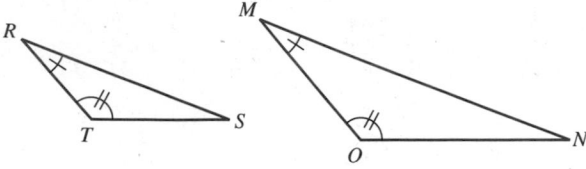

12. △ _____ ~ △TAC

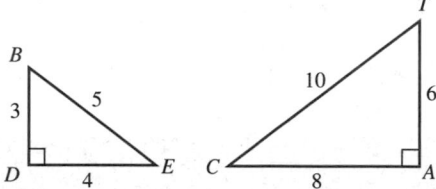

13. Name the six corresponding parts of the congruent triangles shown below.

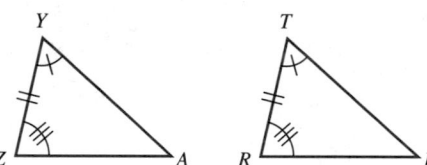

14. Name the six corresponding parts of the congruent triangles shown below.

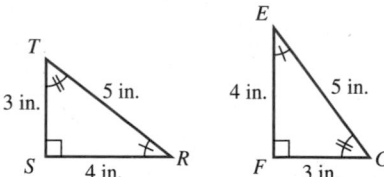

Fill in the blanks.

15. Two triangles are _____ if and only if their vertices can be matched so that the corresponding sides and the corresponding angles are congruent.

16. SSS property: If three _____ of one triangle are congruent to three _____ of a second triangle, the triangles are congruent.

17. SAS property: If two sides and the _____ between them in one triangle are congruent, respectively, to two sides and the _____ between them in a second triangle, the triangles are congruent.

18. ASA property: If two angles and the _____ between them in one triangle are congruent, respectively, to two angles and the _____ between them in a second triangle, the triangles are congruent.

Solve each proportion.

19. $\dfrac{x}{15} = \dfrac{20}{3}$

20. $\dfrac{5}{8} = \dfrac{35}{x}$

21. $\dfrac{h}{2.6} = \dfrac{27}{13}$

22. $\dfrac{11.2}{4} = \dfrac{h}{6}$

Fill in the blanks.

23. Two triangles are similar if and only if their vertices can be matched so that corresponding angles are congruent and the lengths of corresponding sides are _____.

24. If the angles of one triangle are congruent to corresponding angles of another triangle, the triangles are _____.

25. Congruent triangles are always similar, but similar triangles are not always _____.

26. For certain application problems, similar triangles and _____ can be used to find lengths that would normally be difficult to measure.

NOTATION

Fill in the blanks.

27. The symbol ≅ is read as "___ _____ ___."

28. The symbol ~ is read as "___ _____ ___."

29. Use tick marks to show the congruent parts of the triangles shown below.

$$\angle K \cong \angle H \qquad \overline{KR} \cong \overline{HJ} \qquad \angle M \cong \angle E$$

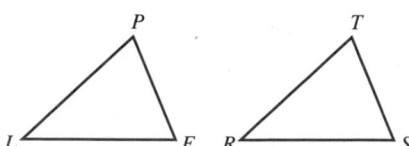

30. Use tick marks to show the congruent parts of the triangles shown below.

$$\angle P \cong \angle T \qquad \overline{LP} \cong \overline{RT} \qquad \overline{FP} \cong \overline{ST}$$

GUIDED PRACTICE

Name the six corresponding parts of the congruent triangles.
See Objective 1.

31. $\overline{AC} \cong$ ____

$\overline{DE} \cong$ ____

$\overline{BC} \cong$ ____

$\angle A \cong$ ____

$\angle E \cong$ ____

$\angle F \cong$ ____

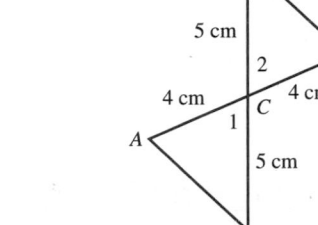

32. $\overline{AB} \cong$ ____

$\overline{EC} \cong$ ____

$\overline{AC} \cong$ ____

$\angle D \cong$ ____

$\angle B \cong$ ____

$\angle 1 \cong$ ____

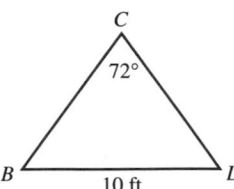 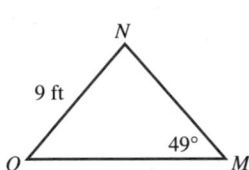

33. Refer to the figure below, where $\triangle BCD \cong \triangle MNO$.

  a. Name the six congruent corresponding parts of the triangles. **See Example 1.**
  b. Find m($\angle N$).
  c. Find m($\overline{MO}$).
  d. Find m($\overline{CD}$).

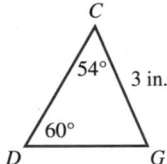

34. Refer to the figure below, where $\triangle DCG \cong \triangle RST$.

  a. Name the six congruent corresponding parts of the triangles. **See Example 1.**
  b. Find m($\angle R$).
  c. Find m($\overline{DG}$).
  d. Find m($\overline{ST}$).

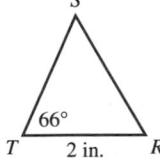

Determine whether each pair of triangles is congruent. If they are, tell why. **See Examples 2 and 3.**

35.

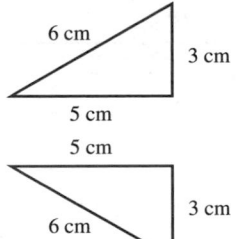

36.

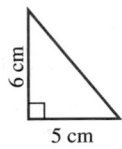

37.

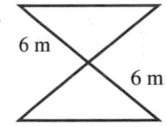

38.

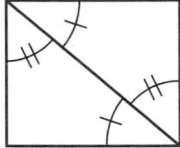

39. Refer to the similar triangles shown below. **See Example 4.**

  a. Name 3 pairs of congruent angles.
  b. Complete each proportion.

$$\frac{LM}{HJ} = \frac{}{JE} \qquad \frac{MR}{JE} = \frac{}{HE} \qquad \frac{}{HJ} = \frac{LR}{HE}$$

  c. We can write the answer to part b in a more compact form:

$$\frac{LM}{} = \frac{MR}{} = \frac{}{HE}$$

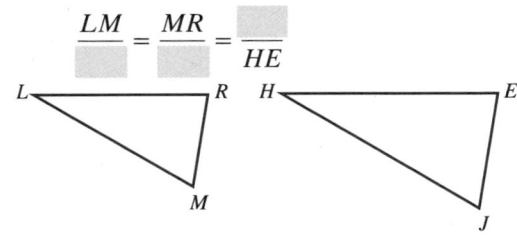

40. Refer to the similar triangles shown below. **See Example 4.**

  a. Name 3 pairs of congruent angles.
  b. Complete each proportion.

$$\frac{WY}{DF} = \frac{}{FE} \qquad \frac{WX}{} = \frac{YX}{FE} \qquad \frac{}{EF} = \frac{WY}{DF}$$

  c. We can write the answer to part b in a more compact form:

$$\frac{}{DF} = \frac{YX}{} = \frac{WX}{}$$

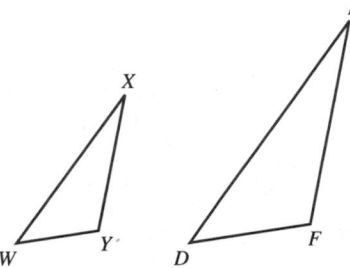

Tell whether the triangles are similar. See Example 5.

41.
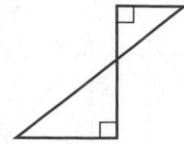

42.
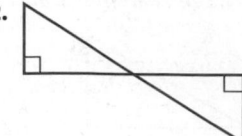

43.
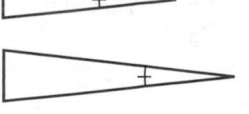

44.
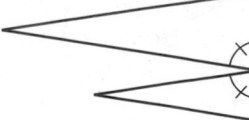

45.
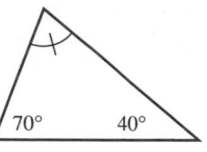

46.
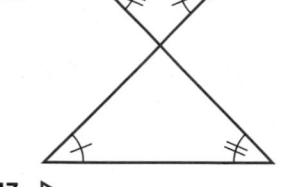

47.
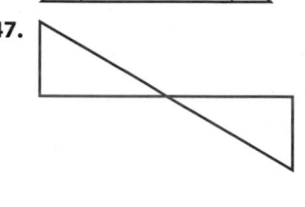

48.
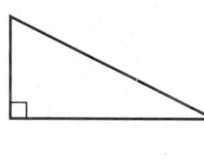

49. $\overline{XY} \parallel \overline{ZD}$
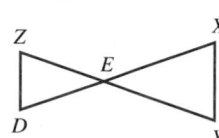

50. $\overline{QR} \parallel \overline{TU}$
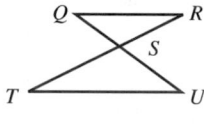

51.
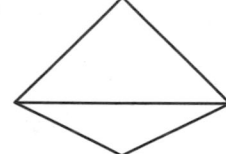

52.
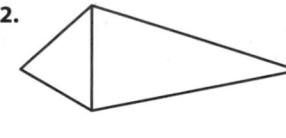

In Problems 53 and 54, △MSN ~ △TPR. *Find* x *and* y.
See Example 6.

53.
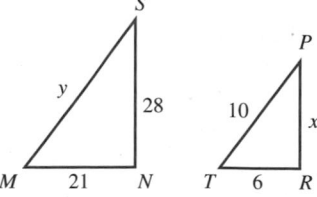

54.
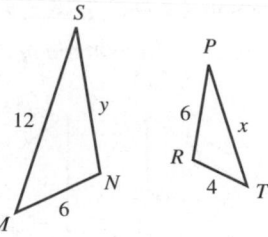

In Problems 55 and 56, △MSN ~ △TPN. *Find* x *and* y.
See Example 6.

55.
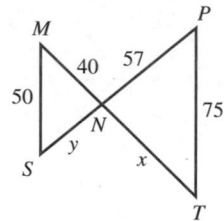

56.
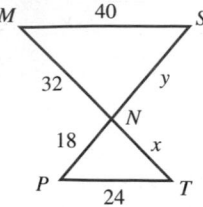

TRY IT YOURSELF

Tell whether each statement is true. If a statement is false, tell why.

57. If three sides of one triangle are the same length as the corresponding three sides of a second triangle, the triangles are congruent.

58. If two sides of one triangle are the same length as two sides of a second triangle, the triangles are congruent.

59. If two sides and an angle of one triangle are congruent, respectively, to two sides and an angle of a second triangle, the triangles are congruent.

60. If two angles and the side between them in one triangle are congruent, respectively, to two angles and the side between them in a second triangle, the triangles are congruent.

Determine whether each pair of triangles are congruent. If they are, tell why.

61.
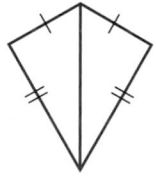

62.
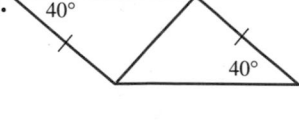

63.
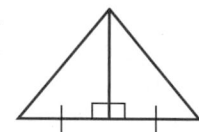

64.
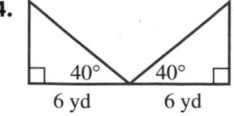

65. $\overline{AB} \parallel \overline{DE}$

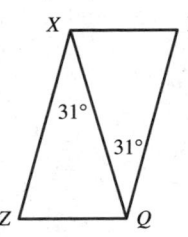

66. $\overline{XY} \parallel \overline{ZQ}$

67.

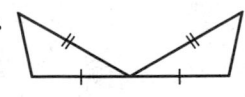

68.

In Problems 69 and 70, $\triangle ABC \cong \triangle DEF$. Find x and y.

69.

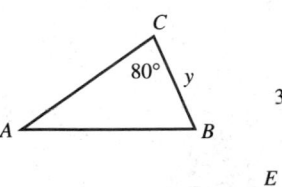

70.

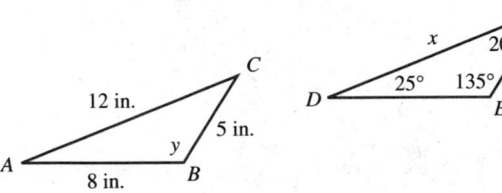

In Problems 71 and 72, find x and y.

71. $\triangle ABC \cong \triangle ABD$

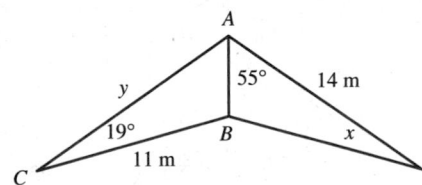

72. $\triangle ABC \cong \triangle DEC$

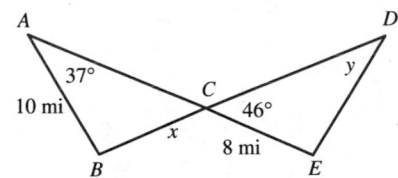

In Problems 73–76, find x.

73.

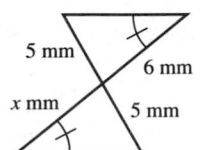

74.

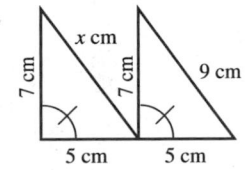

75.

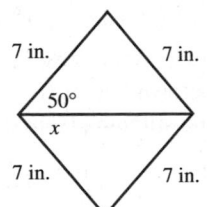

76.

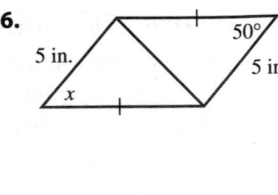

77. If $\overline{DE}$ in the figure below is parallel to $\overline{AB}$, $\triangle ABC$ will be similar to $\triangle DEC$. Find x.

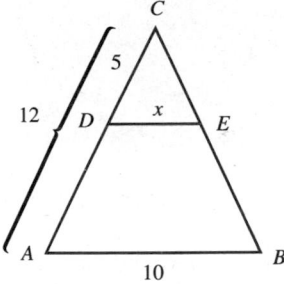

78. If $\overline{SU}$ in the figure below is parallel to $\overline{TV}$, $\triangle SRU$ will be similar to $\triangle TRV$. Find x.

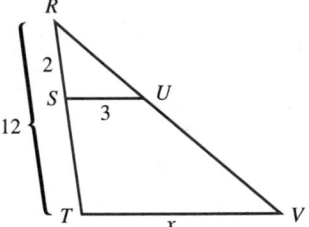

79. If $\overline{DE}$ in the figure below is parallel to $\overline{CB}$, $\triangle EAD$ will be similar to $\triangle BAC$. Find x.

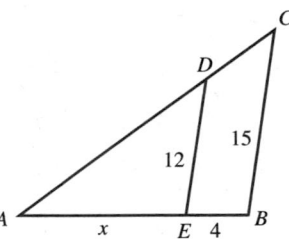

80. If $\overline{HK}$ in the figure below is parallel to $\overline{AB}$, $\triangle HCK$ will be similar to $\triangle ACB$. Find x.

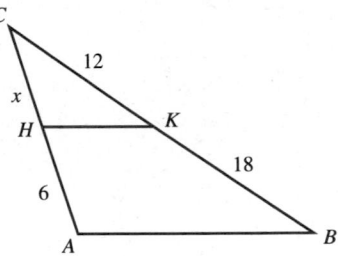

APPLICATIONS

81. SEWING The pattern that is sewn on the rear pocket of a pair of blue jeans is shown below. If $\triangle AOB \cong \triangle COD$, how long is the stitching from point A to point D?

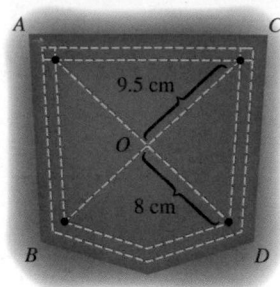

82. CAMPING The base of the tent pole is placed at the midpoint between the stake at point A and the stake at point B, and it is perpendicular to the ground, as shown below. Explain why $\triangle ACD \cong \triangle BCD$.

83. A surveying crew needs to find the width of the river shown in the illustration below. Because of a dangerous current, they decide to stay on the west side of the river and use geometry to find its width. Their approach is to create two similar right triangles on dry land. Then they write and solve a proportion to find w. What is the width of the river?

from Campus to Careers

Surveyor

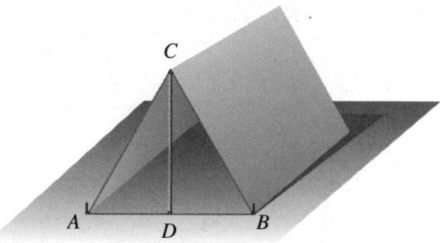

© iStockphoto.com/Lukaz Laska

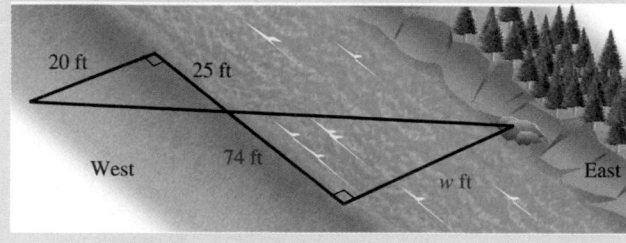

84. HEIGHT OF A BUILDING A man places a mirror on the ground and sees the reflection of the top of a building, as shown below. Find the height of the building.

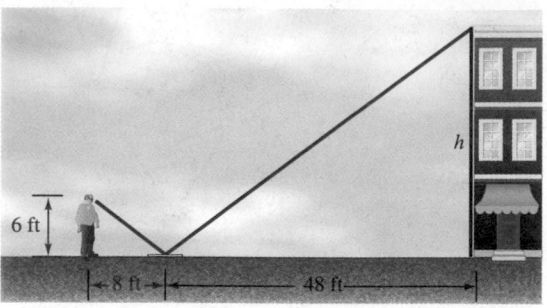

85. HEIGHT OF A TREE The tree shown below casts a shadow 24 feet long when a man 6 feet tall casts a shadow 4 feet long. Find the height of the tree.

86. WASHINGTON, D.C. The Washington Monument casts a shadow of $166\frac{1}{2}$ feet at the same time as a 5-foot-tall tourist casts a shadow of $1\frac{1}{2}$ feet. Find the height of the monument.

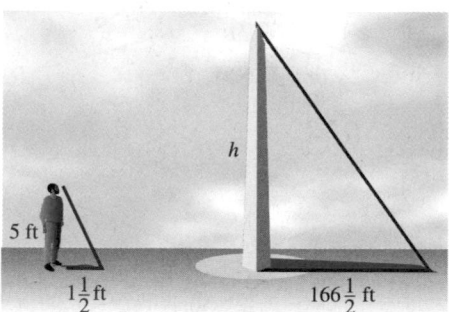

87. HEIGHT OF A TREE A tree casts a shadow of 29 feet at the same time as a vertical yardstick casts a shadow of 2.5 feet. Find the height of the tree.

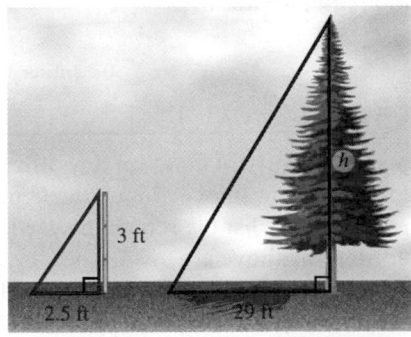

3 ft

2.5 ft   29 ft

88. GEOGRAPHY The diagram below shows how a laser beam was pointed over the top of a pole to the top of a mountain to determine the elevation of the mountain. Find *h*.

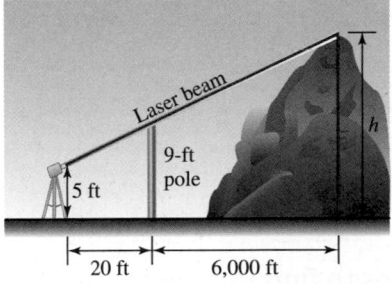

Laser beam

5 ft

9-ft pole

h

20 ft   6,000 ft

89. FLIGHT PATH An airplane ascends 200 feet as it flies a horizontal distance of 1,000 feet, as shown in the following figure. How much altitude is gained as it flies a horizontal distance of 1 mile? (*Hint*: 1 mile = 5,280 feet.)

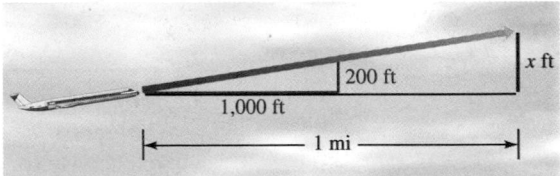

200 ft

1,000 ft

x ft

1 mi

▌ WRITING

90. Tell whether the statement is true or false. Explain your answer.

  a. Congruent triangles are always similar.

  b. Similar triangles are always congruent.

91. Explain why there is no SSA property for congruent triangles.

▌ REVIEW

Find the LCM of the given numbers.

92. 16, 20   **93.** 21, 27

Find the GCF of the given numbers.

94. 18, 96   **95.** 63, 84

SECTION 9.6

Quadrilaterals and Other Polygons

Objectives

1 Classify quadrilaterals.

2 Use properties of rectangles to find unknown angle measures and side lengths.

3 Find unknown angle measures of trapezoids.

4 Use the formula for the sum of the angle measures of a polygon.

Recall from Section 9.3 that a polygon is a closed geometric figure with at least three line segments for its sides. In this section, we will focus on polygons with four sides, called *quadrilaterals*. One type of quadrilateral is the *square*. The game boards for Monopoly and Scrabble have a square shape. Another type of quadrilateral is the *rectangle*. Most picture frames and many mirrors are rectangular. Utility knife blades and swimming fins have shapes that are examples of a third type of quadrilateral called a *trapezoid*.

1 Classify quadrilaterals.

A **quadrilateral** is a polygon with four sides. Some common quadrilaterals are shown below.

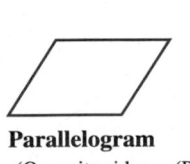 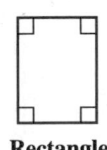 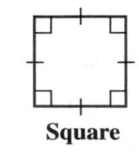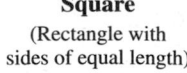 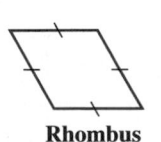 

Parallelogram
(Opposite sides parallel)

Rectangle
(Parallelogram with four right angles)

Square
(Rectangle with sides of equal length)

Rhombus
(Parallelogram with sides of equal length)

Trapezoid
(Exactly two sides parallel)

We can use the capital letters that label the vertices of a quadrilateral to name it. For example, when referring to the quadrilateral shown on the right, with vertices A, B, C, and D, we can use the notation quadrilateral $ABCD$.

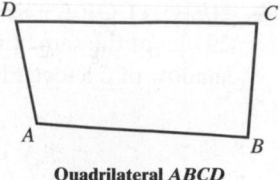

Quadrilateral $ABCD$

> **The Language of Algebra**   When naming a quadrilateral (or any other polygon), we may begin with any vertex. Then we move around the figure in a clockwise (or counterclockwise) direction as we list the remaining vertices. Some other ways of naming the quadrilateral above are quadrilateral $ADCB$, quadrilateral $CDAB$, and quadrilateral $DABC$. It would be unacceptable to name it as quadrilateral $ACDB$, because the vertices would not be listed in clockwise (or counterclockwise) order.

A segment that joins two nonconsecutive vertices of a polygon is called a **diagonal** of the polygon. Quadrilateral $ABCD$ shown below has two diagonals, $\overline{AC}$ and $\overline{BD}$.

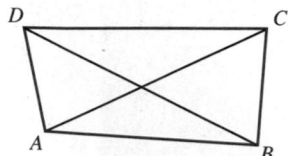

2 Use properties of rectangles to find unknown angle measures and side lengths.

Recall that a **rectangle** is a quadrilateral with four right angles. The rectangle is probably the most common and recognizable of all geometric figures. For example, most doors and windows are rectangular in shape. The boundaries of soccer fields and basketball courts are rectangles. Even our paper currency, such as the $1, $5, and $20 bills, is in the shape of a rectangle. Rectangles have several important characteristics.

Properties of Rectangles

In any rectangle:

1. All four angles are right angles.
2. Opposite sides are parallel.
3. Opposite sides have equal length.
4. The diagonals have equal length.
5. The diagonals intersect at their midpoints.

EXAMPLE 1   In the figure, quadrilateral $WXYZ$ is a rectangle. Find each measure:

a. m($\angle YXW$)   **b.** m($\overline{XY}$)   **c.** m($\overline{WY}$)   **d.** m($\overline{XZ}$)

Strategy We will use properties of rectangles to find the unknown angle measure and the unknown measures of the line segments.

WHY Quadrilateral $WXYZ$ is a rectangle.

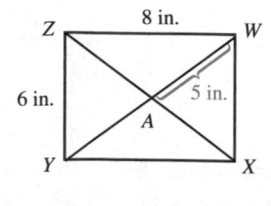

Solution

a. In any rectangle, all four angles are right angles. Therefore, $\angle YXW$ is a right angle, and m($\angle YXW$) = 90°.

b. $\overline{XY}$ and $\overline{WZ}$ are opposite sides of the rectangle, so they have equal length. Since the length of $\overline{WZ}$ is 8 inches, m($\overline{XY}$) is also 8 inches.

c. $\overline{WY}$ and $\overline{ZX}$ are diagonals of the rectangle, and they intersect at their midpoints. That means that point A is the midpoint of $\overline{WY}$. Since the length of $\overline{WA}$ is 5 inches, m($\overline{WY}$) is 2 · 5 inches, or 10 inches.

d. The diagonals of a rectangle are of equal length. In part c, we found that the length of $\overline{WY}$ is 10 inches. Therefore, m($\overline{XZ}$) is also 10 inches. ∎

We have seen that if a quadrilateral has four right angles, it is a rectangle. The following statements establish some conditions that a parallelogram must meet to ensure that it is a rectangle.

Self Check 1

In rectangle *RSTU* shown below, the length of $\overline{RT}$ is 13 ft. Find each measure:

a. m($\angle SRU$)

b. m($\overline{ST}$)

c. m($\overline{TG}$)

d. m($\overline{SG}$)

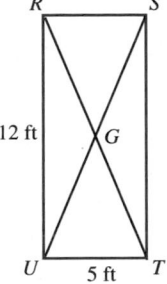

Now Try **Problem 27**

Parallelograms That Are Rectangles

1. If a parallelogram has one right angle, then the parallelogram is a rectangle.

2. If the diagonals of a parallelogram are congruent, then the parallelogram is a rectangle.

EXAMPLE 2 *Construction* A carpenter wants to build a shed with a 9-foot-by-12-foot base. How can he make sure that the foundation has four right-angle corners?

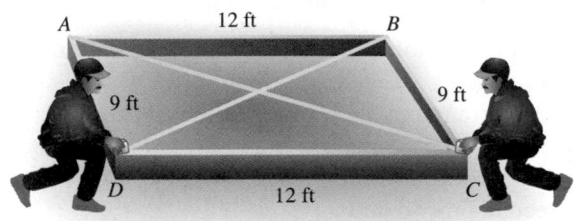

Strategy The carpenter should find the lengths of the diagonals of the foundation.

WHY If the diagonals are congruent, then the foundation is rectangular in shape and the corners are right angles.

Solution The four-sided foundation, which we will label as parallelogram *ABCD*, has opposite sides of equal length. The carpenter can use a tape measure to find the lengths of the diagonals $\overline{AC}$ and $\overline{BD}$. If these diagonals are of equal length, the foundation will be a rectangle and have right angles at its four corners. This process is commonly referred to as "squaring a foundation." Picture framers use a similar process to make sure their frames have four 90° corners. ∎

Now Try **Problem 59**

3 Find unknown angle measures of trapezoids.

A **trapezoid** is a quadrilateral with exactly two sides parallel. For the trapezoid shown on the next page, the parallel sides $\overline{AB}$ and $\overline{DC}$ are called **bases.** To distinguish between the two bases, we will refer to $\overline{AB}$ as the **upper base** and $\overline{DC}$ as the **lower base.** The angles on either side of the upper base are called **upper base angles,** and the angles on either side of the lower base are called **lower base angles.** The nonparallel sides are called **legs.**

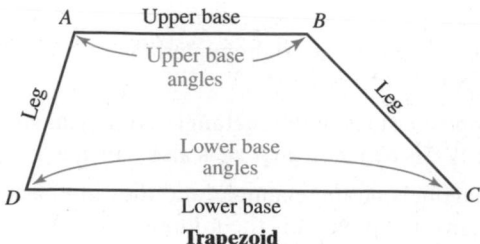

Trapezoid

In the figure above, we can view $\overleftrightarrow{AD}$ as a transversal cutting the parallel lines $\overleftrightarrow{AB}$ and $\overleftrightarrow{DC}$. Since $\angle A$ and $\angle D$ are interior angles on the same side of a transversal, they are supplementary. Similarly, $\overleftrightarrow{BC}$ is a transversal cutting the parallel lines $\overleftrightarrow{AB}$ and $\overleftrightarrow{DC}$. Since $\angle B$ and $\angle C$ are interior angles on the same side of a transversal, they are also supplementary. These observations lead us to the conclusion that *there are always two pairs of supplementary angles in any trapezoid.*

Self Check 3

Refer to trapezoid *HIJK* below, with $\overline{HI} \parallel \overline{KJ}$. Find x and y.

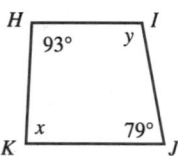

Now Try Problem 29

EXAMPLE 3   Refer to trapezoid *KLMN* below, with $\overline{KL} \parallel \overline{NM}$. Find x and y.

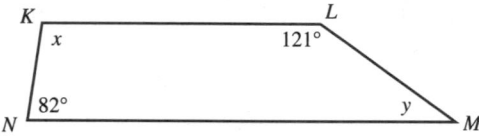

Strategy We will use the interior angles property twice to write two equations that mathematically model the situation.

WHY We can then solve the equations to find x and y.

Solution $\angle K$ and $\angle N$ are interior angles on the same side of transversal $\overleftrightarrow{KN}$ that cuts the parallel lines segments $\overline{KL}$ and $\overline{NM}$. Similarly, $\angle L$ and $\angle M$ are interior angles on the same side of transversal $\overleftrightarrow{LM}$ that cuts the parallel lines segments $\overline{KL}$ and $\overline{NM}$. Recall that if two parallel lines are cut by a transversal, interior angles on the same side of the transversal are supplementary. We can use this fact twice—once to find x and a second time to find y.

$m(\angle K) + m(\angle N) = 180°$	The sum of the measures of supplementary angles is 180°.	$\overset{17}{\cancel{1}}{}^{10}$ $18\cancel{0}$
$x + 82° = 180°$	Substitute x for $m(\angle K)$ and 82° for $m(\angle N)$.	$\underline{-82}$ 98
$x = 98°$	To isolate x, subtract 82° from both sides.	

Thus, x is 98°.

$m(\angle L) + m(\angle M) = 180°$	The sum of the measures of supplementary angles is 180°.	$\overset{7\,10}{18\cancel{0}}$
$121° + y = 180°$	Substitute 121° for $m(\angle L)$ and y for $m(\angle M)$.	$\underline{-121}$ 59
$y = 59°$	To isolate y, subtract 121° from both sides.	

Thus, y is 59°.

If the nonparallel sides of a trapezoid are the same length, it is called an **isosceles trapezoid.** The figure on the right shows isosceles trapezoid *DEFG* with $\overline{DG} \cong \overline{EF}$. In an isosceles trapezoid, *both pairs of base angles are congruent.* In the figure, $\angle D \cong \angle E$ and $\angle G \cong \angle F$.

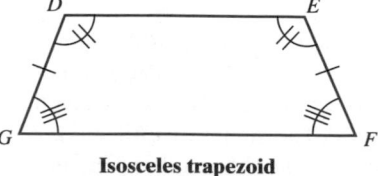

Isosceles trapezoid

EXAMPLE 4   *Landscaping*   A cross section of a drainage ditch shown below is an isosceles trapezoid with $\overline{AB} \parallel \overline{DC}$. Find x and y.

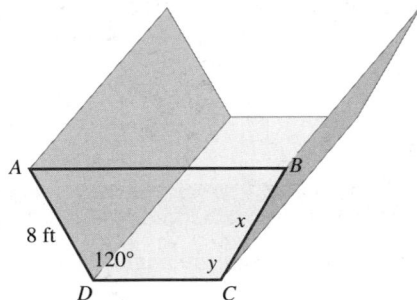

Self Check 4

Refer to the isosceles trapezoid shown below with $\overline{RS} \parallel \overline{UT}$. Find x and y.

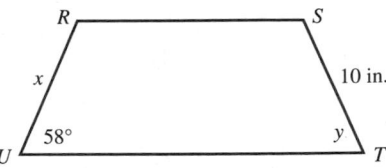

Now Try **Problem 31**

Strategy We will compare the nonparallel sides and compare a pair of base angles of the trapezoid to find each unknown.

WHY The nonparallel sides of an isosceles trapezoid have the same length and both pairs of base angles are congruent.

Solution Since $\overline{AD}$ and $\overline{BC}$ are the nonparallel sides of an isosceles trapezoid, $m(\overline{AD})$ and $m(\overline{BC})$ are equal, and x is 8 ft.
   Since $\angle D$ and $\angle C$ are a pair of base angles of an isosceles trapezoid, they are congruent and $m(\angle D) = m(\angle C)$. Thus, y is 120°. ∎

4 **Use the formula for the sum of the angle measures of a polygon.**

In the figure shown below, a protractor was used to find the measure of each angle of the quadrilateral. When we add the four angle measures, the result is 360°.

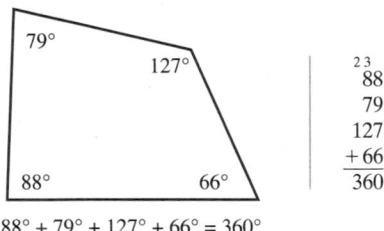

$$88° + 79° + 127° + 66° = 360°$$

   This illustrates an important fact about quadrilaterals: The sum of the measures of the angles of *any* quadrilateral is 360°. This can be shown using the diagram in figure (a) on the following page. In the figure, the quadrilateral is divided into two triangles. Since the sum of the angle measures of any triangle is 180°, the sum of the measures of the angles of the quadrilateral is $2 \cdot 180°$, or 360°.
   A similar approach can be used to find the sum of the measures of the angles of any pentagon or any hexagon. The pentagon in figure (b) is divided into three triangles. The sum of the measures of the angles of the pentagon is $3 \cdot 180°$, or 540°. The hexagon in figure (c) is divided into four triangles. The sum of the measures of the angles of the hexagon is $4 \cdot 180°$, or 720°. In general, a polygon with n sides can be divided into $n - 2$ triangles. Therefore, the sum of the angle measures of a polygon can be found by multiplying 180° by $n - 2$.

Quadrilateral	Pentagon	Hexagon
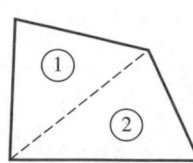	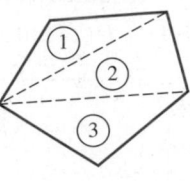	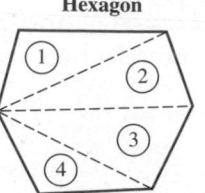
$2 \cdot 180° = 360°$	$3 \cdot 180° = 540°$	$4 \cdot 180° = 720°$
(a)	(b)	(c)

Sum of the Angles of a Polygon

The sum S, in degrees, of the measures of the angles of a polygon with n sides is given by the formula

$$S = (n - 2)180°$$

Self Check 5

Find the sum of the angle measures of the polygon shown below.

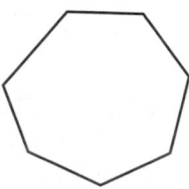

Now Try Problem 33

EXAMPLE 5   Find the sum of the angle measures of a 13-sided polygon.

Strategy We will substitute 13 for n in the formula $S = (n - 2)180°$ and evaluate the right side.

WHY The variable S represents the unknown sum of the measures of the angles of the polygon.

Solution

$S = (\textbf{n} - 2)180°$ *This is the formula for the sum of the measures of the angles of a polygon.*

$S = (\textbf{13} - 2)180°$ *Substitute 13 for n, the number of sides.*

$\quad = (11)180°$ *Do the subtraction within the parentheses.*

$\quad = 1{,}980°$ *Do the multiplication.*

$$
\begin{array}{r}
180 \\
\times 11 \\
\hline
180 \\
1800 \\
\hline
1{,}980
\end{array}
$$

The sum of the measures of the angles of a 13-sided polygon is $1{,}980°$.

Self Check 6

The sum of the measures of the angles of a polygon is $1{,}620°$. Find the number of sides the polygon has.

Now Try Problem 41

EXAMPLE 6   The sum of the measures of the angles of a polygon is $1{,}080°$. Find the number of sides the polygon has.

Strategy We will substitute $1{,}080°$ for S in the formula $S = (n - 2)180°$ and solve for n.

WHY The variable n represents the unknown number of sides of the polygon.

Solution

$S = (n - 2)180°$ *This is the formula for the sum of the measures of the angles of a polygon.*

$\textbf{1,080}° = (n - 2)180°$ *Substitute 1,080° for S, the sum of the measures of the angles.*

$1{,}080° = 180°n - 360°$ *Distribute the multiplication by 180°.*

$1{,}080° + \textbf{360}° = 180°n - 360° + \textbf{360}°$ *To isolate 180°n, add 360° to both sides.*

$1{,}440° = 180°n$ *Do the additions.*

$\dfrac{1{,}440°}{\textbf{180}°} = \dfrac{180°n}{\textbf{180}°}$ *To isolate n, divide both sides by 180°.*

$8 = n$ *Do the division.*

$$
\begin{array}{r}
\overset{1}{1}{,}080 \\
+ \quad 360 \\
\hline
1{,}440
\end{array}
\qquad
\begin{array}{r}
8 \\
180{\overline{)1{,}440}} \\
-1\,440 \\
\hline
0
\end{array}
$$

The polygon has 8 sides. It is an octagon.

SECTION 9.6   STUDY SET

VOCABULARY

Fill in the blanks.

1. A _____ is a polygon with four sides.

2. A _____ is a quadrilateral with opposite sides parallel.

3. A _____ is a quadrilateral with four right angles.

4. A rectangle with all sides of equal length is a _____.

5. A _____ is a parallelogram with four sides of equal length.

6. A segment that joins two nonconsecutive vertices of a polygon is called a _____ of the polygon.

7. A _____ has two sides that are parallel and two sides that are not parallel. The parallel sides are called _____. The legs of an _____ trapezoid have the same length.

8. A _____ polygon has sides that are all the same length and angles that are all the same measure.

CONCEPTS

9. Refer to the polygon below.

   a. How many vertices does it have? List them.

   b. How many sides does it have? List them.

   c. How many diagonals does it have? List them.

   d. Tell which of the following are acceptable ways of naming the polygon.

   quadrilateral *ABCD*

   quadrilateral *CDBA*

   quadrilateral *ACBD*

   quadrilateral *BADC*

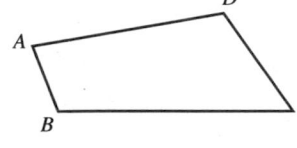

10. Draw an example of each type of quadrilateral.

   a. rhombus   **b.** parallelogram

   c. trapezoid   **d.** square

   e. rectangle   **f.** isosceles trapezoid

11. A parallelogram is shown below. Fill in the blanks.

   a. $\overline{ST} \parallel$ _____   **b.** $\overline{SV}$ ▢ $\overline{TU}$

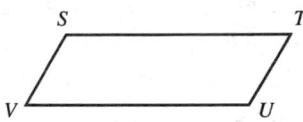

12. Refer to the rectangle below.

   a. How many right angles does the rectangle have? List them.

   b. Which sides are parallel?

   c. Which sides are of equal length?

   d. Copy the figure and draw the diagonals. Call the point where the diagonals intersect point *X*. How many diagonals does the figure have? List them.

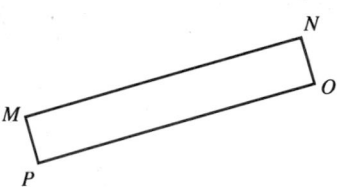

13. Fill in the blanks. In any rectangle:

   a. All four angles are _____ angles.

   b. Opposite sides are _____.

   c. Opposite sides have equal _____.

   d. The diagonals have equal _____.

   e. The diagonals intersect at their _____.

14. Refer to the figure below.

   a. What is m($\overline{CD}$)?   **b.** What is m($\overline{AD}$)?

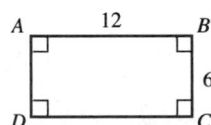

15. In the figure below, $\overline{TR} \parallel \overline{DF}$, $\overline{DT} \parallel \overline{FR}$, and m($\angle D$) = 90°. What type of quadrilateral is *DTRF*?

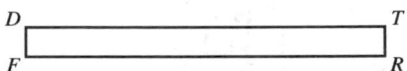

16. Refer to the parallelogram shown below. If m($\overline{GI}$) = 4 and m($\overline{HJ}$) = 4, what type of figure is quadrilateral *GHIJ*?

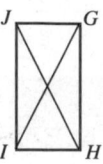

17. a. Is every rectangle a square?

  b. Is every square a rectangle?

  c. Is every parallelogram a rectangle?

  d. Is every rectangle a parallelogram?

  e. Is every rhombus a square?

  f. Is every square a rhombus?

18. Trapezoid *WXYZ* is shown below. Which sides are parallel?

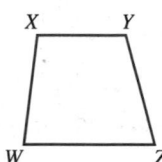

19. Trapezoid *JKLM* is shown below.

  a. What type of trapezoid is this?

  b. Which angles are the lower base angles?

  c. Which angles are the upper base angles?

  d. Fill in the blanks:

  m($\angle J$) = m(⬚)

  m($\angle K$) = m(⬚)

  m($\overline{JK}$) = m(⬚)

20. Find the sum of the measures of the angles of the hexagon below.

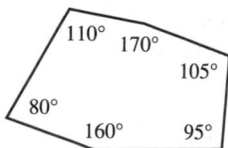

21. What do the tick marks in the figure indicate?

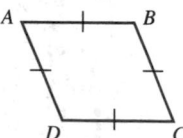

22. Rectangle *ABCD* is shown below. What do the tick marks indicate about point *X*?

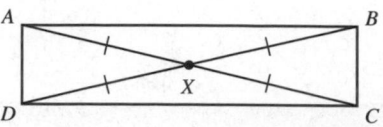

23. In the formula $S = (n - 2)180°$, what does *S* represent? What does *n* represent?

24. Suppose $n = 12$. What is $(n - 2)180°$?

█ **GUIDED PRACTICE**

In Problems 25 and 26, classify each quadrilateral as a rectangle, a square, a rhombus, or a trapezoid. Some figures may be correctly classified in more than one way. **See Objective 1.**

25. a. 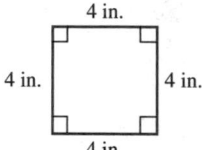   **b.**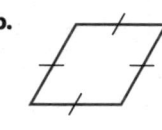

  c.    **d.**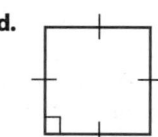

26. a. 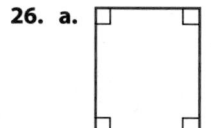   **b.**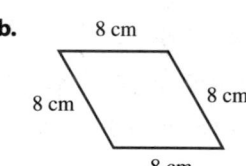

  c. 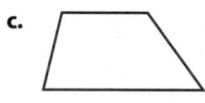   **d.**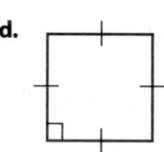

27. Rectangle *ABCD* is shown below. **See Example 1.**

  a. What is m($\angle DCB$)?

  b. What is m($\overline{AX}$)?

  c. What is m($\overline{AC}$)?

  d. What is m($\overline{BD}$)?

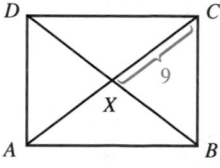

28. Refer to rectangle *EFGH* shown below. **See Example 1.**

  a. Find m($\angle EHG$).   **b.** Find m($\overline{FH}$).

  c. Find m($\overline{EI}$).   **d.** Find m($\overline{EG}$).

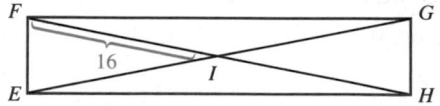

29. Refer to the trapezoid shown below. **See Example 3.**

   a. Find *x*.   **b.** Find *y*.

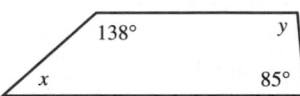

30. Refer to trapezoid *MNOP* shown below. **See Example 3.**

   a. Find m(∠*O*).   **b.** Find m(∠*M*).

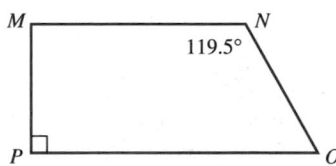

31. Refer to the isosceles trapezoid shown below. **See Example 4.**

   a. Find m($\overline{BC}$).   **b.** Find *x*.

   c. Find *y*.   **d.** Find *z*.

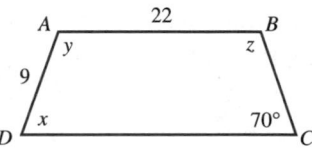

32. Refer to the trapezoid shown below. **See Example 4.**

   a. Find m(∠*T*).

   b. Find m(∠*R*).

   c. Find m(∠*S*).

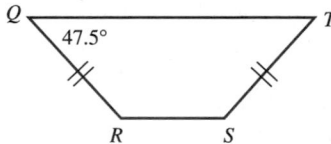

Find the sum of the angle measures of the polygon.
See Example 5.

33. a 14-sided polygon

34. a 15-sided polygon

35. a 20-sided polygon

36. a 22-sided polygon

37. an octagon

38. a decagon

39. a dodecagon

40. a nonagon

Find the number of sides a polygon has if the sum of its angle measures is the given number. **See Example 6.**

41. 540°   **42.** 720°

43. 900°   **44.** 1,620°

45. 1,980°   **46.** 1,800°

47. 2,160°   **48.** 3,600°

| TRY IT YOURSELF

49. Refer to rectangle *ABCD* shown below.

   a. Find m(∠1).

   b. Find m(∠3).

   c. Find m(∠2).

   d. If m($\overline{AC}$) is 8 cm, find m($\overline{BD}$).

   e. Find m($\overline{PD}$).

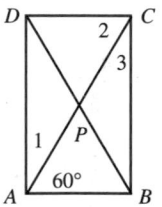

50. The following problem appeared on a quiz. Explain why the instructor must have made an error when typing the problem.

   The sum of the measures of the angles of a polygon is 1,000°. How many sides does the polygon have?

For Problems 51 and 52, find x. Then find the measure of each angle of the polygon.

51.

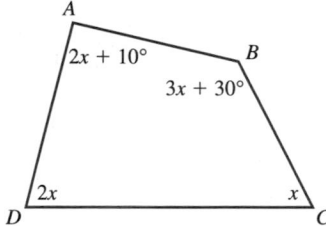

52.

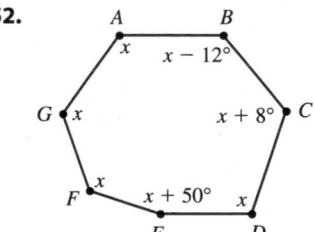

APPLICATIONS

53. QUADRILATERALS IN EVERYDAY LIFE What quadrilateral shape do you see in each of the following objects?

a. podium (upper portion)   **b.** checkerboard

c. dollar bill   **d.** swimming fin

e. camper shell window

54. FLOWCHART A flowchart shows a sequence of steps to be performed by a computer to solve a given problem. When designing a flowchart, the programmer uses a set of standardized symbols to represent various operations to be performed by the computer. Locate a rectangle, a rhombus, and a parallelogram in the flowchart shown to the right.

55. BASEBALL Refer to the figure to the right. Find the sum of the measures of the angles of home plate.

56. TOOLS The utility knife blade shown below has the shape of an isosceles trapezoid. Find x, y, and z.

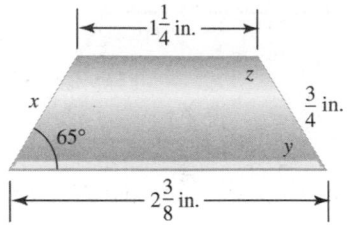

WRITING

57. Explain why a square is a rectangle.

58. Explain why a trapezoid is not a parallelogram.

59. MAKING A FRAME After gluing and nailing the pieces of a picture frame together, it didn't look right to a frame maker. (See the figure to the right.) How can she use a tape measure to make sure the corners are 90° (right) angles?

60. A decagon is a polygon with ten sides. What could you call a polygon with one hundred sides? With one thousand sides? With one million sides?

REVIEW

Write each number in words.

61. 254,309

62. 504,052,040

63. 82,000,415

64. 51,000,201,078

SECTION 9.7
Perimeters and Areas of Polygons

Objectives

1 Find the perimeter of a polygon.

2 Find the area of a polygon.

3 Find the area of figures that are combinations of polygons.

In this section, we will discuss how to find perimeters and areas of polygons. Finding perimeters is important when estimating the cost of fencing a yard or installing crown molding in a room. Finding area is important when calculating the cost of carpeting, painting a room, or fertilizing a lawn.

1 Find the perimeter of a polygon.

The **perimeter** of a polygon is the distance around it. To find the perimeter P of a polygon, we simply add the lengths of its sides.

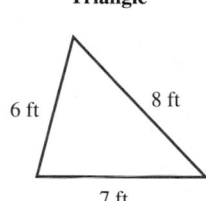

Triangle

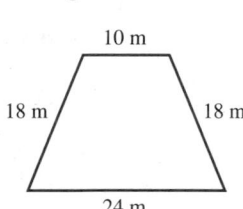

Quadrilateral

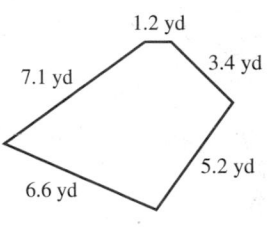

Pentagon

$P = 6 + 7 + 8$
$\quad = 21$

$P = 10 + 18 + 24 + 18$
$\quad = 70$

$P = 1.2 + 7.1 + 6.6 + 5.2 + 3.4$
$\quad = 23.5$

The perimeter is 21 ft.

The perimeter is 70 m.

The perimeter is 23.5 yd.

For some polygons, such as a square and a rectangle, we can simplify the computations by using a perimeter formula. Since a square has four sides of equal length s, its perimeter P is $s + s + s + s$, or $4s$.

Perimeter of a Square

If a square has a side of length s, its perimeter P is given by the formula

$$P = 4s$$

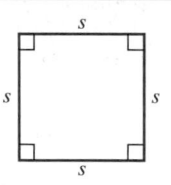

EXAMPLE 1
Find the perimeter of a square whose sides are 7.5 meters long.

Strategy We will substitute 7.5 for s in the formula $P = 4s$ and evaluate the right side.

WHY The variable P represents the unknown perimeter of the square.

Solution

$P = 4s$	This is the formula for the perimeter of a square.
$P = 4(\mathbf{7.5})$	Substitute 7.5 for s, the length of one side of the square.
$P = 30$	Do the multiplication.

$$\begin{array}{r} \overset{2}{7.5} \\ \times\ 4 \\ \hline 30.0 \end{array}$$

The perimeter of the square is 30 meters.

Self Check 1

A Scrabble game board has a square shape with sides of length 38.5 cm. Find the perimeter of the game board.

Now Try Problems 17 and 19

Since a rectangle has two lengths l and two widths w, its perimeter P is given by $l + w + l + w$, or $2l + 2w$.

> ### Perimeter of a Rectangle
>
> If a rectangle has length l and width w, its perimeter P is given by the formula
>
> $$P = 2l + 2w$$
>
>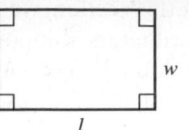

> **Caution!** When finding the perimeter of a polygon, the lengths of the sides must be expressed in the same units.

Self Check 2

Find the perimeter of the triangle shown below, in inches.

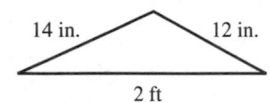

Now Try Problem 21

EXAMPLE 2 Find the perimeter of the rectangle shown on the right, in inches.

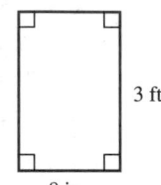

Strategy We will express the width of the rectangle in inches and then use the formula $P = 2l + 2w$ to find the perimeter of the figure.

WHY We can only add quantities that are measured in the same units.

Solution Since 1 foot = 12 inches, we can convert 3 feet to inches by multiplying 3 feet by the unit conversion factor $\frac{12 \text{ in.}}{1 \text{ foot}}$.

$$3 \text{ ft} = 3 \text{ ft} \cdot \frac{\textbf{12 in.}}{\textbf{1 ft}} \quad \text{Multiply by 1:} \frac{12 \text{ in.}}{1 \text{ ft}} = 1.$$

$$= \frac{3 \text{ ft}}{1} \cdot \frac{12 \text{ in.}}{1 \text{ ft}} \quad \text{Write 3 ft as a fraction. Remove the common units of feet from the numerator and denominator. The units of inches remain.}$$

$$= 36 \text{ in.} \quad \text{Do the multiplication.}$$

The width of the rectangle is 36 inches. We can now substitute 8 for l, the length, and 36 for w, the width, in the formula for the perimeter of a rectangle.

$$P = 2l + 2w \quad \text{This is the formula for the perimeter of a rectangle.}$$

$$P = 2(\textbf{8}) + 2(\textbf{36}) \quad \text{Substitute 8 for } l, \text{ the length, and 36 for } w, \text{ the width.}$$

$$= 16 + 72 \quad \text{Do the multiplication.}$$

$$= 88 \quad \text{Do the addition.}$$

$$\begin{array}{r} \overset{1}{36} \\ \times 2 \\ \hline 72 \end{array} \qquad \begin{array}{r} 16 \\ + 72 \\ \hline 88 \end{array}$$

The perimeter of the rectangle is 88 inches. ∎

Self Check 3

The perimeter of an isosceles triangle is 58 meters. If one of its sides of equal length is 15 meters long, how long is its base?

Now Try Problem 25

EXAMPLE 3 *Structural Engineering* The truss shown below is made up of three parts that form an isosceles triangle. If 76 linear feet of lumber were used to make the truss, how long is the base of the truss?

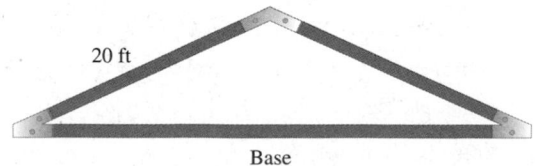

Analyze

- The truss is in the shape of an isosceles triangle.   *Given*
- One of the sides of equal length is 20 feet long.   *Given*
- The perimeter of the truss is 76 feet.   *Given*
- What is the length of the base of the truss?   *Find*

Form We can let b equal the length of the base of the truss (in feet). At this stage, it is helpful to draw a sketch. (See the figure on the right.) If one of the sides of equal length is 20 feet long, so is the other.

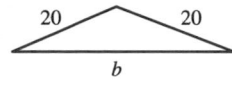

Because 76 linear feet of lumber were used to make the triangular-shaped truss,

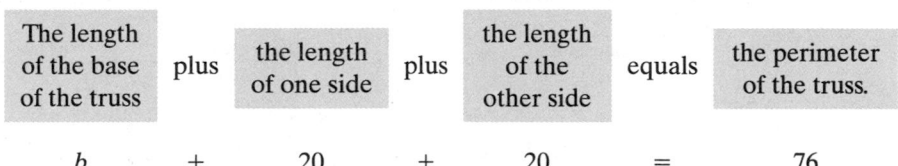

The length of the base of the truss	plus	the length of one side	plus	the length of the other side	equals	the perimeter of the truss.
b	+	20	+	20	=	76

Solve

$$b + 20 + 20 = 76$$
$$b + 40 = 76 \quad \text{Combine like terms.}$$
$$b = 36 \quad \text{To isolate } b, \text{ subtract 40 from both sides.}$$

$$\begin{array}{r} 76 \\ -40 \\ \hline 36 \end{array}$$

State The length of the base of the truss is 36 ft.

Check If we add the lengths of the parts of the truss, we get 36 ft + 20 ft + 20 ft = 76 ft. The result checks.

Using Your CALCULATOR   Perimeters of Figures That Are Combinations of Polygons

To find the perimeter of the figure shown below, we need to know the values of x and y. Since the figure is a combination of two rectangles, we can use a calculator to see that

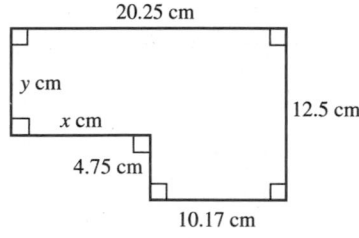

$$x = 20.25 - 10.17 \quad \text{and} \quad y = 12.5 - 4.75$$
$$= 10.08 \text{ cm} \qquad\qquad = 7.75 \text{ cm}$$

The perimeter P of the figure is

$$P = 20.25 + 12.5 + 10.17 + 4.75 + x + y$$
$$P = 20.25 + 12.5 + 10.17 + 4.75 + \mathbf{10.08} + \mathbf{7.75}$$

We can use a scientific calculator to make this calculation.

20.25 $\boxed{+}$ 12.5 $\boxed{+}$ 10.17 $\boxed{+}$ 4.75 $\boxed{+}$ 10.08 $\boxed{+}$ 7.75 $\boxed{=}$ $\boxed{\ 65.5}$

The perimeter is 65.5 centimeters.

2 Find the area of a polygon.

The **area** of a polygon is the measure of the amount of surface it encloses. Area is measured in square units, such as square inches or square centimeters, as shown below.

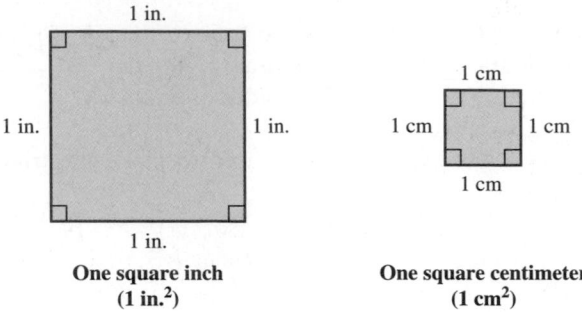

One square inch
(1 in.²)

One square centimeter
(1 cm²)

In everyday life, we often use areas. For example,

* To carpet a room, we buy square yards.
* A can of paint will cover a certain number of square feet.
* To measure vast amounts of land, we often use square miles.
* We buy house roofing by the "square." One square is 100 square feet.

The rectangle shown below has a length of 10 centimeters and a width of 3 centimeters. If we divide the rectangular region into square regions as shown in the figure, each square has an area of 1 square centimeter—a surface enclosed by a square measuring 1 centimeter on each side. Because there are 3 rows with 10 squares in each row, there are 30 squares. Since the rectangle encloses a surface area of 30 squares, its area is 30 square centimeters, which can be written as 30 cm².

This example illustrates that to find the area of a rectangle, we multiply its length by its width.

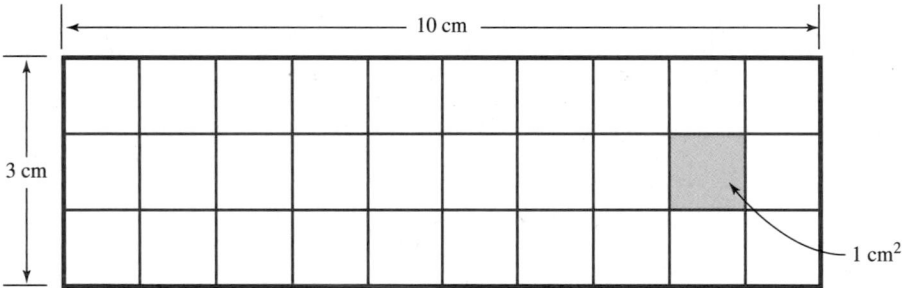

Caution! Do not confuse the concepts of perimeter and area. Perimeter is the distance around a polygon. It is measured in linear units, such as centimeters, feet, or miles. Area is a measure of the surface enclosed within a polygon. It is measured in square units, such as square centimeters, square feet, or square miles.

In practice, we do not find areas of polygons by counting squares. Instead, we use formulas to find areas of geometric figures.

Figure	Name	Formula for Area
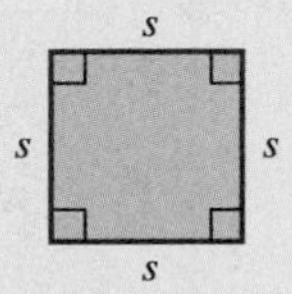	Square	$A = s^2$, where s is the length of one side.
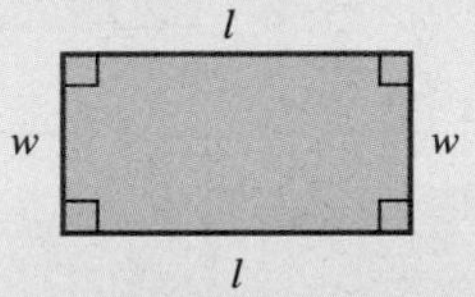	Rectangle	$A = lw$, where l is the length and w is the width.
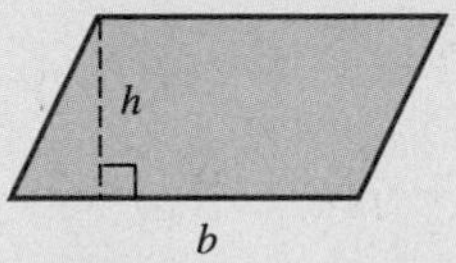	Parallelogram	$A = bh$, where b is the length of the base and h is the height. (A height is always perpendicular to the base.)
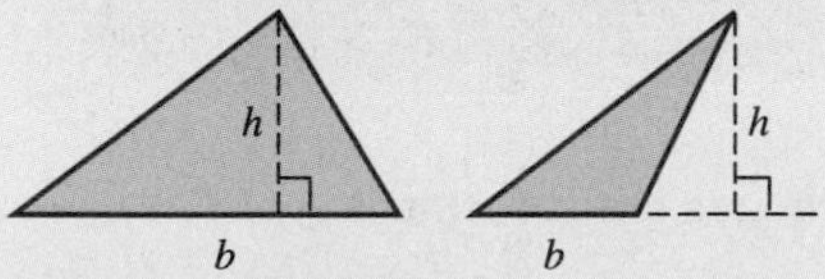	Triangle	$A = \frac{1}{2}bh$, where b is the length of the base and h is the height. The segment perpendicular to the base and representing the height (shown here using a dashed line) is called an **altitude.**
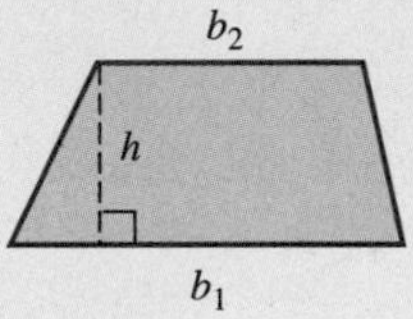	Trapezoid	$A = \frac{1}{2}h(b_1 + b_2)$, where h is the height of the trapezoid and b_1 and b_2 represent the lengths of the bases.

EXAMPLE 4   Find the area of the square shown on the right.

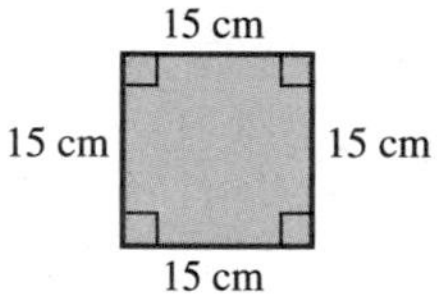

Strategy We will substitute 15 for s in the formula $A = s^2$ and evaluate the right side.

WHY The variable A represents the unknown area of the square.

Solution

$A = s^2$ This is the formula for the area of a square.

$A = 15^2$ Substitute 15 for s, the length of one side of the square.

$A = 225$ Evaluate the exponential expression.

$$\begin{array}{r} 15 \\ \times 15 \\ \hline 75 \\ 150 \\ \hline 225 \end{array}$$

Recall that area is measured in square units. Thus, the area of the square is 225 square centimeters, which can be written as 225 cm².

Self Check 4

Find the area of the square shown below.

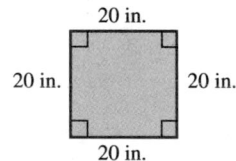

Now Try Problems 29 and 31

EXAMPLE 5   Find the number of square feet in 1 square yard.

Strategy A figure is helpful to solve this problem. We will draw a square yard and divide each of its sides into 3 equally long parts.

WHY Since a square yard is a square with each side measuring 1 yard, each side also measures 3 feet.

Self Check 5

Find the number of square centimeters in 1 square meter.

Now Try Problems 33 and 39

Solution

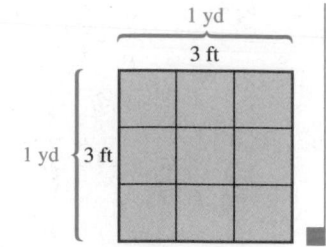

$$1 \text{ yd}^2 = (\mathbf{1 \text{ yd}})^2$$
$$= (\mathbf{3 \text{ ft}})^2 \quad \text{Substitute 3 feet for 1 yard.}$$
$$= (3 \text{ ft})(3 \text{ ft})$$
$$= 9 \text{ ft}^2$$

There are 9 square feet in 1 square yard.

Self Check 6

PING-PONG A regulation-size Ping-Pong table is 9 feet long and 5 feet wide. Find its area in square inches.

Now Try Problem 41

EXAMPLE 6 *Women's Sports*

Field hockey is a team sport in which players use sticks to try to hit a ball into their opponents' goal. Find the area of the rectangular field shown on the right. Give the answer in square feet.

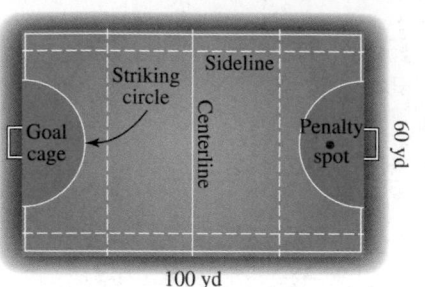

Strategy We will substitute 100 for l and 60 for w in the formula $A = lw$ and evaluate the right side.

WHY The variable A represents the unknown area of the rectangle.

Solution

$$A = lw \qquad \text{This is the formula for the area of a rectangle.}$$
$$A = 100(60) \qquad \text{Substitute 100 for } l, \text{ the length, and 60 for } w, \text{ the width.}$$
$$= 6,000 \qquad \text{Do the multiplication.}$$

The area of the rectangle is 6,000 square yards. Since there are 9 square feet per square yard, we can convert this number to square feet by multiplying 6,000 square yards by $\frac{9 \text{ ft}^2}{1 \text{ yd}^2}$.

$$6,000 \text{ yd}^2 = 6,000 \text{ yd}^2 \cdot \frac{\mathbf{9 \text{ ft}^2}}{\mathbf{1 \text{ yd}^2}} \qquad \text{Multiply by the unit conversion factor: } \frac{9 \text{ ft}^2}{1 \text{ yd}^2} = 1.$$
$$= 6,000 \cdot 9 \text{ ft}^2 \qquad \begin{array}{l}\text{Remove the common units of square yards in the}\\ \text{numerator and denominator. The units of ft}^2 \text{ remain.}\end{array}$$
$$= 54,000 \text{ ft}^2 \qquad \text{Multiply: } 6,000 \cdot 9 = 54,000.$$

The area of the field is 54,000 ft^2.

THINK IT THROUGH *Dorm Rooms*

"The United States has more than 4,000 colleges and universities, with 2.3 million students living in college dorms."

The New York Times, 2007

The average dormitory room in a residence hall has about 180 square feet of floor space. The rooms are usually furnished with the following items having the given dimensions:

- 2 extra-long twin beds (each is 39 in. wide × 80 in. long × 24 in. high)
- 2 dressers (each is 18 in. wide × 36 in. long × 48 in. high)
- 2 bookcases (each is 12 in. wide × 24 in. long × 40 in. high)
- 2 desks (each is 24 in. wide × 48 in. long × 28 in. high)

How many square feet of floor space are left?

EXAMPLE 7 Find the area of the triangle shown on the right.

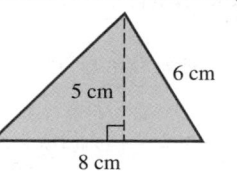

Strategy We will substitute 8 for b and 5 for h in the formula $A = \frac{1}{2}bh$ and evaluate the right side. (The side having length 6 cm is additional information that is not used to find the area.)

WHY The variable A represents the unknown area of the triangle.

Solution

$$A = \frac{1}{2}bh \qquad \text{This is the formula for the area of a triangle.}$$

$$A = \frac{1}{2}(8)(5) \qquad \text{Substitute 8 for } b\text{, the length of the base, and 5 for } h\text{, the height.}$$

$$= 4(5) \qquad \text{Do the first multiplication: } \frac{1}{2}(8) = 4.$$

$$= 20 \qquad \text{Complete the multiplication.}$$

The area of the triangle is 20 cm^2.

Self Check 7
Find the area of the triangle shown below.

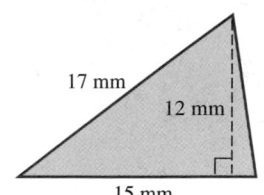

Now Try Problem 45

EXAMPLE 8 Find the area of the triangle shown on the right.

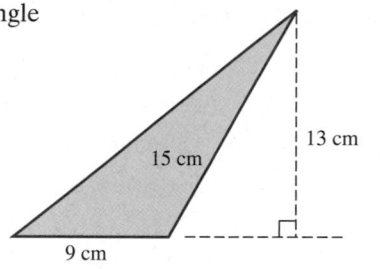

Strategy We will substitute 9 for b and 13 for h in the formula $A = \frac{1}{2}bh$ and evaluate the right side. (The side having length 15 cm is additional information that is not used to find the area.)

WHY The variable A represents the unknown area of the triangle.

Solution In this case, the altitude falls outside the triangle.

$$A = \frac{1}{2}bh \qquad \text{This is the formula for the area of a triangle.}$$

$$A = \frac{1}{2}(9)(13) \qquad \text{Substitute 9 for } b\text{, the length of the base, and 13 for } h\text{, the height.}$$

$$= \frac{1}{2}\left(\frac{9}{1}\right)\left(\frac{13}{1}\right) \qquad \text{Write 9 as } \frac{9}{1} \text{ and 13 as } \frac{13}{1}.$$

$$= \frac{117}{2} \qquad \text{Multiply the fractions.}$$

$$= 58.5 \qquad \text{Do the division.}$$

The area of the triangle is 58.5 cm^2.

$$\begin{array}{r} \overset{2}{13} \\ \times 9 \\ \hline 117 \end{array} \qquad \begin{array}{r} 58.5 \\ 2\overline{)117.0} \\ -10 \\ \hline 17 \\ -16 \\ \hline 1\,0 \\ -1\,0 \\ \hline 0 \end{array}$$

Self Check 8
Find the area of the triangle shown below.

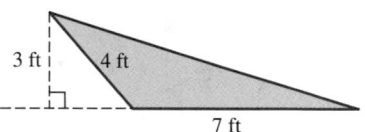

Now Try Problem 49

EXAMPLE 9 Find the area of the trapezoid shown on the right.

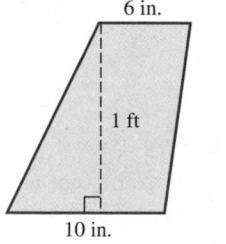

Strategy We will express the height of the trapezoid in inches and then use the formula $A = \frac{1}{2}h(b_1 + b_2)$ to find the area of the figure.

WHY The height of 1 foot must be expressed as 12 inches to be consistent with the units of the bases.

Self Check 9
Find the area of the trapezoid shown below.

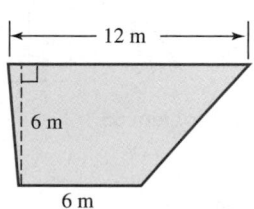

Now Try Problem 53

Solution

$$A = \frac{1}{2}h(b_1 + b_2)$$ This is the formula for the area of a trapezoid.

$$A = \frac{1}{2}(12)(10 + 6)$$ Substitute 12 for h, the height; 10 for b_1, the length of the lower base; and 6 for b_2, the length of the upper base.

$$= \frac{1}{2}(12)(16)$$ Do the addition within the parentheses.

$$= 6(16)$$ Do the first multiplication: $\frac{1}{2}(12) = 6$.

$$= 96$$ Complete the multiplication.

$$\begin{array}{r} \overset{3}{16} \\ \times 6 \\ \hline 96 \end{array}$$

The area of the trapezoid is 96 in².

Self Check 10

The area of the parallelogram below is 96 cm². Find its height.

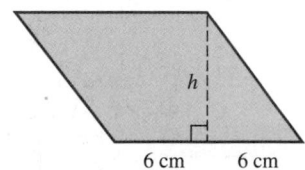

6 cm   6 cm

Now Try Problem 57

EXAMPLE 10   The area of the parallelogram shown on the right is 360 ft². Find the height.

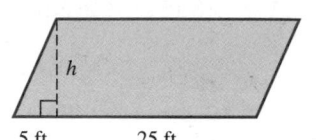

5 ft   25 ft

Strategy To find the height of the parallelogram, we will substitute the given values in the formula $A = bh$ and solve for h.

WHY The variable h represents the unknown height.

Solution From the figure, we see that the length of the base of the parallelogram is

5 feet + 25 feet = 30 feet

$$A = bh$$ This is the formula for the area of a parallelogram.

$$360 = 30h$$ Substitute 360 for A, the area, and 30 for b, the length of the base.

$$\frac{360}{30} = \frac{30h}{30}$$ To isolate h, undo the multiplication by 30 by dividing both sides by 30.

$$12 = h$$ Do the division.

$$\begin{array}{r} 12 \\ 30 \overline{)360} \\ -30 \\ \hline 60 \\ 60 \\ \hline 0 \end{array}$$

The height of the parallelogram is 12 feet.

---

3 **Find the area of figures that are combinations of polygons.**

> **Success Tip** To find the area of an irregular shape, break up the shape into familiar polygons. Find the area of each polygon and then add the results.

Self Check 11

Find the area of the shaded figure below.

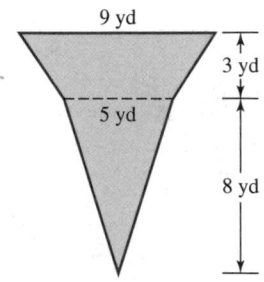

9 yd

3 yd

5 yd

8 yd

Now Try Problem 65

EXAMPLE 11   Find the area of one side of the tent shown below.

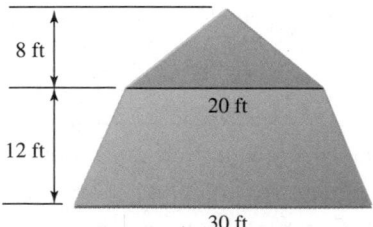

8 ft

20 ft

12 ft

30 ft

Strategy We will use the formula $A = \frac{1}{2}h(b_1 + b_2)$ to find the area of the lower portion of the tent and the formula $A = \frac{1}{2}bh$ to find the area of the upper portion of the tent. Then we will combine the results.

WHY A side of the tent is a combination of a trapezoid and a triangle.

Solution To find the area of the lower portion of the tent, we proceed as follows.

$$A_{\text{trap.}} = \frac{1}{2}h(b_1 + b_2)$$ This is the formula for the area of a trapezoid.

$$A_{\text{trap.}} = \frac{1}{2}(12)(30 + 20)$$ Substitute 30 for b_1, 20 for b_2, and 12 for h.

$$= \frac{1}{2}(12)(50)$$ Do the addition within the parentheses.

$$= 6(50)$$ Do the first multiplication: $\frac{1}{2}(12) = 6$.

$$= 300$$ Complete the multiplication.

The area of the trapezoid is 300 ft².
    To find the area of the upper portion of the tent, we proceed as follows.

$$A_{\text{triangle}} = \frac{1}{2}bh$$ This is the formula for the area of a triangle.

$$A_{\text{triangle}} = \frac{1}{2}(20)(8)$$ Substitute 20 for b and 8 for h.

$$= 80$$ Do the multiplications, working from left to right:
$\frac{1}{2}(20) = 10$ and then $10(8) = 80$.

The area of the triangle is 80 ft².
    To find the total area of one side of the tent, we add:

$$A_{\text{total}} = \mathbf{A_{\text{trap.}}} + A_{\text{triangle}}$$
$$A_{\text{total}} = \mathbf{300\ ft^2} + \mathbf{80\ ft^2}$$
$$= 380\ ft^2$$

The total area of one side of the tent is 380 ft².

EXAMPLE 12   Find the area of the shaded region shown on the right.

Strategy We will subtract the unwanted area of the square from the area of the rectangle.

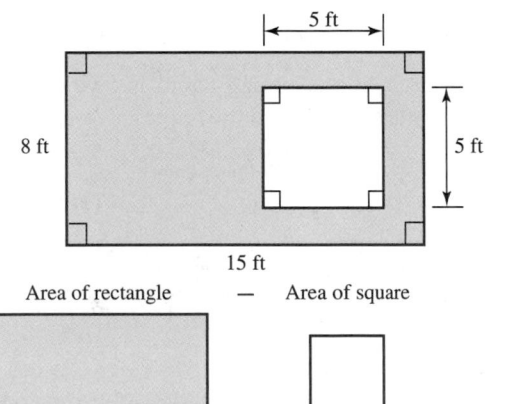

Area of shaded region = Area of rectangle − Area of square

WHY The area of the rectangular-shaped shaded figure does not include the square region inside of it.

Solution

$$A_{\text{shaded}} = lw - s^2$$ The formula for the area of a rectangle is A = lw.
The formula for the area of a square is A = s².

$$A_{\text{shaded}} = \mathbf{15}(8) - \mathbf{5}^2$$ Substitute 15 for the length l and 8 for the width w of the rectangle. Substitute 5 for the length s of a side of the square.

$$= 120 - 25$$

$$= 95$$

$$\begin{array}{r} \overset{4}{15} \\ \times 8 \\ \hline 120 \end{array}$$

$$\begin{array}{r} \overset{11}{\cancel{12}0} \\ -25 \\ \hline 95 \end{array}$$

The area of the shaded region is 95 ft².

Self Check 12

Find the area of the shaded region shown below.

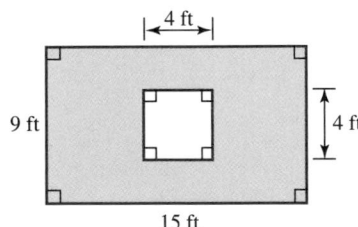

Now Try Problem 69

EXAMPLE 13   *Carpeting a Room*   A living room/dining room has the floor plan shown in the figure. If carpet costs $29 per square yard, including pad and installation, how much will it cost to carpet both rooms? (Assume no waste.)

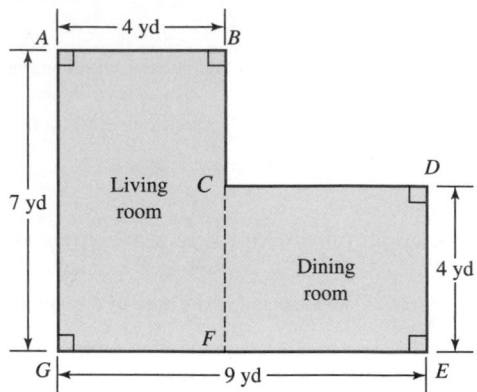

Strategy We will find the number of square yards of carpeting needed and multiply the result by $29.

WHY Each square yard costs $29.

Solution First, we must find the total area of the living room and the dining room:

$$A_{\text{total}} = A_{\text{living room}} + A_{\text{dining room}}$$

Since $\overline{CF}$ divides the space into two rectangles, the areas of the living room and the dining room are found by multiplying their respective lengths and widths. Therefore, the area of the living room is 4 yd · 7 yd = **28 yd²**.

The width of the dining room is given as 4 yd. To find its length, we subtract:

$$\text{m}(\overline{CD}) = \text{m}(\overline{GE}) - \text{m}(\overline{AB}) = 9 \text{ yd} - 4 \text{ yd} = 5 \text{ yd}$$

Thus, the area of the dining room is 5 yd · 4 yd = **20 yd²**. The total area to be carpeted is the sum of these two areas.

$$A_{\text{total}} = A_{\text{living room}} + A_{\text{dining room}}$$
$$A_{\text{total}} = \textbf{28 yd}^2 + \textbf{20 yd}^2$$
$$= 48 \text{ yd}^2$$

$$
\begin{array}{r}
48 \\
\times\,29 \\
\hline
432 \\
960 \\
\hline
1{,}392
\end{array}
$$

Now Try Problem 73

At $29 per square yard, the cost to carpet both rooms will be 48 · $29, or $1,392. ∎

ANSWERS TO SELF CHECKS

1. 154 cm   **2.** 50 in.   **3.** 28 m   **4.** 400 in.²   **5.** 10,000 cm²   **6.** 6,480 in.²   **7.** 90 mm²
8. 10.5 ft²   **9.** 54 m²   **10.** 8 cm   **11.** 41 yd²   **12.** 119 ft²

SECTION 9.7   STUDY SET

VOCABULARY

Fill in the blanks.

1. The distance around a polygon is called the _____.

2. The _____ of a polygon is measured in linear units such as inches, feet, and miles.

3. The measure of the surface enclosed by a polygon is called its _____.

4. If each side of a square measures 1 foot, the area enclosed by the square is 1 _____ foot.

5. The _____ of a polygon is measured in square units.

6. The segment that represents the height of a triangle is called an _____.

CONCEPTS

7. The figure below shows a kitchen floor that is covered with 1-foot-square tiles. Without counting *all* of the squares, determine the area of the floor.

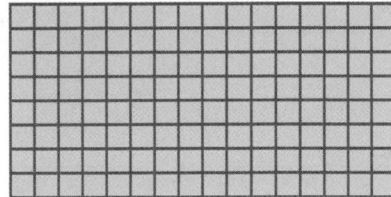

8. Tell which concept applies, perimeter or area.

   a. The length of a walk around New York's Central Park

   b. The amount of office floor space in the White House

   c. The amount of fence needed to enclose a playground

   d. The amount of land in Yellowstone National Park

9. Give the formula for the perimeter of a

   a. square   **b.** rectangle

10. Give the formula for the area of a

   a. square   **b.** rectangle

   c. triangle   **d.** trapezoid

   e. parallelogram

11. For each figure below, draw the altitude to the base *b*.

   a. 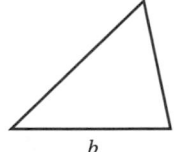   **b.**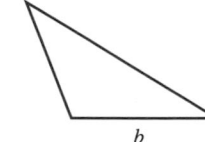

   c. 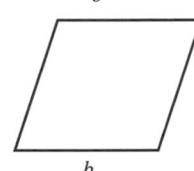   **d.**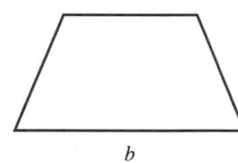

12. For each figure below, label the base *b* for the given altitude.

   a. 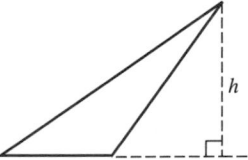   **b.**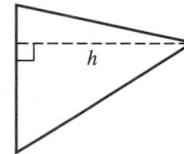

   c. 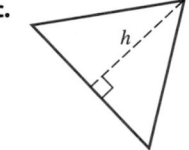   **d.**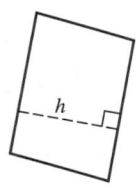

13. The shaded figure below is a combination of what two types of geometric figures?

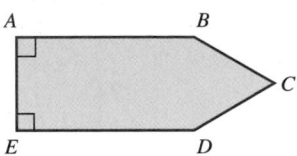

14. Explain how you would find the area of the following shaded figure.

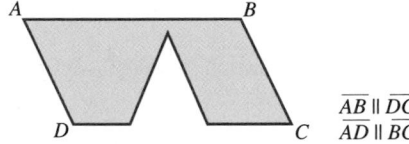

$\overline{AB} \parallel \overline{DC}$
$\overline{AD} \parallel \overline{BC}$

NOTATION

Fill in the blanks.

15. a. The symbol 1 in.^{2} means one _____ _____.

   b. One square meter is expressed as _____.

16. In the figure below, the symbol ⌐ indicates that the dashed line segment, called an *altitude*, is _____ to the base.

GUIDED PRACTICE

Find the perimeter of each square. See Example 1.

17. 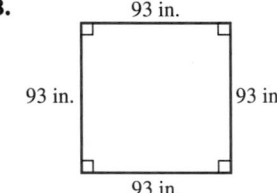   **18.**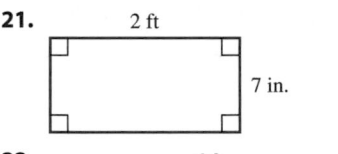

19. A square with sides 5.75 miles long

20. A square with sides 3.4 yards long

Find the perimeter of each rectangle, in inches. See Example 2.

21.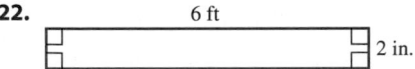

22.

23. 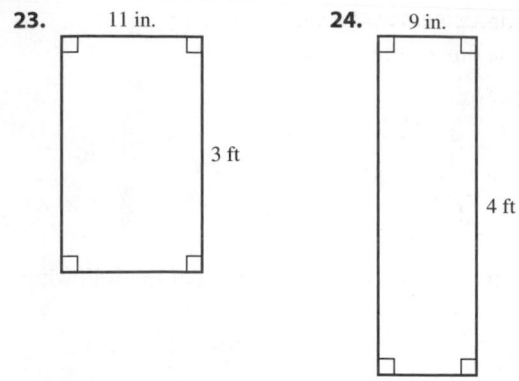 **24.**

Write and then solve an equation to answer each problem.
See Example 3.

25. The perimeter of an isosceles triangle is 35 feet. Each of the sides of equal length is 10 feet long. Find the length of the base of the triangle.

26. The perimeter of an isosceles triangle is 94 feet. Each of the sides of equal length is 42 feet long. Find the length of the base of the triangle.

27. The perimeter of an isosceles trapezoid is 35 meters. The upper base is 10 meters long, and the lower base is 15 meters long. How long is each leg of the trapezoid?

28. The perimeter of an isosceles trapezoid is 46 inches. The upper base is 12 inches long, and the lower base is 16 inches long. How long is each leg of the trapezoid?

Find the area of each square. See Example 4.

29. 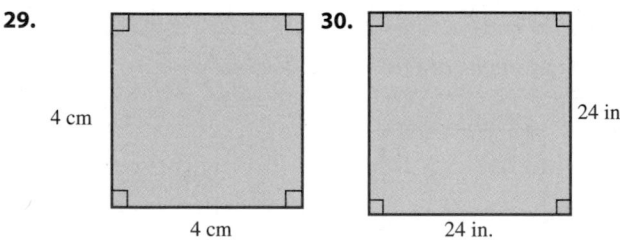 **30.**

31. A square with sides 2.5 meters long

32. A square with sides 6.8 feet long

For Problems 33–40, see Example 5.

33. How many square inches are in 1 square foot?

34. How many square inches are in 1 square yard?

35. How many square millimeters are in 1 square meter?

36. How many square decimeters are in 1 square meter?

37. How many square feet are in 1 square mile?

38. How many square yards are in 1 square mile?

39. How many square meters are in 1 square kilometer?

40. How many square dekameters are in 1 square kilometer?

Find the area of each rectangle. Give the answer in square feet.
See Example 6.

41. 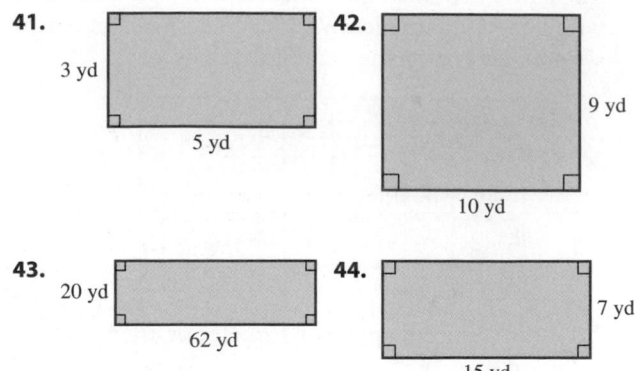 **42.**

43. **44.**

Find the area of each triangle. See Example 7.

45.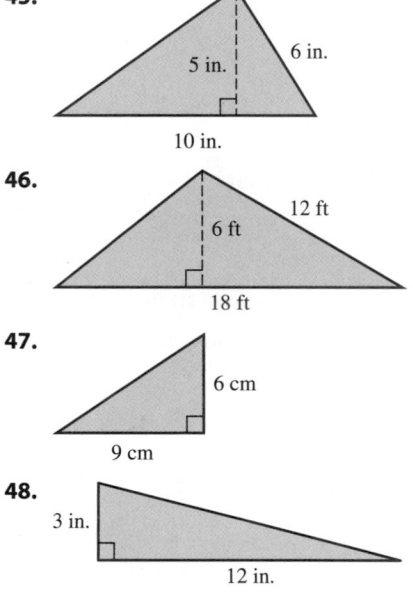

46.

47.

48.

Find the area of each triangle. See Example 8.

49.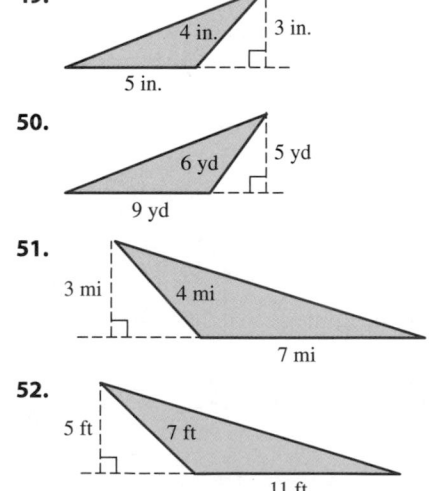

50.

51.

52.

Find the area of each trapezoid. **See Example 9.**

53.
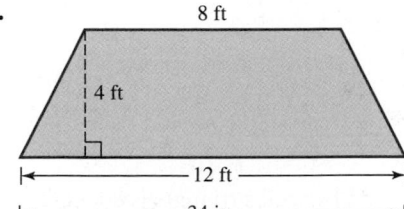
8 ft
4 ft
12 ft

54.
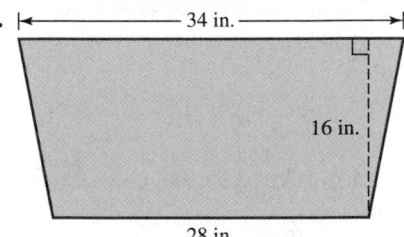
34 in.
16 in.
28 in.

55.
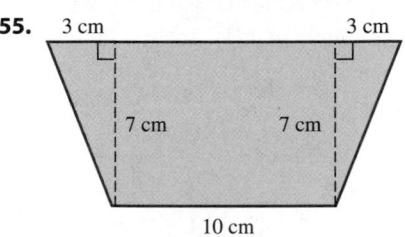
3 cm   3 cm
7 cm   7 cm
10 cm

56.
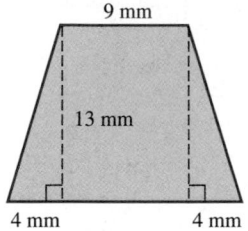
9 mm
13 mm
4 mm   4 mm

Solve each problem. **See Example 10.**

57. The area of a parallelogram is 60 m², and its height is 15 m. Find the length of its base.

58. The area of a parallelogram is 95 in.², and its height is 5 in. Find the length of its base.

59. The area of a rectangle is 36 cm², and its length is 3 cm. Find its width.

60. The area of a rectangle is 144 mi², and its length is 6 mi. Find its width.

61. The area of a triangle is 54 m², and the length of its base is 3 m. Find the height.

62. The area of a triangle is 270 ft², and the length of its base is 18 ft. Find the height.

63. The perimeter of a rectangle is 64 mi, and its length is 21 mi. Find its width.

64. The perimeter of a rectangle is 26 yd, and its length is 10.5 yd. Find its width.

Find the area of each shaded figure. **See Example 11.**

65.
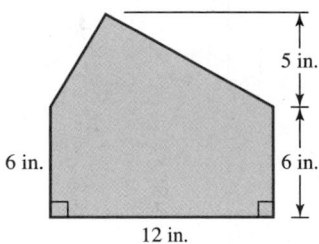
5 in.
6 in.   6 in.
12 in.

66.
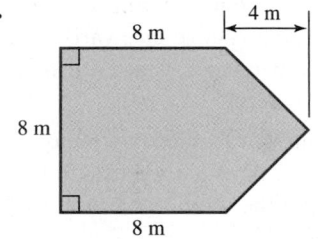
8 m   4 m
8 m
8 m

67.
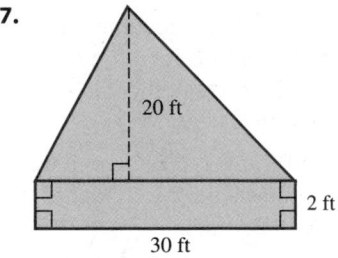
20 ft
2 ft
30 ft

68.
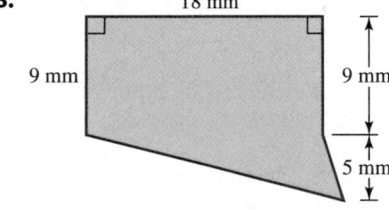
18 mm
9 mm   9 mm
5 mm

Find the area of each shaded figure. **See Example 12.**

69.
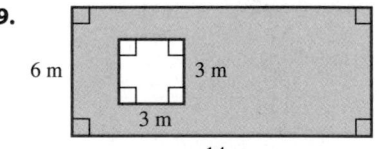
6 m   3 m
3 m
14 m

70.
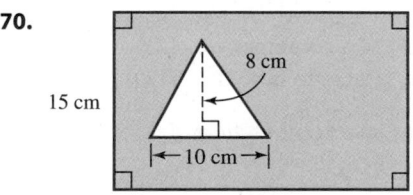
8 cm
15 cm
10 cm
25 cm

71.
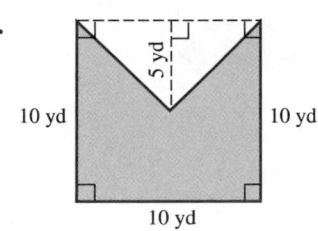
5 yd
10 yd   10 yd
10 yd

72.

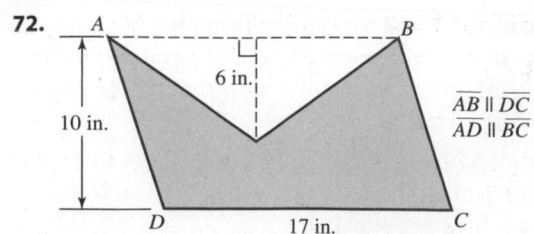

$\overline{AB} \parallel \overline{DC}$
$\overline{AD} \parallel \overline{BC}$

Solve each problem. See Example 13.

73. FLOORING A rectangular family room is 8 yards long and 5 yards wide. At $30 per square yard, how much will it cost to put down vinyl sheet flooring in the room? (Assume no waste.)

74. CARPETING A rectangular living room measures 10 yards by 6 yards. At $32 per square yard, how much will it cost to carpet the room? (Assume no waste.)

75. FENCES A man wants to enclose a rectangular yard with fencing that costs $12.50 a foot, including installation. Find the cost of enclosing the yard if its dimensions are 110 ft by 85 ft.

76. FRAMES Find the cost of framing a rectangular picture with dimensions of 24 inches by 30 inches if framing material costs $0.75 per inch.

TRY IT YOURSELF

Sketch and label each of the figures.

77. Two different rectangles, each having a perimeter of 40 in.

78. Two different rectangles, each having an area of 40 in.^{2}

79. A square with an area of 25 m^2

80. A square with a perimeter of 20 m

81. A parallelogram with an area of 15 yd^2

82. A triangle with an area of 20 ft^2

83. A figure consisting of a combination of two rectangles, whose total area is 80 ft^2

84. A figure consisting of a combination of a rectangle and a square, whose total area is 164 ft^2

Find the area of each parallelogram.

85.

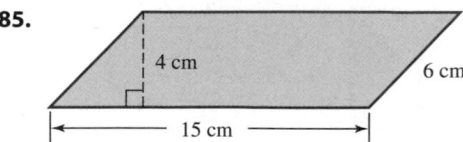

86.

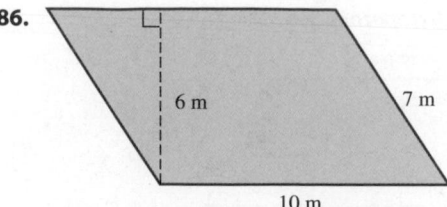

87. The perimeter of an isosceles triangle is 80 meters. If the length of one of the congruent sides is 22 meters, how long is the base?

88. The perimeter of a square is 35 yards. How long is a side of the square?

89. The perimeter of an equilateral triangle is 85 feet. Find the length of each side.

90. An isosceles triangle with congruent sides of length 49.3 inches has a perimeter of 121.7 inches. Find the length of the base.

Find the perimeter of the figure.

91.   **92.**

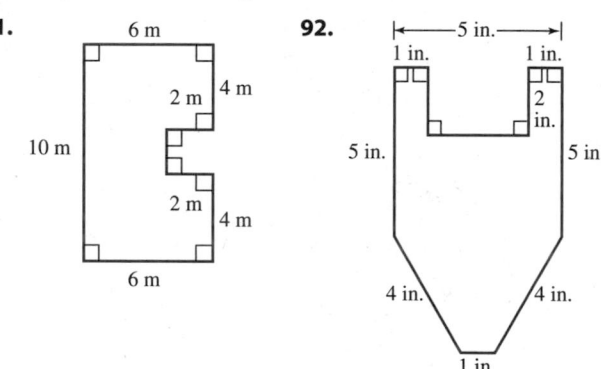

Find x and y. Then find the perimeter of the figure.

93.

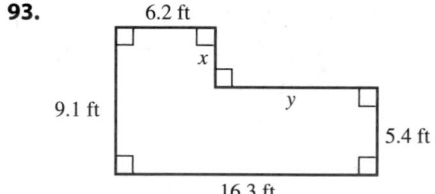

94.

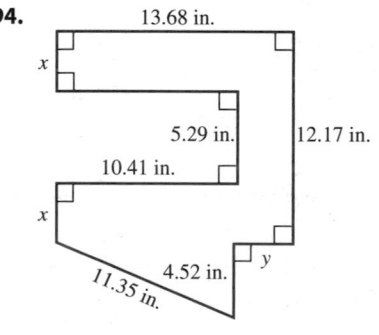

APPLICATIONS

95. LANDSCAPING A woman wants to plant a pine-tree screen around three sides of her rectangular-shaped backyard. (See the figure below.) If she plants the trees 3 feet apart, how many trees will she need?

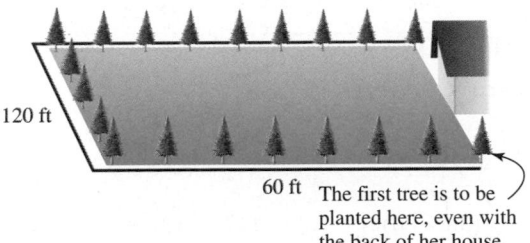

120 ft

60 ft   The first tree is to be planted here, even with the back of her house.

96. GARDENING A gardener wants to plant a border of marigolds around the garden shown below, to keep out rabbits. How many plants will she need if she allows 6 inches between plants?

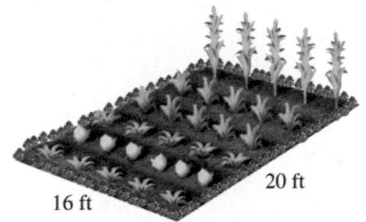

20 ft

16 ft

97. COMPARISON SHOPPING Which is more expensive: a ceramic-tile floor costing $3.75 per square foot or vinyl costing $34.95 per square yard?

98. COMPARISON SHOPPING Which is cheaper: a hardwood floor costing $6.95 per square foot or a carpeted floor costing $37.50 per square yard?

99. TILES A rectangular basement room measures 14 by 20 feet. Vinyl floor tiles that are 1 ft² cost $1.29 each. How much will the tile cost to cover the floor? (Assume no waste.)

100. PAINTING The north wall of a barn is a rectangle 23 feet high and 72 feet long. There are five windows in the wall, each 4 by 6 feet. If a gallon of paint will cover 300 ft², how many gallons of paint must the painter buy to paint the wall?

101. SAILS If nylon is $12 per square yard, how much would the fabric cost to make a triangular sail with a base of 12 feet and a height of 24 feet?

102. REMODELING The gable end of a house is an isosceles triangle with a height of 4 yards and a base of 23 yards. It will require one coat of primer and one coat of finish to paint the triangle. Primer costs

$17 per gallon, and the finish paint costs $23 per gallon. If one gallon of each type of paint covers 300 square feet, how much will it cost to paint the gable, excluding labor?

103. GEOGRAPHY Use the dimensions of the trapezoid that is superimposed over the state of Nevada to estimate the area of the "Silver State."

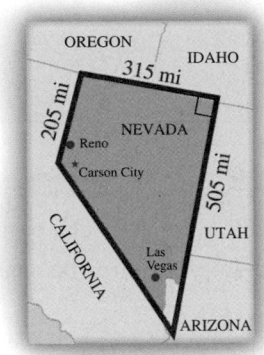

OREGON   IDAHO
315 mi
205 mi   NEVADA
Reno
Carson City   505 mi
CALIFORNIA   UTAH
Las Vegas
ARIZONA

104. SOLAR COVERS A swimming pool has the shape shown below. How many square feet of a solar blanket material will be needed to cover the pool? How much will the cover cost if it is $1.95 per square foot? (Assume no waste.)

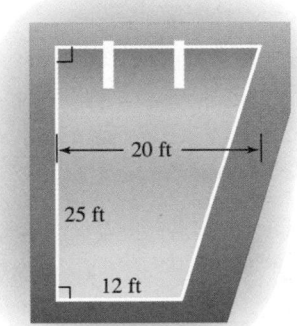

20 ft

25 ft

12 ft

105. CARPENTRY How many sheets of 4-foot-by-8-foot sheetrock are needed to drywall the inside walls on the first floor of the barn shown below? (Assume that the carpenters will cover each wall entirely and then cut out areas for the doors and windows.)

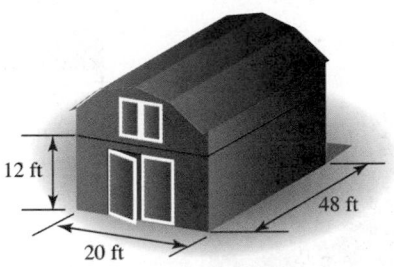

12 ft

48 ft

20 ft

106. CARPENTRY If it costs $90 per square foot to build a one-story home in northern Wisconsin, find the cost of building the house with the floor plan shown below.

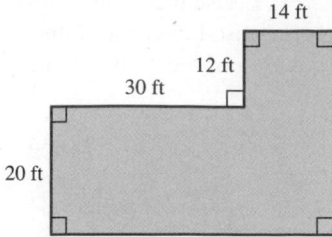

14 ft

12 ft

30 ft

20 ft

107. Explain the difference between perimeter and area.

108. Why is it necessary that area be measured in square units?

109. A student expressed the area of the square in the figure below as 25^2 ft. Explain his error.

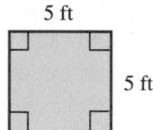

5 ft

5 ft

110. Refer to the figure below. What must be done before we can use the formula to find the area of this rectangle?

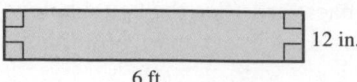

12 in.

6 ft

REVIEW

Simplify each expression.

111. $8\left(\dfrac{3}{4}t\right)$

112. $27\left(\dfrac{2}{3}m\right)$

113. $-\dfrac{2}{3}(3w - 6)$

114. $\dfrac{1}{2}(2y - 8)$

115. $-\dfrac{7}{16}x - \dfrac{3}{16}x$

116. $-\dfrac{5}{18}x - \dfrac{7}{18}x$

117. $60\left(\dfrac{3}{20}r - \dfrac{4}{15}\right)$

118. $72\left(\dfrac{7}{8}f - \dfrac{8}{9}\right)$

Objectives

1 Define circle, radius, chord, diameter, and arc.

2 Find the circumference of a circle.

3 Find the area of a circle.

SECTION 9.8

Circles

In this section, we will discuss the circle, one of the most useful geometric figures of all. In fact, the discoveries of fire and the circular wheel are two of the most important events in the history of the human race. We will begin our study by introducing some basic vocabulary associated with circles.

1 Define circle, radius, chord, diameter, and arc.

> **Circle**
>
> A **circle** is the set of all points in a plane that lie a fixed distance from a point called its **center.**

A segment drawn from the center of a circle to a point on the circle is called a **radius.** (The plural of *radius* is *radii.*) From the definition, it follows that all radii of the same circle are the same length.

A **chord** of a circle is a line segment that connects two points on the circle. A **diameter** is a chord that passes through the center of the circle. Since a diameter D of a circle is twice as long as a radius r, we have

$$D = 2r$$

Each of the previous definitions is illustrated in figure (a) below, in which O is the center of the circle.

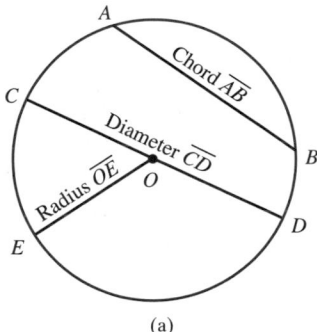 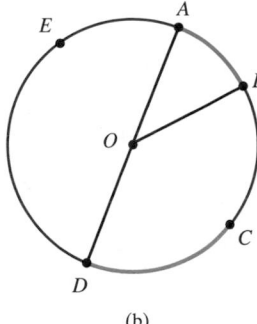

        (a)   (b)

Any part of a circle is called an **arc.** In figure (b) above, the part of the circle from point A to point B that is highlighted in blue is $\overset{\frown}{AB}$, read as "arc AB." $\overset{\frown}{CD}$ is the part of the circle from point C to point D that is highlighted in green. An arc that is half of a circle is a **semicircle.**

> ### Semicircle
>
> A **semicircle** is an arc of a circle whose endpoints are the endpoints of a diameter.

If point O is the center of the circle in figure (b), $\overline{AD}$ is a diameter and $\overset{\frown}{AED}$ is a semicircle. The middle letter E distinguishes semicircle $\overset{\frown}{AED}$ (the part of the circle from point A to point D that includes point E) from semicircle $\overset{\frown}{ABD}$ (the part of the circle from point A to point D that includes point B).

An arc that is shorter than a semicircle is a **minor arc.** An arc that is longer than a semicircle is a **major arc.** In figure (b),

    $\overset{\frown}{AE}$ is a minor arc   and $\overset{\frown}{ABE}$ is a major arc.

> ***Success Tip***   It is often possible to name a major arc in more than one way. For example, in figure (b), major arc $\overset{\frown}{ABE}$ is the part of the circle from point A to point E that includes point B. Two other names for the same major arc are $\overset{\frown}{ACE}$ and $\overset{\frown}{ADE}$.

2   Find the circumference of a circle.

Since early history, mathematicians have known that the ratio of the distance around a circle (the circumference) divided by the length of its diameter is approximately 3. First Kings, Chapter 7, of the Bible describes a round bronze tank that was 15 feet from brim to brim and 45 feet in circumference, and $\frac{45}{15} = 3$. Today, we use a more precise value for this ratio, known as π (pi). If C is the circumference of a circle and D is the length of its diameter, then

$$\pi = \frac{C}{D} \quad \text{where } \pi = 3.141592653589\ldots \quad \text{\small $\frac{22}{7}$ and 3.14 are often used as estimates of } \pi.$$

If we multiply both sides of $\pi = \frac{C}{D}$ by D, we have the following formula.

Circumference of a Circle

The circumference of a circle is given by the formula

$$C = \pi D \quad \text{where } C \text{ is the circumference and } D \text{ is the length of the diameter}$$

Since a diameter of a circle is twice as long as a radius r, we can substitute $2r$ for D in the formula $C = \pi D$ to obtain another formula for the circumference C:

$$C = 2\pi r \quad \text{The notation } 2\pi r \text{ means } 2 \cdot \pi \cdot r.$$

Self Check 1

Find the circumference of the circle shown below. Give the exact answer and an approximation.

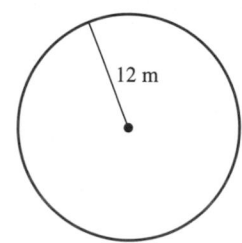
12 m

Now Try Problem 25

EXAMPLE 1   Find the circumference of the circle shown on the right. Give the exact answer and an approximation.

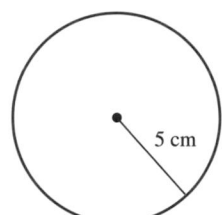
5 cm

Strategy We will substitute 5 for r in the formula $C = 2\pi r$ and evaluate the right side.

WHY The variable C represents the unknown circumference of the circle.

Solution

$C = 2\pi r$	This is the formula for the circumference of a circle.
$C = 2\pi(5)$	Substitute 5 for r, the radius.
$C = 2(5)\pi$	When a product involves π, we usually rewrite it so that π is the last factor.
$C = 10\pi$	Do the first multiplication: 2(5) = 10. This is the exact answer.

The circumference of the circle is exactly 10π cm. If we replace π with 3.14, we get an approximation of the circumference.

$$C = 10\pi$$
$$C \approx 10(\textbf{3.14})$$
$$C \approx 31.4 \qquad \text{To multiply by 10, move the decimal point in 3.14 one place to the right.}$$

The circumference of the circle is approximately 31.4 cm.

Using Your CALCULATOR   Calculating Revolutions of a Tire

When the $\boxed{\pi}$ key on a scientific calculator is pressed (on some models, the $\boxed{\text{2nd}}$ key must be pressed first), an approximation of π is displayed. To illustrate how to use this key, consider the following problem. How many times does the tire shown to the right revolve when a car makes a 25-mile trip?

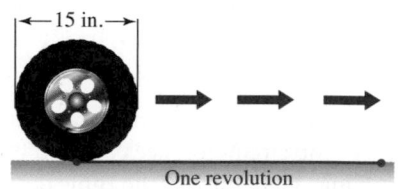
15 in.
One revolution

We first find the circumference of the tire. From the figure, we see that the diameter of the tire is 15 inches. Since the circumference of a circle is the product of π and the length of its diameter, the tire's circumference is $\pi \cdot 15$ inches, or 15π inches. (Normally, we rewrite a product such as $\pi \cdot 15$ so that π is the second factor.)

We then change the 25 miles to inches using two unit conversion factors.

$$\frac{25 \text{ miles}}{1} \cdot \frac{5{,}280 \text{ feet}}{1 \text{ mile}} \cdot \frac{12 \text{ inches}}{1 \text{ foot}} = 25 \cdot 5{,}280 \cdot 12 \text{ inches}$$ The units of miles and feet can be removed.

The length of the trip is $25 \cdot 5{,}280 \cdot 12$ inches.

Finally, we divide the length of the trip by the circumference of the tire to get

$$\frac{\text{The number of}}{\text{revolutions of the tire}} = \frac{25 \cdot 5{,}280 \cdot 12}{15\pi}$$

We can use a scientific calculator to make this calculation.

(25 × 5280 × 12) ÷ (15 × π) = $\boxed{\text{33613.52398}}$

The tire makes about 33,614 revolutions.

EXAMPLE 2   *Architecture*   A Norman window is constructed by adding a semicircular window to the top of a rectangular window. Find the perimeter of the Norman window shown here.

Strategy We will find the perimeter of the rectangular part and the circumference of the circular part of the window and add the results.

WHY The window is a combination of a rectangle and a semicircle.

Solution The perimeter of the rectangular part is

$$P_{\text{rectangular part}} = 8 + 6 + 8 = 22$$ Add only 3 sides of the rectangle.

The perimeter of the semicircle is one-half of the circumference of a circle that has a 6-foot diameter.

$P_{\text{semicircle}} = \dfrac{1}{2}C$ This is the formula for the circumference of a semicircle.

$P_{\text{semicircle}} = \dfrac{1}{2}\pi D$ Since we know the diameter, replace C with πD. We could also have replaced C with $2\pi r$.

$\qquad = \dfrac{1}{2}\pi(6)$ Substitute 6 for D, the diameter.

$\qquad \approx 9.424777961$ Use a calculator to do the multiplication.

The total perimeter is the sum of the two parts.

$P_{\text{total}} = P_{\text{rectangular part}} + P_{\text{semicircle}}$

$P_{\text{total}} \approx 22 + 9.424777961$

$\qquad \approx 31.424777961$

To the nearest hundredth, the perimeter of the window is 31.42 feet.

8 ft   8 ft

6 ft

Self Check 2

Find the perimeter of the figure shown below. Round to the nearest hundredth. (Assume the arc is a semicircle.)

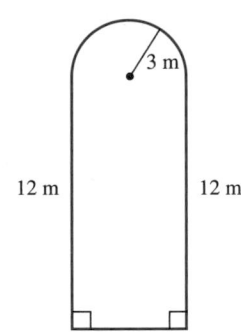

3 m

12 m   12 m

Now Try Problem 29

---

3   Find the area of a circle.

If we divide the circle shown in figure (a) on the following page into an even number of pie-shaped pieces and then rearrange them as shown in figure (b), we have a figure that looks like a parallelogram. The figure has a base b that is one-half the circumference of the circle, and its height h is about the same length as a radius of the circle.

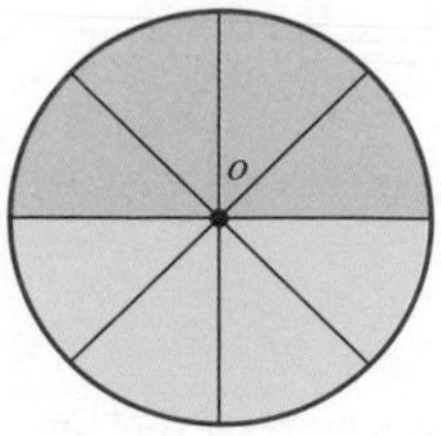

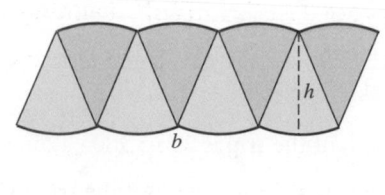

(a)   (b)

If we divide the circle into more and more pie-shaped pieces, the figure will look more and more like a parallelogram, and we can find its area by using the formula for the area of a parallelogram.

$A = bh$

$A = \dfrac{1}{2}Cr$ Substitute $\frac{1}{2}$ of the circumference for b, the length of the base of the "parallelogram." Substitute r for the height of the "parallelogram."

$\quad = \dfrac{1}{2}(2\pi r)r$ Substitute $2\pi r$ for C.

$\quad = \pi r^2$ Simplify: $\frac{1}{2} \cdot 2 = 1$ and $r \cdot r = r^2$.

This result gives the following formula.

Area of a Circle

The area of a circle with radius r is given by the formula

$A = \pi r^2$

Self Check 3

Find the area of a circle with a diameter of 12 feet. Give the exact answer and an approximation to the nearest tenth.

Now Try Problem 33

EXAMPLE 3   Find the area of the circle shown on the right. Give the exact answer and an approximation to the nearest tenth.

Strategy We will find the radius of the circle, substitute that value for r in the formula $A = \pi r^2$, and evaluate the right side.

WHY The variable A represents the unknown area of the circle.

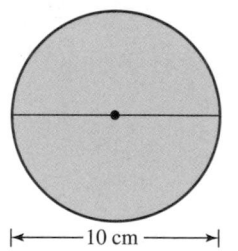

$\longleftarrow$ 10 cm $\longrightarrow$

Solution Since the length of the diameter is 10 centimeters and the length of a diameter is twice the length of a radius, the length of the radius is 5 centimeters.

$A = \pi r^2$ This is the formula for the area of a circle.

$A = \pi(\mathbf{5})^2$ Substitute 5 for r, the radius of the circle. The notation πr^2 means $\pi \cdot r^2$.

$\quad = \pi(25)$ Evaluate the exponential expression.

$\quad = 25\pi$ Write the product so that π is the last factor.

The exact area of the circle is 25π cm². We can use a calculator to approximate the area.

$A \approx 78.53981634$ Use a calculator to do the multiplication: $25 \cdot \pi$.

To the nearest tenth, the area is 78.5 cm².

Using Your CALCULATOR   Painting a Helicopter Landing Pad

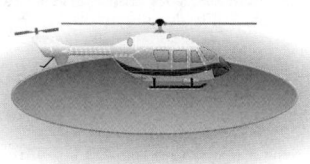

Orange paint is available in gallon containers at $19 each, and each gallon will cover 375 ft². To calculate how much the paint will cost to cover a circular helicopter landing pad 60 feet in diameter, we first calculate the area of the helicopter pad.

$A = \pi r^2$ This is the formula for the area of a circle.

$A = \pi(30)^2$ Substitute one-half of 60 for r, the radius of the circular pad.

$= 30^2\pi$ Write the product so that π is the last factor.

The area of the pad is exactly $30^2\pi$ ft². Since each gallon of paint will cover 375 ft², we can find the number of gallons of paint needed by dividing $30^2\pi$ by 375.

$$\text{Number of gallons needed} = \frac{30^2\pi}{375}$$

We can use a scientific calculator to make this calculation.

$30 \boxed{x^2} \boxed{\times} \boxed{\pi} \boxed{=} \boxed{\div} 375 \boxed{=}$ $\boxed{7.539822369}$

Because paint comes only in full gallons, the painter will need to purchase 8 gallons. The cost of the paint will be 8($19), or $152.

EXAMPLE 4   Find the area of the shaded figure on the right. Round to the nearest hundredth.

Strategy We will find the area of the entire shaded figure using the following approach:

$A_{\text{total}} = A_{\text{triangle}} + A_{\text{smaller semicircle}} + A_{\text{larger semicircle}}$

WHY The shaded figure is a combination of a triangular region and two semicircular regions.

Solution The area of the triangle is

$$A_{\text{triangle}} = \frac{1}{2}bh = \frac{1}{2}(6)(8) = \frac{1}{2}(48) = 24$$

Since the formula for the area of a circle is $A = \pi r^2$, the formula for the area of a semicircle is $A = \frac{1}{2}\pi r^2$. Thus, the area enclosed by the smaller semicircle is

$$A_{\text{smaller semicircle}} = \frac{1}{2}\pi r^2 = \frac{1}{2}\pi(4)^2 = \frac{1}{2}\pi(16) = 8\pi$$

The area enclosed by the larger semicircle is

$$A_{\text{larger semicircle}} = \frac{1}{2}\pi r^2 = \frac{1}{2}\pi(5)^2 = \frac{1}{2}\pi(25) = 12.5\pi$$

The total area is the sum of the three results:

$A_{\text{total}} = 24 + 8\pi + 12.5\pi \approx 88.4026494$ Use a calculator to perform the operations.

To the nearest hundredth, the area of the shaded figure is 88.40 in.².

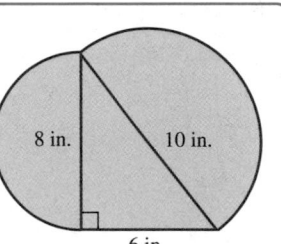

8 in. 10 in. 6 in.

```
        12.5
    2)25.0
      -2
      ___
      05
      -4
      ___
       10
      -10
      ___
        0
```

Self Check 4

Find the area of the shaded figure below. Round to the nearest hundredth.

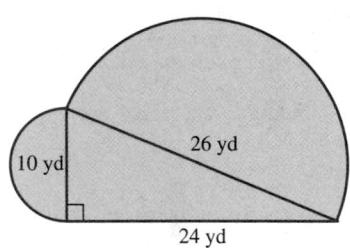

10 yd   26 yd   24 yd

Now Try Problem 37

SECTION 9.8   STUDY SET

VOCABULARY

Fill in the blanks.

1. A segment drawn from the center of a circle to a point on the circle is called a _____.

2. A segment joining two points on a circle is called a _____.

3. A _____ is a chord that passes through the center of a circle.

4. An arc that is one-half of a complete circle is a _____.

5. The distance around a circle is called its _____.

6. The surface enclosed by a circle is called its _____.

7. A diameter of a circle is _____ as long as a radius.

8. Suppose the exact circumference of a circle is 3π feet. When we write $C \approx 9.42$ feet, we are giving an _____ of the circumference.

CONCEPTS

Refer to the figure below, where point 0 is the center of the circle.

9. Name each radius.

10. Name a diameter.

11. Name each chord.

12. Name each minor arc.

13. Name each semicircle.

14. Name major arc $\overarc{ABD}$ in another way.

15. **a.** If you know the radius of a circle, how can you find its diameter?

    b. If you know the diameter of a circle, how can you find its radius?

16. **a.** What are the two formulas that can be used to find the circumference of a circle?

    b. What is the formula for the area of a circle?

17. If C is the circumference of a circle and D is its diameter, then $\frac{C}{D} = $ [ ].

18. If D is the diameter of a circle and r is its radius, then $D = $ [ ] r.

19. When evaluating $\pi(6)^2$, what operation should be performed first?

20. Round $\pi = 3.141592653589\ldots$ to the nearest hundredth.

NOTATION

Fill in the blanks.

21. The symbol $\overarc{AB}$ is read as "____ ____."

22. To the nearest hundredth, the value of π is _____.

23. **a.** In the expression $2\pi r$, what operations are indicated?

    b. In the expression πr^2, what operations are indicated?

24. Write each expression in better form. Leave π in your answer.

    a. $\pi(8)$ **b.** $2\pi(7)$ **c.** $\pi \cdot \dfrac{25}{3}$

GUIDED PRACTICE

The answers to the problems in this Study Set may vary slightly, depending on which approximation of π is used.

Find the circumference of the circle shown below. Give the exact answer and an approximation to the nearest tenth. See Example 1.

25. 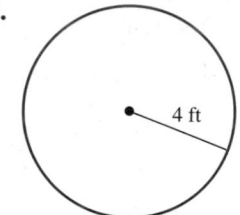   26.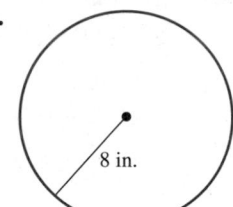

27. 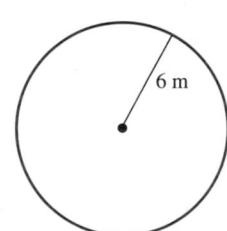   28.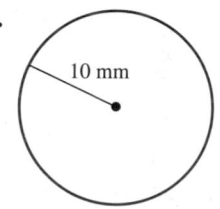

Find the perimeter of each figure. Assume each arc is a semicircle. Round to the nearest hundredth. See Example 2.

29. 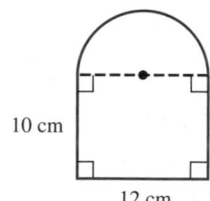   30.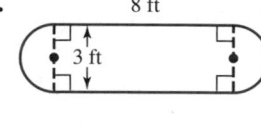

31. 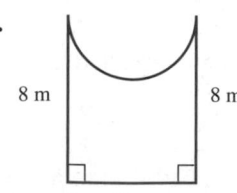   32.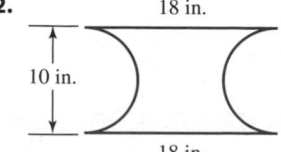

Find the area of each circle given the following information. Give the exact answer and an approximation to the nearest tenth.
See Example 3.

33.

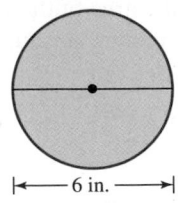

|← 6 in. →|

34.

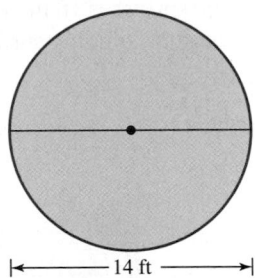

|← 14 ft →|

35. Find the area of a circle with diameter 18 inches.

36. Find the area of a circle with diameter 20 meters.

Find the total area of each figure. Assume each arc is a semicircle. Round to the nearest tenth. **See Example 4.**

37.

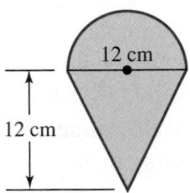

12 cm

12 cm

38.

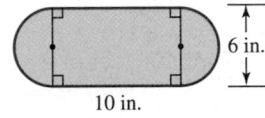

10 in.

6 in.

39.

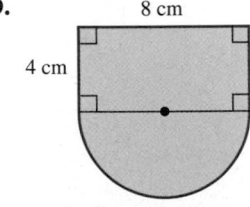

8 cm

4 cm

40.

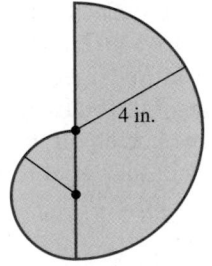

4 in.

▌TRY IT YOURSELF

Find the area of each shaded region. Round to the nearest tenth.

41.   4 in.

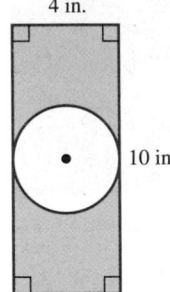

10 in

42.   8 in.

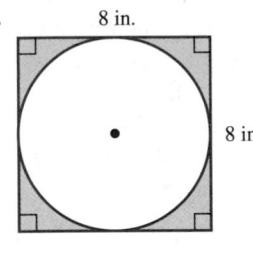

8 in.

43.   *r* = 4 in.

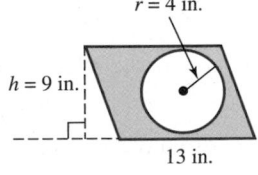

h = 9 in.

13 in.

44.

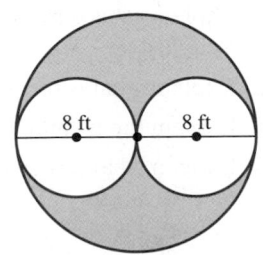

8 ft   8 ft

45. Find the circumference of the circle shown below. Give the exact answer and an approximation to the nearest hundredth.

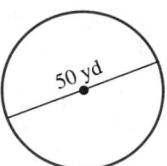

50 yd

46. Find the circumference of the semicircle shown below. Give the exact answer and an approximation to the nearest hundredth.

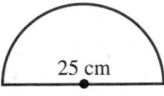

25 cm

47. Find the circumference of the circle shown below if the square has sides of length 6 inches. Give the exact answer and an approximation to the nearest tenth.

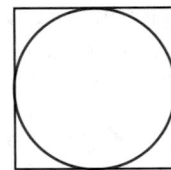

48. Find the circumference of the semicircle shown below if the length of the rectangle in which it is enclosed is 8 feet. Give the exact answer and an approximation to the nearest tenth.

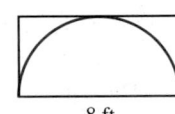

8 ft

49. Find the area of the circle shown below if the square has sides of length 9 millimeters. Give the exact answer and an approximation to the nearest tenth.

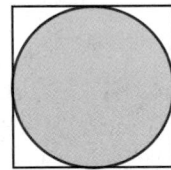

50. Find the area of the shaded semicircular region shown below. Give the exact answer and an approximation to the nearest tenth.

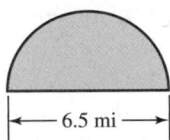

|← 6.5 mi →|

APPLICATIONS

51. Suppose the two "legs" of the compass shown below are adjusted so that the distance between the pointed ends is 1 inch. Then a circle is drawn.

a. What will the radius of the circle be?

b. What will the diameter of the circle be?

c. What will the circumference of the circle be? Give an exact answer and an approximation to the nearest hundredth.

d. What will the area of the circle be? Give an exact answer and an approximation to the nearest hundredth.

52. Suppose we find the distance around a can and the distance across the can using a measuring tape, as shown to the right. Then we make a comparison, in the form of a ratio:

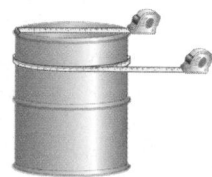

$$\frac{\text{The distance around the can}}{\text{The distance across the top of the can}}$$

After we do the indicated division, the result will be close to what number?

When appropriate, give the exact answer and an approximation to the nearest hundredth. Answers may vary slightly, depending on which approximation of π is used.

53. LAKES Round Lake has a circular shoreline that is 2 miles in diameter. Find the area of the lake.

54. HELICOPTERS Refer to the figure below. How far does a point on the tip of a rotor blade travel when it makes one complete revolution?

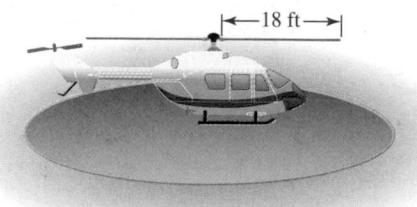

|←—18 ft—→|

55. GIANT SEQUOIA The largest sequoia tree is the General Sherman Tree in Sequoia National Park in California. In fact, it is considered to be the largest living thing in the world. According to the *Guinness Book of World Records*, it has a diameter of 32.66 feet, measured $4\frac{1}{2}$ feet above the ground. What is the circumference of the tree at that height?

56. TRAMPOLINE See the figure below. The distance from the center of the trampoline to the edge of its steel frame is 7 feet. The protective padding covering the springs is 18 inches wide. Find the area of the circular jumping surface of the trampoline, in square feet.

Protective pad

57. JOGGING Joan wants to jog 10 miles on a circular track $\frac{1}{4}$ mile in diameter. How many times must she circle the track? Round to the nearest lap.

58. CARPETING A state capitol building has a circular floor 100 feet in diameter. The legislature wishes to have the floor carpeted. The lowest bid is $83 per square yard, including installation. How much must the legislature spend for the carpeting project? Round to the nearest dollar.

59. ARCHERY See the figure on the right. Find the area of the entire target and the area of the bull's eye. What percent of the area of the target is the bull's eye?

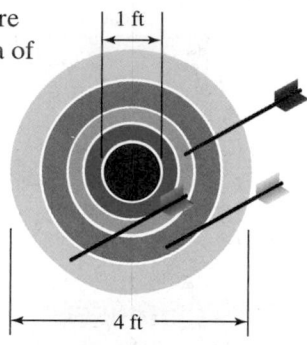

1 ft

4 ft

60. LANDSCAPE DESIGN See the figure on the right. How many square feet of lawn does not get watered by the four sprinklers at the center of each circle?

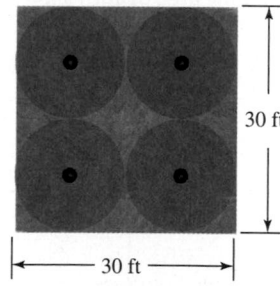

30 ft

30 ft

WRITING

61. Explain what is meant by the circumference of a circle.

62. Explain what is meant by the area of a circle.

63. Explain the meaning of π.

64. Explain what it means for a car to have a small *turning radius*.

REVIEW

65. Write $\frac{9}{10}$ as a percent.

66. Write $\frac{7}{8}$ as a percent.

67. Write 0.827 as a percent.

68. Write 0.036 as a percent.

69. UNIT COSTS A 24-ounce package of green beans sells for $1.29. Give the unit cost in cents per ounce.

70. MILEAGE One car went 1,235 miles on 51.3 gallons of gasoline, and another went 1,456 on 55.78 gallons. Which car got the better gas mileage?

71. How many sides does a pentagon have?

72. What is the sum of the measures of the angles of a triangle?

SECTION 9.9
Volume

Objectives

1 Find the volume of rectangular solids, prisms, and pyramids.

2 Find the volume of cylinders, cones, and spheres.

We have studied ways to calculate the perimeter and the area of two-dimensional figures that lie in a plane, such as rectangles, triangles, and circles. Now we will consider three-dimensional figures that occupy space, such as rectangular solids, cylinders, and spheres. In this section, we will introduce the vocabulary associated with these figures as well as the formulas that are used to find their volume. Volumes are measured in cubic units, such as cubic feet, cubic yards, or cubic centimeters. For example,

- We measure the capacity of a refrigerator in cubic feet.
- We buy gravel or topsoil by the cubic yard.
- We often measure amounts of medicine in cubic centimeters.

1 Find the volume of rectangular solids, prisms, and pyramids.

The **volume** of a three-dimensional figure is a measure of its capacity. The following illustration shows two common units of volume: cubic inches, written as in.^{3}, and cubic centimeters, written as cm^3.

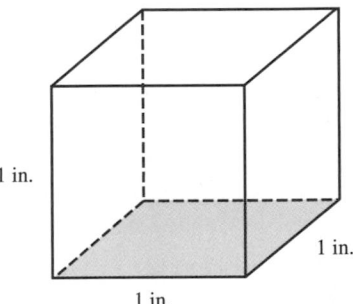

1 cubic inch: 1 in.^{3}

1 in.

1 in.

1 in.

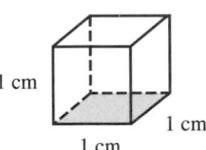

1 cubic centimeter: 1 cm^3

1 cm

1 cm

1 cm

The volume of a figure can be thought of as the number of cubic units that will fit within its boundaries. If we divide the figure shown in black below into cubes, each cube represents a volume of 1 cm^3. Because there are 2 levels with 12 cubes on each level, the volume of the prism is 24 cm^3.

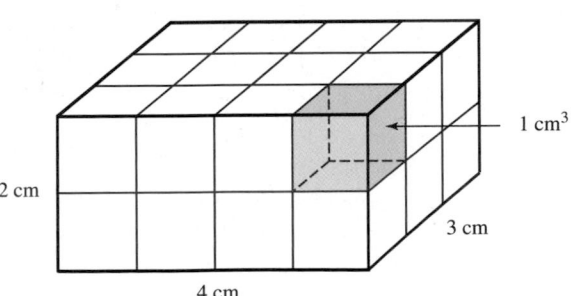

1 cm^3

2 cm

3 cm

4 cm

Self Check 1
How many cubic centimeters are
in 1 cubic meter?

Now Try Problem 25

EXAMPLE 1   How many cubic inches are there in 1 cubic foot?

Strategy A figure is helpful to solve this problem. We will draw a cube and divide
each of its sides into 12 equally long parts.

WHY Since a cubic foot is a cube with each side measuring 1 foot, each side also
measures 12 inches.

Solution The figure on the right helps us
understand the situation. Note that each level
of the cubic foot contains $12 \cdot 12$ cubic inches
and that the cubic foot has 12 levels. We can
use multiplication to count the number of
cubic inches contained in the figure. There are

$$12 \cdot 12 \cdot 12 = 1,728$$

cubic inches in 1 cubic foot. Thus, $1 \text{ ft}^3 = 1,728 \text{ in.}^3$.

Cube	Rectangular Solid	Sphere
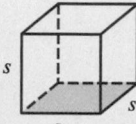	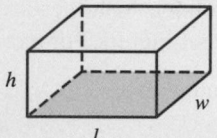	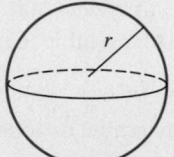
$V = s^3$	$V = lwh$	$V = \frac{4}{3}\pi r^3$
where s is the length of a side	where l is the length, w is the width, and h is the height	where r is the radius

Prism	Pyramid
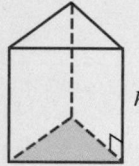 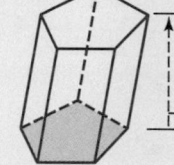	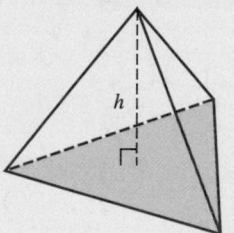 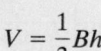
$V = Bh$	$V = \frac{1}{3}Bh$
where B is the area of the base and h is the height	where B is the area of the base and h is the height

Cylinder	Cone
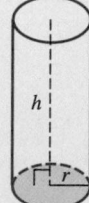 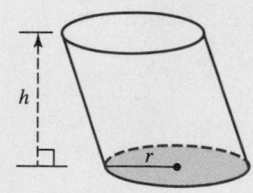	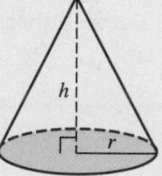 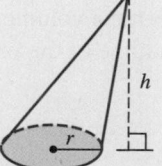
$V = Bh$ or $V = \pi r^2 h$	$V = \frac{1}{3}Bh$ or $V = \frac{1}{3}\pi r^2 h$
where B is the area of the base, h is the height, and r is the radius of the base	where B is the area of the base, h is the height, and r is the radius of the base

In practice, we do not find volumes of three-dimensional figures by counting cubes. Instead, we use the formulas shown in the table on the preceding page. Note that several of the volume formulas involve the variable B. It represents the area of the base of the figure.

> **Caution!** The height of a geometric solid is always measured along a line perpendicular to its base.

EXAMPLE 2 *Storage Tanks* An oil storage tank is in the form of a rectangular solid with dimensions 17 feet by 10 feet by 8 feet. (See the figure below.) Find its volume.

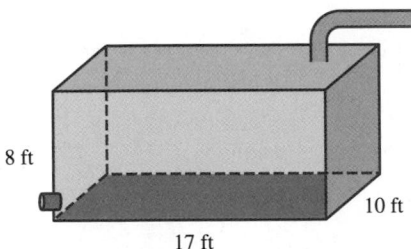

8 ft
10 ft
17 ft

Strategy We will substitute 17 for l, 10 for w, and 8 for h in the formula $V = lwh$ and evaluate the right side.

WHY The variable V represents the volume of a rectangular solid.

Solution

$V = lwh$ This is the formula for the volume of a rectangular solid.

$V = 17(10)(8)$ Substitute 17 for l, the length, 10 for w, the width, and 8 for h, the height of the tank.

$\quad\ = 1{,}360$ Do the multiplication.

$$\begin{array}{r} \overset{5}{1}70 \\ \times\ \ 8 \\ \hline 1{,}360 \end{array}$$

The volume of the tank is 1,360 ft³.

EXAMPLE 3 Find the volume of the prism shown on the right.

Strategy First, we will find the area of the base of the prism.

WHY To use the volume formula $V = Bh$, we need to know B, the area of the prism's base.

Solution The area of the triangular base of the prism is $\frac{1}{2}(6)(8) = 24$ square centimeters. To find its volume, we proceed as follows:

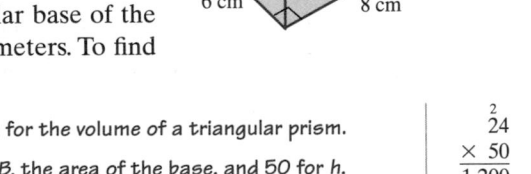

10 cm
50 cm
6 cm
8 cm

$V = Bh$ This is the formula for the volume of a triangular prism.

$V = 24(50)$ Substitute 24 for B, the area of the base, and 50 for h, the height.

$\quad\ = 1{,}200$ Do the multiplication.

$$\begin{array}{r} \overset{2}{2}4 \\ \times\ 50 \\ \hline 1{,}200 \end{array}$$

The volume of the triangular prism is 1,200 cm³.

> **Caution!** Note that the 10 cm measurement was not used in the calculation of the volume.

Self Check 2

Find the volume of a rectangular solid with dimensions 8 meters by 12 meters by 20 meters.

Now Try Problem 29

Self Check 3

Find the volume of the prism shown below.

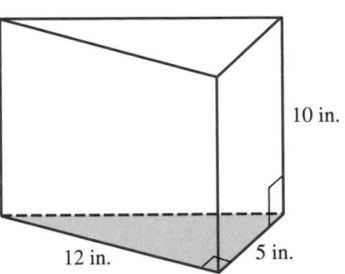

10 in.
12 in.
5 in.

Now Try Problem 33

Self Check 4

Find the volume of the pyramid shown below.

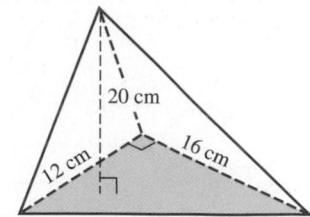

Now Try Problem 37

EXAMPLE 4 Find the volume of the pyramid shown on the right.

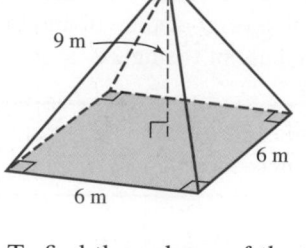

Strategy First, we will find the area of the square base of the pyramid.

WHY The volume of a pyramid is $\frac{1}{3}$ of the product of the area of its base and its height.

Solution Since the base is a square with each side 6 meters long, the area of the base is $(6\text{ m})^2$, or 36 m². To find the volume of the pyramid, we proceed as follows:

$$V = \frac{1}{3}Bh \qquad \text{This is the formula for the volume of a pyramid.}$$

$$V = \frac{1}{3}(36)(9) \qquad \text{Substitute 36 for } B \text{, the area of the base, and 9 for } h \text{, the height.}$$

$$= 12(9) \qquad \text{Multiply: } \frac{1}{3}(36) = \frac{36}{3} = 12.$$

$$= 108 \qquad \text{Complete the multiplication.}$$

$$\begin{array}{r} 12 \\ \times 9 \\ \hline 108 \end{array}$$

The volume of the pyramid is 108 m³.

---

2 **Find the volume of cylinders, cones, and spheres.**

Self Check 5

Find the volume of the cylinder shown below. Give the exact answer and an approximation to the nearest hundredth.

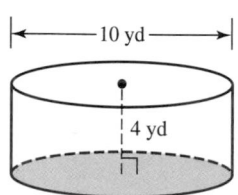

Now Try Problem 45

EXAMPLE 5 Find the volume of the cylinder shown on the right. Give the exact answer and an approximation to the nearest hundredth.

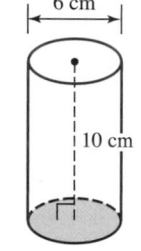

Strategy First, we will find the radius of the circular base of the cylinder.

WHY To use the formula for the volume of a cylinder, $V = \pi r^2 h$, we need to know r, the radius of the base.

Solution Since a radius is one-half of the diameter of the circular base, $r = \frac{1}{2} \cdot 6\text{ cm} = 3\text{ cm}$. From the figure, we see that the height of the cylinder is 10 cm. To find the volume of the cylinder, we proceed as follows.

$$V = \pi r^2 h \qquad \text{This is the formula for the volume of a cylinder.}$$

$$V = \pi(3)^2(10) \qquad \text{Substitute 3 for } r \text{, the radius of the base, and 10 for } h \text{, the height.}$$

$$V = \pi(9)(10) \qquad \text{Evaluate the exponential expression: } (3)^2 = 9.$$

$$= 90\pi \qquad \text{Multiply: } (9)(10) = 90. \text{ Write the product so that } \pi \text{ is the last factor.}$$

$$\approx 282.7433388 \qquad \text{Use a calculator to do the multiplication.}$$

The exact volume of the cylinder is 90π cm³. To the nearest hundredth, the volume is 282.74 cm³.

EXAMPLE 6 Find the volume of the cone shown on the right. Give the exact answer and an approximation to the nearest hundredth.

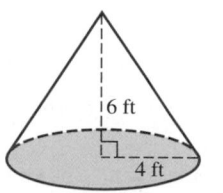

Strategy We will substitute 4 for r and 6 for h in the formula $V = \frac{1}{3}\pi r^2 h$ and evaluate the right side.

WHY The variable V represents the volume of a cone.

Solution

$$V = \frac{1}{3}\pi r^2 h \qquad \text{This is the formula for the volume of a cone.}$$

$$V = \frac{1}{3}\pi (4)^2 (6) \qquad \text{Substitute 4 for } r, \text{ the radius of the base, and 6 for } h, \text{ the height.}$$

$$= \frac{1}{3}\pi (16)(6) \qquad \text{Evaluate the exponential expression: } (4)^2 = 16.$$

$$= 2\pi (16) \qquad \text{Multiply: } \frac{1}{3}(6) = 2.$$

$$= 32\pi \qquad \text{Multiply: } 2(16) = 32. \text{ Write the product so that } \pi \text{ is the last factor.}$$

$$\approx 100.5309649 \qquad \text{Use a calculator to do the multiplication.}$$

The exact volume of the cone is 32π ft³. To the nearest hundredth, the volume is 100.53 ft³.

Self Check 6

Find the volume of the cone shown below. Give the exact answer and an approximation to the nearest hundredth.

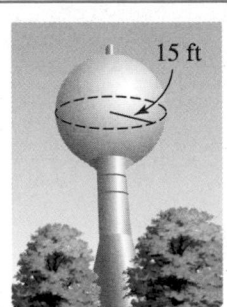

5 mi

2 mi

Now Try Problem 49

EXAMPLE 7 **Water Towers** How many cubic feet of water are needed to fill the spherical water tank shown on the right? Give the exact answer and an approximation to the nearest tenth.

Strategy We will substitute 15 for r in the formula $V = \frac{4}{3}\pi r^3$ and evaluate the right side.

WHY The variable V represents the volume of a sphere.

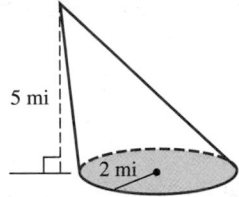

15 ft

Self Check 7

Find the volume of a spherical water tank with radius 7 meters. Give the exact answer and an approximation to the nearest tenth.

Now Try Problem 53

Solution

$$V = \frac{4}{3}\pi r^3 \qquad \text{This is the formula for the volume of a sphere.}$$

$$V = \frac{4}{3}\pi (15)^3 \qquad \text{Substitute 15 for } r, \text{ the radius of the sphere.}$$

$$= \frac{4}{3}\pi (3{,}375) \qquad \text{Evaluate the exponential expression: } (15)^3 = 3{,}375.$$

$$= \frac{13{,}500}{3}\pi \qquad \text{Multiply: } 4(3{,}375) = 13{,}500.$$

$$= 4{,}500\pi \qquad \text{Divide: } \frac{13{,}500}{3} = 4{,}500. \text{ Write the product so that } \pi \text{ is the last factor.}$$

$$\approx 14{,}137.16694 \qquad \text{Use a calculator to do the multiplication.}$$

$$\begin{array}{r} {}^{1\,3\,2} \\ 3375 \\ \times \quad 4 \\ \hline 13{,}500 \end{array}$$

The tank holds exactly $4{,}500\pi$ ft³ of water. To the nearest tenth, this is 14,137.2 ft³.

Using Your CALCULATOR   Volume of a Silo

A silo is a structure used for storing grain. The silo shown on the right is a cylinder 50 feet tall topped with a dome in the shape of a hemisphere. To find the volume of the silo, we add the volume of the cylinder to the volume of the dome.

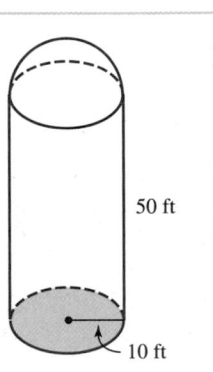

50 ft

10 ft

$$\textbf{Volume}_{\text{cylinder}} + \textbf{Volume}_{\text{dome}} = (\textbf{Area}_{\text{cylinder's base}})(\textbf{Height}_{\text{cylinder}}) + \frac{1}{2}(\textbf{Volume}_{\text{sphere}})$$

$$= \pi r^2 h + \frac{1}{2}\left(\frac{4}{3}\pi r^3\right)$$

$$= \pi r^2 h + \frac{2\pi r^3}{3} \qquad \text{Multiply and simplify: } \frac{1}{2}\left(\frac{4}{3}\pi r^3\right) = \frac{4}{6}\pi r^3 = \frac{2\pi r^3}{3}.$$

$$= \pi(10)^2 (50) + \frac{2\pi(10)^3}{3} \qquad \text{Substitute 10 for } r \text{ and 50 for } h.$$

We can use a scientific calculator to make this calculation.

$\boxed{\pi}\ \boxed{\times}\ 10\ \boxed{x^2}\ \boxed{\times}\ 50\ \boxed{+}\ \boxed{(}\ 2\ \boxed{\times}\ \boxed{\pi}\ \boxed{\times}\ 10\ \boxed{y^x}\ 3\ \boxed{)}\ \boxed{\div}\ 3\ \boxed{=}$

$\boxed{17802.35837}$

The volume of the silo is approximately 17,802 ft^3.

ANSWERS TO SELF CHECKS

1. 1,000,000 cm^3 **2.** 1,920 m^3 **3.** 300 in.^{3} **4.** 640 cm^3 **5.** 100π yd^3 ≈ 314.16 yd^3
6. $\frac{20}{3}\pi$ mi^3 ≈ 20.94 mi^3 **7.** $\frac{1,372}{3}\pi$ m^3 ≈ 1,436.8 m^3

SECTION 9.9   STUDY SET

VOCABULARY

Fill in the blanks.

1. The _____ of a three-dimensional figure is a measure of its capacity.

2. The volume of a figure can be thought of as the number of _____ units that will fit within its boundaries.

Give the name of each figure.

3.

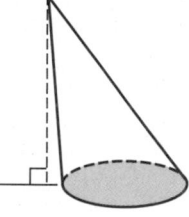

4.

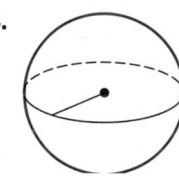

5.

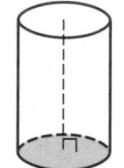

6.

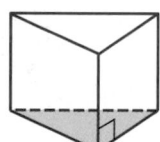

7.

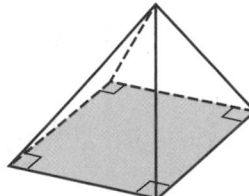

8.

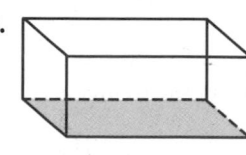

CONCEPTS

9. Draw a cube. Label a side *s*.

10. Draw a cylinder. Label the height *h* and radius *r*.

11. Draw a pyramid. Label the height *h* and the base.

12. Draw a cone. Label the height *h* and radius *r*.

13. Draw a sphere. Label the radius *r*.

14. Draw a rectangular solid. Label the length *l*, the width *w*, and the height *h*.

15. Which of the following are acceptable units with which to measure volume?

ft^2	mi^3	seconds	days
cubic inches	mm	square yards	in.
pounds	cm^2	meters	m^3

16. In the figure on the right, the unit of measurement of length used to draw the figure is the inch.

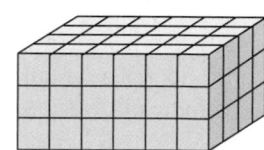

a. What is the area of the base of the figure?

b. What is the volume of the figure?

17. Which geometric concept (perimeter, circumference, area, or volume) should be applied when measuring each of the following?

    a. The distance around a checkerboard

    b. The size of a trunk of a car

    c. The amount of paper used for a postage stamp

    d. The amount of storage in a cedar chest

    e. The amount of beach available for sunbathing

    f. The distance the tip of a propeller travels

18. Complete the table.

Figure	Volume formula
Cube	
Rectangular solid	
Prism	
Cylinder	
Pyramid	
Cone	
Sphere	

19. Evaluate each expression. Leave π in the answer.

    a. $\dfrac{1}{3}\pi(25)6$

    b. $\dfrac{4}{3}\pi(125)$

20. a. Evaluate $\dfrac{1}{3}\pi r^2 h$ for $r = 2$ and $h = 27$. Leave π in the answer.

    b. Approximate your answer to part a to the nearest tenth.

NOTATION

21. a. What does "in.^{3}" mean?

    b. Write "one cubic centimeter" using symbols.

22. In the formula $V = \dfrac{1}{3}Bh$, what does B represent?

23. In a drawing, what does the symbol ⌐ indicate?

24. Redraw the figure below using dashed lines to show the hidden edges.

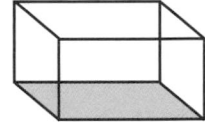

GUIDED PRACTICE

Convert from one unit of measurement to another.
See Example 1.

25. How many cubic feet are in 1 cubic yard?

26. How many cubic decimeters are in 1 cubic meter?

27. How many cubic meters are in 1 cubic kilometer?

28. How many cubic inches are in 1 cubic yard?

Find the volume of each figure. See Example 2.

29.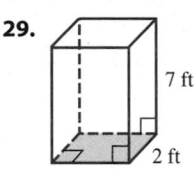

30.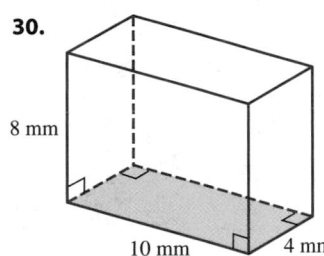

31.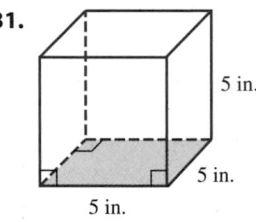

32.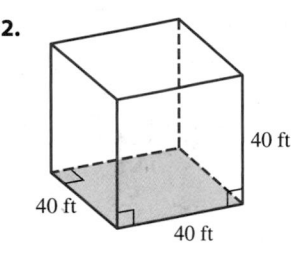

Find the volume of each figure. See Example 3.

33.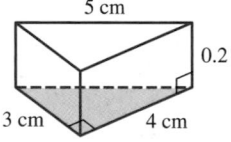

34.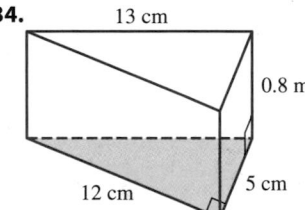

35.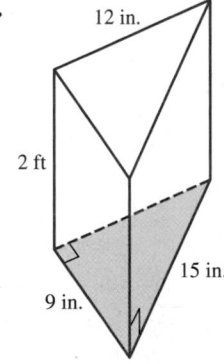

36.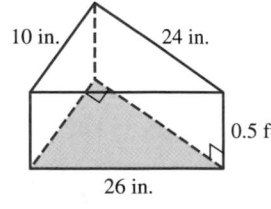

Find the volume of each figure. **See Example 4.**

37.

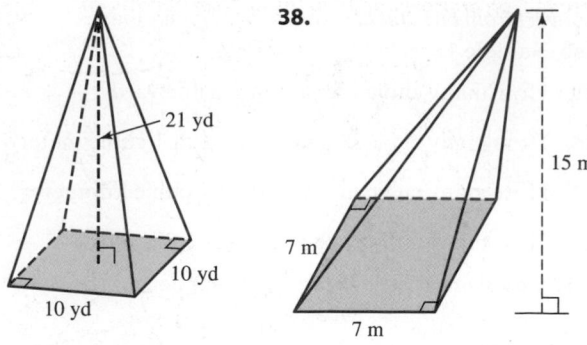

21 yd

10 yd

10 yd

38.

15 m

7 m

7 m

39.

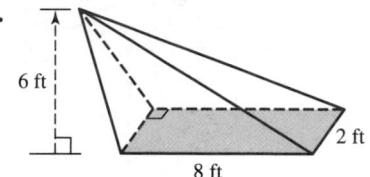

6 ft

2 ft

8 ft

40.

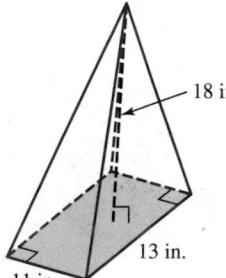

18 in.

13 in.

11 in.

41.

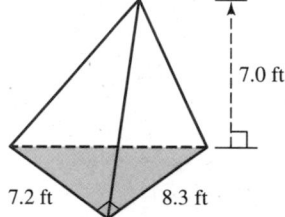

7.0 ft

7.2 ft   8.3 ft

42.

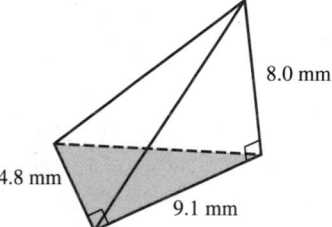

8.0 mm

4.8 mm

9.1 mm

43.

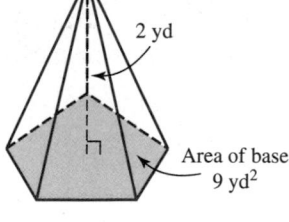

2 yd

Area of base
9 yd²

44.

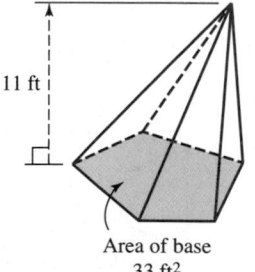

11 ft

Area of base
33 ft²

Find the volume of each cylinder. Give the exact answer and an approximation to the nearest hundredth. **See Example 5.**

45.

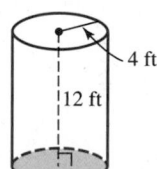

4 ft

12 ft

46.

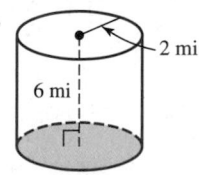

2 mi

6 mi

47.

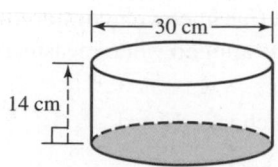

30 cm

14 cm

48.

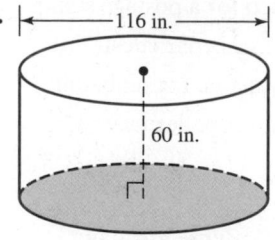

116 in.

60 in.

Find the volume of each cone. Give the exact answer and an approximation to the nearest hundredth. **See Example 6.**

49.

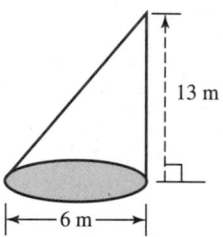

13 m

6 m

50.

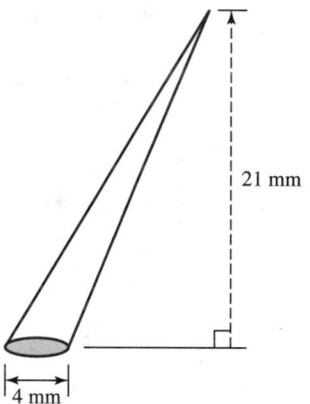

21 mm

4 mm

51.

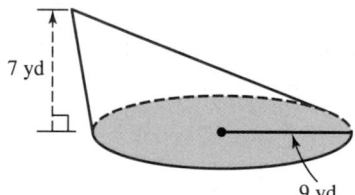

7 yd

9 yd

52.

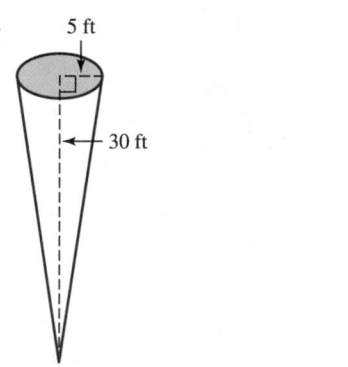

5 ft

30 ft

Find the volume of each sphere. Give the exact answer and an approximation to the nearest tenth. **See Example 7.**

53.
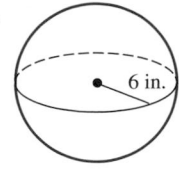
6 in.

54.
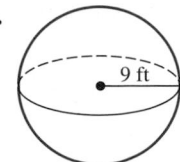
9 ft

55.
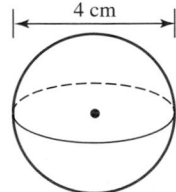
4 cm

56.
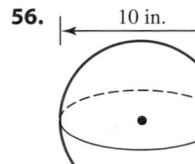
10 in.

TRY IT YOURSELF

Find the volume of each figure. If an exact answer contains π, approximate to the nearest hundredth.

57. A hemisphere with a radius of 9 inches
   (*Hint:* a **hemisphere** is an exact half of a sphere.)

58. A hemisphere with a diameter of 22 feet
   (*Hint:* a **hemisphere** is an exact half of a sphere.)

59. A cylinder with a height of 12 meters and a circular base with a radius of 6 meters

60. A cylinder with a height of 4 meters and a circular base with a diameter of 18 meters

61. A rectangular solid with dimensions of 3 cm by 4 cm by 5 cm

62. A rectangular solid with dimensions of 5 m by 8 m by 10 m

63. A cone with a height of 12 centimeters and a circular base with a diameter of 10 centimeters

64. A cone with a height of 3 inches and a circular base with a radius of 4 inches

65. A pyramid with a square base 10 meters on each side and a height of 12 meters

66. A pyramid with a square base 6 inches on each side and a height of 4 inches

67. A prism whose base is a right triangle with legs 3 meters and 4 meters long and whose height is 8 meters

68. A prism whose base is a right triangle with legs 5 feet and 12 feet long and whose height is 25 feet

Find the volume of each figure. Give the exact answer and, when needed, an approximation to the nearest hundredth.

69.
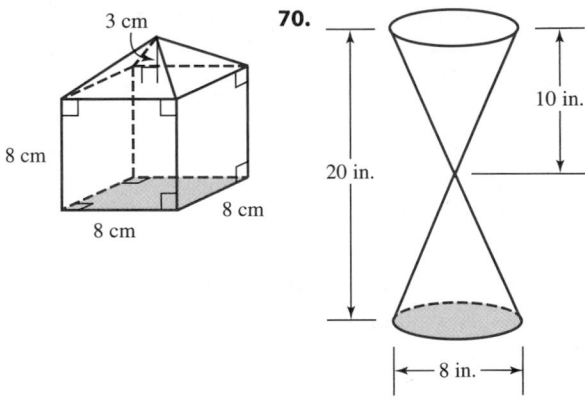
3 cm
8 cm
8 cm
8 cm

70.
10 in.
20 in.
8 in.

71.
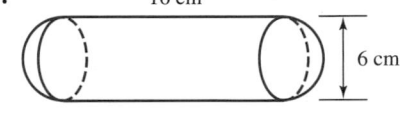
16 cm
6 cm

72.
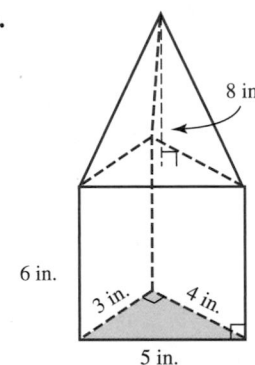
8 in.
6 in.
3 in.
4 in.
5 in.

APPLICATIONS

Solve each problem. If an exact answer contains π, approximate the answer to the nearest hundredth.

73. SWEETENERS A sugar cube is $\frac{1}{2}$ inch on each edge. How much volume does it occupy?

74. VENTILATION A classroom is 40 feet long, 30 feet wide, and 9 feet high. Find the number of cubic feet of air in the room.

75. WATER HEATERS Complete the advertisement for the high-efficiency water heater shown below.

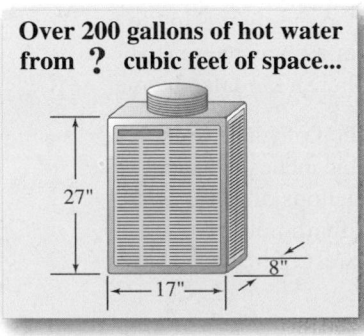

Over 200 gallons of hot water from ? cubic feet of space...

27"
17"
8"

76. REFRIGERATORS The largest refrigerator advertised in a JC Penny catalog has a capacity of 25.2 cubic feet. How many cubic inches is this?

77. TANKS A cylindrical oil tank has a diameter of 6 feet and a length of 7 feet. Find the volume of the tank.

78. DESSERTS A restaurant serves pudding in a conical dish that has a diameter of 3 inches. If the dish is 4 inches deep, how many cubic inches of pudding are in each dish?

79. HOT-AIR BALLOONS The lifting power of a spherical balloon depends on its volume. How many cubic feet of gas will a balloon hold if it is 40 feet in diameter?

80. CEREAL BOXES A box of cereal measures 3 inches by 8 inches by 10 inches. The manufacturer plans to market a smaller box that measures $2\frac{1}{2}$ by 7 by 8 inches. By how much will the volume be reduced?

81. ENGINES The *compression ratio* of an engine is the volume in one cylinder with the piston at bottom-dead-center (B.D.C.), divided by the volume with the piston at top-dead-center (T.D.C.). From the data given in the following figure, what is the compression ratio of the engine? Use a colon to express your answer.

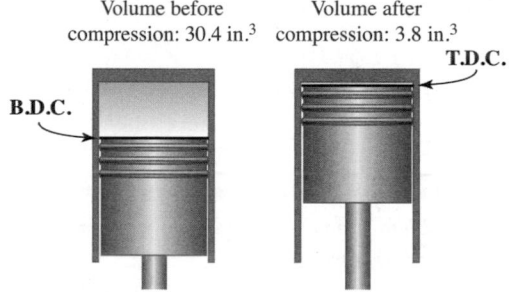

Volume before compression: 30.4 in.³   Volume after compression: 3.8 in.³

82. GEOGRAPHY Earth is not a perfect sphere but is slightly pear-shaped. To estimate its volume, we will assume that it is spherical, with a diameter of about 7,926 miles. What is its volume, to the nearest one billion cubic miles?

83. BIRDBATHS

— 30 in. —

a. The bowl of the birdbath shown on the right is in the shape of a hemisphere (half of a sphere). Find its volume.

b. If 1 gallon of water occupies 231 cubic inches of space, how many gallons of water does the birdbath hold? Round to the nearest tenth.

84. CONCRETE BLOCKS Find the number of cubic inches of concrete used to make the hollow, cube-shaped, block shown below.

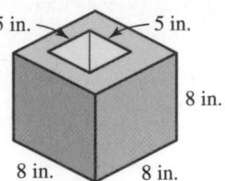

5 in.   5 in.   8 in.   8 in.   8 in.

WRITING

85. What is meant by the *volume* of a cube?

86. The stack of 3×5 index cards shown in figure (a) forms a right rectangular prism, with a certain volume. If the stack is pushed to lean to the right, as in figure (b), a new prism is formed. How will its volume compare to the volume of the right rectangular prism? Explain your answer.

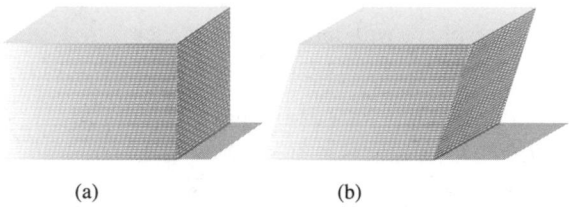

(a)   (b)

87. Are the units used to measure area different from the units used to measure volume? Explain.

88. The dimensions (length, width, and height) of one rectangular solid are entirely different numbers from the dimensions of another rectangular solid. Would it be possible for the rectangular solids to have the same volume? Explain.

REVIEW

89. Evaluate: $-5(5 - 2)^2 + 3$

90. BUYING PENCILS Carlos bought 6 pencils at $0.60 each and a notebook for $1.25. He gave the clerk a $5 bill. How much change did he receive?

91. Solve: $-x = 4$

92. 38 is what percent of 40?

93. Express the phrase "3 inches to 15 inches" as a ratio in simplest form.

94. Convert 40 ounces to pounds.

95. Convert 2.4 meters to millimeters.

96. State the Pythagorean equation.

STUDY SKILLS CHECKLIST

Know the Vocabulary

A large amount of vocabulary has been introduced in Chapter 9. Before taking the test, put a checkmark in the box if you can define and draw an example of each of the given terms.

☐ Point, line, plane

☐ Line segment, midpoint

☐ Ray, angle, vertex

☐ Acute angle, obtuse angle, right angle, straight angle

☐ Adjacent angles, vertical angles

☐ Complementary angles, supplementary angles

☐ Congruent segments, congruent angles

☐ Parallel lines, perpendicular lines, a transversal

☐ Alternate interior angles, interior angles, corresponding angles

☐ Polygon, triangle, quadrilateral, pentagon, hexagon, octagon

☐ Equilateral triangle, isosceles triangle, scalene triangle

☐ Acute triangle, obtuse triangle

☐ Right triangle, hypotenuse, legs

☐ Congruent triangles, similar triangles

☐ Parallelogram, rectangle, square, rhombus, trapezoid, isosceles trapezoid

☐ Circle, arc, semicircle, radius, diameter

☐ Rectangular solid, cube, sphere, prism, pyramid, cylinder, cone

CHAPTER 9   SUMMARY AND REVIEW

SECTION 9.1   Basic Geometric Figures; Angles

DEFINITIONS AND CONCEPTS	EXAMPLES
The word **geometry** comes from the Greek words *geo* (meaning Earth) and *metron* (meaning measure). Geometry is based on three undefined words: **point, line,** and **plane.**	 **Point** • A — Points are labeled with capital letters. **Line** $\overleftrightarrow{BC}$ 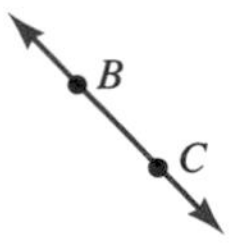 — We can name a line using any two points on it. **Plane** *EFG* 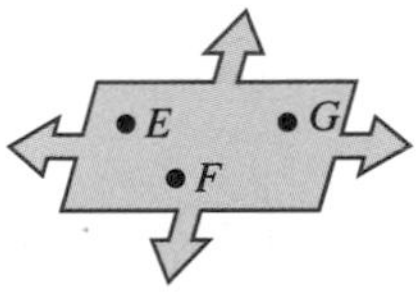 — Floors, walls, and table tops are all parts of planes.

A **line segment** is a part of a line with two endpoints. Every line segment has a **midpoint,** which divides the segment into two parts of equal length.

The notation m($\overline{AM}$) is read as "the measure of line segment $\overline{AM}$."

When two line segments have the same measure, we say that they are **congruent.** Read the symbol $\cong$ as "is congruent to."

A **ray** is a part of a line with one **endpoint.**

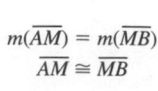   Line segment $\overline{AB}$ Ray $\overrightarrow{CD}$

$m(\overline{AM}) = m(\overline{MB})$
$\overline{AM} \cong \overline{MB}$

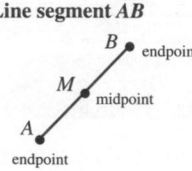 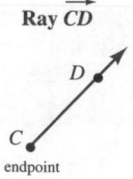

An **angle** is a figure formed by two rays (called **sides**) with a common endpoint. The common endpoint is called the **vertex** of the angle.

We read the symbol $\angle$ as "angle."

The angle below can be written as $\angle BAC$, $\angle CAB$, $\angle A$, or $\angle 1$.

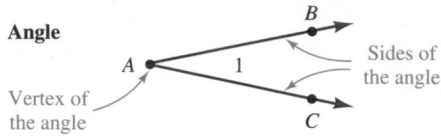

When two angles have the same measure, we say that they are **congruent.**

A **protractor** is used to find the measure of an angle. One unit of measurement of an angle is the **degree.**

The notation m($\angle DEF$) is read as "the measure of $\angle DEF$."

Congruent angles

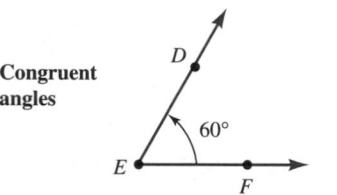 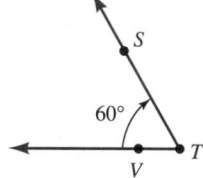

Since $m(\angle DEF) = m(\angle STV)$, we say that $\angle DEF \cong \angle STV$.

An **acute angle** has a measure that is greater than 0° but less than 90°. An **obtuse angle** has a measure that is greater than 90° but less than 180°. A **straight angle** measures 180°.

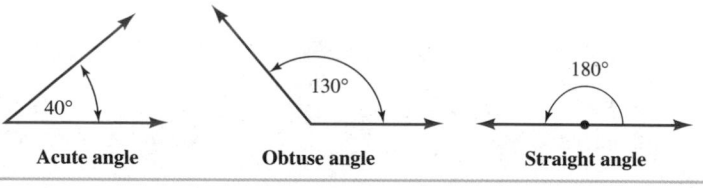

Acute angle   Obtuse angle   Straight angle

A **right angle** measures 90°.

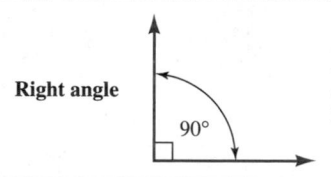

Right angle

A ⌐ symbol is often used to label a right angle.

Two angles that have the same vertex and are side-by-side are called **adjacent angles.**

Two angles with degree measures of x and 21° are adjacent angles, as shown here. Use the information in the figure to find x.

Adjacent angles

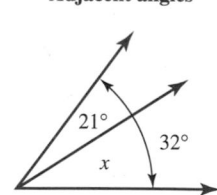

We can use the algebra concepts of variable and equation to solve many types of geometry problems.

The *sum* of the measures of the two adjacent angles is 32°:

$x + 21° = 32°$ The word *sum* indicates addition.

$x + 21° - \mathbf{21°} = 32° - \mathbf{21°}$ Subtract 21° from both sides.

$x = 11°$ Do the subtraction.

Thus, x is 11°.

When two lines intersect, pairs of nonadjacent angles are called **vertical angles.**	**Vertical angles**
Vertical angles are congruent (have the same measure).	Refer to the figure below. Find x and $m(\angle XYZ)$. Since the angles are vertical angles, they have equal measures. $3x + 20° = 2x + 70°$ Set the expressions equal. $3x + 20° - \mathbf{2x} = 2x + 70° - \mathbf{2x}$ Eliminate 2x from the right side. $x + 20° = 70°$ Combine like terms. $x = 50°$ Subtract 20° from both sides. Thus, x is 50°. To find $m(\angle XYZ)$, evaluate the expression $3x + 20°$ for $x = 50°$. $3x + 20° = 3(\mathbf{50°}) + 20°$ Substitute 50° for x. $= 150° + 20°$ Do the multiplication. $= 170°$ Do the addition. Thus, $m(\angle XYZ) = 170°$.
If the sum of two angles is 90°, the angles are **complementary.** If the sum of two angles is 180°, the angles are **supplementary.**	**Complementary angles** **Supplementary angles** $63° + 27° = 90°$ $146° + 34° = 180°$
We can use algebra to find the complement of an angle.	Find the complement of an 11° angle. Let x = the measure of the complement (in degrees). $x + 11° = 90°$ The sum of the angles' measures must be 90°. $x = 79°$ To isolate x, subtract 11° from both sides. The complement of an 11° angle has measure 79°.
We can use algebra to find the supplement of an angle.	Find the supplement of a 68° angle. Let x = the measure of the supplement (in degrees). $x + 68° = 180°$ The sum of the angles' measures must be 180°. $x = 112°$ To isolate x, subtract 68° from both sides. The supplement of a 68° angle has measure 112°.

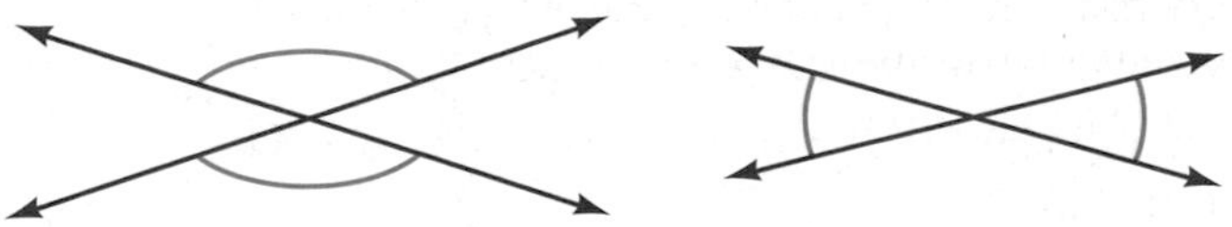

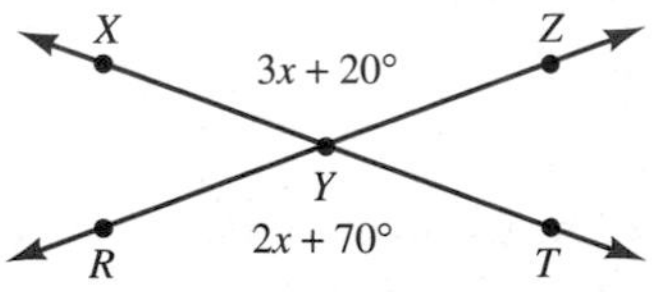

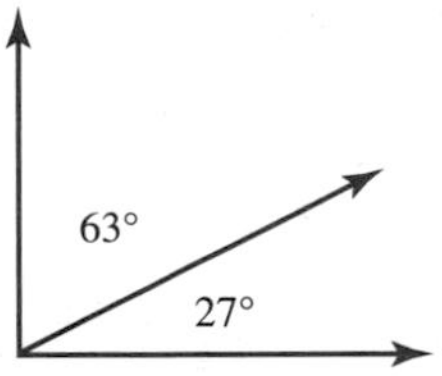

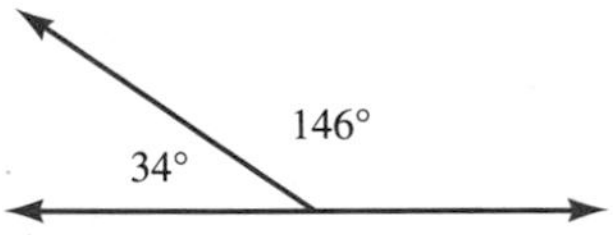

REVIEW EXERCISES

1. In the illustration, give the name of a point, a line, and a plane.

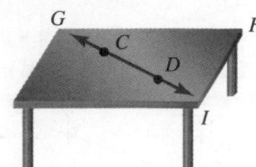

2. a. In the figure below, find m($\overline{AG}$).
 b. Find the midpoint of $\overline{BH}$.
 c. Is $\overline{AC} \cong \overline{GE}$?

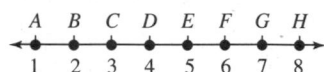

3. Give four ways to name the angle shown below.

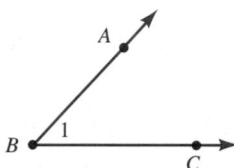

4. a. Is the angle shown above acute or obtuse?
 b. What is the vertex of the angle?
 c. What rays form the sides of the angle?
 d. Use a protractor to find the measure of the angle.

5. Identify each acute angle, right angle, obtuse angle, and straight angle in the figure below.

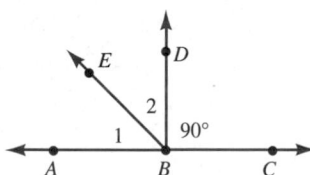

6. In the figure above, is $\angle ABD \cong \angle CBD$?

7. In the figure above, are $\overrightarrow{AC}$ and $\overrightarrow{AB}$ the same ray?

8. The measures of several angles are given below. Identify each angle as an acute angle, a right angle, an obtuse angle, or a straight angle.
 a. m($\angle A$) = 150°
 b. m($\angle B$) = 90°
 c. m($\angle C$) = 180°
 d. m($\angle D$) = 25°

9. The two angles shown here are adjacent angles. Find x.

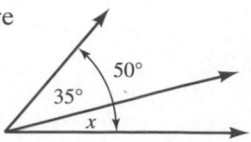

10. Line AB is shown in the figure below. Find y.

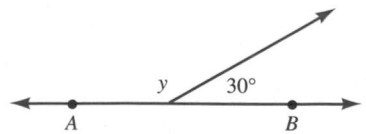

11. Refer to the figure on the right.
 a. Find m($\angle 1$).
 b. Find m($\angle 2$).

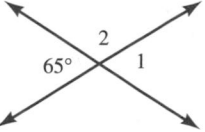

12. Refer to the figure below.
 a. What is m($\angle ABG$)?
 b. What is m($\angle FBE$)?
 c. What is m($\angle CBD$)?
 d. What is m($\angle FBG$)?
 e. Are $\angle CBD$ and $\angle DBE$ complementary angles?

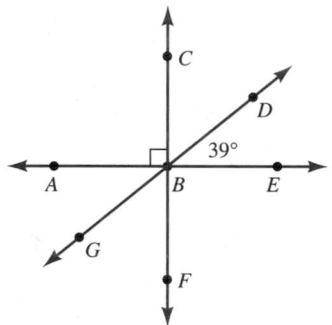

13. Refer to the figure.
 a. Find x.
 b. What is m($\angle HFI$)?
 c. What is m($\angle GFH$)?

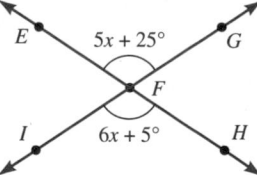

14. Find the complement of a 71° angle.

15. Find the supplement of a 143° angle.

16. Are angles measuring 30°, 60°, and 90° supplementary?

SECTION 9.2   Parallel and Perpendicular Lines

DEFINITIONS AND CONCEPTS	EXAMPLES
If two lines lie in the same plane, they are called **coplanar.** **Parallel lines** are coplanar lines that do not intersect. We read the symbol ∥ as "is parallel to." **Perpendicular lines** are lines that intersect and form right angles. We read the symbol ⊥ as "is perpendicular to."	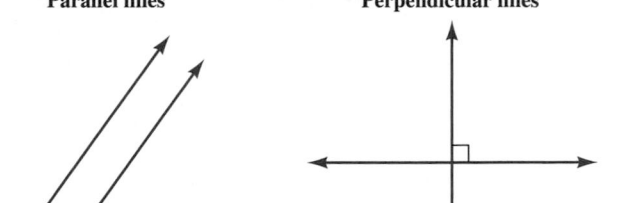
A line that intersects two coplanar lines in two distinct (different) points is called a **transversal.** When a transversal intersects two coplanar lines, four pairs of **corresponding angles** are formed. If two parallel lines are cut by a transversal, *corresponding angles are congruent* (have equal measures).	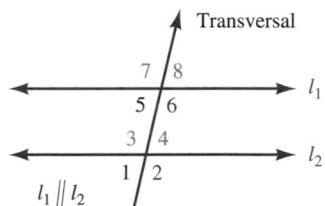
When a transversal intersects two coplanar lines, two pairs of **interior angles** and two pairs of **alternate interior angles** are formed. If two parallel lines are cut by a transversal, *alternate interior angles are congruent* (have equal measures). If two parallel lines are cut by a transversal, *interior angles on the same side of the transversal are supplementary.*	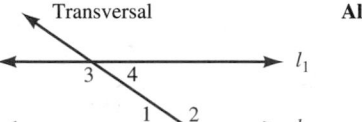
We can use algebra to find the unknown measures of corresponding angles.	In the figure, $l_1 \parallel l_2$. Find x and the measure of each angle that is labeled. 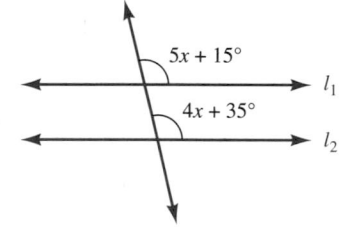 Since the lines are parallel, and the angles are corresponding angles, the angles are congruent. $5x + 15° = 4x + 35°$ The angle measures are equal. $x + 15° = 35°$ Subtract 4x from both sides. $x = 20°$ To isolate x, subtract 15° from both sides. Thus, x is 20°. To find the measures of the angles labeled in the figure, we evaluate each expression for $x = 20°$. $5x + 15° = 5(\mathbf{20°}) + 15°$ $4x + 35° = 4(\mathbf{20°}) + 35°$ $\qquad\qquad = 100° + 15°$ $\qquad\qquad = 80° + 35°$ $\qquad\qquad = 115°$ $\qquad\qquad = 115°$ The measure of each angle is 115°.

We can use algebra to find the unknown measures of interior angles.

In the figure, $l_1 \parallel l_2$. Find x and the measure of each angle that is labeled.

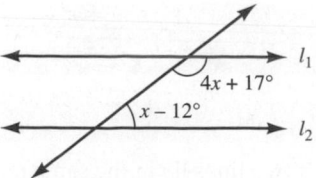

Since the angles are interior angles on the same side of the transversal, they are supplementary.

$$4x + 17° + x - 12° = 180°$$ The sum of the measures of two supplementary angles is 180°.

$$5x + 5° = 180°$$ Combine like terms.

$$5x = 175°$$ Subtract 5° from both sides.

$$x = 35°$$ Divide both sides by 5.

Thus, x is 35°. To find the measures of the angles in the figure, we evaluate the expressions for $x = 35°$.

$$4x + 17° = 4(\mathbf{35°}) + 17°$$ $$x - 12° = \mathbf{35°} - 12°$$
$$= 140° + 17°$$ $$= 23°$$
$$= 157°$$

The measures of the angles labeled in the figure are 157° and 23°.

REVIEW EXERCISES

17. a. Lines l_1 and l_2 shown in figure (a) below do not intersect and are coplanar. What word describes the lines?

    b. In figure (a), line l_3 intersects lines l_1 and l_2 in two distinct (different) points. What is the name given to line l_3?

    c. What word describes the two lines shown in figure (b) below?

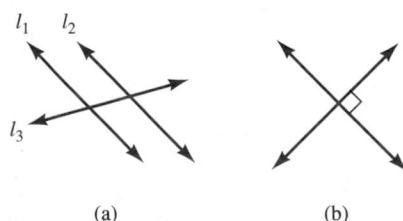

            (a)   (b)

18. Identify all pairs of alternate interior angles shown in the figure below.

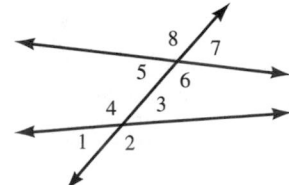

19. Refer to the figure in Problem 18. Identify all pairs of corresponding angles.

20. Refer to the figure in Problem 18. Identify all pairs of vertical angles.

21. In the figure below, $l_1 \parallel l_2$. Find the measure of each angle.

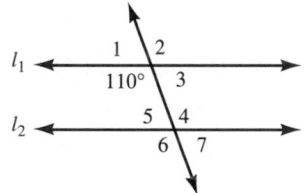

22. In the figure on the right, $\overline{DC} \parallel \overline{AB}$. Find the measure of each angle that is labeled.

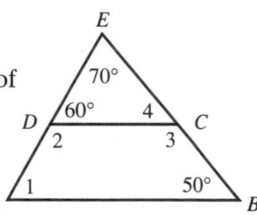

23. In the figure below, $l_1 \parallel l_2$.

    a. Find x.

    b. Find the measure of each angle that is labeled.

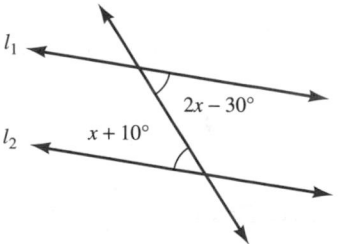

24. In the figure below, $l_1 \parallel l_2$.
   a. Find x.
   b. Find the measure of each angle that is labeled.

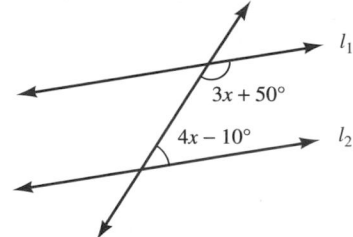

25. In the figure below, $\overline{AB} \parallel \overline{DC}$.
   a. Find x.
   b. Find the measure of each angle that is labeled.

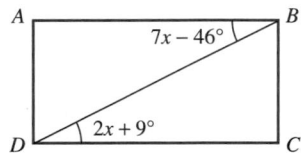

26. In the figure below, $\overline{EF} \parallel \overline{HI}$.
   a. Find x.
   b. Find the measure of each angle that is labeled.

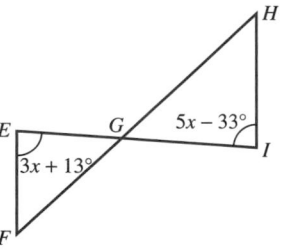

SECTION 9.3 Triangles

DEFINITIONS AND CONCEPTS	EXAMPLES

A **polygon** is a closed geometric figure with at least three line segments for its sides. The points at which the sides intersect are called **vertices.** A **regular polygon** has sides that are all the same length and angles that are all the same measure.

The number of vertices of a polygon is equal to the number of sides it has.

Polygon   Regular polygon

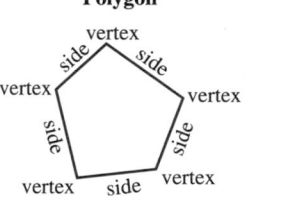 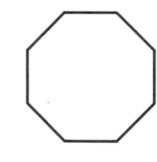

Classifying Polygons

Number of sides	Name of polygon	Number of sides	Name of polygon
3	triangle	8	octagon
4	quadrilateral	9	nonagon
5	pentagon	10	decagon
6	hexagon	12	dodecagon

Quadrilateral (4 sides) Hexagon (6 sides) Octagon (8 sides)

 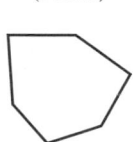 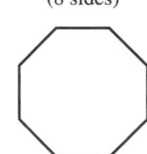

A **triangle** is a polygon with three sides (and three vertices).

Triangles can be classified according to the lengths of their sides.

Tick marks indicate sides that are of equal length.

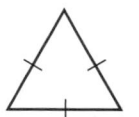 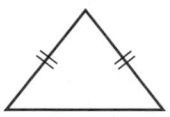 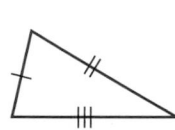

Equilateral triangle (all sides of equal length)   Isosceles triangle (at least two sides of equal length)   Scalene triangle (no sides of equal length)

Triangles can be classified by their angles.	 **Acute triangle** (has three acute angles) **Obtuse triangle** (has an obtuse angle) **Obtuse triangle** (has an obtuse angle)
The longest side of a right triangle is called the **hypotenuse,** and the other two sides are called **legs.** The hypotenuse of a right triangle is always opposite the 90° (right) angle. The legs of a right triangle are adjacent to (next to) the right angle.	**Right triangle**
In an isosceles triangle, the angles opposite the sides of equal length are called **base angles.** The third angle is called the **vertex angle.** The third side is called the **base.** **Isosceles Triangle Theorem:** If two sides of a triangle are congruent, then the angles opposite those sides are congruent. **Converse of the Isosceles Triangle Theorem:** If two angles of a triangle are congruent, then the sides opposite the angles have the same length, and the triangle is isosceles.	**Isosceles triangles**

The **sum of the measures of the angles** of any triangle is 180°. We can use algebra to find unknown angle measures of a triangle.	Find the measure of each angle of $\triangle ABC$. The sum of the angle measures of any triangle is 180°: $\quad x + 3x - 25° + x - 5° = 180°$ $\quad\quad\quad\quad\quad 5x - 30° = 180°$ Combine like terms. $\quad\quad\quad\quad\quad\quad\quad 5x = 210°$ Add 30° to both sides. $\quad\quad\quad\quad\quad\quad\quad\quad x = 42°$ Divide both sides by 5. To find the measures of $\angle B$ and $\angle C$, we evaluate the expressions $3x - 25°$ and $x - 5°$ for $x = 42°$. $\begin{array}{l\|l} 3x - 25° = 3(\mathbf{42}°) - 25° & x - 5° = \mathbf{42}° - 5° \\ \quad\quad\quad = 126° - 25° & \quad\quad\quad = 37° \\ \quad\quad\quad = 101° & \end{array}$ Thus, $m(\angle A) = 42°$, $m(\angle B) = 101°$, and $m(\angle C) = 37°$.
We can use algebra to find unknown angle measures of an isosceles triangle.	If the vertex angle of an isosceles triangle measures 26°, what is the measure of each base angle? If we let x represent the measure of one base angle, the measure of the other base angle is also x. (See the figure.) Since the sum of the measures of the angles of any triangle is 180°, we have $\quad\quad x + x + 26° = 180°$ $\quad\quad\quad 2x + 26° = 180°$ On the left side, combine like terms. $\quad\quad\quad\quad\quad 2x = 154°$ To isolate $2x$, subtract 26° from both sides. $\quad\quad\quad\quad\quad\quad x = 77°$ To isolate x, divide both sides by 2. The measure of each base angle is 77°.

REVIEW EXERCISES

27. For each of the following polygons, give the number of sides it has, tell its name, and then give the number of vertices that it has.

a.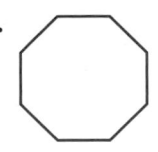

b.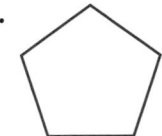

c.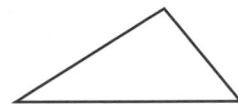

d.

e.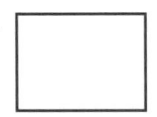

f.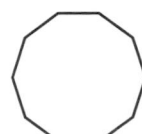

28. Classify each of the following triangles as an equilateral triangle, an isosceles triangle, a scalene triangle, or a right triangle. Some figures may be correctly classified in more than one way.

a.

b.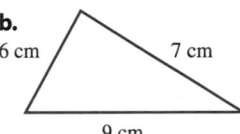

c.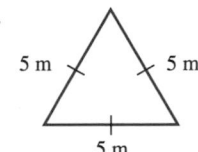

d.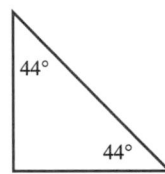

29. Classify each of the following triangles as an acute, an obtuse, or a right triangle.

a.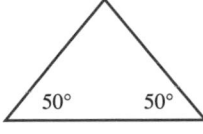

b.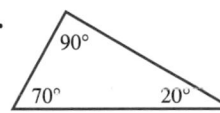

c.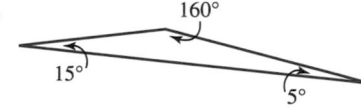

d.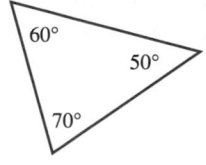

30. Refer to the triangle shown here.

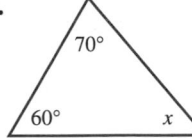

a. What is the measure of $\angle X$?

b. What type of triangle is it?

c. What two line segments are the legs?

d. What line segment is the hypotenuse?

e. Which side of the triangle is the longest?

f. Which side is opposite $\angle X$?

In each triangle shown below, find x.

31.

32.

33. In $\triangle ABC$, $m(\angle B) = 32°$ and $m(\angle C) = 77°$. Find $m(\angle A)$.

34. For the triangle shown below, find x. Then determine the measure of each angle of the triangle.

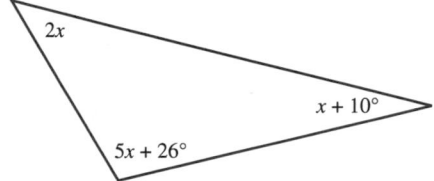

35. One base angle of an isosceles triangle measures 65°. Find the measure of the vertex angle.

36. The measure of the vertex angle of an isosceles triangle is 68°. Find the measure of each base angle.

37. Find the measure of $\angle C$ of the triangle shown here.

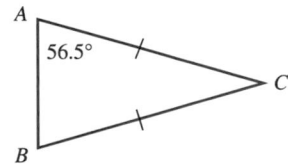

38. Refer to the figure shown here. Find $m(\angle C)$.

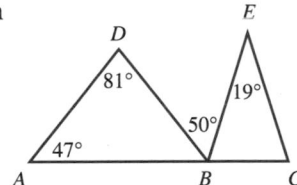

SECTION **9.4**   **The Pythagorean Theorem**

DEFINITIONS AND CONCEPTS	EXAMPLES
Pythagorean theorem If a and b are the lengths of the legs of a right triangle and c is the length of the hypotenuse, then $$a^2 + b^2 = c^2$$ 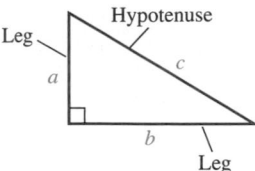 $a^2 + b^2 = c^2$ is called the **Pythagorean equation.**	Find the length of the hypotenuse of the right triangle shown here. We will let $a = 6$ and $b = 8$, and substitute into the Pythagorean equation to find c. 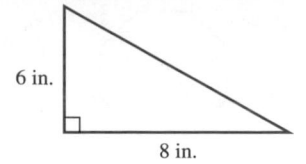 $a^2 + b^2 = c^2$ This is the Pythagorean equation. $6^2 + 8^2 = c^2$ Substitute 6 for a and 8 for b. $36 + 64 = c^2$ Evaluate the exponential expressions. $100 = c^2$ Do the addition. $c^2 = 100$ Reverse the sides of the equation so that c^2 is on the left. To find c, we must find a number that, when squared, is 100. There are two such numbers, one positive and one negative; they are the square roots of 100. Since c represents the length of a side of a triangle, c cannot be negative. For this reason, we need only find the positive square root of 100 to get c. $c = \sqrt{100}$ The symbol $\sqrt{\ }$ is used to indicate the positive square root of a number. $c = 10$ Because $10^2 = 100$. The length of the hypotenuse of the triangle is 10 in.
When we use the Pythagorean theorem to find the length of a side of a right triangle, the solution is sometimes the square root of a number that is not a perfect square. In that case, we can use a calculator to *approximate* the square root.	The lengths of two sides of a right triangle are shown here. Find the missing side length. We may substitute 9 for either a or b, but 11 must be substituted for the length c of the hypotenuse. If we substitute 9 for a, we can find the unknown side length b as follows. $a^2 + b^2 = c^2$ This is the Pythagorean equation. $9^2 + b^2 = 11^2$ Substitute 9 for a and 11 for c. $81 + b^2 = 121$ Evaluate each exponential expression. $81 + b^2 - 81 = 121 - 81$ To isolate b^2 on the left side, subtract 81 from both sides. $b^2 = 40$ We must find a number that, when squared, is 40. Since b represents the length of a side of a triangle, we consider only the positive square root. $b = \sqrt{40}$ This is the exact length. The missing side length is exactly $\sqrt{40}$ feet long. Since 40 is not a perfect square, we use a calculator to approximate $\sqrt{40}$. To the nearest hundredth, the missing side length is 6.32 ft.

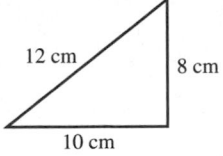

The converse of the Pythagorean theorem:
If a triangle has sides of lengths a, b, and c, such that $a^2 + b^2 = c^2$, then the triangle is a right triangle.

Is the triangle shown here a right triangle?

We must substitute the longest side length, 12, for c, because it is the possible hypotenuse. The lengths of 8 and 10 may be substituted for either a or b.

$$a^2 + b^2 = c^2 \qquad \text{This is the Pythagorean equation.}$$
$$8^2 + 10^2 \overset{?}{=} 12^2 \qquad \text{Substitute 8 for } a, \text{ 10 for } b, \text{ and 12 for } c.$$
$$64 + 100 \overset{?}{=} 144 \qquad \text{Evaluate each exponential expression.}$$
$$164 = 144 \qquad \text{This is a false statement.}$$

Since $164 \neq 144$, the triangle is not a right triangle.

REVIEW EXERCISES

Refer to the right triangle below.

39. Find c, if $a = 5$ cm and $b = 12$ cm.

40. Find c, if $a = 8$ ft and $b = 15$ ft.

41. Find a, if $b = 77$ in. and $c = 85$ in.

42. Find b, if $a = 21$ ft and $c = 29$ ft.

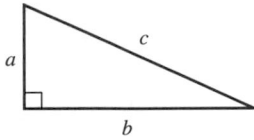

The lengths of two sides of a right triangle are given. Find the missing side length. Give the exact answer and an approximation to the nearest hundredth.

43.

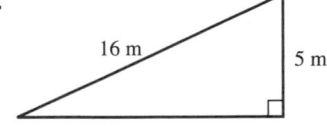

44.

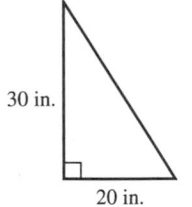

45. HIGH-ROPES ADVENTURE COURSES A builder of a high-ropes adventure course wants to secure a pole by attaching a support cable from the anchor stake 55 inches from the pole's base to a point 48 inches up the pole. See the illustration in the next column. How long should the cable be?

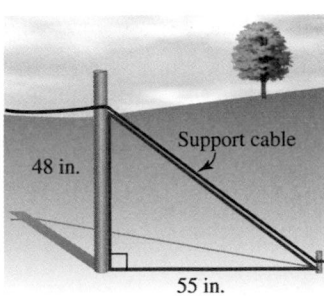

46. TV SCREENS Find the height of the television screen shown. Give the exact answer and an approximation to the nearest inch.

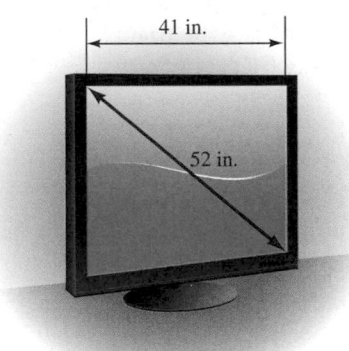

Determine whether each triangle shown here is a right triangle.

47.

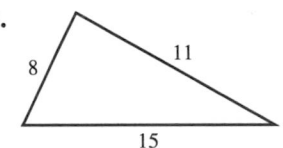

48.

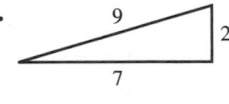

DEFINITIONS AND CONCEPTS	EXAMPLES
If two triangles have the same size and the same shape, they are **congruent triangles.**	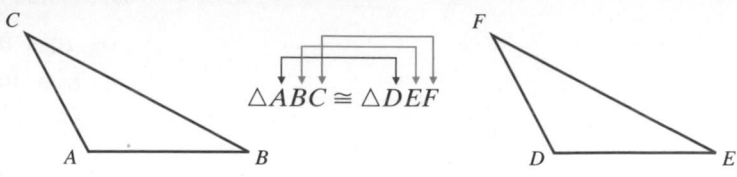
Corresponding parts of congruent triangles are congruent (have the same measure).	There are six pairs of congruent parts: three pairs of congruent angles and three pairs of congruent sides. • $m(\angle A) = m(\angle D)$ • $m(\overline{BC}) = m(\overline{EF})$ • $m(\angle B) = m(\angle E)$ • $m(\overline{AC}) = m(\overline{DF})$ • $m(\angle C) = m(\angle F)$ • $m(\overline{AB}) = m(\overline{DE})$
Three ways to show that two triangles are congruent are: 1. The **SSS property** If three sides of one triangle are congruent to three sides of a second triangle, the triangles are congruent.	$\triangle MNO \cong \triangle RST$ by the SSS property. 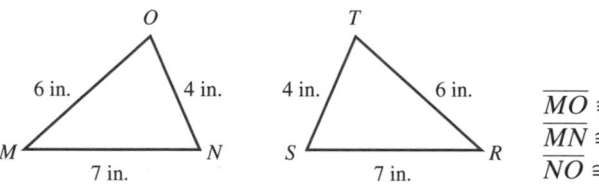 $\overline{MO} \cong \overline{RT}$ $\overline{MN} \cong \overline{RS}$ $\overline{NO} \cong \overline{ST}$
2. The **SAS property** If two sides and the angle between them in one triangle are congruent, respectively, to two sides and the angle between them in a second triangle, the triangles are congruent.	$\triangle DEF \cong \triangle XYZ$ by the SAS property. 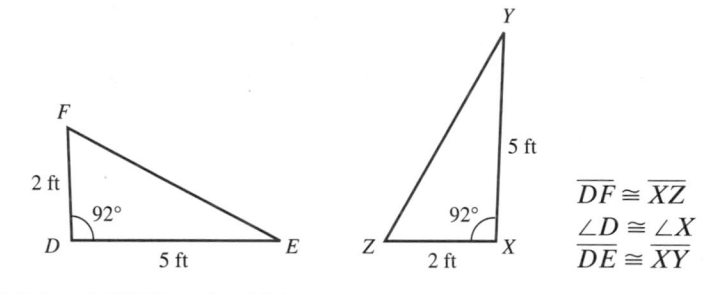 $\overline{DF} \cong \overline{XZ}$ $\angle D \cong \angle X$ $\overline{DE} \cong \overline{XY}$
3. The **ASA property** If two angles and the side between them in one triangle are congruent, respectively, to two angles and the side between them in a second triangle, the triangles are congruent.	$\triangle ABC \cong \triangle TUV$ by the ASA property. 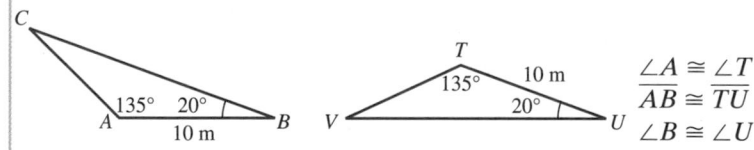 $\angle A \cong \angle T$ $\overline{AB} \cong \overline{TU}$ $\angle B \cong \angle U$
Similar triangles have the same shape, but not necessarily the same size. We read the symbol $\sim$ as "is similar to." **AAA similarity theorem** If the angles of one triangle are congruent to corresponding angles of another triangle, the triangles are similar.	$\triangle EFG \sim \triangle WXY$ by the AAA similarity theorem. 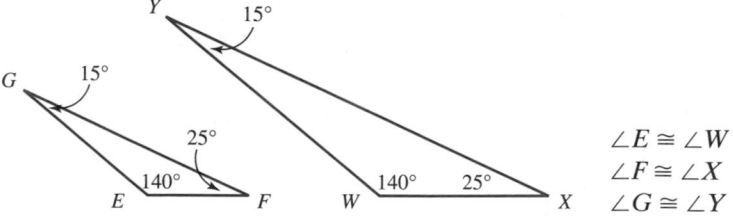 $\angle E \cong \angle W$ $\angle F \cong \angle X$ $\angle G \cong \angle Y$

Property of similar triangles

If two triangles are similar, all pairs of corresponding sides are in proportion.

Similar triangles are determined by the tree and its shadow and the man and his shadow. Since the triangles are similar, the lengths of their corresponding sides are in proportion.

LANDSCAPING A tree casts a shadow 27 feet long at the same time as a man 5 feet tall casts a shadow 3 feet long. Find the height of the tree.

If we let h = the height of the tree, we can find h by solving the following proportion.

The height of the tree → $\dfrac{h}{5} = \dfrac{27}{3}$ ← The length of the tree's shadow
The height of the man → ← The length of the man's shadow

$3h = 5(27)$ Find each cross product and set them equal.

$3h = 135$ Do the multiplication.

$\dfrac{3h}{3} = \dfrac{135}{3}$ To isolate h, divide both sides by 1.3.

$h = 45$ Do the division.

The tree is 45 feet tall.

REVIEW EXERCISES

49. Two congruent triangles are shown below. Complete the list of corresponding parts.

   a. $\angle A$ corresponds to _____ .

   b. $\angle B$ corresponds to _____ .

   c. $\angle C$ corresponds to _____ .

   d. $\overline{AC}$ corresponds to _____ .

   e. $\overline{AB}$ corresponds to _____ .

   f. $\overline{BC}$ corresponds to _____ .

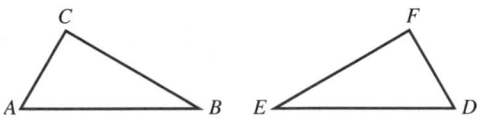

50. Refer to the figure below, where $\triangle ABC \cong \triangle XYZ$.

   a. Find m($\angle X$).

   b. Find m($\angle C$).

   c. Find m($\overline{YZ}$).

   d. Find m($\overline{AC}$).

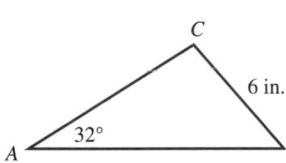

Determine whether the triangles in each pair are congruent. If they are, tell why.

51.

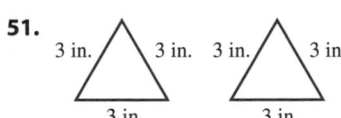

52.

53.

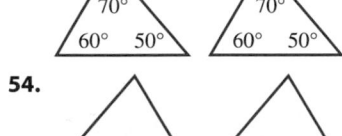

54.

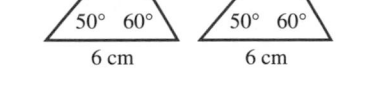

Determine whether the triangles are similar.

55.   **56.**

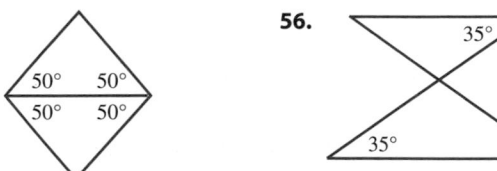

57. In the figure below, $\triangle RST \sim \triangle MNO$. Find x and y.

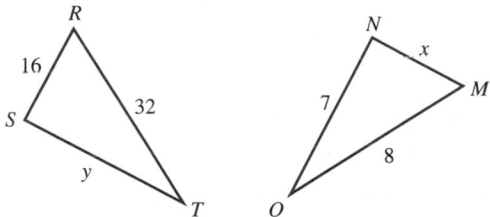

58. HEIGHT OF A TREE A tree casts a 26-foot shadow at the same time a woman 5 feet tall casts a 2-foot shadow. What is the height of the tree? (*Hint:* Draw a diagram first and label the side lengths of the similar triangles.)

SECTION 9.6   Quadrilaterals and Other Polygons

DEFINITIONS AND CONCEPTS	EXAMPLES
A **quadrilateral** is a polygon with four sides. Use the capital letters that label the vertices of a quadrilateral to name it. A segment that joins two nonconsecutive vertices of a polygon is called a **diagonal** of the polygon.	**Quadrilateral *WXYZ*** 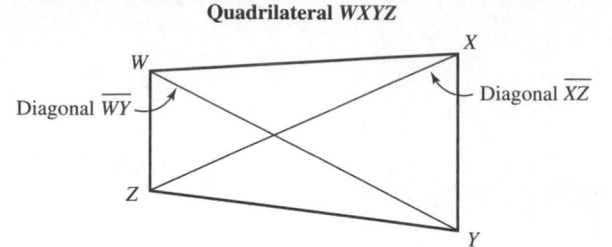
Some special types of quadrilaterals are shown on the right.	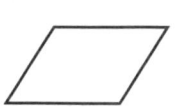 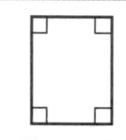 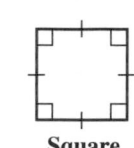 **Parallelogram** **Rectangle** **Square** (Opposite sides parallel) (Parallelogram with four right angles) (Rectangle with sides of equal length) 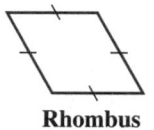  **Rhombus** **Trapezoid** (Parallelogram with sides of equal length) (Exactly two sides parallel)
A **rectangle** is a quadrilateral with four right angles.	**Rectangle *ABCD*** 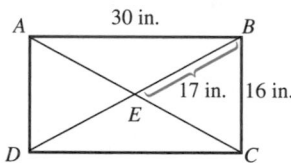
Properties of rectangles: 1. All four angles are right angles. 2. Opposite sides are parallel. 3. Opposite sides have equal length. 4. Diagonals have equal length. 5. The diagonals intersect at their midpoints.	1. $m(\angle DAB) = m(\angle ABC) = m(\angle BCD) = m(\angle CDA) = 90°$ 2. $\overline{AD} \parallel \overline{BC}$ and $\overline{AB} \parallel \overline{DC}$ 3. $m(\overline{AD}) = 16$ in. and $m(\overline{DC}) = 30$ in. 4. $m(\overline{DB}) = m(\overline{AC}) = 34$ in. 5. $m(\overline{DE}) = m(\overline{AE}) = m(\overline{EC}) = 17$ in.
Conditions that a parallelogram must meet to ensure that it is a rectangle: 1. If a parallelogram has one right angle, then the parallelogram is a rectangle. 2. If the diagonals of a parallelogram are congruent, then the parallelogram is a rectangle.	Read Example 2 on page 877 to see how these two conditions are used in construction to "square a foundation." 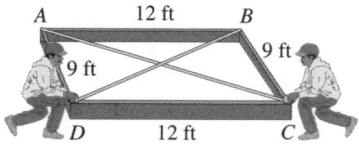

A **trapezoid** is a quadrilateral with exactly two sides parallel.

The parallel sides of a trapezoid are called **bases**. The nonparallel sides are called **legs**.

If the legs (the nonparallel sides) of a trapezoid are of equal length, it is called an **isosceles trapezoid.**

In an isosceles trapezoid, both pairs of **base angles** are congruent.

Trapezoid ABCD

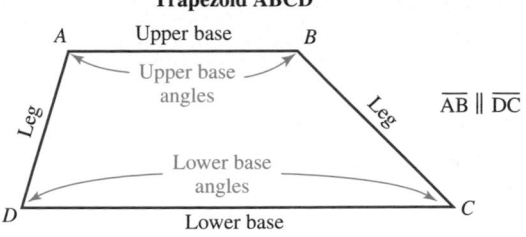

$\overline{AB} \parallel \overline{DC}$

The sum S, in degrees, of the measures of the angles of a polygon with n sides is given by the formula

$$S = (n - 2)180°$$

Find the sum of the angle measures of a hexagon.

Since a hexagon has 6 sides, we will substitute 6 for n in the formula.

$$S = (n - 2)180°$$
$$S = (6 - 2)180° \quad \text{Substitute 6 for } n, \text{ the number of sides.}$$
$$= (4)180° \quad \text{Do the subtraction within the parentheses.}$$
$$= 720° \quad \text{Do the multiplication.}$$

The sum of the measures of the angles of a hexagon is 720°.

We can use the formula $S = (n - 2)180°$ to find the number of sides a polygon has.

The sum of the measures of the angles of a polygon is 2,340°. Find the number of sides the polygon has.

$$S = (n - 2)180°$$
$$\mathbf{2{,}340}° = (n - 2)180° \quad \text{Substitute 2,340° for } S. \text{ Now solve for } n.$$
$$2{,}340° = 180°n - 360° \quad \text{Distribute the multiplication by 180°.}$$
$$2{,}340° + \mathbf{360}° = 180°n - 360° + \mathbf{360}° \quad \text{Add 360° to both sides.}$$
$$2{,}700° = 180°n \quad \text{Do the addition.}$$
$$\frac{2{,}700°}{\mathbf{180}°} = \frac{180°n}{\mathbf{180}°} \quad \text{Divide both sides by 180°.}$$
$$15 = n \quad \text{Do the division.}$$

The polygon has 15 sides.

REVIEW EXERCISES

59. Classify each of the following quadrilaterals as a parallelogram, a rectangle, a square, a rhombus, or a trapezoid. Some figures may be correctly classified in more than one way.

a.

b.

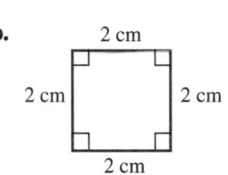

2 cm, 2 cm, 2 cm, 2 cm

c.

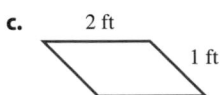

2 ft, 1 ft

d.

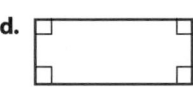

e.

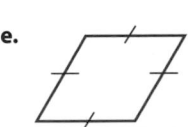

f.

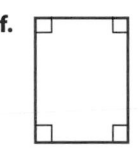

60. The length of diagonal $\overline{AC}$ of rectangle $ABCD$ shown below is 15 centimeters. Find each measure.

a. m($\overline{BD}$)

b. m($\angle 1$)

c. m($\angle 2$)

d. m($\overline{EC}$)

e. m($\overline{AB}$)

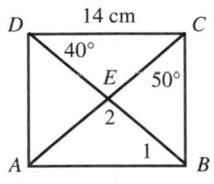

61. Refer to rectangle $WXYZ$ below. Tell whether each statement is true or false.

a. m($\overline{WX}$) = m($\overline{ZY}$)

b. m($\overline{ZE}$) = m($\overline{EX}$)

c. Triangle WEX is isosceles.

d. m($\overline{WY}$) = m($\overline{WX}$)

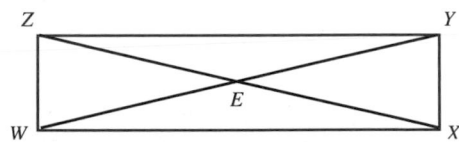

62. Refer to isosceles trapezoid *ABCD* below. Find each measure.

   a. m(∠*B*)

   b. m(∠*C*)

   c. m($\overline{CB}$)

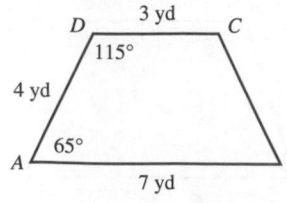

63. Find the sum of the angle measures of an octagon.

64. The sum of the measures of the angles of a polygon is 3,240°. Find the number of sides the polygon has.

SECTION 9.7   Perimeters and Areas of Polygons

DEFINITIONS AND CONCEPTS	EXAMPLES

The **perimeter** of a polygon is the distance around it.

Figure	Perimeter Formula
Square	$P = 4s$
Rectangle	$P = 2l + 2w$
Triangle	$P = a + b + c$

Find the perimeter of the triangle shown below.

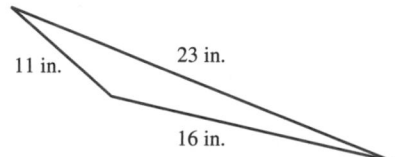

$P = a + b + c$ This is the formula for the perimeter of a triangle.

$P = 11 + 16 + 23$ Substitute 11 for *a*, 16 for *b*, and 23 for *c*.

$\quad = 50$ Do the addition.

The perimeter of the triangle is 50 inches.

The **area** of a polygon is the measure of the amount of surface it encloses.

Figure	Area Formulas
Square	$A = s^2$
Rectangle	$A = lw$
Parallelogram	$A = bh$
Triangle	$A = \frac{1}{2}bh$
Trapezoid	$A = \frac{1}{2}h(b_1 + b_2)$

Find the area of the triangle shown here.

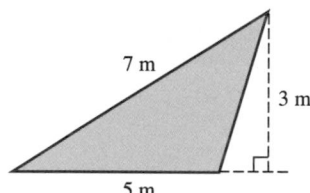

$A = \frac{1}{2}bh$ This is the formula for the area of a triangle.

$A = \frac{1}{2}(5)(3)$ Substitute 5 for *b*, the length of the base, and 3 for *h*, the height. Note that the side length 7 m is not used in the calculation.

$\quad = \frac{1}{2}\left(\frac{5}{1}\right)\left(\frac{3}{1}\right)$ Write 5 as $\frac{5}{1}$ and 3 as $\frac{3}{1}$.

$\quad = \frac{15}{2}$ Multiply the numerators. Multiply the denominators.

$\quad = 7.5$ Do the division.

The area of the triangle is 7.5 m².

To find the perimeter or area of a polygon, all the measurements must be in the **same units.** If they are not, use unit conversion factors to change them to the same unit.

To find the perimeter or area of the rectangle shown here, we need to express the length in inches.

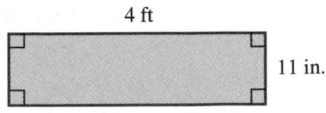

4 ft

11 in.

$$4 \text{ ft} = \frac{4 \text{ ft}}{1} \cdot \frac{12 \text{ in.}}{1 \text{ ft}}$$ *Convert 4 feet to inches using a unit conversion factor.*

$$= 4 \cdot 12 \text{ in.}$$ *Remove the common units of feet in the numerator and denominator. The unit of inches remain.*

$$= 48 \text{ in.}$$ *Do the multiplication.*

The length of the rectangle is 48 inches. Now we can find the perimeter (in inches) or area (in in.^{2}) of the rectangle.

If we know the area of a polygon, we can often use algebra to find an unknown measurement.

The area of the parallelogram shown here is 208 ft^2. Find the height.

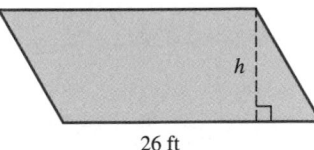

h

26 ft

$$A = bh$$ *This is the formula for the area of a parallelogram.*

$$208 = 26h$$ *Substitute 208 for A, the area, and 26 for b, the length of the base.*

$$\frac{208}{26} = \frac{26h}{26}$$ *To isolate h, undo the multiplication by 26 by dividing both sides by 26.*

$$8 = h$$ *Do the division.*

The height of the parallelogram is 8 feet.

To find the area of an irregular shape, break up the shape into familiar polygons. Find the area of each polygon, and then add the results.

Find the area of the shaded figure shown here.

We will find the area of the lower portion of the figure (the trapezoid) and the area of the upper portion (the square) and then add the results.

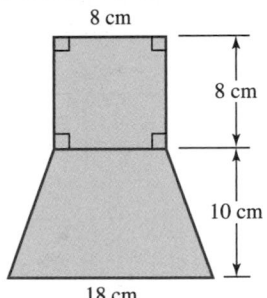

8 cm

8 cm

10 cm

18 cm

$$A_{\text{trapezoid}} = \frac{1}{2}h(b_1 + b_2)$$ *This is the formula for the area of a trapezoid.*

$$A_{\text{trapezoid}} = \frac{1}{2}(10)(8 + 18)$$ *Substitute 8 for b_1, 18 for b_2, and 10 for h.*

$$= \frac{1}{2}(10)(26)$$ *Do the addition within the parentheses.*

$$= 130$$ *Do the multiplication.*

The area of the trapezoid is 130 cm^2.

$$A_{\text{square}} = s^2$$ *This is the formula for the area of a square.*

$$A_{\text{square}} = 8^2$$ *Substitute 8 for s.*

$$= 64$$ *Evaluate the exponential expression.*

The area of the square is 64 cm^2.

The total area of the shaded figure is

$$A_{\text{total}} = A_{\text{trapezoid}} + A_{\text{square}}$$
$$A_{\text{total}} = \mathbf{130\ cm^2} + \mathbf{64\ cm^2}$$
$$= 194\ cm^2$$

The area of the shaded figure is 194 cm².

To find the area of an irregular shape, we must sometimes use subtraction.	To find the area of the shaded figure below, we subtract the area of the triangle *from* the area of the rectangle.

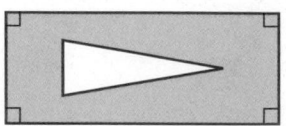

$$A_{\text{shaded}} = A_{\text{rectangle}} - A_{\text{triangle}}$$

REVIEW EXERCISES

65. Find the perimeter of a square with sides 18 inches long.

66. Find the perimeter (in inches) of a rectangle that is 7 inches long and 3 feet wide.

Find the perimeter of each polygon.

67.

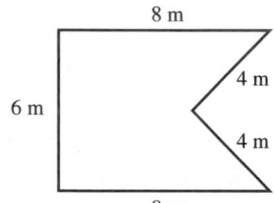

68.

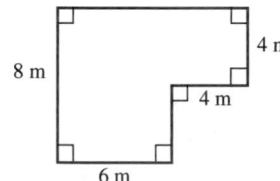

69. The perimeter of an isosceles triangle is 107 feet. If one of the congruent sides is 24 feet long, how long is the base?

70. a. How many square feet are there in 1 square yard?

    b. How many square inches are in 1 square foot?

Find the area of each polygon.

71.

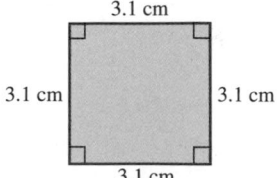

72.

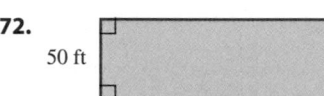

73.

74.

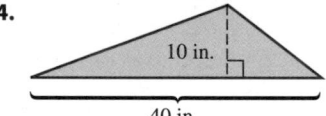

75.

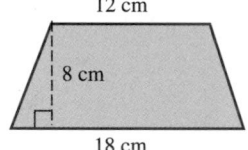

76.

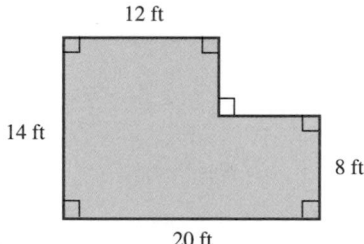

77.

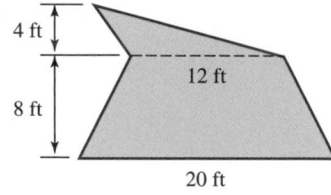

78.

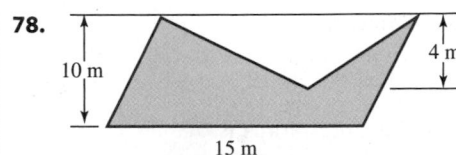

10 m   4 m
15 m

79. The area of a parallelogram is 240 ft². If the length of the base is 30 feet, what is its height?

80. The perimeter of a rectangle is 48 mm and its width is 6 mm. Find its length.

81. FENCES A man wants to enclose a rectangular front yard with chain link that costs $8.50 a foot (the price includes installation). Find the cost of enclosing the yard if its dimensions are 115 ft by 78 ft.

82. LAWNS A family is going to have artificial turf installed in their rectangular backyard that is 36 feet long and 24 feet wide. If the turf costs $48 per *square yard,* and the installation is free, what will this project cost? (Assume no waste.)

SECTION 9.8   Circles

DEFINITIONS AND CONCEPTS	EXAMPLES
A **circle** is the set of all points in a plane that lie a fixed distance from a point called its **center.** The fixed distance is the circle's **radius.** A **chord** of a circle is a line segment connecting two points on the circle. A **diameter** is a chord that passes through the circle's center. Any part of a circle is called an **arc.** A **semicircle** is an arc of a circle whose endpoints are the endpoints of a diameter.	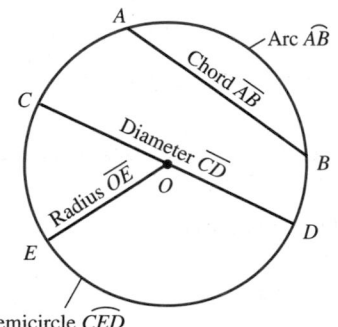 Arc $\overset{\frown}{AB}$, Chord $\overline{AB}$, Diameter $\overline{CD}$, Radius $\overline{OE}$, Semicircle $\overset{\frown}{CED}$
The **circumference** (perimeter) of a circle is given by the formulas $C = \pi D$ or $C = 2\pi r$ where $\pi = 3.14159\ldots$	Find the circumference of the circle shown here. Give the exact answer and an approximation. 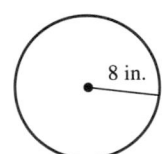 8 in. $C = 2\pi r$ This is the formula for the circumference of a circle. $C = 2\pi(8)$ Substitute 8 for r, the radius. $C = 2(8)\pi$ Rewrite the product so that π is the last factor. $C = 16\pi$ Do the first multiplication: 2(8) = 16. This is the exact answer.
If an exact answer contains π, we can use 3.14 as an approximation, and complete the calculations by hand. Or, we can use a calculator that has a pi key $\boxed{\pi}$ to find an approximation.	The circumference of the circle is exactly 16π inches. If we replace π with 3.14, we get an approximation of the circumference. $C = 16\pi$ $C \approx 16(3.14)$ Substitute 3.14 for π. $C \approx 50.24$ Do the multiplication. The circumference of the circle is approximately 50.2 inches. We can also use a calculator to approximate 16π. $C \approx 50.26548246$

The **area** of a circle is given by the formula

$$A = \pi r^2$$

Find the area of the circle shown here. Give the exact answer and an approximation to the nearest tenth.

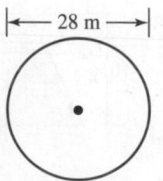

Since the diameter is 28 meters, the radius is half of that, or 14 meters.

$A = \pi r^2$	This is the formula for the area of a circle.
$A = \pi(\mathbf{14})^2$	Substitute 14 for r, the radius of the circle.
$= \pi(196)$	Evaluate the exponential expression.
$= 196\pi$	Write the product so that π is the last factor.

The exact area of the circle is 196π m². We can use a calculator to approximate the area.

$$A \approx 615.7521601 \quad \text{Use a calculator to do the multiplication.}$$

To the nearest tenth, the area is 615.8 m².

To find the area of an irregular shape, break it up into familiar figures.

To find the area of the shaded figure shown here, find the area of the triangle and the area of the semicircle, and then add the results.

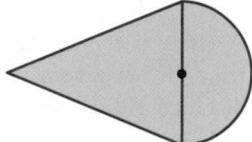

$$A_{\text{shaded figure}} = A_{\text{triangle}} + A_{\text{semicircle}}$$

REVIEW EXERCISES

83. Refer to the figure.

    a. Name each chord.

    b. Name each diameter.

    c. Name each radius.

    d. Name the center.

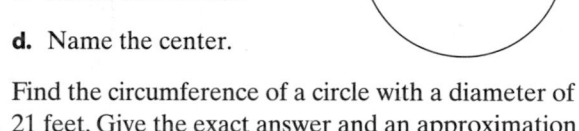

84. Find the circumference of a circle with a diameter of 21 feet. Give the exact answer and an approximation to the nearest hundredth.

85. Find the perimeter of the figure shown below. Round to the nearest tenth.

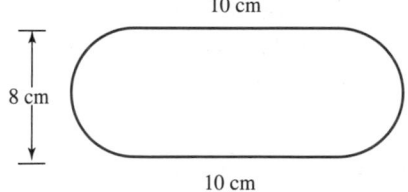

86. Find the area of a circle with a diameter of 18 inches. Give the exact answer and an approximation to the nearest hundredth.

87. Find the area of the figure shown in Problem 85. Round to the nearest tenth.

88. Find the area of the shaded region shown on the right. Round to the nearest tenth.

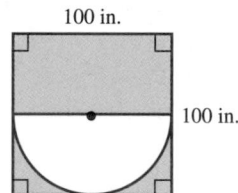

SECTION **9.9**   **Volume**

DEFINITIONS AND CONCEPTS	EXAMPLES

The volume of a figure can be thought of as the number of **cubic units** that will fit within its boundaries.

Two common units of volume are cubic inches (in.^{3}) and cubic centimeters (cm^3).

1 cubic inch: 1 in.^{3}   **1 cubic centimeter: 1 cm^3**

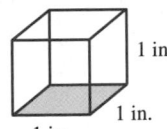 1 in.
1 in.
1 in.

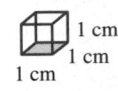 1 cm
1 cm
1 cm

The **volume** of a solid is a measure of the space it occupies.

Figure	Volume Formula
Cube	$V = s^3$
Rectangular solid	$V = lwh$
Prism	$V = Bh$*
Pyramid	$V = \frac{1}{3}Bh$*
Cylinder	$V = \pi r^2 h$
Cone	$V = \frac{1}{3}\pi r^2 h$
Sphere	$V = \frac{4}{3}\pi r^3$

*B represents the area of the base.

CARRY-ON LUGGAGE The largest carry-on bag that Alaska Airlines allows on board a flight is shown on the right. Find the volume of space that a bag that size occupies.

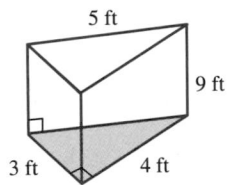 Width: 17 in.
Height: 10 in.
Length: 24 in.

$V = lwh$ This is the formula for the volume of a rectangular solid.

$V = 24(17)(10)$ Substitute 24 for l, the length, 17 for w, the width, and 10 for h, the height of the bag.

$= 4{,}080$ Do the multiplication.

The volume of the space that the bag occupies is 4,080 in.^{3}.

Caution! When finding the volume of a figure, only use the measurements that are called for in the formula. Sometimes a figure may be labeled with measurements that are not used.

Find the volume of the prism shown here.

The area of the triangular base of the prism is $\frac{1}{2}(3)(4) = 6$ square feet. (The 5-inch measurement is not used.) To find the volume of the prism, proceed as follows:

5 ft   9 ft   3 ft   4 ft

$V = Bh$ This is the formula for the volume of a prism.

$V = 6(9)$ Substitute 6 for B, the area of the base, and 9 for h, the height.

$= 54$ Do the multiplication.

The volume of the triangular prism is 54 ft^3.

The letter B appears in two of the volume formulas. It represents the area of the base of the figure.

Note that the volume formulas for a pyramid and a cone contain a factor of $\frac{1}{3}$.

Cone: $V = \frac{1}{3}\pi r^2 h$

Pyramid: $V = \frac{1}{3}Bh$

Find the volume of the pyramid shown here.

Since the base is a square with each side 5 centimeters long, the area of the base is $5 \cdot 5 = 25$ cm^2.

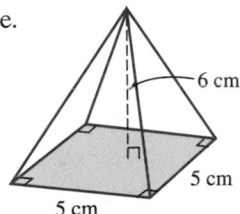 6 cm
5 cm
5 cm

$V = \frac{1}{3}Bh$ This is the formula for the volume of a pyramid.

$V = \frac{1}{3}(25)(6)$ Substitute 25 for B, the area of the base, and 6 for h, the height.

$= 25(2)$ Multiply the first and third factors: $\frac{1}{3}(6) = 2$.

$= 50$ Complete the multiplication by 25.

The volume of the pyramid is 50 cm^3.

Note that the volume formulas for a cone, cylinder, and sphere contain a factor of π.

Cone $V = \frac{1}{3}\pi r^2 h$

Cylinder $V = \pi r^2 h$

Sphere $V = \frac{4}{3}\pi r^3$

Find the volume of the cylinder shown here. Give the exact answer and an approximation to the nearest hundredth.

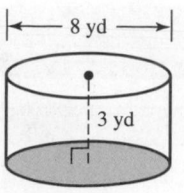

Since a radius is one-half of the diameter of the circular base, $r = \frac{1}{2} \cdot 8$ yd $= 4$ yd. To find the volume of the cylinder, proceed as follows:

$V = \pi r^2 h$ This is the formula for the volume of a cylinder.

$V = \pi(4)^2(3)$ Substitute 4 for r, the radius of the base, and 3 for h, the height.

$V = \pi(16)(3)$ Evaluate the exponential expression.

$= 48\pi$ Write the product so that π is the last factor.

≈ 150.7964474 Use a calculator to do the multiplication.

The exact volume of the cylinder is 48π yd^3. To the nearest hundredth, the volume is 150.80 yd^3.

If an exact answer contains π, we can use 3.14 as an approximation, and complete the calculations by hand. Or, we can use a calculator that has a pi key $\boxed{\pi}$ to find an approximation.

Find the volume of the sphere shown here. Give the exact answer and an approximation to the nearest tenth.

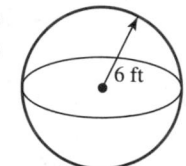

$V = \frac{4}{3}\pi r^3$ This is the formula for the volume of a sphere.

$V = \frac{4}{3}\pi(6)^3$ Substitute 6 for r, the radius of the sphere.

$= \frac{4}{3}\pi(216)$ Evaluate the exponential expression.

$= \frac{864}{3}\pi$ Multiply: $4(216) = 864$.

$= 288\pi$ Divide: $\frac{864}{3} = 288$.

≈ 904.7786842 Use a calculator to do the multiplication.

The volume of the sphere is exactly 288π ft^3. To the nearest tenth, this is 904.8 ft^3.

REVIEW EXERCISES

Find the volume of each figure. If an exact answer contains π,
approximate to the nearest hundredth.

89.
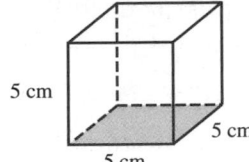
5 cm
5 cm
5 cm

90.
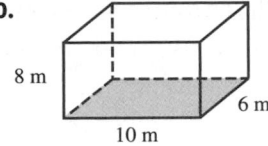
8 m
6 m
10 m

91.
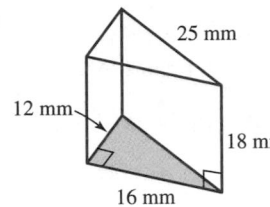
25 mm
12 mm
18 mm
16 mm

92.
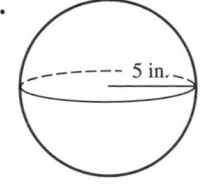
5 in.

93.
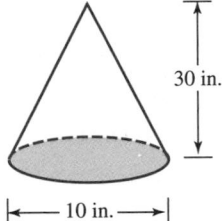
30 in.
10 in.

94.
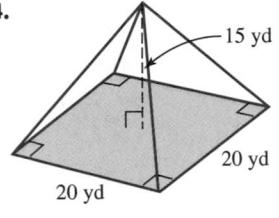
15 yd
20 yd
20 yd

95.
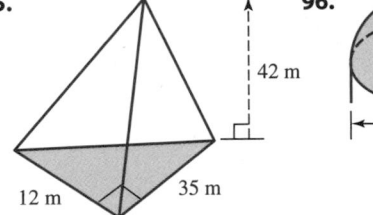
42 m
12 m
35 m

96.
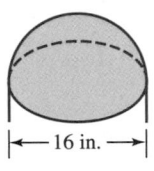
16 in.

97. FARMING Find the volume of the corn silo shown below. Round to the nearest one cubic foot.

10 ft
16 ft
ALLEN
FARMS

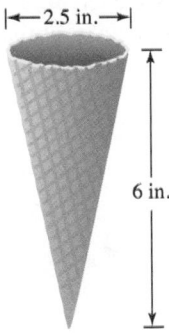
2.5 in.
6 in.

98. WAFFLE CONES Find the volume of the ice cream cone shown above. Give the exact answer and an approximation to the nearest tenth.

99. How many cubic inches are there in 1 cubic foot?

100. How many cubic feet are there in 2 cubic yards?

CHAPTER 9 TEST

1. Estimate each angle measure. Then tell whether it is an acute, right, obtuse, or straight angle.

a.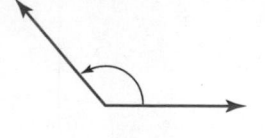

b.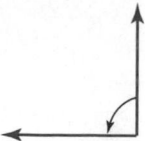

c.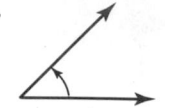

d.

2. Fill in the blanks.

a. If $\angle ABC \cong \angle DEF$, then the angles have the same _____.

b. Two congruent segments have the same _____.

c. Two different points determine one _____.

d. Two angles are called _____ if the sum of their measures is 90°.

3. Refer to the figure below. What is the midpoint of $\overline{BE}$?

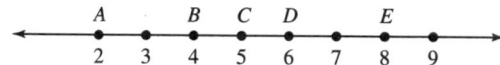

4. Refer to the figure below and tell whether each statement is true or false.

a. $\angle AGF$ and $\angle BGC$ are vertical angles.

b. $\angle EGF$ and $\angle DGE$ are adjacent angles.

c. $m(\angle AGB) = m(\angle EGD)$.

d. $\angle CGD$ and $\angle DGF$ are supplementary angles.

e. $\angle EGD$ and $\angle AGB$ are complementary angles.

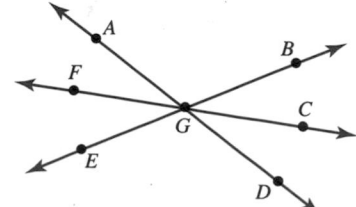

5. Find x. Then find $m(\angle ABD)$ and $m(\angle CBE)$.

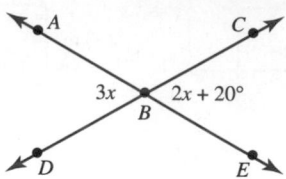

6. Find the supplement of a 47° angle.

7. Refer to the figure below. Fill in the blanks.

a. l_1 intersects two coplanar lines. It is called a _____.

b. $\angle 4$ and ____ are alternate interior angles.

c. $\angle 3$ and ____ are corresponding angles.

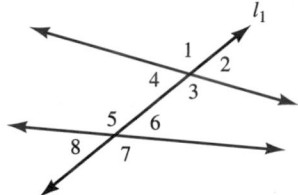

8. In the figure below, $l_1 \parallel l_2$ and $m(\angle 2) = 25°$. Find the measures of the other numbered angles.

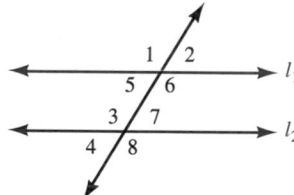

9. In the figure below, $l_1 \parallel l_2$. Find x. Then determine the measure of each angle that is labeled in the figure.

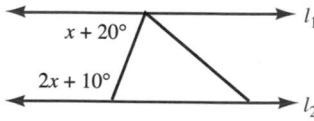

10. For each polygon, give the number of sides it has, tell its name, and then give the number of vertices it has.

a.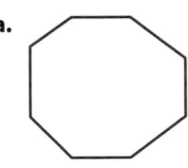

b.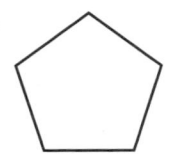

c.

d.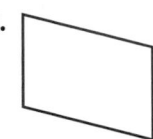

11. Classify each triangle as an equilateral triangle, an isosceles triangle, or a scalene triangle.

a.

b.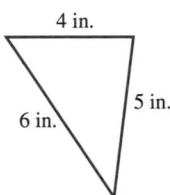

c.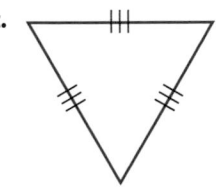

d.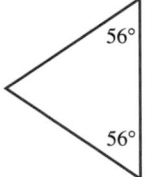

12. Find x.

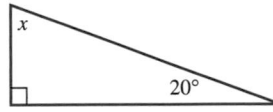

13. The measure of the vertex angle of an isosceles triangle is 12°. Find the measure of each base angle.

14. Refer to rectangle *EFGH* shown below.

  a. Find m($\overline{HG}$).   **b.** Find m($\overline{FH}$).

  c. Find m($\angle FGH$).   **d.** Find m($\overline{EH}$).

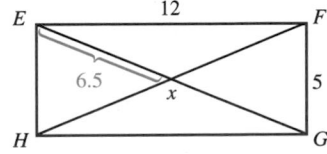

15. Refer to isosceles trapezoid *QRST* shown below.

  a. Find m($\overline{RS}$).   **b.** Find x.

  c. Find y.   **d.** Find z.

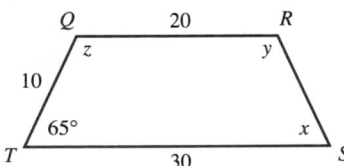

16. Find the sum of the measures of the angles of a decagon.

17. Find the perimeter of the figure shown below.

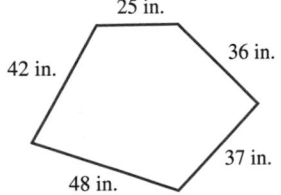

18. The perimeter of an equilateral triangle is 45.6 m. Find the length of each side.

19. Find the area of the shaded part of the figure shown below.

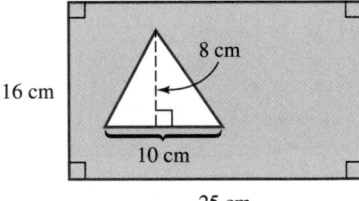

20. DECORATING A patio has the shape of a trapezoid, as shown on the right. If indoor/outdoor carpeting sells for $18 a *square yard* installed, how much will it cost to carpet the patio?

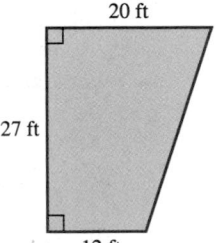

21. How many square inches are in one square foot?

22. Find the area of the rectangle shown below in square inches.

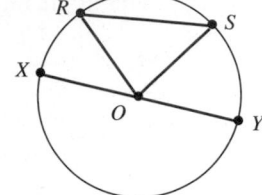

10 ft

1 in.

23. Refer to the figure below, where O is the center of the circle.

   a. Name each chord.

   b. Name a diameter.

   c. Name each radius.

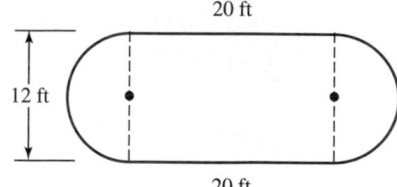

24. Fill in the blank: If C is the circumference of a circle and D is the length of its diameter, then $\frac{C}{D} = \boxed{}$.

In Problems 25–27, when appropriate, give the exact answer and an approximation to the nearest tenth.

25. Find the circumference of a circle with a diameter of 21 feet.

26. Find the perimeter of the figure shown below. Assume that the arcs are semicircles.

20 ft

12 ft

20 ft

27. HISTORY Stonehenge is a prehistoric monument in England, believed to have been built by the Druids. The site, 30 meters in diameter, consists of a circular arrangement of stones, as shown below. What area does the monument cover?

28. See the figure below, where $\triangle MNO \cong \triangle RST$. Name the six corresponding parts of the congruent triangles.

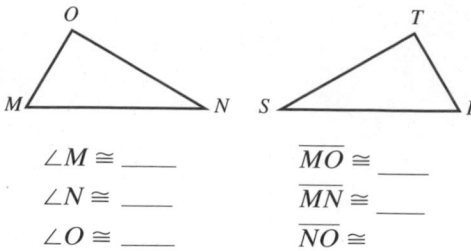

$\angle M \cong$ _____ $\overline{MO} \cong$ _____

$\angle N \cong$ _____ $\overline{MN} \cong$ _____

$\angle O \cong$ _____ $\overline{NO} \cong$ _____

29. Tell whether each pair of triangles are congruent. If they are, tell why.

   a.

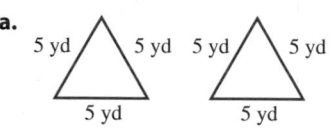

5 yd   5 yd   5 yd   5 yd

5 yd   5 yd

   b.

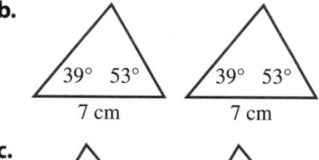

39°   53°   39°   53°

7 cm   7 cm

   c.

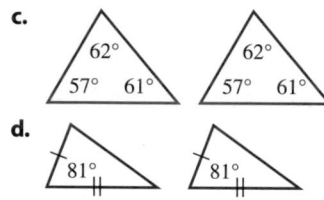

62°   62°

57°   61°   57°   61°

   d.

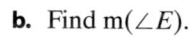

81°   81°

30. Refer to the figure below, in which $\triangle ABC \cong \triangle DEF$.

   a. Find $m(\overline{DE})$.   **b.** Find $m(\angle E)$.

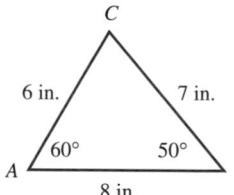
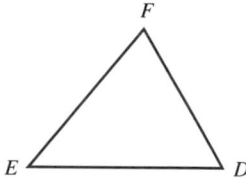

31. Tell whether the triangles in each pair are similar.

   a.   **b.**

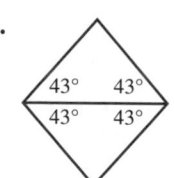
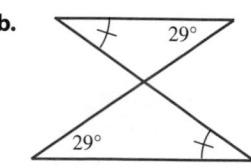

43°   43°   29°

43°   43°   29°

32. Refer to the triangles below. The units are meters.

  a. Find x.   **b.** Find y.

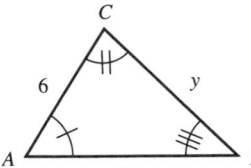
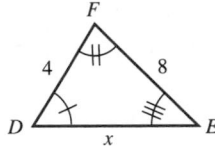

33. SHADOWS If a tree casts a 7-foot shadow at the same time as a man 6 feet tall casts a 2-foot shadow, how tall is the tree?

34. Refer to the right triangle below. Find the missing side length. Approximate any exact answers that contain a square root to the nearest tenth.

  a. Find c if $a = 10$ cm and $b = 24$ cm.

  b. Find b if $a = 6$ in. and $c = 8$ in.

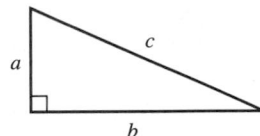

35. TELEVISIONS To the nearest tenth of an inch, what is the diagonal measurement of the television screen shown below?

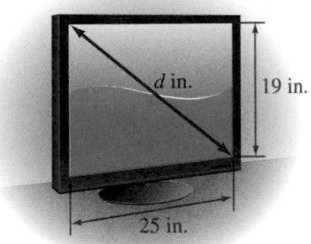

36. How many cubic inches are there in 1 cubic foot?

Find the volume of each figure. Give the exact answer and an approximation to the nearest hundredth if an answer contains π.

37.
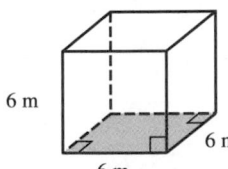

38.
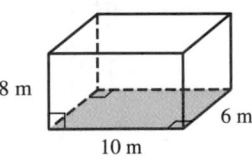

39.
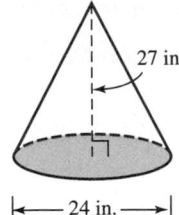

40.
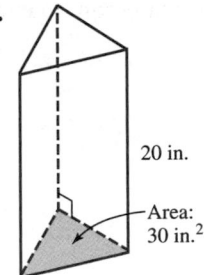

41.
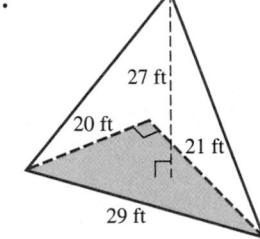

42.
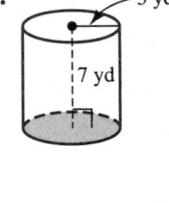

43.
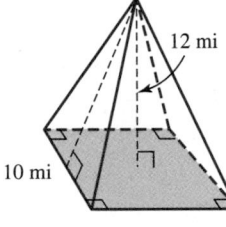

44.
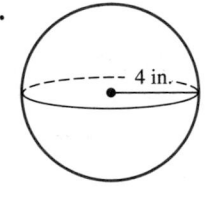

45. FARMING A silo is used to store wheat and corn. Find the volume of the silo shown below. Give the exact answer and an approximation to the nearest cubic foot.

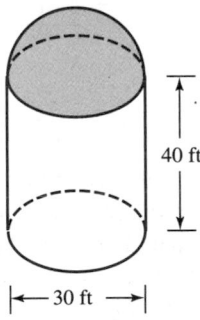

46. Give a real-life example in which the concept of perimeter is used. Do the same for area and for volume. Be sure to discuss the type of units used in each case.

CHAPTERS 1–9   CUMULATIVE REVIEW

1. Write 104,052,005 in words. [Section 1.1]

2. Add: $257 + 99{,}085 + 4{,}101 + 33$ [Section 1.2]

3. Multiply: $\begin{array}{r} 196 \\ \times\ 78 \end{array}$ [Section 1.3]

4. Divide: $34\overline{)2{,}006}$ [Section 1.4]

5. **a.** Find the factors of 32. [Section 1.5]
   b. Find the prime factorization of 140. [Section 1.5]

6. Find the LCM and the GCF of 35 and 45. [Section 1.6]

7. Evaluate: $46 + 3[5^2 - 4(9 - 5)]$ [Section 1.7]

8. Solve each equation and check the result.
   a. $x + 33 = 91$ [Section 1.8]
   b. $24m = 312$ [Section 1.9]

9. Is the statement $-72 > -73$ true or false? [Section 2.1]

10. Perform each operation, if possible.
    a. $-17 + 8$ [Section 2.2]
    b. $-14 - (-2)$ [Section 2.3]
    c. $-3(72)$ [Section 2.4]
    d. $\dfrac{-60}{0}$ [Section 2.5]

11. Evaluate: $\dfrac{2 + 3[5 - (1 - 10)]}{\left|2[(-2)^3 + 2] + 10\right|}$ [Section 2.6]

12. Solve $-2(5) = \dfrac{y}{-3} + 3$ and check the result.
    [Section 2.7]

Write an equation and solve it to answer the following question.
[Section 2.7]

13. BANKING After she made deposits of $55 and $80, a student's account was still $82 overdrawn. What was her checking account balance before the deposits?

14. Translate to mathematical symbols: 18 less than twice the width w. [Section 3.1]

15. Evaluate $b^2 - 4ac$ for $a = -1$, $b = -5$, and $c = -2$.
    [Section 3.2]

16. Simplify: $-4(6u)(-2)$ [Section 3.3]

17. PING-PONG Write an algebraic expression that represents the perimeter (in feet) of the Ping-Pong table shown here.
    [Section 3.4]

x ft $(x + 4)$ ft

18. Solve $6(2j + 6) + 4j = 4(j - 30)$ [Section 3.5]

Write an equation and solve it to answer the following question.
[Section 3.6]

19. TAX REFUNDS After receiving their tax refund, a husband and wife split the refunded money equally. The husband then gave $250 of his money to charity, leaving him with $395. What was the amount of their tax refund check?

20. Simplify: $\dfrac{60x^5}{108x^3}$ [Section 4.1]

Perform each operation. Simplify, if possible.

21. $\dfrac{4}{5} \cdot \dfrac{15}{16}$ [Section 4.2]

22. $-\dfrac{7a}{4b^2} \div \left(-\dfrac{21a}{8b}\right)$ [Section 4.3]

23. Subtract $\dfrac{3}{4}$ from $\dfrac{4}{n}$. [Section 4.4]

24. $\dfrac{7}{10} + \dfrac{3}{14}$ [Section 4.4]

25. TIRES The road surface "footprint" of a sport truck tire is approximately rectangular, as shown here. If the width of the tire is 6 inches, what is the area of the tire "footprint"?
    [Section 4.5]

$7\tfrac{1}{2}$ in.

26. Subtract: $140\dfrac{5}{6} - 129\dfrac{4}{5}$ [Section 4.6]

27. Simplify the complex fraction: $\dfrac{-\dfrac{1}{2} + \dfrac{7}{8}}{\dfrac{3}{4} - \dfrac{1}{2}}$ [Section 4.7]

Write an equation and solve it to answer the following question. [Section 4.8]

28. CAR REPAIRS Two-fifths of the cars that an automobile repair shop serviced last year had transmission problems. If the shop repaired 140 transmissions, how many cars did they service that year?

Solve each equation and check the result.

29. $-\dfrac{5}{6}x = 10$ [Section 4.8]

30. $\dfrac{12}{5} + \dfrac{n}{2} = \dfrac{n}{10}$ [Section 4.8]

31. Round 9.50966 to the nearest thousandth. [Section 5.1]

32. Use a check to determine whether the subtraction shown below is correct. [Section 5.2]

$$
\begin{array}{r}
451.3 \\
-\ 89.8 \\
\hline
361.5
\end{array}
$$

33. GEOMETRY The length of a rectangle is 2.72 feet and the width is 3.81 feet. Find its area. [Section 5.3]

34. Divide: $\dfrac{14.637}{-4.1}$ [Section 5.4]

35. Write $\dfrac{11}{12}$ as a decimal. Use an overbar in your answer.

[Section 5.5]

36. Evaluate: $-4\sqrt{81} + 2\sqrt{4}$ [Section 5.6]

37. Solve $3(y - 1.1) + 3.2 = 2.3$ and check the result.

[Section 5.7]

38. Write the ratio *21 inches to 3 feet* in simplest form.

[Section 6.1]

39. MAKING BROWNIES A recipe for brownies calls for 4 eggs and $1\frac{1}{2}$ cups of flour. If the recipe makes 15 brownies, how many cups of flour will be needed to make 260 brownies? [Section 6.2]

40. Make each conversion.

  a. 50 pounds to ounces [Section 6.3]

  b. 500 milliliters to liters [Section 6.4]

  c. 35 centimeters to inches [Section 6.5]

41. Write $\dfrac{27}{50}$ as a decimal and as a percent. [Section 7.1]

42. 36 is 24% of what number? [Section 7.2]

43. Find the total cost of a $196.56 purchase if the sales tax rate is 4.3%. [Section 7.3]

44. Estimate 20% of 6,700. [Section 7.4]

45. Find the simple interest on a loan of $1,250 borrowed at 10% for 3 months. [Section 7.5]

46. Find the mean, median, mode, and range of the values listed below. [Section 8.2]

  73   5   38   45   11   36   27   35   45

47. Find the coordinates of each point shown in red.
[Section 8.3]

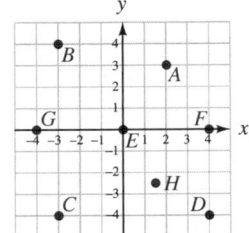

48. Is $(9, -3)$ a solution of $x + 5y = -7$? [Section 8.3]

Graph each equation. [Section 8.4]

49. $y = 3x + 1$ 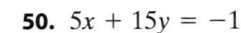   **50.** $5x + 15y = -15$

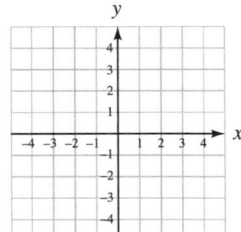   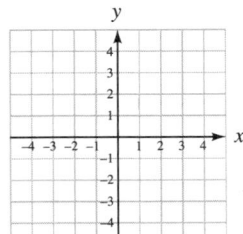

51. First find x. Then find $m(\angle ABD)$ and $m(\angle DBE)$.
[Section 9.1]

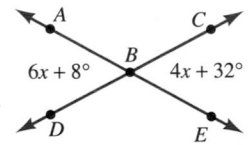

52. The measure of one base angle of an isosceles triangle is 76°. Find the measure of the vertex angle.
[Section 9.3]

53. The length of the hypotenuse of a right triangle is 41 yards and the length of one leg is 9 yards. Find the length of the other leg. [Section 9.4]

54. Find the area of a circle with diameter 10 centimeters. Round to the nearest tenth. [Section 9.8]

Exponents and Polynomials

© Robert E. Daemmrich/Getty Images

from *Campus to Careers*

Police Officer

People depend on the police to protect their lives and property. The job can be dangerous because police officers must arrest suspects and respond to emergencies. The daily activities of police officers can vary greatly depending on their specialty, such as patrol officer, game warden, or detective. Regardless of their duties, they must write reports and maintain records that will be needed if they testify in court.

In **Problem 51** of **Study Set 10.2,** you will see how police officers can compute the stopping distance of a car.

JOB TITLE:
Police Officer

EDUCATION: For many departments, two years of college or a college degree may be required. Physical education courses are helpful. Foreign language skills are desirable.

JOB OUTLOOK: Excellent—employment opportunities are expected to grow 11 percent through 2016.

ANNUAL EARNINGS: Median annual salary in 2007 was $50,330. Earnings often exceed their salary because of payments for overtime.

FOR MORE INFORMATION:
www.bls.gov/oco/ocos160.htm

Objectives

1. Identify bases and exponents.

2. Multiply exponential expressions that have like bases.

3. Raise exponential expressions to a power.

4. Find powers of products.

Multiplication Rules for Exponents

In this section, we will use the definition of exponent to develop some rules for simplifying expressions that contain exponents.

1 Identify bases and exponents.

Recall that an **exponent** indicates repeated multiplication. It indicates how many times the base is used as a factor. For example, 3^5 represents the product of five 3's.

$$\text{Exponent} \rightarrow \overbrace{3^5 = 3 \cdot 3 \cdot 3 \cdot 3 \cdot 3}^{5 \text{ factors of } 3}$$
Base

In general, we have the following definition.

Natural-Number Exponents

A natural-number* exponent tells how many times its base is to be used as a factor.

For any number x and any natural number n,

$$x^n = \overbrace{x \cdot x \cdot x \cdot \,\cdots\, \cdot x}^{n \text{ factors of } x}$$

*The set of natural numbers is $\{1, 2, 3, 4, 5, \dots\}$.

Expressions of the form x^n are called **exponential expressions.** The base of an exponential expression can be a number, a variable, or a combination of numbers and variables. Some examples are:

$$10^5 = 10 \cdot 10 \cdot 10 \cdot 10 \cdot 10$$
The base is 10. The exponent is 5. Read as "10 to the fifth power."

$$y^2 = y \cdot y$$
The base is y. The exponent is 2. Read as "y squared."

$$(-2s)^3 = (-2s)(-2s)(-2s)$$
The base is −2s. The exponent is 3. Read as "negative 2s raised to the third power" or "negative 2s cubed."

$$-8^4 = -(8 \cdot 8 \cdot 8 \cdot 8)$$
Since the − sign is not written within parentheses, the base is 8. The exponent is 4. Read as "the opposite (or the negative) of 8 to the fourth power."

When an exponent is 1, it is usually not written. For example, $4 = 4^1$ and $x = x^1$.

Caution! Bases that contain a − sign *must* be written within parentheses.

$$(-2s)^3 \leftarrow \text{Exponent}$$
Base

EXAMPLE 1 Identify the base and the exponent in each expression:

a. 8^5 **b.** $7a^3$ **c.** $(7a)^3$

Strategy To identify the base and exponent, we will look for the form ▦▪.

WHY The exponent is the small raised number to the right of the base.

Solution

a. In 8^5, the base is 8 and the exponent is 5.

b. $7a^3$ means $7 \cdot a^3$. Thus, the base is a, not $7a$. The exponent is 3.

c. Because of the parentheses in $(7a)^3$, the base is $7a$ and the exponent is 3. ▪

Self Check 1

Identify the base and the exponent:
a. $3y^4$
b. $(3y)^4$

Now Try Problems 13 and 17

EXAMPLE 2 Write each expression in an equivalent form using an exponent: **a.** $b \cdot b \cdot b \cdot b$ **b.** $5 \cdot t \cdot t \cdot t$

Strategy We will look for repeated factors and count the number of times each appears.

WHY We can use an exponent to represent repeated multiplication.

Solution

a. Since there are four repeated factors of b in $b \cdot b \cdot b \cdot b$, the expression can be written as b^4.

b. Since there are three repeated factors of t in $5 \cdot t \cdot t \cdot t$, the expression can be written as $5t^3$. ▪

Self Check 2

Write as an exponential expression:
$(x + y)(x + y)(x + y)(x + y)(x + y)$

Now Try Problems 25 and 29

2 Multiply exponential expressions that have like bases.

To develop a rule for multiplying exponential expressions that have the same base, we consider the product $6^2 \cdot 6^3$. Since 6^2 means that 6 is to be used as a factor two times, and 6^3 means that 6 is to be used as a factor three times, we have

$$6^2 \cdot 6^3 = \underbrace{6 \cdot 6}_{\text{2 factors of 6}} \cdot \underbrace{6 \cdot 6 \cdot 6}_{\text{3 factors of 6}}$$

$$= \underbrace{6 \cdot 6 \cdot 6 \cdot 6 \cdot 6}_{\text{5 factors of 6}}$$

$$= 6^5$$

We can quickly find this result if we keep the common base 6 and add the exponents on 6^2 and 6^3.

$$6^2 \cdot 6^3 = 6^{2+3} = 6^5$$

This example illustrates the following rule for exponents.

Product Rule for Exponents

To multiply exponential expressions that have the same base, keep the common base and add the exponents.

For any number x and any natural numbers m and n,

$$x^m \cdot x^n = x^{m+n} \qquad \text{Read as "x to the mth power times x to the nth power equals x to the m plus nth power."}$$

Self Check 3

Simplify:

a. $7^8(7^7)$

b. x^2x^3x

c. $(y-1)^5(y-1)^5$

d. $(s^4t^3)(s^4t^4)$

Now Try Problems 33, 35, and 37

EXAMPLE 3   Simplify:

a. $9^5(9^6)$ **b.** $x^3 \cdot x^4$ **c.** y^2y^4y **d.** $(c^2d^3)(c^4d^5)$

Strategy In each case, we want to write an equivalent expression using one base and one exponent. We will use the product rule for exponents to do this.

WHY The product rule for exponents is used to multiply exponential expressions that have the same base.

Solution

a. $9^5(9^6) = 9^{5+6} = 9^{11}$ Keep the common base, 9, and add the exponents. Since 9^{11} is a very large number, we will leave the answer in this form. We won't evaluate it.

Caution! Don't make the mistake of multiplying the bases when using the product rule. Keep the *same* base.

$$9^5(9^6) \neq 81^{11}$$

b. $x^3 \cdot x^4 = x^{3+4} = x^7$ Keep the common base, x, and add the exponents.

c. $y^2y^4y = y^2y^4y^1$ Write y as y^1.

$= y^{2+4+1}$ Keep the common base, y, and add the exponents.

$= y^7$

d. $(c^2d^3)(c^4d^5) = (c^2c^4)(d^3d^5)$ Use the commutative and associative properties of multiplication to group like bases together.

$= (c^{2+4})(d^{3+5})$ Keep the common base, c, and add the exponents. Keep the common base, d, and add the exponents.

$= c^6d^8$

Caution! We cannot use the product rule to simplify expressions like $3^2 \cdot 2^3$, where the bases are not the same. However, we can simplify this expression by doing the arithmetic:

$$3^2 \cdot 2^3 = 9 \cdot 8 = 72 \qquad 3^2 = 3 \cdot 3 = 9 \text{ and } 2^3 = 2 \cdot 2 \cdot 2 = 8.$$

Recall that *like terms* are terms with exactly the same variables raised to exactly the same powers. To add or subtract exponential expressions, they must be like terms. To multiply exponential expressions, only the bases need to be the same.

$x^5 + x^2$ These are not like terms; the exponents are different. We cannot add.

$x^2 + x^2 = 2x^2$ These are like terms; we can add. Recall that $x^2 = 1x^2$.

$x^5 \cdot x^2 = x^7$ The bases are the same; we can multiply.

3 Raise exponential expressions to a power.

To develop another rule for exponents, we consider $(5^3)^4$. Here, an exponential expression, 5^3, is raised to a power. Since 5^3 is the base and 4 is the exponent, $(5^3)^4$ can be written as $5^3 \cdot 5^3 \cdot 5^3 \cdot 5^3$. Because each of the four factors of 5^3 contains three factors of 5, there are $4 \cdot 3$ or 12 factors of 5.

12 factors of 5

$$(5^3)^4 = 5^3 \cdot 5^3 \cdot 5^3 \cdot 5^3 = \overbrace{5 \cdot 5 \cdot 5 \cdot 5 \cdot 5 \cdot 5 \cdot 5 \cdot 5 \cdot 5 \cdot 5 \cdot 5 \cdot 5}^{} = 5^{12}$$

$5^3 \qquad 5^3 \qquad 5^3 \qquad 5^3$

We can quickly find this result if we keep the common base of 5 and multiply the exponents.

$$(5^3)^4 = 5^{3 \cdot 4} = 5^{12}$$

This example illustrates the following rule for exponents.

Power Rule for Exponents

To raise an exponential expression to a power, keep the base and multiply the exponents.

For any number x and any natural numbers m and n,

$$(x^m)^n = x^{m \cdot n} = x^{mn}$$ Read as "the quantity of x to the mth power raised to the nth power equals x to the mnth power."

The Language of Algebra   An exponential expression raised to a power, such as $(5^3)^4$, is also called a *power of a power.*

EXAMPLE 4   Simplify: **a.** $(2^3)^7$ **b.** $[(-6)^2]^5$ **c.** $(z^8)^8$

Strategy In each case, we want to write an equivalent expression using one base and one exponent. We will use the power rule for exponents to do this.

WHY Each expression is a power of a power.

Solution

a. $(2^3)^7 = 2^{3 \cdot 7} = 2^{21}$ Keep the base, 2, and multiply the exponents. Since 2^{21} is a very large number, we will leave the answer in this form.

b. $[(-6)^2]^5 = (-6)^{2 \cdot 5} = (-6)^{10}$ Keep the base, -6, and multiply the exponents. Since $(-6)^{10}$ is a very large number, we will leave the answer in this form.

c. $(z^8)^8 = z^{8 \cdot 8} = z^{64}$ Keep the base, z, and multiply the exponents.

Self Check 4

Simplify:
a. $(4^6)^5$
b. $(y^5)^2$

Now Try Problems 49, 51, and 53

EXAMPLE 5   Simplify: **a.** $(x^2x^5)^2$ **b.** $(z^2)^4(z^3)^3$

Strategy In each case, we want to write an equivalent expression using one base and one exponent. We will use the product and power rules for exponents to do this.

WHY The expressions involve multiplication of exponential expressions that have the same base and they involve powers of powers.

Solution

a. $(x^2x^5)^2 = (x^7)^2$ Within the parentheses, keep the common base, x, and add the exponents: $2 + 5 = 7$.

$\qquad = x^{14}$ Keep the base, x, and multiply the exponents: $7 \cdot 2 = 14$.

b. $(z^2)^4(z^3)^3 = z^8z^9$ For each power of z raised to a power, keep the base and multiply the exponents: $2 \cdot 4 = 8$ and $3 \cdot 3 = 9$.

$\qquad = z^{17}$ Keep the common base, z, and add the exponents: $8 + 9 = 17$.

Self Check 5

Simplify:
a. $(a^4a^3)^3$
b. $(a^3)^3(a^4)^2$

Now Try Problems 57 and 61

4 Find powers of products.

To develop another rule for exponents, we consider the expression $(2x)^3$, which is a *power of the product* of 2 and x.

$$(2x)^3 = 2x \cdot 2x \cdot 2x \qquad \text{Write the base 2x as a factor 3 times.}$$
$$= (2 \cdot 2 \cdot 2)(x \cdot x \cdot x) \qquad \text{Change the order of the factors and group like bases.}$$
$$= 2^3 x^3 \qquad \text{Write each product of repeated factors in exponential form.}$$
$$= 8x^3 \qquad \text{Evaluate: } 2^3 = 8.$$

This example illustrates the following rule for exponents.

Power of a Product

To raise a product to a power, raise each factor of the product to that power.
  For any numbers x and y, and any natural number n,

$$(xy)^n = x^n y^n$$

Self Check 6

Simplify:
a. $(2t)^4$
b. $(c^3 d^4)^6$

Now Try Problems 65 and 69

EXAMPLE 6   Simplify: **a.** $(3c)^4$ **b.** $(x^2 y^3)^5$

Strategy In each case, we want to write the expression in an equivalent form in which each base is raised to a single power. We will use the power of a product rule for exponents to do this.

WHY Within each set of parentheses is a product, and each of those products is raised to a power.

Solution

a. $(3c)^4 = 3^4 c^4 \qquad$ Raise each factor of the product 3c to the 4th power.
$$= 81c^4 \qquad \text{Evaluate: } 3^4 = 81.$$

b. $(x^2 y^3)^5 = (x^2)^5 (y^3)^5 \qquad$ Raise each factor of the product $x^2 y^3$ to the 5th power.
$$= x^{10} y^{15} \qquad \text{For each power of a power, keep each base, } x \text{ and } y, \text{ and multiply the exponents: } 2 \cdot 5 = 10 \text{ and } 3 \cdot 5 = 15.$$

Self Check 7

Simplify: $(4y^3)^2 (3y^4)^3$

Now Try Problem 73

EXAMPLE 7   Simplify: $(2a^2)^2 (4a^3)^3$

Strategy We want to write an equivalent expression using one base and one exponent. We will begin the process by using the power of a product rule for exponents.

WHY Within each set of parentheses is a product, and each product is raised to a power.

Solution

$$(2a^2)^2 (4a^3)^3 = 2^2 (a^2)^2 \cdot 4^3 (a^3)^3 \qquad$$ Raise each factor of the product $2a^2$ to the 2nd power. Raise each factor of the product $4a^3$ to the 3rd power.

$$= 4a^4 \cdot 64a^9 \qquad$$ Evaluate: $2^2 = 4$ and $4^3 = 64$. For each power of a power, keep each base and multiply the exponents: $2 \cdot 2 = 4$ and $3 \cdot 3 = 9$.

$$= (4 \cdot 64)(a^4 \cdot a^9) \qquad$$ Group the numerical factors. Group the factors that have the same base.

$$= 256a^{13} \qquad$$ Do the multiplication: $4 \cdot 64 = 256$. Keep the common base a and add the exponents: $4 + 9 = 13$.

$$\begin{array}{r} \overset{1}{6}4 \\ \times \; 4 \\ \hline 256 \end{array}$$

The rules for natural-number exponents are summarized as follows.

Rules for Exponents

If m and n represent natural numbers and there are no divisions by zero, then

Exponent of 1	**Product rule**	**Power rule**
$x^1 = x$	$x^m x^n = x^{m+n}$	$(x^m)^n = x^{mn}$

Power of a product

$$(xy)^n = x^n y^n$$

ANSWERS TO SELF CHECKS

1. a. base: y, exponent: 4   **b.** base: $3y$, exponent: 4   **2.** $(x + y)^5$ **3. a.** 7^{15} **b.** x^6
c. $(y - 1)^{10}$ **d.** $s^8 t^7$ **4. a.** 4^{30} **b.** y^{10} **5. a.** a^{21} **b.** a^{17} **6. a.** $16t^4$ **b.** $c^{18} d^{24}$
7. $432 y^{18}$

SECTION 10.1   STUDY SET

VOCABULARY

Fill in the blank.

1. Expressions such as x^4, 10^3, and $(5t)^2$ are called _____ expressions.

2. Match each expression with the proper description.

$(a^4 b^2)^5$ $(a^8)^4$ $a^5 \cdot a^3$

   a. Product of exponential expressions with the same base
   b. Power of an exponential expression
   c. Power of a product

CONCEPTS

Fill in the blanks.

3. a. $(3x)^4 = \boxed{} \cdot \boxed{} \cdot \boxed{} \cdot \boxed{}$

   b. $(-5y)(-5y)(-5y) = \boxed{}$

4. a. $x = x^{\boxed{}}$ **b.** $x^m x^n = \boxed{}$
   c. $(xy)^n = \boxed{}$ **d.** $(a^b)^c = \boxed{}$

5. To simplify each expression, determine whether you add, subtract, multiply, or divide the exponents.

   a. $b^6 \cdot b^9$
   b. $(n^8)^4$
   c. $(a^4 b^2)^5$

6. To simplify $(2y^3 z^2)^4$, what factors within the parentheses must be raised to the fourth power?

Simplify each expression, if possible.

7. a. $x^2 + x^2$ **b.** $x^2 \cdot x^2$

8. a. $x^2 + x$ **b.** $x^2 \cdot x$

9. a. $x^3 - x^2$ **b.** $x^3 \cdot x^2$

10. a. $4^2 \cdot 2^4$ **b.** $x^3 \cdot y^2$

NOTATION

Complete each solution to simplify each expression.

11. $(x^4 x^2)^3 = (\boxed{})^3$
   $= x^{\boxed{}}$

12. $(x^4)^3 (x^2)^3$
   $= x^{\boxed{}} \cdot x^6$
   $= x^{\boxed{}}$

GUIDED PRACTICE

Identify the base and the exponent in each expression.
See Example 1.

13. 4^3 **14.** $(-8)^2$

15. x^5 **16.** $\left(\dfrac{5}{x}\right)^3$

17. $(-3x)^2$ **18.** $(2xy)^{10}$

19. $-\dfrac{1}{3}y^6$ **20.** $-x^4$

21. $9m^{12}$ **22.** $3.14r^4$

23. $(y + 9)^4$ **24.** $(z - 2)^3$

Write each expression in an equivalent form using an exponent.
See Example 2.

25. $m \cdot m \cdot m \cdot m \cdot m$
26. $r \cdot r \cdot r \cdot r \cdot r \cdot r$
27. $4t \cdot 4t \cdot 4t \cdot 4t$
28. $-5u(-5u)(-5u)(-5u)(-5u)$
29. $4 \cdot t \cdot t \cdot t \cdot t \cdot t$
30. $5 \cdot u \cdot u \cdot u$
31. $a \cdot a \cdot b \cdot b \cdot b$
32. $m \cdot m \cdot m \cdot n \cdot n$

Use the product rule for exponents to simplify each expression.
Write the results using exponents. See Example 3.

33. $5^3 \cdot 5^4$ **34.** $3^4 \cdot 3^6$

35. $a^3 \cdot a^3$ **36.** $m^7 \cdot m^7$

37. bb^2b^3 **38.** aa^3a^5

39. $(c^5)(c^8)$ **40.** $(d^4)(d^{20})$

41. $(a^2b^3)(a^3b^3)$ **42.** $(u^3v^5)(u^4v^5)$

43. $cd^4 \cdot cd$ **44.** $ab^3 \cdot ab^4$

45. $x^2 \cdot y \cdot x \cdot y^{10}$ **46.** $x^3 \cdot y \cdot x \cdot y^{12}$

47. $m^{100} \cdot m^{100}$ **48.** $n^{600} \cdot n^{600}$

Use the power rule for exponents to simplify each expression.
Write the results using exponents. See Example 4.

49. $(3^2)^4$ **50.** $(4^3)^3$

51. $[(-4.3)^3]^8$ **52.** $[(-1.7)^9]^8$

53. $(m^{50})^{10}$ **54.** $(n^{25})^4$

55. $(y^5)^3$ **56.** $(b^3)^6$

Use the product and power rules for exponents to simplify each
expression. See Example 5.

57. $(x^2x^3)^5$ **58.** $(y^3y^4)^4$
59. $(p^2p^3)^5$ **60.** $(r^3r^4)^2$
61. $(t^3)^4(t^2)^3$ **62.** $(b^2)^5(b^3)^2$
63. $(u^4)^2(u^3)^2$ **64.** $(v^5)^2(v^3)^4$

Use the power of a product rule for exponents to simplify each
expression. See Example 6.

65. $(6a)^2$ **66.** $(3b)^3$
67. $(5y)^4$ **68.** $(4t)^4$
69. $(3a^4b^7)^3$ **70.** $(5m^9n^{10})^2$
71. $(-2r^2s^3)^3$ **72.** $(-2x^2y^4)^5$

Use the power of a product rule for exponents to simplify each
expression. See Example 7.

73. $(2c^3)^3 (3c^4)^2$ **74.** $(5b^4)^2(3b^8)^2$
75. $(10d^7)^2(4d^9)^3$ **76.** $(2x^7)^3(4x^8)^2$

TRY IT YOURSELF

Simplify each expression.

77. $(7a^9)^2$ **78.** $(12b^6)^2$

79. $t^4 \cdot t^5 \cdot t$ **80.** $n^4 \cdot n \cdot n^3$

81. $y^3y^2y^4$ **82.** y^4yy^6

83. $(-6a^3b^2)^3$ **84.** $(-10r^3s^2)^2$

85. $(n^4n)^3(n^3)^6$ **86.** $(y^3y)^2(y^2)^2$

87. $(b^2b^3)^{12}$ **88.** $(s^3s^3)^3$

89. $(2b^4b)^5 (3b)^2$ **90.** $(2aa^7)^3 (3a)^3$

91. $(c^2)^3 (c^4)^2$ **92.** $(t^5)^2 (t^3)^3$

93. $(3s^4t^3)^3(2st)^4$ **94.** $(2a^3b^5)^2(4ab)^3$

95. $x \cdot x^2 \cdot x^3 \cdot x^4 \cdot x^5$ **96.** $x^{10} \cdot x^9 \cdot x^8 \cdot x^7$

APPLICATIONS

97. ART HISTORY Leonardo da Vinci's drawing
relating a human figure to a square and a circle is
shown. Find an expression for the area of the square
if the man's height is $5x$ feet.

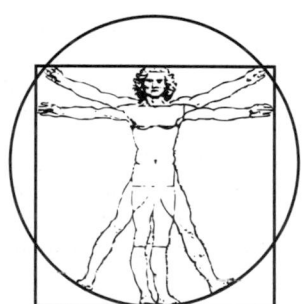

98. PACKAGING Find an expression for the volume of the box shown below.

6x in.

6x in.

6x in.

WRITING

99. Explain the mistake in the following work.

$$2^3 \cdot 2^2 = 4^5 = 1{,}024$$

100. Explain why we can simplify $x^4 \cdot x^5$, but cannot simplify $x^4 + x^5$.

REVIEW

101. JEWELRY A lot of what we refer to as gold jewelry is actually made of a combination of gold and another metal. For example, 18-karat gold is $\frac{18}{24}$ gold by weight. Simplify this ratio.

102. After evaluation, what is the sign of $(-13)^5$?

103. Divide: $\dfrac{-25}{-5}$

104. How much did the temperature change if it went from $-4°F$ to $-17°F$?

105. Evaluate: $2\left(\dfrac{12}{-3}\right) + 3(5)$

106. Solve: $-10 = x + 1$

107. Solve: $-x = -12$

108. Divide: $\dfrac{0}{10}$

SECTION **10.2**

Introduction to Polynomials

Objectives

1 Know the vocabulary for polynomials.

2 Evaluate polynomials.

1 Know the vocabulary for polynomials.

Recall that an **algebraic term,** or simply a **term,** is a number or a product of a number and one or more variables, which may be raised to powers. Some examples of terms are

$$17, \quad 5x, \quad 6t^2, \quad \text{and} \quad -8z^3$$

The *coefficients* of these terms are $17, 5, 6$, and -8, in that order.

> **Polynomials**
>
> A **polynomial** is a single term or a sum of terms in which all variables have whole-number exponents and no variable appears in the denominator.

Some examples of polynomials are

$$141, \quad 8y^2, \quad 2x + 1, \quad 4y^2 - 2y + 3, \quad \text{and} \quad 7a^3 + 2a^2 - a - 1$$

The polynomial $8y^2$ has one term. The polynomial $2x + 1$ has two terms, $2x$ and 1. Since $4y^2 - 2y + 3$ can be written as $4y^2 + (-2y) + 3$, it is the sum of three terms, $4y^2, -2y$, and 3.

We classify some polynomials by the number of terms they contain. A polynomial with one term is called a **monomial.** A polynomial with two terms is called a **binomial.** A polynomial with three terms is called a **trinomial.** Some examples of these polynomials are shown in the table below.

Monomials	Binomials	Trinomials
$5x^2$	$2x - 1$	$5t^2 + 4t + 3$
$-6x$	$18a^2 - 4a$	$27x^3 - 6x + 2$
29	$-27z^4 + 7z^2$	$32r^2 + 7r - 12$

Self Check 1

Classify each polynomial as
a monomial, a binomial, or
a trinomial:
a. $8x^2 + 7$

b. $5x$

c. $x^2 - 2x - 1$

Now Try Problems 11, 13, and 17

EXAMPLE 1   Classify each polynomial as a monomial, a binomial, or a trinomial: **a.** $3x + 4$ **b.** $3x^2 + 4x - 12$ **c.** $25x^3$

Strategy We will count the number of terms in the polynomial.

WHY The number of terms determines the type of polynomial.

Solution
a. Since $3x + 4$ has two terms, it is a binomial.
b. Since $3x^2 + 4x - 12$ has three terms, it is a trinomial.
c. Since $25x^3$ has one term, it is a monomial.

The monomial $7x^3$ is called a **monomial of third degree** or a **monomial of degree 3,** because the variable occurs three times as a factor.

- $5x^2$ is a monomial of degree 2.   Because the variable occurs two times as a factor: $x^2 = x \cdot x$.
- $-8a^4$ is a monomial of degree 4.   Because the variable occurs four times as a factor: $a^4 = a \cdot a \cdot a \cdot a$.
- $\frac{1}{2}m^5$ is a monomial of degree 5.   Because the variable occurs five times as a factor: $m^5 = m \cdot m \cdot m \cdot m \cdot m$.
- 8 is a monomial of degree 0.   The degree of a nonzero constant is 0.

We define the degree of a polynomial by considering the degrees of each of its terms.

Degree of a Polynomial

The **degree of a polynomial** is equal to the highest degree of any term of the polynomial.

For example,

- $x^2 + 5x$ is a binomial of degree 2, because the degree of its term with largest degree (x^2) is 2.
- $4y^3 + 2y - 7$ is a trinomial of degree 3, because the degree of its term with largest degree $(4y^3)$ is 3.
- $\frac{1}{2}z + 3z^4 - 2z^2$ is a trinomial of degree 4, because the degree of its term with largest degree $(3z^4)$ is 4.

Self Check 2

Find the degree of each
polynomial:
a. $3p^3$

b. $17r^4 + 2r^8 - r$

c. $-2g^5 - 7g^6 + 12g^7$

Now Try Problems 23, 25, and 29

EXAMPLE 2   Find the degree of each polynomial:
a. $-2x + 4$ **b.** $5t^3 + t^4 - 7$ **c.** $3 - 9z + 6z^2 - z^3$

Strategy We will determine the degree of each term of the polynomial.

WHY The term with the highest degree gives the degree of the polynomial.

Solution
a. Since $-2x$ can be written as $-2x^1$, the degree of the term with largest degree is 1. Thus, the degree of the polynomial $-2x + 4$ is 1.

b. In $5t^3 + t^4 - 7$, the degree of the term with largest degree (t^4) is 4. Thus, the degree of the polynomial is 4.

c. In $3 - 9z + 6z^2 - z^3$, the degree of the term with largest degree $(-z^3)$ is 3. Thus, the degree of the polynomial is 3.

2 Evaluate polynomials.

When a number is substituted for the variable in a polynomial, the polynomial takes on a numerical value. Finding this value is called **evaluating the polynomial.**

EXAMPLE 3 Evaluate each polynomial for $x = 3$:

a. $3x - 2$ **b.** $-2x^2 + x - 3$

Strategy We will substitute the given value for each x in the polynomial and follow the order of operations rule.

WHY To *evaluate a polynomial* means to find its numerical value, once we know the value of its variable.

Solution

a. $3x - 2 = 3(3) - 2$ Substitute 3 for x.

$\qquad\quad = 9 - 2$ Multiply: 3(3) = 9.

$\qquad\quad = 7$ Subtract.

b. $-2x^2 + x - 3 = -2(3)^2 + 3 - 3$ Substitute 3 for x.

$\qquad\qquad\quad = -2(9) + 3 - 3$ Evaluate the exponential expression.

$\qquad\qquad\quad = -18 + 3 - 3$ Multiply: −2(9) = −18.

$\qquad\qquad\quad = -15 - 3$ Add: −18 + 3 = −15.

$\qquad\qquad\quad = -18$ Subtract: −15 − 3 = −15 + (−3) = −18.

Self Check 3

Evaluate each polynomial for $x = -1$:

a. $-2x^2 - 4$

b. $3x^2 - 4x + 1$

Now Try Problems 35 and 45

EXAMPLE 4 *Height of an Object* The polynomial $-16t^2 + 28t + 8$ gives the height (in feet) of an object t seconds after it has been thrown into the air. Find the height of the object after 1 second.

Strategy We will substitute 1 for t and evaluate the polynomial.

WHY The variable t represents the time since the object was thrown into the air.

Solution

To find the height at 1 second, we evaluate the polynomial for $t = 1$.

$-16t^2 + 28t + 8 = -16(1)^2 + 28(1) + 8$ Substitute 1 for t.

$\qquad\qquad\qquad = -16(1) + 28(1) + 8$ Evaluate the exponential expression.

$\qquad\qquad\qquad = -16 + 28 + 8$ Multiply: −16(1) = −16 and 28(1) = 28.

$\qquad\qquad\qquad = 12 + 8$ Add: −16 + 28 = 12.

$\qquad\qquad\qquad = 20$ Add.

At 1 second, the height of the object is 20 feet.

Self Check 4

Refer to Example 4. Find the height of the object after 2 seconds.

Now Try Problems 47 and 49

ANSWERS TO SELF CHECKS

1. a. binomial **b.** monomial **c.** trinomial **2. a.** 3 **b.** 8 **c.** 7

3. a. −6 **b.** 8 **4.** 0 ft

SECTION 10.2   STUDY SET

VOCABULARY

Fill in the blanks.

1. A _____ is a single term or a sum of terms in which all variables have whole number exponents and no variable appears in the denominator.

2. A polynomial with one term is called a _____.

3. A polynomial with three terms is called a _____.

4. A polynomial with two terms is called a _____.

CONCEPTS

5. How many terms does each of the following polynomials have?

   a. $x^2 + 3x$

   b. $3a^4$

   c. $-4r^2 + 9r + 11$

6. Fill in the blank so that $10c^{\square}$ has degree 3.

7. Fill in the blank. The degree of a polynomial is equal to the _____ degree of any term of the polynomial.

8. What is the degree of each term of the polynomial $4x^3 + x^2 - 7x + 5$?

NOTATION

Complete each solution.

9. Evaluate $3a^2 + 2a - 7$ for $a = 2$.

$$3a^2 + 2a - 7 = 3(\square)^2 + 2(\square) - 7$$
$$= 3(\square) + \square - 7$$
$$= 12 + 4 - 7$$
$$= \square - 7$$
$$= 9$$

10. Evaluate $-q^2 - 3q + 2$ for $q = -1$.

$$-q^2 - 3q + 2 = -(\square)^2 - 3(\square) + 2$$
$$= -(\square) - 3(-1) + 2$$
$$= -1 + \square + 2$$
$$= \square + 2$$
$$= 4$$

GUIDED PRACTICE

Classify each polynomial as a monomial, a binomial, or a trinomial. See Example 1.

11. $3x^2 - 4$

12. $5t^2 - t + 1$

13. $17e^4$

14. $x^2 + x + 7$

15. $25u^2$

16. $x^2 - 9$

17. $q^5 + q^2 + 1$

18. $4d^3 - 3d^2$

19. $81x^3 - 27$

20. $125m^3 - 8$

21. $4c^2 - 8c + 12$

22. $16n^4 - 8n^2 + n$

Find the degree of each polynomial. See Example 2.

23. $5x^3$

24. $3t^5 + 3t^2$

25. $2x^2 - 3x + 2$

26. $\frac{1}{2}p^4 - p^2$

27. $2m$

28. $7q - 5$

29. $25w^6 + 5w^7$

30. $p^6 - p^8$

31. $a^2 - 9$

32. $b^2 - 25$

33. $-m + 3m^4 + 5m^2$

34. $-r + 6r^8 - r^7$

Evaluate each polynomial for the given value. See Example 3.

35. $3x + 4$ for $x = 3$

36. $5n - 10$ for $n = 6$

37. $2x^2 + 4$ for $x = -1$

38. $9r^2 + 12$ for $r = -3$

39. $\frac{1}{2}x - 3$ for $x = -6$

40. $-\frac{1}{2}x^2 - 1$ for $x = 2$

41. $0.5t^3 - 1$ for $t = 4$

42. $0.75a^2 + 2.5a + 2$ for $a = 0$

43. $\frac{2}{3}b^2 - b + 1$ for $b = 3$

44. $\frac{3}{2}n^2 - n + 2$ for $n = 2$

45. $-2s^2 - 2s + 1$ for $s = -1$

46. $-4r^2 - 3r - 1$ for $r = -2$

APPLICATIONS

The height h (in feet) of a ball shot straight up with an initial velocity of 64 feet per second is given by the equation $h = -16t^2 + 64t$. Find the height of the ball after the given number of seconds.

47. 0 second

48. 1 second

49. 2 seconds

50. 4 seconds

51. The number of feet that a car travels before stopping depends on the driver's reaction time and the braking distance, as shown below. For one driver, the stopping distance d is given by the equation

from Campus to Careers
Police Officer

$$d = 0.04v^2 + 0.9v$$

where v is the velocity (speed) of the car. Find the stopping distance when the driver is traveling at 30 mph.

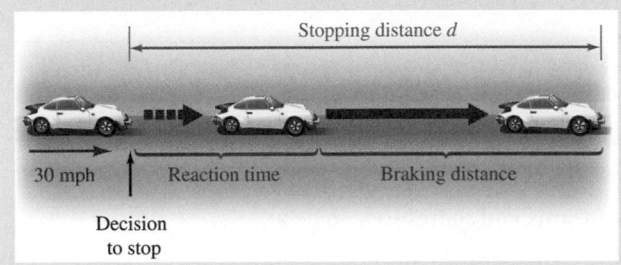

Stopping distance d

30 mph   Reaction time   Braking distance

Decision to stop

In Problems 52–54, refer to Problem 51. Then find the stopping distance for each of the following speeds.

52. 50 mph

53. 60 mph

54. 70 mph

WRITING

55. Explain how to find the degree of the polynomial $2x^3 + 5x^5 - 7x$.

56. Explain how to evaluate the polynomial $-2x^2 - 3$ for $x = 5$.

REVIEW

Perform the operations.

57. $\dfrac{2}{3} + \dfrac{4}{3}$

58. $\dfrac{36}{7} - \dfrac{23}{7}$

59. $\dfrac{5}{12} \cdot \dfrac{18}{5}$

60. $\dfrac{23}{25} \div \dfrac{46}{5}$

Solve each equation.

61. $x - 4 = 12$

62. $4z = 108$

63. $2(x - 3) = 6$

64. $3(a - 5) = 4(a + 9)$

SECTION 10.3
Adding and Subtracting Polynomials

Objectives

1 Add polynomials.

2 Subtract polynomials.

Polynomials can be added, subtracted, and multiplied just like numbers in arithmetic. In this section, we show how to find sums and differences of polynomials.

1 Add polynomials.

Recall that like terms have exactly the same variables and the same exponents. For example, the monomials

$3z^2$ and $-2z^2$ are like terms   Both have the same variable (z) with the same exponent (2).

However, the monomials

$7b^2$ and $8a^2$ are not like terms   They have different variables.

$32p^2$ and $25p^3$ are not like terms   The exponents of p are different.

Also recall that we use the distributive property in reverse to simplify a sum or difference of like terms. We **combine like terms** by adding their coefficients and keeping the same variables and exponents. For example,

$$2y + 5y = (2 + 5)y \qquad \text{and} \qquad -3x^2 + 7x^2 = (-3 + 7)x^2$$
$$= 7y \qquad\qquad\qquad\qquad = 4x^2$$

These examples suggest the following rule.

> **Adding Polynomials**
>
> To add polynomials, combine their like terms.

Self Check 1

Add: $7y^3 + 12y^3$

Now Try Problem 15

EXAMPLE 1   Add: $5x^3 + 7x^3$

Strategy We will use the distributive property in reverse and add the coefficients of the terms.

WHY $5x^3$ and $7x^3$ are like terms and therefore can be added.

Solution

$$5x^3 + 7x^3 = 12x^3 \quad \text{Think: } (5 + 7)x^3 = 12x^3.$$

Self Check 2

Add:

$\dfrac{1}{9}a^3 + \dfrac{2}{9}a^3 + \dfrac{5}{9}a^3$

Now Try Problem 21

EXAMPLE 2   Add: $\dfrac{3}{2}t^2 + \dfrac{5}{2}t^2 + \dfrac{7}{2}t^2$

Strategy We will use the distributive property in reverse and add the coefficients of the terms.

WHY $\frac{3}{2}t^2$, $\frac{5}{2}t^2$, and $\frac{7}{2}t^2$ are like terms and therefore can be added.

Solution
Since the three monomials are like terms, we add the coefficients and keep the variables and exponents.

$$\frac{3}{2}t^2 + \frac{5}{2}t^2 + \frac{7}{2}t^2 = \left(\frac{3}{2} + \frac{5}{2} + \frac{7}{2}\right)t^2$$

$$= \frac{15}{2}t^2 \qquad \text{To add the fractions, add the numerators and keep the denominator: } 3 + 5 + 7 = 15.$$

To add two polynomials, we write a + sign between them and combine like terms.

Self Check 3

Add:
$5y - 2$ and $-3y + 7$

Now Try Problem 25

EXAMPLE 3   Add: $2x + 3$ and $7x - 1$

Strategy We will reorder and regroup to get the like terms together. Then we will combine like terms.

WHY To add polynomials means to combine their like terms.

Solution

$$(2x + 3) + (7x - 1) \qquad \text{Write a + sign between the binomials.}$$

$$= (2x + 7x) + (3 - 1) \qquad \text{Use the associative and commutative properties to group like terms together.}$$

$$= 9x + 2 \qquad \text{Combine like terms.}$$

The binomials in Example 3 can be added by writing the polynomials so that like terms are in columns.

$$
\begin{array}{r}
2x + 3 \\
+\ 7x - 1 \\
\hline
9x + 2
\end{array}
\qquad \text{Add the like terms, one column at a time.}
$$

EXAMPLE 4   Add: $(5x^2 - 2x + 4) + (3x^2 - 5)$

Strategy We will combine the like terms of the trinomial and binomial.

WHY To add polynomials, we combine like terms.

Solution

$(5x^2 - 2x + 4) + (3x^2 - 5)$

$\quad = (5x^2 + 3x^2) + (-2x) + (4 - 5)$ Use the associative and commutative properties to group like terms together.

$\quad = 8x^2 - 2x - 1$ Combine like terms. ■

The polynomials in Example 4 can be added by writing the polynomials so that like terms are in columns.

$$\begin{array}{r} 5x^2 - 2x + 4 \\ +\ 3x^2 \quad\ \ - 5 \\ \hline 8x^2 - 2x - 1 \end{array}$$ Add the like terms, one column at a time.

Self Check 4

Add:
$(2b^2 - 4b) + (b^2 + 3b - 1)$

Now Try Problem 29

EXAMPLE 5   Add: $(3.7x^2 + 4x - 2) + (7.4x^2 - 5x + 3)$

Strategy We will combine the like terms of the two trinomials.

WHY To add polynomials, we combine like terms.

Solution

$(3.7x^2 + 4x - 2) + (7.4x^2 - 5x + 3)$

$\quad = (3.7x^2 + 7.4x^2) + (4x - 5x) + (-2 + 3)$ Use the associative and commutative properties to group like terms together.

$\quad = 11.1x^2 - x + 1$ Combine like terms. ■

The trinomials in Example 5 can be added by writing them so that like terms are in columns.

$$\begin{array}{r} 3.7x^2 + 4x - 2 \\ +\ \ 7.4x^2 - 5x + 3 \\ \hline 11.1x^2 -\ \ x + 1 \end{array}$$ Add the like terms, one column at a time.

Self Check 5

Add:
$(s^2 + 1.2s - 5) + (3s^2 - 2.5s + 4)$

Now Try Problem 31

2 Subtract polynomials.

To subtract one monomial from another, we add the opposite of the monomial that is to be subtracted. In symbols, $x - y = x + (-y)$.

EXAMPLE 6   Subtract: $8x^2 - 3x^2$

Strategy We will add the opposite of $3x^2$ to $8x^2$.

WHY To subtract monomials, we add the oppostie of the monomial that is to be subtracted.

Solution

$8x^2 - 3x^2 = 8x^2 + (-3x^2)$ Add the opposite of $3x^2$.

$\quad = 5x^2$ Add the coefficients and keep the same variable and exponent. Think: $[8 + (-3)]x^2 = 5x^2$ ■

Self Check 6

Subtract: $6y^3 - 9y^3$

Now Try Problem 39

Recall from Chapter 1 that we can use the distributive property to find the opposite of several terms enclosed within parentheses. For example, we consider $-(2a^2 - a + 9)$.

$$-(2a^2 - a + 9) = -\mathbf{1}(2a^2 - a + 9)$$ Replace the − symbol in front of the parentheses with −1.

$$= -2a^2 + a - 9$$ Use the distributive property to remove parentheses.

This example illustrates the following method of subtracting polynomials.

Subtracting Polynomials

To subtract two polynomials, change the signs of the terms of the polynomial being subtracted, drop the parentheses, and combine like terms.

Self Check 7

Subtract:
$(3.3a - 5) - (7.8a + 2)$

Now Try **Problem 43**

EXAMPLE 7   Subtract: $(3x - 4.2) - (5x + 7.2)$

Strategy We will change the signs of the terms of $5x + 7.2$, drop the parentheses, and combine like terms.

WHY This is the method for subtracting two polynomials.

Solution

$$(3x - 4.2) - (5x + 7.2)$$
$$= 3x - 4.2 - 5x - 7.2$$ Change the signs of each term of $5x + 7.2$ and drop the parentheses.

$$= -2x - 11.4$$ Combine like terms: Think: $(3 - 5)x = -2x$ and $(-4.2 - 7.2) = -11.4$.

The binomials in Example 7 can be subtracted by writing them so that like terms are in columns.

$$
\begin{array}{r}
3x - 4.2 \\
-(5x + 7.2)
\end{array}
\quad\longrightarrow\quad
\begin{array}{r}
3x - 4.2 \\
+ \underline{-5x - 7.2} \\
-2x - 11.4
\end{array}
$$
     Change signs and add, column by column.

Self Check 8

Subtract:
$(5y^2 - 4y + 2) - (3y^2 + 2y - 1)$

Now Try **Problem 47**

EXAMPLE 8   Subtract: $(3x^2 - 4x - 6) - (2x^2 - 6x + 12)$

Strategy We will change the signs of the terms of $2x^2 - 6x + 12$, drop the parentheses, and combine like terms.

WHY This is the method for subtracting two polynomials.

Solution

$$(3x^2 - 4x - 6) - (2x^2 - 6x + 12)$$
$$= 3x^2 - 4x - 6 - 2x^2 + 6x - 12$$ Change the signs of each term of $2x^2 - 6x + 12$ and drop the parentheses.

$$= x^2 + 2x - 18$$ Combine like terms: Think: $(3 - 2)x^2 = x^2$, $(-4 + 6)x = 2x$, and $(-6 - 12) = -18$.

The trinomials in Example 8 can be subtracted by writing them so that like terms are in columns.

$$3x^2 - 4x - 6$$
$$-(2x^2 - 6x + 12) \longrightarrow + \frac{-2x^2 + 6x - 12}{x^2 + 2x - 18}$$

Change signs and add, column by column.

ANSWERS TO SELF CHECKS

1. $19y^3$ 2. $\frac{8}{9}a^3$ 3. $2y + 5$ 4. $3b^2 - b - 1$ 5. $4s^2 - 1.3s - 1$ 6. $-3y^3$
7. $-4.5a - 7$ 8. $2y^2 - 6y + 3$

SECTION 10.3   STUDY SET

VOCABULARY

Fill in the blanks.

1. If two algebraic terms have exactly the same variables and exponents, they are called _____ terms.

2. Because the exponents on x are different, $3x^3$ and $3x^2$ are _____ terms.

CONCEPTS

Fill in the blanks.

3. To add two monomials, we add the _____ and keep the same _____ and exponents.

4. To subtract one monomial from another, we add the _____ of the monomial that is to be subtracted.

Determine whether the monomials are like terms. If they are, combine them.

5. $3y, 4y$ 6. $3x^2, 5x^2$

7. $3x, 3y$ 8. $3x^2, 6x$

9. $3x^3, 4x^3, 6x^3$ 10. $-2y^4, -6y^4, 10y^4$

11. $-5x^2, 13x^2, 7x^2$ 12. $23, 12x, 25x$

NOTATION

Complete each solution.

13. $(3x^2 + 2x - 5) + (2x^2 - 7x)$

$= (3x^2 + \boxed{}) + (2x - \boxed{}) + (-5)$

$= \boxed{} + (-5x) - 5$

$= 5x^2 - 5x - 5$

14. $(3x^2 + 2x - 5) - (2x^2 - 7x)$

$= 3x^2 + 2x - 5 \boxed{} 2x^2 \boxed{} 7x$

$= (\boxed{} - \boxed{}) + (2x + 7x) - 5$

$= x^2 + 9x - 5$

GUIDED PRACTICE

Add. See Example 1.

15. $4y + 5y$ 16. $-2x + 3x$

17. $8t^2 + 4t^2$ 18. $15x^2 + 10x^2$

Add. See Example 2.

19. $\frac{1}{8}a + \frac{3}{8}a + \frac{5}{8}a$ 20. $\frac{1}{4}b + \frac{3}{4}b + \frac{1}{4}b$

21. $\frac{2}{3}c^2 + \frac{1}{3}c^2 + \frac{2}{3}c^2$ 22. $\frac{4}{9}d^3 + \frac{1}{9}d^3 + \frac{3}{9}d^3$

Add. See Example 3.

23. $3x + 7$ and $4x - 3$

24. $2y - 3$ and $4y + 7$

25. $2x^2 + 3$ and $5x^2 - 10$

26. $-4a^2 + 1$ and $5a^2 - 1$

Add. See Example 4.

27. $(5x^3 - 42x) + (7x^3 - 107x)$

28. $(-43a^3 + 25a) + (58a^3 - 10a)$

29. $(3x^2 + 2x - 4) + (5x^2 - 17)$

30. $(5a^2 - 2a) + (-2a^2 + 3a + 4)$

Add. See Example 5.

31. $(2.5a^2 + 3a - 9) + (3.6a^2 + 7a - 10)$

32. $(1.9b^2 - 4b + 10) + (3.7b^2 - 3b - 11)$

33. $(3n^2 - 5.8n + 7) + (-n^2 + 5.8n - 2)$

34. $(-3t^2 - t + 3.4) + (3t^2 + 2t - 1.8)$

35. $3x^2 + 4x + 5$ 36. $2x^2 - 3x + 5$
   $+ \underline{2x^2 - 3x + 6}$ $+ \underline{-4x^2 - x - 7}$

37. $-3x^2 \qquad - 7$ 38. $4x^2 - 4x + 9$
   $+ \underline{-4x^2 - 5x + 6}$ $+ \underline{\qquad 9x - 3}$

Subtract. See Example 6.

39. $32u^3 - 16u^3$

40. $25y^2 - 7y^2$

41. $18x^5 - 11x^5$

42. $17x^6 - 22x^6$

Subtract. See Example 7.

43. $(4.5a + 3.7) - (2.9a - 4.3)$

44. $(5.1b - 7.6) - (3.3b + 5.9)$

45. $(7.2x^2 - 3.1x) - (9.4x^2 + 6.8x)$

46. $(3.7y^3 + 9.8y^2) - (2.4y^3 - 1.1y^2)$

Subtract. See Example 8.

47. $(2b^2 + 3b - 5) - (2b^2 - 4b - 9)$

48. $(3a^2 - 2a + 4) - (a^2 - 3a + 7)$

49. $(5p^2 - p + 71) - (4p^2 + p + 71)$

50. $(10m^2 - m - 19) - (6m^2 + m - 19)$

51. $3x^2 + 4x - 5$
$- (-2x^2 - 2x + 3)$

52. $3y^2 - 4y + 7$
$- (6y^2 - 6y - 13)$

53. $-2x^2 - 4x + 12$
$- (10x^2 + 9x - 24)$

54. $25x^3 - 45x^2 + 31x$
$- (12x^3 + 27x^2 - 17x)$

TRY IT YOURSELF

Perform the operations.

55. $(30x^2 - 4) - (11x^2 + 1)$

56. $(5x^3 - 8) - (2x^3 + 5)$

57. $(7y^2 + 5y) + (y^2 - y - 2)$

58. $(4p^2 - 4p + 5) + (6p - 2)$

59. $(3x^2 - 3x - 2) + (3x^2 + 4x - 3)$

60. $(4c^2 + 3c - 2) + (3c^2 + 4c + 2)$

61. $(m^2 - m - 5) - (m^2 + 5.5m - 75)$

62. $(3.7y^2 - 5) - (2y^2 - 3.1y + 4)$

63. $(t^2 - 4.5t + 5) - (2t^2 - 3.1t - 1)$

64. $(a^4 - 5.1a^3 + 1.1a) - (3a^4 - 6.7a^3 + 0.1a)$

65. $-3x^2 + 4x + 25.4$
$+ \quad 5x^2 - 3x - 12.5$

66. $-6x^3 - 4.2x^2 + 7$
$+ -7x^3 + 9.7x^2 - 21$

67. $3s^2 + 4s^2 + 7s^2$

68. $-2a^3 + 7a^3 - 3a^3$

69. $\frac{1}{3}b^4 + \frac{2}{3}b^4 - \frac{5}{3}b^4$

70. $\frac{4}{5}n^6 - \frac{1}{5}n^6 - \frac{2}{5}n^6$

71. $z^3 + 6z^2 - 7z + 16$
$+9z^3 - 6z^2 + 8z - 18$

72. $3x^3 + 4x^2 - 3x + 5$
$+3x^3 - 4x^2 - x - 7$

73. $(-4h^3 + 5h^2 + 15) - (h^3 - 15)$

74. $(-c^5 + 5c^4 - 12) - (2c^5 - c^4)$

75. $0.6x^3 + 0.8x^4 + 0.7x^3 + (-0.8x^4)$

76. $1.9m^4 - 2.4m^6 - 3.7m^4 + 2.8m^6$

77. $(12.1h^3 - 9.9h^2 + 9.5) + (7.3h^3 - 1.2h^2 - 10.1)$

78. $(7.1a^2 + 2.2a - 5.8) - (3.4a^2 - 3.9a + 11.8)$

79. $4x^3 - 3x + 10$
$- (5x^3 - 4x - 4)$

80. $3x^3 + 4x^2 + 12$
$- (-4x^3 + 6x^2 - 3)$

APPLICATIONS

81. BILLIARDS Billiard tables vary in size, but all tables are twice as long as they are wide.

   a. If the billiard table is x feet wide, write an expression that represents its length.

   b. Write an expression that represents the perimeter of the table.

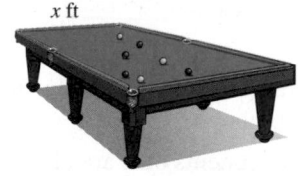

x ft

82. GARDENING Find a polynomial that represents the length of the wooden handle of the shovel.

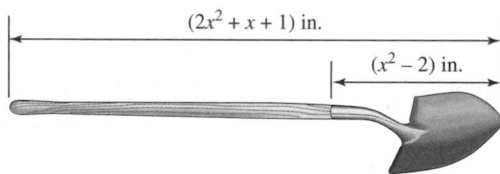

$(2x^2 + x + 1)$ in.

$(x^2 - 2)$ in.

83. READING BLUEPRINTS

   a. What is the difference in the length and width of the one-bedroom apartment shown below?

   b. Find the perimeter of the apartment.

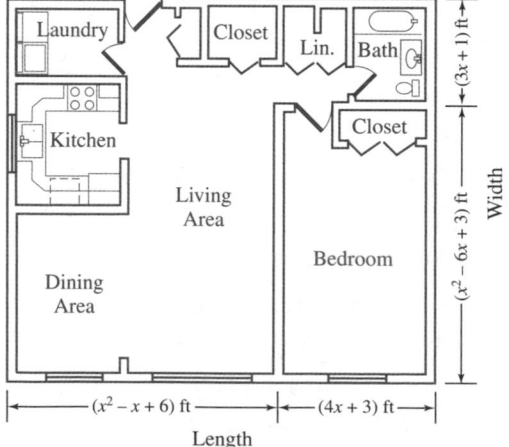

84. PIÑATAS Find the polynomial that represents the length of the rope used to hold up the piñata.

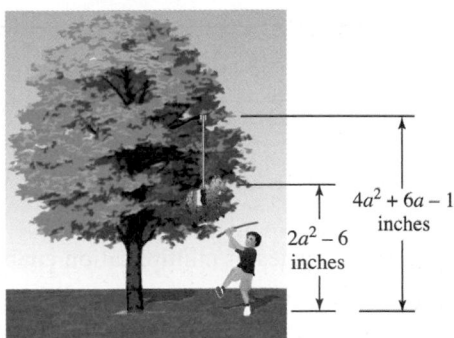

$4a^2 + 6a - 1$ inches

$2a^2 - 6$ inches

WRITING

85. What are *like terms*?

86. Explain how to add two polynomials.

87. Explain how to subtract two polynomials.

88. When two binomials are added, is the result always a binomial? Explain.

REVIEW

89. BASKETBALL SHOES Use the following information to find how much lighter the Kevin Garnett shoe is than the Michael Jordan shoe.

Nike Air Garnett III	Air Jordan XV
Synthetic fade mesh and leather.	Full grain leather upper with woven pattern.
Sizes $6\frac{1}{2}$–18	Sizes $6\frac{1}{2}$–18
Weight: 13.8 oz	Weight: 14.6 oz

90. AEROBICS The number of calories burned when doing step aerobics depends on the step height. How many more calories are burned during a 10-minute workout using an 8-inch step instead of a 4-inch step?

Step height (in.)	Calories burned per minute
4	4.5
6	5.5
8	6.4
10	7.2

Source: *Reebok Instructor News* (Vol. 4, No. 3, 1991)

SECTION 10.4

Multiplying Polynomials

Objectives

1 Multiply monomials.

2 Multiply a polynomial by a monomial.

3 Multiply binomials.

4 Multiply polynomials.

We now discuss how to multiply polynomials. We will begin with the simplest case—finding the product of two monomials.

1 Multiply monomials.

To multiply $4x^2$ by $2x^3$, we use the commutative and associative properties of multiplication to reorder and regroup the factors.

$$(4x^2)(2x^3) = (4 \cdot 2)(x^2 \cdot x^3) \quad \text{Group the coefficients together and the variables together.}$$

$$= 8x^5 \quad \text{Simplify: } x^2 \cdot x^3 = x^{2+3} = x^5.$$

This example suggests the following rule.

Multiplying Two Monomials

To multiply two monomials, multiply the numerical factors (the coefficients) and then multiply the variable factors.

Self Check 1

Multiply: $-7a^3 \cdot 2a^5$

Now Try Problems 13 and 15

EXAMPLE 1   Multiply: **a.** $3y \cdot 6y$ **b.** $-3x^5(2x^5)$

Strategy We will multiply the numerical factors and then multiply the variable factors.

WHY The commutative and associative properties of multiplication enable us to reorder and regroup factors.

Solution

a. $3y \cdot 6y = (3 \cdot 6)(y \cdot y)$ Group the numerical factors and group the variables.

$\qquad\qquad = 18y^2$ Multiply: $3 \cdot 6 = 18$ and $y \cdot y = y^2$.

b. $(-3x^5)(2x^5) = (-3 \cdot 2)(x^5 \cdot x^5)$ Group the numerical factors and group the variables.

$\qquad\qquad\quad = -6x^{10}$ Multiply: $-3 \cdot 2 = -6$ and $x^5 \cdot x^5 = x^{5+5} = x^{10}$.

2   Multiply a polynomial by a monomial.

To find the product of a polynomial and a monomial, we use the distributive property. To multiply $x + 4$ by $3x$, for example, we proceed as follows:

$$3x(x + 4) = 3x(x) + 3x(4)$$ Use the distributive property.

$$\qquad\qquad = 3x^2 + 12x$$ Multiply the monomials: $3x(x) = 3x^2$ and $3x(4) = 12x$.

The results of this example suggest the following rule.

Multiplying Polynomials by Monomials

To multiply a polynomial by a monomial, multiply each term of the polynomial by the monomial.

Self Check 2

Multiply:
a. $3y(5y^3 - 4y)$

b. $5x(3x^2 - 2x + 3)$

Now Try Problems 17 and 19

EXAMPLE 2   Multiply: **a.** $2a^2(3a^2 - 4a)$ **b.** $8x(3x^2 + 2x - 3)$

Strategy We will multiply each term of the polynomial by the monomial.

WHY We use the distributive property to multiply a monomial and a polynomial.

Solution

a. $2a^2(3a^2 - 4a)$

$\qquad = 2a^2(3a^2) - 2a^2(4a)$ Use the distributive property.

$\qquad = 6a^4 - 8a^3$ Multiply: $2a^2(3a^2) = 6a^4$ and $2a^2(4a) = 8a^3$.

b. $8x(3x^2 + 2x - 3)$

$\qquad = 8x(3x^2) + 8x(2x) - 8x(3)$ Use the distributive property.

$\qquad = 24x^3 + 16x^2 - 24x$ Multiply: $8x(3x^2) = 24x^3$, $8x(2x) = 16x^2$, and $8x(3) = 24x$.

3 **Multiply binomials.**

The distributive property can also be used to multiply binomials. For example, to multiply $2a + 4$ and $3a + 5$, we think of $2a + 4$ as a single quantity and distribute it over each term of $3a + 5$.

$$(2a + 4)(3a + 5) = (2a + 4)3a + (2a + 4)5$$
$$= (2a + 4)3a + (2a + 4)5$$
$$= (2a)3a + (4)3a + (2a)5 + (4)5 \qquad \text{Distribute the multiplication by 3a and by 5.}$$
$$= 6a^2 + 12a + 10a + 20 \qquad \text{Multiply the monomials.}$$
$$= 6a^2 + 22a + 20 \qquad \text{Combine like terms.}$$

In the third line of the solution, notice that each term of $3a + 5$ has been multiplied by each term of $2a + 4$. This example suggests the following rule.

Multiplying Binomials

To multiply two binomials, multiply each term of one binomial by each term of the other binomial, and then combine like terms.

We can use a shortcut method, called the **FOIL method,** to multiply binomials. FOIL is an acronym for **F**irst terms, **O**uter terms, **I**nner terms, **L**ast terms. To use the FOIL method to multiply $2a + 4$ by $3a + 5$, we

1. multiply the **F**irst terms $2a$ and $3a$ to obtain $6a^2$,
2. multiply the **O**uter terms $2a$ and 5 to obtain $10a$,
3. multiply the **I**nner terms 4 and $3a$ to obtain $12a$, and
4. multiply the **L**ast terms 4 and 5 to obtain 20.

Then we simplify the resulting polynomial, if possible.

$$(2a + 4)(3a + 5) = 2a(3a) + 2a(5) + 4(3a) + 4(5)$$

$$= 6a^2 + 10a + 12a + 20 \qquad \text{Multiply the monomials.}$$
$$= 6a^2 + 22a + 20 \qquad \text{Combine like terms.}$$

The Language of Algebra   An *acronym* is an abbreviation of several words in such a way that the abbreviation itself forms a word. The *acronym* FOIL helps us remember the order to follow when multiplying two binomials: First, Outer, Inner, Last.

EXAMPLE 3   Multiply: **a.** $(x + 5)(x + 7)$ **b.** $(3x + 4)(2x - 3)$

Strategy We will use the FOIL method.

WHY In each case we are to find the product of two binomials, and the FOIL method is a shortcut for multiplying two binomials.

Self Check 3

Multiply:
a. $(y + 3)(y + 1)$
b. $(2a - 1)(3a + 2)$

Now Try Problems 21 and 23

Solution

a.
$$(x + 5)(x + 7) = x(x) + x(7) + 5(x) + 5(7)$$
$$= x^2 + 7x + 5x + 35 \quad \text{Multiply the monomials.}$$
$$= x^2 + 12x + 35 \quad \text{Combine like terms.}$$

b.
$$(3x + 4)(2x - 3) = 3x(2x) + 3x(-3) + 4(2x) + 4(-3)$$
$$= 6x^2 - 9x + 8x - 12 \quad \text{Multiply the monomials.}$$
$$= 6x^2 - x - 12 \quad \text{Combine like terms.}$$

Self Check 4

Find: $(5x + 4)^2$

Now Try Problem 25

EXAMPLE 4   Find: $(5x - 4)^2$

Strategy We will write the base, $5x - 4$, as a factor twice, and perform the multiplication.

WHY In the expression $(5x - 4)^2$, the binomial $5x - 4$ is the base and 2 is the exponent.

Solution

$$(5x - 4)^2 = (5x - 4)(5x - 4) \quad \text{Write the base as a factor twice.}$$

$$= 5x(5x) + 5x(-4) + (-4)(5x) + (-4)(-4)$$
$$= 25x^2 - 20x - 20x + 16 \quad \text{Multiply the monomials.}$$
$$= 25x^2 - 40x + 16 \quad \text{Combine like terms.}$$

> **Caution!** A common error when squaring a binomial is to square only its
> first and second terms. For example, it is incorrect to write
>
> $$(5x - 4)^2 = (5x)^2 - (4)^2$$
> $$= 25x^2 - 16$$
>
> The correct answer is $25x^2 - 40x + 16$.

4 Multiply polynomials.

To develop a general rule for multiplying any two polynomials, we will find the product of $2x + 3$ and $3x^2 + 3x + 5$. In the solution, the distributive property is used four times.

$$(2x + 3)(3x^2 + 3x + 5)$$
$$= (2x + 3)3x^2 + (2x + 3)3x + (2x + 3)5 \quad \text{Distribute.}$$

$$= (2x + 3)3x^2 + (2x + 3)3x + (2x + 3)5$$
$$= (2x)3x^2 + (3)3x^2 + (2x)3x + (3)3x + (2x)5 + (3)5 \quad \text{Distribute.}$$
$$= 6x^3 + 9x^2 + 6x^2 + 9x + 10x + 15 \quad \text{Multiply the monomials.}$$
$$= 6x^3 + 15x^2 + 19x + 15 \quad \text{Combine like terms.}$$

In the third line of the solution, note that each term of $3x^2 + 3x + 5$ has been multiplied by each term of $2x + 3$. This example suggests the following rule.

Multiplying Polynomials

To multiply two polynomials, multiply each term of one polynomial by each term of the other polynomial, and then combine like terms.

EXAMPLE 5 Multiply: $(7y + 3)(6y^2 - 8y + 1)$

Strategy We will multiply each term of the trinomial, $6y^2 - 8y + 1$, by each term of the binomial, $7y + 3$.

WHY To multiply two polynomials, we must multiply each term of one polynomial by each term of the other polynomial.

Solution

$$(7y + 3)(6y^2 - 8y + 1)$$

$$= 7y(6y^2) + 7y(-8y) + 7y(1) + 3(6y^2) + 3(-8y) + 3(1)$$
$$= 42y^3 - 56y^2 + 7y + 18y^2 - 24y + 3 \quad \text{Multiply the monomials.}$$
$$= 42y^3 - 38y^2 - 17y + 3 \quad \text{Combine like terms.}$$

Caution! The FOIL method cannot be applied here—only to products of two binomials.

Self Check 5

Multiply:
$(3a^2 - 1)(2a^4 - a^2 - a)$

Now Try Problem 29

It is often convenient to multiply polynomials using a **vertical form** similar to that used to multiply whole numbers.

Success Tip Multiplying two polynomials in vertical form is much like multiplying two whole numbers in arithmetic.

$$
\begin{array}{r}
347 \\
\times\ \ 25 \\
\hline
1\,735 \\
+\ 6\,940 \\
\hline
8{,}675
\end{array}
$$

EXAMPLE 6 Multiply using vertical form:

a. $(3a^2 - 4a + 7)(2a + 5)$ **b.** $(6y^3 - 5y + 4)(-4y^2 - 3)$

Strategy First, we will write one polynomial underneath the other and draw a horizontal line beneath them. Then, we will multiply each term of the upper polynomial by each term of the lower polynomial.

WHY *Vertical form* means to use an approach similar to that used in arithmetic to multiply two whole numbers.

Self Check 6

Multiply using vertical form:
a. $(3x + 2)(2x^2 - 4x + 5)$
b. $(-2x^2 + 3)(2x^2 - 4x - 1)$

Now Try Problem 33

Solution

a. Multiply:

$$
\begin{array}{r}
3a^2 - 4a + 7 \\
\times \qquad 2a + 5 \\
\hline
15a^2 - 20a + 35 \\
6a^3 - 8a^2 + 14a \\
\hline
6a^3 + 7a^2 - \ 6a + 35
\end{array}
$$

    Multiply $3a^2 - 4a + 7$ by 5.

    Multiply $3a^2 - 4a + 7$ by $2a$.

    In each column, combine like terms.

b. With this method, it is often necessary to leave a space for a missing term to vertically align like terms.

Multiply:

$$
\begin{array}{r}
6y^3 - 5y + 4 \\
\times \qquad -4y^2 - 3 \\
\hline
-18y^3 \qquad\quad + 15y - 12 \\
-24y^5 + 20y^3 - 16y^2 \\
\hline
-24y^5 + \ 2y^3 - 16y^2 + 15y - 12
\end{array}
$$

    Multiply $6y^3 - 5y + 4$ by -3.

    Multiply $6y^3 - 5y + 4$ by $-4y^2$.

    Leave a space for any missing powers of y.

    In each column, combine like terms.

ANSWERS TO SELF CHECKS

1. $-14a^8$ **2. a.** $15y^4 - 12y^2$ **b.** $15x^3 - 10x^2 + 15x$ **3. a.** $y^2 + 4y + 3$ **b.** $6a^2 + a - 2$
4. $25x^2 + 40x + 16$ **5.** $6a^6 - 5a^4 - 3a^3 + a^2 + a$ **6. a.** $6x^3 - 8x^2 + 7x + 10$
b. $-4x^4 + 8x^3 + 8x^2 - 12x - 3$

SECTION 10.4   STUDY SET

VOCABULARY

Fill in the blanks.

1. $(2x^3)(3x^4)$ is the product of two _____.

2. $(2a - 4)(3a + 5)$ is the product of two _____.

3. In the acronym FOIL, F stands for ____ terms, O for ____ terms, I for ____ terms, and L for ____ terms.

4. $(2a - 4)(3a^2 + 5a - 1)$ is the product of a _____ and a _____.

CONCEPTS

Fill in the blanks.

5. To multiply two polynomials, multiply ____ term of one polynomial by ____ term of the other polynomial, and then combine like terms.

6. Label each arrow using one of the letters F, O, I, or L. Then fill in the blanks.

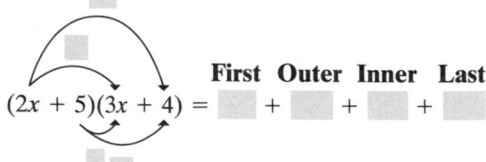

First Outer Inner Last

$(2x + 5)(3x + 4) = \boxed{} + \boxed{} + \boxed{} + \boxed{}$

7. Simplify each polynomial by combining like terms.

   a. $6x^2 - 8x + 9x - 12$

   b. $5x^4 + 3x^2 + 5x^2 + 3$

8. a. Add: $(x - 4) + (x + 8)$

   b. Subtract: $(x - 4) - (x + 8)$

   c. Multiply: $(x - 4)(x + 8)$

NOTATION

Complete each solution.

9. $(9n^3)(8n^2) = (9 \cdot \boxed{})(\boxed{} \cdot n^2) = \boxed{}$

10. $7x(3x^2 - 2x + 5) = \boxed{}(3x^2) - \boxed{}(2x) + \boxed{}(5)$

                  $= \boxed{} - 14x^2 + 35x$

11. $(2x + 5)(3x - 2) = 2x(3x) - \boxed{}(2) + \boxed{}(3x) - \boxed{}(2)$

                  $= 6x^2 - \boxed{} + \boxed{} - 10$

                  $= 6x^2 + \boxed{} - 10$

12.

$$
\begin{array}{r}
3x^2 + \ 4x - 2 \\
\times \qquad\quad 2x + 3 \\
\hline
\boxed{} + 12x - 6 \\
6x^3 + \ 8x^2 - \ 4x \\
\hline
\boxed{} + 17x^2 + \boxed{} - 6
\end{array}
$$

GUIDED PRACTICE

Multiply. See Example 1.

13. $(3x^2)(4x^3)$

14. $(-2a^3)(3a^2)$

15. $(3b^2)(-2b)$

16. $(3y)(-y^4)$

Multiply. See Example 2.

17. $2x^2(3x^2 + x)$

18. $4b^3(2b^2 - 2b)$

19. $2x(3x^2 + 4x - 7)$

20. $3y(2y^2 - 7y - 8)$

Multiply. See Example 3.

21. $(a + 4)(a + 5)$

22. $(y - 3)(y + 5)$

23. $(3x - 2)(x + 4)$

24. $(t + 4)(2t - 3)$

Square each binomial. See Example 4.

25. $(2x + 3)^2$

26. $(2y + 5)^2$

27. $(9b - 2)^2$

28. $(7m - 2)^2$

Multiply. See Example 5.

29. $(2x + 1)(3x^2 - 2x + 1)$

30. $(x + 2)(2x^2 + x - 3)$

31. $(x - 1)(x^2 + x + 1)$

32. $(x + 2)(x^2 - 2x + 4)$

Multiply. See Example 6.

33. $x^2 - x + 1$
         $\times \underline{\qquad x + 1}$

34. $4x^2 - 2x + 1$
         $\times \underline{\qquad 2x + 1}$

35. $4x^2 + 3x - 4$
         $\times \underline{\qquad 3x + 2}$

36. $5r^2 + r + 6$
         $\times \underline{\qquad 2r - 1}$

TRY IT YOURSELF

Perform the operations.

37. $(2a + 4)(3a - 5)$

38. $(2b - 1)(3b + 4)$

39. $-p(2p^2 - 3p + 2)$

40. $-2t(t^2 - t + 1)$

41. $(-2x^2)(3x^3)$

42. $(-7x^3)(-3x^3)$

43. $4x + 3$
         $\times \underline{\ x + 2}$

44. $5r + 6$
         $\times \underline{\ 2r - 1}$

45. $(2x - 3)^2$

46. $(2y - 5)^2$

47. $3q^2(q^2 - 2q + 7)$

48. $4v^3(-2v^2 + 3v - 1)$

49. $\left(-\dfrac{2}{3}y^5\right)\left(\dfrac{3}{4}y^2\right)$

50. $\left(\dfrac{2}{5}r^4\right)\left(\dfrac{3}{5}r^2\right)$

51. $(x + 2)(x^2 - 3x + 1)$

52. $(x + 3)(x^2 + 3x + 2)$

53. $2a^2 + 3a + 1$
         $\times \underline{\ 3a^2 - 2a + 4}$

54. $3y^2 + 2y - 4$
         $\times \underline{\ 2y^2 - 4y + 3}$

55. $(x + 6)(x^3 + 5x^2 - 4x - 4)$

56. $(x - 8)(x^3 - 4x^2 - 2x - 2)$

57. $(3n + 1)(3n - 1)$ **58.** $(5a + 4)(5a - 4)$

59. $(r^2 - r + 3)(r^2 - 4r - 5)$

60. $(w^2 + w - 9)(w^2 - w + 3)$

61. $(5t - 1)^2$ **62.** $(6a - 3)^2$

63. $3x(x - 2)$ **64.** $4y(y + 5)$

APPLICATIONS

65. GEOMETRY Find a polynomial that represents the area of the rectangle (*Hint:* Recall that the area of a rectangle is the product of its length and width).

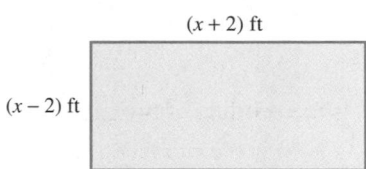
$(x + 2)$ ft
$(x - 2)$ ft

66. SAILING The height h of the triangular sail is $4x$ feet, and the base b is $(3x - 2)$ feet. Find a polynomial that represents the area of the sail. (*Hint:* The area of a triangle is given by the formula $A = \frac{1}{2}bh$.)

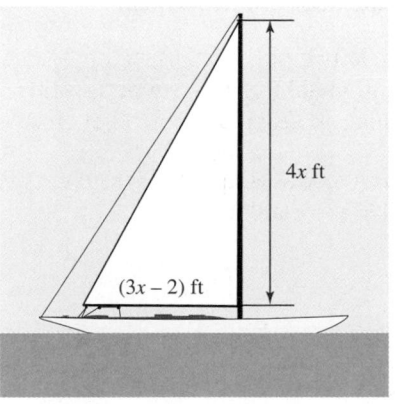
$4x$ ft
$(3x - 2)$ ft

67. STAMPS Find a polynomial that represents the area of the stamp.

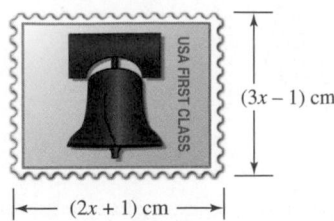

(3x − 1) cm

(2x + 1) cm

68. PARKING Find a polynomial that represents the total area of the van-accessible parking space and its access aisle.

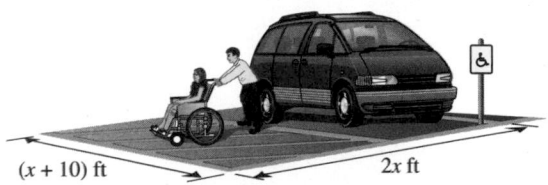

(x + 10) ft   2x ft

69. TOYS Find a polynomial that represents the area of the Etch-A-Sketch.

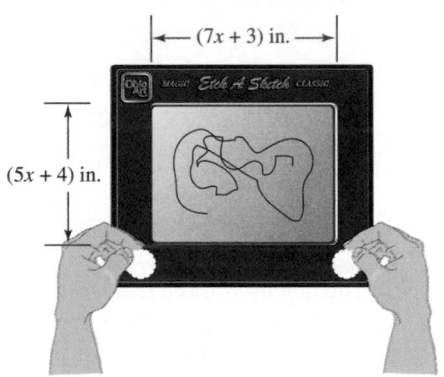

(7x + 3) in.

(5x + 4) in.

70. PLAYPENS Find a polynomial that represents the area of the floor of the playpen.

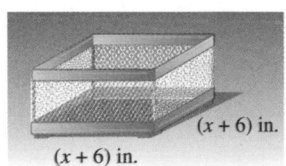

(x + 6) in.

(x + 6) in.

WRITING

71. Explain how to multiply two binomials.

72. Explain how to find $(2x + 1)^2$.

73. Explain why $(x + 1)^2 \neq x^2 + 1^2$. (Read $\neq$ as "is not equal to.")

74. If two terms are to be added, they have to be like terms. If two terms are to be multiplied, must they be like terms? Explain.

REVIEW

75. THE EARTH It takes 23 hours, 56 minutes, and 4.091 seconds for the Earth to rotate on its axis once. Write 4.091 in words.

76. TAKE-OUT FOOD The sticker shows the amount and the price per pound of some spaghetti salad that was purchased at a delicatessen. Find the total price of the salad.

Joan's Spaghetti Salad
303 Foothill Plaza
Plaza Deli
0.78   3.95
NET WT. LB.   PRICE/ LB. Ⓢ   (TOTAL PRICE Ⓢ)

77. Write $\frac{7}{64}$ as a decimal.

78. Write $-\frac{6}{10}$ as a decimal.

79. Evaluate: $56.09 + 78 + 0.567$

80. Evaluate: $-679.4 - (-599.89)$

81. Evaluate: $\sqrt{16} + \sqrt{36}$

82. Divide: $103.6 \div 0.56$

CHAPTER 10 SUMMARY AND REVIEW

DEFINITIONS AND CONCEPTS	EXAMPLES
An **exponent** indicates repeated multiplication. It tells how many times the **base** is to be used as a factor. Exponent ⟶ n factors of x $x^n = x \cdot x \cdot x \cdot \;\cdots\; x$ Base ⟶	Identify the base and the exponent for each given expression. $2^6 = 2 \cdot 2 \cdot 2 \cdot 2 \cdot 2 \cdot 2$ 2 is the base and 6 is the exponent. $(-xy)^3 = (-xy)(-xy)(-xy)$ Because of the parentheses, $-xy$ is the base and 3 is the exponent. $5t^4 = 5 \cdot t \cdot t \cdot t \cdot t$ The base is t and 4 is the exponent. $8^1 = 8$ The base is 8 and 1 is the exponent.
Rules for Exponents: If m and n represent integers, **Product rule:** $x^m x^n = x^{m+n}$ **Power rule:** $(x^m)^n = x^{m \cdot n} = x^{mn}$ **Power of a product rule:** $(xy)^m = x^m y^m$	Simplify each expression: $5^2 5^7 = 5^{2+7} = 5^9$ Keep the common base, 5, and add the exponents. $(6^3)^7 = 6^{3 \cdot 7} = 6^{21}$ Keep the base, 6, and multiply the exponents. $(2p)^5 = 2^5 p^5 = 32p^5$ Raise each factor of the product $2p$ to the 5th power.

To simplify some expressions, we must apply two (or more) rules for exponents.	Simplify: $(c^2 c^5)^4 = (c^7)^4$ Within the parentheses, keep the common base, c, and add the exponents: $2 + 5 = 7$.
	
	  $= c^{28}$ Keep the base, c, and multiply the exponents: $7 \cdot 4 = 28$.
	
	 Simplify: $(t^2)^4(t^3)^3 = t^8 t^9$ For each power of t raised to a power, keep the base and multiply the exponents: $2 \cdot 4 = 8$ and $3 \cdot 3 = 9$.
	
	  $= t^{17}$ Keep the common base, t, and add the exponents: $8 + 9 = 17$.

REVIEW EXERCISES

1. Identify the base and the exponent in each expression.
   a. n^{12} b. $(2x)^6$
   c. $3r^4$ d. $(y - 7)^3$

2. Write each expression in an equivalent form using an exponent.
   a. $m \cdot m \cdot m \cdot m \cdot m$ b. $-3 \cdot x \cdot x \cdot x \cdot x$
   c. $a \cdot a \cdot b \cdot b \cdot b \cdot b$ d. $(pq)(pq)(pq)$

3. Simplify, if possible.
   a. $x^2 \cdot x^2$ b. $x^2 + x^2$
   c. $x \cdot x^2$ d. $x + x^2$

4. Explain each error.
   a. $3^2 \cdot 3^4 = 9^6$
   b. $(3^2)^4 = 3^6$

Simplify each expression.

5. $7^4 \cdot 7^8$

6. $mmnn^2$

7. $(y^7)^3$

8. $(3x)^4$

9. $(6^3)^{12}$

10. $-b^3 b^4 b^5$

11. $(-16s^3)^2 s^4$

12. $(2.1x^2 y)^2$

13. $[(-9)^3]^5$

14. $(a^5)^3 (a^2)^4$

15. $(2x^2 x^3)^3$

16. $(m^2 m^3)^2 (n^2 n^4)^3$

17. $(3a^4)^2 (2a^3)^3$

18. $x^{100} \cdot x^{100}$

19. $(4m^3)^3 (2m^2)^2$

20. $(3t^4)^3 (2t^5)^2$

SECTION 10.2   Introduction to Polynomials

DEFINITIONS AND CONCEPTS	EXAMPLES
A **polynomial** is a single term or a sum of terms in which all variables have whole-number exponents and no variable appears in a denominator.	Polynomials: $32, \quad -5x^2 y^3, \quad 7p^3 - 14q^3, \quad 4m^2 + 5m - 12$

A polynomial with exactly one term is called a **monomial**. A polynomial with exactly two terms is called a **binomial**. A polynomial with exactly three terms is called a **trinomial**.	*Monomials* $3x^2$ $-12m^3 n^2$	*Binomials* $2y^3 + 3y$ $87t - 25$	*Trinomials* $3p^2 - 7p + 12$ $4p^2 q^3 - 8p^2 q^2 + 12p^2 q$

The **degree of a polynomial** is equal to the highest degree of any term of the polynomial.	*Polynomial* $\qquad$ *Degree of the polynomial* $7m^3 - 4m^2 + 5m - 12 \qquad\qquad 3$ $3a^3 - 2a^2 + a^4 \qquad\qquad 4$

To **evaluate a polynomial** for a given value, substitute the value for the variable and follow the order of operations rule.	Evaluate $3x^2 - 4x + 2$ for $x = 2$. $3x^2 - 4x + 2 = 3(2)^2 - 4(2) + 2 \quad$ Substitute 2 for each x. $ = 3(4) - 4(2) + 2 \quad$ Evaluate $(2)^2$ first. $ = 12 - 8 + 2 \quad$ Multiply. $ = 6$

REVIEW EXERCISES

Classify each polynomial as a monomial, a binomial, or a trinomial.

21. $3x^2 + 4x - 5$

22. $3t^2$

23. $2x^2 - 1$

24. $\frac{1}{2}d^5 + \frac{3}{2}d^3 + \frac{5}{2}d$

Give the degree of each polynomial.

25. $3x^2 + 2x^3$

26. $3t^4 - 4t^2 - 3$

27. $3q^2 - 4q^5$

28. $0.2a - 4.5a^5 + 1.3a^3$

29. Evaluate $2t^2 + t - 2$ for $t = -3$.

30. WATER BALLOONS Some college students launched water balloons from the balcony of their dormitory on unsuspecting sunbathers. The height h in feet of the balloons at a time t seconds after being launched is given by the polynomial

$$h = -16t^2 + 12t + 20$$

What was the height of the balloons 1 second after being launched?

SECTION 10.3   Adding and Subtracting Polynomials

DEFINITIONS AND CONCEPTS	EXAMPLES
To **add polynomials,** combine their like terms.	Add: $(4x^2 + 9x + 4) + (3x^2 - 5x - 1)$ $= (4x^2 + 3x^2) + (9x - 5x) + (4 - 1)$ *Group like terms.* $= 7x^2 + 4x + 3$ *Combine like terms.*
To **subtract two polynomials,** change the signs of the terms of the polynomial being subtracted, drop the parentheses, and combine like terms.	Subtract: $(8a^3 - 4a) - (-3a^3 + 9a)$ $= 8a^3 - 4a + 3a^3 - 9a$ *Change the sign of each term of $-3a^3 + 9a$ and drop the parentheses.* $= 11a^3 - 13a$ *Combine like terms.*

REVIEW EXERCISES

Add.

31. $3x^3 + 2x^3$

32. $\dfrac{1}{2}p^2 + \dfrac{5}{2}p^2 + \dfrac{7}{2}p^2$

33. $(3x - 1) + (6x + 5)$

34. $(3x^2 - 2x + 4) + (-x^2 - 1)$

35. $\begin{array}{r} 5x - 2 \\ + \ 3x + 5 \\ \hline \end{array}$

36. $\begin{array}{r} 3x^2 - 2x + 7 \\ + \ -5x^2 + 3x - 5 \\ \hline \end{array}$

Subtract.

37. $16p^3 - 9p^3$

38. $4y^2 - 9y^2$

39. $(2.5x + 4) - (1.4x + 12)$

40. $(3z^2 - z + 4) - (2z^2 + 3z - 2)$

41. $\begin{array}{r} 5x - 2 \\ -(3x + 5) \\ \hline \end{array}$

42. $\begin{array}{r} 3x^2 - 2x + 7 \\ -(-5x^2 + 3x - 5) \\ \hline \end{array}$

SECTION 10.4   Multiplying Polynomials

DEFINITIONS AND CONCEPTS	EXAMPLES
To **multiply two monomials,** multiply the numerical factors (the coefficients) and then multiply the variable factors.	Multiply: $(5p^6)(2p^5) = (5 \cdot 2)(p^6 \cdot p^5)$ *Group the coefficients together and the variables together.* $= 10p^{11}$ *Think: $5 \cdot 2 = 10$ and $p^6 \cdot p^5 = p^{6+5} = p^{11}$.*
To **multiply a monomial and a polynomial,** multiply each term of the polynomial by the monomial.	Multiply: $3r^2(2r^4 + 7r^2 - 4)$ $= 3r^2(2r^4) + 3r^2(7r^2) + 3r^2(-4)$ *Distribute the multiplication by $3r^2$.* $= 6r^6 + 21r^4 - 12r^2$ *Multiply the monomials.*

To **multiply two binomials,** use the *FOIL method:*

F: First
O: Outer
I: Inner
L: Last

Multiply: $(3m + 4)(2m - 5) = 3m(2m) + 3m(-5) + 4(2m) + 4(-5)$

$= 6m^2 - 15m + 8m - 20$ Multiply the monomials.

$= 6m^2 - 7m - 20$ Combine like terms.

To **multiply two polynomials,** multiply each term of one polynomial by each term of the other polynomial and then combine like terms.

Multiply: $(a - 3)(6a^2 - 4a + 1)$

$= a(6a^2) + a(-4a) + a(1) - 3(6a^2) - 3(-4a) - 3(1)$

$= 6a^3 - 4a^2 + a - 18a^2 + 12a - 3$ Multiply the monomials.

$= 6a^3 - 22a^2 + 13a - 3$ Combine like terms.

REVIEW EXERCISES

Multiply.

43. $3x^2 \cdot 5x^3$

44. $(3z^2)(-2z^2)$

45. $2x^2(3x + 2)$

46. $-5t^3(7t^2 - 6t - 2)$

47. $(2x - 1)(3x + 2)$

48. $(5t + 4)(7t - 6)$

49. $(3x + 2)(2x^2 - x + 1)$

50. $(2r - 3)(3r^2 + 2r - 3)$

51.
$$\begin{array}{r} 5x^2 - 2x + 3 \\ \times \quad\quad 3x + 5 \\ \hline \end{array}$$

52.
$$\begin{array}{r} 3x^2 - 2x - 1 \\ \times \quad\quad 5x - 2 \\ \hline \end{array}$$

Square each binomial.

53. $(x + 2)^2$

54. $(8a - 3)^2$

CHAPTER 10   TEST

Fill in the blanks.

1. In the exponential expression 7^5, 7 is the _____ and 5 is the _____.

2. Expressions such as x^4, 10^3, and $(5t)^2$ are called _____ expressions.

3. A _____ is a term or a sum of terms in which all variables have whole-number exponents and no variable appears in a denominator.

4. A _____ is a polynomial with exactly one term. A _____ is a polynomial with exactly two terms. A _____ is a polynomial with exactly three terms.

5. The _____ of the term $3x^7$ is 7 because x appears as a factor 7 times: $3 \cdot x \cdot x \cdot x \cdot x \cdot x \cdot x \cdot x$.

6. To _____ the polynomial $x^2 - 2x + 1$ for $x = 6$, we substitute 6 for x and follow the order of operations rule.

7. $(b^3 - b^2 - 9b + 1) + (b^3 - b^2 - 9b + 1)$ is the sum of two _____.

8. $(b^2 - 9b + 11) - (4b^2 - 14b)$ is the _____ of a trinomial and a binomial.

9. In the acronym FOIL, F stands for _____ terms, O for _____ terms, I for _____ terms, and L for _____ terms.

10. $(2a - 4)(3a^2 + 5a - 1)$ is the product of a _____ and a _____.

11. $(2x + 3)^2$ is the _____ of a binomial.

12. To multiply two polynomials, multiply _____ term of one polynomial by _____ term of the other polynomial, and then combine like terms.

13. Identify the base and the exponent of each expression.
    a. 6^5
    b. $7b^4$

14. Simplify each expression, if possible.
    a. $a^2 + a^2$ b. $a^2 \cdot a^2$
    c. $a^2 + a$ d. $a^2 \cdot a$

Simplify each expression.

15. $h^2 h^4$

16. $(m^{10})^2$

17. $b^2 \cdot b \cdot b^5$

18. $(x^3)^4 (x^2)^3$

19. $(a^2 b^3)(a^4 b^7)$

20. $(12a^9 b)^2$

21. $(2x^2)^3 (3x^3)^3$

22. $(t^2 t^3)^3$

Classify each polynomial as a monomial, a binomial, or a trinomial.

23. $5x^2 + 4x$

24. $-\dfrac{3}{4}t^{15}$

25. $-3x^2 - 2x + 3$

26. $x - 8$

Give the degree of each polynomial.

27. $3t^4 - 2t^3 + 5t^6 - t$

28. $7q^7 + 5q^5 - 8q^2$

Evaluate each polynomial.

29. $3x^2 - 2x + 4$ for $x = 3$

30. $-2r^2 - r + 3$ for $r = -1$

Perform the operations.

31. $(2.1p^2 - 2p - 2) - (3.3p^2 - 5p - 2)$

32. $(-2x^3)(4x^2)$

33. $(3x^2 + 2x) + (2x^2 - 5x + 4)$

34. $(2x - 5)(3x + 4)$

35. $\quad 3d^2 - 3d + 7.2$
    $-(-5d^2 + 6d - 5.3)$

36. $3y^2(y^2 - 2y + 3)$

37. $\quad 4x^2 - 5x + 5$
    $+ 3x^2 + 7x - 7$

38. $(2x - 3)(x^2 - 2x + 4)$

39. FILTERS The length of one side of the square furnace filter shown below is $(x + 4)$ in. Find the perimeter of the filter.

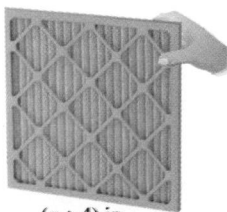

$(x + 4)$ in.

40. Explain what is wrong with the following work:
    $$5^4 \cdot 5^3 = 25^7$$

1. **USED CARS** The following ad appeared in *The Car Trader*. (O.B.O. means "or best offer.") If offers of $8,750, $8,875, $8,900, $8,850, $8,800, $7,995, $8,995, and $8,925 were received, what was the selling price of the car? [Section 1.1]

> 1969 Ford Mustang. New tires
> Must sell!!!! $10,500 O.B.O.

2. Round 2,109,567 to the nearest thousand. [Section 1.1]

3. Add: $458 + 8,099 + 23,419 + 58$ [Section 1.2]

4. Subtract: $35,021 - 23,999$ [Section 1.2]

5. **PARKING** The length of a rectangular parking lot is 204 feet and its width is 97 feet. [Section 1.3]
    a. Find the perimeter of the lot.
    b. Find the area of the lot.

6. Divide: $1,363 \div 41$ [Section 1.4]

7. a. Prime factor 220. [Section 1.5]
    b. Find all the factors of 12. [Section 1.5]

8. a. Find the LCM of 16 and 24. [Section 1.6]
    b. Find the GCF of 16 and 24. [Section 1.6]

9. Evaluate: $\dfrac{(3 + 5)^2 + 2}{2(8 - 5)}$ [Section 1.7]

10. Solve each equation and check the result.
    a. $y - 81 = 243$ [Section 1.8]
    b. $81y = 243$ [Section 1.9]

11. a. Write the set of integers. [Section 2.1]
    b. Simplify: $-(-3)$ [Section 2.1]

12. Perform the operations.
    a. $-16 + 4$ [Section 2.2]
    b. $16 - (-4)$ [Section 2.3]
    c. $-16(4)$ [Section 2.4]
    d. $\dfrac{-16}{-4}$ [Section 2.5]
    e. -4^2 [Section 2.4]
    f. $(-4)^2$ [Section 2.4]

13. **OVERDRAFT PROTECTION** A student forgot that she had only $30 in her bank account and wrote a check for $55 and used her debit card to buy $75 worth of groceries. On each of the two transactions, the bank charged her a $20 overdraft protection fee. Find the new account balance. [Section 2.3]

14. Evaluate: $10 - 4|6 - (-3)^2|$ [Section 2.6]

15. Solve $-x + 2 = 13$ and check the result. [Section 2.7]

16. Solve $4 + \dfrac{x}{-5} - 6 = -1$ and check the result.
    [Section 2.7]

Write and solve an equation to answer the following question.
[Section 2.7]

17. **ACCOUNTING** Because of bad economic times, Acme corporation lost $63 million in 2009. Just one year before, the corporation made a very large profit. If Acme lost a total of $17 million in this two-year span, how much profit did Acme make in 2008?

18. See the illustration below. [Section 3.1]
    a. Let k represent the length (in inches) of the key. Write an algebraic expression that represents the length of the match (in inches).
    b. Let m represent the length (in inches) of the match. Write an algebraic expression that represents the length of the key (in inches).

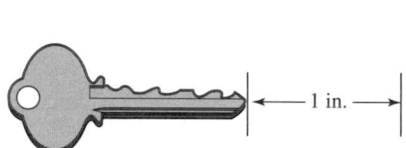

19. Translate into mathematical symbols: 5 less than a number. [Section 3.1]

20. **AIRLINES** Find the distance traveled by a jet if it travels for 3 hours at 475 miles per hour. [Section 3.2]

21. a. Simplify: $-10(-10a)$ [Section 3.3]
    b. Simplify: $-(-4y + 6)$ [Section 3.3]
    c. Simplify: $-3(-6y - 8) - 12 - 4(5 - y)$
       [Section 3.4]

22. **SHOPPING** What is the value (in cents) of x coupons, each of which gives the shopper 50¢ off? [Section 3.6]

Solve each equation and check the result. [Section 3.5]

23. $5r - 24 = r + 5r + 2r$
24. $2(4a + 8) = -3(2 - 3a) - 3a$

Write and solve an equation to answer the following question.
[Section 3.6]

25. WISHING WELLS A city park employee collected
600 cents in nickels, dimes, and quarters at the bottom
of a wishing well. There were 10 nickels, and a
combined total of 25 dimes and quarters. How many
dimes and quarters were at the bottom of the well?

26. Simplify: $\dfrac{35a^2}{28a}$ [Section 4.1]

27. **a.** Write $\dfrac{3}{8}$ as an equivalent fraction with

    denominator 48. [Section 4.1]

    b. What is the reciprocal of $\dfrac{9}{8}$? [Section 4.3]

    c. Write $7\dfrac{1}{2}$ as an improper fraction. [Section 4.5]

28. GRAVITY Objects on the moon weigh only one-
sixth as much as on Earth. If a rock weighs 54 ounces
on the Earth, how much does it weigh on the moon?
[Section 4.2]

Perform the operations.

29. $-\dfrac{5}{77}\left(\dfrac{33}{50}\right)$ [Section 4.2]

30. $\dfrac{15b^3}{16c^5} \div \dfrac{45b}{8c}$ [Section 4.3]

31. $\dfrac{3}{4} - \dfrac{3}{5}$ [Section 4.4]

32. $\dfrac{m}{9} - \dfrac{4}{5}$ [Section 4.4]

33. $-\dfrac{6}{25}\left(2\dfrac{7}{24}\right)$ [Section 4.5]

34. $45\dfrac{2}{3} + 96\dfrac{4}{5}$ [Section 4.6]

35. Simplify: $\dfrac{7 - \dfrac{2}{3}}{4\dfrac{5}{6}}$ [Section 4.7]

36. PET MEDICATION A pet owner was told to use an
eye dropper to administer medication to his sick
kitten. The cup shown below contains 8 doses of the
medication. Determine the size of a single dose.
[Section 4.3]

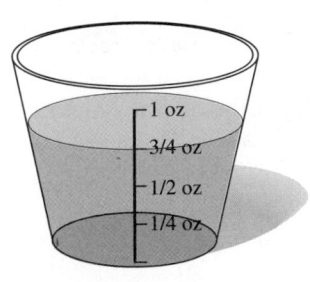

37. BAKING A bag of all-purpose flour contains
$17\dfrac{1}{2}$ cups. A baker uses $3\dfrac{3}{4}$ cups. How many cups
of flour are left? [Section 4.6]

38. Evaluate: $a + b^2c$ for $a = \dfrac{3}{4}$, $b = -\dfrac{1}{3}$, and $c = \dfrac{5}{4}$.
[Section 4.7]

Solve each equation and check the result. [Section 4.8]

39. $\dfrac{x}{12} + \dfrac{2}{3} = \dfrac{x}{6}$

40. $\dfrac{2}{3}q - 1 = -6$

Write and solve an equation to answer the following question.
[Section 4.8]

41. COFFEE DRINKERS Three-fifths of 275 students
surveyed said they started their morning with a cup of
coffee. Of the 275 students, how many would this be?

42. **a.** Round the number pi to the nearest ten thousandth:
    $\pi = 3.141592654. \ldots$ [Section 5.1]

    b. Place the proper symbol ($>$ or $<$) in the blank:
    154.34 ⬜ 154.33999. [Section 5.1]

43. **a.** Write 6,510,345.798 in words. [Section 5.1]

    b. Write 7,498.6461 in expanded notation. [Section 5.1]

Perform the operations.

44. $3.4 + 106.78 + 35 + 0.008$ [Section 5.2]

45. $5{,}091.5 - 1{,}287.89$ [Section 5.2]

46. $-8.8 + (-7.3 - 9.5)$ [Section 5.2]

47. $-5.5(-3.1)$ [Section 5.3]

48. $\dfrac{0.0742}{1.4}$ [Section 5.4]

49. $\dfrac{7}{8}(9.7 + 15.8)$ [Section 5.5]

50. PAYCHECKS If you are paid every other week,
your *monthly gross income* is your gross income
from one paycheck times 2.17. Find the monthly
gross income of a secretary who earns $1,250 every
two weeks. [Section 5.3]

51. Perform each operation in your head.

    a. $(89.9708)(10{,}000)$ [Section 5.3]

    b. $\dfrac{89.9708}{100}$ [Section 5.4]

52. Estimate the quotient: $9.2\overline{)18{,}460.76}$ [Section 5.4]

53. Evaluate $\dfrac{(-1.3)^2 + 6.7}{-0.9}$ and round the result to the
nearest hundredth. [Section 5.4]

54. Write $\dfrac{2}{15}$ as a decimal. Use an overbar. [Section 5.5]

55. Evaluate each expression. [Section 5.6]

   a. $2\sqrt{121} - 3\sqrt{64}$

   b. $\sqrt{\dfrac{49}{81}}$

56. Graph each number on the number line. [Section 5.6]

$$\left\{ -4\frac{5}{8}, \sqrt{17}, 2.89, \frac{2}{3}, -0.1, -\sqrt{9}, \frac{3}{2} \right\}$$

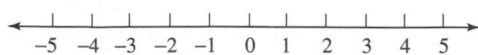

57. Solve $1.7y + 1.24 = -1.4y - 0.62$ and check the result. [Section 5.7]

Write and solve an equation to answer the following question. [Section 5.8]

58. DECORATIONS A mother has budgeted $20 for decorations for her daughter's birthday party. She decides to buy a tank of helium for $15.15 and some balloons. If the balloons sell for 5 cents apiece, how many balloons can she buy?

59. Write each phrase as a ratio (fraction) in simplest form. [Section 6.1]

   a. 3 centimeters to 7 centimeters

   b. 13 weeks to 1 year

60. COMPARISON SHOPPING A dry-erase whiteboard with an area of 400 in.² sells for $24. A larger board, with an area of 600 in.², sells for $42. Which board is the better buy? [Section 6.1]

61. Solve the proportion: $\dfrac{x}{14} = \dfrac{13}{28}$ [Section 6.2]

62. INSURANCE CLAIMS In one year, an auto insurance company had 3 complaints per 1,000 policies. If a total of 375 complaints were filed that year, how many policies did the company have? [Section 6.2]

63. SCALE DRAWINGS On the scale drawing below, $\frac{1}{4}$-inch represents an actual length of 3 feet. The length of the house on the drawing is $6\frac{1}{4}$ inches. What is the actual length of the house? [Section 6.2]

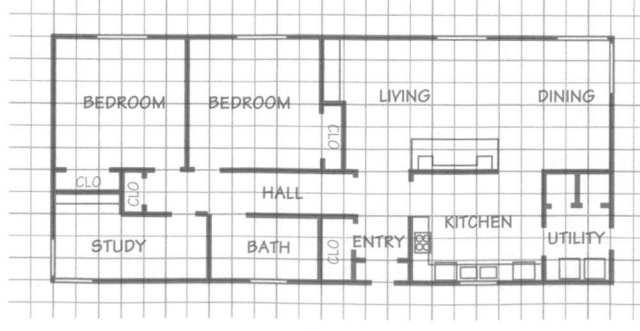

Scale $\frac{1}{4}$ in. : 3 ft

64. Make each conversion. [Section 6.3]

   a. Convert 168 inches to feet.

   b. Convert 212 ounces to pounds.

   c. Convert 30 gallons to quarts.

   d. Convert 12.5 hours to minutes.

65. Make each conversion. [Section 6.4]

   a. Convert 1.538 kilograms to grams.

   b. Convert 500 milliliters to liters.

   c. Convert 0.3 centimeters to kilometers.

66. THE AMAZON The Amazon River enters the Atlantic Ocean through a broad estuary, roughly estimated at 240,000 m in width. Convert the width to kilometers. [Section 6.4]

67. COOKING What is the weight of a 10-pound ham in kilograms? [Section 6.5]

68. Convert 75°C to degrees Fahrenheit. [Section 6.5]

69. Complete the table. [Section 7.1]

Percent	Decimal	Fraction
57%		
	0.001	
		$\frac{1}{3}$

70. Refer to the figure on the right. [Section 7.1]

   a. What percent of the figure is shaded?

   b. What percent is not shaded?

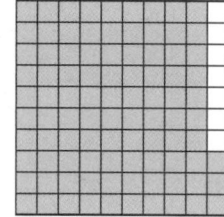

71. What number is 15% of 450? [Section 7.2]

72. 24.6 is 20.5% of what number? [Section 7.2]

73. 51 is what percent of 60? [Section 7.2]

74. CLOTHING SALES Find the amount of the discount and the sale price of the coat shown below. [Section 7.3]

Men's Open Range Coat
Save 25%
Regularly $820⁰⁰
Winter Coats on Sale!
Genuine leather

75. SALES TAX If the sales tax rate is $6\frac{1}{4}$%, how much sales tax will be added to the price of a new car selling for $18,550? [Section 7.3]

76. COLLECTIBLES A porcelain figurine, which was originally purchased for $125, was sold by a collector ten years later for $750. What was the percent increase in the value of the figurine? [Section 7.3]

77. TIPPING Estimate a 15% tip on a dinner that cost $135.88. [Section 7.4]

78. PAYING OFF LOANS To pay for tuition, a college student borrows $1,500 for six months. If the annual interest rate is 9%, how much will the student have to repay when the loan comes due? [Section 7.5]

79. FREEWAYS Refer to the pictograph below to answer the following questions. [Section 8.1]

Freeway Traffic
Average number of vehicles daily

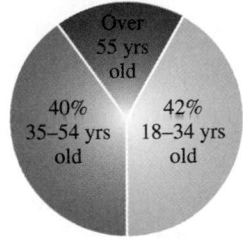

I-405 Los Angeles

I-5 Seattle

I-95 New York

I-94 Minneapolis   = 50,000 vehicles

Source: www.skyscraperpage.com

a. Estimate the number of vehicles that travel the I-405 Freeway in Los Angles each day.

b. Estimate the number of vehicles that travel the I-95 Freeway in New York each day.

c. Estimate how many more vehicles travel the I-5 Freeway in Seattle than the I-94 Freeway in Minneapolis each day.

80. VEGETARIANS The graph below gives the results of a recent study by *Vegetarian Times*. [Section 8.1]

Survey Results: Ages of Adult
Vegetarians in the United States, 2008

Over 55 yrs old

40% 35–54 yrs old

42% 18–34 yrs old

Source: *Vegetarian Times*

a. According to the study, what percent of the adult vegetarians in the United States are over 55 years old?

b. The study estimated that there were 7,300,000 adult vegetarians in the United States. How many of them are 35 to 54 years old?

81. SPENDING ON PETS Refer to the bar graph below to answer the following questions. [Section 8.1]

a. In what category was the most money spent on pets? Estimate how much.

b. Estimate how much money was spent on purchasing pets.

c. Estimate how much more money was spent on vet care than on grooming and boarding.

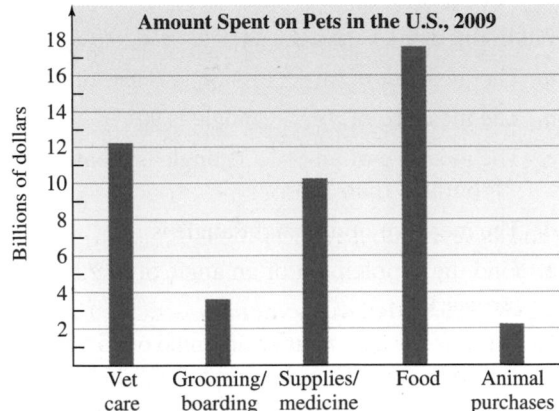

Source: American Pet Products Organization

82. TABLE TENNIS The weights (in ounces) of 8 Ping-Pong balls that are to be used in a tournament are as follows: 0.85, 0.87, 0.88, 0.88, 0.85, 0.86, 0.84, and 0.85. Find the mean, median, and mode of the weights. [Section 8.2]

83. Graph each point: $(-4, -3), (1.5, 1.5), \left(-\frac{7}{2}, 0\right), (0, 3.5)$ [Section 8.3]

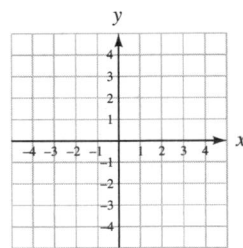

84. Is $(-2, 1)$ a solution of $3x - y = -8$? [Section 8.3]

Graph each equation. [Section 8.4]

85. $y = -2x$

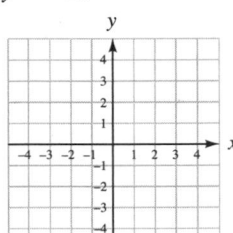

86. $2x - 3y = 12$

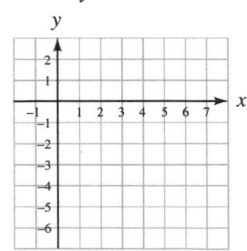

87. $x = 4$

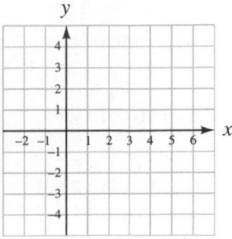

88. Fill in the blanks. [Section 9.1]

   a. The measure of an _____ angle is less than 90°.

   b. The measure of a _____ angle is 90°.

   c. The measure of an _____ angle is greater than 90° but less than 180°.

   d. The measure of a straight angle is _____.

89. a. Find the supplement of an angle of 105°.
      [Section 9.1]

   b. Find the complement of an angle of 75°.
      [Section 9.1]

90. Refer to the figure below, where $l_1 \parallel l_2$. Find the measure of each angle. [Section 9.2]

   a. m($\angle 1$)   **b.** m($\angle 3$)

   c. m($\angle 2$)   **d.** m($\angle 4$)

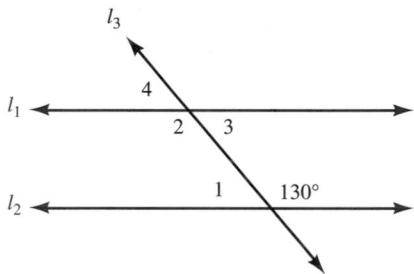

91. Refer to the figure below, where $AB \parallel DE$ and m($\overline{AC}$) = m($\overline{BC}$). Find the measure of each angle. [Section 9.3]

   a. m($\angle 1$)   **b.** m($\angle C$)

   c. m($\angle 2$)   **d.** m($\angle 3$)

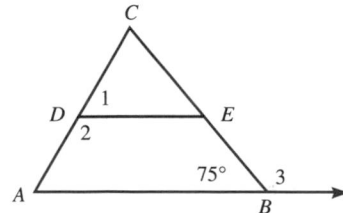

92. JAVELIN THROW Refer to the illustration below. Determine x and y. [Section 9.3]

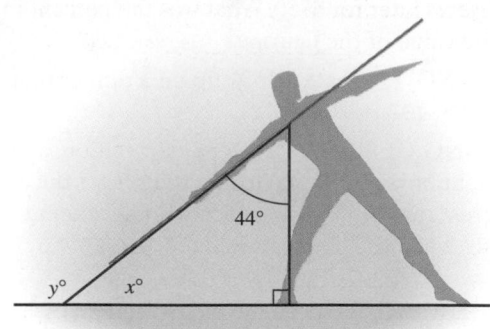

93. If the vertex angle of an isosceles triangle measures 34°, what is the measure of each base angle? [Section 9.3]

94. If the legs of a right triangle measure 10 meters and 24 meters, how long is the hypotenuse? [Section 9.4]

95. Determine whether a triangle with sides of length 16 feet, 63 feet, and 65 feet is a right triangle. [Section 9.4]

96. SHADOWS If a tree casts a 35-foot shadow at the same time as a man 6 feet tall casts a 5-foot shadow, how tall is the tree? [Section 9.5]

97. Find the sum of the angles of a pentagon. [Section 9.6]

98. Find the perimeter and the area of a square that has sides each 12 meters long. [Section 9.7]

99. Find the area of a triangle with a base that is 14 feet long and a height of 18 feet. [Section 9.7]

100. Find the area of a trapezoid that has bases that are 12 inches and 14 inches long and a height of 7 inches. [Section 9.7]

101. How many square inches are in 1 square foot? [Section 9.7]

102. Find the circumference and the area of a circle that has a diameter of 14 centimeters. For each, give the exact answer and an approximation to the nearest hundredth. [Section 9.8]

103. Find the area of the shaded region shown below, which is created using two semicircles. Round to the nearest hundredth. [Section 9.8]

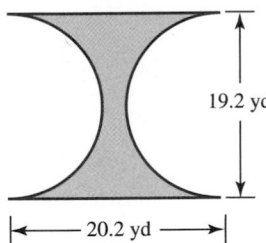

19.2 yd

|← 20.2 yd →|

104. ICE Find the volume of a block of ice that is in the shape of a rectangular solid with dimensions 15 in. × 24 in. × 18 in. [Section 9.9]

105. Find the volume of a sphere that has a diameter of 18 inches. Give the exact answer and an approximation to the nearest hundredth. [Section 9.9]

106. Find the volume of a cone that has a circular base with a radius of 4 meters and a height of 9 meters. Give the exact answer and an approximation to the nearest hundredth. [Section 9.9]

107. Find the volume of a cylindrical pipe that is 20 feet long and has a radius of 1 foot. Give the exact answer and an approximation to the nearest hundredth. [Section 9.9]

108. How many cubic inches are there in 1 cubic foot? [Section 9.9]

Simplify each expression. [Section 10.1]

109. $s^4 \cdot s^5$

110. $(a^5)^7$

111. $(y^5)^2(y^4)^3$

112. $(2b^3c^6)^3$

113. Classify $3x^2 - 7x + 1$ as a monomial, a binomial, or a trinomial. Then give its degree. [Section 10.2]

114. Evaluate $0.5t^3 - t^2 - 4t$ for $t = 4$. [Section 10.2]

Perform the operations.

115. $(5x^2 - 2x + 4) - (3x^2 - 5)$ [Section 10.3]

116. $(6.2a^3 + 7.1a^2 - 4.1a) + (3.8a^3 - 4.1a)$ [Section 10.3]

117. $-3h^9(-5h)$ [Section 10.4]

118. $-3p(2p^2 + 3p - 4)$ [Section 10.4]

119. $(3x + 5)(2x - 1)$ [Section 10.4]

120. $(2y - 7)^2$ [Section 10.4]

Inductive and Deductive Reasoning

Objectives

1. Use inductive reasoning to solve problems.
2. Use deductive reasoning to solve problems.

SECTION I.1
Inductive and Deductive Reasoning

To reason means to think logically. The objective of this appendix is to develop your problem-solving ability by improving your reasoning skills. We will introduce two fundamental types of reasoning that can be applied in a wide variety of settings. They are known as *inductive reasoning* and *deductive reasoning*.

1 Use inductive reasoning to solve problems.

In a laboratory, scientists conduct experiments and observe outcomes. After several repetitions with similar outcomes, the scientist will generalize the results into a statement that appears to be true:

- If I heat water to 212°F, it will boil.
- If I drop a weight, it will fall.
- If I combine an acid with a base, a chemical reaction occurs.

When we draw general conclusions from specific observations, we are using **inductive reasoning.** The next examples show how inductive reasoning can be used in mathematical thinking. Given a list of numbers or symbols, called a *sequence*, we can often find a missing term of the sequence by looking for patterns and applying inductive reasoning.

Self Check 1

Find the next number in the sequence $-3, -1, 1, 3, \ldots$.

Now Try Problem 11

EXAMPLE 1 Find the next number in the sequence $5, 8, 11, 14, \ldots$.

Strategy We will find the *difference* between pairs of numbers in the sequence.

WHY This process will help us discover a pattern that we can use to find the next number in the sequence.

Solution
The numbers in the sequence $5, 8, 11, 14, \ldots$ are increasing. We can find the difference between each pair of successive numbers as follows:

$8 - 5 = 3$ Subtract the first number, 5, from the second number, 8.

$11 - 8 = 3$ Subtract the second number, 8, from the third number, 11.

$14 - 11 = 3$ Subtract the third number, 11 from the fourth number, 14.

The difference between each pair of numbers is 3. This means that each number in the sequence is 3 greater than the previous one. Thus, the next number in the sequence is $14 + 3$, or 17.

EXAMPLE 2 Find the next number in the sequence $-2, -4, -6, -8, \ldots$.

Strategy The terms of the sequence are decreasing. We will determine how each number differs from the previous number.

WHY This type of examination helps us discover a pattern that we can use to find the next number in the sequence.

Self Check 2

Find the next number in the sequence −0.1, −0.3, −0.5, −0.7, ...

Now Try Problem 15

Solution

Since each successive number is 2 less than the previous one, the next number in the sequence is −8 − 2, or −10.

This number is 2 less than the previous number.	This number is 2 less than the previous number.	This number is 2 less than the previous number.

−2 ,   −4 ,   −6 ,   −8 , ... ■

Self Check 3

Find the next letter in the sequence B, G, D, I, F, K, H, ...

Now Try Problem 19

EXAMPLE 3   Find the next letter in the sequence A, D, B, E, C, F, D, ...

Strategy We will create a letter–number correspondence and rewrite the sequence in an equivalent numerical form.

WHY Many times, it is easier to determine the pattern if we examine a sequence of numbers instead of letters.

Solution

The letter A is the 1st letter of the alphabet, D is the 4th letter, B is the 2nd letter, and so on. We can create the following letter–number correspondence:

Letter		Number	
A	→	1	Add 3.
D	→	4	Subtract 2.
B	→	2	Add 3.
E	→	5	Subtract 2.
C	→	3	Add 3.
F	→	6	Subtract 2.
D	→	4	

The numbers in the sequence 1, 4, 2, 5, 3, 6, 4, . . . alternate in size. They change from smaller to larger, to smaller, to larger, and so on.

We see that 3 is added to the first number to get the second number. Then 2 is subtracted from the second number to get the third number. To get successive numbers in the sequence, we alternately add 3 to one number and then subtract 2 from that result to get the next number.

Applying this pattern, the next number in the given numerical sequence would be 4 + 3, or 7. The next letter in the original sequence would be G, because it is the 7th letter of the alphabet.

Self Check 4

Find the next shape in the sequence below.

Now Try Problem 23

EXAMPLE 4   Find the next shape in the sequence below.

Strategy To find the next shape in the sequence, we will focus on the changing positions of the dots.

WHY The star does not change in any way from term to term.

Solution

We see that each of the three dots moves from one point of the star to the next, in a counterclockwise direction. This is a circular pattern. The next shape in the sequence will be the one shown here.

EXAMPLE 5   Find the next shape in the sequence below.

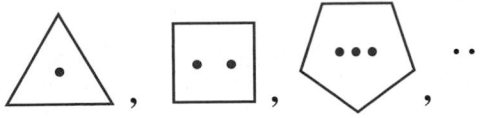

Strategy To find the next shape in the sequence, we must consider two changing patterns at the same time.

WHY The shapes are changing and the number of dots within them are changing.

Solution
The first figure has three sides and one dot, the second figure has four sides and two dots, and the third figure has five sides and three dots. Thus, we would expect the next figure to have six sides and four dots, as shown to the right.

Self Check 5

Find the next shape in the sequence below.

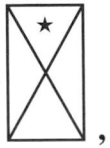

Now Try Problem 27

2 Use deductive reasoning to solve problems.

As opposed to inductive reasoning, deductive reasoning moves from the general case to the specific. For example, if we know that the sum of the angles in any triangle is 180°, we know that the sum of the angles of △ABC shown in the right margin is 180°. Whenever we apply a general principle to a particular instance, we are using deductive reasoning.

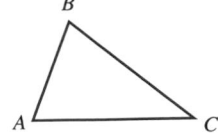

A deductive reasoning system is built on four elements:

1. **Undefined terms:** terms that we accept without giving them formal meaning
2. **Defined terms:** terms that we define in a formal way
3. **Axioms** or **postulates:** statements that we accept without proof
4. **Theorems:** statements that we can prove with formal reasoning

Many problems can be solved by deductive reasoning. For example, suppose a student knows that his college offers algebra classes in the morning, afternoon, and evening and that Professors Anderson, Medrano, and Ling are the only algebra instructors at the school. Furthermore, suppose that the student plans to enroll in a morning algebra class. After some investigating, he finds out that Professor Anderson teaches only in the afternoon and Professor Ling teaches only in the evening. Without knowing anything about Professor Medrano, he can conclude that she will be his algebra teacher, since she is the only remaining possibility.

The following examples show how to use deductive reasoning to solve problems.

EXAMPLE 6   *Scheduling Classes*   An online college offers only one calculus course, one algebra course, one statistics course, and one trigonometry course. Each course is to be taught by a different professor. The four professors who will teach these courses have the following course preferences:

1. Professors A and B don't want to teach calculus.
2. Professor C wants to teach statistics.
3. Professor B wants to teach algebra.

Who will teach trigonometry?

Strategy We will construct a table showing all the possible teaching assignments. Then we will cross off those classes that the professors do not want to teach.

Now Try Problem 31

WHY The best way to examine this much information is to describe the situation using a table.

Solution

The following table shows each course, with each possible instructor.

Calculus	Algebra	Statistics	Trigonometry
A	A	A	A
B	B	B	B
C	C	C	C
D	D	D	D

Since Professors A and B don't want to teach calculus, we can cross them off the calculus list. Since Professor C wants to teach statistics, we can cross her off every other list. This leaves Professor D as the only person to teach calculus, so we can cross her off every other list. Since Professor B wants to teach algebra, we can cross him off every other list. Thus, the only remaining person left to teach trigonometry is Professor A.

Calculus	Algebra	Statistics	Trigonometry
A̶	A	A	A
B̶	B	B̶	B̶
C̶	C̶	C	C̶
D	D̶	D̶	D̶

Self Check 7

USED CARS Of the 50 cars on a used-car lot, 9 are red, 31 are foreign models, and 6 are red, foreign models. If a customer wants to buy an American model that is not red, how many cars does she have to choose from?

Now Try Problem 35

EXAMPLE 7   *State Flags*   The graph below gives the number of state flags that feature an eagle, a star, or both. How many state flags have neither an eagle nor a star?

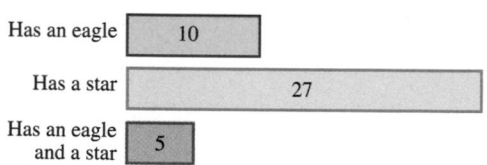

Strategy We will use two intersecting circles to model this situation.

WHY The intersection is a way to represent the number of state flags that have both an eagle and a star.

Solution

In figure (a) on the following page, the intersection (overlap) of the circles shows that there are 5 state flags that have both an eagle and a star. If an eagle appears on a total of 10 flags, then the red circle must contain 5 more flags outside of the intersection, as shown in figure (b). If a total of 27 flags have a star, the blue circle must contain 22 more flags outside the intersection, as shown.

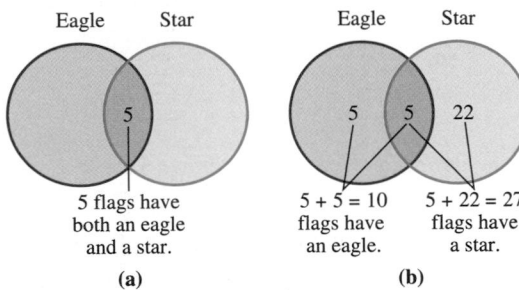

(a)   (b)

From figure (a), we see that 5 + 5 + 22, or 32 flags have an eagle, a star, or both. To find how many flags have neither an eagle nor a star, we subtract this total from the number of state flags, which is 50.

$$50 - 32 = 18$$

There are 18 state flags that have neither an eagle nor a star.

ANSWERS TO SELF CHECKS

1. 5   **2.** −0.9   **3.** M   **4.**   **5.**    **7.** 16

APPENDIX I   STUDY SET

VOCABULARY

Fill in the blanks.

1. _____ reasoning draws general conclusions from specific observations.

2. _____ reasoning moves from the general case to the specific.

CONCEPTS

Tell whether the pattern shown is increasing, decreasing, alternating, or circular.

3. 2, 3, 4, 2, 3, 4, 2, 3, 4, . . .

4. 8, 5, 2, −1, . . .

5. −2, −4, 2, 0, 6, . . .

6. 0.1, 0.5, 0.9, 1.3, . . .

7. a, c, b, d, c, e, . . .

8.  , , , , . . .

9. ROOM SCHEDULING From the chart, determine what time(s) on a Wednesday morning a practice room

in a music building is available. The symbol X indicates that the room has already been reserved.

	M	T	W	Th	F
9 A.M.	X		X		X
10 A.M.	X	X			X
11 A.M.		X	X	X	

10. COUNSELING QUESTIONNAIRE A group of college students were asked whether they were taking a mathematics course and whether they were taking an English course. The results are displayed below.

a. How many students were taking a mathematics course and an English course?

b. How many students were taking an English course but not a mathematics course?

c. How many students were taking a mathematics course?

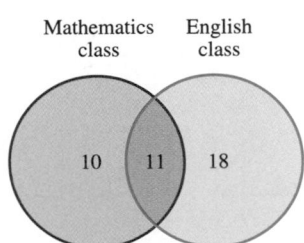

GUIDED PRACTICE

Find the number that comes next in each sequence.
See Example 1.

11. $1, 5, 9, 13, \ldots$

12. $11, 20, 29, 38, \ldots$

13. $5, 9, 14, 20, \ldots$

14. $6, 8, 12, 18, \ldots$

Find the number that comes next in each sequence.
See Example 2.

15. $15, 12, 9, 6, \ldots$

16. $81, 77, 73, 69, \ldots$

17. $-3, -5, -8, -12, \ldots$

18. $1, -8, -16, -25, -33, \ldots$

Find the letter that comes next in each sequence.
See Example 3.

19. E, I, H, L, K, O, N, $\ldots$

20. C, H, D, I, E, J, F, $\ldots$

21. c, b, d, c, e, d, f, $\ldots$

22. z, w, y, v, x, u, w, $\ldots$

Find the figure that comes next in each sequence.
See Example 4.

23.

24.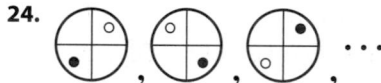

25.

26.

Find the figure that comes next in each sequence.
See Example 5.

27.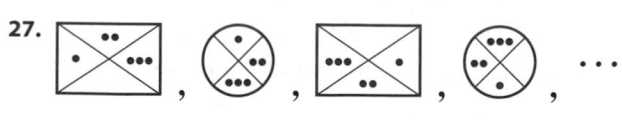

28.

29.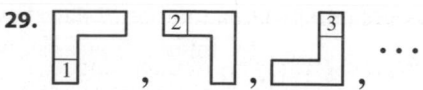

30.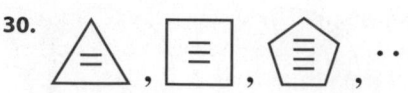

What conclusion can be drawn from each set of information?
See Example 6.

31. TEACHING SCHEDULES A small college offers only one biology course, one physics course, one chemistry course, and one zoology course. Each course is to be taught by a different adjunct professor. The four professors who will teach these courses have the following course preferences:

  1. Professors B and D don't want to teach zoology.

  2. Professor A wants to teach biology.

  3. Professor B wants to teach physics.

Who will teach chemistry?

32. DISPLAYS Four companies will be displaying their products on tables at a convention. Each company will be assigned one of the displays shown below. The companies have expressed the following preferences:

  1. Companies A and C don't want display 2.

  2. Company A wants display 3.

  3. Company D wants display 1.

Which company will get display 4?

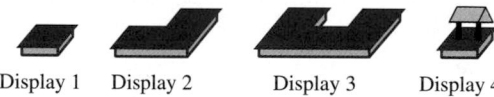

Display 1   Display 2   Display 3   Display 4

33. OCUPATIONS Four people named John, Luis, Maria, and Paula have occupations as teacher, butcher, baker, and candlestick maker.

  1. John and Paula are married.

  2. The teacher plans to marry the baker in December.

  3. Luis is the baker.

Who is the teacher?

34. PARKING A Ford, a Buick, a Dodge, and a Mercedes are parked side by side.

  1. The Ford is between the Mercedes and the Dodge.

  2. The Mercedes is not next to the Buick.

  3. The Buick is parked on the left end.

Which car is parked on the right end?

Use a circle diagram to solve each problem. **See Example 7.**

35. EMPLOYMENT HISTORY One hundred office managers were surveyed to determine their employment backgrounds. The survey results are shown below. How many office managers had neither sales nor manufacturing experience?

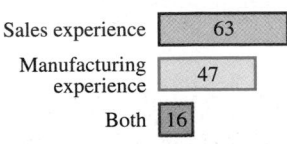

36. PURCHASING TEXTBOOKS Sixty college sophomores were surveyed to determine where they purchased their textbooks during their freshman year. The survey results are shown below. How many students did not purchase a book at a bookstore or online?

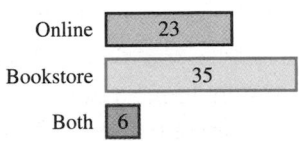

37. SIBLINGS When 27 children in a first-grade class were asked, "How many of you have a brother?" 11 raised their hands. When asked, "How many have a sister?" 15 raised their hands. Eight children raised their hands both times. How many children didn't raise their hands either time?

38. PETS When asked about their pets, a group of 35 sixth-graders responded as follows:

- 21 said they had at least one dog.
- 11 said they had at least one cat.
- 5 said they had at least one dog and at least one cat.

How many of the students do not have a dog or a cat?

▌TRY IT YOURSELF

Find the next letter or letters in the sequence.

39. A, c, E, g, . . .   **40.** R, SS, TTT, . . .

41. Z, A, Y, B, X, C, . . .   **42.** B, N, C, N, D, . . .

Find the missing figure in each sequence.

43.

44.

Find the next letter in the sequence.

45. C, B, F, E, I, H, L, . . .

46. d, h, g, k, j, n, . . .

Find the next number in the sequence.

47. $-7, 9, -6, 8, -5, 7, -4, \ldots$

48. $2, 5, 3, 6, 4, 7, 5, \ldots$

49. $9, 5, 7, 3, 5, 1, \ldots$

50. $1.3, 1.6, 1.4, 1.7, 1.5, 1.8, \ldots$

51. $-2, -3, -5, -6, -8, -9, \ldots$

52. $8, 5, 1, -4, -10, -17, \ldots$

53. $6, 8, 9, 7, 9, 10, 8, 10, 11, \ldots$

54. $10, 8, 7, 11, 9, 8, 12, 10, 9, \ldots$

55. ZOOS In a zoo, a zebra, a tiger, a lion, and a monkey are to be placed in four cages numbered from 1 to 4, from left to right. The following decisions have been made:

   1. The lion and the tiger should not be side by side.

   2. The monkey should be in one of the end cages.

   3. The tiger is to be in cage 4.

In which cage is the zebra?

56. FARM ANIMALS Four animals—a cow, a horse, a pig, and a sheep—are kept in a barn, each in a separate stall.

   1. The cow is in the first stall.

   2. Neither the pig nor the sheep can be next to the cow.

   3. The pig is between the horse and the sheep.

What animal is in the last stall?

57. OLYMPIC DIVING Four divers at the Olympics finished first, second, third, and fourth.

   1. Diver B beat diver D.

   2. Diver A placed between divers D and C.

   3. Diver D beat diver C.

In which order did they finish?

58. FLAGS A green, a blue, a red, and a yellow flag are hanging on a flagpole.

   1. The only flag between the green and yellow flags is blue.

   2. The red flag is next to the yellow flag.

   3. The green flag is above the red flag.

What is the order of the flags from top to bottom?

APPLICATIONS

59. JURY DUTY The results of a jury service questionnaire are shown below. Determine how many of the 20,000 respondents have served on neither a criminal court nor a civil court jury.

Jury Service Questionnaire

997	Served on a criminal court jury
103	Served on a civil court jury
35	Served on both

60. ELECTRONIC POLL For the Internet poll shown below, the first choice was clicked on 124 times, the second choice was clicked on 27 times, and both the first and second choices were clicked on 19 times. How many times was the third choice, "Neither" clicked on?

Internet Poll	You may vote for more than one.
What would you do if gasoline reached $5.50 a gallon?	⦿ Cut down on driving ⦿ Buy a more fuel-efficient car ⦿ Neither
	Number of people voting 178

61. THE SOLAR SYSTEM The graph below shows some important characteristics of the nine planets in our solar system. How many planets are neither rocky nor have moons?

Rocky planets	4
Planets with moons	7
Rocky planets with moons	2

62. WORKING TWO JOBS Andres, Barry, and Carl each have a completely different pair of jobs from the following list: jeweler, musician, painter, chauffeur, barber, and gardener. Use the facts below to find the two occupations of each man.

1. The painter bought a ring from the jeweler.
2. The chauffeur offended the musician by laughing at his mustache.
3. The chauffeur dated the painter's sister.
4. Both the musician and the gardener used to go hunting with Andres.
5. Carl beat both Barry and the painter at monopoly.
6. Barry owes the gardener $100.

WRITING

63. Describe deductive reasoning and inductive reasoning.

64. Describe a real-life situation in which you might use deductive reasoning.

65. Describe a real-life situation in which you might use inductive reasoning.

66. Write a problem in such a way that the diagram below can be used to solve it.

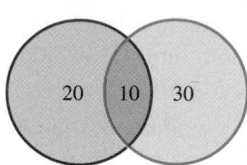

Roots and Powers

n	n^2	$\sqrt{n}$	n^3	$\sqrt[3]{n}$	n	n^2	$\sqrt{n}$	n^3	$\sqrt[3]{n}$
1	1	1.000	1	1.000	51	2,601	7.141	132,651	3.708
2	4	1.414	8	1.260	52	2,704	7.211	140,608	3.733
3	9	1.732	27	1.442	53	2,809	7.280	148,877	3.756
4	16	2.000	64	1.587	54	2,916	7.348	157,464	3.780
5	25	2.236	125	1.710	55	3,025	7.416	166,375	3.803
6	36	2.449	216	1.817	56	3,136	7.483	175,616	3.826
7	49	2.646	343	1.913	57	3,249	7.550	185,193	3.849
8	64	2.828	512	2.000	58	3,364	7.616	195,112	3.871
9	81	3.000	729	2.080	59	3,481	7.681	205,379	3.893
10	100	3.162	1,000	2.154	60	3,600	7.746	216,000	3.915
11	121	3.317	1,331	2.224	61	3,721	7.810	226,981	3.936
12	144	3.464	1,728	2.289	62	3,844	7.874	238,328	3.958
13	169	3.606	2,197	2.351	63	3,969	7.937	250,047	3.979
14	196	3.742	2,744	2.410	64	4,096	8.000	262,144	4.000
15	225	3.873	3,375	2.466	65	4,225	8.062	274,625	4.021
16	256	4.000	4,096	2.520	66	4,356	8.124	287,496	4.041
17	289	4.123	4,913	2.571	67	4,489	8.185	300,763	4.062
18	324	4.243	5,832	2.621	68	4,624	8.246	314,432	4.082
19	361	4.359	6,859	2.668	69	4,761	8.307	328,509	4.102
20	400	4.472	8,000	2.714	70	4,900	8.367	343,000	4.121
21	441	4.583	9,261	2.759	71	5,041	8.426	357,911	4.141
22	484	4.690	10,648	2.802	72	5,184	8.485	373,248	4.160
23	529	4.796	12,167	2.844	73	5,329	8.544	389,017	4.179
24	576	4.899	13,824	2.884	74	5,476	8.602	405,224	4.198
25	625	5.000	15,625	2.924	75	5,625	8.660	421,875	4.217
26	676	5.099	17,576	2.962	76	5,776	8.718	438,976	4.236
27	729	5.196	19,683	3.000	77	5,929	8.775	456,533	4.254
28	784	5.292	21,952	3.037	78	6,084	8.832	474,552	4.273
29	841	5.385	24,389	3.072	79	6,241	8.888	493,039	4.291
30	900	5.477	27,000	3.107	80	6,400	8.944	512,000	4.309
31	961	5.568	29,791	3.141	81	6,561	9.000	531,441	4.327
32	1,024	5.657	32,768	3.175	82	6,724	9.055	551,368	4.344
33	1,089	5.745	35,937	3.208	83	6,889	9.110	571,787	4.362
34	1,156	5.831	39,304	3.240	84	7,056	9.165	592,704	4.380
35	1,225	5.916	42,875	3.271	85	7,225	9.220	614,125	4.397
36	1,296	6.000	46,656	3.302	86	7,396	9.274	636,056	4.414
37	1,369	6.083	50,653	3.332	87	7,569	9.327	658,503	4.431
38	1,444	6.164	54,872	3.362	88	7,744	9.381	681,472	4.448
39	1,521	6.245	59,319	3.391	89	7,921	9.434	704,969	4.465
40	1,600	6.325	64,000	3.420	90	8,100	9.487	729,000	4.481
41	1,681	6.403	68,921	3.448	91	8,281	9.539	753,571	4.498
42	1,764	6.481	74,088	3.476	92	8,464	9.592	778,688	4.514
43	1,849	6.557	79,507	3.503	93	8,649	9.644	804,357	4.531
44	1,936	6.633	85,184	3.530	94	8,836	9.695	830,584	4.547
45	2,025	6.708	91,125	3.557	95	9,025	9.747	857,375	4.563
46	2,116	6.782	97,336	3.583	96	9,216	9.798	884,736	4.579
47	2,209	6.856	103,823	3.609	97	9,409	9.849	912,673	4.595
48	2,304	6.928	110,592	3.634	98	9,604	9.899	941,192	4.610
49	2,401	7.000	117,649	3.659	99	9,801	9.950	970,299	4.626
50	2,500	7.071	125,000	3.684	100	10,000	10.000	1,000,000	4.642

Think It Through (page 9)

1. c **2.** b **3.** e **4.** d **5.** a

Study Set Section 1.1 (page 10)

1. digits **3.** standard **5.** expanded **7.** inequality

9.

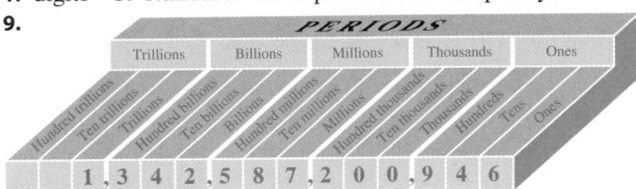

PERIODS

Trillions | Billions | Millions | Thousands | Ones

1, 3 4 2, 5 8 7, 2 0 0, 9 4 6

11. a. forty **b.** ninety **c.** sixty-eight **d.** fifteen

13.

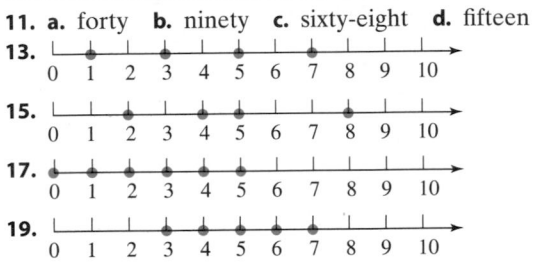

15.

17.

19.

21. braces **23. a.** 3 tens **b.** 7 **c.** 6 hundreds **d.** 5

25. a. 1 hundred million **b.** 7 **c.** 9 tens **d.** 4

27. ninety-three **29.** seven hundred thirty-two

31. one hundred fifty-four thousand, three hundred two

33. fourteen million, four hundred thirty-two thousand, five hundred **35.** nine hundred seventy billion, thirty-one million, five hundred thousand, one hundred four

37. eighty-two million, four hundred fifteen **39.** 3,737

41. 930 **43.** 7,021 **45.** 26,000,432 **47.** 200 + 40 + 5

49. 3,000 + 600 + 9 **51.** 70,000 + 2,000 + 500 + 30 + 3

53. 100,000 + 4,000 + 400 + 1

55. 8,000,000 + 400,000 + 3,000 + 600 + 10 + 3

57. 20,000,000 + 6,000,000 + 100 + 50 + 6

59. a. > **b.** < **61. a.** > **b.** < **63.** 98,150

65. 512,970 **67.** 8,400 **69.** 32,400 **71.** 66,000

73. 2,581,000 **75.** 53,000; 50,000 **77.** 77,000; 80,000

79. 816,000; 820,000 **81.** 297,000; 300,000 **83. a.** 79,590

b. 79,600 **c.** 80,000 **d.** 80,000 **85. a.** $419,160

b. $419,200 **c.** $419,000 **d.** $420,000 **87.** 40,025

89. 202,036 **91.** 27,598 **93.** 10,700,506 **95.** Aisha

97. a. the 1970s, 7 **b.** the 1960s, 9 **c.** the 1960s, 12

d. the 1980s

99.

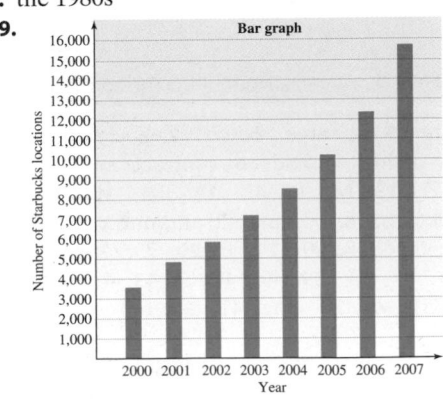

Bar graph

101. a.

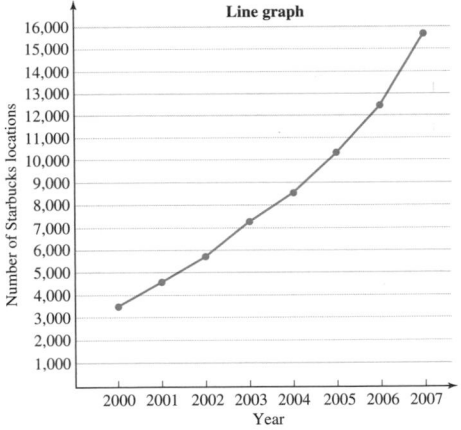

Line graph

DON SMITH
1234 MILL STREET
HILLDALE, CA
7155
DATE *March 9, 2010*
Payable to *Davis Chevrolet* $ 15,601.00
Fifteen thousand six hundred one and 00/100 DOLLARS
FIRST FEDERAL BANK
195 JEFFS STREET
HILLDALE, CA
Memo _____ *Don Smith*

b.

JUAN DECITO
24 ARBOR LANE
ARGENTO, CA
4251
DATE *Aug. 12, 2010*
Payable to *DR. ANDERSON* $ 3,433.00
Three thousand four hundred thirty-three and 00/100 DOLLARS
FIRST FEDERAL BANK
195 JEFFS STREET
HILLDALE, CA
Memo _____ *Juan Decito*

103. 1,865,593; 482,880; 1,503; 269; 43,449

105. a. hundred thousands **b.** 980,000,000; 9 hundred millions + 8 ten millions **c.** 1,000,000,000; one billion

Study Set Section 1.2 (page 29)

1. addend, addend, sum **3.** commutative **5.** estimate

7. rectangle, square **9.** square **11.** minuend, subtrahend, difference **13.** related **15. a.** commutative property of addition **b.** associative property of addition **c.** associative property of addition **d.** commutative property of addition

17. 4, 3, 7 **19.** left, right **21.** parentheses, first **23.** 17, 29

25. 38 **27.** 689 **29.** 461 **31.** 8,937 **33.** 33 **35.** 137

37. 37,500 **39.** 1,020,000 **41.** 88 ft **43.** 68 in.

45. 376 mi **47.** 186 cm **49.** 7,642 **51.** 2,562 **53.** 8,457

55. 6,483 **57.** 51,677 **59.** 44,444 **61.** correct

63. incorrect **65.** 66,000 **67.** 50,000 **69.** 29 **71.** 608

73. 15,907 **75.** 2,901 **77.** 56,460 **79.** 65 **81.** 65

83. 19,929 **85.** 197 **87.** 979 **89.** 303 **91.** 30,000

93. 48,760 **95.** 91 ft **97.** 79,787,000 visitors **99.** $28,800

101. 196 in. **103.** 384 ft **105.** 1,420 lb **107.** 1,495 mi

109. 6,034,093 magazines **111.** 1,764°F **113. a.** $39,565

b. $1,322 **119. a.** 3,000 + 100 + 20 + 5 **b.** 60,000 + 30 + 7

121. a. 5,370,650 **b.** 5,370,000 **c.** 5,400,000

Study Set Section 1.3   (page 44)

1. factor, factor, product   3. commutative, associative
5. square   7. a. $4 \cdot 8$ b. $15 + 15 + 15 + 15 + 15 + 15 + 15$
9. a. 3   b. 5   11. a. area   b. perimeter   c. area
d. perimeter   13. $\times, \cdot, (\ )$ 15. $A = l \cdot w$ or $A = lw$
17. 105   19. 272   21. 3,700   23. 750   25. 1,070,000
27. 512,000   29. 2,720   31. 11,200   33. 390,000
35. 108,000,000   37. 9,344   39. 18,368   41. 408,758
43. 16,868,238   45. 1,800   47. 135,000   49. 18,000
51. 400,000   53. 84 in.^{2} 55. 144 in.^{2} 57. 1,491
59. 68,948   61. 7,623   63. 0   65. 1,590   67. 44,486
69. 8,945,912   71. 374,644   73. 9,900   75. 2,400,000
77. 355,712   79. 166,500   81. 72 cups   83. 204 grams
85. 3,900 times   87. 63,360 in.   89. 77,000 words
91. $73,645,500   93. 72 entries   95. no   97. 18 hr
99. $1,386 per night   101. 84 tablets   103. 54 ft^2
105. 1,260 mi, 97,200 mi^2 109. 20,642

Study Set Section 1.4   (page 59)

1. dividend, divisor, quotient; divisor, quotient, dividend;
dividend, divisor, quotient   3. long   5. divisible   7. a. 7
b. 5, 2   9. a. 1   b. 6   c. undefined   d. 0   11. a. 2   b. 6
c. 3   d. 5   13. 37; 333   15. a. 0, 5   b. 2, 3   c. sum
d. 10   17. $\div, \overline{)\ }, -$ 19. 5, 9, 45   21. 4, 11, 44
23. $7 \cdot 3 = 21$ 25. $6 \cdot 12 = 72$ 27. 16   29. 29
31. 325   33. 218   35. 504   37. 602   39. 39 R 15
41. 21 R 33   43. 47 R 86   45. 19 R 132   47. 2, 3, 4, 5, 6, 10
49. 3, 5, 9   51. none   53. 2, 3, 4, 5, 6, 10   55. 70   57. 22
59. 9,000   61. 50   63. 4,325   65. 6   67. 8 R 25   69. 160
71. 106 R 3   73. 509   75. 3,080   77. 5   79. 23 R 211
81. 30 R 13   83. 89   85. 7 R 1   87. 625 tickets
89. 27 trips   91. 2 cartons, 4 cartons   93. 9 times, 28 ounces
95. 14,500 lb   97. $105   99. 5 mi   101. 13 dozen
103. 9 girls   105. $4,344, $3,622, $2,996   111. 3,281
113. 1,097,334

Study Set Section 1.5   (page 70)

1. factors   3. prime   5. prime   7. base, exponent
9. 45, 15, 9; 1, 3, 5, 9, 15, 45   11. yes   13. a. even, odd
b. 0, 2, 4, 6, 8, 10, 12, 14, 16, 18   c. 1, 3, 5, 7, 9, 11, 13, 15, 17, 19
15. 5, 6, 2; 2, 3, 5, 5   17. 2, 25, 2, 3, 5, 5   19. a. base: 7,
exponent: 6   b. base: 15, exponent: 1   21. 1, 2, 5, 10
23. 1, 2, 4, 5, 8, 10, 20, 40   25. 1, 2, 3, 6, 9, 18   27. 1, 2, 4, 11,
22, 44   29. 1, 7, 11, 77   31. 1, 2, 4, 5, 10, 20, 25, 50, 100
33. $2 \cdot 4$ 35. $3 \cdot 9$ 37. $7 \cdot 7$ 39. $2 \cdot 10$ or $4 \cdot 5$ 41. $2 \cdot 3 \cdot 5$
43. $3 \cdot 3 \cdot 7$ 45. $2 \cdot 3 \cdot 9$ or $3 \cdot 3 \cdot 6$ 47. $2 \cdot 3 \cdot 10$ or $2 \cdot 2 \cdot 15$
or $2 \cdot 5 \cdot 6$ or $3 \cdot 4 \cdot 5$ 49. 1 and 11   51. 1 and 37   53. yes
55. no, $(9 \cdot 11)$ 57. no, $(3 \cdot 17)$ 59. yes   61. $2 \cdot 3 \cdot 5$
63. $3 \cdot 13$ 65. $3^2 \cdot 11$ 67. $2 \cdot 3^4$ 69. 2^6 71. $3 \cdot 7^2$
73. $2^2 \cdot 5 \cdot 11$ 75. $2 \cdot 3 \cdot 17$ 77. 2^5 79. 5^4 81. $4^2(8^3)$
83. $7^7 \cdot 9^2$ 85. a. 81   b. 64   87. a. 32   b. 25   89. a. 343
b. 2,187   91. a. 9   b. 1   93. 90   95. 847   97. 225
99. 2,808   101. 1, 2, 4, 7, 14, 28, $1 + 2 + 4 + 7 + 14 = 28$
103. 2^2 square units, 3^2 square units, 4^2 square units
109. 125 band members

Study Set Section 1.6   (page 81)

1. multiples   3. divisible   5. a. 12   b. smallest   7. a. 20
b. 20   9. a. two   b. two   c. one   d. 2, 2, 3, 3, 5, 180

11. a. two   b. three   c. 2, 3, 108   13. a. 2, 3, 5   b. 30
15. a. GCF   b. LCM   17. 4, 8, 12, 16, 20, 24, 28, 32
19. 11, 22, 33, 44, 55, 66, 77, 88   21. 8, 16, 24, 32, 40, 48, 56, 64
23. 20, 40, 60, 80, 100, 120, 140, 160   25. 15   27. 24   29. 55
31. 28   33. 12   35. 30   37. 80   39. 150   41. 315   43. 600
45. 72   47. 60   49. 2   51. 3   53. 11   55. 15   57. 6
59. 14   61. 1   63. 1   65. 4   67. 36   69. 600, 20
71. 140, 14   73. 2,178; 22   75. 3,528; 1   77. 3,000; 5
79. 204, 34   81. 138, 23   83. 4,050; 1   85. 15,000 mi,
22,500 mi, 30,000 mi, 37,500 mi, 45,000 mi   87. 180 min or 3 hr
89. 6 packages of hot dogs and 5 packages of buns
91. 12 pieces   93. a. $7   b. 1st day: 4 students, 2nd day:
3 students, 3rd day: 9 students   99. 11,110   101. 15,250

Study Set Section 1.7   (page 92)

1. expressions   3. parentheses, brackets   5. inner, outer
7. a. square, multiply, subtract   b. multiply, cube, add,
subtract   c. square, multiply   d. multiply, square
9. multiply, square   11. the fraction bar, the numerator and
the denominator   13. quantity   15. 4, 20, 8   17. 9, 36, 16, 20
19. 47   21. 13   23. 38   25. 36   27. 24   29. 12
31. a. 33   b. 15   33. a. 43   b. 27   35. 100   37. 512
39. 64   41. 203   43. 73   45. 81   47. 3   49. 4   51. 6
53. 5   55. 16   57. 4   59. 5   61. 162   63. 27   65. 10
67. 3   69. 5,239   71. 15   73. 25   75. 22   77. 53   79. 2
81. 1   83. 25   85. 813   87. 49   89. 11   91. 191   93. 34
95. 323   97. undefined   99. 14   101. 192   103. 74
105. $3(7) + 4(4) + 2(3)$, $43   107. $3(8 + 7 + 8 + 8 + 7)$, 114
109. brick: $3(3) + 1 + 1 + 3 + 3(5)$, 29;
aphid: $3[1 + 2(3) + 4 + 1 + 2]$, 42
111. $2^2 + 3^2 + 5^2 + 7^2 = 4 + 9 + 25 + 49 = 87$
113. 79°   115. 31 therms   117. 300 calories
119. a. 125   b. $11,875   c. $95
125. two hundred fifty-four thousand, three hundred nine

Study Set Section 1.8   (page 102)

1. equation, =   2. solve   5. equivalent   7. a. $x + 6$
b. neither   c. no   d. yes   9. a. same   b. c 11. a. add
b. subtract   13. 5, 5, 50; 50, 45, 50   15. is possibly equal to
17. yes   19. yes   21. no   23. no   25. 10   27. 70   29. 3
31. 61   33. 3   35. 5   37. 1,700, 425, jar; jar, addition, 1,700,
x; 1,700, 425, 425, 1,275; 1,275; 1,700   39. 45   41. 4   43. 13
45. 75   47. 740   49. 339   51. 9   53. 10   55. 1   57. 56
59. 84   61. 105   63. 4   65. 12   67. 8   69. 47   71. She
will need to borrow $248,000.   73. 50 Cent earned $150
million in 2008.   75. The earplugs reduce the noise level by
29 decibels.   77. The reading must increase by 25 units to
cause the system to shut down.   79. The gas station was
going to charge her $219.   81. Jimmy Boyd was 12 years
old when he had the number 1 song.   91. 325,780
93. 90   95. 3

Study Set Section 1.9   (page 110)

1. solve   3. isolate   5. a. same   b. cb 7. a. x b. x
9. a. multiply   b. divide   c. add   d. subtract
11. 5, 5, 45, 45, 9, 45   13. 14   15. 42   17. 384   19. 341
21. 1   23. 10   25. 3, $318,500, amount; amount, x, 318,500;
3, 3, 955,500; $955,500; $318,500   27. 1   29. 75   31. 2
33. 10   35. 3,000   37. 50   39. 49   41. 3   43. 4
45. 4,020   47. 1,251   49. 30   51. 141   53. 6

55. Before the course, Alicia could read 133 words per minute.   **57.** The initial cost estimate was $54 million.   **59.** 49 rows will need to be used.   **61.** The shelter received 32 calls each day after being featured on the news.   **63.** The scale would register 55 pounds.   **65.** The average life span of a guinea pig is 8 years.   **71.** 48 cm   **73.** $2^3 \cdot 3 \cdot 5$ **75.** 72   **77.** 0

Chapter 1 Review   (page 114)

1. a. 6   **b.** 7   **c.** 1 billion   **d.** 8   **2. a.** ninety-seven thousand, two hundred eighty-three   **b.** five billion, four hundred forty-four million, sixty thousand, seventeen   **3. a.** 3,207   **b.** 23,253,412   **4.** 61,204   **5.** $500,000 + 70,000 + 300 + 2$ **6.** $30,000,000 + 7,000,000 + 300,000 + 9,000 + 100 + 50 + 4$

7.
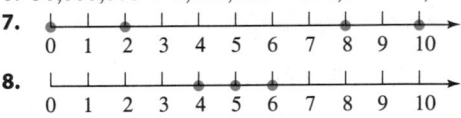

8.

9. $>$ **10.** $<$ **11. a.** 2,507,300   **b.** 2,510,000   **c.** 2,507,350   **d.** 3,000,000   **12. a.** 970,000   **b.** 1,000,000

13. a.
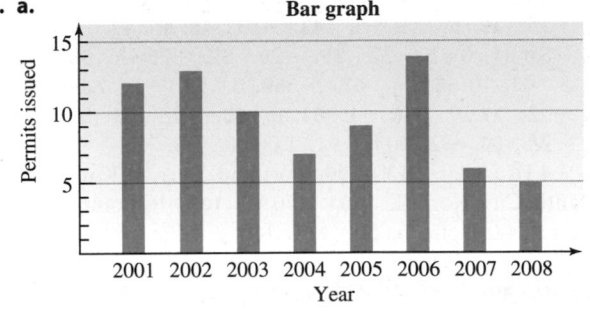

Bar graph

b.
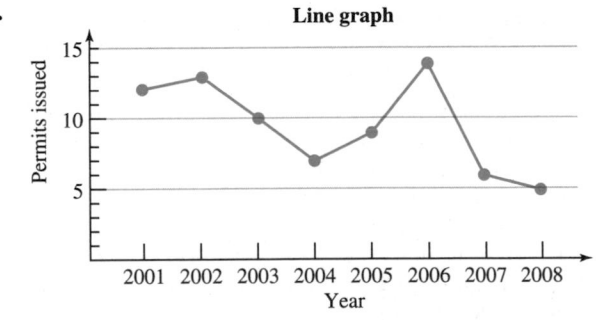

Line graph

14. Nile, Amazon, Yangtze, Mississippi-Missouri, Ob-Irtysh   **15.** 463   **16.** 59   **17.** 6,000   **18.** 50   **19.** 12,601   **20.** 152,169   **21.** 59,400   **22. a.** $61 + 24$ **b.** $(9 + 91) + 29$ **23.** 227,453,217 passengers   **24.** no   **25.** 14,661   **26.** 779,666   **27.** $1,324,700,000   **28.** 2,746 ft   **29.** 61   **30.** 217   **31.** 505   **32.** 2,075   **33.** incorrect   **34.** $12 + 8 = 20$ **35.** 160,000   **36.** 3,041,092 square miles   **37.** $13,445   **38.** 54 days   **39.** 423   **40.** 210   **41.** 720,000   **42.** 9,263   **43.** 1,580,344   **44.** 230,418   **45.** 2,800,000   **46. a.** $5 \cdot 7$ **b.** $2t$ **c.** mn **47. a.** 0   **b.** 7   **48. a.** associative property of multiplication   **b.** commutative property of multiplication   **49.** 32 cm²   **50.** 6,084 in.²   **51. a.** 2,555 hr   **b.** 3,285 hr   **52.** 330 members   **53.** Santiago   **54.** 14,400 eggs   **55.** 18   **56** 37   **57.** 307   **58.** 19 R 6   **59.** 0   **60.** undefined   **61.** 42 R 13   **62.** 380   **63.** $40 \cdot 4 = 160$ **64.** It is not correct.   **65.** It is divisible by 3, 5, and 9.   **66.** 4,000   **67.** 16; 25   **68.** 34 cars   **69.** 1, 2, 3, 6, 9, 18   **70.** 1, 3, 5, 15, 25, 75

71. $2 \cdot 10$ or $4 \cdot 5$ **72.** $2 \cdot 3 \cdot 9$ or $3 \cdot 3 \cdot 6$ **73. a.** prime   **b.** composite   **c.** neither   **d.** neither   **e.** composite   **f.** prime   **74. a.** odd   **b.** even   **c.** even   **d.** odd   **75.** $2 \cdot 3 \cdot 7$ **76.** $3 \cdot 5^2$ **77.** $2^2 \cdot 5 \cdot 11$ **78.** $2^2 \cdot 5 \cdot 7$ **79.** 6^4 **80.** $5^3 \cdot 13^2$ **81.** 125   **82.** 121   **83.** 784   **84.** 2,700   **85.** 9, 18, 27, 36, 45, 54, 63, 72, 81, 90   **86. a.** 24, 48   **b.** 1, 2   **87.** 12   **88.** 12   **89.** 45   **90.** 36   **91.** 126   **92.** 360   **93.** 140   **94.** 84   **95.** 4   **96.** 3   **97.** 10   **98.** 15   **99.** 21   **100.** 28   **101.** 24   **102.** 44   **103.** 42 days   **104. a.** 8 arrangements   **b.** 4 red carnations, 3 white carnations, 2 blue carnations   **105.** 45   **106.** 23   **107.** 243   **108.** 4   **109.** 32   **110.** 72   **111.** 8   **112.** 8   **113.** 1   **114.** 3   **115.** 28   **116.** 9   **117.** 77   **118.** 60   **119.** no   **120.** yes   **121.** 9   **122.** 31   **123.** 340   **124.** 133   **125.** 9   **126.** 14   **127.** 120   **128.** 5   **129.** The couple needed to borrow $97,250.   **130.** The doctor originally had 185 patients.   **131.** 4   **132.** 3   **133.** 21   **134.** 14   **135.** 21   **136.** 36   **137.** 315   **138.** 425   **139.** The week before, the company received 182 orders.   **140.** The chain cost $128.

Chapter 1 Test   (page 132)

1. a. whole   **b.** inequality   **c.** area   **d.** parentheses, brackets   **e.** prime   **f.** equation   **g.** solution   **h.** equality

2.
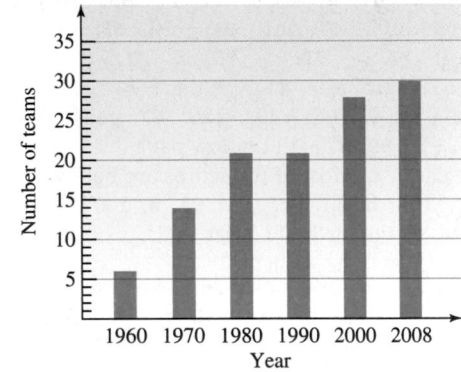

3. a. 1 hundred   **b.** 0   **4. a.** seven million, eighteen thousand, six hundred forty-one   **b.** 1,385,266   **c.** $90,000 + 2,000 + 500 + 60 + 1$ **5. a.** $>$ **b.** $<$ **6. a.** 35,000,000   **b.** 34,800,000   **c.** 34,760,000

7.
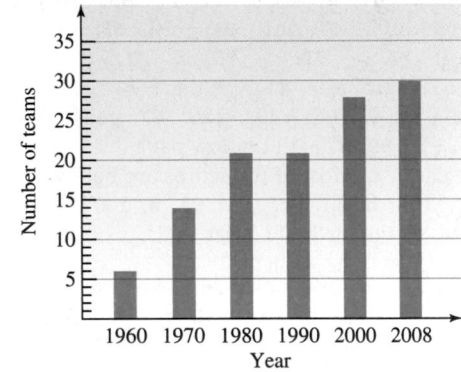

8. $248, 248 + 287 = 535$ **9.** 225,164   **10.** 942   **11.** 424   **12.** 41,588   **13.** 72   **14.** 114 R 57, $(73 \cdot 114) + 57 = 8,379$ **15.** 13,800,000   **16.** 250   **17.** 43,000   **18.** 2,168 in.   **19.** 529 cm²   **20. a.** 1, 2, 3, 4, 6, 12   **b.** 4, 8, 12, 16, 20, 24   **c.** $8 \cdot 5$ **21.** $2^2 \cdot 3^2 \cdot 5 \cdot 7$ **22.** 4,933 tails   **23.** 96 students   **24.** 4,085 ft²   **25.** 414 mi   **26. a.** associative property of multiplication   **b.** commutative property of addition   **27. a.** 0   **b.** 0   **c.** 1   **d.** undefined   **28.** 90   **29.** 72   **30.** 6   **31.** 4   **32. a.** 40 in.   **b.** rice: 5 boxes, potatoes: 4 boxes   **33.** It is divisible by 2, 3, 4, 5, 6, and 10.   **34.** 58   **35.** 29   **36.** 762   **37.** 44   **38.** 1   **39.** yes   **40.** To solve an equation means to find all the values of the variable that, when substituted into the equation, make a true statement.   **41.** 99   **42.** 30   **43.** 11   **44.** 81   **45.** At this time, the college has 2,080 parking spaces.   **46.** The sound intensity of the band is 114 decibels.   **47.** There were 72 students in the class.   **48.** She needs to borrow $14,750.

Think It Through (page 139)

$4,621, $1,073, $3,325

Study Set Section 2.1 (page 143)

1. Positive, negative 3. graph 5. absolute value
7. a. −225 b. −10 sec c. −3° d. −$12,000 e. −1 mi
9. a. The spacing is not uniform. b. The numbering is not uniform. c. Zero is missing. d. The arrowheads are not drawn. 11. a. −4 b. −2 13. a. −7 b. 8
15. a. 15 > −12 b. −5 < −4
17.

Number	Opposite	Absolute value
−25	25	25
39	−39	39
0	0	0

19. a. −(−8) b. |−8| c. 8 − 8 d. −|−8|
21. a. greater, equal b. less, equal
23.
25.
27.
29.
31. < 33. < 35. > 37. > 39. true 41. true
43. false 45. false 47. 9 49. 8 51. 14 53. 180
55. 11 57. 4 59. 102 61. 561 63. −20 65. −6
67. −253 69. 0 71. > 73. < 75. > 77. <
79. −52, −22, −12, 12, 52, 82 81. −3, −5, −7
83. −31 lengths 85. 0, 20, 5, −40, −120 87. peaks: 2, 4, 0; valleys: −3, −5, −2 89. a. −1 (1 below par)
b. −3 (3 below par) c. Most of the scores are below par.
91. a. −20° to −10° b. 40° c. 10° 93. a. 200 yr
b. A.D. c. B.C. d. the birth of Christ
95.

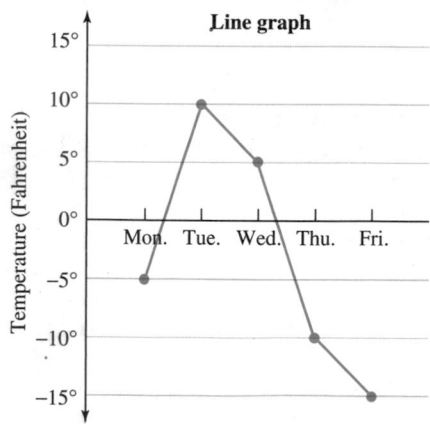

105. 23,500 107. 761
109. associative property of multiplication

Think It Through (page 152)

decrease expenses, increase income, decrease expenses, increase income, increase income, increase income, decrease expenses, decrease expenses, increase income, decrease expenses

Study Set Section 2.2 (page 156)

1. like 3. identity 5. Commutative
7. a. |10| = 10, |−12| = 12 b. −12 c. 2 9. subtract, larger 11. a. yes b. yes c. no d. no 13. a. 0 b. 0
15. −18, −19 17. 5, 2 19. −9 21. −10 23. −62
25. −96 27. −379 29. −874 31. −3 33. 1
35. −22 37. 48 39. 357 41. −60 43. 7 45. −4
47. −10 49. 41 51. 3 53. −6 55. 3 57. −7
59. 9 61. −562 63. 2 65. 0 67. 0 69. −2 71. −1
73. −3 75. −1,032 77. −21 79. −8,348 81. −20
83. 112°F, 114°F 85. a. −15,720 ft b. −12,500 ft
87. a. −9 ft b. 2 ft above flood stage 89. 195°
91. 5, 4% risk 93. 3,250 m 95. ($967) 103. a. 16 ft
b. 15 ft² 105. 2 · 5³

Study Set Section 2.3 (page 166)

1. opposite, additive 3. value 5. opposite 7. −3, 6
9. change 11. a. 3 b. −12 13. +, 6, 9
15. a. −8 − (−4) b. −4 − (−8) 17. −3, 2, 0
19. −2, −10, 6, −4 21. −7 23. −10 25. 9 27. 18
29. −18 31. −50 33. a. −10 b. 10 35. a. 25 b. −25
37. −15 39. 9 41. −2 43. −10 45. 9 47. −12
49. −8 51. 0 53. 32 55. −26 57. −2,447 59. 43,900
61. 3 63. 10 65. 8 67. 5 69. 3 71. −1 73. −9
75. −22 77. 9 79. −4 81. 0 83. −18 85. 8
87. −25 89. −2,200 ft 91. 1,066 ft 93. −8
95. −4 yd 97. −$140 99. Portland, Barrow, Kansas City, Atlantic City, Norfolk 101. 470°F 103. 16-point increase
109. a. 24,090 b. 6,000 111. 156

Study Set Section 2.4 (page 176)

1. factor, factor, product 3. unlike 5. Associative
7. positive, negative 9. negative 11. unlike/different
13. 0 15. a. 3 b. 12 17. a. base: 8, exponent: 4
b. base: −7, exponent: 9 19. 6, −24 21. −15 23. −18
25. −72 27. −126 29. −1,665 31. −94,000 33. 56
35. 7 37. 156 39. 276 41. 1,947 43. 72,000,000
45. 90 47. 150 49. −384 51. −336 53. −48 55. −81
57. 36 59. 144 61. −27 63. −32 65. 625 67. 1
69. 49, −49 71. 144, −144 73. −60 75. 0 77. −64
79. −20 81. −18 83. 60 85. −48 87. −8,400,000
89. −625 91. 144 93. 1 95. −120 97. −2,000 ft
99. a. high: 2, low: −3 b. high: 4, low: −6
101. a. −402,000 jobs b. −423,000 jobs c. −581,000 jobs
d. −528,000 jobs 103. −324°F 105. −$1,200
107. −18 ft 109. −$215,718 115. 2, 3, 5, 7, 11, 13, 17, 19, 23, 29 117. 43 R 3

Study Set Section 2.5 (page 184)

1. dividend, divisor, quotient; dividend, divisor, quotient
3. by, of 5. a. −5(5) = −25 b. 6(−6) = −36
c. 0(−15) = 0 7. a. positive b. negative 9. a. 0
b. undefined 11. a. always true b. sometimes true
c. always true 13. −7 15. −4 17. −6 19. −8
21. −22 23. −39 25. −30 27. −50 29. 2 31. 5
33. 9 35. 4 37. 16 39. 21 41. 40 43. 500
45. a. undefined b. 0 47. a. 0 b. undefined 49. 3
51. −17 53. 0 55. −5 57. −5 59. undefined
61. −19 63. 1 65. −20 67. −1 69. 10 71. −24

73. −30 **75.** −4 **77.** −542 **79.** −1,634 **81.** −$35 per week **83.** −1,010 ft **85.** −7° per min **87.** −6 (6 games behind) **89.** −$15 **91.** −$17 **99.** 211 **101.** associative property of addition **103.** no

Study Set Section 2.6 (page 192)

1. order **3.** inner, outer **5. a.** square, multiplication, subtraction **b.** multiplication, cube, subtraction, addition **c.** subtraction, multiplication, addition **d.** square, multiplication **7.** parentheses, brackets, absolute value symbols, fraction bar **9.** 4, 20, −20, −28 **11.** −8, −1, −5, −14 **13.** −10 **15.** −62 **17.** 15 **19.** 12 **21.** −12 **23.** −80 **25.** −72 **27.** −200 **29.** 4 **31.** 28 **33.** 17 **35.** 71 **37.** 21 **39.** 50 **41.** −6 **43.** −12 **45. a.** 12 **b.** 5 **47. a.** 60 **b.** 14 **49.** −2 **51.** −3 **53.** −770 **55.** −5,000 **57.** −7 **59.** 1 **61.** 17 **63.** −21 **65.** 19 **67.** −7 **69.** 12 **71.** −14 **73.** −11 **75.** 2 **77.** −5 **79.** −3 **81.** −5 **83.** 166 **85.** 0 **87.** −14 **89.** 112 **91.** 22 **93.** 8 **95.** +3 **97.** −400 points **99.** 19 **101.** −$8 million **103.** It's better to refer to the last four years, because there was an average budget surplus of $16 billion. **105. a.** 90 ft below sea level (−90) **b.** $600 lost (−600) **c.** −400 ft **111. a.** −3 **b.** −4 **113.** no

Study Set Section 2.7 (page 203)

1. solve **3.** check **5. a.** multiplication by −2 **b.** addition of −6 **c.** division by −5 **d.** subtraction of 4 **7. a.** add 9 to both sides **b.** divide both sides by −8 **9.** same **11.** subtracting, dividing **13.** −13, 7, 7; −6, −13, −6 **15. a.** −10 · x **b.** x ÷ (−8) **17.** −9 **19.** 5 **21.** −24 **23.** −52 **25.** 17 **27.** 5 **29.** −8 **31.** 5 **33.** 27 **35.** −77 **37.** −14 **39.** 58 **41.** −4 **43.** −8 **45.** 10 **47.** −9 **49.** 6 **51.** −52 **53.** −4 **55.** −1 **57.** 0 **59.** −3 **61.** 15 **63.** −6 **65.** −14 **67.** −1 **69.** −3 **71.** −8 **73.** −9 **75.** −14 **77.** −6 **79.** −2 **81.** −495 **83.** −2 **85.** 54 **87.** −7 **89.** −120, −75, feet; raised, addition; x, −75; x, 120, 120, 45; 45, −75 **91.** In 2007, Crocs made $168 million in profit. **93.** The company gained 34 points of market share in five years. **95.** His checking account balance before the deposit was −$175. **97.** Detroit had −53 yards rushing that day. **99.** The Roman Empire lasted for 503 years. **101.** In the second quarter of 2009, Continental Airlines lost $3 million (−$3 million). **105.** 5 · 5 · 5 · 5 · 5 · 5 **107.** 0 **109.** 9, 218 **111.** 26, 058

Chapter 2 Review (page 208)

1. {..., −5, −4, −3, −2, −1, 0, 1, 2, 3, 4, 5, ...} **2. a.** −$1,200 **b.** −10 sec **3.** −33 ft
4. a.

$$-4 \quad -3 \quad -2 \quad -1 \quad 0 \quad 1 \quad 2 \quad 3 \quad 4$$

b.

$$-4 \quad -3 \quad -2 \quad -1 \quad 0 \quad 1 \quad 2 \quad 3 \quad 4$$

5. a. > **b.** < **6. a.** false **b.** true **7. a.** 5 **b.** 43 **c.** 0 **8. a.** −8 **b.** 8 **c.** 0 **9. a.** −12 **b.** 12 **c.** 0 **10. a.** negative **b.** the opposite **c.** negative **d.** minus

11.

Position	Player	Score to par
1	Helen Alfredsson	−12
2	Yani Tseng	−9
3	Laura Diaz	−8
4	Karen Stupples	−7
5	Young Kim	−6
6	Shanshan Feng	−5

12. a. 1998, $60 billion **b.** 2000, $230 billion **c.** 2004, −$420 billion **13.** −10 **14.** −9 **15.** 32 **16.** 73 **17.** 0 **18.** 0 **19.** −8 **20.** −3 **21.** 10 **22.** 8 **23.** −4 **24.** −20 **25.** −76 **26.** −31 **27.** −374 **28.** 3,128 **29. a.** 11 **b.** −4 **30. a.** yes **b.** yes **c.** no **d.** no **31. a.** −100 ft **b.** −66 ft **32.** 136°F **33.** opposite **34. a.** −9 − (−1) **b.** −6 − (−10) **35.** −3 **36.** −21 **37.** 4 **38.** −6 **39.** −112 **40.** −8 **41.** −37 **42.** 30 **43.** 16 **44.** −24 **45.** −4 **46.** 22 **47.** 6 **48.** −8 **49.** −62 **50.** 103 **51.** 75 **52. a.** −77 **b.** 77 **53.** −225 ft **54.** 180°, 140° **55.** 44 points **56.** −$80 **57.** −14 **58.** −376 **59.** 322 **60.** 25 **61.** −25 **62.** −204 **63.** −68,000,000 **64.** 30,000,000 **65.** −36 **66.** −36 **67.** 120 **68.** 100 **69.** 450 **70.** 48 **71.** −260, −390 **72.** −540 ft **73.** −125 **74.** −32 **75.** 4,096 **76.** 256 **77.** negative **78.** In the first expression, the base is 9. In the second expression, the base is −9. −81, 81 **79.** −3, 5, −15 **80.** The answer is incorrect: 18(−8) ≠ −152 **81.** −5 **82.** −2 **83.** 8 **84.** −8 **85.** 10 **86.** 1 **87.** −50 **88.** 400 **89.** 23 **90.** −17 **91.** 0 **92.** undefined **93.** −32 **94.** 5 **95.** −2 min **96.** −4,729 ft **97.** −22 **98.** 4 **99.** 40 **100.** 8 **101.** 41 **102.** 0 **103.** −13 **104.** 32 **105.** 12 **106.** −16 **107.** −4 **108.** −34 **109.** −1 **110.** −4 **111.** 5 **112.** 55 **113.** 2,300 **114.** −2 **115.** 10 **116.** 32 **117.** −50 **118.** 5 **119.** 42 **120.** −25 **121.** −3 **122.** −7 **123.** −1 **124.** 0 **125.** 2 **126.** −2 **127.** In 2006, Foot Locker made $253 million in profit. **128.** The candidate gained 29 points in eight weeks.

Chapter 2 Test (page 219)

1. a. integers **b.** inequality **c.** absolute value **d.** opposites **e.** base, exponent **f.** solve **g.** check **2. a.** > **b.** < **c.** < **3. a.** true **b.** true **c.** false **d.** false **e.** true **4.** Poly
5.

$$-5 \quad -4 \quad -3 \quad -2 \quad -1 \quad 0 \quad 1 \quad 2 \quad 3 \quad 4 \quad 5$$

6. a. −3 **b.** −145 **c.** −1 **d.** −32 **e.** −3 **7. a.** −13 **b.** −1 **c.** 191 **d.** −15 **e.** −150 **8. a.** −70 **b.** 292 **c.** 48 **d.** 54 **e.** −26,000,000 **9.** 5(−4) = −20 **10. a.** −8 **b.** −8 **c.** 9 **d.** −34 **e.** −80 **11. a.** −12 **b.** 18 **c.** 4 **d.** −80 **12. a.** commutative property of addition **b.** commutative property of multiplication **c.** adding **13. a.** undefined **b.** −5 **c.** 0 **d.** 1 **14. a.** 16 **b.** −16 **15.** 1 **16.** −27 **17.** −34 **18.** 88 **19.** 6 **20.** 48 **21.** −24 **22.** 58 **23.** −72°F **24.** $203 lost (−203) **25.** 154 ft **26.** −350 ft **27.** −15 **28.** −$60 million **29.** −40 **30.** 16 **31.** −15 **32.** 0 **33.** −5 **34.** −38 **35.** 2 **36.** −18 **37.** Her account balance before the deposit was −$244. **38.** The weight of the people that boarded the elevator on the second floor was 250 pounds.

Chapters 1–2 Cumulative Review (page 221)

1. a. 7 millions   **b.** 3   **c.** 7,326,500   **d.** 7,330,000
2. CRF Cable
3.

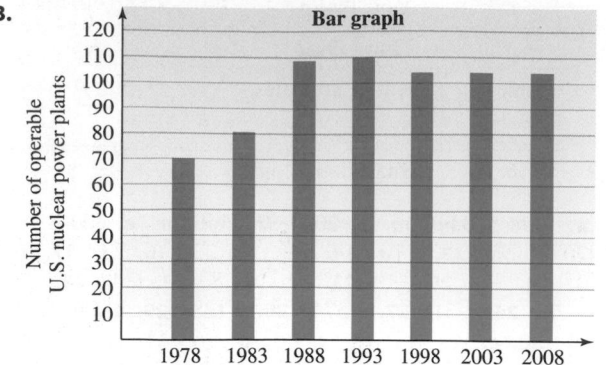

Source: allcountries.org and *The World Almanac and Book of Facts,* 2009

4. 360   **5.** 1,854   **6.** 24,388   **7.** 3,806   **8.** 4,684   **9.** 37,777
10. 1,432   **11.** no   **12.** 65 wooden chairs   **13.** 11,745
14. 5,528,166   **15.** 21,700,000   **16.** 864 tennis balls
17. 104 ft, 595 ft^2 **18.** 25; 144; 10,000   **19.** 87 R 5   **20.** 13
21. 467   **22.** 28   **23.** yes   **24.** 10 times, 20 ounces
25. 60 rolls   **26.** 1, 2, 3, 6, 9, 18   **27. a.** prime number, odd
number   **b.** composite number, even number   **c.** neither,
even number   **d.** neither, odd number   **28.** $2^3 \cdot 3^2 \cdot 7$
29. 11^4 **30.** 175   **31.** 24   **32.** 30   **33.** 6   **34.** 27   **35.** 38
36. 10   **37.** 2   **38.** 41 mph   **39.** yes   **40. a.** no   **b.** yes
c. no   **d.** no   **41.** 13   **42.** 53   **43.** 27   **44.** 24   **45.** There
are 8,835 Dunkin' Donut shops.   **46.** The capacity of Sun
Devil Stadium is 75,000 people.
47. a.

$$\begin{array}{ccccccc} \bullet & \circ & \bullet & \bullet & \bullet & \bullet & \circ \\ -3 & -2 & -1 & 0 & 1 & 2 & 3 \end{array}$$

b.

$$\begin{array}{ccccccc} \circ & \bullet & \bullet & \bullet & \bullet & \bullet & \circ \\ -4 & -3 & -2 & -1 & 0 & 1 & 2 \end{array}$$

48. −3   **49.** 21   **50.** −$79   **51.** −273° Celsius
52. −$55,000   **53.** −37   **54.** 70   **55.** −3   **56.** 4
57. 129   **58.** 1   **59.** −23   **60.** 0   **61.** −4   **62.** −3
63. −100 ft   **64.** −$4,000,000   **65.** 5   **66.** −9
67. 24   **68.** −36   **69.** The account balance before the
deposit was −$735.   **70.** It must be heated 346°F.

Study Set Section 3.1 (page 231)

1. variable   **3.** addition, subtraction   **5. a.** $10 + x$
b. $3t - 2$ (answers may vary)   **7. a.** ii.   **b.** iii.   **c.** iv.   **d.** i.
9. $12 - h$ **11.** 10, 20, 30, 10d; multiply   **13. a.** $8x$ **b.** $5t$
c. $\dfrac{10}{g}$ **15.** $l + 15$ **17.** $50x$ **19.** $\dfrac{w}{l}$ **21.** $P + \dfrac{2}{3}p$
23. $k^2 - 2,005$ **25.** $2a - 1$ **27.** $\dfrac{1,000}{n}$ **29.** $2p + 90$
31. $3(35 + h + 300)$ **33.** $p - 680$ **35.** $4d - 15$
37. $2(200 + t)$ **39.** $|a - 2|$ **41.** $0.1d$ or $\dfrac{1}{10}d$ **43.** $f - 2$
45. $6s$ **47.** $\dfrac{p}{15}$ **49.** $t + 2$ **51.** $\dfrac{h}{4}$ **53.** $450 - x$
55. w = the width of the rectangle (in inches); $w + 6$ = the
length of the rectangle (in inches)   **57.** g = the number of
quarts of coolant originally in the radiator; $g - 3$ = the
number of quarts of coolant that are left in the radiator
59. v = the area of Vermont (in square miles);

$50v + 380$ = the area of Alaska (in square miles)
61. s = the number of calories in a scoop of ice cream;
$2s + 100$ = the number of calories in a slice of pie
63. a. $b + 30$ **b.** $e - 30$ **65. a.** $s + 11$ **b.** $w - 11$
67. x = the age of the ATM; $x + 11$ = the age of the digital
clock; $x - 15$ = the age of the camcorder   **69.** x = the age of
the Empire State Building; $x + 18$ = the age of the
Woolworth Building; $x - 21$ = the age of the United Nations
Building   **71.** $60m$ sec   **73.** $12f$ in.   **75.** $\dfrac{y}{100}$ centuries
77. $\dfrac{e}{12}$ dozen   **79.** three-fourths of r **81.** 50 less than t
83. the product of $x, y,$ and z **85.** twice m, increased by 5
87. a. $7x$ hours   **b.** $365x$ hours   **89. a.** $\dfrac{s}{12}$ dollars
b. $\dfrac{s}{52}$ dollars   **91.** x = the number of votes received by
Nixon; $x + 118,550$ = the number of votes received by
Kennedy   **93.** let x = the age of Apple; $x + 80$ = the age
of IBM; $x - 9$ the age of Dell   **95.** $500, 500 + x, 500 - x$
97. $5,000 - x$ **103.** −10   **105.** −4
107. $\{\ldots, -3, -2, -1, 0, 1, 2, 3, \ldots\}$ **109.** −5

Think It Through (page 245)

4, 6; 6, 9; 8, 12; 10, 15

Section 3.2 (page 246)

1. expression   **3.** substitute   **5.** Celsius   **7. a.** $s = p - d$
b. $p = r - c$ **c.** $r = c + m$ **9.** 5, 25, 45   **11. a.** x = the
length of part 1; $x - 40$ = the length of part 2; $x + 16$ = the
length of part 3 (answers may vary)   **b.** part 2: 20 in.; part 3:
76 in.   **13. a.** x = the weight of a Honda Element;
$2x - 340$ = the weight of an H2 Hummer; $x - 1,720$ = the
weight of a Smart Fortwo car (answers may vary)
b. H2 Hummer: 6,400 lb; Smart Fortwo car: 1,650 lb   **15.** 27
17. −4   **19.** 16   **21.** −17   **23.** −51   **25.** 2   **27.** 144
29. 6   **31.** $165   **33.** $190   **35.** $150   **37.** $8,200
39. 1,650 mi   **41.** 96 mi   **43.** 15°C   **45.** −20°C   **47.** 64 ft
49. 256 ft   **51.** 239   **53.** 25 lb   **55.** −6   **57.** 4   **59.** −30
61. 23   **63.** −3   **65.** 4   **67.** 3   **69.** 65   **71.** 44   **73.** −21
75. −26   **77.** 25   **79.** undefined   **81.** −5   **83.** −270
85. 21   **87.** 6,166; 6,744   **89.** speedometer: rate; odometer:
distance; clock: time; $d = rt$ **91.** 30°, 15°, −5°   **93.** 16, 16 ft;
64, 48 ft; 144, 80 ft; 256, 112 ft   **95.** 32 therms
107. 17, 37, 41   **109.** 7   **111.** division by 3   **113.** 3

Section 3.3 (page 257)

1. simplify   **3.** terms   **5.** removed   **7. a.** 4, 36
b. associative property of multiplication
9. $x(y + z) = xy + xz$ **11.** sign, −1, −   **13. a.** $24x$
b. $24 + 6x$ **15.** −5, −35   **17.** 9, 9, 45y   **19. a.** x
b. $x + 5$ **c.** $10y - 15$ **d.** $5x$ **21.** $12x$ **23.** $40y$ **25.** $100t$
27. $45a$ **29.** $-63xy$ **31.** $-16rs$ **33.** $-30xy$ **35.** $-30br$
37. $4x + 4$ **39.** $7b + 14$ **41.** $27e - 27$ **43.** $6q - 21$
45. $-6h - 10$ **47.** $-40y - 60$ **49.** $-16q + 32$
51. $-35g + 5$ **53.** $20s + 12$ **55.** $90t + 54$ **57.** $-x + 5$
59. $-5d - 8$ **61.** $24d + 42$ **63.** $21q + 140$ **65.** $-24 - 6d$
67. $9t - 108$ **69.** $24t - 18$ **71.** $60h + 20$
73. $9z + 9x - 15$ **75.** $-16a - 32b + 48$ **77.** $3w + 4$

79. $18x + 19$ **81.** $-x - 3$ **83.** $-4t - 5$ **85.** $78c + 18$
87. $12s$ **89.** $-35q$ **91.** $-36c + 42$ **93.** $9x - 21y + 6$
95. $-40h$ **97.** $80c$ **99.** $48t + 32$ **101.** $-8e$
103. $-5x + 4y - 1$ **105.** $2(4x + 5)$ **107.** $-3(4y + 2)$
109. $3(4 - 7t - 5s)$ **111.** $(-4 - 3x)5$ **119.** 5
121. multiplication, division, subtraction, addition **123.** $>$
125. carpeting, painting

Section 3.4 (page 265)

1. term **3.** coefficient **5.** implied **7.** combined
9. $3, -10, 8$ **11.** $8, m$ **13. a.** unlike **b.** unlike **c.** unlike
d. like **15.** $2, 3, 5$ **17.** $2, 5x$ **19. a.** the perimeter of a
rectangle **b.** 2 times the length **c.** 2 times the width
21. a. true **b.** true **c.** true **d.** false **23.** $3x^2, -9x, 4$
25. $5, 5t, -8t, -1$ **27.** $-35a$ **29.** $9mn, -6n$ **31. a.** term
b. factor **33. a.** factor **b.** term **35.** $5, 1, -12$
37. $1, -27$ **39.** $1, -1, 1, 10$ **41.** $-1, 6, -1, 5$ **43.** $8x, 2x$
45. none **47.** $-3k^3, k^3; 6k, -3k$ **49.** $12a, 15a; -8, 1$
51. $15t$ **53.** $50b$ **55.** $-9x$ **57.** $-4d$ **59.** $2s^2$ **61.** $-14e^3$
63. does not simplify **65.** $8z$ **67.** $38a$ **69.** does not
simplify **71.** s **73.** $39a^2$ **75.** $-m$ **77.** $15r$
79. $5x^2 + 16x + 6$ **81.** $y^2 - 10y - 4$ **83.** $2m + 3$
85. $3x - 11$ **87.** 46 ft **89.** 148 yd **91.** $-7x^3$
93. $10y - 28$ **95.** $11t + 12$ **97.** $4x$ **99.** $2t + 8$
101. $-50x$ **103.** does not simplify **105.** $8x^2 - 4x - 9$
107. 0 **109.** $-7r + 11R$ **111.** $-2y^3$ **113.** $-3s + 23$
115. a. $(2d + 15)$ mi **b.** $2b + 30$ **117.** $(4x + 8)$ ft **119.**
$288 **121.** 36 ft, 48 ft, 60 ft, 72 ft, 84 ft **127.** 2 **129.** 16

Section 3.5 (page 273)

1. solve **3.** check **5.** simplify **7. a.** $5x, 3x; 5t, 3t; 5h, 3h$
b. $5t = 3t + 8$ **9. a.** combine like terms: $6x = 36$
b. distribute the multiplication by 5: $5x + 5 = 15$
c. combine like terms: $11x - 5 = 2x + 4$ **d.** distribute the
multiplication by -3 and 2: $-3x + 12 = 2x + 2$
11. $1, d, 4$ **13. a.** $2t - 8$ **b.** -4 **c.** -12 **d.** no
15. $3x, 3, 3, -9; -9, -9, -45, 18, -27, -9$ **17.** $9, 45, 45, 45,$
$5x, 5, 5, 10; 10, 1, 5, 10$ **19.** yes **21.** no **23.** 6 **25.** 3
27. 306 **29.** 257 **31.** -8 **33.** -4 **35.** -2 **37.** 7 **39.** 0
41. -1 **43.** 8 **45.** -13 **47.** -10 **49.** 6 **51.** 37
53. -7 **55.** -30 **57.** 3 **59.** -3 **61.** 10 **63.** 1
65. -28 **67.** 42 **69.** -4 **71.** 735 **73.** 2 **75.** 0
77. -11 **79.** -8 **81.** -5 **83.** 5 **85.** 0 **87.** -12 **89.** 4
91. 2 **93.** 26 **101.** -16 **103.** -3 **105.** 5 **107.** positive

Section 3.6 (page 283)

1. Analyze, equation, Solve, conclusion, Check **3.** division
5. addition **7.** Number **9.** $5x$ **11.** $g - 100$ **13.** $3m$
15. $2w$ **17. a.** 9 **b.** $9 - d$ **19.** $88, 10,$ first-class;
first-class; multiply, $10, 10x$; $88, 11x, 11, 11, 8; 8; 10, 88$
21. It will take 17 months for him to reach his goal.
23. Last year, 7 scholarships were awarded. This year,
13 scholarships were awarded. **25.** There were 10 nickels
and 15 dimes in the piggy bank. **27.** $30, 24, 5x, 4(9 - x)$
29. $h, 18, 18h; 40 - h, 20, 20(40 - h)$ (answers may vary)
31. The number is 8. **33.** The number is 4. **35.** She must
take 4 more sessions. **37.** She has made 6 payments.
39. It will take 9 months to reach the goal. **41.** The
freighter was 21 miles from port. **43.** The father left a total

of $420,500 to his sons. **45.** The monthly rent for the
apartment was $975. **47.** The premium gas tank holds
400 gallons. **49.** There were 6 minutes of commercials.
51. The width of the room is 10 feet. **53.** The width of the
court is 27 feet and the length is 78 feet. **55.** He sold 6 pairs
of dress shoes and 3 pairs of athletic shoes. **57.** He worked
14 hours at the regular rate and 6 hours going up and down
stairs. **59.** He has 18 movie star autographs and
12 television celebrity autographs. **65.** the associative
property of addition **67.** -100 **69.** addition **71.** $2^3 \cdot 5^2$

Chapter 3 Summary and Review (page 289)

1. Brandon is closer by 250 mi. **2.** $h + 7$ **3.** $n - 5$
4. $7x$ **5.** $\dfrac{6}{p}$ **6.** $s + (-15)$ **7.** $2l$ **8.** $D - 100$ **9.** $r + 2$
10. $\dfrac{45}{x}$ **11.** $100 - 2s$ **12.** $|2 - a^2|$ **13.** five hundred less
than m (answers may vary) **14. a.** $(n + 4)$ in. **b.** $(b - 4)$ in.
15. $\dfrac{c}{6}$ **16.** $1{,}000 - x$ **17.** $x + 1$ **18.** $\dfrac{p}{8}$ **19.** $x =$ the
number of hours driven by the wife, $2x =$ the number of
hours driven by the husband **20.** $w =$ the width,
$w + 3 =$ the length **21.** $x =$ the weight of the volleyball
(in ounces), $2x + 2 =$ the weight of the NBA basketball
(in ounces) **22.** $x =$ the age of *To Kill a Mockingbird*,
$x + 6 =$ the age of *The Lord of the Rings*, $x - 9 =$ the
age of *The Godfather* **23.** $12x$ **24.** $\dfrac{d}{7}$
25. a. $h =$ the height of the wall, $h - 5 =$ the length
of the upper base, $2h - 3 =$ the length of the lower base
b. upper base: 5 ft, lower base: 17 ft **26.** $-1{,}000$ means the
sod farm is short $1{,}000$ ft^2 to fill the city's order. **27.** 12
28. -8 **29.** 100 **30.** 64 **31.** 100 **32.** -4 **33.** The sale
price is $278. **34.** The retail price is $15,230.
35. The profit the store made its first month was $4,915.
36. $130, 114, 6x, 55(t + 1)$ **37.** The pool is $2°C$ warmer.
38. The wrench will fall 144 ft. **39.** 24 yr **40.** 4 **41.** $-10x$
42. $42xy$ **43.** $60de$ **44.** $32s$ **45.** $2e$ **46.** $49xy$ **47.** $84k$
48. $100t$ **49.** $4y + 20$ **50.** $-30t - 45$ **51.** $-21 - 21x$
52. $-12e + 24x + 3$ **53.** $48w - 24$ **54.** $-36x - 36$
55. $-6t + 4$ **56.** $-5 - x$ **57.** $-6t + 3s - 1$ **58.** $5a + 3$
59. $8x^2, -7x, 9$ **60.** $-15y$ **61.** $16ab, -6b$
62. $4x, -3, 5x, -7$ **63.** $5, -4, 8$ **64.** $7, 3, 1, -1$
65. $1, 1, -1, 6$ **66.** $-5,125$ **67.** factor **68.** term **69.** term
70. factor **71.** yes **72.** no **73.** yes **74.** no **75.** $7x$
76. does not simplify **77.** $-3z$ **78.** $5x$ **79.** $-12y$
80. $w^2 - 5$ **81.** $-46d + 2a$ **82.** $10y + 15h - 1$
83. $10a^2 - 11a + 6$ **84.** $29w$ **85.** $13y + 48$ **86.** $-5t + 22$
87. $3x + 8$ **88.** $-50f + 73$ **89.** $3x + 3$ **90.** 194 ft
91. not a solution **92.** it is a solution **93.** -18 **94.** 8
95. 305 **96.** -3 **97.** 15 **98.** -3 **99.** 2 **100.** -4
101. -9 **102.** -2 **103.** They can rent the hall for 7 hours.
104. It will take 6 hr to lower the temperature to $29°F$.
105. It cost $32 to rent the trailer. **106.** She runs 9 miles
and she walks 6 miles. **107.** The attendance on the first day
was 2,200 people. The attendance on the second day was
4,400 people. **108.** The width of the parking lot is 25 feet and
the length is 100 feet. **109.** $10, 60; 25, 175; 1, x; 5, 5(n + 25)$
110. There were 15 $3 drinks and 35 $4 drinks sold.

Chapter 3 Test (page 299)

1. **a.** Variables **b.** distributive **c.** like **d.** solve
e. coefficient **f.** expressions **g.** substitute **h.** equation
i. combined **j.** check **2. a.** $2h - 1,000$ **b.** $3,700$
3. $56 - c$ **4.** $t =$ the length of the trout, $t + 10 =$ the length
of the salmon; or $s =$ the length of the salmon, $s - 10 =$ the
length of the trout **5. a.** $r - 2$ **b.** $3xy$ **c.** $x + 100$
d. $\left|\dfrac{x}{-9}\right|$ **6.** $10d$ **7. a.** -12 **b.** 26 **c.** 1 **d.** 23
8. The distance traveled is 165 mi. **9.** The profit is $37,000.
10. It would be 56 ft short of hitting the ground. **11.** The
mean meter reading is 1. **12.** The project requires 250 ft of
edging. **13.** The temperature was 15°C. **14. a.** $25x + 5$
b. $-42 + 6x$ **c.** $-6y - 4$ **d.** $6a + 9b - 21$ **e.** $8a - 120$
f. $36r + 54$ **15. a.** factor **b.** term **16. a.** $11x$ **b.** $12e$
c. $5x^2$ **d.** $30y$ **e.** $-7x$ **f.** $9y$ **g.** $-72ab$ **h.** $-280m$
17. **a.** $8x^2, -x, -6$ **b.** $8, -1, -6$ **18. a.** $-28y + 10$
b. $-3t$ **c.** $-6y - 3$ **d.** $9m^4 + 23m^3$ **19. a.** $10k$¢
b. $20(p + 2)$ dollars **20.** not a solution **21.** -9 **22.** -3
23. 4 **24.** -10 **25.** -10 **26.** -4 **27.** 4 **28.** 0
29. Each classroom session is 3 hr long. **30.** Each day, there
were 8 hr of local shows and 16 hr of national shows.
31. The developer donated 44 acres of land to the city.
32. The width of the frame is 24 in. and the length is 48 in.
33. We simplify expressions and we solve equations.
34. $5x \cdot 2 = 10x$; $5x$ and 2 are not like terms and therefore
cannot be combined.

Chapter 1–3 Cumulative Review (page 301)

1. 3,290,057,000 barrels **2.** 50,000 **3.** 54,604 **4.** 4,209
5. 23,115 **6.** 87 **7. a.** $683 + 459 = 1,142$
b. $\dfrac{5}{0}, \dfrac{0}{5}$ (answers may vary); division by 0 **8.** 2011
9. $4 \cdot 5 = 5 + 5 + 5 + 5 = 20$ **10.** $10,912$ in.^{2}
11. The car had 186 oil changes.
12. **a.** $1, 2, 3, 6, 9, 18$ **b.** $27 = 3 \cdot 9$ **c.** $2 \cdot 3^2$
13. $2, 3, 5, 7, 11, 13, 17, 19, 23, 29$ **14. a.** 315 **b.** 4
15. **a.** 22 **b.** 37 **16.** the addition property of equality
17. 500 **18. a.** 6 **b.** 5 **c.** false
19.

$$\overset{\longleftarrow}{\underset{-4\ -3\ -2\ -1\ \ 0\ \ 1\ \ 2\ \ 3\ \ 4}{\mid\ \ \mid\ \ \bullet\ \ \bullet\ \ \bullet\ \ \bullet\ \ \mid\ \ \bullet\ \ \bullet}}\overset{\longrightarrow}{}$$

20. $-21 - (-73)$ **21. a.** -20 **b.** 30 **c.** 125 **d.** 5
22. $1,630; 575$ **23.** $1,100°F$
24. **a.** $-3^2 = -(3 \cdot 3) = -9; (-3)^2 = (-3)(-3) = 9$
b. the commutative property of multiplication **25.** -5
26. 429 **27.** 7 **28.** -10 **29.** 6 **30.** 17 **31.** -87 **32.** -4
33. **a.** $h + 12$ **b.** $w - 4$ **b.** $\dfrac{1,000}{x}$ **34. a.** $(26 - x)$ in.
b. $25q$¢ **35.** $12f$ in. **36.** 36 **37.** 220 **38. a.** $10x - 35$
b. $-5t + 7$ **39. a.** $24t$ **b.** $-48yz$ **40.** $4, -2, 1, -1$
41. $x + 3; 3x$ (answers may vary) **42. a.** $-b + 1$
b. $2x + 19$ **43.** -1 **44.** -2 **45.** -4 **46.** -13
47. The students spend 175 min in lecture and 125 min in lab
each week. **48.** The width is 10 ft and the length is 50 ft.

Study Set Section 4.1 (page 314)

1. fraction **3.** proper, improper **5.** equivalent
7. simplest **9.** equivalent fractions: $\dfrac{2}{6} = \dfrac{1}{3}$
11. **a.** improper fraction **b.** proper fraction **c.** proper
fraction **d.** improper fraction **13.** 5 **15.** numerators
17. $\dfrac{-7}{8}, -\dfrac{7}{8}$ **19.** $3, 1, 3, 18$ **21.** numerator: 4; denominator: 5
23. numerator: 17; denominator: 10 **25.** $\dfrac{3}{4}, \dfrac{1}{4}$ **27.** $\dfrac{5}{8}, \dfrac{3}{8}$
29. $\dfrac{1}{4}, \dfrac{3}{4}$ **31.** $\dfrac{7}{12}, \dfrac{5}{12}$ **33. a.** 4 **b.** 1 **c.** 0 **d.** undefined
35. **a.** undefined **b.** 0 **c.** 1 **d.** 75 **37.** $\dfrac{35}{40}$ **39.** $\dfrac{12}{27}$
41. $\dfrac{45}{54}$ **43.** $\dfrac{4}{14}$ **45.** $\dfrac{36}{9}$ **47.** $\dfrac{48}{8}$ **49.** $\dfrac{15}{5}$ **51.** $\dfrac{28}{2}$
53. **a.** no **b.** yes **55. a.** yes **b.** no **57.** $\dfrac{2}{3}$ **59.** $\dfrac{4}{5}$ **61.** $\dfrac{1}{3}$
63. $\dfrac{1}{24}$ **65.** in simplest form **67.** $\dfrac{3}{8}$ **69.** in simplest form
71. $\dfrac{10}{11}$ **73.** $\dfrac{5}{9}$ **75.** $\dfrac{6}{7}$ **77.** $\dfrac{17}{13}$ **79.** $\dfrac{5}{2}$ **81.** $\dfrac{35}{12}$ **83.** $-\dfrac{1}{17}$
85. $-\dfrac{6}{7}$ **87.** $-\dfrac{8}{13}$ **89.** $\dfrac{3a}{6a}$ **91.** $\dfrac{45c}{50c}$ **93.** $\dfrac{55}{44a}$ **95.** $\dfrac{42}{45x}$
97. $\dfrac{1}{2a}$ **99.** $\dfrac{x}{4}$ **101.** $\dfrac{4m}{25}$ **103.** $\dfrac{2b^3}{3}$ **105.** $\dfrac{7a}{5b}$ **107.** $-\dfrac{2n^4}{3}$
109. not equivalent **111.** equivalent **113. a.** 32 **b.** $\dfrac{5}{32}$
115. **a.** 16 **b.** $\dfrac{5}{8}$ **117. a.** $28, 22$ **b.** $\dfrac{28}{50} = \dfrac{14}{25}$ **c.** $\dfrac{22}{50} = \dfrac{11}{25}$
119. **a.** 20 **b.** $\dfrac{2}{5}, \dfrac{3}{5}$ **127.** $2,307

Study Set Section 4.2 (page 327)

1. multiplication **3.** simplify **5.** area **7.** numerators,
denominators, simplify **9. a.** negative **b.** positive
c. positive **d.** negative **11. a.** base, height, $\dfrac{1}{2}bh$
b. square **13. a.** $\dfrac{4}{1}$ **b.** $-\dfrac{3}{1}$ **c.** $\dfrac{x}{1}$
15. $7, 15, 2, 3, 5, 5, 24$ **17.** $\dfrac{1}{8}$ **19.** $\dfrac{1}{45}$ **21.** $\dfrac{14}{27}$
23. $\dfrac{24}{77}$ **25.** $-\dfrac{4}{15}$ **27.** $-\dfrac{35}{72}$ **29.** $\dfrac{9}{8}$ **31.** $\dfrac{5}{2}$ **33.** $\dfrac{1}{2}$
35. $\dfrac{1}{7}$ **37.** $\dfrac{1}{10}$ **39.** $\dfrac{2}{15}$ **41.** $\dfrac{3}{10}$ **43.** $\dfrac{4}{7}$ **45.** $-\dfrac{9}{56m}$
47. $\dfrac{5x}{9y^2}$ **49.** w **51.** x **53.** 2 **55.** $-4y$ **57. a.** $\dfrac{9}{25}$ **b.** $\dfrac{9}{25}$
59. **a.** $-\dfrac{1}{36}$ **b.** $-\dfrac{1}{216}$ **61.** $\dfrac{36t^2}{49}$ **63.** $-\dfrac{8a^3}{125}$
65. $\dfrac{15}{32}$ **67.** 9 **69.** 15 ft^2 **71.** 63 in.^{2} **73.** 6 m^2 **75.** 60 ft^2

77.

	$\frac{1}{2}$	$\frac{1}{3}$	$\frac{1}{4}$	$\frac{1}{5}$	$\frac{1}{6}$
$\frac{1}{2}$	$\frac{1}{4}$	$\frac{1}{6}$	$\frac{1}{8}$	$\frac{1}{10}$	$\frac{1}{12}$
$\frac{1}{3}$	$\frac{1}{6}$	$\frac{1}{9}$	$\frac{1}{12}$	$\frac{1}{15}$	$\frac{1}{18}$
$\frac{1}{4}$	$\frac{1}{8}$	$\frac{1}{12}$	$\frac{1}{16}$	$\frac{1}{20}$	$\frac{1}{24}$
$\frac{1}{5}$	$\frac{1}{10}$	$\frac{1}{15}$	$\frac{1}{20}$	$\frac{1}{25}$	$\frac{1}{30}$
$\frac{1}{6}$	$\frac{1}{12}$	$\frac{1}{18}$	$\frac{1}{24}$	$\frac{1}{30}$	$\frac{1}{36}$

79. $-\frac{x}{5}$ **81.** $\frac{21}{128}$ **83.** $\frac{1}{30}$ **85.** $-15x$ **87.** $-\frac{27}{64a^3}$ **89.** a

91. $\frac{8}{3}$ **93.** $-\frac{3}{2}$ **95.** $\frac{2}{9}$ **97.** $-\frac{25}{81}$ **99.** $\frac{2x}{3y}$ **101.** $\frac{5}{6}$

103. 60 votes **105.** 18 in., 6 in., and 2 in.

107. $\frac{3}{8}$ cup sugar, $\frac{1}{6}$ cup molasses

109.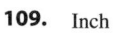

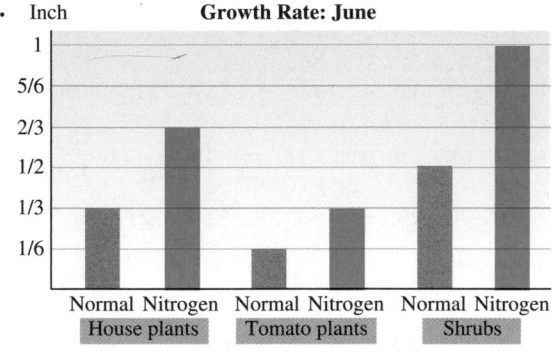

Growth Rate: June

111. 27 ft^2 **113.** 42 ft^2 **115.** 9,646 mi^2 **117.** $\frac{3}{4}$ in.

123. -2 **125.** 23

Study Set Section 4.3   (page 340)

1. reciprocal **3.** quotient **5. a.** multiply, reciprocal
b. $\cdot$, $\frac{3}{2}$ **7. a.** negative **b.** positive **9. a.** 1 **b.** 1
11. 27, 27, 8, 9, 2, 4, 4, 9, 3 **13. a.** $\frac{7}{6}$ **b.** $-\frac{8}{15}$ **c.** $\frac{1}{10}$

15. a. $\frac{8}{11a}$ **b.** $-14b$ **c.** $-\frac{1}{63x}$ **17.** $\frac{3}{16}$ **19.** $\frac{14}{23}$ **21.** $\frac{35}{8}$

23. $\frac{3}{4}$ **25.** 45 **27.** 320 **29.** -4 **31.** $-\frac{7}{2}$ **33.** $\frac{4}{55}$

35. $\frac{3}{23}$ **37.** $\frac{28a}{15}$ **39.** $\frac{66x}{35}$ **41.** $\frac{4a^2b}{3}$ **43.** $\frac{14}{9x^3}$ **45.** $-\frac{5}{3a}$

47. $\frac{3}{m^3}$ **49.** $50x$ **51.** $\frac{18}{x^2}$ **53.** $\frac{2}{3}$ **55.** 1 **57.** $-\frac{5}{8}$

59. $36a^2$ **61.** $\frac{2}{15}$ **63.** $\frac{1}{192}$ **65.** $-\frac{27n}{8}$ **67.** $-\frac{15b^2}{2a}$

69. $-\frac{x}{y^2}$ **71.** $-\frac{1}{64}$ **73.** $\frac{3m}{14}$ **75.** $\frac{8}{15}$ **77.** $\frac{13}{16x}$ **79.** $\frac{2}{9}$

81. $-6y^3$ **83.** $\frac{11}{6}$ **85.** $\frac{15x}{28}$ **87.** $-\frac{5}{2}$ **89.** 4 applications

91. 6 cups **93. a.** 30 days **b.** 15 mi **c.** 25 days

d. route 2 **95. a.** 16 **b.** $\frac{3}{4}$ in. **c.** $\frac{1}{120}$ in.

97. 7,855 sections **105.** is less than **107.** Zero

109.

```
      −2  −1   0        |−4| = 4
 ┿━━┿━━┿━●━●━┿━┿━┿━┿━●━┿
 −5 −4 −3 −2 −1  0  1  2  3  4  5
```

Think It Through   (page 354)

$\frac{7}{20}$

Study Set Section 4.4   (page 354)

1. common **3. a.** numerators, common, Simplify **b.** LCD,
same **5.** $\frac{9}{9}$ **7. a.** once **b.** twice **c.** three times
9. 7, 7, 14, 35, 14, 5, 19 **11.** $\frac{5}{9}$ **13.** $\frac{1}{2}$ **15.** $\frac{4}{15}$ **17.** $\frac{2}{5}$

19. $-\frac{3}{5}$ **21.** $-\frac{5}{21}$ **23.** $\frac{3}{8}$ **25.** $\frac{7}{11}$ **27.** $\frac{10}{21}$ **29.** $\frac{23}{45}$

31. $\frac{1}{20}$ **33.** $\frac{13}{28}$ **35.** $\frac{1}{4}$ **37.** $\frac{1}{2}$ **39.** $-\frac{13}{9}$ **41.** $-\frac{3}{4}$

43. $\frac{19}{24}$ **45.** $\frac{31}{36}$ **47.** $\frac{24}{35}$ **49.** $\frac{9}{20}$ **51.** $\frac{x}{2}$ **53.** $\frac{5c}{7}$ **55.** $\frac{5}{21m}$

57. $\frac{3}{5y}$ **59.** $\frac{3a+10}{15}$ **61.** $\frac{3+8x}{24}$ **63.** $\frac{36-5n}{12n}$

65. $\frac{8d-99}{9d}$ **67.** $\frac{3}{8}$ **69.** $\frac{4}{5}$ **71.** $\frac{11}{12}$ **73.** $\frac{7}{6}$ **75.** $\frac{2}{3}$

77. $\frac{11}{10}$ **79.** $\frac{1}{3}$ **81.** $\frac{9n+8}{12}$ **83.** $\frac{2}{5}$ **85.** $-\frac{11}{20}$ **87.** $-\frac{3}{16}$

89. $\frac{x}{3}$ **91.** $\frac{23}{10}$ **93.** $\frac{5}{12}$ **95.** $\frac{341}{400}$ **97.** $\frac{9}{20}$ **99.** $-\frac{23}{4}$

101. $-\frac{1}{50}$ **103.** $\frac{26-3d}{2d}$ **105.** $-\frac{17}{60}$ **107. a.** $\frac{7}{32}$ in.

b. $\frac{3}{32}$ in. **109.** $\frac{11}{16}$ in. **111. a.** $\frac{3}{8}$ **b.** $\frac{2}{6}=\frac{1}{3}$

c. $\frac{17}{24}$ of a pizza was left **d.** no **113.** $\frac{1}{16}$ lb, undercharge

115. $\frac{7}{10}$ of the full-time students study 2 or more hours a day.

117. no **119. a.** RR: right rear **b.** LR: left rear

123. a. $\frac{3}{8}$ **b.** $\frac{1}{8}$ **c.** $\frac{1}{32}$ **d.** 2

Study Set Section 4.5   (page 368)

1. mixed **3.** improper **5. a.** $5\frac{1}{3}^{\circ}$ **b.** $-6\frac{7}{8}$ in.

7. Multiply, Add, denominator **9.** $-\frac{4}{5}, -\frac{2}{5}, \frac{1}{5}$

11. improper **13.** not reasonable: $4\frac{1}{5} \cdot 2\frac{5}{7} \approx 4 \cdot 3 = 12$

15. a. and, sixteenths **b.** negative, two **17.** 4, 8, 8, 4, 4,
4, 6, 6 **19.** $\frac{19}{8}, 2\frac{3}{8}$ **21.** $\frac{34}{25}, 1\frac{9}{25}$ **23.** $\frac{13}{2}$ **25.** $\frac{104}{5}$

27. $-\dfrac{68}{9}$ **29.** $-\dfrac{26}{3}$ **31.** $3\dfrac{1}{4}$ **33.** $5\dfrac{3}{5}$ **35.** $4\dfrac{2}{3}$ **37.** $10\dfrac{1}{2}$

39. 4   **41.** 2   **43.** $-8\dfrac{2}{7}$ **45.** $-3\dfrac{1}{3}$

47.

$-2\dfrac{8}{9}$ $-\dfrac{1}{2}$ $1\dfrac{2}{3}$ $\dfrac{16}{5}=3\dfrac{1}{5}$

49.

$-\dfrac{10}{3}=-3\dfrac{1}{3}$ $-\dfrac{98}{99}$ $\dfrac{3}{2}=1\dfrac{1}{2}$ $3\dfrac{1}{7}$

51. $8\dfrac{1}{6}$ **53.** $7\dfrac{2}{5}$ **55.** 8   **57.** -10 **59.** $\dfrac{4}{9}$ **61.** $6\dfrac{9}{10}$ **63.** $2\dfrac{1}{3}$

65. $1\dfrac{10}{21}$ **67.** $-13\dfrac{3}{4}$ **69.** $-\dfrac{9}{10}$ **71.** $\dfrac{25}{9}=2\dfrac{7}{9}$ **73.** $2\dfrac{1}{2}$

75. 12   **77.** 14   **79.** -2 **81.** $-8\dfrac{1}{3}$ **83.** $\dfrac{35}{72}$ **85.** $\dfrac{5}{16}$

87. $-1\dfrac{1}{4}$ **89.** $-\dfrac{64}{27}=-2\dfrac{10}{27}$ **91. a.** $3\dfrac{2}{3}$ **b.** $\dfrac{11}{3}$ **93.** $2\dfrac{1}{2}$

95. a. $2\dfrac{2}{3}$ **b.** $-1\dfrac{1}{3}$ **97.** size 14, slim cut   **99.** $76\dfrac{9}{16}$ in.^{2}

101. $42\dfrac{5}{8}$ in.^{2} **103.** 64 calories   **105.** $357¢ = \$3.57$

107. $1\dfrac{1}{4}$ cups   **109.** 600 people   **111.** $8\dfrac{1}{2}$ furlongs

115. 60   **117.** 4

Think It Through   (page 381)

workday: $6\dfrac{2}{3}$ hr; non-workday: $7\dfrac{5}{12}$ hr; $\dfrac{3}{4}$ hr

Study Set   Section 4.6   (page 382)

1. mixed   **3.** fractions, whole   **5.** carry   **7. a.** $76, \dfrac{3}{4}$
b. $76 + \dfrac{3}{4}$ **9. a.** 12   **b.** 30   **c.** 18   **d.** 24   **11.** 5, 5, 21,
35, 31, 35   **13.** $3\dfrac{7}{12}$ **15.** $6\dfrac{11}{15}$ **17.** $-2\dfrac{3}{8}$ **19.** $-3\dfrac{1}{6}$

21. $376\dfrac{17}{21}$ **23.** $714\dfrac{19}{20}$ **25.** $59\dfrac{28}{45}$ **27.** $132\dfrac{29}{33}$ **29.** $121\dfrac{9}{10}$

31. $147\dfrac{8}{9}$ **33.** $102\dfrac{13}{24}$ **35.** $129\dfrac{28}{45}$ **37.** $10\dfrac{1}{4}$ **39.** $13\dfrac{8}{15}$

41. $31\dfrac{14}{33}$ **43.** $71\dfrac{43}{56}$ **45.** $579\dfrac{4}{15}$ **47.** $62\dfrac{23}{32}$ **49.** $11\dfrac{1}{30}$

51. $5\dfrac{11}{30}$ **53.** $9\dfrac{3}{10}$ **55.** $3\dfrac{7}{8}$ **57.** $5\dfrac{2}{3}$ **59.** $10\dfrac{7}{16}$ **61.** $397\dfrac{5}{12}$

63. $-1\dfrac{11}{24}$ **65.** $7\dfrac{1}{2}$ **67.** $-5\dfrac{1}{4}$ **69.** $6\dfrac{1}{3}$ **71.** $53\dfrac{5}{12}$ **73.** $2\dfrac{1}{2}$

75. $-5\dfrac{7}{8}$ **77.** $3\dfrac{5}{8}$ **79.** $4\dfrac{1}{3}$ **81.** $461\dfrac{1}{8}$ **83.** $\dfrac{1}{4}$ **85.** $5\dfrac{1}{4}$ hr

87. $7\dfrac{1}{6}$ cups   **89.** $20\dfrac{1}{16}$ lb   **91.** $108\dfrac{1}{2}$ in.   **93.** $2\dfrac{3}{4}$ mi

95. $48\dfrac{1}{2}$ ft   **97. a.** 20¢ per gallon   **b.** 20¢ per gallon

99. $3\dfrac{1}{4}$ in.   **105. a.** $4\dfrac{3}{4}$ **b.** $2\dfrac{1}{4}$ **c.** $4\dfrac{3}{8}$ **d.** $2\dfrac{4}{5}$

Study Set   Section 4.7   (page 393)

1. operations   **3.** complex   **5.** raising to a power
(exponent), multiplication, and addition

7. $\left(\dfrac{2}{3}-\dfrac{1}{10}\right)+1\dfrac{2}{15}$ **9.** $\dfrac{2}{3}\div\dfrac{1}{5}$ **11.** $\dfrac{23}{4}$ **13.** 3, 6, 2, 2, 2, 5

15. $\dfrac{17}{20}$ **17.** $-\dfrac{1}{6}$ **19.** $-\dfrac{7}{26}$ **21.** $-\dfrac{1}{12}$ **23.** $5\dfrac{13}{30}$ **25.** $2\dfrac{2}{3}$

27. $26\dfrac{1}{4}$ **29.** 18   **31.** $\dfrac{5}{32}$ **33.** $\dfrac{5}{6}$ **35.** $\dfrac{5}{18}$ **37.** $-\dfrac{1}{2}$

39. $\dfrac{50}{13}$ **41.** $\dfrac{25}{26}$ **43.** $-1\dfrac{27}{40}$ **45.** $-1\dfrac{1}{3}$ **47.** 36   **49.** $\dfrac{1}{3}$

51. $\dfrac{31}{45}$ **53.** 5   **55.** $14\dfrac{5}{24}$ **57.** 11   **59.** $-1\dfrac{1}{6}$ **61.** $\dfrac{3}{7}$

63. $\dfrac{3}{10}$ **65.** $44\dfrac{1}{3}$ **67.** $8\dfrac{1}{2}$ **69.** $\dfrac{4}{9}$ **71.** $1\dfrac{37}{70}$ **73.** 3

75. $8\dfrac{4}{15}$ **77.** $91\dfrac{1}{4}$ in.   **79.** yes   **81.** $3\dfrac{1}{4}$ hr   **83.** 9 parts

85. 7 full tubes; $\dfrac{2}{3}$ of a tube is leftover   **87.** 7 yd^2 **89.** 6 sec

95. 2,248   **97.** 20,217   **99.** 1, 2, 3, 4, 6, 8, 12, 24

Study Set   Section 4.8   (page 411)

1. solve   **3.** reciprocal   **5.** Since $25 = 25$ is a true

statement, 40 is the solution of $\dfrac{5}{8}x = 25$.   **7.** 1   **9. a.** $\dfrac{4}{5}p$

b. $\dfrac{1}{4}t$ **11. a.** 6   **b.** 24   **13.** $\dfrac{8}{7},\dfrac{8}{7}, 24, 24$ **15. a.** true

b. false   **c.** true   **d.** true   **17.** $\dfrac{7}{10}$ **19.** $\dfrac{4}{5}$ **21.** $\dfrac{13}{18}$ **23.** $\dfrac{9}{20}$

25. 27   **27.** 70   **29.** $-\dfrac{21}{8}$ **31.** $-\dfrac{45}{16}$ **33.** 27   **35.** 70

37. $\dfrac{25}{9}$ **39.** $\dfrac{43}{11}$ **41.** $-\dfrac{15}{68}$ **43.** $-\dfrac{13}{87}$ **45.** -12 **47.** -16

49. $\dfrac{7}{18}$ **51.** $\dfrac{17}{36}$ **53.** -56 **55.** 126   **57.** 3   **59.** -2

61. $\dfrac{8}{9}$ **63.** 0   **65.** -32 **67.** $\dfrac{1}{3}$ **69.** $-\dfrac{14}{5}$ **71.** $-\dfrac{11}{4}$ **73.** $\dfrac{13}{2}$

75. $\dfrac{5}{12}$ **77.** $-\dfrac{20}{3}$ **79.** -27 **81.** 6   **83.** $\dfrac{10}{9}$ **85.** 24

87. 10   **89. a.** -6 **b.** $\dfrac{24 + 5n}{10}$ **91. a.** $-\dfrac{4}{3}$ **b.** $\dfrac{3x - 4}{12}$

93. $\dfrac{1}{8}$, 32, cars; cars, multiplication, $\cdot, x, 32; 8, 8, 256; 256;$
1, 8   **95.** 20 teeth   **97.** 450 pages   **99.** 27 ft   **101.** 240 in.^{2}
103. 360 min   **105.** 40 players   **107.** 36 homes
113. a. 13,000,000   **b.** 12,600,000   **c.** 12,599,800

Chapter 4 Review   (page 416)

1. numerator: 11, denominator: 16; proper fraction

2. $\dfrac{4}{7},\dfrac{3}{7}$ **3.** The figure is not divided into equal parts.

4. $-\dfrac{2}{3},\dfrac{-2}{3}$ **5. a.** 1   **b.** 0   **c.** 18   **d.** undefined

6. equivalent fractions: $\dfrac{6}{8}=\dfrac{3}{4}$ **7.** $\dfrac{12}{18}$ **8.** $\dfrac{6}{16}$ **9.** $\dfrac{21a}{45a}$

10. $\dfrac{65}{60x}$ **11.** $\dfrac{45}{9}$ **12. a.** no   **b.** yes   **13.** $\dfrac{1}{3}$ **14.** $\dfrac{5x}{12}$

15. $\dfrac{11}{18}$ **16.** $\dfrac{9b^5}{16a^3}$ **17.** in simplest form   **18.** equivalent

19. $\dfrac{7}{24}, \dfrac{17}{24}$ **20. a.** The fraction $\dfrac{5}{8}$ is being expressed as an

equivalent fraction with a denominator of 16. To build the

fraction, multiply $\dfrac{5}{8}$ by 1 in the form of $\dfrac{2}{2}$.   **b.** The fraction $\dfrac{4}{6}$

is being simplified. To simplify the fraction, remove the

common factors of 2 from the numerator and denominator.

This removes a factor equal to 1: $\dfrac{2}{2} = 1$.   **21.** numerators,

denominators, simplify   **22.** $\dfrac{5}{6} \cdot \dfrac{2}{3}$ **23.** $\dfrac{1}{6}$ **24.** $-\dfrac{14}{45}$

25. $\dfrac{5c^2}{12}$ **26.** $-\dfrac{1}{25}$ **27.** $\dfrac{21}{5}$ **28.** $\dfrac{9m}{4}$ **29.** $-x$ **30.** 1

31. $-\dfrac{9}{16}$ **32.** $-\dfrac{125a^3}{8}$ **33.** $-\dfrac{8}{125}$ **34.** $\dfrac{4}{9}$ **35.** 2 mi

36. 30 lb   **37.** 60 in.^{2} **38.** 165 ft^2 **39. a.** 8   **b.** $-\dfrac{12}{11}$ **c.** $\dfrac{1}{5}$

d. $\dfrac{7}{8a}$ **40.** multiply, reciprocal   **41.** $\dfrac{25}{66}$ **42.** $-\dfrac{7}{8}$ **43.** $\dfrac{6m}{5}$

44. $\dfrac{30d}{7}$ **45.** $-\dfrac{3}{2}$ **46.** $\dfrac{8}{5}$ **47.** $-\dfrac{1}{180}$ **48.** 1   **49.** 12 pins

50. 30 pillow cases   **51.** $\dfrac{5}{7}$ **52.** $\dfrac{1}{2}$ **53.** $\dfrac{5x}{4}$ **54.** $-\dfrac{6}{5}$

55. a. $\dfrac{5}{8}$ **b.** $\dfrac{1}{5}$ **56.** 2, 3, 3, 5, 90   **57.** $\dfrac{5}{6}$ **58.** $-\dfrac{31}{40}$

59. $\dfrac{19}{48}$ **60.** $\dfrac{20}{7}$ **61.** $-\dfrac{23}{36}$ **62.** $\dfrac{7}{12}$ **63.** $-\dfrac{23}{6}$ **64.** $\dfrac{47}{60}$

65. $\dfrac{16 + n}{2n}$ **66.** $\dfrac{11x - 36}{9x}$ **67.** $\dfrac{11a - 12}{33}$ **68.** $\dfrac{49 + 8r}{56}$

69. $\dfrac{7}{32}$ in.   **70.** $\dfrac{3}{4}$ **71.** the second hour: $\dfrac{3}{11} > \dfrac{2}{9}$

72. $\dfrac{1}{250}$ **73.** $4\dfrac{1}{4} = \dfrac{17}{4}$

74.

$$-2\dfrac{2}{3} \quad -\dfrac{3}{4} \quad \dfrac{8}{9} \quad \dfrac{59}{24} = 2\dfrac{11}{24}$$

number line from -5 to 5

75. $3\dfrac{1}{5}$ **76.** $-3\dfrac{11}{12}$ **77.** 17   **78.** $2\dfrac{1}{3}$ **79.** $\dfrac{75}{8}$ **80.** $-\dfrac{11}{5}$

81. $\dfrac{53}{14}$ **82.** $\dfrac{199}{100}$ **83.** $2\dfrac{1}{10}$ **84.** $-\dfrac{21}{22}$ **85.** 40   **86.** $2\dfrac{1}{2}$ **87.** 16

88. $-40\dfrac{4}{5}$ **89.** $7\dfrac{9}{16}$ **90.** $6\dfrac{2}{9}$ **91.** $48\dfrac{1}{8}$ in.   **92.** 87 in.^{2}

93. 40 posters   **94.** 9 loads   **95.** $3\dfrac{23}{40}$ **96.** $6\dfrac{1}{6}$ **97.** $255\dfrac{19}{20}$

98. $23\dfrac{32}{35}$ **99.** $83\dfrac{1}{18}$ **100.** $113\dfrac{7}{20}$ **101.** $20\dfrac{1}{2}$ **102.** $34\dfrac{3}{8}$

103. $39\dfrac{11}{12}$ gal   **104.** $\dfrac{5}{8}$ in.   **105.** $\dfrac{8}{9}$ **106.** $\dfrac{19}{72}$ **107.** $8\dfrac{8}{15}$

108. $-3\dfrac{5}{8}$ **109.** $-\dfrac{12}{17}$ **110.** $\dfrac{26}{29}$ **111.** $-\dfrac{2}{5}$ **112.** $\dfrac{63}{17}$

113. $2\dfrac{23}{40}$ **114.** $14\dfrac{1}{16}$ **115.** $8\dfrac{1}{3}$ **116.** $11\dfrac{1}{6}$

117. 5 full tubes, $\dfrac{9}{10}$ of a tube is left over   **118.** 8 in.

119. $\dfrac{3}{10}$ **120.** $\dfrac{7}{12}$ **121.** 99   **122.** 25   **123.** $-\dfrac{15}{7}$ **124.** $\dfrac{27}{4}$

125. $-\dfrac{19}{111}$ **126.** -16 **127.** -56 **128.** -3

129. 330 pages   **130.** 100 minutes

Chapter 4 Test   (page 437)

1. a. numerator, denominator   **b.** equivalent   **c.** simplest
d. simplify   **e.** reciprocal   **f.** mixed   **g.** complex

2. a. $\dfrac{4}{5}$ **b.** $\dfrac{1}{5}$ **3.** $\dfrac{13}{6} = 2\dfrac{1}{6}$

4.

$$-1\dfrac{1}{7} \quad -\dfrac{2}{5} \quad \dfrac{7}{6} = 1\dfrac{1}{6} \quad 2\dfrac{4}{5}$$

number line from -2 to 3

5. yes   **6. a.** $\dfrac{36}{45}$ **b.** $\dfrac{21x}{24x}$ **7. a.** 0   **b.** undefined

8. a. $\dfrac{3}{4}$ **b.** $\dfrac{2n^2}{5}$ **9.** $\dfrac{5}{8}$ **10.** $-\dfrac{3}{20}$ **11.** $\dfrac{12b^4}{a}$ **12.** $\dfrac{11}{20}$

13. $\dfrac{11}{7}$ **14.** $\dfrac{1}{3}$ **15.** $\dfrac{9}{10}$ **16. a.** $-\dfrac{53}{17}$ **b.** $\dfrac{3a^2}{7}$ **17.** $\dfrac{5x - 24}{30}$

18. 40   **19.** $\dfrac{47}{50}$ **20. a.** $9\dfrac{1}{6}$ **b.** $\dfrac{39}{21}$ **21.** $261\dfrac{1}{6}$ **22.** $37\dfrac{5}{12}$

23. $1\dfrac{2}{3}$ **24. a.** Foreman, $39\dfrac{1}{2}$ lb   **b.** Foreman, $5\dfrac{1}{2}$ in.

c. Ali, $\dfrac{1}{4}$ in.   **25.** $\dfrac{8}{9}$ **26.** $\$1\dfrac{1}{2}$ million   **27.** $11\dfrac{3}{4}$ in.

28. perimeter: $53\dfrac{1}{3}$ in., area: $106\dfrac{2}{3}$ in.^{2} **29.** 60 calories

30. 12 servings   **31.** $\dfrac{13}{24}$ **32.** $\dfrac{3}{10}$ **33.** $\dfrac{20}{21}$ **34.** $-\dfrac{5}{3}$ **35.** $\dfrac{8}{3}$

36. $-\dfrac{7}{8}$ **37.** -77 **38.** $\dfrac{1}{2}$ **39.** 210 minutes

40. a. removing a common factor from the numerator and
denominator (simplifying a fraction)   **b.** equivalent fractions
c. multiplying a fraction by a form of 1 (building an
equivalent fraction)

Chapters 1–4 Cumulative Review   (page 439)

1. a. 5   **b.** 8 hundred thousands   **c.** 5,896,600
d. 5,900,000   **2.** hundred billions   **3.** Orange, San Diego,
Kings, Miami-Dade, Dallas, Queens   **4. a.** 450 ft
b. 11,250 ft^2 **5.** 30,996   **6.** 16,544, 16,544 + 3,456 = 20,000
7. 2,400 stickers   **8.** 299,320   **9.** 991, 991 · 35 = 34,685
10. a. 1, 2, 3, 4, 6, 8, 12, 24   **b.** $2 \cdot 3^2 \cdot 5^2$ **11.** 80
12. 21   **13.** 35   **14.** \$156,000   **15.** 65   **16.** 21
17. a. $\{\ldots, -5, -4, -3, -2, -1, 0, 1, 2, 3, 4, 5, \ldots\}$ **b.** true
18. -15 **19.** -324 **20.** 10   **21.** -200 ft
22. $-11°$F per hour   **23.** 16   **24.** -35 **25.** 1   **26.** 2
27. -4 **28.** 52   **29.** The case was raised 68 ft between
observations.   **30.** The pet store made a profit of \$37,639 the

second year.   **31. a.** $x + 15$ **b.** $x - 8$ **c.** $4x$ **d.** $\dfrac{x}{10}$

32. 52   **33. a.** $-15x$ **b.** $28xy$ **34. a.** $6x - 8$
b. $-15x + 10y - 20$ **35.** $5x$ **36.** $7a^2$ **37.** $-x - y$
38. $-4x + 8$ **39.** 4   **40.** 4   **41.** -3 **42.** -8 **43.** She
must complete 21 more shifts.   **44.** The width is 21 ft and the

length is 84 ft.   **45.** $\dfrac{3}{4}$ **46.** $\dfrac{5x^3}{2y}$ **47.** $-\dfrac{4}{5}$ **48.** $\dfrac{1}{2p}$ **49.** $1\dfrac{5}{12}$

50. $\dfrac{20 - 3m}{5m}$ **51.** $\dfrac{11}{16}$ in.   **52. a.** $10\dfrac{5}{7}$ **b.** $-\dfrac{53}{8}$ **53.** $7\dfrac{2}{5}$

54. $6\frac{9}{10}$ **55.** $9\frac{11}{12}$ **56.** $5\frac{11}{15}$ **57.** width: 28 in., height: 6 in.

58. $274\frac{1}{4}$ gal **59.** $3\frac{5}{12}$ ft **60.** $-\frac{3}{64}$ **61.** $\frac{5}{6}$ **62.** $-\frac{2}{49}$

63. $-\frac{14}{15}$ **64.** 4 **65.** -15 **66.** $\frac{8}{3}$ **67.** It would have taken 135 seconds to shave with the older model. **68.** The maximum number of points that a student can earn is 1,000.

Study Set   Section 5.1   (page 453)

1. point **3.** expanded **5.** Thousands, Hundreds, Tens, Ones, Tenths, Hundredths, Thousandths, Ten-thousandths

7. a. 10 **b.** $\frac{1}{10}$ **9. a.** $\frac{7}{10}$, 0.7 **b.** $\frac{47}{100}$, 0.47

11. Whole-number part, Fractional part **13.** ths
15. 79,816.0245 **17. a.** 9 tenths **b.** 6 **c.** 4 **d.** 5 ones
19. a. 8 millionths **b.** 0 **c.** 5 **d.** 6 ones

21. $30 + 7 + \frac{8}{10} + \frac{9}{100}$

23. $100 + 20 + 4 + \frac{5}{10} + \frac{7}{100} + \frac{5}{1,000}$

25. $7,000 + 400 + 90 + 8 + \frac{6}{10} + \frac{4}{100} + \frac{6}{1,000} + \frac{8}{10,000}$

27. $6 + \frac{4}{10} + \frac{9}{1,000} + \frac{4}{10,000} + \frac{1}{100,000}$

29. three tenths, $\frac{3}{10}$

31. fifty and forty-one hundredths, $50\frac{41}{100}$ **33.** nineteen and five hundred twenty-nine thousandths, $19\frac{529}{1,000}$

35. three hundred four and three ten-thousandths, $304\frac{3}{10,000}$

37. negative one hundred thirty-seven hundred-thousandths, $-\frac{137}{100,000}$ **39.** negative one thousand seventy-two and four hundred ninety-nine thousandths, $-1,072\frac{499}{1,000}$ **41.** 6.187

43. 10.0056 **45.** -16.39 **47.** 104.000004 **49.** $>$ **51.** $<$
53. $>$ **55.** $>$ **57.** $<$ **59.** $>$

61.

63.

65. 506.2 **67.** 33.08 **69.** 4.234 **71.** 0.3656 **73.** -0.14
75. -2.7 **77.** 3.150 **79.** 1.414213 **81.** 16.100
83. 290.30350 **85.** $0.28 **87.** $27,842 **89.** -0.7
91. $1,025.78
93.

95. two-thousandths, $\frac{2}{1,000} = \frac{1}{500}$ **97.** $0.16, $1.02, $1.20, $0.00, $0.10 **99.** candlemaking, crafts, hobbies, folk dolls, modern art **101.** Cylinder 2, Cylinder 4 **103.** bacterium, plant cell, animal cell, asbestos fiber **105. a.** $Q3, 2007; $2.75 **b.** Q4, 2006; $-$2.05 **113. a.** $12\frac{1}{2}$ in. **b.** $9\frac{5}{8}$ ft^2

Study Set   Section 5.2   (page 467)

1. addend, addend, addend, sum **3.** minuend, subtrahend, difference **5.** estimate **7.** It is not correct: $15.2 + 12.5 \ne 28.7$ **9.** opposite **11. a.** -1.2 **b.** 13.55
c. -7.4 **13.** 46.600, 11.000 **15.** 39.9 **17.** 8.59 **19.** 101.561
21. 202.991 **23.** 3.31 **25.** 2.75 **27.** 341.7 **29.** 703.5
31. 7.235 **33.** 43.863 **35.** -14.7 **37.** -18.8 **39.** -14.68
41. -6.15 **43.** -66.7 **45.** -45.3 **47.** 6.81 **49.** 17.82
51. -4.5 **53.** -3.4 **55.** 790 **57.** 610 **59.** -10.9
61. -16.6 **63.** 38.29 **65.** 55.00 **67.** 47.91 **69.** 658.04007
71. 0.19 **73.** 4.1 **75.** 288.46 **77.** 70.29 **79.** -14.3
81. -57.47 **83.** 8.03 **85.** 15.2 **87.** 4.977 **89.** 2.598
91. $815.80, $545.00, $531.49 **93.** 1.74, 2.32, 4.06; 2.90, 0, 2.90
95. 2.375 in. **97.** 42.39 sec **99.** $523.19, $498.19 **101.** 1.1°, 101.1°, 0°, 1.4°, 99.5° **103.** 20.01 mi **105. a.** $101.94

b. $55.80 **113. a.** $\frac{73}{60} = 1\frac{13}{60}$ **b.** $\frac{23}{60}$ **c.** $\frac{1}{3}$ **d.** $\frac{48}{25} = 1\frac{23}{25}$

Study Set   Section 5.3   (page 481)

1. factor, factor, partial product, partial product, product
3. a. 2.28 **b.** 14.499 **c.** 14.0 **d.** 0.00026 **5. a.** positive
b. negative **7. a.** 10, 100, 1,000, 10,000, 100,000 **b.** 0.1, 0.01, 0.001, 0.0001, 0.00001 **9.** 29.76 **11.** 49.84 **13.** 0.0081
15. 0.0522 **17.** 1,127.7 **19.** 2,338.4 **21.** 684 **23.** 410
25. 6.4759 **27.** 0.00115 **29.** 14,200,000 **31.** 98,200,000,000
33. 1,421,000,000,000 **35.** 657,100,000,000 **37.** -13.68
39. 5.28 **41.** 448,300 **43.** $-678,231$ **45.** 11.56 **47.** 0.0009
49. 3.16 **51.** 68.66 **53.** 119.70 **55.** 38.16 **57.** 14.6
59. 15.7 **61.** 250 **63.** 66.69 **65.** -0.1848 **67.** 1.69
69. 0.84 **71.** 0.00072 **73.** $-200,000$ **75.** 12.32
77. -17.48 **79.** 0.0049 **81.** 14.24 **83.** 8.6265
85. -57.2467 **87.** -22.39 **89.** -3.872 **91.** 24.48
93. -0.8649 **95.** 0.01, 0.04, 0.09, 0.16, 0.25, 0.36, 0.49, 0.64, 0.81
97. 1.9 in **99.** $74,100 **101.** $95.20, $123.75
103. 0.000000136 in., 0.0000000136 in., 0.00000004 in.
105. a. 2.1 mi **b.** 3.5 mi **c.** 5.6 mi **107.** $102.65
109. a. 19,600,000 acres **b.** 6,500,000,000
c. 3,026,000,000,000 miles **111. a.** 192 ft^2 **b.** 223.125 ft^2
c. 31.125 ft^2 **113. a.** $12.50, $12,500, $15.75, $1,575
b. $14,075 **115.** 136.4 lb **117.** 0.84 in. **125.** $2^2 \cdot 5 \cdot 11$
127. $2 \cdot 3^4$

Think It Through   (page 496)

1. 2.86

Study Set   Section 5.4   (page 496)

1. divisor, quotient, dividend **3. a.** 5.26 **b.** 0.008

5. a. $13\overline{)106.6}$ **b.** $371\overline{)1669.5}$ **7.** $\frac{10}{10}$ **9.** thousandths

11. a. left **b.** right **13.** moving the decimal points in the divisor and dividend 2 places to the right **15.** 2.1 **17.** 9.2
19. 4.27 **21.** 8.65 **23.** 3.35 **25.** 4.56 **27.** 0.46
29. 0.39 **31.** 19.72 **33.** 24.41 **35.** $280 \div 70 = 28 \div 7 = 4$
37. $400 \div 8 = 50$ **39.** $4,000 \div 50 = 400 \div 5 = 80$

41. $15,000 \div 5 = 3,000$ **43.** 4.5178   **45.** 0.003009   **47.** 12.5
49. 545,200   **51.** -8.62 **53.** 4.04   **55.** 20,325.7
57. -0.00003 **59.** -5.162 **61.** 0.1   **63.** 3.5   **65.** 58.5
67. 2.66   **69.** 7.504   **71.** 0.0045   **73.** 0.321   **75.** -1.5
77. -122.02 **79.** -2.4 **81.** 9.75   **83.** 789,150   **85.** 0.6
87. 13.60   **89.** 0.0348   **91.** 1,027.19   **93.** 0.15625
95. 280 slices   **97.** 2,000,000 calculations   **99.** 500 squeezes
101. 11 hr, 6 P.M.   **103.** 1,453.4 million trips   **105.** 0.231 sec
113. a. 5   **b.** 50

Study Set Section 5.5 (page 510)

1. equivalent   **3.** terminating   **5.** $\div$ **7.** zeros   **9.** repeating
11. a. 0.38   **b.** 0.212   **13. a.** $\dfrac{7}{10}$ **b.** $\dfrac{77}{100}$ **15.** 0.5
17. 0.875   **19.** 0.55   **21.** 2.6   **23.** 0.5625   **25.** -0.53125
27. 0.6   **29.** 0.225   **31.** 0.76   **33.** 0.002   **35.** 3.75
37. 12.6875   **39.** $0.\overline{1}$ **41.** $0.58\overline{3}$ **43.** $0.0\overline{7}$ **45.** $0.01\overline{6}$
47. $-0.\overline{45}$ **49.** $-0.\overline{60}$ **51.** 0.23   **53.** 0.49   **55.** 1.85
57. -1.08 **59.** 0.152   **61.** 0.370
63.

65.

67. $<$ **69.** $>$ **71.** $=$ **73.** $<$ **75.** $6.25, \dfrac{19}{3}, 6\dfrac{1}{2}$
77. $-\dfrac{8}{9}, -\dfrac{6}{7}, -0.\overline{81}$ **79.** $\dfrac{37}{90}$ **81.** $\dfrac{19}{60}$ **83.** $\dfrac{3}{22}$ **85.** 1
87. 0.57   **89.** 5.27   **91.** 0.35   **93.** -0.48 **95.** -2.55
97. 0.068   **99.** 7.305   **101.** 0.075   **103.** 0.0625, 0.375,
0.5625, 0.9375   **105.** $\dfrac{3}{40}$ in.   **107.** 23.4 sec, 23.8 sec, 24.2 sec,
32.6 sec   **109.** 93.6 in²   **111.** $7.02   **119. a.** $\{0, 1, 2, 3, 4, 5,$
$6, 7, 8, 9\}$ **b.** $\{2, 3, 5, 7, 11, 13, 17, 19, 23, 29\}$ **c.** $\{\ldots, -3, -2,$
$-1, 0, 1, 2, 3, \ldots\}$

Study Set Section 5.6 (page 519)

1. square   **3.** radical   **5.** perfect   **7. a.** $25, 25$ **b.** $\dfrac{1}{16}, \dfrac{1}{16}$
9. a. 7   **b.** 2   **11. a.** 1   **b.** 0   **13.** Step 2: Evaluate all
exponential expressions and any square roots.
15.

17. a. square root   **b.** negative   **19.** $-7, 8$ **21.** 5 and -5
23. 4 and -4 **25.** 4   **27.** 3   **29.** -12 **31.** -7 **33.** 31
35. 63   **37.** $\dfrac{2}{5}$ **39.** $-\dfrac{4}{3}$ **41.** $-\dfrac{1}{9}$ **43.** 0.8   **45.** -0.9
47. 0.3   **49.** 7   **51.** 16   **53.** -16 **55.** -3 **57.** 20
59. -140 **61.** -48 **63.** 43   **65.** 75   **67.** -7 **69.** -1
71. -10 **73.** $-\dfrac{7}{20}$ **75.** -140 **77.** 9.56   **79.** -1.4
81. 15   **83.** 7   **85.** 1, 1.414, 1.732, 2, 2.236, 2.449, 2.646,
2.828, 3, 3.162   **87.** 3.87   **89.** 8.12   **91.** 4.904   **93.** -3.332
95. a. 5 ft   **b.** 10 ft   **97.** 127.3 ft   **99.** 42-inch screen
109. 82.35   **111.** 39.304

Study Set Section 5.7 (page 531)

1. simplifying   **3.** combined   **5. a.** $4, 3.2, 12.8$
b. associative property of multiplication   **c.** $-6.1, 2, -12.2$
d. commutative property of multiplication   **7. a.** $4.2, 6.3,$
10.5 **b.** $3.6, 5.8, -2.2$ **c.** 2.7   **d.** coefficients   **9.** adding,
dividing   **11.** $2.3, 2.3, 0.6a, 0.6, 0.6, 0.8; 0.8, -1.82, -1.82, 0.8$
13. $12.8t$ **15.** $-56.42m$ **17.** $33.5t$ **19.** $26.4c$
21. $14.8x + 11.1$ **23.** $-11.4m + 16.8$ **25.** $0.06y - 0.564$
27. $3t - 27.5$ **29.** $7.9x$ **31.** $-8.8v$ **33.** $-0.27b^2$
35. $8.67a + 1.44$ **37.** $1.5m + 18.5$ **39.** $2d + 6.8$
41. $18.1y - 12.6$ **43.** $9.1b - 75.6$ **45.** 1.7   **47.** 2.24
49. -4.4 **51.** 7.11   **53.** -28.2 **55.** 0.42   **57.** 1.3
59. -3.9 **61.** -0.7 **63.** 1   **65.** -11 **67.** 2   **69.** 0.8
71. -2.05 **73.** -4.36 **75.** 0.8   **77.** $-0.5x + 3.9$ **79.** -1.1
81. -8.16 **83.** 5   **85.** $-3.7r$ **87.** -21.18 **89.** -2.2
91. -0.1 **93.** 11.5   **95.** 0.4   **97.** 48, 95, 200, signatures;
signatures, 0.95, 48, 200; $0.95x$, 200, 48, 48, 152, 0.95, 0.95, 160;
160; 160, 160, 200   **99.** $8.6 million   **101.** 3.27
103. 200 words   **105.** 2,500 balloons   **107.** 40 hours
109. 22 VHS cassettes   **113. a.** $\dfrac{17}{12} = 1\dfrac{5}{12}$ **b.** $\dfrac{3 + 2x}{3x}$
115. a. $\dfrac{23}{35}$ **b.** $\dfrac{28 - 5n}{35}$

Chapter 5 Review (page 535)

1. a. $0.67, \dfrac{67}{100}$ **b.**

2. a. 7 hundredths   **b.** 3   **c.** 8   **d.** 5 ten-thousandths
3. $10 + 6 + \dfrac{4}{10} + \dfrac{5}{100} + \dfrac{2}{1,000} + \dfrac{3}{10,000}$ **4.** two and three
tenths, $2\dfrac{3}{10}$ **5.** negative six hundred fifteen and fifty-nine
hundredths, $-615\dfrac{59}{100}$ **6.** six hundred one ten-thousandths,
$\dfrac{601}{10,000}$ **7.** one hundred-thousandth, $\dfrac{1}{100,000}$ **8.** 100.61
9. 11.997   **10.** 301.000016   **11.** $<$ **12.** $<$ **13.** $>$ **14.** $>$
15.

16. a. true   **b.** false   **c.** true   **d.** true   **17.** 3,706.082
18. -0.1 **19.** 11.3150   **20.** 0.222228   **21.** $0.67
22. $13   **23.** Washington, Diaz, Chou, Singh, Gerbac
24. Sun: 1.8, Mon: 0.6, Tues: 2.4, Wed: 3.8   **25.** 66.7
26. 28.428   **27.** 1,932.645   **28.** 24.30   **29.** -7.7 **30.** 3.1
31. -4.8 **32.** -29.09 **33.** -25.6 **34.** 4.939   **35. a.** 760
b. 280   **36.** 10.75 mm   **37.** $48.21   **38.** 8.15 in.   **39.** 15.87
40. 0.0068   **41.** -151.9 **42.** 0.00006   **43.** 90,145.2
44. 0.002897   **45.** 0.04   **46.** -10.61 **47.** 0.0001089
48. 115.741   **49. a.** 9,600,000 km²   **b.** 2,310,000,000
50. a. 1,600   **b.** 91.76   **51.** 98.07   **52.** $19.43   **53.** 0.07 in.
54. 68.62 in.²   **55.** 9.3   **56.** 1.29   **57.** -6.25 **58.** 0.053
59. 63   **60.** 0.81   **61.** 0.08976   **62.** -0.00112 **63.** 876.5
64. 770,210   **65.** $4,800 \div 40 = 480 \div 4 = 120$
66. $27,000 \div 9 = 3,000$ **67.** 12.9   **68.** -776.86 **69.** 13.95
70. 20.5   **71.** $8.34   **72.** 0.51 ppb   **73.** 14 servings
74. 9.5 revolutions   **75.** 0.875   **76.** -0.4 **77.** 0.5625
78. 0.06   **79.** $0.\overline{54}$ **80.** $-1.\overline{3}$ **81.** 3.056   **82.** $0.5\overline{7}$ **83.** 0.58

84. 1.03 **85.** > **86.** = **87.** $0.3, \dfrac{10}{33}, 0.\overline{3}$

88.

$-3.\overline{3}$ $-\dfrac{9}{10}$ 1.125 $2\dfrac{3}{4}$

89. $\dfrac{11}{15}$ **90.** $\dfrac{307}{300} = 1\dfrac{7}{300}$ **91.** 93 **92.** 7.305

93. 34.88 in.^{2} **94.** \$22.25 **95.** 5 and -5 **96.** 7, 7

97. 7 **98.** -4 **99.** $\dfrac{8}{13}$ **100.** 0.9

101.

$-\sqrt{16}$ $-\sqrt{2}$ $\sqrt{3}$ $\sqrt{9}$

102. a. 4.36 **b.** 24.45 **c.** 3.57 **103.** -27 **104.** $18\dfrac{1}{3}$

105. 70 **106.** -440 **107.** 8 **108.** 33 in. **109.** 9 and 10
110. Since $(2.646)^2 = 7.001316$, we cannot use an = symbol.
111. $32.2w$ **112.** $-30.4t$ **113.** $10.6y + 15.9$
114. $-20.3x + 58.8$ **115.** $14.1p$ **116.** $-0.12m^2$
117. $2.8a - 12.4$ **118.** $3t - 1.4$ **119.** -18.41 **120.** 5.23
121. -5.34 **122.** 17 **123.** -0.6 **124.** 12
125. \$2.81 per gallon **126.** 8 games

Chapter 5 Test (page 552)

1. a. addend, addend, sum **b.** minuend, subtrahend,
difference **c.** factor, factor, product **d.** divisor, quotient,
dividend **e.** repeating **f.** radical **g.** combined **h.** solve
2. $\dfrac{79}{100}, 0.79$ **3. a.** 1 thousandth **b.** 4 **c.** 6 **d.** 2 tens
4. Selway, Monroe, Paston, Covington, Cadia **5.** 4,519.0027
6. a. $60 + 2 + \dfrac{5}{10} + \dfrac{5}{100}$, sixty-two and fifty-five hundredths,
$62\dfrac{55}{100}$ **b.** $\dfrac{8}{100} + \dfrac{1}{10,000} + \dfrac{3}{100,000}$, eight thousand
thirteen one hundred-thousandths, $\dfrac{8,013}{100,000}$ **7. a.** 461.7
b. 2,733.050 **c.** -1.983373 **8.** \$0.65 **9.** 10.756
10. 6.121 **11.** 0.1024 **12.** 0.57 **13.** 14.07 **14.** 0.0348
15. $1.\overline{18}$ **16.** -0.8 **17.** -2.29 **18. a.** 210
b. $4,000 \div 20 = 400 \div 2 = 200$ **19. a.** 0.567909 **b.** 0.458
20. 61,400,000,000 **21.** 1.25 mi^2
22. 0.004 in. **23.** Saturday, \$23.75
24. 20.825 lb **25.** 10.676 **26. a.** 0.34 **b.** $0.41\overline{6}$
27. 3.588 **28.** 56.86 **29.** -12 **30.** $\dfrac{41}{30}$

31. a.

-0.8 0.375 $0.\overline{6}$

b.

$-\sqrt{9}$ $-\sqrt{5}$ $\sqrt{2}$ $\sqrt{16}$

32. \$5.65 **33.** 37 **34. a.** -1.08 **b.** 2.5625 **35.** 12, 12
36. a. > **b.** < **c.** = **d.** < **37.** 11 **38.** $-\dfrac{1}{30}$
39. a. -0.2 **b.** 1.3 **c.** 15 **d.** $-1\frac{1}{4}$ **41. a.** $14.4t$
b. $-28.2a$ **42. a.** $4.96s$ **b.** $52x - 18.7$ **43.** -7 **44.** 9.18
45. -0.6 **46.** 12 **47.** 0.42 grams **48.** 80 announcements

Chapters 1–5 Cumulative Review (page 555)

1. a. one hundred fifty-four thousand, three hundred two
b. $100,000 + 50,000 + 4,000 + 300 + 2$
2. $(3 + 4) + 5 = 3 + (4 + 5)$ **3.** 16,693 **4.** 102
5. 75,625 ft^2 **6.** 27 R 42 **7.** 1, 2, 4, 5, 10, 20 **8.** $2^2 \cdot 5 \cdot 11$
9. 600, 20 **10.** 4 **11. a.** 266 **b.** 15 **12.** > **13.** -13
14. adding **15.** 83°F increase **16.** -270 **17.** -1
18. $-2,100$ ft **19.** $3(-5) = -15$ **20.** 60 **21.** 42 **22.** -7
23. It must be heated 712°F. **24.** $\dfrac{c}{5}$ **25.** $60h$ min **26.** 219
27. $30x$ **28.** $15x$ **29.** 3 **30.** There are 12 first-class seats.
31. $\dfrac{6}{13}$ **32.** $\dfrac{5}{7}$ **33.** $\dfrac{21}{128}$ **34.** $-\dfrac{3a}{16}$ **35.** $\dfrac{7a + 45}{63}$ **36.** $19\dfrac{1}{8}$
37. $26\dfrac{7}{24}$ **38.** $-\dfrac{1}{3}$ **39.** $\dfrac{7}{64}$ **40.** $11\dfrac{1}{8}$ in. **41.** $\dfrac{3}{4}$ **42.** -36
43. 0.001 in.
44.

$-3\dfrac{1}{4}$ -1.5 $-\dfrac{9}{8}$ 0.75 $\sqrt{4}$ 3.8

45. 1.101 **46.** -8.136 **47.** 0.056012 **48.** 5.6
49. 157.5 in.^{2} **50.** 232.8 **51.** $0.41\overline{6}$ **52.** -2.325 **53.** -6
54. 11.5

Study Set Section 6.1 (page 567)

1. ratio **3.** unit **5.** 3 **7.** 10 **9.** $\dfrac{11 \text{ minutes}}{60 \text{ minutes}} = \dfrac{11}{60}$
11. $\dfrac{13}{9}$, 13 to 9, 13:9 **13.** $\dfrac{5}{8}$ **15.** $\dfrac{11}{16}$ **17.** $\dfrac{5}{3}$ **19.** $\dfrac{7}{4}$ **21.** $\dfrac{2}{3}$
23. $\dfrac{1}{2}$ **25.** $\dfrac{1}{3}$ **27.** $\dfrac{3}{4}$ **29.** $\dfrac{1}{3}$ **31.** $\dfrac{13}{3}$ **33.** $\dfrac{19}{39}$ **35.** $\dfrac{2}{7}$
37. $\dfrac{1}{2}$ **39.** $\dfrac{6}{1}$ **41.** $\dfrac{1}{5}$ **43.** $\dfrac{3}{7}$ **45.** $\dfrac{3}{4}$ **47.** $\dfrac{7}{12}$ **49.** $\dfrac{32 \text{ ft}}{3 \text{ sec}}$
51. $\dfrac{15 \text{ days}}{4 \text{ gal}}$ **53.** $\dfrac{21 \text{ made}}{25 \text{ attempts}}$ **55.** $\dfrac{3 \text{ beats}}{2 \text{ measures}}$
57. 12 revolutions per min **59.** \$5,000 per year
61. 1.5 errors per hr **63.** 320.6 people per square mi
65. \$4 per min **67.** \$68 per person **69.** 1.2 cents per ounce
71. \$0.07 per ft **73. a.** $\dfrac{2}{3}$ **b.** $\dfrac{3}{2}$ **75.** $\dfrac{1}{55}$ **77.** $\dfrac{3}{1}$
79. a. \$1,800 **b.** $\dfrac{4}{9}$ **c.** $\dfrac{1}{3}$ **d.** $\dfrac{1}{18}$ **81.** $\dfrac{1}{1}$ **83.** $\dfrac{1}{20}$
85. $\dfrac{5 \text{ compressions}}{2 \text{ breaths}}$ **87.** $\dfrac{329 \text{ complaints}}{100,000 \text{ passengers}}$ **89. a.** 108,000
b. 24 browsers per buyer **91.** 7¢ per oz **93.** 1.25¢ per min
95. \$4.45 per lb **97.** 440 gal per min **99. a.** 325 mi
b. 65 mph **101.** the 6-oz can **103.** the 50-tablet boxes
105. the truck **107.** the second car **113.** 43,000 **115.** 8,000

Study Set Section 6.2 (page 582)

1. proportion **3.** cross **5.** variable **7.** isolated **9.** true,
false **11.** 9, 90, 45, 90 **13.** Children, Teacher's aides
15. $3 \cdot x, 18, 3, 3, 6, 6$ **17.** $\dfrac{20}{30} = \dfrac{2}{3}$ **19.** $\dfrac{400 \text{ sheets}}{100 \text{ beds}} = \dfrac{4 \text{ sheets}}{1 \text{ bed}}$
21. false **23.** true **25.** true **27.** false **29.** false
31. true **33.** true **35.** false **37.** yes **39.** no **41.** 6

43. 4   **45.** 0.3   **47.** 2.2   **49.** $3\frac{1}{2}$ **51.** $\frac{7}{8}$ **53.** 3,500   **55.** $\frac{1}{2}$

57. 36   **59.** 1   **61.** 2   **63.** $8\frac{1}{5}$ **65.** 180   **67.** 18   **69.** 3.1

71. $\frac{1}{6}$ **73.** $218.75   **75.** $77.32   **77.** yes   **79.** 24 drops

81. 975   **83.** 80 ft   **85.** 65.25 ft = 65 ft 3 in.

87. 2.625 in. = $2\frac{5}{8}$ in.   **89.** $4\frac{2}{7}$, which is about $4\frac{1}{4}$ **91.** 19 sec

93. 31.25 in. = $31\frac{1}{4}$ in.   **95.** $309   **101.** 49.188   **103.** 31.428

105. 4.1   **107.** −49.09

Study Set Section 6.3   (page 596)

1. length   **3.** unit   **5.** capacity   **7. a.** 1   **b.** 3   **c.** 36
d. 5,280   **9. a.** 8   **b.** 2   **c.** 1   **d.** 1   **11.** 1   **13. a.** oz
b. lb   **15. a.** $\frac{1\ ton}{2,000\ lb}$ **b.** $\frac{2\ pt}{1\ qt}$ **17. a.** iv   **b.** i   **c.** ii
d. iii   **19. a.** iii   **b.** iv   **c.** i   **d.** ii   **21. a.** pound
b. ounce   **c.** fluid ounce   **23.** 36, in., 72   **25.** 2,000, 16, oz,
32,000   **27. a.** 8   **b.** $\frac{5}{8}$ in., $1\frac{1}{4}$ in., $2\frac{7}{8}$ in.   **29. a.** 16
b. $\frac{9}{16}$ in., $1\frac{3}{4}$ in., $2\frac{3}{16}$ in.   **31.** $2\frac{9}{16}$ in.   **33.** $10\frac{7}{8}$ in.   **35.** 12 ft
37. 105 ft   **39.** 42 in.   **41.** 63 in.   **43.** $\frac{21}{352}$ mi ≈ 0.06 mi
45. $\frac{7}{8}$ mi = 0.875 mi   **47.** $2\frac{3}{4}$ lb = 2.75 lb   **49.** $4\frac{1}{2}$ lb = 4.5 lb
51. 800 oz   **53.** 1,392 oz   **55.** 128 fl oz   **57.** 336 fl oz
59. $2\frac{3}{4}$ hr   **61.** $5\frac{1}{2}$ hr   **63.** 6 pt   **65.** 5 days   **67.** $4\frac{2}{3}$ ft
69. 48 in.   **71.** 2 gal   **73.** 5 lb   **75.** 4 hr   **77.** 288 in.
79. $2\frac{1}{2}$ yd = 2.5 yd   **81.** 15 ft   **83.** 24,800 lb   **85.** $2\frac{1}{3}$ yd
87. 3 mi   **89.** 2,640 ft   **91.** $3\frac{1}{2}$ tons = 3.5 tons   **93.** 2 pt
95. 150 yd   **97.** 2,880 in.   **99.** 0.28 mi   **101.** 61,600 yd
103. 128 oz   **105.** $4\frac{19}{20}$ tons = 4.95 tons   **107.** 68 quart cans
109. $71\frac{7}{8}$ gal = 71.875 gal   **111.** 320 oz
113. $6\frac{1}{8}$ days = 6.125 days   **117. a.** 3,700   **b.** 3,670
c. 3,673.26   **d.** 3,673.3

Study Set Section 6.4   (page 610)

1. metric   **3. a.** tens **b.** hundreds **c.** thousands   **5.** unit,
chart   **7.** weight   **9. a.** 1,000   **b.** 100   **c.** 1,000
11. a. 1,000   **b.** 10   **13. a.** $\frac{1\ km}{1,000\ m}$ **b.** $\frac{100\ cg}{1\ g}$
c. $\frac{1,000\ milliliters}{1\ liter}$ **15. a.** iii   **b.** i   **c.** ii   **17. a.** ii   **b.** iii
c. i   **19.** 1, 100, 0.2   **21.** 1,000, 1, mg, 200,000   **23.** 1 cm,
3 cm, 5 cm   **25. a.** 10, 1 millimeter   **b.** 27 mm, 41 mm,
55 mm   **27.** 156 mm   **29.** 280 mm   **31.** 3.8 m   **33.** 1.2 m
35. 8,700 mm   **37.** 2,890 mm   **39.** 0.000045 km
41. 0.000003 km   **43.** 1,930 g   **45.** 4,531 g   **47.** 6 g

49. 3.5 g   **51.** 3,000 mL   **53.** 26,300 mL   **55.** 3.1 cm
57. 0.5 L   **59.** 2,000 g   **61.** 0.74 mm   **63.** 1,000,000 g
65. 0.65823 kL   **67.** 0.472 dm   **69.** 10   **71.** 0.5 g
73. 5.689 kg   **75.** 4.532 m   **77.** 0.0325 L   **79.** 675,000
81. 0.0000077   **83.** 1.34 hm   **85.** 6,578 dam   **87.** 0.5 km,
1 km, 1.5 km, 5 km, 10 km   **89.** 3.43 hm   **91.** 12 cm, 8 cm
93. 0.00005 L   **95.** 3 g   **97.** 3,000 mL   **99.** 4   **101.** 3 mL
107. $0.\overline{8}$ **109.** $0.0\overline{7}$

Think It Through   (page 617)

1. 216 mm × 279 mm   **2.** 9 kilograms   **3.** 22.2 milliliters

Study Set Section 6.5   (page 620)

1. Fahrenheit, Celsius   **3. a.** meter   **b.** meter   **c.** inch
d. mile   **5. a.** liter   **b.** liter   **c.** gallon   **7. a.** $\frac{0.03\ m}{1\ ft}$
b. $\frac{0.45\ kg}{1\ lb}$ **c.** $\frac{3.79\ L}{1\ gal}$ **9.** 0.30 m, m   **11.** 0.035, 1,000, oz
13. 10 in.   **15.** 34 in.   **17.** 2,520 m   **19.** 7,534.5 m
21. 9,072 g   **23.** 34,020 g   **25.** 14.3 lb   **27.** 660 lb
29. 0.7 qt   **31.** 1.3 qt   **33.** 48.9°C   **35.** 1.7°C   **37.** 167°F
39. 50°F   **41.** 11,340 g   **43.** 122°F   **45.** 712.5 mL
47. 17.6 oz   **49.** 147.6 in.   **51.** 0.1 L   **53.** 39,283 ft
55. 1.0 kg   **57.** 14°F   **59.** 0.6 oz   **61.** 243.4 fl oz
63. 91.4 cm   **65.** 0.5 qt   **67.** 10°C   **69.** 127 m   **71.** −20.6°C
73. 5 mi   **75.** 70 mph   **77.** 1.9 km   **79.** 1.9 cm
81. 411 lb, 770 lb   **83. a.** 226.8 g   **b.** 0.24 L   **85.** no
87. about 62°C   **89.** 28°C   **91.** −5°C and 0°C
93. the 3 quarts   **99.** $\frac{29}{15}$ **101.** $\frac{4}{5}$ **103.** 8.05   **105.** 15.6

Chapter 6 Review   (page 623)

1. $\frac{7}{25}$ **2.** $\frac{15}{16}$ **3.** $\frac{2}{3}$ **4.** $\frac{3}{2}$ **5.** $\frac{1}{3}$ **6.** $\frac{7}{8}$ **7.** $\frac{4}{5}$ **8.** $\frac{3}{1}$
9. $\frac{7}{8}$ **10.** $\frac{5}{4}$ **11.** $\frac{1}{12}$ **12.** $\frac{1}{4}$ **13.** $\frac{16\ cm}{3\ yr}$ **14.** $\frac{\$3}{5\ min}$
15. 30 tickets per min   **16.** 15 inches per turn
17. 32.5 feet per roll   **18.** 3.2 calories per piece
19. $2.29 per pair   **20.** $0.25 billion per month
21. $\frac{37}{32}$ **22.** $7.75   **23.** 1,125 people per min
24. the 8-oz can   **25. a.** $\frac{20}{30} = \frac{2}{3}$ **b.** $\frac{6\ buses}{100\ cars} = \frac{36\ buses}{600\ cars}$
26. 2, 54, 6, 54   **27.** false   **28.** true   **29.** true   **30.** true
31. false   **32.** false   **33.** yes   **34.** no   **35.** 4.5   **36.** 16
37. 7.2   **38.** 0.12   **39.** $1\frac{1}{2}$ **40.** $3\frac{1}{2}$ **41.** $\frac{1}{3}$ **42.** 1,000
43. 192.5 mi   **44.** 300   **45.** 12 ft   **46.** 30 in.   **47. a.** 16
b. $\frac{7}{16}$ in., $1\frac{1}{2}$ in., $1\frac{3}{4}$ in., $2\frac{5}{8}$ in.   **48.** $1\frac{1}{2}$ in.   **49.** $\frac{1\ mi}{5,280\ ft} = 1$,
$\frac{5,280\ ft}{1\ mi} = 1$ **50. a.** min   **b.** sec   **51.** 15 ft   **52.** 216 in.
53. $5\frac{1}{2}$ ft = 5.5 ft   **54.** $1\frac{3}{4}$ mi = 1.75 mi   **55.** 54 in.
56. 1,760 yd   **57.** 2 lb   **58.** 275.2 oz   **59.** 96,000 oz
60. $2\frac{1}{4}$ tons = 2.25 tons   **61.** 80 fl oz   **62.** $\frac{1}{2}$ gal = 0.5 gal
63. 68 c   **64.** 5.5 qt   **65.** 40 pt   **66.** 56 c   **67.** 1,200 sec

68. 15 min   **69.** $8\frac{1}{3}$ days   **70.** 360 min   **71.** 108 hr

72. 86,400 sec   **73.** $\frac{21}{176}$ mi $\approx$ 0.12 mi   **74.** $20\frac{1}{4}$ tons = 20.25 tons

75. $484\frac{2}{3}$ yd   **76.** 100   **77. a.** 10, 1 millimeter

b. 19 mm, 3 cm, 45 mm, 62 mm   **78.** 4 cm

79. a. $\frac{1\text{ km}}{1,000\text{ m}} = 1, \frac{1,000\text{ m}}{1\text{ km}} = 1$ **b.** $\frac{1\text{ g}}{100\text{ cg}} = 1, \frac{100\text{ cg}}{1\text{ g}} = 1$

80. 5 places to the left   **81.** 4.75 m   **82.** 8,000 mm

83. 165,700 m   **84.** 678.9 dm   **85.** 0.05 kg   **86.** 8 g

87. 5.425 kg   **88.** 5,425,000 mg   **89.** 1.5 L   **90.** 3.25 kL

91. 40 cL   **92.** 1,000 dL   **93.** 1.35 kg   **94.** 0.24 L   **95.** 50 g

96. 1,000 mL   **97.** 164 ft   **98.** Sears Tower   **99.** 3,107 km

100. 198 cm   **101.** 850.5 g   **102.** 33 lb   **103.** 22,680 g

104. about 909 kg   **105.** about 2.0 lb   **106.** LaCroix

107. about 159.2 L   **108.** 221°F   **109.** 25°C   **110.** 30°C

Chapter 6 Test   (page 638)

1. a. ratio   **b.** rate   **c.** proportion   **d.** cross   **e.** tenths, hundredths, thousandths   **f.** metric   **g.** Fahrenheit, Celsius   **2.** $\frac{9}{13}$, 9:13, 9 to 13   **3.** $\frac{3}{4}$ **4.** $\frac{1}{6}$ **5.** $\frac{2}{5}$ **6.** $\frac{6}{7}$

7. $\frac{3\text{ feet}}{2\text{ seconds}}$ **8.** the 2-pound can   **9.** 22.5 kwh per day

10. $\frac{15\text{ billboards}}{50\text{ miles}} = \frac{3\text{ billboards}}{10\text{ miles}}$ **11. a.** no   **b.** yes

12. yes   **13.** 15   **14.** 63.24   **15.** $2\frac{1}{2}$ **16.** 0.2   **17.** $3.43

18. 2 c   **19. a.** 16   **b.** $\frac{5}{16}$ in., $1\frac{3}{8}$ in., $2\frac{3}{4}$ in.   **20.** introduce, eliminate   **21.** 15 ft   **22.** $8\frac{1}{3}$ yd   **23.** 172 oz   **24.** 3,200 lb

25. 128 fl oz   **26.** 115,200 min   **27. a.** the one on the left   **b.** the longer one   **c.** the right side   **28.** 12 mm, 5 cm, 65 mm   **29.** 0.5 km   **30.** 500 cm   **31.** 0.08 kg   **32.** 70,000 mL   **33.** 7.5 g   **34.** the 100-yd race   **35.** Jim   **36.** 0.9 qt   **37.** 42 cm   **38.** 182°F   **39.** A scale is a ratio (or rate) comparing the size of a drawing and the size of an actual object. For example, 1 inch to 6 feet (1 in.:6 ft).   **40.** It is easier to convert from one unit to another in the metric system because it is based on the number 10.

Chapters 1–6 Cumulative Review   (page 640)

1. a. five million, seven hundred sixty-four thousand, five hundred two
b. 5,000,000 + 700,000 + 60,000 + 4,000 + 500 + 2
2. a. 186 to 184   **b.** Detroit   **c.** 370 points   **3.** 69,658
4. 367,416   **5.** 20 R3   **6.** 1, 2, 3, 5, 6, 10, 15, 30   **7.** $2^3 \cdot 3^2 \cdot 5$
8. 140, 4   **9.** 81   **10. a.** 45   **b.** 17,100   **11.** <   **12.** −4
13. 15 shots   **14.** −9, 9   **15. a.** −8   **b.** undefined   **c.** −8
d. 0   **e.** 8   **f.** 0   **16.** 30   **17.** −5,000   **18.** The candidate gained 21 points over the last three months.   **19.** −2

20. −6   **21. a.** $w - 29$ **b.** $\frac{10}{m^3}$ **22.** −1   **23. a.** $-90a$
b. $20x - 10y + 35$ **24. a.** $-6x$ **b.** $x + 10y$ **25.** 4
26. −2   **27.** The width of the lawn is 18 feet and the length is 54 feet.   **28.** $-\frac{4}{5}$ **29.** $\frac{54a}{60a}$ **30.** 59,100,000 sq mi

31. $A = \frac{1}{2}bh$ **32.** $\frac{1}{b}$ **33.** $\frac{9}{20}$ **34.** $\frac{19}{15}$ **35.** $\frac{40 + 7m}{8m}$

36. $\frac{31}{32}$ in.   **37.** $6\frac{9}{10}$ **38.** 34 strips   **39.** $\frac{3}{4}$ hp

40. $-\frac{26}{15} = -1\frac{11}{15}$ **41.** −32   **42.** 10

43. 1,600 tickets were sold for the concert.   **44.** >

45.

$$-3.2 \quad -1\frac{3}{4} \quad -0.5 \quad \frac{11}{8} = 1\frac{3}{8} \quad 2.25 \quad \sqrt{9}$$

(number line from −5 to 5)

46. −17.64   **47.** −23.38   **48.** 250   **49.** 458.15 lb

50. 0.025   **51.** 12.7   **52.** $0.08\overline{3}$ **53.** $9.95   **54.** 23

55. 14.6   **56.** 120   **57.** 3.3   **58.** −73.5   **59.** She must get 800 signatures to earn $60.   **60.** $\frac{1}{5}$ **61.** the 94-pound bag

62. false   **63.** 202 mg   **64.** 15   **65. a.** 960 hr   **b.** 4,320 min   **c.** 480 sec   **66.** 2.5 lb   **67.** 2,400 mm   **68.** 0.32 kg   **69. a.** 1 gal   **b.** a meterstick   **70.** 36 in.

Study Set Section 7.1   (page 653)

1. Percent   **3.** 100, simplify   **5.** right   **7.** percent
9. 84%, 16%   **11.** 107%   **13.** 99%   **15. a.** 15%   **b.** 85%
17. $\frac{17}{100}$ **19.** $\frac{91}{100}$ **21.** $\frac{1}{25}$ **23.** $\frac{3}{5}$ **25.** $\frac{19}{1,000}$ **27.** $\frac{547}{1,000}$

29. $\frac{1}{8}$ **31.** $\frac{17}{250}$ **33.** $\frac{1}{75}$ **35.** $\frac{17}{120}$ **37.** $\frac{13}{10}$ **39.** $\frac{11}{5}$

41. $\frac{7}{2,000}$ **43.** $\frac{1}{400}$ **45.** 0.16   **47.** 0.81   **49.** 0.3412

51. 0.50033   **53.** 0.0699   **55.** 0.013   **57.** 0.0725   **59.** 0.185
61. 4.6   **63.** 3.16   **65.** 0.005   **67.** 0.0003   **69.** 36.2%
71. 98%   **73.** 171%   **75.** 400%   **77.** 40%   **79.** 16%
81. 62.5%   **83.** 43.75%   **85.** 225%   **87.** 105%

89. $16\frac{2}{3}\% \approx 16.7\%$ **91.** $166\frac{2}{3}\% \approx 166.7\%$

93. $\frac{157}{5,000}, 3.14\%$ **95.** $\frac{51}{125}, 0.408$ **97.** $\frac{21}{400}, 0.0525$

99. $2.33, 233\frac{1}{3}\% \approx 233.3\%$ **101.** 91%   **103. a.** 12%
b. 24%   **c.** 4% (Alaska, Hawaii)   **105. a.** 0.0775   **b.** 0.05
c. 0.1425
107. torso: 27.5%

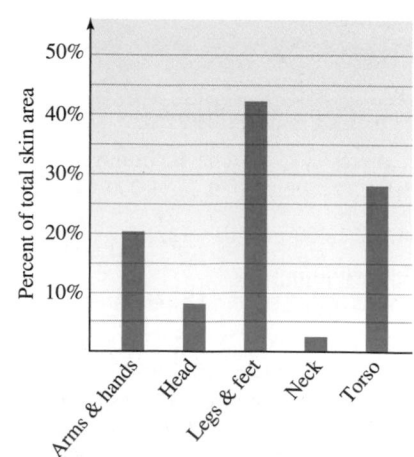

109. a. $\dfrac{5}{64}$ **b.** 0.078125 **c.** 7.8125% **111.** $33\dfrac{1}{3}\%, \dfrac{1}{3}, 0.\overline{3}$

113. a. $\dfrac{13}{15}$ **b.** $86\dfrac{2}{3}\% \approx 86.7\%$ **115. a.** $\dfrac{1}{4}\%$ **b.** $\dfrac{1}{400}$

c. 0.0025 **117.** 0.27% **123. a.** 34 cm **b.** 68.25 cm²

Think It Through   (page 673)

36% are enrolled in college full time, 43% of the students work less than 20 hours per week, 10% never

Study Set Section 7.2   (page 673)

1. sentence, equation **3.** solved **5.** part, whole **7.** cross **9.** Amount, base, percent, whole **11.** 100% **13. a.** 0.12 **b.** 0.056 **c.** 1.25 **d.** 0.0025

15. a. $x = 7\% \cdot 16, \dfrac{x}{16} = \dfrac{7}{100}$ **b.** $125 = x \cdot 800, \dfrac{125}{800} = \dfrac{x}{100}$

c. $1 = 94\% \cdot x, \dfrac{1}{x} = \dfrac{94}{100}$ **17. a.** $5.4\% \cdot 99 = x, \dfrac{x}{99} = \dfrac{5.4}{100}$

b. $75.1\% \cdot x = 15, \dfrac{15}{x} = \dfrac{75.1}{100}$ **c.** $x \cdot 33.8 = 3.8, \dfrac{3.8}{33.8} = \dfrac{x}{100}$

19. 68 **21.** 132 **23.** 17.696 **25.** 24.36 **27.** 25% **29.** 85% **31.** 62.5% **33.** 43.75% **35.** 110% **37.** 350% **39.** 30 **41.** 150 **43.** 57.6 **45.** 72.6 **47.** 1.25% **49.** 65 **51.** 99 **53.** 90 **55.** 80% **57.** 0.096 **59.** 44 **61.** 2,500% **63.** 107.1 **65.** 60 **67.** 31.25% **69.** 43.5 **71.** 12K bytes = 12,000 bytes **73. a.** $20.75 **b.** $4.15 **75.** 2.7 in. **77.** yes **79.** 5% **81.** 120 **83.** 13,500 km **85.** $1,026 billion **87.** 24 oz **89.** 30, 12 **91.** 40,000%

93.

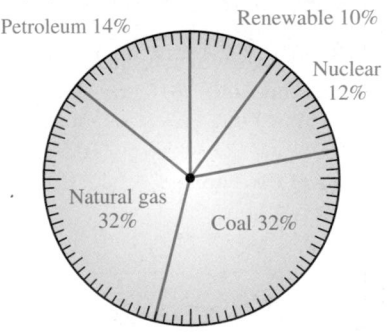

95. 32%, 43%, 13%, 6%, 6%   **2007 Federal Income Sources**

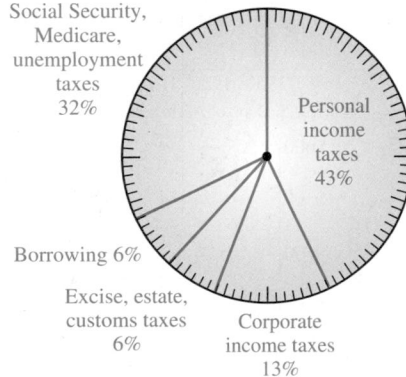

103. 18.17 **105.** 5.001 **107.** 0.008

Think It Through   (page 687)

1. 1970–1975, about a 75% increase
2. 2000–2005, about a 15% decrease

Study Set Section 7.3   (page 690)

1. commission **3. a.** increase **b.** original **5.** purchase price **7.** sales **9. a.** $64.07 **b.** $135.00 **11.** subtract, original **13.** $3.71 **15.** $4.20 **17.** $70.83 **19.** $64.03 **21.** 5.2% **23.** 15.3% **25.** $11.40 **27.** $168 **29.** 2% **31.** 4% **33.** 10% **35.** 15% **37.** 20% **39.** 10% **41.** $29.70, $60.30 **43.** $8.70, $49.30 **45.** 19% **47.** 14% **49.** $53.55 **51.** $47.34, $2.84, $50.18 **53.** 8% **55.** 0.25% **57.** $150 **59.** 8%, 3.75%, 1.2%, 6.2% **61.** 5% **63.** 31% **65.** 152% **67.** 36% **69.** 12.5% **71. a.** 25% **b.** 36% **73.** $2,955 **75.** 1.5% **77.** 90% **79.** $12,000 **81. a.** $7.99 **b.** $31.96 **83.** 6% **85.** $349.97, 13% **87.** 23%, $11.88 **89.** $76.50 **91.** $187.49 **97.** −50 **99.** 3 **101.** 13

Study Set Section 7.4   (page 701)

1. Estimation **3.** two **5.** 2 **7.** 4 **9.** 10, 5 **11.** 2.751, 3 **13.** 0.1267, 0.1 **15.** 405.9 lb, 400 lb **17.** 69.14 min, 70 min **19.** 70 **21.** 14 **23.** 2,100,000 **25.** 200,000 **27.** 4 **29.** 12 **31.** 820 **33.** 20 **35.** $9 **37.** $4.50 **39.** $18 **41.** $1.50 **43.** 8 **45.** 72 **47.** 12 **49.** 5.4 **51.** 180 **53.** 230 **55.** 6 **57.** 18 **59.** 7 **61.** 70 **63.** 12,000 **65.** 1.8 **67.** 0.49 **69.** 12 **71.** 164 students **73.** $60 **75.** $6 **77.** $7.50 **79.** $30,000 **81.** 320 lb **83.** 210 motorists **85.** 220 people **87.** 18,000 people **89.** 3,100 volunteers

95. a. $\dfrac{4}{3} = 1\dfrac{1}{3}$ **b.** $\dfrac{1}{3}$ **c.** $\dfrac{5}{12}$ **d.** $\dfrac{5}{3} = 1\dfrac{2}{3}$

Study Set Section 7.5   (page 710)

1. interest **3.** rate **5.** total **7. a.** $125,000 **b.** 5% **c.** 30 years **9. a.** 0.07 **b.** 0.098 **c.** 0.0625 **11.** $1,800 **13. a.** compound interest **b.** $1,000 **c.** 4 **d.** $50 **e.** 1 year **15.** $I = Prt$ **17.** $100 **19.** $252 **21.** $525 **23.** $1,590 **25.** $16.50 **27.** $30.80 **29.** $13,159.23 **31.** $40,493.15 **33.** $2,060.68 **35.** $5,619.27 **37.** $10,011.96 **39.** $77,775.64 **41.** $5,300 **43.** $198 **45.** $5,580 **47.** $46.88 **49.** $4,262.14 **51.** $10,000, $7\dfrac{1}{4}\% = 0.0725, 2$ yr, $1,450 **53.** $192, $1,392, $58 **55.** $19.449 million **57.** $755.83 **59.** $1,271.22 **61.** $570.65 **63.** $30,915.66 **65.** $159,569.75

71. $\dfrac{1}{2}$ **73.** $\dfrac{29}{35}$ **75.** $8\dfrac{1}{3}$ **77.** −36

Chapter 7 Review   (page 714)

1. $39\%, 0.39, \dfrac{39}{100}$ **2.** $111\%, 1.11, \dfrac{111}{100}$ **3.** 61% **4. a.** 54% **b.** 46% **5.** $\dfrac{3}{20}$ **6.** $\dfrac{6}{5}$ **7.** $\dfrac{37}{400}$ **8.** $\dfrac{1}{500}$ **9.** 0.27 **10.** 0.08 **11.** 6.55 **12.** 0.018 **13.** 0.0075 **14.** 0.0023 **15.** 83% **16.** 162.5% **17.** 5.1% **18.** 600% **19.** 50% **20.** 80% **21.** 87.5% **22.** 6.25% **23.** $33\dfrac{1}{3}\% \approx 33.3\%$ **24.** $83\dfrac{1}{3}\% \approx 83.3\%$ **25.** $91\dfrac{2}{3}\% \approx 91.7\%$ **26.** $166\dfrac{2}{3}\% \approx 166.7\%$ **27. a.** 0.972 **b.** $\dfrac{243}{250}$ **28.** 63% **29. a.** 0.0025 **b.** $\dfrac{1}{400}$ **30.** $6\dfrac{2}{3}\% \approx 6.7\%$ **31. a.** amount: 15, base: 45 percent: $33\dfrac{1}{3}\%$ **b.** Amount, base, percent **32. a.** 0.13 **b.** 0.071 **c.** 1.95 **d.** 0.0025

e. $\frac{1}{3}$ **f.** $\frac{2}{3}$ **g.** $\frac{1}{6}$ **33. a.** $x = 32\% \cdot 96$ **b.** $64 = x \cdot 135$

c. $9 = 47.2\% \cdot x$ **34. a.** $\frac{x}{96} = \frac{32}{100}$ **b.** $\frac{64}{135} = \frac{x}{100}$

c. $\frac{9}{x} = \frac{47.2}{100}$ **35.** 200   **36.** 125   **37.** 1.75%   **38.** 2,100

39. 121   **40.** 30   **41.** 600   **42.** 5,300%   **43.** 0.6 gal methane
44. 68   **45.** 87%   **46.** $5.43

47.

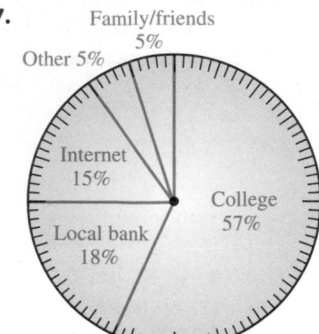

48. 139,531,200 mi^2
49. $3.30, $63.29
50. 4%   **51.** $40.20
52. 4.25%   **53.** $100,000
54. original   **55.** 18%
56. 9.6%
57. a. purchase price
b. sales tax
c. commission rate
58. a. sale price
b. original price
c. discount

59. $180, $2,500, 7.2%   **60.** 5%   **61.** 3.4203, 3
62. 86.87, 90   **63.** 4.34 sec, 4 sec   **64.** 1,090 L, 1,000 L
65. 12   **66.** 120   **67.** 140,000   **68.** 150   **69.** 3   **70.** 10
71. 350   **72.** 1,000   **73.** 60   **74.** 2   **75.** $36   **76.** $7.50
77. about 12 fluid oz   **78.** about 120 people   **79.** 200
80. $30,000   **81.** $6,000, 8%, 2 years, $960   **82.** $27,240
83. $75.63   **84.** $10,308.22   **85. a.** $116.25   **b.** 1,616.25
c. $134.69   **86.** $2,142.45   **87.** $6,076.45   **88.** $43,265.78

Chapter 7 Test   (page 732)

1. a. Percent   **b.** is, of, what, what   **c.** amount, base

d. increase   **e.** Simple, Compound   **2. a.** $61\%, \frac{61}{100}, 0.61$

b. 39%   **3.** $199\%, \frac{199}{100}, 1.99$ **4. a.** 0.67   **b.** 0.123

c. 0.0975   **5.** 0.0006   **b.** 2.1   **c.** 0.55375   **6. a.** 25%
b. 62.5%   **c.** 112%   **7. a.** 19%   **b.** 347%   **c.** 0.5%
8. a. 66.7%   **b.** 200%   **c.** 90%

9. a. $\frac{11}{20}$ **b.** $\frac{1}{10,000}$ **c.** $\frac{5}{4}$ **10. a.** $\frac{1}{15}$ **b.** $\frac{3}{8}$ **c.** $\frac{2}{25}$

11. a. $3\frac{1}{3}\% = 3.3\%$ **b.** $177\frac{7}{9}\% = 177.8\%$ **12.** 6.5%

13. 250%   **14.** 93.7%   **15.** 90   **16.** 21   **17.** 134.4   **18.** 7.8
19. a. 1.02 in.   **b.** 32.98 in.   **20.** $26.24   **21.** 3%   **22.** 23%
23. $35.92   **24.** 11%   **25.** $41,440   **26.** $9, $66, 12%
27. $6.60, $13.40   **28. a.** two, left   **b.** one, left   **29. a.** 80
b. 3,000,000   **c.** 40   **30.** 100   **31.** $4.50   **32.** 16,000 females
33. $150   **34.** $28,175   **35.** $39.45   **36.** $5,079.60

Chapters 1–7 Cumulative Review   (page 735)

1. a. six million, fifty-four thousand, three hundred forty-six
b. $6,000,000 + 50,000 + 4,000 + 300 + 40 + 6$ **2.** 239
3. 42,156   **4.** 23,100   **5.** 64 ft^2 **6.** 15 R6   **7. a.** 1, 2, 4, 5, 8, 10, 20, 40   **b.** $2 \cdot 3 \cdot 7^2$ **8.** 120, 6   **9.** 15   **10.** $2,106
11. a. 184   **b.** 9   **12.** >   **13.** 0   **14.** −$135   **15.** −36, 36
16. a. undefined   **b.** 0   **c.** 0   **d.** 14   **17.** 9   **18.** −1,900
19. −2   **20.** The company gained 36 points of market share in five years.   **21.** $2t − 16$ **22.** 15°C   **23. a.** $32m$
b. $8d + 16$ **24. a.** x **b.** $2l + 2w$ **25.** 3   **26.** It will take her 11 months to reach the goal.   **27.** $\frac{4}{11}$ **28.** $\frac{2}{3}$ **29.** $\frac{36}{45}$

30. −60   **31.** 650 in.^{2} **32.** $-\frac{5}{21a}$ **33.** $-\frac{3}{4}$ **34.** $\frac{24}{35}$

35. $\frac{28 + 2m}{7m}$ **36.** $\frac{7}{6}$ **37.** $\frac{1}{12}$ lb   **38.** −30   **39.** $35\frac{3}{4}$ in.

40. $20\frac{5}{18}$ **41.** $-\frac{5}{6}$ **42.** The total number of pages in the

telephone book is 525.   **43.** −30   **44.** 8   **45. a.** 452.03
b. 452.030   **46.** −5.5   **47.** $731.40   **48.** 0.27   **49.** $0.7\overline{3}$
50. −29   **51.** 4   **52.** 0.6   **53.** It took 3.5 hours of labor to

repair the car.   **54.** $\frac{5}{6}$ **55.** 4   **56.** It will take her 9 minutes.

57. 40 days   **58.** 2.4 m   **59.** 14.3 lb   **60.** $29\%, \frac{29}{100}; 0.473,$

$\frac{473}{1,000}; 87.5\%, 0.875$ **61.** 125   **62.** 0.0018%   **63.** 78%
64. $428, $321, $107, 25%   **65. a.** $12   **b.** $90.18   **66.** $1,450

Study Set Section 8.1   (page 748)

1. (a)   **3.** (c)   **5.** (d)   **7.** axis   **9.** intersection
11. pictures   **13.** bars, edge, equal   **15.** about 500 buses
17. $10.70   **19.** $4.55 ($21.85 − $17.30)   **21.** fish, cat, dog
23. no   **25.** yes   **27.** about 10,000,000 metric tons
29. 1990, 2000, 2007   **31.** 4,000,000 metric tons   **33.** seniors
35. $50   **37.** Chinese   **39.** no   **41.** 62%   **43.** 1,219,000,000
45. 493   **47.** 2002 to 2003; 2004 to 2005; 2005 to 2006;
2007 to 2008   **49.** 2001 and 2003   **51.** 2005 to 2006;
a decrease of 14 resorts   **53.** 1   **55.** B   **57.** 1   **59.** Runner 1
was running; runner 2 was stopped.   **61. a.** 27   **b.** 22
63. $16,168.25   **65. a.** $9,593.75   **b.** $6,847.50   **c.** $2,746.25
67. 2000; about 3.2%   **69.** increase; about 1%   **71.** it
increased   **73.** D   **75.** reckless driving and failure to yield
77. reckless driving   **79.** about $440   **81.** no   **83.** the
miner's   **85.** the miners   **87.** about $42   **89.** about $30
91. 11%   **93.** 21%
95.

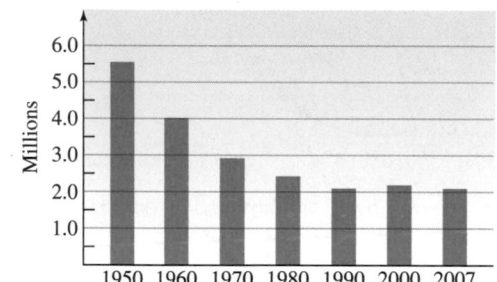

Source: U.S. Dept. of Agriculture

97.

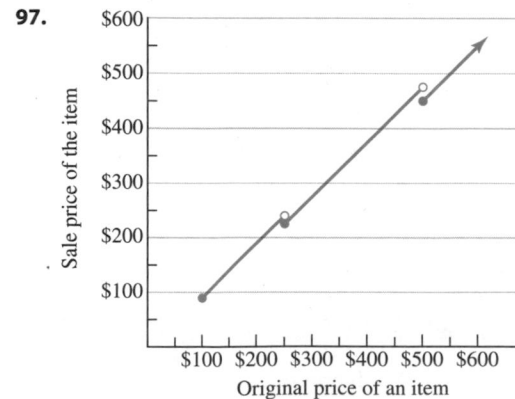

101. 11, 13, 17, 19, 23, 29   **103.** 0, 4

Think It Through   (page 762)

**Median Annual Earnings of Full-Time Workers
(25 years and older) by Education**

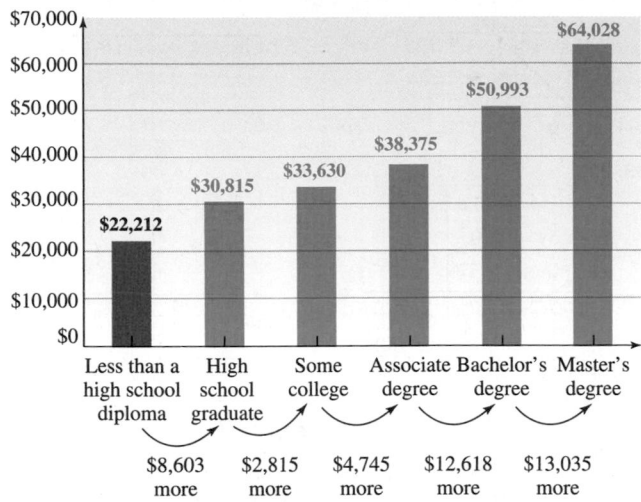

Source: Bureau of Labor Statistics, Current Population Survey (2008)

Study Set Section 8.2   (page 763)

1. mean   **3.** mode   **5.** the number of values   **7. a.** an even number   **b.** 6 and 8   **c.** 6, 8, 14, 7   **d.** 15, 4, 11   **9.** 8

11. 35   **13.** 19   **15.** 5.8   **17.** 9   **19.** 5   **21.** 17.2   **23.** $\frac{5}{8}$

25. 9   **27.** 44   **29.** 2.05   **31.** 1   **33.** 3   **35.** -6 **37.** 22.7

39. bimodal: $\frac{1}{3}, \frac{1}{2}$ **41. a.** 82.5   **b.** 83   **43. a.** 2,670 mi

b. 89 mi   **45. a.** \$11,875   **b.** 125   **c.** \$95   **47. a.** 65¢
b. 60¢   **c.** 50¢   **d.** 55¢   **49.** 61°   **51.** 2.23 GPA
53. 2.5 GPA   **55.** median and mode are 85   **57.** same
average (56); sister's scores are more consistent
59. 22.525 oz, 25 oz, 17.3 oz   **61.** 6.8, 6.9, 1.9
63. 5 lb, 4 lb, 10 lb   **69.** 65%   **71.** 42   **73.** 62.5%   **75.** 43.5

Study Set Section 8.3   (page 774)

1. variables   **3.** satisfies   **5.** table   **7.** origin
9. x, y
11.

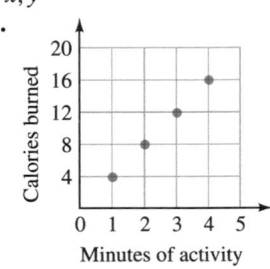

13. a. one   **b.** two

15. a. right   **b.** upward   **17.** origin, right, down
19. A: x-axis, B: quadrant I, C: quadrant III, D: x-axis and
y-axis, E: quadrant II, F: quadrant IV   **21.** x, y **23.** yes
25. 2, 8, 8, 8, 6, 3, 3, 2   **27.** yes   **29.** no   **31.** no   **33.** yes
35. no   **37.** yes   **39.** yes   **41.** yes   **43. a.** 8   **b.** 3

45. a. -1 **b.** 14   **47. a.** 15   **b.** $\frac{3}{5}$ **49. a.** $\frac{4}{5}$ **b.** -8

51. 8, 8; 6, 6; 12, 12   **53.** $-5, -5; 4, 4; -10, -10$
55.

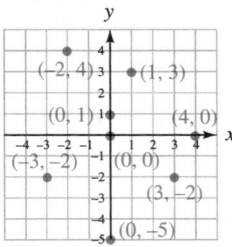

57.

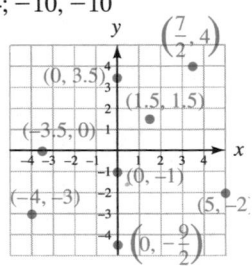

59. $A(2, 4), B(-3, 3), C(-2, -3), D(4, -3), E(3, 0), F(0, -1)$

61. $A(-3, -4), B(2.5, 3.5)$ or $B\left(2\frac{1}{2}, 3\frac{1}{2}\right), C(-2.5, 0)$ or

$C\left(-2\frac{1}{2}, 0\right), D(2.5, 0)$ or $D\left(2\frac{1}{2}, 0\right), E(3, -3), F(0, 3)$

63. yes   **65.** no   **67.** Rockford $(5, B)$, Forreston $(2, C)$,
Harvard $(7, A)$, intersection $(5, E)$ **69. a.** $(2, -1)$ **b.** no
c. yes   **d.** no   **71.** New Delhi, Kampala, Coats Land,
Reykjavik, Buenos Aires, Havana
73.

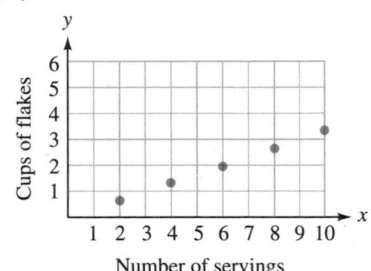

81. -16 **83.** -25 **85.** 21   **87.** -3

Study Set Section 8.4   (page 789)

1. two   **3.** graph   **5.** x-intercept   **7.** x, y; Plot/graph; line,
check   **9. a.** $(0, 1)$ **b.** $(-2, 0)$ **c.** yes   **11.** $0, x$
13. a. horizontal   **b.** vertical   **15. a.** $(-2, 4), (0, 4), (2, 4)$
(answers may vary)   **b.** $(-1, 2), (-1, 0), (-1, -2)$ (answers
may vary)   **17.** 0, 0, 8, 2, 2, 4, 4   **19.** $-3, (-2, -3); 3, (0, 3);$
9, (2, 9)   **21.** $1, (-2, 1); -3, (0, -3); -7, (2, -7)$
23.

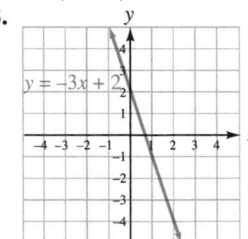

25.

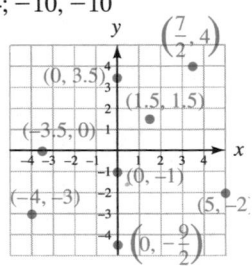

27.

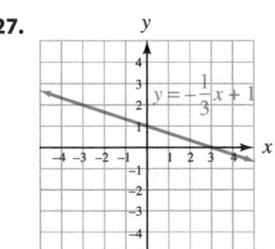

29.

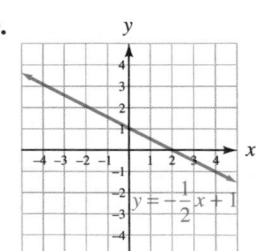

31.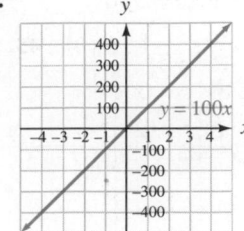

33.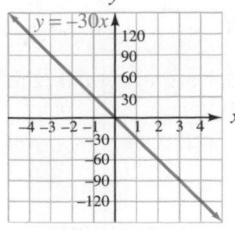

35. $2, -4, 4$

37. $4, 5, \dfrac{16}{5} = 3\dfrac{1}{5}$

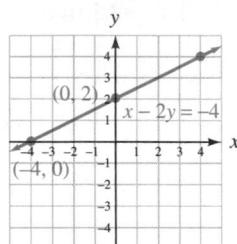

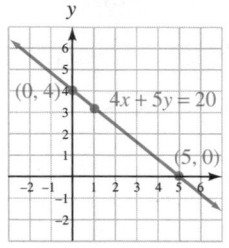

39.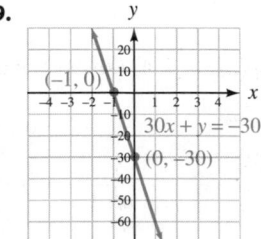

41.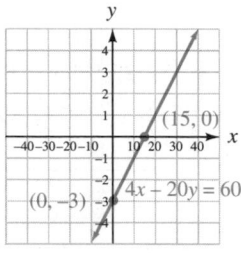

43.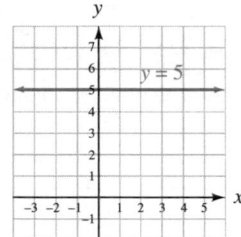

45.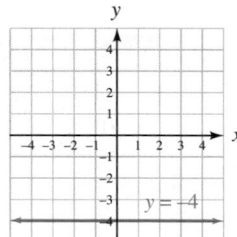

47.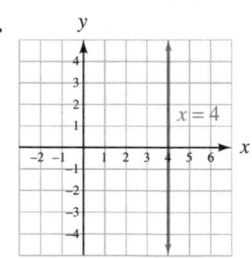

49.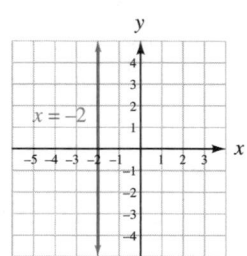

51. x-intercept: $(8, 0)$, y-intercept: $(0, 8)$

53. x-intercept: $(25, 0)$, y-intercept: $(0, 20)$

55.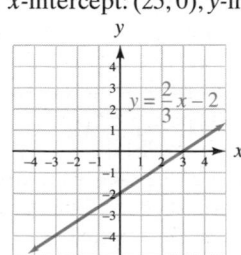

57.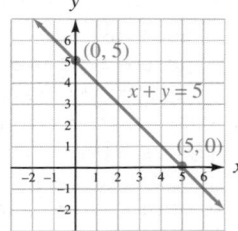

59.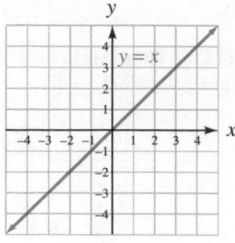

61.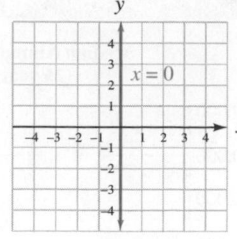

63.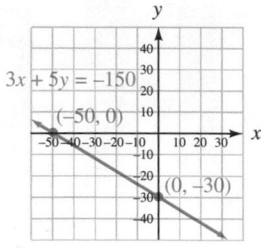

65.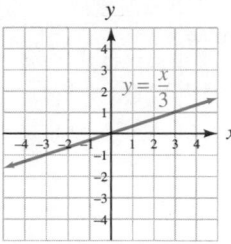

67.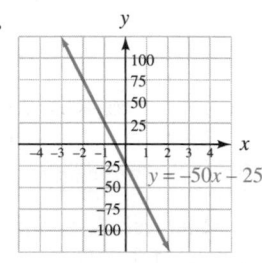

69.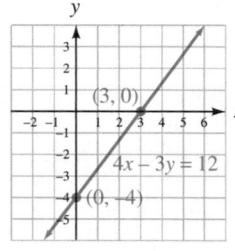

71.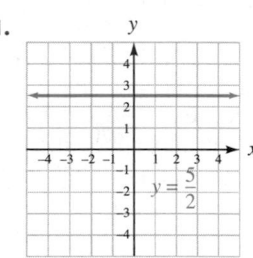

73. $22.50

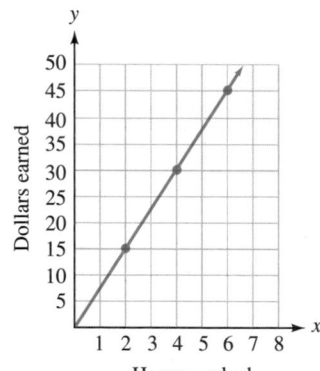

75. $2, 4, 6, 8, 10$

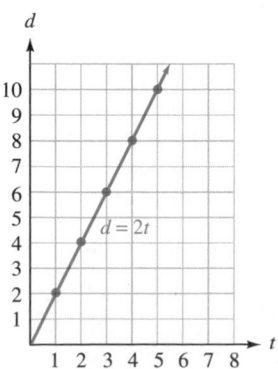

77.

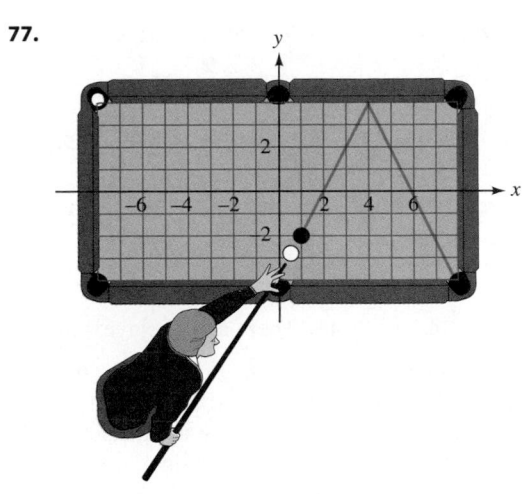

85. $2^2 \cdot 3^2 \cdot 5$ **87.** -5 **89.** $\dfrac{1}{4}$ sec

Chapter 8 Review   (page 797)

1. a. $-18°$ **b.** $-71°$ **2. a.** 30 mph   **b.** 15 mph   **3.** 20
4. about 59   **5.** Germany and India; about 17   **6.** about 35
7. about 29%   **8.** men; about 15% more   **9.** women
10. No, I would not date a co-worker (31% to 29%)
11. about 4,100 animals   **12.** the Columbus Zoo; about
7,250 animals   **13.** about 3,000 animals   **14.** about
12,500 animals   **15.** oxygen   **16.** 4%   **17.** 13.5 lb
18. 166 lb   **19.** about 3,000 million eggs   **20.** about
3,050 million eggs   **21.** 2007; about 2,950 million eggs
22. about 5,750 million eggs   **23.** between 2006 and 2007
24. between 2007 and 2008   **25.** about 290 million more
eggs   **26.** about 500 million more eggs   **27.** 60   **28.** 180
29. 160
30.

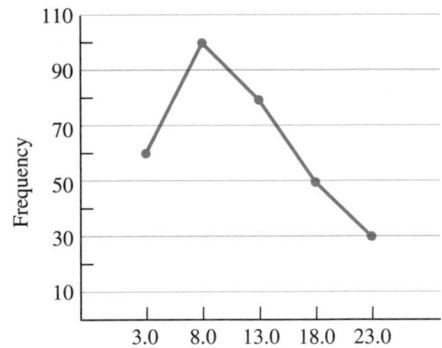

31. yes   **32.** median   **33.** 1.2 oz   **34.** 1.138 oz, 0.5 oz
35. 7.3 microns, 7.2 microns, 6.9 microns, 1.3 microns
36. 32 pages per day   **37.** $20   **38.** 2.62 GPA   **39.** yes
40. no   **41.** $-3, 4$ **42.** $-5, -5; 3, 3; 4, 4$
43.

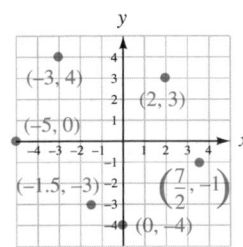

44. $A(4, 3), B(-3, 3), C(-4, 0), D(-1.5, -3.5)$ or
$D\left(-1\dfrac{1}{2}, -3\dfrac{1}{2}\right), E(2.5, -1.5)$ or $E\left(2\dfrac{1}{2}, -1\dfrac{1}{2}\right), F(0, 0)$ **45.** III
46. second column, third row from the bottom
47.

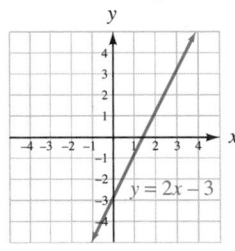

48.

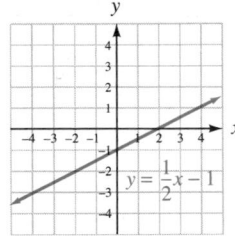

49.

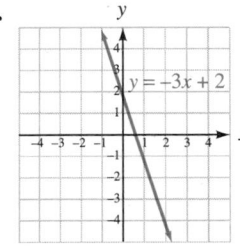

50.

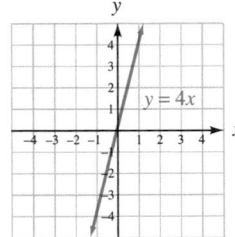

51. $(-3, -2), (-1, 0), (0, 1), (1, 2), (2, 3), (3, 4)$ (answers may
vary)   **52.** $(-3, 0), \left(0, 2\dfrac{1}{2}\right)$ or $(0, 2.5)$
53.

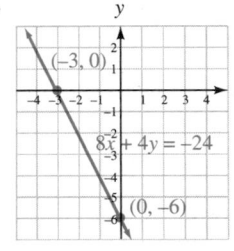

54.

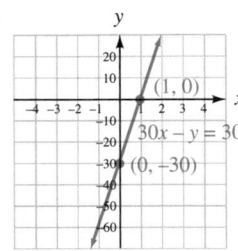

55.

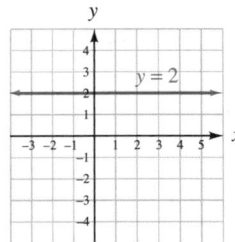

56.

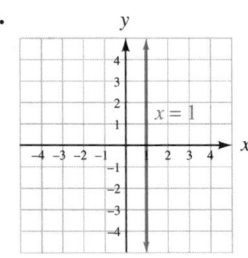

Chapter 8 Test   (page 810)

1. a. axis   **b.** mean   **c.** median   **d.** mode   **e.** central
2. a. 563 calories   **b.** 129 calories   **c.** about 8 mph
3. a. love seat; 150 ft   **b.** 50 feet more   **c.** 340 ft   **4. a.** 75%
b. 14.1%   **c.** lung cancer   **d.** prostate cancer; 32.7%
5. a. about 38 g   **b.** about 15 g   **6. a.** 17%   **b.** 529,550
7. a. about 27,000 police officers   **b.** 1989; about 26,000
police officers   **c.** 2000; about 41,000 police officers
d. about 5,000 police officers   **8. a.** bicyclist 1
b. Bicyclist 1 is stopped, but is ahead in the race. Bicyclist 2 is
beginning to catch up.   **c.** time C   **d.** Bicyclist 2 never lead.
e. bicyclist 1   **9. a.** 22 employees   **b.** 30 employees
c. 57 employees   **10. a.** 7.5 hr   **b.** 7.5 hr   **c.** 5 hr   **d.** 17 hr
11. 3 stars   **12.** 3.36 GPA   **13.** mean: 4.41 million; median:
4.25 million; mode: 4.25 million; range: 1.46 million

14. Of all the existing single-family homes sold in May of 2009, half of them sold for less than \$172,900 and half sold for more than \$172,900.   **15.** yes   **16.** no   **17.** $-2, 4, -1$
18. $1, (0, 1); 2, (3, 2); 0, (-3, 0)$
19. $(30, 32), (30, 34), (31, 34), (38, 30)$
20.

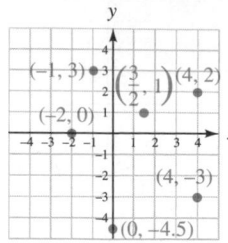

21. $A(0, 0), B(2.5, 3.5)$ or $B\left(2\frac{1}{2}, 3\frac{1}{2}\right), C(-3, -2), D(0, -2),$
$E(4, 0), F(-5, 5)$ **22.** III

23.

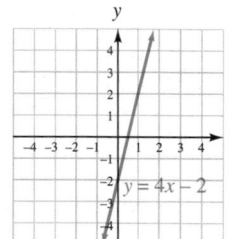

24.

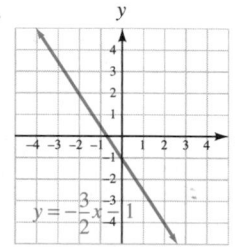

25. a. $(-3, 0)$ **b.** $(0, -1)$ **26.** $(-3, -1), (0, 0), (3, 1)$
(answers may vary)

27.

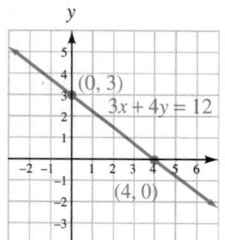

28.

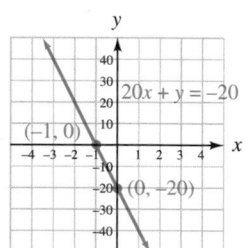

29.

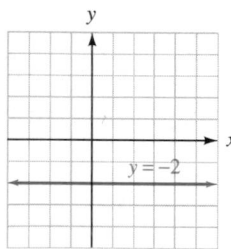

30.

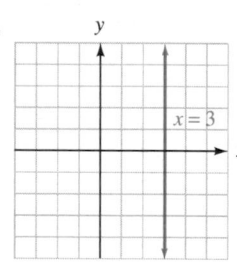

Chapters 1–8 Cumulative Review   (page 814)

1. fifty-two million, nine hundred forty thousand, five hundred fifty-nine; $50,000,000 + 2,000,000 + 900,000 + 40,000 + 500 + 50 + 9$ **2.** 60,000   **3.** 54,604   **4.** 3,209
5. 27,336   **6.** 23   **7.** $1,683 + 459 = 2,142$ **8.** 40 in., 84 in.²
9. a. $1, 2, 3, 4, 6, 9, 12, 18, 36$ **b.** $2, 3, 5, 7, 11, 13, 17, 19, 23, 29$
c. $2^2 \cdot 3^2$ **10. a.** 24   **b.** 4   **11.** 35   **12.** 9   **13.** 179   **14.** 3
15.

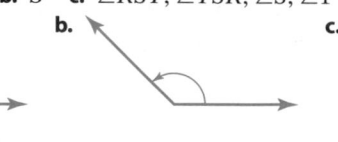

16. a. 6   **b.** 5   **c.** false   **17. a.** -30 **b.** -30 **c.** -175
d. 7   **18.** 1,100°F   **19.** -5 **20.** 429   **21.** -4 **22.** -200
23. 36   **24.** The temperature fell 28° overnight.   **25.** $x^2 - 4$
26. 225 mi   **27. a.** $72t$ **b.** $-15x + 10y + 50$ **28. a.** $4x$

b. $-3x + 23$ **29.** -4 **30.** His score on the interview was 47 and his score on the written part was 51.   **31. a.** $\frac{5}{9}$

b. $\frac{3a^2}{2}$ **32. a.** 0   **b.** undefined   **33.** $\frac{8}{35}$ **34.** $\frac{4}{63n}$ **35.** $-\frac{1}{6}$

36. $\frac{19}{20}$ **37.** $\frac{8x - 9}{72}$ **38.** 160 minutes are spent in lecture

each week.   **39.** $-\frac{21}{20} = -1\frac{1}{20}$ **40.** $6\frac{3}{4}$ in.   **41.** $10\frac{5}{8}$

42. $-\frac{3}{8}$ **43.** $\frac{12}{5}$ **44.** $-\frac{25}{4}$ **45.** There are 15 players on the
team.   **46.** 428.91   **47.** \$1,815.19   **48. a.** 345   **b.** 0.000345
49. 145.5   **50.** -0.744 **51.** 745   **52.** 0.01825   **53.** $0.\overline{72}$
54. 75   **55.** 3   **56.** The business can make up to 2,500 copies each month without exceeding the budget.

57. $\frac{2}{3}$ **58.** \$59.95   **59.** $\frac{1}{7}$ **60.** 128 fl oz   **61.** 6.4 m

62. 19.8°C   **63.** $\frac{3}{100}, 0.03; 2.25, 225\%; \frac{41}{1,000}, 4.1\%$

64. 17%   **65.** 24.36   **66.** 57.6   **67.** \$7.92   **68.** 16%
69. \$12   **70.** \$3,312   **71.** \$13,159.23   **72. a.** 7%   **b.** 5,040
73. a. 2008; 36   **b.** 2007 to 2008; an increase of 16 deaths
c. 2008 to 2009; a decrease of 8 deaths   **74.** mean: 3.02;
median: 3.00; mode: 2.75; range: 1.79   **75.** yes   **76.** 4
77.

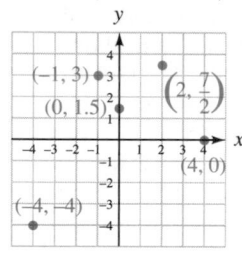

78.

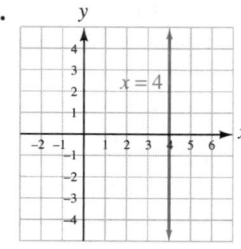

79.

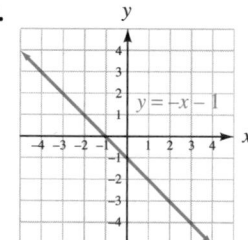

80.

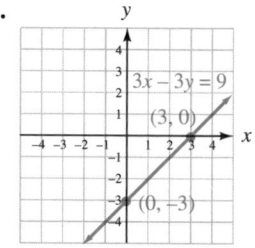

Study Set Section 9.1   (page 828)

1. point, line, plane   **3.** midpoint   **5.** angle   **7.** protractor
9. right   **11.** 180°   **13.** Adjacent   **15.** congruent
17. 90°   **19. a.** one   **b.** line

21. a. $\overrightarrow{SR}, \overrightarrow{ST}$ **b.** S **c.** $\angle RST, \angle TSR, \angle S, \angle 1$
23. a.   **b.**   **c.**

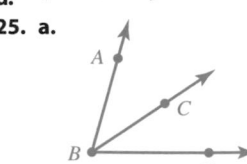

d.

25. a.   **b.**

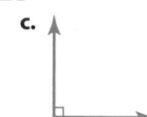

c. 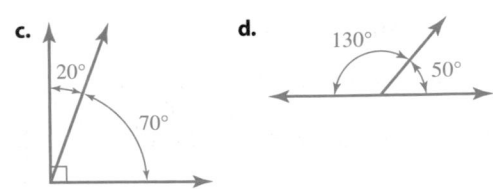 **d.**

27. congruent **29. a.** false **b.** false **c.** false **d.** true
31. true **33.** false **35.** line **37.** ray **39.** angle
41. degree **43.** congruent
45. a. *T* **b.** **c.**

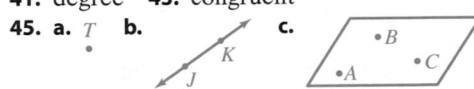

47. a. 2 **b.** 3 **c.** 1 **d.** 6 **49.** 50° **51.** 25° **53.** 75°
55. 130° **57.** right **59.** acute **61.** straight **63.** obtuse
65. 10° **67.** 27.5° **69.** 70° **71.** 65° **73.** 30°, 60°, 120°
75. 25°, 115°, 65° **77.** 60° **79.** 75° **81. a.** true
b. false, a segment has two endpoints **c.** false, a line does
not have an endpoint **d.** false, point *G* is the vertex of the
angle **e.** true **f.** true **83.** 40° **85.** 135° **87. a.** 50°
b. 130° **c.** 230° **d.** 260° **89. a.** 66° **b.** 156° **91.** 141°
93. 1° **95. a.** about 80° **b.** about 30° **c.** about 65°
97. a. 27° **b.** 30° **103.** $\frac{23}{12}$ or $1\frac{11}{12}$ **105.** $\frac{1}{10}$

Study Set Section 9.2 (page 839)

1. coplanar, noncoplanar **3.** Perpendicular **5.** alternate
7. a. **b.**

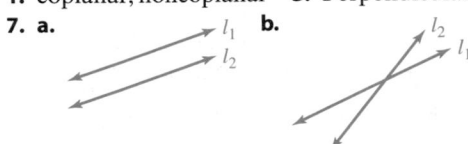

9. a. **b.**

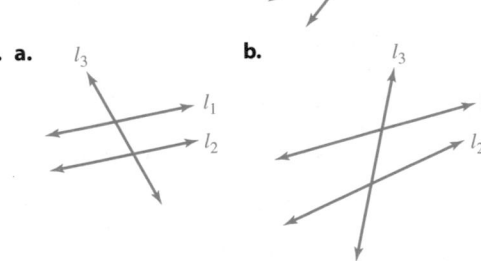

11. corresponding **13.** interior **15.** They are perpendicular.
17. right **19.** perpendicular **21. a.** ∠1 and ∠5, ∠4 and
∠8, ∠2 and ∠6, ∠3 and ∠7 **b.** ∠3, ∠4, ∠5, and ∠6
c. ∠3 and ∠5, ∠4 and ∠6 **23.** m(∠1) = 130°, m(∠2) = 50°,
m(∠3) = 50°, m(∠5) = 130°, m(∠6) = 50°, m(∠7) = 50°,
m(∠8) = 130° **25.** ∠1 ≅ ∠*X*, ∠2 ≅ ∠*N* **27.** 12°, 40°, 40°
29. 10°, 50°, 130° **31. a.** 50°, 135°, 45°, 85° **b.** 180° **c.** 180°
33. vertical angles: ∠1 ≅ ∠2; alternate interior angles:
∠*B* ≅ ∠*D*, ∠*E* ≅ ∠*A* **35.** 40°, 40°, 140° **37.** 12°, 70°, 70°
39. The plummet string should hang perpendicular to the
top of the stones. **41.** 50° **43.** The strips of wallpaper
should be hung on the wall parallel to each other, and they
should be perpendicular to the floor. **45.** 75°, 105°, 75°
53. 72 **55.** 45% **57.** yes **59.** $\frac{1}{3}$

Study Set Section 9.3 (page 849)

1. polygon **3.** vertex **5.** equilateral, isosceles, scalene
7. hypotenuse, legs **9.** addition

11. a. **b.** **c.** **d.**

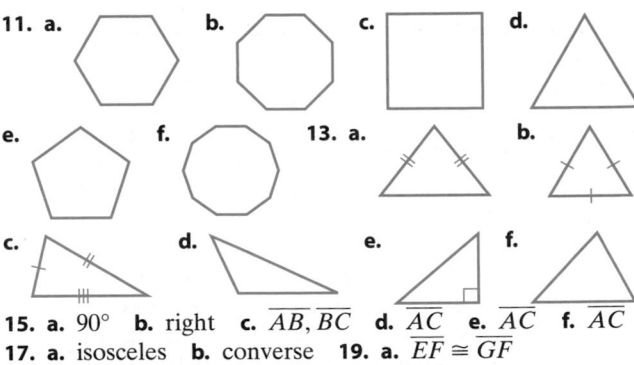

e. **f.** **13. a.** **b.**

c. **d.** **e.** **f.**

15. a. 90° **b.** right **c.** $\overline{AB}, \overline{BC}$ **d.** $\overline{AC}$ **e.** $\overline{AC}$ **f.** $\overline{AC}$
17. a. isosceles **b.** converse **19. a.** $\overline{EF} \cong \overline{GF}$
b. isosceles **21.** triangle **23.** $\overline{AB} \cong \overline{CB}$
25. a. 4, quadrilateral, 4 **b.** 6, hexagon, 6
27. a. 7, heptagon, 7 **b.** 9, nonagon, 9
29. a. scalene **b.** isosceles **31. a.** equilateral **b.** scalene
33. yes **35.** no **37.** 55° **39.** 45° **41.** 50°; 50°, 60°, 70°
43. 20°; 20°, 80°, 80° **45.** 68° **47.** 9° **49.** 39° **51.** 44.75°
53. 28° **55.** 73° **57.** 90° **59.** 45° **61.** 90.7° **63.** 61.5°
65. 12° **67.** 52.5° **69.** 39°, 39°, 102° or 70.5°, 70.5°, 39°
71. 73° **73.** 75° **75. a.** octagon **b.** triangle **c.** pentagon
77. As the jack is raised, the two sides of the jack remain the
same length. **79.** equilateral **85.** 22 **87.** 40% **89.** 0.10625

Study Set Section 9.4 (page 859)

1. hypotenuse, legs **3.** Pythagorean **5.** a^2, b^2, c^2
7. right **9. a.** $\overline{BC}$ **b.** $\overline{AB}$ **c.** $\overline{AC}$ **11.** 64, 100, 100
13. 10 ft **15.** 13 m **17.** 73 mi **19.** 137 cm
21. 24 cm **23.** 80 m **25.** 20 m **27.** 19 m
29. $\sqrt{11}$ cm ≈ 3.32 cm **31.** $\sqrt{208}$ m ≈ 14.42 m
33. $\sqrt{90}$ in. ≈ 9.49 in. **35.** $\sqrt{20}$ in. ≈ 4.47 in. **37.** no
39. yes **41.** 12 ft **43.** 25 in. **45.** $\sqrt{16,200}$ ft ≈ 127.28 ft
47. yes, $\sqrt{1,288}$ ft ≈ 35.89 ft **53.** no **55.** no **57.** no **59.** no

Study Set Section 9.5 (page 869)

1. Congruent **3.** congruent **5.** similar **7. a.** No, they are
different sizes. **b.** Yes, they have the same shape.
9. *PRQ* **11.** *MNO* **13.** ∠*A* ≅ ∠*B*, ∠*Y* ≅ ∠*T*,
∠*Z* ≅ ∠*R*, $\overline{YZ} \cong \overline{TR}$, $\overline{AZ} \cong \overline{BR}$, $\overline{AY} \cong \overline{BT}$
15. congruent **17.** angle, angle **19.** 100 **21.** 5.4
23. proportional **25.** congruent **27.** is congruent to
29.

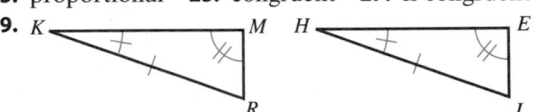

31. $\overline{DF}, \overline{AB}, \overline{EF}, ∠D, ∠B, ∠C$ **33. a.** ∠*B* ≅ ∠*M*,
∠*C* ≅ ∠*N*, ∠*D* ≅ ∠*O*, $\overline{BC} \cong \overline{MN}$, $\overline{CD} \cong \overline{NO}$, $\overline{BD} \cong \overline{MO}$
b. 72° **c.** 10 ft **d.** 9 ft **35.** yes, SSS **37.** not necessarily
39. a. ∠*L* ≅ ∠*H*, ∠*M* ≅ ∠*J*, ∠*R* ≅ ∠*E* **b.** *MR, LR, LM*
c. *HJ, JE, LR* **41.** yes **43.** not necessarily **45.** yes
47. not necessarily **49.** yes **51.** not necessarily **53.** 8, 35
55. 60, 38 **57.** true **59.** false: the angles must be between
congruent sides **61.** yes, SSS **63.** yes, SAS **65.** yes, ASA
67. not necessarily **69.** 80°, 2 yd **71.** 19°, 14 m **73.** 6 mm
75. 50° **77.** $\frac{25}{6} = 4\frac{1}{6}$ **79.** 16 **81.** 17.5 cm **83.** 59.2 ft
85. 36 ft **87.** 34.8 ft **89.** 1,056 ft **93.** 189 **95.** 21

Study Set Section 9.6 (page 881)

1. quadrilateral **3.** rectangle **5.** rhombus **7.** trapezoid, bases, isosceles **9. a.** four; A, B, C, D **b.** four; $\overline{AB}, \overline{BC}, \overline{CD}, \overline{DA}$ **c.** two; $\overline{AC}, \overline{BD}$ **d.** yes, no, no, yes **11. a.** $\overline{VU}$ **b.** $\parallel$ **13. a.** right **b.** parallel **c.** length **d.** length **e.** midpoint **15.** rectangle **17. a.** no **b.** yes **c.** no **d.** yes **e.** no **f.** yes **19. a.** isosceles **b.** $\angle J, \angle M$ **c.** $\angle K, \angle L$ **d.** $\angle M, \angle L, \overline{ML}$ **21.** The four sides of the quadrilateral are the same length. **23.** the sum of the measures of the angles of a polygon; the number of sides of the polygon **25. a.** square **b.** rhombus **c.** trapezoid **d.** rectangle **27. a.** $90°$ **b.** 9 **c.** 18 **d.** 18 **29. a.** $42°$ **b.** $95°$ **31. a.** 9 **b.** $70°$ **c.** $110°$ **d.** $110°$ **33.** $2,160°$ **35.** $3,240°$ **37.** $1,080°$ **39.** $1,800°$ **41.** 5 **43.** 7 **45.** 13 **47.** 14 **49. a.** $30°$ **b.** $30°$ **c.** $60°$ **d.** 8 cm **e.** 4 cm **51.** $40°$; $m(\angle A) = 90°$, $m(\angle B) = 150°$, $m(\angle C) = 40°$, $m(\angle D) = 80°$ **53. a.** trapezoid **b.** square **c.** rectangle **d.** trapezoid **e.** parallelogram **55.** $540°$ **61.** two hundred fifty-four thousand, three hundred nine **63.** eighty-two million, four hundred fifteen

Think It Through (page 890)

about 108 ft^2

Study Set Section 9.7 (page 894)

1. perimeter **3.** area **5.** area **7.** $8 \text{ ft} \cdot 16 \text{ ft} = 128 \text{ ft}^2$ **9. a.** $p = 4s, p = 2l + 2w$

11. a. 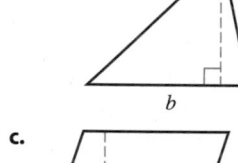 **b.**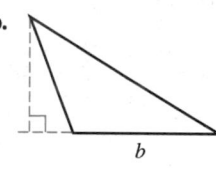

c. 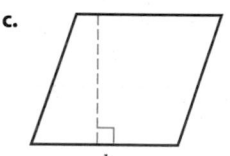 **d.**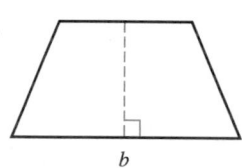

13. a rectangle and a triangle **15. a.** square inch **b.** 1 m^2 **17.** 32 in. **19.** 23 mi **21.** 62 in. **23.** 94 in. **25.** 15 ft **27.** 5 m **29.** 16 cm^2 **31.** 6.25 m^2 **33.** 144 in.^{2} **35.** 1,000,000 mm^2 **37.** 27,878,400 ft^2 **39.** 1,000,000 m^2 **41.** 135 ft^2 **43.** 11,160 ft^2 **45.** 25 in.^{2} **47.** 27 cm^2 **49.** 7.5 in.^{2} **51.** 10.5 mi^2 **53.** 40 ft^2 **55.** 91 cm^2 **57.** 4 m **59.** 12 cm **61.** 36 m **63.** 11 mi **65.** 102 in.^{2} **67.** 360 ft^2 **69.** 75 m^2 **71.** 75 yd^2 **73.** \$1,200 **75.** \$4,875 **77.** length 15 in. and width 5 in.; length 16 in. and width 4 in. (answers may vary) **79.** sides of length 5 m **81.** base 5 yd and height 3 yd (answers may vary) **83.** length 5 ft and width 4 ft; length 20 ft and width 3 ft (answers may vary) **85.** 60 cm^2 **87.** 36 m **89.** $28\frac{1}{3}$ ft **91.** 36 m **93.** $x = 3.7$ ft, $y = 10.1$ ft; 50.8 ft **95.** $80 + 1 = 81$ trees **97.** vinyl **99.** \$361.20 **101.** \$192 **103.** 111,825 mi^2 **105.** 51 sheets **111.** $6t$ **113.** $-2w + 4$ **115.** $-\frac{5}{8}x$ **117.** $9r - 16$

Study Set Section 9.8 (page 906)

1. radius **3.** diameter **5.** circumference **7.** twice **9.** $\overline{OA}, \overline{OC}, \overline{OB}$ **11.** $\overline{DA}, \overline{DC}, \overline{AC}$ **13.** $\overset{\frown}{ABC}, \overset{\frown}{ADC}$ **15. a.** Multiply the radius by 2. **b.** Divide the diameter by 2. **17.** π **19.** square 6 **21.** arc AB **23. a.** multiplication: $2 \cdot \pi \cdot r$ **b.** raising to a power and multiplication: $\pi \cdot r^2$ **25.** 8π ft ≈ 25.1 ft **27.** 12π m ≈ 37.7 m **29.** 50.85 cm **31.** 31.42 in. **33.** 9π in.$^2 \approx 28.3$ in.^{2} **35.** 81π in.$^2 \approx 254.5$ in.^{2} **37.** 128.5 cm^2 **39.** 57.1 cm^2 **41.** 27.4 in.^{2} **43.** 66.7 in.^{2} **45.** 50π yd ≈ 157.08 yd **47.** 6π in. ≈ 18.8 in. **49.** 20.25π mm$^2 \approx 63.6$ mm^2 **51. a.** 1 in. **b.** 2 in. **c.** 2π in. ≈ 6.28 in. **d.** π in.$^2 \approx 3.14$ in.^{2} **53.** π mi$^2 \approx 3.14$ mi^2 **55.** 32.66π ft ≈ 102.60 ft **57.** 13 times **59.** 4π ft$^2 \approx 12.57$ ft^2; 0.25π ft$^2 \approx 0.79$ ft^2; 6.25% **65.** 90% **67.** 82.7% **69.** 5.375¢ per oz **71.** five

Study Set Section 9.9 (page 914)

1. volume **3.** cone **5.** cylinder **7.** pyramid

9. 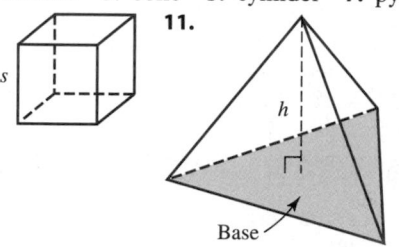 **11.** 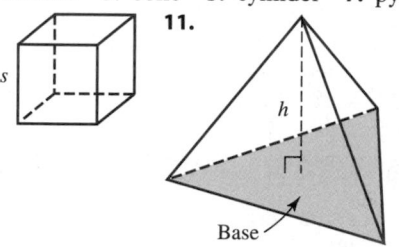 **13.**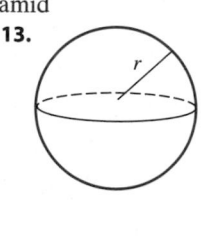

Base

15. cubic inches, mi^3, m^3 **17. a.** perimeter **b.** volume **c.** area **d.** volume **e.** area **f.** circumference **19. a.** 50π **b.** $\frac{500}{3}\pi$ **21. a.** cubic inch **b.** 1 cm^3 **23.** a right angle **25.** 27 **27.** 1,000,000,000 **29.** 56 ft^3 **31.** 125 in.^{3} **33.** 120 cm^3 **35.** 1,296 in.^{3} **37.** 700 yd^3 **39.** 32 ft^3 **41.** 69.72 ft^3 **43.** 6 yd^3 **45.** 192π ft$^3 \approx 603.19$ ft^3 **47.** $3,150\pi$ cm$^3 \approx 9,896.02$ cm^3 **49.** 39π m$^3 \approx 122.52$ m^3 **51.** 189π yd$^3 \approx 593.76$ yd^3 **53.** 288π in.$^3 \approx 904.8$ in.^{3} **55.** $\frac{32}{3}\pi$ cm$^3 \approx 33.5$ cm^3 **57.** 486π in.$^3 \approx 1,526.81$ in.^{3} **59.** 423π m$^3 \approx 1,357.17$ m^3 **61.** 60 cm^3 **63.** 100π cm$^3 \approx 314.16$ cm^3 **65.** 400 m^3 **67.** 48 m^3 **69.** 576 cm^3 **71.** 180π cm$^3 \approx 565.49$ cm^3 **73.** $\frac{1}{8}$ in.$^3 = 0.125$ in.^{3} **75.** 2.125 **77.** 63π ft$^3 \approx 197.92$ ft^3 **79.** $\frac{32,000}{3}\pi$ ft$^3 \approx 33,510.32$ ft^3 **81.** 8:1 **83. a.** $2,250\pi$ in.$^3 \approx 7,068.58$ in.^{3} **b.** 30.6 gal **89.** -42 **91.** -4 **93.** $\frac{1}{5}$ or 1:5 **95.** 2,400 mm

Chapter 9 Review (page 919)

1. points C and D, line CD, plane GHI **2. a.** 6 units **b.** E **c.** yes **3.** $\angle ABC, \angle CBA, \angle B, \angle 1$ **4. a.** acute **b.** B **c.** $\overrightarrow{BA}$ and $\overrightarrow{BC}$ **d.** $48°$ **5.** $\angle 1$ and $\angle 2$ are acute, $\angle ABD$ and $\angle CBD$ are right angles, $\angle CBE$ is obtuse, and $\angle ABC$ is a straight angle. **6.** yes **7.** yes **8. a.** obtuse angle **b.** right angle **c.** straight angle **d.** acute angle **9.** $15°$ **10.** $150°$ **11. a.** $m(\angle 1) = 65°$ **b.** $m(\angle 2) = 115°$ **12. a.** $39°$ **b.** $90°$ **c.** $51°$ **d.** $51°$ **e.** yes **13. a.** $20°$ **b.** $125°$ **c.** $55°$ **14.** $19°$ **15.** $37°$

16. No, only two angles can be supplementary.
17. a. parallel **b.** transversal **c.** perpendicular
18. $\angle 4$ and $\angle 6$, $\angle 3$ and $\angle 5$ **19.** $\angle 1$ and $\angle 5$, $\angle 4$ and $\angle 8$, $\angle 2$ and $\angle 6$, $\angle 3$ and $\angle 7$ **20.** $\angle 1$ and $\angle 3$, $\angle 2$ and $\angle 4$, $\angle 5$ and $\angle 7$, and $\angle 6$ and $\angle 8$ **21.** $m(\angle 1) = m(\angle 3) = m(\angle 5) = m(\angle 7) = 70°$; $m(\angle 2) = m(\angle 4) = m(\angle 6) = 110°$
22. $m(\angle 1) = 60°$, $m(\angle 2) = 120°$, $m(\angle 3) = 130°$, $m(\angle 4) = 50°$
23. a. $40°$ **b.** $50°, 50°$ **24. a.** $20°$ **b.** $110°, 70°$ **25. a.** $11°$
b. $31°, 31°$ **26. a.** $23°$ **b.** $82°, 82°$ **27. a.** 8, octagon, 8
b. 5, pentagon, 5 **c.** 3, triangle, 3 **d.** 6, hexagon, 6
e. 4, quadrilateral, 4 **f.** 10, decagon, 10 **28. a.** isosceles
b. scalene **c.** equilateral **d.** isosceles **29. a.** acute
b. right **c.** obtuse **d.** acute **30. a.** $90°$ **b.** right
c. $\overline{XY}, \overline{XZ}$ **d.** $\overline{YZ}$ **e.** $\overline{YZ}$ **f.** $\overline{YZ}$ **31.** $90°$ **32.** $50°$
33. $71°$ **34.** $18°$; $36°, 28°, 116°$ **35.** $50°$ **36.** $56°$ **37.** $67°$
38. $83°$ **39.** 13 cm **40.** 17 ft **41.** 36 in. **42.** 20 ft
43. $\sqrt{231}$ m ≈ 15.20 m **44.** $\sqrt{1,300}$ in. ≈ 36.06 in.
45. 73 in. **46.** $\sqrt{1,023}$ in. ≈ 32 in. **47.** not a right triangle
48. not a right triangle **49. a.** $\angle D$ **b.** $\angle E$ **c.** $\angle F$
d. $\overline{DF}$ **e.** $\overline{DE}$ **f.** $\overline{EF}$ **50. a.** $32°$ **b.** $61°$ **c.** 6 in.
d. 9 in. **51.** congruent, SSS **52.** congruent, SAS
53. not necessarily congruent **54.** congruent, ASA
55. yes **56.** yes **57.** 4, 28 **58.** 65 ft **59. a.** trapezoid
b. square **c.** parallelogram **d.** rectangle **e.** rhombus
f. rectangle **60. a.** 15 cm **b.** $40°$ **c.** $100°$ **d.** 7.5 cm
e. 14 cm **61. a.** true **b.** true **c.** true **d.** false
62. a. $65°$ **b.** $115°$ **c.** 4 yd **63.** $1,080°$ **64.** 20 sides
65. 72 in. **66.** 86 in. **67.** 30 m **68.** 36 m **69.** 59 ft
70. a. 9 ft^2 **b.** 144 in.^{2} **71.** 9.61 cm^2 **72.** 7,500 ft^2
73. 450 ft^2 **74.** 200 in.^{2} **75.** 120 cm^2 **76.** 232 ft^2
77. 152 ft^2 **78.** 120 m^2 **79.** 8 ft **80.** 18 mm **81.** $3,281
82. $4,608 **83. a.** $\overline{CD}, \overline{AB}$ **b.** $\overline{AB}$ **c.** $\overline{OA}, \overline{OC}, \overline{OD}, \overline{OB}$
d. O **84.** 21π ft ≈ 65.97 ft **85.** 45.1 cm
86. 81π in.^{2} ≈ 254.47 in.^{2} **87.** 130.3 cm^2 **88.** 6,073.0 in.^{2}
89. 125 cm^3 **90.** 480 m^3 **91.** 1,728 mm^3
92. $\dfrac{500}{3}\pi$ in.^{3} ≈ 523.60 in.^{3} **93.** 250π in.^{3} ≈ 785.40 in.^{3}
94. 2,000 yd^3 **95.** 2,940 m^3 **96.** $\dfrac{1,024}{3}\pi$ in.^{3} $\approx 1,072.33$ in.^{3}
97. 1,518 ft^3 **98.** 3.125π in.^{3} ≈ 9.8 in.^{3} **99.** 1,728 in.^{3}
100. 54 ft^3

Chapter 9 Test   (page 942)

1. a. $135°$, obtuse **b.** $90°$, right **c.** $40°$, acute
d. $180°$, straight **2. a.** measure **b.** length **c.** line
d. complementary **3.** D **4. a.** false **b.** true **c.** true
d. true **e.** false **5.** $20°$; $60°, 60°$ **6.** $133°$
7. a. transversal **b.** $\angle 6$ **c.** $\angle 7$ **8.** $m(\angle 1) = 155°$,
$m(\angle 3) = 155°$, $m(\angle 4) = 25°$, $m(\angle 5) = 25°$, $m(\angle 6) = 155°$,
$m(\angle 7) = 25°$, $m(\angle 8) = 155°$ **9.** $50°$; $110°, 70°$
10. a. 8, octagon, 8 **b.** 5, pentagon, 5 **c.** 6, hexagon, 6
d. 4, quadrilateral, 4 **11. a.** isosceles **b.** scalene
c. equilateral **d.** isosceles **12.** $70°$ **13.** $84°$ **14. a.** 12
b. 13 **c.** $90°$ **d.** 5 **15. a.** 10 **b.** $65°$ **c.** $115°$ **d.** $115°$
16. $1,440°$ **17.** 188 in. **18.** 15.2 m **19.** 360 cm^2
20. $864 **21.** 144 in.^{2} **22.** 120 in.^{2} **23. a.** $\overline{RS}, \overline{XY}$
b. $\overline{XY}$ **c.** $\overline{OX}, \overline{OR}, \overline{OS}, \overline{OY}$ **24.** π **25.** 21π ft ≈ 66.0 ft
26. $(40 + 12\pi)$ ft ≈ 77.7 ft **27.** 225π m^2 ≈ 706.9 m^2
28. $\angle R, \angle S, \angle T$; $\overline{RT}, \overline{RS}, \overline{ST}$ **29. a.** congruent, SSS
b. congruent, ASA **c.** not necessarily congruent
d. congruent, SAS **30. a.** 8 in. **b.** $50°$

31. a. yes **b.** yes **32. a.** 6 m **b.** 12 m **33.** 21 ft
34. a. 26 cm **b.** $\sqrt{28}$ in. ≈ 5.3 in. **35.** $\sqrt{986}$ in. ≈ 31.4 in.
36. 1,728 in.^{3} **37.** 216 in.^{3} **38.** 480 m^3
39. $1,296\pi$ in.^{3} $\approx 4,071.50$ in.^{3} **40.** 600 in.^{3} **41.** 1,890 ft^3
42. 63π yd^3 ≈ 197.92 yd^3 **43.** 400 mi^3
44. $\dfrac{256}{3}\pi$ in.^{3} ≈ 268.08 in.^{3} **45.** $11,250\pi$ ft^3 $\approx 35,343$ ft^3

Chapters 1–9 Cumulative Review   (page 946)

1. one hundred four million, fifty-two thousand, five
2. 103,476 **3.** 15,288 **4.** 59 **5. a.** 1, 2, 4, 8, 16, 32
b. $2^2 \cdot 5 \cdot 7$ **6.** 315; 5 **7.** 73 **8. a.** 58 **b.** 13 **9.** true
10. a. -9 **b.** -12 **c.** -216 **d.** undefined **11.** 22
12. 39 **13.** Her checking account balance before the
deposit was $-$217. **14.** $2w - 18$ **15.** 17 **16.** $48u$
17. $4x + 8$ **18.** -13 **19.** The amount of the tax refund
check was $1,290. **20.** $\dfrac{5x^2}{9}$ **21.** $\dfrac{3}{4}$ **22.** $\dfrac{2}{3b}$ **23.** $\dfrac{16 - 3n}{4n}$
24. $\dfrac{32}{35}$ **25.** 45 in.^{2} **26.** $11\dfrac{1}{30}$ **27.** $\dfrac{3}{2}$ **28.** The shop
serviced 350 cars last year. **29.** -12 **30.** -6 **31.** 9.510
32. It is correct: $361.5 + 89.8 = 451.3$. **33.** 10.3632 ft^2
34. -3.57 **35.** $0.91\overline{6}$ **36.** -32 **37.** 0.8 **38.** $\dfrac{7}{12}$
39. 26 cups **40. a.** 800 oz **b.** 0.5 L **c.** about 13.7 in.
41. 0.54, 54% **42.** 150 **43.** $205.01 **44.** 1,400
45. $31.25 **46.** mean: 35; median: 36; mode: 45; range: 68
47. $A(2,3), B(-3,4), C(-3,-4), D(4,-4), E(0,0), F(4,0), G(-4,0),$
$H(1.5, -2.5)$ or $H\left(1\dfrac{1}{2}, -2\dfrac{1}{2}\right)$ **48.** no

49.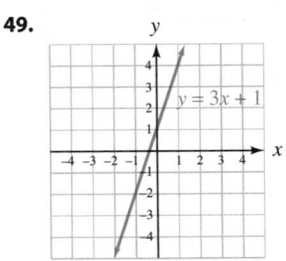
50.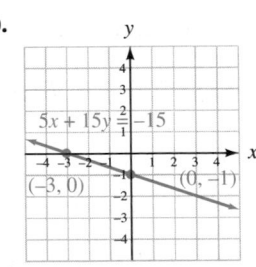

51. $12°, 80°, 100°$ **52.** $28°$ **53.** 40 yards **54.** 78.5 cm^2

Study Set Section 10.1   (page 955)

1. exponential **3. a.** $3x, 3x, 3x, 3x$ **b.** $(-5y)^3$ **5. a.** add
b. multiply **c.** multiply **7. a.** $2x^2$ **b.** x^4 **9. a.** doesn't
simplify **b.** x^5 **11.** $x^6, 18$ **13.** base 4, exponent 3
15. base x, exponent 5 **17.** base $-3x$, exponent 2
19. base y, exponent 6 **21.** base m, exponent 12
23. base $y + 9$, exponent 4 **25.** m^5 **27.** $(4t)^4$ **29.** $4t^5$
31. $a^2 b^3$ **33.** 5^7 **35.** a^6 **37.** b^6 **39.** c^{13} **41.** $a^5 b^6$
43. $c^2 d^5$ **45.** $x^3 y^{11}$ **47.** m^{200} **49.** 3^8 **51.** $(-4.3)^{24}$
53. m^{500} **55.** y^{15} **57.** x^{25} **59.** p^{25} **61.** t^{18} **63.** u^{14}
65. $36a^2$ **67.** $625y^4$ **69.** $27a^{12} b^{21}$ **71.** $-8r^6 s^9$ **73.** $72c^{17}$
75. $6,400d^{41}$ **77.** $49a^{18}$ **79.** t^{10} **81.** y^9 **83.** $-216a^9 b^6$
85. n^{33} **87.** 6^{60} **89.** $288b^{27}$ **91.** c^{14} **93.** $432s^{16} t^{13}$
95. x^{15} **97.** $25x^2$ ft^2 **101.** $\dfrac{3}{4}$ **103.** 5 **105.** 7 **107.** 12

Study Set Section 10.2   (page 960)

1. polynomial **3.** trinomial **5. a.** 2 **b.** 1 **c.** 3
7. highest **9.** 2, 2, 4, 4, 16 **11.** binomial **13.** monomial
15. monomial **17.** trinomial **19.** binomial **21.** trinomial
23. 3 **25.** 2 **27.** 1 **29.** 7 **31.** 2 **33.** 4 **35.** 13 **37.** 6
39. -6 **41.** 31 **43.** 4 **45.** 1 **47.** 0 ft **49.** 64 ft

51. 63 ft **53.** 198 ft **57.** 2 **59.** $\frac{3}{2} = 1\frac{1}{2}$ **61.** 16 **63.** 6

Study Set Section 10.3   (page 965)

1. like **3.** coefficients, variables **5.** yes, $7y$ **7.** no
9. yes, $13x^3$ **11.** yes, $15x^2$ **13.** $2x^2, 7x; 5x^2$ **15.** $9y$

17. $12t^2$ **19.** $\frac{9}{8}a$ **21.** $\frac{5}{3}c^2$ **23.** $7x + 4$ **25.** $7x^2 - 7$

27. $12x^3 - 149x$ **29.** $8x^2 + 2x - 21$ **31.** $6.1a^2 + 10a - 19$
33. $2n^2 + 5$ **35.** $5x^2 + x + 11$ **37.** $-7x^2 - 5x - 1$
39. $16u^3$ **41.** $7x^5$ **43.** $1.6a + 8$ **45.** $-2.2x^2 - 9.9x$
47. $7b + 4$ **49.** $p^2 - 2p$ **51.** $5x^2 + 6x - 8$
53. $-12x^2 - 13x + 36$ **55.** $19x^2 - 5$ **57.** $8y^2 + 4y - 2$
59. $6x^2 + x - 5$ **61.** $-6.5m + 70$ **63.** $-t^2 - 1.4t + 6$

65. $2x^2 + x + 12.9$ **67.** $14s^2$ **69.** $-\frac{2}{3}b^4$ **71.** $10z^3 + z - 2$

73. $-5h^3 + 5h^2 + 30$ **75.** $1.3x^3$ **77.** $19.4h^3 - 11.1h^2 - 0.6$

79. $-x^3 + x + 14$ **81. a.** $2x$ ft **b.** $6x$ ft **83. a.** $(6x + 5)$ ft

b. $(4x^2 + 26)$ ft **89.** 0.8 oz

Study Set Section 10.4   (page 972)

1. monomials **3.** first, outer, inner, last **5.** each, each
7. a. $6x^2 + x - 12$ **b.** $5x^4 + 8x^2 + 3$ **9.** $8, n^3, 72n^5$
11. $2x, 5, 5; 4x, 15x; 11x$ **13.** $12x^5$ **15.** $-6b^3$
17. $6x^4 + 2x^3$ **19.** $6x^3 + 8x^2 - 14x$ **21.** $a^2 + 9a + 20$
23. $3x^2 + 10x - 8$ **25.** $4x^2 + 12x + 9$
27. $81b^2 - 36b + 4$ **29.** $6x^3 - x^2 + 1$ **31.** $x^3 - 1$
33. $x^3 + 1$ **35.** $12x^3 + 17x^2 - 6x - 8$
37. $6a^2 + 2a - 20$ **39.** $-2p^3 + 3p^2 - 2p$ **41.** $-6x^5$
43. $4x^2 + 11x + 6$ **45.** $4x^2 - 12x + 9$

47. $3q^4 - 6q^3 + 21q^2$ **49.** $-\frac{1}{2}y^7$ **51.** $x^3 - x^2 - 5x + 2$

53. $6a^4 + 5a^3 + 5a^2 + 10a + 4$
55. $x^4 + 11x^3 + 26x^2 - 28x - 24$ **57.** $9n^2 - 1$
59. $r^4 - 5r^3 + 2r^2 - 7r - 15$ **61.** $25t^2 - 10t + 1$
63. $3x^2 - 6x$ **65.** $(x^2 - 4)$ ft^2 **67.** $(6x^2 + x - 1)$ cm^2
69. $(35x^2 + 43x + 12)$ in.^{2} **75.** four and ninety-one
thousandths **77.** 0.109375 **79.** 134.657 **81.** 10

Chapter 10 Review   (page 975)

1. a. base n, exponent 12 **b.** base $2x$, exponent 6
c. base r, exponent 4 **d.** base $y - 7$, exponent 3
2. a. m^5 **b.** $-3x^4$ **c.** a^2b^4 **d.** $(pq)^3$ **3. a.** x^4 **b.** $2x^2$
c. x^3 **d.** does not simplify **4. a.** Keep the base 3, don't
multiply the bases. **b.** Multiply the exponents, don't add
them. **5.** 7^{12} **6.** m^2n^3 **7.** y^{21} **8.** $81x^4$ **9.** 6^{36}
10. $-b^{12}$ **11.** $256s^{10}$ **12.** $4.41x^4y^2$ **13.** $(-9)^{15}$ **14.** a^{23}
15. $8x^{15}$ **16.** $m^{10}n^{18}$ **17.** $72a^{17}$ **18.** x^{200} **19.** $256m^{13}$
20. $108t^{22}$ **21.** trinomial **22.** monomial **23.** binomial
24. trinomial **25.** 3 **26.** 4 **27.** 5 **28.** 5 **29.** 13

30. 16 ft **31.** $5x^3$ **32.** $\frac{13}{2}p^2$ **33.** $9x + 4$

34. $2x^2 - 2x + 3$ **35.** $8x + 3$ **36.** $-2x^2 + x + 2$
37. $7p^3$ **38.** $-5y^2$ **39.** $1.1x - 8$ **40.** $z^2 - 4z + 6$
41. $2x - 7$ **42.** $8x^2 - 5x + 12$ **43.** $15x^5$ **44.** $-6z^4$
45. $6x^3 + 4x^2$ **46.** $-35t^5 + 30t^4 + 10t^3$ **47.** $6x^2 + x - 2$
48. $35t^2 - 2t - 24$ **49.** $6x^3 + x^2 + x + 2$
50. $6r^3 - 5r^2 - 12r + 9$ **51.** $15x^3 + 19x^2 - x + 15$
52. $15x^3 - 16x^2 - x + 2$ **53.** $x^2 + 4x + 4$
54. $64a^2 - 48a + 9$

Chapter 10 Test   (page 979)

1. base, exponent **2.** exponential **3.** polynomial
4. monomial, binomial, trinomial **5.** degree **6.** evaluate
7. polynomials **8.** difference **9.** first, outer, inner, last
10. binomial, trinomial **11.** square **12.** each, each
13. a. base: 6, exponent: 5 **b.** base: b, exponent: 4
14. a. $2a^2$ **b.** a^4 **c.** does not simplify **d.** a^3 **15.** h^6
16. m^{20} **17.** b^8 **18.** x^{18} **19.** a^6b^{10} **20.** $144a^{18}b^2$
21. $216x^{15}$ **22.** t^{15} **23.** binomial **24.** monomial
25. trinomial **26.** binomial **27.** 6 **28.** 7 **29.** 25 **30.** 2
31. $-1.2p^2 + 3p$ **32.** $-8x^5$ **33.** $5x^2 - 3x + 4$
34. $6x^2 - 7x - 20$ **35.** $8d^2 - 9d + 12.5$
36. $3y^4 - 6y^3 + 9y^2$ **37.** $7x^2 + 2x - 2$
38. $2x^3 - 7x^2 + 14x - 12$ **39.** $(4x + 16)$ in. **40.** Keep the
common base 5, and add the exponents. Do not multiply the
common bases to get 25.

Chapter 1–10 Cumulative Review   (page 980)

1. \$8,995 **2.** 2,110,000 **3.** 32,034 **4.** 11,022 **5. a.** 602 ft
b. 19,788 ft^2 **6.** 33 R10 **7. a.** $2^2 \cdot 5 \cdot 11$ **b.** 1, 2, 3, 4, 6, 12
8. a. 48 **b.** 8 **9.** 11 **10. a.** 324 **b.** 3
11. a. $\{\ldots, -3, -2, -1, 0, 1, 2, 3, \ldots\}$ **b.** 3 **12. a.** -12
b. 20 **c.** -64 **d.** 4 **e.** -16 **f.** 16 **13.** $-\$140$
14. -2 **15.** -11 **16.** -5 **17.** In 2008, Acme made
\$46 million in profit. **18. a.** $k + 1$ **b.** $m - 1$ **19.** $x - 5$
20. 1,425 mi **21. a.** $100a$ **b.** $4y - 6$ **c.** $22y - 8$
22. $50x¢$ **23.** -8 **24.** -11 **25.** There were 5 dimes and

20 quarters at the bottom of the wishing well. **26.** $\frac{5a}{4}$

27. a. $\frac{18}{48}$ **b.** $\frac{8}{9}$ **c.** $\frac{15}{2}$ **28.** 9 oz **29.** $-\frac{3}{70}$ **30.** $\frac{b^2}{6c^4}$

31. $\frac{3}{20}$ **32.** $\frac{5m - 36}{45}$ **33.** $-\frac{11}{20}$ **34.** $142\frac{7}{15}$ **35.** $\frac{38}{29} = 1\frac{9}{29}$

36. $\frac{3}{32}$ fl oz **37.** $13\frac{3}{4}$ cups **38.** $\frac{8}{9}$ **39.** 8 **40.** $-\frac{15}{2}$

41. 165 students said they started their morning with a cup of
coffee. **42. a.** 3.1416 **b.** $>$ **43. a.** six million, five
hundred ten thousand, three hundred forty-five and seven
hundred ninety-eight thousandths

b. $7,000 + 400 + 90 + 8 + \frac{6}{10} + \frac{4}{100} + \frac{6}{1,000} + \frac{1}{10,000}$

44. 145.188 **45.** 3,803.61 **46.** -25.6 **47.** 17.05
48. 0.053 **49.** 22.3125 **50.** \$2,712.50 **51. a.** 899,708
b. 0.899708 **52.** $18,000 \div 9 = 2,000$ **53.** -9.32 **54.** $0.1\overline{3}$

55. a. -2 **b.** $\frac{7}{9}$

56.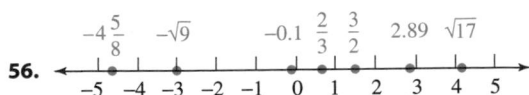

57. -0.6 **58.** She can buy 97 balloons.   **59. a.** $\dfrac{3}{7}$ **b.** $\dfrac{1}{4}$

60. the smaller board   **61.** $6\dfrac{1}{2} = 6.5$ **62.** The company had 125,000 policies.   **63.** 75 ft   **64. a.** 14 ft

b. 13.25 lb $= 13\dfrac{1}{4}$ lb   **c.** 120 quarts   **d.** 750 min

65. a. 1,538 g   **b.** 0.5 L   **c.** 0.000003 km   **66.** 240 km

67. about 4.5 kg   **68.** 167°F   **69.** $0.57, \dfrac{57}{100}, 0.1\%, \dfrac{1}{1,000},$

$33\dfrac{1}{3}\%, 0.\overline{3}$ **70. a.** 93%   **b.** 7%   **71.** 67.5   **72.** 120

73. 85%   **74.** \$205, \$615   **75.** \$1,159.38   **76.** 500%
77. \$21   **78.** \$1,567.50   **79. a.** 380,000 vehicles
b. 295,000 vehicles   **c.** 90,000 vehicles   **80. a.** 18%
b. 2,920,000   **81. a.** food: about \$17.5 billion   **b.** about
\$2.2 billion   **c.** about \$8.5 billion   **82.** mean: 0.86 oz,
median: 0.855 oz, mode: 0.85 oz

83. 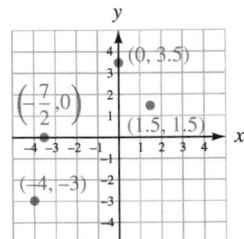   **84.** no

85. 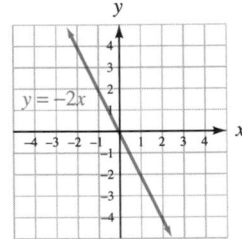   **86.**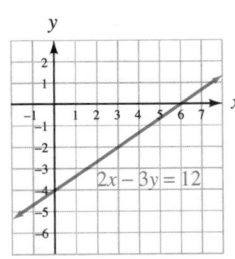

87.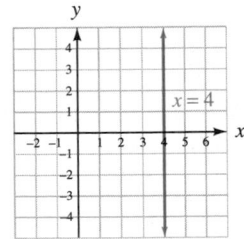

88. a. acute   **b.** right   **c.** obtuse   **d.** 180°
89. a. 75°   **b.** 15°   **90. a.** 50°   **b.** 50°   **c.** 130°   **d.** 50°
91. a. 75°   **b.** 30°   **c.** 105°   **d.** 105°   **92.** 46°, 134°
93. 73°   **94.** 26 m   **95.** yes   **96.** 42 ft   **97.** 540°
98. 48 m, 144 m^2 **99.** 126 ft^2 **100.** 91 in.^{2} **101.** 144 in.^{2}
102. circumference: 14π cm $\approx$ 43.98 cm, area:
49π cm^2 $\approx$ 153.94 cm^2 **103.** 98.31 yd^2 **104.** 6,480 in.^{3}
105. 972π in.^{3} $\approx$ 3,053.63 in.^{3} **106.** 48π m^3 $\approx$ 150.80 m^3
107. 20π ft^3 $\approx$ 62.83 ft^3 **108.** 1,728 in.^{3} **109.** s^9 **110.** a^{35}
111. y^{22} **112.** $8b^9c^{18}$ **113.** trinomial; 2   **114.** 0
115. $2x^2 - 2x + 9$ **116.** $10a^3 + 7.1a^2 - 8.2a$ **117.** $15h^{10}$
118. $-6p^3 - 9p^2 + 12p$ **119.** $6x^2 + 7x - 5$
120. $4y^2 - 28y + 49$

Study Set Appendix I   (page A-5)

1. Inductive   **3.** circular   **5.** alternating   **7.** alternating
9. 10 A.M.   **11.** 17   **13.** 27   **15.** 3   **17.** -17 **19.** R   **21.** e
23.    **25.**   **27.**   **29.**

31. D   **33.** Maria   **35.** 6 office managers   **37.** 9 children
39. I   **41.** W   **43.**   **45.** K   **47.** 6   **49.** 3

51. -11 **53.** 9   **55.** cage 3   **57.** B, D, A, C
59. 18,935 respondents   **61.** 0

INDEX